THE GREAT
CONTEMPORARY
ISSUES

SCIENCE IN THE TWENTIETH CENTURY

Theodore Lownik Library
Illinois Benedictine College
Lisle, Illinois 60532

THE GREAT CONTEMPORARY ISSUES

OTHER BOOKS IN THE SERIES

DRUGS
   Introduction by J. Anthony Lukas
THE MASS MEDIA AND POLITICS
   Introduction by Walter Cronkite
CHINA
   O. Edmund Clubb, *Advisory Editor*
LABOR AND MANAGEMENT
   Richard B. Morris, *Advisory Editor*
WOMEN: THEIR CHANGING ROLES
   Elizabeth Janeway, *Advisory Editor*
BLACK AFRICA
   Hollis Lynch, *Advisory Editor*
EDUCATION, U.S.A.
   James Cass, *Advisory Editor*
VALUES AMERICANS LIVE BY
   Garry Wills, *Advisory Editor*
CRIME AND JUSTICE
   Ramsey Clark, *Advisory Editor*
JAPAN
   Edwin O. Reischauer, *Advisory Editor*
THE PRESIDENCY
   George E. Reedy, *Advisory Editor*
POPULAR CULTURE
   David Manning White, *Advisory Editor*
FOOD AND POPULATION: THE WORLD IN CRISIS
   Sen. George S. McGovern, *Advisory Editor*
THE U.S. AND THE WORLD ECONOMY
   Leonard Silk, *Advisory Editor*

THE GREAT CONTEMPORARY ISSUES

SCIENCE IN THE TWENTIETH CENTURY

The New York Times

ARNO PRESS

NEW YORK/1976

WALTER SULLIVAN

Advisory Editor

509.04
S416

Copyright 1906, 1910, 1911, 1913, 1915, 1917, 1918, 1919, 1920, 1921, 1922, 1923, 1924, 1925, 1926, 1927, 1928, 1929, 1930, 1931, 1932, 1933, 1934, 1935, 1936, 1937, 1938, 1939, 1940, 1941, 1942, 1943, 1944, 1945, 1946, 1947, 1948, 1949, 1950, 1951, 1952, 1953, 1954, 1955, 1956, 1957, 1958, 1959, 1960, 1961, 1962, 1963, 1964, 1965, 1966, 1967, 1968, 1969, 1970, 1971, 1972, 1973, 1974, 1975 by The New York Times Company.

Copyright © 1976 by The New York Times Company.
Library of Congress Cataloging in Publication Data
Main entry under title:

Science in the twentieth century.

(The Great contemporary issues)
Collection of articles from the New York Times.
Bibliography: p. 381
Includes index.
1. Science-History. I. Sullivan, Walter.
II. New York times. III. Series.

Q125.S43434   509'.04   75-24607
ISBN 0-405-06672-4

Manufactured in the United States of America by Arno Press, Inc.

The editors express special thanks to The Associated Press, United Press International, and Reuters for permission to include in this series of books a number of dispatches originally distributed by those news services.

A HUDSON GROUP BOOK
Produced by Morningside Associates
Edited by Gene Brown

Contents

Publisher's Note About the Series	vi
Introduction, by Walter Sullivan	vii

1. Physics and Astrophysics — 1
- Matter and Energy — 2
- Pursuit of a Unified Field Theory — 64
- Atomic Energy: Promise and Threat — 80

2. Astronomy and Planetary Science — 91
- Probing the Universe — 92
- Exploring the Solar System — 123

3. Earth Sciences — 173
- Geology — 174
- Continental Drift — 187
- Oceanography — 203
- The Atmosphere — 216
- Climate — 231

4. Life — 237
- Genesis — 238
- Genetics — 250
- Manipulating Life — 283
- Evolution — 296

5. Science and Society — 327
- Funding and Organizing Research — 328
- Science and Politics — 346
- Religion and Science — 374

Suggested Reading — 381

Index — 383

Publisher's Note About the Series

It would take even an accomplished speed-reader, moving at full throttle, some three and a half solid hours a day to work his way through all the news The New York Times prints. The sad irony, of course, is that even such indefatigable devotion to life's carnival would scarcely assure a decent understanding of what it was really all about. For even the most dutiful reader might easily overlook an occasional long-range trend of importance, or perhaps some of the fragile, elusive relationships between events that sometimes turn out to be more significant than the events themselves.

This is why "The Great Contemporary Issues" was created—to help make sense out of some of the major forces and counterforces at large in today's world. The philosophical conviction behind the series is a simple one: that the past not only can illuminate the present but must. ("Continuity with the past," declared Oliver Wendell Holmes, "is a necessity, not a duty.") Each book in the series, therefore has as its subject some central issue of our time that needs to be viewed in the context of its antecedents if it is to be fully understood. By showing, through a substantial selection of contemporary accounts from The New York Times, the evolution of a subject and its significance, each book in the series offers a perspective that is available in no other way. For while most books on contemporary affairs specialize, for excellent reasons, in predigested facts and neatly drawn conclusions, the books in this series allow the reader to draw his own conclusions on the basis of the facts as they appeared at virtually the moment of their occurrence. This is not to argue that there is no place for events recollected in tranquility; it is simply to say that when fresh, raw truths are allowed to speak for themselves, some quite distinct values often emerge.

For this reason, most of the articles in "The Great Contemporary Issues" are reprinted in their entirety, even in those cases where portions are not central to a given book's theme. Editing has been done only rarely, and in all such cases it is clearly indicated. (Such an excision occasionally occurs, for example, in the case of a Presidential State of the Union Message, where only brief portions are germane to a particular volume, and in the case of some names, where for legal reasons or reasons of taste it is preferable not to republish specific identifications.) Similarly, typographical errors, where they occur, have been allowed to stand as originally printed.

"The Great Contemporary Issues" inevitably encompasses a substantial amount of history. In order to explore their subjects fully, some of the books go back a century or more. Yet their fundamental theme is not the past but the present. In this series the past is of significance insofar as it suggests how we got where we are today. These books, therefore, do not always treat a subject in a purely chronological way. Rather, their material is arranged to point up trends and interrelationships that the editors believe are more illuminating than a chronological listing would be.

"The Great Contemporary Issues" series will ultimately constitute an encyclopedic library of today's major issues. Long before editorial work on the first volume had even begun, some fifty specific titles had already been either scheduled for definite publication or listed as candidates. Since then, events have prompted the inclusion of a number of additional titles, and the editors are, moreover, alert not only for new issues as they emerge but also for issues whose development may call for the publication of sequel volumes. We will, of course, also welcome readers' suggestions for future topics.

Introduction

At no time in human history have the horizons of our knowledge concerning nature expanded with such awesome rapidity as during the past century. The discoveries, the brilliant insights, the misinterpretations, the bitter controversies have come in such quick succession that, for most of us immersed in the problems of the present, they are easily forgotten.

To freshen memories and provide insights into the present as well as the past, an effort is made in this book to present the most significant or telling developments as reported, on a day-to-day basis, by *The New York Times*. Instantaneous views of the passing scene often lack perspective — although a number of articles by such writers as Albert Einstein present a broader view. But seeing events through contemporary eyes offers special advantages. We easily forget "how things were." For this reason and because of their historical interest, accounts of findings that proved spurious have been included as well as those of lasting significance.

Some of the articles received little "play" — were treated editorially as of little consequence. Yet from hindsight certain articles prove remarkably farsighted. A prime example is that describing the predictions of Sir Oliver Lodge in 1919. Sir Oliver was both a distinguished physicist and a mystic who had served as president of the British Association for the Advancement of Science.

According to *The Times* account of September 18, 1919, conventional forms of energy were fast becoming exhausted. "The great source in the future, he thought, would be atomic energy, as this supply was inexhaustible." While it was still inaccessible, he made "a startling statement" in this regard. While only radium had been found to give off such energy, "it must not be thought that radium was the only matter that gave forth atomic energy. Such energy was present in most substances but was usually latent and we had not so far the means of bringing it into force. There were millions of foot tons of energy per ounce in substances at present inaccessible, enough energy to raise the whole German fleet from the bottom of the sea to the top of the Scottish mountains." At the end of World War I the High Sea Fleet of the Imperial German Navy after its surrender had been interned in the anchorage north of Scotland known as Scapa Flow while the Allies debated its disposition. On June 21, 1919, the skeleton German crews scuttled most of the ships and, when Sir Oliver made his comments three months later, they rested firmly on the bottom.

With regard to the possibility of releasing atomic energy, the article continued, "He hoped future investigations would lie in this direction, but he hoped, also, that the human race would not discover how to use this energy until it had brains and morality enough to use it properly, for, if the discovery were made before its time, and by the wrong people, this very planet would be unsafe." One cannot but wonder who today speaks with such prescience and hope that he is listened to.

The articles within each section are presented roughly in chronological order, but in some cases are grouped by subject. Those that deal with a variety of subjects, as in reports on a large, many-faceted meeting, have in some cases been edited to eliminate material of minor interest.

As reflected in the pages that follow, the past century marked the evolution of scientific research from experimentation that typically could be performed on a tabletop by one or two men to the present era when there is heavy dependence on mammoth devices. The latter include the telescope atop Mount Palomar in California, whose mirror is 200 inches in diameter, the particle accelerator at the Fermi National Accelerator Laboratory in Batavia, Ill., that is four miles in circumference, rockets that tower higher than the Statue of Liberty and a radio-radar antenna so large it fills an entire valley in Puerto Rico. While these devices are costly, they have enabled scientists to look far deeper into the atom and much farther out into space than ever before.

Hidden within the pages of this book are a variety of detective stories whose clues are provided by the day-to-day accounts of progress in particular fields of research. There are also false leads and when we read how wrong our predecessors sometimes were, one cannot but wonder what false doctrines lie within the current orthodoxy.

The first section of the chapter on "Physics and Astrophysics," dealing with "Matter and Energy," begins with the discoveries of X-rays and radioactivity at the turn of the century. Following Sir Oliver Lodge's forecasts regarding atomic energy is a statement, published in *The Times* the following day, reprimanding him for usurping the role of H. G. Wells as a science fiction writer. Nevertheless Frederick Soddy of the University of Glasgow, an authority on chemistry and radio-activity, reiterated the atomic energy concept in a lecture published under the not-so-modest heading: SCIENCE ON ROAD TO REVOLUTIONIZE ALL. That was in 1928. Professor Soddy had made the news eight years earlier by refusing to work on a War Office chemical warfare project. (See the account in the Science and Politics section of the final chapter.)

In retrospect the confirmation in 1919 of Einstein's prediction that stars near the sun would appear out of place during an eclipse was among the most significant landmarks in the history of physics, providing dramatic confirmation of his "general" theory of relativity — the one chiefly concerned with gravity, including the effect of solar gravity on passing light waves. The press accounts — LIGHTS ALL ASKEW IN THE HEAVENS — show that the impact was immediate and dramatic. Charles Lane Poor, professor of celestial mechanics at Columbia University, related Einstein's theory to the general period of upheaval following World War I. He is quoted as commenting that "the entire world has been in a state of unrest, mental as well as physical. It may well be that the physical aspect of the unrest, the war, the strikes, the Bolshevik uprisings, are in reality the visible objects of some underlying deep mental disturbance, world-wide in character." This spirit of unrest, he continued, had invaded science: "There are many who would have us throw aside the well-tested theories upon which have been built the entire structure of modern scientific and mechanical development in favor of psychological speculations and fantastic dreams about the universe." After reading about Einstein's theory and the idea of a fourth dimension he said he felt, as one senator stated his feeling after a "celebrated" dinner in Washington, "as if I had been wandering with Alice in Wonderland and had tea with the Mad Hatter."

By the 1930's such skepticism regarding the "new physics" had largely vanished from the scientific community and *The Times* reprinted an entire scientific communication from Einstein — one in which, despite his lack of enthusiasm for some aspects of quantum mechanics and its intrinsic elements of uncertainty, he and two colleagues argued that the uncertainty principle makes it impossible to determine with complete precision either the past path of a particle or its future behavior. The argument leads, he and his colleagues wrote, to "the remarkable conclusion" that this uncertainty also applies to larger-scale events, such as the precise times at which a camera shutter opens and closes.

In 1933 William L. Laurence, who had joined *The Times* three years earlier and later became Science Editor, described at length and with special elegance a lecture by Niels Bohr on the theory of complementarity that had grown out of the uncertainty principle. One of its central consequences is that physical phenomena that can be described in terms of momentum and energy can just as well be described in terms of wavelength and frequency.

This was a period of major developments in physics, as reflected in the news columns of that time: the discovery of the positron, the prediction and observation of the neutrino, and development of the mass spectrometer. Physicists were also moving toward the release of atomic energy. On February 2, 1932, it was reported that Cockcroft and Walton had transmuted hydrogen into helium. It is noteworthy that many of these stories originated in Britain, where much new ground in physics was being broken at that time.

Rereading the excited reports of 1939 — on splitting of the uranium atom, on proposals for uranium isotope separation and on the possibility of atomic bombs powered by a chain reaction — one is struck at how many ideas that later became "secret" had been widely publicized. At a meeting of the American Physical Society Niels Bohr told how bombardment of a small amount of uranium 235 would initiate an explosive chain reaction sufficient "to blow up a laboratory and the surrounding country for many miles." The same report noted the skepticism of some physicists as to the possibility of separating uranium 235, which constitutes less than one per cent of raw uranium. It was reported, however, that Lars Onsager of Yale University, later to become a Nobel laureate, described an apparatus whereby such isotopes might be separated in the gaseous state. (It was a thermal diffusion process that later proved impractical.) The chain reaction could be initiated by a single "slow" neutron penetrating the nucleus of a uranium 235 atom. A nuclear explosion sufficient to "wreck an area as large as New York City" could result.

A little more than four months later Germany invaded Poland and World War II began. On May 5, 1940, William L. Laurence published an extended discussion of atomic energy possibilities. "Germany is seeking it," said part of the headline. But then there was a sudden halt in such accounts. John J. O'Neill, Science Editor of the *New York Herald Tribune* and president of the

National Association of Science Writers, charged that censorship of all news on the subject represented "a totalitarian revolution against the American people." "Can we trust our politicans and war makers with a weapon like that? The answer is no," he said. Nevertheless, he added, the politicians have "taken over control" of those working in the atomic energy field "and are driving them to develop it for war uses."

Ultimately Laurence was brought into the project so that, once the bombs had been dropped, he could explain what had happened in lay language. His reporting, including a description of the first explosion at Alamagordo and an eyewitness account of the Nagasaki bombing, won for him his second Pulitzer Prize.

The postwar years were also marked by great moments, such as the overthrow of parity — the discovery that, except perhaps in very subtle ways, events on the atomic scale do not necessarily occur with equal frequency in right-handed and left-handed manners. On January 15, 1957, Columbia University held a press conference to make the announcement. Harold M. Schmeck, Jr., a new science writer on the paper, went to it but saw Laurence among the reporters — "Atomic Bill," as he was widely known by then — and assumed he would cover. Bill, however, as Science Editor, was now a sage and columnist rather than reporter. Back in the office elevator he waved to Schmeck and wished him luck with the story and Harold realized that one of the most difficult tasks in science writing had fallen on his inexperienced shoulders. He did a fine job, published the next day.

Another landmark was the discovery of a particle called the omega-minus, announced in February 1964, for its existence had been predicted by a new theory organizing the fundamental particles into families. There was a gap in the structure that could only be filled by a particle with certain characteristics and they proved to be those of the omega-minus. The theory had been developed by Murray Gell-Mann and Yuval Ne'eman. More recent has been the discovery of a very heavy but short-lived particle (called the psi at Stanford University and the J particle at Brookhaven National Laboratory). The finding was made almost simultaneously at both places.

There were also, however, premature or spurious announcements. The report of October 13, 1963, that the long-sought W particle or "intermediate vector boson" had been found proved unfounded. The particle would be the "vector" or embodiment of the force responsible for radioactivity — the so-called "weak" force. On June 27, 1966, various research centers called a press conference at Columbia University to announce the results of an elaborate experiment at Brookhaven National Laboratory that indicated an astonishing asymmetrical decay process (decay of the neutral eta meson) in which the resulting positive particles were more energetic than the negative ones. It, too, proved spurious.

There was the prolonged — and inconclusive — effort of Robert H. Dicke of Princeton University to show that the version of general relativity proposed by himself and Carl H. Brans was closer to the truth than that of Einstein. And there was the futile attempt by Joseph Weber of the University of Maryland to convince his scientific peers that he had measured gravity waves impinging on the earth.

Lurking in the minds of theorists is a deep-seated conviction that nature must be symmetrical, that it should not favor right over left, or vice versa, that in the universe as a whole positively charged material should balance that with negative charge — and even, perhaps, that the amount of antimatter should equal that of matter. There is also a belief that all natural phenomena should fit into a single theoretical structure that explains, for example, why the proton is 1836 times heavier than the electron, yet of equal (though opposite) electric charge. Or why the muon exists as a seemingly heavy electron. It should also spell out a rational relationship between the four basic forces of nature — the two forces that act at a distance and are familiar to everyone (gravity and electromagnetism) and the two forces that are effective only over tiny distances — that which binds together the particles of the atomic nucleus — the nuclear or "strong" force — and the "weak" force already mentioned.

Seeking such unified theories has been an obsession of such great physicists as Einstein (see his own exposition, published February 3, 1929, and a number of subsequent articles in the section on pursuit of unified field theories) and Werner Heisenberg, inventor of the uncertainty principle. None have been generally accepted. However one theory, proposed by Hideki Yukawa of Japan in 1950, was in a different category. In 1934 he had predicted the existence of particles intermediate in mass between electrons and particles of the nucleus (protons and neutrons). They were discovered two years later and named mesons. His effort in 1950 was to fit the mesons into a more complete theoretical framework. A more recent step toward unification has been evidence to support proposals of Steven Weinberg of Harvard University and others that the weak and electromagnetic forces may be closely related.

The post-war development of atomic energy for both military and civilian goals is extensively covered including early, nongovernmental reports indicating the wide extent of fallout from nuclear explosions in the atmosphere. These were followed by bitter debate as to the hazards of such fallout, culminating in the accord barring atmospheric tests, reported July 26, 1963.

The discoveries made by sky-watchers, as reflected in the chapter on "Astronomy and Planetary Science," rival in importance those made by the physicists. As bigger and bigger telescopes came into operation the vast scope of the universe became apparent. Edwin Hubbell realized that the Milky Way is not the only "universe" or galaxy of stars, but that other "island universes" exist far beyond it. An account on October 6, 1948, told of the birth of a new science — radio astronomy — whereby, as Laurence put it, astronomers can listen to "the music of the spheres." Radio antennas were able to spot sources of intense radio emission that were then shown by optical astronomers to coincide with peculiar blue, star-like objects. These were the quasars — objects that seem to be shining with extraordinary brilliance and are thus visible at far greater distances than anything else.

This was followed by discovery of the pulsars, taken at first to be beacons constructed by distant civilizations, and of the all-pervading "glow" in the sky that seems to be a residue of the fireball flash in which the universe was born. These and subsequent discoveries, particularly in terms of intense x-ray emissions from some parts of the sky, made more plausible the extraordinary concept of "black holes," largely attributable to J. Robert Oppenheimer.

The possibility of radio communication with civilizations in other planetary systems began to be seriously considered. In 1971 a conference on this subject, sponsored jointly by the American and Soviet academies of science was held in seclusion in Soviet Armenia. An on-the-scene account was published September 13, 1971.

To the man in the street the exploration of the solar system — including the close-up photography of planets by space-craft and the Apollo landings on the moon — was without question the most easily appreciated development of the period in science and technology. The progression went from firm reports in 1906 of canals on Mars, to the Mariner 4 and Mariner 9 missions that sent close-up pictures of that planet in July, 1969, and through much of 1972, showing no evidence for canals. The evolution of modern rocketry that finally carried men to the moon began with Robert H. Goddard's "moon rocket" tests reported on July 21, 1929.

The "Earth Sciences" chapter deals with the revolution in our understanding of our own planet that has occurred since the start of this century. It came about in part through the development of new tools, such as determining rock ages from their radioactive constituents. A section on continental drift deals with a subject to which this writer devoted much attention (see the book review of November 3, 1974). Wegener's proposal in this regard was described in *The Times* on March 25, 1923, and it was put to a vote at a heated meeting of the British Association, reported on September 2, 1950. The result was diplomatically described as a tie.

New tools made possible the exploration of the oceans and the atmosphere of those otherwise inaccessible realms. Echo sounders began to probe the bottom of the sea and heroic balloon flights, like that in 1935 by Captains Albert Stevens and Orvil Anderson, were made, although they are now almost forgotten. The pair reached a height close to 14 miles.

A 1906 account opens the chapter on life and discusses the long-standing debate as to whether or not life could arise — or ever have arisen — from inanimate matter. The "news peg" in this case was the report of an experiment in which radium chloride was sprinkled on sterile bouillon, resulting in the appearance of "germ-like" objects. It was asserted in a commentary that the little critters "are really not in the liquid at all, but are running riot through the brains of their discoverer and his disciples." The articles that follow recount the slow progression from viewing the spontaneous origin of life as an improbability to considering it having been a virtual certainty. Other articles cover such topics as the famous Miller experiment on spontaneous synthesis in a primitive atmosphere, the discovery through radio astronomy, that many of the basic constituents of life have evolved in space, that amino acids — the building blocks of all living protein — had been found in a meteorite.

The spectacular developments in genetics are traced from early efforts to find the chemical basis for heredity, to the announcement that Watson and Crick in England had shown molecules of the critical substance — DNA or deoxyribonucleic acid — to be formed in a double helix, like two interwoven spiral staircases. By early 1962 enough of the chemistry of heredity and its control of protein synthesis had been deduced to justify a special journalistic effort and on February 2 *The Times* published an extensively illustrated article, prepared by John Osmundsen of the science staff and covering almost one full page.

In 1967, when this writer visited the newly-formed Institute of Molecular Biology in Moscow he was told that the creation of the institute came about in part because of that article. It was argued by proponents of the institute that, as the article indicated, this was a new field of basic importance — and it was one in which the Soviet Union was lagging seriously.

"Manipulating Life" shows how this controversial field evolved from seemingly innocuous efforts to understand why some creatures can regenerate severed limbs whereas others cannot. Highlights include the dramatic achievement of Dr. John Furdon in England, who has been able to produce identical frogs by introducing nuclei from body cells into egg cells whose nucleus has been destroyed (reported October 7, 1968). By 1974 the techniques of gene transplantation and genetic manipulation had reached the stage where it was feared that microorganisms against which human beings have no resistance and which are immune to existing drugs might be created in laboratory experiments and escape to attack the populace. Efforts of world biologists to tighten rules on such research is reported.

The ''Evolution'' section includes a sympathetic review of the fifth edition of Darwin's *Origin of Species* in 1871, followed by an indignant review in *The Times* of London. The articles that follow trace the gradual extension of man's history to millions of years before the present, starting with early searches for the ''missing link,'' the discovery of Peking Man and exposure of the Piltdown Man hoax.

The chapter on ''Science and Society'' reminds us that the involvement of national governments in research began as early as World War I. There followed attempts in the Soviet Union to mobilize science for that country's political and economic goals. A by-product, tragic for Soviet science, was the decision of the government to make the unscientific doctrines of Trofim D. Lysenko the official genetics.

In the United States, following World War II, a historic decision was made to remove control of atomic energy from the Army, creating the Atomic Energy Commission (1947). This was followed in 1950 with creation of the National Science Foundation and, in 1957, the post of White House science advisor. The post was abolished by President Nixon but there were indications that it might be restored under President Ford.

''Science and Politics'' begins with an almost incredible account, from May 20, 1906, of the creation by the Assistant Secretary of Agriculture of a ''Heredity Commission,'' drawn from members of the American Breeders Association, to improve the quality of the American people. The project was to be one of ''encouraging the increase of families of good blood, and of discouraging the vicious elements in the cross-bred American civilization.'' Whatever became of this commission and its goals is not clear.

Accounts of activity that many regarded as the misuse of science in developing particularly insidious weapons begin with an article on German chemical warfare research published in 1918. One reads later of Soddy's refusal to take part in such work in Britain. Then there are accounts of the flight of scientists from Nazi Germany, followed after World War II by the agonizing confrontations of American scientists with the manipulators of anticommunist hysteria, culminating in the Oppenheimer case. On April 1, 1950, the newspaper told how 3,000 copies of *Scientific American* were burned on orders of the Atomic Energy Commission because that issue contained an article on the hydrogen bomb considered too revealing. Finally, there was the blacklisting of researchers, including one Nobel laureate, by the National Institutes of Health, presumably because they were outspoken against the Vietnam War and a nationwide work stoppage of scientists. One forgets what a turbulent century this has been.

The section on ''Religion and Science'' traces the battle of fundamentalists like William Jennings Bryan against Darwinism which, he said, ''makes the Bible a scrap of paper.'' The arrest of John Thomas Scopes was reported on May 7, 1925, and his subsequent trial for teaching evolution (at which Bryan and Clarence Darrow confronted one another) was reported. A quarter century later a Papal encyclical sought to bring the doctrine of the Catholic Church more into line with scientific beliefs on evolution and the age of the earth (reported November 23, 1951).

The record of scientific progress reported in these pages is perforce incomplete. Key steps in some cases passed unnoticed at the time. Developments in medicine and related fields have been relegated to another volume. And, because of space limitations, only a fraction of what was reported over the years could be included.

Walter Sullivan

CHAPTER 1

Physics and Astrophysics

Technicians remove a "hot" target after bombardment in a cyclotron.

Courtesy The New York Times.

MATTER AND ENERGY

ELECTRICITY AND X RAYS

THE SUNLIGHT CANNOT PRODUCE THESE PHENOMENA.

By Means of Electric Currents an Artificial Aurora Borealis Is Produced—Early Experiments with the Rays—Prof. Roentgen's Theory of a Change in the Rays—Proof that They Come from the Anode—How Crookes Tubes Are Made.

Many conflicting statements have been published as to the nature and source of the X rays. The cause of these lies in the existence of a large number of very ambitious men in the scientific field, whose knowledge and natural capacity are too limited to give them the position before the world for which they crave. Whenever any discoveries are made that attract unusual attention many of them eagerly seek cheap notoriety, either by publishing their views on the subject or by making experiments and talking of the results.

It is owing to the work of these men that so many statements have been made to the effect that X rays are to be found in light obtained from any source, and also that ordinary sunlight will photograph objects covered by opaque substances. The first announcement of this kind was that X rays could be obtained from the ordinary arc light; then some one else came forward with the statement that the incandescent

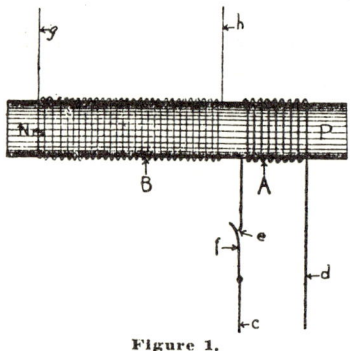

Figure 1.

light would answer just as well as the arc light. These statements were followed by another claiming that ordinary sunlight was as good a source of X rays as any other.

Among the first to clearly demonstrate that the photographs obtained by means of the arc or sunlight were not the same as those obtained by means of the X rays was Prof. W. M. Stion. The method used by Prof. Stion consisted simply in photographing on the same plate objects that are opaque to ordinary light, but transparent to the X rays, and other objects transparent to light and opaque to X rays.

The X rays cannot be called light rays in the ordinary sense of the word, because there is very little light about them. They are produced by the action of electric currents, but are not electrical. The form of electric energy used for their development is different from that used for ordinary purposes; it is different from that used for electric lighting, for operating trolley cars electric elevators, &c. The difference, however, is only in the pressure of the current and the steadiness of the flow. This can be made clear by a simple explanation.

An electric impulse is impelled through a wire by a force that is known in electrical science as electromotive force, and is measured in volts. It is similar to pressure in connection with water. If the hand is placed on a stream of water flowing from a nozzle with a low pressure it will not pro-

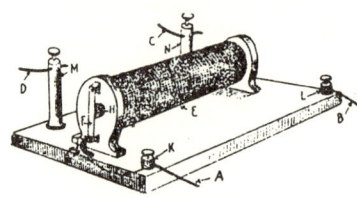

Figure 2.

duce any unpleasant effects; but if the pressure is suddenly increased the hand will be forced away with considerable violence, and perhaps pain and injury. This is due to the fact that the velocity of the stream is increased with the pressure, and, therefore, it acts more like a solid than a liquid.

If the pressure is increased sufficiently the force of the stream may be made such as to enable it to pierce a board or even a plank.

In order to make a current jump a considerable distance without using an enormously high pressure, the terminals of the wire between which it is to jump are placed within a tube from which the air is exhausted, so as to form nearly a perfect vacuum. Tubes of this kind are made by numerous manufacturers, and generally receive the name of the maker or designer. Those most commonly used are called Crookes tubes, as they are patterned after the design and made in conformance with the directions of Prof. Crookes, who originated them. But even with these tubes the electrical pressure required to make a current jump over a space of say two inches is very great, many thousand times as great as that used for incandescent lighting. To obtain such currents a special apparatus is used. It is known as an induction coil, because a comparatively low pressure current passing through it will induce a current of very high pressure to flow in a separate wire wound outside of the one through which the energizing current passes.

The principle of construction and operation of such an apparatus are both shown in the sample diagram, Fig. 1, in which N P is a bar formed of a large number of small iron wires packed within a tube. The lines c and d represent wires leading from an electric battery and forming a coil, A, around the bar. B represents a similar coil of wire, but composed of a much greater number of turns than A. Its ends, g and h, serve to convey the current generated in it (in the manner that will presently be explained) to any other apparatus in which it is to be used. The wire in both coils—that is, A and B—is made of copper, and is covered with cotton or silk, the latter material being used in all but the cheapest grade of machines. The object of this covering is to insulate the wire so that the electric current may not escape either into the iron bar or from one turn of the coil to another.

The action of the apparatus is as follows: The wires c and d are connected with the electric battery, the current flows through the coil A, and this converts the bar N P into a magnet for the time being. While the strength as a magnet of the bar is building up, a current of electricity is induced in the coil B and may be taken off by the wires g and h. The strength of the magnet N P builds up almost instantly, and therefore the current induced in B is only an instantaneous impulse, as no current is induced after B is magnetized to its maximum point. In order to obtain a continuous current, or rather a continuous succession of electric impulses, in the coil B, it is necessary that the current circulating through the coil A be stopped and started with great frequency. This is accomplished by using a spring, as shown at f, which is made to vibrate. At each vibration it strikes the end of the wire at e, and so long as the contact lasts a current flows through A. It ceases during that portion of each vibration of f when the contact with e is broken. In this way there is induced in the coil B a continuous succession of impulses.

The electrical pressure of the impulses induced in the coil B is in proportion to the number of turns of wire it contains, as compared with the number of turns in A. Thus, if the latter has ten turns, and the former 1,000, the pressure in the former (B) will be 100 hundred times as great as in the latter. By making B of a great many thousand turns, the pressure induced in it will be enormous.

Fig. 1, as explained in the foregoing, represents the principle of operation of an induction coil, but the construction used in practice differs from this diagram in the fact that the coil A is wound either under or over the coil B, and the spring F is so arranged that it may be kept vibrating automatically.

Fig. 2 shows the form of construction generally used for induction coils. The spool E has for a core the magnet for

N P of Fig. 1, which in this figure is only seen at one end, where it is marked H. The current from the battery comes in through the wire A to the post K, which is connected with the contact piece G through the wire shown by the dotted line. The spring F is connected with one end of the primary coil, which is marked A, in Fig. 1, and which in this figure is wound on the spool E next to the core H. When the current is off the force of the spring F keeps its lower end against the contact G. If the wires A and B are then connected with a battery the current will pass from A to G and, F, and through the coil around E to B, and thus back to the battery.

As soon, however, as the current passes around E the iron core becomes magnetized, and its end H attracts the spring F, thus breaking the contact at G. The current is

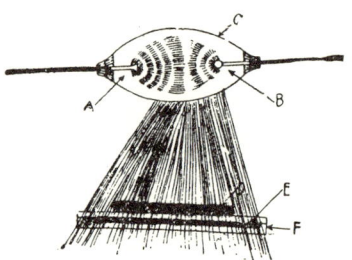

Figure 3.

then interrupted, and H loses magnetism, and therefore ceases to attract F. As a result, the latter swings back and comes in contact with G, thus re-establishing the current. The distance through which F swings, or, more properly speaking, vibrates, is very short, not much more than the thickness of a sheet of paper. Therefore, its motion is very rapid. In elaborate induction coils, such as are used for experimenting with the X rays, the spring F is not used, as it is desirable to have an apparatus of such construction that the rapidity with which the current is broken may be changed so as to determine what effect changes in this direction may have. The device used in place of F is a wheel, which is kept revolving by a small electric motor. This wheel is called a commutator. By changing its velocity the number of interruptions of the currents per second may be varied.

The current derived from the secondary coil of the induction coil, marked B in Fig. 1, is used in connection with a Crookes tube to develop the cathode and X rays. A Crookes tube, one form of which is shown in Fig. 3, is simply a glass globe, which may be of almost any shape desired. It has two wires fastened into it, at or near its opposite ends. It is made perfectly airtight, and before being sealed up as much of the air within it as is possible is exhausted, so as to form a nearly perfect vacuum. When the two wires, A and B, are connected with the wires C and D of Fig. 2, an electric discharge passes through the tube.

If the vacuum is sufficiently perfect, and the pressure of the current great enough, the ends of the wires A and B will become luminous and the space between them will be filled with a glow, which very much resembles the light of a well-defined aurora borealis. The amount of light given out by these tubes is very small, and can only be seen to good advantage in a dark room. If the tube is so constructed that it can be kept connected with an air pump while the current is passing through it, it is found that by increasing the vacuum the color of the light can be changed. The glow around the ends of the wires A and B is separated into bands with dark streaks between them. The change in the vacuum causes these bands to change their width, as well as the intensity and color of the light, and also to increase and decrease in number and change their shape and position. The appearance of these changing phenomena is very similar, but on a very small scale, to the changes noticed in a well-defined display of the northern lights.

The terminal A is called the cathode, and B the anode. For a number of years it has been known that the cathode could be made to emit invisible rays that would become visible if they impinged on certain substances, and also that they were capable of producing photographic effects. But this radiation, which has been called "cathode rays," could only produce these effects within the tube, the glass forming a barrier beyond which the rays could not pass. It was found that, while the rays would not pass through glass, they would pass freely through aluminium. Therefore, tubes were made with a thin sheet of this metal inserted in the glass, opposite the cathode. In this way the cathode rays were obtained outside of the tube, but their action was feeble.

Prof. Röntgen discovered that by properly adjusting the degree of vacuum in the tubes, it was possible to obtain a radiation that would pass through the glass, although not as well as through the aluminium, and thus made it possible to obtain very marked effects outside of the tube. These rays, known as X rays, differ from cathode rays. It was supposed by Prof. Röntgen that they came from the surface of the tube—that is, that the cathode rays in passing into the glass, or at the main surface, were in some way changed so as to be able to proceed in their course, and produce effects in the space surrounding the tube. The investigations of Prof. Rowland of Johns Hopkins University and one or two others lead to the belief that this conclusion of Prof. Röntgen was not correct, and that the real source of the X rays is the anode, or, in other words, that they are anode rays.

The way in which photographs are obtained with X rays is as follows: A photographic plate, E, shown in Fig. 3, is inclosed in a case made of ebonite or any other substance that is opaque to ordinary light. The object to be photographed is placed on the case, as shown at D. If this object is made of a substance that is opaque to X rays, it will cast a shadow on the photographic plate. As none of the X rays will reach the plate within the boundaries of the shadow, only the parts outside will be affected. Therefore, when the plate is developed the outline of the object D will be clearly defined. If D is not wholly opaque to the X rays, but allows some to pass through it, the shadow shown in the photograph will not be very dark. If the object, in addition to being somewhat transparent to the X rays, is of varying thickness, those parts that are the thinnest will be reproduced on the plate by the lightest shadows, and the thickest parts will be the darkest. Thus, not only the outline, but, to a considerable extent, the general shape of the object, will be defined.

If the photographic plate holder F is removed and a board, covered on its under side with certain chemicals, is put in its place, that portion of the board not covered by the shadow of the object D will become slightly luminous, or fluorescent, as it is called. If the object is composed of substances that are not equally opaque to X rays, there will be a certain amount of fluorescence within the shadows, so that the outline of the parts of different degrees of transparency may be easily defined. If the object D is the hand or arm of a man, the bones can be defined by the fact that the part of the board covered by their shadows will not be so luminous. If the fingers are moved, their shadow on the board will move. Therefore the motion of the bones may be seen by the eye, and in his way we can say that we can see through the body or any portion of it and observe the parts while in motion. But we do not in reality see through the body itself. We see only the shadow cast upon the board by the X rays that radiate from the tube and through the body.

April 12, 1896

RADIUM.

Much popular interest has of late been excited by the mention in cable dispatches and mail advices from England of experiments with the newly discovered metal radium, which has such surprising properties that Lord KELVIN was moved to say of it that it threatened to overthrow the law of the correlation of forces. It seems already to have overturned, or at least unsettled, the accepted theory of light, and when the experiments of Profs. MENDELEIFF, YEGOROFF, and BORGMAN of the Electro Technical Institute at St. Petersburg are completed, the results may be to give us a new science and a new nomenclature.

Radium is a rare metal, and extremely difficult to procure. It is a constituent of pitchblende, which is found in many places, but only in a very small way. All that has so far been segregated has come from a mine in Cornwall. A ton of pitchblende carries about 15½ grains of radium, and it is difficult to extract. This quantity, a gram by the metric scale, is at present estimated to be worth about $2,000, and a kilogram (2.2046 pounds) is theoretically worth somewhere in the neighborhood of $2,000,000. It has many curious and as yet inexplicable properties, and also entails many dangers to those who handle it carelessly. Prof. WILLIAM CROOKES, in describing it recently, said: " Probably if half a kilogram were in a bottle on that table it would kill us all. It would almost certainly destroy our sight and burn our skins to such an extent that we could not survive. The smallest bit placed on one's arm would produce a blister which it would need months to heal." This would seem to indicate that it emits something more than light. Heat and actinic energy must make up a large part of its radiation. It also emits electrons with a velocity so great that, according to Prof. CROOKES, " one gram is enough to lift the whole of the British fleet to the top of Ben Nevis; and I am not quite certain that we could not throw in the French fleet as well." This is popular rather than scientific, but it warrants the conclusion that radium will always be a labbratory metal, and that efforts to recover it in large quantities will not meet with much commercial encouragement. Perhaps the universal solvent might have been found long ago if there had been anything to keep it in.

Radium was discovered by M. and Mme. CURIE in France, after they had familiarized the remarkable properties of uranium and polonium. Its influence upon the development of electrical science promises to be very important. A quarter of a century ago it might have been said that electricity had a property somewhat resembling inertia. Now it looks as if Sir OLIVER LODGE had a substantial basis for the assertion that the inertia of matter will have to be explained electrically, since there is no inertia but electricity. This concept of electric inertia was first expressed in a magnificent mathematical paper by J. J. THOMPSON of Cambridge in 1881, when it was regarded as a mathematical curiosity. Radium promises to establish the Thompson hypothesis, and in so doing will possibly open the way to new and yet more important discoveries in a field in which the future doubtless has many surprises in store for even the present generation.

February 22, 1903

NEW PRIMARY ELEMENT

Another Extraordinary Discovery by Prof. and Mme. Curie.

Chemical Congress Astonished at Properties of Polonium — The Sterilization of Water with Ozone.

BERLIN, June 7.—At to-day's session of the Chemical Congress Prof. W. Markwald of Berlin showed the electro-chemical and physical section a smudge of dark powder on a piece of paper, which was the first time that any of the eminent scientists present had seen the metal polonium, discovered by Prof. and Mme. Curie of Paris.

Its discoverers doubted whether polonium was a primary element or related to bismuth, but Prof. Markwald demonstrated that it was indeed a primary element. He exhibited a bit of the metal, weighing 15-100 of a grain, which was produced from two tons of uranium at a cost of $75. It is more thinly distributed in uranium than xenon, the most rarified gas, is in the atmosphere.

Prof. Markwald proceeded to give a marvelous exhibition of the powers of his speck of polonium. It intercepted a strong current of electricity passing through the air from the generator to the receiver, the air ceasing to be a conductor for the flashes. The room was then darkened, and pieces of barium, platinum, and zincblende placed near the polonium glowed with a bright greenish light.

The assemblage of chemists was thrilled with astonishment. It appeared to be a miracle.

In the section of organic preparations, Prof. Proskauer of Berlin read a paper on the sterilization of drinking water with ozone and on ozone water works, the German electricians having succeeded in cheaply producing a concentrated solution of ozone. The speaker, with Profs. Ohlmüller and Prall of the Imperial Health Office, made exhaustive experiments with the solution in purifying water. The experiments included tests with water artificially impregnated with the deadliest disease germs, like typhus, cholera, and dysentery. Such water was pumped through the so-called " ozonizing tower," and then rigidly analyzed. All the germs were found to be killed, whereas the ordinary method of sand filters left the germs living. Moreover, the water was greatly improved in quality through the increase of oxygen from the ozone.

Prof. Proskauer said the ozonizing plant was cheaper than the sand filtering system, usuially used in city water works, hence the time had come for the general introduction of ozone plants. The town of Wiesbaden, added the professor, already had one of these plants, which sterilized 250 cubic meters of water hourly.

June 8, 1903

MYSTERY OF RADIUM.

Sir William Ramsay Describes Its Nature.

[By Sir William Ramsay, the Discoverer of Helium.]

The story of the discovery of radium is full of interest, and my readers may pardon me even if it is again told; for it forms the first chapter in a volume of which many have still to be written.

M. Henri Becquerel, prompted by a hint from the celebrated mathematician M. Poincaré, discovered that the compounds of uranium, a somewhat rare metal, as well as the metal itself, were capable of impressing a photographic plate wrapped up in black paper, or otherwise protected from light. It was also found that such salts, placed near a charged electroscope, discharged it, the gold leaves falling together. An electroscope, it may be explained, is a metal box with glass sides; through a hole in the lid a wire passes, the stopper which closes the hole and supports the wire being made of sulphur, or sealing wax, or some other material which does not conduct electricity. From the end of the wire are suspended two pieces of gold leaf, hanging down so as to be visible through the glass sides of the box.

GOLD LEAVES IN THE BOX.

If a piece of sealing wax is rubbed, so as to excite it electrically, and if the projecting end of the wire is touched with the rubbed sealing wax, a small charge of electricity is given to the wire, and through it to the gold leaves, so that they repel each other, and fly apart, making a figure like an inverted "V." If the wire is touched with the finger the electric charge is conducted away through the body, and the leaves swing back into their original position.

This effect of discharging was found to be produced when a salt or mineral containing uranium was placed inside the box. Mme. Curie, a Polish lady, living in Paris, noticed that the rate at which the gold leaves fell together was more rapid with certain uranium minerals (specimens of pitchblende) than could be accounted for by the uranium oxide in the mineral; she therefore separated the mineral into its groups of constituents—uranium, iron, lead, barium, bismuth, &c., (for the mineral contains all these and many other elements,) and tested each group as to its power of discharge. At first she thought that she had traced the discharging power to the bismuth group, and attributed it to an element which she named "polonium," after her native country.

OTHER ELEMENTS.

This discovery has not been disproved, but it appears that the amount of polonium obtainable is exceedingly small, and difficult to separate from bismuth. Subsequently Mme. Curie discovered another element of the barium group, possessing enormous powers of discharge, and to this element, which occurs in relatively greater amount, she gave the name "radium."

It is an undoubted element in the sense in which that term is generally used; its salts resemble closely those of barium, and its spectrum has been observed by M. Demarçay, Prof. Runge, and Sir William Crookes. Its atomic weight has been determined by Mme. Curie as 225; the atomic weight of uranium is the highest known—240, and there is some evidence from its spectrum that radium may have even a higher atomic weight—over 250—and that the sample analyzed by Mme. Curie may not have been quite free from barium, of which the atomic weight is only 137.

While these researches were in progress M. Curie and Dr. Schmidt discovered simultaneously that another element, thorium, of which the atomic weight is 232, also possesses the power of discharging an electroscope, and, moreover, that if air be led over salts of thorium, the air acquires and retains for a short time discharging power.

FURTHER DISCOVERIES.

The subject was taken up by Prof. Rutherford of Montreal and by Mr. Frederic Soddy, who then worked in his laboratory; and they found that if the "active" air were cooled by passing it through a tube cooled with liquid air it lost its "activity," the active portion remaining in the cold tube. On warming the tube the active portion was carried forward, and with it the discharging power. They also found that a similar "emanation," or gas, was evolved from salts of radium, possessing a much more permanent discharging power. While the "emanation" from thorium salts "decayed" in a few minutes, that from radium salts lasted a month. It, too, was condensible when cooled; it was luminous, and imparted temporary luminosity to objects which it touched. (" excited activity.")

The fact that a radium salt is always hotter than its surroundings, discovered by the Curies, implies that radium is continually losing energy; and if the radium salt be dissolved in water some of this energy is expended in decomposing a portion of the water into oxygen and hydrogen gases. Prof. Rutherford and H. T. Barnes have recently shown that "more than two-thirds of the heating effect is not due to the radium at all, but to the radioactive emanation which it produces from itself." In November, 1902, Messrs. Rutherford and Soddy concluded from their experiments on the emanations from radium and thorium that they are "inert gases, analogous in nature to the members of the argon family," and also they threw out the surmise, "whether the presence of helium in minerals and its invariable association with uranium and thorium may not be connected with the radio-activity."

NATURE OF ELEMENTS.

Now, I had the good fortune to discover helium in 1895; it is one of the argon gases, and is contained in certain minerals, and when Mr. Soddy came to work with me in the early Summer of this year, we tested the truth of this surmise, and we were rewarded by success. The fresh emanation from radium does not show the spectrum of helium, but as it "decays," helium is produced in minute bt ever-increasing quantity.

We can help ourselves by an analogy. Very complicated compounds of carbon and hydrogen can be produced; one containing 30 atoms of carbon and 62 atoms of hydrogen is known. But one of, say, 200 atoms of carbon and 402 of hydrogen would almost certainly fall to bits; it would split up and give out heat. The supposition appears reasonable that just as there is a limit to the possible number of atoms in such compounds, (for the molecules or groups of atoms fall apart by their own weight,) so there may be a limit to the atomic weight of an element.

Those elements with high atomic weight, such as thorium, uranium, and radium, are apparently decomposing into elements with low atomic weight; in doing so they give off heat, and also possess the curious property of radio-activity. What these elements are is unknown, except in one case; one of the products of the decomposition of the emanation from radium is helium.

Can the process be reversed? No one knows. But as gold is an element of high atomic weight, it may be confidently stated that if it is changing, it is much more likely that it is being converted into silver and copper than that it is being formed from them. At this stage, however, speculation is futile. It is certain that further experiment will lead to more positive knowledge of the nature of the elements, and of the transformations which at least some of them are undergoing.—William Ramsay in London Mail.

December 23, 1903

THE ENERGY OF RADIUM.

The romance stage of radium literature has passed and we have finally reached the stage of critical investigation tempered by a well-grounded skepticism that this curious mineral has given us a new theory of matter or upset any of the established axioms of modern science. No doubt the romance was vastly more interesting than the truth will ever be made, but that is beside the fact.

Lord Kelvin, for whom no one will claim infallibility of observation or conclusion, is still the greatest living authority in physics by reason of the fact that he is more often right and less often wrong than any other investigator. His study of radium leads him to regard with some impatience what he characterizes as the English explanation of its energy. This has been presented by Prof. J. J. Thomson of Cambridge in a theory which may be briefly stated as assuming that the energy of radium comes wholly from within. Lord Kelvin holds exactly the opposite view and affirms with confidence that it comes wholly from without. The discussion cannot be profitably followed in the columns of a newspaper, but if one not himself a learned physicist of international reputation may venture to pass judgment on the case as it stands, it is difficult to escape the conclusion that Lord Kelvin has the better of the argument. He cannot bring himself to believe that so enormous an amount of energy as radium displays can be evolved from within. One reason for doubting it is that if endowed with such a store of energy it could scarcely hold together in the form recognized as material and tangible. Hence, as its activity is exerted superficially, Lord Kelvin is sure that radium has the power of drawing its energy from some external source and transforming its manifestations into the familiar phenomena of radio-activity. The strength of Prof. Thomson's position is found in the number of new phenomena which are at the moment without satisfactory explanation, and concerning which any explanation if sufficiently fanciful commands respect and attention. Meanwhile, however, the more conservative students of nature who apply to the law of the conservation and correlation of force the excellent rule to prove all things and hold fast to that which is good are more and more convinced that radium does not, and later discoveries will not, invalidate the great generalization upon which Herbert Spencer built the synthetic philosophy and which has been the cornerstone of scientific thought for more than half a century.

August 1, 1904

PHYSICS AND ASTROPHYSICS

NEW FORCE PREDICTED BY SIR OLIVER LODGE

Atomic Energy Inexhaustible, but Would Imperil This Planet if Misused.

Copyright, 1919, by The New York Times Company.
Special Cable to THE NEW YORK TIMES.

LONDON, Sept. 17.—A great future for atomic energy was outlined by Sir Oliver Lodge today. He was lecturing on "Principles of Energy" at Midland Institute, Birmingham, in connection with the James Watt centenary memorial meetings. Molecular energy, he said, was beginning to show signs of exhaustion, and he instanced such energy as that supplied by coal. It was for scientific man to pursue his studies in other directions.

The great source in future, he thought, would be atomic energy, as this supply was inexhaustible, but at present it was in a sense inaccessible as they had not yet discovered means to make it accessible. Wireless telephony was the first instance of the utilization of the atomic properties of matter.

Sir Oliver made a startling statement in regard to the extraordinary power of energy such as that which emanates from radium, and said it must not be thought that radium was the only matter that gave forth atomic energy. Such energy was present in most substances but was usually latent and we had not so far the means of bringing it into force. There were millions of foot tons of energy per ounce in substances at present inaccessible, enough energy to raise the whole German fleet from the bottom of the sea to the top of the Scottish mountains. He hoped future investigations would lie in this direction, but he hoped, also that the human race would not discover how to use this energy until it had brains and morality enough to use it properly, for, if the discovery were made before its time, and by the wrong people, this very planet would be unsafe.

September 18, 1919

Perhaps the Ghosts Warned Him.
When Sir OLIVER LODGE expresses an opinion on matters purely scientific and lying in the domain of the demonstrable, he must be heard, of course, with the respect earned by his attainments and achievements, both of which are well known. But Sir OLIVER, to the regret of many of his cordial appreciators, of late years has shown more than enough interest in phenomena not beyond the field of science, indeed—no phenomena are that—but beyond the sort of investigation on which alone definite conclusions can be based. He is not alone among scientific men in doing this, but companions of anything like his own eminence in science can be counted on less than the fingers of one hand. His conclusions may be true, but they have no scientific support whatever, and unfortunately it is as a man of science that he demands their acceptance.

And now he is talking about atomic energy as an inexhaustible and enormous source of power, and expressing a fear lest that source be tapped before men are virtuous enough to be trusted with the increased potency for evil as well as for good that it would give them. These apprehensions somehow are remindful of Sir OLIVER's excursions into the Land of Spirits, and of the "evidence" he has brought back from that most misty region. Of course, atomic energy is real, and its amount he does not exaggerate, but—well, why did not Sir OLIVER permit its exploitation for destructive purposes to remain in the very competent hands of H. G. WELLS? For Mr. WELLS, too, is a man of science, in a way, but it is a way unlike what ought to be the way of Sir OLIVER, and his responsibilities are much lighter.

September 19, 1919

SCIENCE ON ROAD TO REVOLUTIONIZE ALL EXISTENCE

By Frederick Soddy, M. A., F. R. S.

> The following article was delivered by Prof. Soddy as a citizen's lecture in the Digbeth Institute, Birmingham, England, on Sept. 16.
> When it is remembered that the author is a scientist of international reputation, important and astounding in the extreme is the language which he uses to suggest the revolution in man's existence which will follow the finding of the secret of the artificial transmutation of the elements, a goal now before science as the result of the discovery of radio-activity.
> Prof. Soddy, who was born in 1877, is lecturer in physical chemistry and radio-activity at the University of Glasgow. Educated at Oxford, he was demonstrator in chemistry at McGill University, Montreal, from 1900 to 1902. He was the Wilde lecturer of the Manchester Literary and Philosophical Society in 1904, the same year going to Western Australia as university extension lecturer. In 1905-06 he was President of the English Röntgen Society. He has studied scientific investigation under Sir William Ramsay. He is the author of several works on radio-activity.

SEVENTEEN years ago not a single fact was known or discovery made on the subject of "The Evolution of Matter." Philosophical theories of the essential unity of all matter, and of the possible transformation of one kind into another, have come down to us from the ancients. Actual attempts at such transformation or transmutation constituted the pseudo-science of alchemy in the Middle Ages, which the modern science of chemistry has displaced and discredited. But until 1896 there was no hint that around and about us there was going on, at the present time, a slow, continuous transformation of certain of the elements into others.

This process is not a transmutation, certainly, in so far as that word implies the artificial change of one element into another at will. But in every other sense it is a veritable transmutation of the elements going on naturally and spontaneously, capable of being experimentally demonstrated and theoretically correlated with other departments of knowledge. Till 1896 the universal experience of physical and chemical science was that in all kinds of changes then known, and in all processes that had been studied, the elements remained essentially unchanged and unchangeable.

Might Transmute.

The efforts of the mediaeval alchemist toward transmutation seem to have been largely dominated with the idea that at a very high temperature one element, such as lead, might be changed into another, such as gold. But modern chemistry in this direction, by the aid of the electric furnace and in every other direction, has infinitely more powerful methods of experiment than were known to the alchemists. Still the elements resisted all efforts to decompose them or to change them into others, and the universal experience of science until 1896 has been that in all processes and changes to which it is possible to subject matter the elements remain essentially unchanged.

All matter is made up out of certain elementary substances, of which the number known is under 100, and the atoms or smallest parts of these elements, "the foundation stones of the natural universe," as Clerk Maxwell called them, to whatever treatment they may be subjected, "remain unbroken and unworn." This conclusion, though now challenged, still remains unshaken with regard to all the processes and changes known up to 1896, but since that time, thanks to the discovery of the entirely new phenomenon of radioactivity, new natural processes have come to light in which certain of the elements have been proved to be in course of change.

Found With Uranium.

The discovery of radioactivity was made by Becquerel with the element uranium, the element with the heaviest and most complex atom known, and shortly afterward it was found

FREDERICK SODDY, F.R.S.

that thorium, the element with the next heaviest and most complex atom, showed similar properties. These elements and all their compounds give out rays of a new kind, allied to the X-rays rather than to light in their power of penetrating opaque metals. Like the X-rays, these rays are invisible to the unaided eye. They are studied by three methods—the photographic method, which depends on their power of darkening a sensitive film in the same way as light; the fluorescent method, in which the rays cause certain fluorescent substances to glow; and by the electric or ionization method, dependent on the power of the rays to render gases conductors of electricity.

The use of these methods led to the discovery, mainly by M. and Mme. Curie, that in a natural state in minerals containing uranium and thorium these elements are associated with minute quantities of other extremely powerfully radioactive elements, radium, polonium, actinium, &c. So that it is possible to concentrate the radioactivity of large quantities of mineral, and to obtain minute preparations so powerfully active that some of the effects may be demonstrated to large audiences. With a few milligrams, for example, of a pure radium compound, the rays can be made visible to an audience through a thick plate of lead by means of a fluorescent screen.

The explanation of the phenomenon, due to Rutherford and Soddy in 1902, is that the atoms of the radio-elements are in process of spontaneous disintegration, in which small parts of the original atoms are expelled with enormous velocity as alpha-rays or alpha-particles, the residue of the atom forming an entirely new element, which usually is itself unstable and proceeds to undergo a definite sequence of similar successive disintegrations, sometimes alpha-rays, sometimes beta-rays being expelled at each change. The alpha-particles have been proved to be atoms of the element helium expelled from the atom of the parent element with a velocity varying from 9,000 to 13,000 miles per second.

The discovery that radium continuously generates helium, by Ramsay and Soddy in 1903, was the first direct proof that these radioactive processes are veritable transmutations. Their essential characteristic is their complete immunity from all external interference. The radioactive changes proceed at the same rate at the lowest as at the highest temperatures attainable, and are not in the least affected by the most powerful influences it is possible to bring to bear upon them.

The energy given out in these changes is, weight for weight of matter concerned, of the order of a million times as great as that given out in any previous change known. The energy that would be liberated in the course of the complete change of an ounce of radium, a process which would require some thousands of years, is equal to that furnished by ten tons of coal during combustion. If it could be made to disintegrate suddenly and explosively, instead of gradually, an ounce of radium would produce as much effect as many tons of the most powerful detonator known. So far all attempts to accelerate or retard the rate of disintegration have been failures, though many such experiments have been tried.

Results Would Be Vast.

Success would mean the accomplishment of a veritable transmutation in the alchemist's sense, and from the point of view of the energy liberated would be of infinitely more importance than the mere change of one element into another. Conversely, from the standpoint of modern chemistry, it is quite impossible to regard the radio-elements as differing entirely from the non-radioactive elements in regard to the amount of energy held within the atom, and which would be evolved if the atom were broken up. If it were possible to accelerate the rate of change of a radio-element, probably the same means would suffice to break up other elements at present entirely non-radioactive and unchanging, and to obtain from them at will quantities of energy of the same order as those obtainable from radium, and a million times greater than any at present utilized.

Considering how completely the face of the globe has been transformed by virtue of the proper utilization and control of the energy liberated in the combustion of fuel, it is not too much to say that the general solution of the problem of the artificial transmutation of the elements would absolutely revolutionize the whole condition of existence. Already it is becoming evident that these newly discovered processes play a large part in cosmical evolution. They alone, of all known processes, are at once powerful and enduring enough to supply the wealth of energy dissipated so prodigally throughout the universe over apparently endless periods of time. In comparison with the enormous amounts of energy stored up within the ultimate atoms of matter, and which would be liberated and rendered available if these could be artificially broken up at will, the sources at present available to man are insignificant indeed.

Yet their proper utilization has rendered the world of to-day barely recognizable as the successor of that of only a century ago. Physical science at least places no limit to the upward path of progress, and no end to the power and ascendancy over nature to which the race may in due course attain.

The progress and application of science resembles the increase of a capital sum under compound interest, slow and almost insignificant at first, but increasing always at an increasing rate without limit, changing and ever more rapidly rechanging the whole aspect of human affairs. Even now the pace is so rapid and the acceleration of the rate of progress so marked that the social and political systems of a less strenuous age are everywhere bursting like a chrysalis with the pressure of the new life within, while for the future, science promises an endless vista of new powers, new opportunities, and new thought. Yet this flood of progress, at least in one aspect, and that probably the most fundamental aspect of all, may be comprehended as a single whole and traced to a single origin. That origin is the general replacement of animate by inanimate sources of energy.

Unscientific man, the highest type of animal, naturally worshipped physical prowess and force; combined himself in communities to multiply his individual powers; enslaved the lower animals and the lower orders of his own species to carry on the heavy labor of the world; adjusted his courses to the favoring wind and the flowing tide; seized on the passing opportunities and the fluctuating currents of the time; and developed into a skilled master mariner of the well-pulled oar and the well-trimmed sail.

Scientific man drives straight on like a steamship. Beasts of burden, slaves, and other devices for increasing and multiplying the physical powers of the individual he has outgrown. Clans, communities, nationalities have outlived their primary purpose, and, in consequence, frontiers and territorial possessions are changing their political significance. The oars have been shipped and the sails, to catch the varying winds of circumstance, have been furled forever.

In short, man has secured the control of larger and purely inanimate sources of energy wherewith to carry on the main work of the world. The energy evolved by coal and fuel during combustion, and by the "white fuel" of foaming waterfalls, is transformed, subdivided, and retransformed in a thousand ways, until in one form or other it enters and lightens the labors of every enterprise and of every home.

Science Pressing Forward.

In almost any large power house at any time there may be seen a single machine doing the work of perhaps ten thousand horses or of a considerable army of laborers, energizing possibly a whole city, under the absolute guidance and control of a single artisan. Physical force, we are still sometimes told, is the ultimate arbiter of all human affairs, the final court beyond which no appeal even yet is possible. We spend well over a third of our revenue yearly on our war services in the attempt to maintain this archaic political philosophy in face of the limitless resources of inanimate energy which science is harnessing. In a very few years we shall be spending it all on this purpose if science continues to develop in the way it assuredly will.

The overwhelming march of science unaccompanied by any proportionate advance in the older departments of thought and philosophy, can only have one end. And as if to clinch the argument, there comes back the revelation from these distant horizons of science, to which our attention has been directed to-night, that the sources of inanimate energy which mankind controlled during the Victorian era, and which is coursing like fire in the veins of the present century, are but a secondary and insignificant offshoot of the primary tide. The main stream which vivifies and rejuvenates the whole universe passes by our very doors, and to its ultimate control and utilization it is now legitimate to aspire.

September 28, 1913

ALCHEMISTS' GOAL REACHED BY BRITON?

Paris Matin Says Sir Ernest Rutherford Has Discovered Transmutation.

RAMSAY MADE LIKE CLAIM

But British Chemist Died Without Making Full Reports of His Experiments.

PARIS, Dec. 8.—Sir Ernest Rutherford, since 1907 Langworthy Professor and Director of Physical Laboratories at the University of Manchester, England, has solved the riddle of the transmutation of matter, the secret sought by the ancient alchemists, according to the Matin.

Professor Rutherford as early as 1903 had divided the atom, or had at least extracted the Alpha particle from it, thus proving that the atom was not, as had been supposed, indivisible. At that time he was known as the greatest expert on radioactivity. The news published in Le Matin evidently shows that his experiments in the transmutation of so-called elements have been carried further than the non-conclusive experiments announced by Sir William Ramsay in 1907 and 1913.

The transmutation of matter, supposing the "matter" to be either physical mixtures or chemical compounds is no secret, but the production of a new element or the change from one to another has been the "riddle."

As early as the Spring of 1907 private advices received at Johns Hopkins University from Sir William Ramsay, who died in 1916, were interpreted to mean that the distinguished chemist of Cambridge University had succeeded in making the segregation of one element from another and the production of copper by the synthetic or combination process from the elements sodium, lithium and potassium. The process was said to include the treatment of these elements by radium.

Sir William did not then impart his formula and so left his critics with the alternatives of believing that copper was not an element, but a compound, or that it had actually been made from other elements, in which case the feat, attempted by alchemists and never heretofore accomplished by modern chemistry had been performed—the transmutation of elements.

The full report of the discovery was printed in the transactions of the Chemical Society of August, 1907. There the experiments were described in great detail, with the single conclusion which defied all tests—that Sir William had actually changed the element copper to the element lithium. The Lancet headed an article on the subject: "Modern Alchemy: Transmutation Realized," and was criticised by Lord Kelvin and others for so doing.

On June 1, 1913, Sir William delivered a lecture on the transmutation of elements at the Chemical Institute of Rome in the presence of King Victor Emmanuel. Here he adduced two fresh instances of transmutation, one of which he called conclusive, the other less so. He said:

"I introduced dry hydrogen into a tube of which the electrodes consisted of plates of aluminium. The anode was covered with a small coating of sulphur, and I subjected it to the action of the cathode discharge for five or six hours. I examined the gases extracted with a pump and did not succeed in finding either neon or helium, but exclusively argon mixed with hydrogen. Of the presence of argon I. entertain no doubt. * * *

"The last experiment I made refers to the action of cathode rays on selenium in an atmosphere of hydrogen. Having absorbed the condensable gases with cold charcoal, I removed the hydrogen as far as possible with a pump. I then drove the gases of the carbon by heating it in a vapor of boiling sulphur, the gas of which in volume did not exceed a few hundredths of a cubic millimeter. It showed lines of hydrogen and mercury, and, though very weakly, the characteristic yellow and green lines of kypton * * * but I admit that I am still not absolutely certain about it."

Sir Ernest Rutherford is one of the best known physicists in the world, holding degrees from more than a dozen universities in the old and new worlds, and having received medals of honor from a number of institutions of higher learning.

He has devoted much attention in recent years to radioactivity and has written several books on the phenomena of radioactive substances and their radiations.

December 9, 1919

LIGHTS ALL ASKEW IN THE HEAVENS

Men of Science More or Less Agog Over Results of Eclipse Observations.

EINSTEIN THEORY TRIUMPHS

Stars Not Where They Seemed or Were Calculated to be, but Nobody Need Worry.

A BOOK FOR 12 WISE MEN

No More in All the World Could Comprehend It, Said Einstein When His Daring Publishers Accepted It.

Special Cable to THE NEW YORK TIMES.
LONDON, Nov. 9.—Efforts made to put in words intelligible to the non-scientific public the Einstein theory of light proved by the eclipse expedition so far have not been very successful. The new theory was discussed at a recent meeting of the Royal Society and Royal Astronomical Society. Sir Joseph Thomson, President of the Royal Society, declares it is not possible to put Einstein's theory into really intelligible words, yet at the same time Thomson adds:

"The results of the eclipse expedition demonstrating that the rays of light from the stars are bent or deflected from their normal course by other aerial bodies acting upon them and consequently the inference that light has weight form a most important contribution to the laws of gravity given us since Newton laid down his principles."

Thompson states that the difference between theories of Newton and those of Einstein are infinitesimal in a popular sense, and as they are purely mathematical and can only be expressed in strictly scientific terms it is useless to endeavor to detail them for the man in the street.

"What is easily understandable," he continued, "is that Einstein predicted the deflection of the starlight when it passed the sun, and the recent eclipse has provided a demonstration of the correctness of the prediction.

"His second theory as to the anomalous motion of the planet Mercury has also been verified, but his third prediction, which dealt with certain sun lines, is still indefinite."

Asked if recent discoveries meant a reversal of the laws of gravity as defined by Newton, Sir Joseph said they held good for ordinary purposes, but in highly mathematical problems the new conceptions of Einstein, whereby space became warped or curled under certain circumstances, would have to be taken into account.

Vastly different conceptions which are involved in this discovery and the necessity for taking Einstein's theory more into account were voiced by a member of the expedition, who pointed out that it meant, among other things, that two lines normally known as parallel do meet eventually, that a circle is not really circular, that three angles of a triangle do not necessarily make the sum total of two right angles.

"Enough has been said to show the importance of Einstein's theory, even if it cannot be expressed clearly in words," laughed this astronomer.

Dr. W. J. S. Lockyer, another astronomer, said:

"The discoveries, while very important, did not, however, affect anything on this earth. They do not personally concern ordinary human beings; only astronomers are affected. It has hitherto been understood that light traveled in a straight line. Now we find it travels in a curve. It therefore follows that any object, such as a star, is not necessarily in the direction in which it appears to be astronomically.

"This is very important, of course. For one thing, a star may be a considerable distance further away than we have hitherto counted it. This will not affect navigation, but it means corrections will have to be made."

One of the speakers at the Royal Society's meeting suggested that Euclid was knocked out. Schoolboys should not rejoice prematurely, for it is pointed out that Euclid laid down the axiom that parallel straight lines, if produced ever so far, would not meet. He said nothing about light lines.

Some cynics suggest that the Einstein theory is only a scientific version of the well-known phenomenon that a coin in a basin of water is not on the spot where it seems to be and ask what is new in the refraction of light.

Albert Einstein is a Swiss citizen, about 50 years of age. After occupying a position as Professor of Mathematical Physics at the Zurich Polytechnic School and afterward at Prague University, he was elected a member of Emperor William's Scientific Academy in Berlin at the outbreak of the war. Dr. Einstein protested against the German professors' manifesto approving of Germany's participation in the war, and at its conclusion he welcomed the revolution. He has been living in Berlin for about six years.

When he offered his last important work to the publishers he warned them there were not more than twelve persons in the whole world who would understand it, but the publishers took the risk.

November 10, 1919

LIGHT AND LOGIC.

British scientists seem to have been seized with something like an intellectual panic when they heard of photographic verification of the Einstein theory, but they are slowly recovering as they realize that the sun still rises—apparently—in the east and will continue to do so for some time to come. "Perhaps the greatest achievement in the history of human thought" was the phrase used by the President of the Royal Society in describing to his parishioners what some cisatlantic skeptics have declared will turn out to be no more than an illustration of the somewhat well known fact that a coin lying at the bottom of a glass of water is not exactly where it appears to be. This explanation of the deflection of starlight in passing the sun rests, of course, on the assumption that it was not a gravitational attraction, but merely refraction in passing through gases of different density; but it seems to have been overlooked in London.

Even supposing, however, that light does not travel in straight lines—that is no reason why thought cannot travel in straight lines. Eminent men of science in their first alarm at the prospect of their gravitational universe collapsing about them declared that, if light rays are deflected by gravitation, space has its limits, and all straight lines are really curved and come back ultimately to their starting point. These gentlemen may be great astronomers, but they are sad logicians. Critical laymen have already objected that scientists who proclaim that space comes to an end somewhere are under some obligation to tell us what lies beyond it; and Euclid would hardly have admitted that a theoretical straight line must be curved merely because a line of light is curved, any more than because a road which appeared as straight on a large-scale map really had a few bends in it.

The Einstein theory, we are told, also involves the conception of the universe as four-dimensional, with time as the fourth dimension: a view which seems to have commended itself to science very recently, though H. G. Wells used it as the basis of an entertaining romance a quarter of a century ago. But this still fails to explain why our astronomers should appear to think that logic and ontology depend on the shifting views of astronomers. Dr. Einstein has perhaps proved that the stars are not where they appear to be, a fact known to any one who has watched the constellations rise and set. From the view that the stars are not where they seem to the layman, but are where they seem to the astronomers, to the recent demonstration that they are not where the astronomers think they are, is not so wide a step as an astronomer may think. It remains probable that the stars are somewhere, and if they are not it will require another science than astronomy to prove it. Speculative thought was highly advanced long before Anaxagoras got into trouble for suggesting that the sun might be as large as the Peloponnesus, and much of the thought of that day is useful still. A sense of proportion ought to be useful to mathematicians and physicists, but it is to be feared that British astronomers have regarded their own field as of somewhat greater consequence than it really is.

November 16, 1919

JAZZ IN SCIENTIFIC WORLD

Prof. Charles Lane Poor of Columbia Explains Prof. Einstein's Astronomical Theories.

WHEN is space curved? When do parallel lines meet? When is a circle not a circle? When are the three angles of a triangle not equal to two right angles?

Why, when Bolshevism enters the world of science, of course!

It is thus that Charles Lane Poor, Professor of Celestial Mechanics at Columbia University, explains the extraordinary cable announcements from London about Professor Albert Einstein's theories, which some suppose to have been verified by observations of the recent total eclipse of the sun. These observations were assumed to show that the rays of stars were deflected as they passed the sun, which led to the Q. E. D. that they were subject to the attraction of the sun, that is to gravitation; and from this premise it was easy to jump to the conclusion that Sir Isaac Newton's theory had been knocked to smithereens.

Well, Sir Isaac, after he saw the apple fall in his garden at Woolsthorpe, and evolved therefrom his theory of gravitation, couldn't prove it for a long time. He made his calculations from a wrong estimate of the radius of the earth; and it was not until years later, when another scientist had corrected the figure for the radius, that he was able to give the gravitational principle to a shocked and incredulous world. Once the incredulity had evaporated in the light of proof, and the theory had become an established fact, it still was not immune from mistaken attack, as Professor Poor points out.

"For some years past," Professor Poor said the other day, after reading the cable dispatches about the Einstein theory, "the entire world has been in a state of unrest, mental as well as physical. It may well be that the physical aspects of the unrest, the war, the strikes, the Bolshevist uprisings, are in reality the visible objects of some underlying, deep mental disturbance, worldwide in character. This mental unrest is evidenced by the widespread intent in social problems, by the desire, on the part of many, to throw aside the well-tested authors of Governments in favor of radical and untried experiments.

"This same spirit of unrest has invaded science, and today there is just as great a conflict in the realm of scientific thoughts as there is in the realm of political and social life. There are many who would have us throw aside the well-tested theories upon which have been built the entire structure of modern scientific and mechanical development in favor of psychological speculations and fantastic dreams about the universe.

"Whenever a new observation is made which apparently does not directly fit into the old-time theories these modern disciples of scientific unrest rush into some weird explanation, involving psychological speculations as to the constitution of matter or our fundamental concepts of mathematics.

"The eclipse observations reported to have been made on May 29 last are a case in point. If these observations are as reported (and such seems unquestionably to be the case), then these explanations, under present accepted theories, may be difficult, but such observations certainly do not warrant the acceptance of the speculations of Einstein.

"It may be that history is merely repeating itself. When Newton's theory of universal gravitation was given to the world in 1685 it was received with incredulity, especially among scientists on the Continent of Europe. Observations were adduced which these scientists asserted, proved the fallacy of the Newtonian ideas. One by one these observations were shown to be in harmony with the law, to be direct consequences of it.

"Nearly one hundred years later (1770) Euler, one of the greatest mathematicians of the age, who had devoted a lifetime to developing and perfecting the Newtonian theory, in discussing the observed motion of the moon, wrote:

"'There is not one of its equations about which any uncertainty prevails, and it now appears to be established by indisputable evidence that the secular inequality in the moon's motion cannot be produced by the forces of gravitation.'

"The essay in which this statement was made appeared during a time of profound mental and political unrest, such as now pervades the world. It won the prize of the Paris Academy of Sciences. To explain this peculiar motion of the moon, the greatest scientists of that age adopted theories involving a resisting medium in space, or introduced a time element into gravitation. Yet only a few years later Laplace found a full and complete explanation in certain intricate relationships between the motion of the moon and the varying shape of the earth's orbit, which had been overlooked by Euler and his followers, and found that this motion was a direct result of the forces of gravitation.

"Now, the so-called Einstein theories, or rather speculations, are such as completely to overthrow not only the law of gravitation, but the fundamental conceptions on which all geometry and physics rest. And to sustain such a complete overturning of the entire basis on which scientific thought has been built, two—just two—observed facts are quoted: the motion of the perihelion of Mercury and certain displacements of stars when photographed near the sun.

"There is no need to go outside the law of gravitation to explain the motion of Mercury's perihelion. The explanation may well be in some term of the most complicated formulas which the mathematicians have overlooked or in some distribution of matter near the sun which the astronomer has hitherto failed to properly note. As a matter of fact, in order to make their equations usable, the mathematical observer assumes that the sun is a perfect sphere and that the space between the sun and the planets is empty. Yet both these assumptions are known to be false; the well-known sun spots and the many photographs of its corona prove the sun to be not perfectly spherical and to be surrounded by an irregular and changeable mass of matter. The real trouble is that the mathematicians have not yet been able to introduce the effects of these into their equations and to deduce their possible effects upon the motion of Mercury.

"The displacements of the stars noted in the recent eclipse photographs may be a phenomenon analogous to the refraction of light. All rays of light, when they pass from one medium to another, from air to glass, for example, are bent or refracted. Upon this principle are based the ordinary eyeglass, or the telescope. When the rays from the stars enter the earth's atmosphere they are bent and travel in curved paths. Now, the sun is surrounded by an envelope of gases of irregular shape and of varying densities, an envelope which certainly extends to the orbit of the earth, and probably millions of miles beyond. Would it not be in accord with all known laws of optics if the rays of light from distant stars were bent and refracted when passing through such an envelope?

"The fact that such a bending effect has now been measured is of great scientific importance, and the results may change some of the hitherto accepted ideas as to the density and distribution of matter near the sun, but I fail to see how such an observation can prove the existence of a fourth dimension, or can overthrow the fundamental concepts of geometry.

"I have read various articles on the fourth dimension, the relativity theory of Einstein and other psychological speculation on the constitution of the universe; and after reading them I feel as Senator Brandegee felt after a celebrated dinner in Washington. 'I feel,' he said, 'as if I had been wandering with Alice in Wonderland and had tea with the Mad Hatter.'"

November 16, 1919

EINSTEIN EXPOUNDS HIS NEW THEORY

It Discards Absolute Time and Space, Recognizing Them Only as Related to Moving Systems.

IMPROVES ON NEWTON

Whose Approximations Hold for Most Motions, but Not Those of the Highest Velocity.

INSPIRED AS NEWTON WAS

But by the Fall of a Man from a Roof Instead of the Fall of an Apple.

Copyright, 1919, by The New York Times Company
Special Cable to THE NEW YORK TIMES.

BERLIN, Dec. 2.—Now that the Royal Society, at its meeting in London on Nov. 6, has put the stamp of its official authority on Dr. Albert Einstein's much-debated new "theory of relativity," man's conception of the universe seems likely to undergo radical changes. Indeed, there are German savants who believe that since the promulgation of Newton's theory of gravitation no discovery of such importance has been made in the world of science.

When THE NEW YORK TIMES correspondent called at his home to gather from his own lips an interpretation of what to laymen must appear the book with the seven seals, Dr. Einstein himself modestly put aside the suggestion that his theory might have the same revolutionary effect on the human mind as Newton's theses. The doctor lives on the top floor of a fashionable apartment house on one of the few elevated spots in Berlin—so to say, close to the stars which he studies, not with a telescope, but rather with the mental eye, and so far only as they come within the range of his mathematical formulae; for he is not an astronomer but a physicist.

It was from his lofty library, in which this conversation took place, that he observed years ago a man dropping from a neighboring roof—luckily on a pile of soft rubbish—and escaping almost without injury. This man told Dr. Einstein that in falling he experienced no sensation commonly considered as the effect of gravity, which, according to Newton's theory, would pull him down violently toward the earth. This incident, followed by further researches along the same line, started in his mind a complicated chain of thoughts leading finally, as he expressed it, "not to a disavowal of Newton's theory of gravitation, but to a sublimation or supplement of it."

When he read in the message from THE TIMES requesting the interview a reference to Dr. Einstein's statement to his publishers on the submission of his last book that not more than twelve persons in all the world could understand it, coupled with the editor's request that Dr. Einstein put his theory in terms comprehensible to a larger number than twelve, the doctor laughed good-naturedly, but still insisted on the difficulty of making himself understood by laymen.

"However," he said, "I am trying to talk as plainly as possible. To begin with the difference between my conception and Newton's law of gravitation: Please imagine the earth removed, and in its place suspended a box as big as a room or a whole house, and inside a man naturally floating in the centre, there being no force whatever pulling him. Imagine, further, this box being, by a rope or other contrivance, suddenly jerked to one side, which is scientifically termed 'difform motion,' as opposed to 'uniform motion.' The person would then naturally reach bottom on the opposite side. The result would consequently be the same as if he obeyed Newton's law of gravitation, while, in fact, there is no gravitation exerted whatever, which proves that difform motion will in every case produce the same effects as gravitation.

"I have applied this new idea to every kind of difform motion and have thus developed mathematical formulas which I am convinced give more precise results than those based on Newton's theory. Newton's formulas, however, are such close approximations that it was difficult to find by observation any obvious disagreement with experience.

"One such case, however, was presented by the motion of the planet Mercury, which for a long time baffled astronomers. This is now completely cleared up by my formulas, as the Astronomer Royal, Sir Frank Dyson, stated at the meeting of the Royal Society.

"Another case was the deflection of rays of light when passing through the field of gravitation. No such deflections are explicable by Newton's theory of gravitation.

"According to my theory of difform motion, such deflections must take place when rays pass close to any gravitating mass, difform motion then coming into activity.

"The crucial test was supplied by the last total solar eclipse, when observations proved that the rays of fixed stars, having to pass close to the sun to reach the earth, were deflected the exact amount demanded by my formulas, confirming my idea that what so far has been regarded as the effect of gravitation is really the effect of difform motion. Elaborate apparatus and the closest and most indefatigable attention to the difficult task enabled that English expedition, composed of the most talented scientists, to reach those conclusions.

"Why is your idea termed the theory of relativity?" asked the correspondent.

"The term relativity refers to time and space," Dr. Einstein replied. "According to Galileo and Newton, time and space were absolute entities, and the moving systems of the universe were dependent on this absolute time and space. On this conception was built the science of mechanics. The resulting formulas sufficed for all motions of a slow nature; it was found, however, that they would not conform to the rapid motions apparent in electrodynamics.

"This led the Dutch professor, Lorenz, and myself to develop the theory of special relativity. Briefly, it discards absolute time and space and makes them in every instance relative to moving systems. By this theory all phenomena in electrodynamics, as well as mechanics, hitherto irreducible by the old formulae —and there are multitudes—were satisfactorily explained.

"Till now it was believed that time and space existed by themselves, even if there was nothing else—no sun, no earth, no stars—while now we know that time and space are not the vessel for the universe, but could not exist at all if there were no contents, namely, no sun, earth, and other celestial bodies.

"This special relativity, forming the first part of my theory, relates to all systems moving with uniform motion; that is, moving in a straight line with equal velocity.

"Gradually I was led to the idea, seeming a very paradox in science, that it might apply equally to all moving systems, even of difform motion, and thus I developed the conception of general relativity which forms the second part of my theory.

"It was during the development of the formulas for difform motions that the incident of the man falling from the roof gave me the idea that gravitation might be explained by difform motion."

"If there is no absolute time or space, supposedly forming the vessel of the universe," the correspondent asked, "what becomes of the ether?"

"There is no ether, as hitherto conceived by science, which is proved by the well known experiment of the celebrated American savant, Michelson, showing that no influence by the motion of the earth on the ether is perceptible through change in velocity of light, such as ought to be produced if the old conception were true."

"Are you yourself absolutely convinced of the correctness of this revolutionary theory of relativity, or are there still any reservations?"

"Yes, I am," Dr. Einstein answered. "My theory is confirmed by the two crucial cases mentioned before. But there is still one test outstanding, namely, the spectroscopic. According to my theory, the lines of the spectra of fixed stars must be slightly shifted through the influence of gravitation exerted by the very stars from which they emanate. So far, however, the results of the examinations have been contradictory; but I have no doubt of final confirmation, even through this test."

Just then an old grandfather's clock in the library chimed the mid-day hour, reminding Dr. Einstein of some appointment in another part of Berlin, and old-fashioned time and space enforced their wonted absolute tyranny over him who had spoken so contemptuously of their existence, thus terminating the interview.

December 3, 1919

EINSTEIN ABSORBED BY QUANTA THEORY

Partly Atones for Upsetting Newton on Gravity by Developing Sir Isaac's Lead.

STUDIES HAVE LINK HERE

Dr. Sommerfeld Follows Langmuir's Research on the Atom—Thinks Eclipse Will Uphold Relativity.

Dr. Albert Einstein, propounder of the relativity theory, is much occupied at present with the study of electrons and with the quanta theory, according to Dr. Arnold Sommerfeld, Professor of Mathematical Physics at the University of Munich, who is in this city on his way to the University of Wisconsin, where he will hold the Carl Schurz Memorial Professorship this Winter.

In working on the quanta theory, which is the theory that light and other radiations do not consist of vibrations or waves only, but of flying particles of energy, Einstein is making partial reparations to Newton, Dr. Sommerfeld pointed out, for the damage which he did to the Newtonian theory of gravity. Newton, who was the first profound investigator of light, held that light consisted of flying particles or corpuscles which produced the sensation of light when they struck the eyes. He was overruled in favor of the wave theory.

"The wave theory explains most of the behavior of light, but not all of it," said Dr. Sommerfeld. "Planck's researches on heat indicated that light did not consist only of waves spreading out uniformly in all directions, but that it consisted partly at least of concentrated particles or quanta of energy. X-ray and other researches have indicated this even more strongly. Newton's theory was that light consisted of flying particles, but his theory was undeveloped and bears no great resemblance to the quanta theory. Einstein is doing much work in this field.

"This is closely connected with the study of the structure of the atom and I follow with great interest the work of your physicist Langmuir at Schenectady, although Bohr's theory and mine differ from his. Langmuir holds that the electrons of an atom are in a fixed position, one to another. We hold that they are in regular motion with relation to each other."

Thinks Eclipse Will Uphold Einstein.

Dr. Sommerfeld said that he entertained no doubt that Einstein's theory would be fully confirmed by the observations of the eclipse next Thursday which will be made by American, English and Dutch-German astronomical parties in Australia and the Indian Ocean. He said that he was convinced that eventually the third prediction of Einstein would also be confirmed by Dr. Charles E. St. John at the Mount Wilson observatory.

This prediction was that a slight difference would be found between a light originating on the sun and a similar light produced on earth, the difference being due to the retarding effect which, under the relativity theory, the sun's mass should have on the vibration of the atoms producing light. He said that physicists at Bonn had shown that only six of the thousands of lines in the sun's spectrum could be safely used for the experiment, because disturbances to which other wave lengths of light are subject would make it impossible to detect the Einstein effect among the other lines. Dr. Sommerfeld added that Professor Birge of the University of California had narrowed the number of test lines from six to two.

"There is no doubt," said Dr. Sommerfeld, "that among other excellent observatories the one at Mount Wilson is the chosen forum before which this important question will be decided; partly owing to its excellent staff and partly owing to its unrivaled equipment.

"It seems to me that the problems of atomic structure and quanta will necessarily be of greater interest than the relativity theory to the physicist of the future, inasmuch as they afford him the opportunities of comparing theory with experiment, whereas the theory of relativity will probably occupy the attention chiefly of mathematicians and philosophers, and perhaps astronomers." Speaking of the condition of science in Germany since the war, Professor Sommerfeld said:

"Experimental work is, of course, much hampered, owing to the high cost of apparatus and the difficulties that stand in the way of young research students who cannot afford to spend longer time at the university than is absolutely necessary.

"Even men who before the war judged themselves comparatively wealthy are now reduced to dire poverty and have to seek any available means of replenishing their incomes. Notwithstanding these obstacles there are a number of institutes which succeed in carrying on good work, as, for example, in the realm of spectroscopy and electronic collisions. work that has greatly stimulated the scientists of your own country.

Quanta Theory Interests Germans.

"On the other hand, theoretical investigations have never stood in such high favor and have never attracted so many workers as at the present time. A possible reason for this is the great popularity of the theory of relativity, works about which already form a library in themselves. On the other hand, studies in the theory of quanta have played a part almost equally important in the German universities. It is a matter of satisfaction to a German professor like myself that these two important branches of modern physics have emerged in the last decades out of the minds of fellow-countrymen.

"I am happy to express how great is the interest of German scientists in the present good work that is being done in America, especially in the experimental and technical branches of physics. Owing to the enormous cost of foreign books and periodicals, the German scientist is, unfortunately, almost entirely dependent on external sources, and in many cases on the good-will of American colleagues. My own institute is fortunate in having been particularly well treated in this respect, and, although no German hopes that the present economic situation will be reversed, we do trust that it will be sufficiently improved before long to bring the prices of journals within reach of German purchasers."

September 18, 1922

SCIENTISTS WITNESS SMASH-UP OF ATOMS

View Extraordinary Photographs of Crash of Projectiles at 20,000 Miles an Hour.

ONE BEHAVES STRANGELY

Hydrogen Atom, Hit by Helium Atom, Travels in Direction From Which Blow Came.

CIVILIZATION'S REAL TASK

Carnegie Institute Head Asserts Energy Must Be Maintained to Prevent Degeneration.

Special to The New York Times.
CAMBRIDGE, Mass., Dec. 27.—Extraordinary photographs of the smash-up of atoms were displayed to chemists of the American Association for the Advancement of Science here today in the course of a session at which many of the chief American investigators of atoms and the electrons that compose them were present.

New calculations as to the size of different atoms were laid before the chemical section by Dr. Theodore W. Richards of Harvard. A new periodic table of elements was explained by Dr. W. D. Harkins of the University of Chicago, which arranges the atoms and subvarieties of atoms called isotopes. Strictly speaking, there are no longer a mere group of ninety-two atoms, according to the old perodic table, but a total of several hundred. For practical purposes, however, the old list of ninety-two holds good, because the new atoms, or isotopes, have no characteristics of their own except minute differences in mass. For instance, lead is never found as a single element, but is a combination of different varieties of lead atoms, each having apparently the same physical and chemical natures, except for the slight difference in mass.

Photographs of two new types of atomic collisions were showed by Dr. Richards. These were taken by the Wilson and Shimidzu methods of photographing an alpha particle, otherwise known as double charged helium nucleus, during its flight through damp air.

One He Can't Explain.

"I have not been able to explain this one," said Dr. Richards, showing a picture of the helium atom striking a hydrogen or nitrogen atom.

This showed a helium atom traveling in a straight line after sideswiping a hydrogen atom, while the hydrogen atom which had been hit went backward instead of forward. Instantly on being struck the atom had started to travel backward in the direction from which the blow came. The atom which had delivered the blow in the meantime continued ahead on a straight course.

The atomic projectile made a speed of 20,000 miles an hour, and in other cases drove its targets in all directions except backward. Sometimes, in the case of a head-on collision, the projectile itself rebounded, but its target was knocked forward.

This unusual effect occurred only once in 10,000 photographs of crashing atoms, and thousands more are to be taken, Dr. Harkins said, to see whether repetitions of the effect throw any further light on the behavior of an atom.

The other type of photograph which is not understood resembles a follow shot in billiards. The projectile, after striking another atom, proceeds in almost the same path. The target atom also moves forward in almost the same path. There were several photographs of this effect.

Photographs Are Vivid.

The atom itself is about a million times too small to be seen with the naked eye, but the high-speed alpha particle, or helium nucleus, which is used as a projectile, is able to produce effects which can be seen and photographed vividly. The particles are bombarded from radium or some other radio-active substances into an enclosure which is full of air supersaturated with water. All dust particles have been pumped out previously so that there is nothing for the water to condense on and form drops.

When there is no dust for it to deposit itself on, water vapor will throw itself as drops on the trail of an alpha particle. Air molecules are electrically charged or transformed into ions by the speeding atom, and the water vapor will seize on an ion when dust particles are absent.

A fine mist is thus set up along the trail of the projectile and along the secondary trail made by the target after it has been knocked forward or to one side at high speed. Under a strong light these moisture lines can be beautifully photographed. The head-on collisions have been shown to knock some of the lighter atoms to pieces, thus producing the artificial disintegration of atoms first shown by Sir Ernest Rutherford. This rarely occurs, however, because of the smallness of the bodies involved. The alpha particle, or double charged helium nucleus, occupies only about one-millionth part of the space of the ordinary atom, which consists of one or more outer rings of atoms as well as the nucleus. The nucleus weighs 1,740 times as much as the outer shell, but is less than one-millionth as big.

The very small pellets has to penetrate the outer shell of another atom and hit the nucleus before the break-up of the atom can take place. Direct hits on this bull's-eye are seldom made. The helium nucleus is so small that on its flight through an inch of air it passes on the average 500,000 atoms before it hits one. Photographs and even motion pictures of these atomic collisions are being taken by the thousand in the hope of getting a good picture of a head-on collision resulting in a complete smash-up of the target and an artificial transmutation of the target atoms into hydrogen and helium.

The helium atom is apparently made of four hydrogen atoms, but the atom of helium has less mass than four atoms of hydrogen. This would mean that mass was lost, but that would conflict with the theory that mass is indestructible. But what has happened is that the lost mass is transformed into energy and energy is mass.

The new measurements which Dr. Richards gave for atoms was based on a study of their compressibility. Usually when atoms unite they pull toward each other and occupy a smaller space then they did when they existed separately. Dr. Richards sought to determine how much different elements could be compressed by their own force in these combinations.

Previous measurements of the atom had made the average about a hundred millionth part of an inch. The measurements of Dr. Richards does not disturb these dimensions greatly, but is supposed to give them with greater refinement.

Isotopes, or the variations in the atoms of the same element, were discussed at greater length by Dr. Robert S. Mulliken, of the "Isotope Laboratory" of the University of Chicago, who said that no laboratory method had yet been discovered of separating isotopes so completely that the isotopes could be individually studied. So far the quantities of isotopes which have been isolated have been too small for examination.

"While isotopes are very much like each other in their behavior," he said, "there are indications that this agreement is not quite complete and they will be studied for slight differences in their properties."

December 28, 1922

DR. BOHR EXPOUNDS THEORY OF ATOMS

Begins a Series of Lectures at Yale on His Study of Revolving Electrons.

LIKENED TO SOLAR SYSTEM

He Pictures the Atom With Nucleus Corresponding to Sun, and Electrons to Planets.

Special to The New York Times.
NEW HAVEN, Nov. 6.—Dr. Nils Bohr, Professor of Physics at the University of Copenhagen and winner of the Nobel Prize in Physics for 1922, began a series of six lectures at Yale University today in explanation of his theory of the structure of the atom, which has been accepted by many scientists as the most plausible hypothesis yet put forward. These scientists believe that Dr. Bohr's study of the revolving electrons inside the atom and his theory of the similarity of the atom to the solar system have constituted a considerable advance on the path toward solving the riddle of the universe.

The title which Dr. Bohr has given to his course of lectures is "The Atom and the Natural System of the Elements." He will lecture tomorrow and on Thursday afternoons and on Tuesday, Wednesday and Thursday afternoons next week. Today's lecture was largely introductory. He discussed the old ideas of the atom and showed how various scientists had laid the foundation, with one discovery after another, for new ideas which require the rejection of many of the old teachings in mechanics and electro-dynamics. He showed how the use of a "picture" of the atom itself, as the result of recent discoveries, had revolutionized the study of physics, and had made it possible to begin the study of the universe from the smallest particles of matter, instead of working down from large masses to the minute.

Dr. Bohr, using diagrams on a blackboard, pictured the atom according to his theory and emphasized its resemblance to the solar system, with the nucleus at the centre of the atom representing the sun, and with electrons revolving around the nucleus in the place of the planets.

Theory of "Explosive Atoms."

The Danish scientist also explained in part the theory of "explosive atoms," throwing off minute projectiles traveling at a speed of several thousand miles a second, though they are too small to be seen. Delicate modern machinery, he said, had enabled scientists to determine the changes in the nucleus made by these explosions, which he regarded as an important advance in the study of physics.

Dr. Bohr said that he would confine himself in his lectures to a discussion of the natural system of the atoms—such features of the problem as showed relationships between the chemical elements. He traced the history of the atomic system from Mendeleeff to the present, showing by charts and diagrams how many investigators had come closer and closer to the solution of the problem.

According to Dr. Bohr, the scientist Balmer first discovered, in the study of the spectrum of hydrogen, that the spectrum lines followed a very simple formula, and first found indications of the simplicity of hydrogen atoms. Rydberg then examined the spectra of other elements, the lecturer went on, and found remarkable resemblances to the hydrogen spectrum. The investigation of spectra by X-ray, he continued, had helped to simplify the problem further.

After this summary of the history of his subject, Dr. Bohr went on:

"The object of these lectures will be to tell what light has been thrown on this system of the elements by development of the atomic theory. A list of what has been done in the last twenty years is really a remarkable one. We assumed twenty years ago that we might be able to work down some day to a picture of the atom from larger bodies of matter; but now we have been presented with a very definite picture of the atom from other sources.

"All doubt regarding the existence of the atom has disappeared. Furthermore, we have learned how to count atoms with great accuracy. Experiments in the discharge of electricity through gases have led to the discovery of the electron as the common constituent of atoms of all elements. The atom of electricity has been isolated, and we have been able to understand the discharge of electricity through gases.

"Next the discovery of radioactivity and the atomic theory which I have sketched here (pointing to the diagram on the blackboard). We now have a picture of the atom which we believe—I think all physicists now believe—is just as real as any of the natural phenomena that we are in the habit of discussing."

"Open Structure" of the Atom.

Explaining the diagram, which consisted of a big dot in the centre surrounded by two rings, on each of which were placed four smaller dots, at great distances from one another, Professor Bohr explained that the principal feature of the structure of the atom, according to his theory, was its openness. In this respect, he pointed out, it bore a "very close resemblance to the solar system."

The theory of "explosive atoms" was touched upon by Dr. Bohr in a discussion of the radioactive properties of the atom. He said that the nucleus had a diameter of only one-hundred-thousandth part of an atom, but that the refinement of instruments of measurement by Rutherford had made it possible not only to determine the number of electrons in an atom but even to measure the changes in a nucleus following "explosions" with great accuracy.

Dr. Bohr said that fundamental difficulties in the way of understanding the atom existed in the old conception of mechanics and electro-dynamics. He pointed out that, although the atom was similar to the solar system in some respects, it was dissimilar in others; therefore, the ordinary laws of mechanics and electro-dynamics could not be used in the study of the atom.

Under the old laws of science, he went on, it was entirely impossible to explain the stability of the atom in spite of "explosions." If the ordinary laws held good for the atom, he declared, electrons would fall into the nucleus "and destroy the life of the atom," very much as if the earth or some other planet fell into the sun and destroyed the life of the universe.

Beginning with his lecture tomorrow, Dr. Bohr said, he would explain the laws which he had worked out in his theory of the structure of the atom to account for its stability.

"I will show in what ways we shall have to make departures from the ordinary ideas of mechanics and electrodynamics," he said. "This is something that could not have been learned before, because we have been obliged in the past to form our laws on large masses of matter and have never before been able to experiment on such a small scale. But the ability to experiment on a small scale is the characteristic feature of the development of the atomic system.

Problem of Physics Inverted.

"Now that we are actually familiar with the structure of the atom, the problem of physics has been inverted. Our object now is to find the fundamental laws that govern the workings of such systems on a small scale, and I will try to tell tomorrow how we have advanced in this direction."

He added that he would describe his new hypothesis based on the quantum theory, to explain the stability of the atomic system according to his views.

Dr. Bohr was introduced this afternoon by Dr. James R. Angell, President of Yale University, who described the scientist as "the winner of the blue ribbon of modern science," referrring to the Nobel Prize.

Dr. Bohr's course is part of the Silliman lectures, an annual feature at Yale. His lectures are delivered in the Sterling Chemical Laboratory, which has a capacity of 300. The hall was crowded this afternoon with students, professors and those of the general public specially interested in the subject. Experts from electrical companies from New York and elsewhere were present.

Professor John Zeleny, Chairman of the Physics Department of Yale University, made the following comment this evening on Dr. Bohr's lectures:

"The discovery that all atoms are made up of negatively charged electrons surrounding a positively charged nucleus has given us the possibility of explaining the properties of all substances by the interaction of these two simple constituents. A closer study of the problem has revealed the fact that the classical laws, which have proved adequate for the treatment of problems dealing with gross matter are not able to account fully for the processes within the atomic structure. To explain the new facts, assumptions have to be made which it is not possible at present to reconcile with the older views.

"It is one of Professor Bohr's great achievements that he had the daring to disregard these older views and to make a new formulation which in a simple way leads to the explanation of many of the properties of the chemical elements. The new formulation and the classical laws are each admitted to be incomplete pictures of a reality the deeper meaning of which is not yet understood. We are fortunate in this course of Silliman lectures to have the new ideas presented to us by their great apostle."

November 7, 1923

Dr. Millikan Discovers Strange New Rays, 10 Miles Above Earth, Coming From the Void

MADISON, Wis., Nov. 10 (AP).—Scientists attending the National Academy of Sciences here are seeking a name for the powerful new rays discovered by Dr. R. A. Millikan of the California Institute of Technology. It is probable that the rays, still in an early stage of development and offering as yet little basis for a theory, will become known as Millikan rays, in honor of the man who first found authentic evidence of their existence.

The new rays are "in-coming"; that is, they form outside or finite space. In this fact, and in the fact that they originate in the passing of atoms over atoms, scientists proclaim something novel in rays.

What the force collects in the way of electrons or other elements will be matter for years of investigation. Uses for the rays are not yet known. In fact, so little is known of their character that the discoverer is not indulging in theories as to what the future will bring to science when they have been further studied.

The power character of the rays, if it continues to increase, might have a disastrous result to life upon earth, although scientists discussing the rays today do not believe this would happen.

The reason for this lies in the statements of Dr. Millikan that while the frequency of the "in-coming" rays is great and their action is going on all the time, there is not enough of them to produce any tangible result on earth at this time. The tests were made more than ten miles above the surface of the earth by instruments attached to balloons.

The sun, it was stated, has no effect on the atomic action producing the rays.

The rays are believed by Dr. Millikan to have extraordinary absorbing power. Discoveries of the scientist are the result of research since the close of the World War, following out theories first advanced by German scientists on the possible existence of other rays in space.

Scientists attending the academy believe the discovery of Dr. Millikan to be one of the outstanding scientific achievements of the year.

In describing his discovery yesterday Dr. Millikan said that the new ray had 1,000 times the frequency of the X-ray, with ionization the same at all times of the day or night and of 10,000,000-volt variety. The rays are due to atoms passing over to other atoms, with the sun having no effect on the action, he said. They appear throughout space and bombard the earth from all directions at all times.

Determining the velocity of light to a minute degree of accuracy is a possibility which may be established early next year, Dr. A. A. Michelson, head of the Department of Physics of the University of Chicago, said. He has come within "plus or minus" twenty miles a second of establishing the true velocity, the result of experiments with octagon mirrors in two mountains in the Sierra Nevadas.

November 11, 1925

CALLS LIGHT WAVE AN ATOMIC SHUDDER

Energy Is Transformed Into Ether Vibration, Asserts Dr. Millikan at Yale.

OUTLINES RADIATION TESTS

Pasadena Scientist Holds Recent Experiments There Disprove Old Periodicity Theory.

Special to The New York Times.

NEW HAVEN, Conn., April 9.—Physical science has had to be completely revised, Professor Robert A. Millikan, director of the Norman Bridge Laboratories of Pasadena, declared in an address at the Sheffield Scientific School in the Lee De Forest course last evening.

He described a group of new experiments in the spectroscopy of the extreme ultra-violet, which has been under intensive study in the Norman Bridge Laboratory, in which, he said, it was definitely shown that two different electrons can simultaneously jump from two different levels to two energy levels and in that act emit a single monochromatic ether wave or ray of light.

"In other words," Dr. Millikan said, "the nineteenth century conception that the periodicity directly observed in a monochromatic ether wave corresponds with and has its origin in a vibrating body within the atom, which vibrates in synchronism with the emitted ether wave, must apparently forever be discarded.

"All the evidence of spectroscopy of the past few years justifies the assertion that the atom in the act of emitting radiation simply transforms the energy of an atomic shudder into a monochromatic.

"We are completely unable at present to form any mechanical picture at all of how this is done. In other words, the nature of the act by which an atom in the sun sends out its characteristic frequency is as yet incomprehensible to us.

"But when we turn to the wide field of modern science we have something which is brand new under the sun, conceptions of which preceding races never dreamed. The conceptions of Galileon and Newtonian mechanics, introduced about 1500, not only revolutionized the material world but they wrought the most profound changes in the philosophy and religion of mankind.

"We built up around these ideas a view of the physical world which we thought only thirty years ago was incapable of further change, at least in the fundamental elements, but within the last thirty years practically all the conceptions which underlay all the thinking of the late nineteenth century have had to be revised. Scarcely one of the generalizations then current is now regarded as of universal validity."

The change in conceptions concerning the nature of radiant energy was the field which Dr. Millikan took for particular discussion in detailing the group of new experiments.

Dr. Millikan is a winner of the Nobel Prize for Physics.

April 10, 1926

DETAILS CONCEPTS OF QUANTUM THEORY

Heisenberg of Germany Gives Exposition Before British Scientists.

PUTS STRESS ON "ENERGY"

Is Mathematical Principle of Revolutionary Effect Comparable Only to Relativity.

By WALDEMAR KAEMPPFERT.
Copyright, 1927, by The New York Times Company.
By Wireless to THE NEW YORK TIMES.

LEEDS, England, Sept. 1.—Of thirty addresses delivered today before the various sections of the British Association for the Advancement of Science, one of the most important was that of a young German, Dr. W. Heisenberg. Fully 200 mathematical physicists listened to his brief exposition of a conception which will make it necessary to modify belief in what we are pleased to call "common sense" and "reality."

The layman without a knowledge of higher mathematics, listening to Dr. Heisenberg and those who discussed his conclusions, would have decided that this particular section of the British Association is composed of quiet and polite but determined lunatics, who have created a wholly illusory mathematical world of their own. The conception is that they and their kind alone have a proper view of "reality"; the rest of us live in a dream world fashioned by ill-understood words.

To explain the quantum theory and its modification by Dr. Heisenberg and others is even more difficult than explaining relativity. It is much like trying to tell an Eskimo what the French language is like without talking French. In other words, the theory cannot be expressed pictorially and mere words mean nothing. One is dealing with something that can be expressed only mathematically.

The consequences, however, are startling. Elections and atoms cease to have any reality as things that can be detected by the senses directly or indirectly. Yet we are convinced the world is composed of them.

Action Supersedes Substance.

In the new mathematical universe events are more important than substances, and energy more important than matter. All mental pictures we have formed of bodies moving through space are thrown into confusion. So simple a conception as a baseball flying from the pitcher to the batter turns out to be obscure, doubtful and even ridiculous.

Planck, the originator of the quantum theory, Heisenberg, Schroedinger and De Broglie have shown that the whole science of mechanics must be rewritten. And when it is rewritten, no one but a mathematician will be able to understand it. The scientific world is faced with an upheaval as great as that brought about by Einstein.

* * * * *

September 2, 1927

TO PRESS RESEARCH ON LIGHT AS MATTER

Professor Compton, Nobel Prize Winner, Sees Recoil of Electrons as Test of Theory.

Special to The New York Times.

CHICAGO, Nov. 19.—Professor Arthur H. Compton of the Physics Department of the University of Chicago, joint winner with Professor C. T. R. Wilson of Cambridge, England, of the Nobel Prize in Physics, has announced that he would carry his experiments a step further in developing the theory that light waves are really "light bullets," acting like projectiles, and that light rays consist of particles of matter.

Professor Compton, who is 35 years old, was eight years ago research physicist for the Westinghouse Lamp Company, but gave up the chance of a profitable commercial career to devote his attention solely to research.

It was Einstein, Professor Compton pointed out, who gave the world ten years ago the theory of light as consisting of streams of particles. In testing this Professor Compton discovered what has come to be called the "Compton effect," which shows that X-rays bouncing off a substance in their path lose some of their energy and become longer in wave length, wave length being in inverse ratio to energy. He explained it this way:

"One's finger placed in the path of an X-ray becomes a source of scattered or reflected rays. We should expect, if X-rays were waves, that the scattered wave lengths are the same length as the waves that produced them. Spectra, however, prove that a part of the X-rays is of distinctly greater wave length than the parent primary rays, and this has been called the 'Compton effect.' This observed change in wave length is directly contrary to the prediction of the wave theory.

"When the X-ray is considered as a particle which is deflected when it collides with the matter through which it passes, then the electron which it strikes must recoil. This was the next part of our experiment to prove that the electron which was hit flew off at the appropriate angle. Professor Wilson of Cambridge University, the other winner of the prize, had developed a method for photographing the trails left by individual electrons passing through the air. With this we found indisputable evidence of the existence of the recoil electrons.

"The obvious interpretation is that X-rays, light rays and radio rays—for they are all the same kind of thing in varying intensity—are streams of particles. Light waves are nothing more than the successive sheets of light particles which may be compared with sheets of drops in a rainstorm."

November 20, 1927

CREATION CONTINUES, MILLIKAN'S THEORY

Cosmic Rays Herald 'Birth of the Elements,' Scientist Says Experiments Tend to Show.

BORN IN STARS AND NEBULAE

Electrons Unite There to Form Helium and Others—Warns to Await More Proof.

Special to The New York Times.

PASADENA, Cal., March 17.—Discovery of evidence tending to show that the process of creation is now going on in the heavens and that the earth, instead of being a disintegrating world, as has long been believed, is a continuously changing and evolving one, was announced by Dr. Robert A. Millikan of the California Institute of Technology last night at a meeting of California Institute Associates. The meeting was held at the home of Mr. and Mrs. Albert B. Roddock in San Marino.

Dr. Millikan's announcement followed his new measurements of cosmic rays, which he has not yet completed. He said that these investigations showed that the cosmic rays, instead of being spread widely, consisted of definite bands of color like the light from a neon lamp.

They had frequencies identical with those which would result from the loss of mass in accordance with the equation of Einstein, thus upholding the latter's theory, the scientist said.

Warning to Await Proof.

At the same time Dr. Millikan issued a warning against accepting his announcement as one of fact until it had been proved by further tests, but he declared that his discoveries were the first indication that the creative process was actually going on now and that ordinary elements were being formed continually from electrons.

Dr. Millikan classified the cosmic rays which have been the subject of his measurements as "announcements of the birth of the elements." His statement said, in part:

"Through new and more precise measurements on cosmic rays than those heretofore made, Millikan and Cameron have just succeeded in bringing forth quantitative evidence that those rays represent the precise amount of energy which should, according to Einstein's equation showing the relation of mass to energy, be emitted in the form of ether waves when the primordial positive and negative electrons unite to create helium atoms and other light atoms such as oxygen and silicon, magnesium and iron.

Tests in Mountain Lakes.

"Millikan and Cameron have investigated these rays through experiments in high mountain lakes, both in California and in Bolivia, and Millikan and Bowen have studied them with the aid of self-recording electroscopes sent up by sounding balloons which reached nine-tenths of the way to the top of the earth's atmosphere.

"The results obtained in such investigations during the past eight months constitute the first indubitable evidence that the cosmic rays on which they have been experimenting. Instead of being spread like white light over a considerable spectral region, consist of bands of definite frequency, or color, like the light from a neon lamp or from a Cooper-Hewitt mercury arc.

"The general spectral region, however, in which these bands are found, corresponds to frequencies 100,000,-000,000 times greater than those emitted by the aforementioned lamps. This is why these cosmic radiations are powerful enough to penetrate 200 feet down into a mountain lake before they are completely absorbed.

Sees Four Main Radiations.

"The rays brought to light by this most recent work correspond to four main radiations extending over a spectral region three octaves wide and having frequencies identical with those which are computed theoretically from the loss of mass which would occur in accordance with the foregoing equation of Einstein, first, when the helium atom is created out of the nucleus of the hydrogen atom (the positive electron) two negative electrons acting as the binding agents; second, when oxygen and nitrogen atoms are created out of hydrogen; third, when silicon and magnesium are so produced, and fourth, when the atom of iron is born.

"Hydrogen and helium are extraordinarily abundant gasses, while the four elements—oxygen, magnesium, silicon and iron—are the most abundant elements found in meteorites and a not unlike percentage of the earth. The agreement between the observed and computed frequencies is so good as to make it highly improbable that it represents an accidental coincidence.

Calculations Detailed.

"The quantitative nature of the agreements obtained is illustrated as follows: While the atomic weight of hydrogen is 1.00778, the atomic weight of helium is 4.00054; when helium is created by the union of four hydrogen atoms an amount of matter disappears which is equal to four times 1.00778.

"The difference—namely, .03058 grams—must, according to Einstein's equation (MC 2-E) go off in the form of radiant energy when the helium atom is formed, and the appearance of this amount of energy in the form of a monochromatic ether wave would give that ether wave the penetrating power which is represented by an absorption coefficient numerically equal to .305.

"This is within a few per cent. of the absorption coefficient directly observed by Millikan and Cameron for the most conspicuous band in their cosmic ray spectrum.

"There is, further, a philosophic argument which supports the results of this observation. We have long known that all elements have a structure which indicates that they are exact multiples of the mass of the positive electron, which is the nucleus of the hydrogen atom.

Expectation of Building Up.

"We have also known for thirty years that in the radio-active process the heavier atoms are disintegrating into lighter ones. It is, therefore, to be expected that somewhere in the universe the building-up process is going on to replace the tearing-down process represented by radio activity.

"Up to the present, however, no evidence had ever been found that this building-up or creative process is going on now. The present experiments constitute the first discovery of such evidence.

"It must be taken with some reserve and must be subjected to further critical analysis and further experimental tests. But, so far as they go, these experiments are at least indications, and the first indications, that all about us, either in the stars, the nebulae or in the depths of space, the creative process is going on, and that the cosmic rays which have been studied for the past few years constitute the announcements broadcast through the heavens of the birth of the ordinary elements out of positive and negative electrons.

Hypothesis Held Plausible.

"When it is remembered that the positive electron is the nucleus of the hydrogen atom, and that all the spectroscopic survey of the heavens shows the extraordinary abundance everywhere of hydrogen; and when we reflect that we have known for fifteen years that all the elements have weights that are practically exact multiples of the weight of the hydrogen atom as it appears in the structure of helium, the foregoing conclusion that the process of atom-building out of positive and negative electrons (the latter have a mass that is negligible in comparison with the former) is now going on gains additional plausibility.

"If it is confirmed it will constitute new proof that this is a changing, dynamic and continuously evolving world instead of a static or a merely disintegrating one.

"Further qualitative support for the validity of the foregoing evidence is derived from the fact that so far as we can now see there are no sort of nuclear changes which could take place powerful enough to produce the observed cosmic rays except those herewith suggested.

"Putting together, then, the quantitative and the qualitative evidence, we may have some confidence in the conclusion that the heretofore mysterious cosmic rays, which unceasingly shoot through space in all directions, are the announcements sent out through the ether of the birth of the elements."

March 18, 1928

4,500 Battle in Museum to See Einstein Film; Police Quell Stampede After 8 Guards Fail

A stampede of 4,500 persons struggling to get into the main lecture hall of the American Museum of Natural History to witness a free showing of a motion picture on the Einstein theory last evening got beyond control of the museum's uniformed guards and necessitated the calling of reserves from the West Sixty-eighth Street police station.

Although many of the glass cases containing exhibits were endangered and the clothing of many of the participants was torn in the crush, the police finally restored order after twenty minutes of hard work without serious damage and with no injuries reported. No arrests were made.

The Amateur Astronomers' Association had sent out 1,500 invitations to its members in this city to attend the showing of the picture at 8 o'clock last night. As that hour approached a throng, estimated by the police at 4,500, gathered at the main entrance of the museum. At the request of Miss M. Louise Rieker, secretary of the association, J. B. Fahlk, superintendent of the building, sent eight uniformed guards among the crowd to inform them that preference would be given ticket holders.

As the word spread the crowd began indignantly hooting the guards. A sudden rush from those behind sent the forefront of the throng against a grilled iron gate leading from the main lobby into the large exhibit room at the front of the museum, where a collection of Alaskan Indian data is housed.

The gate was broken down by the pressure and the milling crowd poured into the passageway, about eight feet wide, leading through the exhibit room to the main lecture hall, where the picture was to be shown. They buffeted the glass exhibit cases lining the passage on either side, but in some inexplicable way none was damaged.

A huge mahogany door blocked the stampede at the end of the passageway, but not for long. The accumulated pressure from behind burst it from its hinges, and the crowd swarmed into the lecture hall. The museum attachés were helpless before the throng, but a telephone message brought the police reserves under Captain Edmund Meade.

When the police had finally brought order many of the crowd had left, but enough remained to require two showings of the picture.

January 9, 1930

ELECTRON THEORIES WIDENED IN DUALITY

Conflict of Particle and Wave Views Explained at Meeting of Philosophical Society.

From a Staff Correspondent of The New York Times.

PHILADELPHIA, Pa., April 24.— The two hundred and fourth annual meeting of the American Philosophical Society, founded by Benjamin Franklin and the oldest learned society in America, was opened here this afternoon with a program of discussion which ranged from the structure of the atom to railroad consolidation, from the League of Nations to Silurian fossils.

In the society's ancient red brick home alongside Independence Hall, Dr. C. J. Davisson, physicist of the Bell Telephone Laboratories of New York, discussed the latest developments in the effort to discover the ultimate constitution of the matter.

Recent studies of the atom, Dr. Davisson said, had cast doubt upon the recently accepted theory of the atom as an infinitely small planetary system, in which electrons were arranged about a nucleus as the earth and its sister worlds were arranged about the sun. In its place, he said, there had come a conception of the atom as continuous and vibrating in more than three dimensions.

Conflict of Electron Theories.

Dr. Davisson's discussion of this new theory was incidental to an attempt to reconcile two apparently unreconcilable conceptions of the electron.

These two theories present electrons as behaving as particles and as waves. Dr. Davisson found ample evidence to support each of these theories, but held that scientists need not be bothered by their apparent conflict. Inability to conceive this duality of apparently irreconcilable properties, he added, was only another evidence of the limitation imposed by nature on man's understanding.

Dr. Davisson agreed with the thesis expressed last year by Dr. Francis X. Dercum, president of the society, that our thinking processes were dependent on the nature of our neutral protoplasm.

It was conceivable, he said, that other forms of stimulation existed in our environment for which we have evolved no receptors whatever. Our conception of our environment and of the processes going on within it, therefore, was imperfect and incomplete and must forever remain so.

Inadequacy of Planetary Idea.

Dr. Davisson sketched the compelling evidence for each conception of the electron. The corpuscular, or particle, theory, Dr. Davisson recalled, evolved from the observations made more than thirty years ago of the beautiful phenomena in highly exhausted electrical discharge tubes, due primarily to a radiation proceeding from the cathodes.

Years of experimentation piled up more evidence of the corpuscular nature of electrons and even made visible the tracks pursued by individual electrons in traversing a gas arc, Dr. Davisson went on. An elaborate theory based on this conception of the atom was built up to explain the electrical and optical properties of matter.

But in certain cases this theory proved inadequate, Dr. Davisson said, and this dissatisfaction "has led to a new and remarkably successful conception of the atom from which the corpuscular electron as an essential feature has disappeared."

"The planetary system of electrons conceived by Bohr is replaced by a medium continuous, though inhomogeneous, capable of natural vibrations," he continued.

"The fact that these vibrations take place in general in a space of more dimensions than three and that we have no idea what it is that vibrates makes visualization of atomic processes a discouraging enterprise, and yet this is less disturbing to the theoretical physicist than might be supposed.

"He has outgrown the ambition of Lord Kelvin. He no longer tries to devise a mechanical model of every phenomenon. It has been discovered, in fact, that a certain esthetic pleasure is derived from dealing in calculations with symbols which evoke no mental picture whatever. In certain quarters a fetish has been made of this mental attitude."

Dr. Davisson went on to list experiments which support the conception that electrons are waves, and concluded:

"It used to be said that physicists regarded light as a wave phenomenon on Mondays, Wednesdays and Fridays and as a corpuscular phenomenon on the other days of the week. This statement must now be extended to include electrons and be modified, I think, to state that he regards light and electrons as both waves and particles on all days of the week.

"And it might be added that familiarity with this duality of properties is dulling his sense of its paradoxical nature."

April 25, 1930

PHYSICS AND ASTROPHYSICS

DR. COMPTON 'SEES' INSIDE THE ATOM

Tells Ithaca Scientists Electrons Are Diffused in It Like Raindrops in Clouds.

REVELATION MADE BY X-RAY

Professor Says the World May Never Get Nearer 'View' Than Beam-Refraction Gives.

From a Staff Correspondent of The New York Times.

ITHACA, June 21.—Experiments which "come the closest of any yet performed" to showing men of science "what the atom looks like" were described here today by Dr. Arthur H. Compton, Nobel prize winner. Dr. Compton made the revelation before 200 members of the American Physical Society at the last meeting of their three-day convention at Cornell University in his paper on the subject of "X-ray Scattering and the Structure of Atoms."

"We are now at the final stage of this study. Not that the final stage is yet complete, but the general, large scale structure of the atom is now fairly definitely known," he said in the paper.

The atom as we now "see" it, according to Dr. Compton, is not at all the miniature solar system, in which the electron, the minutest negative charge of electricity, revolved in a fixed orbit around the proton, or the nucleus, consisting of the minutest positive charge. Nor is the electron a diffuse cloud of negative electricity, as Schrodinger's theory would have it. Rather it is composed of discreet particles of matter, somewhat diffused like raindrops in a cloud, "a probability cloud after the manner of Heisenberg."

"If you look at an atom," said Dr. Compton, "you will find it spherical in shape, the size of the spheres differing with the different elements. The electron surrounds the nucleus like a nebular haze, the nebula being thicker toward the centre than it is toward the circumference. We no longer find it convenient to say that it is a definite particle revolving in a fixed orbit around the nucleus. Rather it is like a cloud of raindrops, diffused through the sphere of the atom."

Similar to Moon Halo

"Several weeks ago," he began, "I noticed a beautiful halo around the moon. Half an hour later the halo was visibly smaller in diameter, and it was no surprise when a few hours later rain began to fall.

"The interpretation of such halos, as due to the diffraction of the moonlight by droplets of water suspended in the air, is well known. The larger the droplets the smaller the angle of diffraction necessary for the appropriate phase difference between the rays coming from the two sides of the drop. So by observing the diameter of the halo we can estimate the size of the water drops which cause it. A shrinking halo means a growing drop and hence probable rain.

"In a very similar manner it is possible to find the size of molecules and atoms in a gas, by observing the diffraction halos produced when they are traversed by a beam of X-rays. We substitute X-rays for the moonlight, and instead of the rain drops or vapor, we use helium, neon, argon, or some other gas. The X-rays make diffraction halos on the gas, just as the moonlight makes it on the raindrops, and just as we can estimate the size and position of the rain drops, we similarly can estimate the size and position of the electrons."

Accuracy Now Possible.

"For many years it has been possible by this method to make rough estimates of the sizes of the atoms, but only very recently has the theory of the process become well understood, and the experimental technique become sufficiently developed to give us precise information regarding the electron distributions in atoms."

There were two methods, Dr. Compton said, for studying the diffraction of X-rays. One was by means of crystals, in which, after taking into account the interference occurring between the various atoms in the space lattice, there emerges the diffraction pattern due to the atom itself.

The second method was the study of the scattering of X-rays by amorphous substances, in which we are really limited to the scattering produced by gases, since only in this case can we neglect the effect of interference by neighboring atoms.

"We have been working on this problem for the past twenty years," said Dr. Crompton, "and now we are at last in a position to tell something definite about the distribution of electrons in atoms and molecules. We have gone from one type of atomic theory to another, and it may perhaps appear too bold to say that the particular theory now in vogue has any finality.

Many Theories Are Discarded.

"Since 1911 many atomic theories have been proposed and discarded. One by one the vortex ring atom of Kelvin, the positively charged jelly of Thomson, the minute solar systems of Rutherford, Bohr and Sommerfeld, as well as the tiny atoms of Crehore, the ring electron atoms of Parson, and the cubic atom of Lewis and Langmuir, have given way to more promising successors.

"We replace even Schrodinger's diffuse cloud of negative electricity by a probability cloud of electrons after the manner of Heisenberg. It now appears, however, that the only one of these many proposals which can account for the observed X-ray diffraction haloes is that of Heisenberg."

After reviewing the difficulties encountered in the study of X-ray diffraction by crystals and by gas, and the recent methods devised for overcoming the difficulties, so that now the method of study by gases especially affords us precise information hitherto lacking about the distribution of electrons in atoms, Dr. Compton concluded as follows:

"We may accordingly say with some confidence that the aspect of the problem of atomic structure which is concerned with the distribution of the electrons in atoms is finding a satisfactory solution.

Revelation Was Predicted.

"It is a relief to note that a theory is at hand which affords a reasonable interpretation of the electron distributions which the experiments show.

"In a bulletin of the National Research Council, published in 1922, having experiments of this character in mind, I had the temerity to predict that within ten years the electron's positions in the lighter atoms would probably be known as reliably as were the positions of the atoms in certain crystals. I believe that prediction is now verified. For this information regarding electron positions in atoms is based upon precisely the same principles as is, for example, our information regarding the position of the oxygen atoms in a

"I suppose it would be fair to say that experiments such as these come the closest of any yet performed to showing us what the atom looks like. For, after all, is not seeing an object in a diffraction phenomenon similar to these under discussion? And when we thus look at the atom we find it composed of electrons diffusely distributed, as shown you in the figures on the screen."

* * * * *

June 22, 1930

SCIENTISTS ACCLAIM NEW ATOM THEORY

Sir Oliver Lodge Says Dr. Dirac Has Important Contribution to Solving Mystery of Matter.

SUMMARY OF HIS FINDINGS

Cambridge Savant Says There Must Be Times When Kinetic Energy of Electron Is Negative.

Special Cable to THE NEW YORK TIMES.

LONDON, Sept. 9.—Dr. P. M. Dirac, a young Cambridge physicist, was acclaimed today for his new atomic theory, which, in the opinion of some of the foremost scientists of Britain, upsets all present conceptions of space and matter.

Without accepting all the implications of his theory, physicists here at the meeting of the British Association for the Advancement of Science admitted he had given a shock to their accepted ideas.

Sir Oliver Lodge, as enthusiastic as his youthful confrère despite his seventy-eight years, called it "a most important contribution" toward solving the mystery of matter and said Dr. Dirac had for the first time discovered the nature of the proton.

As for Dr. Dirac, wherever he went today he was lionized for propounding a theory which his elders yesterday found so difficult to understand. In the simplest language he could summon, Dr. Dirac prepared the following summary of his findings today, admitting that they upset accepted theories, in "a suggestion which seems to explain something which is not understood."

"It is believed all matter is built up from the two elementary kinds of particles, the electron and the proton," he said. "Recent theoretical work seems to suggest that these two kinds of particles are not independent and that actually there is only one fundamental kind of particle in nature.

"The quantum theory of the electron, combined with the principle of relativity, shows there must be states for the electron in which its kinetic energy is negative—and is less, the faster the particle moves in addition to the usual state in which its energy is positive.

"To give a physical meaning to these negative energy states, we must assume that they are nearly all occupied by electrons with just one electron in each state, in accordance with the exclusion principle. We can then interpret the unoccupied negative energy state as protons. They will appear to us as things with a positive energy and also a positive charge.

"There are certain difficulties in the theory which have not yet been removed. They are, firstly, the great difference in the masses of the proton and the electron and, secondly, the fact that the theory predicates that electrons and protons will annihilate one another at a rate which is much too great to be correct. These difficulties are perhaps due to the fact that the interaction between electrons and protons has not yet been properly taken into account."

This was Sir Oliver Lodge's comment today: "Dr. Dirac has propounded a new theory of matter and space. It may be true, and I think it is. It goes more fundamentally than ever before into the secrets of the atom, and it has always been a puzzle to know what the proton was. Dr. Dirac has discovered it is a negative electron that behaves as if it were positively charged. He has found what importance should be attached to something that has hitherto been regarded in science as nonsensical. He is not afraid to conduct his researches into the nonsensical and has told us the effects of them, and they are very valuable."

Dr. F. E. Smith, president of the section of mathematical and physical sciences, said Dr. Dirac's theory was a radical departure from present ideas and would make people think in new channels.

"The departure from the present theory of the atom may lead to an explanation of gravity," Dr. Smith said. "Dr. Dirac has sought to go beyond the horizon of our knowledge. These physical scientists have a more exciting life than Columbus."

Professor Leonard Jones of Bristol University said Dr. Dirac's theory was a brilliant attempt to remove an outstanding theoretical difficulty, but he added that it is only partly successful. "There appears to be no reason why the electron should not have large negative energies as well as positive energies and why the electron should not switch from one to the other," Professor Jones said.

* * * * *

September 10, 1930

EINSTEIN ADVANCES UNCERTAINTY IDEA

In Joint Letter With Two California Scientists, He Holds the Past Cannot Be Gauged.

POINTS TO CAMERA SHUTTER

No Knowing When It Opens or Closes, He States in Quantum Mechanics Discussion.

EXPLAINS IN EXPERIMENT

He Shows Impossibility of Calculating Past or Future Velocity and Energy of Particles.

MINNEAPOLIS, Minn., March 21 (Science Service).—Dr. Albert Einstein has concluded that past events of any sort cannot be described with precise certainty.

This extension of the principles of the new physics is contained in a letter to the editor of The Physical Review, journal of the American Physical Society. Dr. Einstein, jointly with Professor Richard C. Tolman and Dr. Boris Podolsky of the California Institute of Technology, wrote this communication just before he left Pasadena, to return to Germany.

The letter, entitled "Knowledge of Past and Future in Quantum Mechanics," will be published in the next issue of The Physical Review.

Not only does Dr. Einstein conclude that there is an uncertainty in the description of what has happened in the submicroscopic world with which the most recent theories of physics usually deal but he also applies this principle of uncertainty to such everyday happenings as the opening and closing of a shutter on a camera, showing that one cannot know just when the shutter opens or closes.

"It is of special interest to emphasize the remarkable conclusion that the principles of quantum mechanics would actually impose limitations on the localization in time of a macroscopic phenomenon such as the opening and closing of a shutter," Dr. Einstein and his colleagues write.

Development of Older Theory.

The idea that it is impossible to predict the exact path of an object in the future was advanced some two years ago by a young German physicist, Professor W. Heisenberg. This principle of uncertainty has had an influence on the philosophy as well as the practice of science comparable with the idea of relativity introduced by Dr. Einstein.

As the opening paragraph of the Einstein - Tolman - Podolsky letter states:

"It is well known that the principles of quantum mechanics limit the possibilities of exact prediction as to the future path of a particle. It has sometimes been supposed, nevertheless, that the quantum mechanics would permit an exact description of the past path of a particle."

Dr. Einstein laid one of the foundations of the quantum theory, building on the work of Professor Max Planck. The Einstein paper of 1905 applied the quantum theory of energy to light and electricity. The quantum idea that energy is not continuous but in packets or gobs, like matter, has been one of the most fruitful conceptions of the new physics.

Now Dr. Einstein adds the latest building block to our conception of matter and energy by declaring that the past as well as the future is uncertain.

Dr. Einstein's associates in his new pronouncement are on the staff of the California Institute of Technology at Pasadena, where he worked during his recent stay in America. Professor Tolman is a leading authority on thermodynamics, and is noted for his theory of a non-static universe. Dr. Podolsky is a young physicist, Russian-born, but now an American citizen. He was a National Research fellow in physics for several years.

Letter Expounding Principle.

Dr. Einstein's letter reads:

"It is well known that the principles of quantum mechanics limit the possibilities of exact prediction as to the future path of a particle. It has sometimes been supposed, nevertheless, that the quantum would permit an exact description of the past path of a particle.

"The purpose of the present note is to discuss a simple, ideal experiment which shows that the possibility of describing the past path of one particle would lead to predictions as to the future behavior of a second particle of a kind not allowed in the quantum mechanics.

"It will hence be concluded that the principles of quantum mechanics actually involve an uncertainty in the description of past events which is analogous to the uncertainty in the prediction of future events. And it will be shown for the case in hand that this uncertainty of the description of the past arises from a limitation of the knowledge that can be obtained by measurement of momentum.

Arrangement of the Experiment.

"Consider a small box B, as shown in the figure, containing a number of identical particles in thermal agitation and provided with two small openings which are closed by the shutter S. The shutter is arranged to open automatically for a short time and then close again, and the number of particles in the box is so chosen that cases arise in which one particle leaves the box and travels over the direct path SO to an observer at O, and a second particle travels over the longer path SRO through elastic reflection at the ellipsoidal reflector R. The box is accurately weighed before and after the shutter has opened, in order to determine the total energy of the particles which have left, and the observer at O is provided with means for observing the arrival of particles, a clock for measuring their time of arrival and some apparatus for measuring momentum.

"Furthermore, the distance SO and SRO are accurately measured beforehand—the distance SO being sufficient so that the rate of the clock at O is not disturbed by the gravitational effects involved in weighing the box, and the distance SRO being very long in order to permit an accurate re-weighing of the box before the arrival of the second particle.

Paradoxical Results.

"Let us now suppose that the observer at O measures the momentum of the first particle as it approaches along the path SO, and then measures its time of arrival. Of course, the latter observation, made, for example, with the help of gamma-ray illumination, will change the momentum in an unknown manner.

"Nevertheless, knowing the momentum of the particle in the past and hence also its past velocity and energy, it would seem possible to calculate the time when the shutter must have been open from the known time of arrival of the first particle and to calculate the energy and velocity of the second particle from the known loss of energy content of the box when the shutter opened.

"It would then seem possible to predict beforehand both the energy and the time of arrival of the second particle, a paradoxical result, since energy and time are quantities which do not commute in quantum mechanics.

"The explanation of the apparent paradox must lie in the circumstance that the past motion of the first par-

Diagram, Explained in Letter, Illustrates Attempt to Calculate Energy and Velocity of One Particle in Past and Another in Future.

ticle cannot be accurately determined as was assumed. Indeed, we are forced to conclude that there can be no method for measuring the momentum of a particle without changing its value.

"For example, an analysis of the method of observing the Doppler effect in the reflected infra-red light from an approaching particle shows that, although it permits a determination of the momentum of the particle before and after collision with the light quantum used, it leaves an uncertainty as to the time at which the collision with the light quantum takes place.

Past and Future Uncertainty.

"Thus in our example, although the velocity of the first particle could be determined both before and after interaction with the infra-red light, it would not be possible to determine the exact position along the path SO at which the change in velocity occurred as would be necessary to obtain the exact time at which the shutter was open.

"It is hence to be concluded that the principles of the quantum mechanics must involve an uncertainty in the description of past events, which is analogous to the uncertainty in the prediction of future events.

"It is also to be noted that although it is possible to measure the momentum of a particle and follow this with a measurement of position, this will not give sufficient information for a complete reconstruction of its past path, since it has been shown that there can be no method for measuring the momentum of a particle without changing its value.

"Finally, it is of special interest to emphasize the remarkable conclusion that the principles of quantum mechanics would actually impose limitations on the localization in time of a microscopic phenomenon such as the opening and closing of a shutter."

(Copyright, 1931, By Science Service.)

March 22, 1931

COMPTON TO STRIVE FOR ATOMIC ENERGY

He Will Supervise Experiment at University of Chicago That Will Take Many Years.

TELLS ITS PRIMARY AIMS

One Will Be to Raise Voltage of Electrons in X-Rays to That Approaching Sun's.

FACES GREAT OBSTACLES

But Declares Success Would Create a Source of Power That Would Change Civilization.

An extensive study of the problem of releasing atomic energy, the solution of which would create a limitless reservoir of power and would bring undreamed-of changes to civilization, will begin at the University of Chicago shortly, Dr. Arthur H. Compton, Nobel Prize physicist of that institution, announced here yesterday. Dr. Compton will supervise the experiments, which are to continue for several years.

A primary goal of the study will be to produce an extremely high voltage in electrons in X-rays. If this voltage can be raised to between 10,000,000 and 20,000,000 volts pressure it is likely that the experiment will be successful, and the door pointing the way to the release of atomic energy will have been thrown open, in the opinion of Dr. Compton.

The achievement of this tremendous voltage, which has never been attained in any laboratory on earth, would help to approach the high temperatures existing normally on the sun, where atomic energy is continuously being released, Dr. Compton explained.

Temperatures Inside the Sun.

"On the sun, the electron and proton particles of the atom constantly coalesce to produce the photon, which is radiated away into space," Dr. Compton said. "The temperatures in the sun's interior are probably as high as 40,000,000 degrees and it is quite impossible to achieve any such temperature here. But it is possible that the essential

PHYSICS AND ASTROPHYSICS

characteristics of the vital action of coalescence may be due to the high speed of electrons, and in that case it is not a vain hope to anticipate making electrons and protons coalesce on earth to release the huge energies which they contain."

Because of the purely experimental stage of the work, Dr. Compton said he is unwilling to divulge publicly now the exact nature of the experiments to be undertaken at the University of Chicago. He explained that it would be years before the undertaking could succeed, and that "tremendous" technical difficulties would have to be overcome. Among these technical difficulties, he said, were the production of a tube capable of withstanding the great pressure it is hoped to create, and the arrangement of new and intricate equipment of many kinds.

The combination of the electron and proton, the two invisible particles that pursue their orbits inside the tiny atom, is one of three possible ways known for the release of atomic energy. The experiments at the University of Chicago are to attack the coalescense problem, as it seems the most likely to find a solution, Dr. Compton explained. The other two methods, he said, consist of a study of radioactivity and the cosmic rays.

'If the key to harnessing the power of atomic energy is ever discovered, our present civilization will undergo a very radical change," Dr. Compton said. "There is enough atomic energy in a teaspoonful of ordinary water to provide all the energy to run New York City, with all of its transit systems, factories and the life of the metropolis in general."

Dr. Compton revealed the forthcoming study of atomic energy in an address on "Do Things Have a Beginning and an End?" at the College of the City of New York, Twenty-third Street and Lexington Avenue. More than two thousand persons heard Dr. Compton's talk, the last of a series on the "Nature of Things."

Discusses Age of the Earth.

Touching on the age of the earth, Dr. Compton asserted that radioactive material give an approximately accurate basis for concluding that our globe is about 2,000,000,000 years old. The oldest minerals found on the earth are over 1,500,000,000 years old, he said. Life in some form began on this planet about 1,000,000,000 years ago he added.

"The best modern evidence of science points to the belief that man is no more than 2,000,000 years old," he said.

The sun probably attained the ripe age of some 5,000,000,000,000 years, Dr. Compton asserted. So vast, however, is the amount of energy on the sun that, although the earth's parent body is radiating away 400,000 tons of matter a second, there has been no noticeable decrease in the solar energy.

"If the sun's energy were other than radioactive or atomic, our parent globe would be completely burned up in less than 10,000 years," he said. The noted physicist said the best evidence tends to show that the universe began about 1,000,000,000,000,000 years ago.

March 28, 1931

RELEASING THE ATOM'S ENERGY

To the Editor of The New York Times:

The reasons which prompt people to write to you are legion and it will be no surprise to you to read a letter written because of a most unusual fear.

THE TIMES published an article discussing certain experiments to be conducted at the University of Chicago by Dr. Arthur H. Compton, Nobel prize physicist of that institution. As I understand the nature of the experiments to be conducted, it is intended to release the energy locked up in the atom, through the force of a ten or twenty million voltage to be raised in electrons in X-rays.

If the laboratory were to succeed in obtaining necessary voltage and found that it could resolve the atom into energy, I fully realize that a wonderful discovery would have been made—one whose possibilities can hardly be visualized by man as now limited and hedged in by present difficulties in securing energy in any form. There is, however, another possibility which comes to mind in contemplating such experiments, for we know that every invention and every discovery has made new problems for man to solve, and new difficulties for man to surmount.

In releasing the energy of the atom we invade the realm of the unknown. Should a danger confront us in that realm we might not be in a position to battle it. What if the atom whose energy man has released should turn into a "Golem" would destroy man? What if the explosion of the energy in one atom would automatically have the same effect on the other atoms surrounding it? And what if, after Dr. Compton had experimented with an atom in his laboratory one morning, we were to find all our erstwhile peaceful neighbors suddenly rise in mutiny and make it hot for us?

Dr. Compton believes that on the sun temperatures of 40,000,000 degrees are found. What if our tiny friends here on earth are capable of raising our temperature to a similar degree of warmth? I doubt if any one would welcome the change—unless it be a comparatively few Eskimos.

Let us have experiments, by all means, for it is only through them that science will ever advance. In the hands of an expert the most dangerous weapon may be safe. But I hope in every experiment our friends the scientists make they will endeavor first to make provision for every conceivable possibility. They will have to be particularly circumspect in this matter.

For all we know in the 2,000,000 years which Dr. Compton claims man has existed on earth there may have been other experiments which just got out of the experimenter's hands in some prehistoric laboratory and resulted in the complete destruction of everything on earth.

PHILIP H. LEIB.
New York, June 1, 1931.

June 7, 1931

New Microscope Records Workings of Atom; Photographs Electrons, Millikan Announces

By The Associated Press.

ROME, Oct. 13.—The invention of a microscope for observing and measuring the velocity of electrons was announced today by Professor Robert A. Millikan of the California Institute of Technology, who is a winner of the Nobel prize, at the meeting here of fifty world-famous physicists.

How scientists detect the antics of the nuclei of atoms also was demonstrated dramatically at the meeting.

The instrument announced by Professor Millikan consists of an X-ray microscope, which he termed a multiple crystal spectrometer. He credited its invention to Professors Jesse Dumond and Harry Kirkpatrick of the California Institute of Technology at Pasadena.

Dr. Millikan projected photographs made with the machine, and described it as giving the first evidence of the inner workings of the dynamic, instead of the static, atom.

The apparatus consists of fifty spectroscopes arranged in an arc, along which sweeps an X-ray. The photographs showed the activities of two electrons in beryllium, a hard, silver-white metallic element.

The electron is approximately 1,700 times smaller than the atom, yet the invention caught it in action. Dr. Millikan explained that it had been perfected this Summer and that Professors Dumond and Kirkpatrick had examined only the one metallic element. Further researches are now in progress.

The nuclei, the movements of which were demonstrated, are conceived for the sake of convenience as miniature suns, immensely heavy, about which revolve a cloud of electrons, the whole forming an atom.

Photographs of the trails which the nuclei of atoms leave behind when ejected were exhibited by Professor S. Gouldsmit of the University of Michigan. Professor Neils Bohr, Danish Nobel prize winner and author of the solar system conception of atoms, showed how he approached the study of the protons, that is, the nuclei, of atoms by observing the action of electrons outside the nucleus.

October 14, 1931

BIG GENERATOR OPENS WAY TO SMASH ATOM

Scientist Learns How to Build 20,000,000-Volt Apparatus, Hinting at Transmutation.

Dr. Van de Graaff's Device Uses Static Electricity—Operator to Sit in Sphere Terminal.

Special to The New York Times.

PRINCETON, N. J., Nov. 5.—A simple and inexpensive method of building generators capable of developing 15,000,000 to 20,000,000 volts, which experts regard as finally opening the way toward the realization of the age old alchemists' dream of transmutation of the elements, has been perfected at the Palmer Physical Laboratory, Princeton University, by Dr. Robert J. Van de Graaff, it was announced here today.

So far Dr. Van de Graaff has constructed a generator capable of developing 1,500,000 volts. This apparatus will have its first public demonstration in New York next Tuesday at a dinner of the newly-organized American Institute of Physics, at which Dr. Arthur H. Compton of the University of Chicago, Nobel Prize winner in physics, will be the principal speaker.

The same method employed in building this generator, however, can be employed to build generators ten times the voltage and more, it was said, and ultimately can be used to build generators capable of developing 50,000,000 volts.

One to Be Built at Once.

A generator capable of developing 10,000,000 volts, more than four times the highest voltage ever attained, will be constructed immediately by Dr. Van de Graaff under the auspices of the Massachusetts Institute of Technology in the institute's airship dock at its research field station on the estate of Colonel Edward H. R. Green, Round Hill, South Dartmouth, Mass. The voltage of the generators, experiments indicate, are limited only by the size of the apparatus.

The terminals of the new 1,500,000-volt generator consist of two brass spheres, each two feet in diameter. The spheres are supported and insulated from the ground by glass rods, 1¼ inches in diameter and 5 feet high. In each section a belt conveyor, operated by a motor at the base of the supporting rod, and running over a pulley within the sphere, conveys the electric charge to the sphere.

The charge is produced on the silk belt by "spraying," a method technically known as "corona" or "brush" discharge. While the voltage of the charge "sprayed" upon the belt is comparatively low the sphere becomes charged with higher and higher voltages by storing up the low voltages conveyed by the belt. When fully charged one terminal takes a positive charge of 750,000 volts while the other takes a negative charge of the same voltage, making a total difference of potential of 1,500,000 volts, at which point there is a discharge like a lightning flash, with a spark three feet long jumping from one terminal to the other.

In the 10,000,000-volt generator the terminals will be from ten to fifteen feet in diameter, mounted on towers twenty feet high. The operator of the apparatus will sit within one of the spheres, where he will be safer than in any other place. His body, like the spheres, will be charged up to 5,000,000 volts, but as he will not be connected in any way with the earth the charge will have no effect on him.

While present high-voltage generators are very costly and elaborate, it was said that Dr. Van de Graaff's apparatus would require only a few hundred dollars for construction. The 1,500,000-volt generator cost only $90.

New Tool to Smash Atom.

The new generator is expected to supply science with a new tool with which to bombard and smash up the atom. At present this is done on a small scale by employing the alpha particles from radium as the "cannon." But whereas the supply of alpha particles from radium is small and extremely costly, the new generators are expected to supply alpha particles cheaply and in much larger amounts.

"The amount of alpha particles from this source," said Dr. Karl T. Compton, president of Massachusetts Institute of Technology, according to The Associated Press, "will be so enormously larger than that from radium, that the experiment opens up the possibility of transmutation of the elements on a commercial scale. America's leading experts in high voltage X-ray sources state that there is, on the horizon, no other comparably promising source of electric power current at tremendous voltages."

When the apparatus is completed a big vacuum tube may be fixed between the spheres to transform these gigantic voltages into equally gigantic X-rays. The gamma-rays from such a tube, equivalent to great quantities of radium, will be available for medical experiments on cancer.

Dr. Van de Graaff was graduated from the University of Alabama in 1922 and studied at the Sorbonne, Oxford and Princeton. Dr. Compton, who, while at the physics department at Princeton, encouraged Dr. Van de Graaff in his work, has appointed the latter a research associate at M. I. T.

November 6, 1931

NEW 'SCALE' WEIGHS MINUTE ATOM NUCLEI

Hidden Secrets of Science to Be Revealed by Machine Built by Dr. K. T. Bainbridge.

ELEMENTS TO BE ANALYZED

Device Hailed as Research Tool of Immense Value in Study of Isotopes' Variation.

WILL CHECK ON RELATIVITY

High-Precision Mass-Spectrograph to Add to Radiation Data of Einstein and Millikan.

A "scale" which weighs the nuclei of atoms of masses less than a trillionth of a trillionth of an ounce has been built at the Bartol Research Foundation at Swarthmore, Pa., it was announced yesterday by Dr. Henry A. Barton, director of the American Institute of Physics.

The apparatus, a new precision mass-spectrograph, is the work of Dr. Kenneth T. Bainbridge, a National Research Fellow at the Bartol laboratories, under the direction of Dr. W. F. G. Swann, physicist and lecturer. Dr. Bainbridge will describe his instrument before the meeting of the American Physical Society at Cambridge, Mass., on Feb. 25 to 27.

With the new scale scientists will be able actually to weigh not only the atoms of the various elements, as well as their nuclei, but also the different isotopes of which the elements are known to be composed. Isotopes are atoms which are identical chemically but which differ in their weights.

Until now there has been only one such instrument of high precision in the world, that of Dr. F. W. Aston in England, who was knighted recently for discoveries he made with it. Its ability to furnish information concerning atomic nuclei — now in the centre of research activities—brings to the American world of physics, Dr. Barton said, "a research tool of immense value."

Will Check Millikan and Einstein.

It was Sir J. J. Thomson who first discovered that neon, instead of being composed of atoms all of which have the same atomic weight, consists of a mixture of atoms which, though chemically of the same constitution, are nevertheless different in their atomic weights. Dr. Aston, with his new device, proved that most elements consist of a mixture of identical atoms of different weight, or isotopes, but many elements have not yet been analyzed from this point of view. Dr. Bainbridge's instrument is expected to help fill this gap.

Not only will the new instrument show the number of different isotopes an element is composed of; it also will equip science with the first direct experimental data with which to check the theory of Dr. Robert A. Millikan that the cosmic rays originate through the constant creation of heavier elements out of lighter ones in the interstellar spaces. In this connection it also will provide further data with which to check certain deductions of Einstein's general theory of relativity.

One of the most puzzling discoveries of modern physics is the phenomenon that atoms, when they combine to form an element, weigh less in combination than the total of their combined weights. For example, one atom of hydrogen, consisting of one electron and one proton, has an atomic weight of 1.0077. An atom of helium consists of four atoms of hydrogen. It should therefore weigh four times as much. Yet the atomic weight of helium is exactly 4, the fraction of .0306 somehow mysteriously getting lost in the combination. The fractional loss is known technically as the "packing fraction."

This loss of mass, which goes contrary to one of the principal tenets of physics, the law of conservation of matter, is explained by relativity as due to the conversion of part of the matter into radiant energy. Dr. Millikan is convinced that the cosmic rays, most powerful radiation known to science, are the result of this conversion of lost matter into energy in the far-off spaces, and that lost matter can be accounted for only by the formation of heavier elements out of lighter ones.

"Packing Fraction" to Be Weighed.

The instrument of Dr. Bainbridge will be able to weigh, with a precision of one part in 10,000, the amount of lost matter, or the "packing frac-

WEIGHS 'ATOMS' NUCLEI.

Dr. Kenneth Bainbridge, Builder of New Mass-Spectrograph.

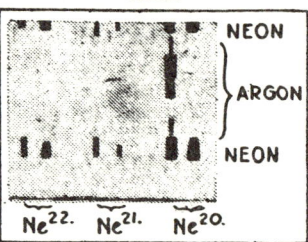

This photographic plate, on which bombarding atoms have left their marks, is measured by Dr. Bainbridge to determine the atoms' weight. Top and bottom strips reveal three kinds of neon atoms, the four others argon atoms.

tion," resulting through the formation of a relatively simple element out of a still simpler one, and thus will be able to determine experimentally whether the amount of matter lost is sufficient to account for radiations as powerful as the cosmic ray. The theory of relativity has given to science the formula which fixes the amount of radiation produced by a given amount of matter.

In this respect Dr. Bainbridge will have the advantage over Dr. Millikan, Dr. Aston and other researchers in the field through the recent discovery of the new isotope of hydrogen, which weighs approximately twice as much as the ordinary hydrogen atom. The new hydrogen atom, the existence of which was reported two months ago by Professor Harold C. Urey and Dr. G. M. Murphy of Columbia University and Dr. F. G. Brickwedde of the United States Bureau of Standards, is supposed to consist of one electron and two protons, one proton more than the ordinary hydrogen atom.

Dr. Millikan had to base his conclusions on the building of rather complicated nuclei, about the nature of which science as yet has very little direct information. With the new hydrogen Dr. Bainbridge will have at his disposal for the first time a substance in which two of the simplest "bricks of matter" are united into one. The two "bricks" in combination are expected to weigh less than the total weight of the two "bricks" individually, due to the loss of the "packing fraction."

On determining the "packing fraction" of the double-weight hydrogen atom, it will be possible to figure out the amount of radiant energy that should result from it. This in turn will be compared with the penetrability and wave length of the actually observed cosmic ray.

"This will be a simple matter for Dr. Bainbridge," Dr. Barton declared, "as soon as he receives a suitable sample of the new hydrogen. In fact, his apparatus works best for the lighter elements like hydrogen, and the paradoxical situation arises that single atoms of elements like gold and mercury are actually almost, but not quite, too heavy to be separated and weighed conveniently."

In the case of many elements, the isotopes of which have as yet undiscovered, science has no idea of the weight of their atoms, Dr. Barton added, even though their "atomic weights" have been accurately determined by chemists. The situation is like knowing the average weight of each orange in a bag of a dozen oranges, easily found by weighing the bagful, but not knowing the weight of any one of the individual oranges in it. The new scale makes it possible to weigh the individual "oranges."

Huge Magnet Used in Device.

The apparatus consists of one of the largest electromagnets in the United States, with a special steel frame weighing 3,500 pounds. The magnet winding consists of 3,000 turns of heavy wire containing 750 pounds of copper. The current supplied comes from the 1,000 ampere-hour storage battery of the Bartol Foundation, sources other than large-capacity storage batteries not being steady enough.

February 13, 1932

PHYSICS AND ASTROPHYSICS

DISCOVERS NEUTRON, EMBRYONIC MATTER

Dr. James Chadwick Describes It as Halfway Between Electricity and Helium.

FIRST STEP IN EVOLUTION

Scientists Say Results Rank With Finding of Electron, Proton and X-Ray.

Special Cable to THE NEW YORK TIMES.

LONDON, Feb. 27.—Dr. James Chadwick, working in Cavendish Laboratory at Cambridge, has discovered the neutron—one of the ultimate particles of nature. The discovery, first made public today, was hailed by scientists here as the most important achievement in experimental physics since Lord Rutherford demonstrated the nuclei structure of the atom in 1911.

The particle Dr. Chadwick discovered, according to The Manchester Guardian's scientific correspondent, appears to consist of a proton and an electron bound together, and hence without an electric charge. It was therefore named the neutron, representing the first step in the evolution of the elements out of electricity and may well be the material of the famous cosmic rays, which might be streams of neutrons.

Some physicists here are of the opinion that it is one of the fundamental discoveries in physics, ranking with the discovery of the electron, the proton and the X-ray. A question also has been raised as to the practical use of the discovery—whether it will result in entirely new experiments, as the X-ray has done.

The neutron represents the first step in the evolution of matter, the first step in the building up of the common materials of everyday life out of primeval electrons and protons. It is the embryonic form of ordinary matter, growing but not yet born.

Scientists Hail Discovery.

"The discovery is of the greatest interest and importance—possibly the greatest since the artificial disintegration of the atom," declared Lord Rutherford, director of the Cavendish laboratory. "It already has afforded a number of examples of unexpected types of atomic disintegration and offers a promising approach to a number of important problems."

The neutron's principal properties bear out a prediction by Lord Rutherford twelve years ago in a lecture before the Royal Society, when he discussed possible types of matter then unknown.

The discovery recalls Dr. Chadwick's first researches as a student at Manchester University. In 1912, at the age of 21, he discovered that substances struck by the heavy particles ejected from radium emit a penetrating wave radiation. He has now shown that the same kind of particles may also cause neutrons to be emitted.

Later, Dr. Chadwick studied at Berlin and Cambridge and became one of Lord Rutherford's chief lieutenants during a period of radioactivity research at Manchester. Always distinguished by exceptional thoroughness, now he has added special inspiration to his gifts.

Completed in Fortnight.

Those who heard him describe the latest research, which was completed within a fortnight, said to be a record for an important discovery in modern physics, were profoundly impressed by his mastery of detail and theory.

During the World War, Dr. Chadwick was interned at Ruhleben, where he improvised a small research laboratory in camp and passed his time experimenting. C. D. Ellis, a fellow prisoner, who had been interested in physics, revived his interest and after the war went to Cambridge, becoming one of the chief figures at the Cavendish laboratory. Dr. Chadwick and Mr. Ellis collaborated with Lord Rutherford in the recent revision of Lord Rutherford's treatise on radioactive substances. When Lord Rutherford left Manchester for Cambridge, Dr. Chadwick went along as his right-hand man.

Hailed by American Scientists.

BALTIMORE, Feb. 27 (AP).—The discovery in Cavendish Laboratory, Cambridge University, London, by James Chadwick of the "neutron," a particle so minute it carries no electrical charge and hailed as the greatest find in the scientific world since the electron, the proton and the X-ray, was announced today by The Baltimore Sun in a dispatch from The Manchester Guardian.

The "neutron" was described as one of the ultimate particles in nature, so tiny it would take 200,000,000,000,000,000,000,000 (200 septillions) to make a mass weighing an ounce. Neutrons are not waves, the dispatch said, but particles. But they have, as particles, hitherto unknown powers of penetration.

The ultimate substance of the world is electricity, and the simplest element built of electricity is helium. The neutron is half way between electricity and helium.

Neutrons are so penetrative and elusive because they have no electric charge. The neutron consists of a proton and an electron bound closely together. It is conjectured that the neutron may be the unit of magnetism, for it presumably is a doublet, as magnets are, with both a north and a south pole, the dispatch continued. The discovery of Dr. Chadwick, assistant director of radio-activity at the laboratory, arose out of investigations by Bothe of Giessen, Millikan, Irene Curie (daughter of Mme. Curie) and her husband, M. Joliot, and Webster, who had experimented with the properties emitted by beryllium when bombarded by radiations from the radioactive substance known as polonium.

Physicists at the Johns Hopkins University, when told of the discovery, hailed the achievement as one of the first importance.

February 28, 1932

ATOM FILM THREATENS STRUCTURAL THEORIES

Dr. Millikan Says Photographs Shown in Pasadena May Be of Utmost Importance.

PASADENA, Cal., Oct. 1 (AP).—A single photograph perplexed noted scientists today and amazement threatened to upset accepted theories regarding the structure of the atom.

The photograph, one of 10,000 made by Dr. Carl D. Anderson, a graduate of the California Institute of Technology, shows the tracks left in vapor by particles blasted out of the hearts of atoms by cosmic rays.

The probable interpretation of this track, according to scientists who viewed the photograph at a meeting of the Astronomy and Physics Club last night, is that it was made by a positively charged atomic particle of such small mass that it is entirely out of line with the previous atomic conceptions.

For many years physicists have based their picture of the atom on the theory that the mass of a proton, one of the positively charged particles in the nucleus or heart of the atom, was more than 1,000 times that of an electron, one of the negatively charged particles which fly about the nucleus.

Dr. Anderson warned against premature conclusion, declaring that if other photographs will show similar tracks he may have "something to talk about."

Dr. Robert A. Millikan, who was present, stated the results of Dr. Anderson's photograph may prove of the utmost importance in the scientific realm.

Other scientists said the photograph indicated the mass of the proton has shrunk to that of an electron, and what was first thought to be a part of the proton actually is a mass of unidentified, neutrally charged material in the nucleus.

October 2, 1932

ATOM TORN APART, YIELDING 60% MORE ENERGY THAN USED

But Two British Scientists Succed Only Once in Each 10,000,000 Bombarded.

BATTERED WITH PROTONS

Hydrogen Atoms Are Thus Transmuted Into Helium—Conservation Theory Seen Upset.

Special Cable to THE NEW YORK TIMES.

LONDON, May 1.—The atom has not only been split but an element has been transmuted into another element—atoms of hydrogen have been turned into atoms of that rare and commercially valuable gas, helium. At the same time the energy of part of an atom has been released in a quantity 60 per cent greater than the amount used to produce this phenomenon, although it occurred only once in each 10,000,000 atoms bombarded.

This is the claim of two young scientists, Dr. J. D. Cockroft and Dr. E. T. S. Walton, working in the Cavendish laboratory at Cambridge University, of which laboratory Lord Ernest Rutherford is the director.

Lord Rutherford is one of the founders of the modern atomic theory of physics and was the first to propound the theory that atoms were composed of electrons, or negative charges of electricity, and protons, of positive units or particles of electricity. He received the Nobel Prize in Chemistry in 1908.

Lord Rutherford tonight described the result of the three years' work of Dr. Cockroft and Dr. Walton as a "discovery of great importance." It is difficult yet, he added, to say to what the discovery might lead.

"Thus far," he said, "the experiments have not yielded anything which would be of immediate commercial value. I have seen it stated that the discovery means that we shall be able to produce an immense accumulation of added electrical energy for commercial purposes. We cannot claim that for our experiments thus far, for the simple reason that for every particle of additional energy obtained it requires millions of particles to make it effective.

"The experiments, however, are of great scientific interest and are likely to be powerful agents in extending our knowledge of the atom."

Dr. Cockroft explained tonight that he and Dr. Walton had concentrated on an atom of hydrogen.

"We found," he said, "that at 120,000 volts some atoms we were bombarding by protons began to build up into helium. These helium atoms came out with energies of the order of 100 to 160 times that of the particles we were firing into them. We only got these results in the last fortnight after nearly three years of work."

The experiments were made in the high-voltage laboratory which Lord Rutherford developed. The experimenters employed voltages of between 120,000 and 600,000 to send millions of protons per second through a vacuum tube at a speed of nearly 4,000 miles a second.

It was found that the bombardment of different elements by the particles split light elements and transformed them into matter. For every atom split several millions of particles were required.

It had long been recognized by Lord Rutherford and others that effects similar to those caused by the bombardment of the atom with Alpha particles would be produced if the atoms could be bombarded by very swift protons. The swiftness of the stream of protons depended on the voltage applied to the vacuum tube, but on average only one in 10,000,000 particles which bombarded the elements was effective, the experimenters said, in bringing about disintegration.

Explains Electrical Aspect.

What led to the belief that the operation of splitting an atom of hydrogen into helium might ultimately produce added electrical energy for commercial purposes was the fact that the helium atoms came out with greater energy than those producing them. Dr. Cockroft explained the position thus:

"In one sense it is true that by this means we were turning 120,000 volts into about 190,000 volts. But only one particle of hydrogen breaks up for every 10,000,000 we employ to bombard it.

"In other words, we were producing from these atoms 100 to 160 times of known energy, but only once in 10,000,000 times. Therefore, it would be only strictly true to say we were turning 100,000 volts into 160,000 volts if every atom broke up."

There was general agreement with Dr. Cockroft, therefore, that while the discovery was of immense scientific importance it was not of immediate practical value, although an economic process of adding considerable value to the original electrical charge might eventually be discovered along these lines.

Dr. Cockroft is 34 years old. Dr. Walton, who is still in his twenties, went to Cambridge from King's College, Dublin.

Sir Leonard Hill, director of research of the London Light and Electrical Clinic, said:

"This discovery is the beginning of something far bigger and more important than the layman might imagine. For the first time in history we have got more energy out of something than we put in it.

"Every schoolboy learns the law of the conservation of energy—briefly, that we cannot 'make' energy. To all intents and purposes this law may be considered broken. Wireless valve [tube] experts should study this discovery closely."

According to The Daily Herald, when lithium was bombarded with hydrogen proton "shells," Alpha particles were driven out, and this revealed to the experimenters that the atom had definitely been split. Actually, lithium had been changed into another element.

Similarly, in the case of aluminum, Lord Rutherford believes it was changed into magnesium. The question is now asked: Can lead or some cheap metal be changed into gold? Has the alchemist's dream come true?

Professor Harkins Hails Finding.

Special to THE NEW YORK TIMES.

CHICAGO, May 1.—William Draper Harkins, Professor of Physical Chemistry at the University of Chicago and the first man to make calculations of the vast energies that could be released by the transmutation of hydrogen atoms into helium atoms, said tonight that the experiments of Dr. Cockroft and Dr. Walton at Cambridge University were of great importance.

"If the results of their work have been correctly reported," he qualified, "they have done something that was never done before and have taken an important step forward."

"From the description I have," he said, "these experimenters did not split the atom—they made atoms of one element from the atoms of another element."

In 1915 Professor Harkins calculated that if the atoms in one pound of hydrogen were built into helium atoms energy equal to that produced by the burning of 10,000 tons of coal would be released—enough to send a battleship around the world. He referred, as did Lord Rutherford, to the fact that laboratory experiments so far had required far more energy than they produced.

Atoms Disintegrated Before.

This is not the first time that an atom has been "split" in laboratory experiments. The first to accomplish this was Lord Rutherford himself, who more than ten years ago succeeded in transmuting nitrogen into carbon by bombarding the nitrogen with Alpha particles from radium.

About a year ago Bothe, a German physicist, bombarded the nucleus of the beryllium atom with Alpha particles from polonium and succeeded in producing carbon. In doing so Bothe also succeeded in producing a radiation more penetrating than the hardest natural Gamma rays from radium, the artificial rays coming very near to having the characteristics of cosmic rays.

Smashing, splitting or disintegrating the atom into its component parts of negative electrons and positive protons, both of which are regarded as the ultimate units of matter, is the most stupendous problem engaging the attention of experimental physicists today. It is the modern equivalent of the alchemist's search for "the philosopher's stone," with which it was hoped to achieve the transmutation of base metals into gold.

To smash the atom would mean, only theoretically for the present, of course, two achievements, which, should they ever be put into practical use, would result in revolutionizing life upon earth. For it would mean not only the realization of the age-old dream of the alchemists to transform one element into another but the opening up for the use of man of the practically limitless energy known to be securely locked up inside the atom.

May 2, 1932

FINDS COSMIC RAYS HAVE ODD PARTICLE

British Scientist Says That They Consist in Part of 'Positive Electrons.'

CREDITS DR. C. D. ANDERSON

Dr. P. M. S. Blackett Says American Should Not Be Robbed of Leads in Discovery.

Special Cable to THE NEW YORK TIMES.

LONDON, Feb. 16.—A new theory of cosmic rays was put forward today by P. M. S. Blackett, young Cambridge physicist, who holds they consist, partly at least, of a new kind of particle, which he called a "positive electron."

Before the Royal Society Dr. Blackett advanced evidence of the existence of a particle having a positive charge of proton but a mass like that of an electron. His conclusions were based on eighteen months of work in the Cavendish Laboratory at Cambridge University in making cosmic rays photograph themselves.

Great Energy Used.

Shooting into an expansion chamber with energy estimated at 100,000,000 to 300,000,000 volts, the particles were made to pass through two lead disks, thus causing expansion and leaving behind automatic photographs of their passage Altogether 500 photographs and many tracks on them have been studied by Dr. Blackett and his colleague, G. Occhialini. Theirs is believed to be the most extensive and most systematic research yet made into the properties of penetrating radiation.

The tracks on the photographic plates shown today were of an astonishing variety and complexity. Some were almost straight, some in circles, and some had been made by weaker particles which had struck the lead counter and had been deflected. Sometimes twenty tracks crossed a single plate, giving the appearance of a shower of particles and seeming to radiate from a single point or a number of points.

Analyzing the nature of the particles that produced the tracks, Dr. Blackett concluded that some were every-day negative electrons, but that others were particles with a positive charge but with a mass comparable to electrons rather than protons. These positive electrons, he concluded, seemed to be produced during nuclear collisions, giving rise to showers of particles.

The radiating points were not visible on his photographs, but he showed several slides with tracks not quite parallel and with radiating points apparently a few centimeters outside the photographic plate.

The scientists listening to Dr. Blackett were impressed by his findings, although in some respects they are not entirely new. Last Autumn Dr. Robert A. Millikan's assistant, Dr. Carl D. Anderson, advanced the very tentative conclusion, based on a few random photographs, that cosmic rays included positive particles which could not be protons.

Dr. Blackett asserted today his researches confirmed Dr. Anderson's findings completely, although they threw little or no light on the origin of cosmic rays.

"I have no idea what this discovery will mean," Dr. Blackett admitted after his lecture, saying he was anxious that Dr. Anderson should not be robbed of the credit of first having suggested the idea. "If it has upset scientists' theories, then the theorists must revise them," he added. "Our job was to find the particles and we did."

Fit Atomic Theory, He Says.

Incidentally Dr. Blackett said his discoveries fit "extraordinarily well" with the atomic theory, with which the young Cambridge scientist, P. M. Dirac, caused a sensation before the British Association for the Advancement of Science in 1930.

Professor Charles T. Wilson, inventor of the Wilson cloud chamber and winner of the Nobel prize, said after listening to Dr. Blackett today:

"It is an amazing achievement. It may be that it is not one phenomenon but many."

Lord Rutherford, under whose guidance Dr. Blackett made his experiments, said:

"There is no doubt this new method has given very valuable data for extending our knowledge of penetrating radiation and should make possible a considerable extension of the work being done in America. In general there seems to be strong evidence of the existence of a light positive particle corresponding to the electron. But the whole phenomenon is exceedingly complex and a great deal of work will have to be done on it."

February 17, 1933

JEKYLL-HYDE MIND ATTRIBUTED TO MAN

'Complementarity,' New Theory of Knowledge, Is Presented by Prof. Niels Bohr.

ALL THINGS DUAL IN ASPECT

We Can Know Only One at Any One Time—Scientists Hail Theory as Revolutionary.

By WILLIAM L. LAURENCE.

Special to THE NEW YORK TIMES.

CHICAGO, June 22.—The Theory of Complementarity, a new theory of knowledge, based on the concepts of modern atomic physics, was presented today for the first time before an American audience by Professor Niels Bohr of the University of Copenhagen, originator of the Bohr model of the atom and Nobel Prize winner.

The new theory is expected by some of the leading scientists here, attending the Summer meeting of

PHYSICS AND ASTROPHYSICS

the American Association for the Advancement of Science, to take its place alongside the relativity and quantum theories as one of the revolutionary developments of modern scientific thought.

Complementarity is the outgrowth of relativity, quantum mechanics, and the Heisenberg principle of uncertainty. The new theory expands the uncertainty principle beyond the realm of atomic physics, where it had been primarily applicable, to include man's entire relation to the world around him and to all processes of knowing and thinking.

Briefly, and in non-technical language, Professor Bohr, after a lifetime of contemplation of both the ponderables and the imponderables of the physical and mental world, has come to discover an inherent essential duality in the nature of things, as they relate to man's ability to know them. The paradox of this duality lies in the fact that the Jekyll-Hyde nature of all things is essentially contradictory, with both aspects being true at different times, but with only one aspect being true at any one given time.

Duality Held Inescapable.

In other words, the very process of knowing one aspect of nature makes it impossible for us to know the other aspect. We can know only one side of its nature at any one time. There is a definite discontinuity in all things partaking of existence and of knowledge, so that when one thing is true this very truth perforce makes another thing non-existent as far as any possible knowledge on our part is concerned.

This contradictory duality is inescapable because it lies at the very heart of things.

It is wrong, according to this theory, to say that there are either free will and determinism, causality or chance. Both are essential parts of one and the same reality, the convex and concave sides of the same sphere. Both are true at different times, but it is never possible to know both at one and the same time. When you are inside the sphere the sphere is concave and it is never possible to experience its convex aspect. In that case it is impossible to have any knowledge of the convex aspect. When one gets outside the sphere only the convex side becomes the reality.

Dr. Bohr told the story of how he and Einstein had recently joined intellects to find a way somehow to get around the troublesome principle of uncertainty, according to which it is possible to determine either the position or the velocity of an electron but absolutely impossible to determine both the position and the velocity at the same time.

Thwarted by Uncertainty.

After wrestling with the problem for many days the two super-minds believed for a time that they had accomplished their ends. Jubilantly they began working out the details of their solution only to find themselves in another intellectual vicious circle. After believing that they had succeeded in killing the monster of "uncertainty" it was there, Mephistopheles-like in another form, mocking at them.

This finally convinced both Bohr and Einstein that the "uncertainty principle" is not the result of a lack in our knowledge but is an inherent part of the very mechanism of knowledge. Trying to get around "uncertainty" in nature would be, to use a homely unscientific phrase, like trying to preserve the hole while at the same time wanting to eat the doughnut.

The uncertainty principle, one of the most startling developments in modern physics, was first announced by Professor Werner Heisenberg of Leipzig in 1925, and created as much of a sensation in scientific circles as did the original announcement of the theory of relativity.

Science, depending for its results on accurate measuring instruments and on observation, discovered an inherent weakness in its fundamental structure. It found that in order to observe anything you must have light to see it by. Yet it found to its consternation that light does something to the object observed, changing it somehow, so that actually one could never hope to observe any object at any particular position as it really is before the light has affected it.

Points to Effect of Light.

The effect of the light on an observed object does not mean merely the change in the color of the object. Light, it was found by Professor Arthur H. Compton of the University of Chicago, another Nobel Prize winner, acts in the manner of a small bullet when played on an electron, so that it collides with it and changes its momentum and its energy.

By the intricate laws of quantum mechanics theoretical physics has devised means which make it possible to determine either the position of an electron or its velocity. But the "principle of uncertainty" decrees that one cannot determine both the position and the velocity of an electron at the same time.

In his address Dr. Bohr first pointed out that our recently gained knowledge of atoms and molecules demanded fundamental changes in our general philosophical outlook.

"We have been forced to recognize," Professor Bohr said, "that we must modify not only all our concepts of classical physics but even the ideas we use in every day life—such as our ideas of space and time.

"Indeed, the features which are so characteristic of atomic phenomena are of such a nature that they cannot be analyzed in ordinary mechanical and electro-dynamic concepts, and we have to renounce a description of phenomena based on the concept of cause and effect.

Must Apply Classical Ideas.

"However, since all measurements must ultimately be interpreted on the basis of classical ideas, we cannot desert or modify these ideas but are compelled to apply them also in our description of atomic phenomena.

"We are thus led to a description of the universe which is 'complementary,' that is, in quantum mechanics the application of any classical concept will invariably exclude the simultaneous use of other classical concepts.

"Logically speaking, it is this peculiar 'complementarity' which leaves room for the atomic phenomena, and the principle of uncertainty."

The apparent contradiction between classical physics and modern atomic physics, according to Professor Bohr, does not mean that both one and the other are wrong or that one must be discarded in favor of the other.

On the contrary, Dr. Bohr holds, this realization of the "complementarity" of our knowledge furnishes us with the long-looked-for bridge with which to link the contradictory concepts of classical and atomic physics.

This "bridge" is our recognition as a fact, or rather as an axiom, that we can apply any classical concept to any modern concept of the quantum theory or of relativity, but that we cannot apply any more than one classical concept at the time.

The difficulty in the apparent inability of modern thought to reconcile the old concepts with the new lies in the fact, Dr. Bohr holds, that we tried to apply all the old concepts to all the new concepts simultaneously. If we do it one concept at a time the apparent contradictions disappear.

Dr. Compton Presides.

Dr. Compton, whose work on the collision of photons, such as light and X-rays with electrons, led to the formulation of the "uncertainty principle," and thus directly to the Bohr theory of complementarity, presided at the meeting, which was held in the Illinois Host Building of the Century of Progress Exposition.

After the meeting Dr. Compton remarked that Dr. Bohr's address was a "brilliant presentation, such as one only of Dr. Bohr's calibre was capable of, which brings into harmony the electron theory, the quantum theory and the theory of relativity."

June 23, 1933

NEW 'WATER' BARES SECRETS OF ATOM

Heavy Hydrogen Gives 'Mirror' to Habits of the Molecules, Professor Urey Reveals.

RESEARCH FUNDS GRANTED

Columbia Trustees Provide for Additional Facilities for Making Costly Liquid.

A new magic "mirror" for gazing into the wonderland of primeval matter, through which science may for the first time behold the "dance of the molecules," and "tag" the specific atoms composing them, was described last night before the New York section of the American Chemical Society.

Professor Harold C. Urey of Columbia University, co-discoverer of "heavy water," which has a hydrogen atom twice the mass of ordinary hydrogen, told the chemists how he had employed the new hydrogen, named by him deuterium (from the Greek, meaning "second"), in the rôle of a chemical "peeping Tom," for peering through infinitesimal "atomic windows" hitherto barred to the gaze of man.

Last September Dr. Frank H. Spedding of the University of California told the American Chemical Society how he used the heavy hydrogen in the construction of a "scale of light beams" with which to "weigh" electrons and protons.

Professor Urey also revealed a new process for the making of "heavy water" at the Columbia University laboratories on a larger scale than heretofore possible and at a tenth of the present cost. But even at the reduced price the precious water will still be worth $6,000 a pound.

Columbia already has half a pint of this water, which looks like ordinary water, but boils at 215 degrees, freezes at 39 degrees and kills small plants and animals.

Trustees Aid Research.

Dr. Nicholas Murray Butler, president of Columbia, announced yesterday that the trustees of the university had made increased financial provision for research on the "heavy water." As a result, Professor Urey said, a new laboratory plant has been completed at Columbia which will yield more "heavy water" per day than any other plant in existence.

The new magic "mirror," it was explained, enables the scientist for the first time to obtain greater certainty in regard to fundamental notions as to the behavior of molecules.

For instance, it was not known whether two molecules of hydrogen (each molecule containing two atoms), colliding and separating again, retained their original composition, or whether, in the mad rush, they exchanged partners. Use of the telltale heavy hydrogen molecule, however, has solved this problem.

In Professor Urey's experiments, each molecule collided with ten billion others per second in a vessel, yet the atoms, it was found, did not change partners. Basing his results on theories of molecules built up during the last ten years, Professor Urey was able to predict this reaction last February.

Before the discovery of the new type of hydrogen it was impossible to test chemical reaction theories, especially in solutions, because the compounds, mixed with water, usually contained some of the same kind of hydrogen as the water.

For example, no one could tell what happened when sugar, which contains carbon, oxygen and hydrogen, was first dissolved in water and then recovered in its original form through evaporation. The question was whether the sugar left some of its own oxygen or hydrogen in the liquid and exchanged them for the oxygen and hydrogen in the original water.

Using the heavy water as the solvent, Professor Urey said, is like labeling the atoms red and green, for, when the evaporation process is complete it is easy to decide whether the hydrogen in the sugar is all "light," as it was originally, or whether it had collected some of the "heavy" hydrogen from the "heavy water." Experiments in Europe have shown, Professor Urey told, that the atoms of compounds do "wander about," because the restored sugar did contain heavy hydrogen.

Ordinary Water Reduced.

Ordinary water contains 5,000 parts of the light hydrogen, of atomic weight one (named by Dr. Urey protium, from the Greek, meaning "first"), to one part of the heavy hydrogen, of atomic weight two. The Ohio Chemical Company reduced 4,000 gallons of ordinary water to 150 gallons. In this concentrate the ratio of the heavy to the light hydrogen was 200 to 1, or one-half of one per cent. Columbia University bought fifty gallons of this concentrate and reduced it still further by a process of electrolysis, devised by Dr. Urey and Dr. E. W. Washburn, of the United States Bureau of Standards. Columbia's present supply of the precious liquid in various degrees of purity is slightly under one half a pint. In two or three months the supply will amount to 400 grams, or about one pound, of the highly concentrated liquid.

Professor Urey also offered a possible explanation why heavy water kills small plants and animals. The heavy water, he said, may ionize differently, and thus have a different electrical effect on living things. More important still, it has a different speed in chemical reactions, and thus is possibly disturbing "the nice balance of chemical reactions taking place in living organisms."

The work at Columbia is in charge of Dr. Urey, Professor Victor K. La Mer, who presided last night, and Professor J. Enrique Zanetti.

Dean Frank C. Whitmore, of the Pennsylvania State College, was another speaker at the meeting. Dr. J. M. Weiss, president of Weiss & Downs, Inc., consulting chemists, was elected chairman of the New York section to succeed Dr. La Mer.

December 9, 1933

New Neutron Ray More Powerful Than Either the X-Ray or Radium

New Development at the University of California Has Possibilities in Medicine and for Building Heavy Chemical Elements From Lighter.

Copyright, 1934, by The Associated Press.

BERKELEY, Calif., Jan. 16.—A new ray more penetrating than either X-rays or radium, a ray made of neutrons, the most disruptive substance known to science, has been developed at the University of California.

The ray is a flow of 10,000,000 neutrons a second, coming from a lead window. It is invisible. Although just a baby in the ray class, and potentially the most dangerous ever produced, it has possibilities both for experiments in medicine and for building up the heavy chemical elements out of the lighter.

Neutrons are ultimate particles of matter discovered in England less than three years ago. They differ from the other fundamental particles because they have no electrical charge. This "zero charge" is held to explain their penetration.

They find no obstacle in the heavy "wall" of electricity guarding the nuclei of the atoms of all elements. They shoot through undeflected unless they make a direct hit on a material particle in an atom. If they hit, their weight, 1,800 times the mass of an electron, causes more spectacular results than other types of rays.

Neutrons have been available only in small quantities, their action confined to interiors of vacuum tubes. The California apparatus yields them in great numbers, and for the first time, so far as reported, brings them outside the tube.

This neutron ray was developed by Professor E. O. Lawrence and

Associated Press Photo.
DEVELOPS NEW RAY.
Professor Ernest O. Lawrence.

Dr. M. Stanley Livingston. It is made in the magnetic field of a magnet as high and wide as a barn door, weighing eighty-five tons. The south pole of this magnet rises, flat topped, as high as a stove from the floor. Inverted directly above is its twin, the magnet's north pole.

Into the open space between these two poles the scientists slide a metal disc resembling a covered frying pan, its interior a vacuum. Inside the vacuum pan lies what looks like a large, flat, round, brass pill-box, called the "merry-go-round."

The pill-box really stands still, but charged particles inside circle around at fabulous speeds—50,000 miles or more a second during incubation of this newest ray.

The rotation is due to opposite sides of the pill-box acting as alternating poles. They alternately pull the particles toward them, while the immense magnetic field keeps the motion going in rings. Pretty timing of alternations causes a complete merry-go-round of particles.

As ammunition, a stream of deutons is fed by tube into the vacuum pill-box. Deutons are nuclei of the recently discovered heavy hydrogen atoms. They are the richest source yet found for neutrons, being almost entirely a mixture of neutrons and protons.

Passing through a spray of electrons, these deutons acquire electrical charges and begin circling. With only 12,000 to 20,000 volts of current to the pill-box alternating poles, the deutons reach speeds having the energies of 2,000,000 volts.

They emerge from the rim of the box and smash themselves utterly against a metal target. From the impact come clouds of neutrons and protons. But protons cannot penetrate much lead. So a lead window filters out the protons, leaving a pure spray of neutrons to emerge from the "frying pan" tube.

Paraffin Proves Presence.

As they are invisible, the first job was to prove their presence. Paraffin did this. By a strange quality, neutrons penetrate heavy elements, like lead, much easier than light ones. Least of all do they pierce hydrogen, the lightest element. Paraffin contains much hydrogen. The neutrons, striking the paraffin in a chamber outside the lead window, hit hydrogen nuclei, smash them, and, with the recoil, set free hydrogen protons. These latter are visible in an ionization chamber. In a second, 5,000 recoil protons appear.

This means 10,000,000 neutrons a second spraying the paraffin. Theoretically the present apparatus, by using a larger "race track," can multiply this ray to 100,000,000 neutrons a second—or tenfold.

These neutrons can pierce the human body with greater ease than X-rays. They also penetrate shields which ordinarily screen X-rays from doing indiscriminate harm. The body, being largely water, is full of hydrogen. While hydrogen is the best stopper for neutrons, it is also the stuff in which neutrons produce very frequent atomic explosions.

Experiments of Dean George B. Pegram and John R. Dunning at Columbia University upon neutrons penetrating into water have indicated that such a ray as the new one at California will go more effectively than X-rays into the deep portions of the human body.

German Tests Not Favorable.

Use of X-rays on deep-buried cancer is limited, for one thing, by the amount of burning the skin will stand. For the X-rays burn the surface more than the parts deep below. It has been suggested that neutrons will not have extra-surface burning effect, but will scour equally at all depths.

On the other side of this prospectus are experiments in Germany by Nora Feightinger, and elsewhere by other scientists, indicating that neutrons may not be more effective on human diseased tissues than X-rays.

X-rays function by setting off a flow of electrons in the flesh. The electrons do the burning. Neutrons, instead of starting the electrons, burst the atomic nuclei and set off particles much heavier than electrons.

These heavier particles might act differently on human tissue. The experiments causing doubt were made on tiny protozoan animals, and not with neutrons, but with alpha radiation, a type of particle resembling those which neutrons would set into action.

For the important scientific job of smashing atoms, to learn how matter is made, neutrons are the best projectiles.

But in addition, Drs. Lawrence and Livingston have found building possibilities. The light metals lithium and aluminum were both made heavier by neutron bombardment. In each case a neutron, instead of breaking a nucleus, joined it, producing a heavier atomic weight isotope of the metal.

January 17, 1934

PHYSICS AND ASTROPHYSICS

Radioactivity Is Produced Artificially By Daughter of Curies and Her Husband

Copyright, 1934, by Science Service.

LONDON, Feb. 1.—Artificial radioactivity has been produced for the first time by Professor F. Joliot and Mme. Irene Curie-Joliot, the famous Paris physicists who are husband and wife. Mme. Curie-Joliot is the daughter of the discoverers of radium.

This achievement has stirred interest at the famous Cavendish Laboratory, Cambridge, where experiments attempting confirmation are in progress. There is hope that through artificial radioactivity medically useful radiation will be produced.

The artificial radioactivity produced by the Joliots consists of obtaining positrons, or positive electrons, from the bombardment of boron with alpha particles. The important fact is that the activity or disintegration produced continues for many minutes after the bombardment is stopped. Boron decays exponentially to 30 per cent in fifteen minutes. Similarly, artificial radioactivity proceeds in aluminum and magnesium. The decay period in aluminum is four minutes.

Lord Rutherford, the famous Cambridge physicist, said:

"It is remarkable that the life of the unstable atom produced is as long as it is. We do not know whether the atoms so far made artificially radioactive are typical or whether other unstable atoms which may be produced will have a longer or shorter life.

"The discovery of the Joliots shows how little we really know about radioactivity."

The mechanism of the artificial radioactivity of boron is interpreted to mean that a boron atom and the helium nucleus or alpha particle unite to form a neutron and an unstable nitrogen atom of weight thirteen which, in turn, changes to a carbon atom of weight thirteen with the release of a positron. The positron is the new particle discovered in 1932 at Pasadena, Calif.

February 2, 1934

PREDICTS FINDING OF FIVE-PART ATOM

Dr. Langer of California Tech Says New Particle Will Clarify Nucleus Theories.

'NEUTRINO' A LIKELY NAME

Its Mass, or Energy, Is Put at 6,000,000 Volts as Against Neutron's 1,006,000,000.

Special to THE NEW YORK TIMES.

PASADENA, Calif., March 7.—Dr. R. M. Langer, physicist of the California Institute of Technology, predicts the discovery within a few months of a new atomic particle destined to complete the picture of the nucleus of the atom and eliminate present discrepancies in nuclear theories.

Speaking at a conference of distinguished scientists in Norman Bridge Laboratory last night, he said that a mathematical analysis of the results of world-wide experiments indicated that the mass of the neutron was "1.006."

As it is impossible directly to weigh the neutron, its mass is expressed in terms of its energy. Thus the figure 1.006 would mean that the neutron's mass or energy is 1,006,000,000 volts of electricity.

In addition to the atomic particles known as the proton, neutron, electron and positron, Dr. Langer predicted that a fifth "building block of matter" would soon be discovered. He said that it was first suggested two years ago at the institute by Dr. Pauli of Switzerland.

Dr. Langer said that the new fundamental particle, which would probably be named the "neutrino," or little neutron, as suggested a few weeks ago by an Italian scientist, would resemble the neutron in carrying neither a positive nor negative electric charge, but would differ in possessing a much smaller mass, amounting in terms of energy to about 6,000,000 volts.

University of California physicists recently completed experiments which fixed the mass of the neutron at about 5,000,000 volts less than that set by Dr. Langer, who pointed out that Dr. C. C. Lauritsen and Richard Crane had suggested that this discrepancy might have resulted from a carbon film covering the substances used in the experiments.

The highest mass found for the neutron, 1,012,000,000 volts, resulted from the work of Curie-Joliot in France, this being 6,000,000 volts higher than Dr. Langer's estimate.

Proof of the existence of a "neutrino," Dr. Langer said, would "solve most of the differences in nuclear pictures."

March 8, 1934

Italian Produces 93d Element By Bombardment of Uranium

Professor Enrico Fermi, Academician, Uses Neutrons Formed by Decomposition of Berylium Under the Action of Alpha Particles of Radium.

Wireless to THE NEW YORK TIMES.

ROME, June 4.—Speaking at a meeting of the Academy of Lincei, which was attended by King Victor Emmanuel, Senator Mario Corbino, who enjoys international reputation as a physicist, announced that the Italian Academician, Enrico Fermi, had produced artificially a new chemical element. The new element was produced by subjecting uranium to a bombardment of neutrons produced by decomposition of berylium under the action of alpha particles of radium.

Uranium, as is known, hitherto had the highest atomic weight of all the elements and occupied the ninety-second place in the Mendelieff scale. The new element occupies ninety-third place. Its production is, therefore, of enormous importance, because all the new elements which have hitherto been found since the discovery of uranium have merely filled in some gaps in the range of elements between hydrogen and uranium.

The new element produced by Professor Fermi is completely outside the range of elements hitherto known to exist on the earth. It opens the possibility that the number of elements which can be produced by artificial means is infinite.

The new element is radioactive and falls in its proper place in the Mendelieff table, having properties akin to manganese and cerium. It is an unstable element, as thirteen minutes are sufficient for 50 per cent of its atoms to decompose. This explains why it is not found in nature, since it can exist permanently only under the influence of the special forces which are necessary for its creation.

"New Youth" in Old Material.

By The Associated Press.

ROME, June 4.—Senator Mario Corbino told his distinguished audience at the Lincei Academy today that "the study of the nucleus, now in its beginning, aims at the ambitious plan of giving back to the earth the youthfulness of its materials."

"One can now affirm," he declared, "that this ambitious design has been converted into reality." Professor Enrico Fermi's discovery, he added, "is in effect a manifestation of new youth communicated by nuclear collision to old established material."

The new element was described as the hardest element known to man.

Dozen Artificial Elements.

More than a dozen artificial elemental physicists both here and abroad since the end of January this year by methods similar to those employed by Professor Fermi.

The pioneering work in the artificial creation of new elements that had not been found in nature was done by Irene Curie, daughter of the discoverers of radium, and her husband, F. Joliot of the famous Curie Radium Institute in Paris. It was this young scientific team that first announced the production of artificial radioactive elements on Jan. 31, this year.

Since the original Curie-Joliot announcement the artificial creation of new radioactive elements has been duplicated by the workers in the famous Cavendish Laboratory, University of Cambridge, England, and in the American physical laboratories of the University of California and the California Institute of Technology.

Professor Fermi's work is, therefore, an extension of the original work of the Curie-Joliot experiments and of the British and American scientists. The Italian physicist's contribution is the first application of the discovery of the French scientists to the heaviest of the known natural elements, uranium, thus extending the Periodic Table of the ninety-two elements of which physicists and chemists believe the material universe to be constituted at present.

Joliots Pioneers in Field.

Curi and Joliot first produced their artificial new elements by bombarding the elements boron, magnesium and aluminum with alpha particles from polonium, a radioactive element discovered by the mother of Irene Curie. In doing so they produced radioactive forms of nitrogen, silicon and phosphorus, for which they proposed special names, radionitrogen, radiosilicum and radiophosphorus.

The Cavendish Laboratory workers corroborated the result of the Joliots by bombarding carbon with high velocity protons, or hearts of hydrogen atoms. The California physicists used deutons, or hearts of hydrogen of mass two, in their experiments, in which they bombarded more than a dozen of the lighter elements and produced new artificial radioactive elements in each case.

Professor Fermi, although only thirty-two years old, is one of the best known of Italian physicists. Last Summer he attended the meeting of the American Association for the Advancement of Science at Chicago as one of the foreign guests of the association and the A Century of Progress Exposition.

June 5, 1934

ENERGY MULTIPLIED 200,000,000 TIMES

Yield From Atom Increased by Columbia Men to That Ratio Over Force Applied.

'SLOW NEUTRONS' USED

Control of 'Bullets' Described at Capital Science Session— Key to Vast Power Seen.

By WILLIAM L. LAURENCE.
Special to THE NEW YORK TIMES.

WASHINGTON, April 26.—The release of atomic energy 200,000,000 times greater than the amount of energy applied was reported here today by physicists of Columbia University before the American Physical Society. The development was hailed as the greatest advance so far toward the practical utilization of the inconceivably large quantities of energy known to be locked in the heart of atoms.

Neutrons slowed to travel with energies of one-fortieth of a volt were shot straight into the hearts of lithium atoms with such accuracy and destructive force that the lithium atom yielded up energy corresponding to 5,000,000 volts, a yield of 200,000,000 per cent on the original investment.

This is a million times greater than the largest amount of atomic energy obtained by other methods, it was reported.

The work was carried on at the Columbia University physics department by Drs. J. R. Dunning, G. B. Pegram, G. A. Fink and D. P. Mitchell. Dr. Dunning presented the report.

Hint Provided by Fermi.

The first hint of a means for "taming" the neutron was obtained last December by Professor Enrico Fermi of Rome, Italy, who succeeded several months ago in artificially creating Element 93.

Ordinary neutrons travel with tremendous speeds and were found to be absolutely "wild." Professor Fermi tried to establish that by slowing them down, they might be controlled.

In "firing" neutrons at atoms he catapulted them through paraffin. This reduced their energy and their speed.

Taking the clue from Professor Fermi, the Columbia physicists proceeded even more earnestly to tame the neutron by the process of slowing it down.

Drs. Dunning and Pegram further discovered—and this was very startling indeed—that the hearts of the atoms literally expanded enormously in their diameters when a "slow neutron" was approaching them.

This "atomic heart expansion" was found to vary greatly with the atoms of the various elements.

The heart of hydrogen, it was found, becomes eight times as large for a slow neutron as it is for a fast one; the heart of lithium becomes thirty times as large, while that of boron becomes 500 times as large.

Target Is Enlarged.

The amount of the enlargement of "the atom heart" means, of course, the corresponding enlargement in the same ratio of the effective target.

The nucleus of the atom seems to act as a "bullseye magnet" for the slow neutron. It literally seems to draw the "slow neutron" straight to its very centre. And in doing so it literally "commits suicide" and yields up all its vast energy in one terrific burst.

The most startling display of "suicidal mania" was found in the case of the cadmium atom. When a slow neutron advances toward it the cadmium nucleus becomes enlarged 10,000 times in its diameter as compared with its size on the approach of a fast neutron. The chances of a slow neutron hitting the nucleus of the cadmium atom, as compared with the chances of a fast neutron, are as 10,000 to 1.

Science at last has found a "key for the cupboard," as Sir Arthur Eddington called it, of the vast store of energy locked up within the atom.

The Columbia scientists have also undertaken to determine if "cooling" of "slow neutrons" has an effect. The results are so recent, obtained only a few days ago, that their report was only preliminary.

More Effective When Cold.

They found that "cold slow neutrons" are even more effective as atomic "range-finders" than "hot slow neutrons."

Using lithium as a target they found that the percentage of direct hits for "cold neutrons" was, respectively, 10, 4 and 9 per cent greater than the direct hits, large as they were, obtained by the use of "hot neutrons."

To cool off the neutrons the scientists made them pass through an atmosphere of liquid air.

While a neutron will penetrate three to four centimeters of lead, it is stopped by a thin plate of cadmium. This discovery promises to be of considerable importance in the field of health.

This absorption power of cadmium for neutrons makes it possible, Dr. Dunning stated, to make tubes of cadmium for "canalizing" the neutrons—that is, to make them travel in a directional beam through a slit, as is done in the case of X-rays and light.

All that science needs now, Dr. Dunning declared, is a larger source for obtaining neutrons. A few minutes after he had said this Dr. Ernest O. Lawrence of the University of California revealed that his new method for splitting the deuteron, heart of the double-weight hydrogen atom, into its component parts, one proton and one neutron, increases the neutron source used by Dr. Dunning and his colleagues a thousand times.

ATOMIC RESEARCH REVEALS PARTICLE

Harvard Evidence of 'Neutrino,' 1,800 Times Lighter Than Neutron, Given to Scientists.

DETECTED BY HUGE SCALE

'Law-Defying' Isobars Found by Device Weighing Million-Million-Millionths of Gram.

By WILLIAM L. LAURENCE.
Special to THE NEW YORK TIMES.

ST. LOUIS, Jan. 3.—The first experimental evidence of the existence of another fundamental particle of matter, which scientists have been hunting for the past few years in the primordial atomic jungle without success, was reported today before the meeting of the American Association for the Advancement of Science by Professor Kenneth T. Bainbridge and Dr. Edward B. Jordan of the Research Laboratory of Physics, Harvard University.

The newest of nature's building blocks, one of the two "atomic ghosts" that have been haunting modern physicists and could not be laid to rest, is named the neutrino. It resembles the neutron in that it has no electric charge, but it differs from it considerably in mass, being about 1,850 times lighter than the neutron.

This first glimpse of the elusive neutrino, which has baffled scientists so much that some doubted it could exist, was made possible by the construction at Harvard by Dr. Bainbridge and Dr. Jordan of a new "atomic scale" for the weighing of individual atoms, known as a mass-spectograph.

Fine Distinctions in Weight.

This new scale, the most sensitive and accurate of its kind ever built, which in itself weighs several tons, actually weighs the individual atoms of the elements of the order of a million-million-millionths of a gram.

So delicate is the scale that it can detect the difference in weight between an atom of hydrogen of mass 1 and an atom of hydrogen of mass 2. The atom of the double-weight hydrogen, known as deuterium, has a mass almost, but not quite, double that of ordinary hydrogen atoms, but the difference is so infinitesimal in terms of grams as to be practically meaningless to the human mind.

Similarly, this supersensitive atom scale can detect the difference in weight between an atom of oxygen, of atomic weight 16, and a molecule of methane gas, which is composed of one atom of carbon, of weight 12, and four atoms of hydrogen, each of atomic weight 1, the total thus being also 16. Here, too, the actual difference is so slight as to be inconceivable.

Source of Neutrino Evidence.

The evidence for the neutrino comes as a product of another important Harvard discovery made by means of the new atom scale.

It has become known in recent years that each of the eighty-nine of the ninety-two chemical elements that have so far been actually discovered has what is known as isotopes, a species of identical twins of the atomic world. These isotopes occupy the same place in the periodic table of elements, yet differ slightly in their atomic weight.

For example, oxygen stands at No. 8 in the periodic table, yet it has been found that there are at least three oxygen "twins," or isotopes, one with atomic weight 16, a second of atomic weight 17, and a third of atomic weight 18.

Similarly there are two hydrogen twins, both occupying place No. 1 on the periodic table, yet the first has an atomic weight of 1, while the second weighs twice as much.

The difference is accounted for by the contents of the nucleus, or core, of the atom, in which the mass of the atom is concentrated.

Oxygen of atomic weight 16, for instance, has a nucleus composed of eight protons and eight neutrons, while oxygen of atomic weight 17 has a nucleus of eight protons and nine neutrons, the one extra neutron accounting for the additional unit in its mass.

Each added unit in atomic weight is thus accounted for by the addition of an extra neutron in the atom's nucleus.

The units of electric charge in the atom, contributed, respectively, by the evenly balanced number of negatively charged electrons in the outer shell of the atom and positively charged protons in the inner nucleus, are exactly the same for all the twins of each element.

So far 250 different chemical twins, or isotopes, of the same electric charge, but different in weight, have been discovered.

Isobars of Different Elements.

Within recent times, however, another species of chemical twins have been discovered to complicate the already intricate picture of the atom.

These new species are known as isobars. Unlike isotopes, they do not belong to the same chemical elements.

Instead, they were either neighbors, or second-door neighbors, on the periodic table and therefore differ in their atomic charges. Yet, strange to say, they have the same atomic weights.

This applies only to certain of the isotopes, or second-door neighbors, of some of the elements. That is, some of the isotopes of two different elements have the same atomic weight. Such isotopes of neighbor or second-door neighbor elements are known as the isobars.

There was no trouble in understanding the isobars of elements that were only second-door neighbors. But, according to modern

April 27, 1935

24

PHYSICS AND ASTROPHYSICS

theoretical physics, isobars of elements that are real neighbors could not be stable and therefore must be radioactive.

Here comes one of the important discoveries made at Harvard with the latest atom-scale, which led indirectly to the evidence for the existence of the neutrino.

Dr. Bainbridge and Dr. Jordan found with their new device six isobars of neighbors on the periodic table of elements. Yet, to their surprise, they violated all laws of modern physics by refusing to be radioactive, or unstable. Instead, they have been found to be as stable as any other well-behaved ordinary isotope.

The "law-defying" isobars belong to the isotopes of the "respectable" elements cadmium, indium, tin, antimony and tellurium.

Now these elements are all real neighbors on the periodic table. Cadmium occupies a place at No. 48; indium is at No. 49; tin, at No. 50; antimony, at No. 51, and tellurium at No. 52.

Cadmium has an atomic weight of 112; indium, 114; tin, 118; antimony, 121, and tellurium, 127. These weights, of course, refer only to respective isotopes of these elements.

Dr. Bainbridge and Dr. Jordan, with their new atomic scale, found an isotope of cadmium and indium, next-door neighbors, with the same atomic weight of 113; an isotope of indium and tin, also next-door neighbors, of atomic weight 115; an isotope of antimony and tellurium, two other next-door neighbors, of atomic weight 123.

Properties of "Neutrino."

Since these isotopes are all next-door neighbors and yet have the same atomic weight, they are thus true isobars of the type that the theoretical laws of physics decree to be unstable and radioactive.

Yet they have been found to be very stable and well-behaved. It is as though one were to discover a lion that refuses to eat meat.

The best possible explanation for the tameness of these newly discovered "ferocious beasts of the atom jungle," Dr. Bainbridge and Dr. Jordan find, is that there must be some strong force to play the rôle of "lion tamer."

This rôle exactly fits a particle that has no electric charge and has a mass somewhat less than the combined separate weights of the electron and the positron, two of the fundamental units of matter.

Because of its similarity to the neutron in its lack of an electric charge, and because of its being much lighter in weight than the neutron, the new particle which fits the picture is named the "neutrino," or little neutron.

The latest addition to the atomic world brings the number of elemental particles up to five. Science now has the electron with a negative charge; the positron, which has the same weight as the electron but is positively charged; the proton, which has a positive charge but weighs 1,842 times as much as the electron or the positron; the neutron, which has no charge, and a weight of the same order as that of the proton, and, finally, the neutrino, which has no charge and a weight less than that of the combined separate weights of the electron and the positron.

January 4, 1936

Atom Explosion Frees 200,000,000 Volts; New Physics Phenomenon Credited to Hahn

By The Associated Press.

WASHINGTON, Jan. 28.—American scientists heard today of a new phenomenon in physics—explosion of atoms with a discharge of 200,000,000 volts of energy.

Theoretical physicists attending a meeting sponsored by the Carnegie Institution of Washington and George Washington University said that Dr. Enrico Fermi of the University of Rome told yesterday that this had been accomplished by Dr. G. Hahn of Berlin.

The report so stirred the limited circle of scientists with facilities to carry on such experiments that work on attempts to duplicate Dr. Hahn's accomplishment has begun at the Carnegie institution's terrestrial magnetism laboratory and at Columbia University.

Scientists at the meeting said the discovery was comparable in significance to the original discovery of radioactivity thirty years ago.

They said that it was too soon to discuss possible applications of the new 200,000,000-volt force, which is thirty times more powerful than radium, but pointed to the fact that radium is now the most efficient weapon used for the treatment of cancer. Like radium, it may be twenty or twenty-five years before the phenomenon could be put to practical use and it might not be practical at all, they said.

Dr. Fermi related that Dr. Hahn bombarded a synthetic element known as "ekauranium" with neutrons, the slow-moving particles of the atom, and produced barium, the substance used in making X-ray pictures of the stomach and intestines.

The only way that this could occur, according to physicists, would be for the ekauranium atom to split apart to form barium and the rare element masyrium.

In causing such a split a force of 200,000,000 volts would be generated since atoms are held together by electrical forces many hundred times more powerful than the force of gravity which holds the stars, planets, sun, earth and moon in their orbits.

January 29, 1939

VAST ENERGY FREED BY URANIUM ATOM

Split, It Produces 2 'Cannonballs,' Each of 100,000,000 Electron Volts

HAILED AS EPOCH MAKING

New Process, Announced at Columbia, Uses Only 1-30 Volt to Liberate Big Force

The splitting of a uranium atom into two parts, each consisting of a gigantic atomic "cannonball" of the tremendous energy of 100,000,000 electron-volts, the greatest amount of atomic energy so far liberated by man on earth, was announced here yesterday by the Columbia University Department of Physics in a statement by Dean George P. Pegram of the Columbia Graduate Faculties.

The splitting of the uranium atom, it was said, constitutes an entirely new atomic process, the possibility of which did not even occur to any of the world's atom smashers. This new process, it was added, "yields the largest conversion of mass into energy that has yet been obtained by terrestrial methods."

Small Force Used for Splitting

One of the most startling phenomena in this newly discovered atomic process is the relatively small amount of energy necessary to liberate the enormous amounts developed through the splitting of the uranium atom. The uranium atom is split by means of neutrons, that is, neutral atomic particles carrying no electrical charge. These neutron bullets travel with energies of only one-thirtieth of a volt. Yet they produce two atomic "cannonballs" of a total of 200,000,000 electron-volts, representing an energy 6,000,000,000 times greater.

The Columbia announcement came under dramatic circumstances, following word received here last week of startling developments along similar lines in physical laboratories in Germany and Denmark. Two European Nobel Prize winners in physics, both of whom are now in this country, took a prominent part in the work. They are Professor Enrico Fermi of Rome, Italy, now at Columbia University, and Professor Niels Bohr of Copenhagen, Denmark, now at the Institute for Advanced Study at Princeton, N. J.

It was Professor Fermi who first fired neutron bullets into uranium, the heaviest element found in nature. Professor Fermi believed he had succeeded in creating an element heavier than uranium, which, being unstable, disintegrated into an isotope (twin of an element) of radium.

The work was continued at the Kaiser Wilhelm Research Institute for Chemistry at Berlin-Dahlem, Germany, by Dr. Lise Meitner and Professor Otto Hahn, who had been working together for many years. Dr. Meitner was discharged last year for racial reasons and she went to Stockholm, Sweden.

Professor Hahn continued his research with a new associate, Dr. F. Strassmann. On checking on Dr. Fermi's work they found to their great amazement that the uranium bombarded with neutrons, instead of disintegrating, as Professor Fermi thought, into an isotope of radium, a close neighbor of uranium on the atomic scale and nearly of the same atomic weight, formed the much lighter element barium. Uranium has an atomic weight of 238, that of radium is 225, whereas the atomic weight of barium is only 137.

Professor Hahn and Dr. Strassmann reported their startling observations on Jan. 6 without offering any theory to explain the new phenomenon. Never before had it been observed, or even suspected, that an element so far removed on the atomic table (uranium occupies No. 92 on the Periodic Table of Elements, while barium occupies No. 56) and so much lighter could be created from another element so much heavier.

Process Analyzed in Sweden

The exiled Dr. Meitner, in Stockholm, was continuing this work in collaboration with Dr. R. Frisch, a colleague of Dr. Bohr at the Institute of Theoretical Physics of the University of Copenhagen. When the work of their German colleagues came to their attention they came to the conclusion that they were here dealing with a new atomic process. They were the first to realize that what was happening was the actual splitting of the uranium atom, of atomic weight 238, into two lighter atoms, barium, of atomic weight 137, and possibly masurium, of atomic weight 97, or krypton, of atomic weight 82.

While the creation of the barium has been determined by physicochemical tests, the identity of the second element split off the uranium is still undetermined.

Dr. Frisch told of his and Dr. Meitner's findings to Dr. Bohr, who, in turn, told Dr. Fermi on his arrival in this country recently. Dr. Fermi computed that if heavy uranium was split in two an enormous amount of energy, approximately 200,000,000 electron volts, must be liberated. He proposed last week that the experiment be performed at Columbia, where the new 150,000-pound cyclotron (atom-smasher) had just been installed.

The experiment was undertaken last Wednesday. Protons (nuclei of hydrogen atoms), catapulted with energies of 10,000,000 volts, were hurled at lithium atoms. This liberated neutrons of approximately the same order of energies. These neutrons, in turn, were slowed down to one-thirtieth of an electron-volt and fired at the uranium atoms.

Those who participated in the Columbia experiments with Professor Fermi were Professor John R. Dunning, Dr. G. Norris Glasoe, Dr. Eugene T. Booth, Dr. Herbert L. Anderson and Professor Francis G. Slack of Vanderbilt University.

Word of these experiments spread and on Friday the physicists at the Carnegie Institution in Washington corroborated the Columbia results.

January 31, 1939

REVOLUTION IN PHYSICS

Great news came out of the physical laboratories of Columbia University and the Kaiser Wilhelm Institute the other day. Slow neutrons were hurled at uranium. Out came two complete atoms, the one barium, the other still to be identified. In addition, the energy released, which is of the order of a hundred million volts, far exceeds that of the neutron that does the shattering. Nothing like transmutation of matter or this conversion of mass into energy has ever been obtained before. A new chapter in physics remains to be written, with consequences that cannot be foreshadowed.

The spontaneous disintegration of uranium, for instance, proceeds at so regular a rate that it serves as a kind of clock to measure the age of the earth. What if the accepted theory of disintegration is wrong? Obviously, the whole problem of the earth's age, which has engaged geophysicists for many years, must be re-examined. And what of the energy of the sun and the stars? Latterly it has been accounted for by assuming that under terrific temperature and pressure matter is annihilated, a process which has nothing in common with ordinary combustion and which entails the complete conversion of mass into heat, light, X-rays and the like. Now we behold uranium giving off tremendous amounts of energy without annihilation. Sir James Jeans and Sir Arthur Eddington will have to revise the theories which they have evolved to explain why the sun and the stars shine.

Lastly, the possibility of harnessing the energy of the atom crops up again. Rutherford, Millikan and other distinguished physicists did their best in late years to discourage speculation on the subject, because bombardment was so inefficient that more energy was expended on the atom than ever came out of it. Now the picture is changed. Though it is still necessary to fire hundreds of millions of bullets at uranium before a single hit is scored, the amount of energy released is enormous. Romancers have a legitimate excuse for returning to Wellsian utopias where whole cities are illuminated by energy in a little matter—this time uranium.

February 3, 1939

6 ELEMENTS FOUND IN URANIUM ATOM

Physicists Bare Discovery of Greatest Amount of Energy Liberated Thus Far

REPORT WIDELY HAILED

Professors Bohr and Fermi, at Columbia Meeting, Tell of Atomic 'Cannon Ball'

The creation of a half dozen of the heavier elements out of uranium, accompanied by the release of tremendous quantities of atomic energy, were reported yesterday at Columbia University by two of the world's leading physicists, Professor Niels Bohr of Denmark and Professor Enrico Fermi of Italy, both Nobel Prize winners and pioneers in the unravelling of the nature of the nucleus which hides the secret of the constitution of matter.

Professor Bohr is now at the Institute for Advanced Study, Princeton, N. J., and Professor Fermi is at Columbia. Both men, who were not scheduled on the program, held their audience of 300 distinguished physicists fascinated by their reports on what is being hailed as "the most sensational discovery in modern physics since the discovery of radioactivity more than forty years ago."

This refers to the discovery made about a month ago in Europe that by bombarding uranium with a slow neutral particle (neutron) of an energy of only one-thirtieth of a volt, the uranium atom is split into two heavy elements, each constituting a gigantic radioactive atomic "cannonball" of 100,000,000 volts each. This is the greatest amount of atomic energy so far to be liberated by man on earth.

Only One Named Till Now

Until now only one element, barium, had been definitely identified as one of the halves of the split uranium atom. Yesterday it was reported that the smash-up of the uranium yields at different times a number of other heavy elements not suspected before. These are krypton, strontium, tellurium, iodine and xenon.

Uranium, the last and heaviest on the table of elements, has an atomic number (which corresponds to the number of positively charged electrical particles in the nucleus) of 92. The atomic numbers of the elements created by the uranium split are: krypton, 36; strontium, 38; tellurium, 52; iodine, 53; xenon, 54, and barium, 56. Some of these elements are not transmuted directly from the uranium, but are break-down products after the first split.

The work on the newest "fountain of atomic energy" is going on feverishly in many laboratories both here and in Europe, it was reported by Professors Bohr and Fermi. It constitutes the biggest "big game hunt" in modern physics, opening up a new milestone in man's mastery over the elements and marking the most important step yet made by science toward the transmutation of the elements and the utilization of the vast stores of energy locked up within the nuclei of atoms.

The new method for the release of atomic energy and the transmutation of the elements is regarded as the nearest approach yet to be made to the finding of a modern version of the "Philosophers' Stone" of the alchemists. Its discussion by two of the greatest authorities in the field came as a surprise to the physicists attending the joint meeting of the American Physical Society and the Optical Society of America, as the official program gave no hint of the event.

Research Is Cited

So recent is the discovery and so feverish is the research going on that most of it still remains to be correlated and explained, Professor Bohr said. The find has been so startling that it has left the scientists in a state of breathless wonder, and the general feeling prevails that physics is now on the eve of epoch-making discoveries.

A compact direct-current electrostatic generator that produces the most powerful X-rays yet to be produced, at a potential of 1,250,000 volts, was described at the meeting by Dr. John G. Trump and Professor Robert J. Van de Graaff of the Massachusetts Institute of Technology.

This powerful new tool, to be used for engineering and scientific research and for the treatment of malignant diseases (cancer), is only one-tenth the size of a 1,000,000 volt generator also designed by the Boston Tech scientists and installed at the Huntington Memorial Hospital in Boston since 1937.

February 25, 1939

VISION EARTH ROCKED BY ISOTOPE BLAST

Scientists Say Bit of Uranium Could Wreck New York

WASHINGTON, April 29 (P).—Tempers and temperatures increased visibly today among members of the American Physical Society as they closed their Spring meeting with arguments over the probability of some scientist blowing up a sizable portion of the earth with a tiny amount of uranium, the element which produces radium.

Dr. Nils Bohr of Copenhagen, a colleague of Dr. Albert Einstein at the Institute for Advanced Study, Princeton, N. J., declared that bombardment of a small amount of the pure Isotope 235 of uranium with slow neutron particles of atoms would start a "chain reaction" or atomic explosion sufficiently great to blow up a laboratory and the surrounding country for many miles.

Many physicists declared, however, that it would be difficult, if not impossible, to separate Isotope 235 from the more abundant Isotope 238. The Isotope 235 is only 1 per cent of the uranium element.

Dr. L. Onsager of Yale University described, however, a new apparatus in which, according to his calculations, the isotopes of elements can be separated in gaseous form in tubes which are cooled on one side and heated to high temperatures on the other.

Other physicists argued that such a process would be almost prohibitively expensive and that the yield of Isotope 235 would be infinitesimally small. Nevertheless, they pointed out that, if Dr. Onsager's process of separation should work, the creation of a nuclear explosion which would wreck as large an area as New York City would be comparatively easy. A single neutron particle, striking the nucleus of a uranium atom, they declared, would be sufficient to set off the chain reaction of millions of other atoms.

April 30, 1939

MESON CONFIRMED AS ATOMIC BINDER

Prof. Bethe of Cornell Finds Mathematical Basis for Attraction Within Nucleus

SOURCE OF FORCES TRACED

Elementary Particle 200 Times Electron's Mass Transmits Essential Nuclear Energy

Special to THE NEW YORK TIMES.

ITHACA, N. Y., Jan. 13—Mathematical confirmation for the first time that the forces holding the nucleus of the atom together are transmitted by the recently found meson, the elementary particle, was presented today at Cornell University by Professor Hans Bethe.

The hypothesis envisaging the function of the meson was made by Yukawa in Japan in 1935, before the meson was actually discovered. He postulated the existence of a particle about 100 to 200 times the mass of the electron and showed that the most important features of nuclear forces could be explained if they were transmitted by such a particle.

The meson was first recorded two years ago in cosmic radiation by Anderson and Neddermeyer of the University of California and simultaneously by Street and Stevenson of Harvard.

Ever since the discovery of the electron scientists have been finding new particles inside the atom. They first attributed the force that holds the atom together to the electric attraction between the electrons and protons. This electrical force was soon found too small, however, to account for the tremendous attraction in the nucleus.

Likened to Light Quanta

Professor Bethe's findings confirm the role of the meson as the binding force within the atomic nucleus.

"The meson transmits the energy inside the nucleus from one particle to another in a manner similar to that in which light quanta transmit the energy from one atom to other atoms," he states. "But, while light quanta have no mass, the meson has a mass about 200 times that of an electron."

According to the Einstein equivalence of mass and energy, a tremendous energy is required to shake a particle of such a large mass loose from the atomic nucleus. Based on mass, it can be calculated that 80-100 million electron volts would be required for the emission of a meson from the nucleus.

Such energies are not yet available in nuclear physics laboratories, the highest energy as yet obtained being about 16,000,000 electron volts from the latest cyclotron built by Professor Lawrence at the University of California. But physicists hope that in the near future they will actually produce the enormous energies required to unlock the door for the meson.

According to the Cornell physicist there are three different kinds of mesons, each having a different electrical charge. In cosmic radiation experimenters have found mesons which were either positively or negatively charged.

"To explain nuclear forces," Professor Bethe states, "it is necessary to assume also the existence of a third kind of meson which is electrically neutral."

Action Upon Simple Nuclei

Professor Bethe investigated the consequence of the meson theory of nuclear forces for the structure of simple atomic nuclei. He found that the deuteron, the nucleus of a heavy hydrogen atom, should have the shape of a football rotating about its long axis. Earlier theories of the nucleus had predicted a spherical shape for the deuteron.

The football shape is exactly what had been found experimentally a few months before by Professor Rabi at Columbia University. His experiments gave a certain value for the deviation from the spherical shape. Bethe's calculations yield exactly the same value on theoretical grounds. Thus mathematical confirmation has been obtained for the hypothesis that the nuclear forces are actually transmitted by mesons.

Professor Bethe, author of an authoritative work on nuclear physics, was born in Strasbourg, Alsace-Lorraine. After studying in the German universities of Munich and Frankfort, he received the Rockefeller Foundation Fellowship in 1930. This fellowship brought him in contact with the Nobel Prize winners, Rutherford, Fermi and Bohr.

Last year he received the Morrison Prize of the New York Academy of Science for his paper explaining the tremendous heat given off by the sun.

He came to the Cornell physics department in 1935 and was promoted to a full professorship in 1937.

January 14, 1940

VAST POWER SOURCE IN ATOMIC ENERGY OPENED BY SCIENCE

Relative of Uranium Found to Yield Force 5 Million Times as Potent as Coal

GERMANY IS SEEKING IT

Scientists Ordered to Devote All Time to Research—Tests Made at Columbia

By WILLIAM L. LAURENCE

A natural substance found abundantly in many parts of the earth, now separated for the first time in pure form, has been found in pioneer experiments at the Physics Department of Columbia University to be capable of yielding such energy that one pound of it is equal in power output to 5,000,000 pounds of coal or 3,000,000 pounds of gasoline, it became known yesterday.

The discovery was announced in the current issue of The Physical Review, official publication of American physicists and one of the leading scientific journals of its kind in the world.

Professor John R. Dunning, Columbia physicist, who headed the scientific team whose research led to the experimental proof of the vast power in the newly isolated substance, told a colleague, it was learned, that improvement in the methods of extraction of the substance was the only step that remained to be solved for its introduction as a new source of power. Other leading physicists agreed with him.

A chunk of five to ten pounds of the new substance, a close relative of uranium and known as U-235, would drive an ocean liner or an ocean-going submarine for an indefinite period around the oceans of the world without refueling, it was said, for such a chunk would possess the power-output of 25,000,000 to 50,000,000 pounds of coal, or of 15,000,000 to 30,000,000 pounds of gasoline.

Uranium ore, in which the U-235 also is present, is found in the Belgian Congo, Canada, Colorado, England and Germany, in relatively large amounts. It is 1,000,000 times more abundant than radium, with which it is associated in pitchblende ores.

Tested With Atom-Smasher

Until about two months ago not even an infinitesimal drop of the substance had been isolated in pure form and the task of doing so appeared hopeless from a practical point of view. Toward the end of February a minute fraction of a gram was isolated at the University of Minnesota Physics Department, under the direction of Professor Alfred O. Nier. The sample was rushed at once to Columbia University, where Professor Dunning, in collaboration with Dr. E. T. Booth and Dr. Aristid V. Grosse, submitted it to tests with the Columbia 150-ton cyclotron (atom-smasher).

The sample, however, was so small that the results, while striking, served merely to stimulate the scientists at Columbia and Minnesota to further efforts. So fast has the work progressed since the beginning of March, the report in the Physical Review says, that the yield has been increased 200-fold.

Such an increase in two months has given new hope that a process for isolating the substance in larger quantities, in grams and pounds instead of millionths of a gram, will be found in the not too distant future. While scientists refuse to make predictions, it is not impossible that a few months or a year hence may see the realization of this quest.

Industrial Laboratories Aid

The fact that industrial laboratories also have taken up the quest and are lending to their university colleagues the vast experimental resources at their disposal is revealed in the same issue of The Physical Review, in a report from the research laboratories at the General Electric Company by Dr. K. H. Kingdon and Dr. H. C. Pollock, also signed by Professor Dunning and Dr. Booth. The report reveals that the G. E. scientists also have set up an apparatus similar to that of Professor Nier and in their turn have separated a relatively large sample of the U-235. This sample was submitted also to experimental tests at Columbia and corroborated the results obtained from the University of Minnesota samples.

The main reason why scientists are reluctant to talk about this development, regarded as ushering in the long dreamed of age of atomic power and, therefore, as one of the greatest, if not the greatest, discovery in modern science, is the tremendous implications this discovery bears on the possible outcome of the European war, it was explained.

The news has leaked out, through highly reliable channels, that the Nazi government had heard of the research in American laboratories and had ordered its greatest scientists to concentrate their energies on the solution of this problem. Every German scientist in this field, physicists, chemists and engineers, it was learned, have been ordered to drop all other researches and devote themselves to this work alone. All these research workers, it was learned, are carrying on their tasks feverishly at the laboratories of the Kaiser Wilhelm Institute at Berlin.

The American scientists, it was said, are in the dark as to what their German colleagues are doing and what progress, if any, they have made. It is believed, however, that the American scientists are in the lead, as Germany does not possess the powerful cyclotrons of American laboratories, and these machines are necessary for carrying out the most effective experiments in studying the energies within the nuclei (cores) of atoms.

However, it was asserted that while cyclotrons are a prime requisite for determining the amount of energy contained in the new substance, the apparatus necessary for its isolation in small quantities was relatively simple and inexpensive, so that the Germans, on learning of the American research, no doubt could duplicate it.

On the other hand it is not believed that this particular apparatus will ever be useful in separating U-235 on a large scale. New plans are being made to isolate the substance on a practical scale, but the plans and the designs for these will be kept a secret to be given only to the United States Government, to do with as it sees fit.

A startling discovery about the new power source, also made as a result of the Columbia experiments, is the simplicity of the method of liberating its vast energy. All that is needed to put it to work running motors and steamships is to place it in a tank of water and keep it supplied with a constant flow of cold water.

Left by itself the substance would be inactive. As soon as it touched water of ordinary temperature it would automatically start to liberate its energy. The water would be turned into steam and the steam would drive powerful turbines. The new water supplied would keep the process going indefinitely. To stop it, all that would be necessary would be to cut off the water supply.

Thus the process would be the nearest practical approach to a form of perpetual motion, for as long as the U-325 would be supplied with water it would keep on liberating its energy until exhausted.

Teriffic Explosive Power

It was figured out, by way of another example, that one pound of the U-235 contains as much energy as 15,000 tons (30,000,000 pounds) of TNT, or 300 carloads of fifty tons each. If this one pound of U-235 exploded within 1/10,000ths of a second, as does ordinary TNT, the pressure produced would be on the order of 100,000,000,000 atmospheres (ten to the seventeenth power dynes to a square centimeter), about 1,000,000 times the pressure produced by TNT or by nitroglycerin.

On the other hand, it was estimated by an explosion expert, that the explosion of such a pound of U-235 would produce a crater much less than 300 feet in radius and probably only seventy-five feet in radius.

The largest amount of explosives ever to have been exploded was 6,000 tons of a mixture of ammonium nitrate and ammonium sulphate in Oppau, Germany, about fourteen years ago. That explosion was accidental, as the mixture was supposed to be a fertilizer. The resulting crater was about 600 feet wide and 250 feet deep.

The U-235 is what is known as an isotope, or chemical twin of ordinary uranium. Even its existence was not known until a few years ago, and its properties had been unsuspected until a short time ago.

Up to a few months ago it was believed to exist in minute amounts in association with ordinary uranium, the proportion of the two having been regarded as being of the order of 1,000 parts of uranium to one part of the U-235, or about one-tenth of 1 per cent.

Suspicion first was cast on the possible nature of the U-235 as a great energy source, on purely theoretical grounds, by Professor Niels Bohr, Noble Prize winning physicist of the University of Copenhagen, Denmark, who carried on his researches last Summer at the Institute for Advanced Study at Princeton, N. J., and by Dr. John A. Wheeler of Princeton University.

Dane Speeded Tests Here

Professor Bohr was one of the

REPORT ON NEW SOURCE OF POWER

Scientists at Columbia University with cyclotron recording machine. Left to right: Dr. E. T. Booth, research physicist; Dr. J. R. Dunning, Professor of Physics, and Dr. A. V. Grosse, a John Simon Guggenheim Research Fellow.

first to learn of the discovery in Germany, in January, 1939, that when an ordinary sample of uranium, containing mixtures of three chemical twins, is bombarded with neutrons (fundamental atomic constituents carrying no electric charge) the uranium produces among the debris of its atoms the much lighter element barium.

Immediately communicating these results to his colleague, Professor Enrico Fermi, another Nobel Prize winning physicist, at Columbia University, and to other leading physicists, the true nature of the results obtained in Germany soon was determined. Repeating the German experiments at Columbia, Johns Hopkins, the Carnegie Institution of Washington and other laboratories, the physicists came upon the discovery that they were witnessing here for the first time a release of the binding energy within atoms on a scale greater than ever before.

What was happening, they discovered, was a splitting of the heavy uranium atom into two parts, one of which was barium of atomic weight 137, thus accounting for the barium observed in the German experiments by Professor Otto Hahn and Dr. Lise Meitner. In the process of the splitting (fission) of the uranium atom by the neutrons, the binding energy holding the uranium atom together was liberated to the extent of 200,000,000 electron volts.

It was then for the first time (it took place in the early Spring and Summer, 1939) that attention was called to the neglected and unsung uranium isotope of atomic weight 235. Basing their reasoning on observations that atoms of even atomic weight are inclined to be less stable than atoms of odd atomic weights, Professor Bohr and Dr. Wheeler presented the theory that it was the minute fraction of U-235 in the uranium sample that was responsible for the release of the 200,000,000 electron volts when the uranium atom was split by a neutron.

The reasoning by Dr. Bohr and Dr. Wheeler was as follows: When uranium 238 is hit by a neutron that enters its nucleus, the atomic weight of the uranium is increased to 239, an odd-numbered weight, and hence a stable, non-exploding atom. On the other hand, when uranium 235 is hit by a neutron that enters its nucleus the atomic weight becomes 236, an even-numbered atomic weight, and hence a non-stable, exploding atom.

This theory split the physicists into two camps, one agreeing with Dr. Bohr and Dr. Wheeler and the other disagreeing. The Columbia experiments have settled the question definitely in favor of the Bohr hypothesis, according to the report.

Another question that until now had remained unsettled, and upon which the crux of the whole matter depended, was whether an atom of uranium, once hit by a neutron and split into two, would release other neutrons from its nucleus and thus start a "chain-reaction," in the manner of a fire-cracker, that would keep the process regenerat-

Dr. Alfred O. Nier, associated with them in the discovery.

Times Wide World

PHYSICS AND ASTROPHYSICS

ing by itself, without any further need for neutron-bullets from outside sources. Unless such a process took place naturally there would be no hope of putting the uranium 235 to work on a practical scale.

This all-important question also has been settled by the Columbia experiments in favor of the "firecracker hypothesis," the experiments establishing definitely for the first time that only one neutron, slowed down by water to travel with very low energy (one-fortieth of a volt), behaves in the manner of a trigger that sets off the process of energy-liberation from the uranium 235, it was reported.

Moreover, it was pointed out, even this one "trigger" neutron is not necessary to be supplied from any apparatus. The air is full of minute amounts of radium that constantly liberate neutrons by hitting atoms in the air. In addition, there are the omnipresent, all-penetrating cosmic rays that constantly bombard the earth from outer space. These powerful radiations also liberate neutrons from air atoms, and in fact play a part in starting off the trigger action in neon signs, the starting of which needs free electrons that are supplied by the radiations in the air produced by the radium and cosmic rays.

A building such as the Empire State, it was asserted, contains scattered radium in minute amounts that would equal a whole gram, worth about $25,000 in current prices. Without such radiations in the atmosphere, it was explained, neon lamps would have to be supplied with free electrons from an outside source and would be much more expensive.

Starting Energy to Work

To start the "Philosopher's Stone" in the U-235 to work, it was explained further, all that is necessary is one neutron traveling at low energy, the lower the better. Now, neutrons, as they come out of the cores of atoms, travel with high energy, and it is therefore necessary to slow them down.

Fortunately, protons, the cores of hydrogen atoms, constituting two-thirds of the volume of water, have the power to make neutrons yield up their high energies and to slow them down to almost no energy at all. Hence, all that is necessary to start the U-235 liberating its great energy is to place it in an environment of ordinary water.

The process then becomes automatic and self-regenerating, it was explained. A neutron liberated by a cosmic ray hitting atoms in the air, for example, is slowed down by the water surrounding the U-235. This splits an atom of the substance into two parts, liberating 200,000,000 volts of atomic binding-energy. In doing so it also liberates other neutrons from the nucleus of the U-235. These neutrons, in turn, are slowed down also as they hit the water, and again split another U-235 atom. The process then continues as long as there are atoms left, and there are 2,500 billion billion atoms of U-235 to the gram, and 453.72 grams make one pound.

Not only is the energy-liberating process automatic and self-regenerating, it was explained, but it also is self-regulating. The energy liberated from the atoms heats up the water so that it turns into steam. When all the water supplied has been turned into steam, there is nothing left to slow down the fast-traveling neutrons, and fast neutrons just go through the uranium without breaking up its atoms and releasing its energy. This brings the whole process to a stop until more cool water is supplied.

As one leading physicist explained it, "the colder the water the better the reaction. The reaction is self-limiting because heat (generated by the split atoms) speeds up the neutrons, and the faster the neutrons the less the reaction."

"The faster you feed in the cold water," the scientist added, "the faster the water will come out red hot on the other side, because more neutrons will be slowed down, and thus more atoms split and more energy is liberated. Thus the process is admirably suited for power purposes."

Another significant discovery, of practical bearing on the question of isolating the substance in usable amounts, was the redetermination of the relative abundance of the U-235 by Professor Nier. Whereas it had been believed as recently as last year that it existed in a ratio of only 1,000 to 1, compared with ordinary uranium of atomic weight 238, Professor Nier found that the ratio is much smaller, 139 pounds of ordinary uranium containing one pound of the U-235. This finding alone has increased automatically the amount of the new "Philosopher's Stone" by more than seven times and therefore makes its isolation seven times easier.

Germany, it was asserted, may regret her act of having sent into exile Dr. Lise Meitner, who, with Professor Hahn, made the first observations that led to the discovery of the fountain-head of atomic energy that German scientists are so feverishly working to harness. Soon after her exile, when she settled at Stockholm, Sweden, Dr. Meitner revealed the results of her work with Professor Hahn to colleagues of Professor Bohr, who at once communicated it to his colleagues in America. Had Germany then realized the importance of the findings it is highly probable, it was said, that she would have kept it a strict military secret and possibly later would have surprised the world with it.

In addition to the uranium of mass 238 and the newly isolated uranium of atomic weight 235 there is a third and much rarer type of uranium, of atomic weight 234. This isotope exists in the ratio of 1 to 17,000, compared with ordinary uranium 238.

Neither uranium 238 nor uranium 234, small amounts of which also have been separated, has been found to liberate energy on being bombarded with slow neutrons. Uranium 238 responds to fast neutrons, but the process in this case would be impracticable, it was pointed out. The Columbia experiments have demonstrated also for the first time, Dr. Dunning and his associates report, that the uranium of mass 234 plays no part in the energy-liberating process.

Five-Pound Mass Necessary

Because of the nature of the neutrons, even the slow-traveling ones, it was explained further, it is necessary to have a mass of at least five pounds, and possibly as high as twenty, to make the process work on a practical scale. In a smaller amount even low energy neutrons would escape into the open without splitting the initial "trigger-atom" that sets off the process. To start the process it is necessary for the neutron to remain inside the mass, so that it would enter the nucleus of an atom to start the splitting process.

However, it was said, it would not be necessary to obtain a mass of five to twenty pounds of pure U-235 to start the process. A concentration of 10 to 50 per cent would be sufficient. In other words, a five-pound mass of uranium mixture, that contained half to two-and a half pounds of the U-235, would be sufficient for use as a prime motive power for submarines, and for other sources of power.

Such a mass, it was explained, "would make the most powerful cyclotron puny by comparison, and would provide neutron radiations thousands of times greater than that produced by any cyclotron. By comparison with such a mass, a cyclotron would be a mere plaything, and the mass would be much less expensive than a cyclotron.

The power from the U-235, it was added, could be applied in many other useful ways. It would provide the most powerful source of neutron rays that might possibly be used in the treatment of cancer, as neutrons are much more powerful than either X-rays or radium. The neutron from the U-235 could be used also for the creation of artificially radioactive elements more powerful than radium. They could even be used for making gold out of mercury, but scientists expressed themselves as disdainful of such uses for their newest creative tool.

One of the scientists explained the process of the energy-liberation from U-235 by comparing it with the burning of coal. Whereas coal uses oxygen to liberate its energy, he explained, the U-235 uses slow neutrons for the same purpose. The process of combustion in the case of the U-235, he added, is, atom for atom, 100,000,000 times as effective as is the case in the combustion of coal. However, as the atomic weight of the uranium is 235, compared with 16 for the oxygen and 12 for the carbon, there are fewer uranium atoms to a given weight than there are oxygen and carbon atoms. This reduces the energy relations of the U-235, compared with coal, to a ratio of 5,000,000 to 1.

There are several new methods being considered for increasing the yield of the new substance to large-scale amounts. But as to this, scientists greet the questioner with a profound silence.

May 5, 1940

WRITER CHARGES U. S. WITH CURB ON SCIENCE

Tells Housatonic Session Work on Uranium 235 Is Censored

FALLS VILLAGE, Conn., Aug. 13 (AP)—John J. O'Neill, president of the National Association of Science Writers, charged tonight that the government had clapped "a censorship" on laboratories developing an element, which, if contained in a ten-pound bomb "would blast a hole twenty-five miles in diameter and more than a mile deep, and would wreck every structure within 100 miles."

Mr. O'Neill, who is the science editor of The New York Herald Tribune, asserted in a prepared address at the Housatonic Valley Conference that the Administration "is staging a totalitarian revolution against the American people."

He said that scientists had recently discovered the method of releasing energy from the uranium atom, and after attributing terrific destructive power to a ten-pound missile of uranium 235, asked:

"Can we trust our politicians and war makers with a weapon like that? The answer is no. Nevertheless, our politicians have taken over control of the scientists who have been working on the application and control of this discovery and are driving them to develop it for war uses.

"They have clapped a censorship in the scientific laboratories where this work is being done and no scientist dares to discuss what he is doing."

August 14, 1941

Dec. 2, 1942— The Birth of the Atomic Age

Story of the great experiment which first released the energy that runs the universe.

By WILLIAM L. LAURENCE

Mr. Laurence, science writer for THE NEW YORK TIMES, *served as special consultant to the War Department to study the development of the atomic bomb and to explain this new force to the public. He was present when the first atomic bomb was exploded at Alamogordo, New Mexico, and saw from a plane the explosion of the bomb dropped on Nagasaki, which brought about Japan's surrender. He also witnessed the explosion at Bikini.*

TOMORROW will be the first official birthday anniversary of the atomic age, commemorating that fateful December day when man lighted the first atomic fire on this planet, the first fire that did not have its origin in the sun. For it was on that day, Dec. 2, 1942, at the gloomy squash court underneath the west stands of Stagg Field on the University of Chicago campus, that man succeeded at last in operating an atomic furnace, the energy of which came from the vast cosmic reservoir supplying the sun and the stars with their radiant heat and light—the nucleus of the atoms of which the material universe is constituted.

To understand what took place four years ago tomorrow it is necessary to review briefly the events that led up to it. It was known that to produce an atomic bomb it would be necessary to start a chain reaction in a mass of atoms. As one atom split, it would have to set off the trigger that would split a second atom.

Early in 1939 came the epoch-making discovery that when uranium atoms were split by bombarding them with neutrons —so called because they are electrically neutral parts of atoms—each atom gave off at least one more neutron. This neutron, it was reasoned, could in turn split another atom of uranium, and the process could go on indefinitely under the proper conditions. In theory all the essentials for a chain reaction were fulfilled. If such a reaction were started, calculations showed, it would liberate vast amounts of energy—3,000,000 times more energy than that given out by equal masses of coal, 20,000,000 times more explosive force than equal masses of TNT. In 1939 this meant the threat of the most destructive weapon the world had ever dreamed of.

Soon came another important discovery. It was established that the uranium that undergoes fission was not the ordinary abundant type of the element, of atomic weight 238, but the much rarer type, having the atomic weight 235. (In nature uranium always contains one part of U.235 to 140 parts of U.238.)

Why, it may be asked, didn't ordinary uranium explode?

First, because U.238 simply absorbs a good portion of the neutrons liberated in the process of splitting U.235.

Second, because a number of impurities present in a natural mixture of uranium also have a strong appetite for neutrons.

Third, because many of the neutrons escape from the surface of the mass of uranium like steam bubbling off hot water.

IT was therefore recognized that before a chain reaction could be achieved, it would be necessary to do three things: First of all, the U.235 would have to be separated from the U.238; second, all the other impurities would have to be removed; third, it would be necessary to get a "critical mass" of the concentrated U.235, that is, a size large enough to retain most of the neutrons within the system. This critical mass, the minimum necessary to start an atomic explosion, was soon determined on theoretical grounds to be somewhere between one and one hundred kilograms.

How could these three objectives be achieved? U.238 and U.235 could not be separated by chemical means because they are different forms (isotopes) of the same element, and the physical means then available were so slow that it would have taken one thousand separation devices one thousand years to produce about thirty grams. Obviously, another approach had to be found—and quickly. The Germans were known to be at work on an atomic bomb and there was then good reason to fear that they would develop it first.

THE solution suggested itself independently and almost simultaneously to Prof. Enrico Fermi and Dr. Leo Szilard, both of whom were then working at Columbia University. It had by that time been established that U.238 would absorb neutrons only if they were moving rather fast. The slow-speed, low-energy neutrons would not penetrate the nucleus. But it was exactly those slow neutrons that split U.235. What was needed was some method of slowing down neutrons.

Elements that slow down neutrons are known as "moderators." One of the most efficient of these is "heavy water," in which the hydrogen is double the weight of ordinary hydrogen. But the only plant in the world producing "heavy water" on a large scale was situated in Norway and was being worked by the Nazis in their atomic bomb project. (This plant, by the way, was destroyed in 1943 by members of the Norwegian underground in one of the great epics of the war, as a result of which the Nazis were left far behind in the race.)

Another "moderator" is graphite, the soft carbon used in pencils. Drs. Fermi and Szilard decided that graphite, because of its availability, would be the most suitable moderator for their purpose. It was their idea to build a huge spherical lattice of graphite bricks in which small lumps of the natural uranium mixture would be imbedded at regular intervals. They named the structure a "pile," a name that has stuck to the gigantic descendants of the original, at Oak Ridge, Tenn., and at Hanford, Wash.

IF such a structure were large enough, they reasoned, most of the neutrons, born through the fission of U.235, would remain inside the "pile," to produce further fissions and thus maintain a chain reaction. The neutrons would move slowly because the graphite moderator would slow them down.

Two formidable obstacles stood in their way. To build such a pile it was first of all necessary to get large quantities of metallic uranium in a form purer than any uranium then in existence; otherwise the impurities would absorb neutrons and stop the reaction. Only very small quantities of metallic uranium were being produced at that time, and these were of doubtful purity. Second, the graphite bricks had to be of a similar degree of purity, if they were to be

PHYSICS AND ASTROPHYSICS

free of neutron-absorbing substances. No graphite of such purity was then in existence.

EARLY in 1942 the researches on the uranium-graphite pile which had been carried on at Columbia University were shifted to the University of Chicago under the direction of Prof. Arthur H. Compton. Having learned that the Westinghouse Company had been producing small amounts of metallic uranium, Dr. Compton telephoned Dr. Harvey C. Rentschler, Westinghouse research director.

"How soon can Westinghouse supply three tons of metallic uranium?" Dr. Compton casually asked.

Dr. Rentschler was aghast. The total output of pure uranium metal up to that time had been a few grams. On being informed that uranium was necessary for a vital secret war project, he went to work. By November, 1942, the three tons were delivered.

Meantime other companies entered the picture, and new and simpler processes for purifying uranium ore and graphite were developed. By Nov. 7, 1942, a total of 12,400 pounds of pure uranium metal had been collected at the west stands of the Chicago squash court, and many more tons of highly purified uranium oxide, as well as tons of the purest graphite ever produced. The stage was set.

ACTUAL work on the first self-sustaining chain-reaction pile began on Nov. 7, 1942. Many preliminary experiments had enabled the pile-builders to figure out the proper shape and dimensions that would yield the most efficient results. The structure was to be a sphere. The graphite was cut in square bricks and built up in layers. At the corners of the graphite bricks in each alternate layer were placed the uranium lumps, those of the pure metal being placed in the center of the pile. A timber framework resting on the squash court floor supported the structure, which was planned to consist of sixteen layers.

Success or failure hinged on the "multiplication factor" of the neutrons. If, for example, each neutron that split an atom of U.235 liberated another neutron which in turn would split another atom, the fission process would maintain itself. In that case the multiplication factor, designated by the scientists as "K," would be equal to one. In other words, the effective birth rate would be equal to the death rate and the chain-reaction would go on.

IF, on the other hand, only ninety-nine fission-producing neutrons were born for every hundred that caused fission, then the multiplication factor would be less than one and the reaction would die out. If the multiplication factor was slightly greater than one so much the better, for this would give an added safety factor. The scientists referred to such a multiplication factor as the "Great God K."

As the pile grew precautions had to be taken to prevent a runaway chain-reaction, which would have caused untold catastrophe. In any mass of uranium, fission occurs spontaneously as stray neutrons smash into the parts of U.235. Therefore an elaborate system of controls was devised. These consisted of a series of boron steel rods and strips of cadmium inserted through slots in the pile. Boron and cadmium have an enormous capacity for devouring neutrons, so that the number of neutrons in the pile could be controlled by either pushing or pulling the boron rods and cadmium strips in and out of the pile. From the very beginning the cadmium strips and boron rods were placed in "retard" position to make sure that "Great God K" did not make a surprise appearance.

BY the night of Dec. 1, 1942, eleven layers of graphite-uranium bricks had been piled up. Late that evening there were signs that the goal was near. But Dr. Fermi, with true scientific imperturbability, decided to call it a day. Early the next morning the atomic "bricklayers" were back on the job. It was one of the coldest days of the winter. The squash court was badly heated. But the bricklayers worked on, oblivious of the gloom and cold.

Dr. Walter H. Zinn, then on leave from the College of the City of New York, was master of ceremonies on that Dec. 2. Present were Drs. Fermi, Szilard and Compton; Drs. Samuel K. Allison, Herbert L. Anderson, George Weil, Eugene Wigner, Norman Hilberry, Volney C. Wilson and John Marshall. There was one young woman in the group, Leona Woods, who later became Mrs. John Marshall.

Present also was Dr. Crawford H. Greenewalt, a member of the board of directors of the du Pont Company, which later built and operated the giant piles at Oak Ridge, Tenn., and at Hanford, Wash. It was largely on the basis of what Dr. Greenewalt saw later that afternoon that the du Pont Company agreed to undertake to build these plants.

As the twelfth layer was completed everyone present became aware that one of the great moments in history was near. As the cadmium strips were cautiously pulled out, the instruments registering the multiplication rate of the neutrons began clicking louder and louder. These clicks were the heralds of the atomic age.

AS the clicks grew more frequent, extra precautions were decided on. Two of the young physicists in the group, Dr. Alvin C. Graves of the University of Texas and Herald V. Lichtenberger of Millikan College, Decatur, Ill., were selected to serve in what their colleagues called the "suicide brigade." They stood silently on a high platform overlooking the pile, each holding a bucket filled with a solution of cadmium, ready to pour it should the "Great God K" show any sign of becoming rambunctious. For two hours they stood thus tensely waiting for a signal that never came, hoping the while that human nerves and muscles would be equal to the task.

Slowly all the cadmium strips but one were pulled out. Then, as the "suicide brigade" stood on the alert, the last one was pulled out to the proper calculated distance.

THE scientists had previously figured out that if the number of neutrons per second reached a count of more than 1,600 it would mean a multiplication factor greater than one. Tensely and silently they stood around the neutron counters. Click, click, click. Twelve hundred, fourteen hundred, sixteen hundred. Sixteen hundred and one. The atomic "baby" had emitted its first lusty cry.

Dr. Greenewalt rushed back to a conference room in Eckhert Hall, where his colleagues had been debating since morning as to whether the du Pont Company should go into the building and operation of the fantastic plants suggested by the Manhattan District.

"Gentlemen," Dr. Greenewalt said, his eyes popping, "there is no need for further discussion."

At the same time Dr. Compton held a short long-distance telephone conversation with President James Bryant Conant of Harvard.

"The Italian navigator has arrived in the New World and found the continent much smaller than he thought it was," said Dr. Compton.

"I hope the natives received him kindly," replied Dr. Conant.

In its final appearance that day the "pile," actually a man-made model of a living star, was a three-quarter complete sphere, flat at the top, a shape geometers know as an oblate spheroid, resembling a giant doorknob. A thirteenth layer was added for luck, and the scientists called it a day.

THE atomic power output that day was at the rate of only one-half watt, which corresponds to the splitting of one-half of a millionth of a gram of U.235 per hour. The force of fission had been so well controlled that there was no danger at all of fire or explosion. Later the rate was increased to 200 watts. There was still a long way to go to the giant plants with an output of millions of kilowatts, at the same time transmuting the useless U.238 into huge quantities of fissionable plutonium, the man-made element that destroyed Nagasaki. Furthermore, what had been demonstrated was a controlled chain-reaction with slow neutrons, whereas for use as an explosive it is necessary to produce an uncontrolled chain-reaction with fast neutrons.

But while there was still much work ahead, the road was clear. It was a straight line from then on to New Mexico, Hiroshima, Nagasaki and victory. It marked the end of an era and the beginning of a new one, with incalculable potentialities for good and for evil.

December 1, 1946

FIRST ATOMIC BOMB DROPPED ON JAPAN

NEW AGE USHERED

Day of Atomic Energy Hailed by President, Revealing Weapon

HIROSHIMA IS TARGET

'Impenetrable' Cloud of Dust Hides City After Single Bomb Strikes

By SIDNEY SHALETT
Special to THE NEW YORK TIMES.

WASHINGTON, Aug. 6—The White House and War Department announced today that an atomic bomb, possessing more power than 20,000 tons of TNT, a destructive force equal to the load of 2,000 B-29's and more than 2,000 times the blast power of what previously was the world's most devastating bomb, had been dropped on Japan.

The announcement, first given to the world in utmost solemnity by President Truman, made it plain that one of the scientific landmarks of the century had been passed, and that the "age of atomic energy," which can be a tremendous force for the advancement of civilization as well as for destruction, was at hand.

At 10:45 o'clock this morning, a statement by the President was issued at the White House that sixteen hours earlier—about the time that citizens on the Eastern seaboard were sitting down to their Sunday suppers—an American plane had dropped the single atomic bomb on the Japanese city of Hiroshima, an important army center.

Japanese Solemnly Warned

What happened at Hiroshima is not yet known. The War Department said it "as yet was unable to make an accurate report" because "an impenetrable cloud of dust and smoke" masked the target area from reconnaissance planes. The Secretary of War will release the story "as soon as accurate details of the results of the bombing become available."

But in a statement vividly describing the results of the first test of the atomic bomb in New Mexico, the War Department told how an immense steel tower had been "vaporized" by the tremendous explosion, how a 40,000-foot cloud rushed into the sky, and two observers were knocked down at a point 10,000 yards away. And President Truman solemnly warned:

"It was to spare the Japanese people from utter destruction that the ultimatum of July 26 was issued at Postdam. Their leaders promptly rejected that ultimatum. If they do not now accept our terms, they may expect a rain of ruin from the air the like of which has never been seen on this earth."

Most Closely Guarded Secret

The President referred to the joint statement issued by the heads of the American, British and Chinese Governments, in which terms of surrender were outlined to the Japanese and warning given that rejection would mean complete destruction of Japan's power to make war.

[The atomic bomb weighs about 400 pounds and is capable of utterly destroying a town, a representative of the British Ministry of Aircraft Production said in London, the United Press reported.]

What is this terrible new weapon, which the War Department also calls the "Cosmic Bomb"? It is the harnessing of the energy of the atom, which is the basic power of the universe. As President Truman said, "The force from which the sun draws its power has been loosed against those who brought war to the Far East."

"Atomic fission" — in other words, the scientists' long-held dream of splitting the atom—is the secret of the atomic bomb. Uranium, a rare, heavy metallic element, which is radioactive and akin to radium, is the source essential to its production. Secretary of War Henry L. Stimson, in a statement closely following that of the President, promised that "steps have been taken, and continue to be taken, to assure us of adequate supplies of this mineral."

The imagination-sweeping experiment in harnessing the power of the atom has been the most closely guarded secret of the war. America to date has spent nearly $2,000,000,000 in advancing its research. Since 1939, American, British and Canadian scientists have worked on it. The experiments have been conducted in the United States, both for reasons of achieving concentrated efficiency and for security; the consequences of having the material fall into the hands of the enemy, in case Great Britain should have been successfully invaded, were too awful for the Allies to risk.

All along, it has been a race with the enemy. Ironically enough, Germany started the experiments, but we finished them. Germany made the mistake of expelling, because she was a "non-Aryan," a woman scientist who held one of the keys to the mystery, and she made her knowledge available to those who brought it to the United States. Germany never quite mastered the riddle, and the United States, Secretary Stimson declared, is "convinced that Japan will not be in a position to use an atomic bomb in this war."

A Sobering Awareness of Power

Not the slightest spirit of braggadocio is discernable either in the wording of the official announcements or in the mien of the officials who gave out the news. There was an element of elation in the realization that we had perfected this devastating weapon for employment against an enemy who started the war and has told us she would rather be destroyed than surrender, but it was grim elation. There was sobering awareness of the tremendous responsibility involved.

Secretary Stimson said that this new weapon "should prove a tremendous aid in the shortening of the war against Japan," and there were other responsible officials who privately thought that this was an extreme understatement, and that Japan might find herself unable to stay in the war under the coming rain of atom bombs.

It was obvious that officials at the highest levels made the important decision to release news of the atomic bomb because of the psychological effect it may have in forcing Japan to surrender. However, there are some officials who feel privately it might have been well to keep this completely secret. Their opinion can be summed up in the comment by one spokesman: "Why bother with psychological warfare against an enemy that already is beaten and hasn't sense enough to quit and save herself from utter doom?"

The first news came from President Truman's office. Newsmen were summoned and the historic statement from the Chief Executive, who still is on the high seas, was given to them.

"That bomb," Mr. Truman said, "had more power than 20,000 tons of TNT. It had more than 2,000 times the blast power of the British 'Grand Slam,' which is the largest bomb (22,000 pounds) ever yet used in the history of warfare."

Explosive Charge Is Small

No details were given on the plane that carried the bomb. Nor was it stated whether the bomb was large or small. The President, however, said the explosive charge was "exceedingly small." It is known that tremendous force is packed into tiny quantities of the element that constitutes these bombs. Scientists, looking to the peacetime uses of atomic power, envisage submarines, ocean liners and planes traveling around the world on a few pounds of the element. Yet, for various reasons, the bomb used against Japan could have been extremely large.

Hiroshima, first city on earth to be the target of the "Cosmic Bomb," is a city of 318,000, which is—or was—a major quartermaster depot and port of embarkation for the Japanese. In addition to large military supply depots, it manufactured ordnance, mainly large guns and tanks, and machine tools and aircraft-ordnance parts.

President Truman grimly told the Japanese that "the end is not yet."

"In their present form these bombs are now in production," he said, "and even more powerful forms are in development."

He sketched the story of how the late President Roosevelt and Prime Minister Churchill agreed that it was wise to concentrate re-

PHYSICS AND ASTROPHYSICS

search in America, and how great, secret cities sprang up in this country, where, at one time, 125,000 men and women labored to harness the atom. Even today more than 65,000 workers are employed.

"What has been done," he said, "is the greatest achievement of organized science in history.

"We are now prepared to obliterate more rapidly and completely every productive enterprise the Japanese have above ground in any city. We shall destroy their docks, their factories and their communications. Let there be no mistake; we shall completely destroy Japan's power to make war."

The President emphasized that the atomic discoveries were so important, both for the war and for the peace, that he would recommend to Congress that it consider promptly establishing "an appropriate commission to control the production and use of atomic power within the United States."

"I shall give further consideration and make further recommendations to the Congress as to how atomic power can become a powerful and forceful influence toward the maintenance of world peace," he said.

Secretary Stimson called the atomic bomb "the culmination of years of herculean effort on the part of science and industry, working in cooperation with the military authorities." He promised that "improvements will be forthcoming shortly which will increase by several fold the present effectiveness."

"But more important for the long-range implications of this new weapon," he said, "is the possibility that another scale of magnitude will be developed after considerable research and development. The scientists are confident that over a period of many years atomic bombs may well be developed which will be very much more powerful than the atomic bombs now at hand."

Investigation Started in 1939

It was late in 1939 that President Roosevelt appointed a commission to investigate use of atomic energy for military purposes. Until then only small-scale research with Navy funds had taken place. The program went into high gear.

By the end of 1941 the project was put under direction of a group of eminent American scientists in the Office of Scientific Research and Development, under Dr. Vannevar Bush, who reported directly to Mr. Roosevelt. The President also appointed a General Policy Group, consisting of former Vice President Henry A. Wallace, Secretary Stimson, Gen. George C. Marshall, Dr. James B. Conant, president of Harvard, and Dr. Bush. In June, 1942, this group recommended vast expansion of the work and transfer of the major part of the program to the War Department.

Maj. Gen. Leslie R. Groves, a native of Albany, N. Y., and a 48-year-old graduate of the 1918 class at West Point, was appointed by Mr. Stimson to take complete executive charge of the program. General Groves, an engineer, holding the permanent Army rank of lieutenant colonel, received the highest praise from the War Department for the way he "fitted together the multifarious pieces of the vast country-wide jigsaw," and, at the same time, organized the virtually air-tight security system that kept the project a secret.

A military policy committee also was appointed, consisting of Dr. Bush, chairman; Dr. Conant, Lieut. Gen. Wilhelm D. Styer and Rear Admiral William R. Purnell.

In December, 1942, the decision was made to proceed with construction of large-scale plants. Two are situated at the Clinton Engineer Works in Tennessee and a third at the Hanford Engineer Works in the State of Washington.

These plants were amazing phenomena in themselves. They grew into large, self-sustaining cities, employing thousands upon thousands of workers. Yet, so close was the secrecy that not only were the citizens of the area kept in darkness about the nature of the project, but the workers themselves had only the sketchiest ideas—if any—as to what they were doing. This was accomplished, Mr. Stimson said, by "compartmentalizing" the work so "that no one has been given more information than was absolutely necessary to his particular job."

The Tennessee reservation consists of 59,000 acres, eighteen miles west of Knoxville; it is known as Oak Ridge and has become a modern small city of 78,000, fifth largest in Tennessee.

In the State of Washington the Government has 430,000 acres in an isolated area, fifteen miles northwest of Pasco. The settlement there, which now has a population of 17,000, consisting of plant operators and their immediate families, is known as Richland.

A special laboratory also has been set up near Santa Fe, N. M., under direction of Dr. J. Robert Oppenheimer of the University of California. Dr. Oppenheimer also supervised the first test of the atomic bomb on July 16, 1945. This took place in a remote section of the New Mexico desert lands, with a group of eminent scientists gathered, frankly fearful to witness the results of the invention, which might turn out to be either the salvation or the Frankenstein's monster of the world.

Mr. Stimson also gave full credit to the many industrial corporations and educational institutions which worked with the War Department in bringing this titanic undertaking to fruition.

In August, 1943, a combined policy committee was appointed, consisting of Secretary Stimson, Drs. Bush and Conant for the United States; the late Field Marshal Sir John Dill (now replaced by Field Marshal Sir Henry Maitland Wilson) and Col. J. J. Llewellin (since replaced by Sir Ronald Campbell), for the United Kingdom, and C. D. Howe for Canada.

"Atomic fission holds great promise for sweeping developments by which our civilization may be enriched when peace comes, but the overriding necessities of war have precluded the full exploration of peacetime applications of this new knowledge," Mr. Stimson said. "However, it appears inevitable that many useful contributions to the well-being of mankind will ultimately flow from these discoveries when the world situation makes it possible for science and industry to concentrate on these aspects."

Although warning that many economic factors will have to be considered "before we can say to what extent atomic energy will supplement coal, oil and water as fundamental sources of power," Mr. Stimson acknowledged that "we are at the threshold of a new industrial art which will take many years and much expenditure of money to develop."

The Secretary of War disclosed that he had appointed an interim committee to study post-war control and development of atomic energy. Mr. Stimson is serving as chairman, and other members include James F. Byrnes, Secretary of State; Ralph A. Bard, former Under-Secretary of the Navy; William L. Clayton, Assistant Secretary of State; Dr. Bush, Dr. Conant, Dr. Carl T. Compton, chief of the Office of Field Service in OSRD and president of Massachusetts Institute of Technology; and George L. Harrison, special consultant to the Secretary of War and president of the New York Life Insurance Company. Mr. Harrison is alternate chairman of the committee.

The committee also has the assistance of an advisory group of some of the country's leading physicists, including Dr. Oppenheimer, Dr. E. O. Lawrence, Dr. A. H. Compton and Dr. Enrico Fermi.

The War Department gave this supplementary background on the development of the atomic bomb:

"The series of discoveries which led to development of the atomic bomb started at the turn of the century when radioactivity became known to science. Prior to 1939 the scientific work in this field was world-wide, but more particularly so in the United States, the United Kingdom, Germany, France, Italy and Denmark. One of Denmark's great scientists, Dr. Neils Bohr, a Nobel Prize winner, was whisked from the grasp of the Nazis in his occupied homeland and later assisted in developing the atomic bomb.

"It is known that Germany worked desperately to solve the problem of controlling atomic energy."

August 7, 1945

Drama of the Atomic Bomb Found Climax in July 16 Test

Following is the first of a number of articles by a staff member of THE NEW YORK TIMES *who was detached for service with the War Department at its request to explain the atomic bomb to the lay public. He witnessed the first test of the bomb in New Mexico and, on a flight to Nagasaki, its actual use.*

By WILLIAM L. LAURENCE

The Atomic Age began at exactly 5:30 Mountain War Time on the morning of July 16, 1945, on a stretch of semi-desert land about fifty airline miles from Alamagordo, N. M., just a few minutes before the dawn of a new day on this earth.

At that great moment in history, ranking with the moment in the long ago when man first put fire to work for him and started on his march to civilization, the vast energy locked within the hearts of the atoms of matter was released for the first time in a burst of flame such as had never before been seen on this planet, illuminating earth and sky for a brief span that seemed eternal with the light of many super-suns.

The elemental flame, first fire ever made on earth that did not have its origin in the sun, came from the explosion of the first atomic bomb. It was a full-dress rehearsal preparatory to use of the bomb over Hiroshima and Nagasaki — and other Japanese military targets had Japan refused to accept the Potsdam Declaration for her surrender.

The rehearsal marked the climax in the penultimate act of one of the greatest dramas in our history and the history of civilized man—a drama in which our scientists, with the Army Corps of Engineers as director, were working against time to create an atomic bomb ahead of our German enemy.

The collapse of Germany marked the end of the first act of this drama. The successful completion of our task, in the greatest challenge by man against nature so far, brought down the curtain on the second act.

The grand finale came three weeks afterward over the skies of Japan with a swift descent of the curtain on the greatest war in history.

The atomic flash in New Mexico came as a great affirmation to the prodigious labors of our scientists during the past four years, in which they managed to "know the unknowable and unscrew the inscrutable."

It came as the affirmative answer to the until then unanswered question: "Will it work?"

With the flash came a delayed roll of mighty thunder, heard, just as the flash was seen, for hundreds of miles. The roar echoed and reverberated from the distant hills and the Sierra Oscuro Range near by, sounding as though it came from some supramundane source as well as from the bowels of the earth.

The hills said "yes" and the mountains chimed in "yes." It was as if the earth had spoken and the suddenly iridescent clouds and sky had joined in one mighty affirmative answer. Atomic energy —yes.

It was like the grand finale of a mighty symphony of the elements, fascinating and terrifying, uplifting and crushing, ominous, devastating, full of great promise and great forebodings.

I watched the birth of the Era of Atomic Power from the slope of a hill in the desert land of New Mexico, on the northwestern corner of the Alamogordo Air Base, about 125 miles southwest of Albuquerque. The hill, named Compania Hill for the occasion, was twenty miles to the northwest of Zero, the code name given to the spot chosen for lighting the first atomic fire on this planet. The area embracing Zero and Compania Hill, twenty-four miles long and eighteen miles wide, had the code name Trinity.

Caravan of Scientists by Night

I joined a caravan of three buses, three automobiles and a truck carrying radio equipment at 11 P. M. Sunday, July 15, at Albuquerque. There were about ninety of us in that strange caravan, traveling silently and in utmost secrecy through the night on probably as unusual an adventure as any in our day.

With the exception of your correspondent, the caravan consisted of scientists from the highly secret atomic bomb research and development center in the mesas and canyons of New Mexico, twenty-five miles northwest of Santa Fe, where we solved the secret of translating the fabulous energy of the atom into the mightiest weapon ever made by man. It was from there that the caravan set out at 5:30 that Sunday afternoon for its destination, 212 miles to the south.

These were the "mesa-men" on the march, dwellers in the "caves" in the interior of atoms, pioneer explorers of vast new continents in hitherto forbidden realms of the cosmos, builders of the civilization of tomorrow.

Here on trails hallowed by pioneers of other days, who opened new frontiers and did not rest until they conquered a continent, "covered wagons" were rolling again through the night on their way to open still newer frontiers of a continent that has no limits in space.

The caravan wound its way slowly over the tortuous roads overlooking the precipitous canyons of northern New Mexico, passing through Espagnola, Santa Fe and Bernadillo, arriving at Albuquerque at about 10 P. M. Here it was joined by Sir James Chadwick, who won the Nobel Prize and knighthood for his discovery of the neutron, the key that unlocks the atom; Professor Ernest O. Lawrence of the University of California, master atom-smasher, who won the Nobel Prize for his discovery of the cyclotron; Professor Edwin H. McMillan, also of the University of California, one of the discoverers of plutonium, the new atomic energy element, and several others from the atomic bomb center, who, with your correspondent, had arrived during the afternoon.

The night was dark with black clouds and not a star could be seen. Occasionally a bolt of lightning would rend the sky and reveal for an instant the flat semi-desert landscape, rich with historic lore of past adventure. We, too, were headed for adventure, Argonauts on the way to a Golden Fleece richer by far than Jason ever found. We were on the road to the fabled golden Seven Cities of Cibola, sought in vain by Coronado on trails not too far away from the area we were traversing.

We rolled along on U. S. Highway 85, running between Albuquerque and El Paso, through sleeping ancient Spanish-American towns, their windows dark, their streets deserted—towns with music in their names, Las Lunas, Belen, Bernardo, Alamillo, Socorro, San Antonio.

At San Antonio we turned east and crossed "the bridge on the Rio Grande with the detour in the middle of it." We traveled ten and one-half miles eastward on U. S. Highway 380, where we turned south on a specially built dirt road, running for twenty-five miles to the Base Camp at Trinity.

The end of our trail was reached after we had covered about five and one-fifth miles on the dirt road. Here we saw the first signs of life since we had left Albuquerque about three hours earlier, a line of silent men dressed in helmets. A little further ahead a detachment of military police examined our special credentials.

We descended and looked about us. The night was still pitch black save for an occasional flash of lightning in the eastern sky, outlining for a brief instant the range of Sierra Oscuro directly ahead of us. We were in the middle of the New Mexico desert, miles away from nowhere, not a sign of life, not even a blinking light on the distant horizon. This was to be our caravansary until the zero hour.

From a distance to the southeast the beam of a searchlight probed the clouds. This gave us our first sense of orientation. The bombing test site, Zero, was a little to the left of the searchlight beam, twenty miles away. With the darkness and the waiting in the chill of the desert the tension became almost unendurable.

Directions for Observers' Safety

We gathered around in a circle to listen to directions on what we were to do at the time of the "shot," directions read aloud by the light of a flashlight:

At a short signal of the siren at minus five minutes to zero "all personnel whose duties did not specifically require otherwise" were to prepare "a suitable place to lie down on."

At a long signal of the siren at minus two minutes to zero "all personnel whose duties did not specifically require otherwise" were to "lie prone on the ground immediately, the face and eyes directed toward the ground and with the head away from Zero."

"Do not watch for the flash directly," the directions read, "but turn over after it has occurred and watch the cloud. Stay on the ground until the blast wave has passed (two minutes).

"At two short blasts of the siren, indicating the passing of all hazard from light and blast, all personnel will prepare to leave as soon as possible.

"The hazard from blast is reduced by lying down on the ground in such a manner that flying rocks, glass and other objects do not intervene between the source of blast and the individual. Open all car windows.

"The hazard from light injury to eyes is reduced by shielding the closed eyes with the bended arms and lying face down on the ground. If the first flash is viewed a 'blind spot' may prevent your seeing the rest of the show.

"The hazard from ultraviolet light injuries to the skin is best overcome by wearing long trousers and shirts with long sleeves."

David Dow, assistant to the scientific director of the Atomic Bomb Development Center, handed each of us a flat piece of colored glass used by arc welders to shield their eyes. Dr. Edward Teller of George Washington University cautioned us against sunburn. Someone produced sunburn lotion and passed it around.

It looked eerie seeing a number of our highest ranking scientists seriously rubbing sunburn lotion on their faces and hands in the pitch blackness of the night, twenty miles away from the expected flash. These were the men who, more than anybody, knew the potentialities of atomic energy on the loose. It gave one an inkling of their confidence in their handiwork.

The bomb was set on a structural steel tower 100 feet high. Nine miles away to the southwest was the base camp. This was G. H. Q. for the scientific high command, of which Professor Kenneth T. Bainbridge of Harvard University was field commander.

Here were erected barracks to serve as living quarters for the scientists, a mess hall, a commissary, a Post Exchange and other buildings. Here the vanguard of the atomists, headed by Prof. J. R. Oppenheimer of the University of California, scientific director of the atomic bomb project, lived like soldiers at the front, supervising the enormously complicated details involved in the epoch-making tests. Here early that Sunday afternoon gathered Maj. Gen. Leslie R. Groves, Commander in Chief of the Atomic Bomb Project; Brig. Gen. T. F. Farrell, hero of World War I, General Groves' deputy; Prof. Enrico Fermi, Nobel Prize winner and one of the leaders in the project; President James Bryant Conant of Harvard; Dr. Vannevar Bush, Director of the Office of Scientific Research and Development; Dean Richard C. Tolman of the California Institute of Technology, Prof. R. F. Bacher of Cornell, Col. Stafford L. Warren, University of Rochester (N. Y.) radiologist, and a host of other leaders in the atomic bomb program.

At the Base Camp was a dry, abandoned reservoir, about 500 feet square, surrounded by a mound of earth about eight feet high. Within this mound bulldozers dug a series of slit trenches, each about three feet deep, seven feet wide and about twenty-five feet long.

At a command over the radio at zero minus one minute all observers at Base Camp, about 150 of the "Who's Who" in science and the armed forces, lay down "prone on the ground" in their pre-assigned trenches, "face and eyes directed toward the ground and with the head away from Zero."

Three other posts had been established, south, north and west of Zero, each at a distance of 10,000 yards (5.7 miles). These were known, respectively, as South-10,000, North-10,000 and West-10,000, or S-10, N-10 and W-10.

Here the shelters were much more elaborate, wooden structures, their walls reinforced by cement, buried under a massive layer of earth.

S-10 was the control center. Here Professor Oppenheimer, as scientific commander in chief, and his field commander, Professor Bainbridge, issued orders and synchronized the activities of the other sites.

Here the signal was given and a complex of mechanisms was set in motion that resulted in the greatest burst of energy ever released by man on earth up till that time.

No switch was pulled, no button pressed, to light this first cosmic fire on this planet.

At forty-five seconds to zero, set for 5:30 o'clock, young Dr. Joseph L. McKibben of the University of California, at a signal from Professor Bainbridge, activated a master robot that set off a series of other robots. Moving "electronic fingers" writ and moved on, until at last strategically spaced electrons moved to the proper place at the proper split second.

The forty-five seconds passed and the moment was zero.

At our observation post on Compania Hill the atmosphere had grown tenser as the zero hour approached. We had spent the first part of our stay partaking of an early morning picnic breakfast that we had taken along with us. It had grown cold in the desert and many of us, lightly clad, shivered. Occasionally a drizzle came down and the intermittent flashes of lightning made us turn apprehensive glances toward Zero.

We had had some disturbing reports that the test might be called off because of the weather. The radio we had brought along for communication with Base Camp kept going out of order, and when we had finally repaired it some blatant band would drown out the news we wanted to hear.

PHYSICS AND ASTROPHYSICS

We knew there were two specially equipped B-29 Superfortresses high overhead to make observations and recordings in the upper atmosphere, but we could neither see nor hear them. We kept gazing through the blackness.

Suddenly, at 5:29:50, as we stood huddled around our radio, we heard a voice ringing through the darkness, sounding as though it had come from above the clouds:

"Zero minus ten seconds!"

A green flare flashed out through the clouds, descended slowly, opened, grew dim and vanished into the darkness.

The voice from the clouds boomed out again:

"Zero minus three seconds!"

Another green flare came down. Silence reigned over the desert. We kept moving in small groups in the direction of Zero. From the east came the first faint signs of dawn.

And just at that instant there rose from the bowels of the earth a light not of this world, the light of many suns in one.

It was a sunrise such as the world had never seen, a great green super-sun climbing in a fraction of a second to a height of more than 8,000 feet, rising ever higher until it touched the clouds, lighting up earth and sky all around with a dazzling luminosity.

Up it went, a great ball of fire about a mile in diameter, changing colors as it kept shooting upward, from deep purple to orange, expanding, growing bigger, rising as it was expanding, an elemental force freed from its bonds after being chained for billions of years.

For a fleeting instant the color was unearthly green, such as one sees only in the corona of the sun during a total eclipse.

It was as though the earth had opened and the skies had split. One felt as though he had been privileged to witness the Birth of the World—to be present at the moment of Creation when the Lord said: Let There Be Light.

On that moment hung eternity. Time stood still. Space contracted into a pinpoint.

To another observer, Prof. George B. Kistiakowsky of Harvard, the spectacle was "the nearest thing to Doomsday that one could possibly imagine."

"I am sure," he said, "that at the end of the world—in the last milli-second of the earth's existence—the last man will see what we saw!"

A great cloud rose from the ground and followed the trail of the Great Sun.

At first it was a giant column that soon took the shape of a supramundane mushroom. For a fleeting instant it took the form of the Statue of Liberty magnified many times.

Up it went, higher, higher, a giant mountain born in a few seconds instead of millions of years, quivering convulsively.

It touched the multi-colored clouds, pushed its summit through them, kept rising until it reached a height of 41,000 feet, 12,000 feet higher than the earth's highest mountain.

All through this very short but extremely long time-interval not a sound was heard. I could see the silhouettes of human forms motionless in little groups, like desert plants in the dark.

The new-born mountain in the distance, a giant among pigmies against the background of the Sierra Oscuro range, stood leaning at an angle against the clouds, a vibrant volcano spouting fire to the sky.

Roar Reverberations Over Desert

Then out of the great silence came a mighty thunder. For a brief interval the phenomena we had seen as light repeated themselves in terms of sound.

It was the blast from thousands of blockbusters going off simultaneously at one spot.

The thunder reverberated all through the desert, bounced back and forth from the Sierra Oscuros, echo upon echo. The ground trembled under our feet as in an earthquake.

A wave of hot wind was felt by many of us just before the blast and warned us of its coming.

The Big Boom came about 100 seconds after the Great Flash—the first cry of a new-born world. It brought the silent, motionless silhouettes to life, gave them a voice.

A loud cry filled the air. The little groups that hitherto had stood rooted to the earth like desert plants broke into a dance, the rhythm of primitive man dancing at one of his fire festivals at the coming of spring.

They clapped their hands as they leaped from the ground—earthbound man symbolizing a new birth in freedom—the birth of a new force that for the first time gives man means to free himself from the gravitational pull of the earth that holds him down.

The dance of the primitive man lasted but a few seconds, during which an evolutionary period of about 10,000 years had been telescoped. Primitive man was metamorphosed into modern man—shaking hands, slapping each other on the back, laughing like happy children.

The sun was just rising above the horizon as our caravan started on its way back to Albuquerque and Los Alamos. It rose to see a new thing under the sun, a new era in the life of man.

We looked at it through our dark lenses to compare it with what we had seen.

"The sun can't hold a candle to it!" one of us remarked.

September 26, 1945

SCIENTISTS 'CREATE' IN ATOMIC PROJECT

Forming of Element Unknown in Nature From Uranium Epic in Man's History

NEUTRON THE KEY FACTOR

Its Effect in Atom Nucleus Put to Use in Production of the New Plutonium

By WILLIAM L. LAURENCE

When the full details of the development of the atomic bomb can finally be told the story of the creation, production and purification of Element 94, named plutonium, will stand out as one of the great epics of history and as a distinct turning-point in the life of man on earth.

In this achievement our scientists not only have realized the dream of the ages, the transmutation of one element into another; they have accomplished what even the ancient alchemists did not dare dream about. For not only have they succeeded in transmuting one element into another, in relatively enormous quantities, they have created an entirely new element that never before was known to exist in nature, an element that, like U-235, releases enormous amounts of atomic energy.

It is as if nature had become tired after creating uranium, her ninety-second element, and decided to "call it a day." Now comes man and takes up the work where nature left off billions of years ago.

To understand how this was done requires the statement of some elementary facts about the constitution of the nuclei of atoms, in which more than 99.999 per cent of the mass and energy of the material universe is concentrated.

The nuclei are composed of two types of fundamental particles, protons and neutrons. Both have about the same mass, or atomic mass 1. But whereas the proton carries a constant fundamental unit of positive electricity, the neutron, as its name implies, is electrically neutral.

Chadwick's Discovery Basis

The discovery of the neutron in 1932 by Sir James Chadwick, then at the Cavendish Laboratory, Cambridge University, England, for which he won the Nobel Prize and knighthood, ranks with the greatest scientific discoveries of all time. It was this discovery that finally gave man the key to the atom. It opened the way for the transmutation of the elements and for the release of atomic energy. It made possible the atomic bomb and holds the promise for greater things to come. It is the Philosopher's Stone the alchemists looked for in vain.

While much still remains to be learned about the neutron, enough about it has been learned during the past ten years to make it the most useful tool in the study of the atom. It behaves differently under different conditions. Under certain conditions it acts in the manner of a particle that carries both a positive and a negative fundamental unit of electrical charge of equal magnitude. The two electrical charges thus balance one another, making the particle electrically neutral.

By the use of the cyclotron, gigantic atomic "merry-go-round" apparatus, neutrons can be fired at tremendous energies at the nuclei of atoms. Some of these neutrons penetrate the nucleus and remain there. When that happens the atomic weight of the atom is increased by one atomic mass unit. In that event the particular element does not change in identity. It becomes what is known as an "isotope" of the element, slightly heavier in weight, but still possessing the same chemical characteristics.

In many cases, however, the entry of a new neutron produces a cosmic cataclysm inside the nucleus, setting to work mighty forces greater, in proportion, than the eruption of the greatest terrestrial volcano or the most devastating earthquake.

When this atomic eruption takes place it often manifests itself in the emission of a negative charge of electricity, generally referred to as a beta particle, or negative electron. This negative electron comes from one of the neutrons in the nucleus.

Neutron Now Becomes Proton

Since that particular neutron originally had both a negative and a positive charge of electricity, the loss of the negative charge changes its character entirely. It is no longer an electrically neutral particle, for it now has a positive charge of electricity. In other words, the neutron has become a proton.

The atom to which this happens thus gets not only an extra unit of atomic weight but also an extra unit of positive electricity in its nucleus. This brings about a fundamental change in the nature of the atom, creating an entirely different element, since the nature and properties of the various elements depend entirely on the number of positively charged particles —i. e., protons, in their nucleus.

The material universe as we know it is made up of ninety-two elements, beginning with hydrogen, the lightest of the elements, at one end of the atomic table, and ending with uranium, the heaviest of the natural elements, at the other end. The elements are numbered from 1 to 92, each number corresponding to the number of positive electrical particles in the nuclei of their atoms. Thus, hydrogen, with atomic No. 1, has one positive particle, or proton, in its nucleus; helium, atomic No. 2, has two protons; lithium, the third element, has 3; and so on, up to uranium, the ninety-second and last natural element, which has ninety-two protons in the nucleus of its atom.

In addition to the protons, the nuclei of the atoms of most of the elements also contain neutrons. The effect of this is to make their atomic weight greater than the atomic number, this increase in weight depending on the number of neutrons, each neutron, as well as each proton, having an atomic mass value of 1.

For example, carbon occupies atomic No. 6 on the Periodic Table of Elements. This means it contains 6 protons in its nucleus. But it has an atomic weight of 12. This means that in addition to the 6 protons its nucleus also has 6 neutrons.

Similarly with uranium. It occupies atomic No. 92 on the Periodic Table. This means its nucleus contains 92 protons. One form (isotope) of uranium has an atomic weight of 238. Subtracting 92 from 238 gives the number of neutrons in the uranium nucleus as 146.

Another form of uranium known as uranium 235, the spectacular atomic energy element, still occupies No. 92 on the Periodic Table. But its atomic weight is 235. This means that it has the same number of protons in its nucleus as uranium 238 (U-238) but 3 fewer neutrons.

Two Transmutations of U-238

With these facts in mind we are now ready to understand the basic principles of the creation of plutonium, Element 94.

We start with uranium 238. A neutron is fired into its nucleus. This sets in motion a series of cosmic events of tremendous consequences.

The neutron first lodges in the nucleus of the uranium 238 atom. This increases the atomic weight of the atom by 1 mass unit. The element is still uranium, but instead of an atomic weight of 238 it now has a mass of 239. It has 147 neutrons instead of 146.

But this form of uranium 239 has a rather turbulent and hectic existence. Atomic eruptions take place of super-volcanic dimensions. Soon a negative electric charge—namely, a beta particle—comes flying out.

This negative particle is lost by one of the 147 neutrons in the uranium 239 nucleus. As explained earlier, that neutron is now left with a positive charge. It is now a proton. In other words, the uranium 239 now has only 146 neutrons in its nucleus. But instead of its original 92 protons it now has 93.

This means that a new element has been created out of the uranium, Element 93 out of Element 92. Since uranium was named after the planet Uranus, the element beyond uranium was named neptunium, after Neptune, the planet beyond Uranus.

But the volanic eruptions started by the original neutron does not end here. Neptunium, an element that was not known to exist in nature, also has a turbulent short life.

In a short time a negative electron (beta particle) comes shooting out of its nucleus. The same process repeats itself. One of the 146 neutrons in the neptunium nucleus, having lost its negative electrons, becomes transmuted into a proton.

This means that the nucleus of the element now has ninety-four protons and only 145 neutrons. It still has the same atomic weight of 239 but it is now once again an entirely new element, totally different from both uranium 238, its grandparent, and its immediate ancestor, neptunium.

How this transmutation was achieved on a mass-production basis will be told in a subsequent article.

October 4, 1945

ELEMENT 94 KEY TO ATOMIC PUZZLE

Liberation of Enough Neutrons in U-235 Created Plutonium Through a Chain Reaction

MATING OCCURS IN A 'PILE'

Great Heat Energy Generated in Vast 3-in-1 Power Plant— Perils, Precautions Cited

By WILLIAM L. LAURENCE

Plutonium, the man-made atomic energy element, contains, as described in detail in a previous article, two more positive charges in its nucleus than uranium; namely, it has ninety-four protons as compared with uranium's ninety-two. It has an atomic weight of 239, as compared with 238 for the common form of uranium, and thus is heavier than any of the elements found in nature.

The new element is, in fact, a great grandchild of uranium 238 and uranium 235. These two great-grandparents of Element 94, instead of being separated, are left together to mate.

The offspring of this mating is a new isotope of uranium, containing an extra neutron (electrically neutral basic unit of matter) in its nucleus, namely, uranium of atomic weight 239.

This new uranium 239 isotope is unstable. Soon, by a process known as beta-transformation, one of its neutrons loses its negative electrical charge (beta particle, or electron), thus leaving the neutron with only its positive charges, that previously had served to neutralize it. This means the neutron is converted into a proton.

Now the addition of one extra proton to the ninety-two protons in the uranium nucleus transmutes the uranium into Element 93, neptunium, an entirely new element not existing in nature, totally different physically and chemically from uranium.

Neptunium also has a turbulent short life time. By another spontaneous beta-transformation, one of the neutrons in its nucleus also emits a negative electron. This means that another extra proton has been added to the ninety-three protons already in the atom's nucleus—signalizing the birth of another new element—Element 94.

First Produced in California

Element 93 first was produced in the University of California Radiation Laboratory in the spring of 1940 by Prof. Edwin M. McMillan and Dr. Philip H. Abelson. They were ready to begin further work to investigate the possibilities of creating Element 94 when they were called away on another secret war project.

With permission of Drs. McMillan and Abelson the work was taken up by another scientific team at the University of California, Drs. Glenn T. Seaborg, Emilio Segrè, Joseph W. Kennedy, Arthur C. Wahl and Ernest O. Lawrence, with Dr. McMillan participating in the work through correspondence.

An isotope of Element 94, of atomic weight 238, produced by bombarding uranium with deuterons (nuclei of heavy hydrogen), was produced in December, 1940. This form, however, was not useful for atomic energy. The isotope of major interest, plutonium of atomic weight 239, that yields atomic energy in amounts equal to uranium 235, was discovered in March, 1941.

This discovery opened up a host of possibilities of enormous import. Since Element 94 could be produced from ordinary uranium 238, it promised a hundred-fold increase of the total atomic energy available from uranium. Since it was different chemically from uranium, it could be separated from it by chemical means.

The question was whether or not enough neutrons could be liberated from uranium 235, in an ordinary, unseparated natural mixture of uranium, to start what is known as a chain reaction, providing enough neutrons to create plutonium out of ordinary uranium 238.

Columbia Work Cited

This question, one of the most vital in the atomic bomb project, was solved brilliantly by a team of physicists working first at Columbia University and later at the Government Metallurgical Project established at the University of Chicago. This team included Prof. Enrico Fermi, famous Nobel Prize winning physicist; Dr. Leo Szilard, Prof. Walter H. Zinn of the College of the City of New York, Dr. Herbert L. Anderson of Columbia, Dr. B. Feld and Dr. George Weil.

Dec. 2, 1942, was one of the climactic days of the atomic bomb project, and, therefore, one of the historic days in the annals of mankind. On that day, in the handball court underneath the West Stands of Stagg Field, on the University of Chicago campus, Professor Fermi and his team demonstrated that plutonium could be produced in large amounts by a special lattice arrangement of uranium and graphite.

The mating of U-235 and U-238 to produce plutonium is brought about by means of a monumental structure designated as a "Pile." Actually the "Pile" in this case is the first atomic power plant built on earth, generating enormous amounts of atomic energy in the form of heat.

In this structure, atoms by the trillions are ripped asunder, and hosts of new elements are constantly being created. These elements, products of the fission of U-235, and distinct from plutonium, are highly important by-products and will have enormous value in biology, medicine and industry.

Thus the Atomic Pile actually is a three-in-one plant. It creates large quantities of plutonium. It creates a host of valuable new elements. It liberates vast amounts of atomic energy. No means, however, are at present available to produce this energy in a form that could be utilized.

The basic building block of the Atomic Pile is uranium metal, containing both U-238 and U-235, embedded in graphite. The uranium and the graphite are arranged in a geometrical lattice.

The graphite serves as the moderator to slow down the neutrons emitted in the process of the splitting of U-235 atoms by other neutrons. This is essential because slow neutrons are more likely to split U-235 atoms than fast neutrons.

The neutrons escaping from the uranium must pass through the graphite blocks before they can hit another piece of uranium in the lattice. In doing so they collide with graphite atoms and thus are deprived of the greater part of their enegry; i. e., they are slowed down. When they are reduced to a certain slow speed the "cosmic fireworks" begin.

The process here started is a self-perpetuating chain reaction, that may be described as a "cosmic firecracker."

When the slow neutron hits the U-235 nucleus, the phenomenon of nuclear fission takes place, namely, the U-235 atom is split into two nearly equal parts that fly apart with tremendous amounts of kinetic energy in the form of heat. In the process of being split, neu- atom of U-235 emits other neutrons, which, after being slowed down by the graphite, split other

PHYSICS AND ASTROPHYSICS

atoms of U-235 in the uranium mixture.

Each U-235 atom split thus acts as a firecracker, starting off the next U-235 atom and the next, ad infinitum, setting off an atomic conflagration in which enormous amounts of atomic energy are liberated every second.

Chain Reaction Perpetuated

While part of the neutrons liberated from the U-235 go into splitting other U-235 atoms, and thus perpetuate the chain reaction, enough of them go into the nuclei of the U-238 atoms to form plutonium in large amounts. When a certain quantity of the plutonium is formed, the uranium is removed and the plutonium is separated by chemical procedures.

The heat produced by the energy liberated in the fission of the U-235 is equivalent to the burning of millions of pounds of coal. The energy emitted in the form of radiations is many thousands of times greater than that generated by all the radium isolated in the entire world prior to the outbreak of the war.

Such a gigantic quantity of radiation would kill any living thing in its vicinity within a fraction of a second. The Atomic Pile therefore created the greatest problem for protecting human life mankind ever had faced.

To study these new problems, and to find solution for them, a staff of several hundred of the country's leading radiologists and biophysicists was organized. Like all other major problems involved in the atomic bomb project, it was successfully solved.

The heat generated in the Atomic Pile is so enormous that if allowed to accumulate it would result in the greatest of catastrophes. Hence herculean measures had to be devised to carry off the heat at a rate never contemplated before.

To behold this atomic power plant standing there in its silent majesty, so silent that the silence itself could be heard, is one of the most terrifying and awe-inspiring spectacles on earth today.

There is not a sound heard, not the slightest hint that within this huge man-made block titanic cosmic fires are raging such as had never raged on this earth in its present form. One stands before it as though beholding the realization of a vision such as a Michelangelo might have had of a world yet to be, as indescribable as the Grand Canyon of Arizona, Beethoven's Ninth symphony or the "Presence that dwells in the light of setting suns."

Here in the great silence one stands silent in the presence of a new form of creation. It is as though Mother Nature had "called it a day" when she created Element 92 and her "problem child" had taken up the work where she had left off.

One is reassured on seeing the most remarkable system of automatic controls, and controls of controls, devised to keep this man-made Titan from running wild. Left without control for even a few seconds, the giant would break his bonds, a super-Frankenstein on the loose.

Enormous as the mass is, its mechanisms and controls are adjusted with the fineness of the most delicate jewelled watch, and they respond with the sensitiveness of a fine Stradivarius. The slightest deviation from normal behavior and the automatic controls go into operation. They can stop the Titan in his tracks almost instantly.

October 5, 1945

Basic Pulse Beat of Universe Seen In Particle Born Within the Atom

By WILLIAM L. LAURENCE

A cosmic entity pulsating on the borderline of space and time, constantly being created and destroyed within the nuclei of atoms, which has a "lifetime" of only one one-hundredth of a sextillionth of a second, was offered yesterday as a possible solution to one of the fundamental secrets of the universe. The theory was presented at the closing session of the meeting of the American Physical Society at Columbia University.

The new entity, named the neutral meson, the existence of which is postulated on evidence provided by the cosmic rays from interstellar space, was described by Prof. J. Robert Oppenheimer of the University of California, one of the world's most distinguished physicists, under whose direction the atomic bomb was developed at Los Alamos, N. M.

Dr. Oppenheimer's presentation was hailed by leading physicists as a highly challenging contribution to modern physical theory, which, if substantiated by further experimental observations and studies, would provide a new key to man's understanding of his physical universe and of the vast cosmic forces that hold it together.

So far all the evidence available comes from cosmic rays, the most powerful and penetrating radiations known in nature, which constantly bombard the earth from all directions. Their origin is one of the most baffling secrets of nature, the solution of which may provide a key to the very origin of the physical universe.

But intensive investigations during the last thirty years have gradually pushed back the frontier of the unknown, and it was on recent discoveries made through explorations of the farthest regions of the "no-man's land" of the cosmos that Dr. Oppenheimer has based his working hypothesis.

The studies on cosmic rays have brought to light the existence of new fundamental particles of matter, known as mesons. They possess a mass about 200 times the mass of the electron, or about a tenth the mass of the proton or of the neutron. Those actually observed have been found to carry electrical charges, some positive and some negative, the charges being equal to the positive and negative charges carried by the proton and the electron, respectively.

The discovery of the meson led to the formulation of a new concept of the forces that hold the nuclei of the atoms together, and, consequently, of the basic forces of the universe as a whole. According to this concept, known as the "meson theory," the protons and neutrons within the nucleus of atoms constantly give birth to mesons, which die as matter and appear again as energy, thus supplying the enormous adhesive forces that keep the nuclei together despite the tremendous electrical repulsive forces that tend to tear the positively charged protons apart.

Not Observed in Nucleus

However, while positive and negative mesons have been observed as components of the cosmic rays, no one has ever observed them in the nucleus, and the "meson theory" has so far proved unsatisfactory in that it does not fully account for the complex forces within the nucleus.

In an effort to build up a more consistent meson theory that would fit the observed facts, physicists turned to the possibility that, in addition to positive and negative mesons, there exists also a neutral meson, which, like the neutron, carries no electrical charge. But, like that other hypothetical nuclear entity, the neutrino, no one has so far succeeded in actually observing a neutral meson, either in cosmic rays or, a fortiori, in the nucleus.

Professor Oppenheimer presented reasons why, firstly, the neutral meson exists, and, secondly, why it has not so far been found.

The reason is, he explained through a series of mathematical formulae, that the neutral meson has such an extremely short lifetime in the form of matter, its duration being one-hundredth of a sextillionth of a second (one divided by one followed by 23 zeros) within the nucleus.

According to this new concept of Dr. Oppenheimer, matter is constantly being created within the nucleus out of the energy of its protons and neutrons and being converted again into different forms of energy. The extremely short "material" existence of the neutral meson, during which it pulsates and dies, to reappear again as energy, may thus be said to represent the basic "pulse beat" of the material universe.

Because they have charge as well as mass, the positive and negative mesons have a relatively much longer existence in the form of matter.

The primary cosmic ray, as it enters the outer fringes of the atmosphere from interstellar space, is believed to be a proton. These protons, traveling with extremely high energies, sometimes reaching a hundred quadrillion electronvolts, collide with the atoms in the atmosphere. It is as a result of these collisions, in which very little momentum to speak of is transferred by the incident protons, that the observed mesons, charged as well as neutral, are born.

But, Dr. Oppenheimer pointed out, while the positive and negative mesons have a lifetime of two-millionths of a second, and hence can be observed before they decay or are absorbed by near-by nuclei, the neutral meson blows up in one-tenth of a quadrillionth of a second. Thus, while it has a lifetime in the open ten million times longer than it has when it is born within the confines of the nucleus, it is still too short to be observed.

Converted Into Gamma Rays

The neutral meson, on blowing up, Dr. Oppenheimer said, become converted into two gamma rays, highly penetrating radiations similar to X-rays. It is these gamma rays that account for the large component of the so-called soft radiations observed in the cosmic rays, Professor Oppenheimer added.

Dr. Oppenheimer also presented mathematical equations for calculating the number of mesons produced by the primary cosmic ray protons. This number varies with the energy of the incoming protons. At a height of twenty miles, the calculations show, a proton traveling with the energy of a hundred quadrillion electron volts (1 followed by 17 zeros) would produce as many as 20,000 mesons, he said.

While no experiments could ever be devised to observe the neutrino, which is believed to be a component of the meson, the possibility exists, Dr. Oppenheimer said, that means may be devised for observing the neutral meson. Furthermore, he added, the much-sought-for negative proton is being created in the nucleus by an exchange of charges through the medium of mesons, and will be detected when is found out how and under what conditions to look for it.

February 2, 1947

GAME OF 'TENNIS' IN ATOM PROVED

Prof. Lawrence Reports on Beam of Neutrons Produced by Synchro-Cyclotron

REQUIRES 10-FOOT SHIELD

Tests Confirm the Heisenberg Theory of Electrical Paradox in Basic Cosmic Forces

By WILLIAM L. LAURENCE
Special to THE NEW YORK TIMES.

NEW HAVEN, Conn., Oct. 15—A beam of neutron radiation so powerful that it required the erection of a ten-foot concrete wall to protect the scientists working with it has provided the first experimental evidence on the nature of the forces binding the neutrons and protons inside the nuclei of atoms—in other words, the fundamental forces underlying the very existence of the physical universe.

The experiments, opening a new era in nuclear physics, were carried out under the sponsorship of the United States Atomic Energy Commission with the 4,000-ton synchro-cyclotron of the University of California.

They were described here today by Prof. Ernest O. Lawrence, University of California Nobel Prize winning physicist, at the opening of the program arranged in celebration of the centennial of the Sheffield Scientific School, Yale University.

Starts Silliman Lectures

Professor Lawrence, who received his doctorate at Yale in 1925 and is a former member of the Yale faculty, delivered the first of a series of four Silliman Lectures on the general subject "Progress and Promise in the Sciences."

The neutron beam discussed today, the most powerful of its kind so far produced by man, was made possible by the post-war redesigning of the giant 184-inch cyclotron as a synchro-cyclotron, an atom-smashing instrument first proposed independently by Russian and American scientists.

This, Prof. Lawrence said, more than doubled the power of the great accelerator. It was turned on for the first time Nov. 1, 1946, and since then has been producing deuterons (nuclei of heavy hydrogen) of energies up to 200,000,000 electron-volts and alpha particles (helium nuclei) up to 400,000,000 electron volts.

"During the first evening of operation last fall", Professor Lawrence reported, "a sharply defined beam of penetrating radiation was observed from the probe-target in the cyclotron chamber. This extraordinarily interesting phenomenon, of course, was immediately the subject of the preliminary exploratory experiments."

By far the most significant experiments so far with this mighty neutron beam were in its application to the study of the sub-nuclear forces that hold the nuclei of the atoms of the material universe together, despite the fact that the nuclei of all the elements are composed of protons (positively charged fundamental particles) and neutrons (elemental particles electrically neutral).

Since, according to the laws of electricity operating in the every-day world, like electrical charges repel each other, science has been confronted with the profound cosmic mystery why this rule does not seem to apply in the world of the atomic nucleus, although it has been suspected that the neutron plays a vital role in preventing the universe from flying apart in a cloud of hydrogen gas.

In an effort to offer a solution to the mystery, Werener Heisenberg, German Nobel Prize winning physicist, who headed the Nazi version of the Manhattan District during the recent war, proposed in the early Nineteen Thirties a theory that promised some explanation of the manner in which neutrons and protons are held together to form the atomic nucleus.

Tossed Back and Forth

He proposed that "exchange forces" oscillated back and forth between protons and neutrons in the nucleus, providing a balance which permits the particles to be held together in a tight core. The thing that is exchanged, according to the theory, is the electrical charge on the proton which, Heisenberg believed, was tossed back and forth, like a tennis ball, between protons and neutrons.

Thus neutrons and protons were not fixed entities but kept being converted into each other, in incredibly short time intervals, inside the atomic nucleus.

For fifteen years physicists had tried unsuccessfully to obtain experimental evidence to check on the validity of the Heisenberg theory. Their greatest handicap was the lack of means of hitting the nucleus hard enough for the "exchange forces" to be detected with available instruments.

The new hundred million electron-volt neutron beam at last gave the scientists the tool they had been looking for. With it they obtained for the first time the experimental evidence demonstrating that these mysterious "exchange forces" are indeed real.

The universe exists, the powerful invisible light of the neutron beam revealed, by virtue of a cosmic "tennis game" played between neutrons, with a positive electric charge as the "ball" being tossed back and forth from one neutron to another, each neutron being hit by the "ball" becoming a proton for an infinitesimal time interval, after which it tosses the "ball" back to its partner, and so on to the end of time.

Presumably the purpose of the "game" is not to allow the positive charges in the nucleus to come ever within repelling distance of each other.

The experiments were done by bombarding paraffin with the 100,000,000-1 volt neutron beam, emitted from a target of beryllium when it is bombarded with 200,000,000-volt deuterons. Paraffin was used because it contains many hydrogen atoms, which have the simplest of all nuclei, consisting of only one proton.

The experiments lead to the conclusion, Professor Lawrence declared, that both ordinary forces and exchange forces are necessary to describe the interaction of neutrons and protons.

The California physicists used four radiation counters similar to Geiger counters. With only ordinary mechanical laws operating they would expect to find a certain pattern, in which "cue ball" neutrons would be deflected by the protons in the paraffin and emerge at small angles in the same general direction of the neutron beam.

However, the scientists found high energy protons emerging in the pattern they would expect of the neutrons, if only mechanical laws were operating. The only conclusion was that the "exchange force" theory was right.

When the bombarding neutrons struck the protons in the paraffin target a glancing blow, the charge on the proton oscillated back and forth between the two particles.

As the two particles moved apart the neutron acquired the charge of the proton, thereby becoming a proton, while the proton became a neutron.

Professor Lawrence showed drawings for a gigantic synchro-cyclotron, with a circular magnet weighing 12,000 tons. This colossus, he said, would accelerate atomic particles from five to ten billion electron volts, a range at which energy would be converted into matter, creating both negative as well as positive protons. It is to be named a bevatron, an abbreviation for billions of electron volts. It will be capable of enlargement.

October 16, 1947

SCHWINGER STATES HIS COSMIC THEORY

Physicists Awed as Harvard Man of 30 Tells Version of Electrodynamic Forces

By WILLIAM L. LAURENCE
Special to THE NEW YORK TIMES.

PASADENA, Calif., June 24—Leading physicists from many parts of this country and abroad listened with fascination today to a 30-year-old Harvard Professor of Physics expound a new theory through the symbols of higher mathematics, about the interactions of energy and matter, the two basic phenomena in which the physical universe is manifested to man.

The new theory was presented by Dr. Julian Schwinger, whom American physicists regard as the heir-apparent to the mantle of Einstein. While not yet complete, the Schwinger theory outlined today at the opening sessions of the meeting of the American Physical Society is looked upon by topflight theoretical physicists as the most important development in the last twenty years in our basic understanding of the cosmic forces holding the material universe together.

While not as revolutionary and far-reaching, at least in its present stage, as the theory of relativity and the quantum theory, the Schwinger theory about the electrodynamic forces operating within electrons and protons and other fundamental particles of forces operating within electrons and protons (expected later to be extended to mesons and neutrons and other fundamental particles of matter), has already removed some of the most serious stumbling blocks that have stood in the way and has provided theoretical physicists with a new road-map for exploring the vast "jungles" in the interior of the nuclei of atoms.

Supported by Columbia Tests

Dr. Schwinger's theory has already found its first support in experiments carried out at Columbia University. Its further elaboration to include all the present known particles of the material universe awaits further experimental observations

The new branch of physics exploring the electrical and magnetic forces within the nuclei of atoms, where nearly all the matter and energy in the universe is concentrated, is known as quantum electrodynamics. While the theory has had brilliant success in explaining the observed phenomena, it had constantly been confronted with a "cosmic ghost," known to the quantum physicist as "infinite inertia." All mathematical roads of the quantum theory of electrodynamics have led to this inevitable "ghost" of infinite inertia, which of course endowed the electron, and hence the physical universe, with an aspect of unreality, contrary to observed facts.

Theoretical physicists used various stratagems to dodge the ghost of infinite inertia, all the time realizing that the day of reckoning must come. The experiments at Columbia made further dodging no longer possible, and it looked for a time that the entire foundation on which the current theory of the electron is based must come tumbling down.

Schwinger Theory to Rescue

Professor Schwinger's theory comes to the rescue in the nick of time. He succeeded in locating the terms in the theory that produce the "ghost of the infinite" and has shown the way how to get around it in a way that it no longer affects the theory. In his theory, it becomes possible to separate the "ghost energy" from the real energy, so that no longer is reality always mixed up with unreality.

Dr. Schwinger demonstrated that only the terms in the mathematical equations defining finite energies had a correspondence with reality, while the terms defining infinities could never be found in the world of observable phenomena.

Thus far the theory can be applied only to energies up to 100,000,000 volts. It is expected, however, that it will be applicable to higher energies, as more knowledge is obtained through observations of the cosmic rays and production of high-energy particles with the super-cyclotrons now being built.

June 25, 1948

Much Hydrogen Bomb Data Known; Process Involves Fusion of Atoms

Information Has Been Published in Scientific Journals Here and Abroad Since 1935 and Therefore Is Available to Soviet

By WILLIAM L. LAURENCE

While a curtain of official secrecy surrounds the hydrogen super-atomic bomb, fundamental data published in scientific journals in this country, England and Germany as far back as 1935 show that such a bomb is definitely possible, and that Russia, as well as other countries, is therefore fully aware of its potentialities.

This information reveals that such a super-atomic bomb could be designed to be somewhere between 100 to 1,000 times more destructive than present-day atomic bombs, though from a practical point of view, our present stage of nuclear technology would limit its power to the lower figure. The first models, when and if constructed, would probably have no more than ten times the explosive power of the present models.

The nuclear process that makes the hydrogen bomb possible is just the opposite of that responsible for the release of the vast energy in uranium or plutonium bombs. In these elements, the nuclear energy is released by the process known as fission; namely, the splitting of their atoms with neutrons in a self-multiplying chain reaction.

In the hydrogen bomb, the process of energy release is that of fusion; that is the fusion of four hydrogen atoms to form one atom of helium.

Same Process as in Sun

The process of welding four atoms of hydrogen, the lightest element in nature, into one helium atom is, in fact, the same process that is universally accepted as the one going on in the sun, and that enables it to keep on radiating for billions of years without being burned out. The sun is thus, in a sense, a hydrogen bomb, the energy of which makes life on earth possible.

The process of both fission and fusion involves the conversion of a small amount of matter into energy, each gram of matter thus converted yielding 25,000,000 kilowatt-hours of energy—the equivalent of 20,000 tons of TNT. In the fission of uranium or plutonium only one-tenth of 1 per cent, or one gram in a kilogram, is thus converted into energy. In the fusion of hydrogen, as high as eight-tenths of 1 per cent, or eight grams a kilogram, is liberated as energy.

This means that one kilogram of hydrogen transmuted into helium would yield about 200,000,000 kilowatt-hours of energy, or an explosive force equal to that of 160,000 tons of TNT. A hydrogen bomb of ten kilograms would thus be equal to 1,600,000 tons of TNT, or eighty times the power of the Hiroshima bomb, while the fusion of 100 kilograms of hydrogen would yield the power of 800 Hiroshima-type bombs.

Such a bomb, it is estimated, would devastate an area of 300 to 400 square miles, as compared with an area of about ten square miles for the Nagasaki model bomb, and possibly fifteen square miles for the improved Eniwetok type models.

Atomic Bomb Is the Key

The process of fusion of hydrogen into helium has been the subject of speculation in scientific journals here and abroad for nearly fifteen years, and especially since 1939, when Prof. Hans A. Bethe, noted Cornell University physicist, presented his famous hypothesis about the hydrogen fusion process in the sun as the reason for the apparently inexhaustible source of solar energy. However, it was not until the advent of the uranium-plutonium fission bomb that the hydrogen fusion process was brought within the realm of possibility on earth.

The reason is that the hydrogen fusion process in the sun takes place at a temperature of 20,000,000 degrees Centigrade and a pressure on one hundred and sixty billion atmospheres. Since no such temperatures or pressures were regarded as possible under terrestrial conditions, man's attainment of hydrogen fusion was held to be forever beyond his reach.

The atomic bomb turned what had not been even a pipe dream into a practical possibility, for the atomic bomb is veritably "a piece of the sun." The temperature at the center of explosion is estimated at 60,000,000 degrees Centigrade, three times the temperature in the sun's interior, and 10,000 times the temperature of the solar surface. The pressure it creates at the moment of explosion is within the range of the pressures in the interior of the sun. Here then are the temperatures and pressures required to produce hydrogen fusion.

From the foregoing it becomes clear that to produce hydrogen fusion on the earth, it would be first necessary to produce a nuclear explosion by uranium or plutonium fission. This, of course, means that the hydrogen super-bomb would require the uranium or plutonium bomb as a fuse to set it off.

While ordinary hydrogen is the most plentiful element in nature, since all water, which covers three-fourths of the earth's surface, is composed of hydrogen and oxygen, the discussions in scientific journals on fusion deal largely with the rare form of hydrogen known as deuterium. This has an atomic mass twice that of common hydrogen and constitutes about one part in 5,000 of all waters on the earth.

While deuterium, more commonly known as "heavy hydrogen," is much rarer and more difficult to produce than ordinary hydrogen, and while, furthermore, it would release only about one-eighth of the energy released in the fusion of light hydrogen, it offers, theoretically at least, some advantages over ordinary hydrogen.

The principal gain is that instead of fusing four atoms of light hydrogen into one atom of helium, it is necessary to fuse only two atoms of heavy hydrogen. This process, the published scientific data show, would require a temperature of a little above 1,000,000 degrees instead of 20,000,000 degrees required for fusion of four light hydrogen atoms.

The nucleus of the light hydrogen atom is composed of one proton, the fundamental atomic particle carrying one unit of positive electricity. The nucleus of heavy hydrogen consists of one proton and one neutron, the fundamental atomic particle that is electrically neutral but has an atomic mass about equal to that of the proton. The nucleus of the heavy hydrogen atom is known as the deuteron.

When two deuterons are fused together under the high temperature and pressure of a fission bomb explosion two possible nuclear reactions take place. Each involves the transmutation of part of the matter of the two deuterons into vast quantities of energy. In each of these reactions, one of two deuterons is split in half, the neutron and proton being separated from each other.

In one reaction, the split off proton is joined to the second deuteron, thus forming a nucleus of two protons and one neutron. Such a nucleus is that of the light form of helium; namely, helium of atomic mass three.

In the second type of reaction, the liberated neutron, instead of the proton, joins the second deuteron, thus forming a nucleus of one proton and two neutrons. Such a nucleus is an isotope (twin) of hydrogen, with a mass of three atomic units, known as tritium.

Neutrons Would Aid Fission

In the first type of reaction, namely, the formation of helium three, one kilogram (2.2 pounds) of heavy hydrogen would produce 0.74998 kilograms of helium three, 0.24916 kilograms of neutrons and 21,560,000 kilowatt-hours of energy.

While the energy yield is thus less than one-eighth that could be produced by the fusion of light hydrogen, the enormous quantities of neutrons liberated in the process, not available from the light hydrogen, would go far towards making up the difference, since these neutrons would greatly increase the fission of the uranium or plutonium bomb "fusing" the explosion.

In the second type of process, namely, the formation of hydrogen three, two possible further reactions are theoretically possible. One of these is that the vast quantity of protons (about 0.25 kilograms) liberated would join the hydrogen three to form the ordinary helium of atomic mass four (two protons and two neutrons), thus releasing further vast quantities of energy.

The second possibility is that the hydrogen three, an unstable nucleus, would break up, thus releasing further neutrons for increasing the efficiency of the fission process in the uranium or plutonium "fuse."

The foregoing indicates that the light hydrogen bomb would involve the overcoming of many more technical difficulties than the construction of a successful device for the release of vast quantities of nuclear energy by the fusion of nuclei of heavy hydrogen.

While this process would yield no greater energy a given weight than that released by the fission of uranium or plutonium, it offers a number of advantages. The principal one is that uranium or plutonium are limited as to size by what is known as the critical mass, whereas no such limit is required for heavy hydrogen.

This would mean that as much

as a ton (1,000 kilograms) of heavy water could be incorporated with a uranium or plutonium bomb. Such a bomb, if all the atoms of heavy hydrogen were to fuse together, would yield an explosive force equal to about 20,000,000 tons of TNT, or 1,000 times the destructive power of the Hiroshima model bomb.

Prof. George Gamow, noted nuclear physicist of George Washington University, speculated on the possibility of setting off a heavy hydrogen explosion by means of "a powerful electric discharge through a thin wire," which, he states, may produce the high temperature required to start the reaction. If that proves to be the case, the uranium or plutonium "fission fuse" would become unnecessary.

January 18, 1950

ATOM TESTS UPSET UNIVERSE THEORIES

Physicists Are Left Uncertain About the Forces Holding Material Together

By WILLIAM L. LAURENCE
Special to The New York Times.

ROCHESTER, N. Y., Dec. 16—The country's top nuclear physicists, gathered at a conference here today at the University of Rochester, frankly admitted that the results of their experiments with giant atom-smashing machines have upset all their previous theories about the forces that hold the material universe together, with no hope for a satisfactory substitute in sight.

As a result of their latest explorations, the conference revealed, they find themselves hopelessly lost in the dark, impenetrable "jungles" of the "darkest Africa of matter"—the nucleus of the atom, in which the secret of the universe is locked up more tightly than ever before.

The physicists had hoped that by assaulting the citadel of the material universe with atomic artillery of hundreds of millions of volts they would make nature yield its most intimate secrets. Instead, nature hurled them back for a sizable loss, figuratively pulling the rug from under their feet.

The latest experiments, the physicists learned to their dismay, instead of lending further support to their theories, have yielded results that invalidate most of them. Instead of a neat picture of the universe, they now have a gigantic crossword puzzle so confusing that leading physicists expressed the view that it will be impossible to solve it within the lifetime of the youngest of their group.

The principal question physicists are trying to answer is "what holds the universe together?" In other words, "how does the material universe exist at all?" The reason for this question is not difficult to find. The universe consists of atoms, and more than 99.9 per cent of the matter and energy of the atoms of the material universe is concentrated within the atomic nucleus.

Action of Protons

Now the nucleus consists of particles carrying a positive electrical charge (protons), and particles that are electrically neutral (neutrons). Since, as every high school boy knows, like charges repel each other, the protons in the nucleus should repel each other with a tremendous force.

A British physicist has figured out that two pounds of protons at opposite sides of the earth, a pound at each pole, should repel each other with a force of twenty-six tons, yet we know that the protons are held together within the infinitesimally small atomic nucleus with a force millions of times greater than the electrical forces tending to repel them.

Without this great nuclear force, it is obvious that the material universe could not exist at all, and everything in it would evaporate into one cosmic cloud of hydrogen gas, the nucleus of which consists of only one proton, with nothing to repel it. But the fundamental question is: "What is the nature of this cosmic force and how does it arise?"

Prof. Hideki Yukawa, noted Japanese physicist, came forth with a theory several years ago, known as the meson theory, to explain the mystery. There are in the nucleus of the atom, he postulated, particles possessing masses about 200 times the mass of the electron, which provide the "cosmic cement" to hold the protons together against the electrical repelling forces.

A few years later, the meson was actually discovered to exist in nature as a by-product of the cosmic rays that constantly bombard the earth from outer space. Later several types of heavier mesons were discovered, and for a time there was hope that these may account for the cohesive forces within the nucleus.

For his providing what appeared to be the most satisfactory theory to explain the nuclear forces, Professor Yukawa was awarded the Nobel Prize in physics last year. But even before the prize had been awarded to him many serious cracks had appeared in his theory.

These cracks became much wider when physicists learned how to create mesons artificially with their giant atom-smashing machines. The results of the latest experiments appear to have given the coup de grace to the meson theory, as Professor Yukawa sat silently listening to the many contradictory and bewildering results yielded by the latest experiments to give it further support.

ROCHESTER, N. Y., Dec. 16 (UP)—Five hospitals, the University of Rochester Medical School and the Atomic Energy Commission announced today a cooperative plan for distributing radio-isotopes for medical treatment.

The cooperative venture means safety for patients from radioactive atoms and considerable monetary savings for patients and participating hospitals. It works like this:

Radioactive materials from Oak Ridge will be stored and processed by the University Isotope Center at facilities of the medical school and the atomic energy project. After being standardized at the medical school's "hot" laboratory, the radio-isotopes will be measured, taken to the hospitals, administered to patients.

If each hospital maintained a separate isotope program, the Atomic Energy Commission said, it would cost the institution up to $74,000 a year plus an initial investment of perhaps $50,000. Under the cooperative plan, each hospital will save about $10,000 every year and the initial investment for each institution will be only $1,500 to $2,000.

Joe W. Howland, chief of the Division of Medical Services at the University of Rochester Atomic Energy Project, originated the idea. The particpating hospitals are Genesee, Rochester General, Highland, St. Mary's and Strong Memorial.

December 17, 1950

CLUE IS REPORTED TO ATOM MYSTERY

Particle Observed in Cosmic Rays May Turn Out to Be the Negative Proton

By WILLIAM L. LAURENCE
Special to The New York Times.

CHICAGO, Oct. 27—Observation in the cosmic rays of a mysterious atomic particle that may turn out to be the long-sought negative proton, one of the missing building blocks of the universe, was reported here today at the closing sessions of the twentieth anniversary meeting of the American Institute of Physics.

The proton is one of the two basic constituents of the nuclei of all the atoms of which the material universe is made. It carries a constant unit of positive electrical charge, equal and opposite to that of the negatively charged electron, and has a mass about 2,000 times that of the election.

Mass of Neutron Noted

So far only one other particle about equal to the mass of the proton has been found. That is the neutron, the second major component of the atomic nucleus, which carries no electrical charge at all.

Since the electrons, as well as the mesons, atomic particles intermediate in mass between the electron and the proton, all carry either negative or positive charges, the absence of a proton with a negative charge constitutes one of the major mysteries of the cosmos. Scientists the world over have been on the lookout for it for many years and many a "big game hunt" has been organized to track it down in the vast "atomic jungle," only to meet failure.

The first hint that the negative proton, or something closely resembling it, may at last have been found, was obtained in studies at the University of Indiana's cosmic ray Laboratory by Assistant Prof. J. Gordon Retallack.

Evidence of Negative Particle

He reported today that, in the course of his studies on cosmic rays, the mysterious radiations that constantly bombard the earth from all directions, he found evidence of a negative particle that had a mass approximating that of the proton.

The particle appeared, at any rate, he said, to be the heaviest so far found to carry a negative electrical charge. It appeared to be heavier than the heavy negative meson, known as the "tau" meson, he added.

So far, he stated, seventeen photographic tracks of this heavy negatively-charged particle approximating the proton mass had been observed in more than 2,000 photographs.

Physicists at the meeting agreed that Dr. Retallack's observations were still of a preliminary nature and therefore could not be taken as evidence that the negative proton had at last been found.

October 28, 1951

40

SCIENTISTS LEARN OF NEW-TYPE ATOM

M.I.T. Physicist Reports Study of Cosmic Entity Without Nucleus in Its Structure

CLUE TO MYSTERIES SEEN

'Positronium,' With 1,000th the Mass of Hydrogen Unit, Regarded as Searchlight

By WILLIAM L. LAURENCE
Special to THE NEW YORK TIMES.

UPTON, L. I., Dec. 4—Discovery of a new type of cosmic entity, an atom without a nucleus, composed of just one positive and one negative electron revolving around each other in the manner of a double star, was reported here today at the Brookhaven National Laboratory of the Atomic Energy Commission.

The discovery is expected to provide a powerful new searchlight for probing the inner mysteries of the atomic nucleus.

The newly found element is nearly one one-thousandth the mass of the hydrogen atom, lightest of all the elements of which the universe is constituted. It was named "positronium" by its discoverer, Prof. Martin Deutsch, 34-year-old physicist of the Massachusetts Institute of Technology, who outlined the results of his findings so far at the closing sessions of a two-day symposium on nuclear science. The conference was arranged by the professional group on nuclear science of the Institute of Radio Engineers.

All the elements in the universe are composed of three fundamental particles—electrons, protons and neutrons. The electrons, lightest of the three, carry a fundamental unit of negative electricity. The protons carry an equal but opposite charge of positive electricity, but have a mass 1,837 times the mass of the electron. The neutron has a mass slightly greater than that of the proton.

Planet-like Action Involved

The protons and neutrons constitute the inner core, or nuclei, of the atoms of all the elements, about which the negative electrons revolve as do planets around the sun. Hydrogen, the lightest element, has one proton in its nucleus, and one electron revolving around it as its planet. Helium, the second element, has two protons in its nucleus and two planetary electrons. The helium nucleus also contains two neutrons.

Nature has built up its elements one proton at the time, with uranium, the ninety-second and last natural element, containing ninety-two protons in its nucleus, balanced by ninety-two negative electrons revolving around it in planetary orbits.

In addition to the negative electron, its electrical counterpart, an electron with a positive charge, known as a positron, also exists. It was discovered as a component of the cosmic radiations that bombard the earth from outer space. It also has been observed to be emitted by some twins, or isotopes, of certain radioactive elements.

Positrons, however, have very little chance in this part of the universe, in which the negative electron predominates. They annihilate each other, their mass being converted into energy.

Complete Conversion to Energy

The annihilation of matter that takes place when positron meets electron is the only instance so far discovered in which 100 per cent of the mass of material particles is converted into pure energy. This is a thousand times greater than the conversion of mass into energy in the explosion of an atomic bomb.

While this "de-materialization" of positron-electron pairs must inevitably take place by the immutable laws of nature, Professor Deutsch has found that the pair performs a brief "cosmic dance of death" prior to their mutual doom. It is then that they form the new cosmic entity — positronium — an element without a "heart," playing a still mysterious but possibly vital part in the cosmic scheme of things.

Professor Deutsch has found that positronium exists in two forms—ortho-positronium and para-positronium, depending on the orientation of their north and south poles relative to each other. In ortho-positronium, the magnetic poles of the positron and the electron are parallel to each other, the north and south poles in each pointing in the same direction. In para-positronium the magnetic poles of the electron-pair point in opposite directions.

The ortho-positronium, Professor Deutsch found, has much the shortest life—its "de-materialization" taking place within one-tenth of a billionth of a second. The para-positronium, on the other hand, lives a thousand times longer.

Positronium of both types exists in nature in the cosmic radiations, though it has not been observed directly from that source. The positronium studied so far is being produced synthetically by the use of artificially made elements emitting positrons.

December 5, 1951

EXPERIMENTS FOR HYDROGEN BOMB HELD SUCCESSFULLY AT ENIWETOK

FLASH IS DESCRIBED

Letter From Task Force Navigator Says Light Equaled 'Ten Suns'

3 ASSERT ATOLL VANISHED

Many at Scene Nov. 1 Believed That a Hydrogen Bomb Had Been Set Off

By The United Press.

WASHINGTON, Nov. 16—The recent test explosion at Eniwetok was a devastating blast, according to composite eyewitness reports sent back by service men who evidently believed that a hydrogen bomb had been set off.

Service men's letters disclosed that "the bomb" had been transported to San Francisco under heavy guard, where it was loaded on a Navy vessel and placed in a special compartment. The door was welded shut and heavy chains were welded across the door.

Federal Bureau of Investigation agents accompanied "the bomb" aboard ship, and there were more civilian and security personnel aboard than sailors. There was a moment's anxiety when the ship's electronics gear picked up what was thought to be an unidentified submarine, but one letter writer said "nothing came of it."

The ship carried "the bomb" directly to the test island, apparently an atoll some thirty-five miles from Eniwetok in the Marshall group, where most United States atomic tests are held.

The test island apparently was about three miles long and somewhere between one-quarter and one mile wide, although the description of the island varied in the letters.

Vessels Scattered in Area

Vessels of the task force were scattered in an area around the island with the closest stationed about thirty miles from the center of the explosion, the letters said. Several ships apparently were thirty-five miles away. There was no indication whether land observers were closer to the scene.

But the letters clearly showed that the explosion took place on Nov. 1 with zero hour at 7:15 A. M., Eniwetok time.

Aboard the ships the men had donned protective clothing and had been instructed to turn their backs to the island ten seconds before the blast, close their eyes and cover their faces with their arms.

At 7:14 A. M. a voice over the loudspeaker of each ship started counting the seconds. During that time, one observer wrote home, everything was quiet. He said:

"In those last few minutes, especially when they were counting off the seconds, we all grew real tense and silence was so perfect you could hear a pin drop. In those last few seconds, I think everything I've been told ran through my mind. And in the last second, I said a silent prayer * * *."

For six seconds after zero there was silence, no movement, no flash in the sky.

The first sign of the explosion came to the men aboard ship in the form of a flash many times brighter than the sun, followed by a wave of heat across their backs.

Although the men had their backs turned with their arms across their dark glasses, the blinding light was not kept out.

"It would take at least ten suns" to equal the light of the explosion from a distance of thirty-five miles, a navigator wrote.

Ten seconds after zero the men on ship started turning around to face the direction of the blast.

"I could hardly believe my eyes," one wrote. "A flame about two miles wide was shooting five miles into the air. This lasted for about 7.2 seconds. Then we saw thousands of tons of earth being thrown straight into the sky. Then a cloud began to form about twenty seconds after the shot."

"You would swear," another sailor wrote, "that the whole world was on fire. It was really something I'll never forget."

At least three eyewitnesses reported that the test island on which the bomb had been exploded disappeared after the blast.

"About fifteen minutes after shot time, the island on which the bomb had been set off from, started to burn and it turned a brilliant red. It burned for about six hours.

During this time it was gradually becoming smaller," one man wrote. "Within six hours an island that had once had palm trees and coconuts had now nothing. A mile wide island had disappeared."

November 17, 1952

New 'Eyepiece' Peers Into Atom; Gets Different Picture of Nucleus

By LAWRENCE E. DAVIES
Special to THE NEW YORK TIMES.

PALO ALTO, Calif., Aug. 12—Perfection of "the most powerful microscopic equipment ever built" and its yielding of a series of startling facts about the nucleus of the atom were announced today at Stanford University.

Powerful enough to see ten times deeper into the atom than any equipment ever built previously, the university said, the apparatus is able through its "eyepiece," a magnet, to distinguish particles within the atomic nucleus only two one-hundredths of a trillionth of an inch apart, an incomprehensibly minute figure.

The equipment, weighing two and a half tons, makes use of Stanford's atom smasher of the linear accelerator type, which is expected eventually to develop billion-volt energies. In combination with the accelerator, which "shoots" electrons into the nucleus of the target atom, the Stanford scientists are using a high-energy apparatus that scatters the electrons, measures their number and their angles of deflection.

The resulting data already have produced a picture of the atomic nucleus that is new in several respects. The university physicists say new hypotheses will have to be developed to explain some of its aspects.

"We now find considerable space between particles in the nucleus where we once thought a uniformly dense solid existed," said Dr. Robert Hofstadter, the Stanford nuclear physicist who designed the scattering apparatus. "That suggests it may be possible to compress it. But this is pure speculation."

The university's announcement listed the "astounding facts" obtained so far from experimentation with the new "microscope," as follows:

"That [the nucleus of the atom] is not a solid little ball of uniformly-packed particles, as widely believed, but more of a 'cottony' sphere composed of particles.

"That these particles are so densely crowded at the core that it appears solid, but they gradually thin out almost into nothingness toward the limits of the sphere.

"That although the average density is about what was previously predicted, the core is five to ten times denser. It is about 130 trillion [130,000,000,000,000] times denser than water. A mere drop of water of such density would weigh about 2,000,000 tons."

Dr. Hofstadter said hypotheses that must be developed to explain the findings might indicate "utterly unsuspected new paths for investigation." Already he and his associates are building an apparatus three times stronger than the present one and they have on the drawing boards a third one designed to utilize the linear accelerator's eventual billion-volt energy. The one under construction will use a twenty-five-ton magnet.

Dr. Hofstadter has been using energies of 125,000,000 to 150,000,000 volts. The electron "bullets" do not really strike the particles of the nucleus but "pass freely through them and are deflected by the particles' electrical fields," according to the university's explanation.

The magnet in the scattering apparatus has counters and photomultipliers attached to it to analyze the angles of deflection and the numbers of the electrons shot through the nuclei of the target material, which so far has been gold, lead, tantalum or beryllium. The first three are heavy metals with very dense nuclei whereas beryllium is a light metal.

The targets are used in the form of metal foils, in some cases only one-thousandth of an inch thick, or about as thick as 100,000 atoms.

One beauty of the Stanford device is this: It is able to examine the nuclear characteristics of compounds without separating their elements. Thus it can "look" into the nuclei of water, a compound of oxygen and hydrogen, instead of having to have water broken up into its component elements and analyzing their nuclei separately.

"This technique," today's announcement said, "will be particularly useful in examining unseparated isotopes—two or more forms of the same element that differ in atomic weight and other respects. The most delicate comparisons can now be made of isotopic nuclei, which are believed to contain the same number of protons [positively charged particles] but different numbers of neutrons [neutral particles]."

Dr. Hofstadter has had as his associates Harry R. Fechter, a graduate student, and John A. McIntyre of the W. W. Hansen Laboratories on the Stanford campus. An assistant was Eve Wiener, a graduate student who recently was killed in an automobile accident.

The Stanford physicist began the research two years ago under a grant from the research corporation. The project has received continuing aid also from the Joint Program of the Office of Naval Research and the Atomic Energy Commission. The Office of Scientific Research of the Air Force's Air Research and Development Command has financed its later phases.

August 13, 1953

New Atom Particle Found; Termed a Negative Proton

'Nuclear Ghost' Is Created From Energy in Coast Atom-Machine—A.E.C. Hails Discovery, but Defers Evaluation

By WILLIAM L. LAURENCE

A long-sought atomic particle that nature has until now hidden from man has been created artificially in man's most powerful "atom-machine" at the University of California.

Announcement of the achievement, opening a new era in man's quest to understand his universe and the elements of which it is constituted, was made jointly yesterday at the university's radiation laboratory at Berkeley and by the Atomic Energy Commission at Washington.

The A. E. C. said that it was too early yet to evaluate the discovery. However, the commission added, a big step forward had been made in man's efforts to control his physical environment.

The new particle is known as anti-proton, or the negative proton. Its existence has been postulated for nearly a quarter century as the missing link of the atomic world, the counterpart of the proton, one of the two fundamental building blocks of the nuclei of all atoms.

All the fundamental atomic particles so far discovered or created have been found to exist in three varieties—positively charged, negatively charged and electrically neutral. But the proton, heart of the hydrogen atom and one of the two constituents of the nuclei of all atoms, has never been found to carry a negative electrical charge.

The negative proton had become, because of its elusiveness, a "nuclear ghost" that haunted the world's physicists for a gen-

PHYSICS AND ASTROPHYSICS

eration. Its appearance at last in man's most powerful atomic "crucible," in which matter is created of energy, has restored the faith of science in the basic symmetry of nature, every element in it being balanced by a "mirror image," positive being balanced by its negative.

Thus the tiny electron, revolving in orbits around the atomic nuclei in the manner of planets around the sun, comes in two varieties, those carrying a basic unit of negative electricity and those carrying an exactly equal unit of positive electricity, known as the positron.

For reasons as yet unknown this part of the universe consists of negative electrons and positive protons, the proton having a mass nearly 2,000 times that of the electron. It is quite possible, however, scientists point out, that in another part of the universe the exact reverse is true, the atoms there being made of positive electrons revolving around a nucleus of negative protons.

Atom-Machine's Role

The negative proton was created in a multi-billion-volt atom-machine known as the bevatron, in which atomic bullets are speeded up to the enormous energies of billions of volts and then hurled against other atoms as targets. In the course of such a collision, the energy of flying atomic bullets is converted into matter.

In the California experiments the atomic bullets consisted of protons, the positively charged nuclei of hydrogen atoms. These were made to travel in a circular path in the doughnut-shaped giant machine until their energy had reached 6.2 billion electron volts, the highest ever achieved by man.

These 6.2 billion-volt protons were hurled against the nuclei of atoms of copper, which consist of twenty-nine protons and thirty-four to thirty-six neutrons.

When the 6.2 billion-volt proton crashes into one of the neutrons in the copper atom's nucleus about two billion volts of the energy is transformed into matter.

This newly created matter comes out in the form of a pair, a positive proton and its negative "mirror image," the anti-proton. If left together the two annihilate each other within a tenth of a millionth of a second, being transformed once again into radiation of an energy of two billion volts.

However, the California physicists have built a cunning maze through which only the negative proton can pass in less than a tenth of a millionth of a second. When in a vacuum, negative protons were found to be quite stable, lending support to the hypothesis that they may exist in a part of the universe in which there are no positive protons.

Physicist on 'Team'

Creation of the negative proton was achieved by a team of radiation laboratory physicists including Drs. Owen Chamberlain, Emilio Segre, Clyde Wiegand and Thomas Ypsilantis, with the help of Herbert Steiner and the cooperation of Dr. Edward J. Lofgren, physicist in charge of the bevatron.

Dr. Ernest O. Lawrence, director of the radiation laboratory, who won the Nobel Prize for his invention of the cyclotron, described the creation of the negative proton as "a major fundamental achievement in physics." He added that it was not possible now to evaluate it fully.

The discovery of the anti-proton, Dr. Lawrence said, may be a "milestone on the road to a whole new realm of discoveries in high energy physics in the days and years ahead."

There is no known "practical" application of the anti-proton, it was said, though it may lead to fundamental knowledge that would in turn offer man new controls over nature.

The A. E. C., which provided $9,500,000 for the building of the bevatron, declared in a statement issued in Washington yesterday that the anti-proton discovery was a major development that "may inaugurate a new era of nuclear physics."

Dr. Willard F. Libby, acting chairman of the commission, in a separate statement congratulated the scientific team on behalf of the A. E. C.

Dr. Libby's Statement

"This accomplishment," Dr. Libby said, "once more indicates why the commission, on behalf of the American people, has put emphasis on fundamental research in nuclear physics and other scientific fields, and financed costly equipment such as the bevatron for use in such research.

"The discovery of the anti-proton will open the way toward fuller understanding of the basic nuclear processes, which are fundamental to the entire atomic energy program.

"Though there is no immediate practical application for this discovery it is important. All of the applications of atomic energy rest upon the findings of fundamental research. No one can predict at this moment the implications which the discovery of the anti-proton may have for future research and development.

"However, we do know that every step forward in understanding the basic constituents of matter is valuable in man's effort to control and use his physical environment."

The anti-proton, if it could be created in large quantities, would be the most destructive force in nature, for actually the anti-proton is anti-matter, annihilating any matter in its vicinity. Luckily, anti-protons can be produced only in minute amounts at a prohibitively high cost in energy and resources.

The creation of the anti-proton also opens the possibility for the artificial creation of the "mirror-image" of the neutron, namely, the anti-neutron, which have properties opposite to that of the neutron, such as magnetic moment for example. The anti-neutron, if brought together with a neutron, would also bring about mutual annihilation, as is the case with electrons and protons and their "mirror-images."

October 19, 1955

ATOMIC SCIENTISTS FIND 4TH PARTICLE

'Anti-Neutron' Action Tops H-Bomb in Energy

Special to The New York Times.

BERKELEY, Calif., Sept. 14— Four University of California physicists working with the world's largest atom-smasher here have found a strange new particle of matter called the "anti-neutron."

Their discovery of the fourth particle was based on six months of experiments and twenty-five years of theoretical analysis.

The anti-neutron's most striking property, the scientists said, is that when it comes close to an ordinary neutron in the heart of an atom, the two particles annihilate each other.

At that instant the particles release several hundred times more energy than that produced in the reaction of a hydrogen bomb.

Confirmation of the anti-neutron's existence was made by Dr. Bruce Cork, Dr. Oreste Piccione, Dr. William Wenzel and Dr. Glen R. Lambertson. They worked as a team under Dr. Edward Lofgren, who is in charge of the Bevatron, the university's atom-smasher.

Prof. Ernest O. Lawrence, director of the radiation laboratory at Berkeley and inventor of the original Cyclotron, said today that the new discovery had no practical application yet.

But scientists on the campus recalled that when uranium atoms were first split by ordinary neutrons many years ago no practical application was immediately foreseen. Out of that split came the atom bomb and the harnessing of atomic energy for power.

"The value of this work lies in the expanding of our understanding of the nature of matter," Dr. Lawrence said. "This is part of a program in basic nuclear research, and past experience has shown that even fundamental discoveries of this kind bring material progress in ways that are unpredictable."

Neutrons and protons are the two particles that make up the nuclei of atoms. Protons carry a positive electrical charge, but neutrons have no charge.

Scientists have long believed that "anti-matter" particles also exist—particles whose physical properties are directly opposite to those in known matter.

Forces Symmetrical

Existence of the anti-proton was proved last October; the proof announced today completes the picture science can now draw positively as to the symmetrical nature of nuclear forces.

Despite the immense quantity of energy released when an anti-neutron swings close to a neutron, the process cannot be used as a practical source of energy, the team of physicists said.

This is because it takes an even greater amount of energy to produce a single anti-neutron.

The new particle lives for only a few ten-millionths of a second before it swings up against an ordinary neutron to unlock a blast of energy.

These particles have long been believed to exist in cosmic rays, but their natural occurence is so rare that no one has ever analyzed them.

Particles Identified

The scientists used the Bevatron to accelerate a stream of protons and neutrons to fantastic speeds. They set up a chain of magnets and ten instruments similar to Geiger counters in order to identify the particles as they swept past. One out of every 50,000 protons turned into an anti-proton.

The anti-protons could be spotted and counted because they possessed an electrical charge and made the counters react; but the anti-neutrons, with no charge, caused no reaction from the counters.

But a tally showed the scientists that bursts of energy were occurring at the target point of their machine that could not be ascribed to the already counted anti-protons. They kept track and eventually proved that every such burst represented the grapple of a neutron with its mysterious opposite number, the anti-neutron.

September 15, 1956

Basic Concept in Physics Is Reported Upset in Tests

Conservation of Parity Law in Nuclear Theory Challenged by Scientists at Columbia and Princeton Institute

By HAROLD M. SCHMECK Jr.

Experiments shattering a fundamental concept of nuclear physics were reported yesterday by Columbia University.

The concept, called the "principle of conservation of parity," has been accepted for thirty years. It must now be discarded, according to the Columbia scientists.

The principle of parity states that two sets of phenomena, one of which is an exact mirror of the other, behave in an identical fashion except for the mirror image effect.

The principle might be explained this way:

Assume that one motion picture camera is photographing a given set of actions and that another camera simultaneously is photographing the same set of actions as reflected in a mirror.

If the two films are later screened, a viewer would have no way, according to the principle of parity, of telling which of the two was the mirror image. The recently completed experiments indicate that there is a way of determining which of the two images is the mirror image.

In communicating with people in an intelligent civilization on another world, the Columbia report explained, it would be impossible, with the principle of parity in effect, to tell whether or not they and we meant the same thing by right-handed or left-handed. This could be true and still the basic physical laws in both worlds would behave exactly alike. The recent experiments indicate that this is not the case for some weak interactions of sub-atomic particles.

The idea that destroyed this principle originated with two theoretical physicists, Dr. Tsung Dao Lee of Columbia and Dr. Chen Ning Yang of the Institute for Advanced Study at Princeton, N. J. They suggested certain definitive experiments in papers on the subject: "Is Parity Conserved in Weak Interactions?"

The generally accepted belief, which had been a part of nuclear physics since 1925, was that parity should be conserved.

Two sets of experiments suggested by the two theorists showed that this parity was not conserved. A team of four Columbia physicists in collaboration with a member of the Institute for Advanced Study and a team at the National Bureau of Standards carried out the work.

The meeting that released the results of the experiments was held at 2 P. M. yesterday in Columbia's Pupin Physics Laboratories at 119th Street and Broadway. The chairman of the meeting was Dr. I. I. Rabi, Columbia's Nobel Prize-winning physicist.

"In a certain sense," Dr. Rabi commented on the development," a rather complete theoretical structure has been shattered at the base and we are not sure how the pieces will be put together."

Physicists present at the meeting indicated that it might take a long time to evolve a new concept on the basis of the recently achieved results. One scientist said that nuclear physics, in a sense, had been battering for years at a closed door only to find that it is not a door at all but a likeness of a door painted on the wall. Now science is at least in a position to hunt for the true door again, he observed.

K Mesons Led to Doubts

The Columbia theorists were led to doubt the principle of parity because, during the last few years, phenomena had been described in high energy physics that could not be explained by existing theories. This was particularly true of the patterns by which certain sub-atomic particles called K mesons decayed. Nobody was able to formulate a theory to account for both of the two methods of decay that they followed.

Dr. Lee and Dr. Yang suggested that perhaps it would be necessary to give up the principle of parity to gain an explanation of the sub-atomic interactions. They found that certain experiments dealing with particles better known than the K mesons could resolve the puzzle.

One set of experiments, done in a low temperature physics laboratory of the Bureau of Standards, showed that disintegrating nuclei of radioactive Cobalt 60 exhibited a specific "handedness," or spin in a given direction.

The other set of experiments dealt with the decay patterns of pi mesons. These are sub-atomic particles that are better understood than the K mesons. The pi mesons are believed to be largely responsible for the force that holds atomic nuclei together.

The disintegration pattern of the pi meson also showed a definite "handedness."

Scientists contributing to the work in addition to Dr. Lee and Dr. Yang are listed by Columbia as Dr. Ernest Ambler, Bureau of Standards; Dr. Richard L. Gavin of Columbia's Watson Scientific Laboratory; D. D. Hoppes, physicist of the Bureau of Standards; Associate Prof. Leon M. Lederman of Columbia and Associate Prof. Chien Shiung Wu of Columbia.

Three of the scientists, Dr. Lee, Dr. Yang and Dr. Wu, were born in China. Dr. Wu is considered the world's leading woman physicist. Dr. Lee is 30 years old, Dr Yang, 34. All three are married and have children.

Dr. Ambler, 33, was born in Bradford, England, and attended Oxford, where he earned the B. A., M. A. and Ph. D. degrees. Dr. Garwin, 28, a native of Cleveland, received his B. S. at Case Institute of Technology, and his M. S. and Ph. D. at the University of Chicago. Mr. Hoppes, also 28, was born in Liberty, Ind., and received the B. S. at Purdue and the M. S. at Catholic University. He is married and has two children.

Professor Lederman, born in New York in 1922, received the B. S. at City College, and the A. M. and the Ph. D. in physics at Columbia. He is married and the father of two children.

January 16, 1957

6 Scientists 'Trap' New Particle Of Atom After 70,000 Photos

By ROBERT K. PLUMB

American scientists have reached a major landmark in the exploration of inner space.

A team of six at the University of California, after working a year and a half with the nation's most powerful scientific instrument, the Bevatron, has obtained a ghostly picture of the atomic particle called the Xi zero.

The Xi particle has zero electrical charge. So it left no tracks to be photographed in experiments in which known atomic particles were traced as they plunged through a tank of liquid hydrogen.

But the presence of one Xi zero particle has been deduced from ghostly effects in a photograph that shows the motions of known particles to be peculiarly skewed by "something." The something in the photograph is the Xi zero particle.

An Xi zero, according to the new evidence, weighs about 2,570 times as much as an electron, and it has a lifetime of about one ten-billionth of a second.

Seventy thousand photographs were taken to catch one Xi zero in motion. A photograph taken just before Christmas has been identified after rigorous analysis as a genuine Xi zero ghost track. The finding will be published in Physical Review Letters, a publication of the American Physical Society.

Mathematical calculations two years ago by Dr. K. Nishijima, a Japanese physicist, and Dr. Murray Gell-Mann of the California Institute of Technology, predicted that the Xi zero should exist. The new photograph proves it, the University of California team reported.

An Xi zero is one of thirty fundamental particles discovered in cosmic ray or cyclotron studies or inferred from mathematical theories. Four of the thirty particles are stable. Others exist fleetingly for a few millionths of a second or less. With the discovery of the Xi (the Greek letter pronounced "zy") zero, all of the presently

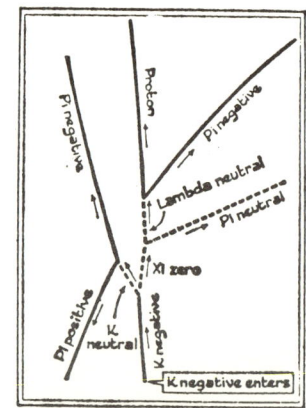

Diagram shows how change in path and nature of other particles illustrated existence of the Xi zero.

PHYSICS AND ASTROPHYSICS

predicated particles making up ordinary matter have been observed. A few predicted particles of anti-matter remain to be seen.

Ordinary matter, which composes the world about us, is made up of atoms consisting of stable particles, of neutrons, protons and electrons. These particles are known to have anti-particles — physical equivalents with opposite charge. Particles and anti-particles annihilate each other. Experiments with anti-particles have led to the speculation that atoms of anti-matter might make up objects that annihilate ordinary objects. Evidence for the existence of an anti-matter universe is now being sought in the heavens by the astronomers.

The fleeting particles and anti-particles, including the Xi zero, have been observed only in powerful smash-ups of atoms by atomic accelerators or cosmic rays. Perhaps these fleeting particles do not make up ordinary matter. But studies of the particles themselves and the way they interact have been a major concern of physicists for more than ten years. At stake is man's understanding of nature.

In the California experiments, done at the university's Lawrence Radiation Laboratory with the support of the Atomic Energy Commission, tracks of charged particles were photographed as the particles moved through a fifteen-inch "bubble chamber."

Charged particles in the liquid-hydrogen-filled bubble chamber set up tracks akin to streams of bubbles in a beer glass. Photographs of the tracks reveal how particles move and interact.

Particles Known to Man

Thirty fundamental particles are known. Thirteen are believed to have anti-particles. They are marked below with asterisks and count as two particles each. Masses are given in terms of mass of an electron. Four particles are stable, others are exceedingly fleet, except the neutron, which has a lifetime of about seventeen minutes. Data adapted from The Scientific American.

PARTICLE.	MASS.	LIFETIME.
Photon	0	Stable
Neutrino*	0	Stable
Electron	1	Stable
Mu meson*	206	Two millionths of second
Pi meson*	273	Two hundred-millionths of second
Pi neutral	264	Less than ten trillionths of second
K meson*	967	One hundred-millionths of second
K_1 meson	973	One ten-billionth of second
K_2 meson	973	Eight hundred-millionths of second
Proton*	1836	Stable
Neutron*	1839	One thousand and ten second
Lambda*	2182	Three ten-billionths of second
Sigma zero*	2326	Less than ten hundred-billionths of second
Sigma positive*	2328	Eight hundred-billionths of second
Sigma negative*	2342	Two ten-billionths of second
Xi negative*	2585	One ten-billionth of second
Xi zero*	?	?

Xi zero, particle just discovered in photographic plate at University of California, has mass 40 per cent greater than proton and lifetime of about one ten-billionth of a second, new studies show.

Picture Traces a 'Ghost' Particle

This is one of the 70,000 photographs taken by scientists at the University of California in an effort to prove the existence of Xi zero atomic particles. Xi particles have zero electrical charge and, therefore, do not leave trails on photographic plate. The apparent reactions of other particles in this photograph have led to deduction that the Xi zero particles do actually exist.

Particles Shot into Target

The Bevatron, the most powerful accelerator in operation outside the Soviet Union, was used to shoot 6,200,000,000-electron-volt particles into a target. The impacts produced negative particles, mostly pi mesons and negative K mesons. Filtering produced a pure secondary beam of K particles.

One photograph among 70,000 produced by the negative K beam showed a K particle entering the chamber. But the track of bubbles suddenly stopped. At this point, the team assumed, the K interacted to produce two neutral particles. The neutral particles left no tracks and did not appear on the photograph.

However, some distance away, two V-shaped tracks started up. Pictures of a "double V," started by neutral particle interaction to produce new charged particles, are common in Bevatron pictures. But in the 70,000th picture, the secondary V's were slightly askew. This observation was established to have been the effect of the never-before-observed Xi zero.

The chain of events was reconstructed from the photograph to be this:

The entering negative K particle hit a proton. The track ended. But a neutral K and a Xi zero were produced. These particles left no tracks of their own. But in a fraction of a billionth of a second the Xi zero decayed into a neutral lambda and a pi zero. The neutral lambda produced the skewed V as it decayed. The other V was produced as the neutral K formed a positive and a negative pi meson.

Only this explanation fits the tracks in the photograph, the group reported. Even if the Xi zero had not been predicted by theory, the evidence would have convinced experimenters that the particle has been found, they said.

The experiments were done by Dr. Luis W. Alvarez, Professor of Physics at the University of California at Berkeley; Dr. Philippe Eberhard, physicist of the Centre National de la Recherche Scientifique de France; Dr. Myron L. Good, physicist at the Lawrence Laboratory; Dr. Harold K. Ticho, Professor of Physics at the University of California, Los Angeles, and two graduate students, William Graziano and Stanley G. Wojcicki.

Searching analysis and a round of conferences by the team brought the conclusion that a Xi zero has been detected. Many particles can be identified by the direction and shape of the tracks they leave in the bubble chamber. The Xi zero remains a ghost, but a documented one.

March 5, 1959

Light Amplification Claimed by Scientist

By JOHN A. OSMUNDSEN

Achievement of the first true amplification of light was claimed here yesterday by a Hughes Aircraft Company scientist.

The feat was said to have been accomplished with an experimental device developed at the company's research laboratories in Culver City, Calif.

Such a device, once perfected, could generate a fine light beam of sufficient intensity to illuminate from the earth small swaths of the moon's surface or to vaporize materials placed in its path. The distance to the moon is 238,840 miles.

A light signal is said to be amplified if its power is increased without changes in its wave length or frequency.

The light would also be of such purity as to permit more precise studies of the structure of matter than are now possible. It also could be used as an information-carrier of unprecedented fidelity.

Thus, new systems for space and earth-bound communications, new tools for scientific research, and conceivably even a death ray—classical in concept—are potential developments that might come from the sort of device described to the press by Dr. Theodore H. Maiman of Hughes' research laboratories. The meeting was held at the Delmonico Hotel.

The light amplifier belongs to a class of devices called "masers." Maser is an acronym for "microwave amplification by stimulated emission of radiation." Microwaves — like radio waves and visible light—are a form of electromagnetic radiation. The Hughes device is an optical maser, or "laser," (the "l" standing for "light") because it amplifies visible electromagnetic radiation (light) rather than the invisible microwave type.

According to the maser principle as conceived in 1955 by Dr. Charles H. Townes of Columbia University, electromagnetic radiation can be amplified from one level of power to a higher one by taking advantage of its interactions with matter.

In the case of the laser, it is light that interacts with a certain group of negatively charged particles called electrons in the crystalline structure of a synthetic ruby.

The electrons involved can occupy one of four energy states, but they spend most of the time in the lowest one.

If the electrons are somehow forced into one of the higher energy states, they absorb energy at a specific frequency. When they "fall" back to the lower energy state, they radiate energy of the same frequency. The electrons can be "pumped" up or made to fall down by sending into the crystal they occupy a signal of electromagnetic energy of that frequency.

The trick is to see that most of the electrons are in a high energy state so that when the electromagnetic signals come in they will fall and radiate, thereby adding to the strength of the incoming signal or amplifying it. It is also necessary to trap the amplified signal in the crystal so that it will bounce back and forth, stimulating more electrons to radiate and add to the amplification.

The Hughes scientists appear to have solved both of these problems with techniques described earlier by Dr. Townes.

To excite the electrons into a high energy state, they flash a burst of green light at the pinkish crystal. This is called "optical pumping."

To trap the amplified signal, they silvered both ends of the ruby. These "mirrors" then reflect the electromagnetic wave back and forth, permitting it to gain power from electron radiation. A small hole in each mirror allows the input signal to gain entrance to the crystal and the output to leave in a fine beam of pure reddish light.

Dr. Maiman reported a power gain of five for the amplified signal over the input. Amplification by a factor of more than 100 is believed to be necessary for applying the laser.

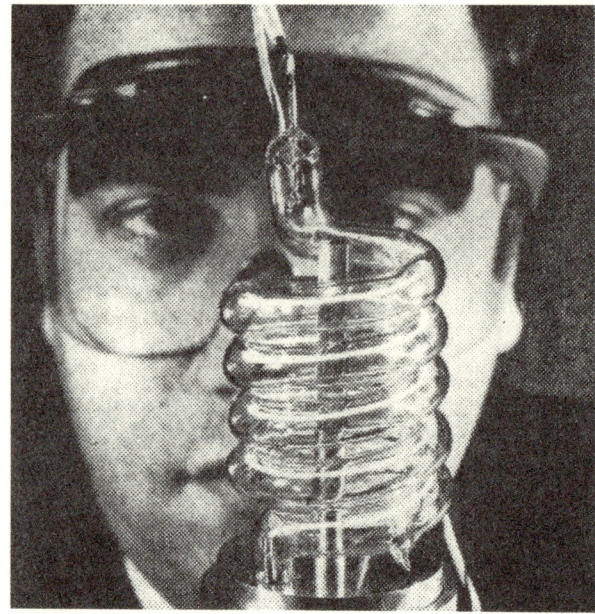

SUPER LIGHT SOURCE: Dr. Theodore H. Maiman of Hughes Aircraft Company, Culver City, Calif., studies new electronic device called a laser. A light source surrounds a rod of synthetic ruby crystal that, when properly activated, can generate a "coherent" beam of light brighter than that emitted by the center of the sun.

July 8, 1960

MOSSBAUER HONORED—

Dr. Rudolf L. Mössbauer, 31-year-old German physicist now a research fellow at the California Institute of Technology, will receive the 1960 Research Corporation Award of $5,000, it was announced last week by Caltech and the Research Corporation. Dr. Mössbauer, on leave from the Institute of Technical Physics in Munich, was cited for discovering a radiation effect that bears his name. The "Mössbauer Effect" is essentially a remarkably accurate yardstick that enables physicists for the first time to measure precisely the effects of natural forces such as gravity, electricity, and magnetism on tiny particles like photons and parts of the nuclei of atoms. What Dr. Mössbauer has done is to find a way to produce and detect gamma rays (similar to X-rays) the wave lengths of which are extremely sharply defined. If this radiation could serve as an atomic "pendulum," the resulting clock would be accurate to within one second in every 3,000,000 years. The effect has already been used to confirm Einstein's prediction that gravity can change the frequency of a light beam. It is now being used in the study of the hitherto little-known internal magnetic and electric fields in isotopes of the rare earth elements. The award will be presented in New York on Jan. 19.

Dr. Rudolf Mossbauer, who won Research Corp. Award.

January 1, 1961

PHYSICS AND ASTROPHYSICS

ELUSIVE NEUTRINO FOUND TO BE TWINS

Brookhaven Discovery May Lead to Clues to Puzzle of the Atomic Nucleus

By WALTER SULLIVAN

An experiment in which, for six months, the world's most powerful atom-smasher fired particles through forty-two feet of armor plate has shown that the most elusive of atomic particles is actually twins.

In the words of one physicist, "it is as though we had discovered two kinds of vacuum."

The particle that has proven to be twins is the neutrino. It has no mass and no electrical charge. Hence it shoots through armor like a bullet through a cloud.

The discovery that there are two such particles almost but not quite identical—may be a crack in the wall that has prevented physicists from seeing the method that lies behind the seeming "madness" of particle physics. The discovery brings to thirty-two the number of "fundamental" particles.

32 Particles Disputed

The ancient Greeks proposed that there was but one invisible and indivisible unit of matter, the "atomos." Many physicists still feel there must be an underlying simplicity. They consider a tally of thirty-two particles intolerable.

The newly completed experiment is, perhaps, most important in that it opens an entirely new avenue of exploration into the nature of matter. Its success is the starting gun for what may prove one of the more dramatic scientific races of the century.

This is the search for the postulated "W" particles that, some believe, embody the force involved in weak interactions between fragments of the atom. Preparations for such an experiment, using the world's two largest atom-smashers, are being rushed here and in Switzerland. Discovery of the particles, in the view of some physicists, would be revolutionary.

Neutrinos are influenced by other particles only in terms of an extremely feeble force— that of the so-called weak interactions. Those of low energy would have to therefore

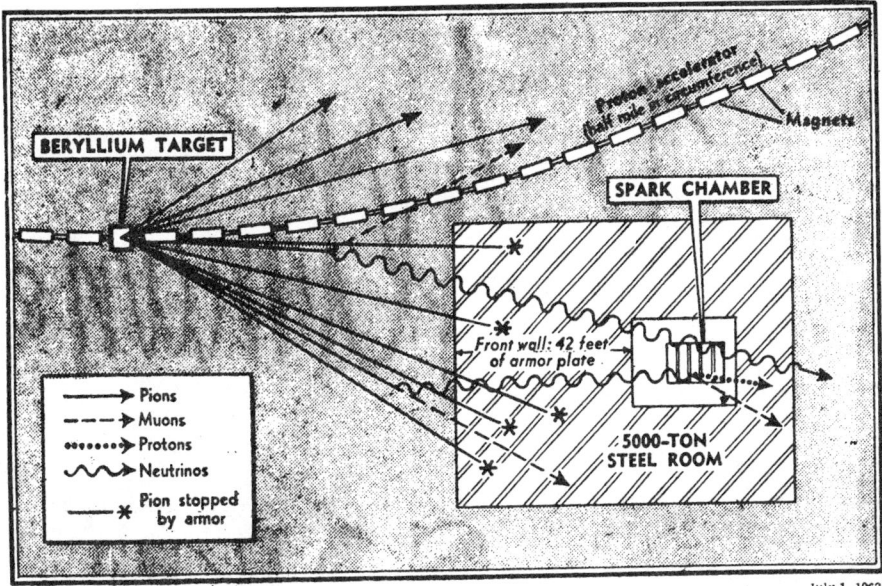

NEUTRINO EXPERIMENT: Protons, entering from the left, hit a strip of beryllium, producing a spray of pions. Before hitting the steel wall, about 10 per cent of the pions decayed into muons and neutrinos. Some of the latter reached the spark chamber and a very few interacted with aluminum atoms there to produce a proton and muon. Their failure to produce electrons demonstrated that there are two kinds of neutrino.

pass, on the average, through ten million miles of lead before hitting a particle squarely enough to interact.

Hence laboratory experiments with neutrinos seemed out of the question until outsized atom-smashers (accelerators) made it possible to produce vast numbers of neutrinos at very high energies. The higher their energy, the more likely they are to interact.

The experiment just completed is the first, using such a beam of high-energy neutrinos. It was conducted on the Alternating Gradient Synchrotron at the Brookhaven National Laboratory in Yaphank, L. I. This is the largest accelerator in existence. It spins particles around an underground path half a mile long. The laboratory is operated for the Atomic Energy Commission by a group of Eastern universities.

Normally an experiment can be completed in a few hours on such a machine, but this one required 800 hours. The allocation of so much time, in view of the urgent demands of other experimenters, is virtually unprecedented and is a measure of the importance ascribed to the neutrino project.

Funds From the A. E. C

Its cost, borne largely by the commission, was more than $1,000,000. It was carried out by a team from Columbia University.

What they have shown is that there is one kind of neutrino that takes part only in reactions involving an electron and another kind that participates only in those involving muons. No other difference between them has been found,

apart from their loyalties to those two particles.

As with other particles, each of the two neutrinos has its mirror-image counterpart, or antiparticle.

In the experiment it was found that a stream of muon-type neutrinos would only interact to produce muons. If all neutrinos were identical, the stream should have produced muons and electrons in equal numbers.

The experimenters believe their discovery may help explain why electrons and muons are identical except in weight. The muon is about 207 times as heavy as the electron. The other basic particles all differ from each other in a number of respects. It is therefore suspected that muons and electrons have some peculiar relationship.

The experiment has also shown that a high-energy neutrino beam can be used to study weak interactions, as had been proposed, independently, by scientists here and in the Soviet Union. They were Dr. Melvin Schwartz of Columbia and Dr. Bruno Pontecorvo, the Italian-born physicist who left England for Russia in 1950.

Weak interactions represent one of the four basic forces of nature. The strongest is that which binds together the nucleus of the atom. When the nucleus is broken apart by a high energy particle, this binding energy is converted into matter in the form of short-lived particles known as pions (pi mesons).

The next strongest force is electromagnetism, or the force of electric charge. It is this that binds atoms and molecules into solids and liquids.

The weakest force of all is gravity. No way has been found to study it, on the atomic level, for it takes a mass of atoms comparable to the entire earth to make a feather fall.

Lying, in strength, between gravity and electromagnetism is the force of the weak interactions. It, for example, controls the emission of electrons by radioactive substances (beta decay). It has also figured in some of the most surprising discoveries of recent years, such as the "overthrow of parity" that, in 1957, won a Nobel Prize for Drs. Tsung Dao Lee and Chen Ning Yang.

Because of the symmetry in nature, it was assumed that particles emerging from weak interactions should come out spinning in either direction. Instead it was found, for example, that neutrinos emerge spinning like a lefthanded screw, whereas antineutrinos have an opposite spin.

Theoretical physicists then took up a suggestion that there might be two as yet undetected particles, plus their corresponding antiparticles, that embody the force of the weak interactions. They would be counterparts of the pions that materialize when the nucleus is smashed. Drs. Lee and Yang christened them "W" particles.

However, one of the difficulties in studying weak interactions is that their effects are hidden by the far greater influences of electric charge and the nuclear force. Because neutrinos ignore these strong forces, but respond to that of weak interactions, they are ideal in searching for "W" particles.

The contestants in this search are Brookhaven and the Columbia group on the one hand and, on the other, an international team at CERN, on the outskirts of Geneva, Switzerland. CERN, a research center formed by a group of European nations, has an accelerator almost identical to that at Brookhaven, except it is slightly smaller.

The participants in the recent Brookhaven experiment included three professors of physics at Columbia: Dr. Schwartz, Dr. Jack Steinberger and Dr. Leon Lederman.

Scientists from Abroad

Other members of the team were Jean-Marc Gaillard, an exchange visitor from the French atomic energy research center at Saclay, near Paris, Konstantin Goulianos, a Fulbright Travel Fellow from Greece, Nariman Mistry of India and Dr. Gordon Danby, a Canadian on the staff at Brookhaven.

The three visitors from overseas were working at Columbia.

With each pulse, the big Brookhaven atom smasher can accelerate 200 billion protons to 33 Bev (billion electron volts). The armor plate, used in the experiment, was provided by the Navy from the deck plates of two obsolete cruisers.

Its role was to keep anything, except neutrinos, from reaching the experimental chamber. However, even armor forty-two feet thick was not enough to keep out charged particles at 33 Bev, and so the proton beam was kept down to 15 Bev.

When the protons of each pulse reached this energy, a beryllium target was flipped into its path. The result was a fan-shaped shower of pions. A pion quickly decays into a muon and neutrino, but since the pions were traveling almost at the speed of light, only 10 per cent could be expected to decay in the seventy feet between the target and the wall of armor.

Beam of Neutrinos

The rest of the pions either missed the wall entirely or smashed into it and vanished. However, a 10 per cent production of neutrinos was enough to generate many millions of such particles per pulse. At this rate, with the detector used, it was possible to observe one or two interactions in each twenty-four hours of operation, pulsing every 1.2 seconds.

Actually, fifty neutrino events were photographed in the detector. Muons were produced in all of them, whereas not a single energetic electron was observed. The detector was a spark chamber, itself an innovation in nuclear experiments. This model consists of ninety aluminum plates, spaced a half inch apart in a chamber filled with neon gas.

The spark chamber is ten feet high, six feet long and four feet wide. The only neutrinos detected were those that penetrated the armor and interacted with the nucleus of an atom in one of the aluminum plates. This produced two or more charged particles—one of them invariably a muon—that sliced through the sandwich array of plates.

These left in their wake chains of ionized neon atoms that provided paths for sparks to jump, like bolts of lightning, between the plates. Since the plates carried alternate electric charges, the result was a chain of sparks, marking the particle path that could be photographed.

Because cosmic rays penetrate the shielding, counters enveloping the apparatus shut it off every time a cosmic ray particle punches into it. Likewise the apparatus was turned on for only a few millionths of a second, following each pulse, when a neutrino reaction was possible.

In this sense, during the six months of the experiment, the spark chamber was active for a total of only three seconds.

One of the major changes planned for the attempt to detect "W" particles is diversion of the main proton beam from its circular track in the accelerator. If the beam, having achieved full power, can suddenly be turned, magnetically, toward the spark chamber, the system will be far more efficient.

This calls for a magnet that can come to full power in an extremely small fraction of a second. CERN is said to have achieved this already. If the proton beam is thus diverted, the shower of pions produced when it hits the target will be aimed squarely at the detector.

Another respect in which CERN is said to be ahead is in building a large spark chamber. It has one three times as big as the one used in the Brookhaven experiment. Columbia is making one five times as big, but it will not be ready until fall and the experiment probably cannot begin until next spring.

July 1, 1962

Physicists Discover The 34th Particle

By JOHN A. OSMUNDSEN

Discovery of the last elementary particle of matter in the list of those that nuclear physicists have expected to find was announced yesterday.

A team of scientists from Yale University and the Brookhaven National Laboratory reported the detection of the particle—called the anti-Xi-zero—in Physical Review Letters, a publication of the American Physical Society, out today.

Detection of the anti-Xi-zero completed the list of 34 elementary particles that nuclear physicists have drawn up to account for various conditions of matter and energy.

The existence of other elementary particles has been conjectured—"predicted' would be too strong," one physicist said yesterday. But the discovery of additional ones would probably serve as much to confuse as to clarify the perplexing state of affairs that exists in particle physics today.

The discovery of the anti-Xi-zero does nothing to make things better. It simply had to be there, and it was.

The chief significance of the anti-Xi-zero is in its confirmation, once again, of the essential symmetry that exists among the constituents of matter.

One aspect of this symmetry is that every elementary particle has an anti-particle, the anti-Xi-zero being the anti-particle of the Xi-zero, which was discovered some time ago.

Anti-particles are fragments of anti-matter, and when they collide with their material counterparts, both are usually annihilated. It was just such matter-anti-matter annihilation, in fact, that produced the anti-Xi-zero.

According to the physicists' report, a beam of anti-protons was siphoned off the main beam of Brookhaven's 30 billion electron volt (bev) Alternating Gradient Synchrotron and shunted into a vessel of liquid hydrogen — essentially a mass of protons.

When anti-protons and protons collide, they either bounce off each other in billiard-ball fashion if they are going very slowly, or they annihilate one another if they are going fast, with the consequent production of other elementary particles.

Sometimes, proton-anti-proton collisions produce particles that are lighter than either of them, such as mesons, in addition to pure energy.

Heavier Than Protons

If the anti-protons are going extremely fast, however, some of their energy of motion is converted — in accordance with Einstein's concept of the equivalence of mass and energy—into matter that makes up new particles that are heavier than protons. This is what was done to produce the anti-Xi-zero, which weighs more than the protons that produced it.

The new particle was extremely hard to find, partly because proton-anti-proton collisions at high energy can produce 100 or more different outcomes and partly because the anti-Xi-zero was invisible under conditions in which the scientists had to look for it.

The reason for this is that the new particle has no electrical charge on it. It is a particle's charge that makes the telltale track of bubbles in the liquid hydrogen bubble chamber by which scientists can observe an event.

Two things, essentially, told the Yale and Brookhaven scientists that an anti-Xi-zero had been created in their 20-inch bubble chamber.

First, a study of the two tracks emanating from the point at which an anti-proton had smashed into a proton revealed that the annihilation had produced a negative pi meson and a Xi-minus particle. To make a balance between energy and mass in and out of the collision, a third particle would have had to have been produced: The invisible anti-Xi-zero.

Sure enough, when the scientists looked at the trackless space a little beyond the collision point of the proton and anti-proton they found another bubble track indicating the decay of the third particle produced.

From an analysis of the decay products of that particle and of the invisible path it must have taken through the bubble chamber, the scientists concluded that it had been the anti-Xi-zero they were looking for.

The rarity of such an event and the difficulty of detecting it was evidenced by the 300,000 photographs that the scientists had to pore over since September, 1961, to find the first one. They now plan to raise the power of the anti-proton beam and make other alterations that will enable them to "collect" more anti-Xi-zeros to determine their properties.

Although the scientists want more information about the anti-Xi-zero, the single one they found was enough to fill in the list of expected elementary particles. The list—as most but not all physicists see it — is made up of the following particles and, of course, their anti-particles:

The photon, two kinds of neutrinos, the electron, the mu meson, two pi mesons, two K particles, the proton, the neutron, the lambda particle, three sigma particles and two Xi particles.

The authors of the report on the anti-Xi-zero particle were: Charles Baltay, Jack Sandweiss and Horace D. Taft of Yale; B. Brian Culwick, William B. Fowler, Joshua K. Kopp, Robert I. Louttit, James R. Sanford, Ralph P. Shutt, David L. Stonehill, Robert Stump, Alan M. Thorndike and Medford S. Webster of Brookhaven, which is at Upton, L. I.

August 15, 1963

PHYSICS AND ASTROPHYSICS

Research Indicates New Atom Particle

By WALTER SULLIVAN

The first of an entirely new family of atomic particles appears to have been observed.

They represent, in material form, one of the four fundamental forces of nature. Two of these forces are gravity and magnetism. Although the two others are less well known, they concern the behavior of atomic particles and relate to the very nature of matter.

It is one of these forces that is embodied in the new family of particles. The latter had been predicted, but their lifetime is so short that they have hitherto defied all attempts to confirm their existence.

They are thought to have been detected in an experiment that took several years to prepare and execute at CERN, the International Atomic Research Center near Geneva.

In the project, nuclei of hydrogen atoms were accelerated virtually to the speed of light by one of the world's most powerful atom smashers. They then crashed into metal and some of the fragments flew through a wall more impenetrable than any yet constructed by man. The survivors interacted in a series of detectors to produce what are thought to be the new particles.

Analogous to Pions

In a sense they are analogous to the pions (pi-mesons) that embody one of the fundamental forces — that which binds together the nucleus of the atom. The pion force is often called the "nuclear glue" or the force of strong interactions.

Within the nucleus of every atom in the universe this glue seems to exist only as a force. However, when the nucleus is shattered by natural radiation or an atom smasher, for a brief instant this force materializes into pions. The latter then decay into other particles.

The new family of particles similarly embodies the so-called force of weak interactions. This is the force that ejects electrons from decaying radioactive material. It also makes the charged pion a short-lived particle, splitting it into a muon (mu-meson) and neutrino.

Of the two other basic forces, electro-magnetism is manifested by the photon, a weightless particle that is a "piece of light." The force of gravity is exerted by a hypothetical particle, the graviton.

The new bits of matter, because of their relationship to weak interactions, are described by some as "W" (for "weak") particles. Others call them "intermediate bosons."

The discovery of W particles is not inconsistent with recent statements that all particles in the established "organization chart" of matter have been discovered. The W's, like the graviton and another postulated group, the omega particles, are "outsiders."

The search for W particles grew out of suggestions made several years ago by Dr. Melvin Schwartz of Columbia University in New York and Dr. Bruno Pontecorvo, the Italian-born physicist now working in the Soviet Union. Preliminary results were described at the International Conference on Elementary Particles, held in Siena, Italy, earlier this month.

An announcement issued by CERN during the conference described the W particles as "new, really fundamental" fragments of the nucleus. However, it said results of the search would not be announced until "all data from the experiment have been fully analyzed."

Other Particles Postulated

Nevertheless, word of the results is spreading through the scientific community. The particle thought to have been identified is the positively charged W. There should also be a negative W and two neutral W's, one of them a so-called antiparticle.

The apparent observation of the W particle seems to mean that the European Organization for Nuclear Research, which operates CERN, has won its race with the Brookhaven National Laboratory at Upton, L. I. Both centers were working toward the experiment. They possess the two most powerful atom-smashers in the world, each of them a proton accelerator a half mile in circumference.

Success depended on achievement of a number of technological feats and the use of a steel wall 82 feet thick, detectors weighing 18 and 30 tons, and hundreds of thousands of photographs.

Many of the protons used in the CERN experiment were probably once part of the water in Lake Geneva. Hydrogen extracted from this water was ionized, forming the hydrogen nuclei, or protons. These were accelerated to 24.8 billion electron volts, then were snatched from the accelerator and fired at a metal target.

Electrically charged fragments of nuclei shattered by the impact were guided by a magnetic horn to prevent their escape before some of them could decay and produce neutrinos. The latter are ghost-like particles that fly through matter like a rocket through a cloud. They apparently have no mass.

Beyond the horn an 82-foot gap allowed time for particles to decay into neutrinos and other fragments. Then a wall 82 feet thick stopped all particles except the neutrinos. The wall was constructed of steel from the Swiss national strategic stockpile.

Behind the wall was a succession of detectors. It was hoped that, in passing through them, a few of the neutrinos, despite their ethereal quality, would hit atomic nuclei squarely enough to leave their telltale marks.

The first detector was a bubble chamber, or tank containing a ton of liquid Freon. Particles generated by a neutrino collision in some cases left trains of bubbles that could be photographed. The response of such particles to a powerful magnetic field of 27,000 gauss, which is a unit of magnetism aided in their identification.

Some 332,000 photographs were taken in an effort to capture such events. It was said to be the first time a bubble chamber was used for neutrino detection. In line beyond the bubble chamber were two spark chambers, the first a sandwich array of aluminum and brass plates weighing 18 tons.

Since the plates are alternately of opposite charge, the passage of an electrically charged particle through the array caused sparks to leap from plate to plate along the particle track.

While W particles are among the theoretical by-products of neutrino collisions, their expected lifetime is too short for direct observation. It is less than 100 millionth of a billionth of a second, which is barely enough time for the particle to travel the width of an atom, even when it is going at nearly the speed of light.

Hence the W particles, which decay into various combinations of others particles, could only be known by their children.

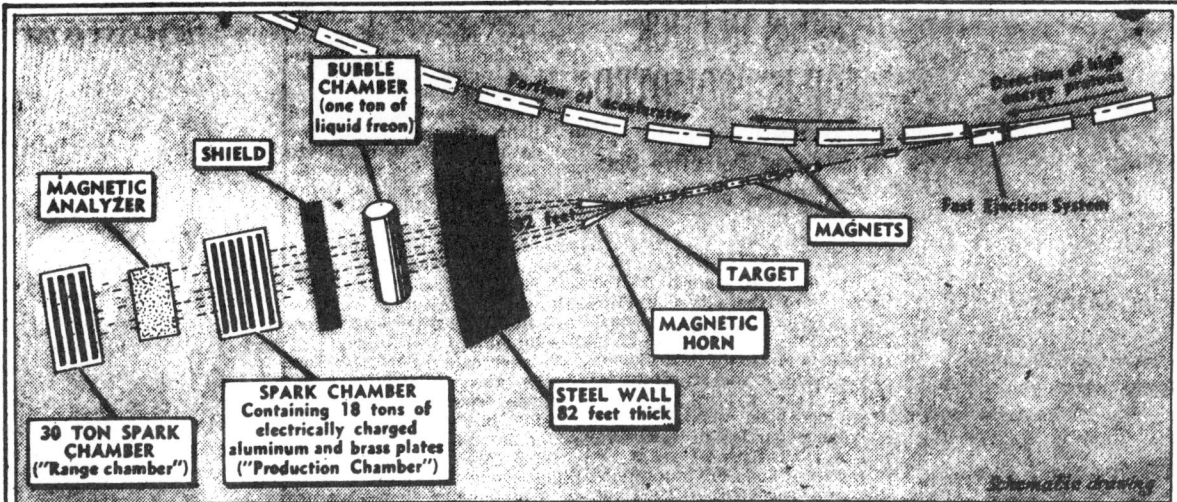

The New York Times   Oct. 13, 1963

ATOMIC DISCOVERY: Detection of "W" particles depended on the extraction from an atom smasher of protons traveling at 99.948 per cent of speed of light. This was done by fast ejection system. Protons struck metal target. Then magnetic horn guided scattered fragments toward the bubble chamber. The steel wall stopped all particles except ghostly neutrinos, which then generated "W" particles in the bubble chamber or spark chambers.

Among the expected products of neutrino collisions were a W particle and a negative muon. The W particle in some cases would then decay into a positive muon and neutrino. This almost instantaneous sequence should reveal itself by leaving two observable muon tracks in the spark chamber.

It was possible, however, that in all the cases where such pairs of tracks were observed, one of the tracks was actually left by a pion, rather than a muon. This possibility had to be eliminated to demonstrate the presence of W particles.

Tracks Are Analyzed

The CERN experimenters sought to do this as follows: The pion is far more likely to interact with other particles than is the muon. Hence the shorter tracks in each pair were the more likely to be pions. These were analyzed statistically in the 76 possible muon pairs generated in the two spark chambers.

The experimenters calculated how many interactions, on the average, should have occurred within a fixed distance along the observed tracks if they were left by pions. In the 32 track pairs photographed in the first chamber it was calculated that 14 interactions should have occurred, yet only 5 were observed. In the second chamber, where 44 pairs were photographed, the figure for pions should have been 16 interactions, and only 4 were observed.

It is chiefly this result that has encouraged the scientists to believe they were seeing pairs of muons produced by W particles. The bubble chamber observations disclosed only one event in which a W particle may have figured.

The experiments indicated that the mass of the W particle, while greater than that of the proton, is not more than 1.5 times the proton mass. It also left no doubt as to the correctness of last year's observation at Brookhaven that there are two kinds of neutrino, related to the two kinds of weak interaction: pion decay and the radioactive emission of electrons.

October 13, 1963

PHYSICISTS UPHOLD CHANGE IN THEORY

3 Teams Confirm Discovery on Behavior of Particles

By WALTER SULLIVAN

It has been found that certain atomic particles, identical except in being of opposite electrical charge, do not behave in an equal although opposite manner.

This discovery, which shakes the foundations of present-day physics, has been confirmed by three teams of physicists, working independently. It culminates a 10-year period in which a variety of asymmetries have been discovered where nature was assumed to be "impartial" and symmetrical.

The earlier findings concerned the so-called "weak interactions" between atomic particles. Although important to physicists, these do not figure prominently in the world about us.

The new discovery relates to the electromagnetic interactions that shape our lives and our world. They control electric, magnetic, chemical and atomic reactions.

The most definitive of the experiments was carried out by physicists from the Nevis Laboratories of Columbia University at Irvington-on-Hudson and the State University of New York at Stony Brook, L. I. The participants from Columbia were led by Dr. Paolo Franzini of Colombia. His wife, Dr. Juliet Lee-Franzini, headed the group from the State University of New York. The experiment is described in today's issue of Physical Review Letters.

The newly found absence of perfect symmetry in certain electromagnetic interactions is subtle—which is why it was not seen before. It is not yet clear whether it affects all such interactions. Experiments to explore this possibility are being planned.

The practical implications of the discovery are uncertain; but in the past any basic finding of this sort has led to important and sometimes historic developments.

Even though very subtle, the absences of symmetry in particle behavior, discovered during the last decade, are thought to have some deep meaning regarding the nature of matter and of the universe.

The situation can be likened to what long prevailed regarding the orbit of the planet Mercury. Very precise measurements and calculations showed a drift in the low point, or perihelion, of the orbit that remained unexplained. It was finally accounted for by the general theory of relativity.

Not Mirror Images

One implication of the new discovery is that the two realms of "matter" and "antimatter" are not, as previously thought, perfect mirror images of one another.

In recent years it has been found that for every particle of matter—electron, proton, meson and so forth—there is, in the catalogue of atomic fragments, an "antiparticle" identical in mass and lifetime, but opposite in electric charge or magnetic properties.

When particles of matter and antimatter meet, they annihilate one another, leaving only a burst of gamma rays. Because our world is dominated by matter, whenever a particle of antimatter (for example, an antiproton) appears it immediately encounters a particle of matter (such as a proton) and vanishes. Some scientists, however, believe that there may be distant worlds of antimatter where everything is opposite, like the looking-glass world visited by the heroine of "Alice in Wonderland."

It had been thought that all physical laws and reactions would look the same to ourselves and to "antipeople" in an "antiworld." It now appears that, if we ever communicate with such a world, we can ask them to conduct an experiment that will show whether theirs is a world of matter or antimatter. Such communication would be possible because light and radio waves look the same in terms of both matter and antimatter.

Symmetry Is Found

The exploration of subatomic particles over the years has brought to light a wide range of behavior patterns that are symmetrical. They seemed as much in the scheme of things as the idea that one cannot make something out of nothing.

One can, however, convert energy into matter. And one can reverse the process, converting mass into energy (as in an atomic bomb). This represents one of the symmetries (time reversal) whereby a reaction, at least in principle, can be made to run in either direction.

The first hint that such symmetries were not universal came a decade ago with the investigation of parity. This concerns a form of symmetry in which the geometry of particle interactions (all else being equal) is righthanded or lefthanded with equal frequency.

Put another way, the world of physical reactions should appear the same, whether viewed directly or through a mirror that reverses "right" and "left." In 1956 it was shown that this was not so in the weak interactions (such as the emission of electrons by radioactive cobalt).

Theory Is Violated

This grieved those who believed nature should be equal and impartial. Then it was observed that if the "mirror" was modified, symmetry still prevailed. The modified "mirror" not only switched right and left but also converted particles into antiparticles and vice versa (a process known as "charge conjugation" and represented by the letter C).

Then, in 1964, another instance of nature's partiality was discovered. Dr. Val L. Fitch and his colleagues at Princeton University found that in a small but significant percentage of cases a particle (the K-2 meson of zero electric charge) decayed into other particles in a manner that violated even the modified symmetry.

Physicists then had to fall back on their last bastion, a form of symmetry, whose "mirror" takes into account not only charge conjugation (C) and parity (P) but the direction of time's flow (T). The reversability of time means that a reaction of particles that began in the past and achieved certain end products can be reversed to produce the original particles. This has been considered a basic law of physics.

It turned out that this three-way "mirror" (known as CPT invariance) resolved the dilemma posed by the Princeton discovery. CPT invariance is a pillar of physics. It is so woven into relativity theory, for example, that few if any believe it can be wrong.

It says that a "mirror" must do three things to show a world of interactions that occur with equal frequency as those seen in front of the "mirror." The "mirror" must:

1. Reverse the image (P).
2. Reverse matter and antimatter (C).
3. Reverse the time flow (T).

Final Solution Absent

Although the application of this rule to the Princeton findings prevented a collapse of physical laws, it did not explain why the asymmetries were being found. There were reasons to suspect that the effect might be one that operated most strongly in the electromagnetic interactions.

Dr. Tsung Dao Lee, who shared a Nobel Prize for his role in the overthrow of parity, and others suggested a way to test this hypothesis.

It involved observing a great

number of events in which a newly discovered particle, the neutral eta meson, decayed into three pi mesons (one positive, one negative and one neutral). The neutral eta is a rare type of atomic fragment that has no features to label it as a particle or antiparticle—a factor that, for certain reasons, made it ideal for such a test.

The three groups of experimenters who have explored this reaction have, in all cases, found that the positive particles came out of the reaction on the average with more energy than the negative ones. The margin of difference lay, in general, between 5 and 9 per cent, which is a far stronger effect than that seen in the weak interactions.

Six Schools Involved

A compilation of data on 1,300 events was made by researchers at the University of California at Berkeley, Columbia, Purdue, Wisconsin and Yale. Another study was carried out by Dr. E. C. Fowler of Duke University. The report published today concerns the most ambitious test to date and the one to which the least margin of error is attributed.

It was carried out with the world's most powerful atom smasher, at Brookhaven National Laboratory near Upton, L. I.

Protons accelerated to great energy within the Brookhaven machine were fired into beryllium, producing a shower of about 60 varieties of subatomic particles. Magnets extracted positive pi mesons that were directed into a tank of liquid deuterium (heavy hydrogen).

Collisions with deuterium nuclei sometimes produced the very-short-lived eta mesons of zero charge and about a third of these decayed in the desired manner.

Roughly 435,000 photographs were taken and, although labor-saving devices were used, it took 300 girls four months to select 80,000 of these pictures as suitable for measurement.

Data on particle tracks from each interraction were punched into business-machine cards and run through a computer. In this way 1,441 events were selected as representing the sought-after decay process.

Positive Found Stronger

It was found that the positive particles were more energetic than the negative ones, more often than vice versa, by a margin of 7.2 per cent. The estimated range of uncertainty was less than 3 per cent.

Because the decaying particle in this experiment can be viewed as both matter and antimatter, its outcome would be the same in an antiworld. It would not be reversed, even though everything else was.

In such a world the proton (actually antiproton) would have an opposite charge to what it has on earth, but there would be no such reversal in the asymmetrical outcome of the experiment. Thus, the looking-glass world of antimatter is not quite a mirror image of our own.

This imbalance in nature, some suspect, may be related to the imbalance in our part of the universe where matter dominates. A balanced universe should be in equilibrium, not only in terms of electric charge, but in having equal amounts of matter and antimatter.

Hence, some scientists have postulated half the galaxies, or star systems, of the universe are of antimatter. The new discovery may indicate that this is not so—that, in fact, nature is peculiarly lopsided.

June 27, 1966

Discovery of New Particle Called 'Crucial Test' of Theory

By WALTER SULLIVAN

Detection of the new subatomic particle, the omega-minus, was described by its discoverers yesterday as a "crucial test" of a theory that could mark a turning point in particle physics.

It may play a role in bringing order out of the chaos of subatomic particles comparable to that played by the periodic table of elements. The latter was devised about 1870 by the great Russian chemist Dmitri Ivanovich Mendeleyev.

By arranging the elements in parallel columns, according to their atomic weights, Mendeleyev showed that they fell into groups with common properties. Furthermore, there were obvious gaps in the table and these enabled him not only to predict the discovery of new elements, but their properties as well.

Moreover, the fact that the elements fell into such a striking arrangement showed that there was an inner symmetry in the structures of these atoms—a symmetry that was yet to be discovered.

This is what now has happened on the much more fundamental level of the atomic particles. In Mendeleyev's case the discovery of his three predicted elements, gallium, scandium and germanium, persuaded the scientific world of the validity of his hypothesis. The omega-minus performs the same role in that its properties are peculiar and would have been unexpected but for the concept known as "the eightfold way."

This theory was proposed independently, early in 1961, by Dr. Murray Gell-Mann at the California Institute of Technology and Dr. Yuval Ne'eman of the Imperial College of Science and Technology in London. Dr. Ne'eman, strange to say, was a newcomer to physics. He was a colonel in the Israeli Army who suddenly decided to turn physicist.

The theory did not come to them out of the blue. Others had recognized various symmetries and relationsnips between the sub-atomic particles. Likewise, Mendeleyev's predecessors had experimented with groupings of elements. In both cases it was dramatic success in prediction that demonstrated the validity of the theory.

As Dr. Maurice Goldhaber, director of the Brookhaven National Laboratory, Upton, L. I., put it at a news conference on Friday, "Most people smiled when they spoke about the eightfold way." They do not smile any longer.

It was with the world's most powerful accelerator, or atom smasher, at Brookhaven that the omega-minus was discovered. The experiment is described in the issue of Physical

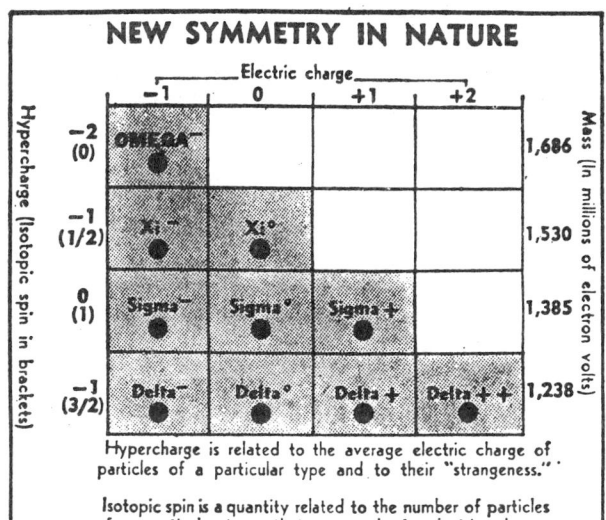

The New York Times — Feb. 23, 1964

The discovery of the Omega minus particle, which was predicted on the basis of this symmetrical arrangement of lighter particles, strongly supports the validity of the theory behind such groupings of atomic particles. It opens the way for prediction of new particles and a better understanding of their diversity. Mass is expressed, as customary, in terms of equivalent energy values. The symbols following each particle's name indicate its electric charge. Figures shown in the column on left are values, peculiar to particle physics, that are fundamental to the theory.

Review Letters dated Feb. 24 and made public yesterday. The scope and importance of the experiment is illustrated by the fact that the report is signed by 33 participants.

The proliferation of particles produced when the nuclei of atoms are broken apart by high energy bombardment has been a source of dismay to physicists. The number of such particles, including some that, because of their extremely short lifetimes, are called "resonances," is at least 82. The discovery of the omega-minus, however, has been a cause for rejoicing.

The eightfold way was suggested by a form of algebra developed in the last century by the Norwegian Sophus Lie. Part of this algebra deals with eight components and this seemed to Dr. Gell-Mann and Dr. Ne'eman to be applicable to eight "conserved quantities" characteristic of the various particles.

These quantities in general cannot be explained in terms of familiar concepts in that they involve mathematical relationships. For example, one of them, "isotopic spin," is derived from the electric charges characteristic of a particular group of particles.

Some particles have a positive charge, some a negative one and some have no charge at all—a zero charge. One kind has two positive charges. Isotopic spin is an arbitrary number equal to the number of charge states characteristic of a given particle (plus, minus, etc.) minus one, divided by two.

For example, a group of particles known as pions appear in three charge states: positive, negative and zero. Their isotopic spin is therefore three minus one, divided by two, which is one.

The other quantities of the eightfold way are similarly derived. Some of them, as stated in an article by Dr. Gell-Mann and others in the February issue of Scientific American, have not yet even been named.

However it was found that application of the hypothesis to eight particles, some of which had hitherto seemed quite unrelated, showed them to be, in a sense, variations of the same particle, differing only in energy levels (or mass) and electric charge. The eightfold nature of this symmetry seemed futher justification for the title of the theory.

Further eight-part groupings were identified. However, there was one misfit group of particles, called by Dr. Gell-Mann and his colleagues the deltas, with a mass of 1,238 million electron volts. It is customary in physics to describe the mass of a particle in terms of its equivalent value in energy, mass and energy being interchangeable in nuclear reactions.

The delta particles, discovered in 1952, appear in four electric states: plus, minus, zero and double-plus. This is the only group of particles with four such states. The eight-part patterns were no more than three units wide and hence could not accommodate four particles. Thus it was proposed that a ten-unit pattern, formed as a triangle or pyramid, was needed to accommodate the Deltas.

The four deltas would form the base Three sigma particles would form the next layer. The discovery of two xi particles with a mass of 1,530 fit the third layer. Hence, at the 1962 high energy physics conference in Geneva Dr. Gell-Mann urged that a search be made for the particle needed to crown this pyramid.

He described the characteristics that it must have to fit this slot and named it the omega-minus. Its properties were in part based on the work of Dr. S. Okubo of the Universities of Rochester and Tokyo.

Meaning Still Obscure

Its discovery, in the words of Dr. Goldhaber, "forms the capstone in a building which was so far held together only by the bold imagination of Dr. Gell-Mann and Dr. Ne'eman"

Like the periodic table of elements and the complex arrangements of lines in the spectrums of light emitted by atoms, the structures of the eightfold way clearly have some inner significance. However this meaning, in the last-named case, is still obscure.

The eightfold way is described by Dr. Gell-Mann in terms of a new set of ten particle names that many physicists consider more logical than the old system. Grouped together under each of these ten names are those heavy particles (baryons) with the same isotopic spin and hypercharge.

Hypercharge, like isotopic spin, is one of the quantities that determine the patterns of the eightfold way. It is related to the average electric charge of a group of particles and to their "strangeness." The latter is a property of particles so named because it seemed strange when first discovered.

When particles are defined in terms of these quantities, some with the same name have widely varying masses. It is the assumption of Dr. Gell-Mann and others that these differences in weight are actually differing energy levels in the same particle. In high energy reactions, such as those required to produce these short-lived fragments, energy is converted into matter and vice versa.

Search at Brookhaven

The search for the omega-minus began several months ago at several laboratories. At Brookhaven the basic tool was a high-energy beam of protons, or hydrogen nuclei. These smashed into a tungsten target generating a multitude of fragments including a variety known as negative K mesons.

With a 400-foot array of magnets and electrostatic separators, these were extracted and directed into an 80-inch bubble chamber. This device, filled with liquid hydrogen at its boiling point, is the world's largest of its kind. The mesons occasionally struck hydrogen nuclei in this chamber, and researchers hunted for the tell-tale decay products of the encounter that theoretically might produce the omega-minus.

The latter would decay, almost immediately, into other particles that could be identified. Thus the omega-minus would be known by its children. On Jan. 31 one such decay mode was observed and, since then, a second form has been detected.

Dr. Ralph P. Shutt was in over-all charge of the experimental team with Dr. Nicholas Samios responsible for conduct of the experiment. Dr. William B. Fowler was in charge of the bubble chamber and Dr. Medford S. Webster headed the group that produced the K mesons.

February 23, 1964

Physicists' New Theory Seeks To Explain Nature of Matter

By WALTER SULLIVAN

A team of physicists has come forth with a scenario for the behavior of matter that may be the closest to the truth yet devised.

It arranges the 100 or more known fragments of matter in tight patterns. Furthermore, it gives added plausibility to the view that there may be only a few basic particles—as yet undiscovered—from which all matter is built.

Since the days of ancient Greece, scientists have suspected that matter may be built of a few indivisible particles.

The new theory does not settle this issue finally, but it leaves open the possible existence in nature of basic building blocks, or "quarks." These could be either real fragments or mathematical entities smaller than the tiny electron.

The theory, presented yesterday at the annual meeting of the American Physical Society, involves concepts that even a number of physicists find difficult to follow. However, those close to the effort to find a unified theory for fragments of the atomic nucleus greeted it as a major advance toward bringing order out of what has seemed to be chaos.

Of particular interest, they pointed out, is a modification of the theory that enables it to stand up when viewed in terms of Einstein's theory of relativity.

The efforts of the theory builders in past years had consisted of painting neat, symmetrical pictures, in mathematical terms. They then held them up to nature to see if they fit. A notable success of this effort was the "eightfold way," developed in 1961 largely by Dr. Murray Gell-Mann of the California Institute of Technology and Dr. Yuval Ne'eman of the Imperial College of Science and Technology in London.

The new theory was presented to the Physical Society, at the Statler-Hilton Hotel, by Dr. Abraham Pais of the Rockefeller Institute.

Early work on the theory was done last summer at Brookhaven National Laboratory, near Upton, L. I., by Dr. Pais, Dr. Feza Guersey of Turkey and Dr. Luigi Radicati of Italy. Other contributors included Dr. M. A. Baqui Beg, Dr. Virenda Singh of India and Dr. Bunji Sakita, a Japanese at the University of Wisconsin.

Like the "eightfold way" the new concept is based on a branch of mathematics known as symmetry group theory. The workings of a symmetry group can be likened to a performance in which the roles of the actors are interchangeable.

This form of mathematics was applied to problems of nuclear physics as early as 1937 by Dr. Eugene Wigner, who later won a Nobel Prize for his demonstration of the symmetries in this field of science. At

PHYSICS AND ASTROPHYSICS

first there were only two variables—two "actors" in the play—and the symmetry group was referred to as SU-2.

When more factors were brought in, it became SU-4. The eightfold way, which lumps particles into groups of eight and ten, that is, on SU-3 transformations.

In the new theory, known as SU-6, there are six actors. The result is that the 100-odd particles are organized into very large groups—one of them numbering 70. The middleweight particles fall into a group of 35. The heaviest-type particles are lumped together in a group of 56.

This theory predicts some particle properties that have already been observed. Perhaps its most dramatic success has been in predicting the ratio between the magnetism of the proton and the magnetism of the neutron, the two stable particles that compose the nucleus of the atom. It has been known for 30 years that the ratio is about .68 but no one knew why. The theory says it should be .667.

Dr. Beg and Dr. Pais have proposed a modification of the new theory. In effect, it doubles the number of actors, making a total of 12 to allow for the effects of relativity. Earlier theories were applicable within a static framework. However, when particles move at close to the speed of light, the effects of relativity become apparent and it was there that earlier theories weakened.

The doubling of actors has been proposed independently by Dr. Abdus Salam, director of the new center of theoretical physics set up by the International Atomic Energy Agency in Trieste, along with Dr. R. Delbourgo and Dr. J. Strathdee.

The theory appears to go a long way toward displaying the symmetry and beauty that physicists believe underlie their science. Yet it makes no pretense of explaining the reason for the patterns. That most fundamental of all truths is not yet in sight.

January 28, 1965

Huge Blast of 1908 Laid to Antimatter

Scientists Ponder Explosion That Felled Many Trees

By WALTER SULLIVAN

In 1908 the earth may have been hit by an "antirock"—a meteorite composed of antimatter—accounting for what was perhaps the most violent natural explosion ever observed.

This is the burden of a report published yesterday in the British journal Nature by three scientists, one of them a Nobel laureate. Confirmation of their suggestion would constitute the first evidence that antimatter exists in substantial form anywhere in the universe.

The hypothesis has been supported, to some extent, by an analysis of tree rings formed during, before and after the year of the explosion. It also fits a peculiar aspect of the blast, which took place high in the air over Siberia, namely that it apparently produced no mushroom cloud.

If a mass of antimatter plunged into the atmosphere, the antimatter would be annihilated leaving no cloud such as that produced by an atomic weapon or chemical explosion.

Particles of antimatter are observed when atoms are smashed in the laboratory, but normally they survive for only a fraction of a second.

For every atomic particle known to physics—protons, neutrons, electrons and so forth—there is a corresponding antiparticle. The latter, in its electric charge or other characteristics, is the opposite of its counterpart in ordinary matter.

When an antiparticle encounters matter, such as an atom of the earth's atmosphere, the antiparticle is annihilated in a reaction in which its entire mass may be converted into energy. Thus a comparatively small chunk of antimatter could produce an explosion of vast dimensions.

Some of the trees knocked down in 1908 in a blast that is subject of scientific report

30 Million Tons of TNT

The 1908 explosion, referred to as the Tunguska meteorite after a nearby river, took place in the air at a height that has been estimated at three miles. Its effects were comparable to those of a nuclear weapon with a yield equivalent to that from 30 million tons of TNT.

The site has been examined by a succession of Soviet expeditions, the most recent in 1961. The effects are still evident, more than a half century later, because trees were blown down 20 miles or more from "ground zero"—the spot directly below the explosion. Their dried trunks still lie along lines radiating from this spot.

The authors of yesterday's report were Dr. Clyde Cowan of the Catholic University of America in Washington, D.C., co-discoverer of the elusive atomic particle known as the neutrino, and C. R. Atluri and Dr. Willard F. Libby, both of the University of California, Los Angeles.

Won Nobel Prize

Dr. Libby won the Nobel prize in chemistry for his discovery that the radioactive form of carbon, carbon 14, can be used as a tool for measuring ages. This technique is a pillar of modern archeology.

The report is couched in extremely cautious terms, stressing the uncertainties in many figures used in the calculations. The authors do not contend that the meteorite was in fact what they call an "antirock." They argue, instead, that, in view of their findings, the hypothesis is sufficiently plausible to merit further investigation.

As one of the authors put it privately, "The proposal is so reminiscent of science fiction that it had to be handled gingerly." The authors hope it will encourage others to examine tree rings in many parts of the world for the unusually high levels of carbon 14 to be expected in wood formed shortly after the blast.

The New York Times — May 30, 1965

Cross marks site of blast

Carbon 14 is normally formed when high energy particles from space, known as cosmic rays, rain onto the atmosphere. However, it was calculated that an antimatter explosion would create enough additional atoms of carbon 14 to produce a worldwide enrichment of this radioactive substance.

In the same way, it is estimated that all nuclear explosions to date have raised the carbon 14 level in the air about 1 per cent. This carbon is "inhaled" by plants the world over in the form of carbon dioxide and becomes a part of their substance.

In the study reported yesterday, a 300-year-old Douglas fir from Arizona and an oak tree from near Los Angeles were analyzed. Wood was stripped from a number of annual rings from 1873 to 1933. In both trees the highest content of carbon 14 was from wood formed in 1909, the year after the explosion.

There is extensive evidence that the blast produced a fireball of sufficient intensity to ignite fire in many places and scald people at great distances.

For example, a farmer, S. B. Semenov, was sitting on the steps of his house at Vanavara, some 40 miles away, when the blast occurred. "My shirt was almost burned on my body," he said afterward.

He dropped his eyes, but when he raised them again the fireball had vanished. Then the blast arrived, hurling him from the steps and knocking him unconscious briefly. His windows were broken and his barn damaged.

A neighbor who was facing in the opposite direction first knew of the blast when his ears began to burn painfully.

The immediate area of the event was sparsely inhabited by Tungus tribesmen, one of whom had a herd of 500 reindeer. The deer and scrubby, sub-arctic forests were destroyed by fire and the tribesman's samovar, apparently of tin, was melted.

A few days ago a report on the Soviet expedition of 1961 became available. It appears in an English-language version of Meteoritica, the Soviet journal on meteorites. The issue was translated by Spectrum Translation and Research, Inc., and published here by Taurus Press.

The report supports the idea that the explosion aloft was intense enough to scald live twigs throughout the region below. The branches of trees that survived the event have been dissected and those that were twigs in 1908 show scalding on their upper sides.

Had they been burned by forest fires that spread from one spot, the burns would presumably have been on the under side.

The most widely accepted explanation of the explosion in recent years has been that a comet head plunged into the atmosphere and exploded before it struck the earth. This would account for the absence of a crater at the site.

White Nights

It would also explain the strange "white nights" observed for two months afterward in Europe and western Siberia. For a time it was possible to read a newspaper even at midnight. This has been attributed to an effect produced by the tail of the supposed comet.

After a Soviet expedition to the area in 1958, it was reported that unusual radioactivity had been found there, leading to a suggestion that the explosion might have been nuclear, long antedating the first man-made blast of this sort. Perhaps, it was suggested in the Soviet Union, the device was delivered by the vehicle of some distant civilization.

This suggestion is dismissed as "fantastic," and as "based on factual material of questionable competence," in the report in Meteoritica by K. P. Florenskiy. One goal of the 1961 expedition, he says, was to search for fragments of the "meteorite" and to explore the peculiarly rapid growth reported for trees in the vicinity after the explosion.

This last phenomenon had led some to suspect that the explosion showered the region with material that acted as a strong fertilizer.

The researchers studied 95 forest plots, laid out across the region in a giant cross and scattered in other sections. Their conclusion is that the rapid growth by surviving trees was simply because of reduced competition from other vegetation.

However, the search for meteorite fragments was more of a problem. Among the many puzzles of the "meteorite" fall has been the failure of successive expeditions to find any fragments that could be attributed to such an object. It seemed strange that anything so large could disintegrate and leave no trace of its material.

The problem is made difficult by the steady rain of meteoritic dust onto the earth. Virtually any sample of surface soil, ocean sediment or polar ice, if examined closely, reveals particles of such dust, often composed of magnetite. These were found also at the Tunguska site, but not in sufficient quantity to represent convincing samples of a meteorite.

In his report Mr. Florenskiy, though citing no such observation, argues that the explosion must have produced a mushroom cloud that was carried downwind. The particles, he says, would be thinly spread over a wide area.

He says that the richest deposit of magnetite particles was found 50 miles north-northwest of ground zero, and he assumes that the wind was in that direction.

"It is our opinion," he says, "that the above relationship is not accidental, although it may be of inadequate statistical certainty." He follows this with a plea for a more thorough search and analysis.

If the particles did in fact come from the object that struck in 1908, it presumably was not an antirock, because an antirock would have left no residue.

If, on the other hand, tree rings of 1909 throughout the northern hemisphere show a high level of carbon 14, it would seem more likely that the earth did in fact encounter a chunk of antimatter.

May 30, 1965

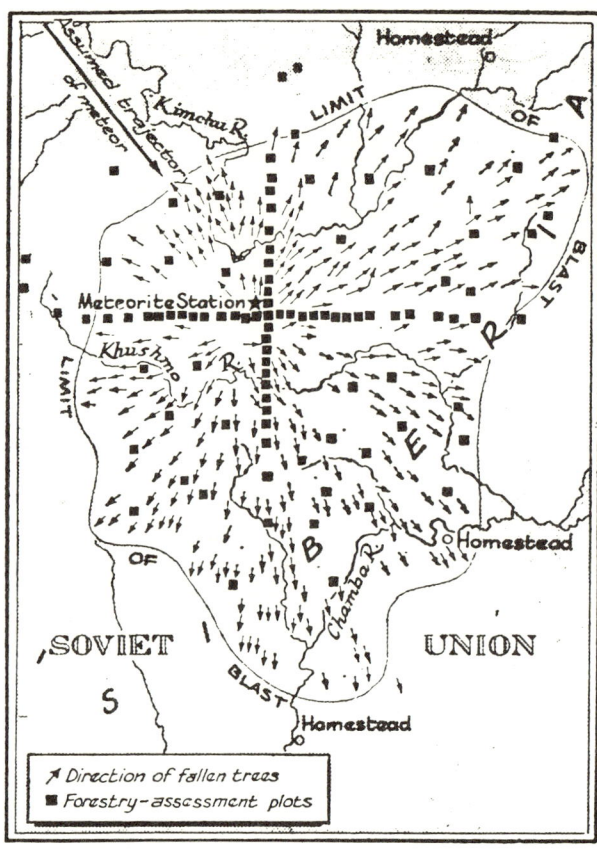

This map of the area where the Tunguska "meteorite" exploded, some 500 miles north of Lake Baikal in Siberia, is based on results of Soviet expedition in 1961. The arrows indicate directions in which trees were felled. The black squares designate forest plots examined for evidence that explosion enriched ground, as was reported earlier.

Physicists Produce Antimatter Particles In a Complex Form

By HAROLD M. SCHMECK Jr.

Are there somewhere antiworlds populated, perhaps, by antipeople?

The question seems like fantasy, but the answer could conceivably be yes in the light of research just reported from Columbia University and Brookhaven National Laboratory.

Using the nation's largest particle accelerator, or atom smasher, an international team of physicists has produced antimatter in a form more complex than ever before. The accomplishment shows that antimatter can exist in aggregations of particles, not only as isolated subatomic particles.

The discovery, therefore, suggests there is nothing in the fundamentals of nuclear physics to bar the possibility that, somewhere in the universe, antimatter may be aggregated into elements, compounds, suns and planets just as the fundamental

PHYSICS AND ASTROPHYSICS

particles of matter form those things in the universe with which man is familiar.

The physicists, who have reported their research in a scientific journal published today, have demonstrated the existence of antideuterons. Each of these is a complex made up of two different antiparticles.

The antideuteron is the first compound of antiparticles ever observed. Its existence had been predicted on theoretical grounds. Its actual production is therefore important because it confirms part of the body of theory on which all man's understanding of the nature of matter rests.

A deuteron is the nucleus of the atom of a form of hydrogen called deuterium or heavy hydrogen. This is a naturally occurring substance found in the oceans and wherever else hydrogen exists. Of every 10,000 atoms of hydrogen two are in the form of deuterium.

The deuteron — the nucleus of the deuterium atom — consists of one proton and one neutron held together by the nuclear force. These are the two fundamental subatomic particles from which all the nuclei of elements are built.

One Proton in Hydrogen

The nucleus of an atom of ordinary hydrogen, the lightest element, consists of one proton. The nuclei of all heavier elements consist of combinations of protons and neutrons.

The antideuteron consists of one antiproton and one antineutron held together, the experiments suggest, by the same nuclear force that binds a proton to a neutron in the nucleus of an atom of ordinary heavy hydrogen. It is the nuclear force that binds together the sub-units of any atomic nucleus.

According to physicists' understanding of the symmetries of physics, whenever a proton is manufactured from energy, an antiproton is produced with it. The particle and tis antiparticle are identical in most respects, but are opposite in electric charge and magnetic characteristics. When particle and antiparticle meet they annihilate each other.

Antimatter is simply a combination of antiparticles in the same sense that matter is a combination of subatomic particles.

Antiproton Predicted

Although the existence of antiparticles was predicted in 1927, the antiproton was not found until 1955. The antineutron was observed shortly afterward.

The senior author of the report describing the production of antideuterons is Dr. Leon M. Lederman, 42-year-old professor of physics and director of Columbia's Nevis Cyclotron Laboratory.

In a further statement made public by the university to explain the research, Dr. Lederman said the nature of the nuclear force binding together the fundamental particles in a nucleus had been studied intensively for the last 30 years and many of its characteristics were known. The fact that an antideuteron can exist, he said, means that the properties of the nuclear force are closely mirrored in the world of antiparticles.

This is a point considered important in understanding the physical laws that govern the universe.

Identical Relationships

The existence of the antideuteron was predicted by one of the most fundamental symmetry theorems of nuclear physics, the so-called TCP symmetry, Dr. Lederman continued. The details of this are impossible to explain precisely in laymen's terms but, in effect, it means that the laws of physics apply identically to a particle and its mirror image provided the flow of time, either forward or backward, and the relationship of the particle to its antiparticle are also taken into account.

It is in supporting the fundamental concept of symmetry that the discovery of the antideuteron is most important, the physicist said.

The research has been supported by the Atomic Energy Commission. The report is in the June 14 issue of Physical Review Letters, a journal devoted to important new discoveries or topics of special current interest in physics. It is published by the American Physical Society, a member of the American Institute of Physics.

In addition to Dr. Lederman, the authors are David Dorfan, 24, a graduate assistant from South Africa; John Eades, 26, a research scientist from Liverpool; Wonyong Lee, 34, assistant professor of physics, from Korea, and Samuel Ting, 29, instructor in physics, from Taiwan. They are engaged in research at Columbia and at Brookhaven National Laboratory, Upton, L. I.

Huge Accelerator Used

The antideuterons were discovered during lengthy experiments in which Brookhaven's huge accelerator, the AGS synchrotron, was used to hurl protons, at almost the speed of light and energy of 30 billion electron volts, at a target of beryllium.

The antideuterons were detected with the aid of a specially built research device called a high-transmission mass analyzer. The major elements of this are 16 magnets, each weighing more than 20 tons, arranged in a pattern stretching 300 feet from the target. Its purpose, in effect, is to sort out and identify the antideuterons in the debris of collisions between the high-energy protons and nuclei of atoms in the target.

Brief mention of the accomplishment was made earlier this year by Dr. Maurice Goldhaber, director of Brookhaven National Laboratory, in testimony to the Joint Atomic Energy Committee of Congress. The details had not been published before their appearance in Physical Review Letters. The current report noted that the experimental data agreed, within 3 per cent, with the expected electric charge and mass characteristics of the antideuteron.

With demonstration that the antideuteron exists, Dr. Lederman said, it is now shown that an antiworld is quite conceivable in terms of contemporary nuclear physics.

"It is not possible now," he declared, "to disprove the grand speculation that these antiworlds could be populated by thinking creatures — perhaps now excited by the discovery of deuterium."

Ever since the first discovery of antiparticles, physicists have been looking for evidence of antimatter in the cosmos at large. Scientists have recently speculated that the great meteorite that hit Siberia in 1908 with cataclysmically destructive results might have been a piece of antimatter.

The concept of antiworlds has even invaded literature. A poem published in The New Yorker magazine several years ago, for example, described an imaginary confrontation on a bit of matter "dark and stellar" between a noted American physicist and his antimatter counterpart, "Dr. Edward Anti-Teller."

The poem described Dr. Anti-Teller's world as conforming so rigorously to the symmetry of physics that the inhabitants kept "macassars," instead of antimacassars, on their chairs.

The piece ended with the physicist and antiphysicist reaching out to shake hands. They touched and, true to the concept that matter and antimatter annihilate on contact, "the rest was gamma rays."

June 14, 1965

Scientists Will Test Challenge To Einstein's Relativity Theory

By WALTER SULLIVAN

The report that observations of the sun had brought into question Albert Einstein's general theory of relativity has awakened intense interest in experiments that might verify the challenge. Several such tests are in preparation.

The challenge to the theory was presented Friday by Dr. Robert H. Dicke, professor of physics at Princeton University, based on his observations of the sun last summer.

He reported that he had found the sun to be flattened at its poles and fat around its equator, much like the earth. This shape would affect the distribution of the sun's gravity in nearby space, thus altering the orbit of Mercury, the planet nearest the sun.

The behavior pattern of Mercury's orbit, known as perihelion rotation, has long been known from observations. It was attributed for the most part to the gravitational influence of other planets, but at least in part it was believed caused by relativity. This residual effect matched the predictions of Dr. Einstein so precisely that it has been regarded as the clearest demonstration of the validity of his theory.

May Need Re-examination

If, as Dr. Dicke believes, the shape of the sun accounts for part of the effect, the general relativity theory must be re-examined.

In the last edition of The Times yesterday it was inadvertently stated that "Albert Einstein's theory of relativity" had been challenged. Earlier editions correctly identified the theory as the "general theory of relativity." The better-known "special theory of relativity," which establishes the relationship between energy and mass, was not involved.

There is little or no doubt that gravity performs, in qualitative terms, what Einstein said it did. It bends or subtracts energy from light, and it can shift the orbit of a planet that moves within a strong gravitational field. Mercury orbits in a region where the gravity of the sun is very strong.

The question concerns the

validity of Einstein's theory as detailed in his field equations. There are other theories, such as the so-called scalar-tensor theories developed by Dr. Dicke and others, that predict such effects, but they are basically different in philosophy and the results, in numerical terms, are different.

Perhaps the most precise laboratory tests of general relativity to date are those carried out by Dr. Robert V. Pound, professor of physics at Harvard University, and his colleagues.

They have exploited an effect discovered in Germany in the late nineteen-fifties by Dr. Rudolph L. Mössbauer. He found that gamma rays of extremely precise wavelengths can be generated by embedding within a crystal material that emits such rays.

Dr. Pound and his colleagues used this effect to test whether gravity alters wavelengths in the manner predicted by relativity. The theory said that gravity "pulling back" on the rays, should stretch their wavelengths.

The Mössbauer effect was sensitive enough to detect this variation between the bottom and top of a 70-foot tower at Harvard. In this month's issue of Physics Today, Dr. Dicke says these experiments have confirmed predictions based on relativity to within 1 per cent.

The trouble is, he says, that the outcome of the experiment would be the same, regardless of whether Dr. Einstein's theory or the scalar-tensor theory was correct.

Other projected tests may settle the matter. Two of them seek to assess the role of gravity in bending light.

The most dramatic apparent confirmation of Dr. Einstein's theory occurred during an eclipse in 1919 with the observation of a star as its light skirted the sun on its way toward the earth.

Seemed Out of Place

The bending effect made it seem out of place in the sky, much as the tip of an oar seems displaced when one looks down into the water. However, sufficiently precise eclipse measurements of this effect have not been possible to date.

One of the new experiments, devised by Dr. Irwin I. Shapiro of the Massachusetts Institute of Technology, will use a radar beam to follow Mercury as that planet passes behind the sun. The radar beam should be bent, just as is a beam of light.

The other was developed by Dr. Henry A. Hill at Wesleyan University in Middletown, Conn., a former associate of Dr. Dicke. He has designed a device that can "see" stars even when they are close to the sun. With this he, too, hopes to test the light-bending effect. He said yesterday in a telephone interview that he hoped to begin observations within a few weeks from a mountaintop in Arizona.

Finally, a group at Stanford University under Dr. William Fairbank is preparing a "top" that can be orbited in an earth satellite. The precession or "wobbling" of this top's spin should be affected by the spinning of the nearby earth and by other effects, making it possible to test which is correct: Dr. Einstein's theory or one of the rival formulations.

January 29, 1967

Cavern Detectors Explore High-Energy Particles

By WALTER SULLIVAN

Initial observations with giant detectors built inside Utah's Wasatch Mountains to peer into realms of physics hitherto beyond reach have produced results astounding to atomic theorists.

Physicists at the University of Utah were measuring the rain of high-energy particles called muons — a by-product of cosmic rays — that constantly strike the earth from the sky.

An analysis of the relative intensities of those particles that have sufficient speed and energy to penetrate the cavern showed the experimenters that a considerable portion of the particles was being generated by a previously unobserved and still unexplained process.

The new observations are grand in the scope of their experimental apparatus, far-reaching in the theoretical implications seen by a number of leading physicists and awesome in the energies involved.

"What we have seen so far in the experiment is only a hint of the vast new discoveries yet to be made in the field of elementary particles," said Dr. Haven E. Bergeson, a member of the University of Utah research team.

"We are certain now that at very high energies, new interactions are taking place that were undreamed of before."

Theorists at various universities are trying to account for the observations in the Utah cavern. Some believe the observed muons may be by-products of the hypothetical, but never observed, "intermediate boson," which would be an entirely new form of matter.

The four basic forces of nature are those responsible for gravity, electromagnetism, the binding together of the atomic nucleus and the so-called "weak interactions" between atomic particles.

Two of these are manifest in particles that have been observed. The electromagnetic force appears as the photon, or "bit" of light. The force holding the nucleus of the atom together becomes manifest in the pion.

On the other hand, a particle called the graviton has been postulated as the manifestation of gravity, but has never been observed.

Likewise, it has been proposed, but never verified, that the intermediate boson, or W-particle, plays a similar role in the weak interactions that, for example, produce certain forms of radioactivity.

Many hours have been spent on the world's largest atom smashers in a vain search for the intermediate boson. Its discovery would add a new pillar to the structure of modern physics.

The muon observations in Utah were made with a million-dollar apparatus crowded into a chamber carved out of the heart of a mountain southeast of Salt Lake City.

It is reached through three miles of tunnel in an abandoned mine and includes giant tanks of water, four miles of electrified pipe cut into 37-foot lengths and stacked around the tanks, and a maze of electronic gear.

The rain of muons onto the earth is an indirect consequence of the rain of cosmic rays onto the high atmosphere. Such rays are actually particles, some of which carry more speed and energy than any fragments of matter observed elsewhere in nature.

When cosmic rays strike particles in the air, a shower of secondary particles (kaons and pions) result.

If these secondaries survive long enough before hitting another particle of air, they decay to produce muons.

However, this decay into muons is less likely to take place when the incoming particles are plunging straight into the lower atmosphere. The reason is that, in penetrating lower and denser layers of air, the parent particles are likely to be destroyed before they have time to decay.

But particles coming into the air obliquely stay up where the air is thin — and collisions are unlikely — until they have had time to give birth to a muon. The result is that more muons have been observed coming from low in the sky than from the zenith. Hence, the lower part of the sky is "brighter" in muons than is the case overhead.

To test the effectiveness of the Utah apparatus in "seeing," whence the muons are coming, it was first operated on the university campus.

Its observations, processed by computer, clearly showed the "muon light" of the eastern sky and the outline of the Wasatch Mountains below it. However, the observations made inside the mountain near Park City have not shown the expected "brightening" in the lower sky.

This indicated that the observations were being swamped by muons produced by some process other than the one previously observed at lower energies.

The muons reaching the detectors in the cavern must go through some 2,000 feet of rock, as well as the snows of the Treasure Mountain ski area above.

Thus, only those of extremely high energy — with more than 1,000 billion electron volts — survive. This is far more than the 70 billion electron volts achieved in man-made atom smashers.

It appears that at such high energies, the sky-brightness rule does not apply.

On this basis, the experimenters reported in Physical Review Letters that a "new process" seems to be at work. The authors include Dr. Jack W. Keuffel, professor of physics at the University of Utah, and two assistant professors, Drs. Bergeson and Richard O. Stenerson.

In three weeks of observing,

PHYSICS AND ASTROPHYSICS

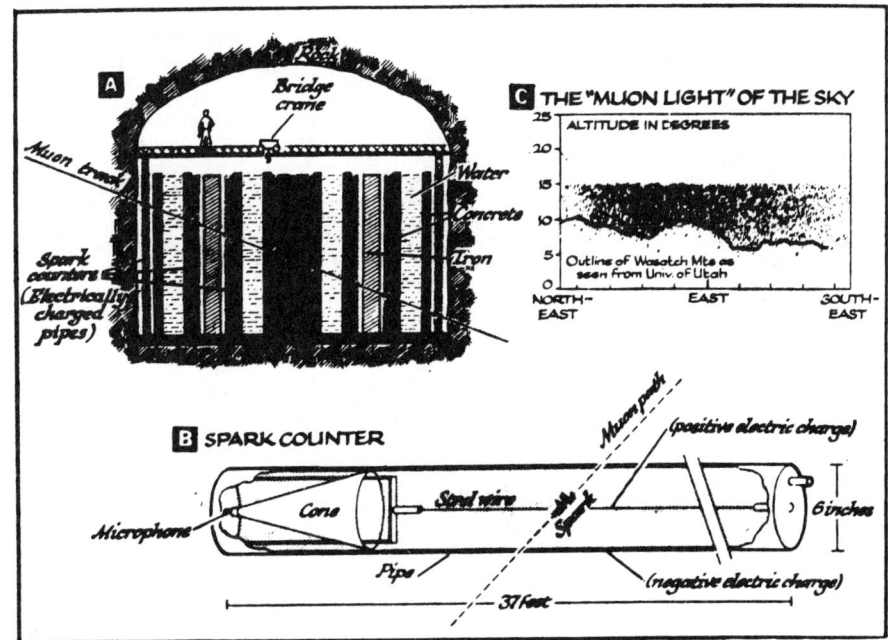

Sparks triggered by muons passing through electrically charged pipes of massive Utah cavern detector trace path like that in (A). Detector is activated only when particle has sufficient energy to penetrate two water tanks. Man on crane indicates scale. One of the 600 pipes is shown in (B). Sound of spark is recorded by microphone at left. (C) is computer-generated plot of source of muons. It also shows Wasatch Mountain profile.

they recorded 14,500 muons energetic enough to penetrate the cavern.

The Utah physicists emphasized in telephone interviews that they were by no means certain it was the long-sought intermediate boson that accounted for their observations.

The Utah apparatus was built with a grant from the National Science Foundation to detect particles that travel through the entire earth at the speed of light. They are known as neutrinos since they carry no electric charge.

Because of their strange properties, neutrinos almost never interact with atoms. They fly right through them as a rocket might fly through the solar system.

On the rare occasions when they hit a particle within the nucleus of an atom squarely enough to interact, they produce, as a by-product, very high energy muons.

Every now and then it is thought that one of the neutrinos that has plunged through the earth generates a muon that streaks upward into the Utah chamber.

The detection of such muons would be the first proof that neutrinos are, in fact, shooting through the earth like bullets through a smoke screen.

Only half the detectors of the Utah array are in operation as yet and it was in testing this equipment that the startling discovery was made.

January 26, 1968

Evidence of Gravity Waves Reported

By WALTER SULLIVAN

Detectors 600 miles apart are recording simultaneously what appear to be gravitational waves impinging upon—and passing through—the earth.

If the observations, announced yesterday by the University of Maryland, have been interpreted correctly, a new chapter in man's observation of the universe has been opened.

In essence, the development would enable man to view the universe from a fresh perspective. Among other things, it may help him to determine whether the universe is infinite or finite, to decide which of two rival theories of gravity is correct and to explain why 90 per cent of the universe seems to be "missing."

Present knowledge of phenomena beyond the earth has been derived almost entirely from a single kind of wave, that of electromagnetic radiation (embracing light waves, radio waves and X-rays.)

A preliminary study of the recordings by Dr. Joseph Weber at the University of Maryland, who designed and carried out the experiment, has persuaded him that a previously unobserved phenomenon, which releases vast quantities of energy, is taking place.

His detectors will have to be rearranged before the location can be narrowed down. However, he believes it to lie in the same general region of the Milky Way Galaxy, or star system, as the sun and earth. The indicated direction is away from the turbulent core of the galaxy.

A different proposal is that the observed gravitational waves are lapping back and forth through the universe as an aftermath of its explosive birth, more than 10 billion years ago. A further residue of this primordial "big bang" is a glow that seems to pervade the universe, primarily at radio wave lengths.

This glow figured in another announcement, made yesterday by the National

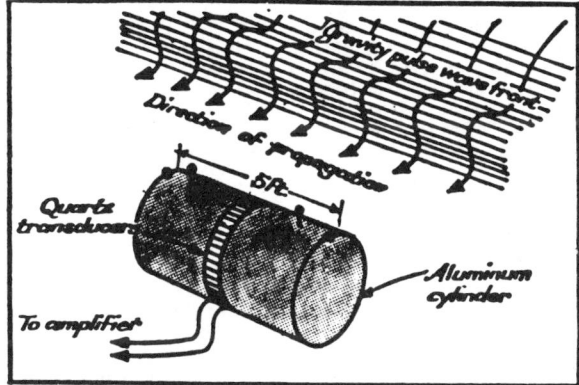

Detector of gravity waves consists of solid aluminum cylinder suspended from a bridge cushioned against earth tremors. Cylinder oscillates when a gravity wave front parallel to its long axis passes through it. Oscillation is then amplified, and the electric impulses are recorded.

Dr. Joseph Weber with model of the device he used.

Radio Astronomy Observatory.

It reported that the glow was illuminating dust clouds in distant space whose composition seems suitable for the synthesis of organic compounds.

The report suggested that "primitive life forms" might have evolved there.

The discovery of gravitational waves would help complete the parallelism between the behavior of the two long-range forces in nature: gravity and electromagnetism. Dr. Albert Einstein, in his General Theory of Relativity published in 1916, predicted the discovery of gravitational waves analogous to electromagnetic waves, such as those of radio.

Gravitational waves would be shed by an asymmetrical, spinning object just as radio waves can be generated by an oscillating or spinning electric charge. However, physicists had despaired of ever observing gravitational waves because they would be so weak.

A dozen years ago Dr. Weber began developing a detection system, even though calculations showed that it could only detect waves generated by a cataclysmic event, such as a supernova (the collapse and death of a star).

Such events occur in this part of the Milky Way only once every few centuries. Dr. Weber's detectors, however, are seeing what appear to be gravitational shock waves at least once a week. They are seen almost daily if one accepts indications of lesser magnitude.

If his observing technique can be refined, according to theorists, it should be possible to illuminate some of the most basic problems in science.

Gravity wave recordings may test the various theories advanced to explain pulsers and quasars—two recently discovered classes of celestial objects both of which generate powerful radio and light emissions.

It should be possible to discriminate between the two contending theories of general relativity. That derived by Dr. Einstein depended on a form of mathematics known as tensor calculus, but a rival version also makes use of scalar calculus.

So far, gravitational effects that might discriminate between them have been too subtle to provide a clear answer.

The gravity pulses may provide a clue to the whereabouts of the "missing matter" in the universe. As noted by Dr. Gart Westerhout, a noted Dutch astronomer, in commenting on Dr. Weber's observations, the motions of stars within our galaxy show that they are under the gravitational influence of 40 per cent more matter than can be seen.

Furthermore, if the picture of the entire universe derived from Dr. Einstein's calculations is correct, more than 90 per cent of its matter is missing. Some believe that part of this may be in the form of "black holes."

These are hypothetical spots in the sky where a very large star has died. With no more heat to support its structure, the star has contracted, crushed together by its own enormous weight until even its atoms could not withstand the pressure.

Such an object, theorists say, would be so dense that its gravity would not permit the escape of any light to testify to its presence. Nor could the light of a star beyond it pass through such an object. The latter would be evident only as a "black hole" in the sky.

One proposed explanation for the gravity pulses being observed is that massive objects are falling into such a black hole somewhere near by.

The missing matter question is of philosophical, as well as of scientific, importance since it bears on the nature and destiny of the universe.

If the total amount of material in the universe can be determined, this should indicate whether it is infinite or finite and whether it is destined to expand forever or fall back together.

The current expansion can be likened to a stone thrown high and slowing as gravity seeks to pull it back to earth. Either the stone has sufficient velocity to escape the earth, despite its slowing, or it is destined to fall back.

The expansion of the universe appears to be slowing, but it is not clear whether it will ultimately stop and fall back together again. Some cosmologists believe the universe has no beginning and no end, but oscillates between periods of expansion and collapse.

Colleagues Skeptical

A year ago Dr. Weber reported the possible detection of gravitational waves. The expected strength of such waves is so weak that skeptical colleagues suspected he might be recording earthquakes or other local phenomena.

The detectors in his work were all in the vicinity of the university campus at College Park, Md. Elaborate arrays of seismic detectors were installed and have shown that earthquakes do not have any effect on the gravity detectors.

Another fear was that some sort of electric impulses, as from a lightning flash, might enter the recording circuits, imprinting simultaneous pulses on all the detectors.

This possibility has been eliminated because one detector has a built-in lag of 11 seconds in recording a gravity event, whereas it would respond immediately to an electric event. The 11-second lag is observed in a number of the events.

However, to eliminate any possibility of a local effect, Dr. Weber six months ago began operating another detector at Argonne National Laboratory near Chicago. Its readings are carried into the Maryland laboratory by telephone line.

The simultaneous recordings 600 miles apart have convinced Dr. Weber that gravity waves are, in fact, being seen.

In the June 16 issue of Physical Review Letters he reports that during an 81-day period there were more than 17 "significant" two-detector coincidences, five three-detector coincidences and three events that showed up on all four detectors (three in Maryland and one in Illinois).

Controversy Expected

He said the probability of two triple-detector coincidences seen on March 20 being accidental was such that this would occur only once in 70 million years of observation. It was therefore "certain," he said, that all of the coincidences could not be accidental.

However, the university, in making its announcement, acknowledged that his findings would be challenged by some. The results promise, it said, "to open wide a vigorous controversy in modern physics."

Each detector consists of a massive cylinder of solid aluminum, suspended inside a vacuum tank from a bridge that is cushioned against local shocks. The elastic properties and dimensions of the cylinder are such that, according to calculations based on the Einstein equations, it should respond to gravitational waves where frequency is about 1,660 cycles a second.

If the wave front is roughly parallel to the axis of the cylinder, the latter should oscillate.

The frequecy was chosen because the collapse of a spinning star should generate a burst of gravitational waves sweeping past 1,660 cycles. Oscillations of the cylinder are sensed by piezoelectric crystals around its waist. The resulting electric impulses are amplified and recorded.

As set forth in relativity theory, the medium through which the gravitational wave moves at the speed of light is the sea of space and time that fills the universe. The geometry of this space-time is curved in a subtle way, and is distorted briefly as the wave passes by. It is this distortion that sets the cylinder to oscillating.

The detectors are all oriented east and west. At any one time, they scan a pole-to-pole segment of the heavens over the meridian of the eastern United States on this side of the earth and India on the far side.

Assumption Noted

It is assumed that gravitational waves pass easily through the earth.

As the earth turns on its axis, the observations sweep around the entire heavens. The finding that the gravity pulses occcur primarily at certain times in this sweep implies one or more primary sources. By reorienting some of the detectors, Dr. Weber explained a few days ago, it should be possible to narrow down these sources.

A comparison of arrival times of each pulse in Illinois and Maryland should give additional clues to their direction.

Dr. Weber has been making continuous, high-precision observations of the earth's gravity in Maryland, looking for changes that might indicate oscillations of the entire earth in response to gravitational waves.

The results so far are negative, but he hopes gravity instruments of greater sensitivity can be installed on the earth and moon. These could be used in tandem to provide far more precise information on directions of the waves.

June 15, 1969

PHYSICS AND ASTROPHYSICS

Subatomic Tests Suggest A New Layer of Matter

By WALTER SULLIVAN

A number of physicists believe that, through a variety of atomic experiments, they have begun opening the door to the innermost sanctum of matter.

In the first, and probably most important, of these experiments, conducted at the Stanford Linear Accelerator in Menlo Park, Calif., evidence has been found of internal components within the proton and neutron—once considered indivisible building blocks of the universe.

Dr. Wolfgang K. H. Panovsky, director of the center, and his staff recently declared jointly that the results "appear to have uncovered another layer of matter."

Specifically these findings suggest the presence, in proton and neutrons, of points of electric charge that, in several respects, resemble the elusive and long-sought quarks.

In 1964 Dr. Murray Gell-Mann of the California Institute of Technology pointed out that characteristics of the multitude of heavier subatomic particles, discovered in atom-smashing experiments, could be explained in terms of smaller building blocks that he called quarks.

An intensive discussion is under way, here and abroad, as to the meaning of the new observations. But there is widespread belief that a new level, within the atom, has been penetrated. And some scientists hope the new findings will lead to an understanding of the basic forces within atomic nuclei and, perhaps, within the nuclear particles themselves.

Such forces would dwarf any with which we have contact in daily life, such as gravity and magnetism.

In a recent assessment for the White House science staff, Dr. Victor F. Weisskopf, professor of physics at the Massachusetts Institute of Technology and former head of the American Physical Society, wrote:

"Not much is understood yet, but it seems most probable that we are touching here the most fundamental questions of nature and the universe."

"We are faced," he contin-

United Press International
Dr. Wolfgang K. H. Panofsky

ued, "with a realm of entirely new phenomena, with a way of behavior of matter which was completely unknown and unexpected before, phenomena which may some day be of use in practical applications."

Earlier Breakthrough

Dr. Weisskopf likens the situation to that, a half-century ago, when it became evident that the atom as a whole has internal structure—a compact nucleus surrounded by a cloud of electrons. The discovery led to an understanding of the chemical properties of the elements and revolutionized both chemistry and physics.

It was subsequently shown that the nucleus itself has structure, being formed of clustered neutrons and protons. Now it appears that even those "elementary" particles have some sort of internal components.

While follow-up experiments are under way in laboratories throughout the world, the initial observation of such internal structure was made at Stanford. The two-mile accelerator there produces by far the most powerful electron beam in the world.

When the electrons hit their target, they are traveling at more than 99 per cent of the speed of light and carry an energy of 20 billion electron volts.

Whereas the interior of the proton previously seemed amorphous, like a snowball, Dr. Sidney Drell, who works with the Stanford accelerator, now likens it to jam—with seeds.

Lines of Attack

A variety of tools for physics research are being used to explore these findings. Two of these lines of attack are modern versions of classic experiments that, early in this century, helped lay the foundations of modern physics.

One, used by Ernest Rutherford, showed that the atom is not a formless glob, but has internal structure. Using those electrons from radioactive material that streamed through a pinhole in a shield, he bombarded atoms and observed the extent to which the electrons were deflected by electric fields within each atom.

This bending, or "scattering," of the electron paths was, in some cases, far too great to have been produced by a diffuse distribution of electrical energy throughout the atom. Rather, energy—or electric charge—seemed to be concentrated at a point in the atom's center. Subsequently, with the development of ways for artificially accelerating electrons, it became possible to produce beams of higher and higher energy. This, in turn, has made it possible to "see" smaller and smaller objects within the atom.

Wave Length Patterns

This is because the electrons in the beams have a wave-like motion whose wave length decreases as the energy goes up. The shorter the wave length, the smaller the objects that can be detected by the beam.

Thus, it became possible to show that the nucleus is formed of particles (protons and neutrons). Now the Stanford accelerator beam is so powerful that it can detect objects one-fortieth the radius of the proton.

It was this that made possible the discovery of what seem concentrations of electric charge inside such particles.

The other classic experiment, now being repeated in a far more sophisticated manner, was that performed before World War I by Robert A. Millikan to determine the nature of electric charges. He sought to find out whether the strength of such charges varies in a smooth, uniform manner.

Millikan measured the strength of electric charges carried by falling drops of oil. He did so by observing the response of the drop as it fell through an electric field of known strength.

Millikan's Experiment

He found that, instead of a smooth variation of intensity, the charges were all simple, whole-number multiples (such as 2, 3, 4, etc.) of a basic charge. He took the basic charge to be that of a single electron.

Only once did he see a fractional charge—one with a value less than that of an electron—and he assumed it was a spurious observation.

In the current experiments, scientists at Stanford University, in Palo Alto, are using a modern version of the experiment to see if fractional charges may, in fact exist. If they do, this would be strong support for the quark theory, which holds that quarks would have charges that would be such fractions of the electron charge.

In the tests at Stanford an attempt is made to reduce to zero the electric charge on a niobium sphere. Niobium is a metal that becomes super conductive at very low temperatures. If an extra quark lurks within the niobium, it should make it impossible to bring the total charge down to zero by electrical means. Electrical charges, as opposed to quark charges, occur only in simple whole numbers.

Last September Dr. Arthur F. Hebard of Stanford reported preliminary results of these experiments to an international conference in Japan. In a paper coauthored by Dr. William M. Fairbank, professor of physics at Stanford, Dr. Hebard said the findings seem to indicate a residual charge equal approximately to one-third of the total electron charge.

Test for Quarks

However, he added, "before this data can be taken as evidence for quarks, further measurements on the first sphere [niobium sphere] will be made under different experimental conditions and the validity of the results checked by measurements on other spheres."

Such tests are now under way.

The finding of a minus-one-third charge would be strongly suggestive of quarks because, in Dr. Gell-Mann's theory, the proton is formed of three quarks, one of which has a charge of minus-one-third. Its two other constituents would be quarks, each with a plus-two-thirds charge. The net charge would therefore be one (as is observed for the proton).

The neutron would also be formed by three quarks: two with minus-one-third charge and one with a plus-two-thirds charge, to make a net charge of zero. The particles of intermediate weight, known as mesons, would be formed of two quarks.

Dr. Gell-Mann, who conceived of quarks (an Israeli, Dr. Yuval Ne'man, made a similar proposal at about the same time), has long argued that they may not be particles that could have an independent existence. In a recent telephone interview, he suggested two other possibilities.

Subatomic 'Currents'

They could, he said, simply manifest forces, or "currents," moving rapidly about within the nuclear particles. Or they may be stationary constituents that exist only within the particle and hence can never be observed outside it.

He feels the recent observations at Stanford have not eliminated any of these possibilities.

His colleague at Caltech, Dr. Richard P. Feynman, (they are both Nobel Prize winners) has also proposed that the nuclear particles contain constituent units that he calls partons. They may, he believes, only be stable when bound together inside the proton or neutron.

An analogue, in this respect, is the neutron itself, which survives indefinitely as long as it is bound inside the atomic nucleus. Yet on its own the neutron decays, within about 12 minutes, into a proton, electron and neutrino.

When liberated from within a proton, neutron or meson, the quarks may vanish so fast no one has detected them.

One possibility, discussed by Dr. Weisskopf in a recent interview, is that the quarks are very massive, but are bound together inside nuclear particles so firmly that their great mass is concealed.

'Negative Energy'

The effect would be comparable to that which makes the nucleus appear to weigh less than the sum of all its parts. The protons and neutrons forming such a nucleus seem to weigh less when bound inside of it than when at liberty because the binding force represents "negative energy." This in-pulling energy reduces the weight of each particle.

Some physicists are wondering whether the reason quarks have not been produced at the energy levels now available in accelerators is that they are extremely massive since the available energy contributes to the mass of many particles produced.

And if the quarks are so massive, the energy needed to conceal this mass within the nuclear particles is awesome. The binding energy, according to some theorists, could be represented by a "sea" of quarks and antiquarks existing as a force field, rather than as real particles.

If there are quarks, it is assumed there must be their mirror-image counterparts — antiquarks. For it has been found that every other atomic particle has its antimatter counterpart, opposite in electric charge or other features, but identical in mass.

The possibility that quarks are very massive and could therefore be produced only in high-energy interactions has focused interest on the new machines that will be far more powerful than any hitherto available.

The Stanford search for tiny objects within protons was initiated following theoretical calculations by Dr. J. D. Bjorken of that laboratory suggesting they might be observable.

In 1968, hardly a year after the machine went into operation, evidence for such objects began to appear and it can now be tentatively deduced that these point charges are fractional—that is, less than the total charge of the proton—and that they have the magnetic properties predicted for quarks (expressed by the term "spin one-half").

Dr. Richard Taylor, leader of the experimental group at the accelerator, says his desk is piled "18 inches high" with papers sent to him by theorists seeking to explain the Stanford results without recourse to subunits within the nuclear particles.

However experimenters in a number of laboratories believe they are seeing evidence for such internal structure, and theorists are devising names to describe their own brands.

Theory Supported

Among the laboratories whose findings have been interpreted, at least by some, as consistent with the existence of quarks (or partons) is one at Frascati, near Rome. A machine there produces head-on collisions between beams of electrons and their antimatter counterparts, positrons.

Another line of attack has been taken at Brookhaven National Laboratory on Long Island where Dr. Leon Lederman of Columbia University, and his colleagues, have been pounding uranium with a beam of very high energy protons.

At the internatonal nuclear research center, known as CERN, near Geneva, proton beams are colliding head-on, producing interactions that are also seen, by some theorists, as indicative of internal structure within the proton.

Dr. Lederman shares the excitement of many physicists over the new discoveries. As bigger and bigger atom smashers were built, he said last week, some researchers had begun to wonder whether they were creating effects that had little meaning in nature.

They asked: "Are we just making puzzles for ourselves?" It now appears that the new experiments are demonstrating in the laboratory, processes characteristic of such puzzling celestial discoveries as pulsars and quasars.

Pulsars seem, in effect, to be gigantic atomic nuclei. Quasars appear to be explosively releasing energy at a rate that may now become explicable, for it appears that unsuspected energy sources may lie within nuclear particles.

While Dr. Gell-Mann believes it is too early to draw firm conclusions from all these observations, he says "an immensely exciting grand synthesis" is within reach. It may come within a couple of years —or not for 15.

But when all these observations fall into place, he adds, the outcome will be as revolutionary as the discovery of quantum behavior on the atomic level that is the basis of present-day physics.

April 25, 1971

Tests Find Changes In Atomic Behavior

By WALTER SULLIVAN

Experiments with the world's most powerful atom smasher, at Batavia, Ill., have shown that five different atomic fragments get "bigger" when accelerated to high energy and that they begin to lose some of their differing characteristics at these hugh energies.

For example, the two building blocks of the atomic nucleus—the proton and neutron —differ in that the proton carries a positive electric charge whereas the neutron does not and is slightly heavier. Yet in the highest energy experiments it was found that these two types of particle behave in the same way.

The findings are considered important because they relate to the strongest force in nature —the "nuclear" force that binds together particles of the nucleus. No comprehensive theory now exists to explain the actions of this force.

The experimenters believe their findings may cast light on the behavior of the nuclear force and help pave the way to a theory explaining its actions.

The findings are to be described to the International Conference on High Energy Physics currently under way at Imperial College in London. It is the 17th of what were originally known as the Rochester conferences because the first was held at Rochester, N. Y.

The papers were summarized here last week at a briefing by leaders of the 16-member team that did the experiments at the Fermi National Accelerator Laboratory west of Chicago.

Determination of Size

It is the nuclear force that determines the apparent size or "total cross-section," of interacting particles.

To determine a cross-section a beam composed of one type of particle is fired through material containing a similar or different type of particle. The probability that the impinging particles will interact defines the cross-section.

In essence it indicates how closely, at a given beam energy, a beam particle must come to a particle in the target material to interact.

As noted by the experimenters, one would expect that, as the energy of the beam increased, the particles would have to come closer to one another to interact. That is, their cross-sections would decrease.

This was, in fact, the case up to beam energies of about 50 billion electron volts. But then a reversal of the trend was observed.

The first hint of this came in 1971 from experiments at Serpukhov in the Soviet Union. Last year, at the European nuclear research center near Geneva, the effect was clearly seen in head-on collisions between colliding proton beams.

The new experiments probed more deeply into the nature of this phenomenon in that six different types of beam were fired into two kinds of target material: liquid hydrogen, whose nuclei consist of protons, and deuterium, whose nuclei contain both protons and neutrons.

The beams were formed of protons, their antimatter twins

PHYSICS AND ASTROPHYSICS

Leaders of a team of 16 experimenters explaining results to reporters. From the left are Dr. Thaddeus F. Kycia of Brookhaven National Laboratory; Dr. Rodney L. Cool of Rockefeller University and Dr. Winslow F. Baker of Fermi Laboratory, scene of experiments.

(antiprotons), pi mesons of positive and negative charge and positive and negative K mesons.

A Secondary Beam

Mesons, which are very short-lived, interact with protons and neutrons under control of the nuclear force. They were generated at the Fermi accelerator by firing the proton beam of that machine at a beryllium target. This produced a secondary beam of mesons, protons and antiprotons.

Two types of target were placed in this secondary beam. One contained liquid hydrogen whose atoms contain a simple proton nucleus. The other target contained deuterium—a form of heavy hydrogen with both a proton and neutron in its nucleus, making it possible to observe interaction with neutrons as well as protons.

Of primary interest were the relative trends in cross-section with the various types of beam. As Dr. Winslow F. Baker of the Fermi Laboratory, one of the experimenters, put it, "a new simplicity" seems to be emerging at the energy goes up. Particles belongng to the same famly, but differing from one another at low energy, behave more and more alike.

According to Dr. Rodney L. Cool of Rockefeller University, another member of the experimental team, not only did the proton and neutron "look substantially the same" under high energy bombardment by these various beams, but this was also true of particles and their antimatter twins. Such a trend was predicted in 1958 by a Russian theorist, Isaak Y. Pomeranchuk. It was seen, as well, in particles that differ from one another in a subtle property known as "strangeness."

Of the six types of particle beam employed, only the antiproton beam failed to show a rise in cross-section. Instead the cross-section droped steadily from energies of 50 to 100 billion electron volts. It then leveled of, however, implying that above 200 billion volts it would begin to rise.

Validity Not Proved

The fact that the meson beam cross-sections were two-thirds of the proton-on-proton bombardments was "compatible," Dr. Cool said, with the theory that particles like protons are formed of fragments, known as quarks. But, he aded, this did not prove validity of the proposition.

It was noted at the briefing that two of the four basic fores—gravity and electromagnetism—have already been described in comprehensive theories.

Gravitational theory is routinely employed in space flight calculations. Electromagnetic theory lies at the basis of all electrical applications. The fourth force, in addition to the nuclear force, is the "weak" force manifested in radioactivity.

While the nuclear force figures in the release of nuclear energy, in the words of Dr. Cool, no "general theory of nuclear interactions" exists.

The usefulness of such a theory, he pointed out, cannot be predicted before it has been formulated, any more than, a century ago, it would have been possible to predict the applicability of James Clerk Maxwell's electromagnetic equations.

The experimental team was led by Dr. Winslow F. Baker of the Fermi laboratory, Dr. Cool and Dr. Thaddeus F. Kycia of Brookhaven National Laboratory near Upton, Long Island.

July 2, 1974

New and Surprising Type Of Atomic Particle Found

By WALTER SULLIVAN

Experiments conducted independently on the East and West Coasts have disclosed a new type of atomic particle.

Its properties are so unexpected that there are differing views as to how it might fit into current theories on the elementary nature of matter.

The experiments were done at the Stanford Linear Accelerator in Palo Alto, Calif., by a team under Dr. Burton Richter and at the Brookhaven National Laboratory in Upton, L.I., by a group under Dr. Samuel C. C. Ting of the Massachusetts Institute of Technology.

In a statement yesterday, the two men said:

"The suddenness of the discovery coupled with the totally unexpected properties of the particle are what make it so exciting. It is not like the particles we know and must have some new kinds of structure.

"The theorists are working frantically to fit it into the framework of our present knowledge of the elementary particle. We experimenters hope to keep them busy for some time to come."

Some scientists believe that the new particle will prove to be the long-sought manifestation of the so-called weak force—one of the four basic forces in nature. The others are gravity, electromagnetism and the strong force that binds together the atomic nucleus.

It is also suspected that the particle may be related to a recently developed theory equating two of those forces — electromagnetism and the weak force— as manifestations of the same phenomenon. However, the properties of the newly discovered particle are not those predicted for either of those roles.

That a major discovery had been made became evident last Monday during a conversation at the Stanford Linear Accelerator between Dr. Wolfgang K. H. Panofsky, director of that two-mile-long device, and Dr. Ting.

"I'd like to talk a little physics," said Dr. Ting, as recalled by Dr. Panofsky. He then told of recent experimental results obtained by himself and his colleagues.

"We just can't believe them," he said. However, Dr. Panofsky said that essentially the same observations had been made in his own laboratory.

News of the discovery has created a sensation in the world of physics, and preparations are being made at CERN—the international European Nuclear Research Center outside Geneva—to try to duplicate the discovery.

One possibility under discussion is that the new particle falls into a class, predicted by some theorists, that would display a combination of properties termed "charm." These properties would be distinct from those, known as strangeness, characterizing another family of particles. In the latter case, the name derived from what seemed the strange manner in which they form and decay.

However, it was found that they fell into a pattern that made possible predictions as to how each would behave. The same would be true of those displaying charm.

The new particle is one of the heaviest known. It was detected at the Stanford Linear Accelerator in experiments in which electrons and their positively charged counterparts, positrons, were collided head on.

When the collision energy reached 3.105 billion electron volts there was, according to yesterday's joint announcement, a "sudden enormous increase" in the number of heavy particles produced. This indicated the production of a particle whoes mass was equivalent to that energy.

Such a particle would be three and a half times as heavy as the proton.

Existence Is Implied

At Brookhaven a proton beam with an energy of 30 billion electron volts, impinging on stationary protons, produced a large number of electron-positron pairs at 3.1 billion electron volts. This, too, implied the existence of such a particle. The fact that similar results emerged from such different experiments is seen as strong confirmation for the finding.

One of the chief surprises is the long life of the particle. Even though it decays on the average in 100 billionths of a billionth of a second, so heavy a particle would be expected to decay 1,000 times that fast.

The findings are to be reported in the Dec. 2 issue of Physical Review Letters.

Participants at Brookhaven included Drs. U. J. Becher and Min Chen of M.I.T. and Y. Y. Lee of Brookhaven. At Palo Alto, they included Drs. Roy F. Schwitten and Rudolf R. Larsen of that laboratory and Drs William Chinowsky, Gerson Goldhaber and George H. Trilling of the Lawrence Berkeley Laboratory of the University of California.

The idea that the weak force — that responsible for radioactive decay — must be transmitted by a particle of some sort derives from the knowledge that the other forces are expressed in this way. For the electromagnetic force, it is the photon — or light wave. For the nuclear force, it is the pi-meson. For gravity, is is the hypothetical gravition.

Because of vain attempts to find such a particle transmitting the weak force, it has been suspected that it is very heavy — more than 10 times the mass of the proton and three times the mass of the new particle.

The latter bears no electric charge and is believed to have the properties described by physicists as spin one and negative parity.

November 17, 1974

Physicists Still Not Sure if New Particles Have 'Charm'

BY WALTER SULLIVAN

After almost six months of hectic research in which the world's largest experimental devices have been used to probe some of the tiniest constituents of matter, the nature of the new class of particles discovered last November has become more perplexing than ever.

The possibility remains that the particles indicate the existence of a subtle property of matter known as charm. A few particles have been fleetingly observed that may be "charmed." But other experiments designed to test the charm hypotheses have failed to find particles predicted by the theory.

As stated in the current issue of the CERN Courier, issued by the European Organization for Nuclear Research near Geneva, the discoveries of recent months "have opened up completely new interpretations about the fundamental components of nature and their behavior."

Largest in Europe

The organization's research center outside Geneva is the site of Western Europe's largest atom smasher. Virtually all research on the new particles depends on the very large smashers, or particle accelerators, including the four-mile ring of the Fermi National Accelerator Laboratory at Batavia, Ill., the world's largest.

The present state of efforts to understand the new particles was described last week by physicists from European and American laboratories at the spring meeting of the American Physical Society.

Charm is a hypothetical property of atomic particles that would be conserved when those particles interact with other heavy particles, much as an electric charge is conserved. In charge conservation, for example, if the net electric charge of particles entering into a reaction is minus one, the net charge of those that emerge from the reaction must also be minus one.

The existence of the property called charm was suggested in 1964 by Drs. James D. Bjorken, now at Stanford University, and Sheldon L. Glashow, now at Harvard University. They saw it as an explanation for certain patterns of particle interactions. When the new particles were discovered in November they were hailed as the first evidence for "charmonium."

Quarks and Antiquarks

The particles of charmonium would be formed from the mating of two subparticles known as quarks. Because one was a so-called antiquark, the properties of charm would be neutralized within charmonium and not evident to outside observation.

Since physicists like to assign odd names to theoretical concepts two researchers at the Fermi Laboratory have facetiously proposed "panda," rather than "charm" for this property. They are M. B. Einham and C. Quigg.

"We chose this name because of the panda's well-known shyness and tendency to stay among his own kind," they have written. The new particle would then be called "pandamonium" which, as noted in the CERN Courier, "is a fair reflection of its impact on the world of high energy physics."

The first of the new particles was discovered almost simultaneously at Brookhaven National Laboratory near Upton on Long Island, and at the Stanford Linear Accelerator in Palo Alto, Calif. Brookhaven named it the J particle. The California group calls it the psi particle. The latter team then found a second, slightly heavier particle that decays into the first one.

Both are uncharged electrically but the charm hypothesis predicted a whole family of particles, including some with electric charge (both positive and negative) and the outward manifestations of charm.

Dr. Samuel C. C. Ting of the Massachusetts Institute of Technology, led the group at

Brookhaven that made the original discovery there. An intense beam of high energy protons (hydrogen nuclei) was fired into a tank of liquid hydrogen to produce proton-proton collisions.

Last week he reported that a continuation of such experiments had not revealed the predicted sister particles, although the data have not been entirely processed.

A completely different type of Brookhaven experiment has, however, produced a single event in which a seemingly charmed particle was formed.

A beam of neutrinos was fired into a tank filled with liquid hydrogen to see what happened when the neutrinos interact with protons.

Neutrinos are elusive particles that have no observable mass and no electric charge. They therefore interact with protons only when they come extremely close to them. In this series only 100 neutrino events were detected in 62,000 photographs and the analysis of one suggested that a particle behaving as though "charmed" had briefly been formed.

Existence Confirmed

Its mass would be roughly 2.426 billion electron volts (abbreviated Gev.) The mass or weight of such particles is normally expressed in terms of the energy that would be released if they were entirely converted to energy. The masses of the particles discovered in November are 3.1 and 3.7 Gev. Their existence was quickly confirmed here and abroad.

Researchers at the organization's base in Switzerland, firing neutrinos into Gargamelle, a bubble chamber filled with heavy liquid, have also observed an event which, as stated by the CERN Courier, is "a strong candidate for the production of a charmed partcle."

However other experiments there, at the Fermi laboratory and at Stanford have not observed phenomena predicted by the charm theory.

For example, it had been expected that the heavier psi particle would decay through a succession of energy levels, giving off characteristic gamma rays at each stage.

Unsuccessful Effort

At last week's meeting, Dr. Robert Hofstadter, a Nobel laureate in physics, reported on unsuccessful efforts at Stanford to observe these gamma ray emissions. Just as the characteristic colors, or spectral lines, emitted by atoms as they change their energy states, reflect their internal structure, one would expect to find analogous emissions from the psi particles, he said.

Thus, Dr. Hofstadter said, the theories advanced so far "do not seem consistent with what we are finding." He expressed confidence, however, that spectral lines of some sort would eventually be detected.

Dr. I. Peruzzi of the Italian nuclear center at Frascati reported that in experiments, like those at Stanford, where beams of electrons and positrons collide head on, no evidence was found for any particles in the range from 1.1 to 3.0 Gev.

Dr. Roy Weinstein of Northeastern University in Boston told of efforts at the Fermi Laboratory to detect charmed particles with a mass of about 2 Gev, produced in pairs and thus producing a "bump" in the data at about 4 Gev. A small bump of this sort has, in fact, been observed, he reported

Evidence for a particle witry a less sharply defined mass has been seen as 4.2 Gev in the Stanford electron-positron collisions, according to Dr. Charles C. Morehouse.

Dr. David Cline of the University of Wisconsin reported discovery of what is tentatively being called the Y particle that seems completely different from the psi of J particles. It was observed at the Fermi Laboratory in a small percentage of high energy neutrino collisions decaying in a manner never before observed (into two muons and a neutrino.) Its mass would lie between 2 and 4 Gev and its interactions would be controlled by the "weak" force responsible for radioactivity.

The psi of J particles are believed to answer to the "strong" force that binds the atomic nucleus. They seem to be "cousins" of light waves in that they behave like vector particles. That is, they are the transmitters of a force—in this case the "strong" force. Light waves transmit the electromagnetic force.

Because of this relationship it was suspected that the n^ **particles would be produced by aiming extremely energetic light waves (gamma rays) at a solid target. This has been done at Stanford, the Fermi Laboratory, Cornell University and in West Germany.** The results, support the view that the new particles are related to the nuclear force.

Puzzling Discoveries

A varety of puzzling discoveries were reported. Dr. Ting found it "very strange" that J particle production at 20 Gev is only one-tenth what it is at 30 Gev. The Northeastern group reported that production by a beam of particles known as pions was five times higher than by a proton beam. "No theoretical explanations" exist, they said.

Dr. Michael J. Tannenbaum o fRockefeller niversity in New York told of a strange phenomenon first observed in the Soviet Union and subsequently at CERN, where he was a visiting experimenter, and elsewhere. The proton beams that collide head-on at CERN produce, in effect, the highest energy collisions in any laboratory. The psi of J particles are seen as well as electron-positron pairs that shoot off at right angles to the colliding beams.

However, once in roughly every 10,000 events only one electron flies off, violating one of the basic laws of physics: that heavy particle collisions cannot produce single light particles, such as the electron.

"Something re'ly interesting" seems to be occurring, he said. "We are now confronted with all these mysteries to solve," he told a standing-room - only - audience," and I think it is a wonderful time."

May 4, 1975

PURSUIT OF A UNIFIED FIELD THEORY

EINSTEIN EXPLAINS HIS NEW DISCOVERIES

In a Simplified Discussion of "Field Theories, Old and New," the Eminent Scientist Shows the Meaning Of His Latest Contribution for Gravitation, Electro-Magnetism and Our Ideas of Time and Space

Wireless to THE NEW YORK TIMES.
Copyright, 1929, by The New York Times Company.
All Rights Throughout the World Reserved.

By ALBERT EINSTEIN.
BERLIN.

WHILE physics wandered exclusively in the paths prepared by Newton, the following conception of physical reality prevailed: Matter is real, and matter undergoes only those changes which we conceive as movements in space. Motion, space and also time are real forms. Every attempt to deny the physical reality of space collapses in face of the law of inertia. For if acceleration is to be taken as real, then that space must also be real within which bodies are conceived as accelerated.

Newton saw this with perfect clarity, and consequently he called space "absolute." In his theoretical system there was a third constituent of independent reality—the motive forces acting between material particles, such forces being considered to depend only on the position of the particles. These forces between particles were regarded as unconditionally associated with the particles themselves and as distributed spatially according to an unchanging law.

The physicists of the nineteenth century considered that there existed two kinds of such matter, namely, ponderable matter and electricity. The particles of ponderable matter were supposed to act on each other by gravitational forces under Newton's law, the particles of electrical matter by Coulomb forces also inversely proportional to the square of the distance. No definite views prevailed regarding the nature of the forces acting between ponderable and electrical particles.

The Old Theory of Space.

Mere empty space was not admitted as a carrier for physical changes and processes. It was only, one might say, the stage on which the drama of material happenings was played. Consequently Newton dealt with the fact that light is propagated in empty space by making the hypothesis that light also consists of material particles interacting with ponderable matter through special forces. To this extent Newton's view of nature involved a third type of material particle, though this certainly had to have very different properties from the particles of the other forms of matter. Light particles had, in fact, to be capable of being formed and of disappearing. Moreover, even in the eighteenth century it was already clear from experience that light traveled in empty space with a definite velocity, a fact which obviously fitted badly into Newton's theoretical system, for why on earth should the light particles not be able to move through space with any arbitrary velocity?

It need not, therefore, surprise us that this theoretical system, built up by Newton with his powerful and logical intellect, should have been overthrown precisely by a theory of light. This was brought about by the Huygens-Young-Fresnel wave theory of light which the facts of interference and diffraction forced on stubbornly resisting physicists. The great range of phenomena, which could be calculated and predicted to the finest detail by using this theory, delighted physicists and filled many fat and learned books. No wonder then that the learned men failed to notice the crack which this theory made in the statue of their eternal goddess. For, in fact, this theory upset the view that everything real can be conceived as the motion of particles in space. Light waves were, after all, nothing more than undulatory states of empty space, and space thus gave up its passive rôle as a mere stage for physical events. The ether hypothesis patched up the crack and made it invisible.

The ether was invented, penetrating everything, filling the whole of space, and was admitted as a new kind of matter. Thus it was overlooked that by this procedure space itself had been brought to life. It is clear that this had really happened, since the ether was considered to be a sort of matter which could nowhere be removed. It was thus to some degree identical with space itself; that is, something necessarily given with space. Light was thus viewed as a dynamical process undergone, as it were, by space itself. In this way the field theory was born as an illegitimate child of Newtonian physics, though it was cleverly passed off at first as legitimate.

To become fully conscious of this change in outlook was a task for a highly original mind whose insight could go straight to essentials, a mind that never got stuck in formulas. Faraday was this favored spirit. His instinct revolted at the idea of forces acting directly at a distance which seemed contrary to every elementary observation. If one electrified body attracts or repels a second body, this was for him brought about not by a direct action from the first body on the second, but through an intermediary action. The first body brings the space immediately around it into a certain condition which spreads itself into more distant parts of space, according to a certain spatio-temporal law of propagation. This condition of space was called "the electric field." The second body experiences a force because it lies in the field of the first, and vice versa. The "field" thus provided a conceptual apparatus which rendered unnecessary the idea of action at a distance. Faraday also had the bold idea that under appropriate circumstances fields might detach themselves from the bodies producing them and speed away through space as free fields; this was his interpretation of light.

Maxwell then discovered the wonderful group of formulae which seems so simple to us nowadays and which finally built the bridge between the theory of electro-magnetism and the theory of light. It appeared that light consists of rapidly oscillating electro-magnetic fields.

After Hertz, in the '80s of the last century, had confirmed the existence of the electro-magnetic waves and displayed their identity with light by means of his wonderful experiments, the great intellectual revolution in physics gradually became complete. People slowly accustomed themselves to the idea that the physical states of space itself were the final physical reality, especially after Lorentz had shown in his penetrating theoretical researches that even inside ponderable bodies the electro-magnetic fields are not to be regarded as states of the matter, but essentially as states of the empty space in which the material atoms are to be considered as loosely distributed.

Dissatisfied with Dual Theory.

At the turn of the century physicists began to be dissatisfied with the dualism of a theory admitting two kinds of fundamental physical reality: on the one hand the field and on the other hand the material particles. It is only natural that attempts were made to represent the material particles as structures in the field, that is, as places where the fields were exceptionally concentrated. Any such representation of particles on the basis of the field theory would have been a great achievement, but in spite of all efforts of science it has not been accomplished. It must even be admitted that this dualism is today sharper and more troublesome than it was ten years ago. This fact is connected with the latest impetus to developments in quantum theory, where the theory of the continuum (field theory) and the essentially discontinuous interpretation of the elementary structures and processes are fighting for supremacy.

We shall not here discuss questions concerning molecular theory, but shall describe the improvements made in the field theory during this century.

These all arise from the theory of relativity, which has in the last six months entered its third stage of development. Let us briefly examine the chief points of view belonging to these three stages and their relation to field theory.

The first stage, the special theory of relativity, owes its origin principally to Maxwell's theory of the electro-magnetic field. From this, combined with the empirical fact that there does not exist any physically distinguishable state of motion which may be called "absolute rest," arose a new theory of space and time. It is well known that this theory discarded the absolute character of the conception of the simultaneity of two spatially separated events. Well known is also the courage of despair with which some philosophers still defend themselves in a profusion of proud but empty words against this simple theory.

On the other hand, the services rendered by the special theory of relativity to its parent, Maxwell's theory of the electro-magnetic field, are less adequately recognized. Up to that time the electric field and the magnetic field were regarded as existing separately even if a close causal correlation between the two types of field was provided by Maxwell's field equations. But the special theory of relativity showed that this causal correlation corresponds to an essential identity of the two types of field. In fact, the same condition of space, which in one coordinate system appears as a pure magnetic field, appears simultaneously in another coordinate system in relative motion as an electric field, and vice versa. Relationship of this kind displaying an identity between different conceptions, which therefore reduce the number of independent hypotheses and concepts of field theory and heighten its logical self-containedness are a characteristic feature of the theory of relativity. For instance, the special theory also indicated the essential identity of the conceptions' inertial mass and energy. This is all generally known and is only mentioned here in order to emphasize the unitary tendency which dominates the whole development of the theory.

We now turn to the second stage in the development of the theory of relativity, the so-called general theory of relativity. This theory also starts from a fact of experience which till then had received no satisfactory interpretation: the equality of inertial and gravitational mass, or, in other words, the fact known since the days of Galileo and Newton that all bodies fall with equal acceleration in the earth's gravitational field. The theory uses a special theory as its basis and at the same time modifies it: the recognition that there is no state of motion whatever which is physically privileged—that is, that not only velocity but also acceleration are without absolute significance—forms the starting point of the theory. It then compels a much more profound modification of the conceptions of space and time than were involved in the special theory. For even if the special theory forced us to fuse space and time together to an invisible four-dimensional continuum, yet the Euclidean character of the continuum remained essentially intact in this theory. In the general theory of relativity this hypothesis regarding the Euclidean character of our space-time continuum had to be abandoned and the latter given the structure of a so-called Riemannian space. Before we attempt to understand what these terms mean let us recall what this theory accomplished.

It furnished an exact field theory of gravitation and brought the latter into a fully determinate relationship to the metrical properties of the continuum. The theory of gravitation, which until then had not advanced beyond Newton, was thus brought within Faraday's conception of the field in a necessary manner; that is, without any essential arbitrariness in the selection of the field laws. At the same time gravitation and inertia were fused into an essential identity. The confirmation which this theory has received in recent years through the measurement of the deflection of light rays in a gravitational field and the spectroscopic examination of binary stars is well known.

The characteristics which especially distinguish the general theory of relativity and even more the new third stage of the theory, the unitary field theory, from other physical theories are the degree of formal speculation, the slender empirical basis, the boldness in theoretical construction and, finally, the fundamental reliance on the uniformity of the secrets of natural law and their accessibility to the speculative intellect. It is this feature which appears as a weakness to physicists who incline toward realism or positivism, but is especially attractive, nay, fascinating, to the speculative mathematical mind. Meyerson in his brilliant studies on the theory of knowledge justly draws a comparison of the intellectual attitude of the relativity theoretician with that of Descartes, or even of Hegel, without thereby implying the censure which a physicist would read into this.

However that may be, in the end experience is the only competent judge.

Yet in the meantime one thing may be said in defense of the theory. Advance in scientific knowledge must bring about the result that an increase in formal simplicity can only be won at the cost of an increased distance or gap between the fundamental hypothesis of the theory on the one hand and the directly observed facts on the other hand. Theory is compelled to pass more and more from the inductive to the deductive method, even though the most important demand to be made of every scientific theory will always remain: that it must fit the facts.

We now reach the difficult task of giving to the reader an idea of the methods used in the mathematical construction which led to the general theory of relativity and to the new unitary field theory.

The Problem Stated.

The general problem is: Which are the simplest formal structures that can be attributed to a four-dimensional continuum and which are the simplest laws that may be conceived to govern these structures? We then look for the mathematical expression of the physical fields in these formal structures and for the field laws of physics—already known to a certain approximation from earlier researches—in the simplest laws governing this structure.

The conceptions which are used in this connection can be explained just as well in a two-dimensional continuum (a surface) as in the four-dimensional continuum of space and time. Imagine a piece of paper ruled in millimeter squares. What does it mean if I say that the printed surface is two-dimensional? If any point P is marked on the paper, one can define its position by using two numbers. Thus, starting from the bottom left-hand corner, move a pointer toward the right until the lower end of the vertical through the point P is reached. Suppose that in doing this one has passed the lower ends of X vertical (millimeter) lines. Then move the pointer up to the point P passing Y horizontal lines. The point P is then described without ambiguity by the numbers X Y (coordinates). If one had used, instead of ruled millimeter paper, a piece which had been stretched or deformed the same determination could still be carried out; but in this case the lines passed would no longer be horizontals or verticals or even straight lines. The same point would then, of course, yield different numbers, but the possibility of determining a point by means of two numbers (Gaussian coordinates) still remains. Moreover, if P and Q are two points which lie very close to one another, then their coordinates differ only very slightly.

When a point can be described by two numbers in this way, we speak of a two-dimensional continuum (surface).

Riemannian Metric.

Now consider two neighboring points P, Q on the surface and a little way off another pair of points P', Q'. What does it mean to say that the distance PQ is equal to the distance P' Q'? This statement only has a clear meaning when we have a small measuring rod which we can take from one pair of points to the other and if the result of the comparison is independent of the particular measuring rod selected. If this is so, the magnitudes of the tracts PQ, P' Q' can be compared. If a continuum is of this kind we say it has a metric. Of course, the distance of the two points PQ must depend on the coordinate differences (dx, dy). But the form if this dependence is not known à priori. If it is of the form:

$$ds^2=g_{11}dx^2+2g_{12}dxdy+g_{22}dy^2$$

then it is called a Riemannian metric.

If it is possible to choose the coordinates so that this expression takes the form: $ds^2=dx^2+dy^2$ (Pythagoras's theorem), then the continuum is Euclidean (a plane).

Thus it is clear that the Euclidean continuum is a special case of the Riemannian. Inversely, the Riemannian continuum is a metric continuum which is Euclidean in infinitely small regions, but not in finite regions. The quantities g_{11}, g_{12}, g_{22} describe the metrical properties of the surface; that is, the metrical field.

By making use of empirically known properties of space, especially the law of the propagation of light, it is possible to show that the space - time continuum has a Riemannian metric. The quantities g_{11}, &c., appertaining to it determine not only the metric of the continuum but also the gravitational field. The law governing the gravitational field is found in answer to the question: Which are the simplest mathematical laws to which the metric (that is the g_{11}, &c.) can be subjected? The answer was given by the discovery of the field laws of gravitation, which have proved themselves more accurate than the Newtonian law. This rough outline is intended only to give a general idea of the sense in which I have spoken of the "speculative" methods of the general theory of relativity.

Expanding the Theory.

This theory having brought together the metric and gravitation would have been completely satisfactory if the world had only gravitational fields and no electro-magnetic fields. Now it is true that the latter can be included within the general theory of relativity by taking over and appropriately modifying Maxwell's equations of the electro-magnetic field, but they do not then appear like the gravitational fields as structural properties of the space-time continuum, but as logically independent constructions. The two types of field are causally linked in this theory, but still not fused to an identity. It can, however, scarcely be imagined that empty space has conditions or states of two essentially different kinds, and it is natural to suspect that this only appears to be so because the structure of the physical continuum is not completely described by the Riemannian metric.

The new unitary field theory removes this fault by displaying both types of field as manifestations of one comprehensive type of spatial structure in the space-time continuum. The stimulus to the new theory arose from the discovery that there exists a structure between the Riemannian space structure and the Euclidean, which is richer in formal relationships than the former, but poorer than the latter. Consider a two-dimension Riemannian space in the form of the surface of a hen's egg. Since this surface is embedded in our (accurately enough) Euclidean space, it possesses a Riemannian metric. In fact, it has a perfectly definite meaning to speak of the distance of two neighboring points P, Q on the surface. Similarly it has, of course, a meaning to say of two such pairs of points (PQ) (P'Q'), at separate parts of the surface of the egg, that the distance PQ is equal to the distance P'Q'. On the other hand, it is impossible now to compare the direction PQ with the direction P' Q'. In particular it is meaningless to demand that P'Q' shall be chosen parallel to PQ. In the corresponding Euclidean geometry of two dimensions, the Euclidean geometry of the plane, directions can be compared and the relationship of parallelism can exist between lines in regions of the plane at any distance from one another (distant parallelism). To this extent the Euclidean continuum is richer in relationships than the Riemannian.

A Mathematical Discovery.

The new unitary field theory is based on the following mathematical discovery: There are continua with a Riemannian metric and distant parallelism which nevertheless are not Euclidean. It is easy to show, for instance, in the case of three-dimensional space, how such a continuum differs from a Euclidean.

First of all, in such a continuum there are lines whose elements are parallel to one another. We shall call those "straight lines." It also has a definite meaning to speak of two parallel straight lines as in the Euclidean case. Now choose two such parallels E_1L_1 and E_2L_2 and mark on each a point P_1, P_2.

On E_1L_1 choose in addition a point Q_1. If we now draw through Q_1 a straight line Q_1-R parallel to the straight line P_1, P_2, then in Euclidean geometry this will cut the straight line E_2L_2; in the geometry now used the line Q_1-R and the line E_2L_2 do not in general cut one another. To this extent the geometry now used is not only a specialization of the Riemannian but also a generalization of the Euclidean geometry. My opinion is that our space-time continuum has a structure of the kind here outlined.

The mathematical problem whose solution, in my view, leads to the correct field laws is to be formulated thus: Which are the simplest and most natural conditions to which a continuum of this kind can be subjected? The answer to this question which I have attempted to give in a new paper yields unitary field laws for gravitation and electro-magnetism.

PROFESSOR ALBERT EINSTEIN'S newest work, a comprehensive theory fusing electro-magnetism and gravitation in a single law, was presented to the Prussian Academy of Sciences last week. The whole world has followed with keenest interest the latest contribution to science of the eminent author of the relativity theory. In the accompanying article, Professor Einstein himself explains his new work in a form as simple as the subject will allow. The title of his article is "Field Theories, Old and New."

In the first part of the article Professor Einstein traces the development of the theories with which he deals, from the time of Newton through Faraday and Maxwell, whose work it is now believed Professor Einstein has brought to a culmination.

In the second part of the article, Professor Einstein deals with the new geometry he has evolved—a geometry described as between Riemannian geometry and the familiar Euclidean geometry. He states his conclusion in mathematical formulae. The study of this part of the article will naturally be profitable in proportion to the reader's familiarity with physics and mathematics.

A glossary on this page gives the meaning of the technical terms used by Professor Einstein.

GLOSSARY OF EINSTEIN'S TERMS

SOME of the terms employed by Dr. Einstein in his article on this page will be recognized by those who recall their elementary physics; others will not be understood except by students of higher mathematics. For the convenience of all readers, the following glossary has been compiled of important terms, names and theories mentioned by Dr. Einstein.

CONTINUUM—A line, straight or curved, is a one-dimension continuum. A surface, flat or curved, is a two-dimension continuum. Space is a three-dimension continuum. Einstein's space time is a four-dimension continuum.

COULOMB FORCES—Coulomb, a French physicist (1736-1806), found that two electrified particles attract or repel each other with a force which is directly proportional to the product of their charges and inversely proportional to the square of the distance between them. Such forces are called Coulomb forces.

DEDUCTIVE METHOD—Establishing particular facts from general principles or truths.

DIFFRACTION—A deviation of the rays of light from a straight line when they are partially cut off by an obstacle or when they pass near the edges of an opening.

ELECTRO-MAGNETIC FIELD—A portion of space in which electric and magnetic forces exist.

ETHER—In physics, ether is a supposed medium which fills all space and through which radiant energy of all kinds—including radio waves, light waves, X rays, cosmic rays—is propagated.

EUCLID—A Greek mathematician (about 350-300 B. C.) called the "father of geometry."

FARADAY—English physicist (1791-1861) and discoverer of electromagnetic induction.

FOURTH DIMENSION—Fourth dimension of space is an assumed dimension whose relation to the recognized dimensions—length, breadth and thickness—is analogous to that borne by any one of them to the other two.

GALILEO—An Italian physicist and astronomer (1564-1642), inventor of the telescope and discoverer of the moons of Jupiter and the laws of falling bodies.

GAUSSIAN COORDINATES—Gauss, a German mathematician (1777-1855). In studying the properties of curved surfaces he used coordinates to latitude and longitude on the surface of a sphere. Such coordinates are called Gaussian coordinates.

GRAVITATIONAL FIELD—A portion of space across which heavy bodies attract each other.

HEGEL—A German philosopher (1770-1831).

HERTZ—A German physicist (1857-1894) who discovered the propagation of electromagnetic waves.

HUYGENS - YOUNG - FRESNEL WAVE THEORY OF LIGHT—Huygens, a Dutch mathematician (1629-1695), Young, an English physicist (1773-1829), and Fresnel, a French physicist (1788-1827), founded the theory that light is propagated by waves.

INDUCTIVE METHOD—The scientific method which attempts to obtain general laws from particular cases.

INERTIA—In physics, inertia is that property of matter by virtue of which it persists in its state of rest or of uniform motion in a straight line, unless some force changes that state.

INTERFERENCE—In physics, interference is the term used to describe the effect of waves in neutralizing or in reinforcing each other.

LORENTZ—A Dutch physicist who developed the theory of electrons.

MAXWELL'S FIELD EQUATIONS—Maxwell, a Scottish physicist (1831-1879), laid down the electromagnetic theory of light. This theory predicted the effects afterward observed by Hertz. The equations of Maxwell's theory are called the Maxwell field equations.

METRIC—A term used to describe a mathematical system of measurement.

NEWTON'S LAW—Newton's law of universal gravitation asserts that every particle of matter attracts every other particle of matter with a force which is directly proportional to their masses and inversely proportional to the square of the distance between them.

PONDERABLE MATTER—Matter that has weight.

QUANTUM THEORY—In atomic physics the quantum theory was founded in 1900 by Planck, a German mathematical physicist. This theory may be briefly characterized by saying that it considers atomic phenomena as essentially discontinuous phenomena.

RIEMANNIAN SPACE AND RIEMANNIAN METRIC—If the properties of two-dimensional space can be described by the formula given in the article, viz.:

$$ds^2 = g_{11}dx^2 + 2g_{12}dxdy + g_{22}dy^2$$

the space is said to be a Riemannian space and to have a Riemannian metric. If the two-dimensional space, however, can be described by the simple formula, $ds^2 = dx^2 + dy^2$, it is said to be a Euclidean space and to possess Euclidean metric. Riemann was a German mathematician (1826-1866).

February 3, 1929

Einstein Recasts Field Theory on Another Basis; Says It Guarantees Equations' Compatability

Wireless to THE NEW YORK TIMES.

BERLIN, April 4.—Professor Albert Einstein has submitted to the Prussian Academy of Sciences a treatise supplementing his New Field Theory, recently published. He calls this latest work "The Unitary Field Theory and the Hamilton Principle," and in the introduction to it, he says:

"It shows that the field equations of the unitary field theory based upon the Riemann metric and far-parallelism can be derived from a Hamilton principle. This method has this advantage over those used by me before, namely, that it gives equations whose compatibility is a priori guaranteed."

Dr. Albert Einstein issued his latest theory early this year, causing the usual mathematical and physical furore.

Briefly, it says there is only one substance, "the field," and only one universal physical law. In other words, it reduces to one formula the basic laws of relativistic mechanics and electricity.

April 5, 1929

EINSTEIN ANNOUNCES A NEW FIELD THEORY

He Introduces a Vector of 5 Components Into 4-Dimensional Space-Time Continuum.

A preliminary announcement by Professor Albert Einstein of the completion by him, in collaboration with Dr. Walter Mayer, his assistant, of part of his work on a new unified field theory, supplanting the one announced by him in 1929, upon which he had spent more than ten years of work, was made public yesterday by the Josiah Macy Jr. Foundation of 565 Park Avenue, which last year created a fellowship to provide a competent collaborator to Dr. Einstein in his research work.

Unified field theory is a term widely applied to represent the theory advanced by Einstein, according to which there is but a single background to all material activity—one unified field.

Before Einstein a material object was commonly conceived of as existing in space, time, a gravitational field, and an electromagnetic field, each object thus having four different backgrounds. Einstein's special theory of relativity amalgamated space and time into one, space-time, while the general theory of relativity, with its Riemannian geometry, further absorbed the gravitational field into space-time. Thus Einstein reduced three of the four backgrounds to one. The unified field theory goes a step further by including the electromagnetic field into the synthesis.

Old Unitary Theory Abandoned.

Einstein's new theory will be published in the near future, according to the announcement, probably in Pasadena, in connection with his investigations last Winter while in California. The Einstein statement was submitted in the president's report at the annual meeting of the board of trustees of the Macy Foundation.

Einstein's preliminary announcement does not go into the details of his new theory, confining itself to a general, brief statement of the mathematical lines of procedure followed by him and Dr. Mayer. It contains, however, the frank admission that his older unitary field theory, which was based on the introduction of the theory of distant parallelism in Riemannian geometry, had been abandoned by him when he found, after a year's further work, that it was a "striving in the wrong direction."

Instead of the theory of distant parallelism Einstein, with Dr. Mayer, has worked out a new unitary field theory on new mathematical concepts, based on the theory of Theodore Kaluza, promulgated in 1921, which Einstein had formerly regarded as "not acceptable."

Kaluza's theory rests on the assumption that the physical space-time continuum is five-dimensional instead of four-dimensional, as had been previously considered. By postulating a fifth dimension he was enabled to obtain field laws which agree in first approximation to the known field laws of both electricity and gravitation.

Einstein objected to this theory at first on the grounds that he considered it "anomalous to replace the four-dimensional continuum by a five-dimensional one "only to find it necessary subsequently to tie up artificially one of these dimensions in order to account for the fact that the fifth dimension does not manifest itself in the physical world of space-time. In other words, Einstein found it objectionable to introduce a fifth dimension the reality of which was not on a par with the other four dimensions.

The new theory, Professor Einstein says, "formally approximates Kaluza's theory without being exposed to the objection just stated." This was accomplished, he adds, "by the introduction of an entirely new mathematical concept."

Until now, Einstein explains, it has been believed that one can introduce into a space of, for example, four dimensions, given vectors or vector-fields of no more than four components. In other words, a given vector is regarded by mathematicians as having as many components as the dimensionality of the space with which it is associated. Only two vector-components, it is held, can be introduced into three-dimensional space, only three vector-components into three-dimensional space, and similarly with higher dimensions.

This restriction, Einstein declares, appears not to be necessary. He and Dr. Mayer have found that a vector of five components can be introduced into the space-time continuum of only four dimensions. It is on this finding that new theory is based on.

Comment of Professor Wills.

Professor A. P. Wills of the Department of Mathematical Physics, Columbia, when asked to comment on Dr. Einstein's statement, said:

"The new concept involves apparently a generalization of the vector idea, in connection with four-dimensional Riemannian geometry. In ordinary three-dimensional space, quantities such as displacement, velocity, force, electric field intensity, are known as vector quantities. The specification of such a quantity requires the use of a set of three numbers—one for length, the other two for direction. These numbers are the components of the vectors.

"A region of space, with each point of which is associated a vector, is called a vector-field. Examples of such are the gravitational, electric and magnetic fields of the physicist.

"The usual concept of a vector in a space of, for example, four dimensions, is a set of four numbers, components of the vector, which transform in passing from one coordinate system to another in accordance with a definite rule. Geometric visualization of vector quantities in space greater than three dimensions is not possible. But the rules for the transformation of the numbers specifying such vectors are the same as for three-dimensional space.

"In the generalization by Einstein of the vector idea the notion of a vector in space of any number of dimensions is such as to permit a vector's possessing a number of components different from the dimensionality of the space. Specifically, Einstein states that he has been successful in introducing a vector of five components into the four-dimensional space-time continuum."

Professor Wills said he felt that further comment on the latest Einstein theory should be deferred until its publication in full.

Einstein's Statement.

Einstein's statement is as follows:

"Ever since the formulation of the general relativity theory in 1915 it has been the persistent effort of theoreticians to reduce the laws of the gravitational and electromagnetic fields to a single basis. It could not be believed that these fields correspond to two spatial structures which have no conceptual relation to each other. Thus arose the theories of Weyl and Eddington, which, however, have been abandoned by their authors; the theory of Kaluza and also the theory of distant parallelism. After we both had worked more than a year on the further development of the last theory we reached the conclusion that we were striving in the wrong direction and that the theory of Kaluza, while not acceptable, was nevertheless nearer the truth than the other theoretical approaches.

"Kaluza's theory rests on the assumption that the physical space-time continuum is five-dimensional (instead of, as formerly, four-dimensional) in which the empiric four-dimensionality of the physical continuum can be accounted for by the hypothesis that the physical variables are independent of the coordinate

$$x_5$$

By postulating a Riemannian metric in five dimensions, Kaluza reaches field laws which agree in first approximation with the known field laws of gravitation and electricity.

"Among the considerations which question this theory stands in the first place the following: It is anomalous to replace the four-dimensional continuum by a five-dimensional one and then subsequently to tie up artificially one of these five dimensions in order to account for the fact that it does not manifest itself.

"We have succeeded in formulating a theory which formally approximates Kaluza's theory without being exposed to the objection just stated. This is accomplished by the introduction of an entirely new mathematical concept which may be described as follows:

"Until now it has been believed that one can introduce into a space of 'n' dimensions only vectors, or vector-fields, of which the number of components agree with the number of dimensions of that space. It appears, however, that this restriction is not necessary. It has its origin in the 'anschauliche' (outwardly apparent) significance of those vectors responsible for the formulation of the vector concept.

"We have been successful in introducing into space

$$R_n$$

of 'n' dimensions, vectors

$$a^i \; (i=1\ldots m)$$

of 'm' components, and in deriving a calculus of such vectors and tensors which is essentially no more complicated than the well-known absolute calculus.

"Our theory arises quite readily from consideration of five vectors (five components) in the four-dimensional continuum. There follows from that a 'five-curvature' of space which is analogous to the Riemannian curvature and which bears a similar relationship to the laws of the unitary field that the Riemannian curvature does to the relativistic equations of the gravitational field alone.

"This theory does not yet contain the conclusions of the quantum theory. It furnishes, however, clues to a natural development, from which we may anticipate further results in this direction. In any event, the results thus far obtained represent a definitive advance in knowledge of the structure of physical space."

October 26, 1931

NEW THEORY LINKS MATTER TO ENERGY, UNITING IN CREATION

Conversion of Radiation Into Matter, and Vice Versa, Shown in Mathematical 'Bridge.'

WORK OF J. R. OPPENHEIMER

By WILLIAM L. LAURENCE.
Special to THE NEW YORK TIMES.
CAMBRIDGE, Mass., Dec. 28.—A new mathematical theory which builds an intellectual "bridge" linking the material and the non-material, and explains how something possessing no dimensions may assume three dimensional existence, was presented here today before the American Association for the Advancement of Science.

The new "bridge" was described before a distinguished audience of scientists from all over the country, at a symposium on nuclear physics at Massachusetts Institute of Technology, by Professor J. R. Oppenheimer of the University of California.

The mathematical bridge was fashioned to enable science to explain the recently discovered, startling phenomenon, observed at the laboratories of California Institute of Technology, in which non-material gamma rays from thorium C, when fired at terrific speed at the heart of an atom, somehow disappear and pairs of material particles, each composed of a positive and a negative electron, come flying out in their place.

The phenomenon led a number of leading scientists to the conclusion that here, for the first time, man was witnessing the creation of matter out of radiant energy.

Dr. Oppenheimer's mathematics, which dovetails with the findings in the laboratory, seeks to fashion a new fundamental unity in nature, supplementing the mathematical unities introduced by Maxwell, who united electricity, magnetism and light, and by Einstein, who added gravitation to the others.

The Oppenheimer theory, if found acceptable, would now link in a new union the worlds of non-material, non-dimensional radiation and material, three-dimensional matter.

Dr. Oppenheimer's theory constitutes an extension of the theory of Dr. P. A. M. Dirac, British scientist who shared this year's Nobel Prize in physics.

In 1931, more than a year before discovery of the positron by Dr. Carl D. Anderson of California Institute of Technology, Professor Dirac predicted existence of this positive particle of matter, of a mass equal to that of the electron.

At that time Dr. Dirac also presented mathematical reasons why radiant energy should be transformed into the energy of positive and negative electrons.

Dr. Dirac's predictions were fulfilled sooner than was expected by scientists, yet this theory contained a number of paradoxes which ran contrary to experience. Professor Oppenheimer's theory obviates the difficulties of the Dirac theory.

According to the Oppenheimer formulae, applied within certain limits, 2 per cent of any given quantity of electro-magnetism will be converted into matter in the form of pairs of electrons and positrons.

These limits, however, are very small indeed, being 10 to minus 11th power centimeters and 10 to the minus 21st power in seconds. In other words, a space in centimeters represented by the numeral 1 divided by 11 ciphers and a period of time in which one second is divided by 1 followed by 21 ciphers.

* * * * *

December 29, 1933

Einstein in Vast New Theory Links Atom and Stars in Unified System

A Pattern Is Envisaged by Him and Dr. Rosen in Structure of Space and Matter That May Harmonize Relativity Rules and Quantum Conception, Hitherto Unreconciled by Scientists.

By WILLIAM L. LAURENCE.

Soaring over a hitherto unscaled "mathematical mountain-top," Dr. Albert Einstein, climber of "cosmic Alps," reports having sighted a new pattern in the structure of space and matter.

From his new mathematical vantage point "outside of space" he has had a glimpse, Dr. Einstein believes, of the "Promised Land of Knowledge," in which the atom and the universe are embraced for the first time in one comprehensive system, with a new "solid" bridge at last spanning the yawning chasm hitherto separating the two worlds governed, respectively, by the rules of relativity and the quantum theory.

Einstein's latest daring flight through the vastnesses of his space-time universe, the fashioning of which gained for him the title of one of the eight "universe-makers" of history, is described by him in the current issue of The Physical Review, published by the American Physical Society and the American Institute of Physics.

To make this "flight," which may take its place as another epoch-making achievement, Einstein has built a new type of "space-ship," entirely of mathematical concepts and designed for penetrating the furthermost reaches of space and time.

Returning from his first "test trip" on this latest breed of Pegasus, Einstein has brought back with him some startling new "pictures of space," "photographed" with the highly intricate mathematical "lenses" of a "cosmic camera" designed in collaboration with Dr. N. Rosen at the Institute for Advanced Study at Princeton.

The new "space-photographs" present an entirely new, radically different vision of reality. Space, instead of being one, is seen as composed of "two identical sheets joined by many bridges." A new particle of matter "without gravitating mass," namely, an electrical atom which weighs nothing, is found to be, according to the mathematics of Einstein, "the most natural electrical particle." Electricity and mass are found not to be related, but appear as independent constants in nature. And the atoms, fundamental building blocks of the universe, are visioned as cosmic "bridges" linking the "two identical sheets of space."

What the new Einstein "space-photographs" indicate, according to Drs. Einstein and Rosen, is the possibility, stupendous in its scientific implications, that a method at last has been found to develop an all-embracing physical and mathematical theory which would include the macrocosm and the microcosm, the universe as a whole as well as the atom, and which would bridge the wide gulf now separating two of the greatest achievements of the human intellect—the theory of relativity and the quantum theory.

"In spite of its great success in various fields," Drs. Einstein and Rosen assert, "the present theoretical physics is still far from being able to provide a unified foundation on which the theoretical treatment of all phenomena could be based.

"We have a general relativistic theory of macroscopic phenomena, which, however, has hitherto been unable to account for the atomic structure of matter and for quantum effects; and we have a quantum theory, which is able to account satisfactorily for a large number of atomic and quantum phenomena, but which, by its very nature, is unsuited to the principle of relativity.

"Under these circumstances it does not seem superfluous to raise the question as to what extent the method of general relativity provides the possibility of accounting for atomic phenomena.

"It is to such a possibility that we wish to call attention in the present paper in spite of the fact that we are not yet able to decide whether this theory can account for quantum phenomena."

The Relativity Theories.

The original theory of relativity, known as the special theory, deals with space and time, and shows that the two are really one, space-time, time being conceived as the fourth dimension. The expanded theory, known as the general theory of relativity, brings gravitation into the fold, making it merely another aspect of the space-time continuum.

Though this synthesis, in one comprehensive system, of space, time and gravitation was described by Bertrand Russell as being probably "the greatest synthetic achievement of the human intellect up to the present time," it left several large gaps which Einstein and many other outstanding intellects have tried to bridge ever since.

To the great dismay of men of science, the universe as it stands today in the light of modern knowledge is divided into two air-tight compartments, each of which is governed by a distinctly separate set of rules, each one valid in its own domain but not applicable in that of the other.

Thus a very anomalous situation exists in the state of modern scientific theory and practice. On the one hand, relativity explains the phenomena of the universe as a whole but tells us nothing about the structure of the atom and the still smaller parts of which the atom is composed. In fact, the very existence of the atom, as far as relativity is concerned, has come to be regarded by many theoretical physicists as an anomaly in the gravitational field, a sort of cosmic "hobo" that must be tolerated but really "does not belong."

On the other hand, the quantum theory makes the atom "king." In it, "all the world's an atom" and there is no other world, as far as the quantum theory is concerned, for the laws that govern the atom and the doings within it fail entirely to explain the phenomena of the relativity-governed universe with its myriads of stars, constellations and galaxies.

In what is known as the "unified field theory" Einstein and others have attempted to bring electricity and magnetism into the fold of space-time and gravitation. This has met with no real success so far.

Bridge Seen for the Gap.

In the present work he reports the first steps in the development of a method, which, he believes, opens the way for bridging the gap between the universe of the stars and the universe of the atom.

The first step has been made, he announces. Whether the next ones will follow logically he confesses he does not yet know.

But, Drs. Einstein and Rosen add, "the publication of this theoretical method is justified, in our opinion, because it provides a clear procedure, characterized by a minimum of assumptions, the carrying out of which has no other difficulties to overcome than those of a mathematical nature."

Drs. Einstein and Rosen set themselves to find out, as a preliminary to attacking the main problem, whether or not it would be possible to modify the field law of gravitation and also the field law of gravitation and electricity, "without any essential change." They credit Dr. Walter Mayer, Dr. Einstein's collaborator, with having shown them the way to do it.

By introducing this modification while at the same time refusing to accept the point of view that the atom is an anomaly or a singularity in the gravitational or in the electrical field, the scientists came upon many strange mathematical adventures.

They found, to their surprise, that "the four-dimensional space is described mathematically by two congruent parts, or 'sheets'"; next they were startled to find a connection between the two "sheets," which they decided to call a "bridge."

On closer examination, the "bridge" turned out to be a "spatially finite" material particle. This particular "spatially finite bridge" was further identified by them as being the now familiar neutron, and possibly the as yet unseen neutrino, which are elementary particles having mass but no electric charge.

"With this conception one is able to understand the atomistic character of matter, as well as the fact that there can be no particles of negative mass," they report.

View in Another Direction.

Heading their "space-ship" in a different direction and focusing their "cosmic camera" on another part of space, Drs. Einstein and Rosen were amazed once again to "see" four-dimensional space as composed of two congruent layers, again connected by a "spatially finite bridge."

This time, however, the "materials" of which the "bridge" was made were found to be entirely different, exactly the reverse of the first bridge. Whereas the first had mass but no electric charge, the second had electric charge but no mass, a "ghost atom."

Thus there are three, instead of only two, fundamental types of material particles, or atoms, if Einstein's latest "space-photographs" are later corroborated by other means, as have been his earlier space-time-gravitation concepts of relativity.

In addition to the neutron and the as yet unseen neutrino, which have mass but no electrical charge; and the electron, proton and positron, which have both mass and electrical charge, there appears to exist, according to Einstein, a third type of elementary "material" particle, an "electric atom" which has a unit of electrical charge but no mass.

If the electrical charge may be described as the "soul" of an atom, the first type of atom could be said to be a "body" without a "soul"; the second type would have both "body" and "soul," while the third

PHYSICS AND ASTROPHYSICS

fundamental type of building block of reality—so far existing only in Einstein's mathematical "picture"—would be a sort of "ghost atom," a "soul" without a "body."

While Einstein does not speculate on the nature of this "ghost atom," it is reasonable to assume that, since electricity is either positive or negative, this "ghost atom" could be either a negative electric charge without mass or a positive electric charge without mass.

These "space-photographs" are all that have been taken so far, but they mark the end of the first reel in what promises to be one of the most exciting "transcosmic flights" yet taken by man.

Unified Universe in View.

So far Drs. Einstein and Rosen have "seen" only two types of "bridges" connecting their double-layer space. But on these they already are building a new foundation to carry what they hope will be a completely unified universe. In this universe the duality now existing between the laws governing galaxies and atoms will be eliminated, and the present state of "anarchy" in the atom, where, according to the quantum theory, the law of cause and effect does not prevail, will be reduced to the same deterministic, immutable law and order that govern the universe of relativity.

Each type of particle, possessing exclusively either mass or electric charge, according to the new picture of reality, constitutes a different "bridge" over the two congruent sheets of space. An electron, a proton, or a positron, having both mass and electric charge, would thus each constitute a "two-bridge" phenomenon across the chasm of the "double-sheet space."

"One might expect," the paper asserts, "that processes in which several elementary particles take part correspond to regular solutions of the field equations with several bridges between the two equivalent sheets corresponding to physical space.

"Only by investigations of these solutions will one be able to determine the extent to which the theory accounts for the facts. For the present, one cannot even know whether regular solutions with more than one bridge exist at all.

"In favor of the theory one can say that it explains the atomistic character of matter as well as the circumstance that there exist no negative neutral masses, and that in principle it can claim to be complete.

"On the other hand, one does not see a priori whether the theory contains the quantum phenomena (the laws governing the interactions and energy-states of atoms and their constituent parts).

"Nevertheless one should not exclude a priori the possibility that the theory may contain the quantum phenomena."

It may turn out that later "space-photographs" will give as faithful a reproduction of the quantum phenomena as they have already given of the atomistic phenomena, they add.

"In any case," Drs. Einstein and Rosen conclude, "here is a possibility for a general relativistic theory of matter which is logically completely satisfying and which contains no new hypothetical elements."

July 5, 1935

NEW MATHEMATICS LINKS TWO WORLDS

Cartan of Paris Offers System Promising Bridge Between Stars and Atom.

ACCLAIMED AT HARVARD

Leaders at Tercentenary Also Hail Contribution by Carnap in the Field of Logic.

By WILLIAM L. LAURENCE
Special to THE NEW YORK TIMES.

CAMBRIDGE, Mass., Sept. 1.—An important extension of modern mathematical methods, which promises to bridge the gap between the universe of the stars and the world of the atom, embracing at last the relativity and quantum theories, was presented here today before the Harvard tercentenary conference of arts and sciences.

The new mathematics, which extends further the mathematics that made possible Einstein's theory of relativity, was presented here for the first time by Professor Elie Joseph Cartan, mathematician, of the University of Paris. His paper dealt with "the extension of tensor analysis to non-affine geometries," which unifies several modern mathematical developments into a new synthesis.

Another contribution, dealing with "truth in mathematics and logic," and offering a "new form of logic," was presented by Professor Rudolf Carnap, philosopher, formerly of the universities of Vienna and Prague, who will join the faculty of the University of Chicago next month.

These adventures in the highest realms of mind, mathematics and logic, were described before a meeting held jointly with the American Mathematical Society, the Mathematical Association of America and the Association for Symbolic Logic.

The contributions were described as epoch-making by authorities present, who at the same time admitted that the material presented was of such a profound and abstruse nature that months of concentrated effort would be required to comprehend and digest it.

Hopes to Unify Two Systems

It was the development in recent times of the mathematical system known as "tensor analysis" that made it possible for Einstein to develop his theory of relativity. On the other hand, the theory of the new quantum mechanics, developed by Dirac, Heisenberg and others, was the outgrowth of an entirely different system of mathematics.

Professor Cartan has now extended the system of tensor analysis in such a way that for the first time it embraces systems of mathematics applicable to the quantum theory. Thus he hopes that at last a way has been opened for solving one of the greatest problems in modern science, the unification of the two great systems of thought which alone make the physical universe understandable.

The search for a unified principle underlying his universe has been one of man's greatest intellectual cravings since the earliest days. From the Greek philosopher Thales down to the present it has been man's intuitive belief, amounting to a conviction, that the world he lives in, from the smallest atom to the totality of the stars and galaxies, is governed by a set of universal laws applicable throughout the entire realm.

There have been times in man's intellectual history when philosophers believed that they had achieved the goal in the search for the underlying, all-embracing principle governing all things. Then came modern science, with the extension of man's knowledge far out in space and deep within the atom, which shattered his dream of universal monism.

For he found, to his disappointment, that the set of mathematical laws by which he could gain an insight into the infinitesimal but infinite universe of the atom was totally different from the principles which embraced the totality of the new stars and galaxies.

Science has thus found itself faced with this vexing duality, a universe made up of atoms which has to have two sets of principles to explain it, one set (relativity) for the whole, and another set (quantum mechanics) for the building blocks (atoms) out of which the whole is constituted.

Discussion by Experts Awaited

Thus any extension of present mathematical systems that promises to supply the missing link between the two dual systems is regarded as of the highest importance to science.

Einstein, Eddington, Levi-Civita and several others of the world's outstanding mathematical physicists, are attending the tercentenary conference, and are expected to discuss the latest development in man's search for unity in consultations which might yield historic results. Dr. Cartan stated after the lecture that he expected to discuss the possibilities of his new mathematics with Einstein here.

Only the introduction to Dr. Cartan's paper is quotable for the general reader. He then plunged into such deep mathematical waters that even many experts present confessed they found it difficult to follow.

"It is unnecessary to recall the great services which tensor analysis has rendered to geometry and to mathematical physics," Dr. Cartan began. "Every one is aware that Einstein's general theory of relativity might not have been conceived had this admirable instrument of research not been created, under the name of 'absolute differential calculus,' by G. Ricci and T. Levi-Civita.

Difficulties of Adaptation Cited

"In its original form tensor analysis is marvelously adapted to Riemannian geometry and especially to the geometries with affine connections which have sprung up in the wake of the great movement created by the discovery of general relativity.

"This movement, it is true, did not stop at tensor analysis, and the geometries with projective, or conformal, connections, and others, have been added to Riemannian geometry, just as the classical projective, or conformal, geometries, and, in a general way, the different geometries of Klein, have been added to that of Euclid.

"But it is rather difficult to adapt tensor analysis to these generalized geometries, and the attempts to do so often give the impression of being artificial and of seeking principally to preserve the external form of the theory of Ricci and Levi-Civita in cases where it is contrary to the nature of the facts.

"I would like to show, in this lecture, the point of view which might be assumed in order to build up a general theory of tensor analysis; that is, a theory which would be applicable to the present state of differential geometry."

Later Dr. Cartan said:

"The principal application, one could almost say the only one, which makes a generalization of the absolute differential calculus indispensable is related to the theory of generalized spaces, and particularly to the necessary and sufficient condition for applicability (of geometric identity of structure) of two spaces of the same nature."

Dr. Carnap States Problem

Professor Carnap first described the work of Dr. K. Goedel, Vienna mathematician, who showed that methods of modern mathematics were "incomplete." He then described his own method, which, he stated, would eliminate present difficulties in the logic of mathematics.

The first problem is to establish what is the nature of the truth of mathematical and logical theorems, Dr. Carnap stated. To do so, we must distinguish between what is known as factual truth and formal truth.

A statement of empirical science is true or false according to whether the possible fact asserted by it exists or not. A statement in the language of mathematics and logic does not assert a fact. Such a statement is true or false "according as the rules of the language system state it to be true; that is, capable of analysis, or false, namely, contradictory to the rules."

The second problem, Dr. Carnap said, is how to construct a system of mathematics, i. e., how to define the concept "true mathematical statement."

The customary method of demonstration, he said, consists of laying down some primitive postulates and some rules of inference, with a finite number of premises. For example, the system known as principia mathematica. According to the results of Dr. Goedel, however, such a system is always incomplete.

"Therefore, we are compelled," Dr. Carnap said, "to apply rules of a new kind, referring to infinite classes of premises. By rules of this kind the concept of the truth of universal statements concerning integers can easily be defined.

"However, for statements concerning real numbers, i. e., classes of integers, certain difficulties arise."

September 2, 1936

New Einstein Theory Gives A Master Key to Universe

Scientist, After 30 Years' Work, Evolves Concept That Promises to Bridge Gap Between the Star and the Atom

By WILLIAM L. LAURENCE

Albert Einstein, whose theory of relativity provided the formula that revealed the existence of atomic energy and offered mankind new visions of the material universe, has developed, after more than thirty years of arduous labors, a mathematical concept that is expected to lead to new and much deeper insights into the cosmos.

The new theory, described by Einstein as a "generalized theory of gravitation," attempts to interrelate all known physical phenomena into one all-embracing intellectual concept, thus providing one major master key to all the multiple phenomena and forces in which the material universe manifests itself to man.

In his special theory of relativity, published in 1905, Einstein proved by mathematics that space and time, rather than being two separate entities, were actually united in one four-dimensional continuum. Out of this intellectual synthesis emerged the discovery that matter and energy were both interchangeable, matter being "frozen" energy, while energy was matter in a fluid state.

In his general theory of relativity, published in 1916, Einstein proved, again by mathematics, that gravitation and inertia were equivalent, thus bringing space, time, matter, energy, gravitation and inertia into one all-embracing intellectual concept. This accomplishment was described by Bertrand Russell as "the greatest intellectual achievement of all time," and caused George Bernard Shaw to refer to Einstein as one of the "eight universe-builders in history."

However, there still remained one of the greatest cosmic forces that could not be brought into the unified structure, the all-pervading force of electromagnetism, which permeates the cosmos at large and the atoms of which the cosmos is constituted. It is this force that Einstein believes he has at last succeeded in bringing into an all-embracing cosmic concept, known among scientists as a "Unified Field Theory." This means that the gravitational field and the electromagnetic field, the two major "fields" in which the material universe manifests itself, can at last be viewed as being two manifestations of one united cosmic entity.

The new Einstein concept is to be published as a new chapter in the third edition of his famous book, "The Meaning of Relativity," originally issued in 1922. The second edition came out in 1944. The new edition will be published in February by the Princeton University Press.

Typewritten Text on View

A typewritten text of the new chapter was displayed yesterday by the Princeton University Press at the opening sessions of the annual meeting of the American Association for the Advancement of Science at the Statler Hotel.

Just as the original theory of relativity had to be checked to determine whether it conformed with established physical facts, the new theory also must await experimental corroboration, Dr. Einstein points out. In the introduction to his new paper he says:

"I shall present an attempt at the solution of this problem [the unification of gravitation with electromagnetism], which appears to me highly convincing, although, due to mathematical difficulties, I have not yet found a way to confront the results of the theory with experimental evidence."

The idea of developing an all-encompassing field theory, uniting the field of gravitation with the electromagnetic field, has been a major goal of physics during the last thirty years, engaging the attention not only of Einstein but of many other great men of science. Over the course of the centuries, beginning with scientists in China, India and Greece, vast stores of knowledge have been accumulated by experimentation and observation, but there has been no single theory to explain and describe them all.

For example, it is known that there are various types of elementary particles of which the atoms are constituted, such as electrons, protons and neutrons, but no theory could explain why there are only a few specific types instead of

DEVELOPS NEW THEORY

Albert Einstein

many. The equations of Clerk Maxwell describe electromagnetic fields caused by moving electrons, yet they do not explain why all electrons and protons have the same charge. Physicists know that bodies produce gravitational fields, but no one knows why they produce them.

Einstein set himself the enormous task to bring all these different manifestations of the cosmos under one all-embracing theory from which all the known phenomena could be deduced. Such a comprehensive theory not only provides a profounder understanding of the universe around us; it generally make possible the prediction of entirely new and unforeseen phenomena.

For example, until about the second half of the nineteenth century, magnetism, electricity and light were regarded as three different phenomena. Along came the Scottish mathematical physicist, Maxwell, and proved by his famous mathematical equations that all three were actually one. Since light was known to be propagated in waves, the identification of light with electromagnetism revealed the existence of electromagnetic waves. This led to a search for such waves, with the result that the German physicist, Hertz, discovered Hertzian waves, now better known as radio waves. In other words, radio and television were first discovered by Maxwell's mathematical formulae.

Similarly, the mathematical synthesis by Einstein of space and time led to the unification of matter and energy, a concept that found its most spectacular verification when the atomic bomb exploded over New Mexico and Japan, forty years after the promulgation of the original theory of relativity, and that promises the utilization of the vast stores of the energy within the nuclei of atoms for the benefit of mankind. The general theory of relativity, in turn, predicted then unknown phenomena about the cosmos, which have since been verified by astronomers and cosmologists.

It took about a quarter-century before Maxwell's mathematical waves were translated into actual radio waves. Similarly, it took more than twenty years before Einstein's mathematical formulae could be transformed into physical realities. Hence it must be stressed that many years may pass before

PHYSICS AND ASTROPHYSICS

The principal equations in the Einstein text which may offer important clues to the major mysteries of the cosmos.

The New York Times — Dec. 27, 1949

Einstein's newest mathematical concepts can be expected to find their counterpart in the world of actualities.

Experimented Evidence to Come

In presenting his relativity theories Einstein made predictions that, he then said, would either prove or disprove the correctness of his concepts. In this case, however, he is frank to admit that he has "not yet found a practical way to confront the results of the theory with experimental evidence." However, the fact that he regards his formulae as "highly convincing" indicates his belief that he has at last found the "key to the cosmos" he has been seeking for more than half of his seventy years, though he above everyone, realizes that the "key" must first be tested against the cosmic lock before it can be definitely known whether it "fits."

Einstein's latest work promises to bridge at last gap that now separates the infinite universe of the stars and galaxies and the equally infinite universe of the atom, which at present are widely separated, one being explained by relativity while our knowledge of the other rests on the quantum theory, of which Einstein was also one of the major architects.

While Einstein mentions the quantum theory only in passing in his present paper, it is known that he intends to crown his life's work as a "universe builder" by bringing the relativity and the quantum theories, the two major pillars on which man's basic understanding of the universe rests, into one all-embracing, comprehensive system. The present work is regarded as a major step in that direction.

December 27, 1949

YUKAWA EXPANDS HIS MESON THEORY

Japanese Nobel Prize Winner Offers New Explanation of the Nature of Atom Fractions

'COSMIC GHOST' SEEN LAID

A Particle Is Conceived That Has Charge and Mass but Lacks 'Infinite Energy'

By WILLIAM L. LAURENCE

Dr. Hideki Yukawa of Kyoto University, Japan, winner of the Nobel Prize in physics in 1949, presents in the current issue of The Physical Review, published yesterday, a new general theory on the nature of the elementary particles of the nuclei of atoms.

His theory, he believes, will lead to a better understanding of the fundamental structure of matter and of the cosmic forces, operating within the nuclei of atoms, responsible for the very existence of the material universe.

Dr. Yukawa, now a visiting professor at Columbia University, won world renown in 1934 when, at the age of 28 as a lecturer at Osaka University, he predicted the existence of a then unknown type of elementary particle of matter, now known as the meson.

The meson, he predicted at the time, had a mass intermediate between the electron, negatively charged fundamental particle, and the proton, positively charged constituent of the atomic nucleus.

Meson's Mass Calculated

Dr. Yukawa even calculated that the mass of the then purely hypothetical meson was about 200 times that of the electron.

Dr. Yukawa's hypothesis was verified two years later when Dr. Carl Anderson of the California Institute of Technology, who won the Nobel Prize for his discovery of the positron, discovered the meson as a product of the cosmic rays. About a year ago mesons were produced artificially with the giant synchro-cyclotron at the University of California.

The meson originally discovered by Dr. Anderson had a mass about 200 times the electron mass, just as was predicted by Dr. Yukawa. About two years ago, a meson of about 400 electron masses was discovered by a group of British scientists. The Yukawa meson has since been named the "pi" meson, while the British meson is known as the "mu" meson.

Dr. Yukawa's original hypothesis and its later verification led to the development of what is known today as the meson theory, the only available, though still not fully satisfactory, theory to explain the forces and interactions between the fundamental particles in the atomic nucleus.

For his formulation of the meson theory Dr. Yukawa received the Nobel Prize in physics in 1949.

However, the meson theory, while fruitful, is still far from complete, posing many questions for each one it answers. It is to place the meson theory on a more firm foundation that Dr. Yukawa has worked out his newest theory, the first part of which was published yesterday.

"This first part," Dr. Yukawa said, "provides the framework for a new concept of the elementary particles. It does not yet contain the picture." This will come later, he said.

Any theory explaining the nature of the elementary particles, Dr. Yukawa pointed out, must satisfy both the relativity and the quantum theories, the two basic pillars on which all our understanding of the material universe is based.

A particle that conforms to the demands of both the special relativity and the quantum theories turns out to possess the properties of charge and mass, but, strangely enough, cannot have any dimension in space, being merely what is known as a "point in a local field."

A Scientific Stumbling Block

This particle without dimensions in space has become one of the greatest stumbling blocks in working out a satisfactory theory to explain the nature of all elementary particles and the forces holding them together. For such a dimensionless particle leads to the inevitable conclusion that all fundamental constituents of matter have an infinite self-energy.

This troublesome infinity of energy has been haunting theoretical physicists like a "cosmic ghost," lending to all their theorizing an aspect of unreality.

It is this hitherto seemingly unsurmountable difficulty that Dr. Yukawa believes he has overcome after two years of effort. His new theory, he said, makes it possible to conceive of elementary particles that possess not only charge and mass but also dimensions in space, while at the same time they fulfill the requirements of the relativity and quantum theories.

Such particles, he declared, do not have infinite energy, thus conforming with observed facts and "driving away the cosmic ghost."

"This problem of infinity," he said, "is a disease that must be cured. I am very eager to be healthy."

In his next paper, he said, he will deal with the interaction of such particles of finite energy, in what he terms "non-local fields."

A colleague of Dr. Yukawa at Columbia said the new theory, if proved successful, would "lead to a much more realistic description of the physical universe than previous theories."

January 25, 1950

Einstein Offers New Theory To Unify Laws of the Cosmos

$$g_{ik;l} = 0, \quad \Gamma_i = 0$$
$$R_{ik} = 0, \quad R_{ik,l} + R_{kl,i} + R_{li,k} = 0$$

Einstein's latest equations for a United Field Theory. These formulas are known as tensors. They are highly condensed mathematical shorthand, representing relationships between the forces of gravitation and electromagnetism in their relationship to space, time and physical forces.

By WILLIAM L. LAURENCE

Albert Einstein, named by George Bernard Shaw as one of the eight "Universe Builders" in recorded history, has returned from a three-year sojourn on the lonely summit of his scientific Sinai with a new set of laws for the cosmos.

These laws, embodied in a few mathematical formulas, will, he believes, reduce the physical universe in its totality to a few simple, fundamental concepts that will unify all its multifarious and seemingly unrelated manifestations into one all-embracing intellectual synthesis.

He calls this all-embracing concept, which he has been seeking with the consecrated devotion of a high priest of science for more than half of his seventy-four years, a Unified Field Theory. He hopes it will provide the master key to the many mansions of the physical universe, and ultimately permit admittance to its very sanctum sanctorum.

Dr. Einstein's latest concepts of the fundamental laws governing the cosmos are published today by the Princeton University Press as an appendix to the fourth edition of his famous book, "The Meaning of Relativity," originally published in 1922. The appendix, headed "Generalization of Gravitation Theory," is a radically revised version of an appendix to the third edition of the book, published in 1950.

His concept of 1950, he says, left one serious difficulty to be solved. This "last step in the theory," he adds, "has been fully overcome in the last few months."

In his quest for a new understanding of the fundamental laws governing the cosmos Dr. Einstein has searched for simple unifying principles underlying the multifarious phenomena in which the material universe manifests itself. In his special and general theories of relativity, published in 1905 and 1916, respectively, which brought about the greatest intellectual revolution since Newton, he united matter and energy, space and time, gravitation and inertia, all considered as individual and unrelated entities, into one all-embracing cosmic concept.

This synthesis, wrote Bertrand Russell in 1924, "is probably the greatest synthetic achievement of the human intellect up to the present time."

"It sums up the mathematical and physical labors of more than 2,000 years," he continued. "Pure geometry from Pythagoras to Riemann, the dynamics and astronomy of Galileo and Newton, the theory of electromagnetism as it resulted from the researches of Faraday, Maxwell and their successors, all are absorbed with the necessary modifications, in the theories of Einstein."

However, having achieved the intellectual synthesis of matter and energy, space and time, gravitation and inertia, Dr. Einstein was confronted with two concepts that defied unification. The material universe, as the theory of relativity conceives it, is composed of two major fields—the field of graviation and the electromagnetic field.

It was this synthesis of the gravitational field and the field of electromagnetism that Dr. Einstein has been seeking, a will-o'-the-wisp that kept eluding him. Three years ago, he had it almost within his grasp, having overcome all the obstacles but one. He now is convinced that he finally has overcome this last obstacle and thus has attained the crowning achievement of his life's work.

"The solution of the problem," Dr. Einstein writes, "appears to me to be highly convincing, although, due to mathematical difficulties, I have not yet found a practical way to confront the results of the theory with experimental evidence."

It took more than forty years before the Einstein synthesis of matter and energy as fundamentally two different manifestations of a single cosmic entity could be demonstrated experimentally on a large scale. This was done on the desert of New Mexico in the early morning of July 16, 1945, when the explosion of the first atomic bomb converted one gram of matter into the explosive energy equivalent to 20,000 tons of TNT. Similarly, it took several years before the predictions of the general theory of relativity could be verified by observations of a total eclipse of the sun.

Behind Dr. Einstein's quest for a Unified Field Theory, which most of present-day physicists regard as unattainable, lies one of the greatest intellectual schisms in the history of science. Dr. Einstein believes the physical universe is one continuous field, like an endless stream. The majority of modern-day physicists, on the other hand, champion the quantum theory, which holds that the universe is discontinuous, made up of particle and quanta (atoms) of energy.

According to the quantum theory, the physical universe is dual in nature, everything in it being both particle and wave. The quantum theory, of which Dr. Einstein himself was one of the principal founders, has as one of its keystones the Heisenberg Uncertainty Principle, according to which it is impossible to predict individual events, so that all knowledge is based on probability and thus at best can be only statistical in nature. The Uncertainty Principle has led, furthermore, to the universal acceptance by present-day physicists (with the exception of Dr. Einstein) that there is no causality or determinism in nature.

Dr. Einstein alone has stood in majestic solitude against all these concepts of the quantum theory. Admitting that it has had brilliant successes in explaining many of the mysteries of the atom and the phenomena of radiation, which no other theory has succeeded in explaining, he maintains that the theory of discontinuity and uncertainty, of the duality of particle and wave, and of a universe not governed by cause and effect, is an incomplete theory, and that eventually laws will be found showing a continuous, non-dualistic universe, governed by immutable laws, in which individual events are predictable.

"I cannot believe," he says, "that God plays dice with the cosmos!"

He admits freely that his present field theory does not find room in the universe for the atom and its component particles, which he describes as "singularities in the field." But he is equally confident that the field theory is the only approach to a well-ordered universe, and that eventually it will find room for the "enfant terrible" of the cosmos—the atom and the vast forces within it.

"I must explain," he writes, "why I have gone to so much trouble to arrive at this result.

"The contemporary physicist cannot, without such an explanation, appreciate this; for he is convinced, as a result of the successes of the probability-based quantum mechanics, that one must abandon the goal of complete descriptions of real situations in a physical theory.

"One cannot keep side by side the concepts of field and particle as elements of the physical description. The field concept requires freedom from singularities, while the particle concept [of the quantum theory, as elementary concept], is a singularity in the field. The field concept, however, seems inevitable, since it would be impossible to formulate general relativity without it. And general relativity is the only means to avoid the 'no-thing' inertial system.

"For this reason I see in the present situation no possible way other than a pure field theory, which then, however, has before it the gigantic task of deriving the atomic character of energy.

"We are separated by an as yet insurmountable barrier from the possibility of confronting the theory with experiment. Nevertheless, I consider it unjustified to assert, a priori, that such a theory is unable to cope with the atomistic character of energy."

When Mr. Shaw in 1930 named Dr. Einstein as one of the eight "makers of the universe" he listed the seven others as Pythagoras, Aristotle, Ptolemy, Copernicus, Galileo, Kepler and Newton.

March 30, 1953

PHYSICS AND ASTROPHYSICS

Einstein's Cosmos Equations Solved

Czech Refugee Finds Electromagnetism Is Basis of Universe

By WILLIAM L. LAURENCE

A mathematical map that promises to open roads through the intellectual jungles of the cosmos, which Prof. Albert Einstein has been seeking to penetrate for more than thirty years, was outlined yesterday at Indiana University, Bloomington, Ind., by Prof. Vaclav Hlavaty, one of the world's select group of experts on the highly involved mathematics of multi-dimensional space.

Professor Hlavaty (pronounced La'vaty), a refugee from Communist Czechoslovakia, is a member of Indiana's Graduate Institute for Applied Mathematics.

In working out his pioneer map, of which much detail still remains to be filled in, Professor Hlavaty has achieved solutions to the equations in Professor Einstein's 1953 mathematical model of the universe, which Professor Einstein himself described as "gropings in the dark" and Erwin Schroedinger, one of the world's top mathematical physicists, described as "next to impossible" of solution.

The solutions, Professor Hlavaty said, lend themselves to physical interpretation, and this opens the possibility at least to devise experiments for testing Professor Einstein's latest Unified Field Theory, which Professor Einstein admitted he had so far been unable to do.

The solutions of the Einstein equations reveal, Dr. Hlavaty said, that electromagnetism is the basis of the universe. This goes one step further than Professor Einstein had anticipated when he sought to unify gravitation and electromagnetism, the two principal forces in which the universe manifests itself, under one set of mathematical laws.

Outgrowth of Relativity Theory

According to Dr. Hlavaty's solution of the Einstein equations, electromagnetism is the basis of all cosmic forces—gravitation, as well as matter and energy, being built up out of the all-pervading all-embracing electromagnetic field.

The electromagnetic field, Dr. Hlavaty's solutions of the Einstein equations reveal, grows logically out of the geometrical properties of the four-dimensional space-time manifold of the Einstein General Relativity Theory.

According to the original version of that theory, only gravitation grew out of the geometry of the four-dimensional curved space-time continuum, electromagnetism somehow not being accounted for in the theory. Furthermore, gravitation, according to the original theory published in 1916, depended on the presence of matter, no gravitation being possible without matter.

Dr. Hlavaty's solutions have revealed the need for drastic changes in the original concepts of general relativity. Instead of two prime cosmic fields, gravitation and electromagnetism, flowing parallel to each other like two mighty streams originating from one source, there is now only one great ocean of electromagnetism, out of which and into which flow the streams of gravitation, matter and energy.

Thus instead of unifying gravitation and electromagnetism as offspring of a common ancestor, the latter becomes the progenitor of the former.

New Possibilities Held Opened

The solution show, Dr. Hlavaty said, that it is possible to have gravitation without matter, a concept not permitted under the original general relativity theory. They also reveal that it is possible for space to exist without gravity and without matter, yet it still would be governed by the electromagnetic field.

In announcing his latest version for a Unified Field Theory last March, Dr. Einstein had stated that he "had not yet found a practical way to confront the results of the theory with experimental evidence." Professor Hlavaty's solutions, Dr. Einstein said yesterday, open the possibility for just such confrontation.

"'If this is true,'" Dr. Hlavaty quoted Dr. Einstein as having said, "'this is the most important thing.'" Professor Einstein, according to Dr. Hlavaty, had told him that "your theory is a definite improvement of my theory." However, Dr. Hlavaty added, "Einstein is a genius while I am only a mathematician."

The corroboration by experiment of Professor Einstein's formula in his special relativity theory of 1905, which revealed that matter and energy were equivalent, and that one gram of matter was the equivalent of 25,000,000 kilowatt-hours of energy, laid the foundation for the development of the atomic bomb and the promise of vast industrial power from atomic energy.

Radio's Development Recalled

Similarly the confirmation by experiment of James Clerk Maxwell's equations, which revealed that light was electromagnetic in nature, led to the discovery of radio waves and to the age of radio and television. Many other similar instances in the history of science may be cited.

Hence the opening of the possibility of devising physical experiments for testing Professor Einstein's latest equations, as elaborated on by Professor Hlavaty, justifies expectations of further revolutionary discoveries to result if the experiments prove the theory correct.

Successful tests may lead to the discovery of cosmic forces at present entirely unsuspected, just as the existence of atomic energy was unknown some fifty odd years ago. They may also lead to the utilization and harnessing of known forces such as the universal gravitation.

Professor Hlavaty hopes that his method may at last build a bridge linking the two major theories of the universe—the relativity and the quantum.

The former deals with the enormous scale of the forces interacting in the universe at large through all the ages, while the latter deals with the minute scale of forces interacting within the infinitesimal spaces of the nuclei of atoms in fraction of fractions of millionths of a second.

At present, there is no link between the theory relating to the universe of the galaxies in the four-dimensional space-time manifold and the universe of the nuclei of atoms of which the galaxies are constituted. The universe, according to relativity, is governed by cause and effect, whereas, according to the quantum theory, chance governs all, a concept that led Professor Einstein to remark that he "could not believe that God played dice with the cosmos."

Spinors Used in Solution

Professor Hlavaty said that his solutions for the equations of the Einstein Unified Field Theory make use of spinors, a mathematical tool also used in quantum mechanics. This leads him to believe that the spinor theory will serve as the first arch of a possible bridge linking the unified field theory, which also embraces relativity, with the quantum theory. He is now attempting to formulate such a bridge.

Professor Schroedinger, now at the Institute for Advanced Study in Dublin, Ireland, wrote that he was "full of admiration for the skillfulness with which Dr. Hlavaty handled this obstreperous problem," according to a member of the University of Indiana mathematics faculty.

Born in Czechoslovakia, Dr. Hlavaty served on the faculty of Charles University, Prague (where Professor Einstein had also served at one time), from 1927 to 1948. He was elected to the Czech Parliament in 1946, after he had been asked to enter politics by the late President Eduard Benes, because so many statesmen had been killed by the Nazis during their occupation of the country. He barely escaped with his life when the Communist putsch took place.

Escaping to France, he served as Professor of Mathematics at the Sorbonne for one term in 1948. In the fall of that year he joined the Indiana University mathematics department. The Graduate Institute for Applied Mathematics, of which he now is a member, was founded in 1950.

"It took two years of an upset stomach," Professor Hlavaty said yesterday, "before I came upon the right solutions of Einstein's equations. It was the best detective story I have ever experienced."

July 30, 1953

Two Theories Offered as Clues to All Matter

Table for Particles and Basic Formula Are Discussed

By ROBERT K. PLUMB

Two ideas that might give man a firmer understanding of the nature of the physical world about him were reported by scientists yesterday.

In Berlin, at a session celebrating the 100th anniversary of the birth of the physicist Max Planck, Prof. Werner Heisenberg presented what he called a "suggestion for the basic equation of matter."

The equation was reported to have been formulated from three measurement units of modern physics and some assumptions about symmetry of the universe that date back to Plato.

In New York last night, at a lecture arranged by the New York Academy of Sciences, 2 East Sixty-third Street, Dr. John J. Grebe, director of nuclear and basic research for the Dow Chemical Company, Midland, Mich., proposed a "periodic table for fundamental particles" that he said might help explain the "structure of the universe."

Both Professor Heisenberg and Dr. Grebe entered modest disclaimers that their efforts might reveal to man exactly what it is he stands on or looks at. But both efforts were identified as attempts to further understand the structure of the matter that composes the universe.

This is an ancient philosophic problem that has been made more complicated instead of easier, although much more precise, by discoveries recently made in the gigantic accelerators with which modern physics breaks up pieces of atoms to see what they are made of.

The most powerful accelerator performing scientific tests of matter now is the Bevatron of the University of California in Berkeley. An accelerator twice as powerful has been built in the Soviet Union, but it is reported not yet ready for experiments. Under construction here is a bigger machine to be in operation in two years.

The Berkeley Bevatron accelerates atomic particles to energies up to 6,200,000,000 electron volts (6.2 BEV) and smashes them into targets. In the smash-up some of the energy put into the machine is transmuted to matter in a relation the reverse of that expressed by Einstein's expression $E=Mc^2$.

Matter is destroyed in the Berkeley experiments, anti-matter is created, and a series of "strange" particles, not all of which have been seen in nature, have been obtained. The "strange" particles have been studied as they break up—some of them in a fraction of a millionth of a second. Their modes of decay and their masses have been measured many times.

28 Particles Arranged

The report here last night by Dr. Grebe, a physical chemist, was an attempt to explain how twenty-eight of these so-called "fundamental" particles might be arranged into a table akin to the table of the chemical elements that Mendeleyev achieved in the nineteenth century. Dr. Heisenberg's report in Berlin was reported to be an attempt to fit the known particles into one mathematical system in which the known characteristics of the particles and their observed actions could be accounted for and predicted.

Professor Heisenberg received the Nobel Prize in 1932 for the "uncertainty principle," a rule of physics that holds that events in the sub-atomic world cannot be observed without destroying them.

The efforts by Dr. Heisenberg and Dr. Grebe were not related, although they have a common ultimate goal: understanding of the structure of matter. Physicists and mathematicians studying evidence from accelerators and from cosmic rays up to this point have been in the position of one who looks at the unassembled pieces of a jig-saw

Dr. John J. Grebe and part of his periodic table for fundamental particles, possible key to matter's structure.

This other equation is believed close to one discussed by Dr. Werner Heisenberg in Berlin, on structure of matter.

The New York Times

puzzle. The pieces do not make sense. Only when they are assembled can a rational grasp of the story the pieces tell be attained.

3 Measurement Units Used

Professor Heisenberg's theory, according to reports from Congress Hall in West Berlin, encompasses three units of measurement. They were the "quantum of action" that Planck found to be a small indivisible unit essential in calculations about energy, the constant velocity of light that Einstein believed to be a universal constant, and a new Heisenberg concept, a measure such as the diameter of a simple atomic center.

Details of the equation connecting these three through known laws were not disclosed publicly by Professor Heisenberg yesterday. However, his mathematics has been the subject of intensive correspondence among theoretical physicists. The trend of Professor Heisenberg's work, it is reported, has been clear for some time. From this information, physicists in the United States tend to doubt the universal application of his equations and to doubt that they may sometime result in a "unified field theory in the sense of the efforts of Einstein," which Professor Heisenberg has declared to be his goal.

In a classical sense, a unified field theory would be an equation in which all forces and bodies in the universe could be stated in one simple proposition. Weak forces such as gravity and strong forces such as those that bind atoms together would both be accounted for in a unified field theory. The behavior of all types and sizes of matter that can be liberated from natural atoms, or created in accelerators, would be explained.

This is hard, it is pointed out, because many new particles have been found recently in accelerators as well as in cosmic ray studies. Cosmic rays bombard the earth with particles a billion times more energetic than the best particles produced in accelerators by man. More complete smash-ups have been found in cosmic ray studies, although accelerators provide large quantities of high energy particles for repeatable scientific studies. Thus the discovery of a new particle often requires that the theory be changed.

Up to this time, physicists in this nation have said, Professor Heisenberg has been vague about precisely how his theories would take into account some of the new "strange" particles. In fact, one objection to the Heisenberg suggestion, voiced by several, is that "strangeness cannot be put into a degenerate vacuum." These words mean that the complex mathematical model earlier proposed by Heisenberg is not capable of explaining all that is observed in accordance with rules that suggest how matter must behave.

How this model might have been improved or modified in Professor Heisenberg's latest work was not clear here last night. His ideas have met a strong current of doubt, however, doubt that was not mitigated by his acknowledged earlier accomplishments.

Here at the Academy of Sciences, speaking before a special audience of the academy's sections on mathematics and engineering, physics and chemistry, Dr. Grebe suggested that the twenty-eight accelerator-produced new particles might all be multiples or combinations made up of two simple particles, the electron, with a negative charge, and the positively charged electron, or positron, a recent entry in problems of nuclear physics.

(The positive-charged electron, the anti-neutron and the anti-proton, all recently produced, might join together to form in theory at least "anti-matter" which would cause a mutual annihilation when it came near ordinary matter made up of negative electrons and ordinary protons and neutrons.)

Dr. Grebe suggested that pairs of charged electrons and positrons might be the common denominator of all the neutral particles. The mass of the particles would vary as suggested in Einstein's relativity as the particles moved faster.

He found between fundamental particles (whose properties are known) unique relations that can be expressed as a ratio of pi divided by 4. (Pi is a constant, the ratio of the circumference of a circle to its diameter.) Dr. Grebe reported that the masses of many of the known fundamental particles appear to be simple multiples of electron masses when related to pi.

Further, Dr. Grebe suggested that gravity itself might be an electromagnetic force accountable in electromagnetic terms.

"It will then be easy to understand why gravity cannot be shielded and yet operates with the laws of electromagnetic attraction," he said. "For still more complete and realistic explanation, one can express the entire process with electro-static equations of forces.

Picture Becomes Familiar

"The whole picture of the conversion of energy to matter by pair formation known for thirty years and the annihilation of matter into energy so ably demonstrated at Berkeley during the last few years becomes familiar because of our long association with the production and decay of elemental charges. It gives one great confidence to know that the only divisible matter we have dealt with appear to be the electron and positron, 1836.12 times lighter in mass that that paragon of stability, the proton, the nucleus of hydrogen.

"I know that there have been a number of attempts to explain the structure of matter in terms of electrons and positrons as the building blocks. They have been discarded. The same might happen to this theory. However, the mathematical relations discovered cannot help but remain and be a useful step forward."

These views were echoed by Dr. Lyle Borst, chairman of the New York University Physics Department and chairman of the New York Science Academy's section of mathematics and engineering. Dr. Borst said:

"Dr. Grebe has gone so far beyond anyone else—he has taken such a tremendous leap into the unknown—that it is not possible to predict at this time where this will lead. It may result in reconciliation of the field and quantum theories. It may supplant them both. It may, as Dr. Grebe pointed out in his paper, be discarded. But, even if discarded, it will have contributed to the eventual explanation of the structure of matter."

April 26, 1958

Physicists Offer New Theories on Gravity Waves and Atomic Particles

By ROBERT K. PLUMB

Shirt-sleeved physicists sat cross-legged in the aisles of the crowded New Yorker Hotel grand ballroom yesterday afternoon to hear two reports that appear to be of classic significance. In the first paper, Prof. Paul A. M. Dirac of Cambridge University proved mathematically that gravitational waves existed and that they could carry energy. "One might find ways of observing them in the future," he said. In the second paper, Dr. Robert E. Marshak proposed a universal rule that he believed might explain the interactions among the twenty-nine mysterious and fleeting elementary particles that physicists have caught in cosmic ray and cyclotron experiments over the last quarter of a century. Many major achievements in man's attempts to understand the nature of the material world constitute the background against which the physicists listened to two reports:

Physicists know about four forces. A universal rule to account for gravitational forces was laid down by Sir Isaac Newton three centuries ago. It says the same law applies to the fall of an apple, the motion of the earth about the sun, or the motion of a satellite about the earth.

Electromagnetic forces were unified a century ago by James Clerk Maxwell. His universal rule explained much about the propagation of light and led to the discovery of radio transmission.

Dr. Marshak proposed a universal rule to explain the weak forces that cause spontaneous disintegration (called decay) of some of the twenty-nine "strange particles" found when atoms are smashed.

A fourth force, nuclear force, which binds atoms together under ordinary circumstances, remains without a "universal rule" upon which science can build.

Nuclear Force Is Strongest

The strongest force of the four is nuclear force. If it were equal to one, the next strongest force, electromagnetic force, would be weaker by a factor of 100 in analogous conditions. The "weak" forces with which Dr. Marshak dealt would be weaker still by a factor of ten followed by thirteen zeros.

Gravitational forces in the ordinary sense of attraction between masses (such as the apple and the satellite and the earth) would be weaker still by a factor of ten followed by more than forty zeros.

Professor Dirac won the Nobel Prize in Physics in 1933 with Erwin Schroedinger for a mathematical suggestion that positive electrons must exist. These were found in 1936 by Dr. Carl D. Anderson and V. G. Hess. This discovery of anti-matter completely changed science's view of the nature of the atom.

Professor Dirac, who has been at the Institute for Advanced Study in Princeton, N. J., since September, has given several mathematical reports on the existence of gravitational waves. The equations are understood by few.

Uses Einstein Theory

As Professor Dirac's ideas were interpreted by colleagues here for the American Physical Society's annual meeting, he has applied mathematics fruitful to quantum mechanics to gravity fields.

He starts with the Einstein Theory of General Relativity, drops out coordinates from the space-time system proposed by Einstein, and demonstrates that a moving mass should produce gravitational waves, just as a moving electrical body produces electromagnetic waves. The gravitational field carries energy away from the moving mass.

Gravity is a very weak force —although it is usually multiplied by large masses such as

The New York Times

Dr. Robert E. Marshak with formulas and other data he used to illustrate his proposed universal rule to explain the weak forces that cause the spontaneous disintegration (or decay) among the twenty-nine particles that are found when atoms are smashed.

that of the earth. The earth itself moving about the sun at a speed one one-thousandth that of the speed of light would create only one gravitational wave in a year. The earth-created wave would have a length of one light year.

Gravitational waves are so weak, according to interpretations of Professor Dirac's idea, that ordinary electromagnetic waves are stronger by a factor of ten followed by forty zeros. This is ten thousand trillion trillion trillion.

Professor Dirac proposed that gravitational wave units be called gravitons. Einstein suggested the existence of gravitational waves, but now it has been established that they can carry energy, so they might be detected.

Present-day experimental techniques cannot observe gravitational waves, nevertheless some experiments have been proposed. One physicist suggested that the present Federal budget could not stand to pay for an experiment, but another said that the Federal income tax would merely have to be raised a little. Other, possibly easier and cheaper, means of detecting gravitons are being studied.

Importance Is Cited

Professor Dirac said he believed that his postulation at this time was in the same category as his postulation of positive electrons a quarter of a century ago. Another leading theoretical physicist said he believed it was likely to be one of the most important of the century.

Dr. Marshak's universal theory of weak interactions was evolved at the University of Rochester. He worked with Dr. E. C. G. Sudarshan of the Tata Institute of Fundamental Research in Bombay, India. Dr. Sudarshan, now at Harvard, was a graduate student of Dr. Marshak's.

As he outlined his theory, physicists now have a list of twenty-nine particles and anti-matter particles that have been carefully studied in accelerator and in cosmic ray researches. Six of these are now classified as "leptons," seven are classified as "bosons" and sixteen heavier particles are called "baryons." Another fundamental particle is the light unit, or photon, which is not included in the list of particles with weak interactions.

The known particles and their anti-matter companions unite and break up in ways for which there has been up to this time no explanation.

Existence Is Varied

Some of the particles exist only a fraction of a billionth of a second. Some exist for 1,000 seconds. Some have no mass at rest, others weigh several times as much as an electron. The decay of what is now known as a K meson into either two or three pi mesons led Dr. Chen Ning Yang and Tsung Dao Lee of Columbia to question in 1956 whether parity was conserved in weak interactions. Later experiments gave a negative answer.

In the absence of a universal rule to explain the reactions and the relations between twenty-five elementary particles that decay and four that are stable, many physicists have resorted to what amounts to nearly a botanic approach—they have attempted to classify the particles into groups with common characteristics.

Dr. Marshak, in explaining his theory at a press conference, said that the theory held that parity broke down in all decay processes involving weak forces, that charge conjugation broke down in all decay processes and that time reversal was maintained.

He said the theory could predict the decays of the particles and the modes of decay. It provides, he said, a complete mathematical explanation for weak interactions.

Prof. Paul A. M. Dirac, who proved mathematically that gravitational waves exist.

The new theory, called "V-A", for vector and axial-vector interaction, is in accordance with important recent experimental results in studying interactions of the particles, he said.

Sometimes nuclear forces — much bigger than the weak forces — do enter into particle interactions, Dr. Marshak said. But nuclear forces appear to change the V minus A only a small amount and not by the enormous amount that might be expected. Corrections for nuclear forces can be put into the V-A expression, Dr. Marshak said.

In his formal report, Dr. Marshak pointed out that the new theory of weak interactions provided a universal rule for three of the four forces. A theory of nuclear interactions is missing. And the ultimate quest is for a theory that will unite the four forces into a single master theory.

Perhaps more than twenty-nine elementary particles will eventually be found, he pointed out. But the theory consolidates what is now known, he said.

Mystery Remains

A mystery remains, Dr. Marshak reported. More modes of decay are known than there are particles and anti-particles. Not all the possible decays take place. Some decays do not take place although they are consistent with conservation laws. The new theory explains why decays take place, but it does not explain why other decays do not take place.

"V-A correlates an enormous amount of experimental data in a rather successful way," Dr. Marshak told the packed and hushed ballroom.

Earlier, Dr. Marshak had reported that the new theory provided one explanation for a series of phenomena that were observed in the elementary particles. Just as Newton's gravitational theory explained the motion of the apple and the motion of the earth—established that they follow the same universal law—so the new idea gathers the mysterious particles together, he noted.

About 3,500 physicists and physics teachers are attending the annual four-day meeting of the American Physical Society here.

January 31, 1959

Particle Tests Seem to Confirm Goal of Unified Physics Theory

By WALTER SULLIVAN

As a result of experiments conducted in Switzerland and apparently confirmed in Illinois, a number of physicists believe that two of the basic forces of nature—electromagnetism and the less-well-known "weak" force responsible for radioactivity—may be expressions of the same phenomenon.

This would be an important step toward the long-sought goal of a "unified field theory," relating the four seemingly diverse forces that control all processes in physics, chemistry and biology.

While the significance of the findings is still being debated, they are in line with the predictions of a controversial attempt to formulate a theory linking the two forces. They also run counter to the predictions of what, until now, had been the generally accepted theory.

The original experiments were done at the European Nuclear Research Center, known as CERN, near Geneva. Fifty scientists from laboratories in Belgium, Britain, France, Germany and Italy took part. An announcement from the center describes the results as "astonishing."

Two Forces Are Familiar

Two of the four recognized forces in nature act at a distance and are familiar to almost everyone. One is gravity, which makes things fall toward a massive object, like the earth; the other is electromagnetism, which drives electric motors and, because it is the force that binds atoms to one another, lies at the basis of chemistry.

The two other forces act only over the short distances involved in close encounters between atomic particles. One is the extremely powerful force that binds together particles of the atomic nucleus—the so-called "strong" force. The other manifests itself by ejecting particles from an atom in what is called radioactive decay. That is the "weak" force.

PHYSICS AND ASTROPHYSICS

Until now all four forces have seemed isolated from one another, not only in terms of the wide gaps in their relative strengths but also in the theoretical framework used to describe each of them.

Gravity is a million billion billion times weaker than the weak forces. The latter, in low energy encounters, is 10 billion times weaker than electromagnetism, which in turn is a thousand times weaker than the strong force.

The CERN findings concerning a possible identity between the weak force and electromagnetism were supported by experiments conducted in a different manner at the National Accelerator Laboratory near Batavia, Ill., using the world's most powerful atom smasher.

Six Weeks of Added Tests

An additional six-week run of tests is being undertaken at Batavia and a new, more revealing set of experiments is also planned at CERN. Other laboratories, as well, are prepared to test the findings.

It was a century ago this year that James Clerk Maxwell published his classic "Treatise on Electricity and Magnetism." It showed that electricity, magnetism and the speed of light were all manifestations of the same phenomenon — electromagnetism. Upon this theoretical foundation, much of modern science was built.

If it is now shown that electromagnetism and the weak force (including radioactivity) are also manifestations of the same basic phenomenon, the finding will be of historic importance.

The new observations have evoked excitement on both sides of the Atlantic because they had been predicted by the controversial effort to bring the weak and electromagnetic forces into a single theoretical framework. While the history of this theory is complex, it is largely associated with Dr. Steven Weinberg of Harvard University.

In an account of the CERN experiment in the Oct. 4 issue of the New Scientist, Dr. Fred Bullock, who led those participants from University College, London, cites a statement Dr. Weinberg made 18 months ago when he was at the Massachusetts Institute of Technology.

'It Smells Right'

"Right now there's not a grain of experimental evidence that this general idea is right," said Dr. Weinberg. "But it solves so many theoretical problems all at once that it smells right."

"It appears," comments Dr. Bullock, "that Weinberg's olfactory sense was functioning with exceptional efficiency."

The importance of the recent work, Dr. Bullock adds, "is that it provides dramatic confirmation of a unified field theory which describes the weak and electromagnetic interactions as different aspects of the same force law."

Assessing the findings in a recent editorial, the British journal Nature is not quite so unequivocal. It says, however, that if the theory is confirmed, it will in fact represent such a unification. Other theorists in recent interviews also said that the findings, while conforming to the Weinberg prediction, did not necessarily prove his formulation of the theory to be correct.

The theory had predicted that the weak force, like the electromagnetic force, would sometimes produce a "neutral current" in interactions between atomic particles. It is this unexpected behavior of the weak force that has apparently been observed.

The term "current" refers to the nature of transformations that occur during collisions between very light, or weightless, particles ("leptons") and heavier particles ("hadrons"). The leptons include the electron (which is negatively charged), the positron (similar, but positively charged), the muon (similar to the electron but heavier) and the elusive neutrino.

The latter apparently has no mass, no electric charge, and always travels at the speed of light.

The two particles of the atomic nucleus — proton and neutron—are both hadrons.

Shower of Particles

When a lepton bangs into a hadron the resulting shower of particles always contains a lepton as well as one or more hadrons. If the electric charge on the lepton that impinges is identical to the charge on the resulting lepton, this represents a "neutral current."

On the other hand, when the resulting lepton differs in electric charge, the transformation is a "charged current." Similar terminology is used for the hadrons.

If the impinging lepton is a neutrino, its interaction with the hadron is controlled only by the weak force. In such a case, according to conventional theory, the reaction should always be through a charged current. In other words, whereas the impinging lepton (a neutrino) is neutral, the emerging lepton should be charged.

On the other hand, if the initial lepton is charged (for example a positron), the electromagnetic force is in control and a similarly charged particle should emerge, constituting a neutral current. This seemed to define a fundamental difference between the weak and electromagnetic forces.

Theory Prediction

The new theory predicted that the weak force could also produce a neutral current, and confirmation has now been reported.

This had never been observed because neutrino experiments are so difficult. Since neutrinos only interact when they impinge very squarely on a hadron, they can go through the entire earth 100 billion times with only a fifty-fifty chance of interacting.

Hence neutrino experiments are possible only with a machine that can generate enormous numbers of high energy neutrinos and fire them into a very large tank of dense fluid. The accelerator at CERN can produce pulses of a billion neutrinos (or antineutrinos) every two seconds, firing them into a giant tank, or "bubble chamber," of fluid.

The beam is purged of all particles except neutrinos by firing it through a steel wall many yards thick, formed from the Swiss national strategic reserve of steel. (At Batavia the beam traverses more than a half-mile of dirt.)

Other Reactions Recorded

The bubble chamber is known as Gargamelle. As noted in the Nature editorial, Gargamelle was Gargantua's mother. "She had a dreadful and Rabelaisian labour before bringing her great child into the world," it said. "In this case the labor has also been long, not because the offspring was caught up in her internal workings but because the data were embedded in background."

The "background" consisted of other reactions generated, for example, by neutrinos hitting the chamber walls. Some mimicked the neutral current phenomenon but are not believed to have had an important affect on the results.

Gargamelle was filled with liquid freon but in the next round of experiments will be filled with propane. This, it is expected, will show more unequivocally the nature of neutrino-proton interactions.

The Batavia experiments not only seem to have confirmed the neutral current findings but also to have shown another similarity between electromatic and weak force phenomena. The neutrino beam—even more intense and energetic than the one at CERN—was fired into

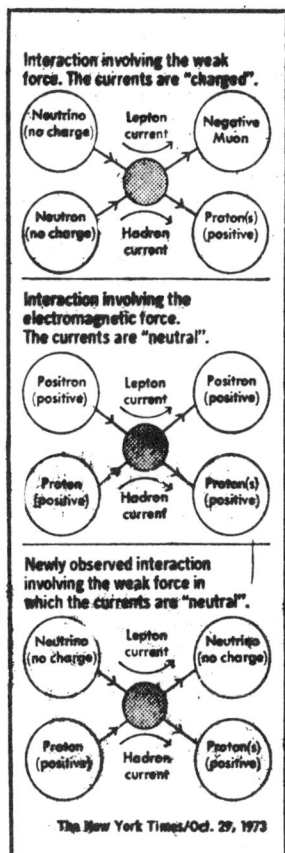

Two Types of Particles in Collision

The experiments that have suggested a relationship between two fundamental forces, electromagnetism and the so-called "weak" force of radioactive decay, involved an analysis of collisions between two types of atomic particles.

They were light-weight or weightless ones, such as neutrinos, positrons and muons (known collectively as leptons) and heavier ones, such as protons and neutrons (known collectively as hadrons.)

When the impinging lepton was a neutrino, with no electric charge (top diagram), the force controlling the outcome was the weak force and it was assumed that the resulting lepton would always differ in electric charge.

When the impinging lepton was charged, as with the positron in the middle diagram, the electromagnetic force was in control and the resulting lepton was identical in charge. Since there was no change in the charge, this was a "neutral" current typical of electromagnetic reactions.

Experiments in Switzerland and Illinois indicate that neutrino reactions, involving the weak force, may also produce a "neutral" current (bottom diagram), and this suggests a similarity between the two forces.

The New York Times/Oct. 29, 1973

77

100 tons of a substance resembling mineral oil.

Detectors then recorded the spray of particles that emerged when an extremely high energy neutrino hit a proton in the fluid. By recording the manner in which particles scattered off protons under this neutrino bombardment, it was possible to "look" at the internal structure of the proton in terms of the weak force.

Earlier experiments with the Stanford Linear Accelerator in California, in which protons were bombarded with electrons, had shown proton structure in terms of the electromagnetic force.

Proton 'Looks' the Same

This revealed points of electric charge within the proton—a discovery that led many physicists to believe the proton is formed of constituents that some call "partons."

Now, as noted last week by Dr. Laurence Sulak at Batavia, it appears that the proton "looks" the same in terms of the weak force, revealed by neutrino bombardments, as it did under electron bombardment. This again suggests a relationship of the two controlling forces.

The Batavia experimenters are led by Drs. Carlo Rubbia of Harvard, Alfred K. Mann of the University of Pennsylvania and David Cline of the University of Wisconsin.

A peculiar feature of the weak force is that it becomes stronger when interacting at higher energy. Thus in the Batavia experiments the weak force, instead of being 10 billion times weaker than electromagnetism, is only 10 times weaker. This has aroused curiosity as to what will happen at energies sufficiently high to equate the two forces.

The Weinberg theory has roots in earlier proposals by Abdus Salam, now head of the international Center for Theoretical Physics in Trieste, and others.

As formulated by Dr. Weinberg in 1967, the equations led to certain improbable predictions, involving infinities. Gerhard 't Hooft in Utrecht and Benjamin W. Lee at the State University of New York in Stony Brook helped weed out these difficulties.

In the CERN and Batavia experiments the ratio of neutral current neutrino events was about one to four. Dr. Bullack, commenting on the findings, said: "It is gratifying, even surprising at this early stage, that the observed events occur at much the rate required by the new theory."

Efforts to incorporate the forces of nature into a unified theory were attempted by theorists of such stature as Albert Einstein and Werner Heisenberg. However, according to specialists in the field, this is the first time confirmation by experiment has been possible.

October 30, 1973

Experimental Findings Challenge Accepted Theories on Atomic Physics and Cause Confusion in Science

By WALTER SULLIVAN

Experiments with new and more powerful research tools, some of them described for the first time last week at a meeting of the American Physical Society in Washington, have thrown into turmoil science's efforts to comprehend nature on its most basic level.

In the words of Dr. Wolfgang K.H. Panofsky, president of the society, who is director of the Stanford Linear Accelerator in California, the new findings there and elsewhere have led to "a state of maximum confusion" in the world of physics.

Yet he and others believe that these problems can be resolved by further pursuing the new avenues of experimentation that have produced the recent surprises about the behavior of atomic particles.

While the physicists in Washington were contemplating these findings, they were also assessing a bold effort by Dr. Steven Winberg of Harvard University, who addressed a meeting of the Optical Society of America also being held in the capital, to develop a unified theory on the forces of nature.

Dr. Weinberg's approach represents a departure from the dominant philosphy of physics in that he suspects the really fundamental laws of nature are not, in all cases, directly manifest. They may instead be only imperfectly demonstrated by observed phenomena.

Unified Theory

His effort is to bring into a unified theory three of the four forces that control all events in nature. These forces differ so radically in their strength, as well as, in many respects, in their behavior, that such a unification has been elusive.

Two of the forces are directly evident to our senses: gravity and electromagnetism, the latter controlling all electrical and magnetic phenomena. The other forces operate only on an atomic scale: The "strong" force that binds together particles of the atomic nucleus and the "weak" force that controls such nuclear reactions as radioactivity.

Earlier Dr. Weinberg and Dr. Abdus Salam had sought to show a link between electromagnetism and the "weak" force and some experimental results have given support to such a link. Now Dr. Weinberg believes the "strong" force can also be brought into this theoretical framework.

Dr. Panofsky, commenting on the turmoil consequent to the new, surprising experimental findings, said that physicists were being forced to consider such "crackpot" ideas as the possibility that the electron has internal structure of some sort. So far, although it displays mass (that is, weight) and an electric charge, the electron has appeared to be infinitely small.

Basis for Technology

It has been demonstrated, for example, that it can be no wider than a millionth of a billionth of a centimeter. All of physics and a highly developed electronics technology have been based on the assumption that the electron is a dimensionless point-source with a negative charge.

Another explanation of the recent findings, according to Dr. Panofsky, might be that the constituents of larger particles, such as the proton and neutron, could themselves have internal complexities.

Those contituents, sometimes called quarks or parons, have never been directly observed. Recent experimental results were described last week as inconsistent with the concept that such internal components are "hard," point-like objects.

At the Physical Society meeting, 13 out of 15 papers at one entire session were devoted to the experiments with colliding beams that were responsible for much of the present ferment. These experiments have been carried out jointly by scientists from the University of California at Berkeley and the Stanford Linear Accelerator at Palo Alto, Calif., using a circular device, called SPEAR, that is adjacent to and is fed by the two-mile-long, straight-line accelerator.

Head-On Collisions

Electrons from the Stanford machine, the most powerful electron accelerator in existence, are injected into the SPEAR ring to circulate in one direction and positrons, likewise produced by the accelerator, are injected to circulate in the opposite direction.

In this way head-on collisions between electrons and positrons are achieved at energies higher than any heretofore produced in such collisions.

The positron is identical with the electron except that it carries a positive instead of negative electric charge. It thus constitutes a particle of "antimatter."

So much energy is involved in the collisions—already up to 5.4 billion electron volts—that heavy particles, or "hadrons," sometimes emerge from the tiny "fireball" that results. In an atomic explosion matter is converted into energy, but in this case it is the other way around.

In so straightforward an experiment it was expected that the results would be predictable in terms of present physics theory. From experiments at lower energies it was shown, as expected, that in some collisions hadrons (such as various kaons, pions, protons and their antiparticles) were formed. In other collisions the products were muons.

The muons are not very much heavier than electrons and are so similar that their existence as separate entities has long been a puzzle. On theoretical grounds it was expected that the ratio of hadron production to muon production would remain relatively stable as the energy of the collisions was increased.

If the ratio of hadron to muon remained about 2 to 3, this would support the quark concept. Instead, as the energy was increased, the ratio steadily increased until at 5.4 billion electron volts it was 6 to 1.

Possibly, Dr. Panofsky said, this is an effect limited to that energy range. But, he added, if the ratio continues to rise, as the energy of the collisions is increased (eight billion electron volts are expected shortly), "then there is something terribly wrong" with present theory.

Equally puzzling is the similarity of the products of electron-positron collisions, at these extremely high energies, to those observed at CERN, the European research center near Geneva, in head-on collisions between protons.

Protons interact with one another in terms of the "strong" force that binds the atomic nucleus. Electrons, on the other hand, do so in terms of electromagnetism, or the "weak" force.

In view of these fundamental differences, it has been astonishing that, at high energies, the two types of collision produce the same mixture of hadrons.

The idea that there are "hard," pointlike constituents inside the larger particles, such as protons, had arisen from a variety of experiments and this had led to specific predictions.

For example, if one throws a baseball to graze a tree, the harder the throw, the less likely that the ball will bounce off the tree at a sharp angle. Likewise, in experiments where a beam of muons is fired at protons (which are the nuclei of hydrogen atoms), the greater the energy of the beam, the less likely should it be that the muons scatter off points of charge inside the protons with strong transverse momentum.

A muon beam experiment, conducted at 150 billion electron volts, has indicated a departure from such "scaling" behavior at lower energies. It was performed with the world's most powerful proton accelerator, at the National Accelerator Laboratory in Batavia, Ill., which differs from the Stanford device in many respects.

The results so far have been compared only with those of other experimenters but a more definitive comparison is expected from a recent rerun of the Batavia test at 56 billion electron volts.

That the forces of nature may not be irreconcilably divorced from one another was argued by Dr. Weinberg in a lecture and was elaborated upon at a news conference. He suspects there is only "one kind of interaction" in nature. The concept, he added, is "an extraordinarily simple view" of the physical laws.

Those laws are manifested, many believe, by particles that, like little bridges, perform the interaction when two other particles meet. If the interaction is electromagnetic, this bridging particle is the photon —a wave of light or radio energy.

If the encounter involves the strong force, the vector particle is a meson. For gravity it is assumed to be the hypothetical graviton and for the weak force, the suspected "W" particle.

The latter, Dr. Weinberg said, must be at least 50 times heavier than the proton. Yet the photon is weightless. How, then, could the forces represented by the photon and "W" particle be woven into a single theory, with all the symmetries that seem intrinsic to nature on that basic level?

The answer, he believes, is that superficially some of these symmetries are broken and that nature, as we observe it, is but an imperfect representation of its own underlying laws. He likens the concept to Plato's images on the cave wall, which were but imperfect imitations of reality.

April 29, 1974

ATOMIC ENERGY: PROMISE AND THREAT

Japanese Reports Doubted

OAK RIDGE, Tenn., Aug. 30 (AP)—Japanese reports of deaths from radioactive effects of atomic bombing are pure propaganda in the opinion of Maj. Gen. Leslie R. Groves, commanding general of the Manhattan District.

Studies by scientists in this country do not bear out the death reports, General Groves said at a press conference here today.

"While the people of the United States would be committing suicide if work on the atomic energy were not continued," General Groves said, "I believe commercial uses of this force are probably decades away."

"The atomic bomb is not an inhuman weapon," General Groves told workers and military personnel in a surprise visit to the project yesterday.

"I think our best answer to anyone who doubts this is that we did not start the war, and if they don't like the way we ended it, to remember who started it," he said.

August 31, 1945

World-Wide Waves Of Atom Bomb Hinted

By The United Press.

SCHENECTADY, N. Y., Oct. 29—Dr. Chauncey G. Suits, vice president and research director of the General Electric Company, said today that there were indications that radio-activity released by atomic-bomb explosions had been felt around the world.

He said that a major American film company had found its films being "fogged" some time after the atomic bombs had been dropped. He said that the film companies suspected that radio-activity had been carried into the stratosphere and had made its way around the world.

Reached by telephone in Rochester, N. Y., Thomas J. Hargrave, president of the Eastman Kodak Company, said that some film had been spotted by radio-activity about a month after a test atomic bomb had been fired in New Mexico but that this had not recurred after two bombs had been dropped on Japan.

October 30, 1945

FILM SPOTS TRACE VAST A-BOMB RANGE

Radioactive Particles Spread Over Australia-Sized Area, Eastman Studies Show

By WALTER S. SULLIVAN

The Eastman Kodak Company made public yesterday its findings on long-range radioactivity effects of atomic bomb explosions. The findings, together with those of other observers, indicate that temporarily radioactive by-products of a single bomb spread in the course of a few days over an area about the size of Australia.

The Kodak findings are based on several months of investigation into the source of radioactivity detected last fall in strawboard stiffeners used between photographic films. Early in the war cardboard stocks had become contaminated by waste from radium instrument dial factories, and Kodak began using strawboard instead.

Strawboard Becomes Radioactive

A routine test of this new material in the fall of 1945 revealed that strawboard made three to eight weeks after the bomb blast in New Mexico showed radioactivity sufficient to fog film. The two plants making this strawboard were in the Middle West, 400 to 500 miles apart.

The radiations showed up after three weeks of exposure as specks of fogging, numbering ten to several hundred on an X-ray film 14 by 17 inches. In many cases the radiation penetrated several layers of film. The points on the strawboard opposite the specks were punched out in quantity and the resulting material was analyzed to determine the nature and probable source of the radioactivity.

Extensive tests showed that beta rays, but no alpha rays, were given off, indicating that the particles in the strawboard were not naturally radioactive, but artificially radioactive elements such as result from an atomic bomb explosion. The mean "half-life" of this activity was found to be about thirty days.

The Kodak scientists concluded that radioactive particles from the New Mexico explosion probably were spread by stratospheric winds and precipitated into the Middle West by rain. The two plants making strawboard are supplied by different watersheds, but sediment taken from the river water used in both plants showed contamination, whereas the straw, fresh from the fields, did not.

This spreading of the particles, as with the world-wide diffusion of dust clouds after volcanic explosions, is supported by the reports of observers in Arizona and Maryland. Lieut. Comdr. A. W. Coven of the Electrical Engineering Department at the Naval Academy was taking observations with a gamma ray counter in the days after the New Mexico test. Less than a day after the explosion he observed an increase in the radio-activity of air in Maryland, which rose to a peak of almost twice normal two and a half days after the explosion.

The Coast and Geodetic Survey observatory in Tucson, Ariz., also was measuring the conductivity of the air—a direct index of radio-activity. It first observed the increase about five days after the New Mexico test and an increase to twice normal by the eighth day. This was only about 25 per cent above the maximum for that time of year. There were light rains in Tucson during these days. Similar observations in this period in Peru and western Australia showed no unusual increase.

Tolerance to Radioactivity

In any case, the atmosphere has so little radioactivity that when doubled it is still very slight. The safe exposure for the human body has been agreed upon by the International Protection Commission, a group of leading radiologists from various countries who set standards of exposure for operators of X-ray and other radiating equipments. The maximum daily safe load, they say, is about ten thousand times the normal radiations of air. This, plus the evidence that one bomb apparently doubles the radioactivity over a wide area, probably is responsible for the rumor that ten thousand bombs in a single barrage would raise the air's radioactivity beyond the tolerance of all life on earth.

Research physicists in radiology at Memorial Hospital in New York point out that even if radio-activity increased in direct proportion to the number of bombs exploded on one continent, the standards of the International Protection Commission cannot be directly applied, since the radiations of atomic bomb by-products may be of a far less potent type than those, for example, of an equivalent amount of radium.

There has been speculation, as well, about the influence of these radioactive clouds on weather and ionization of the upper air. Victor Hess, Nobel prize winner for his pioneering work in cosmic rays, said several months ago that the tests at Bikini Atoll might even bring North America a year of continuous rain. F. W. Reichelderfer, chief of the United States Weather Bureau, believes the effect on weather will be negligible. Condensation of raindrops, he says, will occur on hygroscopic particles such as salt long before it does so on ions.

Long-range radiotelephone experts of RCA and the Bell Laboratories report that their records showed no unusual broadcasting conditions after the bomb explosions. They point out that the E-Layer, lowest level of ions affecting radio transmission, is about seventy miles above the earth and seems to be beyond the reach of the upward thrust of the bombs, whose clouds mushroom at about eight miles.

Army, Navy and civilian scientists will make elaborate observations during the Bikini experiments this summer. They will seek to determine the extent of radioactive diffusion and ionization and its effect on radio transmission. Existing information thus will be greatly enlarged, but it is unlikely that it will entirely settle the argument as to the cumulative effect of many bombs falling on one continent.

May 23, 1946

NEW ATOMIC POWER PLANT PRESAGES PEACETIME USES; IT UTILIZES FAST NEUTRONS

RATE IS REGULATED

Vast Possibilities Opened by Controlled Release of Plutonium Force

SIZE OF POWER UNIT CUT

Adaptation for Propulsion of Ships and Locomotives Now Held Feasible

By WILLIAM L. LAURENCE

Successful operation of an atomic power plant of revolutionary design, using for the first time fast, instead of slow, neutrons for liberating nuclear energy from plutonium at a controlled rate, was announced yesterday by Dr. Norris E. Bradbury, director of the Los Alamos Scientific Laboratory at Los Alamos, N. M., where the atomic bomb was developed.

The announcement, regarded as the most important of its kind since President Truman revealed the existence of the atomic bomb, was released by the United States Atomic Energy Commission in Washington.

The new plant, it is revealed, "has been operated successfully at low power since November, 1946." The initial proposals for its construction were made in December, 1945, shortly after completion of the Los Alamos laboratory's wartime job.

The new plant, the announcement states, is unique in two respects: It is the first to employ the fission of the man-made element, plutonium, instead of normal, unseparated uranium. It is also the first to use fast neutrons, thus eliminating the need for a moderator to slow the neutrons down.

Energy Liberated Slowly

These two unique features open up new vistas in the use of atomic energy for power, of enormous potentialities for peacetime as well as military uses. In the words of the announcement, the new plant is "in a sense a controlled version of the atomic bomb" which also utilizes a fast neutron chain-reaction, the principal difference being that in the bomb the reaction is allowed to go uncontrolled, whereas in the new power unit the fast neutrons are kept in check by a system of controls, so that the energy is liberated at a steady, instead of an explosive rate.

It is also revealed that the "heart of the new atomic power plant, known as a 'fast reactor,' is a small vessel containing a critical mass of the nuclear explosive, plutonium, which emits neutrons of high energy." Since a "critical mass" of plutonium is the amount used in the atomic bomb, which, with all its auxiliaries, could be carried in a B-29, it becomes obvious that the active material in the new atomic chain-reacting pile is very much smaller in its dimensions than the nuclear reactors now producing plutonium at Hanford, Wash.

This great reduction in the dimensions of a nuclear reactor brings it for the first time within the realm of practical use as a power plant for the propulsion of vessels and possibly large locomotives. Since it will still need heavy shielding against the vast quantities of radiation emitted during the fission process, it will still be impractical for use in automobiles or airplanes. But it does open up the possibility of a practical power source for the propulsion of long-range pilot-less airplanes, and, of course, of long-range rockets.

Two Factors Reduce Size

The only major obstacle still in the way of such practical power applications is the fact that the new unit, like the older ones, operates at lower power. An experimental high-power nuclear reactor is now being built at Oak Ridge, Tenn., and is expected to be in operation within about two years.

The gigantic chain-reacting atomic "piles" at Hanford use natural uranium, composed of a mixture of the two isotopes (twins) of the element—uranium 238 and uranium 235. Of these two types of uranium only the uranium of atomic weight 235 is utilizable for the release of atomic or rather nuclear energy.

The ratio of uranium 238 to uranium 235 is 140 to 1, which means that for every pound of fissionable material releasing nuclear power 140 pounds of non-fissionable material must also be included in the pile. Since plutonium has the same fissionable properties as uranium 235 its use in the new nuclear power plant thus means a reduction in size by a factor of 140.

Great as this reduction factor is, it is relatively small in comparison with another factor that reduces the size to even smaller dimensions. This second size-reducing factor comes directly as a result of the utilization of fast neutrons, which, as stated, eliminates the need of vast quantities of material, such as graphite, used as moderators for slowing down the neutron's enormous initial speeds from thousands of miles a second to only one mile a second.

By a fortunate act of nature fast neutrons cannot be used in the chain-reacting pile using natural, unseparated uranium, since uranium 238 absorbs a large proportion of the fast neutrons liberated in the fission of uranium 235, without producing further fission, thus making a chain reaction impossible. Were it not for this fact, the production of atomic bombs would be a relatively simple process, requiring only the purification of natural uranium. Were this the case, the Nazis also would have had atomic bombs and civilization as we know it might have been destroyed in a cataclysmic Goetterdaemmerung.

With slow neutrons, on the other hand, it is possible to arrange a controlled process so that enough are left to maintain a chain reaction, while those absorbed by the uranium 238 convert it into plutonium.

In the new reactor, the announcement states, the "high energy or fast neutrons are caught by other atoms of plutonium before they have been slowed down by colliding with any materials except the nuclear fuel. * * * The Los Alamos reactor utilizes no diluting material, and thus shares with the bomb the property of using the neutrons from fission almost as soon as they are emitted.

"The fast reactor's energy release proceeds at a rate which can be set easily by the operators. The rate can be kept constant.

"The energy generated produces no more heat in the core of the reactor than is given off by an ordinary kitchen oven, but since this heat is generated in a relatively small region, special cooling provisions are required to prevent overheating of the center of the reactor.

"While the over-all energy release is comparatively small, the concentration is intense. During operation the reactor produces a very large quantity of neutrons and other radiations. The effect of the neutrons and the radiations on other materials are studied by means of many plugs and outlets. Experiments are also carried on to determine the properties, formation and consumption of the neutrons and radiations.

"A thick shielding wall of concrete and steel prevents the escape of the radiations. Instruments for gauging every phase of the experiments are contained in a panel in a room adjoining the fast reactor."

Dr. Bradbury said the fast reactor "gives a more intense source of fast neutrons than physicists heretofore have been able to obtain, except during the brief time of the test of the first atomic bomb in the New Mexico desert July 16, 1945."

"It is hoped," he added, "that such a source will make possible the study of fast neutrons' chain reactions in more detail, and thus be another step toward finding the best type of chain reactor for the production of useful power."

Clue to Scientists' Death

The announcement also provides a clue to the nature of the experiments which led to the tragic death of Dr. Louis Slotin. The "original design, testing and construction of the reactor," it is revealed, "were undertaken by a group working with the late Dr. Louis Slotin, victim of a radiation accident at the Los Alamos Laboratory in May, 1946, and Dr. Philip Morrison, now associated with Cornell University."

Supervising the operation of the fast reactor are a young husband-and-wife team of Los Alamos scientists, Drs. David B. and Jane Hamilton Hall, both of whom obtained their doctor's degrees at the University of Chicago in 1942. They have been associated with the atomic energy program since its inception. The engineer in charge is Robert I. Howes of Santa Fe, N. M., who obtained his electrical engineering degree at the University of Wisconsin in 1934.

In announcing the operation of the fast reactor Dr. Bradbury paid tribute to the teamwork which made it possible.

"It is characteristic of the complex and novel construction of the new plutonium reactor," he said, "that essential contributions to it have been made by almost all parts of the laboratory. The chemists and metallurgists prepared the plutonium. Expert machinists fabricated the many parts of unusual materials. Craftsmen erected the heavy and special structures of the reactor and its shield. Designers and engineers worked out the many mechanical devices for control of the plant. Theorists solved the mathematical problems underlying the whole design."

August 30, 1947

CATARACTS IN EYES LAID TO BOMBINGS

A. E. C. Reports First Evidence of Delayed Effects Found at Hiroshima and Nagasaki

Special to The New York Times.

WASHINGTON, June 17—The United States Atomic Energy Commission announced today that eye cataracts had been turning up in recent months as the first evidence of delayed effects among the survivors of the atomic bombings of Hiroshima and Nagasaki.

The commission released data accumulated by the Atomic Bomb Casualty Commission of the National Research Council, which is studying the long-range effects on the Japanese who have apparently recovered from the acute or immediate effects of the bombings.

"Following the discovery that radiation similar to that released in an atomic bomb burst had caused cataracts to form in the eyes of research workers in this country, a preliminary ophthalmic survey was started at Hiroshima last year," today's announcement said. "This survey, led by Dr. David G. Cogan and Dr. S. Forrest Martin of the Harvard Medical School, revealed ten cases of cataracts believed to have been caused by the atomic bomb.

"Subsequent examination of 1,000 persons, most of whom were within 3,000 feet of the point above which the bomb exploded, has led to the discovery of about forty certain cases of radiation cataract and an additional forty suspected cases.

"A full-scale opthalmological study is now under way at Hiroshima and a survey of survivors at Nagasaki is also planned. Annual follow-ups will be made of a statistically significant sample of all persons known to have been within a certain radius of the bomb's hypocenter. More frequent follow-ups of all known and suspected cases of cataract are planned. The eye studies are under the direction of Dr. S. J. Kimura."

Meanwhile the Atomic Bomb Casualty Commission is continuing its over-all survey of the medical and genetics effects of the bombings on the populations of the two cities. The A. E. C. said its findings would "have important significance for scientists and for military and civil defense planning in the United States."

The casualty group has accumulated data on 150,000 persons in the bombed areas, and this is being reported in scientific literature that is made available to the Defense Department, the National Security Resources Board, the United States Public Health Service and other interested agencies.

A medical follow-up of births in the two cities has already covered about 35,000 births, but it has been estimated that at least 200,000 births "must be studied in order to detect small changes in the frequencies of congenital and inherited abnormalities."

June 18, 1950

NEW ATOMIC PLANT 'BREEDS' OWN FUEL; CIVIL USE IS NEARER

Dean Says Reactor Can Create Fissionable Material Out of Common Form of Uranium

LOWER COST NOW IS GOAL

Commission Head Tells Edison Institute Time Has Come to Let Industry Enter Field

Special to The New York Times.

ATLANTIC CITY, June 4—The Atomic Energy Commission has developed a new atomic power plant that can produce new fuel "at least" as rapidly as fuel is consumed in operating the plant.

This new "milestone in the development of atomic energy in this country" was announced here today by Gordon Dean, retiring chairman of the Atomic Energy Commission, in a speech at the closing session of the Edison Electric Institute.

"It is a development," Mr. Dean said, "which holds out the promise of making a civilian atomic power industry even more feasible and attractive in the long range than it has hitherto appeared to be."

The fuel that operates an atomic power plant is uranium 235, the fissionable form of uranium, which constitutes only about 1 per cent of the total uranium in natural deposits. Part of the enormous cost of producing atomic power is in isolating this natural fuel from the elements with which it occurs.

In the new power plant, as Mr. Dean described it here, the burning of uranium 235 not only produces the heat to operate the power plant, but changes non-fissionable uranium into fissionable plutonium at the same time "at a rate that is at least equal to the rate at which uranium 235 is being consumed."

The significance of this, he explained, "is that it is now possible for mankind ultimately to utilize all of the uranium that can be extracted from the earth's surface for atomic fuel, whether it is fissionable or not in its natural state."

Process Is Called 'Breeding'

Scientists have known for a long time, the commission chairman said, that the process, known technically as "breeding," is possible not only for non-fissionable uranium but also for thorium, another relatively plentiful element. But they were never sure, he explained, that as much or more new fuel would be produced as was burned.

The demonstration that this was possible, Mr. Dean said, was made by Dr. Walter Zinn, Dr. Harold Lichtenberger and other scientists of the Argonne National Laboratory near Chicago, using the atomic reactor at the Reactor Testing Station in Arco, Idaho. That is the reactor that in December, 1951, produced the first useful atomic power by using the heat produced by burning uranium 235 to operate a steam turbine.

The success of the experiment with uranium "suggests," the chairman explained, that similar breeding may be possible with thorium, as the scientists have suspected. Thorium was not used in the experiment, however, and therefore he said "I do not wish to imply that its susceptibility to breeding has been proved."

"We must take care," he warned, "to see that this encouraging development is kept in its proper perspective. This news does not mean that economic power from atomic fuels is here. It does not mean that over night we have suddenly obtained all the fissionable material we want or need. It does not mean that uranium can now be regarded as a virtually costless fuel."

Mr. Dean specified two important technical limitations of the new process:

1. "Before the newly created fuel can be extracted and put to use, it must go through a chemical separation process which is currently one of the most expensive aspects of the atomic energy business.
2. "Breeding is a slow process, and a reactor may have to operate for five years or longer before it succeeds in yielding as much new fuel as was initially invested in it."

Policy Is at Crossroads

Nevertheless, the result of this and other developments in the field of atomic energy, Mr. Dean said, "is to bring us to the crossroads in atomic power policy."

"The last remaining technical obstacle," he asserted, "is to learn how to build atomic power plants so cheaply that the power they produce will be competitive with that from conventional fuels. The policy problem that faces us is how this cost-cutting job can best be done."

The Atomic Energy Commission has come to the conclusion, he said, that the time has arrived when "the present Federal monopoly should be relaxed to permit wider participation in the power reactor program." He based this position on two factors:

1. "There is good reason to believe that a demand outside of the Government's own sphere of operations is developing for atomic reactors. It is a commercial demand, an economic demand and a civilian demand. It seems only reasonable to expect that people outside of the Government be given a chance to work toward the development of reactors to meet this demand.
2. "The job ahead is a developmental one—a cost-cutting one —the kind of a job that can be done best by skilled people competing with other skilled people who are working toward the same or a similar goal."

As a start in this direction, the Atomic Energy Commission, he explained, has recommended that the Atomic Energy Act of 1946, which established the Government monopoly in the atomic field, be amended to permit:

1. The ownership and operation of nuclear power facilities by groups other than the commission;
2. The lease or sale of fissionable material under safeguards adequate to assure the national security, and
3. The use and transfer of fissionable or by-product materials by the owners of reactors, subject to purchase by the commission or regulation by the commission in the interest of health and safety.

The chairman urged that those legal changes be accompanied by changes in the practices of the Atomic Energy Commission to permit:
1. The granting of more liberal patent rights;
2. A progressively more liberalized information policy in the power reactor field; and
3. The performance in commission laboratories of the research and development work in the power field that is deemed warranted in the national interest.

"A few people," Mr. Dean said, "have already labeled these policy recommendations as 'the atomic giveaway program.' That is simply not true.

"This all boils down to a question of fair play and common sense. I think it is obvious that we cannot expect private concerns to come in and spend millions of dollars without getting some benefits, and I think it is obvious that private concerns as well as the Government must be in this power program. Otherwise we can never have real competition, the catalytic agent of progress, and we will always have a program limited in size and scope by the range of the Federal Government's imagination and vision, which is not always as broad as it might be."

In an interview before his speech the chairman said that when he retired on June 30 he planned to take a two months' vacation, and then do "some writing."

He said he was in complete agreement with the statement made to the convention's 3,000 delegates by Walter H. Sammis, president of the Ohio Edison Company, who declared he would "endeavor to further the cooperative efforts between the electric utility industry and the Atomic Energy Commission in the development and utilization of atomic power for the general welfare of all the people."

June 5, 1953

2,000 JOIN PAULING IN BOMB TEST PLEA

He Lists Backers of World Ban — Other Scientists Dispute Radiation Views

By GLADWIN HILL
Special to The New York Times.

PASADENA, Calif., June 3— Dr. Linus Pauling, Nobel Prize biochemist, said today that 2,000 scientists had joined him in urging an international agreement to stop testing of nuclear bombs because of the radiation dangers.

The plea was immediately questioned by a number of other scientists.

Dr. Pauling, head of the division of chemistry and chemical engineering at the California Institute of Technology, made public a list of signers of a petition which, he said, would be presented to Congress.

The expression was the latest in a running international controversy over whether current nuclear test explosions by the United States, Great Britain and the Soviet Union are innocuous or have dire radiation hazards.

The petition asserted that "each added amount of radiation causes damage to the health of human beings all over the world and causes damage to the pool of human germ plasma such as to lean to an increase in the number of seriously defective children that will be born in future generations."

Views Stir Clashes

Dr. Pauling's left-of-center political views have made him a figure of occasional earlier controversy. He has made repeated statements that nuclear tests were causing widespread predisposition to such ailments as leukemia, a blood cancer. These contentions have been contradicted by other scientists of equal standing.

The signers of the petition made public today included Dr. L. H. Snyder of the University of Oklahoma, president of the American Association for the Advancement of Science; Dr. H. J. Muller, Indiana University geneticist and a 1946 Nobel Prize winner, and Dr. Joseph Erlanger of Washington University, St. Louis, who won a 1944 Nobel Prize for work in physiology and medicine. Dr. Pauling received the Nobel Prize in 1954 for studies of molecular structures.

Other signers were Dr. Ralph E. Lapp, physicist and writer on atomics; Dr. Edward U. Condon, former head of the Federal Bureau of Standards, and Dr. Harlow Shapely, Harvard astronomer.

Dr. Pauling said that there were about forty members of the National Academy of Sciences on the roster.

At least one Atomic Energy Commission scientist was in the group. He is David L. Hill, physicist at the A. E. C. Los Alamos Laboratory in New Mexico. Associates said that Mr. Hill wrote speech material on nuclear topics last year for Senator Estes Kefauver, Democratic candidate for Vice President.

The list made public today had the names of 134 New Yorkers, ninety-six of them living in New York City.

Many prominent scientists were conspicuous by their absence from the list.

Dr. Joel H. Hildebrand, head of the Chemistry Department at the University of California, challenged the petition's assertion about hazards as "not a true indication of the dangers in the absence of quantitative comparisons with natural radiation and current X-ray usage."

Dr. Kenneth S. Pitzer, dean of the University of California College of Chemistry, formerly an A. E. C. research director, said that he felt the risks from nuclear tests were "much smaller than the risks we take in everyday living."

Dr. George Beadle, head of the biology department of the California Institute of Technology, declared that, while "I agree there is urgent need for reviewing the situation, both scientifically and morally * * * I feel that scientists ought to confine their public statements to their own fields, or to make it clear that they are speaking not as experts but are expressing private opinions."

On this point the petition said that "as scientists we have knowledge of the dangers involved and therefore a special responsibility to make those dangers known."

Dr. Pauling said the petition grew out of the reaction to a speech he made at Washington University in St. Louis May 15. He listed more than a score of persons he said had "collaborated in the formulation of the appeal." All the signers of the petition acted only as individuals, he said, and no organization was involved.

He said that the document would be given to Representative Chet Holifield, Democrat of California, chairman of the subcommittee of the Joint Congressional Committee on Atomic Energy that is holding hearings on the radiation problem.

June 4, 1957

Associated Press Wirephoto
ANNOUNCES APPEAL: Dr. Linus Pauling, who yesterday released paper, signed by 2,000 scientists, asking end of nuclear tests.

TELLER SEES PERIL IF ATOM TESTS END

Tells Senators Move Might Sacrifice Millions in a 'Dirty' Nuclear War

By ALLEN DRURY
Special to The New York Times.

WASHINGTON, April 16— Dr. Edward Teller told the Senate Disarmament subcommittee today that if the United States should suspend nuclear testing now "we may be sacrificing millions of lives in a 'dirty' nuclear war later."

Dr. Teller, one of the leading scientists in this nation's development of the hydrogen bomb, said that if the United States yielded to the Soviet campaign to stop nuclear tests, it would be stopping at a place where it knew much about the military uses of nuclear power but not enough about its peaceful uses.

Further, he said, to stop now would be to abandon what he described as "our only advantage" over the Soviet, the lead in nuclear weapons.

"I believe the Russians are developing their war machine faster than we are," he said.

Skeptical on Inspection

Dr. Teller was skeptical of proposals for international inspection in the event the Soviet Union and the Western powers reached an agreement to stop producing and testing atomic and nuclear weapons. He said he doubted that such inspection could possibly be effective.

The scientist, one of the chief supporters of the Atomic Energy Commission in its opposition to any suspension of tests by the United States at this time, said that inspection would probably not work at all without the complete cooperation of the Russians.

He said that if the Russians experimented sufficiently with

small nuclear weapons, "they will find out more and more, a plethora of methods to make these tests disappear" among the many earthquakes that agitate portions of the globe.

Under questioning by Senator Hubert H. Humphrey, Democrat of Minnesota, the subcommittee chairman, Dr. Teller said he could not reveal the advice he had given Harold E. Stassen at the London disarmament negotiations last year.

He said Mr. Stassen, then President Eisenhower's disarmament adviser, had placed his scientific aides "under deeper secrecy than I have been forced to accept at any other time." Mr. Stassen has resigned and is now seeking the Republican nomination for Governor of Pennsylvania.

Dr. Teller denied that radioactive fall-out was endangering the world's population. He said scientists had not yet been able to prove that blood and bone diseases would be caused by the amount of fall-out from tests to date.

UPHOLDS TESTS: Dr. Edward Teller, physicist.
Associated Press

He admitted there might be "many thousands" of mutations as a result of radioactivity's effect on genetics. But, he argued, mutations were not necessarily all bad, and they might be caused by a number of things, including medical X-rays.

He estimated that the recent Soviet tests, described by Senator Humphrey as "the dirtiest ever," had perhaps put enough radioactivity into the world's atmosphere to shorten human lives by much less than one day's time.

Advice on Atom Victims
Special to The New York Times

PHILADELPHIA, April 16— It might be unwise for people exposed to atom bomb radiation to attempt to walk out of the contaminated region afterward, a scientist said here today.

The speaker was Dr. Thomas J. Haley of the Atomic Energy Project, School of Medicine, University of California at Los Angeles. He addressed a session of the Federation of American Societies for Experimental Biology at the Penn Sherwood Hotel here.

Experiments with animals show that heavy single doses of radiation interfere seriously with the ability to withstand fatigue, Dr. Haley declared.

From this evidence he reasoned that a person who had been heavily hit by radiation during a nuclear blast might well be advised to get under cover and to stay as quiet as possible for several days. The effort involved in trying to get away from the "hot area" immediately might prove fatal, he asserted.

In another paper at the federation meeting a physiologist said there was no physical basis for the old idea that "if you stop thinking about it the pain will go away."

This speaker, Dr. Leon C. Greene of the Navy's Aviation Medical Acceleration Laboratory at Johnsville, Pa., said there appeared to be no adaptive body mechanisms that caused the sensation of pain to disappear gradually under an unchanging stimulus.

April 17, 1958

U. N. ATOM PANEL AGREES FALL-OUT IS PERIL TO MAN

DIVIDES ON TESTS

Group Bars Soviet Bid for Halt—Genetic Threat Noted

By WALTER SULLIVAN

The United Nations Scientific Committee on the Effects of Atomic Radiation has found unanimously that fall-out from nuclear weapons tests is a hazard to mankind.

The only point of disagreement in the fifteen-nation committee was whether or not to call for "immediate cessation" of the tests, as proposed by the Soviet representative. Such a recommendation was opposed by the majority, including the United States, on the ground that it lay beyond the jurisdiction of the committee.

The report followed two and a half years of study.

The cessation of tests and other measures to control radiation dangers, the majority view stated, "involve national and international decisions which lie outside the scope of its work."

Unequivocal on Danger

Nevertheless the committee was unequivocal in its depiction of the danger to generations yet unborn, as well as to those alive today, of radioactive fall-out. "Even the smallest amounts of radiation are liable to cause deleterious genetic, and perhaps also somatic, effects," it said.

Somatic effects are injuries to the body as a whole, including cancer, leukemia, prenatal damage and shortening of life.

The report, throughout, emphasized the smallness of radiation doses resulting from fallout. Even if weapons tests continued, according to one method of computation, the radiation hazard superimposed upon that from natural sources would not be substantially greater than the hazard from the natural causes.

Evidence Held Inconclusive

The committee's conclusions were based on indications that any added radiation exposure, no matter how slight, might be injurious. This appeared to be most likely in the field of heredity, although in this area, as in most others touched on by the reports, the evidence was not found to be conclusive.

Nevertheless the committee said that, at the present stage of research, it appeared that "even the smallest doses" could cause mutations that would undermine heredity. This, it said, laid grave responsibilities on the governments of the world.

The knowledge that man's actions can damage his genetic inheritance, the report said, "clearly emphasizes the responsibilities of the present generation, particularly in view of the social consequences laid on human populations by unfavorable genes."

An increased mutation rate, it said, would probably lower the average intelligence quotient and reduce the life span of the human race.

Once the genes, which carry the fruits of evolution from parent to child, have been altered, there is no changing them back.

The committee drew a distinction between perils undertaken voluntarily and those imposed without consent on all the peoples of the world. Fall-out, the report said, involves dangers that "by their very nature are beyond the control of the exposed persons."

The report noted that "some hazards are implicit in almost all technological advances," but the radiation exposures in research, in industry, in X-ray examination and radiation therapy are "for the benefit of mankind and can be controlled."

It recommended that measures be taken to minimize such exposure. "The committee concludes that all steps designed to minimize irradiation of human populations will act to the benefit of human health," the report said.

"Such steps," it continued, "include the avoidance of unnecessary exposure resulting from medical, industrial and other procedures for peaceful uses on the one hand and the cessation of contamination of the environment by explosions of nuclear weapons on the other."

Although attempts have been made to develop "clean" bombs producing a minimum of fall-out, it is generally agreed that all above-ground nuclear explosions produce some of this airborne, radioactive debris.

The committee was established by the United Nations General Assembly in 1955 as a consequence of widespread alarm over the danger from fall-out and other man-made radiation.

A large part of the committee's attention has been devoted to possible hazards due to cumulative exposure and accidents in the growing use of atomic power for industrial purposes.

Five meetings were held between March 1956, and June of this year. "Full and unrecorded discussions centered on a blackboard," the report said. The committee secretariat worked during the interim periods organizing the voluminous reports from various nations.

The committee consisted of scientists from Argentina, Australia, Belgium, Brazil, Britain, Canada, Czechoslovakia, Egypt (later the United Arab Republic), France, India, Japan, Mexico, Sweden, the Soviet Union and the United States. Its task was to make recommendations for the study of the problem and "evaluate" reports from the various nations.

Many Called Doomed

One of the most controversial aspects of the fall-out problem has been the view of some scientists that large numbers of people are being doomed to death by this phenomenon. Dr. Linus C. Pauling of the California Institute of Technology, a Nobel Prize winner in chemistry, estimated in May that carbon 14 from the bomb tests to that date would produce about 1,000,000 seriously defective children and about 2,000,000 embryonic and neonatal deaths.

In approaching this problem, the committee noted many unknowns, including the population of the earth, the minimum dose with adverse effects and the rate of future testing.

If the world population is 3,000,000,000, if even the slightest dose may induce leukemia, and if the bomb tests are stopped this year, the committee estimated that less than 25,000 to 150,000 cases of leukemia, a usually fatal illness, could be ascribed to fall-out.

This would be the ultimate total. It placed the annual natural incidence at 150,000 cases, with 15,000 cases a year due to natural radiation and from 400 to 2,000 cases a year due to fall-out.

On the other hand, if it requires a dose of 400 or more rems to induce leukemia, the committee said there would probably be no cases due to fall-out, regardless of whether the tests were halted. The rem is an arbitrary unit of radiation exposure. It stands for "roentgen equivlent mammal." The report also mentions mrem, which indicates a thousandth of a rem.

A threshold of 400 rems is high in comparison, for example, to the dose of about three rems from natural sources over a thirty-year period.

So far as genetic defects are concerned, it was estimated that, in a world population of 5,000,000, there would be from 700,000 to 3,000,000 natural occurrences a year, with from 25,000 to 1,000,000 due to natural radiation. If tests were continued until the rising rate of fall-out radiation leveled to a state of equilibrium, there would be 500 to 40,000 additional annual occurrences, the report said.

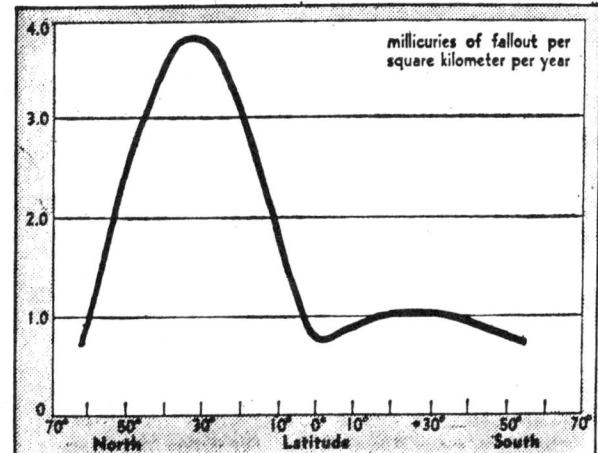

The New York Times   Aug. 11, 1958

FALL-OUT PATTERN: The curved line shows the variation in strontium 90 fall-out with latitude in 1956 and 1957. The three low points are in the latitudes of Anchorage, Alaska (left), the Equator (center) and Cape Horn (right). The peak is in the latitude of Savannah, Ga. New York City is at Lat. 41 deg. N. Rio de Janeiro is in the latitude of the lesser maximum. The horizontal figures represent latitude. The vertical figures are millicuries of radioactive fall-out per square kilometer per year.

The estimated dosages also varied widely. Diet was an important factor. In Japan, with a rice diet and low calcium content in the soil, strontium 90, a primary fall-out hazard, is more readily absorbed into crops and thence into the body.

If weapons tests continue at a high level until equilibrium is reached in about 100 years, the dosage received by bone marrow in a seventy-year lifetime in such countries would be about 17 rem, compared to 7 rem from natural sources. The marrow dose is significant as a possible cause of bone cancer.

On the other hand in a milk-drinking country like the United States, the lifetime total would be only about 2.8 rem under these circumstances.

A "genetically significant" dose is delivered to the reproductive organs — largely from caesium 137. Over a thirty-year period under the given circumstances, the dose would only be .12 rem, according to the report, compared to 3 rem from natural radiation.

The committee outlined its efforts to find clear evidence of radiation effects from small doses. It reported that the death rate of radiologists in the United States, compared with that of other medical men, suggested such an effect. Yet a similar study in Britain failed to support this view.

The report noted that natural radiation — from rocks, from volcanic sources, etc.—in some parts of the world was ten times normal. It urged that studies be made of the populations there as well as of uranium miners. It has been reported that there is a marked incidence of lung cancer in some of those working in such mines.

Another suggested approach was a study of populations living at high altitudes or high geomagnetic latitudes, where cosmic rays were especially intense. Also, a study was proposed of those who have received radiation treatment.

One of the problems has been the delay in appearance of observable effects. Cancer due to radiation or chemical agents may not develop for from ten to twenty years. Leukemia may not appear for from five to ten years, the report said.

While radiation is definitely thought to cause both types of disease, the report noted that malignant tumors resulting from even severe exposure were infrequent. The more likely result of total body irradiation, it said, was leukemia. This was observed in the study of radiologists.

Genetic Limit Studied

In discussing genetic effects, the report described current efforts to determine whether or not there was a lower limit of radiation, below which no damage would be done. The evidence so far does not suggest such a limit, the report said, but much work remains to be done in this field.

It said that, while genetic damage reduced fertility, even a doubling of the present mutation rate was unlikely to lead to extermination of the human race.

The lack of fundamental knowledge of normal cell structure, the report said, is "the major factor" limiting progress in radiobiology.

While the committee dealt largely in world-wide averages, it noted that fall-out tended to be concentrated, by wind patterns, in the middle latitudes north and south of th Equator. Because most bomb tests are in the Northern Hemisphere, the dose there is by far the heaviest.

In a comparison of strontium 90 fall-out from 1954 through 1957 in Argentina, Britain, Japan, Sweden and two American cities—Pittsburgh and New York the committee found the highest level consistently in New York. In terms of natural radiation, nevertheless, all the figures were low.

Rise in Carbon Isotope

Also made public yesterday was a working paper prepared for the committee by a group of United Nations specialists. The committee did not have time to incorporate its findings in its report. The paper revised the methods used to estimate fall-out doses and effects.

It also dealt with the effects of carbon 14. The proportion of this radioactive isotope in carbon of the lower atmosphere is thought to have increased from 5 to 10 per cent as a result of bomb tests. The specialists estimated that a consequent increase of from one-third to two-thirds of a per cent was to be expected in all living matter.

Because this isotope takes 5,600 years to lose half of its radioactivity, the effect of the tests carried out to date, although slight, will continue for 8,000 years, according to the study.

August 11, 1958

A Chronology of Nuclear Explosions

Following is a chronology of nuclear explosions by the United States, the Soviet Union, Britain and France from 1945 to 1963. The United States Atomic Energy Commission says that not all of the nuclear tests by the United States and the Soviet Union have been announced.

1945
July 16—First atomic explosion, experimental bomb, set off at Alamogordo, N. M.
Aug. 6—Hiroshima destroyed.
Aug. 9—Nagasaki destroyed.

1946
July 1—United States tests atomic device at Bikini.
July 25—United States tests atomic device under water.

1948
April—United States tests three nuclear devices at Eniwetok.

1949
Sept. 23—Soviet Union tests its first atomic bomb.

1951
Jan. 27-Feb. 7—United States conducts five tests.
April-May—United States conducts four atomic tests, one under ground.
October—Soviet Union tests two atomic devices.
Oct. 22-Nov. 29—United States conducts seven tests.

1952
April 1-June 5—United States conducts eight tests.
Oct. 3—Britain tests her first nuclear device near the Montebello Islands.
Oct. 31—United States tests one atomic device at Eniwetok.
Nov. 1—United States tests first thermonuclear, or hydrogen, weapon at Eniwetok.

1953
March 17-June 4—United States tests 11 atomic devices.
Aug. 12—Soviet Union tests its first hydrogen bomb.
Aug. 23—Soviet test of an "atomic device."
October—Britain tests two atomic devices.

1954
March 1-May 14—United States tests six nuclear devices.
October—Soviet Union tests a nuclear weapon.

1955
February-May—United States conducts 15 tests, including one under water and one under ground.
August-November—Soviet Union tests four devices and weapons.

1956
March-April—Tests by Soviet Union.
May-July—Thirteen tests by United States.
May 15-June 19—Britain tests two weapons.
August-November—series of Soviet tests resumed, total of seven for year.
September-October—Britain carries out four tests.

1957
Jan. 19-April 16—Soviet conducts a series of weapons tests.
May 17—Britain tests her first hydrogen bomb.
May - October — Twenty-four tests by United States at Nevada range, including one under ground.
August-December—Soviet Union concludes series of 13 tests for year.
September - October — Britain tests three devices.
Nov. 8—Britain explodes hydrogen device, bringing total of tests for year to seven.

1958
February-March — Ten Soviet test shots.
April-July—United States tests 29 devices, including four under ground.
April 28—Britain tests atomic device.
August—United States tests two nuclear missile warheads.
August - September — Britain tests four devices.
August-September — Two high-altitude devices tested by United States.
September - October — United States tests 19 devices, including three under water.
November—Series of 15 Soviet tests.

1960
Feb. 13—France explodes her first atomic bomb in Sahara.
April 1—France explodes nuclear device.
Dec. 27—France sets off third nuclear device.

1961
United States conducts eight under ground tests.
April 25—France sets off fourth nuclear device.
September—Soviet Union resumes testing with 31 shots, including the largest explosion in history and one under water.

1962
April—United States begins series of tests in Pacific. Final American test in atmosphere takes place Nov. 4. Total of 86 devices tested in year.
August—Soviet Union starts series of 40 tests. Shot of Dec. 24 is last known Soviet test in atmosphere.
One test conducted by France. Two underground tests conducted in Nevada by United States and Britain.

1963
June 30—United States reports "inconclusive evidence" of Soviet test.
United States conducts 10 under ground tests.
France conducts one under ground test.
Total shots announced to date: United States, 259; Britain, 21; United States and Britain jointly, 2; Soviet Union, 126; France, 6.

July 26, 1963

U.S., SOVIET AND BRITAIN REACH ATOM ACCORD THAT BARS ALL BUT UNDERGROUND TESTS

TREATY INITIALED

Rusk and Lord Home Will Go to Moscow to Sign Pact

By SEYMOUR TOPPING
Special to The New York Times

MOSCOW, July 25 — The United States, the Soviet Union and Britain concluded today a treaty to prohibit nuclear testing in the atmosphere, in space and under water.

The historic document was initialed at 7:15 P.M., Moscow time, by W. Averell Harriman, Under Secretary of State for Political Affairs; Soviet Foreign Minister Andrei A. Gromyko, and Viscount Hailsham, British Minister for Science.

A communiqué on the initialing said:

"The heads of the three delegations agreed that the test ban treaty constituted an important first step toward the reduction of international tension and the strengthening of peace, and they look forward to further progress in this direction."

Austrian Treaty Recalled

It was noted in the diplomatic community here that the treaty represented the first major East-West accord since the conclusion of the Austrian State Treaty on May 15, 1955. That agreement ended the postwar four-power occupation of Austria.

The United States and Britain agreed at the test-ban talks to further discussion of the Soviet proposal relating to a pact of nonaggression between the North Atlantic Treaty Organization and the Soviet-bloc's Warsaw Pact alliance. The communiqué said this would be done in consultation with the NATO allies "with the purpose of achieving agreement satisfactory to all participants."

PHYSICS AND ASTROPHYSICS

Mr. Harriman, who appeared tired but happy after 10 days of intensive negotiations, said the test-ban treaty would relieve the fears of people all over the world about nuclear contamination of the atmosphere.

He expressed the hope that other nations would adhere to the test ban treaty, which provides for the accession of other members.

Mr. Harriman also announced that Secretary of State Dean Rusk and the Earl of Home, the British Foreign Secretary, would come to Moscow in the near future for the ceremonial signing of the treaty. Foreign Minister Gromyko presumably will sign for the Soviet Union.

Mr. Harriman has requested an appointment with Premier Khrushchev for tomorrow to discuss the results of the conference and to take up the question of civil strife in Laos. He has arranged to leave by air Saturday for Washington to report to President Kennedy and to brief members of Congress on the treaty.

The treaty, after signing, would be subject to parliamentary ratification by the United States Senate, the Supreme Soviet and the British Parliament.

The Western delegates to the three-power talks were unable to persuade Mr. Gromyko to accept international on-site inspection to verify the nature of seismic disturbances. The treaty, therefore, does not cover underground nuclear testing.

The preamble to the treaty pledges the three nations to continue negotiations for the discontinuance of all test explosions for all time.

The second article of the five-article treaty binds the signatories to "refrain from causing, encouraging or in any way participating in" any nuclear explosion in the prohibited environments.

The United States and Britain would thereby be barred from assisting France in her contemplated testing program, which is directed at developing an independent nuclear striking force. The Soviet Union similarly would be restricted in extending any assistance to the nuclear weapons program of Communist China.

Abrogation Clause in Treaty

The treaty is of unlimited duration, but it contains an abrogation clause that already is the subject of controversy in the gathering battle for ratification in the United States Senate.

The clause states: "Each party shall, in exercising its national sovereignty, have the right to withdraw from the treaty if it decides that extraordinary events related to the subject matter of this treaty have jeopardized the supreme interests of its country."

Three months' advance notice of withdrawal is required.

The abrogation clause evidently represented a compromise reached during the closed talks. A United States-British draft treaty proposed last Aug. 27 at the 17-nation disarmament conference in Geneva entailed a more involved procedure for withdrawal.

Under the original Western draft, a signatory would be entitled to summon a conference to study evidence that its national security had been endangered by violations or testing by a nation that was not a party to the treaty. Under the draft, withdrawal would be permissible 60 days after the convening of the requested conference.

Treaty Can Be Amended

The treaty can be amended by a majority of its members. The majority, however, must include the votes of the three original signatories.

The three Governments proclaimed in the preamble that their principal aim was "the speediest possible achievement of an agreement on general and complete disarmament under strict international control in accordance with the objectives of the United Nations."

The preamble said that such an agreement "would put an end to the armaments race and eliminate the incentive to the production and testing of all kinds of weapons, including nuclear weapons."

The Moscow radio broadcast tonight the text of the treaty and the communiqué, which said that the discussions had taken place "in a businesslike, cordial atmosphere."

The initialing of the accord was reported by telephone to Mr. Khrushchev, who was conferring today with Communist leaders assembled here for an economic conference.

The reaction of Soviet officials and ordinary Muscovites to the outcome of the talks was one of pleasure. There were cautious expressions of hope that it might signify a break in the cold war.

Atmosphere Cheerful

The feeling that perhaps a start had been made in arranging a détente between East and West was reflected in the atmosphere in the conference room of the Foreign Ministry's reception mansion.

Correspondents, upon entering the gilded, high-ceilinged room, found the delegates seated about a round table looking somewhat weary after two hours and 45 minutes of discussion prior to the initialing of the treaty.

Behind the banter that the delegates engaged in for the benefit of photographers, there was an air of solemnity and of historic moment.

The United States and Soviet delegates shouldered their way through the crowd and met halfway at the rim of the conference table.

Mr. Harriman grabbed the hand of Mr. Gromyko, whom he has encountered over the years at a series of tension-ridden and fruitless conferences. The tall, 71-year-old diplomat said: "Thank you." Mr. Gromyko replied: "Thank you. Till we meet tomorrow, good-by."

The conference, which marks the first notable success in more than five years of negotiations on a test-ban treaty, ended on this note. Mr. Harriman later said: "It was a businesslike talk. There was goodwill throughout."

The most difficult phase of the negotiations took place during the last four days in defining the link between the conclusion of a test-ban treaty and the Soviet-proposed East-West nonaggression pact.

The Soviet Government, until the final stage of the negotiations, had insisted on a simultaneous signing of the agreements. The Western delegates resisted on the ground that they were not authorized to negotiate a nonaggression agreement since this would have to be done in consultation with the other North Atlantic allies.

The alliance members, especially West Germany, had opposed the form of the Soviet proposal put forward on Feb. 20, which in effect involved the recognition of East Germany.

The compromise, which was worked out last night, was incorporated in the final paragraph of the communique. It said:

"The heads of the three delegations discussed the Soviet proposal relating to a pact of nonaggression between the participants in the North Atlantic Treaty Organization and the participants in the Warsaw Treaty.

"The three Governments have agreed fully to inform their respective allies in the two organizations concerning these talks and to consult with them about continuing discussion on this question with the purpose of achieving agreement satisfactory to all participants."

Mr. Harriman and Lord Hailsham avoided any implied recognition of East Germany by insisting on the term "participants" in the Warsaw Pact rather than member nations, which is the phraseology of the Soviet proposal.

Mr. Gromyko attempted in negotiating the phraseology of the communiqué to bind the United States and Britain as closely as possible to a commitment to negotiate seriously on a nonaggression agreement.

The Soviet Government evidently dropped its insistence on a simultaneous conclusion of a test ban treaty and a nonaggression agreement because of the assurances of Mr. Harriman that the latter would be examined in good faith.

Assurances Given

The Soviet aim in seeking a nonaggression pact has been to obtain Western acceptance of the status quo in Eastern Europe, particularly the division of Germany.

In the exchange of opinions at the three-power talks Mr. Harriman outlined a possible formula for a nonaggression treaty that might be acceptable to the United States and its allies. This would not entail recognition of East Germany.

Such an agreement would bar the use of force in Europe and guarantee the security of West Berlin. It would be realized through parallel declarations by the two military groupings rather than through the signature of a common document.

The process of consultation with the allies on a nonaggression agreement is scheduled to go forward after Mr. Harriman's return to Washington.

Secretary of State Rusk is expected to carry on the discussions of a nonaggression treaty and other measures with Foreign Minister Gromyko when he comes here for the signing of the test-ban treaty.

July 26, 1963

Atomic Age Has Altered the Habitat of Man

BY WALTER SULLIVAN

Twenty years after the first atomic bomb exploded over Hiroshima on Aug. 6, 1945, New York time, studies of fallout residues show that the physical, as well as the political, environment of the world has been permanently altered.

Even if there are no more bomb tests—which seems unlikely—the radioactivity of carbon in the air, the oceans and all living matter will remain twice normal for a prolonged period. It will not sink to its pre-bomb level for thousands of years.

Although large-scale testing

of nuclear weapons in the atmosphere ceased at the end of 1962, the level of strontium 90 on the surface of the earth has continued to rise. According to a Weather Bureau study, it will probably reach its peak this year and then begin to subside.

It has been rising because much of the bomb-produced fallout was injected into the stratosphere, where it has remained suspended, sinking to earth primarily during the spring. This reservoir is thought largely to have been drained by now, but a considerable amount of the bomb-produced strontium is "missing."

From the number of weapons exploded in the air it had been calculated that a certain amount of strontium 90 was added to the environment. Scouting of the upper air by U-2 aircraft, balloon samplers and other devices has enabled specialists to calculate the amount of bomb debris still up there.

Calculations Are Off

These observations, plus ground measurements, indicate that about 90 per cent of the observed strontium has fallen to the ground and only 10 per cent is still aloft. But when these two components are added, they fall far below the estimated total production. Apparently there is a gross error in the calculations or the observations.

The level of strontium is of special importance because it is picked up by the body and incorporated into bones and teeth. This arises from its chemical similarity to calcium.

While the ground burden of strontium 90 is still thought to be rising, the level in milk hit its peak in the spring of 1964.

Milk is the subject of special surveillance by the United States Public Health Service because it constitutes a rapid channel for delivery of fallout to the population. The bomb material falls on fields, is eaten by cows and can reach the consumer within days.

This is particularly critical where iodine 131 is concerned. This bomb product retains its radioactivity for so short a period that it is of no medical significance unless it reaches a person soon after the bomb blast. Its half life is eight days, which means that in such a period it loses half its radioactivity.

Half Life Is 28 Years

By contrast, the half life of strontium 90 is 28 years. That of carbon 14 is roughly 5,730 years. Such radioactive forms of various atoms are produced in the explosion and become at least a temporary part of the environment. Only in special cases have the levels become high enough to be of general medical concern.

Among examples of the spe-

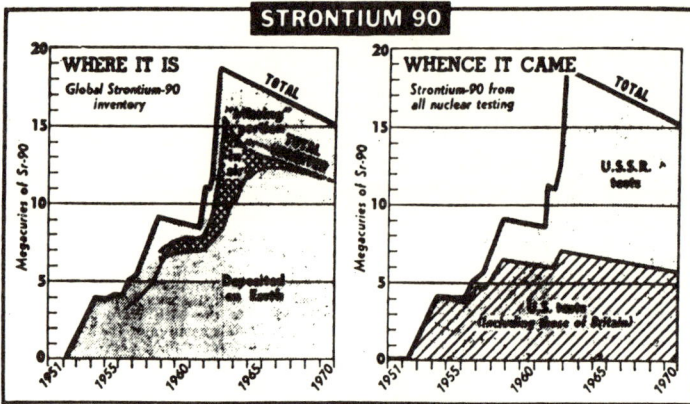

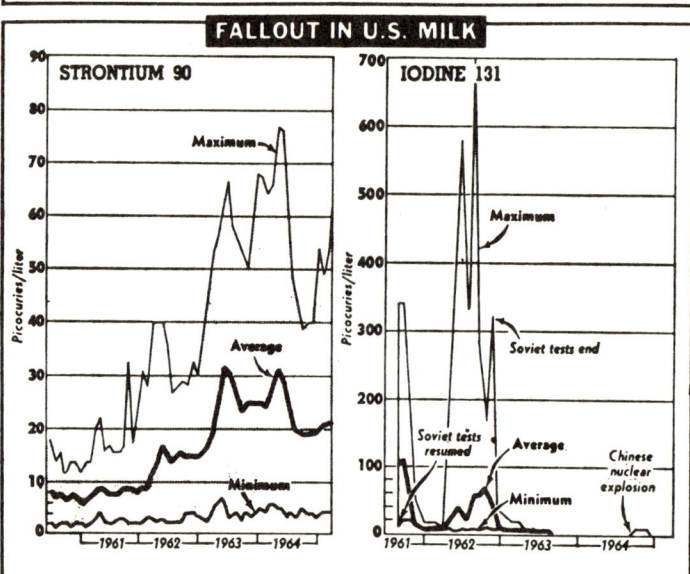

Upper diagrams depict year-to-year distribution of strontium 90. At right is relative contribution of nuclear tests. At left is strontium on ground and still airborne, plus estimated total from all tests. Air and ground samples fall short of estimate. Bottom diagrams show levels of radioactive strontium and iodine in milk. Since iodine decays rapidly, there has been almost none since 1963, except for small amount after Chinese tests.

cial cases are the Alaskan Eskimos who live on caribou that eat reindeer moss. This moss has been comparatively rich in strontium from Soviet tests across the Arctic Ocean. The Eskimos are now under special observation.

Another source of concern has been the exposure of farmers living in valleys near the Nevada test site. On several occasions in recent years radioactive debris has been ejected from shallow underground explosions and been carried by wind beyond the test site.

On occasion, this has contaminated milk on nearby farms to the point where it was necessary to seize the milk and give the farmer a substitute supply.

Within a few days after the Chinese fired their first nuclear explosion last Oct. 17 there was a sharp rise in the radioactivity of thyroid glands of cattle slaughtered in New York State. These glands are monitored by the Public Health Service in cooperation with the United States Department of Agriculture, since radioactive iodine entering the body moves quickly to the thyroid gland.

Rate Rose in November

The rise hit its peak in early November, then sank to normal in December. A similar jump was noted in the nationwide Pasteurized Milk Network of the Public Health Service. It operates 63 stations in the United States, with at least one in each state, plus Puerto Rico and the Canal Zone.

A similar rise was observed after the second Chinese blast, on May 15 of this year. However, the resulting levels were minor compared to those during the intense testing of 1962.

Among the various ways in which the Division of Radiological Health of the Public Health Service keeps an eye on fallout effects is the analysis of bones obtained from the newly deceased or where bone has been removed surgically. The program was begun late in 1961 and has shown strontium levels about half those expected.

This was stated by James G. Terrill, Jr. in his testimony a few weeks ago to a subcommittee of the Congressional Joint Committee on Atomic Energy. Mr. Terrill is deputy chief of the Division of Radiological Health.

Strontium 90 levels in the bones of those less than five years old hit their peak in 1963. Diet studies, however, show that the maximum intake of fallout particles by the population was probably last year.

Students of the world's weather have continued to use fallout as a tool for tracing the movements of air masses.

Special tracers were included in certain of the high altitude explosions. For example, rhodium 102 was used to tag debris from a 1958 shot fired 27 miles above the Pacific. Cadmium 109 tagged fallout from a 1962 shot 250 miles up.

An unplanned tracer was added to the atmosphere when an orbiting reactor accidentally plunged into the atmosphere in April, 1964, dumping plutonium 238 into the air as it burned up. It has been found that it takes about two years for the stratospheric remains of such material to become uniformly distributed over the entire surface of the earth.

Prior to this there was little information on the extent to which there is an exchange of air between the Northern and Southern Hemispheres of the earth.

Likewise, the repeated observation that fallout descends from the stratosphere in spring has led to efforts to explain this phenomenon.

At Pennsylvania State University, Dr. Edwin F. Danielsen proposed that a fold of the stratosphere is sometimes pushed down into the troposphere, the lowest layer of the atmosphere, carrying fallout particles with it.

He thought that the Rocky Mountains might precipitate such folding of the lower boundary of the stratosphere. A project was organized by the Defense Atomic Support Agency the Atomic Energy Commission, the Weather Bureau and others to investigate the proposal.

Air sampling was done by B-50 and B-57 aircraft equipped to filter air en route and measure continuously the radioactivity of the collected material. The flights repeatedly traversed what appeared to be a fold of stratosperic air deep in the troposphere and every time they did so there was a sharp jump in radioactivity.

It is hoped that further studies of this sort will spell out in greater detail the manner in which this apparent folding takes place.

August 7, 1965

PHYSICS AND ASTROPHYSICS

Long-Term Danger In Reactors Feared

By WALTER SULLIVAN
Special to The New York Times

WASHINGTON, April 28—A team of physicists, convened last year by the American Physical Society to assess the safety of American nuclear reactors, has found no reason for "substantial short-term concern," but is critical in terms of long-range prospects.

The study focused on the water-cooled reactors that are the standard energy source in American atomic power plants. Fifty-four are in operation, as well as one gas-cooled reactor. A total of 236 water-cooled reactors have been built or are under construction or projected.

The physicists noted that so far the safety record of such reactors "has been excellent, in that there has been no major release of radioactivity," and said they had uncovered no reasons for substantial short-range concern regarding risk of accidents. But they did express concern for long-term operation of an increasingly large number of such power plants when the likelihood of seemingly "improbable" accidents becomes greater.

The tone of the physicists' report, particularly in dealing with the short-term outlook, contrasted with the views of critics of nuclear power plants such as the Union of Concerned Scientists and environmental organizations such as the Sierra Club.

Attention has been drawn by some critics to mishaps that have occurred, including limited releases or spillages of radioactive material and the recent discovery of cracked piping in several plants.

That the hairline cracks were found was cited by the panel of physicists as evidence of vigilance in the reactor safety effort.

One major source of concern to the panel is the absence of any realistic, full-scale test of what would happen in case of the most serious accident—a core melt-down. If all the cooling systems failed, the reactor core and its fuel rods could heat sufficiently to melt.

The reactors are enclosed in pressure vessels designed to contain the radioactive gases released if melting occurred. If heat or internal pressure became sufficient to rupture the pressure vessel, winds could carry this lethal debris hundreds of miles.

The physicists, chosen to be independent of any connections with the reactor safety program, accepted estimates of the former Atomic Energy Commission that the likelihood of such a casualty was very small. The calculated probability that it would happen to any one reactor in any one year ranged from one in 50,000 to one in 50 million.

The study found, however, that estimates of cancer deaths resulting from such an event should be 50 times higher than those of the A.E.C. It also regarded the A.E.C. estimates of genetic injuries as much too low.

The physicists' higher figure for cancer deaths derived from consideration of two factors not considered by the A.E.C.

One was the long-term effect on populations downwind of radioactive material deposited on the ground. The other was the effect of radioactive debris on organs such as lungs and thyroid gland. In the extreme case considered, the physicists estimated that from 20,000 to 300,000 residents would have thyroid glands damaged from exposure to radioactive iodine.

Although this is less frequently fatal than other such exposures, it would help raise the total cancer deaths to 10,000 to 20,000 in the 10,000 to 20,000 square miles downwind from the plant, the physicists estimated. These estimates assume a population density of 300 a square mile

It was pointed out here today by Dr. Wolfgang K. H. Panofsky, director of the Stanford Linear Accelerator in California, that since the exposed population was very large, these additional cancer deaths would add only 0.1 per cent to the 20 per cent of the exposed population who would be expected to die of cancer normally, according to national statistics.

Since many of the deaths would occur long after the accident, it would be hard to pin down the blame. It would also be difficult to persuade people to leave their homes in the contaminated area if the added risk of their dying of cancer was less than 1 per cent.

Dr. Panofsky presided over a session of the Physical Society's spring meeting here at which results of the study were presented by some of its dozen participants. Dr. Panofsky was chairman of a trio of internationally known physicists who reviewed the study's findings.

The two others were Drs. Hans Bethe of Cornell University and Dr. Victor F. Weisskopf of the Massachusetts Institute of Technology. Like Dr. Panofsky, both are former presidents of the society. The published report was also made public yesterday.

At a news conference before the session at the Shoreham Hotel, Dr. Harold W. Lewis of the University of California at Santa Barbara, chairman of the year-long study, noted that it was focused entirely on the safety of reactors cooled by ordinary water, such as those now in use in this country.

It did not assess other reactor types or the relative merits of other energy sources.

"We studied risks; we didn't study benefits," he said.

Publication of the report followed an assessment last week by the Environmental Protection Agency suggesting that development of breeder reactors be delayed from 4 to 12 years. Such reactors would "breed" new atomic fuel.

Their chain reactions would not only generate power but also would release neutrons with sufficient energy to penetrate the nucleus of Uranium 238, which is useless as fuel. This would convert it into plutonium 239, which can power a reactor — or a bomb. Breeder reactors are already functioning in Europe and the Soviet Union. Opponents of the program fear that, if many are built, enough plutonium will be produced to tempt seizure of the material by terrorists.

The E.P.A. report said that new estimates of national power needs through the year 2020 indicated that the requirements would not be so great as assumed in earlier planning. It proposed, therefore, that the breeder program could be delayed while alternatives were assessed.

Today, Elmer B. Staats, Controller General of the United States, submitted to Congress a report on the primary breeder development effort—that of a liquid metal fast breeder reactor. He called it "our nation's highest priority energy program." While his report is a background paper that makes no recommendations, it finds projections by the Energy Research and Development Administration "optimistic and possibly unrealistic."

The administration, which has acquired many of the functions of the defunct A.E.C., has projected operation of the first breeder reactor in 1987. It has also projected that 186 commercial-size breeders will be operating by the year 2000, and 1,178 by the year 2020.

The cost to the Government of breeder development so far has been close to $2-billion. By 2020 it has been projected to $10.7-billion. However, inquiries within the atomic industry indicated, according to Mr. Staats, "that few utilities would be willing to commit large amounts of capital until they were fairly certain that Lmfbr's [breeder reactors] would be technically and economically viable."

Major Steps Proposed

Members of the Physical Society's study group cited a long delay in carrying out a test designed to provide data needed to assess what would happen if the cooling system of present reactors failed. Much concern has been expressed in recent years about the adequacy of the emergency core cooling systems provided in case of such failures.

The performances of various elements of the reactor system in an emergency have been simulated in computer models. However, the panel found, there has been no such modeling of the system as an interrelated whole.

In view of the large number of water-cooled reactors now operating and being planned, the panel said, "we believe it is important that the reactor safety research program quickly take major steps to bring about a convincing resolution of the uncertainties in [emergency core cooling] performance."

Likewise it was found that operating personnel had not been trained adequately to respond quickly and correctly to unexpected emergencies.

As an example of a "weak point" found in the safety program it was noted that diesel generators, like those at air traffic control centers, were provided at reactor sites in case of a power failure. However, inquiries of the Federal Aviation Administration showed that such generators fail to start up 3 per cent of the time. Two are needed for high reliability.

Among the recommendations of the study were the following:

¶Greater automation of reactor operation to reduce to a minimum the danger of human error.

¶Consideration of controlled venting to prevent explosion of the containment vessel after an accident.

¶Preparations to lessen the effects of a major accident, such as providing the population with iodine pills to reduce uptake of radioactive iodine.

¶Defense measures against sabotage, including provisions for shutdown before invaders could reach their goal.

¶"Careful assessment" of benefits and costs of placing atomic plants underground or in remote "nuclear park" settings.

The study also recommended a major expansion of research on many aspects of reactor safety, including quality control of components and biological effects of an accident.

April 29, 1975

CHAPTER 2

Astronomy and Planetary Science

A radio telescope at the Maryland Point Observatory of the U.S. Naval Research Laboratory.

Courtesy The New York Times.

PROBING THE UNIVERSE

A GREAT SIXTY-INCH REFLECTOR WHICH PHOTOGRAPHS THE STARS

Wonderful Instrument Erected by the Carnegie Institution at Mount Wilson, California.

By Mary Proctor.

At last one of my dreams has been realized, for I have seen the great solar observatory, which has been erected on the top of Mount Wilson. Like the Druids of old who tramped along the roadway leading to the magic circle of stones at Stonehenge, we trudged wearily up the narrow trail, nine miles in length, in order to see the great sun temple of modern days, which is situated at an elevation of 5,886 feet above the level of the surrounding plain.

Our party, consisting of four, started at 6 o'clock on the morning of April 18, although it would have been wiser had we begun the ascent at an earlier hour. We found it fairly easy at first, with a good, broad trail sheltered by rocks and overhanging verdure, but as soon as we emerged from this shadowy retreat the trail became more rugged, and was at times exposed to the fierce glare of sunlight.

We were cheered occasionally by messages inscribed on huge boulders with green paint, such as "Two miles to Orchard Camp," "Stop here for a lunch and cup of coffee." Later on, it was, "One mile to Orchard Camp," and, finally, a huge stone with the suggestive motto: "Now smile," for we were within a short distance of what might be termed the half-way house to the summit of the mountain.

We did smile, under the mistaken impression that Orchard Camp was nestled comparatively close to the Observatory, but soon discovered our mistake when we reached that portion of the trail. The cordial welcome of the proprietor of the "hotel" served somewhat to soften the blow, but even so we were overcome when we learned that we had only traversed three miles of the way, and the best part of the trail as far as grading and steepness were concerned.

From now on, we were told, the grade would be far steeper, so we thought it advisable to rest a while before we continued on our way. It was well we did so, for the ascent was more difficult, and with the hot sun beating down on the trail though it was still only 8 o'clock in the morning, our task was an arduous one before we finally reached the summit.

Stopping occasionally for a brief rest, we saw that a transformation scene was gradually taking place in the San Gabriel Valley below. Its outlines were partially hidden beneath a sea of clouds glistening in the sunlight, forming a marvelous contrast with the deep blue of the sky overhead. Far, far away the distant mountains were outlined in a soft, misty haze, blending with the gray tints of the clouds in their vicinity.

Looking upward we obtained our first glimpse of the observatory, gleaming white against the pines and surrounding shrubbery, and we realized that we were at last nearing our destination. Following the winding trail up to the summit, we finally reached the buildings comprising the greatest solar observatory in the world.

Owing to the kindness of Prof. Walter S. Adams, we were allowed to enter the dome, which contains the wonderful sixty-inch reflector with which such wonderful photographs have been taken of the Milky Way and nebulae, those great clouds of glowing gas from which stars were originally evolved. These photographs, which are exhibited in the Observatory Museum, are of exquisite sharpness and perfection of detail, fulfilling the highest expectations of the Director, Prof. G. E. Hale. The sixty-inch reflector was erected by the Carnegie Institution of Washington, D. C., and located at Mount Wilson, in a district exempt from cyclones and hurricanes, situated above the fogs, with an agreeable climate and an atmosphere wonderfully clear from disturbance.

Becoming profoundly impressed with the work accomplished by the sixty-inch reflector in disclosing the wonders of the heavens, Mr. John Daggett Hooker, one of the most energetic, public-spirited, and prosperous citizens of Los Angeles, conceived the idea of presenting the sum of $50,000 for the erection of a still larger reflector. If a sixty-inch reflector will accomplish so much, what will a 100-inch speculum disclose?

The construction of such an instrument would be the work of a comparatively short time, and a vigorous constitution and years but little past the middle age gave promise that Mr. Hooker might live to explore, through a telescope of this magnitude, the realms far beyond what we now call the outer space. Mr. Hooker entered into communication with the Carnegie Institution, at Washington, and offered to give the sum of $50,000 for the manufacture of a reflector of that size. The proposition was accepted, and there was no delay in carrying on the work.

The order for the casting of the glass disk was given in September, 1906, to the French plate-glass companies at St. Gobain, France, and during the year between the Spring of 1907 and June, 1908, six or eight castings were made. On Dec. 2, 1908, the perfect disk arrived from France and was deposited in the Hooker Building at Pasadena, but upon being unpacked it was immediately seen that the lens was imperfect.

A second attempt was made, and, although not a perfect success, it is hoped by means of grinding and polishing that the disk may be rendered fit for its important task of photographing the wonders of the star depths.

This mammoth lens is about 13 inches thick, and weighs over four tons, and we saw it at the laboratory at Pasadena where it is now undergoing the polishing process. The task of conveying it up the trail and installing it within the Dome, which is to be its abiding place, is not the least of the difficulties to be surmounted before it starts on its labors. The famous trail along which the 60-inch reflector was taken, cost twenty-five thousand dollars in the process of widening.

The mirror was removed on a huge, long, red automobile car, run by a gasoline engine of 40 horse power, connected with a dynamo, which generated the electric current.

Turning our attention from the great reflector, we were allowed the unusual privilege of visiting the canvas-covered building, containing the 5-foot spectroheliograph of the Snow telescope, with which over one thousand photographs of the sun were taken during the year 1909. We were enabled to watch the process of the sun having its picture taken, and later obtained a glimpse of the rainbow-colored band of light, which reveals such marvellous facts concerning the constitution of the sun. These rainbow tinted bands of light which bridge the distance between the sun and the shores of our tiny isle in space, are the special study of the astronomers at the Mount Wilson Observatory.

In learning something of the origin, nature, development, and destiny of our own particular sun, we are becoming better acquainted with the peculiarities of other suns, young and old, great and small, which glitter by millions in the depths of space. With the spectroscope, spectrograph, and spectroheliograph at Mount Wilson, the astronomer detects the nature of the substances contained in the sun and stars, and ascertains the character of the changes going on in these bodies.

Fain would we have lingered on the summit of Mount Wilson until the next day, so as to watch the sun rise, but we had made arrangements to return to Los Angeles that same evening. Retracing our steps downward was a comparatively easy matter, more deftly accomplished by running rather than walking.

In two hours we reached the Orchard Camp, lingering there for a refreshing cup of coffee and sandwiches. The proprietor of the hotel urged us to hasten on our way, as it was growing late and he feared we might be overtaken by the darkness. He also insisted upon giving us a candle to light us on our way, though it was bright daylight at the time and only 6 o'clock, but later on we were sincerely grateful for his gift, especially during the latter part of our downward trip.

Leaving Orchard Camp we hurried down the trail, running the greater part of the way, and when presently we observed the evening star glowing brightly in the sky we knew it boded no good, for it heralded the approach of the darkness of night, which soon enveloped the valley like a shroud. By the fitful gleam of candle light we managed to find our way along the trail. A misstep would surely have been fatal, plunging us into the canyon below.

The roar of the cataract, and the remembrance of the steep cliffs and jagged precipices, which would have been our final resting place, made us specially careful. Occasionally we were compelled to rest a while, for we were footsore and weary after our long tramp. Looking down into the depths of the canyon we felt anything but reassured, so turning our gaze in the direction of the stars, which now glowed with unwonted splendor overhead, we took comfort in their presence.

Finally we came in sight of the lights of the valley of Pasadena, more welcome at that time than the brightest constellation in the sky, and soon reached the end of the trail in safety.

A few evenings later the writer had an interview with John Daggett Hooker, the generous donor of funds for the new reflector, and he showed us some of the photographs taken at Mount Wilson, by means of an ingenious device he has invented, and which he terms an electro-transparency.

It is illuminated with sixteen electric lights, which shine through opalescent glass. In front of this is placed one of the photographs taken with the 60-inch reflector, and the bright light endows the starlike points begemming the photographic plate with a glow somewhat resembling their own inherent light. This creates the illusion that one is in very truth gazing at the stars themselves.

Regions of lucid matter taking form,
Brushes of fire, hazy gleams, clusters
   and beds of worlds,
And beelike swarms of suns and starry
   streams.

No longer is it necessary for the student to endure the discomforts of an all-night watch with the telescope, but by

ASTRONOMY AND PLANETARY SCIENCE

Dome of the 60-Inch Reflecting Telescope at Mount Wilson.

Mounting of the 60-Inch Telescope.

means of photography the wonders of the heavens can be enjoyed at leisure when and where we please. It was the examination of these photographs, revealing the power of the 60-inch reflector, which filled Mr. Hooker with a desire to have a larger reflector made, in which still more stars may be shown, and it is his fervent hope that some day soon his dream may be realized.

From early boyhood, when he was awakened from sleep one night by his father to look at a total eclipse of the moon, he has been an enthusiast regarding astronomy, and now by means of the new reflector in course of construction he hopes that it may be possible to explore new realms far beyond what we now know of the borderland of the stellar universe.

May 21, 1911

GIANT STAR EQUAL TO 27,000,000 SUNS LIKE OURS

As for Our Little Earth, Betelgeuse Is as Big as Trillions of Globes Like It.

DIAMETER 260,000,000 MILES

Michelson Measures Colossus of the Skies Whose Light Comes to Us in 150 Years.

GREAT TRIUMPH OF SCIENCE

By the Famous Physicist Whose Researches Laid the Foundation for Einstein's Theory of Relativity.

Special to The New York Times.

CHICAGO, Dec. 29.—Professor Albert A. Michelson, the noted scientist of the University of Chicago, in a paper read today before the American Physical Society, in conjunction with the annual meeting of the American Association for the Advancement of Science, announced the perfecting of a device for measuring the diameters of stars by interference methods. This is regarded by scientists as a stupendous achievement.

Professor Michelson's paper, which bears the title of "The Application of Interference Methods to Astronomical Measurements," gave the result of the first application of the device to one of the stars in the constellation of Orion, Alpha Orionis, whose distance has already been determined by parallax methods. The common name of this star is Betelgeuse. It has been possible hitherto to determine the distances of some of the nearer stars, though the nearest is trillions of miles away, by measuring their parallax, and the masses of binary stars have been computed by other methods involving mathematical consideration of their observed period. But the method announced by Professor Michelson is most remarkable in being the first which has successfully determined the actual diametrical size of a star.

The result of the measurement of Alpha Orionis, or Betelgeuse, as accomplished by Professor Michelson's method is astounding. Betelgeuse has a diameter, it was discovered, a little more than three hundred times that of the sun, and nearly as large as that of the orbit of the planet Mars. To be explicit the diameter of Betelgeuse is 260,000,000 miles. If it were placed as near to us as the sun its brilliant surface would fill out the whole visible heavens. Compared with the sun in volume it is 27,000,000 times as great. These dimensions make the bodies in our solar sys-

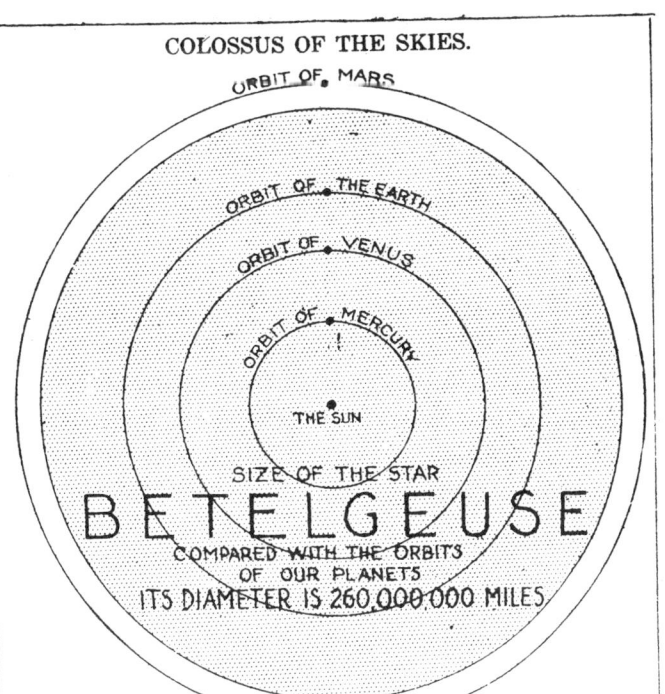

The shaded portion of the diagram shows the size of Betelgeuse compared with the orbits of our planets. As will be seen, the star would nearly fill the orbit of Mars. The sun and the planets as shown here are greatly exaggerated. The sun, for example, if correctly drawn to scale, would be only 1-150th of an inch in diameter. It would take 27,000,000 suns like it to equal Betelgeuse, although the diameter of our sun is 866,000 miles.

DIAGRAM OF ORION,

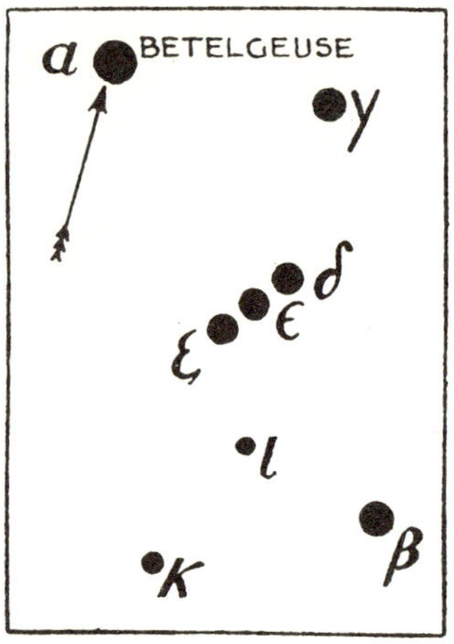

Showing the colossal star Betelgeuse, whose diameter is 260,000,000 miles; one of the most brilliant stars in the sky on a clear Winter evening

tem seem most minute and insignificant and present the conception of celestial bodies of magnitudes hitherto unmeasured and almost beyond comprehension. It would take trillions of globes like our little earth to equal Betelgeuse in size.

Betelgeuse is the northernmost star of the brilliant constellation Orion, lying above the giant's shoulder. This constellation is the finest spectacle of the skies and is conspicuous on any clear evening in Winter. The distance of Betelgeuse from the earth is perhaps 150 "light years," that is the light that strikes your eye when you look up at it started on its journey from the star at the rate of 186,000 miles a second 150 years ago.

Method First Applied at Mount Wilson.

The method used by Professor Michelson was first applied with the eight-foot reflecting telescope at Mount Wilson, in Southern California. The mirror of the telescope was obscured by an opaque cap with two slits adjustable in width and distance apart. When the instrument is focussed on a star, then, instead of an image of the star, there appears a series of interference bands arranged at equal distances apart and parallel to the two slits. When the slits are separated a distance between them will be attained at which the fringes disappear. A simple formula then gives the angle subtended by the star, and when this angle and the distance of the star are known the diameter of the star can be readily determined.

A still more powerful device is obtained by using two adjustable mirrors with an "interferometer" attachment, instead of the telescope and slits. It is this latter device which has been applied with such remarkable success in measuring the diameter of Betelgeuse.

The method of Professor Michelson has also been applied by a member of the staff at the Mount Wilson Observatory to the measurement of the star Capella, which has been known from spectroscopic evidence to be really a double star, though the two stars are so near together as to appear as one through the most powerful telescopes. By means of Professor Michelson's attachment to the eight-foot telescope it was possible to measure the minute angle of .045 second subtended by the two stars and to determine their successive positions as they revolved in their orbits. The calculated and observed results agreed with astonishing precision, the maximum error being .0001 of a second. The magnitude of this exceedingly small angle can be appreciated by the fact that it is roughly equal to that which would be subtended by a pinhead at a distance of more than a thousand miles.

Michelson's Device the Work of Years.

Professor Michelson, generally recognized as the foremost authority in the world on the subject of light, has been working on the principles involved in the mechanism of this device for a number of years. He perfected the idea last Summer, but kept the matter a secret from all except his most intimate associates.

He took the plans out to the Mount Wilson observatory in California last Summer and after some preliminary instructions left his assistants to conduct the tests with the device. Yesterday Professor Michelson received a telegram from the observatory announcing the success of the contrivance in its first experiment.

"There has never before been any means of direct measurement of the stars," said Professor Forrest Moulton, the well-known astronomer of the University of Chicago, in commenting upon the invention.

"The principles of the device are so sound that the figures may be accepted as absolutely accurate. This first test introduces us to a conception of celestial bodies of magnitudes hitherto unmeasured and almost beyond comprehension. This is just another notable achievement in the life of Professor Michelson, which has been devoted to intensive research in the subject of light."

Professor Michelson is a scientist of unique distinction in America and elsewhere. He has made one of the most accurate determinations of the velocity of light and has applied the methods of light interference to numerous delicate physical measurements of lengths and angles, among which was his accurate determination of standards of lengths for the French Government. The famous Michelson-Morley light experiment gave much of the inspiration out of which has grown the Einstein theory of relativity.

A Recipient of the Nobel Prize.

Professor Michelson is one of a very small group of scientists who have been awarded the Nobel prize for their researches.

He has been considered for many years America's leading scientist. He was born at Streino, Germany, in 1852. His primary education was obtained in the public schools of San Francisco. He then sought an appointment to the United States Naval Academy. President Grant found a place for him and he was graduated a midshipman in 1873. After a service of two years at sea he became an instructor in physics and chemistry at the Naval Academy, this really being the start of his career as scientist.

He became so enamored of scientific research that he resigned from the service in 1881, studied two years abroad and then was appointed Professor of Physics in the Case School of Applied Science in Cleveland, holding that position for six years. From 1889 to 1892 he was similarly engaged at Clark University, Worcester, Mass. He was then called to the chair of the department of physics at the University of Chicago, a position which he has ever since held.

The first international recognition of Professor Michelson came to him with his calculation of the velocity of light. More than thirty years ago by a method of his own devising he demonstrated that the earth, instead of slipping through the medium which transmits light, the so-called ether, drags this medium along with it. His experiments along this line resulted in his development of the interferometer, which then was the most efficient instrument devised for analyzing the spectra of incandescent gases and vapors.

Some time later Professor Michelson applied his interference methods to astronomical and spectroscopical measurements and ascertained the diameters of the satellites of the planet Jupiter.

After subsequently specializing in the improvement of scientific instruments Professor Michelson brought out the echelon, which has a solving power many times as great as the largest diffraction grating. It was for this distinguished service rendered to science with his echelon and interferometer that the Nobel Prize for Physics was awarded to him in 1907.

Honored by England, America, Germany.

In 1907 also Professor Michelson received the Royal Society's Copley Prize, the highest scientific honor that can be bestowed in the British Empire. He has received medals and honorary degrees from the Franklin Institute of Philadelphia, the Paris Academy of Sciences, the University of Leipsic, the University of Göttingen and many others.

In 1915 the public was informed that Professor Michelson had ruled 120,000 perfectly straight and parallel lines on a piece of metal 6 inches by 3 inches in area. The metal was a diffraction grating. It split a ray of light into the greatest number of shades of color ever seen. It measured differences between the wave lengths of various colored rays down to a millionth part of a centimeter, or one three-millionth of an inch, recording colors no human eye had ever seen.

Professor Michelson's special work has been the study of light—the light that passes between planets and stars and suns.

Professor Michelson is a brother of Charles Michelson, now the Washington correspondent of The New York World, and of Miss Miriam Michelson of San Francisco, author of "In the Bishop's Carriage," "The Madigans," "A Yellow Journalist," "The Duchess of Suds" and "The Superwoman." Professor Michelson is the father of Albert Heminway Michelson, formerly in the American consular service.

December 30, 1920

Einstein Startles Berlin by Suggesting A Possibility of Measuring the Universe

Special Cable to THE NEW YORK TIMES.

BERLIN, Jan. 28.—Professor Einstein has again startled the scientific world by an argument that it will be possible to prove the universe finite and even to estimate its size in space.

Lecturing to the Prussian Academy of Science, he said that the question of the infinity or otherwise of the universe was a quite sensible one to propound if considered in the light of practical geometry. One possibility of deciding the question, he said, lay in applying the relativity theory to the Newtonian law of reckoning the average velocity of stars as that which they must have to prevent the Milky Way from collapsing on itself through reciprocal action.

Applying certain results of the relativity theory, he came to the following conclusion: If the real velocities of stars (the measurement of which could be carried out) were less than the calculated velocities, then it was proved that real gravitations' great distances were smaller than the gravitational distances demanded by the law of Newton.

From such divergence the finiteness of the universe could be proved indirectly and it was even permitted to estimate its size.

January 29, 1921

New Star Cluster Located by Harvard Widens Universe to Two Quintillion Miles

CAMBRIDGE, Mass., Oct. 10 (Associated Press).—A new outpost of the stellar system has been found. As a result man's knowledge of the limits of the Milky Way has been extended by 50,000 to 100,000 parsecs, or light years. That is, the known stellar system probably has a great diameter of between two quintillions and one hundred quadrillions of miles and two quintillions and four hundred quadrillions of miles.

This represents a newly estimated great diameter for galactic system of 350,000 to 400,000 parsecs. It was only a few years ago that scientists placed the furthermost limits of the Milky Way at 30,000 parsecs.

This latest increase in the stellar system as it is known to man came with observation of photographs of a globular cluster of stars in the constellation Lynx by Dr. Lampland of the Lowell Observatory and by Professor Harlow Shapley and the staff of the Harvard College Observatory. The cluster, of uncommon interest because it is one of the faintest and most distant known, occurs about 60 degrees from the nearest previously known globular clusters and nearly opposite the region in which these clusters are mainly concentrated.

In the official bulletin issued at the Harvard Observatory today regarding this far-flung bunch of stars a slight qualification was made, saying that further observation was being made to justify the present belief. Observatory officials, however, said that for practical purposes it could be assumed that the cluster had been established as typical, and this being true a new boundary for the starry spaces had been found.

This new outpost of the skies is known to the astronomers as N. G. C. 2419. It appears to be 165,000 light years, or 990 quadrillions of miles from the sun, and the distance between the sun and the earth being comparatively small in the larger scheme of astronomy, it would be about the same distance from the earth.

This distance is exceeded by only two or three clusters, the Harvard bulletin said, and these are in far removed parts of the heavens.

The Harvard anouncement went into still greater figures with the statement that "the distance from the centre of the system of known globular clusters is more than 200,000 light years, and the distance separating N. G. C. 2419 and N. G. C. 6517, another faint globular cluster in the opposite part of the sky, is of the order of 350,000 light years."

October 11, 1922

ASTRONOMY AND PLANETARY SCIENCE

FINDS SPIRAL NEBULAE ARE STELLAR SYSTEMS

Dr. Hubbell Confirms View That They Are 'Island Universes' Similar to Our Own.

WASHINGTON, Nov. 22.—Confirmation of the view that the spiral nebulae, which appear in the heavens as whirling clouds, are in reality distant stellar systems, or "island universes," has been obtained by Dr. Edwin Hubbell of the Carnegie Institution's Mount Wilson observatory, through investigations carried out with the observatory's powerful telescopes.

The number of spiral nebulae, the observatory officials have reported to the institution, is very great, amounting to hundreds of thousands, and their apparent sizes range from small objects, almost star-like in character, to the great nebulae in Andromeda, which extends across an angle some 3 degrees in the heavens, about six times the diameter of the full moon.

"The investigations of Dr. Hubbell were made photographically with the 60-inch and 100-inch reflectors of the Mount Wilson observatory," the report said, "the extreme faintness of the stars under examination making necessary the use of these great telescopes. The revolving power of these instruments breaks up the outer portions of the nebulae into swarms of stars, which may be studied individually and compared with those in our own system.

"From an investigation of the photographs thirty-six variable stars of the type referred to, known as Cepheid variables, were discovered in the two spirals, Andromeda and No. 33, of Messier's great catalogue of nebulae. The study of the periods of these stars and the application of the relationship between length of period and intrinsic brightness at once provided the means of determining the distances of these objects.

"The results are striking in their confirmation of the view that these spiral nebulae are distant stellar systems. They are found to be about ten times as far away as the small Magellanic cloud, or at a distance of the order of 1,000,000 light years. This means that light traveling at the rate of 186,000 miles a second has required a million years to reach us from these nebulae and that we are observing them by light which left them in the Pliocene age upon the earth.

"With a knowledge of the distances of these nebulae we find for their diameters 45,000 light years for the Andromeda nebulae and 15,000 light years for Messier 33. These quantities, as well as the masses and densities of the systems, are quite comparable with the corresponding values for our local system of stars."

November 23, 1924

SPEEDS OF NEBULAE AID EINSTEIN BELIEF

High Velocities of Three New Finds, Scientists Say, Are Probably Illusions.

SUGGEST SPACE IS FINITE

One Travels 4,900 Miles a Second, According to Dr. Adams of Mount Wilson Observatory.

Further corroboration of the Einstein theory and of the concept that the universe may be closed, with a definite curvature and a finite volume, has been obtained, scientists hold, by the discovery at Mount Wilson Observatory in California of three nebulae apparently traveling through space at tremendous speeds which are several times the highest velocity ever observed in the heavens. According to the new evidence placed before the American Association for the Advancement of Science by Dr. Walter S. Adams, director of Mount Wilson Observatory, these three celestial bodies are moving away from the earth at speeds of 3,100, 4,600 and 4,900 miles a second.

These high velocities, it is suggested, may be illusions instead of measurements of actual speeds, caused by distortions in light waves that have traveled enormous distances through space. These distortions, it is asserted, may be due to the curvature of space predicted by Einstein.

Dr. Clyde Fisher, Curator of Astronomy at the American Museum of Natural History, said yesterday that the new discoveries are further evidence of the validity of Einstein's ideas "now generally accepted by most scientists." He said they may lead to proof that cosmic space, long held to be illimitable, had definite limits, and that light, instead of traveling on and on indefinitely, may be hedged in by a closed universe.

"The measured velocity is probably not a measure of actual motion but more likely a measure of crumpling space, a relativity effect," in the opinion of Dr. Harlow Shapley, director of the Harvard College Observatory. "One of the deductions from the general theory of relativity is that the space-time universe is finite but unbounded and that very distant objects should show a spurious velocity of recession."

Commenting recently on a previous discovery at Mount Wilson Observatory of heavenly bodies with an apparent velocity of more than a thousand miles a second, Professor Arthur S. Eddington, the famous English astronomer and physicist, declared that tremendous speeds in the skies can be explained only by the view that space is finite though unbounded.

"In such a space, light which has traveled an appreciable part of the way around the world is slowed down in its vibrations, with the result that all spectral lines are displaced toward the red," he said. "Ordinarily we interpret such a red displacement as signifying receding velocity in the line of sight.

"On ordinary grounds it would be hard to explain the striking fact that these spirals seem to move away from us at such tremendous speeds. Why should they shun us like a plague? But the phenomenon is intelligible if what has really been observed is the slowing down of vibrations consequent on the light from these objects having traveled a great part of the way around the cosmos."

June 25, 1929

NEBULA VELOCITIES SUPPORT EINSTEIN

Spectra Observations at Mount Wilson Observatory Back His Cosmology.

EXPANDING UNIVERSE SEEN

Rate of Recession of Distant Stars From the Earth Still a Puzzle for the Astronomers.

Professor Einstein further expounded his theories of an expanding universe before the Prussian Academy of Sciences Wednesday. The doctrine has received support from the observations of Professor E. P. Hubble of Pasadena, Cal., where the spectra of faint nebulae which seem to be receding into space with velocities of 12,000 miles a second have been photographed. This photographic work is described in the following story.

By MILTON U. HUMASON,
Spectroscopist of the Carnegie Institution of Washington Observatory at Mount Wilson.
Copyright, 1931, by The Associated Press.

MOUNT WILSON OBSERVATORY, Cal., June 11.—The 100-inch reflector of the Mount Wilson Observatory has been used during the last two years to photograph spectra of faint nebulae. These photographs have shown that some of the very remote nebulae, at distances greater than 100,000,000 light-years (a light-year is six million of million miles) apparently have velocities in the line of sight as large as 12,000 miles a second.

The direction of the motion is always away from the earth, that is, they appear to be receding, and Dr. Edwin P. Hubble has shown that the velocities become larger as the distance increases.

How are these spectrum photographs obtained and measured?

In order to produce a spectrum, an instrument called a spectograph must be attached to the telescope. In the spectograph is a prism through which the light of the nebula must pass. On emerging from the prism, the light is broken up into colors, each color emerging at a slightly different angle, thereby forming a spectrum.

This spectrum is then photographed with a camera lens, also mounted in the spectrograph. If the spectrum of a faint nebula is to be photographed the scale must be kept as small as possible. This is accomplished in the Mount Wilson spectrograph by the use of a high-speed camera lens designed by Dr. W. B. Rayton of Rochester, N. Y. It has a greater speed and better definition than any other short focus lens heretofore used in a stellar spectrograph.

Essential Focus Called For.

In order to obtain one of these photographs the spectograph is attached to the 100-inch telescope, and this is pointed at the nebula. The light from the nebula must enter the spectograph in exactly the place in order that it will pass correctly through the prism and camera.

If the nebula is faint, the exposure may last twenty to thirty hours.

The astronomer must keep the nebula, little more than a point of light among thousands shown by the telescope, in its correct place by constant watching during the entire exposure. The plate when finished is developed in the observatory.

In the spectrum of a nebula are certain lines which may be seen. Two of the strongest and most important are due to the element calcium. They are known as the H and K lines, and the velocity at which the nebula is traveling through space is determined by measuring the amount of shift or displacement of the calcium lines from their normal position.

In order to measure the amount of shift, it is necessary to have a reference spectrum on the same plate with the spectrum of the nebula. This is usually done by photographing the spectrum of helium directly alongside of the nebular spectrum.

On the photograph is seen the spectrum of the nebula and that produced by a helium tube. The velocity of the nebula is determined by measuring the shift of the lines in the nebula with respect to the lines in the spectrum of helium. If the lines are shifted toward the violet end of the spectrum, it would indicate a velocity of approach. If they are shifted toward the red end of the spectrum, it would indicate a velocity of recession.

In the spectrum of practically every nebula that has been photographed, the lines have been found shifted toward the red. In other words, the nebulae all appear to be moving away from the earth.

Faintest Nebula in Leo.

The faintest nebula whose velocity has been observed is in the constellation of Leo. Dr. Hubble estimates its distance as about 105,000,000 light-years and its visual magnitude at about 15.5. This nebula is therefore 6,300 times fainter than the faintest star which can be seen with the naked eye.

The velocity of this remote nebula he puts at about 12,000 miles a second.

The results of this investigation are observational facts. From them Dr. Hubble has established close relation between apparent faintness of nebulae and red shifts in their spectra. The former are confidently interpreted as distances, but the interpretation of the red shifts as velocities of recession is controversial. For the present, scientists prefer to speak of these velocities as "apparent."

June 12, 1931

STARS NOT 'DYING,' DR. RUSSELL FINDS

Tells Scientists Heat Is the Result of the Creation of Elements by Synthesis.

HOLDS SIZE IS CONSTANT

Recent Discoveries Disprove the Theory of Destruction of Electrons, He Asserts.

From a Staff Correspondent.
Special to The New York Times.

ATLANTIC CITY, Dec. 30.—There is no room for doubt that the synthesis of heavier elements out of lighter ones and hydrogen may actually occur inside the stars and thus liberate the enormous amounts of heat which they are known to generate, Professor Henry Norris Russell, noted Princeton astronomer, told the American Association for the Advancement of Science at the first Hector Maiben lecture tonight.

This synthesis, Professor Russell said, is now made more plausible than the theory that stars produce their heat by the annihilation in their interiors of electrons and protons at tremendous rates, which must eventually result in their becoming dead, burnt cinders in space. The new evidence for the theory of the synthesis of new elements is furnished, he said, by the recent discoveries in atomic physics, particularly the discovery early this year of the neutron, neuter particle of matter.

Dr. Russell further declared that knowledge recently gained in astronomy by the application of the laws of atomic physics shows that the so-called dwarf stars were not necessarily dwarfs because they were very old, as had been generally assumed. There is proof now that some of the dwarf stars were born dwarfs.

Dwarf stars are tremendously compact, with some being estimated at a mass of a ton to the cubic inch, he declared.

Their size formerly was believed to be due to the fact that they were so old that most of their mass had been burned away, but better explanations have come to light, Dr. Russell said.

Ordinary matter, he declared, cannot be compressed in such compact volume. But the matter in the stars is composed largely of ionized particles, or atoms, from which some of the electrons have been taken away. In such a state, matter is much more compressible, without the danger of "jamming." This, he said, would adequately explain the dwarf stars. As proof, he cited the existence of twin stars, "binaries," born at the same time, yet one of which is a sun while the other is a dwarf.

"The hardest problem of all," Dr. Russell said, "is the source of energy which keeps the stars shining. The rate of loss of heat is almost incomprehensibly great.

The only two processes so far suggested which would supply enough heat to last for geological time are the building up of other elements out of hydrogen, or the mutual annihilation of protons and electrons, with transformation of their whole mass into energy. The second process would liberate more than 100 times as much heat as the first; but present theories indicate that it would not happen except at temperatures of many billions of degrees.

"Atomic synthesis of the first type should be possible at much lower temperatures. This was predicted three years ago by Atkinson and Houtermans, and similar occurrences have been observed in the laboratory by Cockcroft and Walton. The theory is still provisional; but it shows that the rate of heat production increases very rapidly with the temperature. In a gas containing hydrogen, oxygen, nitrogen and carbon (all of which are very abundant in the stars) heat should be produced fast enough to keep the stars shining at temperatures of about 20 million degrees. The internal temperatures of the stars of the main sequence appear to be just of this order, and it is probable that they are deriving their heat supply from processes of atomic synthesis of this general nature. What supplies the giants (which must be much cooler inside, unless they have dense cores) is still unknown.

Relation to Mass Held Key.

"If we knew the law governing the heat production exactly, we could work out, theoretically, all the properties of a star. Vogt has shown that, whatever the form of this law (and of the other physical laws involved), stars which are all of the same composition, in every part, must show an exact (though perhaps complicated) relation between mass and luminosity, luminosity and size, and so on.

"The observed relations—especially between luminosity and size for the brighter stars—indicate strongly that this is not the fact, and hence that the stars differ from one another in composition. As they may be composed of more than 200 different elements (counting isotopes) no two of which need behave in the same way as regards heat generation, this is not surprising."

December 31, 1932

'ENERGY ON MOVE' FOUND AMID STARS

Dr. Anderson, Mount Wilson Observer, Says That Space Is Not Empty.

GAS ALSO IS REPORTED

Tenuous Medium, Containing Sodium and Calcium, Floats, He Holds, in Milky Way.

MOUNT WILSON OBSERVATORY, Calif., July 31 (AP).—Gas and radiation, the latter representing "energy on the move," fill the space between the stars, according to Dr. J. A. Anderson, astronomer of the Carnegie Institution's Mount Wilson Observatory, after a study of what most people think of as "empty" space.

But although space between the stars is not empty, he explains, the far vaster reaches of space between the star clouds or galaxies are probably actually very nearly empty, at least so far as can be determined with existing telescopes.

Floating among the stars of our own star cloud, the Milky Way, is a tenuous gaseous medium containing atoms of sodium and calcium, says Dr. Anderson. It also probably contains atoms of other known chemical elements, distributed in about the same proportion as in the atmospheres of the stars.

This gaseous material probably is about equal in mass or weight to the mass of all the stars in our galaxy, he estimates. The best counts indicate that our galaxy contains about 10,000,000,000 stars. There are tremendous spaces between them, occupied by the gaseous medium, however. If each star were considered as a drop of rain one-eighth of an inch in diameter, each drop would have to be at least four miles from every other drop to be in proportion to the distances of the huge stars from each other.

Describes Vastness of Universe.

Going on to give an idea of the vastness of the known universe, Dr. Anderson estimates there are 100,000,000 other galaxies like our own Milky Way within the range of existing telescopes. If each contains at least 10,000,000,000 stars, as does ours, all the stars in the known universe, about 1,000,000,000,000,000,000, would fill a cube about 60,000,000,000 miles square.

Besides the gaseous material in each galaxy, Dr. Anderson says, space contains vast amounts of radiation which is "in transit."

"The stars are continually emitting prodigious amounts of energy in the form of light, both visible and invisible," he states. "Since light travels radially outward from each star at the rate of 186,000 miles a second, at any given instant all the light emitted by a star during the previous year will be found within a sphere of one light-year radius (6 million million miles) having the star as its centre.

"Light emitted two years previously would be found within the next spherical shell of one light-year thickness, and so on for the light emitted in any past time. Since each of the 1,000,000,000,000,000,000 stars in the known universe is emitting light, it is clear that space must contain a great quantity of radiation, all of which is in transit."

Cosmic Radiation Included.

In addition to this ordinary radiation direct from the stars, Dr. Anderson adds, there is the so-called cosmic radiation, also in transit, which apparently is at least a hundred times greater in quantity than the ordinary radiation. Its source is uncertain.

"What happens ultimately to this radiation?" he asks. "Does it travel on forever without any change other than mere attenuation due to distance, or does it slowly become transformed into something else—for example, ordinary matter?"

Some scientists now hold that such a transformation is possible, and there is partial evidence in the laboratory that it can be done. Bigger telescopes and more ingenious astronomers of the future may answer these questions, says Dr. Anderson.

August 1, 1934

New Studies Show 'Cosmic Phoenix' Operating in Main Sequence Stars

BY WILLIAM L. LAURENCE
Special to The New York Times.

PHILADELPHIA, Dec. 26—The cosmic mechanism explaining for the first time the workings in the interiors of the most important stars in the universe, known as the main sequence stars, of which our sun is a relatively minor member, was described today before the American Physical Society, at the opening sessions of the annual convention of the American Association for the Advancement of Science. The sessions, which will embrace reports on advances during the year in all realms of science, will continue seven days.

As a result of their "journeys" into the inconceivably hot interiors of twenty-five of the giant stars of the main sequence group, including such cosmic bodies as Y Cygni, the blue giant that emits 10,000 times as much radiation as our sun and has a mass seventeen times as great, science at last obtains a logical explanation as to the processes that enable these cosmic giants to produce such inconceivable energies for billions of years.

The report was presented by Professor H. A. Bethe of Cornell University, and Dr. R. E. Marshak, of the University of Rochester. Their work was made possible as the result of a WPA project under the direction of Dr. Lyman J. Briggs, director of the United States Bureau of Standards, and Drs. A. N. Lowan and G. Blanch of New York City. With a corps of assistants, they carried out computations of mathematical tables that would have taken individual mathematicians and astrophysicists scores of years to complete.

ASTRONOMY AND PLANETARY SCIENCE

Professor Bethe, discoverer of the mechanism whereby the sun manages to radiate the equivalent of a billion billion dollars' worth of energy every second, at the rate of 1 cent a kilowatt-hour, received the A. Cressy Morrison prize of the New York Academy of Sciences in December, 1938. For further studies into the mechanism of the energy production in the interior of other stars, Dr. Bethe and Dr. Marshak were declared the winners of the same prize two weeks ago in New York. Details of the last prize-winning paper were revealed for the first time here today.

Drs. Bethe and Marshak had reported that they had worked out not only the energy production mechanisms of the main sequence stars but also of another class of stars, stellar bodies much less luminous than the sun and much denser, known as the red dwarfs.

In addition, they have gained new knowledge pointing the way for the first time toward an understanding of a third important group of stars known as the white dwarfs, possessing interior densities as high as 100,000 times that of the interior of the sun.

The white dwarfs are believed to represent a family of dying suns, and our own sun, after it has burnt up most of its hydrogen fuel in the course of the next twelve billion years, will gradually decay into a white dwarf, scientists assert.

The newer studies have revealed, Drs. Bethe and Marshak reported, that the "cosmic phoenix" operating in the interior of the sun also operates with similar effect in the interior of the main sequence stars and in that of the red dwarfs. On the other hand, the studies revealed that the phoenix principle does not operate in the energy production processes of the dying suns.

In the course of their explorations into the interiors of the various families of stars, the scientists stated, they found ranges of internal temperatures different from those hitherto accepted for the red dwarfs. They also said they had succeeded for the first time in obtaining the first approximations of the tremendous temperatures in the interior of the white dwarfs.

According to present concepts of the composition of the sun, that leave out of the calculations the presence of appreciable amounts of helium, Drs. Bethe and Marshak reported, the sun's internal temperature turns out to be 25,700,000 degrees centigrade and its internal density becomes 110 grams a cubic centimeter. Their new studies further show that the sun emits energy at a rate 100 times greater than the rate determined from observations.

At Variance With Eddington

These figures are at considerable variance with the earlier figures based on the studies fifteen years ago by Sir Arthur Eddington, according to which the sun's internal temperature is 19,600,000 degrees centigrade, while its interior density is only 75 grams a cubic centimeter.

The discrepancy between the Eddington temperature-density figures for the interior of the sun and the new Bethe-Marshak figures can be explained, they stated today, by assuming that the composition of the sun contains as much as 30 to 40 per cent of helium, the gas discovered in the sun long before it was found on earth.

The present concept of the sun's composition gives 35 per cent of its substance as hydrogen and the remaining 65 per cent of a mixture of the heavier elements, largely iron, potassium, silicon, magnesium and oxygen. This 65 per cent is generally known as the Russell Mixture, after Professor Henry Norris Russell, Princeton astronomer, who first proposed it.

The newest figures indicate, according to Drs. Bethe and Marshak, that the Russell Mixture constitutes only 25 to 35 per cent of the sun's interior, the remaining 30 to 40 per cent being accounted for by helium.

By assuming the presence of 30 to 40 per cent helium in the interior of the sun, Dr. Marshak pointed out, the temperature for the sun's interior would remain at 19,600,000 degrees centigrade, and the rate of its output of energy would agree with the observed figures of two ergs a gram a second, a rate equivalent to a billion billion dollars worth of energy every second, at the price of only one cent per kilowatt-hour.

The temperature of the sun's surface is only 6,000 degrees centigrade.

Temperatures of Red Dwarfs

On the other hand the red dwarfs, much less luminous than the sun and much denser, in which the "cosmic phoenix" has also been found to operate, seem to have internal temperatures from 2,000,000 to 3,000,000 higher than present figures. The temperature range revealed by the "phoenix" is about 15,000,000 degrees.

Vastly different conditions have been found to exist in the dying suns, or the white dwarfs. The companion to Sirius A, known as Sirius B, belongs to this group. It has the tremendous density, the studies show, of 10,000,000 grams a cubic centimeter, 100,000 times the density of the interior of the sun, while its central temperature has been determined to be 15,000,000 degrees centigrade. While its mass is the same as that of the sun, its radius is only one-hundredth of the sun's radius.

Another member of the family of dying suns studied by Drs. Bethe and Marshak is the white dwarf known as No. 40 Eridani B. Its density is 10,000 times that of the sun, and its temperature has been determined to be 30,000,000 degrees centigrade.

With these new values, Dr. Marshak stated, it becomes possible to calculate what the energy production levels will be.

The observed energy production in this family of stars, Dr. Marshak stated, can be satisfactorily accounted for by assuming that gravitational contraction, the gradual compression of these stars by the tremendous gravitational forces operating within them, manifests itself in the radiations they emanate. Calculations of the amounts of energy produced by such gravitational contraction show a rate of energy production sufficient to last at least a billion years and probably much longer.

These studies, it was asserted, account for the energy production of all the stellar bodies composing the known material universe, with two exceptions. One of these is the family of stars known as the red giants, and the other is the group known as Novae and Super-Novae, the stars that suddenly explode for no known reason, rising from cosmic obscurity practically overnight into the most brilliant stars of the sky and producing radiations equivalent to that of a billion suns, only to vanish again into obscurity in a comparatively short time.

When these two mysteries are cleared up, man will have determined the inner mechanisms of all the stellar bodies that compose the universe he lives in and the tremendous cosmic forces which make them "tick."

Speediest Rotor Described

A new ultra-centrifuge, with a rotor spinning 110,000 times a second, 2,600 times faster than the whirling propeller of the fastest airplane, was described by L. E. Machattie, Dr. Arthur L. Stauffacher, Dr. Leland B. Snoddy and Professor J. W. Beams of the University of Virginia.

The new ultra-centrifuge, operating on the principle of the cream-separator, is expected to find many important uses in the separation of the twins of the elements, isotopes, and in the purification of the ultra-microscopic entities such as viruses and protein molecules.

One of the uses for the new apparatus may be the separation in pure form of Uranium 235 (U-235), the recently discovered element promising to make possible the utilization of the tremendous stores of energy inside the atom. It has been determined within the past year that one pound of U-235, if it could be isolated in pure form, would yield energies from each pound equivalent to the energies released in the burning of 5,000,000 pounds of coal, or 3,000,000 pounds of gasoline. The ultra-centrifuge can also be used in the weighing of molecules.

The possibility that the atom of U-235 may split into three instead of two parts, thus yielding a 10 per cent higher energy than at present believed, was presented at the meeting by Dr. R. D. Present of Purdue University.

December 27, 1940

Astronomers Report Milky Way Two Vast Differing Star Groups

By CHARLES A. FEDERER Jr.
Special to The New York Times.

COLUMBUS, Ohio, Dec. 30—Our huge wheel-shaped system of stars known as the Milky Way galaxy is apparently composed of two vast aggregations of stars which differ from each other in their motions and physical characteristics, so that astronomers are now assigning celestial bodies to one of two groups which they call Population I and Population II.

The far-reaching consequences of this distinction among stars were discussed at a symposium on the relation between the spectral characteristics and the motions of stars held today at the Perkins Observatory of Ohio State and Ohio Wesleyan Universities. Leader in the proposal that stars are either of Population I or Population II is Dr. Walter Baade of Mount Wilson Observatory.

He presented evidence to show that this "class distinction" applies to other galaxies as well as our own, particularly to the great spiral nebula in Andromeda, nearest and most easily studied of the neighboring galaxies and similar to the Milky Way in many respects.

Dr. Baade has found that the spiral arms of the Andromeda system contain mostly Type One objects, and include nebulous dust and gas, whereas the nucleus of that galaxy contains Type Two objects and no dust and gas that can be observed.

Similarly, in our galaxy, stars, clusters, and other objects of Population I seem to lie chiefly in the outer regions, close to the plane of the Milky Way, and extended well out beyond the sun. They form a flattened system in rapid rotation. Our sun and most of the stars in its vicinity are included in Type I.

Type Two stars, on the other hand, are mostly nearer the center of the Milky Way but may extend farther from the plane of the system. They form a more rounded system, apparently in less rapid rotation than the stars of Population I. They also include the globular clusters, long known for their peculiar stellar make-up, for they consist of tens of thousands of stars brighter than the sun packed very closely together.

The velocities of the stars give a clue to whether they belong to Type I or II. The sun requires some 200,000,000 (two hundred million) years to make one revolution around the Milky Way center at its estimated distance of 30,000 (thirty thousand) light years from that center. The orbits of the sun and other Type I stars in its vicinity are generally thought to be nearly circular.

But some stars have very different velocities, even though they are in this region of space, and Dr. A. N. Vyssotsky, of the University of Michigan, characterized them today as possible "émigrés" from the central regions.

Some of them are big red giant stars which have strayed this far out; others are small red dwarfs which have exceedingly high velocities relative to the sun. Their actual orbits around the galactic center may be rather elongated, so that they are only temporarily this far from the center.

In the galactic star clusters, such as the famous Hyades and Pleiades in the constellation of Taurus, the distinction of stellar types also appears.

Dr. R. J. Trumpler, of the University of California, at Berkeley, pointed out that the Pleiades cluster includes stars like the sun, cooler stars and hotter stars, all the way up to bright blue stars. The Hyades, however, have no bright blue members, but there are red giant stars in the Hyades, scores of the open star clusters may be classified according to this simple difference.

Another line of attack has been carried on by Dr. P. C. Keenan of Perkins Observatory, using the spectograph attached to the 69-inch reflector here, sixth largest telescope in the world. By certain minute differences in their spectra, Dr. Keenan is able to distinguish between fast-moving and slow-moving giant yellow stars, assigning the fast-moving ones to population two.

On the basis of their positions in the galaxy, Dr. Cecilia Payne Goposchkin of Harvard Observatory, has assigned intrinsically varying stars to one or the other type. The rapidly pulsating cluster-type variables, originally found in globular star clusters, belong to Type II, whereas the classical cepheid variable stars are of Type I.

December 31, 1947

Palomar Observers Dazzled In First Use of 200-Inch Lens

By WILLIAM L. LAURENCE
Special to THE NEW YORK TIMES.

PALOMAR OBSERVATORY, Calif., June 4—Man took his first blinking glances at the heavens last night through his 200-inch eye in the Hale telescope atop Palomar Mountain and was dazzled by a new radiance from the light of distant stars, four times brighter than any similar star-image he had ever seen before.

Newspaper men and guests attending the dedication ceremonies yesterday afternoon were invited late last night to take a peek at the heavens through the big eye, equivalent in light-gathering power to that of a million human eyes.

This followed an earlier demonstration of the highly intricate mechanism and motions of the telescope and its 137-foot diameter dome, one of the great engineering achievements of modern times.

Under conditions of poor visibility we gazed, through a small eye piece, at the planet Saturn with its three-ring, 42,000-mile-wide "circus," and its satellites. Its image was magnified 700 times, that being the limit set last night by the atmospheric conditions. Magnification, however, was not the purpose for which the big eye was built.

What was unique about our glimpse of Saturn last night was the brightness of its image. It was four times as bright as any image of Saturn ever seen before by human eyes with the most powerful telescope, brighter than a bright moon on a clear night. We could see three of nine satellites, shimmering about her, and at least two of its three concentric rings.

The effulgent light made one blink at first in the darkness, producing the effect of a strong flashlight suddenly brought close to the eye.

Saturn is about 900 million miles distant from the sun, and has a mass ninety-five times that of the earth. It is a gaseous planet with a density only 0.13 that of the earth, by far the least dense of all the planets. Its surface markings usually show a yellowish zone at the Equator and a greenish cap at the Poles.

Of Saturn's nine satellites, the five inner ones move in circular orbits, and nearly in the plane of the rings. The largest, Titan, has a diameter of about 2,600 miles and a brightness 1/1,000th that of the planet.

The outer of the three flat concentric rings (A, B, and C) has a diameter of 171,000 miles, and is 10,000 miles wide. The division between rings A and B, known as the Cassini division, is about 3,000 miles wide. Ring B is about 16,000 miles wide and brighter than A. Ring C is a crepe ring, separated from ring B by A 1,000 miles. The width of the whole ring system is about 42,000 miles.

There is a clear space of 7,000 miles between the inner edge of the ring system and the planet. The thickness of the rings is about ten miles. Their structure is a flock, or swarm, of separate particles.

Last night marked one of the few times when the Hale telescope will be used for viewing any stellar body visually. For, in reality, the giant telescope is a huge camera. The 200-inch mirror is an instrument for gathering light that will enable astronomers to photograph objects a billion light-years from the earth, twice as far as is possible with the 100-inch Mount Wilson telescope.

A light-year is the distance light travels in one year at a speed of 186,000 miles per second, or about six trillion miles.

While the telescope has the light-gathering power of a million human eyes, it does not mean that a human eye with a million times its normal visual power would be able to see as far in space as this telescope will, for the human eye cannot accumulate light as does a photographic plate.

Astronomers use photography for a number of reasons. First, the eye is very limited to what it can actually see, and what it can see is recorded only in the mind of one person. Second, photographic plates can record light that the eye will miss, for they can be exposed to light from an object for any given length of time —hours if necessary. Third, a photograph becomes a permanent measurable record that can be made available and studied by all astronomers.

Telescope's Movement Shown

Equally as impressive and awe-inspiring, producing the feeling of sudden transportation to a strange never-never land where all things move at the same time in all directions, was the operation of the giant telescope itself, which was put through its paces for the benefit of newspapermen and guests.

First we saw the metal covering of the 200-inch mirror open up in the manner of petals of a gigantic flower.

We stood in wonder in the presence of man's biggest artificial eye with which man will be able to see a universe eight times the volume of the universe he could observe until now. It took 180,000 man-hours of labor to give it its parabolic (concave) surface, accurate to within two-millionths of an inch. It weighs 14¾ tons. It is about 24 inches thick at the edges and 20½ inches at the center.

By the pressure of a button at the control desk, the giant telescope, weighing a million pounds, started moving, first from east to west (right ascension) and then from zenith to horizon (declination).

Then the giant dome, 137 feet in diameter, started slowly moving about, in a manner of a merry-go-round, and one was gradually tranported to a fourth-dimensional environment, in which east and west, north and south, merged into one another and then vanished in a strange loss of awareness of space and time.

Controls Highly Elaborate

The telescope has the most elaborate control system ever used by an astronomer. It has an electric remote indicating system of right ascension (moving the yoke from east to west) and declination (moving of the tube from zenith to horizon).

At a control desk located beneath the north bearing pillars, an assistant can control the huge instrument. He can dial any star position desired, press a button and the telescope will move to that position automatically and begin following the object across the skies. This automatic setting system is accurate to less than one second of arc.

In addition to controlling right ascension and declination movements, it also has numerous switches for energizing other devices and also indicators giving the zenith angle of the telescope, position of the wind screen, rates of motion of right ascension and declination, the focus position, sidereal and Pacific standard time. Both talk box and telephone communication with the control desk is provided.

Dummy Telescope Used

By means of a small dummy telescope, rotation of the dome is controlled and synchronized with movement of the telescope, so that the dome slit is always in the proper position.

So friction-free and delicately balanced is the telescope that only a one-twelfth horsepower electric motor is required to move it at celestial rate. For faster movement a two-horsepower motor is used.

Variations in driving rate, caused by atmospheric refraction, and other effects, such as slight deformation of the telescope structure, is calculated by a mechanical computer which automatically adjusts the frequency of the time standard to the proper tracking rate.

The observatory dome is 137 feet in diameter and 135 feet high. It moves on a circular track and has split shutters riding on horizontal rails. The revolving part of the dome is of butt-welded steel construction. Its total weight is 1,000 tons. The entire building is insulated to keep temperature rise to a minimum, so that there will be little difference between inside and outside temperature when the dome is opened for observation.

June 5, 1948

STUDIES REPORTED IN STAR EVOLUTION

Findings of 2 Scientists Held 'Revolutionary Development' at Pasadena Meeting

By WILLIAM L. LAURENCE
Special to The New York Times.

PASADENA, Calif., June 28—Studies of the light of distant galaxies that started its journey to the earth 200,000,000 years ago, traveling at a steady speed of 186,000 miles per second, suggest that we may be seeing for the first time the evolution of stars as the progress from birth to maturity and death.

These studies, which promise to make it possible for man on his insignificant speck of dust to roll back the eons of time and view the cosmos at large as it appeared hundreds of millions of years ago, were outlined here today at the joint meeting of the American Astronomical Society and the Astronomical Society of the Pacific at the California Institute of Technology.

Leading astronomers such as Dr. Edwin Hubble of the Mount Wilson Observatory described them as a "revolutionary development in astronomy."

The investigations, made possible by a new highly sensitive type of photoelectric cell developed by the Radio Corporation of America during the war, were made at the Mount Wilson Observatory by Drs. Joel Stebbins and Albert E. Whitford of the Washburn Observatory of the University of Wisconsin.

The new electric eye, known also as a photomultiplier, makes possible the accurate measurement of light from distant nebulae too faint to be measured by means heretofore available.

Comparisons Were Made

With this new light detector, Drs. Stebbins and Whitford measured for the first time the faint light from distant nebulae at a distance of 200,000,000 light years away, and compared them with the light of nebulae (galaxies composed of hundreds of millions of stars) nearer to us.

These measurements revealed a phenomenon that startled the astronomers. The farther out nebula is in space, the redder is the light emanating from it. In each case it was found that the excess amount of red light was two and a half times greater than could be accounted for in the well-known and still puzzling phenomenon of the "red shift" in the spectra of nebulae.

The red shift effect is observed through the spectra of the light of the distant nebulae, whereas the extra-reddening effect observed by Drs. Stebbins and Whitford has been discovered through actual photographs of the nebulae, taken first in white light and then through filters that eliminate all the colors except red.

The spectrum red shift, namely, the displacement of all the lines in the spectra of the nebulae toward the red end, has been interpreted as meaning that the nebulae are rushing away from us and from each other at the explosive speed of 100 miles per second for each million light years, so that the nebulae at a distance of 200,000,000 light years, for example, would be receding from us at the rate of 20,000 miles per second.

Universe Expands Rapidly

This red shift effect is thus generally interpreted as meaning that the universe is expanding at an explosive rate.

Comparing the amount of the red shift in the spectra of distant nebulae with the color of the nebulae as revealed by the filtered light photographs, Drs. Stebbins and Whitford found to their surprise that the redness of the nebulae was two and a half greater all along the line than could be accounted for by the red shift effect as revealed by the spectrum.

One explanation to the new cosmic mystery is that the intergalactic spaces are filled with cosmic dust that absorb the light and thus produce an effect similar to the reddening of the sun at twilight.

What is regarded as the most logical explanation has been offered by Dr. Martin Schwarzfield of the Princeton University Observatory. According to Dr. Schwarzfield's hypothesis, the distant galaxies are redder than the nearer galaxies because we are seeing them as they were two hundred million years ago, and hence at a younger stage in their evolution than the nearer galaxies.

Now in our galaxy of the Milky Way we observe a few super-giant red stars, such as Antares and Betelgeus. These super-giant red stars are known to lead a very fast life, burning up their energy at a much faster pace than stars such as our sun, and thus growing old much faster.

Explanation of the Light

The reason we find so few of these super-giant red stars in our own immediate vicinity in the cosmos, it is believed, is because it takes much less time for the light of the galaxies in which these stars are found to reach us (since they are nearer to us) and hence we see the galaxies at a stage in their evolution when most of their red stars had been burned up as a result of their fast living. Therefore, the total light of the near galaxies is less red than the light of the distant ones.

The light of the distant galaxies, on the other hand, such as that measured by Drs. Stebbins and Whitford, has taken 200,000,000 years to reach us. This means that we are now seeing these galaxies as they were two hundred million years ago, when they still contained a very large number of the fast-burning super-giant red stars.

If further studies corroborate this hypothesis, man will be able to study his cosmos at its various stages of evolution, just as he can study trees in a forest at their various stages of growth.

The farther he reaches out in space, the younger will be the stage in the evolution of the universe. If he reaches out far enough, he may conceivably see the very beginning of time when his universe was delivered new born in the depths of space, emitting its first "cosmic cry."

June 29, 1948

Radar Yields New World of Sound; Brings 'Music of Spheres' to Earth

By WILLIAM L. LAURENCE
Special to The New York Times.

ITHACA, N. Y., Oct. 5—Radar devices developed during the war for detecting enemy submarines and airplanes have uncovered a new world of sound coming from all around the cosmos. They have provided the first scientific evidence of the existence of something along the lines of the "music of the spheres," postulated by Pythagoras more than 2,500 years ago.

The radar apparatus and techniques have brought to light for the first time a vast range of radio frequencies generated all over the cosmos, the sun, the Milky Way and other galaxies, as well as from spaces where the most powerful telescopes have so far failed to locate any stellar bodies.

The birth of a new science, radio astronomy, offspring of a union of radio engineering and astrophysics, was hailed here today at Cornell University, celebrating its eightieth anniversary this week, where astronomers, physicists, electrical engineers and pioneers in radio communication met in the first conference of its kind to discuss the new techniques for "tuning in on the universe."

The scientists attending the conference, which included representatives from England and Canada, saw two remarkable motion pictures. One was of the aurora borealis, taken by Dr. Carl W. Cartlein, of the Cornell Physics Department, who discussed the possible origin of the spectacular terrestrial "fireworks" display and their probable connection with sunspots, which, in turn, are believed to bear a direct relationship to the "song of the sun."

The second motion picture, the first of its kind, was shown by Prof. Donald H. Menzel of the Harvard College Observatory, under whose direction the picture was taken at the Harvard Coronograph Station at Climax, Col., high in the Rocky Mountains.

The motion pictures, which show the prominences on the face of the sun in their awe-inspiring activities, have revealed motions that, Dr. Menzel said, defy the laws of gravitation and all the known laws of thermodynamics.

The photos show tremendous clouds of gas constantly descending into the sun without being seen to rise from its surface. Where these clouds come from is a new cosmic mystery. Furthermore, they fall with a constant velocity, without being accelerated, as the law of gravitation dictates.

Clouds of High Temperature

Equally mystifying is the temperature of the clouds in the solar prominences. It was determined to reach 35,000 degrees centigrade, whereas the temperature of the surface of the sun is only 6,000 degrees. Where does the energy to supply the enormous temperature come from, when the surface below it is so relatively "cold"? On the face of it, this contradicts all the immutable laws of heat transfer as they exist on earth.

To complicate matters, the corona of the sun, seen during total eclipses, has a temperature reaching a million degrees. Nobody knows where the corona's heat comes from.

The pictures of the prominences covered a gigantic area of the sun 500,000 miles wide and 400,000 miles high, or 200 billion square miles. At one point of the picture the audience of scientists gasped on beholding an arch of cloud, almost a complete circle, with a diameter determined to reach a million miles, equal in size to the diameter of the sun itself. The top of the arch rose to more than half a million miles above the sun's surface.

Another high point in the picture was a solar eruption that resembled in every detail the pattern of the mushroom cloud in the explosion of an atomic bomb, mushroom and all.

Dr. Menzel reported on work done by other groups of scientists which has revealed another brand new solar mystery. It had been believed during the past twenty years that the sun was surrounded by a powerful magnetic field. Recent measurements, however, have determined that the sun has no magnetic field whatsover.

What happened to the sun's magnetic field? There are two possibilities. Either it never had one, meaning that the observations made twenty years ago were not correct or, it had a magnetic field two decades ago which somehow has been lost for no reason that can be given on the basis of our present knowledge.

In either case the mysterious disappearance of the sun's magnetic field knocks one of the major props from under the present theory of the nature of the cosmic rays, which is based on the existence of a magnetic field around the sun. Those who study cosmic rays, Dr. Menzel said, will have to look for a new explanation concerning the so-called latitude phenomenon observed in these powerful radiations bombarding the earth.

Mysterious noises coming from the direction of the Milky Way were first observed more than fifteen years ago by Dr. Karl G. Jansky of the Bell Telephone Laboratories, who tuned in on frequencies of 14.2 megacycles. It was not, however, until the refined radar equipment developed during the war was made available to scientists, that they were able to "tune in on the universe" and to listen in to the "song of the cosmos."

They found that not only does the Milky Way send messages to earth on a frequency of 14.2 megacycles, but that it keeps constantly sending on a wide band of frequencies. Not only the Milky Way, but one can "tune in" on the sun and many other galaxies in the depths of interstellar space and listen in to the "cosmic melody."

Symphonies Heard in Cosmos

The "cosmic melody," according to the radio astronomers, is not "music" to human ears. In the words of Dr. Menzel, it "sounds like a combination of gravel falling on the roof and the howling of wolves." Nevertheless, scientists concede that it may be possible some day to select certain frequencies and combine them in such a way as to make music pleasant to human ears, creating new cosmic harmonies that may counteract the present terrestrial discord. It is conceivable, for example, that instruments may be so arranged as to pick up out of the vast variety of cosmic frequencies a selection of sounds that would constitute a Beethoven symphony. In the true sense of the word, the stars would thus be made to sing, and man may dance to the "music of the spheres."

The radio astronomers are not, however, interested in such things at present. For them the cosmos, instead of being permeated with eternal silences, has suddenly been discovered to be a very noisy cosmos, full of mysterious sounds, a haunted universe forever emitting ghost-like wails.

These wailings have opened to us a new land for exploration by ever inquisitive man. They promise to open up vast regions of the cosmos beyond the reach of the telescope. What man cannot now see of the universe he may be able to hear.

October 6, 1948

GIANT 'EAR' HEARS 'GHOSTS' IN SPACE

Radio Telescope That Taps Sounds of Invisible Stars Is Demonstrated at Cornell

---

COSMIC DUST PENETRATED

Device Cost Only $30,000, as Compared With $6,000,000 for Mount Paloma's 'Eye'

By WILLIAM L. LAURENCE
Special to The New York Times.

ITHACA, N. Y., Oct. 6—A new type of telescope for exploring the cosmos, known as a radio telescope, which will serve as a giant "ear" just as the optical telescope serves as a giant "eye," was demonstrated here today at Cornell University, which is celebrating its eightieth anniversary this week.

Instead of registering the light of distant stars and galaxies invisible to the human eye, the new radio telescope tunes in on the radio waves recently found to be constantly transmitted from celestial bodies.

The instrument was described at a two-day conference on radio astronomy, the new science born as the result of the recent discovery of the existence of a "cosmic radio symphony" of electromagnetic waves of a wide range of frequencies.

The "mirror" of the telescope is a 204-inch saucer-shaped radio reflector and is thus four inches larger than the 200-inch optical mirror of the Hale telescope on Mount Palomar, Calif., the largest in the world.

The cost of the radio telescope was about $30,000, as compared with $6,000,000 for the Mount Palomar "eye." Unlike the optical telescope, it can be operated in cloudy weather, as radio waves penetrate clouds as well as haze.

Reaching "Hot Spots" of Space

Since radio waves of the range of frequencies transmitted by the Milky Way, the sun and other parts of the universe can also penetrate clouds of cosmic dust that make large areas of space opaque to optical telescopes, the new Cornell radio mirror and others being built by other institutions promise to open a new and much wider "window" into the vastnesses of space.

Just as the X-ray penetrates opaque objects impenetrable by visible light rays, the radio mirror will serve as a "celestial X-ray" to penetrate regions now invisible to the telescopic eye.

Preliminary studies with similar devices in England and Australia, it was reported, have already revealed several "hot spots" in the constellations of Cygnus, Orion, Sagittarius, Cassiopeia and about a half-dozen other regions in the sky, where the optical telescope has failed to reveal the existence of any stellar bodies.

These "hot spots" present a new type of stellar "ghosts" that can be heard without being seen. Their light is too faint to be seen by the most powerful seeing telescope or they are "extinct" cold stars which have radiated away their visible light.

Range of "the Optical Window"

The earth's atmosphere is transparent to electro-magnetic radiation near the visible portion of the spectrum. Through this "window," about one decade broad, namely, a band of wavelengths of which the frequencies at one end are ten times greater than frequencies at the other, man has obtained virtually all of his knowledge of the universe.

The optical "window" covers the range of wave lengths from the infra-red through the visible spectrum down to the ultra-violet. The discovery of "cosmic noise" has revealed the existence of a second "window" in the atmosphere which is transparent to a wide range of radio waves.

This radio "window" is about three decades wide and is located in the shorter wave radio region of the spectrum, with wave lengths running from fifty feet down to four inches, corresponding to frequencies of twenty up to 30,000 megacycles a second.

It is through the radio "window" that the information from outside the earth in the form of "cosmic noise," or static is observed by means of the radio telescope.

The result of the cooperative efforts of the mechanical, civil and electrical engineering departments of Cornell, the telescope will be used in a radio astronomy investigation jointly sponsored by Cornell and the Office of Naval Research.

How the Device Operates

Designed to withstand winds up to sixty miles per hour and to track with an angular error of less than one-half a degree, the telescope will see areas of the sky whose diameter varies from about two to thirty degrees, depending on the frequency employed.

In addition to the usual astronomical polar and declination axes, two other rotations are available, one about a vertical axis to facilitate calibration of the antenna, the other the rotation of the 17-foot parabolic reflector about its own axis for polarization studies.

The information from space is received from a sensitive receiver fed by a small antenna at the focal point of the reflector.

The sun was found to radiate at all frequencies of the electro-magnetic spectrum, thus including not only the obvious light and neighboring frequencies, but also the radio portion of the spectrum.

These radio frequencies are too weak to be detected by commercial broadcast receivers, but occasionally present interference in the form of static to the shorter wave bands. This static from the sun and other sources in space, which arrives at the surface of the earth, is the subject of the radio astronomy studies.

October 7, 1948

HYDROGEN PROVED IN HEAVENLY VOIDS

Harvard Scientists Use Giant Radio Listening Post to Detect Its Presence

---

By ROBERT K. PLUMB
Special to The New York Times.

SCHENECTADY, N. Y., June 16—The first physical measurements indicating that hydrogen is present in the "voids" between the stars of our galaxy were reported here this morning at the final session of the summer meeting of the American Physical Society.

The report marks the first time it has been established with certainty that nascent hydrogen fills the cold silent reaches between the planets, the sun and the stars of the Milky Way. Astronomical theory holds that hydrogen should be present in the ether, but it has never before been actually detected.

The report was presented by Dr. H. I. Ewen and Dr. E. M. Purcell, both of Harvard University.

They said they detected atomic hydrogen in space with a giant radio listening post, erected with a horn-type antenna devised to be trained onto the heavens.

"Ear" Tuned to Cosmos' Noise

The giant ear, tuned to the noise of the cosmos, first detected the atomic hydrogen on the night of March 25, 1951. A plot of the frequencies received by the listening ear that night showed a strong line at 1420 megacycles per second. Other investigators have reported that this frequency is that of atomic hydrogen which is almost devoid of energy. If hydrogen is energized (by heat) the frequency of its radiation is higher.

The 1420 megacycle line reached a maximum when the ear was trained to eighteen hours right ascension, which is about in the center of the Milky Way, Dr. Ewen and Dr. Purcell said.

ASTRONOMY AND PLANETARY SCIENCE

Observations of the hydrogen signal, made over a period of several hours, showed that the frequency changed slightly with the passage of time. The Harvard experimenters attributed this to the "doppler" shift caused by the motion of the solar system bearing the measuring apparatus.

The "doppler" effect is a measure of the apparent change in frequency of a signal if the listener or the source is moving. A train whistle seems to change pitch (frequency) when the locomotive moves past.

How the Stars May Form

Analysis of the hydrogen signal indicates that it comes from hydrogen at a temperature somewhere between 10 and 35 degrees above absolute zero, it was said, indicating that this might be the temperature of "space."

Current astronomical theories hold that about half the weight of our galaxy consists of hydrogen particles floating eternally in vast clouds between the stars. Condensation of such a cloud forms a new star, according to one widely-held idea.

Hydrogen has been detected in the visible (light) spectrum of stars, but this is gas contained in the atmosphere of these bodies. About 5 per cent of the galactic hydrogen is concentrated near the surface of hot stars, Dr. Ewen and Dr. Purcell said.

The hydrogen in space, measured by the Harvard group, has an apparent concentration of about one atom for each cubic centimeter of ether. The new evidence indicates that it is widely distributed through the voids.

June 17, 1951

British Astronomer Royal Supports Theory That Creation Is Continuing

By JOHN HILLABY
Special to The New York Times

LONDON, May 23—Before an audience that included some of the most famous mathematicians and physicists in the country, the Astronomer Royal, Sir Harold Spencer Jones, gave what amounted to official approval to a new theory of creation for all universal matter at a meeting of the Royal Institution tonight.

In one sense the meeting was a perfect epilogue to a week made remarkable by the experiments on the giant cosmotron at the Brookhaven Laboratory on Long Island. In the cosmotron the scientists hope to reverse the process of the atomic bomb and make subatomic matter out of energy.

Here the Astronomer Royal explained how cosmic matter (atomic or subatomic hydrogen) could be created out of the energy of the universe. In other words, what will probably take place in a laboratory is already taking place on a limitless scale in the realms of space.

The British theory of universal or continuous creation was first put forward in 1948 by two Cambridge astro-physicists, Thomas Gould and Herman Bondi, who form part of a group called the New Cosmologists. The theory was hotly debated on physical, metaphysical and philosophical grounds largely because of a series of radio talks on the subject last year by Fred Hoyle, an original astronomer with a flair for popular broadcasts.

When the controversy died down it was generally assumed that the Bondi-Gould theory had been flattened by the more conservative and classical astronomers. The suport given to it tonight by the Astronomer Royal was cautious for continuous creation will probably be proved or disproved by work on the most distant galaxies now taking place at Mount Palomar Observatory in California. At the moment, however, Sir Harold says he finds the idea "attractive," which in scientific circles is praise at a high level.

The nub of the continuous creation theory is that it provides a satisfactory answer to the problem of how the stars continually recede without disappearing altogether, and leaving the universe virtually void. The fact that they do recede has been proved by Dr. Edwin Hubble of the Mount Wilson Observatory in Arizona, who showed that light from distant galaxies lowers in pitch or reddens as it get further away.

Sustained Lemaitre Theory

This "law of recession" fitted in neatly with calculations made by Canon Lemaitre, the astronomer at Louvain University, who proved that a static universe was a physical impossibility.

As the stars must have come from somewhere, the previous school of cosmologists here considered that the galaxies must have been concentrated a few thousand million years ago in some quite small region of space. The commencement of expansion, they said, marked the beginning of time and they considered that the universe was running down. It appeared to have a finite life and was finite in extent.

These views were based on the supposition that there must have been a moment in prehistory when everything was suddenly created in a dense mass. Unfortunately, when the astronomers tried to square these views with Albert Einstein's general theory of relativity they had to use questionable mathematical devices like two-time scales and what they called the "secular variation of physical" constants.

As the Astronomer Royal explained tonight, the new cosmological theory of Bondi and Gould is not dependent on these devices. Creation, according to them, did not occur suddenly. It is a continuous process. By estimating the mass of the universe and the speed of recession of the stars they have demonstrated, at least to their own satisfaction, that the universe is in a state of balance. Its density is everywhere the same.

"Creation" Is Continuous

This, they say, is due to the continuous and spontaneous creation of an element or elemental particle to fill, as it were, the voids left by the receding stars and the galaxies.

The creation is brought about by the interchangeability of matter and energy demonstrated by Einstein.

The "matter," furthermore, is atomic hydrogen. This is an assumption, but a reasonable one because hydrogen is the simplest and most abundant universal element and astro-physical evidence indicates that all other elements have been synthesized from hydrogen.

Bondi and Gould now are awaiting more observational evidence from Mount Palomar and other observatories. Instead of a single "miracle of creation" they have substituted a continuous series of miracles, and this "steady state theory of the expanding universe," as they call it, seems to be the best explanation found so far.

May 24, 1952

SCALE OF UNIVERSE SLATED FOR CHANGE

Dr. Shapely Discloses His Measurements at Parley of Astronomical Society

By CHARLES A. FEDERER Jr.
of Harvard College Observatory
Special to The New York Times

AMHERST, Mass., Dec. 30—The dimensions of the universe are being revised drastically. The 200-inch telescope can now see two thousand million light-years into space, and the Andromeda galaxy is at least as large as ours. The Milky Way system no longer has the unique and scientifically undesirable characteristic of being the largest galaxy. The universe is expanding at a slower rate, and is twice as old as previously thought.

All this and more changes of far-reaching significance to astronomy and metaphysics result from a revised scale of distance measurements presented by Dr. Harlow Shapley of Harvard College Observatory to the American Astronomical Society here today.

The results mark the climax of a lifetime of study by Dr. Shapley of the scale of the universe. In 1916 he employed measurements of the distances of the globular clusters to establish the dimensions of the Milky Way system.

He has now crystallized into one distance-scale change the work of many astronomers who in recent years have been indicating with increasing definiteness the need for such a change. He called his results "a note on the revision of the end of the long doubt."

Key to Present Work

The key to Dr. Shapley's present work is the average brightness of the globular clusters in the large and small Magellanic clouds. These clouds are companion galaxies to the Milky Way, located in the far southern sky and invisible from the United States.

On the average, the globular clusters in the clouds have seemed to be intrinsically three or four times fainter than the globular clusters associated with the Milky Way system. This has been on the basis of a distance to the clouds of somewhat more than 75,000 light-years.

Dr. Shapley proposed that the Magellanic clouds now be "placed" 150,000 light-years away, nearly twice as far as formerly supposed. On this basis, the globulars in the clouds and in the Milky Way have the same average brightnesses.

Such a change involves, however, the scale of intrinsic brightnesses of the pulsating stars known as Cepheid variables, which have long been the yardsticks of the universe.

Distances to the magellanic clouds, to the great galaxy in Andromeda, and to other near-by galaxies whose individual stars can be seen, have been established by means of their cepheid variables. Dr. Shapley would now increase their magnitudes on the average nearly four times.

Of course, the observed brightnesses of the cepheids would remain unchanged, and for any one such star to appear as faint as it does, its distance would have to be doubled. Therefore, every galaxy is twice as far away, and the extremely small, faintest galaxies observable with the 200-inch telescope are two thousand million light-years away instead of half that far.

Other Distances Altered

This change will also affect the estimated distances of the classical cepheid variables in the Milky Way, including the star Delta in Cepheus, the prototype from which the cepheid variables were named. But the dimensions of the Milky Way would not be changed. The Milky Way is 100,000 light-years in diameter, with the sun about 25,000 light-years from the center.

At Mount Wilson and Palomar Observatories, Dr. E. P. Hubble has in the past noticed a discrepancy in brightness between the globular clusters in the Andromeda galaxy and those in the Milky Way, just as Dr. Shapley has done for the magellanic clouds. Dr. Hubble's discrepancy is also cleared up by the new distance scale. Messier 31, as the Andromeda galaxy is called, is now to be considered 1,500,000 light-years away.

This famous object, so commonly used to illustrate the nature of our spiral systems, thus must have dimensions twice those previously assigned to it. This makes it equal to or larger than our own system. Its mass is also increased many times.

The galaxies, too, have larger dimensions and masses, and the Milky Way galaxy is not the "largest and best" of the galaxies.

We can figure backward in time to the moment when the universe began its expansion from a compact "point." The new figures mean that to get to its present size the universe started expanding twice as far in the past, about three or four thousand million years ago.

December 31, 1952

DEAD STARS HELD TO LEAVE 'GHOSTS'

Briton Believes They Still Cast Radio Signals Likened to Heavenly Hydrogen Bomb

GREAT CRAB NEBULA IS KEY

900-Year-Old Remains Give Clues to the Existence of Prehistoric Supernovae

Special to THE NEW YORK TIMES.

LONDON, March 26—The theory that long-dead supernovae or exploded stars leave behind radioactive "ghosts" has been propounded by Dr. William H. Ramsey of the physics department of the University of Manchester. The remains of the stars are being sought in the most distant or extra-galactic regions of the universe by astronomers using the new radio telescopes.

The Ramsey theory is based on a series of observations on the Great Crab Nebula, an enormous mass of expanding gas first observed by Chinese astronomers in 1054 A. D. The Crab, like the former stars called Tycho and Kepler, is a dying supernova.

It is believed to be the remains of one of the thermo-nuclear explosions akin to the hydrogen bomb that occur in galaxies at the rate of about one every 500 years. Nobody knows exactly why.

Some supernovae leave visible remains behind. The luminosity of the Crab, for instance, is still 300 times as great as that of the sun although it exploded 900 years ago. Others, like Tycho's "ghost," can be detected only by the intense electromagnetic or radio signals they emit.

Dr. Ramsey set himself the task of analyzing the visible remains of the Crab. From the color of the nebular gas he estimated that the outside temperature was about 7,000 degrees centigrade (12,632 degrees Fahrenheit) but the internal temperature he thought was "at least 50,000 degrees centigrade (90,032 degrees Fahrenheit)."

Light Waves Aid Study

As no known source of energy would enable a mass of dispersed gas to maintain such exceptional brilliance for nine centuries the problem was to decide why the Crab remained so bright.

The clues he found were in the lines of the light waves of the nebula as seen through a spectroscope or prism that enables astronomers to calculate what elements are giving rise to luminosity.

The hydrogen lines, as he expected, were exceptionally weak as the Crab has probably exhausted nearly all its atomic hydrogen fuel. The strongest lines, however, were those of nitrogen and sulphur.

Dr. Ramsey then calculated that as the Crab was in a state of old radio activation it could contain only those radio elements with a relatively slow decay rate. Furthermore, because certain of these elements, the heavy ones, are relatively rare, the common elements in the Crab would have to have an atomic weight of less than seventy.

There are only four possible radio elements with such specifications: beryllium 10, carbon 14, chlorine 36 and potassium 40. Of these four, both beryllium and potassium could not easily be detected through a spectroscope, which fact left only carbon and chlorine as the possible strong luminous components of the Crab.

Two Vital Clues

Radioactive carbon, it is known, decays into nitrogen, and radioactive chlorine decays ultimately into sulphur—the two elements represented by the curious spectral lines in the Crab nebula.

Thus, by a step-by-step process of scientific deduction Dr. Ramsey has provided two vital clues to the existence of prehistoric supernovae or "ghost" stars.

The clues are the presence of nitrogen and sulphur.

Astronomers here are already looking for the remains of other supernovae. On the assumption that the relics should survive for 30,000 years, there should be at least 200 of them in the visible universe.

It has been suggested that the great Loop nebula in Cygnus is one and that the radio star in Cassiopeia is another. These "ghosts," Dr. Ramsey believes, may exist solely as radio signals or as what might be described as the dying glow of a heavenly hydrogen bomb.

March 27, 1953

SCIENTIST EXTENDS COSMIC FRONTIERS

New Data Indicate Universe Is 8 Times Larger, Twice as Old as Had Been Believed

AGE REVISED TO 4 BILLION

Celestial 'Yardsticks' at Fault as 'Big Eye' at Palomar Doubles Its Own Range

By GLADWIN HILL
Special to THE NEW YORK TIMES.

PASADENA, Calif., April 10—Evidence that the visible universe is eight times as large as previously conceived by astronomers, and twice as old, was detailed today by Dr. Walter Baade of the Mount Wilson and Mount Palomar Observatories in California.

The latest indications, he said, are that instead of being 2,000,000,000 years old, the universe is upward of 4,000,000,000. Instead of its farthest reaches being 1,000,000,000 light years away, it seems more likely that they are 2,000,000,000, he added. Such a linear doubling of dimensions, when multiplied three times to get the volume, would represent an eight-fold increase in size.

A light year is the distance that light, traveling at 186,000 miles a second, covers in a year.

The revised dimensions are corollaries of Dr. Baade's finding, from observations with both the 200-inch telescope on Mount Palomar, the world's largest, and the 100-inch telescope on Mount Wilson, that astronomers have been wrong "by a factor of about two" in their traditional calculations of celestial measurements outside the galaxy in which the Earth lies.

He reported this finding last September to the General Assembly of the International Astronomical Union at Rome, the proceedings of which are scheduled to be published soon.

How the Errors Arose

Since then, astronomers in many parts of the world have been reporting corroborative observations. One such item was reported this week by Dr. G. E. Kron of the University of California's Lick Observatory and S. C. B. Gascoigne of the Commonwealth Observatory in Australia.

A noteworthy implication of the revised dimensions is that the "big eye" on Mount Palomar has "seen" 2,000,000,000 light years away instead of the 1,000,000,000 previously announced as its maximum achievement.

The astronomers' errors, Dr. Baade explained in an interview with THE NEW YORK TIMES, arose from a miscalculation of the distances of certain stars traditionally used as "yardsticks" in computing the dimensions of the universe. These were stars of a type called the "Cepheid variables," because they occurred notably in the formation Delta Cephiae and because they "pulsated" or fluctuated in their emission of light.

However, the resultant measurements of the universe produced a number of contradictions and anomalies through the years, in the light of other observations.

A major key to the mystery developed in 1944 when it occurred to Dr. Baade that they might be dealing with two different kinds of stars, with different characteristics that might be misleading as to their distances.

Took Half a Dozen Pictures

In the next few years this was confirmed by observations both from Mount Wilson and with the "big eye," installed in 1948, using special red-sensitive photographic emulsions developed during the war.

Two "populations" of stars were found—the first characterized by

ASTRONOMY AND PLANETARY SCIENCE

"blue-light" stars of very high temperatures, 100,000 times as bright as the sun; and the second consisting predominantly of "red-light" stars of much lower temperatures and only 1,000 times as bright as the sun.

It took no more than a half dozen photographs through the "big eye," Dr. Baade said, to confirm that some of the "yardstick" Cepheids belonged to one group and some to the other, and that their different types of light connoted different scales of distance instead of the presumed uniform scale.

Drs. Lyman Spitzer and Martin Schwarzchild of Princeton University collaborated with Dr. Baade in this work. His principal assistant was Allan Sandage of the Mount Wilson - Mount Palomar staffs. The observatories are operated jointly by the Carnegie Institution and California Institute of Technology.

Dr. Baade said that it would take a couple of years to carry the error factor further and refine it even to one decimal place.

April 11, 1953

BIRTH OF UNIVERSE TRACED TO BLAST

U. S. Scientists Told of Data in Observations on Speed of Receding Galaxies

By WILLIAM L. LAURENCE
Special to The New York Times.

BERKELEY, Calif., Dec. 27—Observations of 800 galaxies show that the material universe was born in a gigantic cosmic explosion some five and a half billion years ago.

Since that original event, data discovered with the world's two largest telescopes indicate, the fragments of the explosion, countless galaxies each made up of billions of giant stars, have been receding from us and from each other at speeds directly proportional to their distance from us— the farther the distance the greater the speed.

The observations were described today at the annual meeting of the American Association for the Advancement of Science.

Studies over the last twenty years have shown that the speed of the recession is 180 kilometers for each million parsecs. The star system farthest in space so far measured was the hydra cluster at a distance of 1,100,-000,000 light-years, roughly 333,-000,000 parsecs.

This cluster, it was found, recedes from us at 60,000 kilometers per second, or one-fifth the speed of light. There is hope, it was reported, that the present equipment will make it possible to extend measurements to objects receding with speeds of 100,000 kilometers per second, one-third the speed of light, or distances of 555,000,000 parsecs.

Studies Began in 1929

Astronomical distances are measured in light-years and parsecs. A light-year is the distance light travels in one year and a parsec is 3.3 light-years. The speed of light is 300,000 kilometers (186,000 miles) a second, so that a light-year is nearly six trillion miles.

The studies were outlined today by Dr. Allan R. Sandage, staff member of the Mount Wilson and Mount Palomar Observatories. There the world's two largest telescopes, the 100 and 200 inch, respectively, are located.

Dr. Sandage collaborated in the twenty-year project with Dr. M. L. Humason of Mount Wilson and Mount Palomar and with Dr. N. U. Mayall of the Lick Observatory of the University of California.

The original observations indicating that the universe was expanding were made in 1929 by the late Dr. Edwin P. Hubble. He observed that the light of distant star systems shifted toward the red end of the spectrum and that the red shift increased with the distance of the light source.

The only interpretation to explain the red shift was that the galaxies were receding from us and from each other at a speed directly proportional to the distance of the galaxy.

Observations Extended

In 1929, however, Dr. Hubble had data on the distances and recessional speeds of only twenty-four galaxies. By 1936 Drs. Hubble and Humason had accumulated data on 106 isolated galaxies and on ten giant clusters in an effort to probe into space as far as possible and to place the observational aspects of the expansion of the universe on a definitive basis.

More data were needed, however. Consequently, an extensive program of observation was begun in 1936 by Dr. Humason at Mount Wilson and Dr. Mayall of Lick to obtain data on the red shifts of many galaxies.

The latest studies have shown, it was reported, that the speed of the recession was about one-third the speed calculated by Dr. Hubble. Rather than receding at the rate of 530 kilometers per million parsecs, as Dr. Hubble had estimated, the new observations indicate that the speed at which the galaxies recede from us and from each other is only 180 kilometers per million parsecs.

Since the speed of light is the maximum speed possible in the universe, this means that the maxium speed at which the galaxies could recede would be reached at about 1,666,000,000 parsecs, or a distance of five and a half billion light years. This corresponds to the present figures about the age of the crust of the earth and the oldest stars in our nearby galaxy.

This would mean also that the radius of the expanding universe was three times greater than it had been believed to be in 1936 and 50 per cent greater than the current figure.

December 28, 1954

RADIO-TELESCOPE TO SCAN COSMOS

Big New British Instrument Designed to Hunt for to Universe's Secrets

By KENNETT LOVE
Special to The New York Times.

JODRELL BANK, England, June 27—A huge radio-telescope that may find a key to the origin of the universe is nearing completion here.

A. B. C. Lovell, Professor of Radio-Astronomy at Manchester University and director of the Jodrell Bank Experimental Station, said yesterday the instrument should be able to plumb the limits of the universe, if the universe has any limits.

He said he expected the telescope to be sensitive enough to pick up radar echoes from an object the size of an airplane as far away as the moon.

The radio-telescope is similar in principle to large optical telescopes, but is designed to receive electro-magnetic radiations having longer wave lengths than those of light.

Stars' Secrets Still Sought

It is not yet known whether all light-emitting stars also emit radio waves. On the other hand, many sources of radio waves from outer space do not emit visible light. They are called radio stars.

Professor Lovell spoke to a group of reporters at the first showing of the radio-telescope to nonscientists. It is the largest steerable radio-telescope in the world. Valued at more than $2,000,000, it is controlled from an instrument room filled with panels and dials.

Reflector 250 Feet Across

Seen from a distance, the silvery painted girders of the instrument, holding aloft a parabolic reflector 250 feet in diameter, tower over the peaceful meadows and trees of the Cheshire countryside like a fantastically enlarged radar scanner.

The reflector, built of sheet steel, weighs 750 tons. It is rotated in a vertical plane by machinery taken from gun turrets of the scrapped battleships Royal Sovereign and Revenge. The entire structure, weighing 2,000 tons, rotates horizontally on steel rails.

The reflector bowl can make a complete vertical rotation in fifteen minutes, turning at the rate of 24 degrees a minute. The structure can rotate horizontally at the rate of 20 degrees a minute. Thus it can track space bodies, including the satellite to be launched by the United States.

H. C. Husband, the consulting engineer who designed the radio-telescope, said it would be ready to start working before October. Workmen crawled around the rim of the reflector 215 feet above ground, fitting the last few steel plates into place.

Radio signals from space are faint and they arrive on the earth's surface in a confusion of varying wave lengths. They are so faint, Professor Lovell said, that the energy of all the space signals received over the entire surface of the earth is only one-millionth the amount required to operate a flashlight.

To Span a Billion Years

The telescope is designed to select wavelengths varying from less than twenty-one centimeters to ten or twenty meters. The reflector focuses the signals on an aerial mast rising sixty-two feet six inches from the center of the bowl.

Professor Lovell said it was difficult to give a figure for the range of the radio-telescope. It depends on the strength of the signal, just as the range of the human ear depends on the volume of sound being received.

But the scientist said he expected the instrument to locate and identify radio-wave sources from 1,000,000,000 to 2,000,000,-000 light-years distant. This means it will pick up signals that started traveling toward the earth one or two billion years ago, the higher figure being close to the estimated age of the earth.

Professor Lovell said the extreme effective range of the 200-inch optical telescope on Mount Palomar, Calif., was about 200,000,000 light-years.

An advantage of the radio-telescope is that it can continue to work in daylight and in foul weather. Professor Lovell said he hoped the new instrument would help to confirm or alter present theories on the origin of the universe.

June 28, 1957

DEVICE WILL AID RADIO TELESCOPE

MASER Amplifier Developed at Harvard to Extend the Range in Space Tenfold

By JOHN H. FENTON
Special to The New York Times

CAMBRIDGE, Mass., Dec. 12—Harvard University scientists have developed a new amplifier they believe will extend the range of radio telescopes tenfold.

The device was operated for the first time Dec. 7 at the Gordon McKay Laboratory of Applied Science, it was announced today.

The amplifier is known as a three-level solid state MASER. The core of the device is a single crystal of potassium cobalticyanide about the size of the outer joint of a man's thumb. The name stands for microwave amplification by stimulated emission of radiation.

In operation, a radio telescope picks up radiation impulses from hydrogen clouds in galaxies of outer space. The MASER amplifies the range beyond that of any known radio telescope.

The amplifier's potential applications include radar systems as well as research in radio astronomy. A MASER-equipped telescope should provide a test of cosmological theories, such as the theory that the universe is expanding, according to Prof. Thomas Gold, Harvard astronomer and authority on radio astronomy.

Professor Gold also said that use of the device should be able to confirm or deny the existance of hydrogen gases between the galaxies, now only suspected.

The Harvard MASER was developed by a group of Harvard scientists, Dr. Nicholaas Bloembergen, McKay Professor of Applied Physics, who proposed the device in 1956; Dr. J. O. Artman, research fellow in applied physics, and Sidney Shapiro, a graduate student. Funds were provided under a contract with the Army, Navy and Air Force and the Harvard Division of Engineering and Applied Physics.

To Use Device in Telescope

Professor Gold said that the college observatory hoped with the assistance of the Division of Engineering and Applied Physics to apply the MASER to its sixty-foot radio telescope at the Agassiz Station at Harvard, Mass., soon. He said that the complicated and experimental nature of the device would prevent its immediate use.

The Harvard MASER is the first to run successfully on the twenty-one-centimeter band, the frequency of emission of radiation from interstellar hydrogen.

The term MASER was coined by Prof. C. H. Townes of Columbia University, who first successfully constructed amplifiers of this type at least in the free world. He used gas instead of a crystal as an operating medium.

The Russians are known to have written papers on MASERS. But there was no information available at Harvard as to whether they have any in operation.

The first operating solid-state MASER of the Harvard type invented by Professor Bloombergen was built at the Bell Telephone Laboratories a year ago.

Similar ones have been successfully built at the Massachusetts Institute of Technology Lincoln Laboratory, which deals with problems of continental defense. None of these operates at twenty-one centimeters.

The Harvard MASER is made up of potassium cobalticyanide and an intentionally introduced impurity of one-half of 1 per cent of potassium chromicyanide. The crystal is kept cold by a bath of liquid helium at 2 degrees Kelvin, a temperature only slightly above absolute zero. (minus 459.6 degrees Fahrenheit).

Electrons of the impurity in three energy levels are used. Amplification is gained by shifting electrons from level to level within the crystal.

Electrons Shift Levels

Whenever an electron falls from one level to another, as a marble on a staircase, it gives off energy at a specific wavelegnth. But if radiation of the same wave-length comes in from the outside and strikes the electron at the lower level, it is kicked back upstairs.

By measuring the amount of radiation created by the electrons bouncing back and forth on the "staircase" scientists are able to interpret the signals that have come from outer space.

Radio astronomy in the twenty-one-centimeter wave band began at Harvard. The radiation from the clouds of hydrogen in space was first recorded in 1951 by two Harvard physicists, Dr. Harold I. Ewen and Prof. Edward M. Purcell, a Nobel Prize winner.

The Harvard College Observatory initiated radio research on the hydrogen cloud in the Milky Way (our own galaxy) in 1953, with a twenty-four-foot radio telescope.

December 13, 1957

A Radio Eye 'Looks' At Heart of Galaxy

By WALTER SULLIVAN

For the first time man has been able to "see" through the dust clouds to the heart of the galaxy of which he is a part with sufficient clarity to distinguish the structure of that region.

The picture, in terms of radio noise on 8,000 megacycles, shows four centers of intensity. The two innermost ones may be twin clusters of stars, similar to those that seem to lie at the centers of some other galaxies.

The mapping has been done by Dr. Frank D. Drake of the National Radio Astronomy Observatory at Green Bank, W. Va. He used the observatory's great, dish-shaped antenna, whose diameter is eighty-five feet. With it, he swept back and forth across the region, charting the intensity of its radio emissions.

As seen from the earth, all four areas lie in a line that is almost identical to the latest estimate of the galactic plane. The latter is, in effect, the center-line of the Milky Way. The clouds of stars forming the galaxy are thought to lie in a spiral formation. This, when viewed from the earth end-on, arches the heavens in what we know as the Milky Way.

The fact that the four areas do not quite lie along the plane, as previously plotted, may require that it be slightly revised.

The galaxy was first mapped by the observatory at Leiden, in the Netherlands, using 1,420 megacycles. This showed a source of intense radio noise in the galactic center. However, because of the low frequency, the picture was too fuzzy to distinguish detailed structure.

Dr. Drake says the radio spectrum of the two center regions shows them to be "hot"—suggesting they may be clusters of stars. The outer spots are comparatively cool. Their contours indicate, Dr. Drake believes, that they may be a ring surrounding the center of the galaxy. When seen in cross section, however, this resembles two spots.

Some astronomers believe the center of the galaxy is more spherical in shape and rotates in the manner of a solid body, whereas the outer, spiral part, in which the sun lies, does not. It is thought at Green Bank that the turbulent transition zone between these two systems may produce a synchrotron-type radio noise and hence account for the "cold" ring.

The center of the galaxy lies about 25,000 light years from the earth in the direction of the constellation Sagittarius. A light year is the distance traveled in one year by light moving at 186,300 miles a second—or about 6,000,000,000,000 miles.

If the twin radio sources at the center are masses of stars, Dr. Drake believes they may be equivalent to a billion suns. As an example of such a phenomenon in another galaxy, he cites the twin regions of intense light at the heart of the spiral nebula in Andromeda. They are believed to be masses of stars.

The two inner radio sources charted by Dr. Drake are about forty light years from the galactic center. The outer sources are some 200 light years from the center.

Dr. Otto Struve, director of the observatory, believes that this discovery, together with the more detailed picture that will soon be provided by larger and better radio telescopes, should help explain the origin of the galaxies and the reason for their spiral formation.

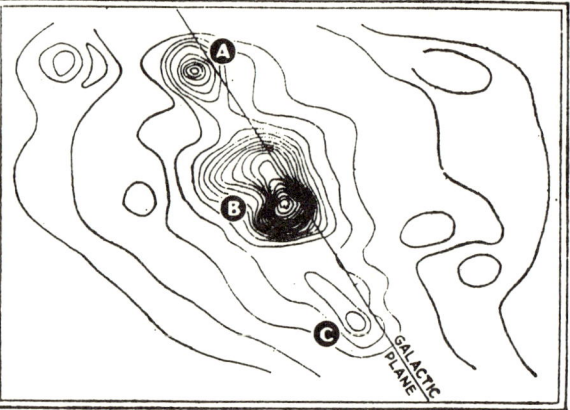

By the National Radio Astronomy Observatory

The center of our galaxy as "seen" in lines of equal radio intensity on 8,000 megacycles. Twin areas of "hot" radio noise come from the center (B), flanked by two areas of "cool" sound (A & C). The center may constitute twin masses of stars. The galactic plane is, in effect, the center-line of Milky Way. Center lies in direction of Sagittarius.

September 26, 1959

ASTRONOMY AND PLANETARY SCIENCE

RADAR ASTRONOMY

Echoes Provide Another 'Window' For Peering Into Space

By WILLIAM L. LAURENCE

For thousands of years astronomy, the oldest of the sciences, could observe the universe only through the "window" of visible light radiated by the sun and the stars. Then, in 1933, an American radio engineer, Karl Jansky, found that the stars and galaxies and the cosmic dust clouds that permeate intergalactic space continually transmit to earth radio waves of definite frequencies. This led to the birth of the new science of radio astronomy. The new science opened up a wide new electromagnetic "window." Through it man gained much new knowledge about the cosmos by tuning in on the radio messages from outer space by means of giant radio telescopes.

Back to Earth

In an article in the August issue of Scientific American, Drs. Von R. Eshleman and Allen M. Peterson, both associate professors of electrical engineering at the Radioscience Laboratory of Stanford University, describe the development of a still newer "window" into space—radar astronomy. In contrast with radio astronomy, which tunes in on radio signals arriving from outer space, the new discipline seeks answers to fundamental questions about the sun, the planets and interplanetary space from echoes of radar signals sent out from the earth.

Radar pulses were first used some thirty-five years ago to probe the ionized layers of the upper atmosphere. In 1946, workers of the United States Army Signal Corps, using more sensitive systems, were the first to detect the return of a radar signal bounced off the moon. Only twelve years later, in 1958, workers in the Lincoln Laboratory of the Massachusetts Institute of Technology employed a radar system 10,000,000 times more sensitive to send a round-trip signal to Venus. And in 1959, the article reports, radar echoes from the sun were detected by the workers at Stanford.

The steps from the moon to Venus and the sun, it is pointed out, "required tremendous increases in sensitivity, but, with a similar additional improvement in equipment, essentially all the solar system can be studied by radar."

"It may well be," Drs. Eshleman and Peterson state, "that before fully instrumented space observatories make their journeys, many of the questions they are designed to answer will be answered by the use of radar in astronomy—that is, by the use of radio signals that man himself chooses to broadcast."

Mars 'Canals'

Among the questions that radar may answer the authors list the following: Are the "canals" of Mars long marshes or simply scars of planetary evolution? Do the clouds that cover the face of Venus conceal oceans, continents and mountain ranges? Is it true that the sun has a highly variable atmosphere of charged particles that extend beyond the orbit of the earth?

Radar studies of Venus, they report, have already brought considerable improvement in estimates of the "astronomical unit," the mean distance between the earth and the sun. Further advances in radar technology, they add, "should make it possible to construct maps of the planets that will present their surfaces in detail approaching that of photographs of the moon." And "because radar is expected to penetrate Venusian or Jovian [Jupiter] clouds as readily as those of the earth, it will give man his first look at the surfaces of these planets."

Radar, they state, will also, in effect, illuminate the dark side of Mercury and distinguish the interplanetary gases that send little or no light or radio energy to earth. And with radar it will be possible to investigate at first hand regions such as the solar corona that cannot be probed by rocket.

Radar astronomy will not displace rocketry in the active exploration of space, the article states, for the two are complementary. Both are confined to operation within the limits of the solar system. While a rocket might be launched into the more distant reaches of space, it would be difficult to maintain radio contact with it much beyond the boundaries of the solar system, and "it will remain quite impossible to secure a radar echo from even the nearest star."

The major drawback to radar astronomy is the expense of powerful enough transmitters, the construction cost of which ranges from $1 to $5 per watt of power capacity. A transmitter that has just been installed at Stanford University fills a barn-sized building with more than fifty tons of electrical equipment, yet puts out only 300,000 watts. The largest transmitters now being built and planned for use in radar astronomy still have a power output averaging less than 10,000,000 watts, an output insignificant, indeed, when compared with that from cosmic sources.

The second most intense radio "star"—two colliding galaxies some 500 million light-years away in the constellation Cygnus—has been found to emit about one trillion trillion trillion watts of electromagnetic energy—which is reduced to a whisper by the time it reaches the earth. The sun has a total power output of more than 100 trillion trillion watts, mostly in the visible and infra-red region of the spectrum, but with as much as ten trillion watts at radio wavelengths.

July 24, 1960

The late Karl Jansky, founder of radio astronomy.

RIVAL COSMOLOGIES

Astronomers Differ Over Whether Universe Is Ageless or Aging

By WALTER SULLIVAN

Is the universe growing older? Or is it ageless, with new galaxies, new stars, new planets—perhaps new life—forever being continuously created.

For the past three weeks this question, marking the difference between the "steady-state" and the "big bang" cosmologic theories, has held the attention of astronomers from many lands. They discussed it first at a small conference in Santa Barbara, Calif., and then at the General Assembly of the International Astronomical Union, which ended at the University of California in Berkeley on Thursday.

In attendance, at one or both meetings, were almost all of the leading figures in the cosmological debate that has been carried on in recent years with increasing fervor. Both sides pointed to recent discoveries and observations which, they feel, support their point of view.

Heart of Controversy

In summing up the controversy Dr. Herman Bondi of Kings College at the University of London set forth the points on which virtually all are agreed:

(1) First, all of the distant galaxies are moving away from us. A galaxy is an assembly of stars, the Milky Way galaxy has a population of stars reckoned in the billions. The galaxies themselves, within range of optical or radio observation,

Werner in The Indianapolis Star
"Well, now, that's interesting."

also probably number in the billions. Expansion of the universe was postulated before it was observed. The receding motion of the galaxies, indicating expansion, has been observed in the stretching, or "reddening" of their light waves, the so-called "red-shift."

(2) The dimmer the galaxy, the greater is the red shift. This indicates that galaxies farthest away are receding fastest.

(3) The distribution of matter throughout the universe is uniform. Optical observations suggesting this have recently received strong confirmation from the observations of radio-astronomy.

(4) The farther we look, the deeper we penetrate into the past. Thus, as Dr. Bondi puts it, "geography turns into history," as astronomers explore the most remote sky areas.

In 1927 Abbé Georges Lemaître, a Belgian astronomer-cleric, published the "big bang" hypothesis to account for the expanding motion of the galaxies. If yourun this motion backwards five billion years, he said, it seems evident that all of the galaxies originated at one point.

In explaining this idea to newsmen at the conference Astronomical Union he described this starting point as "a kind of bottom in space and time." What existed at the beginning was one single entity with no structure. There followed the "beginning of multiplicity" and the flying apart of the components.

Abbé Lemaître denies the allegation of his opponents that his views are tailored to fit the concept of a divine creation. His cosmology differs from the "steady-state" view (the idea that the universe is being continuously created) for one thing in that it permits the existence of superlatives, such as a "largest" galaxy and a "hottest" star, whereas the other cosmology does not.

One of the chief problems in present day astronomy and cosmology arises from the puzzling recent discovery that some stars seem to be roughly twice as old as the "universe." This finding was cited in two of the talks given to the 1,000 astronomers who attended general sessions of the Astronomical Union Conference. One was by Dr. Jan H. Oort of The Netherlands, president of the organization, and the other by Dr. Martin Schwarzschild of Princeton University.

Dr. Schwarzschild noted that, although the beginning of the expansion of the universe is put at thirteen billion years ago, the ages of some star clusters in our galaxy are now estimated at twenty-five billion years. The latter ages are based on theoretical stellar life histories.

To account for this discrepancy, Abbé Lemaître has postulated that the presently observed rate of expansion between the galaxies (or clusters of galaxies) may apply only to the present time. The rate may have been greater in the past. Therefore the true age of the universe may be considerably more than it appears to be from the observed present rate.

The great difficulty in the opposite view—the "steady-state" concept of the universe—is that it theoretically requires the manufacture of new matter from nothing. If the universe is forever expanding, yet forever uniform in density, new matter must form. This would violate the law of the conservation of matter and energy that scientists have long considered established.

Protagonists of the steady-state view, however, point out that relativity has enforced a revision of other seemingly irrevocable laws. Furthermore, according to Dr. Fred Hoyle of Cambridge University, England, who was at Berkeley, it is necessary to produce only one atom of hydrogen in a bucketful of space every ten million years to make up for the expansion.

Looking at the Past

The most widely discussed tests of the rival cosmologies at the California meetings were those based on the assumption that if the universe is aging, we should be able to see it in a more youthful form by looking far enough into the past.

Some, for example, believe the elliptical galaxies are a senile form. And galaxies of other configurations, such as ours is believed to be a typical spiral galaxy which must be younger. But observations are difficult for the most distant galaxies observable visually are mere pinpoints and their structures cannot be determined.

A new approach that has produced results which some regard as damaging to the "steady-state" theory, is the analysis of radio signals from distant objects.

Unfortunately, the raido spectrum is not divided into emission lines whose shift towards the red can be used as a guage of distance. To get around this Dr. Martin Ryle of Cambridge University has for several years been analyzing, in terms of relative strengths, the sources of radio "noise" that dot the sky.

He assumes that the weaker a source the farther it is away. Furthermore the weaker sources should be more numerous, because the field of view expands with distance.

If the universe is thinning out, as required by the "big-bang" concept, one would expect to see a still greater density of radio sources at great distance (far in the past). This is what Dr. Ryle reports he has found.

The Other View

Backers of a "steady-state" universe, however, question whether the weak sources observed by Dr. Ryle are really very far away. Dr. Hoyle notes that only twenty or thirty radio sources have been identified optically so that their distances can be measured. And he cites recent observations made at the University of Manchester with antennas some 100 miles apart.

By a closed television circuit between the two points it has been possible to make phase comparisons of incoming radio waves and thus determine the width of the radio sources. Three sources were found to be very narrow and therefore presumably far away. A large percentage, however, were large and irregularly shaped, indicating that they are comparatively near.

Despite this challenge to Dr. Ryle's results, the "steady state" team admits that it is fighting an up-hill battle. And, Dr. Bondi says, they are going to be fighting it for a long time to come.

August 27, 196

Radio Noise Laid to Star Blasts

Theory Now Gaining Favor Over That of Galaxy Collisions

By WALTER SULLIVAN

Astronomers appear to be abandoning the idea that cataclysmic collisions between galaxies account for the extremely strong radio signals from points in distant space.

Instead it has been proposed that what may be occurring is a chain reaction of star explosions, or supernovae. In this concept millions of stars would blow up within a time span of 150 years.

While this suggestion has evoked much interest, it has not yet been generally accepted. In fact astronomers seem agreed only on their inability to explain the massive flow of radio energy from these spots, far beyond the spiral star clouds of our own galaxy.

These points in the heavens are estimated to generate 10,000,000 times as much radio energy as that produced by the entire galaxy of which the solar system is a part.

For a number of years it has been thought, by many, that the second most powerful radio source, Cygnus-A, consisted of colliding galaxies. The most powerful source of all, Cassiopeia-A, is loud because it is comparatively close. It is thought to be the aftermath of a supernova within our own galaxy.

Mount Wilson Observatory

Photograph of Cygnus-A, taken through world's largest telescope, was thought by astronomers for years to show a collision between two galaxies. The prevailing opinion now is that the supposed collision is an illusion.

When the world's largest telescope, the 200-inch instrument on Mount Wilson, in California, was trained on the spot from which the radio noise was coming, it photographed two galaxies in seeming collision.

However, recent calculations have persuaded a number of astronomers that, even if the star systems were banging into each other full tilt, this would not generate the required energy.

A supernova is, in effect, a massive nuclear explosion. It was recently proposed by Dr. G. R. Burbidge at the Yerkes Observatory of the University of Chicago that a chain reaction of such explosions within the compact heart of a galaxy could produce the required energy. He said that some star systems even appear to have survived more than one such calamity.

The problem was discussed yesterday at a seminar on radio astronomy held at the Statler-Hilton Hotel under the auspices of the American Institute of Physics and the National Association of Science Writers.

Its chairman, Dr. Fred T. Haddock, director of the Radio Astronomy Observatory of the University of Michigan, said that "prevailing opinion," had turned against the collision hypothesis.

The latter was promulgated in 1954 by Drs. Walter A. Baade and Rudolph L. Minkowski of the Mount Wilson-Mount Palomar Observatories.

It was challenged in 1958 by Viktor A. Ambartsumian of the Soviet Union, who argued that the universe was too roomy for such collisions to occur with any significant frequency.

Dr. George B. Field, assistant professor of astronomy at Princeton University, told the seminar that this roominess had developed, in part, from an increase in the distance scale by which the universe is now measured. The distances are far greater than they were thought to be a few decades ago.

Likewise, he said, the "colliding" galaxies of Cygnus-A shine with a peculiarly "hot" light, such as that seen in stellar explosions.

Dr. Philip Morrison, Professor of Physics at Cornell University, discussed the need for a new science of "anticryptography" to devise codes that could be broken easily by an intelligent being. These would be used to establish contact with civilization on other celestial bodies.

January 24, 1962

Magnetic Field in the Milky Way Confirmed by British Scientists

Special to The New York Times

LONDON, Nov. 9—Scientists at the Jodrell Bank Radio Astronomy Observatory announced today that they had succeeded in measuring part of the magnetic field of the Milky Way, the galaxy of which the solar system is a part.

A spokesman said that it was the first time positive evidence had been obtained of the existence of such a magnetic field.

The spokesman at the observatory, which is 25 miles south of Manchester, said that the delicate experiment, involving the measurement of light from a source a thousand light years away, had taken four years.

The Jodrell Bank radio telescope, the world's largest, measured a signal that had a strength of "a million-million-millionth of a watt, or about a million-millionth of the signal picked up by an ordinary television set," the spokesman said. He added that the measurement occupied 2,000 of the station's 20,000 experimental hours over the last four years.

"The magnetic field is believed to be the force which holds the hydrogen in the spiral arms from which the stars eventually form," the spokesman said. "But for years astronomers have had to invoke the existence of a magnetic field to explain its spiral structure."

He said the experiment was accomplished by application of the Zeeman effect. When light passes through a magnetic field it is split into polarized components. By measuring the separation, the spokesman said, the strength of the magnetic field was obtained. The Jodrell Bank scientists found that the magnetic field measured 25-millionths of a gauss. A gauss is the standard unit of magnetic intensity. The magnetic field at the surface of the earth is about half a gauss.

Sir Bernard Lovell, the director of Jodrell Bank, said he was "thrilled with the success of the work."

He said it was a cause of great satisfaction that the decision to pursue the program after initial negative results had been rewarded with results of such high importance.

The leader of the research group was Dr. Rodney D. Davies, a 32-year-old Australian lecturer at the station.

Jodrell Bank is operated and maintained by Manchester University.

November 10, 1962

New Clues to the Size of the Universe

Discovery of 5 Distant Objects Aids Quest for Knowledge

By WILLIAM L. LAURENCE
Special to The New York Times.
NEW YORK.

The dimensions of the universe are among the major mysteries of cosmology. Is it finite or infinite? Will it continue to expand at ever increasing speed until it reaches the velocity of light—the ultimate possible velocity in the universe—or will it stop expanding before it reaches that point? Will it eventually contract to its original dimensions before it exploded, to start a never ending cycle of expansion and contraction, or will it come to a point of no return at which all its energy had run downhill to a state of "heat-death"? These are the fundamental questions to which answers are being sought.

Distances in space are measured in terms of light-years, namely, the distance traveled by light in a year at the velocity of 186,000 miles a second, or six trillion miles. The most powerful optical telescope in existence—the 200-inch Hale reflector on Mount Palomar, Calif.—has penetrated space to a depth of 2,000 million light-years, namely, to a distance that takes light 2,000 million years to traverse. In other words, the light seen now by the 200-inch Mount Palomar reflector started its journey earthward 2,000 million years ago.

That distance appeared to be the limit of penetration by the optical telescope. Last week radio and optical astronomers of the Mount Wilson and Palomar observatories announced the discovery of what probably are the brightest objects in the universe observed so far. Their brilliance, they believe, may make it possible to penetrate two to three times deeper into space than was considered possible.

5 Bodies Recognized

Five of these super-bright celestial bodies have so far been recognized, following a series of remarkable cosmic detective work that may open a new chapter in man's quest for better understanding of the universe around him. The objects are so bright that they had been thought to be nearby stars in our own Milky Way Galaxy, but there is now growing evidence that they are very distant galaxies involved in titanic explosions.

Two of the five objects that have been studied in greater detail than the others appear to be 100 times brighter, intrinsically, then our entire galaxy of 100 billion stars. One of them is the second-most-distant object known. Others may be even farther away.

"Their surprising brightness gives us hope that we can optically identify with present telescopes much more distant objects than was ever thought possible," said Dr. Jesse L. Greenstein, professor of astrophysics at the California Institute of Technology and staff member of the two observatories, which are operated by Caltech and the Carnegie Institution of Washington.

Unable to interpret the spectral patterns of the presumed "radio stars," Drs. Greenstein and Schmidt began to wonder whether they may not be unprecedentedly bright radio galaxies as seen through the curtain of stars in the Milky Way. Working on this assumption they were amazed to find that the spectral pattern suddenly made sense.

Instead of being a radio star, they found that 3C-273 was a vast radio galaxy receding from us at the tremendous rate of one-sixth the speed of light, corresponding to a distance of nearly two billion light-years. The spectral pattern of 3C-48 showed that it was a galaxy receding at the rate of 69,000 miles a second, or more than one-third the speed of light. This indicates that 3C-48 is 3.6 billion light-years away and is thus the second-most-distant object known. It is exceeded in distance only by the radio galaxy 3C-295, which is five to six billion light-years away and is receding at almost half the speed of light. It was identified by Dr. Rudolph Minkowski at Palomar in 1960.

Gave Data on Phenomena

"These spectra," Dr. Greenstein said, "gave information about phenomena that otherwise never could be observed in galaxies. The largest satellite telescopes planned could not find or observe such faint objects."

If the two objects, 3C-273 and 3C-48, are indeed very distant galaxies, as now appears likely, the astronomers calculate that if either of them were placed beside our Milky Way Galaxy, it would be 100 times brighter. While most of their energy is being radiated in the form of light, they are also enormous emitters of energy in the longer radio wavelengths. The total energy each is radiating is equivalent to that of more than a trillion suns.

Attempts are being made to get interpretable spectra of the other three radio sources—3C-196, 3C-286, and 3C-147. It is possible, it was stated, that one or more of them may have a rate of recession greater than 3C-48, and thus may be further away. Indications are, the astronomers say, that all five are, in reality, radio galaxies in the early stages of initial explosion, the biggest explosions known in the universe.

Because of their great brightness and their easily identified ultra-violet spectral lines shifted toward the red end of the spectrum, which serve to determine their distance, the astronomers believe that it may become possible to work optically on objects that are nearly 10 to 12 billion light-years away, a distance believed to be the edge of the visible universe. The study of very distant objects is vital to cosmologists in determining the kind of universe we live in —whether it will continue expanding, stop expanding, or will eventually contract.

The five objects were first discovered by radio astronomers by their radiations of energy in the longer wavelengths, at radio frequencies. They were identified in the third Cambridge (England) Catalogue of Radio Sources as 3C-48, 3C-196, 3C-286, 3C-147, and 3C-273. Radio astronomers at Jodrell Bank in England and then at the Caltech Radio Observatory in Owens Valley focused attention on these five radio sources, which became of special interest because of their very small apparent diameters, smaller than any previously found.

Pinpoints In The Sky

The radio astronomers at Jodrell Bank, which has the world's largest radio telescope, determined that the objects were little more than pinpoints in the sky. The Caltech Radio Observatory's twin 90-foot dishes precisely located four of the objects, making it possible to identify them with optically visible galaxies. Radio astronomers in Australia precisely located the fifth object, 3C-273, when it was occulted several times by the moon. They determined the moment that the radio signal of 3C-273 went silent as the moon passed between it and the earth, and when the signal resumed after the moon had passed by. From these observations and the known position of the moon, they obtained a good position and size of the radio object.

As the radio sources were thus pinpointed, their positions were superimposed by Dr. Thomas A. Matthews of the Caltech Radio Observatory on Palomar photographs of the sky. In this way, objects which appeared optically to be bright stars were located at the sites of the five radio sources. Other significant contributions toward the identification of the five mysterious cosmic objects were made by Drs. Maarten Schmidt and John B. Oke of the two optical observatories.

The five objects were first designated as "radio stars", in contrast to "radio galaxies". A radio galaxy consists of an optically visible galaxy usually framed between two great invisible clouds of gas that radiate vast amounts of energy in the radio frequencies. It has been theorized that a grand explosion of a super star in the galaxy's nucleus ejected the gas clouds into space.

For two years, however, the spectra of the supposed "radio stars" defied interpretation. Spectra separate the lights from an object into patterns of bands that are "fingerprints" of the kind and quantities of atoms in the object. Its chemical composition, motions and temperature are "written" in the spacing and intensities of the bands.

March 26, 1963

STUDY INDICATES WATER IN SPACE

Detection of OH Gives New Radio Astronomy 'Window'

The first indication that water exists in space among the stars has been reported by scientists at the Massachusetts Institute of Technology.

This suggestion arose from their detection of groupings of atoms containing one oxygen and one hydrogen atom each. This grouping is known as the hydroxyl radical, or OH in chemical shorthand.

A radical is a group of atoms that maintains its identity through chemical changes that affect the rest of the molecule that contains it.

The existence in space of the free hydrogen atom—H—was detected by Harvard radio astronomers in 1951. The discovery of OH there thus completes the requirements for the presence of water, HOH or H_2O.

Such interstellar water would not exist as drops but as individual molecules. These would result from collisions of hydroxyl radicals with hydrogen atoms.

Collisions Are Few

This sort of event would be exceedingly rare, because there is, on the average, only one hydrogen atom in every cubic centimeter in space and only a ten millionth as many OH groups, according to the M.I.T. scientists' estimates.

Their detection of hydroxyl radicals means that scientists now have a new way of "seeing" the structure of the Milky Way, the galaxy of which the sun is a star, and the rest of the universe.

Through the emissions at a wavelength of 21 centimeters, from hydrogen atoms in interstellar space, radio astronomers have been able to map in fine detail the spiral structure of the milky way.

This sort of study has now been expanded with the discovery of the new substance to observe with radio telescopes — OH — by Professor Alan H. Barrett of the Institute's Research Laboratory of Electronics, Dr. Marion L. Meeks, Dr. Sander Weinreb and John C. Henry, all of the institute's Lincoln Laboratory.

They did their work, which was supported by the Air Force, with an 84-foot parabolic radio telescope, a digital computer and a radiation measuring device that helped the scientists zero in on the faint OH emissions at two frequencies, 1667.35 and 1665.40 megacycles, in the 18 centimeter wavelength.

The M.I.T. scientists want now to see if the dust and gas clouds in space as seen by means of radio observations on the hydrogen line are the same when viewed according to the hydroxyl radical absorption lines.

Preliminary evidence suggests there is a difference.

Thus, the new work should permit a more accurate assessment of the distribution, abundance and movement of OH, and hence of oxygen, throughout the Galaxy than hitherto possible, adding greatly to knowledge of the structure and dynamics of the cosmos.

November 14, 1963

Behavior of Distant Star-Like Objects Leads Scientists to Question Some Basic Laws of Physics

By WALTER SULLIVAN

Using a spectrum obtained with the world's largest telescope, in California, a leading Soviet astronomer has calculated that a star-like object is by far the most distant thing ever seen.

It is one of the peculiar objects recently discovered to be strong emitters of radio noise and apparently so distant that they must be billions of times brighter than average stars to be visible at all. At least nine of them have been identified in recent months.

Last week world leaders in physics and astronomy met in Dallas in an effort to explain how objects can generate so much energy. Any hypothesis seemed to violate one or more of the rules thought to limit the sizes of stars and the energy that they can produce.

Was it possible, they asked themselves, that the basic laws of physics, as we know them, are deficient? Is there some source of energy, dwarfing that of the hydrogen bomb, with which science is unfamiliar?

The Soviet estimate of the distance was made by Dr. Iosif

ASTRONOMY AND PLANETARY SCIENCE

S. Shklovsky of the Sternberg Astronomical Institute in Moscow. Dr. Shklovsky, a leading theoretician, was invited to the Dallas conference but, for reasons not explained, did not appear. His calculations were based on a spectrum obtained by Dr. Maarten Schmidt with the 200-inch telescope on Mount Palomar and published in the Astrophysical Journal.

Flying Away From Earth

The object in question is known as 3C-286. Dr. Shklovsky, writing in the Soviet Astronomical Circular, has estimated that it is flying away from the earth at 55 per cent of the speed of light. As reported in the Soviet press, he interprets this to indicate a distance of 10 billion light-years.

If correct, this means that light from the object has been traveling toward the earth for 10 billion years at 186,000 miles a second. We would thus be seeing something as it looked and behaved long before the birth of the solar system.

American astronomers who did the observing believe that Dr. Shklovsky's interpretation may be correct, but they are not sure. In any case they would equate the object's recession at 55 per cent the velocity of light with a distance of six billion light years, rather than 10. This would still make it the most distant known object.

Dr. Shklovsky's estimate was based on a rule first recognized in 1929 by Dr. Edwin P. Hubble of the United States. Dr. Hubble found, in his study of galaxies, that the farther away they are, the more their light is shifted toward the red end of the spectrum.

This he attributed to a uniform expansion of the universe. The more distant the object the faster it is flying away from us, the extent of this velocity being shown by the so-called "red-shift" of its spectrum.

If the motion of expansion were reversed into contraction, like a moving picture run backwards, the entire known universe would converge to a single point, indicating an origin some 13 billion years ago. Another by-product of Hubble's Law is that while we can look five or 10 billion years into the past, Nature forbids us to peer further than the theoretical birth of the universe.

This is true because anything that lies more than 13 billion light years away must be receding at more than the speed of light and hence is invisible. In practice astronomers do not believe they will ever be able to see out to this theoretical limit.

The great distance of the objects could indicate that they represent events that occurred early in the history of the universe. The magnitude of the distance enables science to peer back billions of years into the past. It may also mean that these are short-lived events that occur rarely and hence are tainly distributed through the universe.

Strong Signals Emitted

The most distant object previously identified is 3C-295, an object that emits strong radio signals but seems to be a galaxy, rather than a star-like body. According to the California astronomers it is receding at 36 per cent of the speed of light and is some five billion light years distant.

Another distant object, 3C-48, is one of the star-like sources and is thought to be four billion light years away. The designation "3C" refers to the Third Cambridge Catalogue of Radio Sources, prepared at Cambridge University in England.

At the Dallas meeting it was suggested that the equations at the core of Dr. Albert Einstein's General Theory of Relativity may not be entirely adequate to explain what is being seen. One physicist even suggested that atoms may be crushed out of existence in the cores of these strange bodies, releasing vast quantities of energy.

This would violate one of the present fundamental laws of physics. It says that no matter how hard you hit or squeeze an atom you cannot destroy its heavy particles, or "baryons," such as protons and neutrons. The reaction may alter the baryons, but the number left afterward must equal the number that existed at the start.

Theory Doubted

Among those who, on these grounds, doubted the total-annihilation theory was Dr. J. Robert Oppenheimer, director of the Institute for Advanced Study at Princeton, N. J., and one of the early theoreticians in this field.

The suggestion that, with enough pressure, it might be possible to squeeze a nuclear particle entirely into energy was made by Dr. John A. Wheeler, professor of physics at Princeton University and himself an important figure in the history of atomic theory.

In the fusion and fission reactions that produce nuclear energy, be they in bombs or stars, only a very small fraction of the original mass is converted into energy. The reaction in the core of the sun, for example, squeezes the nuclei of four hydrogen atoms into a single helium nucleus. A tiny bit of mass—less than 1 per cent—is left over and becomes energy.

However, the conversion follows the famous Einstein equation $E=MC^2$, meaning that the energy released is equal to the mass of the material being converted multiplied by the speed of light squared. Thus a vast amount of energy is produced by a small bit of matter.

It is thought that the star-like objects, if they derive their energy in the manner of our sun, must be equivalent to 100 million suns, burning their fuel so fast that it lasts only about 100,000 years. This is very fast. The expected life of a star like our sun is some 10 billion years.

Furthermore, as noted by Dr. Martin Schwarzschild, professor of astronomy at Princeton, no assemblage of matter comparable to 100 million suns has ever been observed working "in a concerted manner."

Rhythmical Variation

One of the puzzling features of the newly discovered objects is that the brightness of some varies rhythmically. This is particularly evident in the case of what may be the nearest of them, known as 3C-273, which is thought to be some two billion light years away.

These pulsations quite clearly take place in a 13-year cycle. There is also evidence, less clear, of a subcycle and of occasional flare-ups of great magnitude.

Because 3C-273 is bright enough to be seen, even in the telescopes of amateur astronomers, it has been possible for Dr. Harlan J. Smith of the University of Texas and his co-workers to study about 2,000 photographs of this part of the sky, taken throughout the period since 1886 largely by the Harvard College Observatory.

Using the brightness of stars appearing near 3C-273, in the photographs, for reference, he has been able to chart the systematic changes in its luminosity. The discovery that it pulses at 13-year intervals raises formidable problems.

The visible part of this object may be as much as 3,000 light years wide. In any case it would have to be very large to generate the observed light by processes with which we are familiar.

The puzzle is how anything so big can act in an organized manner, such as pulsating, at ryhthmic intervals shorter than the time for light, or anything else, to travese it.

A number of explanations for the generation of so much energy were suggested at the conference, many of them based on the more bizarre extensions of general relativity. The discussion delved into such situations as that in which an object is so dense that its gravity does not permit anything to escape from it, not even a ray of light.

Such an object would be observable, from without, only through its gravitational pull on other bodies. The concept grew out of the calculations, in 1916, by Karl Schwarzschild, a young German astronomer, father of the professor now at Princeton. The elder Schwarzschild manipulated the Einstein equations to show what would happen if a spherical body of given mass contracted within a certain radius—known now as the Schwarzschild radius.

This radius, for a body with the mass of the earth, is about one inch.

Not only would pressure at the center of such a body become infinite but the normally subtle curvatures of space, produced by gravity, would become so great that nothing could get out. Wavelengths of light and other such forms of radiation would be infinitely shifted toward the red by the gravity of such a dense mass.

Dependent on Symmetry

Such strange results are dependent on perfect symmetry of the body. However, as was pointed out in Dallas, astronomical objects in general seem far from symmetrical. One sees considerable "messiness" through telescopes, and perhaps this shows that the Schwarzschild radius is only a theoretical extreme.

One of the peculiarities of the Einstein equations is that they allow for negative mass. That is, they provide for material that repels other matter in a manner that is the reverse of normal gravitation. The fact that negative gravity has never been observed has led some to argue that the equations are imperfect.

The extreme applications of relativity were a central part of the Dallas discussions because the energy released as clouds of particles fall toward one another at catastrophic velocities seemed to offer an explanation for the observed radiation.

How such a collapse might operate has been the subject of analysis by Dr. Fred Hoyle of Cambridge University in England, Dr. William A. Fowler of the California Institute of Technology and Drs. Geoffrey and Margaret Burbidge at the University of California in La Jolla.

Brevity Poses Difficulty

The difficulty, as pointed out by various speakers, is the calculated brevity of the period from the time when the collapse has proceeded sufficiently to generate the observed radiations to the time when the material vanishes inside the Schwarzschild radius. In the words of Dr. Freeman J. Dyson of the Institute for Advanced Study, "It is all over in about a day."

A number of arguments were built on the idea that asymmetry and other factors would slow the final collapse and prevent the sudden switching off of radiated light and radio waves.

Within the galaxy, or spiral system, of about 100 billion stars, of which the sun is a part, we do not see stars more than about 65 times as heavy as the sun. Stars are in equilibrium between the outward pressure produced by energy within them and the inward pressure of gravity. It has been assumed that anything more than 65 times heavier than the sun would break up.

Dr. Hoyle suggested, as one explanation, that a galaxy could contain many stars of this marginal size, each burning up its nuclear fuel at a rapidly increasing rate until it goes into a runaway reaction that in effect is explosive. A sequence of such explosions, he argued, could account for the throbbing light of some distant objects.

However, the astronomers reported that the strong ultraviolet light from such objects is not that typical of stars, but rather what one would expect from a vast cloud of hydrogen under the influence of a superstar in the center.

Dr. Thomas A. Matthews of the California Institute of Technology suggested that the core of such a system might consist of two extremely heavy bodies, circling each other and slowly coming together. Their approach to one another would release vast amounts of gravitational energy and their rotation might account for the variation in light intensity of the system as a whole.

Powerful Gravity?

The conferees discussed the possibility that the red shift on which the distance estimates have been based might instead be an effect of extremely powerful gravity, such as that of an object contracting toward the Schwarzschild radius. In that case the objects might not be far away at all.

Dr. Jesse L. Greenstein of the Mount Wilson-Palomar Observatories argued that gravity could not account for the red shift because the gas cloud emitting the observed light does not seem strongly squeezed gravitationally. If it were, he said, it would not contain the doubly ionized form of oxygen that shows up in its spectrum. Thus gravity may be very strong deep within these bodies, but not near their radiating surfaces.

Further evidence that the objects are very distant was presented by William Jeffreys 3d of Yale University. He has analyzed the motion of these bodies in relation to nearby stars by examining photographs taken over many years. If a star is part of our galaxy it is expected to travel around the core of the galaxy with all of its brothers. The analysis convinced Mr. Jeffreys that the strange objects are, in effect, motionless and hence lie far outside the galaxy.

The most detailed information concerning the structures of the bodies came from the ingenious use of the moon by Australian radio astronomers as part of their observing equipment. As a receiver they used the huge, 210-foot dish antenna at Parkes, near Sydney.

Whenever the moon crossed the paths traveled by radio waves from one of the distant objects, the Australians made a careful record of changes in strength of the incoming signal. If the source were a pinpoint, the signals would have been cut off by the moon abruptly, as though someone had pulled a switch.

Instead, as described by Dr. Cyril Hazard of the University of Sydney, the intensity dropped in steps. The length of time required for the moon to cut off all signals showed the width of the object. The step-like decline indicated that emissions from various structural units were being cut off, one by one.

When the source emerged from behind the moon, the process was repeated, in reverse.

This furnished such precise information on the location of each source that it was possible for astronomers such as Dr. Matthews and Dr. Allan R. Sandage of the Mount Wilson-Palomar Observatories to search photographs for the star-like object associated with it. So far nine have been identified.

It was proposed that the nearest of them, 3C-273, may have experienced five successive eruptions, producing five concentric shells of trapped, highly energetic particles that generate the observed radio waves.

This object has a peculiar structure. The Australian observations showed twin radio sources and Maarten Schmidt of the Mount Wilson-Palomar Observatories then obtained photographs showing visible objects at both locations. A faint jet of luminous gas reaches 200,000 light years from one toward the other. At least one of the other distant objects also shows such a jet.

This would seem to show that the process, whatever it may be, is not symmetrical.

Some participants in the conference, sponsored by the Southwest Center for Advanced Studies, the University of Texas and Yeshiva University, feel that these objects are the most important discovery in several decades of astronomy.

Dr. Philip Morrison of Cornell University, who made the final summation, said that, in any case, "the sense of wonder and excitement that has been generated is amazing."

December 27, 1963

DISTANT 'QUASAR' FOUND TO PULSATE

Astronomers Study Photos From Heidelberg Archives

By WALTER SULLIVAN
Special to The New York Times

WASHINGTON, April 28 — Early star photographs from the archives of Heidelberg University have confirmed that one of the most distant and awesome objects in the universe pulsates in a regular, 13-year cycle.

The behavior of this recently discovered object, now called a "quasar," is one of the chief puzzles in modern science. Today those attending the annual meeting of the American Physical Society here heard some of the latest efforts at an explanation.

The chief problem is how objects so far away can shine so brightly without violating fundamental laws of physics. Also perplexing is how anything so large can pulsate at a comparatively rapid rate.

It is as though a gigantic bowl of jelly were quivering faster than the travel speed, through it, of any controlling stimulus. Under the circumstances, such organized behavior would appear impossible.

The cyclic behavior of the distant quasar has been found by Dr. Harlan Smith and his colleagues at the University of Texas in Austin.

They have checked its brightness on photographic plates going back almost a century. Today he reported, by telephone, that old photographers sent from Heidelberg had left no doubt about the 13-year cycle.

A quasar, or "quasi-stellar radio source," is so called because it resembles a star, but is so distant that it must be hundreds of billions of times brighter.

Dr. Peter Bergmann of Yeshiva University in New York told today's meeting that the pulsing quasar must have been shining thus for at least a million years.

He based this on the length of a jet of luminous gas that protrudes from one side of the quasar. Even if the gas were ejected at close to the speed of light, it must have taken at least a million years for it to reach its "present" length.

Actually the object, known by its catalog number as 3C-273, appears to be more than a billion light years away.

A light year is the distance traveled by light in a year at 186,000 miles a second. Hence an object a billion light years away is seen as it was a billion years ago.

Dr. Germann discussed the widely held belief that the energy of the quasars is generated by the falling of material into their cores.

He cited the proposal of some that such an object would eventually vanish because it would become so dense that no light or other radiation could escape from it. The effect derives from the influence of gravity on light in the relativity theory.

The earth would have to be compressed to the size of a pingpong ball for its gravity to be sufficiently intense for such a vanishing act.

Only about a dozen quasars have been identified since their existence was recognized, hardly more than a year ago. They were the subject of an international conference in Dallas last December.

April 29, 1964

Finding of Blue Galaxies Backs 'Big Bang' Theory

By WALTER SULLIVAN

The discovery of a "major new constituent of the universe" was reported yesterday by the Mount Wilson and Palomar Observatories in California. It appears that an entire category of peculiar blue objects, hitherto thought to be stars on the outskirts of the Milky Way, the galaxy, or star system, of which the sun is a part, are, in fact, enormous objects of unmatched brilliance that populate the entire universe.

They resemble the strange, recently discovered "quasars," except they are not sources of strong radio emission, and are so numerous that they should enable astronomers to determine the nature of the universe.

A preliminary examination of the evidence, according to Dr. Allan Sandage of Mount Palomar, suggests that the universe is not infinite, but rather a closed system that pulsates.

Expansion Would Stop

This would mean that the currently observed expansion of the universe would ultimately be reversed. The far-flung galaxies would fall back together, converging to be annihilated in an explosion from

ASTRONOMY AND PLANETARY SCIENCE

whose Promethean fire a new universe — and ultimately new life — would be born.

"The clues," said Dr. Sandage, "indicate that our universe is a finite, closed system originating in a 'big bang,' that the expanding universe is slowing down, and that it probably pulsates once every 82 billion years."

Dr. Sandage has named the new objects "quasi-stellar blue galaxies." They seem to be 500 times more plentiful than the quasars. It is estimated that 100,000 of them are within range of the large telescopes that can see stars as dim as the 19th magnitude.

These "blue galaxies" are still very scarce when compared with normal galaxies such as the Milky Way. Dr. Sandage estimates that there is only one blue galaxy for every 100,000 of the normal type of galaxy.

However, like the quasars, the blue galaxies are so extraordinarily brilliant that they can be seen out to distances far beyond any other objects.

Since February, when the search for the blue galaxies began, relative distances have been determined for three and one of them proved to be the second most distant object known. The farthest of all is a quasar known as 3C-9. The actual distance cannot be determined until the geometry of the universe has been established.

Distance Estimated

However, if Dr. Sandage is right about the geometry, 3C-9, or what is left of it, is now 8.7 billion light years away.

Since one light year is the distance that light travels in a year at 186,000 miles a second, these farthest-out objects are seen as they were billions of years ago.

Dr. Sandage believes it is now possible with the 200-inch telescope on Mount Palomar, the world's largest, to look back in time 93 per cent of the way to the last "big bang" explosion.

This volume of space would encompass one-third of the universe, if it is a closed system. It might also extend to the "particle horizon" — the time before condensation of the particles of the universe into structures such as gas clouds, stars and galaxies.

It is this range of vision that promises to reveal definitely the geometry of the universe: Whether it is expanding into infinite space and will continue doing so indefinitely or whether its space is curved in upon itself, limiting the expansion and suggesting a cycle of repeated expansions and contractions.

The discovery of these strange new objects, 100 times as brilliant as an ordinary galaxy, comes on the heels of two other discoveries that bear on the nature of the universe.

One has been the existence of the quasars, which may be galaxies in the process of birth, evolving later into the blue galaxies and, perhaps, finally, into normal galaxies.

Dr. Allan Sandage of the Mount Palomar observatory.

'Flash' Remnants

The other observation, by scientists of the Bell Telephone Laboratories in Holmdel, N. J., was of what some believe to be the primordial flash in which the present universe was born.

Because of the billions of years that have passed since then and the steady expansion of the universe, it had been calculated that the light waves from this flash must have been stretched until they would appear as very short radio waves. It was these predicted radio waves that were reportedly observed.

Not only do these discoveries have great philosophical and scientific implications; it is hard to see how they can fail to influence the creative currents of our time.

The realization, during the Renaissance, that the earth is not central to the cosmos deeply affected the creative geniuses of that time. If Dr. Sandage's preliminary interpretation of the data is correct, and it is established that the universe is ultimately bound to collapse upon itself, this too will have enormous implications.

The cataclysm would not, of course, be imminent. The universe would expand for some 41 billion years, of which about 12 billion have elapsed. Then it would contract for an equal period before the final implosion.

Such stars as our sun have a life span of about 10 billion years, our own being in middle age. As such stars grow old they suddenly expand, vaporizing any nearby planets. Thus it has been known for some time that the solar system is doomed.

As the British mathematician and philosopher, Bertrand Russell has put it:

"All the labors of the ages, all the devotion, all the inspiration, all the noonday brightness of human genius, are destined to extinction in the vast death of the solar system."

It has been argued that, while each inhabited world of the universe is thus mortal, such civilizations may be able to communicate with one another, passing on to young worlds the histories, achievements and wisdom of aging ones.

If, however, the universe periodically collapses, then all life is destroyed and must evolve anew.

A critical feature of the observation is the change in the rate at which the universe is expanding. Dr. Albert Einstein said, in effect, "Give me the deceleration and I will give you the mass and curvature of the universe."

The situation can be likened to the observation of a missile fired into space. The tug of the earth's gravity slows it and, by observation of this deceleration, the strength of gravity, and hence the total mass of the earth, can be estimated.

Furthermore, the rate indicates whether the missile has enough velocity to escape the earth's gravity and sail off endlessly into space or will ultimately fall back to earth.

In the case of the universe, the gravity of all the galaxies appears to counteract their expansion but it has been unclear whether or not they have enough expansion velocity to escape from one another.

Relativity Effect

The picture is further complicated by the effects of relativity.

For example, gravity all through the universe causes light to bend. It was Dr. Einstein who first proposed that gravity would bend light and his prediction was demonstrated in the eclipse of 1919, when the light from a star could be observed as it skirted the sun. But it had long been suspected that the basic geometry of space is curved.

The idea dates back to the late 18th century when the great mathematician, Karl Friedrich Gauss, as a youth, sought to prove that parallel lines will never meet.

His failure to do so led him to suspect that there was something slightly imprecise about the geometry of Euclid that is taught in school.

It seemed possible that, if parallel lines were extended far enough, they might curve enough to meet.

This was pursued during the first half of the 19th century by Nikolai I. Lobachevsky in Russia, G.F.B. Riemann in Germany and Johann Bolyai in Hungary. They developed various geometries suggesting that space is curved.

The question was, is the curvature "positive" like that on the surface of a sphere, or negative and outward bending? The former would mean the universe is closed in upon itself. The latter would indicate that the cosmos is limitless.

There are three possibilities: Either the curvature is positive, negative or space is Euclidian with no curvature. It should be possible to determine which of these is correct by observing the manner in which the rate of expansion, within the universe, has changed over its lifetime.

Expansion Measured

The present expansion rate, as demonstrated by the nearest galaxies, is such that galaxies a million light years apart are withdrawing from each other at about 14 miles a second. Galaxies twice that far apart are separating at twice that speed, and so forth.

If the expansion rate had always been the same, then this pattern would apply all the way out to the limits of observation.

However, several years ago Dr. Sandage reported that the expansion observed at great distances — and therefore far in the past — was faster than at present among nearby galaxies. The new studies have greatly extended the range of these observations. Dr. Sandage's paper in the current issue of The Astrophysical Journal presents expansion rate data from three of the new blue galaxies and 10 of the quasars. The results, he believes, conform with a positively curved, pulsating universe.

They are not conclusive. However, he added, "the evidence is inconsistent with the steady state theory."

The latter proposes that the universe is boundless in time and space, that it has always been and always will be as it is now. This would require the creation of new matter to fill the spaces produced by the constant expansion.

The final answer, Dr. Sandage said, may be forthcoming "within the next few years." He stated this in the announcement made public by the Mount Wilson and Palomar Observatories. The latter are operated jointly by the California Institute of Technology and the Carnegie Institution of Washington.

June 13, 1965

Scientists Trace Birth of Universe With Light Waves

By WALTER SULLIVAN

Five groups, working independently, have identified what they take to be remains of a primordial flash that occurred when the present universe was born.

Because the universe has been expanding ever since this cataclysmic event, light waves from the fireball, being part of the universe, have expanded with it.

In other words, their wave lengths have steadily lengthened until they are no longer in the visible part of the spectrum. They have been stretched into microwaves, like those used in radar.

Variety of Evidence

These emissions apparently filled the universe like a thin gas. In fact, at a conference last week in Miami Beach, where these discoveries were disclosed, the emissions were termed a "photon gas" — a hitherto unobserved phenomenon. Photons are parcels of light, or "light waves."

The Miami Beach conference brought to notice a number of other startling observations relating to the nature of the universe. It was a small gathering of astronomers, physicists and theoreticians from various countries called to consider "the observational aspects of cosmology."

The general chairman and host to the conference was Dr. S. Fred Singer, dean of the School of Environmental and Planetary Sciences at the University of Miami. The meeting dealt with a variety of evidence bearing on such questions as whether the universe was born from nothing, some 13 billion years ago, or whether it was preceded by an endless series of earlier universes.

In the latter theory, each universe was born in an explosive fireball. It then expanded, presumably generating life on some of its planets. Finally, succumbing to the gravitational attraction of all its parts, it fell back together, producing a new explosion and a new universe.

The idea that light from the last fireball might still be observable appears to have first been proposed by Dr. Robert H. Dicke, professor of physics at Princeton University, and his associates, Drs. P. J. E. Peebles, P. G. Roll and D. T. Wilkinson. The concept of an oscillating universe—one that expands, contracts, explodes and expands again—had a philosophical appeal for Dr. Dicke, as it does for many others.

He noted that the present universe is clearly born with little or no heavy elements. The oldest stars seem to be made almost entirely of hydrogen and helium. The heavier elements were synthesized gradually inside the stars and in stellar explosions.

If this universe was born of another, what happened to the elements of the prior universe? Dr. Dicke reasoned that the collapse of the old universe must have produced a fireball hot enough to break up the elements. This would have to exceed 180 billion degrees Fahrenheit.

He proposed that the light of this fireball might still be observable as microwaves and his associates went to work building a special antenna, aimed at the zenith, to look for such a "photon gas."

Then, to their amazement, they discovered that researchers at the Bell Telephone Laboratories, using a huge, horn-shaped antenna built as prototype for the Telstar program, had observed just such a microwave background in the sky. They had been perplexed as to its source.

Last week the Princeton group said it too had seen the effect on a wave length of three centimeters. The Bell Laboratories likewise reported more precise observations at seven centimeters and others at Cambridge University have "seen" the primordial flash at 21 centimeters.

Furthermore the flash seems to explain an old puzzle. In 1938, Dr. Pol Swings in Belgium found evidence, in light that had passed through cyanogen gas in space, that the gas was slightly heated. The cyanogen molecule consists of one carbon and one nitrogen atom. Why such molecules in the cold reaches of space should be heated, even to their lowest energy state, was a mystery.

Now, learning of the microwave evidence, Dr. George Field at the University of California in Berkeley and two scientists at the Institute for Space Studies of the National Aeronautics and Space Administration, Drs. Nicholas Woolf and Patrick Thaddeus, have independently proposed that the heat source is what remains of the flash. It is at a wave length at 2.54 millimeters, on the borderline between infrared light and microwave radio emissions.

All these observations indicate that the residual temperature of the flash is about three degrees above absolute zero. It had been expected that it would be more than 10 degrees.

The lower figure raises a new problem, for to those who believe in the conventional laws of physics it implies that the original explosion produced much helium as well as hydrogen. It could mean that almost a third of the universe is made of helium—far more than present estimates.

This is because, at a certain stage of the explosion, conditions were propitious for the production of helium. If the residual temperature is only three degrees, that implies a comparatively long helium-producing phase.

One solution lies in a modification of relativity theory proposed by Dr. Dicke. This postulates a gradual weakening of gravity during the lifetime of each universe. A by-product of his theory would be to carry the explosion through the helium-producing phase so rapidly that little helium was formed.

Because of observational difficulties, the abundance of helium in the universe is unknown. At the conference Dr. Allan R. Sandage of the Mount Wilson and Palomar Observatories in California welcomed the idea that the basic building material of the universe may have been one-third helium.

This, he pointed out, would resolve an old puzzle: namely that the globular clusters of stars seem much older than the universe itself. The globular clusters look like beautifully symmetrical swarms of bees.

In 1959 their age was estimated as about 24 billion years. This has been reduced to 18 billion, but the age of the universe is thought to be no more than from eight to 13 billion years.

If these clusters contain 30 per cent helium, that would bring their age estimate down within the age of the universe, Dr. Sandage said. On the other hand, Dr. Woolf argued, from observations of grossly varying helium abundances in certain clusters, that most of their helium must have been made inside their stars.

December 20, 1965

Astronomers Hear Signals From Space

By WALTER SULLIVAN

American radio astronomers during the last week have been recording radio signals from beyond the earth that they and their British colleagues believe could be from other civilizations.

They are, however, unwilling to give this idea prominence until all possibilities of a natural origin can be eliminated.

The British, who discovered the signals but could only observe them for about one minute each day, proposed that they might be from neutron stars—stars of high density believed formed by collapsed atoms squeezed together. However, the last week's observations have raised questions concerning that interpretation.

The great bowl-shaped antenna operated by Cornell University at Arecibo, P. R., has been observing one of the sources for three hour daily. The antenna, which fills a bowl-shaped valley is the largest of its kind on this planet.

"It is the most exciting discovery of the past 50 years," said a prominent California astronomer a few days ago. "But don't quote me!"

"Our first thought," said Sir Martin Ryle of Cambridge University, where the observations were first made, "was that this was another intelligence trying to contact us."

"We cannot completely rule that out," Sir Martin said, but he argued in favor of a natural origin.

The British have identified at least four of the sources in space, only one of which has been observed at Arecibo—between Vega and Altair, two stars in the Milky Way. The British have not given out details on the three other sources.

Dr. Frank Drake, director of the Arecibo Ionospheric Observatory, reached by phone in Puerto Rico, said that it should be possible within four or five months to determine whether or not the signals under observation there were coming from a planet in orbit around another star.

Reluctant to Speculate

Although radio astronomers making the observations are reluctant to speculate publicly if the signals were of artificial origin, they are talking privately about the possibility that these sources are navigation beacons or segments of a communication net linking a number of highly advanced civilizations.

Dr. Drake says that the Arecibo observations have confirmed all of the extraordinary features of the signals reported

ASTRONOMY AND PLANETARY SCIENCE

RECEIVING SIGNALS FROM BEYOND EARTH: The Arecibo Ionospheric Observatory, which occupies this valley in Puerto Rico, has been recording mysterious signals.

Associated Press

by the British and have disclosed others:

¶They occur at intervals of 1.337 seconds with a regularity far greater than that of any ordinary timepiece. "They could put WWV out of business!" Dr. Drake said. WWV is the radio station that broadcasts the time signals of the United States Naval Observatory.

¶The intensity of each pulse is highly variable over a period of one minute. The emissions then disappear for three or four minutes, whereupon they reappear for another minute of variable intensity. This cycle is continuous.

¶The Arecibo observations have shown that, at a frequency of 111 megacycles a second the pulses at peak power are one of the strongest radio emissions yet discovered in the sky. The British, unable to observe near that frequency, found the signals very weak.

The signals are inaudible at 40 megacycles, on the bottom side of their range, and can barely be heard at 200 megacycles on its top side.

¶The signals, which when collected by the Arecibo antenna are strong enough to be heard on a loud-speaker, "chirp" like the so-called whistlers. The latter are radio signals generated by distant lightning flashes. Because high-frequency radio waves travel faster through an electron-rich medium, such as the upper atmosphere, the high-frequency part of the signal arrives first. The lowest-frequency part arrives last. The result as heard on a loud-speaker is a swiftly descending whistle or chirp.

The British using estimates of electron density in space, proposed that the source might be roughly 200 light years away. The nearest stars that might be the centers of solar systems like our own are about 11 light years away. That is, their light takes 11 years to reach earth, traveling at 186,-000 miles a second.

Dr. Drake, however, questioned the distance estimate. If the signals traveled through an atmosphere such as our own when leaving their source, the observed chirping effect could be attributed in part to that atmosphere.

This would put the source far closer, however, because if most of the signals were used up escaping an atmosphere, there would be fewer to travel a greater distance in space.

However, the source of the signals lies well beyond the solar system, as it always appears to eminate from the same point in space.

The Cambridge group, describing its discovery in a recent issue of the British scientific magazine, Nature, said: "The remarkable nature of these signals at first suggested an origin in terms of man-made transmissions which might arise from deep space probes, planetary radar or the reflection of terrestrial signals from the moon."

They then found the signals came from fixed points among the stars. The one under observation at Arecibo lies between Vega and Altair, close to the central plain of the Milky Way Galaxy.

So precise is the spacing of the pulses that they reflect the orbital motion of the earth. As the earth moves toward the pulses they seem more closely spaced than when the earth is moving away from them, just as waves strike the bow of a ship at closer intervals when it is steaming upwind than they do when steaming downwind.

It is this effect that may ultimately enable radio astronomers to tell whether the source is itself in orbit around a star. The precise spacing of the pulses also makes them potentially useful for space navigation much as the radio emissions of electronic navigation systems are used by ships and planes on earth.

The Arecibo observatory is timing the pulses with a rubidium vapor clock that, within a week, should show if there is any variability down to one second in 100 million.

The observations at Arecibo so far have been on four frequencies. They show the initial pulse to be 50 thousandths of a second long, but in transit the low frequency part is delayed some 10 seconds more than the high frequency part of the pulse. The relative intensities of these components varies with each pulse.

Dr. Drake said that the variations did not display any properties. However, messages between advanced civilizations might be difficult to identify.

The problem, however, is to explain why the rate is so very precise, but the intensity so variable.

March 10, 1968

A Pulsar Is Discovered at 'Site' of a Neutron Star

Radio Source in Crab Nebula Found Where Astronomers Postulated Celestial 'Cinder'

By WALTER SULLIVAN

A pulsing source of celestial radio emissions, or "pulsar," has been discovered almost exactly where astronomers had predicted the existence of a super-dense "cinder" left by a stellar explosion observed in A.D. 1054.

The remnant of this explosion, a cloud of glowing, turbulent gas, is still visible in telescopes as one of the most dramatic features of the night sky. It is known as the Crab Nebula.

At the time of the original explosion, or supernova, nine centuries ago it was visible day and night as an extremely bright star.

It has been speculated that an extremely dense object, or "neutron star," might be left at the site of this explosion.

The discovery, in this region, of a pulsar has encouraged those who believe the mysterious pulsars are, in fact, neutron stars.

Although at least a dozen pulsars have been detected since the first were discovered last year, none has been clearly identified in optical telescopes. Yet their highly rhythmic signals are strongly received by radio antennas.

Emissions from the new pulsar, plus another source of radio waves near it, were reported on Nov. 6 by scientists who had been using the 300-foot antenna of the National Radio Astronomy Observatory in Green Bank, W. Va.

Now, Cornell University scientists, using the 2,000-foot antenna at Arecibo, P. R., have pinpointed the location.

According to Dr. Frank D. Drake of Cornell, the new pulsar is extraordinary for the very rapid pace of its pulses. They occur every 0.033089 seconds. For many months after the discovery of the first four pulsars all the known ones displayed rhythms between one quarter of a second and two seconds.

It has been suspected that such a rapid and highly rhythmic phenomenon must be associated with the pulsation or spinning of an extremely dense object. The only astronomical object that could spin at such high speed would be a neutron star.

Although there is, as yet, no certain evidence that neutron stars exist, physicists seeking to calculate what happens when a star burns up its atomic fuel have postulated that its atoms may collapse, forming an extraordinarily dense ball of neutrons.

In this process a large portion of the material forming the star would collapse into a sphere from four to ten miles in diameter. One cubic inch of such an object, it is calculated, would weigh about 10 billion tons.

During the collapse the rest of the star's material would be thrown off in a cataclysmic, or supernova, such as that which produced the Crab Nebula.

Conjecture About Star

Astronomers have long wondered whether there might not be a newly formed neutron star in or near the center of that nebula. The receiving system at Arecibo, which scans a patch of sky one degree of arc in width, was aimed at it and detected the high-speed pulsar.

The nebula itself is one twentieth of a degree in width. The Arecibo scanning has shown the pulsar to be within one sixth of a degree of the nebula's center.

According to Dr. Drake the neutron star may not be at the center. It may have been thrown to one side in the explosion.

Estimates of distance to the pulsar, according to the Cornell group, are consistent with the distance ascribed to the Crab Nebula. The pulsar distance can be estimated because the high frequency portion of each of its radio pulses reaches earth ahead of the low frequency portion. The latter is slowed by electrons. The extent of this slowing is an indication of how far the radio waves have traveled since they left the pulsar.

Dr. Drake believes the very fast pulse rate of the new pulsar may indicate that it is comparatively young. However, if the pulse rates slow as the pulsar gets older, this must occur over a very long period, for not even a slight change in pulse rate has been observed in any of the pulsars.

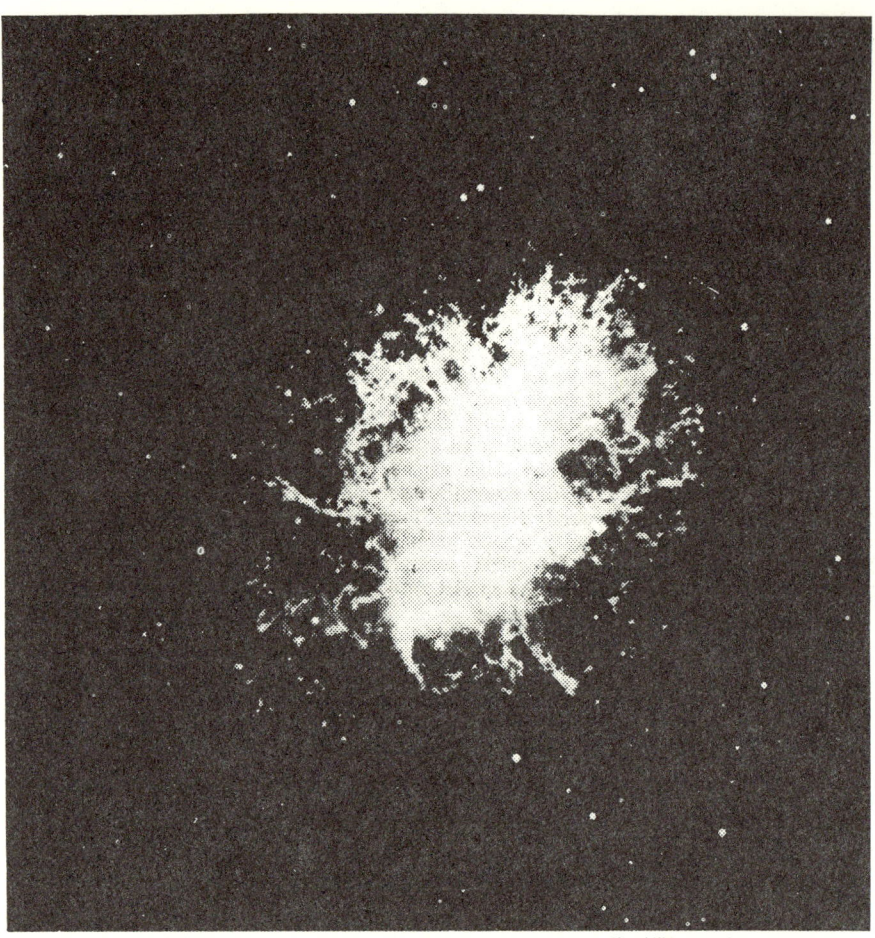

Mount Wilson and Palmer Observatories

SOURCE OF SIGNALS: Highly rhythmic radio signals have been detected from this remnant of a stellar explosion, known as the Crab Nebula. In A.D. 1054, the explosion, or supernova, was visible day and night. Pulses may be from a star formed at that time.

November 20, 1968

An X-Ray Scanning Satellite May Have Discovered a 'Black Hole' in Space

By WALTER SULLIVAN

Observations with an earth satellite named Uhuru, the first vehicle, placed in orbit primarily to scan the heavens at X-ray wave lengths, have revealed what, it is suspected, are one or more of the long-postulated "black holes."

A black hole is the hypothetical remnant of a star that has collapsed into so small and dense an object that light cannot escape from it. The reason for its invisibility would be its extremely intense gravity.

It is known that the gravity of a massive object like the sun increases the wave lengths of light emanating from it. In terms of visible light, the stronger the gravity, the redder the light, the stronger the gravity, the redder the light becomes. In a black hole, gravity is so strong that the wave lengths are infinitely lengthened before they can escape.

Hence the object would be invisible at all wave lengths, including those of radio, visible light and X-rays. The X-ray pulses reported yesterday, it was suspected were generated, perhaps by spinning of the central object outside its most intense gravitational field.

Last Dec. 12, the Uhuru satellite was launched for the National Aeronautics and Space Administration from a movable platform operated by Italian personnel in the Indian Ocean off East Africa. The word "uhuru" means "freedom" in Swahili. The sateillite is No. 42 in the Explorer series.

Equatorial Orbit

Because the platform was on the Equator it was possible to place the satellite in an equatorial orbit by launching it due east. A number of new discoveries are emerging from analysis of its data by a team at American Science and Engineering, Inc., in Cambridge, Mass.

The discovery of a possible black hole was reported yesterday to the American Astronomical Society in Baton Rouge, La., by Dr. Riccardo Giacconi of the Cambridge team and was elaborated on in telephone interviews.

The object is one of a number of previously known sources of X-ray emissions, and is known as Cygnus X-1. The designation means it is X-ray source No. 1 in the constellation Cygnus, the Swan. The new finding is that it is pulsing in a highly rhythmic manner at a rate of at least 15 times a second.

The rate could be double that, but this would not be evident because of the limited sampling rate of the satellite. The X-ray pulses are reminiscent of those from pulsars—objects that emit very rhythmic radio pulses (and, in one case, light and X-ray pulses).

Radio Pulses Absent

However, as noted yesterday by Dr. Herbert Gursky of the Cambridge team, what is extraordinary about this object is the absence of any observable radio pulses. Furthermore, intense efforts over the years to identify a peculiar-looking star in its vicinity have been unsuccessful.

Currently, Dr. Jerome Kristian of the Hale Observatories in California is seeking to find an object in that part of the sky whose light is flickering at 15 times a second. His results so far have been negative, according to Dr. Gursky, but a more thorough search will be possible when the sky is extremely dark, during the next new moon.

It is the apparent absence of any radio or light pulses associated with those at X-ray wave lengths that has led to the suspicion that a black hole is under observation. Pulsars, pulsing generally at slower rates and at radio wave lengths, are believed to be neutron stars.

It is assumed that their pulse rates represent their spin rates, averaging about one second. Neutron stars were presumably formed by the collapse of a burned-out star and differ from the hypothetical black holes in that their density and gravity are not sufficient to make them invisible.

One fear of the group at American Science and Engineering was that the apparent X-ray pulses arose from a defect in the satellite instruments. They therefore examined data from other spots in the sky emitting X-rays and found that most were not pulsing.

This was not, however, true of all of them. One source, in the southern constellation Circinus, seems to be varying in the same manner as Cygnus X-1, although no precise pulse rate has been identified.

Another pulsing source of X-rays, Dr. Gursky said, may be the object in Centaurus reported, early in March, to be "flaring" on a time scale as short as 15 seconds. The variations were reported by Dr. Walter H. G. Lewin of the Massachusetts Institute of Technology, based on observations with a high-flying balloon.

Only by more detailed analysis of the incoming Uhuru data will it be possible to determine whether these other objects are pulsing rhythmically — and whether a number of other such otherwise invisible objects are adrift in space.

Finally, it will be up to the theorists to assess the likelihood that they are, in fact, black holes and, if so, why they are signaling their presence with such powerful X-ray pulses.

April 1, 1971

SATELLITE GAUGES X-RAY STARS' SPIN

Records Sharp Changes in New Types of Objects

By WALTER SULLIVAN
Special to The New York Times

WASHINGTON, April 28 — On April 12 the Uhuru, a compact little satellite in orbit over the equator, recorded what scientists believe were the effects of a cataclysmic event far out in the Milky Way.

What it detected was a sharp change in the spin rate of a new class of celestial object, which is coming to be known as an X-ray star. The change, it was reported here today, must have been caused by some violent event, such as a vast in-fall of material onto the star.

The spin rate of the star, known as Centaurus X-3, has been deducted from its X-ray pulsations. Changes in pulse rates of those heavenly clocks known as pulsars have also been observed, but the extent of the change was one-ten-thousandths as much.

Like the pulsars, the pulsing X-ray stars are thought to be fast-spinning remnants of stars that have burned up their nuclear fuel and have collapsed into cores of extreme density.

However, those studied so far differ from pulsars in several ways. The pulse rates, at least of some, are slower and they have not been observed to emit pulses at any wavelengths except those of X-rays. Pulsars characteristically pulse in radio waves, and one of them, in the Crab Nebula, also pulses in visible light, X-rays and gamma rays.

Gamma rays have shorter wave lengths than X-rays, which in turn, have shorter wave lengths than do the colors of visible light. Radio wave lengths are much longer than those of visible light.

Uhuru was described today at the spring meeting of the American Physical Society at the Sheraton-Park Hotel as the world's first X-ray observatory. Because X-rays from space cannot penetrate the earth's atmosphere, they must be observed from rockets and satellites.

Uhuru was launched last December from an Italian platform in the sea off Kenya. Its name is Swahili for "freedom." The three-foot scientific payload, with an X-ray telescope aimed out each end, is slowly rotating so that the two scopes sweep the heavens.

As explained by Dr. Riccardo Giacconi, chief investigator of the project, the sweep speed was slowed after the surprising discovery that some of the X-ray stars are pulsing. Slowing the sweep increased the observation time on each of them.

Of about a dozen such objects studied so far for possible pulsations, three have shown such variability. However, the first found to be pulsing, in the constellation Cygnus, seems on closer examination to be pulsing at several different rates simultaneously.

Rocket Mission Set

The experimenters hope the nature of this celestial counterpoint will be disclosed by a rocket flight planned by scientists of the Massachusetts Institute of Technology for next Friday. It is to be launched from the White Sands Missile Range in New Mexico.

Because the sampling rate of detectors aboard Uhuru is limited, it cannot sort out all the rhythms, whereas the recording by the rocket would be virtually continuous. In view of the excitement generated by the Uhuru observations, half of the rocket's observation schedule has been reallocated to this target.

Dr. Giacconi is with American Science and Engineering, Inc., in Cambridge, Mass., which designed the experiment. His colleague Dr. Wallace Tucker, sought to explore possible explanations for the X-ray pulsations, noting that three fates have been postulated for burned-out stars.

In each of them, it is believed, gases forming the star, having lost the support of heat generated in the star's core, fall inward toward the core, exerting tremendous pressure. The resulting collapse of the star may produce an object of great density, visible as a so-called "white dwarf."

Or, if the pressure is even greater, the atoms may collapse into a ball of nuclear particles, forming what is known as a neutron star. In such a case a large part of the star's material becomes concentrated in an object only a few miles in diameter, spinning at very high speed. The pulsars are believed to be neutron stars.

However, according to theory, in the most extreme case the collapse continues in a runaway process. This leads to a superdense concentration whose gravity is so strong that nothing — not even light — can escape. It is invisible and would appear in the sky as a "black hole."

While there has been some speculation that the X-ray pulsations may be caused by spinning black holes, Dr. Tucker proposed instead that pulsing X-ray stars may result from the collapse of stars whose original spin rate was so fast that their rate of collapse has been retarded before they reached the most dense stage.

Having stopped short of full collapse, such an object would spin at rates comparable to that observed for Centaurus X-3 — roughly once every five seconds.

While the slowdown observed on April 12 amounted to only a few thousandths of a second, it presumably represented a vast change in spin energy, or "angular momentum," of the object.

The third apparently pulsing X-ray star is Lupus X-1, whose rhythm is not yet clear. However, the intensity of its emissions varies 50 per cent within a third of a second and twofold over a time span of minutes.

At the end of the session, the chairman, Dr. Herbert Friedman of the Naval Research Laboratory, commented, "It is obvious that the whole field exposed by Uhuru is rich with mystery and that we have just begun to scratch the surface."

April 29, 1971

Man Listens for Life on Worlds Afar

By WALTER SULLIVAN

Special to the New York Times

BYURAKAN, U.S.S.R., Sept. 11 — This week, for the first time since life originated on this planet, its most intelligent species met at an international conference to consider the possibility of communication with intelligent life on other worlds.

The Russians revealed that last year they conducted experiments designed to intercept signals from any such distant civilizations. In one series a transcontinental network of stations received a number of simultaneous pulses of suspicious origin, but because of the pattern in which they occurred they are suspected of originating in the atmosphere.

Those attending the conference, sponsored jointly by the Soviet and American Academies of Science, included leading specialists in the origin of life, the development of biological data processing systems, cultural evolution, language, radio astronomy and electronics.

The sessions were held at the astronomical observatory here, one of the most important in the Soviet Union, within sight of snowy Mount Ararat, across the Turkish border.

Because of the sensational nature of the subject and to encourage uninhibited discussion, the conference was held in relative privacy.

There was wide agreement, based on the latest astronomical and biological reasoning, that great numbers of other worlds have probably developed along lines very similar to earth in its early history.

It was also agreed that life might exist on a certain percentage of these worlds and that on some of them such life might have evolved into an intelligent technological species. Since earth's own technology is developing at an almost explosive rate in terms of astronomical time scales, it is calculated that some of these technologies may have reached extraordinary capability.

Such civilizations in various parts of the universe could, it was thought, be trying to make contact with other civilizations. In a resolution adopted almost unanimously at the end of the conference today, it was suggested that the chances of intercepting such signals were sufficient to justify a serious search.

An interim committee was formed to organize a permanent international group to coordinate research in this field. Two areas of search were proposed.

One would concentrate on the several hundred nearest stars and other nearby objects where "astro-engineering" activity by a supercivilization might be modifying celestial bodies.

The other area of search would seek the beacons of any far more distant supercivilizations. In this area an entire galaxy of billions of stars could be scanned at once for any indication of variable emission or signal.

The first attempt to intercept signals came in 1960 at the National Radio Astronomy Observatory in West Virginia. Two nearby stars were monitored on what then seemed a particularly logical wavelength for civilizations to "find one another" on, but the results were negative.

Helping the Other Team

As noted by one Soviet astronomer, the experts here sitting around a U-shaped line of tables were playing a "game" with some other hypothetical team of intelligent beings in some distant world. Only in this case the purpose of the game was not to beat the other team but to help it win.

That is, the goal was to seek out the most logical and economical means of attracting attention and communicating and then to hope that the other team had come to the same conclusion. The game, in this respect, is just beginning. There was no agreement here on the most logical approach.

In the assessment of the likelihood of signals reaching earth from other worlds, two factors emerged as perhaps the least predictable.

One, as pointed out by Dr. Francis H. C. Crick of Cambridge University, the Nobel Prize-winning biologist, is the origin of life.

Recent research has strengthened the argument that on any earth-like planet a rich soup of the chemicals necessary for life's origin is likely to develop. It also appears, Dr. Crick said, that once life has arisen and the process of natural selection has begun, evolution moves forward with almost irresistible momentum.

The uncertainty lies with those critical steps whereby the chemicals of the soup organize themselves into self-reproducing units that develop the characteristics of life. It may have happened only once, here on earth, Dr. Crick said, or it may be a common event throughout the universe. Knowledge is still insufficient for an assessment.

The other area of special uncertainty is the birth of language, an essential prelude to the rise of a civilization.

Dr. Kent Flannery of the University of Michigan, an authority on the development of agriculture, said that cities and other cultural features of civilization have arisen spontaneously in various parts of the earth and that this suggests that the same would happen on other worlds. But too little is known of the nature of speech and language to predict how often they would be likely to arise, he said.

An effort was made to consider every possible means of communication, even the most far-fetched. It is difficult for those familiar only with the limited capabilities of earth, participants said, to guess what means might be used by more advanced societies.

Thus there was discussion—and, in general, a rejection—of the idea of firing hydrogen bombs within a critical part of the magnetic field enveloping a planet, thus releasing a directional pulse of radio energy. It was also proposed that the atmosphere of a planet like Jupiter be used to focus radio emissions from earth into a very narrow beam.

The use of invisible light was also debated. The radio astronomers were more favorable toward radio waves but Dr. Charles H. Townes, of the University of California at Berkeley, a Nobel physicist, argued that the relevant technologies were not sufficiently advanced, and conditions on other planets not sufficiently known, to make such a choice.

"It is hazardous to make absolute decisions at this point," he said, for in 100 years "we will see things quite differently."

Of worlds that have already advanced far beyond earth, he said: "I can see such civilizations monitoring very carefully every form of electromagnetic emission from every star near them." Such emissions include both light and radio waves.

He also proposed that the dispatch of automated messengers to other stars had not been sufficiently explored as a possible means of making contact. Such messengers, he added, could carry flares to attract attention and pictures to tell about the world from which they came.

Dr. Ronald N. Bracewell of Stanford University has recently elaborated his proposal that such messengers may be the most logical means of establishing contact. It was widely agreed at the meeting, however, that signal transmissions by radio or light wave would be by far the most economical way to communicate.

If Everyone Listens

Dr. Frank Drake of Cornell University said that not only are there no clues as to where to look in the sky, but there also is no obvious radio frequency on which to listen.

The frequencies formerly considered "landmarks" have become too numerous with new discoveries, and in terms of optimum propagation over vast distances of space, there is no narrow and therefore "logical" band of frequencies.

Furthermore, he asked, "what if everyone is listening and no one is sending?"

It was suggested that some supercivilizations have built radio beacons powerful enough to be detected in other galaxies. However, an exchange of messages with such a distant world would require millions of years.

Radio waves travel at the speed of light and so an exchange of messages with even a nearby star—say, 10 light years away—would take 20 years. Message exchanges with more distant civilizations, it was said at the conference, would enrich the future generations that received the answers.

However, Dr. Thomas Gold of Cornell cited the comment of a British Member of Parliament: "Why should I do that for posterity? What has posterity ever done for me?"

ASTRONOMY AND PLANETARY SCIENCE

Laws of Universe Are Put Into Question

By WALTER SULLIVAN

Recent astronomical observations have so shaken the foundations of current theory that some physicists are proposing that the laws governing events here and now may not be valid in other regions of space and time.

The observations, for example, have brought into question the reliability of the yardstick used in estimating distances to faraway galaxies.

They have revealed objects that seem to be moving faster than light, contrary to assumed physical law. And they seemingly have shown objects whose energy output defies explanation.

Among the proposals advanced to explain such observations is a theory that atoms were lighter and gravity stronger millions of years ago, when the observed events occurred. They are so distant that it has taken that long for their light to reach the earth.

A more radical suggestion is that matter is entering this universe from other universes, carrying with it the physical "constants" characteristic of those universes. Among such "constants" would be gravity.

Hypotheses of this sort represent a basic departure from the concept, born in the work of Sir Isaac Newton, that the laws controlling the fall of an apple on earth also apply in the most distant parts of the universe.

Interviews with astrophysicists here and abroad have indicated that most of them hope and believe the observations can be reconciled with conventional laws. Yet unconventional explanations have been put forth by internationally known scientists and published in reputable scientific journals.

The yardstick that is now being questioned has provided the chief pillar of the concept that the universe is expanding as a consequence of its origin in a fiery "big bang." It is the so-called red shift in the wave lengths of light from distant galaxies.

As a rule, the dimmer the light from a galaxy—presumably because of its greater distance—the more its characteristic wave lengths of light are lengthened, shifting toward the red end of the spectrum. This red shift, it has been assumed, is caused by motion of the galaxy away from the earth as an aftermath of the "big bang."

The faster it recedes, according to the assumption, the greater this red shift, just as the wave lengths of sound from a horn are lengthened as the vehicle moves away from the observer, lowering its pitch.

However, a number of galaxies have recently been observed with red shifts radically different from those of their seemingly nearest companions. In several cases the differences in their motion, relative to the earth, amount to about 12,500 miles a second.

This has caused scientists to ask if the galaxies are really flying apart at such speed or if the red shifts have been caused in some other way.

Similar doubts have been caused by the observation of apparently distant objects that, according to the red shift yardstick, seem to be 3 billion light-years distant—a light-year is the distance light travels in a year—and flying apart at 10 times the speed of light. This raises the possibility that the red shift might be grossly misleading. If so, the objects could be much nearer and hence moving more slowly.

Radical Proposals

The explanation advanced by Sir Fred Hoyle, recently knighted by Queen Elizabeth, is that atoms have been getting steadily heavier. In the past, when their weight was less, he says, the light they emitted was redder. Thus the red shift, seen in distant galaxies, would be at least in part an indication of their greater youth, in terms of the universe's life history.

Since it has taken the light millions of years to reach us, we see the galaxies as they were when, according to Sir Fred, their atoms were lighter, their light redder and their gravity stronger.

The most radical proposals of recent months concern the possible existence of "black holes" and "white holes."

A generation ago it was deduced, from Albert Einstein's formulations of relativity theory, that the inward pressure of very large accumulations of matter could produce objects approaching infinite density within an infinitely small radius.

The gravitational field of such an object would be so intense that light could not escape from it or pass close by, thus producing a "black hole in the sky."

A year ago, Dr. John A. Wheeler of Princeton University and his associate, Dr. Remo Ruffini, predicted that the first indication of the existence of "black holes" would come from the detection of a massive, yet invisible object circling a visible star. Such an object, they said, would manifest its presence through X-ray emissions and its gravitational influence on the visible companion.

'White Holes' Controversial

The X-rays would be generated as matter was drawn in by the super-powerful gravity of the "black hole" and collided with gas on the "holes's" outer fringes. In recent weeks several objects have been identified that, at least to some extent, match the "black hole" criteria.

The "white hole" concept is far more controversial. A rule considered basic to physics, on the atomic level, is that any process that runs in one direction should just as readily run backwards.

From this it is argued that, if matter can "go down the drain" into a "black hole," perhaps vanishing entirely from our universe's framework of space and time, why should not the opposite process occur?

This has been proposed, for example, by Dr. Robert M. Hjellming of the National Radio Astronomy Observatory in Green Bank, W. Va. He believes that, in this way, it is possible to explain a major problem in astrophysics.

That problem is the nature of the process enabling quasars and the cores of some galaxies to shine with extraordinary brilliance. Quasars are objects that, from their red shifts, seem to be the most distant observable bodies in the sky.

Some 20 years ago the Soviet astronomer Viktor A. Ambartsumian drew attention to seemingly explosive events in the cores of some galaxies and suggested that processes unknown to science might be at work there. This possibility has subsequently been explored by Sir Fred Hoyle and others.

Dr. Hjellming's proposal is that the cores of galaxies and quasars are "white holes" through which matter from other universes, existing in other space-time reference frames, is entering our universe. Such matter would have departed from the other universe via a "black hole."

Conversely, the matter vanishing in a "black hole" of our universe is flowing into another via a "white hole." A modification would be return via the core of a galaxy in our own universe.

Such ideas have been explored, as well, by such Soviet scientists as Dr. Andrei D. Sakharov and I. D. Novikov.

It has been proposed that if some galaxies, via "black hole-white hole" links, are being fed matter carrying the physical constants of another universe, that could explain their discordant red shifts.

Sir Fred Hoyle, co-founder of the so-called steady state concept of the universe, has long argued for the creation of matter to fill gaps in his cosmology. A steady state universe is one that is expanding but eternal and internally unchanging. However, to fill the gaps produced by expansion, new matter must be formed.

One of Sir Fred's earlier suggestions involved antimatter particles — mirror images of particles of matter. Thus the electron, which has a negative electric charge, has an antimatter counterpart with identical mass but a positive charge. Antimatter particles interact and vanish on contact with particles of matter and therefore live only briefly in our matter-dominated world.

Sir Fred's suggestion was that equal amounts of matter and antimatter are formed in the cores of galaxies. This would conform to one of the symmetries of nature, in which a light wave, when converted into matter, produces one particle of matter and one particle of antimatter.

Physicists are troubled by the apparent one-sided nature of our universe, which, according to the symmetries of nature, should contain equal amounts of matter and antimatter. Dr. Hjellming believes his concept of "black hole-white hole" links between universes may be the answer.

He likens the situation to the oft-cited representation of the universe in terms of galaxies sprinkled over the surface of an expanding balloon. The balloon represents the four-dimensional curvature of space and time. As it swells, the galaxies draw farther apart.

In his view, however, there

is a matching universe on the inner surface of this "balloon," linked to the outer one by "white hole-black hole" connections.

Matter "falling" through a "black hole" from this universe comes out as antimatter in the "white hole" of its sister universe, whose antimatter composition balances the domination of our own universe by matter.

Thus the matter of which we and our world is made would formerly have been the antimatter of another universe.

The reaction of many astrophysicists to such speculations is that there is no real evidence that other universes exist or that matter is being formed in the cores of galaxies. They prefer to seek explanations in accepted laws of physics.

But, as one of them put it recently, "rarely in history have theorists questioned so fundamentally the precepts of their time."

January 27, 1972

Star and Pulsar
Discovered 'Waltzing' in Distant Constellation

By WALTER SULLIVAN

Through observations with an earth-orbiting satellite rockets and various ground-based telescopes it has been discovered that an ordinary star and a pulsar, or pulsing object of extreme density are waltzing together through space far out in the constellation Hercules.

The finding, discussed here yesterday, was hailed as of major importance in that it provides a variety of opportunities for further discoveries in what is widely regarded as the most exotic and challenging field of astrophysics.

This is the area dealing with pulsars and other extreme concentrations of matter. Among the latter are the hypothetical "black holes" whose density is so great that light cannot escape their super-strong gravity.

From observation of the star and its companion, visible only as an emitter of X-ray pulses, it has become possible for the first time to determine directly the mass of a pulsar, or neutron star.

The pulsar, though only a few miles in diameter and spinning every one and a quarter seconds, weighs roughly as much as the sun. Such an object, formed of atoms crushed into a tight sphere of neutrons, would be so dense that one cubic inch would weigh billions of tons.

Two-Star System

Binary, or two-star systems have in the past, proved one of the most useful tools in astronomy. They make it possible to determine the mass of one of the two companions if the relative motions of the objects and the mass of the other star are known.

The calculations make use of the laws of motion that determine any form of orbital flight. In this case the two stars are, so to speak, orbiting one another.

The disclosures were made at the Sixth Texas Symposium on Relativistic Astrophysics, being held at the Americana Hotel. The conference is so named because the earliest ones were held in Texas.

Dr. Harvey Tananbaum of American Science and Engineering, the firm in Cambridge, Mass., that has been analyzing X-ray observations from the Uhuru satellite, said the pulse rate of the pulsar was once every 1.2378206 seconds.

This is taken to be its spin rate, and it was found, from January to July of this year, that it speeded up by four and a half millionths of a second.

Tendency to Slow Down

The better-known pulsars, which announce their presence with radio pulses, tend to slow down, rather than accelerate, although Dr. Frank D. Drake of Cornell University reported Wednesday on some erratic changes in this regard.

Apparently the radio pulsars are neutron stars spinning in isolation. Their spinning magnetic fields, not being symmetrical to the spin axis, throw off material in a way that generates the radio pulses.

In this way they lose energy and slow down.

The X-ray pulsars, on the other hand, are circling close to a companion star. As described yesterday by Dr. David Pines of the University of Illinois, the intense gravity of the pulsar draws material from the companion star. As this material falls into the pulsar, it is guided toward the latter's magnetic poles to produce "hot spots" at each of those sites.

It is these that, being offset from the spin axis, beam X-rays toward the earth once per revolution. The infalling material also causes the pulsar to spin faster.

Because the pulses of this X-ray pulsar vanish every 1.7 days, there was a suspicion that it was periodically being eclipsed by a companion star. Close to the inferred site of the pulsar was HZ Hercules, long known to be a variable star.

Hence, as reported by Dr. Neta Bahcall of Princeton University, the Wise Observatory in Israel began close observation of the star and found that it, too, exhibited a brightness cycle of 1.7 days.

Apparently emissions from the pulsar cause the nearest part of its companions to glow brightly as the pulsar flies overhead. The star, as seen from earth, therefore becomes brightest when the pulsar is on this side of it, then dims as the pulsar and its bright region pass behinnd the star.

Dr. William Liller of the Harvard College Observatory has checked photographic plates going back to 1949 and found that the star, for some periods of a few years, showed this cycle of enhanced brilliance, and then failed to do so for a few years.

Superimposed Cycle

It has also been found that a 35-day cycle of varying X-ray intensity is superimposed on the brief pulses. The reasons for this and for the episodic nature of past performance are unknown. However further study has confirmed the "waltzing" companionship of both bodies.

As the pulsar goes around its companion, the effect of this motion in modifying the apparent rate of its pulses is like the changing pitch of a passing horn. The same effect is seen in the wave lengths of light from the visible star, both observations meshing into a coordinated cycle.

This waltz in the sky, said Dr. Pines, represents "an utterly fascinating system about which we are going to learn a very great deal in time to come."

At a session on black holes, Dr. Remo Ruffini of Princeton University reviewed what is known of X-ray sources that seem to be circling other objects (although in general those objects are, themselves, unknown). Those which emit X-ray pulses, he believes, are neutron stars.

They are, at most, only a few times more massive than the sun, which is not enough for collapse into black holes of almost infinite density.

The X-ray sources in two-body systems, which do not pulsate, he believes, are black holes. And since neutron stars are constantly drawing matter from their companions, each may ultimately get massive enough, he said, for collapse into a black hole.

The leading candidate as such an object remains Cygnus X-1, an X-ray source identified as a possible black hole last December.

However Dr. Kip Thorne of the California Institute of Technology, a leading theorist, commented last evening that he was "not at all 100 per cent convinced" that Cygnus X-1 is a black hole.

December 22, 1972

ASTRONOMY AND PLANETARY SCIENCE

Men Report Seeing Edge of the Universe

By WALTER SULLIVAN

Astronomers believe they have seen the edge of the universe.

The announcement last week of the discovery of a quasar more distant than anything previously observed has strengthened this belief, not because it is so distant but because it is not farther away.

The situation can be likened to gazing through a forest of widely scattered trees and finding that none can be seen beyond a certain distance. The implication is that, as Einstein and others believed, the universe is finite.

Beyond its expanding volume, this theory holds, nothing exists—not even space, because, in this concept, over such great distances space curves back upon itself.

Assuming the validity of the method used for estimating distances to quasars and other very remote objects, light from the newly discovered one has probably taken 12 billion years to reach the earth, traveling at 186,000 miles a second.

A Wall to Vision

The nature of quasars is unknown, although some scientists suspect they are galaxies in an early stage of formation. If they are as distant as they seem to be, they shine far more brightly than any other celestial object, both invisible light and light at radio wave lengths.

Because of this intrinsic brightness, some of them would be expected to be visible at considerably greater distances than that of the newly found quasar.

The distance estimates are determined from the observed expansion of the universe. It has been found from a variety of observations that the galaxies, or great star systems, are flying apart much like particles of dust in an expanding cloud. When such expansion is viewed from any one particle in the cloud, the rate at which another particle is receding is directly related to its distance. The faster the rate, the greater the distance.

By determining the rate at which a quasar is flying away from the earth, astronomers can estimate its distance in terms of the length of time its light has been on the way here. The expansion is such that the universe appears to have originated in an enormous explosion of matter—a "big bang"—some 13 billion years ago, and the earliest quasars should be visible far enough away for their light to have been on its way here for that long.

But looking across vast distances — and, hence, far back into time—man can see out only to 12 billion years (in the most distant quasars). Hence, in the words of Dr. Allan R. Sandage of the Hale Observatories in California, in these quasars we are apparently seeing "the edge of the world." To cosmologists, "world" is virtually synonomous with "universe."

For the last year or two, Dr. Sandage said in a telephone interview Friday, it has been suspected that there is some sort of "wall" preventing astronomers from seeing quasars in the region beyond 12 billion years. Using the new, more powerful methods of observing with radio and optical telescopes, such objects should be visible, if they exist.

Now, he said, it is beginning to look as though the "wall" is real.

Detection of the most distant quasar has been reported in the British journal Nature by Dr. R. F. Carswell and Dr. P. A. Strittmatter of the University of Arizona's Stewart Observatory. It is known as OH471 and is apparently receding from the Milky Way, the galaxy of stars within which the earth lies, at a speed of 91 per cent of the speed of light.

Expansion Is Uniform

The previously most distant quasar is calculated to be flying away from earth at 89 per cent of the speed of light. It is assumed from the seemingly uniform expansion of the universe that the speed of recession is a reliable indicator of distance. Such expansion has been thoroughly documented within ranges of shorter distance.

What is now the second-most-distant quasar was detected by Dr. Roger Linds and Dr. Derek Willis of Kitt Peak National Observatory, which, like the Stewart Observatory, is near Tucson, Ariz.

The apparent observation of the edge of the universe is considered by Dr. Sandage and other believers to be further evidence of the validity of the big bang cosmology.

Viewed As Beacons

He and others of this view believe the first billion years in the life of the universe were required for an expanding gas cloud, initially composed chiefly of hydrogen and helium, to form the elements and assemble them into accretions dense enough to initiate nuclear reactions and to shine. Hence, the first billion years was a time of darkness.

When quasars were first discovered a decade ago, it was hoped that they would provide beacons, of standard brightness, on the outer fringes of the universe, indicating the extent to which its expansion is slowing.

From this it could be determined whether the universe is doomed to expand forever or, like a stone thrown into the air, is slowing enough in its expansion so that eventually it will "fall back together" from the gravitational attraction of all its parts.

For quasars to be used for this purpose, they would have to be roughly uniform in intrinsic brightness. As astronomers put it, for such a determination they need a "standard candle" — something whose intrinsic brightness as against apparent brightness at an unknown distance, is known so that they can determine the extent to which the light has been dimmed by the distance it has traveled.

This is not proved to be the case with quasars. Their intrinsic brightness seems to vary by wide margins. Hence, Dr. Sandage, who is one of the world's leading specialists in this field, believes that, for this search, astronomers will have to depend on certain galaxies whose intrinsic brightness is more uniform.

By cataloguing large numbers of such galaxies according to rate of recession and relative distances, determined from the dimming of their light, it should become evident to what extent the rate of expansion was faster in the past—that is, extent of the slowing down.

The speed of recession is determined by measuring the extent to which certain known wave lengths of light from the object have been shifted toward the red, or long-wave, end of the spectrum. Rapid motion away from the observer stretches the wave lengths of light much as the pitch of a horn is lowered when a vehicle is moving away.

The extent of this "red shift" is an indicator of the speed of recession and has become the standard yardstick for distances to other galaxies.

There are some astronomers, apparently a minority, who have questioned the applicability of this yardstick to the quasars. They note that, if the yardstick is valid, the intrinsic brightness of the quasars must vary by remarkably large margins.

They also note that two of the quasars seem linked by a gaseous bridge, although their radically different red shifts would indicate they should be flying apart. Other astronomers have questioned the reality of this bridge.

The yardstick has also been challenged because of its implication that certain very distant objects are flying apart faster than the speed of light—on theoretical grounds, an impossibility. If the red shift yardstick is misleading and these objects are relatively close to earth, estimates of their motion would shrink to more reasonable rates.

In reply, it is argued that these objects may not really be flying apart at all, the observations being an optical effect unrelated to motion.

Attempts have been made to attribute the red shifts to effects other than rapid motion away from the earth—for example, by extremely strong gravity fields. The new observations are cited to challenge this on the ground that the red shifts now being seen are so great that their production by anything other than motion is improbable.

Finally, there is the argument that, in some parts of the universe, or at some times in the past, the laws of physics were different and atoms emitted their characteristic light at wave lengths other than those of today.

This runs so counter to all other observational evidence that scientists tend to consider this the explanation of last resort.

April 8, 1973

MATTER IN SPACE TERMED LIMITED

Scientists Make Estimate Lower Than Expected

Special to The New York Times

PRINCETON, N.J., Aug. 20—Princeton University astronomers say they have determined an estimate for the density of the universe that is much lower than expected.

The estimate, achieved with a telescope aboard the orbiting satellite Copernicus, implies that there is little undiscovered matter in the universe. The finding was announced to coincide with tomorrow's first anniversary of the Copernicus launching.

Dr. John B. Rogerson, professor of astrophysical sciences at Princeton, and Dr. Donald G. York, research staff member, used the telescope to determine the relative abundance of deuterium, a heavy form of hydrogen with twice the weight of a normal hydrogen atom, in interstellar gas.

The scientist found that there is one deuterium atom in interstellar space for every 100,000 atoms of gas.

Deuterium is thought to have been created moments after the "big bang" that is believed to have started the current expansion of the universe. No way has yet been found in which additional deuterium can have been created in significant quantities during the subsequent evolution of the universe.

Little Invisible Mass

Knowing what fraction of hydrogen became deuterium originally, the scientists assumed that all deuterium detected was a relic of the "big bang" and thus calculated the amount of matter created and the average density of the current universe.

"This computed density amounts on the average to one hydrogen atom in each 10 cubic meters," Drs. Rogerson and York explained. "This extremely low density indicates that there is not a relatively large amount of invisible mass in the universe."

Some cosmologists had postulated a huge quantity of undetected matter that would slow down and eventually reverse the expansion of the universe.

The 32-inch Copernicus telescope, designed by members of the Princeton University Observatory, detects the amount of starlight absorbed at ultraviolet wave lengths.

First Measurement

"The deuterium observations were difficult, since the determination of atomic abundance from the amount of light absorbed at characteristic wave lengths requires an absorption that is neither too large nor too small," the astronomers said. "These critical conditions are satisfied at several wave lengths so short that they lie outside the normal operating range of the Copernicus telescope. The instrument sensitivity at these short wave lengths is only about one-tenth of its normal value."

Although interstellar deuterium has been detected in several ways over the last year, this is the first measurement of the over-all deuterium abundance relative to hydrogen. The two Princeton astronomers pointed out that in the normal course of evolution since the "big bang," interstellar gas has been diluted by deuterium-free emissions from the stars.

"The observed abundance is thus less than the original value produced by the 'big bang' and a correction is made to this effect," Dr. Rogerson explained. "Uncertainty as to the exact nature of this correction is a source of uncertainty in the arrived value of the density of the universe. Further, if nature has found a way to create a significant quantity of deuterium since the 'big bang,' then our determination of the density of the universe from the present deuterium abundance is invalid."

August 21, 1973

Quasar's Light Indicates It's Old, Exploding Galaxy

Special to The New York Times

PASADENA, Calif., March 31—Analysis of light from the diffuse halo around the nearest quasar indicates that it consists of a normal and rather ancient galaxy whose core is undergoing a series of catastrophic explosions, according to astronomers of the Hale Observatories here.

The implication is that all quasars, considered by most astronomers to be the brightest and most distantly visible objects in the universe, are such objects.

The discovery that the quasar, known as BL Lacertae is the nearest quasar yet observed was a by-product of the study. As indicated by its designation, the object had long been considered a variable star in the constellation Lacerta, the lizard, of the northern sky. It was assumed to be a member of the same family of stars to which the sun belongs—the Milky Way Galaxy.

A Billion Light Years Away

Instead, it now appears to lie far across the universe, so distant that its light takes one billion years to reach the earth. To generate the visible lights and radio emissions that reveal its presence across the vast distance the object must be massive, yet its brilliance in visible light fluctuates radically even on a time scale as short as a few minutes.

Its radio emissions also vary from month to month. How so huge a "lighthouse" could flicker at so rapid a rate is a puzzle to astrophysicists.

The observations were made with the world's largest operating telescope, the 200-inch reflector on Mount Palomar. A small disk was inserted into the optical system to shut out light from the quasar itself and make possible an analysis of the wavelengths of light from the faint halo around it. This showed the halo to be formed of stars typical of an older galaxy.

Furthermore, the wavelengths were shifted toward the red end of the spectrum in a manner typical of stars a billion light years away — that is, so distant that their light takes that many years to reach us. Previously it had been impossible to estimate the quasars distance because it displays no special lines.

Question About Quasars

A question about quasars, according to Dr. J. B. Oke, who with Dr. James Gunn made the observations, is whether they represent some explosive phase in the early history of a galaxy, or may occur much later in that history. The new observations suggest the latter possibility.

However, as noted by Dr. Maarten Schmidt of the observatories, the further out and, hence, into the past one looks, the more quasars one sees, suggesting they were more common when the universe was young. Such explosive stages "may be a pretty rare event for old galaxies," he was quoted as saying in the observatory announcement.

Drs. Schmidt, Oke and Gunn are also on the faculty of the California Institute of Technology here which operates the observatories with the Carnegie Institution of Washington. The findings are being reported in the April issue of the astrophysical journal.

April 1, 1974

Astronomers 'Film' Biggest Object Yet

By WALTER SULLIVAN

Dutch radio astronomers have "photographed" the largest object yet discovered in the universe. They used a multiple-antenna system of the type that won a Nobel Prize in Physics this week for Sir Martin Ryle, head of a radio astronomy group at Cambridge University in England.

The photograph, an image derived from radio emissions, was made available yesterday. It shows an object so far-flung that light, traveling at 186,000 miles a second, takes 18.6 million years to cross its width. It is thus about three times bigger than the largest previously known component of the universe.

Until now it was believed that clusters of galaxies were the largest units. Some of them are formed of several thousand galaxies, each comparable to the Milky Way Galaxy to which the earth, sun and planets belong. A typical cluster is about three million light years wide. That is, it takes light three million years to cross it, although a few are twice that size.

The newly found structure

by Dr. Joseph H. Taylor of the University of Massachusetts in Amherst, who with Russell A. Hulse found the object during a systematic search for new pulsars.

Discovery of the first pulsars, lies beyond the constellation Leo Minor, far beyond the Milky Way Galaxy and its neighbors. A somewhat smaller one has also been found in the direction of the constellation Lynx. They are believed to have been produced by cataclysmic explosions within a central and relatively brilliant core.

These explosions apparently generated jets of gas that are still flying out in opposite directions. The observed radio emissions are presumably produced by swirling motions within this gas.

The picture has been derived from data collected by the Telescope in the Netherlands, in 1967, has won for Prof. A contour map of radio intensities generated by the object was published several weeks ago by astronomers from the University of Leyden in the British journaaa

The photograph was made available by Dr. Harry van der Laan of that university, whose astronomers use the Westerbork observatory, a national facility. Dr. van der Laan is visiting at the Institute for Advanced Study in Princeton, N.J.

On Tuesday, the day the Nobel awards in physics were announced, another major discovery deriving from work for which the prize was given was described in a presentation at the institute. This was the identification of a pulsar in orbit around an extremely dense object.

The discovery was reported in 1967, has won for Professor Antony Hewish the Nobel Prize that he will share with Sir Martin Ryle.

Dr. Hewish works with Sir Martin at Cambridge University in England. Pulsars, which emit rhythmic radio pulses, are believed to be fast-spinning stars that have collapsed to states of extreme density.

The finding of a pulsar that is whirling in tight circles with another object was hailed by theorists at the Institute for Advanced Study as of major importance. The discovery, made with the giant dish antenna near Arecibo, P.R., was described as opening up a wide range of research opportunities.

"Now, perhaps, for the first time the effects of general relativity will begin to emerge from the noise," one astrophysicist commented. Another said of the studies that can now be made by specialists in various fields, "There is something for everybody."

The opportunities relate to the nature of pulsars, the manner in which their rhythmic radio pulses are generated, and the precise effects of relativity on time.

Because the pulsar, in its calculated orbit, seems to come close to a very massive object, the gravity of that object, according to the predictions of relativity, should make the pulse rate appear slower during each such close passage.

Dr. van der Laan said yesterday that he had sent word of the discovery to the Westerbork observatory, whose antennas last Friday began recording radio emissions from that spot in the sky. The recordings must be analyzed before the results are known.

Because the antenna system is able to map in detail the expending gas cloud left by a stellar explosion, or "supernova," it is hoped these observations will reveal the cloud left by the supernova that gave birth to the orbiting pulsar. This might indicate when the explosion occurred. If the other object of this two-star system was similarly born, its remnant cloud might also appear.

The Westerbork observatory consists of 12 dish antennas on an east-west axis with two more under construction. The observing method, called aperture synthesis and originally devised by Professor Ryle, depends on a technique known as interferometry.

The wave fronts of radio emission from the area under observation, as received by each antenna, are correlated in a manner that indicates with precision the direction from which they came.

By computer analysis it is then possible to construct an image of the source of radio emissions comparable to a photograph taken in visible light. However, the latter, under the best astronomical seeing conditions, shows details about six times smaller than radio images currently generated by the Westerbork system.

The most ambitious such system under development is the Very Large Array Telescope to be built in New Mexico at a projected cost of about $75-million.

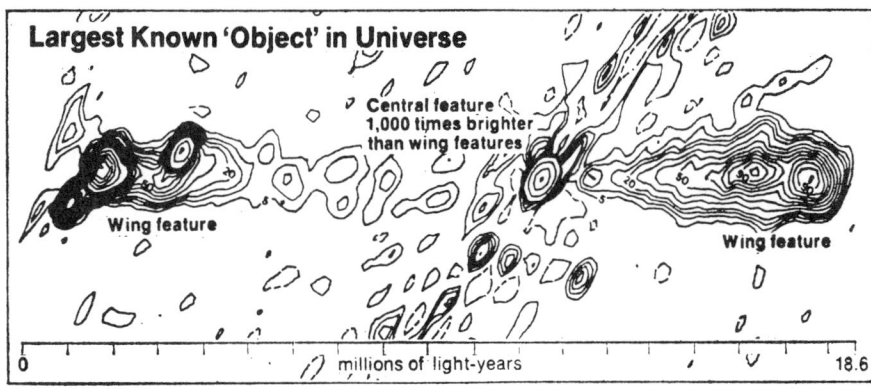
Largest Known 'Object' in Universe
Central feature 1,000 times brighter than wing features
Wing feature
Wing feature
0 — millions of light-years — 18.6
The New York Times/Oct. 17, 1974

October 17, 1974

ANTENNA BEAMS FAR-OUT MESSAGE

Scientists Hope to Reach Hypothetical Civilization in a Cluster of Stars

By WALTER SULLIVAN

There have been scientific experiments whose chances of success were very small, but few can compare in that respect with an effort last Saturday to communicate with a hypothetical civilization far out in space.

The experiment's chief purpose, however, was to dramatize for the inhabitants of earth the capabilities of the resurfaced dish antenna at Arecibo in Puerto Rico, by far the world's largest. The antenna, positioned in a bowl-shaped valley, can be used both for transmission and to receive faint emissions from far out in space.

The space message, which can be converted into a television image, was beamed toward a globular cluster of stars known as Messier 13 near the edge of the Milky Way. It was transmitted in the form of systematic frequency fluctuations at a wave length of 12.6 centimeters.

If any being in the direction of Messier 13 happens some time in the distant future to be observing in the direction of the solar system and at that wave length when the 169-second transmission reaches him, that being will see a signal flesh 10 million times brighter than the sun.

Analysis Required

If that arouses curiosity sufficiently to inspire analysis of the transmission, it will be found that the frequency fluctuations represent 1679 "characters" or "bits" of information in the typical "on-off", binary code.

Those who sent the message hope that any remote observers will note that 1679 is the multiple of two prime numbers—73 and 23—and will guess that an image is being transmitted that can be reconstructed by printing the message either as 73 consecutive lines of 23 "bits" or 23 lines of 73 "bits."

If the first of these steps is taken, with one type of signal printed as a black square and the other as a blank, a picture emerges that is designed to convey meaning.

The picture seeks to indicate that on the third planet out from the sun there exists a civilization, whose beings depend on molecules of DNA deoxyribonucleic acid for the continuity of their species. Also shown is a human figure and a representation of the transmitting antenna.

The transmission was devised by the staff of the National Astronomy and Ionosphere Center, which operates the Arecibo observatory for Cornell Univer-

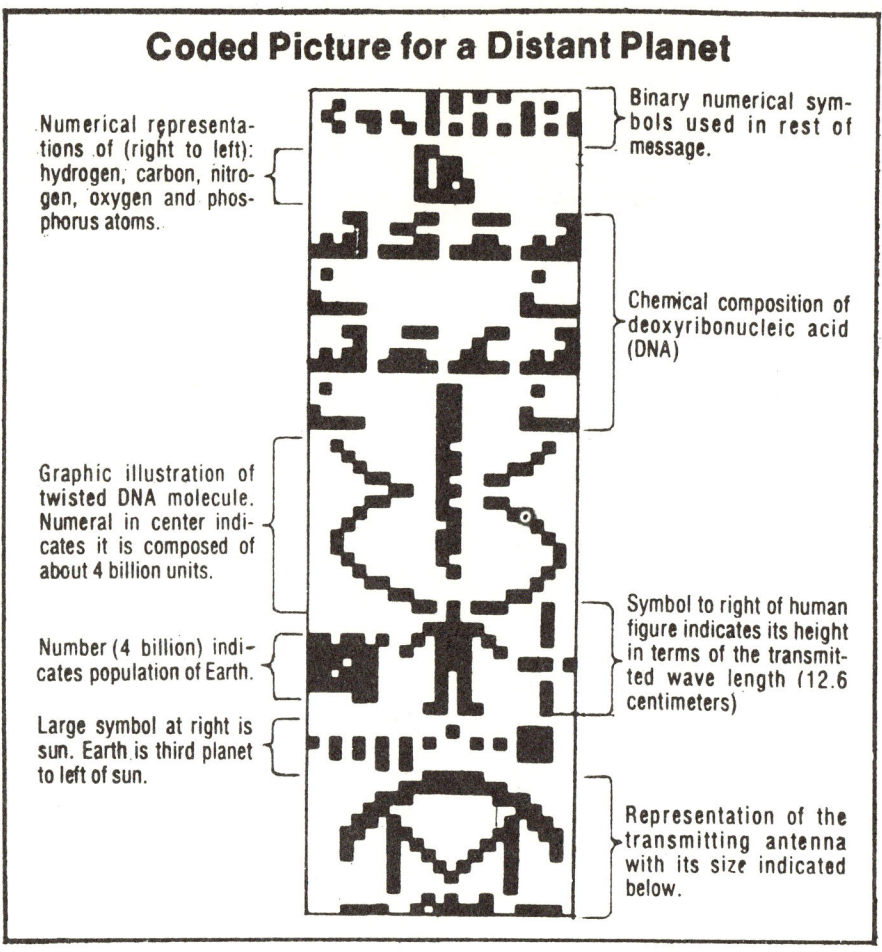

Signals transmitted toward a distant cluster of stars last Saturday, when printed out as 73 lines of 23 units each, form an image intended to convey much information about life on earth. The message was transmitted toward Messier 13, a cluster of stars so distant the signals will take 24,000 years to reach there.

sity and the National Science Foundation. It is believed strong enough to be observed by a comparable antenna anywhere in the Milky Way Galaxy, which is more than 100,000 light years broad. The emissions travel at the speed of light, or 186,000 miles a second.

A Long Time for Delivery

The globular cluster, Messier 13, was chosen because it will fit into the spread of the signal beam when the transmission reaches there, 24,000 years hence. The cluster resembles a swarm of bees and contains about 300,000 stars, some of which may have planets similar to the earth.

As noted by Dr. Frank Drake, director of the Cornell-Arecibo center, no one responsible for the message is likely to learn if it succeeds, since a reply from Messier 13 could not reach the earth for 48,000 years.

Another effort communicate cate with hypothetical distant civilizations, with comparably dim chances for success, was the mounting of a small plaque on Pioneer 10, the spacecraft that flew past Jupiter last December and is on its way out of the solar system, destined to wander indefinitely among the stars.

The plaque, in addition to depicting human figures and other information relating to life on earth, sought to indicate its place and time of origin, using the pulsars as landmarks. Pulsars, extremely dense and fast-rotating objects in space that are thought to have once been stars, are pulsing at distinctive rhythms at radio wave lengths.

In this case such identification was not necessary, since the observer would know from which direction the signals had come. A subtle correction was made to the outgoing signals to allow for slight changes in wave length imposed by the earth's spin and its motion around the sun.

The transmission formed part of ceremonies last Saturday marking completion of a three-year effort to improve the capabilities of the Arecibo antenna system. This included covering the dish, 1,000 feet wide, with a perforated aluminum surface. Suspended 50 stories above the dish is a new transmitter with an output of 450,000 watts.

This can be used for radar probing of the planets and of the upper atmosphere. Its radar sensitivity has been increased 2,000-fold, and the frequencies that can be observed for radio astronomy have been expanded 10-fold.

November 20, 1974

EXPLORING THE SOLAR SYSTEM

CANALS ON MARS.

Further Proof That They Have Actual Material Formation.

From The London Times.

Ever since the presence of canals or channels (canali) on the surface of the planet Mars was first described by the Italian astronomer, G. V. Schiaparelli, in 1877, the question of their character, and even of their real existence, has been keenly debated. On the one hand, they have been accepted as truly material formations, and various hypotheses have been advanced to explain them, such as that they are waterways connecting with oceans, or that they consist of lines of vegetation growing along irrigation works which derive their water from the seasonal melting of the polar snows and are the result of intelligent effort of some sort, or that they are merely great rifts produced in the globe by uneven contraction on cooling. On the other hand, some competent observers have failed to detect the canals at all, while others who have succeeded in seeing them have not agreed with each other in their descriptions of what they saw. In consequence it has been suggested that these canals or channels are of the nature of illusions of vision, and are not the definite features that appear on the drawings, but "rather the result of slight suggestions made to the eye by more or less irregular differences in the minute shadings and color tints on the surface of the planet."

Within the last few months fresh light has been thrown on the question by a piece of work carried out at the Lowell Observatory at Flagstaff, Arizona. The observers there have always been among the most successful in seeing and drawing the canals; and they resolved to supplement their visual observations by an attempt to photograph them in the favorable conditions presented by the planet's opposition in the Spring of this year, when it was comparatively near the earth. The observatory at Flagstaff enjoys exceptional advantages for such a task. It stands 7,250 feet above sea level, and therefore above many of the lower and denser layers of the atmosphere which incommode observatories at smaller altitudes, and it possesses the largest glass in the world mounted at such an elevation. Further the air is intensely dry, and its currents trouble the image less and produce less distortion and obliteration of detail than at lower levels in more humid conditions. Without such advantages the undertaking would be hopeless; even with them it was one of extraordinary difficulty and delicacy, and called for numberless precautions. It must be remembered that the diameter of the planet at the time was only some 15 seconds of arc, and that the view of it to be obtained even with the largest telescopes in favorable conditions is little, if any, better than that obtained of the moon with the naked eye. But to secure the required definition of detail it was usually thought necessary at Flagstaff to use only the central portion of the 24-inch object glass of the telescope, which accordingly was reduced by a diaphragm to an effective diameter of 12 inches; a color screen was employed that allowed only the yellow and orange rays to pass, and the extremely sensitive plates which had to be used permitted exposures of only from six to ten seconds, though during that time the utmost care had to be exercised to insure that the telescope followed the planet smoothly and exactly.

The result was that Mr. Lampland, Prof. Lowell's assistant at Flagstaff, succeeded in obtaining in May and June a number of photographs of the planet at different stages of rotation, which show canals quite distinctly and even present indications of the doubling which has been regarded as still more doubtful than the existence of the canals themselves. When a series of these photographs taken on the same night in close succession is examined the same markings are seen to be repeated from one to another, though naturally some of the pictures are better than others; and comparison of drawings made of the planet's surface by Prof. Lowell himself at about the same dates as, though quite independently of, the photographs shows a close correspondence between the two, canals having the same position and direction being perfectly visible on both drawings and photographs, even to the untrained observer. The defining power of the eye, however, is so much superior to that of the photographic plate that, although some 400 canals and 175 oases have been made out by Prof. Lowell himself, the photographs have so far revealed only about 40 canals and four or five oases. If it be admitted that the photographic plate cannot lie and can yield representations only of things that have a real objective existence, the conclusion to be drawn from the constancy of the markings on the successive photographs of a series and the correspondence between photographs of different series and almost contemporaneous drawings by Prof. Lowell would seem to be that the canals, whatever be their nature and origin, cannot be mere subjective illusions on the part of certain observers, but have an actual existence as material formations on the surface of Mars.

January 2, 1906

COMET GAZERS SEE FLASHES

And Then at 2:30 A. M., the Comet's Tail, 100 Degrees Long, 10 Degrees Wide.

EARLIER LIGHTS AURORAL

Seen in the Northeastern Sky by Miss Proctor and Other Visitors to the Times Tower.

FIRST CONTACT UNSEEN

Results Negative at Yerkes Observatory—Aurora Lights Due to Sun Spots.

WATCHERS AT ALL HOTELS

Throngs Gather on the Roofs in the Hope of Seeing Phenomena—East Side Excited.

Speeding by the earth at a rate which would have carried it from this city to Chicago in about thirty seconds and across the continent in just six seconds over a minute, Halley's comet brushed the earth with its long spreading tail last night, and few of the thousands of sky gazers who thronged Broadway, the hotel roofs, the Times tower, the river bridges, Riverside Drive, and the ferry boats were any the wiser. The earth is supposed to have entered the tail at a point 15,000,000 miles from the comet's head at 10:30 P. M., and to have finished the transit of a million miles about 5 o'clock this morning. The tail and the earth passed each other at a joint speed of forty-three miles a second.

Glimpse of the Tail at 2:30 This Morning.

Between 10:30 and 11:30 P. M. Miss Mary Proctor, the astronomer, and the other watchers on the top of THE TIMES tower saw what appeared to be auroral flashes in the northeastern sky. Miss Proctor remained on watch all night, and at 2:30 this morning, just after the moon had set, discovered a band of light 100 degrees in length stretching from the horizon at the point of sunrise up through the great square of Pegasus and Aquarius to Aquila.

At its widest part, just beneath the first magnitude star, Altair, the width of the band was about ten degrees, and throughout its length it had a brilliance equal to that of the Milky Way, near which it terminated. The path of this band of light was very nearly that along which the comet was last seen and Miss Proctor was convinced that it was the outer boundary of the tail through which the earth was passing. The band of light was still there at 3 o'clock this morning.

Other watchers who have seen the comet and its tail repeatedly during the last two weeks, confirmed Miss Proctor's view.

Unseen by Prof. Jacoby.

Prof. Harold Jacoby, Rutherfurd Professor of Astronomy of Columbia University, stood out on Riverside Drive from 10:30 o'clock until midnight watching for phenomena. At midnight he concluded there would be none, and remarked:

"The fact that the comet's tail has not been visible at the time of its contact with the earth, which must now be chiefly past, is a vindication for the general belief of astronomers that the tail is so thin that its presence near us is undetectable. The wonder of the comet then becomes why so thin and sinuous a body as its tail can be seen so brightly when off in distant space."

It seemed as if everybody was out to see what could be seen. The Times tower proved a vantage point from which an excellent view of the northeastern sky could be had. Early in the evening it was thronged with watchers, among them Miss Mary Proctor, who arrived shortly before 10:15 o'clock, and instantly was surrounded by a group of eager inquirers.

Miss Proctor patiently explained for fifteen minutes what might be expected and what she hoped to see, and then there came what seemed to be the first manifestation of the comet's nearness to us It was a light auroral glow and was observed at exactly 10:30 o'clock. It appeared in the northeastern sky, directly over the Queensboro Bridge, and lasted only for an instant.

Then for an hour, however, came apparent intermittent flashes, sometimes of white and some times of a ruddy hue. The light flared up from beneath a cloud and then was gone.

Once or twice, however, the flashes took definite shape and presented an unusual spectacle of color combinations. One of these which Miss Proctor noted, she described as "resembling an arch of glowing white surmounted by a crest of crimson."

The display occurred above a low lying bank of mist and rose to about five degrees above this. It was not of any considerable breadth, however, and resembled more a splash of color against the dark background of the sky than a wide strip of light.

The light of the moon, which shone brilliantly, interfered seriously with the observations. Its rays dulled possible electrical displays, and what might have been brilliant flashes appeared dull under the moon's counter glamour.

Miss Proctor described the flashes as arcs of fire hovering over the darkness. Some of them arose, fanlike in shape, rays of white light radiating upward to the arcs edge, where a crimson flash completed the effect. The flashes were intermittent between 10:30 and 11:30 o'clock, when they ceased, and the most keen-sighted watchers were unable to catch another glimpse of them.

Aurora Due to Sun Spots.

It is possible that these flashes were due to sun spots, the discovery of which was reported from various observatories yesterday afternoon. The Yerkes Observatory at Williams Bay, Wis., reported an auroral display at 9:30 last night, which allowing for an hours difference in time coincides with its appearance here as seen from the Times Tower. The Yerkes astronomers attribute the display to the new sun spots, not the comet. A familiar effect of sun spots is electrical disturbances in the earth's atmosphere.

On Riverside Drive, near Seventy-sixth Street, where Prof. Jacoby was making his observations, several commented on this more brilliantly lighted portion of the sky.

"The trouble with those folks," said Prof. Jacoby, "is that they don't look at their sky often enough. It's just so every time the moon is full or nearly full, and it has been just so for several nights past."

Astronomer Disappointed.

Prof. Jacoby did not go out less unexpectant than the populace in general that is not skilled in the lore of orbits and gravities and counterpulling astral forces He expected results, and was rather disappointed that there were no visible phenomena.

At 10 o'clock a hurry call from Larchmont arrived over the telephone at Prof. Jacoby's residence, notifying him that peculiar auroral flashes were visible along the eastern horizon.

"Yes," he remarked, "that's exactly where the offshore lighthouses cast their glimmer out to sea at regular intervals. That observer, too, isn't a night hawk in sky parlance. Perhaps he'll learn by the time the next comet comes along."

The crowd that was dense along the bridges and Riverside Drive at 10 o'clock began to dwindle at 11, and was pretty well in doors by midnight. With café and hotel parties, however, it was different. To them it was comet night, and whatever the comet might do for the magnetic pole, and the telegraph wires, and the atmosphere there was no doubt about its influence on the flavor of viands. It made them good, and the music that went with them delicious, even better than on New Year's eve, which comes once a year instead of only once a life time.

Prof. Jacoby enjoyed his comet dinner with professional friends at the club house of the Century Association, 7 West Forty-third Street. At 9:30 o'clock he left the club for his home, where he had an appointment with friends and newspaper reporters.

Prof. Jacoby talked of the passing of a comet while he figured that the earth was experiencing whatever it was going to of this one.

Old Prints of the Comet.

"I have been looking up the old prints of Halley's comet," he said, "for it would be mighty interesting to know just how much the visitor has lost in girth since it was here in 1835. It's a pity photography wasn't invented then.

"All these particles that are swept back from the comet into its tail must surely be lost to it forever. I can't see how it can ever recover them, and since they are thus lost, the comet must diminish in size on each of its visits, so that we can speculate as to a time when it will finally break up in meteoric shower.

"Perhaps there'll be a chance for some one to figure out that matter next time it makes a visit, and then establish himself securely in fame by a prophecy as forceful as that which links the name of this particular visitor with Halley.

"Nobody can do it now, because he can't compare photographs with the showing made in 1835. All that were made then were hand drawings, and after the hand drawings had been reproduced for publication in the rather poor astronomical periodicals of the period they lost much of their value for purposes of measurement.

"All I can say is that from those drawings the comet does not look materially different in 1910 from what it did in 1835. And that brings up an interesting question about that tremendously long tail. If the comet is shedding its core out into that filmy veil all the way through its journey it is hard to figure how it could last so long as it has, with its many periodic returns, never materially more inspiring than at present.

"As we figure from that, and other data about it, that it is perfectly probable comets grow tails only within this particular solar system. A tailess comet off in space on the far end of its orbit is perfectly possible to imagine. When we first locate them they are only a blur. There is no tail, and then the tail begins to grow, increasing constantly as the sun is approached. Maybe, then, the sun makes the tail by making the comet head boil with its heat, or driving the particles off by the friction of its rays.

"To-night the comet parties, except those in the cafés, were disappointed. But to-morrow night they ought not to fare so badly, and after that there should surely be in the western sky each night after sunset a sight glorious enough to satisfy the most fastidious. I shall be keenly interested in observing what the populace does when the comet in its real glory is finally revealed to them."

Gazers All On Roofs.

The streets of New York, except for Riverside Drive and Grand Street, gave little evidence last night that a million or so sky-gazers here were taking a lively interest in the heavens. Had it not been for the frequent glances skyward of pedestrians and clusters of men and women here and there at corners, any one who happened to know nothing about the passage of the earth through the tail of the comet would have detected little of interest in the street throngs.

The reason for this was simple enough. There was no special place in the city where the view of the sky was much better than another, except, of course, on the roof tops. The result was that most of the consistent sky-gazers were not in the streets, but on the roofs of their apartment houses or hotels.

Of course there was comet talk on every hand, in streets and hotels, street cars, and in the homes. Many had thought that Riverside Drive would be crowded early with sky gazers. Some of the astronomers had suggested in the newspapers that the Drive would be an excellent place to witness any atmospheric phenomena which might occur. But the Drive through most of its length was far from crowded.

About 9:15 o'clock the sky gazers on the Drive were put to rout by a sprinkle of rain. It wasn't much, but dark clouds had obscured the stars over New Jersey way, and it looked as though a real rain storm were coming. In a few minutes the Drive was almost deserted. But the patter of rain soon passed, and a few of the sky gazers returned to wait for the "crucial hour," as some of them chose to call it.

The lower east side was probably the liveliest section of the city. All along Grand Street, especially, there was a throng of excitable foreigners, chatting about the comet in their native tongues and gazing at the sky.

Late in the evening some small boys climbed to the roof of 401 Grand Street and sent up a toy balloon with a red light attached. It shot skyward and began drifting eastward at a rapid rate.

"There is the comet!" some one in the street shouted, and in an instant hundreds had taken up the cry, and were eagerly pointing to the speck of light drifting across the sky. On the Williamsburg Bridge some 20,000 people had gathered to watch for whatever phenomena there might be. The bridge crowd soon caught sight of the balloon, and across the bridge ran the shout:

"There is the comet, look!" They were still looking and pointing when the toy balloon passed on out of sight.

Some of the school teachers on the lower east side had a busy time yesterday keeping the children in order. Many of the little ones, especially the foreign born, were sure that the end of the world was imminent. In one of the lower grade classes in Public School 110, at 18 Cannon Street, a photographer took a flash light photograph in the afternoon. The flash so frightened the youngsters that the Principal had to help the teachers quiet them.

About 8:30 o'clock there was a demonstration in front of Old St. Patrick's Cathedral in Mott Street, near Prince. A long procession of Italian children, clad in white, filed past the church chanting the Litany of the Blessed Virgin. Policeman Kirk, on duty there, wanted to know what it was all about. A man who walked at the head of the procession brushed him aside and pointed at the sky. When the procession had passed the church it turned and the children and their elders knelt in front of the old sanctuary and prayed. After awhile other policemen arrived, and they induced the Italians to return to their homes.

In the vicinity of Grant's Tomb most of the benches were occupied, after the sprinkle of rain, until midnight. The watchers there were not disappointed, for most of them had not expected to see anything except an average Spring starlight night. Central Park also was visited during the evening by hundreds of sky gazers.

In the early hours of this morning hundreds of men and women who had attended comet parties were to be seen on their way home, still casting furtive glances at the sky as though half expecting to see something unusual before they retired.

At midnight policemen went through Central Park and drove out the throngs. The exodus at the Fifty-ninth Street entrance caused a jam in the Park circle and about 4,000 people gathered there. For the most part they stayed waiting to see the comet. "Telescope Tom," who has for years been seen at the corner of Twenty-third Street and Fifth Avenue and at the Circle with a telescope, set his apparatus in the Circle and did a land office business.

Chinatown didn't take much stock in the end-of-the-world business. No Chinese astronomer had predicted any such dire fate for the world, and accordingly the residents there went about their business as usual.

Hundreds of comet watchers crowded the roofs of the hotels waiting for the first glimpse of the tail end of the celestial visitor. The majority of the hotels ran express elevators to the roof beginning at 10 o'clock, and many incidents enlivened the night's vigil.

While several hundred men and women were gazing at the sky from the roof of the Waldorf-Astoria, a flashlight was set off unawares by a photographer. The brilliant flash of light and the noise made by the explosion caused the watchers to think that the crack of doom had really arrived, and there was much excitement for a few moments.

Several women screamed as the blinding flash lit up the roof, but their fears were quickly appeased when the cause of the sudden flash was made known.

The sun parlor on the roof of the Waldorf presented a strange scene at midnight. While the throng of watchers outside were being supplemented by the after-theatre crowds, attention was directed to the cozy interior of the sun parlor, for there, unmindful of what might happen, were women playing bridge. Six tables were filled, and the players continued their game until long after midnight.

The roof was illuminated by electric torquiers, and gilt chairs were used for the accommodation of the guests, among whom were Mr. and Mrs. A. J. Drexel, Jr., who recently returned from their honeymoon trip; Sir Herbert and Miss

Marshall, Mr. and Mrs. Henry Watterson, Baron von Bleichroder, Joseph Cannon, Speaker of the House of Representatives, and many others.

A huge telescope was operated in the centre of the roof, and there the watchers stood in line to get a peep at anything that the starry firmament had to offer.

The throng was considerably increased when 200 members of the National Association of Manufacturers, with their wives, came up after their dinner. "Uncle Joe" Cannon, with a big black cigar between his lips, seemed to enjoy the scene.

A large telescope swept the sky from the roof of the Hotel Manhattan, and scores of guests and patrons of the hotel assembled on the roof before 11 o'clock. A number of comet dinner parties were given here, and at midnight the hotel was practically deserted, for everybody was on the roof.

There were throngs on the roof of the Astor and Knickerbocker by 11 o'clock, and folks who supped after the theatre made a rush to get there before the stroke of 12. Four express elevators were taxed to their utmost at the Astor between 11 and 12 o'clock, and all sorts of glasses were brought into use.

The exodus to the roof of the Plaza began about 10 o'clock, and when the watchers were not scanning the sky they were inspecting the wireless telegraph station.

It was generally agreed by the watchers on the roof of the St. Regis and Gotham that the earth was passing through the comet's tail without any visible fuss. There was a large gathering atop the St. Regis, and over at the Gotham the roof was turned into a sort of night camp, with wigwams and tepees, which added to the picturesque setting of the roof.

The Netherland and Savoy had their quota of watchers on the roof, and scores of watchers enjoyed the occasion on the roofs of the Majestic and Ansonia uptown. A meteor luncheon was served on the roof of the Wolcott, and many cosy corners were fitted up beforehand, with soft pillows, easy chairs, and couches, which were placed in convenient angles for the watch parties.

There was a merry party atop the Hoffman House after 11 o'clock. The Holland House had its share of watchers, and the elevators at the Belmont raced up and down for the accommodation of the watchers.

The comet received attention at several of the theatres last night, when it was mentioned in some way or other by singers in the musical shows. De Wolf Hopper called attention to the projected roof parties in his usual curtain speech, and acknowledged that he and his associates intended to go on the roof themselves. At the Astor Theatre the house staff, their friends, and several of the actors from "Seven Days" went to the roof of the theatre after the performance and scanned the heavens with opera glasses.

May 19, 1910

THE COMET.

Considerable additions to the public stock of knowledge about comets have resulted from the untiring scrutiny to which HALLEY's comet has been subjected since it first came within the range of our vision a few weeks ago. As a sort of bull against this and all other comets, it may be remarked that these additions to our knowledge are chiefly negative. Astronomers are so accustomed to the use of minus quantities that they will not balk at a suggestion that addition may be the result of subtraction. In past times men have known, or surmised, a great deal about comets that was not so. There has been a world of speculation as to what would happen if a comet should fall into the earth or into the sun. Prof. CHARLES A. YOUNG used to say that if a comet should fall into the earth it would disturb the inhabitants thereof about as much as the throwing of a featherbed into the ocean would disturb the whales. The results of observations taken at the Jesuit observatories in the Philippines confirm this comforting view. Those patient astronomers, under the most favorable conditions, watched the sun during the time when the head of the comet traversed its disk between 3:30 and 11 A. M. yesterday, and observed no obscuration of any part of the solar surface save that occasioned by a few sunspots, of which notice has been taken elsewhere. This would seem to be a demonstration that the head of the comet is not made up of solid matter; that these wanderers of the skies are throughout their whole extent gaseous, the head being only somewhat more dense than the tail.

As to the comet's tail, it may be described as a mere manifestation without much substance. According to some theories it has no substance. Yet, undeniably, a good many of the earth's people were not a little alarmed about the consequences of the passage of our sphere through that tail. The spectroscope had told us that it contained cyanogen, which is a highly unpleasant gas, so remarkably unpleasant, indeed, that if it had in sufficient quantities been diffused through our atmosphere the people of the earth would have been suffocated. We passed through the tail, or we did not pass through it—the astronomers have not quite settled the point yet—and nothing happened. Inasmuch as astronomers have assured us that the whole tail of HALLEY's comet, with its intolerable millions of miles of length, could be comfortably stowed in a cask, it is not surprising that nothing happened. The so-called auroral displays were very likely connected with the sunspots, and not with the comet. We believe a meteoric stone, or something of the kind, did fall through the roof of a Kansas real estate man's office, but that is liable to happen at any time.

So we know now, so far as knowledge is possible of objects so far away, that the head of a comet is not composed of solid matter, and that the tail of a comet is of such extraordinarily tenuous composition that we can neither hear, taste, feel, nor smell it. So we really have learned something about comets by unlearning what we supposed we knew or had conjectured before.

Ambitious to maintain its place as the leading comet organ, THE TIMES yesterday announced that from the tower of The Times Building that band of light, 100 degrees or more in length, stretching from the horizon to the star Altair, which is the tail of the comet, was still visible in the eastern skies from 2:30 to 3 A. M. Thursday, May 19. It was the opinion of Miss PROCTOR that this part of the tail, seen as the earth was passing through it, was its outer boundary. From this time on the comet will become an adornment of the evening skies. It should be visible soon after sunset, and would be but for the envious moon, which is just now approaching the stage of fullness. However, the moon may be expected to wane as usual, and then the comet can be studied without the inconvenience of keeping late hours. It is rapidly going away from us, however, and will soon become less brilliant. The multitudes of photographs that have been taken of HALLEY's comet upon the occasion of this visit will, no doubt, be preserved, to be used for comparison by astronomers who will study the visitor, with much improved instruments, upon its return seventy-five years hence. The point of chief interest will be to determine whether it has diminished in breadth of head, length of tail, or undergone other changes in appearance.

May 20, 1910

SEEK NEW PLANET OUTSIDE OF NEPTUNE

Astronomers of the World Taking Observations to Detect Possible New Body.

NEPTUNE HAS BEEN WABBLY

Perturbations Like Those That Led to Finding of That Planet Stir Star-Gazers.

Special to The New York Times.

WASHINGTON, Dec. 21.—Since the first week in December the Naval Observatory here has been taking special observations of the planet Neptune as part of a search which observatories all over the world are making for a new planet.

The existence of this new planet is suspected because of perturbations of Neptune, which have been noticed for a quarter of a century. The discovery of the planet Neptune in 1845 as the result of observations made by the British astronomer, Adams, and a French astronomer, La Verrier, acting independently of one another, followed the noting of perturbations of Uranus. This planet was found not to be moving in its calculated orbit, which inspired the belief that there might be another planet. Before the discovery that Neptune was a planet it was thought to be a fixed star.

Just as the perturbations of Uranus led to the discovery of Neptune, so the perturbations of Neptune, which have led some astronomers to suspect that there is still another planet, may result in the finding of a new world.

The observations of Neptune, which began this month, will continue until the end of February. The planet will be nearest the earth on Jan. 31, and the opportunity for observing the planet in its present swing toward the earth will be excellent.

The fact that these special observations were being taken at the Naval Observatory, whose specialty has been the observation of the sun, moon, and planets, was confirmed tonight in an interview with Professor Asaph Hall, Jr., astronomer at that observatory.

"It has been thought by certain astronomers for some time," said Professor Hall, "that there might be a planet outside of Neptune. The place of Neptune as a planet in the solar system was found because it was noticed some years ago that the planet Uranus was not moving in its calculated orbit.

"After Uranus had been watched for about a quarter of a century, it was noticed that the path pursued by that planet was not strictly in accordance with calculations in which the perturbations that had been caused by Jupiter and Saturn had been taken into account. The ultimate result was the discovery of Neptune as a result of the observations made by two astronomers working independently. These were Leverrier, a famous astronomer of the Paris Observatory, and J. C. Adams, a young man at Cambridge, Eng., both of whom calculated that the perturbations of Uranus must be caused by a supposed ulterior planet."

Prof. Hall said that the supposition that there was an extra-Neptunian planet was not a new one and that the matter had been worked upon for some time. Special observations are now being taken to determine whether the supposed new planet can be discovered. At the Naval Observatory several years ago special attention was paid to the matter, when it was discovered that Neptune was not adhering to its calculated orbit. Professor Percival Lowell of Harvard has worked on the matter for some time and has written a book in which he contends that there must be a new extra-Neptunian planet. Prof. Henry Norris Russell of Princeton also has been at work on the suspected new planet.

Professor Hall said that observations at the observatory would be made mainly with the nine-inch meridian circle, which is the instrument it uses principally for determining the position of the planets. Neptune is now coming nearer to the earth, and until Jan. 31 the position of that planet will be favorable all the time for special observations. The observations will continue about a month after that date.

"What we will do here," said Professor Hall, "will be to make a series of very careful observations of Neptune to calculate the position of that planet and then compare those positions with the tables for the purpose of ascertaining whether and to what extent the tables should be corrected. If the perturbation of Neptune turns out to be very great, it might be possible to estimate the possible location of a new planet."

Professor Hall said that the discovery of another planet would be an event of the greatest importance in astronomy, and would be the third planet discovered in modern times. The others were Uranus and Neptune.

The late Professor Simon Newcomb, while on duty here, paid particular attention several years ago to an investigation of the orbits of Neptune and Uranus and supplied the Smithsonian Institution with general tables of their motions and his views on the subject.

December 22, 1919

Say Clouds on Mars Show An Atmosphere Like Ours

Copyright, 1924, by The New York Times Co.
Special Cable to THE NEW YORK TIMES.

PARIS, Aug. 14.—The presence of clouds around the planet Mars has been definitely established by astronomers watching the planet nightly from the summit of the Jungfrau, and the phenomenon is accepted as absolute proof that Mars has an atmosphere probably of the same kind as the earth.

This observation is the first important result of the watch which is being kept on our nearest celestial neighbor during its nearest approach to the earth in many years. Atmospheric conditions have not been altogether favorable for observation, but it is hoped that before Aug. 24, when efforts are to be made to signal with a calcium light reflected off the snow slopes of the mountain through an enormous lens, weather conditions in this part of our planet will have improved.

Astronomers must now, however, count on the possibility of the hypothetical inhabitants of our neighbor being unable to see their signals because of such clouds as have just been discovered.

August 15, 1924

TO EXPLORE SKIES WITH 'MOON ROCKET'

Smithsonian Institution Expects Dr. Goddard's Device to Reach Outer Space.

INSTRUMENTS TO BE MADE

Automatic Sun Camera and Air Sampler Will Complete Rocket— To Float Back by Parachute.

WASHINGTON, July 20 (P).—A day when rockets might be sent as far into the atmosphere as man desires is foreseen by Smithsonian Institution scientists as the result of Dr. R. H. Goddard's "moon rocket" tests in Massachusetts, the last of which alarmed Worcester three days ago.

The institution has spent more than $12,000 in the last twelve years in backing the experiments and Dr. C. C. Abbot, its secretary, today made public from Dr. Goddard's official report the importance of the recent explosive test.

"No such wild project as going to the moon is contemplated," Dr. Abbot said. "We wish to create a method to gather meteorological and atmospheric data in outer space which man cannot reach by aerial navigation, balloons or kites."

To Carry Delicate Instruments.

Delicate instruments will be carried in the completed rocket, which will be equipped with a parachute. When the force of the propellant is expended the rocket, if everything goes as planned, will float gently to earth and the instruments return unharmed.

What the ultimate value of the rocket is to mankind is a question of interesting conjecture to scientists, as it has been throughout the years to fiction writers and inventors with a Jules Verne imagination.

Through the medium of the rocket science seeks to obtain four things—samples of the upper air for chemical analysis, measurements of temperature and pressure in distant space, camera spectographs of the sun beyond the ozone layer which now cuts out the region of the ultraviolet, and measurements at will of the condition of the atmosphere for aviation.

"This last problem, of course, is of great practical interest," Dr. Abbot said. "A rocket which would be set to explore the meteorological condition of the air at any height and to bring back its record to within a mile of the starting point and within an hour of its sending would be a boon to aviation."

Rocket Proposed in 1916.

"Sounding balloons, although they can rise to fifteen or twenty miles, often drift 150 miles from their starting point and may never be recovered with their recording apparatus or only after days or weeks."

Dr. Goddard proposed the development of the rocket to the Smithsonian in 1916. He had already worked on the problem for several years. He showed by mathematics that it is possible for a rocket to carry sufficient high-power explosive to propel itself beyond the atmosphere, which extends upward at least 200 miles, and even to go beyond recovery of the earth's gravitation. It would then become a man-made meteor in outer space.

With automatic stabilizers to insure vertical flight already designed, all that remains to complete the rocket will be the design of automatic apparatus to record meteorological measurements, a camera to photograph the sun's spectrum, and air samplers.

July 21, 1929

NINTH PLANET DISCOVERED ON EDGE OF SOLAR SYSTEM; FIRST FOUND IN 84 YEARS

LIES FAR BEYOND NEPTUNE

Sighted Jan. 21 After 25 Years' Search Begun by Late Percival Lowell.

SEEN AT FLAGSTAFF, ARIZ.

Observatory Staff There Spots It by Special Photo-Telescope —Makes Thorough Check.

ASTRONOMERS HAIL FINDING

The Sphere, Possibly Larger Than Jupiter and 4,000,000,000 Miles Away, Meets Predictions.

By The Associated Press.
FLAGSTAFF, Ariz., March 13.—In the little cluster of orbs which scampers across the sidereal abyss under the name of the solar system there are, be it known, nine instead of a mere eight, worlds.

The presence of a ninth planet in the retinue of the sun, long suspected, was definitely announced here today by Dr. V. M. Slipher of the Lowell Observatory, who headed a group of eminent astronomers whose groupings in the Milky Way with telescopes and cameras located the new-found sphere.

Way out beyond Neptune, tagging bashfully behind its brothers, the new planet's exact whereabouts, size and age are still unknown and it has not even a name.

Its presence was mathematically predicted years ago by the late Dr. Percival Lowell, noted scientist, who founded the observatory here, partly for the very purpose of identifying it. Other astronomers, notably Dr. W. W. Campbell, director of Lick Observatory, verified Dr. Lowell's calculations.

Reward of Long Search.

Today the faith in those calculations was rewarded by an announcement by Dr. Slipher that the new planet had been "sighted" Jan. 21 by an extremely delicate photographic lens developed for the search. Announcement was withheld, Dr. Slipher said, "until we were absolutely sure."

The discovery revealed that the planet is forty-five times as far from the earth as the earth is from the sun. Although its size has not been definitely determined, it is believed it may be bigger than Jupiter, largest member of the solar family, which is 1,200 times larger than the earth; and, said the announcement, it is at least no smaller than the earth. Astronomically the discovery is regarded as the greatest since the location of Neptune, eighth member of the solar system, in 1846.

The astronomers who participated in the discovery are C. O. Lampland, E. C. Slipher, J. C. Duncan, K. P. Williams, E. A. Edwards and T. B. Gill.

Until some one entitled to do so gives the sphere a name, it is to be known as "the trans-Neptunian planet."

First notice of the body was made by C. W. Tombaugh, photographer at the observatory, who saw a tiny spot on one of his plates. With this revelation, investigations were intensified and the scientists soon determined they had come upon the long-sought planet.

Announcement of Discovery.

Announcement in connection with the discovery was made at the observatory as follows:

"The announcement is made by the Lowell Observatory at Flagstaff of the discovery of a celestial body whose rate and path among the stars indicates a new planetary member of the solar family beyond the outermost known planet, Neptune.

"Twenty-five years ago Dr. Percival Lowell, who founded Lowell Observatory, began a mathematical investigation for a planet beyond Neptune. The probability of locating such a body, however, was difficult and involved enormous and intricate computations.

"In 1914 he announced as the result of his calculations the possibility and distance of the predicted body in a large memoir, a Lowell Observatory publication.

"The search of the skies directed by Dr. Lowell's theoretical work was begun by photography in 1905 and has been continued to the present time. The use was made of the best available instruments, the search covering that band of the skies in which the known planet traveled.

"Early last year the Lawrence-Lowell telescope, a highly efficient special instrument for the search, was put in operation. Some weeks ago (Jan. 21) Mr. C. W. Tombaugh detected an object on a plate made with the telescope, which has since been followed carefully here.

"It has been observed photographically with the large Lowell reflector by C. O. Lampland and it has been observed visually with the larger refractor by the various members of the staff.

"All observations indicate the object to be the one which Lowell saw mathematically."

Report to Harvard Observatory.

Special to The New York Times.
CAMBRIDGE, Mass., March 13.—The discovery of the ninth solar planet was reported to the Harvard Observatory here this morning in the following telegram from the Lowell Observatory at Flagstaff:

"Systematic research begun years ago, supplementing Lowell's investigation for a trans-Neptunian planet, has revealed an object which for seven weeks has in rate of motion and path consistently conformed to trans-Neptunian body at the approximate distance he assigned. Fifteenth magnitude. Position March 12 at 3 hours, Greenwich mean time, was seven seconds of time west from Delta Geminorum, agreeing with Lowell's predicted longitude."

Harvard astronomers are checking up on the data, but Professor Harlow Shapley and his assistants have expressed confidence in the accuracy of the discovery and believe that it is the most vital astronomical finding made since Neptune was located in 1846.

Because of its tremendous distance from the sun, the Harvard astronomers believe that the ninth planet receives the light of the sun with a brilliancy at most hardly exceeding that of moonlight. They believe that the newly found body takes at least 330 years to go around the sun.

Percival Lowell was the older brother of President Lowell of Harvard. He was graduated from Harvard in 1876. After graduation he spent several years in business and later traveled in the Far East, especially in Japan, about which he wrote several books.

As a boy he had been greatly interested in astronomy, and this interest returned during his travels. In 1894 he established the Lowell observatory in Flagstaff, under his own direction. He died in 1916.

The Lowell observatory has been in close touch with Harvard University since its foundation, and the 13-inch triplex telescope, the most powerful of its kind, through which the discovery was made, was completed by Robert Lundin of Watertown, Mass., after it had been started by the Rev. Joel Metcalf of Watertown.

Lowell Called It "Planet X."

CAMBRIDGE, Mass., March 13 (AP).—Professor Percival Lowell, in 1915, a year before his death, published his "Memoir on Trans-Neptunian Planet," in which were set forth his elaborate mathematical calculations and prediction of the direction of the newly-found planet. He called it "Planet X."

Fifteen years before the planet was to be discovered, professor Lowell wrote:

"It indicates for the unknown mass between neptune's and the earth's, a visibility of the 12-13 magnitude according to the albedo (reflecting power) and a disk of more than one second in diameter."

Gives Credit to Slipher.

Special to The New York Times.
BOSTON, March 13.—Roger Lowell Putnam of Springfield, Mass., chairman of the board of trustees of the Lowell Observatory, said tonight that discovery of the new planet may safely be credited to Dr. V. M. Slipher, director of the observatory. After the image of the planet had been located on a photographic plate, Mr. Putnam said, Dr. Slipher about three weeks ago found the sphere through the observatory's 24-inch telescope. The observations are to be checked and studied in an effort to determine the size and position of the planet.

CONGRATULATE DISCOVERERS.

Mount Wilson Staff Will Seek Later to Confirm Lowell Finding.

Special to The New York Times.
PASADENA, Cal., March 13.—Congratulations of the staff here were extended today to scientists at Lowell Observatory, Flagstaff, Ariz., by Frederick H. Seares, assistant director of the Mount Wilson Observatory, who recently completed a twenty-year task of cataloguing the stars.

The Mount Wilson astronomers' observations have been confined to stars, nebulae and the sun, but as soon as practical an effort will be made to confirm the Lowell Observatory's discovery through the instruments here, according to Dr. Seth B. Nicholson, sunspot observer of the staff, who formerly devoted much time to planet study.

Just at this time the moon is too bright for best atmospheric conditions in trying to observe the new planet, Dr. Nicholson said.

Tells of Lowell's Work in 1910.

Special to The New York Times.
CHICAGO, March 13.—Professor Philip Fox, director of the Adler Planetarium, now building on the lake front, today hailed the discovery of a new planet beyond Neptune, extending "my congratulations to the astronomers" at the Lowell Observatory.

"I visited the Flagstaff Observatory in 1910," Professor Fox said, "and Mr. Lowell [the late Percival Lowell] told me then that his computations indicated the presence of a body beyond Neptune. Neptune had not moved in accordance with the perturbation, and that led to his conclusion that some other body was present."

As a coincidence to the new discovery, Professor Fox said that Sir Frank Dyson, astronomer royal of England, had recently sent to the Adler Planetarium a reflecting telescope made in 1781 which had been used by Sir William Herschel, the discoverer of Uranus.

Yerkes Observatory Seeking It.

LAKE GENEVA, Wis., March 13 (AP).—Observer O. Struve of the Yerkes Observatory here said today that if the Lowell Observatory had found a new planet beyond Neptune "the discovery will be one of the greatest in the history of astronomy."

Professor George Van Biesbroeck of the observatory has made some observations on the Lowell theory of the existence of another planet and expected to search further for it tonight.

New Planet Compared With Earth and Neptune

Size:
Earth—8,000 miles in diameter.
Neptune—32,000 miles.
New Planet—8,000 or more.

Distance from Sun:
Earth—One astronomical unit.
Neptune—Thirty astronomical units.
New Planet—About fifty units.

Speed of Revolution:
Earth—19 miles a second.
Neptune—3½ miles a second.
New Planet—From 1 to 2 miles a second.

Time of Revolution:
Earth—One year.
Neptune—146 Earth-years (entire revolution not yet observed).
New Planet—Probably 300 to 600 years.

Note—These figures on the new planet are tentative, based upon computations of astronomers here on the Flagstaff announcement.

Find Hailed at Princeton.

Special to The New York Times.

PRINCETON, N. J., March 13.—The announcement of the discovery of the ninth major planet was hailed as "an important and interesting addition to our knowledge of the solar system" by Professor J. Q. Steward of the department of astronomy at Princeton University.

Referring to Professor Percival Lowell's prediction of the discovery being fulfilled, Professor Steward said:

"Lowell may perhaps be termed the posthumous discoverer of this new planet."

Dr. Schlesinger at Yale Comments.

NEW HAVEN, Conn., March 13 (AP).—Declaring that the discovery of a ninth major planet constituted "one of the greatest discoveries in the history of science," Dr. Frank Schlesinger, director of the Yale observatory, today said "it is easy to predict that other major planets will be added to our solar system.

"This strengthens the belief that other planets exist that are fainter, but it will be increasingly difficult to discover them, as the outer ones must be very faint as seen from the earth," he said.

March 14, 1930

Name Pluto Given to Body Believed to Be Planet X

By The Associated Press.

FLAGSTAFF, Ariz., May 24.—Pluto, the title of the Roman gods of the regions of darkness, was announced tonight at Lowell Observatory here as the name chosen for the recently discovered trans-Neptunian body, which is believed to be the long-sought Planet X.

The announcement was made by Roger Lowell Putnam of Springfield, Mass., trustee of the observatory and nephew of the late Dr. Percival Lowell, founder of the observatory, who predicted the existence of Planet X.

"We felt," said Mr. Putnam, "in making our choice of a name for Planet X, that the line of Roman gods for whom the other planets are named should not be broken, and we believe that Dr. Lowell, whose researches led directly to its discovery, would have felt the same way."

May 25, 1930

4 Planets Bathed in 'Coal Damp,' Analysis of Atmospheres Shows

Ammonia Also in Gases Enveloping Jupiter, Saturn, Uranus and Neptune, Science Session Is Told, but No Oxygen, and Life Like That on Earth Is Impossible.

By WILLIAM L. LAURENCE.
Special to The New York Times.

BERKELEY, Calif., June 22.—The constitution of the atmospheres of Jupiter, Saturn, Uranus and Neptune, long a baffling cosmic mystery, has at last been definitely determined after successful reproduction of equivalents of these atmospheres in the laboratory.

This was announced today at the meeting of the American Association for the Advancement of Science in a report by Dr. V. M. Slipher of Lowell Observatory in Arizona, and Dr. Arthur Adel of the Department of Physics of the University of Michigan.

The report was presented by Dr. E. C. Slipher, brother of the co-author of the report. Dr. E. C. Slipher is also an astronomer at Lowell Observatory.

These atmospheres, it has been found, consist of seven parts of hydrogen, one part of carbon and one part of nitrogen. By far the predominating constituent is methane, commonly known as marsh gas, mine gas or coal damp, which consists of four atoms of hydrogen and one atom of carbon. The next main constituent is ammonia, which is composed of three atoms of hydrogen and one of nitrogen. In addition there are large quantities of free hydrogen.

No oxygen exists in the atmospheres of these giant planets, and this, it was pointed out, definitely rules out even the remotest possibility of the existence of life such as we know it.

Were any oxygen to be present, it was pointed out, it would cause a terrific explosion because of the presence of so much hydrogen.

A summary of the accumulating evidence relating to the other planets, it was added, indicates strongly that, as far as our own corner of the universe is concerned, our little earth is the only possible abode of life in the forms in which it is known to us.

The work of Dr. Slipher and Dr. Adel corroborates the pioneer work of two other investigators in the field. Dr. Ruppert Wildt of Goettingen, Germany, was the first to suspect that certain bands in the spectrum of these planets correspond to methane, while Dr. Theodore Dunham Jr. of Mount Wilson Observatory was the first to establish the presence of ammonia.

The study of the planets is much more difficult, it was pointed out, than that of the distant nebulae, because the planets do not furnish their own light, but make themselves known only through the light they borrow from the sun.

Man's "Ladder Into Space."

The spectrum is man's "ladder into space." When a celestial body is incandescent it reveals its nature through the spectrum it sends out.

Because their light is borrowed, the spectra of the planets are difficult to interpret.

The work of identification has been made possible, it was asserted, by the development of highly sensitive photographic plates for the invisible infra red part of the spectrum which enables astronomers to "push the planets further into the red."

It is in these parts of the spectrum that the atmospheres of Jupiter and the other giant planets produce wide intense bands. But until the sufficiently sensitive photographic plates were developed these bands remained invisible to science.

After the bands were at last made visible there was still the problem of identifying them with some element known on earth. To do so it is necessary to reproduce similar bands in the laboratory, for it is in this manner that the existence of known elements in the stars is confirmed.

Similar Bands Produced.

The identification of these bands was not made, Dr. Slipher revealed, until light was passed through the equivalent of 2,000 meters of methane gas. Thus twenty-two of the bands produced by Jupiter and Saturn were reproduced in the laboratory. The spectrum bands in Uranus and Neptune are similar in appearance to the other two.

Dr. Slipher also presented a photographic record of the mysterious eruption that took place on the surface of Saturn last Summer, beginning Aug. 2 and ending Nov. 10.

Only one such eruption on Saturn had been reported before. The earlier one was much shorter in duration, beginning Dec. 7, 1876, and ending Jan. 2, 1877.

The recent Saturn eruption, Dr. Slipher stated, came very suddenly, not a sign of it having been noticed within a few hours before it started. It rapidly rose to a maximum brightness in the week to Aug. 10. Its area became greatest on Sept. 11. Individual spots were most numerous by the middle of September.

At one time, Dr. Slipher reported, the Saturn eruption covered an area 50,000 miles long and 12,000 to 15,000 miles wide.

Comparison to Indicate Size.

"You could have put the earth inside it," he said, "and it would have been no more than a dark spot in that bright band."

Dr. Slipher also displayed a series of recent photographs showing that Jupiter also is subject to great atmospheric disturbances. While nothing is known about what causes these disturbances, it is believed that the cause is the same on both planets.

In the case of Jupiter, Dr. Slipher said, four major disturbances have been reported in this century, beginning in 1906, and coming with a seeming periodicity of eight years.

Coincident with each one of these major disturbances on Jupiter, he added, there have been tremendous auroral displays on earth, but why this should be so is still a mystery.

In fact, he said, "we do not even know whether these four planets have solid surfaces."

Of the other planets in our solar system, Mercury has little or no atmosphere, and indications are that Pluto has none. Venus does have an atmosphere to an appreciable extent, but it is very strongly carbon dioxide, and as yet no oxygen or water vapor has been detected. All these are conditions unfavorable to life as we know it.

Oxygen on Mars in Doubt.

Mars has atmospheric obscuration caused by clouds, and must have water since it has polar caps, it was agreed by Dr. Slipher and Dr. Donald H. Menzel of the Harvard College Observatory. The presence of oxygen on Mars is possible, but still in doubt, and it is also agreed among astronomers that the amount of water in the atmosphere of Mars is much less than that in the earth's atmosphere.

The seasonal changes on Mars furnish strong evidence of the existence of some form of vegetation. The temperature on Mars ranges from 65 to 70 Fahrenheit at midday to minus 140 degrees at night.

There is no question, the scientists added, that human beings transplanted to Mars would just gasp and die. This does not mean, however, that some other form of life might not have been evolved in the millions of years of the planet's existence.

If we did not know that fish lived in the waters we would probably believe that no life could be possible there either, the scientists pointed out.

The asteroids, the planetoids between Mars and Jupiter, the number of which is estimated at 20,000, are also without atmospheres and hence without life, it was added.

New Human-Monkey Link.

Further evidence of the relationship between man and monkey has been obtained by J. K. Van Deventer, research assistant in bacteriology at Stanford University, through studies of the liquefaction of coagulated blood. A report of this work was presented today by Professor W. H. Manwaring of Stanford before the American Society of Naturalists.

Mr. Van Deventer showed that blood clots from human beings and from monkeys, when subjected to the liquefying action of a certain chemical compound, behave alike, and that they differ in behavior from blood clots of all other animals he tested, including the horse, cow, sheep, goat, dog, cat, guinea pig, rat, mouse and chicken.

On the basis of this evidence he concludes that in the nature of his blood fiber man showed a close relationship to other primates. He

ASTRONOMY AND PLANETARY SCIENCE

says: "This finding is typical of recent serological data which indicate that the human body is a taxonomic mosaic of amphibian, lower mammalian and primate immuno-chemical specificities, thinly veneered with anthropoid human colloidal characters."

The liquefying agent was lysin, obtained from parasitic organisms often found in the human body, known as streptococcus hemolyticus.

Groups Hold Elections.

Dr. Bailey Willis, Professor Emeritus of Geology at Stanford, was today elected president of the Pacific division of the association. Dr. William V. Houston of the California Institute of Technology was named a member of the executive committee for five years.

H. E. Burke of the United States Bureau of Entomology was named president of the Pacific Slope branch of the American Association of Economic Entomologists. C. O. Smith of the Citrus Experiment Station, Riverside, was elected president of the American Phytopathological Society.

Professor O. L. Sponsler of the University of California at Los Angeles was named president of the Pacific section of the Botanical Society of America. Dr. T. F. Buhrer of the University of Arizona was chosen president of the Western Society of Soil Sciences.

June 23, 1934

CONTACT WITH MOON ACHIEVED BY RADAR IN TEST BY THE ARMY

Signal Sent From Laboratory in Jersey Is Reflected Back 2.4 Seconds Later

VAST POSSIBILITIES SEEN

Mapping of Planets, Defense Against Bombs in Cosmic Space Are Suggested

By JACK GOULD

The first man-made contact with the moon was achieved on Jan. 10 when the Army Signal Corps beamed a radar signal on it and 2.4 seconds later received an echo reflected by the celestial body, it was announced yesterday. The signal, covering a round-trip distance of an estimated 450,000 miles, was sent out from the Evans Signal Laboratories at Belmar, N. J.

Applications almost beyond immediate comprehension were foreseen as a result of the electronic achievement. New and far more accurate study of the universe, perhaps ultimately resulting in the detailed topographical mapping of distant planets, was anticipated. Detection of enemy missiles flying through cosmic space also was expected to be possible from the new definitive proof that radio waves could penetrate beyond the earth's ionosphere.

The sound that the moon sent back to the earth took the form of a 180-cycle note, or somewhat higher in pitch than the hum to be heard on a home radio receiver when a station is not tuned in. It lasted half a second. The Army also recorded the echo visually on an oscilloscope. There the epic-making peep appeared as a series of jagged, saw-tooth lines.

Army Announces Feat

The official announcement that a radio signal had been bounced off the moon was made by Maj. Gen. George L. Van Deusen, Chief of the Engineering and Technical Service, Office of the Chief Signal Officer, at the annual dinner of the Institute of Radio Engineres at the Hotel Astor.

The first word to reach the public, however, came several hours earlier under circumstances anything but formal. A group of reporters crowded into a small upstairs reception room in the hotel and a quiet, 39-year-old officer, Lieut. Col. J. H. DeWitt, who supervised the experiment, announced what had been done.

As the men who had finally "reached the moon," Colonel DeWitt and his four chief associates in the venture were modest in the extreme. Only upon the reporters' insistence was there any revelation of biographical material on the quintet.

Colonel DeWitt, a former broadcast engineer in Nashville, Tenn., and a "ham" (amateur) radio operator, acknowledged that the results were the climax of his peacetime hobby to put a signal up to the moon. He said he failed in an attempt in 1940.

Jacob Mofsenson, 32, a graduate of City College, who entered the Signal Corps in April, 1942 was even more hesitant, but finally consented to tell his peacetime occupation.

"I was a diamond dealer," he said, with a laugh.

The other principal participants were Dr. E. K. Stodola, 31, a graduate of Cooper Union, who was in charge of research; Dr. Harold Webb, 36, a former teacher of physics and mathematics at West Liberty College, West Liberty, Va., and Herbert Kauffman, 31, who had worked in radio in New Orleans.

Two Conducted First Test

Dr. Webb and Mr. Kauffman were actually the only two at the radar receiving equipment when the first echo came back from the moon, but said they had betrayed no particular emotion at the time over the event.

"We looked for it and got the results," Dr. Webb said.

Work on reaching the moon by radar was started soon after V-J Day, according to Colonel DeWitt, and on Jan. 10 preparations had been completed for a test. On that day, he said, the moon rose at 11:48 A. M. and a few minutes later the initial radar impulse was beamed heavenward on a frequency of 111.6 megacycles. It was at 11:58 A. M., as the scientists remembered it, that the first flick of light appeared on the oscilliscope denoting success.

The tests were continued for five days, three tests being made as the

AN ECHO FROM THE MOON IS RECORDED

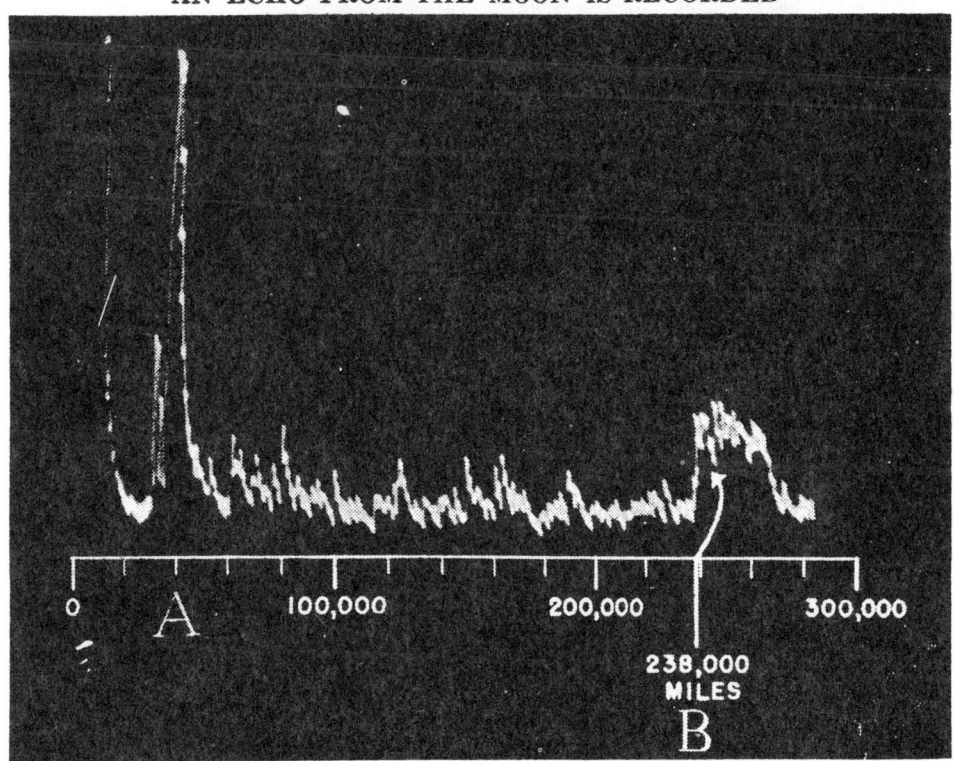

This is the Army's visual representation of the reception of a radar signal reflected by the moon. At "A" is the strong beam sent heavenward from Belmar, N. J. At "B" and the arrow is the record of the same signal as received back from the celestial body. The jagged lines between "A" and "B" are the variations caused by local noise conditions at Belmar. The time it took the signal to make the trip from earth to moon and back was 2.4 seconds. The mean distance from the earth to the moon is 238,857 miles.

moon rose and one as it set. Tests of the moon's "receptiveness" when it was higher in its arc were not possible because of lack of suitable antenna equipment and some days no signal came back, apparently because of propagation characteristics within the earth's atmospheric region.

The peak power of the transmitter was three kilowatts but through use of a special antenna giving a gain of 200 its radiation effectiveness thereby being vastly increased. The strength of the signal received back from the moon was calculated at about three watts.

Colonel DeWitt emphasized the "real trick" in making contact with the moon was not so much in the transmission but in the construction of a receiver of exceptional sensitivity to pick up the feeble echo from the planet. He estimated the sensitivity at .01 microvolts.

The radar waves traveled at the speed of light—186,000 miles a second. The mean distance between the moon and the earth is calculated at 238,857 miles, but the greatest problem for the scientists at Belmar was to allow also for the distance variation involved in the relationship between the speed of the moon and the earth's movement. The moon's speed, it was explained, varies from 750 miles faster than the earth's rotation to 750 miles slower.

Having demonstrated that a signal can reach the moon and return, Colonel DeWitt said the only problem left was the calculation of the time interval. When the echo came back in 2.4 seconds, the scientists were convinced they had achieved their aim "because there was nothing else there but the moon."

To make sure that there might be no error, a small group of scientists, not identified, visited Belmar and verified the conclusions.

"Hello" From Moon Expected

Colonel DeWitt professed a dislike for speculation of a "Buck Rogers or Jules Verne" character, but acknowledged that the Army scientists hoped to increase their transmitter's power so that it could be modulated by voice.

"We should be able to say 'hello' and hear the moon say 'hello' back," he said.

He quickly added: "I hope the moon doesn't answer, 'Good-by.'"

In connection with the announcement in New York, the War Department in Washington issued a statement on the implications of the feat. Maj. Gen. Harry C. Ingles, Chief Signal Officer of the Army, noted that it could have "valuable peacetime as well as wartime applications, although it is impossible at this stage to predict with certainty what these will be."

"One obvious possibility is the radio control of long-range jet or rocket-controlled missiles, circling the earth above the stratosphere," the War Department continued. "The German V-2 missiles already are believed to have reached an altitude of sixty miles.

"The primary significance of the Signal Corps achievement is that this is the first time scientists have known with certainty that a very high frequency radio wave sent out from the earth can penetrate the electrically charged ionosphere which encircles the earth and stratosphere. The several layers of the ionosphere start about thirty-six miles above the surface of the earth and extend to approximately 250 miles.

"On this basis, the V-2 projectiles already have risen above the lower ionopshere levels, and it is now known that radio waves can completely penetrate the ionosphere.

"The new technique will also be valuable for studying the effects of the ionosphere upon radio waves. Scientists already have learned that low and medium frequency waves are reflected by the ionosphere, and these reflections form the 'skywaves' used for long-distance broadcasting. The ionized layers also sometimes distort and bend radio waves, much as a prism distorts light waves.

"Another valuable application may be the provision of new astronomical information. Not only may it be possible to construct detailed topographical maps of distant planets with the aid of radar data, but scientists may be able to determine the composition and atmospheric characteristics of other celestial bodies by this means.

"A less likely application of the new technique will be the possibility of radio control from the earth's surface of 'space ships' venturing thousands of miles from the earth, and the radio reporting of astronomical data electronically computed aboard such vessels."

For the project, which Army officers informally labeled "Diane," the chief deviation from conventional radar application was in the use of a much longer pulse-repetition rate, somewhere between three and five seconds, compared with the usual pulse rate of thousands of times a second. The length of time each pulse of energy existed varied from one-tenth to one-half a second, an "enormously long interval" compared with war-time standards.

It was the factor of the rate and duration of the pulse, plus companion antenna problems, that led Colonel DeWitt and his associates to a relatively cautious approach in speculating on making contacts with Mars and Venus, as different distance ranges present new sets of problems.

Termed "Interesting Tool"

CAMBRIDGE, Mass., Jan. 24 (U.P).—Dr. Harlow Shapley, director of the Harvard College Observatory, described the Army's radar contact with the moon tonight as "an interesting tool in exploring the solar system" and predicted that more startling war-born developments would be revealed within the next few years.

"I believe radar contact with the moon is extremely helpful," he said, "because it will aid us in the study of meteoric material in the vicinity of the earth. I don't think it will ever help us to find another planet."

Indicating that the radar experiments had been known to scientists for at least two years, Dr. Shapley said the importance of this advance was not at all comparable to the atomic bomb. However, he said astronomers were working on other discoveries made during the war, and he predicted they would be far more startling than the radar contact when they were announced.

January 25, 1946

Telemeter Messages

'Electronic Nerve Center' Sends Back Data From Rockets

From White Sands, N. M., and other places rockets loaded with recording apparatus have been sent up to heights of well over a hundred miles. In descriptions of these flights into the stratosphere it is said that information is radioed back to a station on the ground by "telemeter."

What telemetering means Walter Hausz, General Electric engineer, told the American Institute of Electrical Engineers at its recent convention. He called the telemeter "a compact electronic nerve center." During a flight the telemetering system transmits twenty-eight different kinds of facts thirty-five times a second.

After a free fall of sixty to a hundred miles or so it stands to reason that recording instruments are broken. The scientists on the ground are content if films and other records survive the crash. Hence the electronic brain must operate perfectly. It is only by telemetering that the height reached can be determined.

June 26, 1949

DEVICE PLOTS ORBITS FROM 1653 TO 2060

Electronic Machine Calculates the Paths of the 5 Outer Planets to 14 Places

SHOWN TO ASTRONOMERS

Its 12,000 Vacuum Tubes Can Multiply 14-Digit Numbers at Rate of 40 a Second

By CHARLES A. FEDERER Jr.
Of Harvard College Observatory
Special to THE NEW YORK TIMES.

HAVERFORD, Pa., Dec. 28—Machines have accomplished feats of mathematical calculation that man could never hope to complete. The orbits of the five outer planets of the solar system, Jupiter, Saturn, Uranus, Neptune and Pluto, have been calculated to fourteen places for more than a hundred years in the future, to the year 2060.

Three astronomers noted for their work in celestial mechanics, the branch of astronomy that deals with the motions of bodies in the universe in accordance with the laws of gravitation, reported the completion of these orbital calculations to the American Astronomical Society, currently meeting at Haverford College.

They are Dr. W. J. Ecker, director of pure science at International Business Machines; Gerald Clemence, director of the United States Nautical Almanac Office, and Dr. Dirk Brouwer, director of Yale University Observatory. Their cooperative enterprise was supported in part by the Office of Naval Research.

Hitherto the amount of calculation required to get accuracy equal to or better than the observations of the planets themselves has been beyond attempt, but now the selective sequence electronic calculator of International Business Machines at the Thomas J. Watson Computing Bureau in New York City has accomplished the tremendous task in three or four weeks of actual operating time.

ASTRONOMY AND PLANETARY SCIENCE

And Backward, Too

The calculations, for comparison with observations, were also carried back to the year 1653. Data from eclipses of Jupiter's moons observed at that early date are still of value to astronomers.

The problem's mathematical complexity may best be indicated by the language of the mathematician. Solutions of certain types of simultaneous differential equations to the thirtieth order were required, as a total of ninety force factors acting on six bodies were involved. Integration was carried forward from force factor to orbit at forty-day intervals. Each result was required before the next interval could be calculated.

With its 12,000 vacuum tubes, 25,000 relays, and high-capacity memory, the selective sequence electronic calculator can multiply fourteen-digit numbers at the rate of forty a second, and add 100 nineteen-digit numbers in a second. It carries all operations forward in duplicate and constantly compares them, stopping automatically if the two sets fail to agree and trying again.

The new orbits are expected to be adopted internationally, succeeding those theories of Hill and Newcomb of half a century ago. The precise predictions of planetary positions thus made possible will be published in the American Ephemeris and Nautical Almanac and in those of other countries. Therefore, they will have immediate practical value to navigators and geodesists.

Tremendous Influence

As these five planets contain all but 1 per cent of the combined masses of all the planets, they have tremendous disturbing influence on the motions of such lesser bodies as asteroids, comets, and the satellites of the principal planets.

This evening, the annual Russell Lecture was delivered by Dr. Harlow Shapley, director of Harvard College Observatory. He described Harvard's survey of the "inner metagalaxy," comprising those extragalactic nebulae that are brighter than apparent magnitude 17.5. This magnitude corresponds to a distance from us of eighty million light-years for an average-sized galaxy of stars. Some giant galaxies, such as our own Milky Way system, are much larger than average, and they may be included in the Harvard survey to distances four or five times as great.

Dr. Shapley finds about one galaxy in every 100 million million million cubic light-years. For the fraction of the universe available to the 200-inch telescope, he estimates a billion galaxies to be visible, but adds another billion that are hidden by the obscuring dust near the plane of the Milky Way. This total of two billion star systems is probably not more than 1 per cent of the total galaxy-populated universe.

The evolution of star systems appears to proceed from the spiral or pinwheel forms to the ellipsoidal or disk-like forms, in quite the opposite direction to that believed a decade ago. The spiral galaxies, such as the great nebula in Andromeda, are filled with dust and gas from which stars are being born, whereas the ellipsoidal galaxies contain little, if any, dust and gas. They seem to contain no young red giant stars. Less than half of the galaxies are spirals. Nevertheless, Dr. Shapley believes the universe to be in a youthful state.

December 29, 1950

SOVIET FIRES EARTH SATELLITE INTO SPACE

560 MILES HIGH

Visible With Simple Binoculars, Moscow Statement Says

By WILLIAM J. JORDEN
Special to The New York Times.

MOSCOW, Saturday, Oct. 5—The Soviet Union announced this morning that it successfully launched a man-made earth satellite into space yesterday.

The Russians calculated the satellite's orbit at a maximum of 560 miles above the earth and its speed at 18,000 miles an hour.

The official Soviet news agency Tass said the artificial moon, with a diameter of twenty-two inches and a weight of 184 pounds, was circling the earth once every hour and thirty-five minutes. This means more than fifteen times a day.

Two radio transmitters, Tass said, are sending signals continuously on frequencies of 20.005 and 40.002 megacycles. These signals were said to be strong enough to be picked up by amateur radio operators. The trajectory of the satellite is being tracked by numerous scientific stations.

Due Over Moscow Today

Tass said the satellite was moving at an angle of 65 degrees to the equatorial plane and would pass over the Moscow area twice today.

"Its flight," the announcement added, "will be observed in the rays of the rising and setting sun with the aid of the simplest optical instruments, such as binoculars and spy-glasses."

The Soviet Union said the world's first satellite was "successfully launched" yesterday. Thus it asserted that it had put a scientific instrument into space before the United States. Washington has disclosed plans to launch a satellite next spring. Oct. 4."

The Moscow announcement said the Soviet Union planned to send up more and bigger and heavier artificial satellites during the current International Geophysical Year, an eighteen-month period of study of the earth, its crust and the space surrounding it.

Five Miles a Second

The rocket that carried the satellite into space left the earth at a rate of five miles a second, the Tass announcement said. Nothing was revealed, however, concerning the material of which the man-made moon was constructed or the site in the Soviet Union where the sphere was launched.

The Soviet Union said its sphere circling the earth had opened the way to interplanetary travel.

It did not pass up the opportunity to use the launching for propaganda purposes. It said in its announcement that people now could see how "the new socialist society" had turned the boldest dreams of mankind into reality.

Moscow said the satellite was the result of years of study and research on the part of Soviet scientists.

Several Years of Study

Tass said:

"For several years the research and experimental designing work has been under way in the Soviet Union to create artificial satellites of the earth. It has already been reported in the press that the launching of the earth satellites in the U. S. S. R. had been planned in accordance with the program of International Geophysical Year research.

"As a result of intensive work by the research institutes and design bureaus, the first artificial earth satellite in the world has now been created. This first satellite was successfully launched in the U. S. S. R. October four."

The Soviet announcement said that as a result of the tremendous speed at which the satellite was moving it would burn up as soon as it reached the denser layers of the atmosphere. It gave no indication how soon that would be.

Military experts have said that the satellites would have no practicable military application in the foreseeable future. They said, however, that study of such satellites could provide valuable information that might be applied to flight studies for intercontinental ballistic missiles.

The satellites could not be used to drop atomic or hydrogen bombs or anything else on the earth, scientists have said. Nor could they be used in connection with the proposed plan for aerial inspection of military forces around the world.

An Aid to Scientists

Their real significance would be in providing scientists with important new information concerning the nature of the sun, cosmic radiation, solar radio interference and static-producing phenomena radiating from the north and south magnetic poles. All this information would be of inestimable value for those who are working on the problem of sending missiles and eventually men into the vast reaches of the solar system.

Publicly, Soviet scientists have approached the launching of the satellite with modesty and caution. On the advent of the International Geophysical Year last June they specifically disclaimed a desire to "race" the United States into the atmosphere with the little sphere.

The scientists spoke understandingly of "difficulties" they had heard described by their American counterparts. They refused several invitations to give any details about their own problems in designing the satellite and gave even less information than had been generally published about their work in the Soviet press.

Concerning the launching of their first satellite, they said only that it would come "before the end of the geophysical year" —by the end of 1958.

Several weeks earlier, however, in a guarded interview given only to the Soviet press, Alexander N. Nesmeyanov, head of the Soviet Academy of Science, dropped a hint that the first launching would occur "within the next few months."

But generally Soviet scientists consistently refused to boast about their project or to give the public or other scientists much information about their progress. Key essentials concerning the design of their satellites, their planned altitude, speed and instruments to be carried in the small sphere, were carefully guarded secrets.

October 5, 1957

COMETS CALLED SPACE 'ICEBERGS'

In Spite of Fiery Trails They May Not Be Afire, According to Astronomer's Theory

Those comets that blaze fiery trails across our skies may not really be on fire at all. They may, in fact, be icebergs of space and pretty dirty icebergs, at that.

That is the theory propounded by Dr. Fred L. Whipple, director of the Smithsonian Astrophysical Observatory in Cambridge, Mass., according to The Associated Press.

With the Soviet space satellites drawing attention to the heavens, more and more questions are being asked about comets and probably more and more comets will be seen.

Three were reported visible to the naked eye last year—two confirmed by astronomers, the third listed as "possible." Normally, only one or two are reported observed without a telescope.

A layman's chances of discovering a comet are pretty good if he has a telescope. Astronomers want to hear about these discoveries. They would like to know more about these wanderers of the solar system.

Dr. Whipple's "dirty iceberg theory" was formulated only recently.

According to it comets are a collection of frozen gases, ice and bits of spatial dust, dirt and meteors. They form far out in space, millions of miles from the sun.

Pick Up Speed in Space

The bases freeze and collect more celestial debris. Moving through space, they gather speed as they near the sun and the pull of gravity increases.

At the same time, subtle changes begin to occur as the frozen nucleus of the comet warms up. The gases vaporize somewhat and become fluorescent in the rays of the sun. The dust of the nucleus also reflects the light of the sun.

As it first appears on earth, the nucleus (or head) looks like a bright point of light—a little star. Then, as it moves faster and more gases vaporize, the tail appears, a thin stream of fluorescent gases and dust.

There are very few rules for a comet to follow. One inescapable law is that it must revolve around the sun. Some pass close to the sun; some pass far away.

Comets move in elongated ellipses, some of fantastic size. It is estimated that the orbit of a comet sighted in 1864 is so long it may be more than two million years before it returns.

About fifty comets are known to have orbital periods of less than 100 years. Halley's spectacular comet, which took up most of the Southwestern sky when it appeared in 1910, is expected to show up again about 1985. Its tail was so long that the head appeared in some places in early morning while the tail was still visible to other parts of the earth the preceding evening.

Name of Discoverers

Anyone who wants to leave his name to scientific posterity can do so by discovering a comet. A comet is named after the first person to sight and report it.

The first comet visible to the naked eye last year was Arend-Roland, named for its Belgian co-discoverers.
have a slender spike extending from its nose as well as the tail.

The second confirmed comet of 1957 was Mrkos (pronounced Markosh), a bright vagrant recorded by a Czech observer. The possible comet was Encke, which last appeared in 1954 and returns every three and one-third years.

Some comets are highly unusual.

The comet of 1744 had six tails in a great fan. The Great Comet of 1843 had a tail 200,000,000 miles long.

The head of a comet may be 30,000 miles in diameter or it may be a million. Some comets ten times as large as the earth have been seen.

January 6, 1958

U.S. Satellites Find Radiation Barrier

Detect Intense Block to Space Traveler 600 Miles Up

By JOHN W. FINNEY
Special to The New York Times

WASHINGTON, May 1— United States satellites have detected a mysterious band of extremely intense radiation some 600 miles in space.

The radiation, 1,000 times more powerful than had been expected by scientists, raises a new obstacle to manned space flight. Scientists must now start redesigning future space ships to shield human passengers against the radiation.

The radiation is so intense that a space traveler would use up his weekly tolerance dose of radiation in one and a half hours. Scientists believe, however, that the radiation can be reduced to tolerable levels by enclosing the space travelers in a thin protective shield of lead.

The belt of intense radiation, which may stretch for several thousand miles into space, was discovered by the two Army Explorer satellites launched earlier this year. Both satellites were equipped with Geiger counters to measure the radiation, particularly from cosmic rays, in space.

The scientific findings obtained from the two satellites were described in detail for the first time today before scientists of the National Academy of Sciences and the American Physical Society. Later, the results were discussed at a news conference by scientists participating in the International Geophysical Year satellite program.

As a more encouraging sidelight, the satellites demonstrated that it was possible to develop space vehicles whose temperatures could be kept within tolerable limits for humans and confirmed that cosmic dust—or micrometeorites—was not so dense as to pose an undue hazard. The satellites also discovered that space a few hundred miles up was several times more dense than had been supposed.

By coincidence, the findings were disclosed as the first detailed information became available on the experiments performed by the Soviet Union's two satellites.

The findings of the Soviet satellites seemed to correspond closely to the scientific conclusions drawn from the United States satellite and rocket program, such as the higher-than-expected temperatures and density in space. Significantly new in the Soviet report was the effect of prolonged weightlessness on animals—an experiment not yet performed by the United States.

The second Soviet satellite, which was equipped to perform cosmic ray experiments, also apparently encountered the strange layer of radiation. The "mystifying" increase reported in the Soviet report, however, was not so great as that found by the American satellites, nor did the Soviet scientists seem to have so complete a concept of the nature and probable source of the layer of radiation.

'Crash Program' Begun

The United States satellites encountered the intense belt of radiation as they neared the peak altitudes of their orbits. Aside from its implications for space travel, the discovery may shed new scientific light on how the earth's atmosphere is heated up and how aurora borealis (northern lights) is caused.

ASTRONOMY AND PLANETARY SCIENCE

Dr. James A. Van Allen, at a meeting of scientists in Washington, tells of discovery of intense radiation belt.
Associated Press Wirephoto

The discovery of the high-intensity radiation was described by Dr. James A. Van Allen, head of the Department of Physics at the State University of Iowa, who developed the cosmic ray experiments for the two Explorer satellites.

At altitudes of 600 miles and higher, Dr. Van Allen reported, the satellites encountered unexpected radiation "1,000 times as intense as could be attributed to cosmic rays." The radiation became so intense, he said, that at times it "jammed" the Geiger counters so they did not put out any measurements.

The source of the intense radiation layer, he said, is believed to be ionized gas, probably hydrogen, shot out from the sun. In the electrified gas are fairly high-energy electrons, which produce X-rays as they bombard the satellite shell.

Dr. Van Allen offered this still preliminary theory about the radiation layer:

A reservoir of ionized gas, or plasma, apparently stretches from about 600 miles to perhaps 8,000 miles in space. The earth's magnetic field acts as an "umbrella" to hold the layer at least 600 miles away.

The reservoir of ionized gas gradually "drizzles" away into the earth's atmosphere, particularly around the North and South Poles, as the electrons lose energy. Occasionally, the reservoir is replenished by new outbursts of ionized gas from the sun.

Satellite scientists were so startled and impressed by the findings that they are now drafting a "crash program" to launch a satellite within the next eight months that would be designed to probe the nature and intensity of the radiation.

The electrons in the plasma rain on the satellite "like rain on a tin roof" and create X-rays, which go bouncing around inside the satellite. The X-rays are formed in the same manner as in an X-ray tube and apparently have about the same energy as medical X-rays.

Discussing the space-age implications of the satellite findings, Dr. Van Allen said that it was "likely that cosmic rays themselves are not a serious biological hazard, but this new radiation is something to worry about."

In a space ship, he said, the level of radiation could be reduced to about 10 per cent by a shield of lead one millimeter thick. This would represent about 100 pounds of lead shielding for each space passenger.

Dr. Edward Manring of the Air Force Cambridge Research Center, reported that a micrometeorite experiment performed by the two satellites had shown that cosmic dust presented a "small hazard" for future space travel, although over an extended period the dust could erode some of the satellite surfaces. The density of cosmic dust discovered by the satellites was in "fair agreement" with previous estimates, he said.

Dr. G. F. Shilling and Dr. Theodore E. Sterne of the Smithsonian Astrophysical Observatory reported that visual and radio observations of the satellites' orbits had lead to the conclusion that the atmosphere at a height of 220 miles had a density of about two ounces a cubic mile—or fourteen times the density previously predicted. This means the atmosphere at 200 miles is 100,000,000,000 times lighter than at sea level.

May 2, 1958

New Reasons for Comet Tails

The long plume of a comet's tail flows mostly away from the sun, no matter what path the comet follows. Thus the tail, which seems to the layman to be streaming out behind the racing comet, may often be pointing at right angles to the comet's path.

This fact long ago led astronomers to the conclusion that some repulsive force from the sun was producing the tail effect. The most logical conclusion seemed to be that the gases that make up a comet's tail are pushed away from its head by the pressure of light from the sun.

Although this is still believed to be the situation for many comets, calculations have shown that light pressure is not great enough to account for some of the long, straight comet tails stretching out sometimes as far as 100,000,000 miles. This has prompted scientists at the Max Planck Institute for Astrophysics in Munich to seek an alternate explanation.

In an article in the current Scientific American, Ludwig F. Biermann and Rhea Lust of the Institute suggest that particles of matter shot out from the sun may be combing gases back from some comets in the shape of a tail.

Comets normally are frozen chunks of various gases which are held together by their mutual weak gravitational attraction.

When the comet is far away from the sun it has no tail at all. On closer approach to the sun, however, some of the frozen materials are vaporized and surround the nucleus or comet head with an envelope of gas. It is material from this envelope that may eventually become the comet's tail.

There are three kinds of comet tail, according to the article. Those of Type I are long and straight. Tails of Types II and III are curved, are shorter than tails of Type I, and apparently are less complex in structure.

Some comets may change their tails from one type to another. Some have no tails at all and occasionally one will have two tails simultaneously.

To get a clearer idea of the forces at work to produce comet tails astronomers have used photographic techniques to measure the acceleration of matter in the tails. From this type of study the specialists at Max Planck Institute have concluded that comet tails of Type I at least represent forces at work too strong to be explained by the pressure of light alone.

Instead, they have concluded, the tails of Type I must be pushed back from the comet head by a rain of atomic particles ejected by the sun.

This rain, generally believed to originate in the vicinity of sunspots, consists of electrons and charged atomic nuclei which have been stripped of their accompanying electrons.

It is thought that all comets are members of our solar system, rather than chance visitors from the farther reaches of outer space. One astronomer has estimated that as many as 100,000,000,000 comets may form a great cloud out on the fringes of our system 50,000 to 100,000 times as far from the sun as the earth is.

October 5, 1958

SOVIET ROCKET HITS MOON AFTER 35 HOURS

FLAGS IN VEHICLE

Sphere Rams Surface at 7,500 M.P.H.— Moscow Jubilant

By MAX FRANKEL
Special to The New York Times.

MOSCOW, Monday, Sept. 14 —The Soviet Union hit the moon with a space rocket early this morning.

The first object sent by man from one cosmic body to another bore pennants and the hammer-and-sickle emblem of the Soviet Government.

The announcement said steps had been taken to prevent the destruction of the pennants by the impact.

The object was a sphere of unknown size weighing 858.4 pounds. It crashed into the moon at a speed of about 7,500 miles an hour at 2 minutes and 24 seconds after midnight Moscow time. This was 5:02:24 P. M. Sunday in New York.

The time of impact was only 84 seconds later than Soviet scientists had predicted.

Instrument Sphere

The success of the Soviet's moon shot was made known in a jubilant Government announcement at 35 minutes after midnight over the Moscow radio.

The sphere was a hermetically sealed instrument container that had been ejected from the last stage of a multi-stage rocket.

The rocket was launched from Soviet territory at about one o'clock Saturday afternoon Moscow time (6 A. M. in New York).

The container covered a distance of 236,875 miles in about 35 hours.

The impact was not visible from the earth, but the strike was signaled by the sudden end to radio transmissions that were being received here from the container during its space voyage.

[Jodrell Bank in England reported that it had received

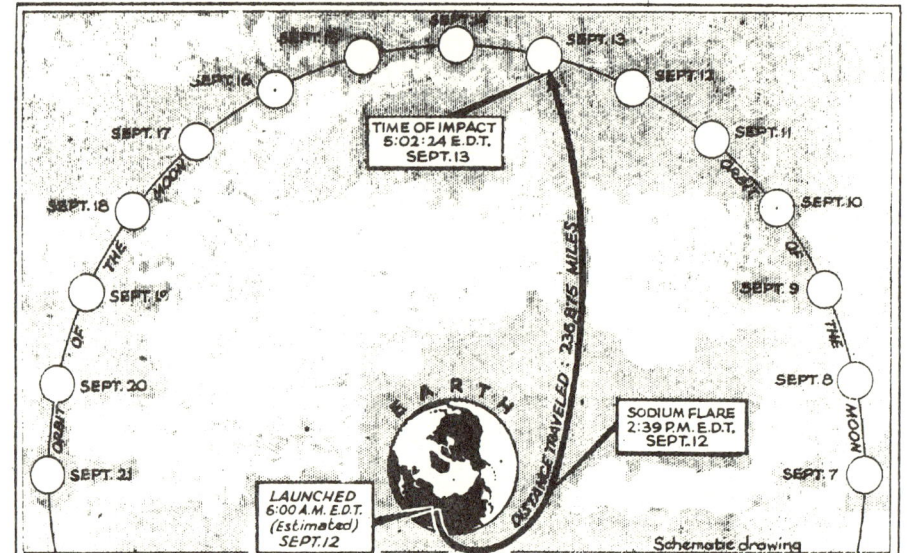

PATH TO THE MOON: Solid line shows path traveled by the Soviet rocket to the moon.

the signals up to the time the rocket hit the moon.]

Fate of Sphere Unknown

The sphere was able to reach the moon's surface because there is little or no atmosphere that would produce friction and burn it up.

It is not known whether it shattered on impact or penetrated the dust that is thought to blanket much of the moon's surface.

Soviet scientists had estimated before the final announcement that the container would hit at a point about 270 miles from the center of the face among three large depressions in the moon's surface known as the Seas of Tranquillity, Serenity and Vapors.

There was no word here on the fate of the last stage of the rocket, which had been flying in space near the container. The container was separated from the rocket segment after they had safely escaped from the earth's gravitational pull in the first hour of flight on Saturday.

The rocket section, which weighed 3,324 pounds without fuel, had been described here as "guided." One scientist said that Soviet experts thus had had "the possibility of correcting its flight."

But there was no official word on whether this actually had been done or how it was to have been done.

The rocket and the container flew at slightly different speeds and in slightly different trajectories toward the moon, according to bulletins issued here yesterday evening.

The rocket, it was disclosed, carried its own radio transmitter. But signals from it, on two frequencies, were reported getting steadily weaker five hours before the moon strike.

Soviet scientists took special measures before the launching to prevent the sphere from contaminating the moon's surface with terrestrial micro-organisms. They did not describe the measures.

Urged by Scientist

Such precautions have been urged repeatedly by scientists so that future studies of the moon will not be thrown off by elements that might have been imported from the earth.

The announcement this morning said special but unspecified measures also had been taken to "preserve" the pennant carried by the container, which bore the inscriptions "Salute to the Union of Soviet Socialist Republics" on one side and the Soviet insignia of a hammer and sickle inside a garland and "September 1959" on the other.

A somber-voiced announcer for the Moscow radio shouted the word "Attention" at thirty-five minutes after midnight, interrupting a program of classical music.

"Moscow speaking," he said and proceeded to read a bulletin from Tass, the official Soviet press agency.

It said the container had made history's first flight from the earth to another cosmic body. The moon strike was an outstanding achievement of Soviet science and engineering and had opened a new page in space research, the announcement added.

The bulletin was read three times. A scientific worker then vaguely described the importance of preventing micro-organisms from the earth from reaching the moon and the program of classical music was resumed.

The moment of impact with the moon was 2 minutes and 30 seconds earlier than Soviet scientists had predicted on Saturday shortly after the rocket was launched.

At 7 P.M. yesterday, five hours before the strike, they forecast the impact for one minute after midnight. This turned out to be a minute and twenty-four seconds too soon.

These projections as well as the entire moon shot operation were evidence of the high degree of accuracy achieved by Soviet rocket experts.

As one scientist pointed out here this morning, the moon shot required infinite care. Although the moon is 2,160 miles in diameter the problem of hitting it with a rocket is equivalent to firing from a moving and rotating platform, the earth, at a moving target, the moon.

ASTRONOMY AND PLANETARY SCIENCE

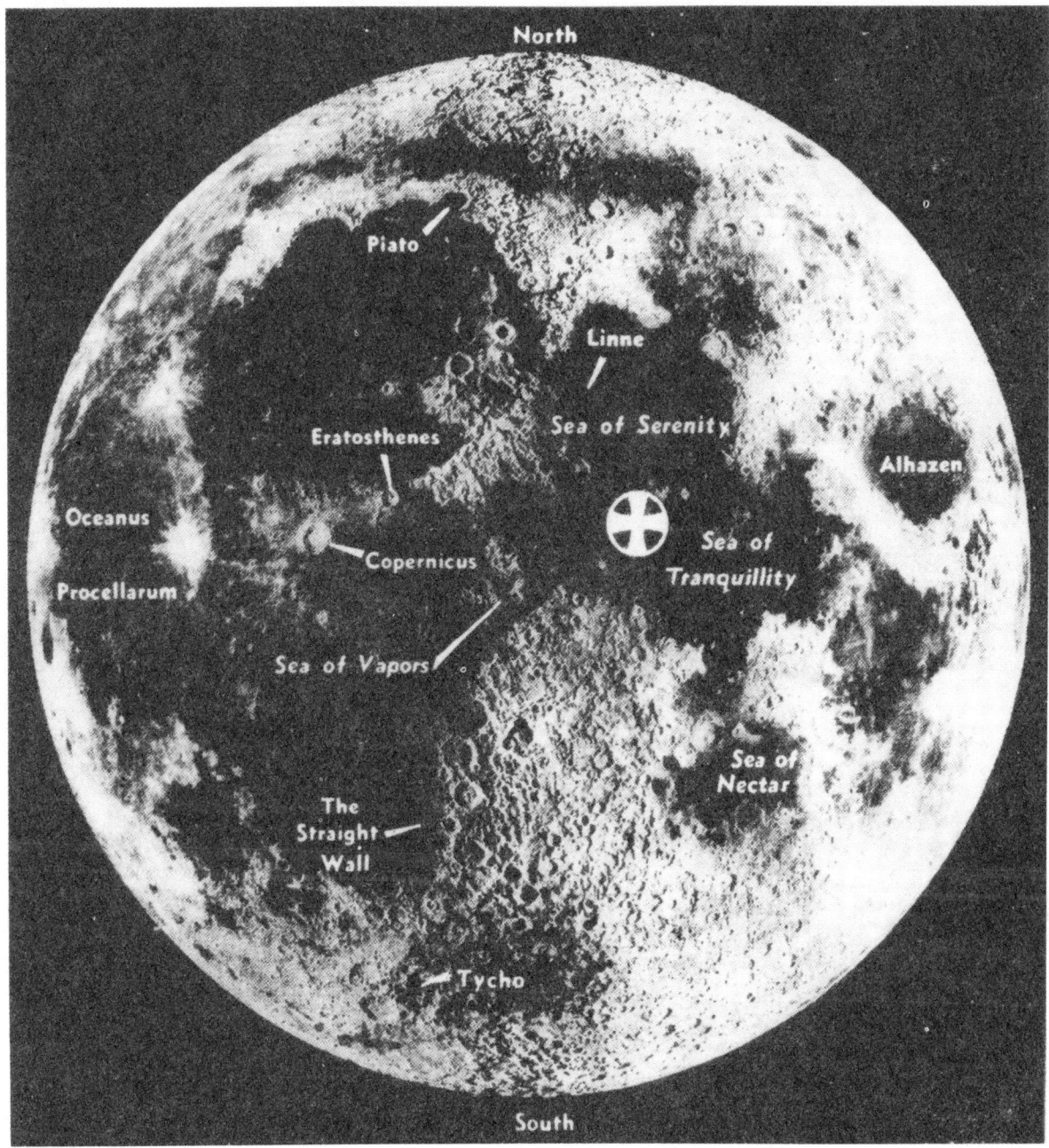

AREA OF IMPACT: Cross near right center shows area where Soviet rocket hit surface of moon. Photograph is a composite picture of face of moon taken by camera of Lick Observatory near San José, Calif. Labeled are craters such as Copernicus at left center and three seas in area of the landing, Tranquillity, Serenity and Vapors.

The New York Times, Sept. 14, 1959

The Soviet press and proclamations issued both at home and for abroad were quick to restate the claim of superiority for Soviet science and by extension the superiority of the Communist system that has supported it.

U. S. Failures Recalled

Some statements also compared the Soviet achievement to last year's moon-shot failures in the United States. Still other commentators contended that the Soviet feat was made possible by rocket fuels and equipment superior to those of the United States.

But most of all, Soviet propaganda seized upon the event as being of special significance to the forthcoming Eisenhower-Khrushchev talks. The Soviet leader will arrive in Washington tomorrow at the dramatic height of world attention to the Soviet moon strike.

The Premier is certain to offer the event as proof of Soviet might, skill and determination to surpass the United States in all other fields of production and technology.

A few minutes before the container struck the moon a special altimeter went into operation inside the sphere. It radioed back data on the angle and speed of approach to the moon, providing a kind of countdown to the moment of impact.

Altimeter Data Sent

Signals from the altimeter were transmitted on a frequency of 183.6 megacycles from one of two radio transmitters in the container.

A second radio also was working well on a frequency of 39.986 megacycles but the impulses it sent on the frequency of 19.993 had become considerably weakened five hours before the impact.

Reception from the radio aboard the last stage of the rocket also had weakened by that time. Its frequencies were 20.003 and 19.997 megacycles.

The fact that one of the radios sent toward the moon was in the last stage of the rocket was made clear in communiqués here only last evening. It also had not been previously clear whether the instrument container was flying in front of or behind the last stage of the rocket.

All communiqués until last evening had spoken only of the rocket. When the final impact was predicted at 7 P. M. yesterday, scientists said they were really talking about the sphere carrying the instruments.

No Photographs

Thus far no pictures or detailed descriptions of either object have been released.

A cartoon in Izvestia pictured a round sphere flying ahead of a blunted rocket section. But the drawings in other papers portrayed pointed rockets streaking into the sky.

The only description of the instrument container offered here has been the statement that it carried pennants, the date of launching and the Soviet Union's insignia.

If it resembles the container sent past the moon by Soviet scientists last January—and apparently it does—then a photograph published by Pravda in January is the best image of the object available thus far.

If they are similar, then the container that struck the moon had a surface of pentagonal stainless steel plates, most of which bore inscriptions that read either "U.S.S.R. September 1959," or simply "U.S.S.R.," with an etching of the Soviet seal, a crosed hammer and sickle surounded by a garland.

Had Inscribed Plate

The final rocket stage of the January shot had also carried a plate reading "Union of Soviet Socialistic Republics" on one side and the seal and date on the other.

The pentagonal segments gave that instrument container the appearance of a seamed volleyball.

The container that hit the moon yesterday was hermetically sealed and filled with an unidentified gas.

The instruments inside were connected with the two radios. They gathered information on the magnetic field of the earth and whether the moon has a magnetic field, the belts of radiation around the earth, the intensity and variation of cosmic rays, the heavy nuclei of cosmic rays, the density of matter in space, and on the number of meteoric particles encountered.

Soviet scientists looked forward especially to the answer on the question of a magnetic field around the moon and on the moon's gravitational pull and therefore its mass.

A leading Soviet space expert, Yevgeni K. Fedorov, said the study of cosmic ray protons and of corpuscular radiation of the sun had been dropped after the January shot because they were of less interest now than radiation zones around the earth.

A study of the radiation belts is crucial to the working out of safety measures to protect future space travelers.

A number of writers in the Soviet press this morning said that it was now clear that the day was not far off when man could be flying toward the moon.

Mr. Fedorov, as well as several other top Soviet scientists, wrote in the newspapers that it was now "especially important to insure extensive international cooperation of scientists, especially the scientists of the Soviet Union and the United States."

There was some speculation here that Premier Khrushchev would bring to the United States some specific proposals for cooperation if not joint ventures into space.

The excitement over the Soviet achievement built up slowly on what was an unseasonably warm and sunny Sunday afternoon in Moscow.

The moon itself, which will be full on Thursday, was easily visible last evening through a partially cloudy sky over the capital.

But there were not many moon gazers in the streets at midnight. Several articles in the papers yesterday had indicated that the moon strike could not be seen from the earth.

Shortly after launching of the multistage rocket on Saturday its final stage achieved a speed of seven miles a second to escape the earth's gravitational pull.

Then the rocket and the container, which was separated from it, slowed down as they coasted toward the moon.

At 7 P. M. last night, Moscow time, they entered the gravitational field of the moon at a speed of 1.44 miles a second. As they approached the moon it exerted an increasing pull and their speed was 2.06 miles a second at the moment of impact.

Log of Moon Trip

The Soviet space vehicles were 95,000 miles from the earth at 10 P. M. Saturday, Moscow time (3 P. M. New York time). At 10 o'clock yesterday morning (3 A. M. in New York) they were 161,250 miles away and at 7 P. M. last night (noon Sunday in New York) they were 201,250 miles from the earth.

Soviet tracking of the last rocket stage was based on radio signals as well as direct observation at one short minute during the flight. This was made possible by the emission from the rocket of a bright yellow sodium cloud at 9:40 P. M. Saturday, Moscow time.

The cloud was not visible in Moscow because of overcast skies, but it was photographed by at least two Soviet observatories, in the Caucasus and in Uzbekistan in Central Asia.

September 14, 1959

BACK OF MOON 'SEEN' FIRST TIME; PHOTO SHOWS FEWER CRATERS THAN FACE HAS

AREAS ARE NAMED

Ground Switch Swung Lens Into Position— Picture Put on TV

By OSGOOD CARUTHERS
Special to The New York Times.

MOSCOW, Tuesday, Oct. 27— The Soviet Union released today what it said was man's first picture of the hidden side of the moon.

Eight of the hazy dark spots shown in the single picture released were promptly given names by a specially appointed committee of the Soviet Academy of Sciences. One of the largest was a depression said to be 187 miles across. It was called the Moscow Sea.

The picture was transmitted by the official press agency, Tass, to its bureaus throughout the world, published in Moscow's two principal newspapers, Pravda and Izvestia, and shown to Soviet viewers over the Moscow television network.

Other Pictures Taken

The picture shown today was one of a "considerable number" that Tass said the Soviet Union's latest lunar rocket had taken twenty days ago as it soared past the far side of the moon.

What does the other side of the moon look like? Here is what the photograph showed in part:

A vast white area with darker shadings covering most of the southern hemisphere and extending halfway up the western quarter of the northern hemisphere. The boundary of this shadow was named by the Soviet scientists the Soviet Mountains.

In the center of the huge, white area was a large irregular indentation, the one that was named the Moscow Sea. Scientists use the word sea to describe the dry depressions on the surface of the moon facing the earth, and the Russians apparently continued the practice for the far side.

In the western sector behind the Soviet Mountains is a group of four round spots, two of which were given names. The other two were said still to be under study for classification, as were ten other clearly defined spots on the unseen surface of the moon.

70 Per Cent Photographed

Photographs published by Pravda also showed the far lunar side was mostly covered by mountains.

[In a broadcast from Moscow Monday night, The Associated Press said, a Soviet scientist described the lunar discovery as follows:

["The unseen part of the moon is considerably more monotonous than the side turned toward the earth. It contains fewer seas and fewer contrasts."

[The scientist was Prof.

ASTRONOMY AND PLANETARY SCIENCE

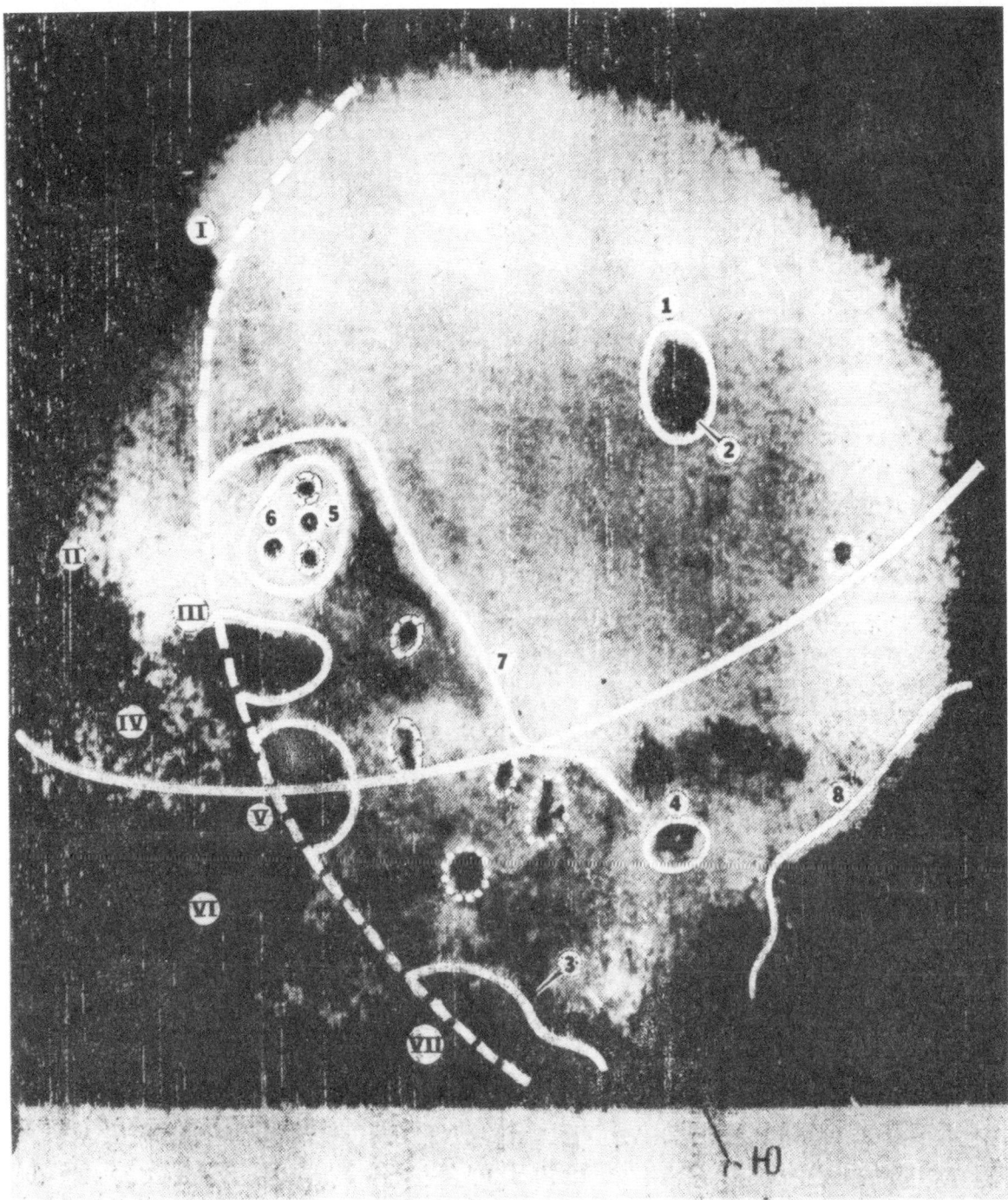

Associated Press Radiophoto

FAR SIDE OF THE MOON as photographed by equipment aboard Soviet vehicle. Picture, released by Tass, has not been retouched except for numerals and lines.

Soviet astronomers identify the long solid line as the moon's equator. Heavy broken line at left separates the part of the moon visible from the earth from the portion that cannot be seen. Solid lines surround objects absolutely established; objects that need more clarity of form are enclosed in heavy dotted lines; fine-dotted lines are around objects now being classified.

The Arabic numerals, as given by Soviet astronomers, are as follows: 1—Moscow Sea, a crater 187 miles in diameter; 2—Astronauts' Bay of Moscow Sea; 3—Continuation of Southern Sea on the moon's face; 4—Crater of the main Tsiolkovsky Hill; 5—Crater of central Lomonosov Hill; 6—Joliot-Curie Crater; 7—Soviet Mountains, and 8—Sea of Dreams.

The Roman numerals designate areas visible from the earth: I—Humboldt's Sea; II—Sea of Crises; III—Marginal Sea; IV—Sea of Waves; V—Smyth's Sea; VI—Sea of Fertility; VII—Southern Sea.

Arrow indicates north pole (top), south pole (bottom).

For those portions not designated by numerals or lines, further processing is now being done.

Aleksandr A. Mikhailov, director of the Pulkovo Observatory.]

An earlier announcement said that the 600-pound cosmic vehicle had succeeded in photographing in bright sunlight 70 per cent of the back of the moon, which is eternally hidden from the earth's view.

The announcement also said that the rocket had automatically developed the pictures it had taken during a forty-minute series of exposures and had transmitted them back to Soviet stations shortly before it had reached its perigee, or nearest point to the earth, on Oct. 18.

Tass also released a picture of the space vehicle that had carried the complex scientific and photographic equipment into its fantastic elliptical trip more than 600,000 miles around the earth and past the moon's orbit.

The vehicle looked like a big ashcan with a rounded bottom and a truncated conical top from which four fixed antennae bristled. In the center of the top was a large lens—the big eye that the Soviet Union asserted had given man his first view of what lay on the other side of the moon.

The Tass bulletin said that an automatic switch controlled from the ground had swung this eye toward the moon when its carrier was zooming through interplanetary space about 37,000 to 43,000 miles from the moon.

According to the picture released, the entire back half of the lunar globe appeared to have been in sunlight when the satellite snapped its three-quarter profile.

The western quarter of the globe in the picture showed areas that already had been seen from the earth and had been given their classical names —Humboldt's Sea, the Sea of Fertility, the Sea of Crisis. This part of the globe also showed the Marginal Sea, part of which extended into newly photographed area, as did Smyth's Sea at the moon's equator, and the Southern Sea at the south of the globe.

In fact, part of the photograph showing the areas seen from the earth had by far the larger masses of dark spots.

The moon television showing also revealed that one of the indentations in the Moscow Sea had been named Astronauts' Bay by the Soviet committee.

One medium-sized spot in the center of the southern hemisphere said to have a crater in its center was called Tsiolkovsky Hill after the Russian whom the Soviet Union has credited with being the father of rocketry.

A huge dark area on the southeastern rim of the globe is called the Sea of Dreams.

The two round spots named in the western sector were named for Frederic Joliot Curie, the late French nuclear scientist, and for Mikhail V. Lomonosov, an eighteenth century Russian scientist.

The picture hsowed the moon with its polar axis leaning at about a 60-degree angle to the left from south to north. Considerably more of the northern hemisphere was visible than was the southern hemisphere.

With the publication of the picture the Russians promptly proclaimed another triumph for their scientists. Although several pictures were said to have been transmitted back to earth by the lunar rocket, it was considered almost certain that the one distributed throughout the world today was one of the best.

The picture of what the Soviet scientists called the automatic interplanetary station had been heavily outlined with a black line by artists. It appeared to be a cylindrical container made up of at least twelve flat panels.

At the top of the cylinder was a thick collar that might have contained solar batteries to operate the radio and other mechanisms. There were other protuberances and "windows" on the sides and top of the huge cannister that were not identified.

The Tass bulletin issued last night said that the lunar rocket's flight beyond the moon and the pictures it had produced had proved that Soviet scientists had solved the problem of orienting, or turning in space, a cosmic vehicle by signals from the earth so that the lens would be pointed at the moon when the camera started operating.

The bulletin also said that the problem of transmitting half-tone photographs of high quality back to earth by television from cosmic distances had been solved.

With the help of a special radio transmission system the satellite sent television photographs through cosmic space as the vehicle approached its closest point to earth. Simultaneously, it transmitted new data that made it possible more accurately to define the vehicle's orbit, Tass said.

It was also predicted by Soviet scientists that the satellite will last for about six months from the time of its launching on Oct. 4 and will have made eleven to twelve orbits around the earth before it re-enters the denser layers of the earth's atmosphere and burns up.

This was the first time that Moscow had so specifically asserted that the rocket would not continue indefinitely its great eliptical orbit around the earth.

The Tass bulletin said that other scientific data about qualities of space, also received from the vehicle's transmitting apparatus, still were being studied by Soviet scientists. The Soviet Union has promised to make public the information it has gathered from its lunar probe as soon as all data have been worked out.

At 8 P. M. Moscow time tomorrow (noon in New York), the bulletin said, the satellite will be 300,564 miles from the center of the earth and will be over a point on the earth's surface with coordinates Lat. 6 degrees 30 minutes N. Long. 38 degrees 6 minutes W.

There was no indication in the bulletin as to whether the space vehicle would continue to transmit messages to the earth.

Theory Is Offered

MOSCOW, Tuesday, Oct. 27 (AP)—The general monotony of the landscape of the moon's far side is "beyond doubt associated with the question of the origin of the configuration of the moon," Professor Mikhailov said today.

"The dark patches of the so-called seas are clearly visible," he went on. "Some of them extend to the other side of the moon."

No Evidence of Life Seen

MOSCOW, Tuesday, Oct. 27 (UPI)—There was no evidence of lunar life in photographs of the moon distributed inside and outside of the Soviet Union.

One photograph released by Tass today showed the entire lunar circumference. It looked like a light gray orange, with black bruise marks.

Tass said the hidden side of the moon had been photographed with a camera having two lenses, with focal distances of 200 and 500 millimeters, in according with a command-signal. The lenses provided simultaneous two-scale photographing, Tass said.

October 27, 195

SOVIET ORBITS MAN AND RECOVERS HIM

187-MILE HEIGHT

Yuri Gagarin, a Major, Makes the Flight in 5-Ton Vehicle

MOSCOW, Wednesday, April 12—The Soviet Union announced today it had won the race to put a man into space. The official press agency, Tass, said a man had orbited the earth in a spaceship and had been brought back alive and safe.

A brief announcement said the first reported space man had landed in what was described as the "prescribed area" of the Soviet Union after a historic flight.

A Moscow radio announcer broke into a program and said in emotional tones:

"Russia has successfully launched a man into space. His name is Yuri Gagarin. He was launched in a sputnik named Vostok, which means "East."

Reports on Landing

Tass said that, on landing, Major Gagarin said: "Please report to the party and Government, and personally to Nikita Sergyevich Khruschev, that the landing was normal, I feel well, have no injuries or bruises."

He landed at 10:55 A.M. Moscow time [2:55 A.M. New York time].

Earlier, the major reported: "Flight is proceeding normally, I feel well."

After orbiting the earth the major applied a braking device, and the vehicle space landed in the Soviet Union, Tass said.

Major Gagarin, 27 years old, is an industrial technician, and married. He was reported to have received pre-flight training similar to that of the astronauts who will man the United States' first space ships.

Soared to 187 Miles

The announcer said the Sputnik reached a minimum altitude

of 175 kilometers (109½ miles) and a maximum altitude of 302 kilometers (187¾ miles).

He said the weight of the Sputnik was 10,395 pounds, or slightly over five tons.

The announcement of the launching came at 2 A. M. New York time.

It said everything functioned normally during the flight.

Constant radio contact was maintained between earth and the sputnik, the Moscow radio said.

The announcer said the duration of each revolution around the earth was 89.1 minutes.

The title of the announcement was "The First Human Flight into the Cosmos."

The radio, which was quoting a Tass press agency statement on the launching, said that Maj. Gagarin "is feeling well" and that "conditions in the cabin are normal."

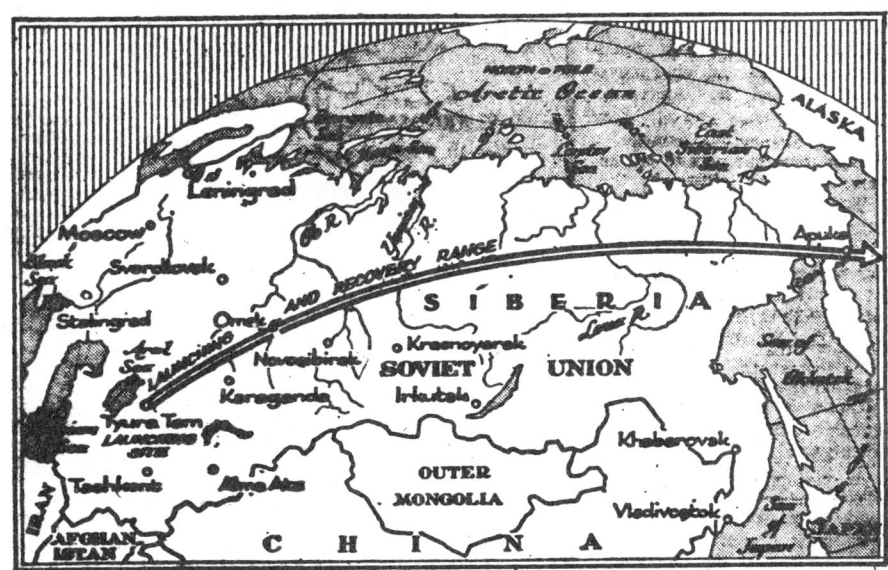

PROBABLE TRACK: Reported path of astronaut's flight from launch site at Tyura Tam.

SOVIET SPACE SUIT: This official photograph, released in 1958, shows a Soviet astronaut testing equipment designed to be used on rocket flights beyond the atmosphere.

As soon as the Moscow announcement was made, Russians began to telephone congratulations to each other.

The first astronaut is a major in the Soviet Air Force and is believed to be a test pilot.

The Tass announcement said that the launching of the multistage space rocket, which carried the Sputnik into orbit, was successful.

After attaining the first escape velocity, it said, and the separation of the last stage of the carrier rocket, the space ship went into free flight on a round-the-earth orbit.

Reports of the launching of a Soviet space man had been reported repeatedly in Moscow for the last twenty-four hours.

The London Daily Worker and other sources had said the Soviet Union had sent a man into space last Friday and had brought him back alive.

Many persons in Moscow were convinced after today's announcement that another flight into space was attempted on Friday and there was speculation that something might have gone wrong.

The announcement of the first flight into space was repeated three times, after which the normal radio program of music was resumed.

The radio also broadcast patriotic songs.

The announcement said the condition of the navigator was being observed by means of radio telemetering devices and television.

Major Gagarin, the announcement went on, withstood satisfactorily the placing of the satellite ship into orbit.

April 12, 1961

Mariner Inspects Venus at Close Range; Radios Data 36,000,000 Miles to Earth

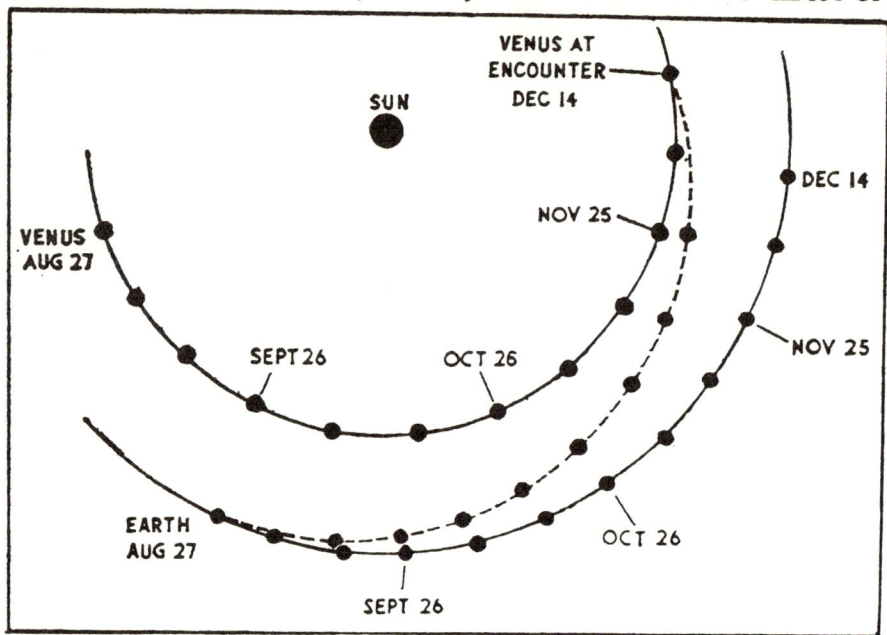

Dotted line shows path traversed by the Mariner 2 satellite from the time of launch until it reached within 21,000 miles of Venus and sent radio reports back to earth.

Information May Tell If Life Could Exist in Planet's Atmosphere

By JOHN W. FINNEY
Special to The New York Times.

WASHINGTON, Dec. 14 — The United States achieved a significant "first" in the exploration of space today by sending a Mariner spacecraft near the planet Venus to take man's first close-up observations of a planet.

For 40 minutes, as it raced toward its near rendezvous with the sister planet of earth, the Mariner 2 spacecraft took instrument readings of the temperature and atmosphere of Venus — readings that may answer the age-old mystery of the nature of the cloud-shrouded planet.

Back from the 447-pound spacecraft, some 36,000,000 miles from earth, came loud and clear radio signals containing information that may resolve the question of whether some form of life can exist on the planet.

The technological and scientific achievement—thus far unparalleled in the five-year history of space exploration — came as a climax to a trip that began 109 days ago on a launching pad at Cape Canaveral, Fla.

The spacecraft, resembling a miniature oil derrick, travelled 180,200,000 miles through the hostile, unknown environment of interplanetary space in a race to overtake the planet. In an impressive display of guidance, the spacecraft passed 21,100 miles from the surface of Venus.

Virtually until the last moment, however, it was uncertain whether the spacecraft, becoming uncomfortably hot as it came ever closer to the sun and suffering from electronic difficulties, would be able to carry out its interplanetary mission.

Twice during the early morning hours, an automatic timer in the spacecraft failed to turn on two radiometers that were to make temperature measurements of the planet. Finally, at 5:45 A.M. Pacific standard time, a ground command to turn on the instruments was flashed from the Goldstone tracking stations in California's Mohave Desert.

The radio command took three minutes to travel the 36,000,000-mile distance to the spacecraft, setting a record in interplanetary communications. Three minutes later, radioed word came back from Mariner 2 that the instruments had been turned on. Some five hours later, the instruments began taking observations of the planet.

Historic Scientific Event

As James E. Webb, head of the National Aeronautics and Space Administration, pointed out at a news conference, it was a "historic scientific event" and an "outstanding 'first'" in space for our country and the free world."

In this "one significant hour" of space exploration, he said, "more may be added to man's knowledge of the planet Venus than has been gained in all the thousands of years of recorded history."

Behind the prevailing jubilation was the fact that the United States had at least beaten the Soviet Union in scoring a spectacular and impressive "first" in the space race. Thus far all the "firsts" had gone to the Soviet Union — the first earth satellite, the first lunar landing of a payload, the first man in space. With Mariner 2, the United States had the honor of obtaining the first direct information in the vicinity of another planet.

Aside from the considerations of international prestige and competition, however, it was the most significant, as well as most spectacular, of the nation's scientific efforts in space thus far. From the information obtained from Mariner 2 it should be possible to end the mystery of whether an oven-hot desert, a dark, hot region of massive dust storms or perhaps oceans and habitable temperatures exist beneath the shroud of clouds hiding the planet against observations from earth.

In a space program that has been plagued by embarrassing mechanical failures, it was also an impressive engineering achievement in designing a spacecraft that could go through intricate mid-course correction maneuvers, endure the conditions in interplanetary space, make observations and radio them back to earth.

Drama Marked Trip

It was the engineering performance of Mariner 2 that lent an air of fingers-crossed anxiety and drama to the trip for with each day there was increasing uncertainty that the spacecraft would be able to perform its mission. As aptly described by space agency scientist Edgar M. Corright at the news conference, the story of Mariner's trip through space "makes the perils of Pauline look like a nursery story."

The spacecraft literally approached its rendezvous with Venus on "a wing and a prayer." In mid-November, one of its two solar panels, providing electrical energy, went off. As the spacecraft approached the planet, its battery and earth sensor temperatures rose to design limits, raising the possibility that the battery would explode or the spacecraft lose its "fix" on earth.

Some 12 hours before the scheduled "flyby," further electronic difficulties developed. Twice — at 11:21 p.m. and 2:41 a.m. Pacific Standard Time — the spacecraft's central computer and sequenser, acting on orders fed into it before launching, tried to turn on the two radiometers. Each time the miniature computer and timer failed.

As the spacecraft came up over the horizon within radio range of the Goldstone tracking station, the decision was made to send the command from the ground. From that point on, everything proceeded smoothly.

The two radiometers — one to measure the microwave radiation, the other to measure the infrared radiation generated by the heat in the planetary surface and atmosphere—were turned on and started their 120 degree scanning motion.

At 10:55 A.M. when the spacecraft was 25,300 miles from the planet, the radiometers obtained the first measurements of Venus. At 11:17 A.M. the spacecraft, approaching the planet from behind and below its orbit passed the "terminator"—the dividing line between the dark and sunlit sides of the planet. By 11:37 A.M. at a distance of 21,700 miles from Venus, the radiometers, peering out the side of the spacecraft, could no longer "see" the planet and the historic experiment was over.

As long as it could be observed only from the earth, Venus has presented a cloud-shrouded riddle. The riddle has only been deepened by uncertainty over what was being measured in the ground-based observations.

Spectrographic observations indicate that the Venusian atmosphere is composed primarily of carbon dioxide, with a small amount of nitrogen and oxygen. But there is uncertainty whether the atmosphere also contains some water vapor, a critical factor in determining the temperature on the planet and the possibility of life.

Observations of the microwave energy generated by the heat on Venus indicate that temperature on or near the surface could be as high as 615 degrees Fahrenheit while infrared measurements show that the temperatures somewhere in the upper atmosphere could be as low as minus 38 degrees Fahrenheit. But there is disagreement over the source and accuracy of these observed temperatures.

'Models' of Conditions

Confronted with these inconclusive and often contradictory observations, scientists have advanced three "models" of conditions on Venus and in its atmosphere:

¶The "greenhouse model" that attributes the high temperatures to the same process that heats the air in the florist greenhouse. Visible sunlight pierces the cloud layer and warms the planet surface. As it heats up, the surface re-radiates infrared energy. However, the infrared energy cannot escape back through the cloud layer, with the result that trapped energy creates oven-hot conditions on the planet. For such trapping of the infrared energy, however, considerable water vapor would be required in the Venusian atmosphere.

¶The "aeolosphere model" in which the surface of the planet is a dark, dusty region with giant dust storms. The surface is heated by the friction of the swirling dust and gas, and the heat is trapped by the dense clouds of dust.

¶The "ionospheric model" that holds that the high temperature observations are being traced to the wrong place. Instead of coming from the planetary surface, this theory holds, the observed radiation comes from energy transmitted by electrons in an extremely dense and energetic ionosphere around Venus. Only this theory offers the slim ray of hope that habitable conditions may exist on Venus. Temperatures below the boiling point could exist, thus raising the possibility of oceans and an environment for some primitive form of life.

Mariner 2 may make it possible to choose among three models — or perhaps for scientists to develop a completely new theory — by making the first direct measurements of the temperature and composition of the Venusian atmosphere.

One of the experiments, the microwave radiometer, is designed to determine whether water vapor exists in the planetary atmosphere, and thus establish the validity of a "greenhouse model."

The radiometer, basically a device for receiving electromagnetic signals such as radio waves, will scan the planetary surface to detect radiation at two wave lengths, 13.5 and 19 millimeters. The first frequency is absorbed by water vapor. The second is unaffected by the presence of water vapor, and thus should be capable of "seeing" through the atmosphere to the surface.

The instrument, therefore, should be able to give a direct reading of temperatures at the planetary surface. Furthermore, by comparing the differences in temperature at the two wave lengths it should be possible to determine whether water vapor exists in the Venusian atmosphere.

The microwave radiometer experiment also may be able to test the "ionospheric model." If a dense ionosphere exists, the energy transmitted by the electrons should be detected at 19-millimeter wave lengths as the radiometer scans the planet.

A companion infrared radiometer experiment is designed to find out if there are any "breaks" in the cloud cover of Venus, and if so how much heat is escaping through the holes into space.

The infrared experiment will observe heat energy in two wave lengths — 8 to 9 microns and 10 to 10.8 microns. In the first wave lengths the atmosphere is transparent except for clouds. In the second wave length the lower atmosphere is hidden by the presence of carbon dioxide.

If there are any appreciable breaks in the clouds, therefore, a substantial difference will be detected in the heat measurements at the two wave lengths.

With its magnetometer experiment, Mariner 2 also should provide indirect information about the interior of the planet, particularly whether it has a fluid center.

December 15, 1962

VENUS DATA POSE MAGNETIC PUZZLE

'Quick Look' Suggests Field Smaller Than Earth's— Cloud Cover Studied

By ROBERT C. TOTH
Special to The New York Times

WASHINGTON, Dec. 19 Space scientists were puzzled today over their tentative analysis of measurements made by the Mariner space probe of the magnetic field around Venus.

A "quick look" at the data, first available on the historic voyage of Mariner, suggests that the earth's sister planet has a smaller magnetic field than earth.

Scientists of the National Aeronautics and Space Administration were extremely cautious, however. They said that a "solar wind" near Venus could have depressed the Venutian field below the 21,500-mile altitude at which Mariner passed the planet.

It was also learned that preliminary analysis of Mariner data indicated Venus has a solid cloud cover rather than one broken by surface winds or other weather conditions there.

If Venus was shown to have a magnetic field like earth, it would have been strong evidence that the planet was formed like man's and probably that it, like earth, still has a hot interior.

How Field Is Created

Current theory is that the earth's field is created in large part by the different rate of rotation of the hot core inside the cold outer crust.

An extremely weak Venutian field would also raise the possibility that Venus spins on its axis much more slowly than earth. Previous long-range observations suggested that Venus rotated about once each 225 days, although other studies indicated a much faster rotation rate.

The complete analysis of the data will not be available for some days. Dr. Edward J. Smith, a N. A. S. A. scientist of the Jet Propulsion Laboratory in California, said his team would be ready to report its findings on the magnetic field at a scientific meeting in California next week.

There is some sentiment within the space agency here to withhold the Mariner results, however, until they are accepted for publication in a scientific journal.

When queried by telephone, Dr. Smith said that if Mariner had passed earth at the same distance, it probably would have detected the magnetic field here.

Mariner's instruments to measure the field were working properly when the 447-pound craft passed Venus Dec. 14 after a 180,000,000 mile flight through space, N.A.S.A. has said.

Venus is closer to the sun and the solar wind of charged particles could have been stronger there than it is near earth, other N.A.S.A. scientists noted.

Jet Propulsion Laboratory officials, in counseling caution on interpreting the tentative analysis, said that "back of the envelope" calculations based on Mariner's temperature readings of Venus had indicated a surface temperature of about zero degrees, plus or minus 100 degrees centigrade.

That result, which would have had enormous implications on the possibility of life on the cloud-covered planet, has turned out to be false after detailed examination of the data.

December 20, 1962

RANGER TAKES CLOSE-UP MOON PHOTOS REVEALING CRATERS ONLY 3 FEET WIDE

Craft Hits Target Area; 4,000 Pictures Sent Back

Details of Lunar Region Seen Thousand Times Clearer Than Before—Feat Hailed as Leap in Knowledge

By RICHARD WITKIN
Special to The New York Times

PASEDENA, Calif., July 31—Ranger 7 radioed to earth today the first close-up pictures of the moon—a historic collection of 4,000 pictures one thousand times as clear as anything ever seen through earth-bound telescopes.

Scientists here were hailing the achievement, which exceeded all expectations, as by far the greatest advance in lunar astronomy since Galileo.

They said the pictures not only represented a great leap in man's knowledge of the moon, but also, on a more practical level, lent encouragement that the lunar surface was suitable for Project Apollo's manned lunar landings.

Taken in 17 Minutes

The still pictures were snapped and transmitted in the last 17 minutes before the spacecraft crashed into an area northwest of the Sea of Clouds.

They meant in effect that the 240,000 mile distance to the moon had been shrunk by man's ingenuity to a mere half-mile in terms of what he could see of its topography. They showed craters three feet in diameter and a foot to a foot and a half deep.

The best earthbound telescopes, handicapped by the shimmering mantle of the atmosphere, can shrink the lunar distance only to 500 miles and reveal features no smaller than about one-mile across.

The startling disclosures of what Ranger 7 had wrought were made at a packed news conference here by a team of scientists headed by Dr. Gerard P. Kuiper of the University of Arizona.

The conference, televised nationally, was held in the auditorium of the Jet Propulsion Laboratory of the National Aeronautics Space Administration.

"This is a great day for science," the eminent astronomer said at the start, "and a great day for the United States.

"What has been achieved is truly remarkable. We have made progress in resolution [clarity of pictures] not by a factor of 10 . . . not by a factor of 100, which would have been remarkable, but by a factor of 1,000."

As a series of ten samples of the Ranger 7 photographs were flashed on a screen, Dr. Kuiper pointed out some of the more interesting features. Among the highlights of his recital and of answers both he and another member of the scientific panel made were these:

¶A few hours' quick study of Ranger 7's massive output had not revealed that there were any totally unforeseen problems on the moon. But the numberless new details opened a region of knowledge that would keep scientists in deep study for three or four years or more.

¶There was evidence that the white rays around some major craters were caused not by light fluffy material tossed up from the moon but by sizable rocks thrown off in the formation of these large craters. The rocks made numerous secondary craters deep enough to represent an extreme hazard for a manned lunar landing in the area. Such areas were to be avoided like poison, Dr. Kuiper said.

¶The tentative impression of the scientific team was that the lunar surface dust or other substance was not thick enough to swallow an astronaut landing craft. Dr. Eugene Shoemaker of the United States Geological Survey at Flagstaff, Ariz., when asked if he would like to step out on the moon, said:

"I don't think I'd be very worried."

¶The only interesting feature noted in the quick first look was a cluster in one of the pictures projected here of very many small craters showing a soft rather than a hard familiar outline.

Earlier, after a look at a few hastily processed Polaroid pictures, Dr. William H. Pickering said:

"They are several times better than any pictures of the moon we have seen before from the point of view of resolution. We will certainly see things on the final pictures that we have never seen before."

Dr. Pickering is director of the Jet Propulsion Laboratory of the National Aeronautics and Space Administration. The laboratory carried out the Ranger project.

Asked about details of the pictures, he answered:

"Well, if you mean were there any little green men, the answer is 'no.'"

Bernard P. Miller, Ranger manager for the Radio Corporation of America, which built the six television cameras, said his first look showed the pictures to be even better than expected.

Others who saw Polaroid samples, somewhat inferior to prints still to come from the finest top quality cameras used primarily, offered such assessments as "excellent" and "extremely clear."

The final prints were being processed with what was described as "tender loving care" in one of the finest Hollywood laboratories, and were expected to be considerably better in detail.

The lunar close-ups promised not only to multiply what man knows of the moon's terrain and its bleak history but also to remove obstacles to firm planning for the first lunar landing by American astronauts. The closest of the shots was snapped and transmitted three-tenths of a second before impact, when Ranger 7 was just one-half mile from the moon.

The still pictures were the first photographs taken from a spacecraft of the side of the moon facing the earth. However, in October, 1959, the Russians photographed the far side of the moon—the side always hidden from the earth—with their Lunik 3 satellite.

But while these provided the first solid though not surprising evidence of what the hidden side looked like, the resolution or clarity was slight. It was much cruder than the resolution of telescopic photographs taken of the near side of the earth.

The Soviet pictures thus contributed rough outlines for the first far-side lunar maps, but no new understanding of the precise nature of the lunar terrain.

Ranger 7 told a different story.

Even the small-sized preliminary Polaroid shots gave assurance that man was on the threshold of new discoveries whose meaning could be enormous.

The best telescopes on earth cannot delineate objects less

than a mile or more across. Hopes were that the Ranger 7 prints would pick up objects a small fraction of that size.

"If the objects are sharply defined, and there is sharp contrast from the shadows, we should see something down to about a few meters, about as big as a Volkswagen," said Harris M. Schurmeier, head of the laboratory's Ranger team.

Can End Some Doubts

The great potential value of such detail for Project Apollo was that it should clear some doubts remaining about whether the two-man "bug" being built for the first manned landings can safely do the job. Its spidery landing legs have been conservatively designed with a wide-open stance and large feet to cope with a variety of lunar topography. It was understood that the bug could tip 15 degrees or more and still be in no danger of upset.

But there has been some concern that the smooth look of the lunar seas might be an illusion because obstructions less than a mile in width could not be distinguished by telescopes forced to look through the shimmering earthy atmosphere.

It has been thought possible that the seas, which have been believed to be the most hospitable places for putting down manned vehicles, might be pocked with sizable rocks and ditches. These could cause a one-leg-up, one-leg-down landing that would topple the vehicle.

It was not expected that the Ranger 7 pictures would give more than inconclusive clues on another critical issue. This has to do with the consistency of the lunar surface.

Is it soft or hard? If a soft layer of dust or pumice, is it deep enough to swallow a landing craft?

The triumph of Ranger 7 abruptly and happily ended a succession of 12 failures in 12 attempts by various United States agencies to put useful equipment on or near the moon in the last six years.

It was also a vindication of the Jet Propulsion Laboratory, which had come in for painful criticism following the failure of Ranger 6 to transmit any pictures last February. Inquiries by NASA and Congress led to numerous design changes and a tightening of the management structure.

The 804-pound Ranger 7 was launched Tuesday at 12:50 P.M., Eastern daylight time.

It ended its elliptical journey of 243,665 miles at 9:25 A.M. today for an elapsed time of 68 hours 35 minutes. The moon, at its zenith in the early morning sky here, was at a line-of-sight distance of 228,000 miles at time of impact.

Perhaps because the laboratory had gone through so much travail in the past, the reaction here when the success of the mission became apparent was especially emotional.

Several hundred officials, newsmen and laboratory employes monitored the progress of Ranger 7 over closed-circuit television from the same auditorium where many of the same people had, last February, suffered what was perhaps the most shattering letdown of the space program so far.

When the moment came to receive pictures from Ranger 6, after everything seemed as shipshape as on Ranger 7, nothing happened.

Leaders Saluted

When the moment came today, the signals were right on time, and it was clear almost immediately that their quality was good. The audience not only cheered but also stood on its feet in a remote salute to top officials who were still in various control rooms.

They stood again and applauded and whistled when the officials finally arrived for a news conference.

Asked about the reaction in the various control rooms here and at Goldstone, Calif., these men repeatedly used a single word: Chaos.

Someone put on the closed circuit TV a sign reading— "Next Stop Mars."

Ranger 7's six slow-scan TV cameras began snapping their pictures and transmitting them about 19 minutes before impact.

The spacecraft altitude above the moon was about 1,000 miles.

They were able to supply some 4,320 still pictures before the craft wound up in a heap of rubble (some lunar explorer may some day mark it appropriately) just a few miles from the bull's eye chosen when the last rocket maneuver was made midway along the flight path on Wednesday. The impact speed was 5,850 miles an hour.

The impact point was in what looks on existing lunar maps (they will have to be refined now) like an extension of the Sea of Clouds, Mare Nubium in the classical Latin. The area was 10.7 degrees south of the lunar equator and 20.7 degrees west of the central north-south line as seen from earth. The nearest prominent moon mark is the crater Parry, to the east and slightly north.

The TV pictures pouring from the two 60-volt transmitters on the spacecraft were gobbled up by two 85-foot dish antennas at Goldstone. They were recorded by three methods, to play things as safely as possible.

Most important, the signals, which looked to human eyes like a single point of light horizontally scanning a television tube, were immediately photographed by a special 35-mm. kinescope camera whose shutter kept open just long enough for one complete scan.

A human eye could not see a picture on the tube because the eye requires several rapid scans to retain a meaningful image.

The succession of 35-mm. negatives were the ones destined for the archives and were promptly packed away in a refrigerator so that what was on them could not possibly be disturbed.

The second recording method used magnetic tapes. After painstaking calibration of the sensitive ground equipment, the tapes were played back and a second set of 35-mm. negative rolls made in the same manner as the first.

It took five and a half hours to complete the calibration and re-run.

The third recording method involved taking quick-look Polaroid pictures, which can be developed in seconds, of the TV tubes as often as thought necessary. This was to give engineers a check from time to time on how the equipment was working so that they could make any needed adjustments.

Engineers here said the grain of the Polaroid film was not as dense or fine as that in the primary 35-mm. cameras, and that the Polaroid lens was not quite as good. In addition, he said, the developing process for the 35-mm. films was more accurate.

The negatives made belatedly from the magnetic tapes were the ones that were to be studied by a team of prominent scientists here and later released to the public. The master set in the refrigerator would not be touched until the other sets had shown technicians how to get the most out of them in processing.

The scientific team was headed by Dr. Gerard P. Kuiper of the University of Arizona. Its other members included Dr. Eugene Shoemaker, of the United States Geological Survey at Flagstaff, Ariz.; Ewen A. Whitaker of the University of Arizona, and Raymond L. Heacock of the Jet Propulsion Laboratory.

Dr. Harold C. Urey of the University of California was also a member but he was not here for the initial analysis. He was reported in Europe.

The magnetic-to-film negatives were flown from Goldstone to Burbank and then rushed to Hollywood. Officials of NASA were secretive about where the laboratory was and, in all, seemed to be guarding the precious film documents like gold bullion.

Two more virtually identical Ranger craft are scheduled to be launched early next year. But they will aim their camera eyes at different potential landing spots for Apollo astronauts.

The invaluable but still limited information from the Rangers will be elaborated on, starting about a year later, through the first "soft" lunar landing of a Surveyor spacecraft. This will put cameras down on the moon, and they will take their time peering around the nearby moonscape.

The Surveyors also will scoop up lunar soil, run it through ingenious mineral analyzers, and radio back to earth what they have found.

Finally, a third class of spacecraft named Lunar Orbiter will fill in the long gaps between the isolated locations photographed by Ranger and Surveyor. They will orbit the moon at about 20 miles altitude and take continuous strips of photographs of lunar terrain.

Ranger 7 rammed into the moon at an angle of about 23½ degrees from the vertical. Its six TV cameras looked out of a small opening at slightly varying angles averaging 38 degrees from the long axis of the spacecraft.

The difference in pointing angle of the camera from the flight path of the spacecraft, which was canted slightly, was about 7 degrees. This meant that the center of successive pictures would not be a single terrain feature but would move gradually across the ground.

Three of the six cameras were equipped with wide-angle lenses and the other three had narrow-angle lenses.

On only two of the six did the scanner run its beam over the entire face of the tube. The scanners on the four others scanned only the center portion, about 1-16th of the total frame.

This made it possible to scan more quickly, about 2-10ths of a second each scan. But it sacrificed much of the available image.

The full-scan mechanisms took 2.5 seconds for each operation but had the advantage of taking in the whole picture.

Today, everything worked flawlessly. The first full-scan pictures began arriving at the isolated desert antennas 80 seconds after the power had been turned on and technicians did their best to contain their great elation until their jobs had been done.

Fourteen minutes before impact, the warm-up command for the partial-scan cameras was issued by an involved sequencing system aboard the spacecraft. Eighty seconds later, its enheartening signals began arriving at Goldstone.

Man's path to the moon had been measurably cleared.

August 1, 1964

FIRST MARS PHOTO IS TRANSMITTED; MARINER SIGNALS INDICATE PLANET LACKS A LIQUID CORE LIKE EARTH'S

OTHER DATA SENT

Sensors Find Scant Radiation Belt and Thin Atmosphere

By WALTER SULLIVAN
Special to The New York Times

PASADENA, July 15 — Mariner 4 has sent to earth the first close-up photograph of Mars.

The picture, transmitted today in an eight-hour broadcast over a distance of 134 million miles, shows the "limb," or rounded edge of Mars, including a vast, desert-like region.

It does not show any of the controversial canals. But this is not necessarily significant, since the view is extremely oblique and covers a region under the noonday sun. Such lighting makes for little contrast.

The picture, the first ever taken of another planet at close range, covers a region between the areas of Mars known as Cebrenia, Arcadia and Amazonis.

Part of the second picture, which should overlap the first, has already been transmitted to earth and it is possible that as many as 22 pictures of the planet will be delivered in the next 10 days.

Officials here at the Jet Propulsion Laboratory, which is in charge of the project for the National Aeronautics and Space Administration, were jubilant.

No Magnetic Field

Meanwhile, scientists associated with the project reported some of their initial findings. These include:

¶Mars has virtually no magnetic field and hence, presumably, no liquid core. This means the planet may differ fundamentally from earth in terrain and the chemical composition of its surface. The radioactivity of its air must be comparatively high, as well as the exposure of its surface to space radiation.

¶Mars has no significant radiation belt. This is good news for those planning the exploration of Mars. Vehicles will be able to orbit the planet for long periods without radiation hazard to their passengers or instruments.

¶The atmosphere of Mars is extremely thin — probably too much so for the use of parachutes or other conventional devices in the gentle landing of instruments and, ultimately, astronauts on its surface.

¶Mars, like the earth, has swept its orbit clear of much of the cosmic dust that would otherwise be adrift there. Since the orbit is elliptical, the distance of the planet from the sun varies from 128 to 155 million miles. It is this region that has been swept comparatively clean.

If Mars lacks a metallic core, it would appear that the planet has never gone through the churning internal processes that have given the earth its layered structure.

This would mean that Mars does not have continents formed of light-weight rocks, and oceanic basins underlain with basaltic rock, in the manner of the earth.

The surface chemistry of such a planet would be very different from that of the earth. Possibly the forces within it that generate volcanoes and have built the landscape of earth are absent or mitigated, for these forces derive from the internal heat of the planet.

Radar signals bounced off Mars have shown it to be remarkably smooth. In fact, one area that was a target for the Mariner camera seems to reflect the radar beam almost as a mirror would.

Mariner 4 was not designed to determine whether or not there is life on Mars. However, the planet's lack of a substantial magnetic field, plus the fact that its air is very thin, means that its surface is probably bombarded with radiation from space 50 times more intense than that striking the earth.

However, this was described by Dr. James A. Van Allen of the State University of Iowa this afternoon as "not a frightening value."

In fact, Dr. William H. Pickering, director of the Jet Propulsion Laboratory, said, "I have always felt we will find some form of life on Mars," and added that he was not discouraged.

The laboratory has carried out the Mariner project for the National Aeronautics and Space Administration.

Both men spoke at a news conference held to discuss preliminary findings from instruments carried past Mars yesterday by Mariner 4.

At the time, the vehicle, as it sailed on beyond the orbit of Mars, was slowly sending its initial picture.

At first the signals were received by the tracking antennas at Johannesburg, South Africa and Madrid. But as the earth turned, the central station of the tracking network at Goldstone, Calif., picked up the signal and Johannesburg finally lost it.

Sent in Form of Numbers

The data came in as numbers indicating the darkness of each spot along the 200 lines that together form the picture. The sequence of numbers strongly suggested that a picture was being received. It was repeatedly announced from the Mariner control center here that "the data look very good."

As the scientists presented their findings, it became clear that there was unanimity of opinion, from a variety of observations, that the planet has no substantial magnetic field.

As Dr. Van Allen put it, facetiously:

"If there are any Martian men, they do not use compass needles."

It is the earth's magnetic field that makes the compass a useful instrument. The shape of the field, which reaches far out into space around the earth, is comparable to what it would be if there were a powerful bar magnet near, though not directly at, the center of the earth.

The magnetism is thought to be generated, at least in part, by the movement of liquid iron withing the earth's core.

Because electrically charged particles are diverted by such a magnetic field, the magnetism acts as a shield, except over the magnetic poles where its force lines are almost vertical.

Dr. Van Allen said the entire atmosphere of Mars was bombarded by a rain of high energy particles, or radiation, comparable to that which only reaches the top of the earth's atmosphere near the polls. The fact that the air of Mars is very thin means much more of this reaches the surface than on earth.

Analysis Awaited

It is hoped that by tomorrow a more precise estimate of the density of the Martian air will become known by analysis of the behavior of radio waves from Mariner 4 as it went behind the planet last night.

During this part of the flight the waves, skirting Mars, came to earth through the Martian atmosphere.

It is reported that the findings are not much out of line with those obtained from study of the absorption of infrared light that has passed through the Martian air.

These observations, made from both California and Texas observatories, indicated that air pressure at the surface of Mars is probably no more than one to three one-hundredths of that on the surface of the earth.

Dr. Van Allen, discoverer of the earth-encircling radiation belts that bear his name, said the search for such belts near Mars was "completely negative."

He showed results of an electron detector displaying no change from the normal interplanetary level during the flyby.

Proton Rise Noted

However, there was a rise in observed protons when the vehicle was 100,000 miles from Mars. This apparently came from an eruption on the sun.

About a dozen such events had been observed during the 228 days of flight, he said. The absence of radiation belts indicated, he added, that Mars has a magnetic moment, or total force, no more than one-thousandth that of the earth.

Dr. John A. Simpson of the University of Chicago, who was in charge of the cosmic ray observations, said his instruments confirmed that an outburst

ASTRONOMY AND PLANETARY SCIENCE

from the sun coincided with the fly-by.

He, too, saw no sign of magnetic influence from the planet and estimated its maximum magnetic moment at less than five-thousands that of the earth.

He noted that an absence of magnetic field would mean that radiation from space would generate far more radioactive atoms within the air of Mars than is typical of the earth's air.

The best known of these atoms is carbon 14, widely used in dating archaeological objects.

Dr. H. S. Bridge of the Massachusetts Institute of Technology told of efforts to measure the "solar wind," particularly in the vicinity of Mars.

The solar wind is the steady outflow of thin, hot, high-velocity gas from the sun. It is diverted from the earth by magnetism, but no such diversion was seen near Mars.

He, too, estimated that magnetic moment of Mars is less than a thousandth that of the earth.

Dr. Leveritt Davis Jr. of the California Institute of Technology told of the Mariner magnetic measurements. It was hoped not only to detect the planet's magnetism but determine its orientation—that is, to find out the location of the magnetic poles of Mars. No such magnetism was found. He estimated the magnetic moment at less than a three-thousandth that of the earth.

Data Contradicted

"Probably," he said there is "essentially no field at all."

W. Merle Alexander of the space agency's Goddard Space Flight Center, reported that a steady rise in cosmic dust had been observed during the first 90 days of flight, as the vehicle moved out through the solar system.

The trend was then reversed. The reason appears to be the action of Mars in sweeping up dust in its vicinity, Mr. Alexander said. The earth apparently has the same effect.

Furthermore, both planets tend to collect the lighter dust particles, so that the remaining dust consists preferentially of large particles.

The effect of Mars was first observed by Mariner when it reached the region swept periodically by the planet when nearest the sun in its elliptical orbit. This was some 37 million miles out from the earth's orbit.

The absence of a radiation belt around Mars contradicts a report by radio astronomers a year ago that the planet emits "hot" radio waves, indicating the presence of such a belt.

The report worried the Mariner engineers, for intense radiation could have "blinded" the electric eye that was to watch the star Canopus and keep the data-transmitting antenna aimed at the earth.

There was talk of loading gyros on the craft to keep it steady if this happened, but it was decided to take a chance and leave them off.

The radiation belt would complicate plans for manned landings and the selection of long-lived orbits for the Voyagers.

These are the large spacecraft that, in a few years, will survey the planet from orbit and, perhaps, send down small, sterile capsules to study its surface.

The absence of a magnetic field and hence, presumably, of a liquid core was not entirely a surprise. Mars has only a tenth the mass of the earth and hence the gravitational pressure and heat in its core should be far less.

The flight of Mariner 2 past Venus two and a half years ago showed that that planet, almost identical in size with the earth, likewise does not have a strong magnetic field, although the instruments were by no means so sensitive as those of Mariner 4.

Probably the reason is that Venus spins extremely slowly. It is thought that the spin of the earth contributes to the liquid flow that generates its magnetism.

Dr. Van Allen pointed out that radio emissions from Jupiter show it to have an "immense" radiation belt. This is presumably because it spins fast and is very large, generating power for magnetism.

Explanation for Redish Hue

Dr. Pickering pointed out that one explanation for the redish hue of Mars may be the presence of limonite, an iron oxide. This suggests that iron is uniformly spread through the planet instead of being largely concentrated in the core, as on earth.

Studies of the gravitational field of Mars, as disclosed by the movements of bodies near it, have suggested its internal composition is uniform. Again this suggests the absence of an iron core. Analysis of the effect of the planet's gravity on Mariner 4 may pin this down more precisely.

Radio signals during the flyby yesterday hinted at trouble in the tape recorder. The reason for the signals is still obscure, but Dr. Pickering said today: "We believe the tape recorder did operate properly and that the data for all 20 pictures will be there."

The actual number of pictures will not be known until all are received, since some may have been exposed at the end of the sequence when the aim of the camera telescope had slipped across to the night side of Mars.

However, the fact that the first picture shows the edge of the planet indicates that the sequence started just as planned and has led to the hope that from 18 to 22 pictures of the illuminated landscape may be obtained.

July 16, 1965

Mariner 4's Final Photos Depict a Moonlike Mars

By WALTER SULLIVAN
Special to The New York Times

WASHINGTON, July 29—A heavy, perhaps fatal, blow was delivered today to the possibility that there is or once was life on Mars.

The Mariner 4 photographs taken of the planet at close range July 14 show a crater-pocked landscape lacking any sign that there has been water erosion there. There is no evidence of river valleys or ocean basins.

Virtually all theories for the origin of life demand the presence of liquid water, preferably in large quantities. While Mars has frosty caps on its poles, it now appears that the planet may never have known rain.

Mars looks far more like the moon than the earth. It shows the scars of ancient collisions with objects whose impact left giant craters.

No water action, no sedimentation, no plowing under of the landscape by mountain-building processes have erased this record of ancient cataclysms, as has been the case on earth. Nor are there other signs of internal dynamics, such as continental masses, ocean basins and mountain chains.

Dr. Robert B. Leighton of the California Institute of Technology summarized the results at the White House ceremony this morning at which the series of pictures was presented to President Johnson and the nation.

The "scientifically startling" discovery that the red planet is covered with craters, Dr. Leighton said, "further enhances the uniqueness of the earth within the solar system."

The President himself linked the seeming uniqueness of man, at least within the solar system, to the urgency of preserving mankind from the holocaust of nuclear war.

"It may be," he said, "it just may be, that life as we know it with its humanity is more unique than many have thought and we must remember this."

However, he expressed optimism.

"I believe it is very clear," he said, "that in this day, when we are reaching out among the stars, that earth's billions will not set their compass by dogmas and doctrines which reject peace and embrace force and rely upon aggression and terror for fulfillment."

While Mariner 4 transmitted 21 full photographs and a fragment of a 22d to earth, it is primarily those from the middle of the sequence that are revealing.

On numbers 7 through 15 it is possible to distinguish 70 craters, ranging in diameter from two or three miles to 75 miles.

For reasons that are still uncertain, the remaining pictures of the series are increasingly under-exposed and it is not possible to see clearly the transition from the daylight to the night portion of the planet. Nor do the final pictures show the South Polar frost cap.

They do, however, display what is thought to be frost inside some craters in the winter hemisphere and possibly several frost-covered peaks. A giant escarpment estimated to be 13,000 feet high runs diagonally across picture 13, which covers an area measuring 140 by 170 miles.

With publication of these, the first close-up pictures of another planet, the centuries of speculation, of dreaming and science fiction have at last come up against reality. The results do not bear directly on the possibility of life in other planetary systems, a possibility many consider high.

They do indicate that there is only one planet in this system that has ever been suitable for life as we know it—the planet earth.

Those who argue for life in other worlds had hoped Mars would demonstrate that life arises quite commonly in the universe. It now appears that Mars is unlikely to provide a direct demonstration of this hypothesis.

May Reveal Process

Nevertheless, as Dr. Leighton pointed out, closer study of the planet may bring to light an important segment of the process whereby life originated on the earth. The evolution of the complex chemicals of life may easily have proceded part way on Mars, then been interrupted and preserved. Such a record has long since been wiped out on earth.

"If the Martian surface is truly in its primitive form," he added, "that surface may prove to be the best — perhaps the only—place in the solar system still preserving clues to original organic development."

In his remarks, the President facetiously expressed relief that no canals or other signs of intelligent life were observed.

"As a member of a generation that Orson Welles scared out of its wits," he said, "I must confess that I am a little bit relieved that your photographs didn't show more signs of life out there."

On Oct. 30, 1938, the actor and producer, Orson Welles, presented a radio adaptation of "The War of the Worlds," a science fiction novel written by H. G. Wells in 1898. It depicted hideous Martians landing in New Jersey so vividly that many radio listeners fled into the streets.

Despite the dim prospects of life on Mars, the Mariner 4 pictures do not directly rule it out. They cover only about 1 per cent of the surface. They show features no less than two or more miles in width. Above all they do not explain the seasonal color changes that have long been cited as the chief evidence for life.

Every spring, areas near the winter pole of Mars begin to darken and this effect moves slowly toward the equator. It is as though some form of vegetation was coming to life, although the direction of this progression—toward the equator—is the reverse of the advance of spring on earth.

Various proposals have been made as to how such darkening could occur by nonliving processes, but none of them has been widely accepted.

As Dr. Bruce C. Murray of the California Institute of Technology pointed out today, this spring darkening is as much of a mystery as ever.

Drs. Murray and Leighton are both on the team that carried out the television scanning.

Among those who cling to the hope that life may yet be found on Mars is Dr. William H. Pickering, director of the Jet Propulsion Laboratory of the California Institute of Technology. It was this laboratory that carried out the Mariner program for the National Aeronautics and Space Administration.

The evidence for life from ground-based observations is "still there," Dr. Pickering told a news conference after the White House ceremony. "And we have to investigate it," he said.

Dr. Leighton said the photographs were not expected to demonstrate the existence or nonexistence of life. Nevertheless, he added: "The search for a fossil record does appear less promising if Martian oceans never existed."

A preliminary study indicates that Mars has few steep slopes, as befits a planet that, unlike the earth, has not been torn and shaken by internal forces. The maximum slopes observed seem no more than 10 degrees.

No shadows have been identified, although they probably would have shown up had the final pictures been clear, since the terrain was then illuminated by a very low sun.

The craters are evident in terms of different intensities of lighting on their various slopes, the side toward the sun being dark but not in shadow. In the last clear shots these dark slopes seem bright, as though perhaps coated with frost.

The craters have rims that rise a few hundred feet above the terrain and central depths of a few thousand feet. These are typical of craters produced by explosive impacts, such as the famous Meteor Crater in Arizona, formed when a meteorite struck with such force that it exploded.

The craters of the moon are generally of this sort, as are those produced by shallow nuclear explosions.

In a burst of enthusiasm, Dr. Leighton said of picture 11 that it "must surely rank as one of the most remarkable scientific photographs of this age." It shows dimly a crater 75 miles wide within which are a number of smaller, more sharply defined craters that obviously are much younger.

This is what one sees repeatedly on the moon. The giant, dim craters are extremely old —perhaps relics of a time when the solar system was more cluttered with large objects than it is today.

The nearness of Mars to the asteroid belt could explain the plethora of craters; the surprise is that they are still so much in evidence.

The asteroid belt lies between the orbits of Mars and Jupiter. It consists of thousands of minor planets and chunks of rock in orbit around the sun. Because it fills a gap in the otherwise systematic spacing of the planets, some believe the asteroids are either the relics of a former planet, or are material that never conglomerated into a planet when the solar system was being formed.

Dr. Leighton said the evidence was "overwhelmingly preponderant" for the formation of the craters by impact. None appears to be volcanic. He said no clouds were "identified" as such, despite the appearance of a cloud-like blur off the edge of the planet in the first picture.

One possible explanation for underexposure of the last pictures is that Mars reflects low-angle sunlight far less efficiently than expected.

The camera system was designed to adjust the exposure of each picture in terms of light intensity recorded on the previous shot. Possibly the system was unable to keep up with the rapidly falling brightness of the terrain.

While the absence of water on Mars was not a surprise, many thought the planet must once have had liquid water and, perhaps, oceans in which life could have thrived. This now seems unlikely.

The possibility of canals is not entirely dead, Dr. Leighton said.

He noted that some enthusiasts believe the pictures were taken in "a very poor place and a very poor season," for seeing the canals first reported by 19th century astronomers.

He cited the quip of a Canadian cartoonist that "one word from Mars would be worth a thousand pictures."

July 30, 1965

Venus Yields Up a Few of Her Mysteries

By WALTER SULLIVAN

It is remarkable that the nearest planet to our earth—the brilliant "evening star" that has inspired poets and lovers from time immemorial—is one of the least understood of our neighbors in the cosmos. Venus remains so despite the fact that much information was learned when a Soviet capsule landed there on Wednesday and an American spacecraft, Mariner 5, flew past the planet on Thursday.

Both vehicles had been launched last June to take advantage of a favorable alignment of the earth and Venus. A preliminary look at the data they sent to earth has given American and Soviet astronomers new confidence in ground-based observations and the validity of their conclusions therefrom. However, there are also surprises and inconsistencies.

Observations from earth had led astronomers to conclude that a large part of the Venus air is composed of carbon dioxide. The broadening of spectral lines had shown that air pressure at the surface was many times greater than on earth. And the paucity of water vapor in the Venus air had led in recent months to the conclusion that the clouds of Venus are formed of dust rather than ice crystals (in the upper part) and water droplets.

Radio observations from earth and from Mariner 2, as it flew past Venus in 1962, had indicated a surface temperature of roughly 800 degrees Fahrenheit. This is the heat level used in the new, self-cleaning kitchen ovens. Boiled-over remnants in the oven turn to fine dust at such a temperature. It seemed that Venus was not very hospitable to life.

Capsule Released

Early Wednesday the Soviet craft, Venus 4, began to fall toward the earth-facing, night side of Venus. As it felt the first whisps of Venus air, the craft cast loose an instrumented sphere. Signals from the craft itself vanished, either when cut off by the electron wake of its initial plunge or when it burned up in the atmosphere.

The egg-shaped sphere was coated with a thick layer of dark material that wore away as it burned. Finally, at a height of about 16 miles, the sphere slowed enough for a lid to open and release a parachute. A parabolic antenna swung out to face

ASTRONOMY AND PLANETARY SCIENCE

WHAT WE LEARNED FROM THE VENUS FLIGHTS

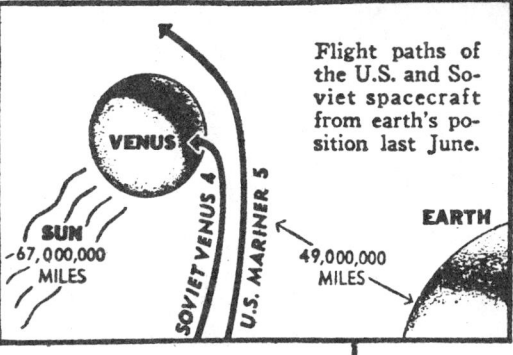

Flight paths of the U.S. and Soviet spacecraft from earth's position last June.

CLOUD COVER Top layer composed of dust particles rather than ice crystals; possibly a second cloud layer, indicated by pause during Venus 4 descent.

ATMOSPHERE Surface pressure is 15 times that on earth; composed mostly of carbon dioxide.

TEMPERATURE Ranges from 104 degrees fahrenheit in clouds to 536 degrees near surface as detected by Soviet capsule during parachute descent.

Soviet Venus 4 measured pressure and temperature during descent.

STRUCTURE Data obtained on planet's magnetic field may indicate whether it has a solid rather than liquid core.

Schematic drawing

The Soviet Union landed a capsule on Venus last week and the U.S. sent one flying past it in an effort to learn more about the mysterious planet. Some of the findings are shown at the right.

directly up—and hence toward Soviet receiving stations. It began to send temperature and pressure readings to earth, as well as an analysis of the air's composition.

It found the air composed almost entirely of carbon dioxide, with about 0.4 per cent oxygen and 1.6 per cent water vapor. No nitrogen was detected, which American scientists found strange.

The air at the start of observations, 16 miles aloft, was at 104 degrees Fahrenheit, which is thought to mean the capsule, by then, was already below the clouds. The last reading was 536 degrees. The pressure, at that point, was 15 to 22 times that at sea level on earth. It was announced yesterday that the capsule was then on the surface.

The surface of Venus is hot enough for pools of molten metal, such as tin, to exist there. However, radar measurements from earth indicate that parts of the planet are mountainous. Temperatures on the mountains and near the poles may be cooler. Hence some Soviet and American scientists still believe there may be primitive forms of life on Venus.

For several minutes during the hour-and-a-half-long descent of the Soviet capsule the pressure and temperature readings hovered at the same level. An updraft may have halted the descent for a while, or the capsule may have passed through some peculiar feature of the Venusian atmosphere. It has been postulated from radar echoes that a second cloud layer exists beneath the one visible from earth.

Measuring the Atmosphere

The mission of America's Mariner 5 seems to have been an unqualified success, even though its public impact was diluted by the prior Soviet landing. Analysis of the results is slow because the most important findings are hidden in subtle changes of radio waves that have passed through the Venus atmosphere.

This type of observation proved highly effective when Mariner 4 flew behind Mars in 1965. Radio signals traveling from the spacecraft to earth pass through successively deeper layers of the target planet's atmosphere. The denser the layer, the more signals are bent and altered, making it possible to reconstruct a rough profile of atmospheric density.

It was expected that the lower, very dense layers of Venus air would bend the signals so much that they would never escape from the planet. However, the observed effect did not seem as great as expected from the high density reported by the Russians.

Mariner 5 also found evidence that Venus has a magnetic field, although a weak one, whereas the Soviet craft detected no local magnetism during its final plunge. However, the Russians were perplexed by a brief magnetic indication early in the craft's approach to that planet.

The question of a magnetic field is of great interest because it bears on some of the chief Venusian mysteries, such as the source of its remarkable heat. The earth's magnetic field is thought to be largely generated by movements of its liquid core and the planet's spinning.

To display a magnetic field like that of the earth, Venus should be spinning and should have a liquid core. From radar measurements it has been estimated that it is hardly spinning at all. It is estimated to make one turn (clockwise, in the direction opposite to the other planets) every 244 days.

Yet ultraviolet photographs of its clouds suggest that the atmosphere rotates every five days. If both observations are correct, the surface must be swept by steady winds of more than 100 miles an hour, which some scientists find implausible.

The mountains, if they exist, were presumably thrust up by heat-generated activity within the planet. Its surface may be seething with volcanoes and it may have a large molten core. Precise measurements of its magnetism will help resolve such questions.

A remarkable feature of the Soviet measurements was pointed out by Dr. S. Ichtiaque Rasool of the Institute for Space Studies of the National Aeronautics and Space Administration. It has been calculated that when the earth was young, all of the carbon dioxide now tied up, chemically, in the earth's rocks was in the atmosphere. Gradually it reacted with water to form carbonates.

This early atmosphere had, it is believed, a surface pressure almost identical to that observed on Venus and a composition that likewise was similar, except for the inclusion of a small amount of nitrogen. Mars, too, is now thought to have an atmosphere at least half of which is carbon dioxide.

The problem, then, is to explain why the earth retained its water and thus solidified its carbon dioxide, whereas Venus apparently did not.

The Russians, in their exhuberance last week, were saying that manned exploration of Venus, perhaps as early as 1980, would resolve such questions. At the moment, however, the problem is to devise unmanned devices that can explore the planet at temperatures higher than those of most kitchen ovens. This is still a severe challenge to the electronics industry. Arming men to do the job will be even more difficult.

October 22, 1967

3 MEN FLY AROUND THE MOON ONLY 70 MILES FROM SURFACE

Astronauts Examine 'Vast, Lonely' Place; Read From Genesis

By JOHN NOBLE WILFORD
Special to The New York Times

HOUSTON, Wednesday, Dec. 25—The three astronauts of Apollo 8 yesterday became the first men to orbit the moon. Early today, after flying 10 times around that desolate realm of dream and scientific mystery, they started their return to earth.

They fired the spacecraft's main rocket engine at 1:10 A.M. to kick them out of lunar orbit and to carry them toward a splashdown in the Pacific Ocean on Friday.

Through the static of 231,000 miles, as Apollo 8 swung around from behind the moon and started for earth, one of the astronauts dispelled any doubts, saying, "Please be informed there is a Santa Claus."

57-Hour Return Trip

It would be a 57-hour return trip from the most far-reaching voyage of the space age thus far—or of any other previous age. The astronauts had seen, as no other men had, the ancient lunar craters, plains and rugged mountains from as close as 70 miles.

At 4:59 A.M. yesterday, about 20 hours before the return trip, Col. Frank Borman of the Air Force, Capt. James A. Lovell Jr. of the Navy and Maj. William A. Anders of the Air Force, swept into an orbit of the moon by firing the spacecraft's main rocket. This occurred after they flew around the leading edge of the moon and were directly behind the earth's only natural satellite.

"We got it! We've got it!" exclaimed a mission commentator of the National Aeronautics and Space Administration as the spacecraft emerged from behind the moon 24 minutes later, and was clearly flying a safe and smooth orbit.

Businesslike Report

The calm and laconic Apollo explorers, however, were all business. Captain Lovell's first message to earth was simply:

"Go ahead, Houston. Apollo 8. Burn complete. Our orbit is 169.1 by 60.5—169.1 by 60.5."

The astronauts flew twice around the moon in the egg-shaped orbit, then dropped to a circular orbit nearly 70 miles above the ancient craters, plains and rugged mountains of the lunar surface.

As they beamed their first live television from orbit on

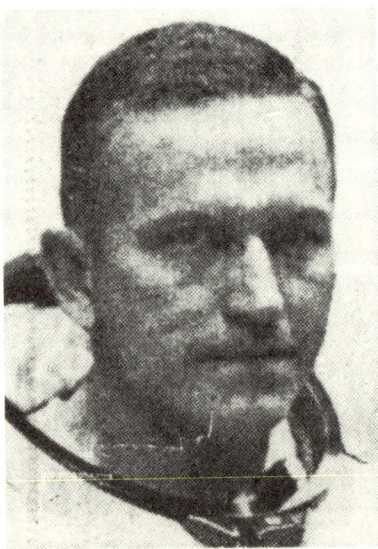
Col. Frank Borman

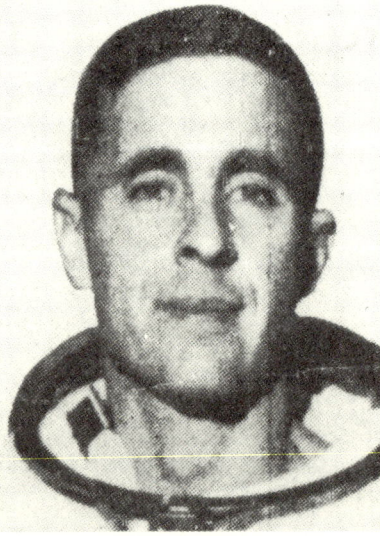
Maj. William A. Anders

Capt. James A. Lovell Jr.

Associated Press

Christmas Eve morning, they described the surface of the moon as a colorless gray, "like dirty beach sand with lots of footprints on it" and said it "looks like plaster of Paris."

At about 9:30 P.M. the astronauts began their second and last television show from lunar orbit. It ran some 30 minutes and showed the bright moon, in a pitch black sky, outside the spacecraft window.

Earth Like on 'Oasis'

Colonel Borman described the moon as a "vast, lonely and forbidding sight," adding that it was "not a very inviting place to live or work."

Captain Lovell saw the earth as a "grand oasis in the big vastness of space."

Major Anders was most impressed by "the lunar sunrise and sunsets."

As the telecast neared its end, Colonel Borman said "Apollo 8 has a message for you." With that, Major Anders began reading the opening verses from the Book of Genesis about creation of the earth.

"In the beginning," Major Anders read, "God created the heaven and the earth.

"And the earth was without form and voice; and darkness was upon the face of the deep . . ."

Captain Lovell then took up with the verse beginning, "And God called the light day, and the darkness He called night."

Colonel Borman closed the reading with the verse that read:

"And God called the dry land Earth; and the gathering together of the water called He Seas: and God saw that it was good."

Sends Holiday Greetings

After that Colonel Borman signed off, saying:

"Good-by, good night. Merry Christmas. God bless all of you, all of you on the good earth."

Glynn S. Lunney, one of the flight directors here, told reporters earlier, "we have a completely 'go' spacecraft."

George M. Low, the spacecraft manager at the Manned Spacecraft Center, said he was "altogether happy" with the mission — the most ambitious and daring thus far in the nation's $24-billion Apollo project to land men on the moon next year.

Although the mission's object was not primarily scientific, Dr. John Dietrich of the space center's geology and geochemistry branch, said that the television pictures and astronaut descriptions had "demonstrated their ability to observe from the spacecraft to a degree I think surprised most of us."

The astronauts' color movies and still pictures, expected to be the most spectacular and most valuable of all the pictures, will be brought back for processing and analysis by scientists. Many of the pictures were taken of a site in the Sea of Tranquility where American astronauts may land next year.

The lunar-orbiting mission, the second manned flight of the Apollo project, is expected to be followed by an earth-orbiting flight in February or March to test the lunar landing vehicle. The first landing on the moon could come as early as next June.

Apollo 8's historic voyage round and round the moon came about 69 hours after the spacecraft was launched by a Saturn 5 rocket at Cape Kennedy, Fla., last Saturday morning.

The spacecraft was falling faster and faster toward the moon's vicinity, having crossed from earth's to the moon's sphere of gravitational influence when the astronauts were more than 214,000 miles away from earth.

To make a fine correction of their aim the astronauts fired the spacecraft's small maneuvering rockets. Ground controllers, fearing some of the Apollo computers' navigation data might be incorrect, radioed a new set of numbers into the instrument's memory unit.

Then, Maj. Gerald Carr of the Marine Corps, the astronaut acting as capsule communicator in the control room, radioed:

"Apollo 8, you are riding the best bird we can find."

"Thanks a lot, troops," Major Anders replied. "We'll see you on the other side."

At 4:49 A.M. as the spacecraft curved behind the moon, the signals died out. Apollo 8 was out of range of the space agency's three deep-space tracking antennas—in California, Spain and Australia.

This was 10 minutes before the spacecraft's 20,500-pound-thrust main rocket was supposed to fire, slowing them down and dropping them into lunar orbit. The rocket is situated in the aft end of the spacecraft's 22-foot-long equipment unit, called the service module. The forward crew compartment, the command module, is only 11 feet long.

Engine Fires on Schedule

If the engine failed to fire, the astronauts would merely loop around the moon's back side, without going into orbit, and then whip back to earth.

But the engine ignited on schedule, at 4:59 A.M., and fired slightly more than four minutes. The engine was pointing toward the moon at an angle so that the firing acted as an explosive brake, slowing the spacecraft from a speed of 5,758 miles an hour to 3,643 miles an hour.

It was 20 minutes more before flight controllers here could know if they had an orbiting spacecraft. They waited in tense silence.

Then came a trickle of data indicating that the spacecraft was emerging and working well. Finally, a long minute later, there was a crackle of sound over the voice communication circuit.

Captain Lovell was talking. He was ever so matter of fact, the pilot first, leaving any poetry for later.

"Go ahead, Houston, Apollo 8," Captain Lovell said. "Burn complete. Our orbit is 109.1 by 60.5."

Happy to Get Message

Amid the jubilation from the control room, Major Carr acknowledged the message:

"Apollo 8, this is Houston. Roger. 169.1 by 60.5. Good to hear your voice."

The numbers were the altitude of their moon orbit given in nautical miles. It translates to about 194.5 statute miles at the highest point, which would be on the front of the moon, and 69.6 miles at the low point, on the back side.

Flight controllers and the astronauts always deal in nautical miles. These figures are translated by reporters into statute miles.

As the astronauts emerged from around the eastern edge of the moon, traveling westward near the equator, the sun was shining high overhead.

Because of its slow rotation, the moon's daytimes and nighttimes both last 14 days. Most of the moon's back side and the eastern edge of the side facing earth are now in sunlight. Much of the center part of the moon's face is partly illuminated now by the earth shine.

The astronauts' first concern was not the moon but a radiator in the spacecraft's cooling system. All the water had evaporated, and had to be replenished.

Then the astronauts looked down to see the crater-scarred moon.

One of the first major lunar features they spotted was the crater Langrenus, one of many craters with peaks rising from the center of their floors. They next flew over the broad plain called the Sea of Fertility.

When asked by ground controllers what "the old moon looks like," Captain Lovell began describing the sights unfolding below:

"The moon is essentially gray, no color. Looks like plaster of Paris or sort of grayish beach sand. We can see quite a bit of detail.

"The Sea of Fertility doesn't stand out as well here as it does back on earth. There's not as much contrast between that and the surrounding craters. The craters are all rounded off. There's quite a few of 'em. Some of them are newer. Many of them look like —especially the round ones— look like they were hit by meteorites or projectiles of some sort.

"Langrenus is quite a huge crater. It's got a central cone to it. The walls of the crater are terraced, about six or seven different terraces on the way down."

Some 'Old Friends' Seen

Captain Lovell went on to describe two craters in the Sea of Fertility — his "old friends" Messier and Pickering. He saw rays of light material extending out from Pickering's rim.

"Now we're coming upon the craters Colombo and Gutenberg," Major Anders said. "We can see the long parallel faults of Gaudibert, and a run through the mare [sea] material right into the highland material."

Faults are cracks in the lunar surface produced by stresses of unknown origin.

By then, Apollo 8 had passed over much of the daytime area of the moon and was reaching the Sea of Tranquility, an even broader plain on the right side of the moon, as seen from earth.

With the sun lower, close to the horizon, the astronauts were able to make out more details, especially enhancing their perception of depth and height on the lunar surface.

This was the way they wanted it. For on the Sea of Tranquility lies one of the five sites being considered for the manned lunar landing planned for next year. The primary purpose of the Apollo 8 mission is to take detailed pictures of the area to help future astronauts steer to their landing.

It's about impossible to miss," Captain Lovell assured flight controllers. "Very easy to pick out."

The spacecraft then flew over the terminator—the point where daylight changes to darkness. The view in that area, they reported, was quite sharp.

But beyond, even with the earthshine, it became more and more difficult to make out any landmarks on the surface.

After moving around the back side and reappearing in

their second orbit, the astronauts aimed their 4.5-pound television camera on the moon and transmitted their first pictures. This began at 7:29 A.M.

At first the light was too bright, with the sun directly overhead. The first crater they looked down on was an unnamed one they called Brand, after Vance D. Brand. He is an astronaut acting as capsule communicator during the night shift.

Throughout their orbit the astronauts called many of the small unnamed craters on the back side by the names of friends, associates and themselves. The craters Borman, Lovell and Anders lay just south of the equator near where the back of the moon ends and the front begins.

Apollo officials said the crater names were in no way official, merely handy ways to identify some nameless features.

During the telecast, Major Anders, who handled the camera, described the scene below:

"The color of the moon looks like a very whitish gray, like dirty beach sand with lots of footprints in it. Some of these craters look like pickaxes striking concrete, creating a lot of fine haze dust."

Captain Lovell then reported seeing "a lot of what appears to be very small new craters that have these little white rays radiating from them."

The moon is believed to be about the same age as earth—some 4.5-billion to 5-billion years old. Since the moon has no atmosphere and no surface water, its face has not undergone the same erosion as the earth's has.

But through the ages it has been peppered with meteorites and rent with volcanic eruptions, presumably the primary sources of the craters.

In the telecast the astronauts took pictures along a path running about 550 miles. A single picture usually covered an area some 175 miles wide.

After about 10 minutes, the astronauts signed off their telecast, promising to resume it on their ninth orbit. The show stopped as their spacecraft passed over Smyth's Sea and two prominent craters, Gilbert and Kaestner.

The astronauts complained again that three of their five spacecraft windows were hazy, clouding somewhat their view of the moon.

An analysis by engineers on the ground indicated that the problem probably lay in the fact, that under the rubberized caulking material used to seal the windows' edges. They suspect that the material creates the fogginess by giving off some traces of contaminating gases. The two clear windows were sealed by different substances.

However, the astronauts reported that the ice that had formed on the center window was melting. This came about because of the warming sunlight being reflected by the moon.

Passing over the Sea of Fertility for the second time, Major Anders took pictures with his 16-mm. movie camera. He described the sea as a darker brown than he had expected.

While in orbit the astronauts shot some 1,200 still pictures, many of them in color, with a 70-mm. Hasselblad camera, as well as movies and the television.

On one orbit Major Anders let the movie camera run for the completed two-hour revolution, taking pictures at one frame a second. He also took pictures of the terminator between day and night.

The subsequent orbits were fairly quiet. The astronauts ate and took a turn at some short naps. They continued to take color pictures, and also practiced sighting landmarks used for navigation.

From the tracking data, flight controllers have noticed that the spacecraft tended to "jiggle" when it flew over the famous crater Copernicus.

This was attributed to the fact, recently noted, that under the moon's surface there are scattered lumps of material of greater density. A slightly greater gravitational tug is exerted at those spots.

The unevenness, however, caused the astronauts no trouble. But the data are being analyzed carefully so that future moonfarers will know what to expect.

It was observed during the 10 orbits that the spacecraft's flight path underwent some slight but puzzling changes. Starting out in a circular orbit of 69.8 miles, Apollo 8 wound up sagging to nearly 68 miles at a low point and rising to nearly 73 miles at a high point.

Flight controllers could not immediately explain this tendency, which was being studied. It might be related to the moon's gravitational anomalies.

When night fell here and a bright quarter moon shone in the clear sky, mission control commented to Apollo 8 that "there is a beautiful moon out there."

Colonel Borman replied, "Now, we were just saying that there's a beautiful earth out there."

"It depends on your point of view," concluded the ground controller.

During the orbital voyage, Colonel Borman, the commander did most of the piloting, while Captain Lovell navigated and Major Anders was cameraman.

At a news conference here, flight controllers were asked what emotional reaction the astronauts experienced on being the first men to fly around the moon and see it from up close. They radioed statements revealed little emotion.

To this Major Carr replied.

"I would certainly say it was one of jubilation, exultation and any other word I can think of that would be synonymous. This is something they have worked many, many months and weeks on. The more craters they saw and recognized the more jubilant they became."

December 25, 1968

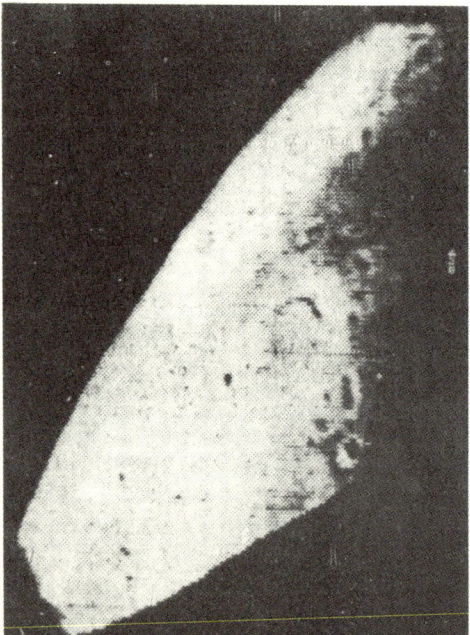 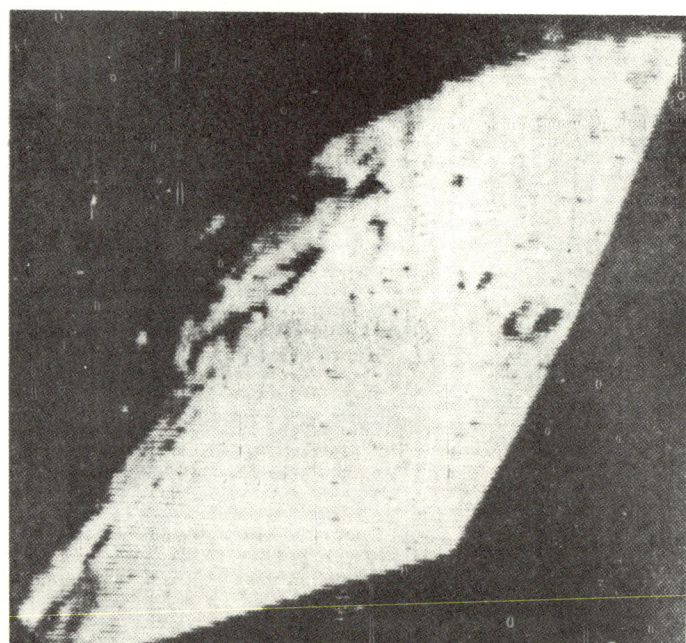

Associated Press

Moon pictures taken through the window of the Apollo 8 spacecraft that were telecast to earth last night. In picture at left of Sea of Crises area, the larger crater is 30 to 40 miles wide. The picture at right was last transmitted.

ASTRONOMY AND PLANETARY SCIENCE

'Proof' of Water on Mars Found, Supporting Idea Life Can Exist

By WALTER SULLIVAN

Astronomers at the McDonald Observatory in Texas said yesterday they had obtained the first "absolutely conclusive proof" that water exists in the atmosphere of Mars.

Although the water content appears to be very low, the observations seem to undermine the arguments of those who say that the air of Mars is so dry that life could not exist there.

Furthermore, the astronomers said, the observations are consistent with the long-postulated migration of water vapor from the martian pole emerging from winter to the pole about to enter winter. This movement was proposed to account for the "wave of darkening" that some have taken as evidence of vegetation.

This darkening of the Martian landscape, as seen through telescopes on earth, begins in spring around the fringes of the polar cap that is melting. The darkening moves toward the Martian equator at about 28 miles a day, as though spring began in the Arctic and moved south.

It has been proposed that melting of polar frost adds enough water vapor to the air to enable plants to flourish, although non-living processes could also account for this color change.

The observations were described yesterday in telephone interviews by Dr. Harlan J. Smith, head of the McDonald Observatory, operated by the University of Texas, and by Dr. Ronald Schorn of the Jet Propulsion Laboratory, operated by the California Institute of Technology in Pasadena. Another member of the observing team was Dr. Stephen Little of the University of Texas.

Since the turn of the century a number of scientists have suspected that the white polar caps of Mars were covered with dry ice (frozen carbon dioxide) rather than frozen water. Dr. Smith believes the new observations show that "at the least an appreciable amount" of water ice must be present in the caps.

Wave Lengths Give Clue

In the Texas observations the observatory's 82-inch reflecting telescope channeled light from Mars into a large spectrograph, which revealed the relative intensities of the light's constituent wave lengths. At several hundred points in the infrared portion of the spectrum there was absorption at wave lengths characteristic of water vapor in the atmosphere of Mars.

Although the observatory is 6,800 feet above sea level in desert country, and although the observing was done on two unusually dry nights this month and last, there was still 20 times more water vapor in the earth's atmosphere overhead than in the Martian atmosphere under observation.

However, the observations were made when Mars was rapidly closing its range from the earth. This relative motion shifted the wave lengths absorbed by Martian water so that they were no longer masked by the effects of water in earth's air.

It was found that air in the southern skies of Mars contained, on the average, enough water to cover the surface to a depth of two-thousandths of an inch. Air over the northern hemisphere was half as moist.

This, according to Dr. Schorn, is consistent with observations made in Texas in 1964-65, which suggested that the northern hemisphere in early spring was comparatively moist. By late spring, when the southern cap was growing rapidly, it was found that the moisture had shifted south.

It is now late spring in the northern hemisphere. The 1964-65 observations were not sufficiently clear-cut, according to Dr. Schorn, to be convincing.

The total humidity of the Martian atmosphere is now estimated, from the Texas observations, to be sufficient to produce one cubic mile of liquid water. The astronomers hope that, by further observations, they will be able to test the hypothesis that there are "oases" or particularly damp regions on Mars where life would find a comparatively hospitable environment.

They have begun obtaining spectra with the observatory's new 107-inch telescope and further studies with a French-designed spectrum analyzer, using so-called interferometry, are planned.

Furthermore, when two Mariner spacecraft fly past Mars on July 31 and Aug. 5 they may obtain "snapshot" observations indicating humidity in various regions. However, Dr. Schorn said, the Mariner instruments will be sensitive enough only to measure the low levels of humidity.

The two shots have been timed to take advantage of the current close approach of the earth and Mars. They will be closest—that is, "in opposition" —on May 31.

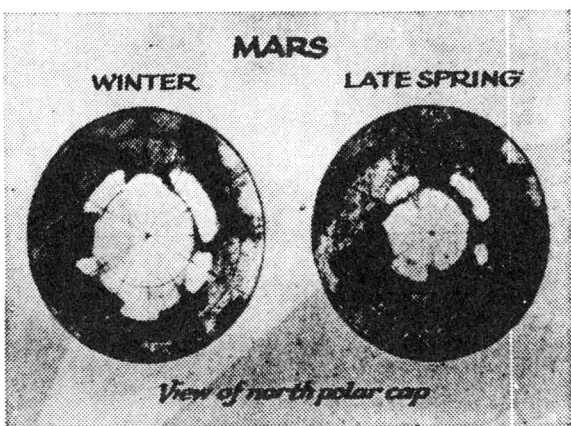

Adapted from Audouin Dollfus   March 25, 1969

As north polar cap of Mars melts in spring, added moisture is believed to cause nearby gray areas to darken. Drawings, based on photographs, appear in "Intelligent Life in the Universe," by I. S. Shklovskii and Carl Sagan.

March 25, 1969

MEN WALK ON MOON

Astronauts Land Module on Lunar Plain; Collect Rocks and Plant American Flag

Powdery Moon Surface Explored Around Craft

By JOHN NOBLE WILFORD
Special to The New York Times

HOUSTON, Monday, July 21—Men have landed and walked on the moon.

Two Americans, astronauts of Apollo 11, steered their fragile four-legged lunar module safely and smoothly to the historic landing yesterday at 4:17:40 P.M., Eastern daylight time.

Neil A. Armstrong, the 38-year-old civilian commander, radioed to earth and the mission control room here:

"Houston, Tranquility Base here. The Eagle has landed."

The first men to reach the moon—Mr. Armstrong and his co-pilot, Col. Edwin E. Aldrin Jr. of the Air Force—brought their ship to rest on a level, rock-strewn plain near the southwestern shore of the arid Sea of Tranquility.

About six and a half hours later, Mr. Armstrong opened the landing craft's hatch, stepped slowly down the ladder and declared as he planted the first human footprint on the lunar crust:

"That's one small step for man, one giant leap for mankind."

His first step on the moon came at 10:56:20 P.M., as a television camera outside the craft transmitted his every move to an awed and excited audience of hundreds of millions of people on earth.

Tentative Steps Test Soil

Mr. Armstrong's initial steps were tentative tests of the lunar soil's firmness and of his ability to move about easily in his bulky white spacesuit and backpacks and under the influence of lunar gravity, which is one-sixth that of the earth.

"The surface is fine and powdery," the astronaut reported. "I can pick it up loosely with my toe. It does adhere in fine layers like powdered charcoal to the sole and sides of my boots. I only go in a small fraction of an inch, maybe an eighth of an inch. But I can see the footprints of my boots in the treads in the fine sandy particles.

After 19 minutes of Mr. Armstrong's testing, Colonel Aldrin joined him outside the craft.

The two men got busy setting up another television camera out from the lunar module, planting an American flag into the ground, scooping up soil and rock samples, deploying scientific experiments and hopping and loping about in a demonstration of their lunar agility.

They found walking and working on the moon less taxing than had been forecast. Mr. Armstrong once reported he was "very comfortable."

And people back on earth found the black-and-white television pictures of the bug-shaped lunar module and the men tramping about it so sharp and clear as to seem unreal, more like a toy and toy-like figures than human beings on the most daring and far-reaching expedition thus far undertaken.

Nixon Telephones Congratulations

During one break in the astronauts' work, President Nixon congratulated them from the White House in what, he said, "certainly has to be the most historic telephone call ever made."

"Because of what you have done," the President told the astronauts, "the heavens have become a part of man's world. And as you talk to us from the Sea of Tranquility it requires us to redouble our efforts to bring peace and tranquility to earth.

"For one priceless moment in the whole history of man all the people on this earth are truly one—one in their pride in what you have done and one in our prayers that you will return safely to earth."

Mr. Armstrong replied:

"Thank you Mr. President. It's a great honor and privilege for us to be here representing not only the United States but men of peace of all nations, men with interests and a curiosity and men with a vision for the future."

Mr. Armstrong and Colonel Aldrin returned to their landing craft and closed the hatch at 1:12 A.M., 2 hours 21 minutes after opening the hatch on the moon. While the

ASTRONOMY AND PLANETARY SCIENCE

The New York Times

CONGRATULATIONS: President Nixon talking with Neil A. Armstrong and Colonel Edwin E. Aldrin Jr. on the moon

third member of the crew, Lieut. Col. Michael Collins of the Air Force, kept his orbital vigil overhead in the command ship, the two moon explorers settled down to sleep.

Outside their vehicle the astronauts had found a bleak world. It was just before dawn, with the sun low over the eastern horizon behind them and the chill of the long lunar nights still clinging to the boulders, small craters and hills before them.

Colonel Aldrin said that he could see "literally thousands of small craters" and a low hill out in the distance. But most of all he was impressed initially by the "variety of shapes, angularities, granularities" of the rocks and soil where the landing craft, code-named Eagle had set down.

The landing was made four miles west of the aiming point, but well within the designated area. An apparent error in some data fed into the craft's guidance computer from the earth was said to have accounted for the discrepancy.

Suddenly the astronauts were startled to see that the computer was guiding them toward a possibly disastrous touchdown in a boulder-filled crater about the size of a football field.

Mr. Armstrong grabbed manual control of the vehicle and guided it safely over the crater to a smoother spot, the rocket engine stirring a cloud of moon dust during the final seconds of descent.

Soon after the landing, upon checking and finding the spacecraft in good condition, Mr. Armstrong and Colonel Aldrin made their decision to open the hatch and get out earlier than originally scheduled. The flight plan had called for the moon walk to begin at 2:12 A.M.

Flight controllers here said that the early moon walk would not mean that the astronauts would also leave the moon earlier. The lift-off is scheduled to come at about 1:55 P.M. today.

Their departure from the landing craft out onto the surface was delayed for a time when they had trouble depressurizing the cabin so that they could open the hatch. All the oxygen in the cabin had to be vented.

Once the pressure gauge finally dropped to zero, they opened the hatch and Mr. Armstrong stepped out on the small porch at the top of the nine-step ladder.

"O.K., Houston, I'm on the porch," he reported, as he descended.

On the second step from the top, he pulled a lanyard that released a fold-down equipment compartment on the side of the lunar module. This deployed the television camera that transmitted the dramatic pictures of man's first steps on the moon.

Ancient Dream Fulfilled

It was man's first landing on another world, the realization of centuries of dreams, the fulfillment of a decade of striving, a triumph of modern technology and personal courage, the most dramatic demonstration of what man can do if he applies his mind and resources with single-minded determination.

The moon, long the symbol of the impossible and the inaccessible, was now within man's reach, the first port of call in this new age of spacefaring.

Immediately after the landing, Dr. Thomas O. Paine, administrator of the National Aeronautics and Space Administration, telephoned President Nixon in Washington to report:

"Mr. President, it is my honor on behalf of the entire NASA team to report to you that the Eagle has landed on the Sea of Tranquility and our astronauts are safe and looking forward to starting the exploration of the moon."

The landing craft from the Apollo 11 spaceship was scheduled to remain on the moon about 22 hours, while Colonel Collins of the Air Force, the third member of the Apollo 11 crew, piloted the command ship, Columbia, in orbit overhead.

"You're looking good in every respect," Mission Control told the two men of Eagle after examining data indicating that the module should be able to remain on the moon the full 22 hours.

Mr. Armstrong and Colonel Aldrin planned to sleep after the moon walk and then make their preparations for the lift-off for the return to a rendezvous with Colonel Collins in the command ship.

Apollo 11's journey into history began last Wednesday from launching pad 39-A at Cape Kennedy, Fla. After an almost flawless three-day flight, the joined command ship and lunar module swept into an orbit of the moon yesterday afternoon.

The three men were awake for their big day at 7 A.M. when their spacecraft emerged from behind the moon on its 10th revolution, moving from east to west across the face of the moon along its equator.

Their orbit was 73.6 miles by 64 miles in altitude, their speed 3,660 miles an hour. At that altitude and speed, it took about two hours to complete a full orbit of the moon.

The sun was rising over their landing site on the Sea of Tranquility.

"We can pick out almost all of the features we've identified previously," Mr. Armstrong reported.

After breakfast, on their 11th revolution, Colonel Aldrin and then Mr. Armstrong, both dressed in their white pressurized suits, crawled through the connecting tunnel into the lunar module.

They turned on the electrical power, checked all the switch settings on the cockpit panel and checked communications with the command ship and the ground controllers. Everything was "nominal," as the spacemen say.

LM Ready for Descent

The lunar module was ready. Its four legs with yard-wide footpads were extended so that the height of the 16½-ton vehicle now measured 22 feet and 11 inches and its width 31 feet.

Mr. Armstrong stood at the left side of the cockpit, and Colonel Aldrin at the right. Both were loosely restrained by harnesses. They had closed the hatch to the connecting tunnel.

The walls of their craft were finely milled aluminum foil. If anything happened so that it could not return to the command ship, the lunar module would be too delicate to withstand a plunge through earth's atmosphere, even if it had the rocket power.

Nearly three-fourths of the vehicle's weight was in propellants for the descent and ascent rockets—Aerozine 50 and nitrogen oxide, which substituted for the oxygen, making combustion possible.

It was an ungainly craft that creaked and groaned in flight. But years of development and testing had determined that it was the lightest and most practical way to get two men to the moon's surface.

Before Apollo 11 disappeared behind the moon near the end of its 12th orbit, mission control gave the astronauts their "go" for undocking—the separation of Eagle from Columbia.

Colonel Collins had already released 12 of the latches holding the two ships together at the connecting tunnel. He did this when he closed the hatch at the command ship's nose. While behind the moon, he was to flip a switch on the control panel to release the three remaining latches by a spring action.

At 1:50 P.M., when communications signals were reacquired, Mission Control asked: "How does it look?"

"Eagle has wings," Mr. Armstrong replied.

The two ships were then only a few feet apart. But at 2:12 P.M., Colonel Collins fired the command ship's maneuvering rockets to move about two miles away and in a slightly different orbit from the lunar module.

"It looks like you've got a fine-looking flying machine there, Eagle, despite the fact you're upside down," Colonel Collins commented, watching the spidery lunar module receding in the distance.

"Somebody's upside down," Mr. Armstrong replied.

What is "up" and what is "down" is never quite clear in the absence of landmarks and the sensation of gravity's pull.

As Mr. Armstrong and Colonel Aldrin rode the lunar module back around to the moon's far side, the rocket engine in the vehicle's lower stage was pointed toward the line of flight. The two pilots were leaning toward the cockpit controls, riding backwards and facing downward.

"Everything is 'go'," they were assured by Mission Control.

Their on-board guidance and navigation computer was instructed to trigger a 29.8-second firing of the descent rocket, the 9,870-pound-thrust throttlable engine that would slow down the lunar module and send it toward the moon on a long, curving trajectory.

The firing was set to take place at 3:08 P.M., when the craft would be behind the moon and once again out of touch with the ground.

Suspense built up in the control room here. Flight controllers stood silently at their consoles. Among those waiting for word of the rocket firing were Dr. Thomas O. Paine, the space agency's administrator, most of the Apollo project officials and several astronauts.

At 3:46 P.M., contact was established with the command ship.

Colonel Collins reported, "Listen, baby, things are going just swimmingly, just beautiful."

There was still no word from the lunar module for two minutes. Then came a weak signal, some static and whistling, and finally the calm voice of Mr. Armstrong.

"The burn was on time," the Apollo 11 commander declared.

When he read out data on the beginning of the descent, Mission Control concluded that it "look great." The lunar module had already descended from an altitude of 65.5 miles to 21 miles and was coasting steadily downward.

Eugene F. Kranz, the flight director, turned to his associates and said, "We're off to a good start. Play it cool."

Colonel Aldrin reported some oscillations in the vehicle's antenna, but nothing serious. Several times the astronauts were told to turn the vehicle slightly to move the antenna into a better position for communications over the 230,000 miles.

"You're 'go' for PDI," radioed Mission Control, referring to the powered descent initiation—the beginning of the nearly 13-minute final blast of the rocket to the soft touchdown.

When the two men reached an altitude of 50,000 feet, which was approximately the lowest point reached by Apollo 10 in May, green lights on the computer display keyboard in the cockpit blinked the number 99.

This signaled Mr. Armstrong that he had five seconds to decide whether to go ahead for the landing or continue on its orbital path back to the command ship. He pressed the "proceed" button.

The throttleable engine built up thrust gradually, firing continuously as the lunar module descended along the steadily steepening trajectory to the landing site about 250 miles away.

"Looking good," Mission Control radioed the men.

Four minutes after the firing the lunar module was down to 40,000 feet. After five and a half minutes, it was 33,500 feet. At six minutes, 27,000 feet.

"Better than the simulator," said Colonel Aldrin, referring to their practice landings at the spacecraft center.

Seven minutes after the firing, the men were 21,000 feet above the surface and still moving forward toward the landing site. The guidance computer was driving the rocket engine.

The lunar module was slowing down. At an altitude of about 7,200 feet, with the landing site still about five miles ahead, the computer commanded control jets to fire and tilt the bug-shaped craft almost upright so that its triangular windows pointed forward.

Mr. Armstrong and Colonel Aldrin then got their first close-up view of the plain they were aiming for. It was then about three and a half minutes to touchdown.

The brownish-gray panorama rushed below them—the myriad craters, hills and ridges, deep cracks and ancient rubble on the moon, which Dr. Robert Jastrow, the space agency scientist, called the "Rosetta Stone of life."

"You're 'go' for landing," Mission Control informed the two men.

The Eagle closed in, dropping about 20 feet a second, until it was hovering almost directly over the landing area at an altitude of 500 feet.

Its floor was littered with boulders.

It was when the craft reached an altitude of 300 feet that Mr. Armstrong took over semimanual control for the rest of the way. The computer continued to have control of the rocket firing, but the astronaut could adjust the craft's hovering position.

He was expected to take over such control anyway, but the sight of a crater looming ahead at the touchdown point made it imperative.

As Mr. Armstrong said later, "The auto-targeting was taking us right into a football field-sized crater, with a large number of big boulders and rocks."

For about 90 seconds, he peered through the window in search of a clear touchdown point. Using the lever at his right hand, he tilted the vehicle forward to redirect the firing of the maneuvering jets and thus shift its hovering position.

Finally, Mr. Armstrong found the spot he liked, and the blue light on the cockpit flashed to indicate that five-foot-long probes, like curb feelers, on three of the four legs had touched the surface.

"Contact light," Mr. Armstrong radioed.

He pressed a button marked "Stop" and reported, "okay, engine stop."

There were a few more cryptic messages of functions performed.

Then Maj. Charles M. Duke, the capsule communicator in the control room, radioed to the two astronauts:

"We copy you down, Eagle."

"Houston, Tranquility Base here. The Eagle has landed."

"Roger, Tranquility," Major Duke replied. "We copy you on the ground. You got a bunch of guys about to turn blue. We are breathing again. Thanks a lot."

Colonel Aldrin assured Mission Control it was a "very smooth touchdown."

The Eagle came to rest at an angle of only about four and a half degrees. The angle could have been more than 30 degrees without threatening to tip the vehicle over.

The landing site, about 120 miles southwest of the crater Maskelyne, is on the right side of the moon as seen from earth. The position: Lat. 0.799 degrees N., Long. 23.46 degrees E.

Although Mr. Armstrong is known as a man of few words, his heartbeats told of his excitement upon leading man's first landing on the moon.

At the time of the descent rocket ignition, his heartbeat rate registered 110 a minute—77 is normal for him—and it shot up to 156 at touchdown.

At the time of the landing, Colonel Collins was riding the command ship Columbia about 65 miles overhead.

Mission control informed the colonel, "Eagle is at Tranquility."

"Yea, I heard the whole thing," Colonel Collins, the man who went so far but not all the way, replied. "Fantastic."

When the Apollo astronauts landed on the Sea of Tranquility, the temperature at their touchdown site was about zero degrees Fahrenheit in the sunlight, even colder in the shade.

During a lunar night, which lasts 14 earth days, temperatures plunge as low as 280 degrees below zero. Unlike earth, the moon, having no atmosphere to act as a blanket, is unable to retain any of the day's warmth during the night.

During the equally long lunar day, temperatures rise as high as 280 degrees. By the time of Eagle's departure from the moon, with the sun higher in the sky, the temperatures there will have risen to about 90 degrees.

This particular landing site was one of five selected by Apollo project officials after analysis of pictures returned by the five Lunar Orbiter unmanned spacecraft.

All five sites are situated across the lunar equator on the side of the moon always facing earth. Being on the equator reduces the maneuvering for the astronauts to get there. Being on the near side of the moon, of course, makes it possible to communicate with the explorers.

July 21, 1969

PROBE FINDS MARS UNLIKE THE EARTH

Planet Is Less Hospitable to Life Than Expected

By WALTER SULLIVAN
Special to The New York Times

PASADENA, Calif., Aug. 2—Mariner 6's probing of the Martian atmosphere and surface has shown the planet to be utterly different from the earth and even less hospitable to life than previously believed.

However, infrared scanners of the surface have revealed areas that may be as warm in the Martian atmosphere as San Francisco. At night, however, they presumably slip into a deep freeze.

One of the most surprising discoveries was the absence of any evidence of nitrogen in the upper atmosphere. Nitrogen is the chief constituent of the air on earth. It comprises 5 per cent of the gas vented by volcanoes on this planet and appears in all life forms.

If the atmosphere of Mars had derived from volcanic activity, as is thought to have been the case on earth, it should show at least 5 per cent nitrogen. None was detected.

Nor was there any evidence of ammonia, which is a compound of nitrogen and hydrogen and is thought to have been one of the basic substances from which the chemistry of life evolved.

Preliminary findings of the flight past Mars by Mariner 6 early Thursday were made public today at the Jet Propulsion Laboratory, which is conducting the Mariner probes for the National Aeronautics and Space Administration. The Jet Propulsion Laboratory is operated by the California Institute of Technology.

Among other findings were the following:

¶Infrared spectra of light reflected from the surface indicated the presence of ice crystals, presumably from an ice fog. This confirms the existence of water on the planet.

¶The dark cap seen near the south pole in the approach photographs appears to be haze over that region, rather than an ice-free area.

¶Part of the infrared spectrum hints at the possible existence of organic material on or near the surface, including hydrocarbons. Such material has been found, for example, in meteorites. However, this is as yet far from being an indication of life.

¶The surface air pressure, as indicated by Mariner 4 in 1965, is comparable to that on earth at heights of 100,000 to 150,000 feet. This is the pressure that has been assumed in planning a system for landing equipment on Mars in 1973.

¶Ultraviolet light from the sun is being reflected from the surface of Mars. Such light would be fatal to most forms of life on earth.

Planet Resembles Moon

Mars, in the close-up Mariner 6 photographs, has even more points of resemblance to the moon than previously believed, according to scientists who have studied the pictures here. In addition to innumerable craters of all sizes and ages, there are long ridge-folds, and the larger craters look strikingly like some on the moon.

Their steep walls have clearly slumped, forming concentric terraces like those of Aristarchus and Copernicus on the moon.

Also, the walls are marked by avalanche chutes like fork marks around the edge of a pie.

However, Dr. Robert P. Sharp, a Cal Tech geologist, said that trying to interpret features on a planet 60 million miles away through remotely acquired photographs was "like a vet looking at an unknown species of elephant through binoculars, trying to tell how he feels by the wrinkles in his skin."

One spot on Mars whose brightness has been observed to vary over the years in earth-based photographs has now been identified in the Mariner photographs as a gigantic crater 300 miles wide. It is known as Nix Olympica.

There is a bright spot in its center—perhaps a central peak

like that in some larger moon craters. It is also surrounded by a whitish band like a gigantic lifesaver.

Mystery Remains

This suggests the possibility that it is a bullseye formation like Mare Orientale (the Eastern Sea) on one side of the moon, whose remarkable concentric ridges have been disclosed by photographs obtained in lunar orbit.

Why the huge Mars crater periodically becomes bright is a mystery that remains to be solved.

Infrared measurements of the surface temperature as the Mariner spacecraft swept across the afternoon side of Mars and across to the nightside showed wide variations. As expected, the areas that looked dark in long-range photographs were the warmest.

Much of the brighter regions were below freezing, though not below zero degrees Fahrenheit. The dark areas showed temperatures as high as 55 degrees in one infrared instrument and 75 degrees in the other.

It was the latter reading that led Dr. George C. Pimentel of the University of California in Berkeley to compare the daylight temperature to that of San Francisco.

The scanners were looking at very large areas and thus obtaining average temperatures. More localized areas could be considerably warmer or colder.

The nightside of Mars, as expected, was deeply frozen.

Atomic Oxygen Seen

Abundances of various elements in the upper atmosphere were determined by looking, at ultraviolet wavelengths, for the fluorescence induced in such elements by sunlight. A small amount of atomic oxygen was seen, and a great deal of carbon dioxide, as expected.

Dr. Charles A. Barth of the University of Colorado, who reported these findings, said that the planet's original quota of nitrogen might still be locked in its interior.

This would imply that Mars has had no history of volcanic activity, which would be difficult to explain in the light of evidence from the Apollo 11 moonquake detectors that implies the moon is active in this respect.

Dr. Pimentel said that his spectrometer had shown no clear evidence of methane, which with ammonia is regarded by many as one of the original substances essential for life's evolution.

Dr. Terry Neugebauer of Cal Tech, in reporting on the temperature measurements, said that those on the nightside of Mars were particularly significant, since that side is never visible from the earth. The earth's orbit lies inside that of Mars, so its sunlit side is always facing the earth.

At 9:15 P.M., Eastern daylight time, Mariner 7 began transmitting the first of 34 pictures it took of Mars today, and there were indications that its views of the red planet, will be considerably more detailed than those of Mariner 6. The first picture was obtained at a range of 1,054,000 miles.

The spacecraft completed sending the first batch of photographs later in the evening.

In tonight's pictures there still appears to be a haze over the south polar ice cap and adjoining the cap is a broad dark region. It is now spring in the southern hemisphere of Mars.

In the past it has been observed that some areas near the poles become darker as the ice cap melts in spring. It has been proposed that this is plant life or surface material responding to an increase in moisture associated with the melting of the ice cap.

Dr. Robert B. Leighton of Cal Tech, who is in charge of the Mariner television program, said today that he and others had suspected that this might be a cap of dry ice—frozen carbon dioxide—rather than water ice.

If there is an abundance of carbon dioxide in the Martian atmosphere, the ice cap could be far more substantial than the thin frost covering of water ice that has been postulated for Mars.

The Mariner 7 temperature readings of the polar cap should tell whether it is water ice, at 32 degrees Fahrenheit, or whether there may be an ice covering and glaciers formed of dry ice at far lower temperatures.

Dr. John E. Naugle, NASA associate administrator for space science and applications, termed the Mariner 6 flight "an unqualified success." He described it as part of a program for visiting all the planets of the solar system by the end of this century.

This would include the orbiting of two Mariner-type vehicles around Mars in 1971 and the orbiting of two larger vehicles in 1973 that could drop life-detection capsules on the planet. The latter project is known as Viking.

Dr. Naugle was asked if, in view of the reduced probability that life exists on Mars, it was worth spending billions to go there. He replied that, if it is in fact found that no native life exists on the planet, it might be possible to implant there certain forms of earth life.

The composition of the earth's atmosphere is known to have been controlled in part by the activity of living organisms, particularly plants.

Hence, some have proposed that Mars might be made habitable, after a prolonged period, by the transplantation there of life forms able to endure its severe environment.

August 3, 1969

Legacy of the Apollo 11 Flight: Verifications and New Discoveries

By WALTER SULLIVAN

One year ago today, with almost half of humanity straining to catch their businesslike, helmet-muffled exchanges, Neil A. Armstrong and Col. Edwin E. Aldrin Jr. lifted off the moon and headed for a rendezvous with their mother ship and the journey back to earth.

Hundreds of millions of persons had watched and listened with high emotion as the bug-shaped module landed on the moon and as the two astronauts excitedly walked like toys on the lunar landscape. Their feat left the United States and the world elated.

With the astronauts of Apollo 11 on their return were 48 pounds of lunar rock and soil. It was the first time that man had reached across the black void of space and plucked specimens from another celestial body.

Even taking into account the political motivations behind Project Apollo—reaching the moon before the Russians—the flight of Apollo 11, during which man walked on the moon, was the most dramatic demonstration yet of his insatiable curiosity.

But what is known today that was unknown before that memorable step on July 20, 1969? And what more can be learned from future study of the moon.

For one thing, the collection and laboratory analysis of carefully collected lunar samples not only reinforced the earlier deductions — such as the extraordinary abundance of titanium in lunar surface material — but also made possible basic new discoveries.

Outstanding among the discoveries has been the extreme age of the moon, its older rocks a billion years more ancient than any found on earth. And the hope that the moon, with its giant craters and other scars, might fill in the earliest chapter in the history of the solar system can now be realized.

The moon is an archive of greater antiquity than any available on earth; few, if any, of the rocks found there seem to have been formed fewer than three billion years ago. Thus, whereas volcanic eruptions and other mountain-building processes have continuously reshaped the earth, the moon has apparently lain dormant over a long period of time.

Report Next January

The specimens brought back were all collected in lunar seas. Apollo 11 landed in the Sea of Tranquility and Apollo 12 set down in the Ocean of Storms.

The Apollo 12 astronauts, who landed on the moon last November, brought back 75 pounds of lunar material; the Apollo 11 crew had returned 48 pounds to the earth.

From Jan. 5 to Jan. 8 of this year, results obtained from examinations of the Apollo 11 samples by more than 500 scientists in nine countries were reported at a conference in Houston. To avoid a scramble for priority in publication, the scientists had been required, as a strict condition for access to the material, to pledge no early publicity.

The same constraint now applies to the Apollo 12 material. Virtually the only information available is that made public by the Lunar Sample Preliminary Examination Team, which took the first look at the specimens. This was mainly done in the quarantined environment of the Lunar Receiving Laboratory at the Manned Spacecraft Center in Houston.

Twenty-eight pounds of Apollo 12 rocks and fine-grained material have been allocated to 139 American scientists and 54 foreign scientists who are to report their conclusions at Houston next January.

So far, the most dramatic of the Apollo 12 findings has been the age and composition of a lemon-sized rock, apparently thrown to the Ocean of Storms by the impact of a meteorite on a distant highland. Its age, from measurements of the extent to which radioactive materials within it have decayed, seems to be 4.6 billion years.

This coincides strikingly with the age of a great many meteorites and is thought to represent the time when solid objects condensed out of the cloud of hot gas and dust around the newly forming sun. This was presumably when the earth and the moon formed, too, but the earth was then subjected to internal cooking.

In this way the earth became "differentiated." That is, heavy material, primarily iron, sank to form the central core and light material rose to form rocks such as the granite of which continents are largely made. Thus, apparently, none of the starting material from which the earth was formed remains.

The lunar seas, resembling great, solidified lakes of lava, suggest that melting occurred on the moon as well, even though the moon apparently does not have a substantial iron core. Moon rocks brought back from both Apollo sites were igneous, or formed by melting.

Furthermore, the oldest, lemon-sized specimen, like the cooked-out crystal rocks on the earth, is relatively rich in potassium, thorium and uranium. But is is evident that the history of the moon has been very different from that of the earth.

No Water or Oxygen

The moon as a whole is not only less dense than the earth but the rocks examined so far have also shown no evidence of contact with substantial amounts of water or oxygen.

Dr. Robert Jastrow, head of the Institute for Space Studies of the National Aeronautics and Space Administration, believes this does not mean water never existed on the moon. Snaking canyons, or "rills," in some areas of the moon seem to indicate some sort of liquid erosion in the past. Perhaps, he said yesterday, the water was "baked out" of the material in lunar seas.

The Apollo samples have shown that the composition of the moon varies greatly from place to place and that landings at a number of sites will be needed to gain a full understanding of this diversity. The two landings have left unresolved such basic problems as the origin of the moon and the role of volcanic activity in shaping its surface.

The great age of some lunar samples seems to rule out the idea that the moon was torn from the earth at an intermediate stage in the earth's history. If the moon was born of the earth, it must have been at the very beginning.

Many of the fruits of Project Apollo are yet to be harvested. When several seismometers are operating at separate locations, it will be possible to pinpoint each event that they record. A laser reflector, left at the Apollo 11 site, is being used by observatories on the earth to keep extremely accurate records of earth-moon distances. These, after a decade or so, may reveal whether or not gravity is weakening as some believe. If so, the moon is very slowly increasing its distance from the earth.

Examination of lunar material that was subjected to bombardment by gas from the sun during various periods of the past should also help outline the history of the earth's parent star.

The Apollo findings, obtained at high cost and high risk, have not so far led to revolutionary discoveries. They have produced one suggestion, concerning the sun, that would be revolutionary if confirmed.

That suggestion is the controversial proposal of Dr. Thomas Gold of Cornell University that glassy splatters atop some lunar rocks were formed on rare occasions when the sun flared up to 100 times its normal brilliance. Such flares, occurring perhaps every few thousand years, would last only a few seconds, long enough to melt the tops of suitably placed lunar rocks and, perhaps, blow off the outer envelop of the earth's atmosphere.

Conventional Explanation

The more conventional explanation for these glassy coatings seen on both Apollo landing missions is that they are splashes of molten rock thrown up by the high velocity impacts of tiny meteorites. Such impacts do not occur on earth because little meteorites are slowed by the atmosphere.

While most of the Apollo results have been reported with scientific seriousness, one in the June 26 issue of Science Magazine is in a lighter vein. It concerns the perplexingly low velocity of sound waves through lunar rock—roughly one-third that in the earth's rocks.

Edward Schreiber of Queens College and Orson L. Anderson of the Lamont-Doherty Geological Observatory sought out various earth materials with a sound velocity comparable to that of the lunar rocks and found that a variety of Swiss, Norwegian, Italian and Wisconsin cheeses fit the bill.

While the lunar rocks are more dense than cheese, they said, this "may readily be accounted for when one considers how much better aged the lunar materials are." Their findings, they added, "lead us to suspect that perhaps old hypotheses are best, after all, and should not be lightly discarded."

July 21, 1970

SOVIET SAYS CRAFT LANDED ON VENUS AND RADIOED DATA

Unmanned Ship Reportedly Touched Down on Dec. 15 After 120-Day Journey

NO SURPRISES REVEALED

Temperature Is 847 to 923F.—Atmospheric Pressure Is 90 Times That of Earth

By BERNARD GWERTZMAN
Special to The New York Times

MOSCOW, Jan. 26—The Soviet Union announced today that its unmanned Venera 7 last month made the first soft landing on Venus's torrid surface and sent back data for more than 20 minutes.

A report published in Izvestia, the Government newspaper, and a summary issued by Tass, the official Soviet press agency, broke a six-week silence on the results of the Soviet Union's latest and most successful Venus probe.

Preliminary data released by Tass on Dec. 15 said nothing about a soft landing and left the impression that Venera 7 had been no more successful than the previous Venera craft that had failed to endure the extreme heat and high pressure of Venus's cloud-shrouded atmosphere, which has a great deal of carbon dioxide.

But today's official report said that "it was the first time that scientific information was relayed directly from the surface of another planet in the solar system."

Another Achievement

Venera 7 clearly was another technological achievement for the Russians, who have recently put the stress in their space program on unmanned projects. It contrasts with the emphasis placed by the United States on its Apollo program and putting men on the moon. Apollo 14 is scheduled to be launched on Sunday.

The scientific information sent back to earth by Venera 7's instrument capsule caused no surprises and tended to fortify the general consensus held by scientists since the Soviet Venera craft and the American Mariner 5 sent back data in recent years on missions that fell short of soft landings.

According to the Soviet report, Venera 7's instruments reported that the surface temperature on Venus was as hot as scientists had been predicting. It was 475 degrees Centigrade, with a margin of error of 20 degrees in either direction. In Fahrenheit, this means that the surface of Venus at the point where Venera 7's instruments lay ranged from 847 to 923 degrees above zero.

Oxygen Impossible

This is more than four times the temperature at which water boils and means that it is impossible for oxygen to exist in any quantity on the planet's surface.

Venera 7's instruments also reported that the atmospheric pressure on the surface was equivalent to 90 atmospheres—or 90 times that of earth at sea level—with a margin of error of 15 atmospheres in either direction.

Because more than 90 per cent of the atmosphere is carbon dioxide, the density of the atmosphere at the surface was about 60 times greater than that of earth, the report said.

The combination of scorching heat, crushing pressure and dense air makes it highly unlikely that any human could long endure on the surface, even in the most technologically advanced life-support systems, experts here said.

Because of the thick, dense clouds that surround the planet, it is impossible to see the surface from earth. And because the space capsule must land on the "night" side—facing earth—to facilitate radio communication, no television pictures were possible.

Comment by Designer

The chief designer of the Venera series said in May, 1969, after the descent of Venera 5 and 6 into the atmosphere, "I am not confident that in the near future it will be possible to receive television pictures of the planet's surface."

Venera 7 was launched on Aug. 17 on its 120-day journey to Venus, the closest planet to earth, and the second closest planet to the sun. On Dec. 15, Tass announced that the 2,590-pound craft had completed its mission and that it transmitted information for 35 minutes as its instrument capsule floated by parachute through the Venusian atmosphere.

Nothing was said, even in subsequent days, about a soft landing, and it was assumed here that Venera 7 had followed the pattern of Venera 4, 5 and 6 — all of which sent data while descending but apparently either burned up or failed to communicate from the surface.

There was no explanation given today as to why the Soviet Union had not announced its latest technological achievement earlier.

The report said it had been ascertained that the craft's capsule had landed on the surface when the radio signal frequency stopped changing.

Signal Volume Reduced

"The signals from the descending probe were received for another 23 minutes after landing," Tass said. "Here the voume of the signal was about a hundred times less than during the descent."

The report said that "a special method" using computers had detected "this weak signal" and allowed it to be deciphered.

The temperature outside the capsule increased as it penetrated the atmosphere, starting about 40 miles from the surface, the report said. At the time of landing, the distance between Venus and the earth was about 36.3 million miles, while the radio signals took 3 minutes 22 seconds to reach the earth.

According to the report, Venera 7's design was similar to that of the last three Venera craft. But it was heavier, to withstand pressures of up to 180 atmospheres and temperatures of up to 530 degrees Centigrade.

Tass called the latest achievement "a successful implementation of the Soviet program" and said it "has pushed back the frontiers of man's knowledge of the nature of this planet which lies closest to earth."

The Venera 7 was on the surface of the planet at the same time as the Soviet moon rover, Lunokhod 1, was sending back information from the lunar surface. This was the first time, the report said, that "earth was receiving information simultaneously from two celestial bodies."

The recent successes of the Soviet Union in unmanned craft — the Luna 16, which brought back lunar samples to earth in September, Luna 17, which deposited the still active Lunokhod on the moon in November, and Venera 7 — underscore the determined thrust of Soviet space research toward unmanned vehicles with manned flight only being used around the earth in preparation for an orbiting space station.

Venera 7 did not receive unusual publicity in the Soviet media tonight. Television did not do more than report the Tass summary. For most Russians, the latest achievement does not arouse any particular interest. Russians by and large are much more interested in the drama of manned flight.

Report Accepted Here
By WALTER SULLIVAN

The Soviet report that Venera 7 landed intact on the planet Venus and transmitted data to earth was viewed here last evening as probably correct.

Several factors supported this view. Moscow reports said data sent from the vehicle during its descent through the atmosphere of Venus showed a steady rise in temperature and atmospheric pressure. Then, suddenly, both these measurements ceased to change.

They remained constant until the signals ceased, some 23 minutes later, presumably because the extreme heat had penetrated the insulated capsule and destroyed its electronics.

Furthermore, the Soviet account noted that the gradual change in frequency of radio signals from the craft, as it slowed in its descent, finally came to an end, indicating that it was no longer falling.

Reason for the Charge

This change in frequency resulted from the reduction in the capsule's speed, relative to the earth. It was analogous to the change in pitch of a horn as a vehicle's speed changes, relative to an observer.

Another explanation of the Soviet account would be a termination of descent because of a strong updraft in the dense Venusian atmosphere. Because the night side of Venus is almost as hot as the day side, in spite of the extremely slow spin rate of the planet, it is assumed that winds in its atmosphere are very strong.

However, such a halting of the parachute-suspended capsule seems unlikely.

Finally, the Russians reported in 1967 that Venera 4, which plunged into the planet's atmosphere on Oct. 18 of that year, had sent data from the surface. This was later shown, by a combination of American measurements, to be incorrect.

The Russians tacitly acknowledged that their craft had gone off the air well above the surface. They are unlikely to risk a repetition of this episode.

January 27, 1971

Heat in Moon Found Surprisingly High

By WALTER SULLIVAN

Data relayed to earth by radio from a hole drilled into the lunar surface last summer by the Apollo 15 astronauts indicate that the moon is a far richer source of heat than had generally been supposed.

The finding of unexpectedly high heat flow from beneath the lunar surface, one of the most surprising of the Apollo program, has persuaded a number of scientists that current theories on how the moon and planets formed have to be revised.

Dr. Marcus E. Langseth of Columbia University's Lamont-Doherty Geological Observatory said that the heatflow determination, while a preliminary result obtained at a single site on the moon, was "one of the more profound and surprising results" of the Apollo missions. Dr. Langseth is responsible for analysis of the lunar heatflow measurements.

Dr. Gerald J. Wasserburg, professor of geology at the California Institute of Technology, who analyzed the returned lunar soil samples, said that henceforth, because of the heatflow finding, "it will be a different ballgame" when one considers how planets were formed.

It is now proposed that during the moon's birth a series of layers were laid down, producing a surface region rich in radioactive elements.

Heat generated by this radioactivity, it is thought, melted much of the top layer when the moon was young and accounts for the newly observed heat flow.

The equipment for measuring the flow of heat from within the moon was installed four months ago in two holes drilled at the Apollo 15 landing site, near Hadley Rille.

The drilling proved frustratingly difficult because the lunar surface material was very dense. The astronauts, Col. David R. Scott and Lieut. Col. James B. Irwin, were impatient to go rock hunting but were persuaded by ground control in Houston, Tex., to persist in drilling until one hole was five feet deep.

Data from sensors in that hole have indicated a flow of heat from the interior measuring 0.8 millionths of a calorie per square centimeter per second. According to Dr. Langseth, the average figure for the earth is roughly twice that, but it varies from region to region. Dr. Langseth is responsible for analysis of the lunar heat-flow measurements.

He said yesterday that based on existing concepts as to the moon's nature he had expected the flow of heat from the lunar interior to be about one-sixth that on earth. However, Dr. Langseth noted that the abundance of radioactive elements found in the returned lunar soil

ASTRONOMY AND PLANETARY SCIENCE

Dr. Marcus E. Langseth

Dr. Gerald J. Wasserburg

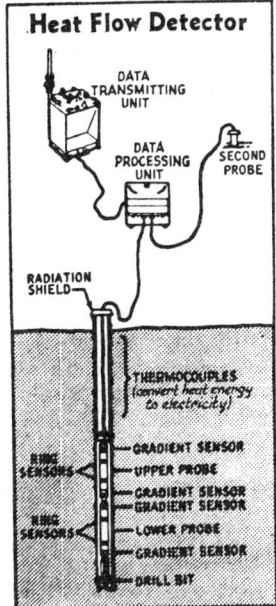

Heat flow data are derived by measuring thermal conductivity of lunar material and temperature differences on a drill rod.

samples by Dr. Wasserburg has made the observed heat flow seem plausible.

"If one asks: 'What's crazy about the moon,'" Dr. Wasserburg said yesterday, "the answer is the very high level of radioactivity in soil brought back from all four Apollo landing sites. If the moon were made throughout of such material," he added, "it would be just a molten puddle."

This is because the radioactivity of such elements—chiefly uranium, thorium and a radioactive form of potassium—generate heat. It is this process on earth that produces the heat released in volcanoes and earthquakes. But it is believed that the earth's history has been very different from that of the moon.

Early in its history the earth became hot enough internally for heavy material (chiefly iron) to sink, forming its core. The lighter material rose to the surface, forming a crust relatively rich in radioactive elements. For a variety of reasons it is believed that the moon never went through such a process.

How then could its original surface have become so rich in radioactive elements? Dr. Wasserburg believes this came about as gravity slowly pulled together into a spherical solid object the dust, gas and other materials that formed the moon. The heat-flow measurement, he says, has now indicated the amount of radioactive material in the resulting surface layer. abundant radioactive elements have been found.

The implication is that layering as a by-product of accretion may have occurred in the formation of other bodies, including smaller planets such as Mars.

The lava-like rock forming lunar seas is not very radioactive and presumably erupted from well below the original surface layer. It is in rocks and dust presumably derived from the original crust that

While an element like uranium is heavy in pure form, it becomes chemically bound into light-weight rocks, such as granite.

The heat flow is determined by gradiant sensors and thermocouples set at intervals along the drill rod and probe. These convert temperature into electrical signals that can be radioed to earth.

November 30, 1971

Data on Mars Indicate It's a Dynamic Planet

By JOHN NOBLE WILFORD
Special to The New York Times

PASADENA, Calif., June 14—A clear and often surprising portrait of Mars as a distinctive, dynamic planet is emerging from the pictures and data transmitted back to earth by Mariner 9, the spacecraft that has been circling Mars twice a day for the last seven months.

Scientists are discovering Mars to be a varied world of sharp relief and many contrasts, a world that has been shaped in part by water and therefore may be or may have been suitable for some forms of life.

They are seeing, as never before, a world of high winds, great temperature extremes, clouds of water vapor as well as of dust, towering volcanos, chasms larger and deeper than the Grand Canyon, glacial terraces, moon-like craters and meandering channels that were probably cut by the flow of water.

And the scientists are concluding that they were wrong when, after Mariners 4, 6 and 7 discovered such bleak vistas, they figured Mars must be a dead world, more like the moon than anything else.

They were misled, they said, because they saw only 10 per cent of the surface, the part that happened to be the older cratered regions, and Mariner 9 is now surveying about 85 per cent in remarkable detail.

At a news conference here, Robert H. Steinbacher, the project scientist, said:

"Mars used to be likened to earth in science fiction, and was compared to the moon after the first pictures from Mariner 4 in 1965. Now we are seeing that Mars has a character all its own. It is not earth-like or moon-like, it is Mars-like."

At the news conference and in a series of interviews, Mariner scientists agreed that the most important discovery thus far involved the growing evidence that water has played an active role in the planet's evolution.

The scientists believe that the south polar ice cap is partly frozen water, not all frozen carbon dioxide (dry ice). They have also detected water vapor in some of the fluffy white clouds capping some of the volcanic mountain peaks.

But Harold Masursky of the United States Geological Survey, leader of the television examination team, said that the most convincing evidence was found in the many photographs showing deep, winding channels that may have once been fast-flowing streams.

"We are forced to no other conclusion," Mr. Masursky said in an interview, "but that we are seeing the effects of water on Mars."

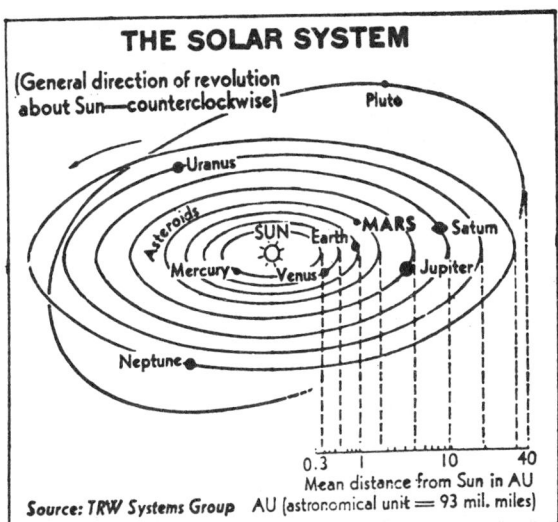

Mars is a comparatively close planetary neighbor. Its average distance from the earth is 48.5 million miles.

Moreover, Mr. Masursky said, some of the water erosion may be fairly recent, geologically speaking — perhaps within the last 50,000 years.

Theories on Water Release

Mr. Masursky and other scientists suggest two explanations for how enough water could be released on the planet to cause the apparent rains and floods.

Since the axis of the planet slowly wobbles, or precesses, as Mars travels around the sun, the polar regions are sometimes exposed to increased doses of sunlight.

This occurs in cycles of 50,000 years. As a result, the polar caps may melt completely, releasing enough moisture into the thin Martian atmosphere to cause heavy rainfalls all over the planet.

Another theory is based on volcanic heat melting water ice stored beneath the Martian surface as permafrost.

The earth's water sprang from volcanic vapors, and the new evidence of past volcanic activity on Mars suggests that the same processes may have been at work there — except that most of the water froze and was trapped below the surface.

Since volcanic activity seems to have extended over a long period of time on Mars, the scientists said that there might have been occasions when the heat thawed the permafrost, producing sufficient water to etch out many of the canyons detected no "cleancut indications of heat sources such as from active volcanoes."

The second province is an equatorial plateau region marked by deep canyons and great cracks, or faults, in the Martian crust. This is evidence of significant tectonic and sinuous "riverbeds" seen in so many of the Mariner pictures.

Questions About Life

"The presence or absence of water is obviously central to our studies," Mr. Masursky said. "Has there been in the past water on Mars, and is it there now? The answer is central to the question of whether Mars has been a suitable place for the development of life."

Without answering directly the question of life on Mars, Mr. Masursky said that Mariner 9's results increased the likelihood that future missions to land on the planet might find some signs of life, or at least fossils of past life.

The United States plans to land two Viking life-detection robots on Mars in 1976. The Soviet Union has indicated that it may attempt such landings as early as 1973.

Mariner 9 has transmitted nearly 7,000 television pictures of Mars since it swung into an orbit of the planet last November, the first spacecraft ever to circle the red planet. It has now completed 428 revolutions, following a course ranging from 1,000 to 1,200 miles from the surface.

The project is directed by the Jet Propulsion Laboratory, which is operated by the California Institute of Technology for the National Aeronautics and Space Administration.

At the news conference, the project scientists displayed a detailed map of the complete Martian equatorial region, encompassing surface features over territory ranging from 30 degrees north to 30 degrees south. In addition, Mariner 9 has photographed much of the south polar region and is now aiming its cameras on the North Pole.

From the photographs, which have been enhanced by a computer removing blemishes and sharpening contrasts, the scientists have identified four "major geological provinces" on Mars.

The first is the volcanic province. The most prominent of the volcanic peaks is Nix Olympica, whose summit Caldera is twice as wide as the volcanoes that formed the Hawaiian Islands.

It is apparently a fairly young feature, since there is little evidence of meteorite impacts in the region. But scientists said that Mariner 9's remote-sensing instruments had detected no "cleancut indications of heat sources such as from active volcanoes."

The second province is an equatorial plateau region marked by deep canyons and great cracks, or faults, in the Martian crust. This is evidence of significant tectonic activity — the cracking and slippage of the crust as in earthquakes — during the recent history of Mars.

The greatest of the canyons in this region is reported to be 10 times the length of the Grand Canyon, and three times deeper. It runs from west to east along the Equator, covering nearly 2,500 miles.

Scientists point to a network of tributaries to the canyon and a delta-like region extending from its eastern end as evidence of possible water erosion.

The third geological province, which seems to extend over half of the planet, is a heavily cratered region that looks much like the broad plains of the moon. One of the impact basins, Hellas, is reported to be larger than any similar basin on the moon. Great sand dunes are also seen in this area.

The cratered terrain is thought to be the most ancient feature on Mars, since many of the craters appear to have been eroded by wind, subsequent impacts and other forces. The three earlier Mariners, which flew by Mars but did not go into orbit, photographed this region.

The fourth province is a spectacular expanse of stair-step terraces and deep grooves radiating from the south polar region. Scientists suspect that glaciers moving out from the polar ice cap gouged out the grooves and piled up rocky debris to form the terraces.

It is expected that the north polar region, when it is completely photographed, will consist of similar terrain — indicating that Mars, like Earth, has had its ice ages and is now in a warmer period.

Other Findings

Other Mariner 9 results include the following:

¶ The North Pole temperature is about 200 degrees below zero Fahrenheit, much colder and dryer than the coldest place on earth, which is Antarctica with 125 degrees below zero. In the more temperate equatorial zone, the temperatures reach 80 degrees above zero in the early afternoon.

¶ Atmospheric winds reach speeds of up to 115 miles an hour even in relatively calm periods. During the dust storms the winds reach 300 miles an hour.

¶ Ozone, probably produced in a reaction between water vapor and solar radiation, has been detected over both polar regions. On earth, ozone in the upper atmosphere screens out the lethal ultra-violet radiation from the sun, making life possible. On Mars, however, the amount of ozone is less than one-hundredth that present in the earth's atmosphere.

¶ Mars bulges at the Equator, is flattened at the poles and is more smoothly circular in between. There are wide variations in the planet's gravitational pull along the Equator, indicating "large stresses at work on the planet's crust."

¶ The south polar cap recedes from about 2,000 miles in diameter in the winter to about 200 miles in the spring. For some reason, the residual cap is not centered at the poles — but is some 100 miles to the side.

Mariner 9 is expected to continue collecting data and taking pictures until it runs out of gas to control its flight position. Flight controllers here estimate that it could remain active until the end of the year, if fuel is carefully husband.

Although Mariner 9 has provided many surprises, Dr. Rudolph A. Hanel of the Goddard Space Flight Center in Greenbelt, Md., one of the project scientists, said:

"We have not seen any sign of life on Mars. We didn't expect to. But we have not seen anything that would exclude life."

But the pendulum of hope, the scientist said, has now swung back toward the prospect that Mars may yet yield evidence that some life exists elsewhere in the solar system.

June 15, 1972

Apollo Rocks Spur 2 Lunar Theories

By WALTER SULLIVAN

As a result of the five manned landings on the lunar surface since Neil A. Armstrong took his "one small step" three years ago, more has been learned of the early history of the moon than the total sum of knowledge accumulated thus far on the comparable early history of the earth.

In the current issue of the British journal Nature it is suggested that some global event on earth about three billion years ago wiped out most of the earlier geologic record.

The knowledge about the moon, still meager and fragmentary, is being extracted from a collection of rocks — chiefly those collected by the Apollo 16 astronauts on the last mission. Some of these specimens are at least a half-billion years older than any ever found on earth.

The lunar samples, and the other findings of the Apollo missions, have not yet answered the age-old question of the moon's origin, but they have narrowed the options and generated two new theories.

ASTRONOMY AND PLANETARY SCIENCE

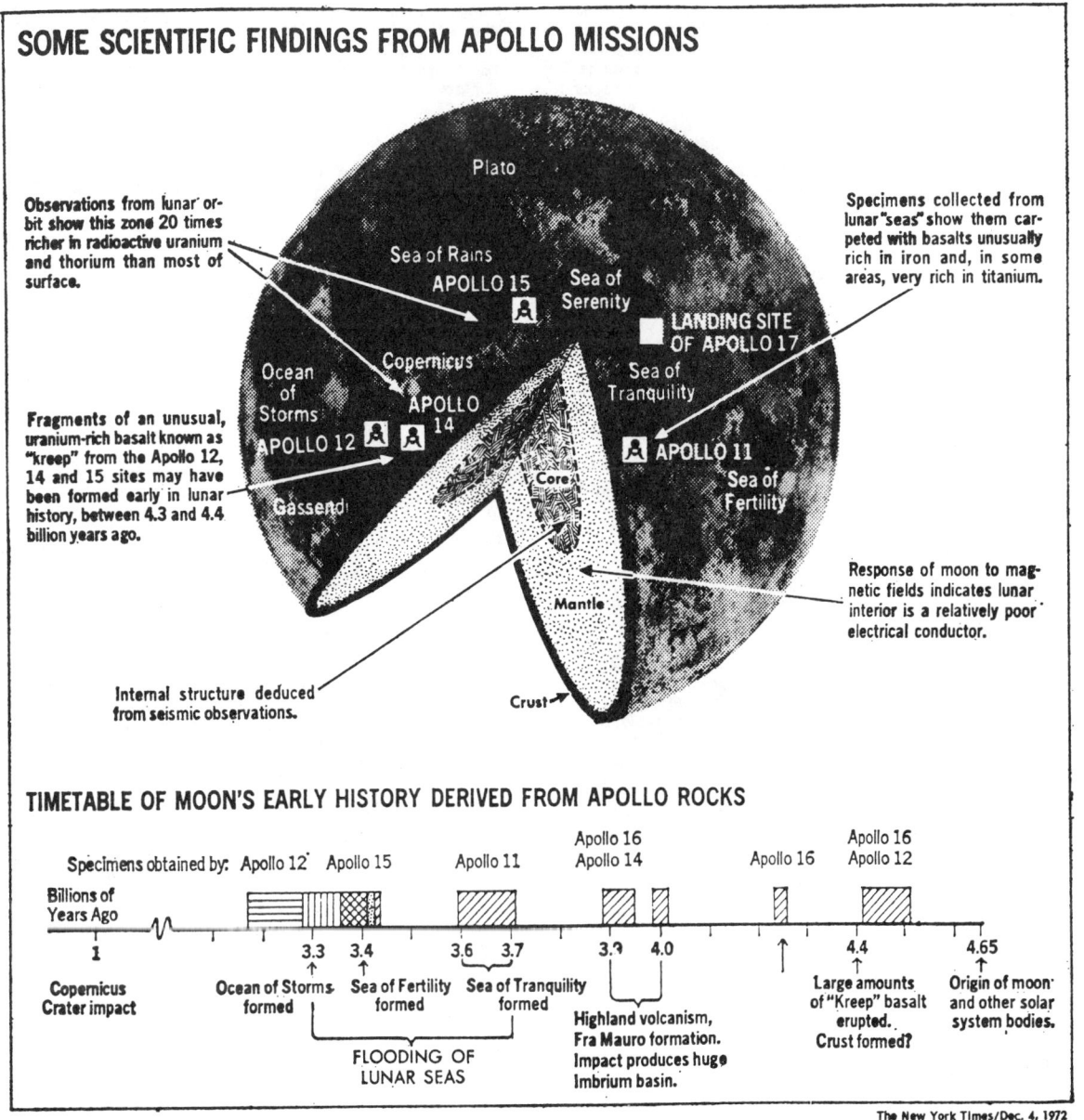

The New York Times/Dec. 4, 1972

'Refractory' Minerals

Both are derived from evidence that the moon is largely made of "refractory" minerals—those that melt only at very high temperatures. These substances would be the first to condense out of a mixture of gaseous compounds such as that thought to have circled the sun as the planets began forming.

One of the new theories holds that the moon formed in an orbit as close to the sun as that of Mercury, the innermost planet. It is proposed that Mercury itself, being so close to the sun, is formed of such materials.

The other hypothesis is that the moon formed at roughly the same distance from the sun as the earth, but in an orbit tilted sharply to the plane of the ecliptic. This plane defines a disk-shaped region within which all the planets except Pluto—the outermost—circle the sun.

Most of the hot gas from which the planets formed is assumed to have lain within this region. Above and below it the low gas pressure would have enabled refractory minerals to condense and, it is proposed, form a moon in an orbit steeply inclined to the ecliptic.

Seeming Contradictions

The moon sampling and data collection have produced seeming contradictions, some of which, it is hoped, can be resolved by further study and particularly by the heat-flow measurements planned for the Apollo 17 mission to be launched Wednesday night.

These measurements of the flow of heat from the lunar interior, to be carried out in two holes drilled eight feet into the surface, should help resolve the dispute as to whether the interior of the moon is hot, but remarkably deficient in iron oxide, or cold and more like the composition of the earth's interior.

None of the Apollo samples brought back so far have shown any evidence of recent volcanic activity, even though photographs from lunar orbit show many features that resemble fresh lava flows. On the airless moon these flows would erode very slowly and may be much older than they look.

However, twice in recent months the impact of a heavy meteorite has jostled the moon enough to produce evidence for a core that is at least partially molten. The first meteorite hit 90 miles north of the Apollo 14 site on May 13, but the biggest, on July 17, plunged into the Moscow Crater, or hit near it, on the far side of the moon.

Impact Recorded

The shock waves of this impact passed through the lunar core en route to the four Apollo landing sites, on the earth-facing side of the moon, where seismometers are in operation.

All these instruments radio their readings to earth, powered by atomic batteries.

The shock waves from these impacts and from "moonquakes" that occur deep inside the moon all suggest partial melting of the lunar core, according to Dr. Gary Latham, who is responsible for the seismic studies.

The so-called shear waves generated by such events do not pass readily through a liquid — a phenomenon that has been used to identify the liquid portion of the earth's core. A shear wave causes the medium through which it passes to oscillate at right angles to the direction in which the wave is traveling. This effect can be simulated by shaking a rope whose far end is secured.

Dr. Latham suspects that the spontaneously occurring moonquakes may be related to invasions of the otherwise solid part, or "mantle," of the lunar interior by hot fluid from the core. The quakes repeatedly originate at precisely defined sites in a zone between 400 and 600 miles below the lunar surface.

So far 33 such sites have been recognized and the locations of six have been pinpointed within the lunar interior.

The quakes tend to occur on that day each month—depending on the lunar orbit—when the moon is closest to the earth and hence is subjected to maximum strain from the earth's gravity. Early in the Apollo-program observations a single "hot spot" was identified that seemed to produce the great majority of quakes, but it has now become less active in relation to other quake centers.

While the seismic observations have pointed to a partially molten core, data obtained with magnetometers on the lunar surface and in lunar orbit have led other experimenters to deduce a relatively cool interior.

Electrical Conductivity

These studies focused on the electrical response of the moon to magnetic fields being carried past by clouds of gas ejected from the sun. The readiness with which these magnetic fields penetrate the moon, as recorded by magnetometers left at the Apollo sites, is an index of the electrical conductivity of the lunar interior.

In a recent summary of the findings, Dr. David Strangway of the Manned Spacecraft Center in Houston said the internal temperatures deduced from the observed conductivity lay between 1600 and 2700 degrees Fahrenheit. Such heat, despite its intensity, would hardly be enough to produce melting.

However, this reasoning assumes a certain iron content of the lunar interior that some believe may be too high. Thus Dr. Don L. Anderson, head of the Seismological Laboratory of the California Institute of Technology, argues that the interior of the moon may be formed of refractory materials, such as the silicates of calcium, aluminum and titanium.

The low electrical conductivity of these substances, even at high temperature, would mean that a core formed from them could be hot and still compatible with the magnetic recordings. Dr. Anderson is author of the idea that the moon formed in an orbit tilted sharply to the ecliptic.

Small Core of Iron

However Dr. Alastair G. W. Cameron of Yeshiva University, proponent of the view that the moon formed just inside the orbit of Mercury, believes it has a small core of relatively pure iron.

He notes that the orbit of Mercury is unusually elliptical and attributes this to an early gravitational battle with the moon. The stronger gravity of Mercury, a full-fledged planet, threw the moon into an orbit that carried it out past the earth, whereas the weaker gravity of the moon left Mercury in its rather lopsided orbit.

Dr. Cameron points out that both these hypotheses leave unanswered the question of how the earth's gravity was able to capture the moon into a relatively circular orbit.

Because of this difficulty the theory is still alive that regards the moon as formed from material in orbit around the evolving earth, like the rings of Saturn. However, the idea that the moon was torn from the upper layers of the earth has virtually died as a consequence of the Apollo findings.

Discovery of Rocks

The discovery of rocks on the moon that are from 4.25 to 4.5 billion years old seems to rule out any such origin unless it occurred immediately after the earth's formation. All bodies of the solar system are thought to have formed about 4.65 billion years ago.

The oldest rocks known on earth have been found by a group from Oxford University in west Greenland. As reported in the Nov. 27 issue of Nature Physical Science, they are granite-like rocks a little older than 3.7 billion years.

According to Dr. Paul Gast of the Manned Spacecraft Center, in Houston, about 4.4 billion years ago there seems to have been an outpouring onto the lunar surface of an unusual type of rock that has been designated "kreep." It is a form of basalt.

Information on the first billion years of lunar history is meager but from age determinations of rock from the lunar "seas" it has been established that most if not all of them were flooded with lava between 3.2 and 3.7 billion years ago.

It had been proposed that the seas, which lie almost entirely on the earth-facing side of the moon, were flooded because of tidal stresses imposed on the moon as it was being captured by the earth's gravity. However, Dr. Gast says the 500-million-year period of their formation, as indicated by the Apollo specimens, is too long for such a process.

December 4, 1972

VENUS UNMASKED BY ASTRONOMERS

Radar Probes Give Man His First Look at Surface— Craters Dot Landscape

By JOHN NOBLE WILFORD
Special to The New York Times

HOUSTON, Aug. 4 — High-resolution radar probes have broken through the thick clouds of Venus and for the first time distinguished features on the planet's surface, which presents a landscape of huge, shallow craters.

Of the dozen craters discovered in man's first look at the Venusian surface, the biggest is 100 miles wide and less than a quarter of a mile deep. Others range in size from 20 to 65 miles wide.

The discoveries have led to the production of the first map of a part of Venus showing discrete features, instead of the blurry shadings of earlier radar maps of the planet. The new map is also the first to include elevation contours of the basically flat surface.

Clouds 13 Miles Thick

The unmasking of Venus, which is concealed by a perpetual bank of clouds some 13 miles thick, was accomplished by a team of radar astronomers at the Jet Propulsion Laboratory at Pasadena, Calif. The results were announced simultaneously there and at the Johnson Space Center here.

Covered in the latest radar scan of Venus is an area along the equator of more than 500,000 square miles, which is about the size of Alaska. The probe achieved a resolution of about six miles, five times better than the last Venus radar experiment in 1970.

In previous radar observations over the last decade, by the Jet Propulsion Laboratory and also the Massachusetts Institute of Technology, more than a sixth of the Venusian surface has been mapped in some fashion. Optical astronomers have yet to see any of the planet's surface through telescopes.

Dr. Richard A. Goldstein, who headed the Jet Propulsion Laboratory team, said, "This area of Venus appears to be as crater-infested as the moon."

This came as no great surprise to planetary scientists. Dr. Carl E. Sagan, director of the Labortory of Planetary Studies at Cornell University, said in a telephone interview that the "same sort of debris that makes holes on the moon should be bombarding Venus."

Dr. Sagan said that the apparent density and size of the craters on Venus suggested that they had been formed by meteorite impacts rather than volcanic eruptions. Volcanic craters are usually smaller.

The radar findings also indicate that meteorites of considerable size can't penetrate the thick, high-pressure atmosphere of Venus without burning up, Dr. Sagan added.

But since there is no water and presumably a low wind velocity at the Venusian surface, he said, it is not clear what forces are causing the craters to erode. The shallowness of the craters suggests that they are eroded.

Dr. Goldstein and his team made the radar sounding of the planet on June 20, 1972— when Venus was last at its closest approach to earth, 26 million miles away. To get the

ASTRONOMY AND PLANETARY SCIENCE

improved resolution, they used two antennas at the Goldstone Tracking Station in the Mojave Desert of California. It took them a year to process the data and make the map.

The radar signals were beamed toward Venus with the 210-foot dish antenna, and the return echoes were received at both the large antenna and an 85-foot antenna. The transmitting power is 400,000 watts. The round-trip signal time is four and a half minutes.

From the slight time-delay variations in the radar bouncebacks, the astronomers were able to detect the relief of the planet and outline the shape and depth of such features as the craters. The signals are timed with hydrogen maser clocks, which are so accurate and precise that they would have an error of one second in one million years.

'Stereo Reception'

In addition, the return signals are analyzed for polarization. If the signals bounce back unpolarized—that is, with their electric fields scrambled —they indicate rough terrain. Unscrambled echoes indicate smooth terrain.

By receiving the return signals at two antennas 14 miles apart, the scientists were better able to distinguish elevation differences at a resolution of about 600 feet.

"This, in effect, gives us stereo reception," Dr. Goldstein said, "and enabled to pinpoint each area touched on Venus. We were able to see depths better."

Dr. Howard C. Rumsey Jr., who developed the computer technique used in producing the map, said that the area surveyed was "basically flat," with elevation variation no greater than 3,300 feet.

Dr. Rumsey is also on the staff of the Jet Propulsion Laboratory, which is operated by the California Institute of Technology for the National Aeronautics and Space Administration.

Early next year, the laboratory's scientists expect to get a much closer radar profile of parts of Venus when Mariner 10 flies by the planet on its way to the first rendezvous with Mercury. The unmanned spacecraft is scheduled to be launched from Cape Kennedy, Fla., in early November.

Mariner 10 is equipped with television cameras and radar. The new radar map is being studied by project scientists to help them to decide where to point Mariner's cameras.

August 5, 1973

Pioneer 10 Passes Jupiter, Will Leave Solar System

Craft Pierces Radiation Belts, Moving Within 81,000 Miles of Giant Planet —Sends the First Close-Up Data

By JOHN NOBLE WILFORD
Special to The New York Times

MOUNTAIN VIEW, Calif., Dec. 3 — A small American spacecraft, Pioneer 10, sped within 81,000 miles of Jupiter tonight, sailing past the glowing giant ball of primordial gases for man's first close-up exploration of the largest planet in the solar system.

After a 21-month, 620-million-mile voyage from earth, the robot spacecraft sailed safely through Jupiter's hazardous radiation belts and transmitted color pictures and volumes of scientific data as it made its closest approach to the planet's cloud tops.

Pictures sent back here by the spacecraft showed the planet growing ever larger, its Great Red Spot bulging like a baleful eye in the Southern Hemisphere and its covering clouds of cold ammonia flowing in bands of many colors. The detail and contrast were reported to be better than in any previous photographs of Jupiter.

Radioed data included measurements of Jupiter's radiation belts, magnetic field, temperatures and atmospheric composition. Scientists reported the first confirmation of helium's existing at Jupiter, a discovery considered important to cosmologists attempting to explain the chemistry of the solar system and the universe.

Pioneer 10, reaching a speed of 23 miles a second (82,800 miles an hour) raced by the planet at 9:25 P.M., Eastern standard time.

The pull of Jupiter's gravity was so great that, as expected, the 570-pound spacecraft was accelerated and slung past the planet on a new trajectory that should make it the first spacecraft ever to leave the solar system altogether.

All indications were that the spacecraft's systems survived the heavy radiation doses without degradation. In fact, radiation levels appeared to reach a peak about 15 minutes before Pioneer 10's closest approach to Jupiter and then to drop.

"We're still alive," Dr. John H. Wolfe, the project's chief scientist, exclaimed.

In some of the pictures transmitted during the rendezvous, Dr. Tom Gehrels of the University of Arizona, the scientist in charge of the spacecraft imaging system, said that he could distinguish the planet's red spot to be apparently " a towering cloud mass" that even seemed to cast a shadow. The picture could help scientists determine the nature of the mysterious red feature.

In a message from the White House, President Nixon congratulated the Pioneer project team for "so impressive a scientific and technical achievement" and for demonstrating that "man's ability to explore the heavens is on the threshold of the infinite."

In the extraordinary event that the spacecraft is discovered by some extraterrestrial civilization, Pioneer 10 bears a plaque with symbols disclosing earth's location and depicts a nude man and woman. The spacecraft is expected to leave the solar system beyond faraway Pluto in 1987 on a galactic course leading somewhere in the direction of the constellation Taurus.

Pioneer 10 traveled farther and faster than any previous manmade object to make its historic rendezvous with Jupiter, one of the most important scientific moments of the 16-year-old space age.

Spacecraft from earth have already explored Venus, Mars and the earth's moon. Mariner is on its way to reconnoiter Mercury, the smallest and inner-most planet. A fleet of four Soviet spaceships is cruising toward Mars for possible landings.

Now, Pioneer 10 has explored what many scientists believe is the most interesting planet in the sun's family— Jupiter. It is 1,300 times bigger than the earth and contains 71 per cent of all the planetary matter in the solar system. From earth-based observations, Jupiter is assumed to be com-

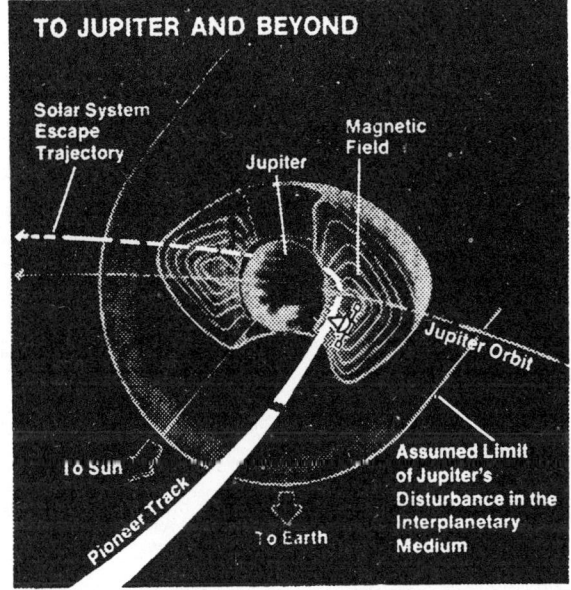

The New York Times/Dec. 4, 1973

Accelerated and put on an altered course by Jupiter's gravity and orbital speed, Pioneer 10 is destined to be the first manmade object to leave the solar system.

Planetary Data

	Mercury	Venus	Earth	Mars	Jupiter	Saturn	Uranus	Neptune	Pluto
Mean Distance from Sun (Millions of Miles)	36	67	93	142	483	886	1,789	2,793	3,663
Diameter (Miles)	3091	7,688	7,926	4,201	88,692	75,059	29,247	27,741	8,718
Mass (Earth = 1)	0.056	0.817	1.000	0.108	318	95.2	14.6	17.3	0.9(?)
Surface Gravity (Earth = 1)	0.36	0.87	1.00	0.38	2.64	1.13	1.07	1.41	(?)
Rotation Period	59 Days	243 Days (retrograde)	*H M S 23 56 04	H M S 24 37 23	H M S 9 50 30	H M S 10 14 0	H M S 10 49 0	H M S 14 (?)	6.39 Days
Moons	0	0	1	2	12	9	5	2	?

*H—Hours M—Minutes S—Seconds

The New York Times/Dec. 4, 1973

posed primarily of a mixture of gases similar to those present at the beginning of the solar system and similar to those out of which life emerged on earth.

For engineers planning future explorations, Jupiter is considered the gateway to the outer planets. By using Jupiter as a gravitational slingshot, they hope to send larger spacecraft out to Saturn and eventually Uranus, Neptune and Pluto.

As Pioneer 10 closed in on Jupiter, it crossed the orbit of Ganymede, the largest of Jupiter's 12 moons and about the size of Mercury. The spacecraft came within 277,000 miles of Ganymede.

Infrared measurements from the earth indicate that Ganymede's surface is coated with water ice. Pioneer's infrared radiometer, an instrument designed by the California Institute of Technology, returned data today showing the satellite's surface temperature to be about 235 degrees below zero Fahrenheit.

About noon today, Pioneer 10 focused its ultra-violet photometer on Jupiter for the second time to make measurements of the planet's helium content and other atmospheric constituents.

"There's a helium glow at Jupiter," Dr. Darrell L. Judge of the University of Southern California, the principal investigator, reported. "It's fairly weak compared to the hydrogen, about 100 times less bright, but it would have been a surprise if there had not been any helium."

Despite its great volume, Jupiter has a low density—one-fourth that of the earth and only slightly greater than that of water. This suggested to scientists long ago that Jupiter must be composed primarily of the lightest elements, hydrogen and helium. Hydrogen had already been discovered through earth-based instruments, but not helium.

Theories of the formation of the solar system and the universe are based on assumed abundances of hydrogen and helium and their ratio to one another. Since Jupiter is so cold and has such strong gravity, it is thought that the planet has lost little of its original gases and therefore should represent the chemistry of the early solar system.

As Pioneer 10 passed closest to Europa, a Jovian satellite about the size of the earth's moon, the spacecraft encountering a steadily increasing amount of radiation — electrons and protons trapped by Jupiter's magnetic field.

Protons are positively charged subatomic particles, and at extremely high energies are considered the most dangerous to the spacecraft's electronics. Electrons are negatively charged particles. As in the earth's Van Allen radiation belts, the trapped particles around Jupiter come from the sun in a flowing stream of interplanetary particles known as the solar wind.

About six hours before the rendezvous, a brief scare shook the control room here at the National Aeronautics and Space Administration's Ames Research Center. Suddenly, the spacecraft's imaging system ceased returning picture data, switching instead to another mode of operation. Had radiation dealt it a spurious command?

A radio command from the control room quickly switched the system back to normal operation. Flight controllers were not sure whether the problem was a random malfunction or somehow related to radiation bombardment. Measurements of Jupiter's radiation belts showed that they were more intense than any naturally occurring radiation in the vicinity of earth.

As a result of the problem, there was a loss of nearly an hour of data, including one picture of Io, another of Jupiter's moons.

The subject of Jupiter's magnetic field and radiation belts is the most puzzling and controversial of the mission. Scientists have been debating the subject ever since Pioneer 10 entered the sphere of Jovian magnetic influence—the magnetosphere — last Tuesday, about 24 hours earlier than had been predicted.

Magnetism Estimates

The early entry led scientists at first to estimate that Jovian magnetism at the surface might be 40 times as strong as the earth's—about 20 gauss, compared with the earth's one-half gauss. A gauss is a unit measure of magnetism. This would have been more than twice the strength predicted by radio astronomers.

But the surprisingly slow build-up in magnetic measurements as Pioneer 10 moved closer to the planet caused a rapid reversal of scientific opinion.

Dr. Edward J. Smith of the Jet Propulsion Laboratory in Pasadena, Calif., chief of the magnetometer experiment, said at a news conference today that the Jovian magnetic field may turn out to be more like 2½ gauss.

In addition, Dr Smith reported evidence suggesting that Jupiter's magnetic field does not appear to conform with most of the predicted models. He said the planet's magnetic axis appeared to be offset from the rotational axis by some 15 degrees.

However, after the spacecraft's closest approach to Jupiter, Dr. Wolfe of the Ames Research Center reported new data indicating "no evidence of a significant offset."

Based on radiation observations, Dr. James A. Van Allen of the University of Iowa, another Pioneer scientist, said today that Jupiter's magnetic field seems to be not only offset but also shaped like a broad, thick disk, not as originally thought, more like the spherical, symmetrical shape of the earth's magnetic field.

The field, Dr. Van Allen said, might be likened in shape to the disk-like rings of Saturn—only invisible, of course. And it must be wobbling as it rotates, like a child's hula hoop in action.

At a news briefing, Dr. Van Allen estimated that the magnetic disk is spun out several million miles but is only some 176,000 miles thick—or four times as thick as Jupiter's radius.

Other project scientists are not so sure. Dr. Wolfe said that Dr. Van Allen's radiation data may support such conclusion, but not the magnetism and thermal plasma measurements.

It may take scientists several days to reach a consensus on the shape and nature of the planet's radiation belts and magnetic field.

This will be important information to astronomers seeking to interpret over long periods of time radio signals emitted by Jupiter. It would also help flight controllers directing future missions to Jupiter and the planets beyond. Pioneer 11, a sister spacecraft, is now halfway to Jupiter for what should be an even closer fly-by.

December 4, 1973

Observations of Kohoutek Appear to Confirm Concept of Comets as 'Snowballs'

By WALTER SULLIVAN

Kohoutek is a giant snowball, coated or impregnated with organic matter and displaying an antitail as well as a tail, according to observations from Skylab and a number of ground observatories.

The detection of water vapor in its tail, announced yesterday by the National Aeronautics and Space Administration, is taken to confirm the concept of comets as "dirty snowballs."

Such a view of cometary composition was proposed several years ago by Dr. Fred L. Whipple, recently retired director of the Smithsonian Astrophysical Observatory in Cambridge, Mass., who was long on the Harvard University faculty. However, his concept has been challenged by some astronomers, notably by those who viewed comets as "flying sand banks."

The discovery that water vapor is being blown off Kohoutek to form its tail appears to support the snowball model. Dr. Whipple said yesterday that he was "extremely gratified" by the finding.

He added, however, that he found even more exciting the discoveries of organic matter and of an antitail.

The organic matter (methyl cyanide and hydrocyanic acid) is of the type detected by radio emissions from distant clouds of dust and gas between the stars. The molecules are among those thought to represent the earliest steps in the evolution of more complex substances characteristic of life.

Their presence in the comet, first reported several weeks ago, was similarly indicated by radio emissions. If they are merely a coating, Dr. Whipple said, they may have been picked up as the solar system sailed through one of the interstellar dust clouds.

In that case, he added, the molecules' characteristic radio emissions should become weaker as the outer layer of the comet boils off. If they impregnate the entire comet, forming part of the material from which it was originally formed, there will be no such drop-off in the emissions.

Hence, Dr. Whipple said, the issue may be settled by the observations now under way.

The antitail has been detected optically by the Skylab astronauts and, through infrared scanning, by Dr. Edward P. Ney and his colleagues at the University of Minnesota. Whereas the tail of Kohoutek, as it recedes from the sun, is blown out ahead of it by solar radiation and gas (the "solar wind"), the antitail points toward the sun.

The antitail presumably is formed of particles, shed by the comet, that are too heavy to be strongly affected by the solar wind. Hence they trail behind the comet instead of being blown ahead of it.

Dr. Stephen P. Maran of the Goddard Space Flight Center in Maryland, who has been coordinating Kohoutek observations, said yesterday of the water detection: "You really have to believe that Whipple is right." The Goddard center is operated by the National Aeronautics and Space Administration.

In 1963, Dr. Maran said, spectral lines that could not readily be identified were seen in reddish light from the tail of a relatively bright comet. More recently these and several other lines were seen by the Asiago and Lick observatories in red light from Kohoutek's tail, but their source was still uncertain.

Now, Dr. Gerhard Herzberg and his co-workers in Ottawa have identified the five observed lines as those produced by ionized water molecules. Dr. Herzberg, who is with the National Research Council of Canada, won a Nobel Prize for his work on the spectra of molecular fragments, known as free radicals.

Dr. Maran said attempts were being made to detect ammonia, which is thought to be a constituent of comets. A few days ago a Convair 990 jet was flown in an effort to detect methane but whatever amount of that gas may be present was insufficient to be recorded.

The unmanned spacecraft, Mariner 10, on the way to Venus, has begun scanning the comet at ultraviolet wave lengths and, on Saturday, will turn on its television camera to provide astronomers with their first view of such an object from elsewhere than on or near the earth.

While Kohoutek is "the best-observed comet in history," Dr. Whipple said, it is admittedly a disappointment to those who predicted a spectacle. "If you want to have a safe gamble," he said, "bet on a horse—not a comet."

January 14, 1974

Mariner 10 Photos Said to Show Nature of the Weather on Venus

By WALTER SULLIVAN
Special to The New York Times

PASADENA, Calif., Feb. 6— Television images of Venus transmitted by Mariner 10 as it flies away from that planet have revealed a system of bands and streaks roughly parallel to the Venusian equator and strikingly reminiscent of those in the clouds of Jupiter.

They are believed to show, for the first time, the nature of weather on earth's nearest planetary neighbor. According to Dr. Verner E. Suomi of the University of Wisconsin, it appears that the winds of Venus constitute a symmetric, spiralling system that carries energy from the equatorial region to the two poles.

Dr. Suomi, an internationally known authority on atmospheric circulation, is a member of the scientific team interpreting the pictures as they come in from Mariner. He pointed out that such a weather system for Venus had been predicted several years ago by Dr. Yale Mintz of the University of California, Los Angeles.

Discovery of the bands came unexpectedly today. Initial pictures taken at close range yesterday as Mariner flew past the planet showed no such features. Mariner is now receding from the planet and its pictures reveal large scale features more clearly than before. As the project manager, W. Eugene Giberson, put it this afternoon, "Yesterday we were looking at the leaves, but today we can see the branches and trees."

Whereas the bands of Jupiter are prominently evident even through a small telescope, those of Venus had never before been observed. They were recorded by the Mariner television cameras at ultraviolet wavelengths that did not readily penetrate the earth's atmosphere. They also were greatly enhanced, photographically, by a computer process that increases the contrast between subtle shadings.

Almost No Spin

The discovery is remarkable in that no two planets of the solar system differ more radically than Venus and Jupiter. The latter is huge, composed largely of hydrogen and other light substances and spins rapidly — once every 10 hours. Its banded cloud structure is clearly related to this fast spin rate.

Venus is no larger than the earth and is comparably dense. It hardly spins at all, rotating once every 243 days in a direction opposite to that of all other major bodies of the solar system. However, in recent years evidence has been found that its clouds rotate much more rapidly.

Observations from earth at ultraviolet wave lengths revealed features that seemed to reappear every four days, suggesting that cloud systems— perhaps "storms"—are circulating the planet at about 200 miles an hour. Such are the maximum velocities in jet streams on earth. The direction of cloud motion on Venus is the same as that of the planet's spin but much faster.

A Puzzling Feature

It is this motion that presumably accounts for the bands, or streaks, some of which when viewed in detail show the whispy elongated quality of cirrus clouds on earth. Why the bands show up only in ultraviolet light is not certain. However, Dr. Suomi believes they lie in the topmost region of the clouds.

They do not however appear to be in the shell of thin mist that was observed in early pictures across the edge, or limb, of Venus yesterday. This mist layer showed up clearly in orange light, but not in the ultraviolet characteristic of the bands.

While it is believed by gaining an understanding of atmospheric circulation on other planets we will better understand weather on earth, the reason for the rapid motion of the Venus upper atmosphere around that planet is puzzling. An enormous amount of energy in the form of angular momentum is involved and, Dr. Suomi pointed out in an interview, it is not clear what originally set it in motion.

The upper winds apparently carry Venusian air, which is primarily carbon dioxide, north and south, spiraling away from the equator to subside at the poles and flow back towards the equator at lower levels.

However the lower atmosphere of Venus is extremely dense, surface air pressure being about 90 times that on earth.

Hence only a moderate flow of this dense air toward the equator is needed to compensate for the winds toward the poles aloft. The Soviet lander Venera 8, recorded only weak winds in contrast to the violent storms once thought to sweep the surface.

Furthermore, the massive Venusian atmosphere stores so much heat that despite the planet's slow rotation its nightside is essentially as hot as the day side. Also, as confirmed yesterday by Mariner's infrared scanner in a sweep across one entire side of Venus, there is little cloud top temperature difference between the equator and poles compared for example, to that on earth.

Whereas pictures of the earth from space typically show great swirling cloud patterns that manifest cyclonic storm systems no such features have been seen so far in the Venus pictures. By contrast, Dr. Suomi said, storms are frequently visible on Jupiter.

Still unexplained are several roughly circular dark regions that appear in the Venus pictures. Dr. Suomi noted that in looking at the dark and bright streaks it was difficult to tell whether the clouds are black or white. He suspects that less prominent patterns at right angles to the streaks manifest wave patterns in the fast-flowing air.

It had been hoped that along the boundary line between day and night on Venus the low sun angle would make visible the layering of the cloud cover. Apparently the top deck of clouds is so dense that even under these circumstances it was impossible to see through it. However, radio probing of the atmosphere in past missions, as well as on this one, plus data from Soviet capsules parachuting onto Venus, have shown that some form of layering exists.

February 7, 1974

Mariner Data Give Hint of Origin of Venus

By WALTER SULLIVAN
Special to The New York Times

PASADENA, Calif., Feb. 7 —Early analysis of the deluge of scientific data sent by Mariner 10 as it flew past Venus on Tuesday has lent support to the view that the manner in which that planet was born and matured differed basically from that of the earth, leaving Venus barren of water from the beginning.

If Venus was formed in a different manner, its present furnace-hot environment cannot be taken, as some have thought, as a preview of the earth's fate.

It appears that the upper air of Venus is constantly being fed by gas blowing out from the sun—the solar wind. In the vicinity of that planet, this wind is intense, because Venus is relatively near the sun. Magnetic detectors on Mariner confirmed that Venus has virtually no magnetic field to fend off the solar wind.

Hence it pounds directly onto the upper atmosphere of Venus. On earth, the solar wind is diverted thousands of miles aloft by the earth's magnetism.

The provisional interpretation of Mariner data indicates that the hydrogen in the upper air of Venus has been derived from the solar wind and not from the breakdown of water erupted from within Venus over millions of years.

Not of Internal Origin

On earth such volcanic eruptions and the breakdown of surface rocks are believed to have provided all the water of the oceans, lakes and atmosphere, as well as the gases of the air. If the hydrogen of Venus is not of internal origin, this would support the hypothesis of some scientists that Venus was formed from a condensing cloud of dust and gas so close to the sun that most of the water in its constituents was blown away before it could become a planet.

The absence of any observable magnetism near Venus supports the view that the earth's own magnetic field is a byproduct of its spin.

Venus hardly spins at all. It takes the planet 243 days to rotate once, relative to the stars, and hence its day is longer than the Venus year. One of the findings, from highly precise tracking of the effect the Venusian gravity had upon the trajectory of the passing spacecraft, has been the extraordinary roundness of the planet. Its flattening at the poles is a hundredth that of the earth's.

The implication is that if Venus once spun faster and therefore was fat around its Equator, like the earth, it was soft enough inside to adjust its shape to an almost perfect sphere after slowing. Its slow rotation also seems to account for the spiraling circulation of its atmosphere shown in the mosaics of television images prepared over the last 24 hours.

Pattern of Air Flow

According to Dr. Verner E. Suomi of the University of Wisconsin, laboratory experiments have shown that the atmosphere of a very slowly rotating planet that is being heated from one side, as by the sun, will develop a relatively simple pattern of upper air flow that spirals smoothly away, to the north and south, from the tropical zone. This "classic Hadley circulation," is evident in the "fabulous" Mariner pictures enthusiastically reported to a press conference here today.

The reference was to the British meteorologist George Hadley who, in 1735, theorized that such a circulation existed on the earth. The laboratory experiments, Dr. Suomi said, show that as the spinrate is increased the circulation acquires an undulating or wavelike form that eventually resembles that seen in photographs of the earth taken from space.

The absence of water emanations from within Venus is implied, though not confirmed, by data from an instrument that scanned the outer edge of the planet's atmosphere as Mariner flew by. It observed the ultraviolet glow from that region recorded at wave lengths typically emitted by ordinary hydrogen, but saw virtually none of those produced by heavy hydrogen (deuterium).

If the hydrogen envelope of Venus had been derived from the breakdown by sunlight of water molecules slowly emerging from the planet—a process that envelopes the earth in a hydrogen cloud—the envelope would be enriched in deuterium since the latter, because of its weight, would be less apt to fly off into space than ordinary hydrogen.

Had the hydrogen come from water contributed by a comet that plunged into Venus, according to this reasoning, it would also be rich in deuterium because of the freezing process whereby comets are thought to have been formed.

On the other hand, if the hydrogen came from the sun, which burns deuterium in its core, virtually no deuterium should be evident. Dr. A. Lyle Broadfoot of Kitt Peak National Observatory in Arizona reported that deuterium had not been detected, but he was cautious about making a final interpretation.

Today's Soviet report of a hydrogen atmosphere around Mercury was greeted with interest in view of the Venus findings. However, Dr. James A. Dunne, project scientist for the Mariner mission, noted that hydrogen would be difficult to see from the earth, since its characteristic ultraviolet glow does not penetrate the atmosphere. It could have been detected from earth orbit.

Dr. Suomi was introduced by Dr. Bruce C. Murray of the California Institute of Technology, leader of the television team, as "the most prominent person in the world on interpretation of the earth's clouds from satellites."

Dr. Suomi said a "bubbling" appearance in the tropical zone of Venus seemed to mark cells of rising gas that were transporting excess solar heat from below to be dissipated toward the poles by the upper wind. Unlike the clusters of cumulus clouds that perform this role in the tropics on earth, however, these cells seem to originate far above the surface of Venus whose lower atmosphere is so dense that it participates only slightly in the wind patterns.

Dr. H. Taylor Howard of Stanford University said that as the spacecraft went behind Venus and its radio transmissions were aimed through the atmosphere those at the shorter or X-band wave length were received until they cut off abruptly when they reached depths less than 33 miles above the surface. It appears, he said, that an extremely dense cloud layer envelopes Venus between heights of 22 and 33 miles.

Data indicated that the top layer is at 37.5 miles. Television views across the edge of the planet showed a succession of three or four haze layers above that.

The Mariner mission is being managed by the Jet Propulsion Laboratory operated here by the California Institute of Technology on behalf of the National Aeronautics and Space Administration.

While Mariner came within a few miles of its target point alongside Venus, the gravity of the planet has thrown it into a trajectory that would carry it past the daylight side of Mercury on March 29.

February 8, 1974

Mercury Found to Have Magnetic Field

By WALTER SULLIVAN
Special to The New York Times

PASADENA, Calif., March 29 —Mariner 10 zoomed in today for the first close-up look at the planet Mercury and sent back television pictures showing an intensely crater-marked surface gouged, torn and blasted by space debris and 4.6 billion years of proximity to the sun.

The surface, criss-crossed here and there by strange-looking valleys of unknown origin, bears many resemblances to those of the moon and Mars. Yet it differs in ways that could ultimately be indicative of Mercury's separate history.

Probably the biggest surprise has been in indications that Mercury has a magnetic field with a strength about 1 per cent that of the earth's. Even so weak a field was unexpected, since it is stronger than any magnetism observed near Venus, Mars or the moon.

The earth's magnetism is believed caused by a dynamo effect from the churning of molten material within the core of a spinning planet. The spin rate of Mercury is so slow that little or no field was expected.

Another surprise has been the presence of a tenuous but appreciable atmosphere formed of helium, neon and argon, with some hydrogen present.

There is apparently no oxygen, carbon, carbon dioxide or nitrogen—the first such planetary atmosphere ever observed.

No "seas" like those of the moon were evident in the approach-phase pictures, although the scarcity of craters in pictures sent after the closest approach was interpreted as possibly indicating a sea-like region. Nor were features typical of Mars observed, such as the mammoth Martian volcano Nix Olympicus, that planet's snaking "river" beds or its great equatorial canyon, indicative of some form of deep ferment.

Yet, as on Mars, the craters seem heavily eroded, even though Mercury, at least now, has no substantial atmosphere and therefore no winds. Much of the surface is chaotically rugged, but some crater floors seem completely smooth except where pocked by an impact more recent (and smaller) than the one that produced the initial crater.

A considerable percentage of the craters have central, ripple-like peaks like those typical of many lunar craters. It is believed, that these were formed by the impact of a meteorite hitting sufficiently hard to produce an explosion. The central peak results from a rebound effect seen also in shallow underground nuclear explosions on earth.

The white spot in the central region of the area under observation during the approach, beginning last Saturday, proved to be a small but very bright crater cut into the wall of a larger, older and darker crater. Both of them have central peaks. From the bright crater there radiates a system of "rays" like those around the large, relatively young craters of the moon.

Scientists of the mission headquarters at the Jet Propulsion Laboratory here noted the striking resemblance of this and several other Mercury craters to large ones on the moon, such as Aristarchus. In some of these Mercury craters, sections of the walls have slumped into a step-like sequence like that around the Copernicus crater of the moon.

The white spot crater is estimated to be 25 miles wide, with the larger one beside it 60 miles across—comparable in width to Aristarchus.

The rays are clearly formed by debris thrown out by the explosive impact that produced the crater. In some cases, the surface is also gouged into valleys that radiate from the crater. However, the origin of other worm-like valleys remains unclear. They seem to be uniformly about five miles wide and form no coherent pattern with regard to one another.

When the pictures are pieced together into mosaics, the significance of some features will probably become clear.

In one picture, for example, a ridge at least 100 miles long can be seen. As noted by Dr. Bruce C. Murray, geologist in charge of the television scanning, such a feature on earth would be attributed to the warping of the surface by deep

Mariner 10 photo of Mercury at distance of 21,700 miles
Jet Propulsion Laboratory via Associated Press

processes. Its significance on Mercury may become clearer when its relation to other features becomes evident.

Particularly striking were small but very bright craters, scattered across the landcape and presumably formed from the relatively recent impacts of objects smaller than those responsible for the larger, older craters. Speculation regarding the processes responsible for crater erosion includes blasting by solar gas—the "solar wind" —constant bombardment by meteoritic particles and "dust storms" during a hypothetical period when the planet had a significant atmosphere.

Because of the solar wind, intense solar heating and radiation bombardment, it is believed that the planet could not retain such an atmosphere for an extended period.

Whereas the pictures sent to earth by this spacecraft as it passed venus on Feb. 5 showed haze layers above the planet's horizon, and haze was also evident in Mariner 9 views of Mars, the horizon of Mercury as seen today through the cameras of Mariner 10 showed a clean edge against a black sky.

Strong Auroral Glows

During the craft's closest approach, when it was within 460 miles of the surface at 4:46 P.M. Eastern daylight time, it was on the nightside collecting scientific data but not recording pictures. The first pictures shown as it emerged on the far side disclosed a surface that seemed less heavily cratered. Throughout the encounter, mission scientists kept up a running commentary on this scanning of the last of the inner planets to come under close scrutiny.

The Jet Propulsion Laboratory is operated by the California Institute of Technology and is conducting the mission on behalf of the National Aeronautics and Space Administration.

According to Dr. Michael B. McElroy of Harvard University, strong auroral glows were observed on the nightside of Mercury at ultraviolet wavelengths typical of helium, argon, hydrogen and probably neon. On the sunlit side, no hydrogen was detected within the sensitivity of the sensors.

This appeared to contradict a recent claim attributed to Dr. Nikolai Kozyrev of the Soviet Union that he had seen evidence of a hydrogen envelope as Mercury passed across the face of the sun last year.

No evidence of argon, carbon dioxide or nitrogen was observed. Thus, in the words of Dr. James C. Fletcher, the NASA administrator, the atmosphere of Mercury "is unlike any atmosphere we have seen anywhere else."

Dr. Fletcher reported that the magnetic field of Mercury was sufficient to deflect the solar wind and produce a "bow shock wave" where the solar wind impinged on the magnetic field at supersonic velocity.

In a congratulatory message, President Nixon said that the mission "marks another historic milestone in America's continuing exploration of the solar system."

By the end of the century, Dr. Fletcher told a news conference, all of the planets should have been examined at close range, and by then, he said, man should regard the entire solar system as his home.

March 30, 1974

Jupiter Mainly Liquid Hydrogen, Data From Pioneer 10 Indicate

WASHINGTON, Sept. 10 (UPI)—The planet Jupiter is a spinning ball of liquid hydrogen with a turbulent interior and raging atmospheric storms. And the mysterious Great Red Spot on the planet may be the vortex of a 25,000-mile-long tornado that has been towering above Jupiter's cloud belt for hundreds of years.

This was the picture of the dense, brightly banded planet that scientists have pieced together from the Pioneer 10 fly-by of Jupiter last December, the National Aeronautics and Space Administration said today.

"At best, Jupiter has only a small rocky core, thousands of miles below the heavily clouded atmosphere," the researchers concluded in a report released by NASA.

The scientists analyzed measurements and pictures radioed back by the Pioneer 10 spacecraft, which flew within 81,000 miles of Jupiter on Dec. 5, 1973.

A sister probe is due to pass within 29,000 miles of Jupiter on Dec. 3 this year.

The space agency said in releasing the report that some of the information appeared to confirm theories about Jupiter while other findings contradicted earlier ideas about the planet.

Some of the major new findings are that Jupiter apparently has a much hotter interior than previously thought, its magnetic field is larger than expected, and its radiation belts are even more intense than predicted by many scientists.

Much of the new picture of Jupiter's interior was based on temperature calculations by Dr. William B. Hubbard of the University of Arizona, using measurements obtained from a Pioneer 10 gravity sensing experiment.

"Jupiter is almost certainly a liquid planet, for it is too hot to solidify, even with its enormous internal pressures of millions of atmospheres," NASA said.

The agency said the atmosphere begins to turn to liquid hydrogen 600 miles below the top of Jupiter's colorful clouds. At this transition zone the temperature is calculated at 3,600 degrees Fahrenheit.

At a depth of 15,000 miles, temperature is estimated at 10,000 degrees and the pressure 90,000 times earth's surface atmospheric pressure.

M1 At a depth of 15,000 miles, the pressure is three million atmospheres and liquid hydrogen turns to liquid metallic hydrogen.

The calculations indicate that the temperature of Jupiter's center is 54,000 degrees—six times the temperature on the surface of the sun.

The temperatures decrease steadily outward from the planet's surface to 184 degrees below zero at the top of Jupiter's clouds.

September 11, 1974

ASTRONOMY AND PLANETARY SCIENCE

Pioneer Photographs Jupiter and Flies On

View of Jupiter taken yesterday by Pioneer 11, as received at NASA's Ames Research Center in Mountain View, Calif. At the upper right is Ganymede, one of four large, or Galilean, satellites of the planet. The spacecraft was about 463,000 miles from Jupiter at the time of transmission. The time on the picture is Pacific Standard.

By WALTER SULLIVAN
Special to The New York Times

MOUNTAIN VIEW, Calif., Dec. 2—Pioneer 11 transmitted the first pictures of Jupiter's south polar region tonight and then, after a perilous journey through what is believed to be the most intense part of that planet's radiation belt, soared into a trajectory that will carry it across the solar system in five years to Saturn.

At 10:24 P.M., Pacific standard time, a hushed and anxious team of scientists awaited the first scheduled transmissions indicating that their spacecraft was still functioning after emerging from behind Jupiter. They came 25 seconds later than expected, and a cheer then went up in the control room of the Ames Research Center here.

Early indications were that all of the vehicle's many systems were operating normally. For 42 minutes, including the moment of closest approach when it came within 26,600 miles of Jupiter, the vehicle's transmissions were blacked out by the planet. It is believed that this period also included a passage through the most intense radiation.

So distant is Jupiter that it took 41 minutes for Pioneer's renewed transmissions to reach earth.

As Pioneer made its final approach to Jupiter tonight it began to do strange things. Heaters were turned on, as though some saboteur had gone aboard. Its antenna feed changed directions, although no command to do so had been sent from earth.

It appears that the spacecraft was flying through a sea of low-energy electrons that built up electric charges of varying strengths in different parts of the vehicle. The resulting sparks injected false commands into the system.

Countermanding commands were hastily transmitted, and the craft emerged from the intense electron sea with only a minor loss of observations.

Earlier as Pioneer plunged toward its perilous loop around that giant planet it cast its electronic eye on Callisto, second largest of the Jovian moons, and revealed what seemed to be a small polar cap like that on Mars.

Callisto is intermediate in size between the planets Mars and Mercury. The composition of the polar cap is uncertain.

Nevertheless Dr. Tom Gehrels of the University of Arizona, who is in charge of the Pioneer 11 imaging project, pointed out today that two observatories on earth had provided spectroscopic evidence for the presence of ice on all four of the large, or Galilean, satellites of Jupiter.

They are so named because of

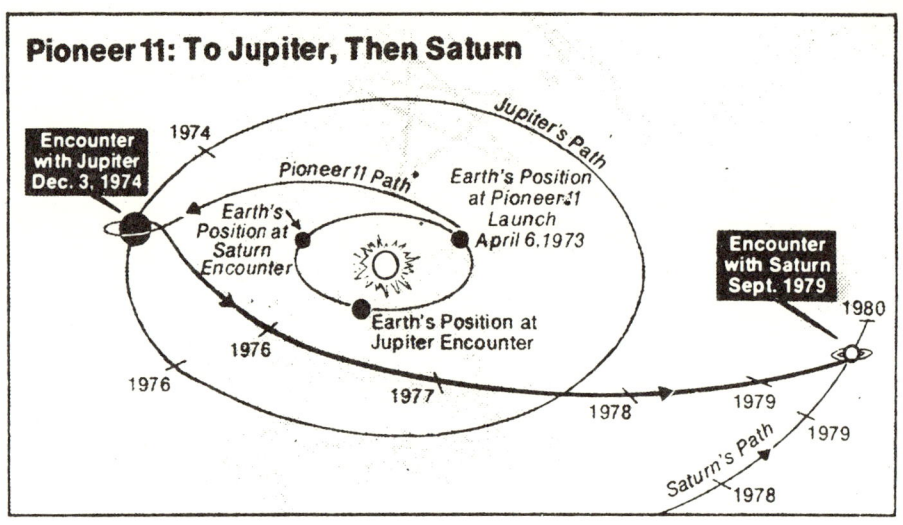

their discovery by Galileo. Callisto is the outermost of the four and thus is presumably less subject to effects of the lethal radiation belt around the Equator of Jupiter in surrounding space. Hence it has been suggested as a potential landing site for a manned mission.

Held by Gravitational Field

All day today Pioneer 11, in the grip of a gravitational field 318 times as powerful as that of the earth, was drawn toward Jupiter prior to making its closest approach earlier tonight. It was then thrown by Jovian gravity back across the solar system toward a 1979 rendezvous with Saturn.

In the control rooms here where high-speed printers poured out a stream of incoming data on the planet, the growing confidence of recent days faded earlier this evening in the face of Pionee's imminent confrontation with Jupiter's inner radiation belt.

It was expected that this would subject the craft to the most intense radiation ever experienced by a space vehicle—enough, many times over, to be fatal to any stowaway on the 570-pound robot explorer.

In the control rooms at the Ames Research Center, one batch of buttons has been covered with a plastic hood and a sign above it warns all not to press the buttons, for that would send a shutdown signal to the spacecraft.

Apparently on last year's Pioneer 10 mission—the only previous visit to Jupiter — someone touched one of the buttons, resulting in the loss of some data.

The Ames Research Center of the National Aeronautics and Space Administration that is conducting the Pioneer missions shares the Navy's Moffett Field with a complex of enormous wind tunnels and other research devices.

The scientists and technicians at the control center became increasingly confident last night as they observed that radiation levels being reported as Pioneer neared Jupiter were relatively low.

Pioneer 10 approached the planet closest to its equatorial plane. But Pioneer 11 made the approach well below that plane, prior to flying under its southern hemisphere behind the planet and back up through the hottest part of the radiation belt to fly over its northern hemisphere. It appears that the most intense radiation is near the equatorial plane, rocking about 10 degrees above and below it as the planet spins.

The data and pictures already received here are of major scientific importance. It is now possible, for example, to see at close range what changes have occurred over the last year in the Jovian weather—in its banded cloud systems.

While there have been significant changes, circulation patterns on Jupiter are remarkably stable. In 1664 Robert Hooke in Britain discovered a red spot on Jupiter and such a feature has been intermittently observed on the planet since then.

It is now widely believed to be a storm, analogous to a hurricane but persisting for centuries rather than days. It is about 15,000 miles wide. Patterns in the white band in which it is imbedded, as seen in Pioneer 10 and 11 pictures, indicate that the Jovian atmosphere is flowing around the spot.

A number of other features visible last year are still there including a prominent plume streaming from east to west in one of the bands north of the Jovian equator. It resembles smoke streaming from a stack in a strong wind. A second such plume is visible on the opposite side of the planet.

Presumably the plume if formed by an upward flow from within a region moving rapidly eastward into an upper layer of atmosphere either moving west or moving east far ess swiftly.

An attempt has been made at the Geophysical Fluid Dynamics Laboratory of the National Oceanic and Atmospheric Administration in Princeton, N. J., to reproduce or "model" the circulation patterns on Jupiter. The results were described at a briefing by Dr. Gareth Williams of that laboratory.

It is believed that the primary driving force of the circulation is heat flowing from the interior of the planet, rather than the heat of sunlight that is relatively weak at the orbit of Jupiter, 480 million miles from the sun. For that same reason, Pioneer does not have wing-like solar panels for power generation but is powered by canisters of radioactive plutonium 238 whose heat generates elecricity.

It is estimated that two and half times as much heat reaches the Jovian atmosphere from within the planet as does from the sun. Because of this, the planet's huge size and rapid spin the circulation pattern is very different from that in the atmospheres of the inner planets. It produces, for example, Jupiter's banded appearance.

The bright bands are believed to be where clouds, heated from below are rising and becoming very cold at their tops. The dark bands that alternate with the bright ones are where subsidence occurs to complete the circulation pattern.

The activity is most intense in the equatorial latitudes of Jupiter, which is where the bands are chiefly prominent. The pattern seems absent in the polar regions that are being viewed at close range for the first time during tonight's encounter.

Eight Bands

There are eight bands, alternately dark and bright, to each side of the Jovian equator. Dr. Williams suspects that the circulation pattern extends to depths that are shallow relative to the enormous volume of Jupiter — a few hundred miles in contrast to a diameter of some 87,000 miles for the entire planet.

The circulation also generates swift eastward flow in the equatorial zones. Dr. Williams said it resembled to some extent equatorial flow in the earth's oceans.

The flow pattern, involving elongated, rolling circulation to a relatively shallow depth, may also be enlightening, according to Dr. Williams, with regard to the slow circulation of semi-molten rock beneath the earth's rigid surface. Such flow is believed responsible for continental movements.

The depth of the circulation patterns on Jupiter is the focus of arguments about whether the red and brown colors on the planet manifest the presence there of organic molecules. It has been proposed that these precursors of life have been synthesized within the Jovian clouds. If, however, the circulation carries this material down to high temperature levels, such molecules would be destroyed. The color in that case might derive from simple sulphur compounds.

December 3, 1974

EXPERTS DISCUSS PIONEER'S FINDINGS

Magnetosphere of Jupiter Is Believed to Measure Nearly 9-Million Miles

By WALTER SULLIVAN
Special to The New York Times

MOUNTAIN VIEW, Calif., Dec. 5 — From evidence collected by Pioneer 11 in the last few days it appears that, if the envelope of magnetically enslaved particles around Jupiter were visible from the earth, it would cover an area of the sky at least four times as large as the moon.

Furthermore, the magnetosphere would be forever changing in extent and shape.

Because Jupiter is the only planet, apart from the earth, known to have a magnetic field embracing such a region of trapped particles, it is hoped that analysis of it will lead to a better understanding of the earth's own field and its magnetosphere.

Scientists who have experiments aboard Pioneer 11, which circled Jupiter Monday night and now is headed toward Saturn, presented preliminary results of their observations at a news conference today. It was held here at the Ames Research Center of the National Aeronautics and Space Administration, which is conducting the Pioneer missions.

It has been said that the earth's magnetic field, the force that controls a compass needle, is the most thoroughly described—and least understood—of all phenomena relating to the planet. The Jovian field, like that of earth, is off center. It also resembles the earth's field in being tilted some 10 degrees relative to the spin axis of the planet.

Reversed Polarity

It differs, however, in that the polarity of its field is reversed. On Jupiter, the north-pointing needle of a compass would point south. It is now known that at irregular intervals, ranging from thousands to millions of years, the earth's field flips over. Jupiter may be in a state of reversed polarity.

The magnetospheres of the earth and of Jupiter are the regions in space within which the planet's magnetic field predominates, holding off the blasts of fast-moving gas or solar wind blowing out from the sun, yet holding in thrall a series of radiation belts around the planet's equator.

However, as Dr. James A. Van Allen, who discovered the earth's radiation belts, pointed out today, the total strength of Jupiter's magnetism is 20,000 times greater than that of the earth. Yet the solar wind at that planet's orbit is 25 times less dense than it is at the earth, which orbits much closer to the sun.

This, Dr. Van Allen suggested, accounts for the enormous size of the Jovian magnetosphere, which from today's observations appears to be close to nine million miles wide.

December 6, 1974

CHAPTER 3

Earth Sciences

The effect of an earthquake in the northern San Fernando Valley on a bridge on the Golden Gate Freeway.

Courtesy The New York Times.

GEOLOGY

TIMES AND PLACES OF EARTHQUAKES

Systematic Observation of Shocks Only 25 Years Old.

EARLIER DATA ENIGMATICAL

Seismological Research Adequately Rewarded Only in Japan — Indifference of the World at Large.

Prof. H. H. Turner, F. R. S., in The London Times.

The occurrence of several disastrous earthquakes and eruptions during the last few months inevitably suggests the question whether all these events may not have a common and determinable origin. To avert any of these disasters, even to modify them in the slightest degree, may be entirely hopeless; but the vaguest foreknowledge of their probable occurrence might be of untold value in saving life and property. Has modern research obtained any clues which enable predictions to be made, or promise that prediction may be possible in the near future? It must be frankly admitted that as yet our knowledge is so slight as to have no commercial value; but still, there are one or two clues in the hands of those working at the subject which may ultimately lead them to more directly useful knowledge. We have learned something of the regions where earthquakes occur, and something of the times when we may specially expect them; and, though the something is in each case a very little, the magnitude of the issues involved lends it interest.

Systematic observation of earthquakes is only about a quarter of a century old, and for fairly complete records of all the shocks occurring in different parts of the globe we can date only from 1892. Before that date information could only be collected on the spot, and was thus frequently lost; but it was realized about 1890 that a series of earthquake observatories, with delicate instruments, could obtain records of shocks in any quarter of the globe, and identify the spot with certainty, even if there were no witnesses of the actual occurrence. From the records of these observatories it appears that there are every year some 30,000 minor shocks of earthquake in different localities, but of these only 60 are "world-shaking" and observable from a great distance. Such numbers indicate immediately that, from one point of view, the San Francisco earthquake cannot be regarded as exceptional; it was only one event out of sixty per annum. What rendered it disastrous was the existence of a great town in the shaken locality. But was the neighborhood known to be a dangerous one? Was it, at any rate, suspected, so that the building of a great city there was an error of judgment? And is it advisable to rebuild the city in the same place? These are questions of the gravest importance; and it is well worth while to review the little knowledge already accumulated with the utmost care to see whether it will give us even provisional answers to them.

Prof. Milne, in the tenth report of the British Association Committee, refers the "world-shaking" earthquakes observed in the six years 1899-1905 to thirteen great earthquake regions, designated by the first thirteen letters of the alphabet. Three of these—I, J, and L—are responsible for only five, three, and two shocks, respectively, and are thus of small importance compared with the others, which average about forty shocks each. Excluding them for the present, the remaining ten regions lie approximately in two rings on the earth's surface, a configuration which is most strikingly apparent when the regions are marked on a globe. The more important ring includes the following seven regions: A, (Alaskan coast,) B, (Californian coast,) C, (West Indies,) D, (Chilean coast,) M, (south of New Zealand,) F, (Krakatoa region,) and E, (Japan.) Its centre is among the conspicuous group of islands which includes Tahiti, and the radius of the ring is about 65 degrees. The other ring has its centre at the opposite point of the earth, which is in the Sahara Desert; and at a radius of 50 degrees from this centre lie regions G, (between India and Madagascar,) H, (the Azores,) and K, (Tashkend.) Now, this is not merely a convenient geographical summary, but a physical fact of vital importance, according to recent researches by Prof. Jeans.

In a remarkable paper read before the Royal Society in 1903 he gave reasons for believing that the earth is by no means a sphere or a spheroid, as we have been accustomed to think, but is of a pear shape.

Under gravitational stress it is continually approaching the spheroidal form—the pear is being crushed into a sphere by its own attraction, and the result is a series of earthquakes. These naturally occur in the weakest places, and if any one will experiment in crushing a pear toward a spherical shape, or even draw a diagram and consider where the weakest points would be, the reasons for the existence of two rings of greatest weakness will readily suggest themselves. The ends of the pear are the centres of these rings, one in Africa, one in the Pacific, and when once this is pointed out the pear shape of the earth is, according to Prof. Sollas, "obvious to mere inspection; it is a geographical fact and not a speculation." Prof. Sollas is indeed responsible for the particular suggestion above sketched, for Prof. Jeans had originally proposed a different axis, which he withdrew in favor of the obvious improvement. The confirmation of Prof. Sollas's view from the distribution of earthquake centres is remarkable. It does not seem, however, quite certain which is the blunt end of the pear; it has been hitherto placed in Africa, but there seem to be several reasons for regarding Africa as the stalk end. This point cannot, however, be dealt with here. The important thing is that there seems to be a real reason for the occurrence of earthquakes in these particular regions, and that they will probably continue to occur there. Prof. Jeans's conclusions have recently been examined by Lord Rayleigh, who announced at the Royal Society only a few weeks ago that he found them generally confirmed, and that we must regard our earth as at present in a state far from stable.

The lessons to be learned from the distribution of earthquakes in space are accordingly tolerably plain in theory, though in practice we may not be able to take advantage of them. If we would be particularly safe from earthquakes we must take up our abode near one of the ends of the pear—either in Africa or in the Pacific. There is also a region of safety between the two dangerous rings—in America generally, for instance, excluding the West, or in Siberia. But the dangerous regions include so vast and so valuable a part of the earth's surface that it is impracticable to leave them unoccupied. Moreover, our knowledge is as yet not specific enough. In the dangerous regions themselves, some parts are much more dangerous than others; for instance, Japan, which is reckoned above as a single region, can be divided into at least fifteen distinct seismic districts. As observations are accumulated we may be able to make similar partitions of the other regions. For the present the general attitude toward earthquakes will probably be similar to that toward other dangers, such as those of travels and voyages for instance; the risk must be incurred. We know that there are at times fatal tornadoes, but other interests are at stake, and we put to sea in the hope that none will occur during our voyage.

We come to the second point, the distribution of earthquakes in time. Are there seasons of special activity such as the recent occurrence of several disasters seems to suggest? Here our knowledge is slighter still, and the observed facts have not yet been co-ordinated by a mathematical investigation. Still there seems to be some evidence in support of the view that exceptional irregularities in the rotation of our earth may be responsible for an increased number of earthquakes at particular times. That the evidence is slight must be attributed to the shortness of the time during which it has been possible to obtain it, and not necessarily to inherent weakness in the evidence itself. The brevity of the earthquake record has been mentioned above; that of irregularities in the earth's rotation is longer; but the discovery that such irregularities existed was made only twenty years ago, though the phenomenon was then traced back through the old observations. The irregularities are systematic in character, and the law governing them is approximately known already; so that, if the presumed connection between them and earthquakes is confirmed, we may be able to predict periods of great earthquake frequency. Such periods would be in some respects analogous to the times of Spring tides.

It is a familiar fact that at new and full moon the tides are much greater than when the moon is at the quarters. The reason is that we have two tide-raising bodies, the moon and the sun, which sometimes act in concert, and then we get large tides; sometimes in opposition, and then we get small tides. If the influence of these two bodies were more nearly equal, instead of the moon being so predominant a partner, we can imagine times when the tides would be barely perceptible. Similarly there are apparently two contributors to the variation in our earth's rotation, which sometimes act in unison and sometimes in opposition. They are more nearly equal in influence than are our moon and sun, and consequently there are times when these two contributors nearly balance one another and the axis of rotation remains almost steady. But in due time the contributors reinforce one another and the axis acquires a considerable "wobble." Each end of the axis then describes a curve composed of wide sweeps and sharp bends, and the evidence seems to be that at the sharp bends we are particularly liable to earthquakes.

The exact statement of the case as given by Prof. Milne in his Bakerian lecture, "Recent Advances in Seismology," delivered before the Royal Society on March 22 last, is as follows:

"In a period of nearly thirteen years (1892 to 1904) I find records for at least 750 world-shaking earthquakes, which may be referred to three periods continuous with each other, and each two-tenths of a year, or seventy-three days' duration. The first period occurred when the pole movement followed an approximately straight line or curve of large radius, the second equal period when it was undergoing deflection or following a path of short radius, and the third when the movement was similar to that of the first period. The numbers of earthquakes in each of these periods taken in the order named were 211, 307, and 232—that is to say, during the period when the change in direction of motion has been comparatively rapid the relief of seismic strain has not only been marked, but it has been localized along the junctions of land blocks and 'and plains, where we should expect to find that the stress due to change in direction of motion was at a maximum. Until the magnitude of these induced stresses has been estimated it would be premature to assume that the frequency under consideration is directly due to change in direction of pole movement, it being quite as likely that both phenomena may result from a general cause."

It is eminently to be desired that a mathematical investigation of the point should be undertaken; but the difficulties are very great, and as yet no one has had the time and courage to attack them. It will be seen, then, that the seismologist is as yet not able to give forecasts of any commercial value, though he is by no means without hope of doing so.

There are, however, some lessons of immediate practical importance which have been learned by seismological study; we may again quote from Prof. Milne's Bakerian lecture:

"At the Imperial University of Tokio a platform was constructed which by means of powerful machinery could be made to reproduce earthquake motion of varying intensity. On this table large models of masonry, wood, and metal designed to resist expected seismic accelerations were tested. This table has been to the builders in Japan what a testing tank in a dockyard has been to constructors of large vessels. The ultimate result of these and other investigations has been to modify and extend the rules and formulae of ordinary construction, and now in Japan, as opportunity presents itself, new types of structure are springing up. These have withstood violent shakings which have materially damaged ordinary types in the neighborhood. While much has thus been done to reduce the loss of life and property, the Japanese Government, stimulated by the results of this experience, has been encouraged to extend its support to seismological investigations in general.

"In 1886 the Chair of Seismology was established at the Imperial University, and since 1892 there has been in existence a Seismological Investigation Committee, which has already issued seventy quarto volumes. At the Central Meterological Observatory in Tokio records are received from nearly 1,500 observing centres."

From these paragraphs it will be seen that there are questions which merit the close study of engineers and architects whose work lies in the dangerous regions, though but little attention has been paid to them except by that wonderful little people who have already taught us more than they learned from us. It is some consolation, doubtless, to reflect that modern seismology owed its origin to Englishmen.

It will, no doubt, be objected to this comparison that an important consideration has been omitted. Seismological questions are of urgent practical importance in Japan, but not in England. That is true, and we all hope that it may remain true; but our guarantee is not absolute. Whether the regions of danger are permanent or shifting is just one of the questions which the whole world is interested in answering, and which can be answered only by patient and laborious research. The British Isles are far from being in a specially safe region; in fact, they lie almost exactly on the smaller dangerous circle above mentioned, through Tashkend, the Azores, and the Indian region; and though earthquake activity seems to be at present limited to these three regions, and so far as it strays in our direction seems to find an outlet rather beyond us, (in the region labeled J by Prof. Milne, between Iceland and the North Cape, where three earthquakes were recorded in six years,) we have no right to assume that this state of things is more than temporary.

During the last year or two, however, more has been done in Europe generally to follow the lead of Japan; international co-operation in seismological work has been organized in Germany, and though the adhesion of some important countries is not yet certain, owing to various difficulties which need not be noticed here, it is hoped that these may be smoothed away in time. If so we may look forward to a welcome strengthening of the corps of workers in seismology, though there is still more than enough work for them all to do.

May 7, 1906

SUN'S PULL ON THE EARTH.

Prof. Michelson Measures Distortion of the Latter's Surface.

CHICAGO, Nov. 30.—The periodic distortions of the earth's surface caused by the sun and moon are being submitted to the most exact measurement in scientific history in experiments at the observatory at Lake Geneva, Wis., under the direction of Albert A. Michelson, professor of the department of physics at the University of Chicago.

Prof. Michelson, whose discoveries won him the Nobel prize in physics in 1907, told of the success of the tests in a paper at the meeting of the American Physical Society at the Ryerson Laboratory yesterday. Physicists assert that his announcement was the most important of the year in their field.

Prof. Michelson has found that the rigidity of the earth is virtually that of steel, and that the surface of solid earth is distorted by the action of the sun and moon about one-fourth as much as water.

The chief apparatus for the experiments is a tube 500 feet long and 8 inches in diameter, half filled with water, and sunk 6 feet in the ground. As the sun and moon draw the water to one end of the tube or the other, the difference in the level is measured with instruments of extreme delicacy. The average change in level between the two ends had been found to be one-thousandth of an inch. The accuracy of the measurement has been carried to 1 per cent. of this fraction, a degree never before achieved.

The tests at Lake Geneva are still in progress, and Prof. Michelson hopes to achieve other results soon.

December 1, 1913

RADIUM UNCOVERS NEW CLUES TO EARTH'S AGE

Scientists Using Radioactive Elements as Time Clock Now Estimate That Certain Rocks Were More Than 1,100 Million Years in the Making

IN pursuing their researches into the age of the earth, scientists have found a "clock" in the radioactive elements, which are constantly engaged in keeping a material register of geological time. The fascinating possibilities of this investigation are revealed in the accompanying article.

By DR. ARTHUR HOLMES,
Of the University of Durham, England.

FOR thousands of years man, with the insatiable curiosity of his kind, has been trying to answer the bold question: How old is the earth? Cicero relates that the venerable priesthood of Chaldea held a belief, based on its system of astrology, that the earth had already existed for over two million years. Zoroaster more modestly estimated the duration of the world's existence at 12,000 years; while the Hebrew chronological tables as interpreted by Archbishop Ussher indicate that the creation of the world took place in the year 4004 B. C.—at 4 o'clock in the morning, according to a profane wag of modern times. Philosophically opposed to these limited ideas of a definite beginning, the old Brahmins of India regarded time and the earth as eternal.

Thus, when the attempt to solve the problem first became a definite scientific aspiration, the rival guesses already lay between infinity and a few thousand years. More than a century ago geologists had come to realize that the earth must have had an inconceivably long history, expressible not in thousands but in millions of years. The record of that history is written in page after page of stratified rocks, and the story disclosed by their study is one of successive changes of life and scene of the most impressive kind. Yet all around us is evidence of the extreme slowness with which similar formations are being accumulated at the present day.

Processes Now Going On.

Here we have a clue to one method of attempting to answer our question. The continents of the present day are continually being worn away by wind and rain, glacier and torrent; and the rivers, more or less heavily charged with alluvial matter, carry the rock waste down to the sea, where the broad ribbons of sand and mud that everywhere fringe the lands are being slowly laid down. The colossal hourglass of rock destruction on the lands and rock formation on the sea-floor runs unceasingly.

The immense systems of strata that have already been built up in this way are shown in order in the accompanying diagram, and it will be seen that their maximum thickness reaches the almost incredible total of 500,000 feet. How long is it since the earliest pages of these volumes of earth history were written? How long since land and sea began their eternal struggle for supremacy? How long since life first appeared and began its mysterious procession through the ages?

If we knew how many years each foot of sedimentary rock took to grow, then these questions could be answered. The "big" trees of California record their ages by annual rings which can be counted when the trees are cut down, and some of the oldest date back more than 4,000 years. Unfortunately, annual rings are found in very few rock formations, and those that do occur show that the formations have generally grown outward like a railway embankment. The first part of the embankment may have been dumped down in a day or two, but a long stretch may have taken months to complete without growing any thicker. For the same reason it is difficult to judge the length of time represented by the whole extent of a formation.

However, to get a preliminary idea of the immensity of time that has to be grasped, we may take the work of the Nile as an example. Since the reign of Rameses II, over three thousand years ago, one foot of sediment has been deposited at Memphis in Lower Egypt every 400 or 500 years. On this scale the accumulated sands and mud in the geological hourglass would represent about 200 million years. We shall see later that even this result is far too low.

Another hour-glass method is based not on the sands of time but on the salt of the sea. Long ago the Royal Astronomer Halley suggested that the oceans had become salt because of their retention of the saline particles annually added to them by the rivers. With prophetic insight he added, "We are therefore furnished with an argument for estimating the duration of all things."

To apply Halley's argument to the age of the oceans two factors are necessary: First, the total amount of salt, or, more accurately, of sodium, that has now accumulated in the ocean waters; and, second, the rate at which sodium is removed from the lands and carried in solution to the sea. These quantities are tolerably well known from analyses of rocks and natural waters in all parts of the world.

Every year more than 8,000 million tons of material are transported by rivers to the sea, and of this material the sands and muds laid down on the sea-floor contain about 0.8 per cent. of sodium. But the parent rocks on the lands themselves had 1.3 per cent. of sodium, and the difference, 0.5 per cent., thus represents the sodium that has passed into solution and become more or less permanently added to the oceans.

Thu we arrive at the rough estimate of 40 million tons as the annual addition of fresh sodium at the present day. The amount of sodium already accumulated is more than 12,000 million million tons and evidently, at the existing rate of supply, this must have taken over 300 million years.

A Steady Hour-Glass.

Unfortunately, these simple geological methods are based on the assumption that the hour-glass of sediment and salt has been running at the same unvarying rate throughout all the vicissitudes of earth history. In recent years it has come to be believed, for many reasons, that the present rates are far higher than those that have prevailed during the average conditions of past ages.

The continents are now more elevated and the lands more widespread than they have generally been, since we are just emerging from one of the world's greatest periods of mountain building and uplift. Only at rare intervals have there been such gigantic ranges as the Alps, Himalayas and Rockies of today, and thus the wearing down of the lands is now exceptionally rapid.

Moreover, human activities have speeded up the rate in all sorts of ways that can never have happened before. The removal of forests, the tilling of soils, engineering excavations and the fumes of chemical works all contribute effectively to the destruction of the lands. As a result of these considerations, and others of a similar kind, we may now feel confident that the estimate of 200 or 300 million years already reached is many times too small, and further than this the geological methods cannot go.

The early geologists were likely to be somewhat casual in their demands on the bank of Time. More than half a century ago Lord Kelvin invaded the mists of speculation, hoping to dispel them with a mathematical wand. He successfully calculated the time that must have elapsed since the earth's crust solidified from its original molten state and he announced that unless the earth were herself generating heat in some unknown way that period could not exceed 40 million years. The long controversy that raged around this result is one of the most familiar of the fierce scientific battles that enlivened Victorian times.

THE RECORD OF THE ROCKS

MAXIMUM THICKNESS		RADIOACTIVE TIME SCALE	LIFE SCALE
4,000 FT. - EUROPE	RECENT & GLACIAL		MAN
34,000 FT. CALIFORNIA	UPPER TERTIARY	33 MILLION YEARS	MAMMALS AND FLOWERING PLANTS
35,000 FT. ITALY AND WYOMING	LOWER TERTIARY	56 MILLION YEARS	
46,000 FT. WESTERN STATES	CRETACEOUS		AGE OF REPTILES AND WELL-KNOWN BIRDS
20,000 FT. ALASKA	JURASSIC		
20,000 FT. ALPS AND U.S.	TRIASSIC		
13,000 FT. AUSTRALIA	PERMIAN	185 MILLION YEARS	AGE OF COAL FORESTS AND AMPHIBIANS
40,000 FT. BRITAIN AND U.S.	CARBONIFEROUS	218 MILLION YEARS	
37,000 FT. BRITAIN	OLD RED SANDSTONE OR DEVONIAN	263 MILLION YEARS	AGE OF FISHES
15,000 FT. BRITAIN	SILURIAN		AGE OF TRILOBITES AND OTHER INVERTEBRATES
40,000 FT. AUSTRALIA	ORDOVICIAN		
40,000 FT. BRITISH COLUMBIA	CAMBRIAN		
AT LEAST 50,000 FT.	UPPER PRE-CAMBRIAN	550 MILLION YEARS	AGE OF PRIMITIVE WATERPLANTS
AT LEAST 30,000 FT.	MIDDLE PRE-CAMBRIAN	845 MILLION YEARS	PRIMITIVE LIFE
		1050 TO 1100 MILLION YEARS	
AT LEAST 90,000 FT. (BASE NOT KNOWN)	LOWER PRE-CAMBRIAN OR ARCHAEAN		THE DAWN OF LIFE
TOTAL AT LEAST 514,000 FT.			

THIS DIAGRAM SHOWS TO SCALE THE MAXIMUM KNOWN THICKNESS OF ALL THE GEOLOGICAL FORMATIONS. THE ACTUAL AGES IN YEARS, READ FROM RADIOACTIVE TIME-KEEPERS, ARE SHOWN ON THE RIGHT.

Despite Kelvin's great authority and apparently unimpeachable reasoning many geologists entertained the forlorn hope that some flaw might yet be found in his method. Their optimism was justified beyond all expectation by three surprising discoveries made in the early years of the present century. The first of these was the discovery of radium by Mme. Curie; the second was the discovery by her husband that radium is constantly giving out heat, and the third was the discovery by Lord Rayleigh that radium is widely distributed throughout all rocks.

With the advent of our knowledge of radioactivity in rocks the suspected flaw in Kelvin's method was revealed, for it was proved that the earth could no longer be regarded merely as a simply cooling body.

The radioactive elements belong to two families having uranium and thorium as their parents, the more familiar radium being one of the daughter elements of the uranium family. Now, although the amounts of these elements in the rocks are very small, yet they are so widespread and so unceasingly active in giving out heat that it has become difficult to believe that the earth can cool down at all except by volcanic processes on a far greater scale than any of which we have experience.

To pursue this fascinating topic further would carry us away from our subject, and it must suffice to say that Kelvin's estimates had to be abandoned with dramatic suddenness, to the great joy of the geologists. If the earth has indeed cooled down from a molten state—and there are no convincing reasons against that traditional view—then the time taken must have been immensely longer than 40 million years.

And now we enter on the most interesting phase of the story. Radioactivity destroyed one argument, but it suggested a new one. It made possible the most elegant and refined method of measuring geological time that has yet been devised. Just as a clock is kept going by the energy of a spring that slowly unwinds itself, giving out its energy as it runs down, so the radioactive elements, uranium and thorium, are maintained by the internal energy of their atoms. As they decay, atom by atom, that energy is liberated in a form that mainly appears as heat.

But while a clock records time in ticks that are heard and pass away, the radioactive elements are engaged in keeping a more material register of time. Uranium and thorium steadily evolve the light gas, helium, and the heavy metal, lead, as stable end products, and these, as they decay no further, necessarily accumulate as time goes on. Every fresh radioactive mineral (such as pitchblende, one of the chief ores of radium), is now regarded as a natural chronometer, registering time by the atoms of helium and lead that are unceasingly produced within it year after year.

Lord Rayleigh measured directly the rate of formation of helium in various minerals, and he was able to show that a gram of uranium generates one cubic centimeter of helium in nine million years and that thorium takes about three times as long to produce the same volume of gas. In the case of minerals from the very old rocks of Ontario Lord Rayleigh found that they contained 600 million to 700 million times as much helium as the amount that could accumulate within them in a single year.

After our experience with sediments and salt, the reader will realize that it is no longer safe to jump to the simple conclusion that these minerals must be 600 or 700 million years old. Once again the true age must be greater than the figure actually reached.

Elements on the Witness Stand.

Helium is a gas, and it has been experimentally demonstrated that as soon as a radioactive mineral is exposed to the air it begins to lose part of the concentrated gas stored within it. When it is powdered for analysis, naturally still more leaks away. Consequently the helium actually found in a mineral can be only a fraction of the total amount generated during the lifetime of the mineral. We must therefore be prepared for ages even higher than 700 million years.

Fortunately lead, the other element produced by the radioactive elements, is much less likely to escape. From the rate of production of helium it can easily be calculated that one gram of uranium generates lead at the rate of one gram in 6,600 million years. Thus, if we know by chemical analysis that the percentage of uranium in a mineral is U, and that of the lead derived from it is Pb, then the time taken for that lead to accumulate is given in millions of years by the formula 6,600 Pb divided by U.

But thorium also produces lead, and as both the radioactive elements are generally present in suitable minerals they must both be taken into consideration. If Th be the percentage of thorium, then the formula for the age in millions of years becomes 6,600 Pb divided by (U+0.37 Th).

Can we be sure that the radioactive clocks have always been recording time at the same steady rate throughout their long history? There is very sound technical evidence to prove that we can. No temperatures or pressures that can be imposed on radioactive substances affect their behavior in the slightest degree, and there is visible record in the rocks themselves that the atoms in the earth's earliest ages behaved in exactly the same way as those of today.

As characteristic examples of time measurements let us consider a few actual results.

In the mines of Gilpin County, Colorado, pitchblende occurs which was crystallized there in early Tertiary times. Its analysis gives an age of 56 million years. In Europe, the famous Joachimsthal mine, from which the silver of the original "thalers" or dollars was taken, now produces pitchblende having an age of 187 million years. This mineral is of special interest, as it dates from a time just after the great coal fields of the carboniferous period had been deposited. It is perhaps 200 million years since most of the world's coal resources were stored up, yet in this extravagant age we shall use up nearly all the available coal in a millionth of the time.

Going back to still earlier ages, we find in Ceylon, Brazil and Central Africa minerals that are all about 550 million years old. Returning to America, the oldest rocks that have so far provided suitable minerals are found in Colorado, Texas and Ontario; all of these have an age of 1,050 to 1,100 million years, and the same age is given by the minerals of similar rocks of Norway, Sweden, India and East Africa.

We seem now to have reached nearly the beginning of the earth's history, but even before these old rocks ascended from the depths there were others already formed at the surface, such as the iron ores mined in the Vermilion Range of Minnesota. Thus the age of the earth must be well over 1,100 million years.

"Sooner or later radioactive clocks in the form of minerals from the very oldest rocks will be orthcoming, and a still nearer approach to the true age will be possible. The next few years are likely to see a very great advance in this fascinating subject, for many American geologists and chemists are now cooperating in the study of radioactive minerals, the National Research Council having appointed an active Committee on the Measurement of Geological Age, under the Chairmanship of Professor A. C. Lane.

Evidence in the Stars.

Although geologists can still find "no vestige of a beginning," astronomical considerations have shown in recent years that the new demands on geological time, so much greater than were thought permissible twenty years ago, are by no means excessive. Dr. Harold Jeffreys has attempted to calculate the age of the solar system itself from the shape of the orbit of Mercury compared with those of the other planets.

He finds that the time required is between 1,000 and 10,000 million years, the most probable figure being about 2,500 million years. And passing far beyond the bounds of thought, Dr. J. H. Jeans has recently suggested that the ancestral sun has been shining through the universe for something like seven or eight million million years. Judged by that appalling standard even the earth is young.

If, for the sake of familiar comparison, we liken the age of the sun to its distance from the earth, then the age of the earth would be represented by the distance from New York to Palm Beach. Man himself has probably already a million years of history behind him, yet on this scale of comparison the length of a train of ten or a dozen coaches would amply express the whole lifetime of humanity. A single human life would be no more than the thickness of a carriage window.

With Keats we have looked through magic casements into the faery lands of the past, and we have traced the earth backward through time until it recedes into an inconceivable remoteness, far beyond the bounds of investigation possible to the geologist. Yet, as Swinburne expressed it, man can "see through the years flowing round him, the law lying under the years," and with Alfred Noyes he can declare:

Here, now, the eternal miracle is renewed;
Now and forever God makes heaven and earth.

June 6, 1926

COMPOSITION OF THE EARTH REVEALED BY QUAKE RECORDS

Seismograph Readings Show the Rock Composition Forty Miles Beneath the Surface and Explain "Floating Mountains"

By W. L. LAURENCE.

IN a report on the most complete earthquake survey yet made in the United States, presented recently before the National Academy of Sciences at Berkeley, Cal., Professor Perry Byerly of the University of California, seismologist, told of the most recent attempts to peek into the interior of the earth and to find out what it is made of.

So far man's knowledge of what the inside of his own planet is composed of has been extremely meager and has been based mostly on conjecture. In fact, we know much less about the interior of the earth a little more than a mile below the surface than we know about the distant nebulae 200,000,000 million light years away. We gaze on these far-off stars through the 100-inch telescope of the Mount Wilson Observatory, study their composition with the spectroscope, even measure their temperature by means of the newly perfected thermocouple. But as for the interior of the earth we have so far not been able to devise any way of probing its mysteries.

Professor Byerly told of the first means devised to penetrate to depths as far as the very "core" of our terrestrial sphere.

Earthquakes in California, he said, are definite indications that the Sierra Nevada Mountains are still growing, being continually pushed up to greater heights by the floor of the Pacific. For this reason, he added, there would of necessity be more earthquakes in California from time to time.

Messages From the Earth.

The new tools devised by science to peep into the insides of the earth, Professor Byerly stated, are the various kinds of waves sent out by earthquakes and registered on seismographs. The earthquake waves are the only tourists which visit the interior of the earth and return to its surface. Each particular kind of earthquake wave writes its own record on the seismographs, and to those versed in that particular alphabet they are just like codes cabled from the very bowels of the earth.

From these messages from the deep, Professor Byerly reported, has it been possible for the first time to determine fairly definitely what the crust of the earth in California consists of to a depth of at least forty miles. There are, at least, three layers in its upper part. The first is an eighteen-mile-thick layer of granite. Below it is a basaltic layer of a thickness of about twenty-one miles. Below that is a third layer of ultra-basaltic rocks of undetermined thickness.

The report describes two different kinds of earthquake-waves which tell the tale of the deep. Fastest of these, and the first to arrive to the recording instrument, are called Primary, or P-waves. The waves to come next, slower than the first, which arrive by another path and are completely different in the nature of their vibrations, are known as secondary, or S-waves. Each one of these tells its own story, and when the two stories are put together there stands revealed a tale never told before.

The Waves Analyzed.

From studying the times of emergence of the various groups of waves at the earth's surface, as well as the type of vibration, it is possible to calculate the path the wave has taken and its speed at various depths. When the type of wave and its speed become known scientists can determine certain properties of the medium through which it has traveled at various depths, such as "elastic constants" and density. From knowledge of these properties in rocks at the surface and experiments on them at high pressure it becomes possible to form some estimate of how they would act at great pressure and temperatures such as must exist in the earth's interior.

The latest studies of Professor Byerly were centred about the earthquake last Thanksgiving Day in the Sierras of East Central California. This earthquake was in itself not a very serious one. No one was hurt, and the greatest damage was the cracking of concrete reservoirs and the breaking of dishes. A large landslide occurred during the earthquake in a mountain canyon near the boundary of Inyo and Fresno Counties, and it so happened that an airplane observer was photographing the region as the landslide was going on. These unusual photographs are now being studied by Professor J. C. Jones of the University of Nevada.

The difference in the time of arrival of the P and S waves at seismographic stations during that earthquake, Professor Byerly stated, allowed a calculation of the depth of the source of the earthquake and also of the thickness of the two layers through which the waves had passed.

The depth of the source was computed to be about two or three miles, while the source itself was located at a point about fifteen miles east of the northwest corner of Inyo County. The thickness of the upper layer was found to be about eighteen miles and that of the second layer about twenty-one miles.

The speed of the P-waves in the upper layer is about three and one-half miles per second and that of the S-waves about two miles per second. These speeds are reasonable for granite, and since granite seems to be the basement rock it is natural to conclude that the upper layer is made of granite.

In the second layer the speed of the P-waves is about four and one-half miles per second, and while no S-wave was observed in this layer it was still possible to compute the speed of S from the known speed of P. This computed speed is two and one-half miles per second, which agrees with the speed fixed by Professor Byerly last Winter for the surface layer under the Pacific Ocean. This layer under the Pacific, it is generally agreed among geologists, is not granitic but basaltic. Hence it was concluded from the similarity of the speeds of the S-waves that the second layer under the Sierras is also basaltic.

The medium lying under the second layer offers a speed of five and one-half miles per second to the P-waves and a speed of three miles per second to the S-waves. From this it is concluded that the third layer is composed of rocks of ultra-basaltic type. The speed of the S-waves in this third layer is the same as that found by Professor Byerly for the second layer under the Pacific.

Core of the Earth.

The S-wave does not go through the "core" of the earth. Nor can it travel through a fluid. From these two facts European seismologists have been led to conclude that the core of the earth is made up of some fluid. Fast P-waves, on the other hand, can be transmitted through the earth's core, and from this seismologists have calculated that it extends from the centre of the earth to about half way from the surface.

Whatever the material of the core, it is evidently different than the outer portions of the earth. The high density has led geophysicists to believe it to be a heavy metal, and since iron is so common in meteorites and on the earth it has been thought probable by many that the core may consist of iron. This core acts as a huge lens to the P-waves, bringing them to a focus similar to light waves. The "image" of the earthquake source, however, would be above the surface of the earth at the antipodes, due to "aberration" of the core lens.

Under the theory of isostasy, now being widely used by geologists to explain the way the crust of the earth is arranged, there is a certain depth below the earth's surface at which the pressures are equalized at all points. At that depth, geologists believe, there is a very slow "flow" of the dense, solid materials of the rocks.

It is as if the earth at that depth were made of a special kind of cold tar or sealing wax. The hardest sealing wax will flow and change its shape if it is subjected to a pressure over a period of several years. Under the theory of isostasy a square-mile chunk of rocks and crustal material, taken any place on the earth from the surface down to this isostatic layer, must have the same weight, or at least strive to have the same weight.

According to this hypothesis mountains actually float on a sea of rock. They and the material below them must, under the theory of isostasy, be made of lighter material than the bottoms of oceans. This idea is well borne out by measurements of the earth gravitational pull, which is greater over the sea than over the land, especially mountains.

Professor Byerly has applied this theory of isostasy to determine whether the earth is still growing in California. In other words, whether the crust there is in isostatic balance.

Through earlier computation he first determined that the thickness of the supposed basaltic layer under the Pacific is about twenty-four miles. This estimate, and the figures he obtained from the Thanksgiving Day earthquake in 1929, allowed him to compare the weight of an area of the Sierra Mountains, from their peaks down to their assumed level of isostatic flow below the crust, with the weight of a similar area of material under the Pacific, taken to the same depth.

The figures corroborated the theory of "floating mountains." The weight of the Sierra column was found to be definitely lighter than the similar area under the ocean in spite of the fact that it sticks out high in the air. From this Professor Byerly concludes that the Sierras must rise still further than they have so far, in order to place the crust in isostatic equilibrium.

This means, Professor Byerly explained, that the earth is still growing in California, that the mountains must rise still higher, and that earth stresses must build up. And as these stresses are relieved from time to time there will, of necessity, be more earthquakes in California.

November 2, 1930

FIND MOON CHANGES EARTH DISTANCES

Scientists Report 63-Foot Shifts Between Continents of America and Europe.

GRAVITATIONAL PULL CITED

Drs. Stetson, Loomis Reveal at Harvard Results of Clock Checking Here and Abroad.

Special to THE NEW YORK TIMES.
CAMBRIDGE, Mass., March 2.— The gravitational pull of the moon apparently creates tides in the solid earth which change the distance between North America and Europe as much as sixty-three feet, according to Dr. Harlan T. Stetson, visiting professor at the Harvard Institute of Geographical Exploration, and Dr. A. L. Loomis of the Loomis Laboratory at Tuxedo Park, N. Y.

The effect of such an earth tide larger than could have been expected was detected when Dr. Stetson and Dr. Loomis found that discrepancies in astronomically checked clocks in Europe and in North America increased and decreased regularly with changes in the moon's position. Clocks are checked astronomically by comparison with the movement of stars across the meridian, a semi-circle running through north and south and a point directly over the observer's head. If discrepancies are found in two clocks, thus checked by the stars, then the position of either one or the other of the stations appears to have moved east or west, thus changing the observer's meridian.

First Discoveries.

The scientists discovered that the discrepancies between American and European clocks, at astronomical stations where the checking was done regularly, moved either east or west of their normal positions. When the position of the moon caused both stations to move apart, the average distance between them might be increased by about thirty-two feet, they found. If the moon caused them to move toward each other, they might be nearer together by the same amount.

United States time signals checked at Washington are broadcast from the naval station at Annapolis, Md.; English time signals, checked at Greenwich, are broadcast from Rugby, and French time signals, checked at Paris, are broadcast from Bordeaux. At specified times, each station picks up the signals of the other two.

Discrepancies noted between Annapolis time signals and those from Rugby were seen to rise and fall with the moon's position. Almost exactly the same curves of rise and fall applied to the transmission between Annapolis and Bordeaux. But between Rugby and Bordeaux, no such relationship was seen. Evidently something takes place over the Atlantic which does not take place between England and France.

Divergencies With the Moon.

The two scientists found that when the moon was north of the equator, the continents were about 32 feet closer together than normally when the moon crossed the meridian, and that they spread apart until, when the moon's hour angle was about 14, the continents were about 32 feet further apart than on the average.

When the moon was south of the equator, exactly the opposite was true, the continents moving together when the moon was crossing the meridian, and traveling apart gradually until the moon's hour angle was about 12.

With the moon on the equator, Drs. Stetson and Loomis discovered that the continents moved back and forth twice during the moon's 24-hour day. The continents never moved as far apart or so close together as when the moon was north or south of the equator.

The scientists point out that as far as the elasticity of rocks is concerned, it is well within the realm of possibility for the continents of Europe and North America to move apart as much as 63 feet. Such a movement would be equivalent to stretching a yard-long rock less than .0004 inch, an amount well within the elastic limit even of solid granite.

March 3, 1935

CALL EARTH'S AGE 2½ BILLION YEARS

Wilkins and Rayton Report to Scientists, at Rochester, on Extensive Study.

'INFANCY' A LONG PERIOD

Time Before Maturity Into Cold Planet Put at 700,000,000 Through Atomic Method.

By WILLIAM L. LAURENCE
Special to THE NEW YORK TIMES.
ROCHESTER, June 16.—A new "birth certificate" for the earth, which fixes the time of its "creation" from the bowels of the flaming sun at a period dating back 2,500,000,000 years, was described here today at the Summer meeting of the American Association for the Advancement of Science, by Drs. T. R. Wilkins and W. M. Rayton of the University of Rochester.

In the last few years studies of the rate of decay of radioactive rocks have furnished science with what have become known as radioactive clocks for determining the age of the earth. However, these clocks lead science back only to

EARTH SCIENCES

the time when the earth had become a solid body, old enough to have formed its ancient rocks.

At that time hundreds of millions of years had already passed from the time when the earth and the other planets in the solar system had been torn, in the form of flaming, tenuous "cosmic infants," from the body of their mother, the sun.

The new "certificate" for the earth's birth gives science the first definite clues as to the time of the actual coming into existence of the earth as a separate cosmic entity. In this respect it may be said to be the first scientific approach toward fixing the period of "creation day" in Genesis, when the earth was separated from the heavens.

Older Method of Telling Age

The older radioactive clocks are based on the fact that the element uranium, the last and heaviest on the table of elements, slowly disintegrates into lead, over the course of millions of years, at a definite given rate.

The lead developed from uranium has a different atomic weight from ordinary lead, and by measuring the amount of uranium-lead present, and knowing the rate at which the transformation takes place, the scientist is enabled to determine how many years have passed from the time the rocks still contained 100 per cent uranium and no lead.

From these determinations, obtained by a number of independent investigators, the age of the most ancient uranium rocks, and thus the age of the solid earth itself, has been found to be no older than 1,852,000,000.

On the other hand, the first "birth certificate" now sets the earth's birthday at a period 2,500,000,000 years back. According to this, therefore, the period of earth's infancy, childhood and adolescence, before it finally matured into a cold, solid planet, lasted something like 700,000,000 years.

Results of Long Research

The new earth birthday is the result of many years of research at the University of Rochester by Dr. Wilkins, who developed many new delicate measuring devices for probing secrets that had been hidden away in the hearts of uranium atoms from the time when the earth came into existence as an individual member of the cosmic family.

These tell-tale atoms may thus be said to constitute the first "eye-witnesses" to the cosmic drama of the creation of the solar system.

At first it was believed that all the uranium atoms were all of one kind. With the discovery within recent years of the existence of isotopes, that is, chemical "twins," which are identical as to the number of units of electrical charge in their nuclei but differ in their atomic weights, evidence began accumulating that uranium also had twins.

The uranium twins, however, are so much alike that it had proved very difficult to tell them apart. In the devising of means for telling them apart, Drs. Wilkins and Rayton reported, lies the key to the original "birth cry" of the earth.

Each of the uranium "twins" gives off its own brand of alpha particles, or hearts of helium atoms, which constitute one of the characteristic radiations of most of the radioactive elements. For, while the alpha particles given off by the various radioactive elements are all the same in their constitution, being composed of two neutrons and two protons, the range of the alpha particles varies in accordance with the decay period of the element from which it is emitted.

Each radioactive element, by constantly giving off emanations and thus radiating away part of its matter, is degenerating into lead at a fixed rate, some relatively fast and others, such as uranium, at a relatively slow rate. The time it takes for radioactive elements to radiate away half of the original energy is known as the half-life time for the element. Thus the half-life time for radium is about 1,600 years, whereas the half-life time for uranium is several million years.

More recently scientists have begun to suspect that there exists a definite relationship between the decay period of a radioactive element and the range of the alpha particles it emits. The evidence pointed to the fact that the longer the range of the alpha particles the shorter the decay period. In other words, the more energetic radioactive elements produce alpha particles of the greater range.

By developing new delicate photographic instruments for recording the tracks of alpha particles and taking thousands of photographs of the alpha particles emitted from uranium, Drs. Wilkins and Rayton discovered that, instead of giving off particles of one range, the uranium registered particles of three different ranges.

The three ranges correspond to three different isotopes of uranium. Two of these were known before, but the existence of the third, named actino-uranium, had only been suspected, and it is this actino-uranium, with its rate of decay, the range of its alpha particles and its relative abundance, that furnishes the key to the original separation of the earth from the sun.

Three Lines of Approach

At least three different lines of evidence, Dr. Wilkins reported, pointed to the existence of actino-uranium with a rate of decay ten times that of the isotope of uranium known as uranium 1. This further suggests that "actino-uranium and uranium 1 existed in some fixed ratio in the sun, where they were being formed when the earth broke away."

"Assuming further that this fixed ratio of these isotopes when in the sun was inversely proportionate to their decay constants—the natural assumption for the amounts of elements in equilibrium with a uniform rate of production—it is shown that the age of the solar system can be calculated directly from the present activity ratios of these isotopes and their decay constants," said the report.

"This gives an age of the solar system of 2,500,000,000 years. It leads further to a curve for the variation of the relative numbers of actinium and radium lead atoms in minerals of various geological ages in splendid agreement with observed values."

According to Dr. Wilkins's evidence, there originally existed in the sun one kilogram of actino-uranium for each ten kilograms of uranium 1 and the same ratio of the two uranium isotopes was handed down to the earth when it was torn away from its parent.

On the other hand, the rate of disintegration of the actino-uranium is ten times greater than that of uranium 1. Measuring the relative amounts of the two found today it thus becomes possible, according to Dr. Wilkins, to figure back to "zero time," when the two started on their path of decay.

Ozone Held to Affect Weather

A new rôle for the ozone, that layer of a rare form of oxygen which surrounds the earth's upper atmosphere like a protective blanket against the more energetic ultra-violet rays from the sun, was described by Dr. Brian O'Brien of the University of Rochester.

The data obtained on the stratosphere flights made by the National Geographic Society and the Army Air Corps, Dr. O'Brien reported, furnish evidence that the ozone, in addition to serving as shield for keeping out those ultra-violet radiations of wave lengths that would prove harmful to life, also serves as a reflecting layer for the heat waves radiating away from the earth, thus preventing the heat furnished by the sun from evaporating.

The ozone thus has an effect on the weather, a role previously not known.

Dr. O'Brien reported on experiments with captive balloons in the United States and in Europe, carried on at two different seasons in two different years, which indicate, he said, that the ozone blanket over Europe lies much closer to the earth than the ozone blanket over North America.

Studies on the internal heat of the earth indicate that at a depth of twenty kilometers the temperature is 500 degrees centigrade and that at the center of the earth the temperature is from 4,000 to 5,000 degrees centigrade, according to a report presented by Dr. Louis B. Slichter of the Massachusetts Institute of Technology.

June 17, 1936

EARTH CRUST AGE FIXED AT PARLEY

Geophysical Group Agrees on 4.5 Billion Years at Meeting in Toronto

By WALTER SULLIVAN
Special to The New York Times

TORONTO, Sept. 9—A number of scientists who have sought in different ways to determine the age of the earth's crust found themselves in agreement today. The figure was 4.5 billion years.

They also agreed on this figure as representing the age of the meteorites that occasionally fall on the earth. This suggested that the earth, the other planets and the meteorites were formed at the same time.

They placed the age of the elements forming these bodies at a minimum of five billion years. Some suggested a maximum age of fifty billion years.

In all cases these figures were based on study of radioactive decay in various elements. The oldest method was to measure the proportions of uranium 238 and uranium 235 that had turned into two isotopes of lead. These forms of uranium each decay at a constant rate that is known with considerable accuracy as a result of nuclear research in recent years.

Attend Assembly

Those who expressed their views on the age of the earth are attending the General Assembly of the International Union of Geodesy and Geophysics. They included Prof. F. G. Houtermans of the University of Berne; Dr. George Wetherel of the Carnegie Institution of Washington, and Prof. G. J. Wasserburg of the California Institute of Technology.

The scientists declined to guess how the earth and the solar system were formed, but they estimated its period of cooling, before formation of the crust, as comparatively "rapid" —somewhere between 50,000 and 500 million years.

A study of angles of impact, evident in meteorite craters, has convinced some of the scientists here that these missiles from space were orbiting within the solar system. It is thought that they were formed at the same time as the earth.

Radioactive age determinations date from early in the century, although they have only recently become widespread.

The trend, in recent decades, has been to extend the earth's birthday further and further into the past. Those at the conference here are convinced that they are now close to the true age of this planet, within a margin of 5 per cent.

September 10, 1957

Latest Picture of 'Solid' Earth

The most up-to-date picture of the composition of the earth was discussed last week by Prof. Richard F. Flint and Prof. Matt S. Walton Jr., Yale geologists.

The earth, Professor Walton said, has an average density of about 5.5 grams per cubic centimeter, while the rock that makes up the earth's crust ranges from about 2.5 to 3.4 grams per cubic centimeter. We now know, Professor Walton added, that the ordinary rock of the surface persists only five to ten miles in depth beneath oceans and twenty to forty miles beneath continents. Next comes the underlying part of the earth's "mantle," which continues to a depth of 1,800 miles. Then comes an outer core extending down to a depth of 3,200 miles from the surface, followed by an inner core of a radius of 800 miles.

According to Prof. Keith E. Bullen, of the University of Sydney, Australia, a world authority on the earth's interior, the "mantle" is in an essentially solid state, while the outer core is in a fluid or molten state, and the inner core is essentially solid, with a rigidity about four times that of steel at ordinary pressure.

Higher Densities

Professor Bullen estimated that the earth's density jumps from 5.5 grams per cubic centimeter at the bottom of the mantle, to 9.5 at the top of the outer core and steadily increases to nearly 12, just outside the inner core. The density at the center of the inner core probably lies between 14½ and 18, Professor Bullen estimates.

Pressure at the bottom of the mantle reaches 1,333,000 atmospheres, while at the center of the earth the pressure is nearly 4,000,000 atmospheres. The pull of gravity stays within 2 per cent of the surface value to a depth of 1,500 miles. The value then rises a little until the bottom of the mantle is reached, then drops steadily to zero at the earth's center. — W. L. L.

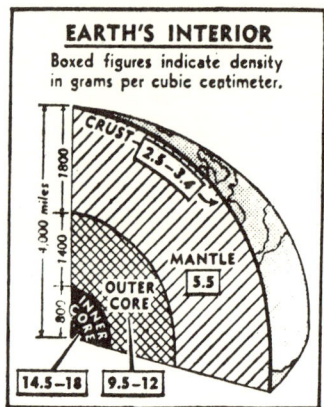

EARTH'S INTERIOR
Boxed figures indicate density in grams per cubic centimeter.

February 23, 1958

Undersea Earth Probe Planned

U. S. Group Studies Feasibility of Boring Into Primordial Crust for Life Secrets

By WALTER SULLIVAN

A committee has been formed by the National Academy of Sciences to prepare a plan for boring below the ocean floor to what is thought to be the primordial crust of the earth, and through it to the mysterious "Moho."

The latter is thought to be the boundary between the crust and whatever lies beneath it. Even in its thinnest areas, the crust is believed to be more than three miles thick. The drilling rig would have to be mounted on a bobbing platform over about three miles of water.

The cost would be great, but sponsors of the project believe the scientific rewards would be even greater. The findings might include a complete history of the earth's climate, the filling of the missing link that makes all of early evolution a mystery, and perhaps clues to the origin of the earth itself.

The chairman of the committee is Dr. Gordon G. Lill, head of the Geophysics Branch of the Office of Naval Research in Washington. The committee held its first official meeting Oct. 3.

Its task is to assemble data for a decision as to whether the plan is feasible. The data include an estimate of the cost, techniques that might be used, possible sites for the hole and what might be learned.

The Moho, as it is popularly known among students of geology, is the Mohorovičić Discontinuity, named for a Yugoslav who identified it from his study of earthquake waves. It is thought to mark the boundary between the earth's crust and the semi-plastic, highly compressed rock beneath it. The crust is thought to consist of two parts. The lower layer, probably of basalt, envelopes the globe. The upper part, of granitic rock surmounted by sedimentary deposits, is limited to the continents.

Hence the ocean floor, where the crust is only a few miles thick, seems to offer easier access to the Moho than the continents, where the crust is twenty-five miles or more deep.

Beneath the crust the earth's interior is subdivided by geologists into three layers. The outer mantle of ultrabasic rock is about 1,800 miles thick. The core, possibly of molten iron and nickel, is thought to be 1,360 miles thick. The inner core, 815 miles in radius, may be a solid form of the same nickel-iron mixture.

Oil companies have developed floating rigs that can drill holes in the ocean floor at water depths as great as 1,500 feet. On land an oil well now being drilled in the Pecos field of Texas is said to have passed four miles.

New Problems Faced

In neither case, however, can the problems be compared to those faced in sinking a well three or four miles into an ocean floor under three miles of water.

The off-shore oil wells are drilled into the oil-bearing deposits of the continental shelf. Even at 1,500 feet the water is shallow compared to the three-mile depth of the true ocean floor.

To the geologist, the continents are not bounded by shorelines, which represent a constantly changing margin, dependent on varying ocean levels. The true boundary, often several hundred miles off shore, is where the continental shelf ends and the bottom drops abruptly, two or three miles, to abysmal depths. From there on seaward, the floor of the ocean is surprisingly flat, except for occasional protrusions.

Upon this floor, presumably since the formation of the oceans, debris has been raining, depositing in successive layers the record of past life and climate—perhaps of past orientations of the earth's magnetic field. Research ships in recent years have been able to bring up cores representing cross-

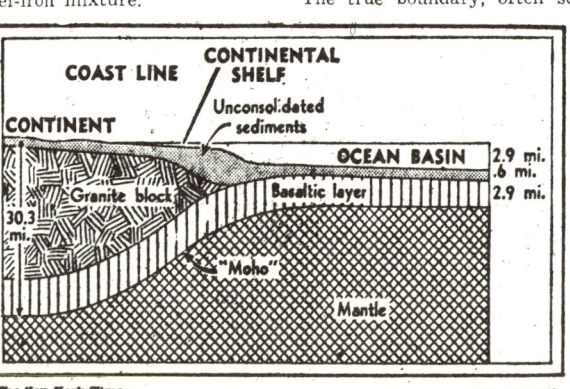

Outer layer of earth's crust can be penetrated more easily from ocean site because of granite barrier under land. Diagram based on drawing by J. Tuzo Wilson of Toronto.
The New York Times — Oct. 20, 1958

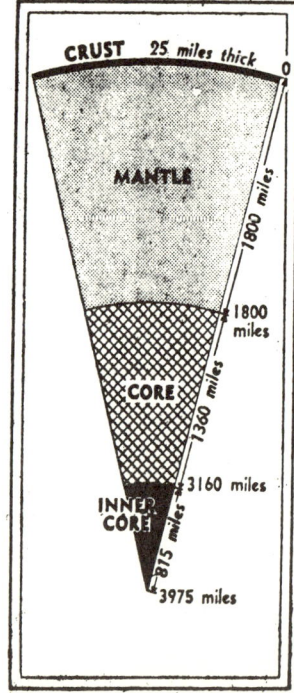

A cross-section of the earth's composition, with heavy line as outer crust.

sections of this sediment to a depth of less than 100 feet.

The full depth of the sediments is thought to be 3,000 or 4,000 feet. They seem to have lain undisturbed. So far as can be learned, no true sea-bottom sediments or sea-bottom rocks have ever been found on continents or islands.

The sedimentary rocks that have given us our knowledge of early life all are thought to represent lake bottoms or deposits on the continental shelves.

For some reason, as yet not understood, the record available in these rocks goes back only to the Cambrian Period. At that time evolution is believed to have reached an advanced stage. Most of the phyla, the basic divisions of the plant and animal kingdoms, are thought to have arisen by then.

There were no vertebrates, but such complex creatures as the trilobites were common. These sea animals, now extinct, had jointed legs and looked somewhat like horseshoe crabs. The ancestry of such animals—the nature of early evolution—is unknown.

It is possible to determine the climate at the time a certain layer of sediment was deposited, because the roster of tiny sea creatures whose shells are found indicates the temperature of the water. Since particles tend to align themselves with the earth's magnetic field as they settle, it is thought by some that ancient locations of the migrating magnetic poles can be determined from sediments. The validity of this, however, is being challenged.

Protagonists of the sea-drilling project hope nevertheless that a cross-section of the sediments will provide a ladder—a framework of time—on which to hang our fragmentary knowledge of the history of life and of climate.

Below the sediments presumably is the surface as it existed when the oceans were formed and the earth took its present shape. Its composition may be a clue to the nature of the moon's surface and to the origin of the earth.

One of the mysteries in the study of earthquake and explosion shocks has been the apparent absence of echoes off the surface beneath the ocean sediments. It has been suggested that it may be owing to its composition—a layer of loose meteoric material, fallen from the skies when the solar system was young.

This may have been the time when the moon, too, was bombarded into its present pock-marked appearance. The meteoric origin of the moon craters is not, however, fully accepted.

In penetrating the oceanic crust, it would be possible for the first time to determine its composition. This is widely thought to be a form of basalt. But how does it differ from the mantle beneath it? Is its origin comparable to that of slag that rises to the surface of molten metal in a smelter?

Or is the rock below Moho chemically similar to that above it, except that immense pressure has changed its molecular state?

Does the radioactivity at these depths, or the heatflow from below them, give any clue to the source of the heat in the core of the earth?

These are some of the questions that advocates of deep drilling believe may be answered.

Site Choice a Problem

They must also find a suitable site. A number of places are known where the bottom seems undisturbed, but the place must also be accessible and free of severe storms. Thus there is an ideal spot near the Fiji Islands, from the bottom point of view, but it lies on the path of typhoons and is extremely remote.

While a figure of $20,000,000 has been mentioned as a possible cost, those connected with the program say extensive investigation will be needed before even a rough estimate can be made.

A year ago a group of four oil companies, known as the Cuss Group, patented an offshore drilling rig that can be operated on a bobbing vessel. The bits are turned by a string of drill pipes flexible enough to allow for tossing of the craft. The hole in the ocean floor is funnel-shaped to simplify the problem of re-seating the portion of the rig that rests on the bottom.

Presumably a device of this type will be used for the deep hole, but it is expected that several test bores will be needed to work out difficulties before attempting a boring to Moho.

October 20, 1958

SATELLITE SHOWS EARTH PEAR-LIKE

Physical Parley Told That Globe Is Stronger Inside Than Formerly Believed

By ROBERT K. PLUMB

The earth is not quite a flattened sphere but is slightly pear-shaped—stem end at the North Pole— it was reported here yesterday at the opening of the American Physical Society's annual meeting.

The discovery that the earth is not the same shape north and south of the equator was made by studies of the orbital flight of the softball-sized Vanguard satellite launched March 17, 1958.

The finding was interpreted by scientists of the National Aeronautics and Space Administration to mean that the inside of the earth is stronger than has been believed. It would have to be to withstand strains imposed by bulges and depressed areas revealed to exist in the new satellite studies.

"We think this is the second most important finding of satellite studies up to this time," said Dr. Robert Jastrow, chief of the theoretical division of the Aeronautics and Space Administration. The most important find, he added, is the discovery of doughnut-shaped fields of radiation surrounding the earth.

Perigee Varied

The studies of Vanguard were discussed at a press conference by Dr. John A. O'Keefe, assistant chief of the space agency's theoretical division. Ann Eckels and R. K. Squires of the agency participated in the study.

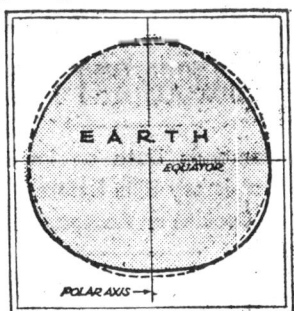

The New York Times — Jan. 29, 1959
NEW VIEW of the earth's shape (heavy line, shaded area, and old (dotted line).

The team found that the height of the nearest approach of the Vanguard satellite to earth (the perigee of the orbit) was lower in the Northern Hemisphere than the Southern Hemisphere. (The perigee of the Vanguard satellite swings completely around the earth in eighty-two days.) The unexpected change in shape of the elliptical orbit was found to be not caused by atmospheric drag, electrical or magnetic forces. This left variations in the earth's "lumpiness" as the explanation.

The earth is known to be flattened some twenty-seven miles at the poles. The newly discovered lumps amount to a fifty-foot rise of the sea level at the North Pole and a fifty-foot lowering at the South Pole, imposed on the known flattening. Outside the polar cap in the Northern Hemisphere the sea level is lower by about twenty-five feet than has been believed and the region in the Southern Hemisphere outside the polar cap is twenty-five feet higher.

'Slightly Pear-Shaped'

"The combined effects of these variations cause the earth to be slightly pear-shaped, with the more narrow end in the Arctic and the broad base in the Antarctic," the space group reported.

The finding will have to be considered for its effects on analyzing gravitational data from future space experiments. Further, the scientists said, it suggests that the hot rock that supports the earth's crust—rock surrounding an inner core, presumably of molten iron—is not as elastic as has been believed.

Dr. Jastrow reported that air-density measurements from American and Soviet satellite and rocket experiments had shown that the amount of residual air at a given distance above the earth varied with the season of the year, the time of day, the sun's activity and the latitude.

The physicists are meeting in the New Yorker Hotel.

January 29, 1959

TEMBLORS MAKE THE EARTH 'RING'

Analyses of '60 Chile Quake Waves Show Pulsations

By WALTER SULLIVAN

When the world is struck a hammer blow by a major earthquake, everything on earth bounces up and down for roughly a month.

The movement is a simultaneous rising and falling through the world, rather than a wave motion. The entire globe expands and contracts in a twenty-minute cycle.

The prolonged life of this effect means that the earth, in this respect, is more resonant than the most perfectly cast bell.

Discovery of the resonance was disclosed here yesterday at a symposium on space physics that formed part of the annual meeting of the American Physical Society, at the Statler-Hilton Hotel.

At the same session it was reported that the earth's magnetic field acts as a great barrier to the hot gases, or plasma, rushing outward from the sun. The result is a "shadow" that extends tens of thousands of miles from the earth, in a direction away from the sun.

The shadow was detected by

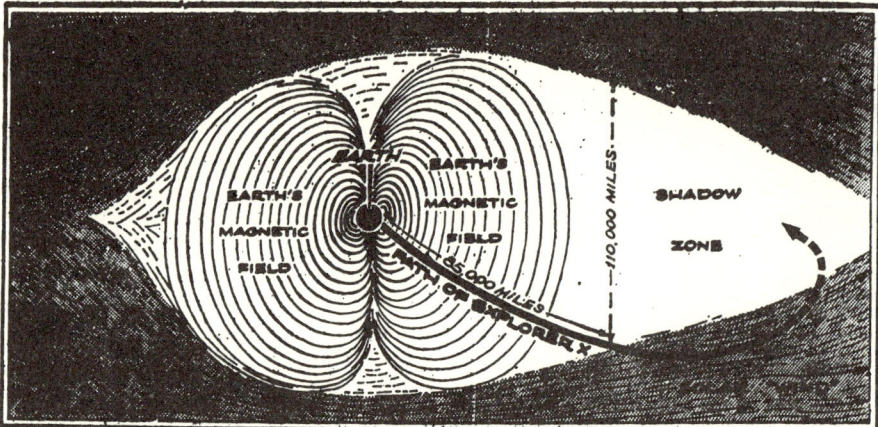

WIND SHADOW: Data from Explorer X indicates that it did not enter the "wind" of hot gases from the sun until it was some 85,000 miles from the earth. Apparently the world's magnetic field (the curved lines enveloping the earth) deflects the "wind" and produces a shadow some 110,000 miles wide and of great length. The vehicle transmitted until it was about 148,000 miles out (solid line). It then curved back and orbited the earth.

Explorer X, a space probe fired last March. It also found that the width of the shadow fluctuates, presumably because magnetism within the solar "wind" occasionally forces the earth's magnetic field to shrink.

Analysis to Chile Quake

The month-long "ringing" of the earth was reported by Dr. Gordon J. F. MacDonald of the University of California at Los Angeles. Its discovery grew out of analysis of shock waves from the Chilean earthquake of 1960, which earlier had indicated similar, but far shorter-lived, free oscillations of the earth. The bouncing motion that he described yesterday amounts to only about a thousandth of an inch.

He said another form of quake-produced oscillation had shown that there was a weak magnetic coupling between the earth's liquid core and the solid mantle that envelopes it. When the mantle wiggles, it tends to pull the core with it.

The Explorer X results were described by Dr. Bruno Rossi of the Massachusetts Institute of Technology, whose laboratory participated in the experiment. The data indicated a thin solar "wind" blowing through space at about 200 miles a second.

He said that Explorer XII, shot 50,000 miles toward the sun last August, had not detected the wind, presumably because it had not passed beyond the shielding influence of the earth's magnetism.

In another paper Dr. Hong-yee Chiu of Yale University and the Institute for Space Studies of the National Aeronautics and Space Administration proposed a fifth of the energy produced by a star, during its lifetime, is in the form of neutrinos.

Peculiar Particles

These are perhaps the most peculiar of all atomic particles, having no weight and no electric charge. They rarely interact with other particles and hence pass through matter with almost no chance of being stopped. Dr. Chiu said that on the average a neutrino must pass through ten billion planets the size of the earth before interacting.

He cited a suggestion by a Soviet astronomer that the universe may originally have been one mass of neutrinos. Their interactions then produced the other particles and elements, as well as the seeming "explosion" of the universe.

Dr. Chiu said that, according to his calculations, stars in their later stages produce large numbers of neutrinos. Because these particles interact so rarely, they may be accumulating in the cosmos.

Hence, he added, there may eventually be nothing left but neutrinos, and, he continued, it may be that "from the neutrino we came and to the neutrino we will return."

A great, universal flow of neutrinos, while suspected, has so far escaped detection, although the particles have been produced and detected in the laboratory, using large atom-smashers.

Balloon Telescope Plan

Dr. Martin Schwarzschild of Princeton University told of difficulties that have delayed about a year his project for lifting a large-sized telescope by balloon.

The three-ton assembly of optical, guiding and television equipment will be by far the largest load ever hoisted by a balloon, he said.

The construction of an adequate balloon has been one of the difficulties, but he said that one was recently inflated with sufficient helium to lift 13,000 pounds, yet did not burst. Another problem has been to produce optics shaped with sufficient precision to exploit the advantages of altitudes. It is calculated that, by surmounting almost all the atmosphere, the telescope will be able to see three times as much detail as has been possible on earth.

The components of the rig have been made, but not all are working properly as yet. Its thirty-six-inch telescope is larger than all but about two dozen such instruments in ground-based observatories of the United States.

When flown, the scope and balloon assembly will be 562 feet high. It should greatly increase our knowledge of the surface of Mars and the other planets.

Eventually a thirty-two-inch telescope is to be flown in a satellite. Its chief task, however, will be to scan the heavens in portions of the ultraviolet spectrum that cannot pierce the atmosphere.

January 26, 1962

SATELLITES DETECT VARIATIONS IN EARTH

BALTIMORE, June 8 (AP)—The earth has four vast bulges and four equally big depressions that affect the flight of satellites, scientists at Johns Hopkins University said yesterday.

Reporting information gained from a series of satellites after analysis at the university's Applied Physics Laboratory in nearby Howard County, the scientists said that the features were each several thousand miles across.

"These highs and lows are not little things," said Dr. Robert R. Newton, supervisor of space research and analysis at the laboratory. "They are as big as the North American continent. We think we found all the big ones and are hoping for more information from other satellites."

The findings developed from information radioed by the Beacon satellites launched between 1961 and 1963. Two newer satellites in the series are now in orbit and are operating satisfactorily.

Radio soundings from six earlier satellites were drawn closer to the earth while passing over the highs and drifted farther off while passing over the lows.

June 9, 1965

Diamonds Linked to Blasts From Within Earth

By WALTER SULLIVAN

An accumulation of evidence is now believed to show that diamonds are brought to the surface by rocket-like eruptions that pierce the entire crust of the earth and roar into the air at supersonic speed — comparable, perhaps, to the noise of 1,000 supersonic transports.

While it has long been suspected that the subsurface columns of cemented fragments, usually a few hundred feet in diameter and known as "pipes," in which diamonds are found are the residue of some form of violent eruption from great depth, the nature of such events was a mystery.

New methods of specimen analysis have now given strong support to the view that the material is blown up from depths of 124 miles or more by high-pressure accumulations of gas in that region. Some of the findings in this research have recently been published in scientific journals and others await publication.

EARTH SCIENCES

One result of this work has been confirmation that a number of formations resembling diamond pipes exist in southeast Utah and northeast Arizona. Known as diatremes, they are packed with material that, it has been shown from analysis, has come from depths as great as 125 miles.

Diamond is a phase of carbon that typically is formed at temperatures and pressures comparable to those below 90 miles.

Dating of Diatremes

Dating of the Utah and Arizona diatremes has shown that they erupted, some 30 million years ago, when North America is thought to have begun pulling away from Europe, moving westward over the earth's deep interior, or "mantle." Evidence has also been found that the African diatremes, or diamond pipes, erupted when South Africa began moving relatively fast, some 120 million years ago.

The diatremes in Utah and Arizona include material identical in most respects to the kimberlite found in the diamond pipes near Kimberley, South Africa. While no diamonds have as yet been found in the Utah-Arizona diatremes, they contain abundant wine-colored garnets of a type rich in chromium and magnesium, typical of diamond pipes elsewhere.

Another chemical analysis has shown that the garnets that were mixed with rough diamonds (apparently bought in London) to simulate a diamond deposit in the great diamond hoax of 1872 came from these diatremes. Today such mixing would have added a realistic touch to the fraud, but in those days the association was unknown.

The stones were sprinkled atop a mesa near the corner of Utah, Wyoming and Colorado to deceive prospective investors. The perpetrators of this fraud apparently believed the garnets were rubies (which do not normally occur with diamonds) and were unaware of their true nature as sought-after clues to the presence of diamond deposits.

Clues to Discovery

In fact garnets of this type have been used in the Soviet Union to backtrack 200 miles along the route where they had been scattered by an advancing ice sheet, leading to discovery of one of the world's richest diamond sources. Some geologists believe similar detective work may finally disclose the source of the diamonds found in glacial deposits from Ohio to Wisconsin.

These were presumably swept down by the advancing ice from a source north of the Great Lakes. So far the only diamond pipe of any consequence discovered in North America has been at Murfreesboro, Ark. It is reportedly no longer sufficiently productive for mining and serves as a tourist attraction.

The 1872 fraud was exposed, in part, by Government surveyors who found the site and, in part, by the city editor of The Times in London. The latter's discovery that rough diamonds had been bought there by Americans under suspicious circumstances broke upon the New York scene in a dispatch to The New York Times on Aug. 30, 1872. By then, however, both Tiffany and Rothschild money had apparently been invested in the bogus scheme.

Now Lowell S. Hilpert of the United States Geological Survey in Salt Lake City has found the site in Colorado and analyzed some of the garnets scattered by the hoaxers, determining their real source in the Arizona diatremes.

'Stuffed in the Ground'

The new theory of diamond eruption has been developed largely by Dr. Thomas R. McGetchin and his students at the Massachusetts Institute of Technology. For a number of years he has been studying the Arizona-Utah diatremes. He likens their eruption to a rocket "stuffed in the ground upside down."

The driving force, he believes, was a combination of gases in their fluid phase—probably chiefly water—under extremely high pressure in that region below the earth's crust known as the upper mantle. When this pressure became great enough to force a passage up through the crust, he believes, the material began fighting its way upward at a relatively modest speed.

But as it neared the surface, according to the theory, the water and other liquefied gases turned to vapor, because of the reduced pressure, and began to drive the eruption at great velocity.

Such rapid decompression would have a chilling effect, as when carbon dioxide turns to "dry ice" on escaping from a fire extinguisher.

This, Dr. McGetchin says, explains an old puzzle, namely why diatremes pierce coal beds without "burning" the coal. In fact coal and limestone are sometimes found unaltered within kimberlite.

Evidence of Eruption

Earlier students of diamond pipes and other diatremes suspected that some form of high velocity eruption was involved because the hodge-podge fragments of rock, cemented together to form kimberlite, tend to be rounded, as though they had been exposed to extreme abrasion.

Some look like beach pebbles, or stream boulders. Those who discovered the diatremes of the Southwest thought they were glacial deposits. Even the diamonds found in other diatremes are rounded, despite being formed of the hardest material in nature.

The explanation seems to be that this material had an extremely rough journey on its long, high velocity trip up from the depths.

Supporting evidence for gas action as the driving force of such eruptions has been found in the discovery of gases inside diamonds. The diamonds serve as little "crystal bottles" containing samples of the gas that surrounded them when they formed in the upper mantle millions of years ago.

The analysis of these gas inclusions has been carried out by Dr. Charles E. Malton of the University of Georgia in Athens. He grinds up a diamond and then determines (with a mass spectrometer) what gases are trapped inside of it. The results, he said in a telephone interview, have been surprising.

For example the gas content of a Brazilian diamond, representing one part per 100,000 of the diamond itself, consisted of 85 per cent water and 0.1 per cent ethyl alcohol, as well as fractions of hydrogen, methane, oxygen, nitrogen and carbon dioxide. A diamond from Kimberley even contained a trace of butene (formed of three carbon and six hydrogen atoms).

The apparent existence of free water in the upper mantle not only supports the view that it can drive rocket-like eruptions, but also conforms to a widely held belief that the presence of water there lowers the melting temperature of upper mantle components. This, it is thought produces a "slushy" layer that permits the

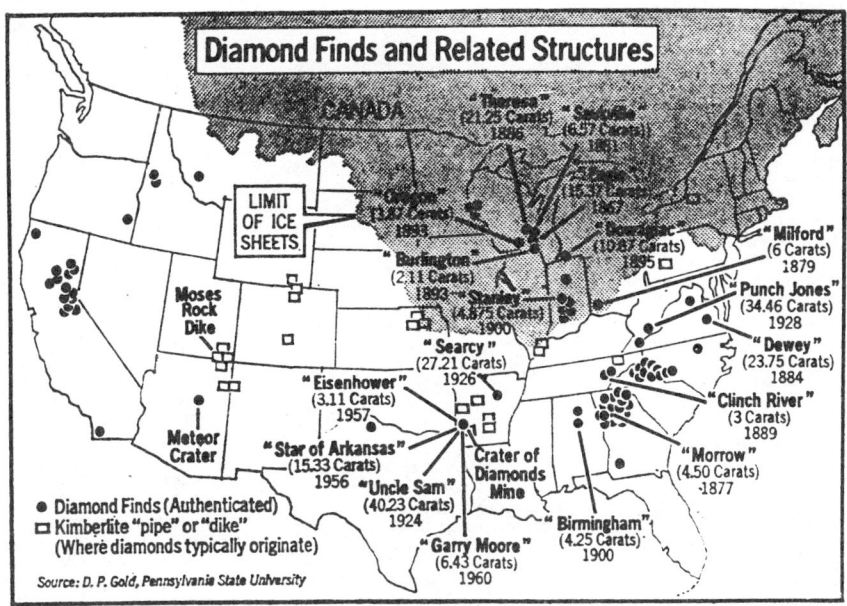

Diamonds have been found in many areas of the United States, but most were of little value. The largest, "Uncle Sam," came from the only known diamond "pipe" of significance, the Crater of Diamonds Mine near Murfreesboro, Ark. Specimens near the furthest advance of the former ice sheets (shaded area on map) were carried from a source north of the Great Lakes. Names, dates and weights of largest finds are shown, as well as Moses Rock Dike in Utah which has been under intensive study.

drifting of great surface plates of the earth.

The analyses of specimens from the diatremes are of great scientific interest since, it turns out, they are derived from a succession of depths well down into the upper mantle. As Dr. McGetchin puts it, the diatremes "are natural Moholes."

The Mohole Project was a plan for drilling several miles through the ocean floor to sample the mantle, which lies there at a relatively shallow depth. It was shelved when the costs multiplied to unacceptable levels.

The use of a device known as an automated electron microprobe has made it possible to determine the composition of tiny inclusions within a garnet or diamond, establishing the temperature and pressure under which the specimens were formed. The pressure then indicates the depth within the earth at which this took place.

The microprobe fires a very narrow beam of electrons at the target material and the resulting X-ray emissions indicate its composition.

Clues to Formation

In his recent presidential address to the Geochemical Society, Dr. Francis R. Boyd of the Carnegie Institution of Washington pointed out that the chemical composition of garnets from the diamond pipes could be used to determine the depth and pressure at which they were formed, so long as the garnets — as often occurs — contain two forms of pyroxene.

In this way it has been possible to show that some South African samples were formed 90 miles down at a temperature of 1,750 degrees Fahrenheit, whereas those formed as deep as 125 miles were at more than 2,500 degrees.

In a similar analysis of material from an elongated feature in southeast Utah, known as Moses Rock Dike, Dr. McGetchin has found that the temperature, at 1,750 degrees, was relatively uniform from the base of the crust, 30 miles down, to 125 miles. The latter depth is thought to be the top of the "slushy" layer of the upper mantle.

In the African pipes a peculiar feature of garnets from the bottom 25 miles of the sequence—the region immediately above the "slushy" layer—is that they are almost all badly sheared, Dr. Boyd said. Yet those from shallower depths are not. He postulates that this shearing occurred, in the bottom layer of the rigid plate, as Africa began to move and that heat generated by the process could have caused the eruptions.

'Peepholes' Into Planets

Dr. McGetchin believes diatreme eruptions probably have also occurred on the moon and Mars, bringing to the surface samples of material from far below. He calls them "peepholes into planetary interiors." The National Aeronautics and Space Administration takes the notion seriously enough to support his work.

He recalls how Ranger 8, the photographic spacecraft that hit the crater Alphonsus, sent pictures showing a chain of little craters on its floor. These he likens to the chain of eruptions that produced Moses Rock Dike, throwing up material from 125 miles down.

Initially these sites on earth looked like "maar" craters, produced by high velocity gas eruptions such as that at Nilahue, Chile, in 1956. An eruption of this kind typically lasts only a few hours and leaves a feature shaped much like Meteor Crater in Arizona.

After a few million years such a feature has eroded until no crater is left. What remains, at the surface, is the throat of the vent, clogged with the material that failed to escape before the eruption faded.

But, as Dr. Boyd points out, a few bucketfuls of this material represents a top-to-bottom sampling of the upper 125 miles of the earth's interior. The new method for determining the depth from which each fragment came, he said in his address, "takes a miscellaneous collection of mantle rocks and puts them back the way they were."

March 20, 1973

Widespread Molten-Rock Shower Eons Ago Indicated

By WALTER SULLIVAN

Evidence has been found that a single catastrophe, whose nature remains uncertain, spread droplets of molten rock over the entire region from Massachusetts as far south as Texas and the Caribbean 35 million years ago.

One hundred ten billion tons of this material, preserved in glassy fragments known as tektites, fell in this event, according to a report in the June issue of Earth and Planetary Science Letters.

Tektites had previously been found in the United States, but all, with a single exception, had been either in Texas or in Georgia. They have now been identified as far south as the Venezuela Basin, south of the Dominican Republic.

Furthermore, it has been shown, by two independent types of age determination, that all of the specimens fell at roughly the same time.

Tektites are glassy fragments, from the width of a hand down to microscopic size, that occur in several well-defined "strewn fields" across the earth. The fragments occur in various characteristic shapes — teardrops, dumbbells, buttons and spheres—but they tend to be streamlined.

That is, they appear to have been sculptured by high-velocity travel through the atmosphere while still in a molten state. Those in each strewn field have chemical characteristics in common and, it has been found, fell in the same event.

An initial step toward recognizing the great extent of the North American strewn field was the discovery in 1971, of a previously unknown tektite specimen in a cabinet that had been used at Columbia University by the late Dr. Arie Poldervaart, professor of geology there.

It was labeled as a tektite from Cuba—remote from any known strewn field—but the circumstances of its collection were not specified. Now it has been found that one layer in a core of sediment extracted from the Caribbean floor midway between Venezuela and the Dominican Republic contains some 6,000 microscopic tektites.

About 1 per cent of these are pitted in a characteristic star-shaped pattern. From the nature of the fossils in this layer of sediment it is known that it accumulated 35 million years ago, and it has been shown that the tiny beads were also formed at that time.

The northernmost limit of this strewn field has been set on the basis of a chance discovery made in 1960 by Dr. John Chase, a zoologist from Ohio Wesleyan University who spends his summers at Martha's Vineyard, off Cape Cod. While hunting for sharks' teeth along the bluffs of Gay Head he found a strange-looking disk of green glass.

Except for a missing corner, it was comparable in shape and size to the palm of a hand. It has been shown, not only that this it a tektite, but also that in age and composition it is part of the same fall recorded in the Georgia, Texas, Cuba and Caribbean specimens.

The report in Earth and Planetary Science Letters is by Billy P. Glass and Ralph N. Baker of the University of Delaware and D. Storzer and G.A. Wagner of the Max Planck Institute for Nuclear Physics at Heidelberg, West Germany.

The tektite ages have been determined both by measurement of their potassium and argon content and by counting the tracks produced in them by the fission, or radioactive decay, of uranium atoms. The potassium-argon method depends on the slow decay of a radiactive form of potassium into argon. Unless the material is heated to high temperature, the argon accumulates and can serve as an index of elapsed time since the specimen cooled and solidified.

In the uranium method the number of fission tracks produced since the glassy material formed serves as a stopwatch. While both methods have been used for the large tektites, only the fission-track technique is applicable to those of microscopic size. They are so small they yield insufficient potassium and argon for the alternate method.

The discoveries make this the second largest strewn field of tektites yet found. The largest covers all of Southeast Asia, the Philippines, Australia, Tasmania

EARTH SCIENCES

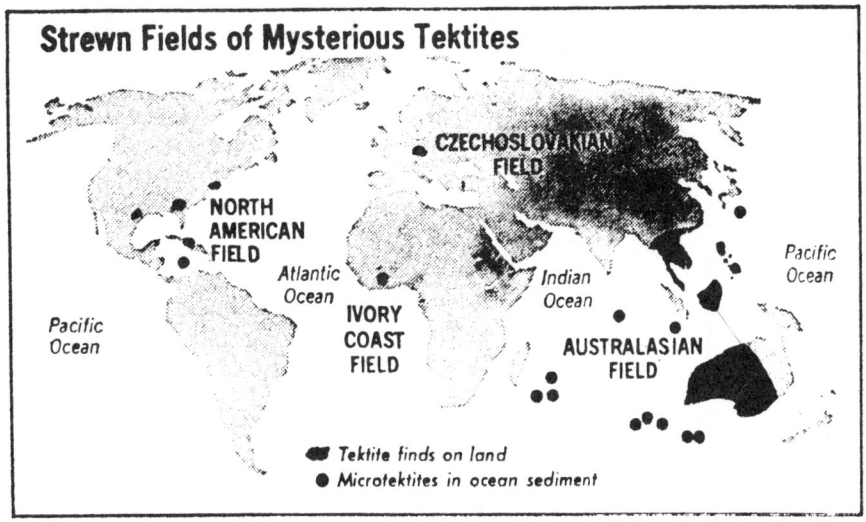

Four tektite fields are currently known

Some Samples of Tektites

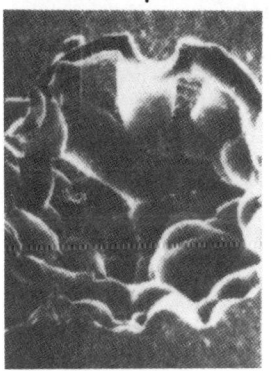 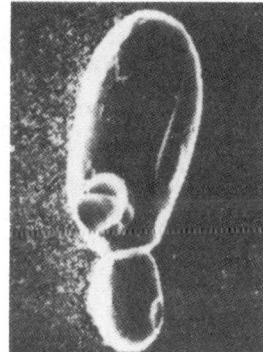 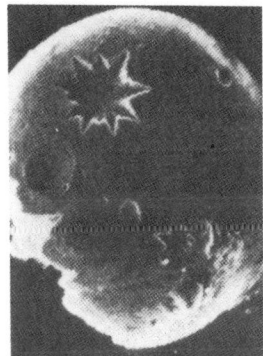

Representative microscopic specimens, as seen in a scanning electron microscope, are the scalloped sphere, left; the bowling-pin configuration, center, and the star-like pit in a sphere, typical of the Caribbean.

and much of the Indian Ocean. It fell 700,000 years ago at a time when the earth's magnetic field made its most recent persistent reversal of polarity.

That is, prior to this event a north-pointing compass needle would have pointed South. Whether the event responsible for the tektites caused this magnetic field reversal (for example, by disruption of the earth's spin) is a subject of current controversy.

Probably the first strewn field to be identified as such was the one that spread glassy fragments across that part of Czechoslovakia close to the junction of the Czech, Austrian and West German borders 15 million years ago.

After gold was discovered in gravel deposits of the Ivory Coast, in West Africa, the local inhabitants observed that smooth, glassy objects were often found with the gold. They called them "agna" and believed they had supernatural powers.

Similar tektites have now been found, in microscopic form, in nearby ocean sediments, and the fall has been put at one million years ago. Some of these tiny beads have a scalloped surface.

Soon after the Caribbean floor was carpeted with microscopic tektites, the tiny sea creatures known as foraminifera began to incorporate them into their shells—perhaps, Dr. Glass suggests, as "status symbols."

It is widely believed that tektites result from a catastrophic impact of some sort. The most obvious would be the fall of a meteorite or comet onto the earth. This is probably the favorite explanation, but, as Dr. Glass pointed out in a recent interview, the chemical composition of the tektites is not similar to that of any rocks that occur in abundance.

The debate on tektite origin is now more than 100 years old. The tektites of Czechoslovakia were described as early as 1788. Charles Darwin speculated on the origin of those in Australia. Now called Darwin Glass locally, they were used by the aborigines in magic rites long before Darwin's arrival.

However, the arguments as to the origin of tektites have been so inconclusive that some pessimists fear that they will not be resolved until the next strewn field is laid down.

July 26, 1973

Quasar Emissions Used in a Quake Prediction Study

By WALTER SULLIVAN

Radio emissions from objects billions of light years away are being used in an effort to record—in fractions of an inch—the kind of warping of the California landscape that is believed to precede major earthquakes.

It is hoped that the emissions, from the most distant objects known, called quasars, will provide a method of earthquake prediction unavailable from other measurements, which reveal only changes in horizontal distance between landmarks.

About 20 quasars are being used a "benchmarks in the sky" to watch for changes in the locations of benchmarks in the earthquake-prone terrain of California. It is now widely believed that a swelling of the landscape occurs prior to major quakes.

Quasars are being used because, in view of their extreme distance from earth, their positions relative to one another within the framework of the sky remain unchanged. Closer sources of radio emission tend to move in that respect.

While the nature of quasars is uncertain, many suspect they are the cores of great star systems or galaxies "burning" so brilliant, both in terms of light and radio waves, that they can be observed billions of light years away. That is, their light has required billions of years to reach the earth.

Accurate Clocks Needed

While the quasar emissions are irregular, the relative arrival times of any particular wave front at two widely spaced antennas can be determined by a method known as interferometry. To this end, extremely precise time recordings are essential and the Cali-

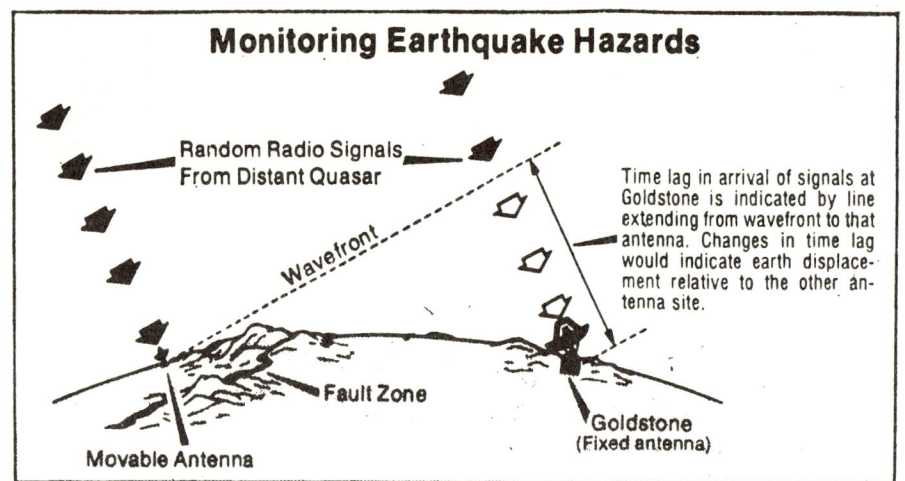

Upper dashed line represents wavefront at time it reaches movable antenna. The wavefront reaches Goldstone later because that antenna is farther from source. Time-lags can be compared with those in earlier measurements to check earth movements.

fornia measurements depend on clocks accurate to one ten-billionth of a second.

It is as though the random radio signals from a quasar were the sounds of a distant gunfight and the arrival times each of gunshot were being monitored at two antennas. The lag in the arrival time at one antenna, relative to the other one, would indicate how much farther away it was from the gunfight.

If, then, a warping of the landscape altered this time lag, the change could be taken as indicative of an impending quake.

A number of quasars in different parts of the sky are being used to develop three-dimensional data on the benchmark locations. One of the antennas, 30 feet in diameter, is movable and is being shifted to a variety of sites.

Measurements on Faults

In this way, measurements can be made across the various faults where earth movements have produced damaging earthquakes in Southern California. The most devastating is the San Andreas Fault. According to Peter F. MacDoran, who is leading the project at the Jet Propulsion Laboratory in Pasadena, Calif., the fault will be spanned by several observation baselines.

It was on that fault, at Fort Tejon north of Los Angeles, that the worst earthquake ever reported in the Western States occurred, in 1857.

To check the reliability of the technique, one baseline has been chosen where no changes are expected. It extends from Goldstone, site of the fixed antenna in the experiment, to southern Nevada. The Goldstone "dish," 210 feet in diameter, is the main antenna of the Jet Propulsion Laboratory's space tracking system.

Lab Run by NASA

The laboratory is operated for the National Aeronautics and Space Administration by the California Institute of Technology. Mr. MacDoran, in a telephone interview yesterday, said that in recent days about nine hours of quasar recording had been done by the portable and fixed antennas.

The immediate goal, ia a measurement accuracy of about four inches, with the expectation that, in a few years, the margin of error will be less than one inch. Some now believe, Mr. MacDoran said, that the region within several miles of an impending earthquake site swells as much as several feet before the event but measurement of such an effect in the past has been close to impossible.

It is planned to make measurements along each baseline at least once a year. The technique, known as ARIES (for Astronomical Radio Interferometric Earth Surveying), depends on instruments initially designed for space exploration.

September 5, 1974

CONTINENTAL DRIFT

NEW THEORY THAT CONTINENTS ARE FLOATING AROUND LIKE SHIPS

SCIENTISTS, especially of England and Germany, have been taking a great interest in the Wegener hypothesis, which holds that the continents are cakes of light material, loosely moored on the globe and floating about. They were all united in one super-continent during the carboniferous period, according to the Wegener hypothesis. The land mass was slipping gradually from east to west. It moved all together for a while. There was no Atlantic Ocean then and New York bordered on Belgium or some other point along what is now the European Coast.

Finally some parts of the mass slipped a little faster than the rest. North and South America began to float away. Australia and the Antarctic Continent broke loose from Africa and India, then separated and drifted to where they are now.

Even the South Sea Islands, Hawaii, the Azores and other isolated bits of land projecting up from the floor of the oceans are not entirely of volcanic origin, according to Dr. Wegener, but crumbs of the crust which stayed behind, instead of drifting away with the larger land masses.

This theory was first put forward by Dr. Alfred Wegener, Professor of Meteorology in the University of Hamburg. Scientific workers usually dislike speculations that cannot be put to the test, but this colossal guess has caused wide discussion in scientific publications and was even the topic for a day's argument by the geological section of the British Association.

That land masses are not anchored fast to the underlying mass has been indicated by observations in many parts of the world, notably at Ukiah in California. Blocks of the continent have been represented as pressing against other blocks in a process of adjustment of the earth's crust, slowly making mountains and valleys and occasionally causing earthquakes. The Federal Government has made appropriations to investigate the alleged "creep" of certain sections of the continent. Interest in this country so far has been mainly in verifying facts, leaving generalizations and guesses as to the causes for a later date.

The Puzzle Continents.

There are several plausible features of the Wegener theory. First, few persons have ever looked at a map of the world without noticing that South America and Africa could neatly be fitted together like pieces in a puzzle. The other continents and islands do not match up so prettily, but this is accounted for by Dr. Wegener on the theory that in being dragged about they were pulled and twisted out of shape, wrinkled into mountains, depressed below sea level. Dr. Wegener gave the name of Gondwanaland to his one big continent.

The strongest apparent evidence lending color to the theory is the fact that records indicate Greenland is, in a way, running wild. It has moved west by almost four-fifths of a mile in thirty-seven years, according to observations made in 1870 and compared to observations in 1907. If this land mass has floated 1,200 meters, as the records indicate, it would show a remarkable looseness between the outer and inner shell of the earth in that part of the world. But rather than accept such an explanation, most scientists have been skeptical of the value of the longitude observations. Professor H. H. Turner,

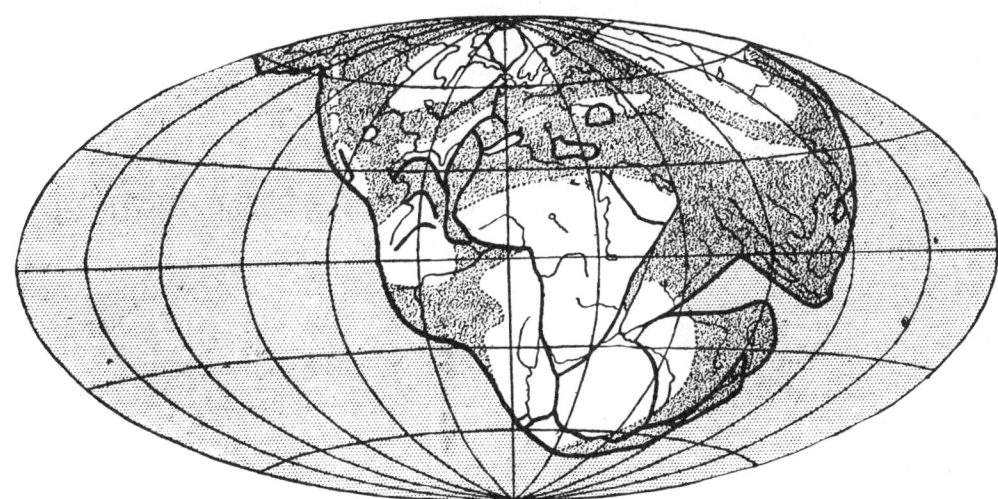

When Earth Was Still One Mass. White Portions Indicate Land Above Water Level, and Heavily Shaded Portions Land Submerged. Border Lines Show Edge of Mass in Carboniferous Period.

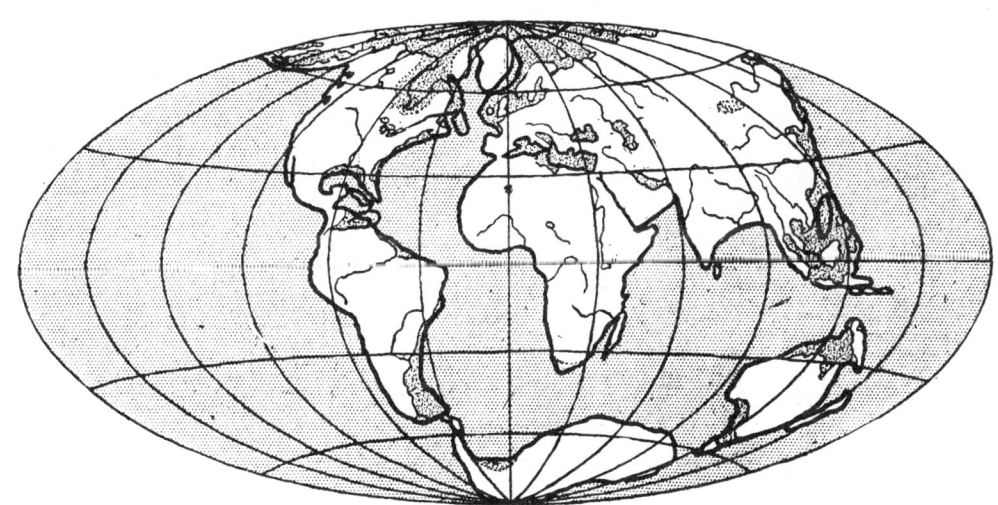

After the Mass Broke Up and Continents Were Formed. Note How the Shore Lines of Africa and South America Could Be Fitted Together. This Represents the Quaternary Era.

who discussed this theory at the meeting of the British Association, advocated a remeasurement of Greenland to see whether it had apparently kept on floating since 1907.

The almost universal distribution of many families of animals, birds, insects and plants, as proved both by fossils and surviving specimens, is another evidence of closer land connections in the past. Tapirs, for instance, apparently once roamed over most of the earth, but survive now only in the Central American regions and on the Malay Peninsula. Fossils of closely related animals in America and Asia have been accounted for on the theory that there was a land connection between Alaska and Siberia.

The rocky mass which forms the exterior of the earth is known to be very light compared to the interior. The first mile or so of the crust is less than one-third of the average weight of the whole mass, the interior having about the weight of iron. The light crust is believed to be lodged in the stratum below about as icebergs are in water. Why the crust should not follow perfectly the rotation of the interior of the earth is a difficult question. Believers in the theory of Dr. Wegener have proposed many explanations. One guess put forward by Dr. A. S. Eddington in a lecture on "The Borderland of Astronomy and Geology" before the Geological Society of London, was that tidal friction, especially resulting from land-locked shallow seas, had slowed down the rate of revolution of the earth.

"The frictional dissipation," he continued, "acts as a break on the earth's rotation, and we now feel confident that the break is a surface-brake applied at certain points on the earth's surface where the favorable conditions exist. The retarding force is transmitted to the earth's interior, and so delays the rotation as a whole; but unless the material is entirely non-plastic there will be a tendency for the outer layers to slip on the inner layers. I do not know how much the material a few hundred miles below the surface would be expected to give under the strains; it may be unappreciable, but I will assume that though small it has some effect.

"We have then the whole crust slipping from east to west over the main part of the interior. Probably it would go very stickily, sometimes arrested by jamming, which would hinder it for a time, and then going more easily. That is helpful in explaining certain astronomical observations. There are irregularities in the motions of heavenly bodies, noticed particularly in the swift-moving moon, but shown also on a smaller scale in the sun and planets, which appear to indicate that our standard timekeeper, the earth, is a little irregular. Now, of course, it is the rotation of the surface of the earth which determines our standard time. I find it difficult to believe that there can be irregular variations in the angular velocity of the earth as a whole; but it seems less difficult if the variations are merely superficial, due to the crust sliding non-uniformly on the interior.

"What interests the geologist more nearly is that the brake is applied only at certain areas on the surface, so that there would be a tendency to crumple the crust, more particularly to the west of these areas.

"I have regarded the crust as fairly mobile from east to west. I suppose the geologists would also like it mobile from north to south in order to have glacial periods in those portions which are now near the Equator. It is not possible to hold out much encouragement for such an idea, because we cannot imagine any force acting from north to south. Still, if the crust, which is being urged by the east-west force of tidal friction, is resisted by obstacles, it may be deflected, finding that, say, a southwest track offers less resistance. In a long enough time almost any displacement may have happened, granting my hypothesis that the connection of the crust to the interior is reasonably plastic. So that I cannot forbid this possible interpretation of glacial periods in the earlier geological times."

As North and South America moved away from the mother land-mass, the Pacific Coast of both continents necessarily bore the brunt of the attack on obstacles in the path of the movement. When the Pacific Coast tended to get stuck to the layer of earth below the crust the weight of the continent forced it on. In the jamming that followed the mountain ranges which line the Pacific Coast from end to end were presumably forced upward. Some of the arguments of the British scientists at the British association meeting were summarized as follows in the British scientific publication Nature by W. B. Wright:

"Mr. W. B. Wright pointed out that a critical comparison of the geological formations on the two sides of the North Atlantic shows on the whole a very remarkable correspondence, both stratigraphical and paleontological, from the Archaean to the Cretaceous, and in particular brings to light certain facts even more strikingly indicative of a former rapprochement between the two continents than any pointed out by Wegener.

Hard to Get Proofs.

"Professor Sollas confessed himself attracted by the theory, but doubtful as to proofs. He was not greatly impressed by arguments based on the similarity of the geological formations on the opposite sides of the oceans, the most remarkable of which was perhaps that cited by Mr. Wright. A certain uniformity is to be expected in rocks derived from the same Archaean base. The explanation on the whole was out of proportion to the points of correspondence cited.

"Dr. Harold Jeffreys stated that the rotational force which could be invoked to explain the movements of the continents was very small and quite insufficient to produce the crumpling up of the Pacific ranges. The ocean floors also presented a difficulty, for, being composed of basaltic rock, they would be less radio-active and therefore stronger than the continental crust. The withdrawal of India northward and its gathering up into the Himalayan folds were, moreover, not easily accounted for.

"Dr. G. C. Simpson thought the theory was a wonderful one from the meteorological point of view, as it explained the marked changes of climate given by the geological record and in particular the eccentric position of the Quaternary ice-sheets with reference to the pole.

"Professor Marshall of Wanganui, New Zealand, pointed out that the movement of that country was to the east and not to the west.

"The President, Professor Kendall, in closing the discussion said he had many years ago examined the question of a land connection across the Atlantic, especially in its bearing upon the distribution of fishes and reptiles. The practical identity of the Old Red fish faunas of the Orkneys and North America seemed to show a very close connection, and the similarity extends to the carboniferous. Divergence, especially in the reptiles, is marked in the Trias and probably complete throughout the Jurassic. Unfortunately, the reptiles require two barriers, one of land to stop the migration of the marine forms and one of sea to inhibit that of the land forms. The evidence adduced by Martin Duncan and marshalled by Gregory proved a connection between Europe and America during the Oligocene. He had long ago found it necessary to abandan a belief in the absolute permanence of ocean basins.

"The discussion as a whole was interesting as bringing out the extreme divergences of opinion produced by viewing the hypothesis from different aspects—astronomical, physical, meteorological and biological—but it becomes very apparent that the surest test of its validity lies in the domain of geology."

March 25, 1923

Britons Weigh Continent Theory; Vote on New Evidence Deadlocked

Wegener Hypothesis on Drifting Land Masses Is Argued in Heated 3-Hour Debate as Scientists Present Their Findings

Special to THE NEW YORK TIMES.

BIRMINGHAM, England, Sept. 1—After three hours of acrimonious debate on new evidence in support of the Wegener theory that continents had broken off from a solid mass and drifted apart, the issue was put to the vote of a crowded British Association meeting here today. The result was a dead heat for and against.

In a hall packed to the window ledges, the scientists argued whether there had once been a solid land mass and whether the Americas, Africa, Australia and Antarctica had slipped their moorings and drifted on a sea of basalt carrying their wild life with them.

Geologists, geographers, zoologists and botanists voted. Most geologists were against the theory and most of the naturalists were for it. The chairman, Dr. W. B. Turrill of the Royal Botanical Gardens at Kew, declined to use his vote to decide.

[The theory of continental rift was first formulated by Prof. Alfred Wegener, German geophysicist and meteorologist, who published his hypothesis in 1920 in "The Origin of Continents and Oceans." Professor Wegener lost his life in November, 1930, during a Greenland expedition in which he attempted to prove his theory.]

It started with a paper by Dr. J. R. F. Joyce, who had done oceanographic and geological surveys in Antarctica. His investigations suggested that the "Scotia Arc," the trail of islands curving from Tierra del Fuego through the Falkland Islands and the South Shetlands back into King Edward VII Land was a sort of hinge on which Antarctica had swung away from its original position in the Indian Ocean. If the arc were straightened and Antarctica replaced, Australia and New Zealand would fall tightly into place in the solid land mass.

Dr. Joyce got support from Prof. W. T. Gordon of London University from his work on Antarctic fossils which he maintained had showed that Antarctica must have originated nearer the Equator because the vegetation found in the fossil remains could not have flourished in a region of four months' daylight.

Dr. H. E. Hinton of the Department of Zoology of Bristol University supported the Wegener theory on the evidence of the similarity of rodents, beetles and primitive reptiles found alike in the African and South American continents. They could not be explained by land bridges or by wind drift or by raft drift, he said, because they could not have carried their own rations for a 4,000-mile ocean crossing.

Prof. J. H. F. Umbgrove, geologist of Delft, The Netherlands, flatly rejected the Wegener theory and among other evidence against it instanced the oceanic sediment of the Atlantic. This is 10,000 feet thick, as shown by seismic tests. The necessary erosion must have taken three to four million years to accumulate and throws out the Wegener time scale, he argued. Another argument against the Wegener theory was the narrow limits of the continental shelf of the American Pacific coast, he suggested. If the continents had moved, they must have, like a bulldozer, pushed and piled up the sea bed ahead. There was no evidence of this, he declared.

Prof. Harold Jeffreys of Cambridge University called on his fellow-scientists to waste no more time on the Wegener theory unless its protagonists could produce proof of the mechanism by which it had happened. He said they had been arguing for thirty-five years against geophysical evidence. Wegener's gross mistake was shown by the fact that a movement of land masses would have been toward the equator and not westward, he said.

Sir John Cockroft, director of the British atomic research establishment, described the atomic power generators of the future. He said British scientists and engineers were carrying out "feasibility tests" on power producers that should lead to a working pilot plant in three to five years. He said it was much too early for anyone to estimate the cost of atomic generators compared with conventional ones, but he was emphatic that atomic energy would not prove prohibitively costly.

The capital outlay on an atomic power station would be about three times that of a present-day electricity generator station, he estimated, but the fuel costs would be only a small fraction and uranium would not need replacement for ten or even thirty years.

Because of improvement in metals and in methods, Sir John said the British atomic pile could now be stepped up to 300 degrees centigrade and 10,000 kilowatts. The waste heat would be employed for central heating of the research station, but the big advances being made and the development of new metals—beryllium and zirconium, which can stand a greater

EARTH SCIENCES

heat and withstand nuclear bombardment—would enable temperatures of at least 450 degrees centigrade to be attained.

Examples of the use of atomic by-products in industrial research were given by Dr. Seligman of the Harwell atomic research station. One that he quoted—carried out by foreign scientists whom he discreetly declined to identify—had been the use of radioactive isotopes in cosmetics to find whether beauty creams actually penetrated the skin. They do not, he asserted.

Another matter of feminine concern was an investigation to improve the quality of nylon. In the production of nylon a fine film of oil is applied to the fibers. Manufacturers wanted to improve this film. Radioactive chemicals were experimentally added to the oil and it was possible to discover and correct the spread of the film.

September 2, 1950

Theory That Continents Wander Is Supported by British Scientist

By WALTER SULLIVAN

After several years of study, one of Britain's leading scientists has found it "highly probable" that the continents have drifted toward and away from each other, covering distances in the thousands of miles.

He believes, for example, that in the last 440,000,000 years North America has moved almost 3,000 miles north and probably has rotated 40 degrees counter-clockwise.

The scientist is Dr. Patrick M. S. Blackett, who won the Nobel Prize in physics in 1948 and, from 1957 to 1958, was president of the British Association for the Advancement of Science. He now heads the physics department at Imperial College of Science and Technology in London.

The idea of continental drift was advanced several decades ago by a German meteorologist, Dr. Alfred L. Wegener, but it has found little favor, particularly in the United States. Dr. Wegener, who led an expedition to Greenland, suggested that continental drift would explain the puzzles of climate change.

Among the more challenging of these puzzles is the existence of vast coal fields within a few hundred miles of the South Pole. The coal beds, rich in massive tree trunks, probably extend to the ice-covered Pole itself. A similar mystery is the presence of ancient coral in northwest Greenland, close to the point on land nearest the North Pole.

Dr. Blackett, at the end of a visit to this country, said last week that the presence of the former forests of Antarctica—which were necessary to produce coal—could be explained in one of three ways: either the crust of the earth has slipped around the spinning top of its molten interior; or the world was once so warm or even-temperatured that trees could grow near the pole; or the continents, individually, have moved.

Checks Magnetic Data

Dr. Blackett has sought to determine which of these hypotheses is correct by examining the magnetic and climatic evidence found by various researchers in the rocks of the different continents. His conclusion is that only the last one fits the available data.

Dr. Blackett originally won world fame for his work in nuclear physics and cosmic rays. He is credited with making the first cloud chamber photograph of the transmutation of an atom. More recently his interest has centered on the magnetism of rocks. It is here that he and others have found evidence that they believe supports the continental drift theory.

Dr. Blackett, in an interview, cited in particular the work of Dr. S. K. Runcorn at Newcastle in England and Dr. John W. Graham of the Department of Terrestrial Magnetism of the Carnegie Institution of Washington.

Such men believe that the traces of magnetism in ancient rocks serve as frozen compass needles pointing toward the magnetic pole as it existed when the rocks were formed. It has been found that the location of the pole, as indicated, for example, by rocks formed at a certain period of India's past, differs from the location determined from European rocks.

Tries to Test Theory

The explanation, according to adherents of the drift theory, is that the continents have moved, with respect to one another, since those rocks were made. Their reasoning is still controversial. Dr. Blackett has sought to test the theory by comparing the magnetic latitude of a continent, as derived from the rocks of a given period, with the latitude indicated by the climate at that time.

He feels that the data are still insufficient to tell where a continent once was, with respect to other land areas. He has simply taken the "dip" or inclination of the magnetic force in the rocks to determine latitude. For clues to ancient climates, he has collected the known information on fossil coral formations, plants, salt deposits and evidences of former ice sheets.

Dr. Blackett has also found a striking similarity between the drift speeds derived from the magnetic and fossil data and those observed along the San Andreas Fault in California and along other fault lines around the rim of the Pacific.

The most rapid drift indicated by magnetic analysis is that of Australia from 300 to 400 million years ago. The continent's mean speed during that period is calculated to have been from three to four inches a year. This is six times the rate attributed to Europe and North America for the last 400 to 500 million years.

Long Journey for India

The longest journey described in Dr. Blackett's analysis has been that of India. The magnetic latitude of that subcontinent during the Permo-Carboniferous Period is calculated to have been near the South Pole. The vegetation that produced India's coal is, in fact, akin to that found in Australia and Antarctica and is not related to the plant life in other coal deposits of the Northern Hemipshere.

More significant, Dr. Blackett believes, is the fact that, during much of this period, from 250 to 300 million years ago, India was extensively covered with ice. He concludes that India has drifted north during the intervening period at about two inches a year. Others have suggested that the crumpling that formed the parallel folds of the Himalayas resulted from the impingement of India on the Asian land mass.

One of the more striking correlations reported by the British scientist concerns the location of Europe during the Devonian Period, some 350 to 400 million years ago.

A study of corals laid down at that time places Paris only about 2 degrees of latitude north of the Equator. Magnetic analysis puts the city some 5 degrees north of the Equator. Likewise, the coal beds of Europe were deposited in a tropical or subtropical climate that existed after the Devonian Period.

The gradual northward movement of the continent in the last 500 million years, as charted by

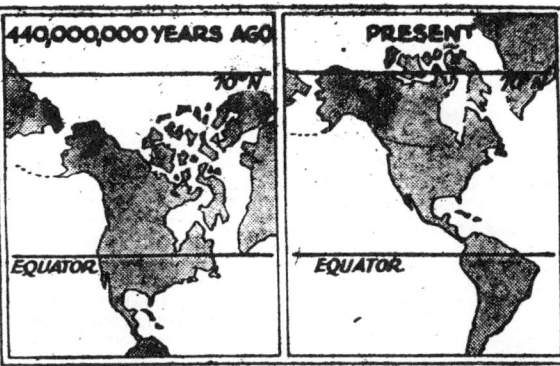

CONTINENTAL DRIFT: The maps illustrate theory of motion by continents. The equatorial location (left) of North America is based on studies by a British scientist.

magnetic studies, appears to be reflected in the composition of its rocks, Dr Blackett believes.

On the bottom are the Old Red Sandstones known to all European students of geology. These rocks were laid down, he suspects, when Europe was in the arid zone south of the Equator. On top of them is coal produced in the rainy tropics, surmounted by New Red Sandstone, deposited as Europe crossed the latitudes of the Sahara.

Dr. Blackett believes that the distribution of coral and salt deposits in ancient times points to zones of climate similar to those of today.

According to him, Europe and North America both appear to have ben near the Equator some 350 million years ago. The coal laid down soon thereafter was the residue of warm-climate forests.

On the other hand, the coal of India, Africa, Australia and South America came from the so-called Glossopteris vegetation thought to have grown in a cool climate.

The northward movement of New York, according to a magnetic study, has been swifter than that of Los Angeles because of the continent's rotation. Thus, it is said that the East Coast parted company with the equator some 250 million years ago, whereas the West Coast has taken 380 million years to make the journey.

Dr. Blackett published an analysis of the magnetic data in the July issue of the Proceedings of the Royal Society. He hopes to present the rest of his evidence in a forthcoming issue. Meanwhile, he has set forth his arguments in several scientific lectures, delivered during his stay in this country.

December 26, 1960

EARTH'S SURFACE RISES AND FALLS

Scientists Say Phenomena Show Plastic Interior

By WALTER SULLIVAN
Special to The New York Times

BERKELEY, Calif., Sept. 1— Observations made this year in the United States and in recent years in Europe indicate that the crust of the restless earth is in constant vertical motion.

The northwestern United States, in recent decades, appears to have risen an average of about two feet. The East Coast has sunk about four inches.

San Jose, Calif., has subsided 11.2 feet since 1912 and its sinking has speeded up in recent years. The level of the city has fallen two feet in the last three years. However, it is exceptional, lying on the west side of the Hayward Fault, the line of active movement in the earth's crust.

In Europe, repeated leveling surveys have shown steady trends of rising or falling. If they persisted, the central European plain would rise into a new mountain range within a few million years. Since there is no sign that mountains are being born, it is suspected that the phenomenon is a wave motion, slowly crossing Europe as waves traverse the oceans.

Continental Drift

The motions, described here at a world assembly of earth scientists, are being cited as evidence that the earth's interior is sufficiently plastic to permit the continents to drift apart.

They have also led to the formation of an international "Commission on Recent Crustal Movements," within the International Union of Geodesy and Geophysics, whose General Assembly completed a two-week session here today.

At the meeting there has been renewed debate between those who believe in continental drift and those who scoff at such an idea.

Adherents of the theory say that relative motions of about an inch a year between the continents could explain such puzzles as the presence of coal in Antarctica and Spitsbergen, showing that both these regions were once warm and vegetated.

The European surveys were cited by Dr. J. Tuzo Wilson, head of the Institute of Earth Sciences at the University of Toronto. He urged that highly precise observations be made in places where relative movements of large lands areas may be measurable.

Even with special earth satellites, he pointed out, it has not been possible to measure transoceanic distances with the required accuracy of a few inches, but this may be possible in certain places where the drift theory suggests movement, such as between Canada and Greenland, between Africa and Madagascar or across the width of Iceland.

The 1963 measurements of vertical movements in the United States were described to the meeting by James B. Small of the United States Coast and Geodetic Survey. He said the results were still somewhat uncertain because of the piecemeal nature of earlier surveys. Leveling adjustments were carried out along two lines running from coast to coast and two North-South lines.

During the next seven years, he said, it is planned to relevel lines extending 20,000 miles along three transcontinental routes and eight North-South routes. The survey would be carried out as fast as possible to obtain an "instantaneous" picture. The land level changes are thought to be independent of the "rebound" that typically follows the melting of an ice sheet.

Parts of Scandinavia and Canada, for example, are still rising at a rate of about one foot every 30 years, having been relieved of a great ice load some 10,000 years ago.

The surveying has also shown the effects of more recent loading. For example the area of Nevada and Arizona bordering Lake Mead sank at a rate of an inch a year during the early period after it was impounded by Hoover Dam.

Nearby Las Vegas, Nev., has sunk 2.2 feet since 1935 because so much water has been pumped out of the ground for city use.

September 2, 1963

Antarctic Rocks Tell of a Vast Land

By WALTER SULLIVAN
Special to The New York Times

BERKELEY, Calif., Dec. 28 —Geologic discoveries of the last 10 years have demonstrated "beyond a doubt" that until comparatively recently the Antarctic continent was not isolated from the rest of the world, it was reported here today.

That region, now sheathed in ice, seems to have once been attached to the steamy subcontinent of India. And although it is now devoid of large land animals, during the period when the rest of the world was inhabited by dinosaurs Antarctica too apparently had its population of reptiles.

Dr. Laurence M. Gould, retiring president of the American Association for the Advancement of Science, said the evidence in Antarctic rocks pointed strongly toward the possibility that that continent was once the "keystone" of a gigantic land mass formed from India and all Southern Hemisphere land areas.

Dr. Gould, geologist and scientific leader of Rear Adm. Richard E. Byrd's first expedition to Antarctica, has in recent years served as chairman of the Committee on Polar Research of the National Academy of Sciences.

In his presidential address to the association this evening he also said that Antarctic waters, acre for acre, were potentially the richest in the world as a food source.

The fish are walled in by the Antarctic "convergence" — an abrupt temperature change where polar waters slip under those of temperate seas. Ninety per cent of Antarctic bottom fish species are found nowhere else because of this barrier.

The idea that the southern lands were once a single continent that broke up and drifted apart is far from new, but it has long been controversial.

Dr. Gould said the discovery of a layer of tillites 900 feet thick in the Horlic Mountains of Antarctica, coupled with similar contemporaneous deposits in South America, India, Africa and Australia, implied that these regions must once have been close together.

All these deposits date from the late part of the carboniferous, or coal-forming, period, 300 million years ago.

Tillite is rock that appears to have been formed from gravel and left by a glacier.

The similarity of both fossil plants and fossil animals from all these areas during past geologic periods also points to their close association, Dr. Gould said.

"In the light of the new information from Antarctica," he asserted, "there is but one explanation, and that is extensive continental glaciation—an explanation hardly consistent with present land relationships."

The hypothetical giant continent, known widely as Gondwana Land, seems to have existed until late Mesozoic time, some 50 million years ago, before it fragmented, he said.

The Mesozoic was the age of dinosaurs. Hence it appears likely that Antarctica was populated by reptiles and, possibly, by early mammals.

Dr. Gould pointed out in an interview that scientists from Ohio State University had found fossil trails in Antarctic rock that they believe were left by ancient reptiles.

The fact that no dinosaur remains have yet been found is not surprising, he said. They are hard to discover even on a continent that is free of ice and has been thoroughly explored.

Dr. Gould told of current efforts to assess the extent of worldwide air pollution by examination of Antarctic snow and ice, both from recent precipitation and from that of a couple of hundred years ago.

There has been some evidence from Greenland ice that the lead content of the air has increased alarmingly in recent years. This is being checked in Antarctica.

December 29, 1965

GEOLOGIST BACKS CONTINENTS' DRIFT

Says Climatic Variation Is a Result of Movement

By JOHN NOBLE WILFORD
Special to The New York Times

WASHINGTON, April 9 — The climate does not really change much, but continents do.

This is the theory of Dr. Warren Hamilton, a scientist with the Department of Interior's Geological Survey. He maintains that a recently completed study of the earth's ancient climates, based on new discoveries of tropical fossil remains in frigid regions, provides "strong confirmation" of the controversial continental drift theory.

The results of his study will soon be published in a number of scholarly journals.

Much of the climatic variation in the past, according to Dr. Hamilton, a geologist, has been caused primarily by the movement of continents across the face of the earth, rather than by worldwide changes in climatic zones.

"Each continent has its own pattern of climatic variations, rather than a pattern shared with all other continents," Dr. Hamilton says. "Thus, it is likely that climatic zones remain relatively unchanged in width and position, and that the continents have drifted through these zones throughout geologic time."

Crust Believed to Expand

A number of scientists have theorized that billions of years ago the earth consisted of one or two continents. Through heat boiling up from the depths of the earth or other causes, the crust may have expanded and eventually split apart, resulting in many separate continents.

The theories first became popular in the late 19th century when scientists made much of the fact that the bulge of South America at Brazil looks as if it once snuggled up to the indentation of the western coast of Africa. What is now Newfoundland, New England and the British Isles may have pulled away from Northern Europe. Earthquake zones and fault lines, such as the San Andreas Fault in California, indicate that the crust is still mobile.

On the basis of a recent study of the earth's ancient climates. Dr. Hamilton sides with the "drift" theorists. He says there have been some fluctuations in climate — thus explaining the glaciers that once extended down as far as the Ohio Valley — but that the primary tropical zones appear to have remained the same.

Fossil Remains in Arctic

Yet, he points out, scientists have found fossil remains of coral reefs dating back 270 million years as far north as Arctic Canada, Greenland and Spitzbergen. These seem to have occurred at the same time that glacial ice caps covered now-tropical parts of India, Australia and Africa.

Neither reefs nor glaciers, he says, could possibly have formed where these lands are now. Dr. Hamilton regards the fossil data as proving that continental drift has occurred throughout the last half-billion years.

Dr. Hamilton, who has led three geological field parties to the Antarctic, says that even that polar region was once tropical and, presumably, in a different location.

Dr. Hamilton declares that the continental drift is still going on as actively as it ever has — at rates of a few feet or a few yards every 100 years.

He postulates that the earth's crust, rather than acting on the basis of internal movement, may be "moving independently of the interior."

April 10, 1966

A Force That Pushes Continents Apart

By WALTER SULLIVAN

The forces that in recent months and years have thrust new islands up through the floor of the North Atlantic, when considered with other new evidence gleaned from the ocean floors, have engendered wide acceptance of the view that our planet is in the throes of dynamic changes.

In the current Polar Record, P. F. Friend of the Scott Polar Research Institute, at Cambridge University in England, assesses the hypothesis that Europe and North America were contiguous 100 or 200 million years ago, then were forced apart to form the Atlantic:

"Recently," he writes, "evidence has accumulated to establish beyond a reasonable doubt that this drift has indeed occurred." The attention of geologists henceforth, he says, "should now be turned from the question of whether drift has occurred to the manner in which it has occurred."

The theory that has evolved in recent years is that hot, plastic rock creeps upward from the bowels of the earth like porridge over a flame, spreads out on the surface and then sinks down in some other region — the edge of the "pot," so to speak. The upwelling occurs along the midocean ridges, as dramatically manifest in the formation of Surtsey and Jolnir islands off Iceland during the last four years. The subsidence begins where material that has flowed outward from the ocean ridges reaches the continental margins.

A large percentage of the world's earthquakes occur either along the midocean ridges or along continental margins where subsidence is thought to be taking place — for example, the coast of Chile, the Aleutians, Japan and other places around the Pacific basin.

Mapped by Computer

As an example of the striking fit between land masses on opposite sides of the North Atlantic, Friend publishes a map drawn by Sir Edward Bullard and his associates at Cambridge University with the aid of a computer. The fit is between the outlines of the opposing continents as determined by their continental shelves — not by coastlines showing on current maps.

The coastline is drawn by the whims of climate that determine sea level. Should polar ice melt, raising worldwide sea levels, Florida and many other coastal areas may vanish. When much of the world's water is tied up in great ice sheets, the oceans recede from the continental shelves that, for example, reach hundreds of miles to seaward from New York. It is the edges of these shelves that divide the world, structurally, into two provinces: continental blocks and ocean basins. Bullard's computer relocated the blocks of Canada, Greenland and Europe to achieve the best fit.

Another of the most recent lines of evidence has come from the magnetic mapping of the ocean floors. The magnetic maps show regions that deviate, either positively or negatively, from the average. This has revealed peculiar striped patterns parallel to virtually all the midocean ridges. What is more remarkable, the patterns seem to be symmetrical on opposite sides of the ridge.

The explanation that has gained a wide following in the last year is that these patterns are caused by rock that has flowed away from both sides of the ridges. As the rock comes out of the depths and cools, it is imprinted with the earth's magnetism existing at that time.

If, as many now believe, the magnetic field of the earth flips over at intervals measured in thousands or millions of years, these reversals will be recorded in the outflowing rock, just as information is imprinted in a tape recorder.

Laboratory experiments indicate that such reversals occur where magnetism is generated within a rotating fluid, like the earth's interior. If such an event occurred, the north needle of a compass would point south.

One of the most clearly defined magnetic patterns on the ocean floor arose from a survey of a limited portion of the ridge region southwest of Iceland. The center line of its symmetrical pattern, marked "A" on the chart above, coincides with the center line of the ridge. It is also roughly aligned with the new islands off Iceland and the regions of most recent volcanic activity on Iceland itself.

Lava Flow Studied

In the chart, based on one published by Dr. James R. Heirtzler

The current volcanic generation of new islands near Iceland, such as Surtsey, in foreground in picture at right, and Jolnir in eruption, is thought to manifest forces that have pushed Europe and America apart. Large computer-drawn map shows how continents might have been joined 100-200 million years ago. Magnetic patterns in inset are discussed below.

and others at the Lamont Geological Observatory of Columbia University, the features marked "B" and "C" show the symmetry of magnetic features on opposite sides of the ridge. The timetable of the more recent magnetic reversals has been determined by study of lava flows in Iceland and elsewhere. The resulting pattern of reversals seems to fit the sequence of magnetic changes imprinted on the ocean floor. The indicated rate of spreading is a few inches per year in the Pacific and about one inch in the Atlantic.

Friend cites evidence collected from cores, or cross sections of the ocean floor, obtained in many parts of the Atlantic. In some areas—presumably younger parts of the ocean — fossils from the Upper Cretaceous Period, some 70 million years ago, are absent. From this it has been concluded that at that time the ocean was only about two thirds its present size.

One of the most remarkable features of the worldwide system of midocean ridges is the existence of deep clefts along their midlines. These rifts are presumably torn open by the upwelling process. Such a rift runs directly across Iceland and it is now appearing in the evolution of Surtsey. The island has continued to erupt and change its geography. Early this year its area was almost one square mile and in March it was still growing at a rate of 2,400 square yards daily.

The evidence for ocean floor spreading has injected new life into the seemingly static science of geology. Much of the world appears to be in flux. At this spring's meetings of the American Geophysical Union some 10 sessions were devoted to the subject and evidence relating to it. Such long-standing mysteries as the "Andesite line" in the Pacific Ocean seem now to be explicable.

This line passes through the islands of the western Pacific. Islands on the continental side of the line have rocks containing andesite — a basaltic rock often rich in minerals. Its name derives from its presence along the Andes, which it has enriched with gold, silver, copper, tin and other ores.

It is now proposed that andesite pushes upward where rock that has spread across the ocean floors is sinking down under the continents. The rock has become hydrated and has picked up continental type material on the way. As it sinks, its lighter fractions push upward, forming the island arcs and volcanoes that rim such oceans as the Pacific.

As one geophysicist put it after a recent discussion of the subject: "It's a revolution. The geology texts will all have to be rewritten."

July 9, 1967

EARTH SCIENCES

A Vertebrate Fossil Is Found in Antarctic

By WALTER SULLIVAN

A fossil fragment of a large amphibian that lived in subtropical forests near the South Pole more than 200 million years ago has been found in Antarctica.

This is the first indication that land vertebrates once inhabited that region. It had previously been thought that Antarctica, isolated by hundreds of miles of stormy seas, had never been reached by such animals.

The discovery supports the view that the continent was once linked to other land masses.

This is because animals of this type are thought to have been unable to travel in salt water. Yet closely related species are known to have lived contemporaneously in South Africa, Australia and even as far away as Spitsbergen, near the North Pole.

Dr. Edwin H. Colbert, curator of vertebrate paleontology at the American Museum of Natural History, at 81st Street and Central Park West, termed the discovery one of the most important fossil finds of the century.

It was he who identified the specimen, a fragment of jaw bone only 2½ inches long. He described the animal as resembling a giant salamander about four feet long.

The specimen, he said, should make "very happy" those who believed the continents were once joined and then drifted apart. The ancient animal belonged to the labyrinthodonts, a major group of extinct amphibians. They are of special interest because they may have been ancestral to all modern land vertebrates, including man.

The Antarctic discovery was made last December by a geology team from the Institute of Polar Studies at Ohio University. The team worked from last November to February in the mountainous country east of the upper Beardmore Glacier and 325 miles from the South Pole. Peter J. Barrett of New Zealand, who led the party, reported from Ohio State, in a telephone interview, that the bone was embedded in a succession of sediment layers 2,500 feet thick.

It lay in what had once been a sediment-filled stream bed

An artist's restoration of a labyrinthodont, which lived more than 200 million years ago
Lois Darling, from "Evolution of the Vertebrates"

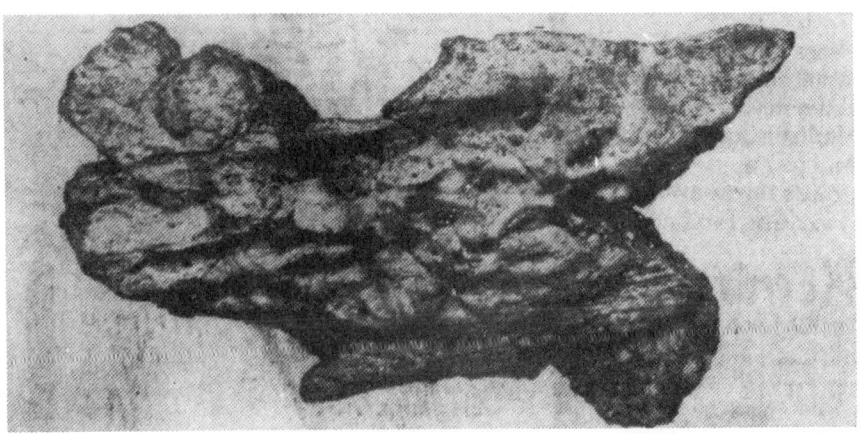

Remains, twice natural size, of jaw of the subtropical amphibian found near South Pole
The American Museum of Natural History

among plant fossils of the early Triassic Period, some 220 million years ago. The ferns and other plants are typical of those that lived along fresh-water streams during that time. The association of the fossil with such material is thought to establish the fact that it was a fresh-water amphibian of the early Triassic.

Findings in recent years have left no doubt that Antarctica was once warm. There has been an argument, however, about whether this was so because the whole world was warmer, or because Antarctica specifically was once in warm latitudes.

When Capt. Robert Falcon Scott of Britain raced the Norwegians to the Pole in 1911-12, he and his companions discovered great beds of coal in the mountain walls lining the Beardmore Glacier.

The fossil vegetation, which includes large tree trunks, is similar to that found in deposits of the same period from such far-flung regions as India, Africa and South America. This led to the theory of an ancient supercontinent, which geologists called Gondwanaland, that joined all these regions.

One view was that they were linked by land bridges that have since been submerged. This view has lost favor because of detailed mapping of the ocean floors. No evidence of such bridges has been found.

The other theory was that Gondwanaland broke apart into fragments that slowly drifted apart. However a number of scientists have rejected this idea of continental drift.

Fossils have been found that show Antarctica was once inhabited by penguins as tall as a small man, but there has hitherto been no evidence that any large land animals roamed the dense forests of that continent. Insects were certainly there, as they are in present-day Antarctica, borne from other continents by winds or birds.

The penguin is not considered a land vertebrate because it spends a large of its life in the ocean.

Vertebrates are animals with spinal cords.

Whatever land animals did exist on Antarctica were annihilated as the ice ages heaped snow on the continent until its ice sheet pushed to the sea in all directions.

Dr. Colbert said that, because of the "very far-reaching implications" of the discovery, he had shown the specimen Monday to Dr. Donald Baird of Princeton University, another authority on fossils of that period. Dr. Baird concurred in his identification.

March 13, 1968

2 Geologists Confirm Continental Drift Theory

Layers of Rock in Africa and South America Found to Match in 3 Areas

By WALTER SULLIVAN

Two University of Georgia geologists believe they have confirmed the previous conjunction of Africa and South America in terms of an intimate geologic "fit" between the two continents.

In the current issue of the journal Science they state that in at least three areas the geologic patterns of the two continents match one another. It is as though the proper joining of two pieces of a jigsaw puzzle had been confirmed because their picture designs matched.

The two scientists present evidence that a 200-mile geologic feature perpendicular to the coast in Brazil continues on the African side in Gabon. They note that two other such "fits" have been shown, leaving, in their view, little doubt that the continents, now thousands of miles apart, were once joined.

The two are Dr. Gilles O. Allard, associate professor in the geology department, and Dr. Vernon J. Hurst, head of the department, at the University of Georgia in Athens.

A striking match has been found in layers dating from before the Jurassic Period, 180 million years ago, until the Middle Cretaceous, some 100 million years ago.

These rocks contained some 30 identical species of freshwater animals (ostracods) and many other indicators of close proximity. Then abruptly in the Middle Cretaceous a thick layer of salt was lain down on both continents.

Presumably at that time, according to the Georgia scientists, salt water first invaded the region as the continents split apart and the Atlantic Ocean began to form. From then on the geologic records of the two regions diverge.

Those who have argued that the continents, over millions of years, have drifted about, altering their relative positions, cited the striking "fit" between the bulge of Brazil and the embayment on the west coast of Africa. They said there should also be a geologic match between the two regions.

In 1967 Dr. Patrick M. Hurley and his colleagues at the Massachusetts Institute of Technology, working with a group from the University of São Paulo in Brazil, demonstrated a match between the ages of rocks at various points along the opposing coasts.

Another match is the appearance of similar mineral formations (mylonite) at Recife and Patos in Brazil and north-central Cameroon in Africa.

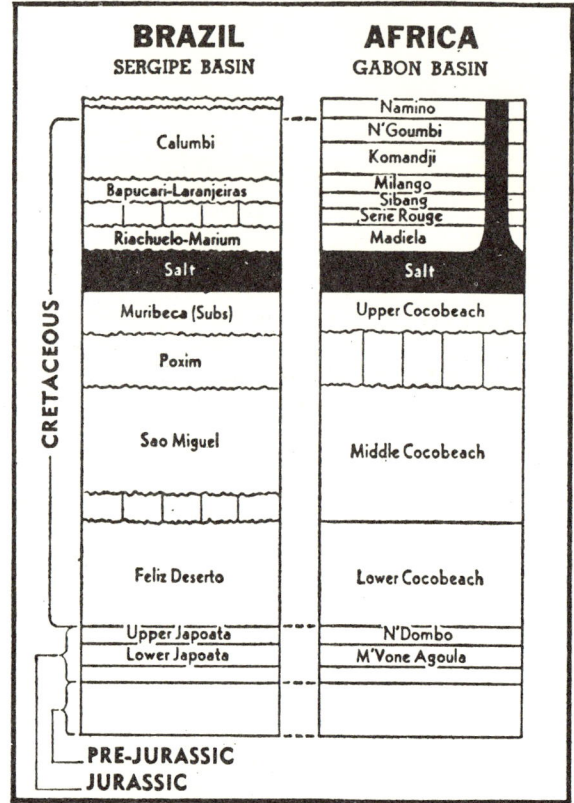

A former link between Africa and Brazil is indicated by similar sequences of older rock layers in Brazil (left) and Gabon (right). However, above salt laid down in mid-Cretaceous Period, 100 million years ago, the layers are no longer similar. The salt apparently marks the birth of the Atlantic Ocean as the continents split.

Rock formations in Africa and Brazil are said to match in terms of age (1), in composition (2) and in the layering and fossils (3).

February 8, 1969

Quake Data Show Drift in Earth Crust

By WALTER SULLIVAN
Special to The New York Times

WASHINGTON, April 22 — Analysis of earthquakes throughout the world has disclosed patterns of horizontal slippage consistent with theories of slow drift by huge plates of the earth's crust.

This was reported here today to the American Geophysical Union by Dr. Lynn R. Sykes of Columbia University's Lamont-Doherty Geological Observatory.

"We can almost say with certainty that continental drift is a reality." Dr. Sykes told newsmen afterwards.

He displayed a map showing the calculated earthquake motions. They indicate, for example, a movement of the Central Pacific basin to the northwest, relative to North America and Asia.

This would account for slippage along the San Andreas Fault that is responsible for most California earthquakes.

The idea that the surface of the earth is subdivided into a number of stable plates that drift about, push each other and sometimes override one another has been suggested by several scientists in the last year or two.

As evidence for the hypothesis Dr. Sykes pointed out that almost all of the world's earthquakes occurred in narrow zones that would be the margins of these hypothetical plates. Material rising along the midocean ridges would add to oceanic plates, pushing them aside.

Downward Flowing Material

Along ocean perimeters, where there are arcs of volcanic islands and deep offshore trenches, the ocean floor material is flowing downward, generating earthquakes to depths as great as 400 miles.

Although the surface motions can be described now with considerable confidence, Dr. Sykes said, it is far from clear what is going on at greater depths to push the plates hither and yon. Presumably, however, they ride on a comparatively soft layer, the "asthenosphere," from 60 to 100 miles down.

Dr. Sykes spoke at a plenary session of the meeting on "Frontiers of Geophysics."

Another speaker, Dr. Arthur E. Maxwell of the Woods Hole Oceanographic Institution in Woods Hole, Mass., reported on the results of drilling into the floor of the South Atlantic.

They show, Dr. Maxwell said, that the ocean there has been spreading at a uniform rate of about one inch a year for the last 70 million years. His estimate was made by dating the deepest sediment brought up at each site from its tiny fossil shells.

The oldest sediment from the hole nearest the ridge was 11 million years old. At four other

EARTH SCIENCES

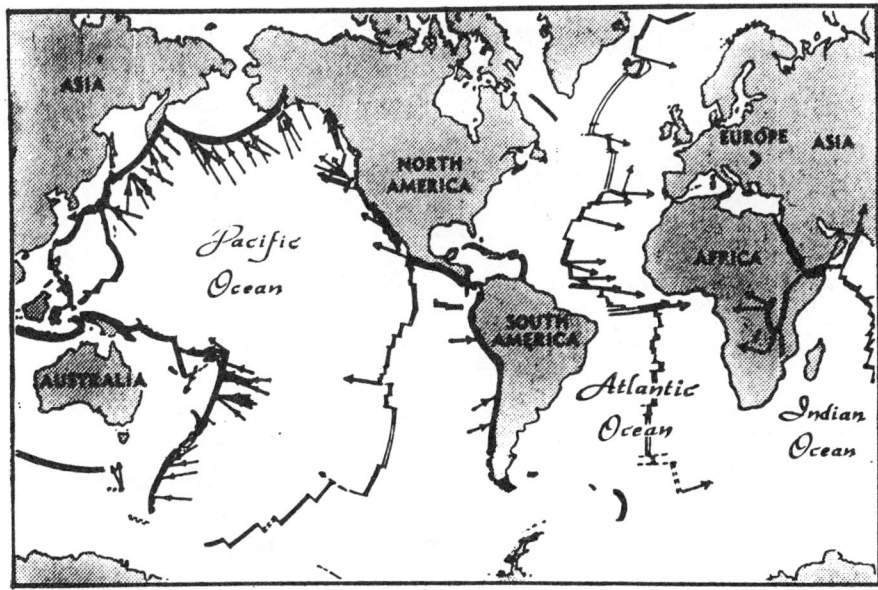

EARTH SLIPPAGE: Motions of large plates of earth's surface, relative to one another, are suggested by the arrows. Each arrow represents horizontal component of earthquake slippage by plate on which arrow is drawn relative to neighboring plate. The hypothetical plates, some 60 miles thick, are outlined by arcs of islands and arclike features (heavy lines) and worldwide system of rifts associated with midocean ridges (double lines). Map was adapted from one displayed by Dr. Lynn R. Sykes of Columbia's Lamont-Doherty Geological Observatory, who believes there may be as many as a dozen plates.

sites increasingly distant to the west the ages were 24, 40, 49 and 67 million years, respectively.

Spreading Rate of Ocean

From this it was possible to reconstruct the spreading rate and estimate that South America broke away from Africa some 150 million years ago, Dr. Maxwell said.

Dr. Maxwell said that, while the fossils found on the mid-Atlantic ridge today are typical of sediments laid down at depths of 6,000 or 7,000 feet, the fossil record indicates that in the past the ridge has lain at twice that depth.

This may mean, he added, that there is "breathing" of the ridge—that it is periodically inflated by material rising from deep in the earth.

Dr. Sykes estimated from earthquake data that the downward slippage of material under the island arcs is as great as six inches a year. This is "extremely high" from a geologic point of view, he said, and it would be possible to confirm such motion by some form of direct measurement.

April 23, 1969

NEW DATA SUPPORT CONTINENT DRIFT

Scientists Say Australia and Antarctica Were Linked

WASHINGTON, May 31 (UPI) —Two scientists said today they had proved that Australia and Antarctica, now 2,000 miles apart, were once joined in a single continent, which broke up millions of years ago.

The findings of Dr. Robert S. Dietz and Walter Sproll of the Environmental Science Services Administration's Atlantic oceanographic laboratoresi in Miami followed computer processing of data obtained by United States and Australian scientists during a global cruise in 1967 of the coast and geodetic survey ship Oceanographer.

The data disclosed a "precise fit" between the two continents, establishing that they used to be one, the scientists said.

The results were further support for the theory, accepted by most earth scientists, that the continents are sliding about on the planet's surface. The rate of drift varies from an inch to more than four inches in various parts of the world.

The continental drift theory, once derided, says the present continents used to be part of one or two supercontinents that were sundered 100 million to 200 million years ago by forces surfacing from the earth's interior.

The scientists said that "one of the last units to be sundered, the Australia-Antarctica split, may have occurred as late as 40 million years ago."

Study of the data, according to the scientists showed that the southeastern end of Australia, including Tasmania, fitted exactly into Antarctica's Ross Sea. The southwestern end of Australia matched Antarctica's Knox Coast. And "between these two end points, the concave Great Bight of Australia fitted snugly against the convex outline of Wilkes Land."

June 1, 1969

Fossil Called Proof of Continent Link

By WALTER SULLIVAN

The discovery in mountains near the South Pole of the fossil remains of a reptilian counterpart of the hippopotamus that lived, as well, in Africa has established "beyond further question" the former joining of all the southern continents, according to a leading authority on the subject.

The key discovery was made Thursday and reported in a message to the National Science Foundation in Washington by Dr. Laurence M. Gould, scientific leader of Adm. Richard E. Byrd's first expedition to Antarctica in 1928.

Dr. Gould and a fellow geologist, Dr. Grover Murray, president of Texas Technological University, visited the site ofter the discovery, and Dr. Gould said they both considered the find "not only the most important fossil ever found in Antarctica but one of the truly great fossil finds of all time."

Dr. Gould and Dr. Murray are members of the National Science Board, which oversees the National Science Foundation.

The fossil was found in the first bed of reptilian and amphibian fossils discovered on the Antarctica continent.

The deposit, apparently an old stream bed, was found a few weeks ago in the Alexandra Range, flanking the

A Lystrosaurus, Triassic creature of the type whose fossil remains were found by officials of National Science Board. Picture is from Edwin H. Colbert's "The Age of Reptiles."

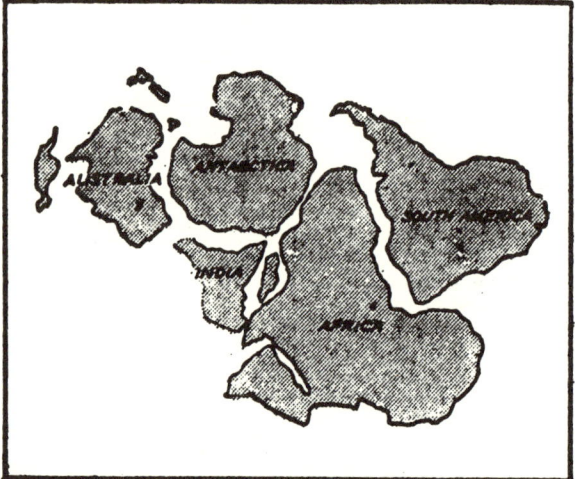

Above map indicates manner in which southern continents were thought to have fitted together hundreds of millions of years ago in a land mass. Scientists later termed the land mass Gondwanaland after a region of India.

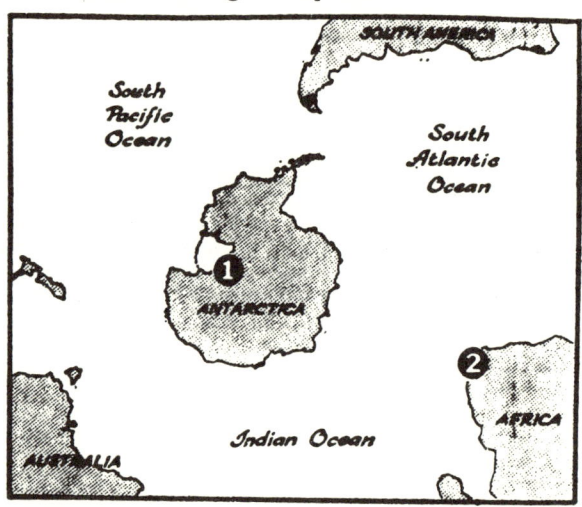

Present relative location of southern continents is indicated above. Fossils of hippo-like reptile discovered Thursday at (1) in Antarctica turned out to be identical with those that were previously found at (2) in South Africa.

Beardmore Glacier on the west. The Beardmore was the route British explorers used in their first attempts to reach the South Pole.

Two years ago, a single fossil fragment was found some 100 miles away in mountains east of the glacier. It was identified as being from a large, salamander-like amphibian that also lived on nearby continents. This was the first fossil hint of a former link between Antarctica and neighboring land masses.

Ever since coal deposits rich in the remains of large trees were found along the Beardmore Glacier, geologists have speculated about why Antarctica was once warm.

One explanation was continental drift—Antarctica was once closer to the equator and drifted to its present position at a speed of a few inches a year.

A striking similarity between the fossil Antarctica vegetation and that from India, South Africa, Australia and South America persuaded a number of scientists that those lands must have been joined during the Permian Period, some 250 million years ago.

The Triassic Period, from which the newly found fossil bed dates, followed the Permian.

Hypothetical Land

To explain the close relationship between the Permian vegetation of India and Africa, geologists conjured up a hypothetical land mass called Gondwanaland (from the Gondwana region of India). Some saw it as a lost continent that once linked the various land masses and then sank into the sea.

Others suggested a joining of the lands before they drifted apart. However, most geologists, until recently, found it hard to believe that the continents could plow their way along the ocean floor like giant barges.

The seeds of the "Gondwana" vegetation, they argued, could have been carried over the oceans by various means without a need for land bridges.

The new find, described by Dr. Gould as "a key index fossil," has been identified by Dr. Edwin H. Colbert of the American Museum of Natural History here. Dr. Colbert, a leading authority on fossils of that period, was with the field party that made the discovery.

The find was a partial skull of Lystrosaurus, considered an index fossil because it occurs in great numbers in some South African locations and has frequently been used to establish the period of a particular deposit there and in Asia.

The skull was unusual in that nostrils and eyes were high on the head, presumably so the animal could see and breathe while wallowing. It is considered inconceivable that such a creature could have migrated across open ocean. Lystrosaurus was two to four feet long.

Also in the fossil bed are the remains of Thecodonts (ancestors of the dinosaurs) and Labyrinthodonts. A jaw fragment of the latter was found in December, 1967. The only living descendents of the Thecodonts, according to yesterday's National Science Foundation announcement, are alligators, crocodiles and, via a more roundabout route of evolution, birds.

The site was discovered by Dr. David H. Elliot of the Institute of Polar Studies at Ohio State University Nov. 23—the first day of field work by his nine-man party. The location is at Coalsack Bluff, so named by New Zealanders because of its coal seams.

A field camp of Jamesway huts had been set up in the area by Navy construction men. The fossil bed is only five minutes from it by helicopter. Dr. Colbert was working in the area with a four-man team in the hopes of finding more vertebrate fossils in addition to the single jaw fragment of 1967.

December 6, 1969

EARTH SCIENCES

Scientists Find Ancient South Pole in the Sahara

By SANDRA BLAKESLEE

A team of earth scientists, dressed in summer shorts and sun hats, searched for the South Pole recently and found it—in the middle of the Sahara.

Dr. Rhodes W. Fairbridge, professor of geology at Columbia University and a member of the team, announced the finding yesterday at a meeting of the American Geophysical Union in Washington.

The expedition, which took scientists from 11 nations to the southeastern corner of Algeria, confirmed what has long been suspected, that the South Pole of 450 million years ago has been slowly edging its way northward, by a sliding action of the earth's crust, to the point where it has arrived today, exposed beneath the desert sun. Inch by inch, the ancient South Pole has traveled a distance of 5,500 miles.

"There is no question about it," Dr. Fairbridge said in an interview. "The territory that was the earth's south polar region in the Upper Ordovician period is now the Central Sahara." The Ordovician geologic period occurred in the Paleozoic era about 500 million years ago. The Taconic mountains in northeastern North America were formed during this period.

The phenomenon of sliding land masses is called the continental drift theory. It holds that, throughout the earth's 4.5-billion-year history, a series of vast plates, or continents, have ben floating over a fluid zone of molten rock about 40 to 60 miles beneath the earth's surface.

Scientists believe that the earth's avis and its magnetic poles have remained in fixed positions throughout the earth's history. So far, however, geologists have not determined the machanism by which this drift phenomenon works

Nevertheless, the evidence that such movements have taken place is indisputable, Dr. Fairbridge said.

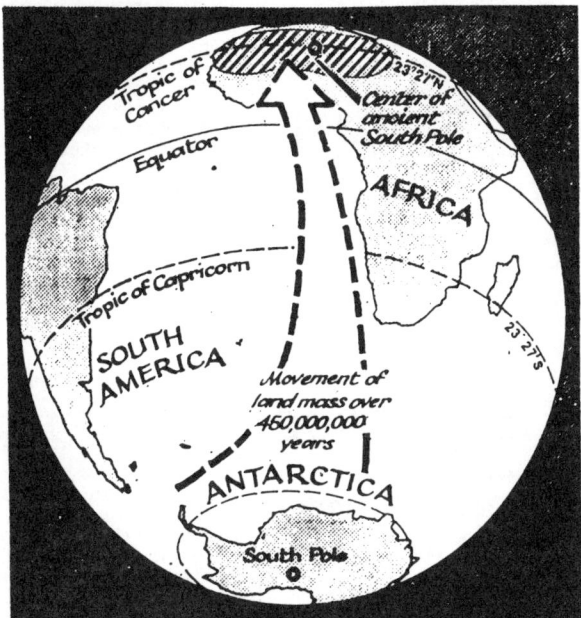

The New York Times   April 21, 1970

Schematic view shows the shift of a portion of the earth's crust from Antarctica to the Sahara region.

Month-Long Search

To accumuiate such evidence, the Australian-born geologist and his colleagues went to the Sahara last January. They spent a month combing the desert from an airplane and a Land Rover and digging in it with picks and shovels.

First, the flat topography showed great parallel grooves, like tractor marks, across hundreds of miles of smooth, pavement-like rock. This indicated that masses of glacial ice, which makes up polar ice caps, hoved across the land.

Second, the rocks held a magnetic history. Lines of magnetism in the rock indicates that it had once been at the South Pole rather than at the North Pole.

Moreover, the sedimentary composition of the rocks pointed to an icy past. Sandy deltas are the likely product of melting ice and soil cracks are the typical results of deep freezing.

Fossil Evidence

Fossil evidence clinched the theory. Trilobites, tiny crab-like animals that disappeared millions of years ago, just after the Qrodvician period, were found imbedded in the

The force of melting ice etched this sandstone in the Sahara, according to scientists who found it and said it showed how the South Pole of 450 million years ago moved north.

rocks. Radioactive dating of the rocks put their age at about 450 million years.

From such evidence, the scientists concluded that the Sahara was once covered by a large body of continental ice.

Although Europe was once covered by glaciers from the North Pole, Dr. Fairbridge said the ice that covered the Sahar did not come from the north.

One reason is that the area, which includes Morocco, Mauretania, Algeria, Niger, Libya and Chad, is only 1,500 miles from the equator — too far south for a northern glacier to come with the earth's axis fixed where it is. Another reason is that the grooves in the stone show that the glaciers moved from the south to the north.

The center of the ancient South Pole is now situated in the border pocket of Algeria, Libya and Niger. Temperatures there reach 137 degrees. Antarctic temperatures can plummet to 137 degrees below zero.

The earth looked quite different 450 million years ago than it does today, Dr. Fairbridge said. South America and Africa were probably one continent. Antarctica and Australia also were probably one continent, situated up near the Equator. Tropical coral deposits have been found in modrn Antarctica.

Scientists first became curious about the flat rocky formations in the Sahara 10 years ago, when Algerian and French geologists searched for oil deposits in the desert. The recent expedition, founded by the Algerian Petroleum Institute, was made to confirm suspicions about continental drift.

Fourteen scientists, from Algeria, Brazil, Denmark, England, France, Holland, Poland, Sweden, the Soviet Union, the United States and West Germany took part in the expedition. A second American on the team was the Rev. Paul Potter of the University of Indiana.

April 21, 1970

Evidence of Moving Earth Plates Is Found

3 Gigantic Sections Charted on Floor of Pacific Ocean

By WALTER SULLIVAN

Oceanographers from the Navy and Princeton University believe discoveries in the Pacific Ocean floor southwest of Central America support the view that the earth's crust is formed of gigantic plates that push and pull at one another.

The theory was formulated two years ago by Dr. W. Jason Morgan of Princeton and Dr. D. P. McKenzie of Cambridge University in England. To test it, the scientists rode the U.S.N.S. De Steiguer, a new research ship, to an area where, from global patterns, it was suspected that three such plates were being pushed apart from a junction point.

If so, it was reasoned, this would be evident in a peculiar, wedge-shaped pattern of undersea mountains, valleys and rock magnetism. The triple junction is where the Galapagos Rise, a ridge system running east and west just north of the Galapagos Islands, meets the north-south East Pacific Rise.

The suspected wedge pattern was found, with the Galapagos Rise as its centerline. Soundings along that rise also revealed the deepest clefts yet discovered in such an oceanic ridge system. According to an announcement yesterday by Princeton, one cleft is deeper than the Grand Canyon, which is about a mile deep.

This valley, which is 30 miles long, lies under three miles of water near the triple junction.

The drifting-plate hypothesis is an elaboration of the theory that the ocean floors are slowly being pushed away from mid-ocean ridges by molten rock welling upward into the ridges. Magnetic "signatures" appear to have been frozen into such rock as it cooled and was pushed away from the ridge.

From these signatures it has been possible, simply through magnetic observations by ships on the ocean surface, to estimate when the ocean floor in each area was laid down. During its 1,500-mile cruise, the De Steiguer sailed back and forth, charting topography and rock magnetism in the junction area.

The resulting pattern indicates ocean floor movement away from both ridges at speeds that vary from one sector to another.

The combined effect of these motions is to push the Pacific Plate, west of the ridges, to the northwest. The so-called Cocos Plate, named for an island off Costa Rica, is being pushed northeast and the Nazca Plate, which is named after a province of Peru, is moving toward the coast of South America.

The existence of a deep ocean trench, paralleling that coast, plus earthquake activity such as that that devasted many Peruvian cities on June 1, is widely believed to result from slow penetration of the Nazca Plate beneath the Pacific coast of South America.

The new findings have been reported by Dr. Kenneth S. Deffeyes of Princeton, leader of the group, and Dr. Leonard Johnson of the Navy Oceanographic Office.

In a premliminary "onboard" report, the scientists said that up-welling of molten rock in the ridge area was also indicated by the appearance, in bottom photographs, of "pillow" lava on the floor of the deep valley. This type of lava, also seen near other midocean ridges, is formed when lava erupts under water.

July 9, 1970

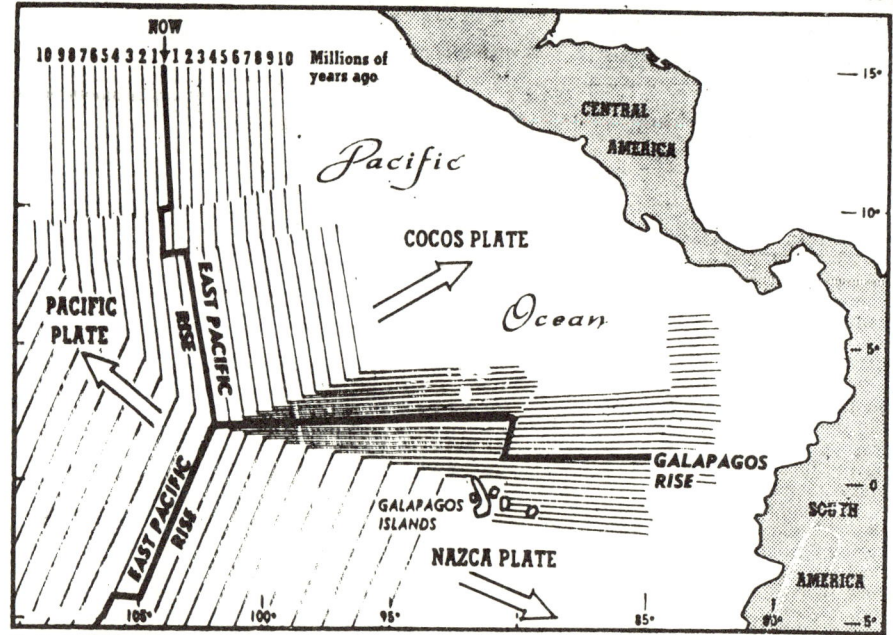

Charting of the Pacific Ocean floor by Navy and Princeton scientists, as shown above, is believed to confirm that three great plates of the earth's crust are being pushed apart southwest of Central America. Such motion could account for the devastating Peruvian earthquake of June 1. Up-welling lava along the ridges, shown as heavy lines, constantly pushes the plates apart. Lighter lines mark increasing ages of ocean floor to either side of the ridges, as indicated by magnetic "signatures" in the rock. The lines are spaced a million years apart, as indicated by numerals at the top.

EARTH SCIENCES

Mountain Birth Linked to Oceans

Giant Shifting Plates in Earth Termed Key to Mystery

By WALTER SULLIVAN

Recent discoveries on the ocean floors are leading to a comprehensive theory on how the far-flung mountain ranges of the world were formed.

The theory explains why the ranges tend to fringe continents. It accounts for the timetable of their formation, from the crumpling of the Appalachians hundreds of millions of years ago to the present emerging mountains.

It offers an explanation for the fact that the world's highest mountains, the Himalayas, are built of ocean floor material.

The theory, known as "global plate tectonics," is part of a revolution in the earth sciences that many specialists in the field consider as important as the Darwinian revolution of a century ago.

While geologists for centuries have studied the folding and upthrusting of mountains from the Alps to the Cascades, there has never been a generally accepted explanation of what thrust them up or why it occurred when and where, in each case, it did so. This theory meets such a need.

It is a consequence of the discovery that the earth's crust is formed of great plates, both continental and oceanic, that move independently, colliding or overriding one another.

Science of Tectonics

Tectonics is the science dealing with processes that have formed the mountains and other surface features of the earth.

Earlier this week a series of maps was made public showing how movements of the earth's crustal plates tore apart an ancient supercontinent some 200 million years ago and transported the fragments to the present locations of the continents.

According to the new theory, encounters between such drifting plates account for all of the world's major mountain systems.

While the theory has evolved in stages, with many contributors, it has now been set forth in detail by two geologists.

They are Dr. John M. Bird, head of the geology department at the State University of New York in Albany, and Dr. John F. Dewey of Cambridge University in England.

Their presentations appear in recent issues of The Journal of Geophysical Research and The Geological Society of America Bulletin.

Elements of the theory are still in dispute. Not all specialists in the field agree with their hypothesis that the contraction of an ancient ocean, the so-called Appalachian Atlantic, led to the formation of the Appalachian Mountains.

Questions About Concepts

Dr. Bird, in turn, does not accept the widely held view that the oceanic plates are being driven apart by material welling up into the mid-ocean ridges.

He suspects, rather, that the plates are being carried on the shoulder of a layer of flowing rock, 100 miles thick, that is part of a deep churning activity within the earth.

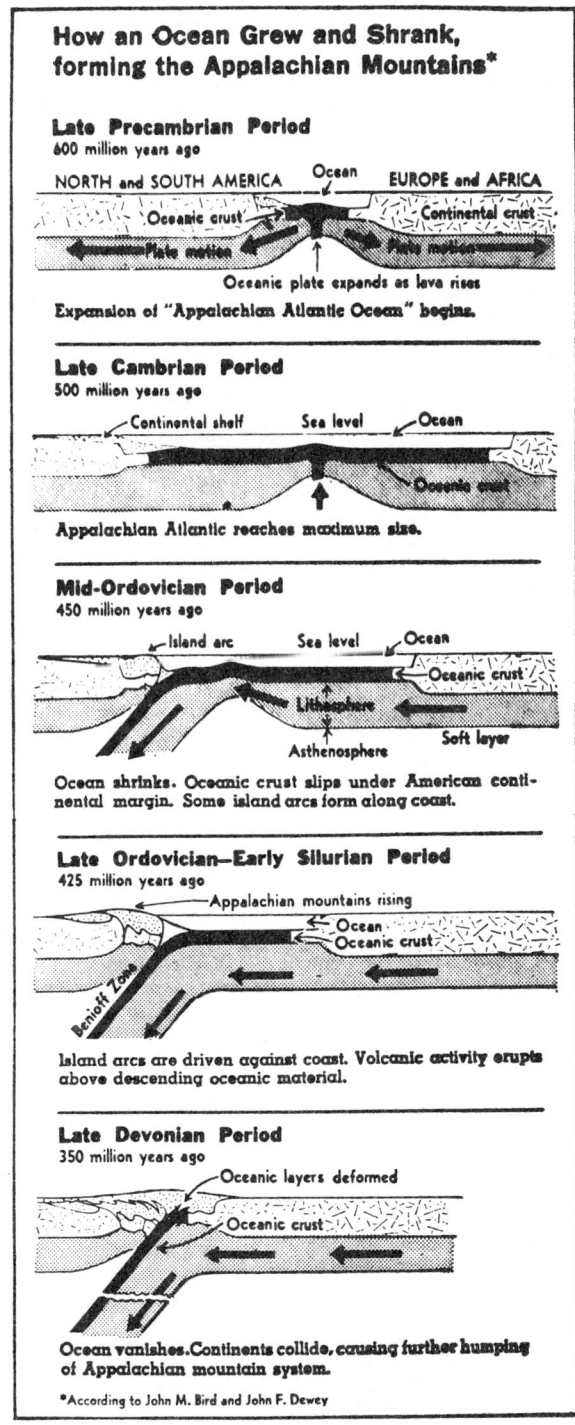

How an Ocean Grew and Shrank, forming the Appalachian Mountains*

Late Precambrian Period — 600 million years ago
Expansion of "Appalachian Atlantic Ocean" begins.

Late Cambrian Period — 500 million years ago
Appalachian Atlantic reaches maximum size.

Mid-Ordovician Period — 450 million years ago
Ocean shrinks. Oceanic crust slips under American continental margin. Some island arcs form along coast.

Late Ordovician—Early Silurian Period — 425 million years ago
Island arcs are driven against coast. Volcanic activity erupts above descending oceanic material.

Late Devonian Period — 350 million years ago
Ocean vanishes. Continents collide, causing further humping of Appalachian mountain system.

*According to John M. Bird and John F. Dewey

The two geologists have sought, through detailed analysis of such encounters, to explain the complex rock structures of various mountain systems. Even in as relatively simple an encounter as that between South America and the Pacific Ocean floor the effects are manifold.

In such a process the ocean plate is warped downward where it meets the offshore edge of the continent. The ocean plate descends, forcing its way under the continental rock in a slope known as the Benioff Zone. This is a region of frequent earthquakes.

Where the plate first bends down there is a trench in the sea bottom and earthquakes near there are of shallow origin. However, the Benioff Zone, as it extends inland, slopes down several hundred miles, as shown by the pinpointing of its earthquakes.

At such depths pressures and temperatures become so great that some components of the oceanic plate melt and force their way to the surface in volcanic eruptions. The penetration of oceanic material also humps the rim of the continent.

Vast Oceanic Plates

There is upwelling of lava into mid-ocean ridges, Dr. Bird says, but it does not provide enough energy to push the plates.

Where an oceanic plate (the entire eastern Atlantic constitutes a single plate) has become thick, cool and heavy near its continental margin, in his view, its own weight may cause it to sink, pulling the rest of the plate slowly behind it.

An analogy, Dr. Bird said in a recent telephone interview, would be a towel that floats on water until one edge becomes soaked and sinks, pulling the rest behind it.

Dr. Bird and Dr. Dewey list four basic types of mountain-building processes: a collision between continents, such as the one that formed the Himalayas when India drifted into Asia; a plate of ocean floor pushing under a continent, as along the West Coast of South America; ocean floor undercutting another plate of ocean floor, forming island arcs like those of Tongas and Marianas in the Western Pacific; and an island arc impinging on a continent, as may have occurred in New Guinea.

Effect of Volcanic Activity

It is this humping and volcanic activity that, it is now believed, has built the world's second highest mountain sys- Similar volcanic activity is also evident in western Mexico and the Alaska Peninsula.

According to Dr. Dewey and

Dr. Bird, it also occurred along the eastern shores of North America as the ancient Appalachian Atlantic shrank.

Off Newfoundland, where the oceanic plate penetrated under an extension of the continent (the continental shelf), volcanic activity helped build an island arc that, as the ocean was squeezed out of existence, was thrown against the continent.

At this stage, some 350 million years ago, the Appalachians may have been the loftiest mountains of the world having, like the Himalayas, been formed by continental collision. Then, at least 150 million years later, the Americas again broke away from Europe and Africa as hot lava welled up into the rupture zone.

Such a process of oceanic birth is now thought to be taking place in the Red Sea.

The process that gave birth to the modern Atlantic rumpled the landscapes of Africa and North America, forming highly symmetrical folds such as those of the Blue Ridge and Pocono Mountains.

When an island arc such as Japan forms off a continental coast enclosing an inland sea, oceanic sediments accumulate in the enclosed basin.

If the island arc is then driven against the continent, those sediments are crumpled. Or, in a collision of two continents, sediments on the intervening continental shelves are folded into mountain ranges.

Formation of Himalaya

The sediments that were heaped to form the Himalayas accumulated on an ocean floor during a 500-million-year period, according to specialists in that region.

There are indications, from submarine earthquakes, that the moving oceanic plate that pushed India against Asia, having been blocked by the impact, is beginning to burrow downward south of India.

Another front where great forces are contending lies between Africa and Eurasia. It appears that Africa has long been pushing north in a plate motion that has formed the Alps.

The most active zone is in the eastern Mediterranean, where, according to Dr. Dewey and Dr. Bird, "a collision of North Africa with Greece and Turkey seems inevitable."

The time is not imminent, the plate movements being in inches per year. However, the compression can already be blamed for the destructive earthquakes of Turkey and the Balkans. Similarly, the burrowing of the Pacific floor under South America has caused the quakes that in recent years have taken thousand of lives in Chile and Peru.

September 2, 1970

Continental Drift Is Traced to an Iceland Volcano

By WALTER SULLIVAN

Intensive international efforts to determine what is slowly pushing the land masses of Europe and America farther apart have shown that, in a broad sense, a large part of the North Atlantic is one vast volcanic structure centered on Iceland.

This emerged at a conference of specialists from both sides of the Atlantic held at the University of Iceland in Reykjavik during the week of July 1 to 7.

Such evidence has been seized upon by those earth scientists who believe a great plume, or upwelling, of hot, relatively plastic rock from a depth of 1,000 miles or more is rising beneath Iceland, making that the most volcanic region on earth and pushing apart the eastern and western halves of the ocean.

However, other explanations were offered at the meeting. It was proposed, for example, that the hot, swollen nature of the North Atlantic floor arises from the slowness of spreading from its centerline.

Perhaps the most striking discovery reported at the meeting was the manner in which, chemically speaking, the second highest volcano in Iceland serves as the focal point of eruptions that extend at least 1,000 miles down the Mid-Atlantic Ridge.

Dr. G. E. Sigvaldason of the Nordic Volcanological Institute in Reykjavik described the analysis of 513 samples of lava collected from all over Iceland. These showed that, in a strikingly uniform manner, those specimens of a potash-rich variety showed less and less potassium content at increasing distances from Kverkfjöll volcano.

The latter rises on the north edge of Iceland's largest ice sheet. Not only does the volcano mark the bullseye center of a symmetrical pattern of changing potassium chemistry, but it does so also for other chemical constituents of Icelandic eruptions, such as titanium and phosphorus.

What, in particular, created a sensation among those at the meeting was the manner in which this trend not only radiates in all directions from Kverkfjöll, but extends under water for 1,000 miles south, down the local sector of the Mid-Atlantic Ridge known as the Reykjanes Ridge.

This implies that something is taking place, centered beneath Kverkfjöll whose influence extends at least 1,000 miles to the south. In fact it was noted by M. H. P. Bott of Durham University in Britain that the floor to the Atlantic slopes up toward Iceland from the Azores—a distance comparable to that from New York to Los Angeles.

The Plastic Layer

This, Dr. Bott said, may be interpreted as a south-to-north increase in temperature of the plastic layer or "asthenosphere," believed to lie beneath the rigid surface of the earth.

It is the softness of this layer that is thought to make possible the motions of great plates of the earth's surface, such as those moving Europe and America away from one another.

If, for some reason, this layer were particularly hot beneath Iceland and the vast region surrounding it, the layer would swell and the sea would become correspondingly shallower.

Such a possibility was reinforced by a report from Dr. Marcus G. Langseth and Gary Zielinski of Columbia University's Lamont-Doherty Geological Observatory. Virtually the entire sea floor, extending about 1,000 miles north and south of Iceland, displays an unusually high rate of heat flow from the earth's interior, Dr. Langseth told the meeting.

This applies to the region north of the Charlie-Gibbs fracture zone, where the Mid-Atlantic Ridge is offset at latitude 53 degrees North, as well as to the Norwegian and Greenland Seas. "Evidence is increasing," Dr. Langseth said, "that Iceland is but the emerged part of a much larger anomalous oceanic region."

Among the meeting participants was Jean-Guy E. Schilling of the University of Rhode Island who had previously reported on the analysis of 9,000 pounds of volcanic specimens dredged by the research ship Trident from the Reykjanes sector of the ridge and from the central zone of Icelandic volcanic activity.

These provided the first indication of a systematic change in lava chemistry as one approaches Iceland from the south. Now, Dr. Schilling told the meeting, the Trident has obtained a comparable collection of specimens along the ridge extending north from Iceland.

While the potassium analysis is incomplete there are other indications (from relative abundances of lanthanum and samarium) of an abrupt cutoff just north of Iceland. An east-west fracture zone beneath the sea there, Dr. Schilling believes, may impede the northward flow of material from the suspected plume beneath Iceland.

Dr. Schilling has reported a change in lava chemistry as one approaches a hypothetical plume beneath the Afar region of Ethiopia that is allegedly pushing Africa and Arabia apart. This is where three rift systems converge—from the Red Sea, the Gulf of Aden and East Africa. An even more pronounced effect, according to Dr. Schilling, centers on the Azores in the central Atlantic. A similar trend in eruption chemistry has been reported as one approaches a hypothetical plume under the Galapagos Islands off Ecuador.

Such modifications in the chemistry of lavas erupted at increasing distances from a suspected plume have been attributed to dilution of the lavas as they traveled away from the plume beneath the rigid plates that form the earth's surface.

Gravity Force Higher

Further evidence for an unusual situation beneath the North Atlantic was presented by J. R. Cochran and Dr. Manik Talwani, director of the Lamont-Doherty Observatory. Through the entire region the local force of gravity tends to be higher than in comparable ocean areas elsewhere.

In presenting an alternate explanation for the North Atlantic findings, Dr. Langseth pointed out that the spreading of sea floor away from midocean ridges provides an efficient escape route for heat from deep within the earth. The heat is carried up into the newly forming rifts by lava and also the melting of the lava before it erupts absorbs heat from the depths.

Hinge Region

Whereas the spreading process in the South Atlantic is rapid, amounting to several inches a year, in the north—the hinge region in the swinging of the Americas away from Eurasia and Africa—the motion is much slower.

As a result, Dr. Langseth said, the heat does not escape so readily, perhaps accounting for the swelling and high heat flow of the North Altantic floor and the eruptions centered in Iceland.

Another explanation was offered by Dr. G. P. L. Walker of the Royal School of Mines in London, who has studied Icelandic geology for years. He noted that properties of the earth's crust in land areas facilitate eruptions from below more readily than the crust beneath deep seas.

If land, or a slightly submerged continental fragment, remained behind in the position of Iceland when the two sides of the Atlantic split apart, this would have favored intense eruptions there. So, as the continents drift away to east and west, "land stays as land," he said, "and perhaps we don't need a plume—a hot spot."

One task of the conference, which was organized by the Geoscience Society of Iceland, was to reconstruct the history of the North Atlantic.

80 Million Years

Dr. Walter C. Pitman 3d of the Lamont-Doherty Observatory proposed that the splitting open of the North Atlantic began about 80 million years ago. At that time, according to his reconstruction, the easternmost part of Siberia, facing the Bering Sea, was attached to Alaska and separated by sea from Asia.

As spreading of the Atlantic pushed North America west, this Alaska-Siberian promontory struck Asia, producing the Verkhoyansk Mountains and other ranges of eastern Siberia.

Meanwhile rifting between Greenland and Scandinavia led to volcanic eruptions that poured lava onto what is now the Greenland coast in the area of Gunnbjorns Fjeld, Greenland's highest mountain, rising 12,139 feet above the sea. According to C. Kent Brooks of the University of Copenhagen the eruptions come close to having been the most voluminous ever identified.

They poured onto bedrock that itself had been lifted by what Dr. Brooks believes was a former plume. The result was a dome 20,000 feet high, he said. Its eastern part was split off, as the continents separated, and subsided, part of it surviving in the Faeroe Islands.

Other continental fragments were left behind and sank beneath the sea in the spreading process, notably a large region known as Rockall Bank. As with the submerged area around the Faeroes, this has been shown to be continental in structure.

From the evidence presented in Reykjavik, the spreading patterns changed. The Jan Mayen Ridge, extending under water from Jan Mayen Island, which is still volcanically active, south toward the Faeroes was the original source of spreading between Norway and Greenland.

Some 30 million years ago the spreading zone apparently moved west to the presently active ridge that then gave birth to Iceland. Meanwhile spreading was taking place between Greenland and North America, forming Davis Strait and the Labrador Sea. This also seems to have pushed Greenland some 100 miles north before the activity died out 13 million years ago.

Dr. Peter R. Vogt of the Naval Oceanographic Office in Washington attributed this to a plume rising off southern Greenland — "Iceland's extinct cousin," he called it. Perhaps, he added, the plume jumped from there to a position beneath Iceland.

Dr. Vogt proposed, as well, that the shallow ridge extending from Greenland through Iceland to the Faeroes once served as a dam impeding the exchange of water between the Atlantic and Arctic Oceans. This ridge, some believe, was formed by the plume that now feeds the volcanoes of Iceland.

As this ridge spread away from Iceland and its foundations cooled, it presumably shrank and subsided. This, according to Dr. Vogt, allowed cold, dense water from the Arctic Ocean to penetrate the Atlantic, altering current patterns in that ocean. Some believe changing currents helped initiate the ice ages that followed.

It was noted that, although the opening of the North Atlantic began at least 60 million years ago, the fossil record indicates that mammals were able to move freely between Europe and America as recently as 40 million years ago. If the ridge described by Dr. Vogt was above water at that time, it could have provided such a link.

Evidence that Iceland itself is being split apart was presented by Dr. Robert W. Decker and Richard Plumb of Dartmouth College, Paul Einarsson of the Lamont-Doherty Observatory and, separately, by a German group from the University of Branuschweig.

Laser Devices

Periodic readings, using lasers and other high-precision distance measuring devices, have shown an apparent widening over the last few years of one or two inches across the Thingvellir rift zone. This is the area of broad fissures northeast of Reykjavik where the Althing began to meet in 930 A.D.—often described as the world's first parliament.

If, Dr. Schilling said, the activity beneath Iceland is fading, as some bellieve, less and less material will rise to fill the gaps formed by this spreading and Iceland will eventually split in two.

While the overwhelming majority of earth scientists now accept the evidence for sea-floor spreading as conclusive, there are still dissenters. Among them are several prominent Soviet scientists. The latter failed to make a scheduled appearance at the meeting, but this was thought unrelated to its subject matter—or to its partial sponsorship by the Science Committee of the North Atlantic Treaty Organization.

The only voice raised in opposition was that of Dr. Mac L. Keith of Pennsylvania State university. He argued that the sea floors are converging on the Mid-Atlantic Ridge instead of flowing in the opposite direction.

In the title of his talk he referred to it as "An Outrageous Hypothesis" and it was evident that this was one of the few aspects of his paper with which his listeners were in general agreement.

July 14, 1974

Continents In Motion

The New Earth Debate.
By *Walter Sullivan.*
Illustrated. 399 pp. New York:
McGraw-Hill Book Company. $17.95.

By SANDRA SCHMIDT ODDO

Down the middle of the Atlantic Ocean there is a ridge, and down the middle of the ridge there is a rift, a

Sandra Schmidt Oddo is a writer and freelance critic.

crack through which molten lava bubbles up. Once there were glaciers in the Sahara. A lot of mountains contain fossils of sea creatures; often the rocks, folded like tablecloths, are youngest on the bottom. The earth's magnetic field, the thing that causes compass needles to point north, occasionally reverses to make them point south. Once there were swamps in Antarctica; once there were seas in the American Middle West. The water under the Red Sea is hotter and saltier than it ought to be. These are facts.

They have been explained, some of them, by theories that the earth shrinks; that it expands; that it is subject to occasional attacks by wandering planets (Venus, ripped loose from Jupiter and careening across the solar system before being bound by the sun); that the poles, which do wobble, go drifting around the globe.

But every once in a while in scientific work the whole unwieldy accumulation of facts tips over, spills across a field of knowledge in a grand, new and unexpected pattern that puts all the facts in order, with vast spaces for more, as yet undiscovered, facts—and that changes man's whole perception of himself and the world he lives in. It happened to Galileo. It happened to Darwin. And it may have happened again in this century, first in 1915 to a man named Alfred Wegener, who couldn't seem to get the idea across, then gradually, since the 1950's, to nearly the whole of geology, biology and perhaps physical science. Rather like describing a mystery story by saying "the butler did it," the new theory can be condensed into this statement: continents move.

The idea that New York was once across the street from Morocco, that Brazil untucked itself from under the bulge of Africa, that India slewed from the bottom of the world and somehow charged headlong into Asia, raising the Himalayas with its impact, is by now more or less familiar to anybody who reads the newspapers. But the continental drift theory is familiar as doings in space are familiar, as briefly interesting results of sketchily understood projects. Now Walter Sullivan, science editor for The New York Times, has done an admirable job, a journalist/scholar's job, an educator's job of arranging all this information clearly, suspensefully and fully for the reader.

He tells the story like a mystery story, clue by clue, investigator by investigator. He sets up difficulties, problems and, just as the reader's intellectual itch becomes unbearable, he provides possible answers. "Continents in Motion" is a staggering assemblage of facts; it is also a history of discovery as revealing as the ways of science and scientists as J. D. Watson's account of unraveling the genetic code.

Sullivan subtitles the book "The New Earth Debate," and carefully fosters in the reader a proper sense of scientific skepticism, the "oh, yeah?" attitude that helps keep discoverers on the straight and narrow. But Sullivan, it is evident eventually, is a true believer.

The theory is so elegant: 10 major plates carrying continents and oceans on their backs, skating around the globe in a grand slow-motion bumper-car game. Ridges rise between the plates and new earth appears, spreading them apart. Where old plates are squeezed, trenches open and the old edges dive back into the mantle, the mysterious dense hot middle of the earth, and volcanic mountains are pushed up as they descend. Or perhaps two plates collide: granite, the stuff of continents, is comparatively light, a froth boiled to the surface in the first seethe of the planet 4.5 billion years ago. It isn't easily plowed under. Since the first split of the first huge supercontinent there have apparently been numerous ruptures and re-suturings. Europe and Africa have banged repeatedly; the Atlantic has opened twice.

More elegance: the material of the world apparently recycles itself in a process taking perhaps 150 million years—the age of the oldest ocean floor. As old plates are swallowed down, metal, minerals, oil may be renewed; more land may be accumulated from sediment, the bodies of uncountable numbers of living things deposited on the ocean floor, scraped off as the plates descend. And all of it powered, perhaps by convection currents in the mantle, or moved—perhaps — by plumes of — what? — sprung from the molten, moving heart of the planet. The mind struggles to comprehend it.

And comes up with awe as the only appropriate reaction. "Uniformitarianism lives!" Sullivan exults, to his own satisfaction (and mine), finally triumphing over the opposing scientific camp which holds that wholesale catastrophe is necessary to explain the earth's anomalies. But uniformitarianism—the idea that all the processes needed to shape the face of the earth are here and at work now with earthquake and volcano, upthrust, inundation, and inch-by-inch drift—is quite catastrophic enough.

Like most orderly arrangements of facts, plate tectonics—the name is less than five years old—also have potential practical applications. If the forces that power earthquakes are understood, perhaps earthquakes can be controlled. Geothermal heat might be (and is being) drawn from areas where contention between plates produces hot spots close to the surface. If plate movement has an important effect on the formation of oil and gas, as it now appears, the right places to look for them are easier to predict. If some part of the process is laying down new deposits of minerals (slowly, too majestically slowly to replace our feverish ravages), it would be useful to know it.

But the uses of plate tectonics are still secondary; continental drift is still reasonably pure science and so the search for answers to all the new questions is still, by and large, a kind of tune-in-next-week adventure.

When any field has a theory as new and roomy as this one to explore, a certain ebullience and excitement is inevitable. Suddenly, there is the quantum leap: geology knows what it *doesn't* know — and where to go to find out. It behooves one to go dig holes in the ocean floor? Fine, let's do it—through typhoons and icebergs. There's something interesting in Africa's Afar triangle? So what if the temperature in summer hits 134 degrees, let's go see. I kept imagining these intrepid scientists—Sullivan sets the stage, but resists doing so—as natives must: a Vermonter scratching his head and saying, "You wouldn't believe it, Min, but there's this fella up on the highway cut doing funny things to the rock." Or an African tribesman trying to tell his friends that he's been hired to help men in Land Rovers look for scratches! in the Sahara. Or, the more serious side, Robert Scott frozen in Antarctica, leaving on his sledge 35 pounds of old rock painfully gathered, hauled week after week until his death—and bearing evidence of long-ago life.

Sullivan's style is lucid and readable. His subject is fascinating. His book is one that makes you think well of yourself for reading it, and of him for writing it. ■

November 3, 1974

OCEANOGRAPHY

NEW DEVICE MEASURES OCEAN DEPTHS BY SOUND

Invention of Government Scientist Has Made Possible First Comprehensive Chart of an Area in the Pacific—Course of Earthquakes and Lost Continents May Be Revealed.

CONTINENTS sunk for ages below the surafce of the ocean, new lands growing up from the depths of the sea, can now be traced and charted by means of a device developed by experts of the United States Navy. By the use of this interesting instrument sound waves are projected even 3½ miles to the ocean's bed. The depth is determined by the interval elapsing before the wave returns and is registered on a delicate scale. Even in shallows and shoal water the ceremony of heaving the lead may now be abandoned, for the sonic depth finder, as the invention is called, will serve all purposes.

Officers at the Navy Department are so impressed with the working of this instrument that they speculate on finding the lost Atlantis, the continent believed to have spread between South America and the African coast, and the new land said to be emerging from the vast Pacific. Through the sonic depth finder hundreds of soundings can be made while the vessel from which the observations are taken is traveling at twelve knots an hour. After these soundings are marked on a chart all that is necessary to map the contours of the ocean's bed is to run lines from sounding to sounding.

The clearest illustration of what the sonic depth finder—taking its title from the Latin word sono, meaning sound—can accomplish is illustrated in a contour chart of the ocean's floor off the California and Mexican coasts, made after recent surveys directed by the navy's hydrographic office, at the request of the Carnegie Institution at Washington. Under orders from Captain Frederic B. Bassett Jr., the navy hydrographer, the destroyers Hull and Corry carried on the soundings, beginning operations in November, 1922, making 5,000 soundings in thirty-eight days.

Leaving San Francisco, the destroyers ran lines straight out from the west coast, ten miles apart, until a depth of 2,000 fathoms—12,000 feet—was reached. This process continued until they had charted the sea floor from San Francisco to Point Conception, near Santa Barbara. They ran lines five miles apart until Point Descano, just south of the Mexican border line, was reached. The average distance between soundings on each line was one to two miles, the distance covered was 5,600 miles, and the area canvassed 34,000 square miles. The Hull and Corry steamed at a steady pace of twelve knots.

Starting the Work.

The sonic depth finder is the development of a Government scientist, Dr. H. C. Hayes of the Experimental Station of the Bureau of Engineering, Navy Department. After months of research he brought the instrument to such perfection that in June, 1922, it was placed on board the destroyer Stewart, then detailed to proceed from Newport to Gibraltar, and on to Manila through the Suez Canal. Dr. Hayes accompanied the Stewart and soundings were constantly made while the destroyer was steaming across the Atlantic, a striking contrast to the old method of stopping and throwing out the lead, a process which usually consumed one to two hours. Depths from 90 to 250 fathoms were registered. So eminently successful was this experiment that the Navy Department determined to make the exhaustive test just completed in the Pacific.

The theory that a sound impulse starting from a point near the surface of the water will be reflected to that point by the sea bottom is the basis upon which the sonic depth finder was designed. Considering the velocity of sound in water to be 4,800 feet, or 800 fathoms, the distance the sound travels to the sea floor and back will be equal to the number of seconds shown on a stop watch times 800 fathoms. The depth would be equal to one-half of this total.

At first stop watches were used, but this involved the human equation and was not dependable. The next development was a constant speed mechanism, which could be set in motion by the outgoing sound impulse, then by means of a delicate relay system brought to rest by the received echo. But this, too, proved impracticable, for the relays could not be kept in adjustment on shipboard, and often the device was stopped by various unexpected sounds coming from under the water. The navy's experts devised an indirect means of measuring the time interval and produced results which were correct within one-half of one per cent. This device depends upon the principle that if sound impulses are sent out at known intervals, some of these will agree with the lapse of time necessary for the sound to travel to the bottom and back to the receiver.

The Navy Department does not wish it understood that the sonic depth finder alone can measure distances greater than forty or fifty fathoms down. For depths such as encountered in mid-ocean and in the paths of ocean currents it is necessary to employ an instrument known as a high-power sound transmitter, amplifying the impulse received. A secret device for reception also is used.

Tracing the Earthquakes.

The Carnegie Institution sought the chart of the Pacific because of its deep interest in the origin of earthquakes. The Pacific has been conducting itself in strange fashion for the last year, and the seismological societies of the United States have been conducting studies of the great ocean and the disturbances on adjoining shores. More knowledge was demanded. For instance, in order to study the earth movements of California it was necessary to know the zones of structural weakness. Some of them can be plainly seen, others can be found only by inference from the adjacent geological structures. Knowing where the zones of weakness are situated, it then becomes comparatively easy to discover the source and direction of underground movements.

Before the Hull and Corry began their work, the Seismology Society of America prepared a fault map of the land area most liable to earthquake disturbances on the Pacific Coast, and the destroyers worked over the sea territory adjoining. As they went on with their soundings and the map grew, it became apparent that, as the earthquake experts had already deduced, many of the earth faults led to the seashore and on beneath the surface of the water. The finished chart represents the configuration of the ocean floor in the region, showing the submerged hills, cliffs, valleys and precipices. It is, in fact, the first successful contour map of a zone of deep-sea soundings ever made.

"It is clear from the contour map that a number of steep slopes or cliffs have been located, some of which may be fault scarfs of considerable elevation," said Captain Bassett in describing the achievement. "The indications are that the chart also locates the so-called continental shelf which is commonly thought of as representing the structural demarcation between a sinking ocean bed and a rising continent.

"The future study of the inaccessible ocean depths appears now to be of high precision. Regions in which changes occur frequently, such as the coast of Chile or the Hawaiian Island group, can be studied with great care and detail and the direction in which future displacement may be expected can be ascertained.

"The clearness with which these contours are delineated by the sonic survey suggests the possibilities of a more effective study of continent building and of the general problem of isostasy than has heretofore been possible. Ancient connections such as are supposed to have existed between South America and Africa, or across the Indian Ocean, can now be worked out almost as positively as upon a land area. The perfection of the sonic depth finder offers unlimited possibilities to tectonic geology and oceanography."

It is known by the navy hydrographers that great changes are going on in the deeps of the oceans. Scientists believe that land is rising near the West Indies and in the region of Malaysia, and even near Hawaii. By means of the sonic depth finder studies of today can be adjusted at any later day. Depths up to 3,200 fathoms, or 19,200 feet, have been measured with the device.

April 8, 1923

Life Discovered 6 Miles Down in the Pacific; Anemones, Bivalves and Crustaceans Found

Special to THE NEW YORK TIMES.

MANILA, July 25—A group of international oceanographers and marine scientists exploring the Mindanao Deep off Mindanao Island reported today finding proof that life exists more than six miles beneath the surface of the sea where hydrostatic pressures exceed 15,000 pounds a square inch.

Ingenious scoops and steel nets lowered by cable from the Danish research vessel Galathea first brought from the sea floor samples of primeval ooze containing bacterial matter from 34,000 feet, just under six miles.

Then, exploring at greater depths in the so-called Mindanao Trench, they trawled seventeen sea anemones, sixty-one sea cucumbers, two bivalves and one crustacean, demonstrating the fairly rich variety of life in the previously unexplored submarine region believed to be the deepest spot in the world's oceans—between six and seven miles below the surface.

The 1,600-ton corvette Galathea, borrowed from the Danish navy and fitted for oceanographic research, docked briefly today at Cebu in the central Philippines for refueling and to report on its findings.

The head of the expedition, Dr. Anton Bruun of Copenhagen University, said that the expedition's work and especially its recent findings would help solve mysteries locked prehistorically under the world's deepest waters.

Members of the expedition include Danish zoologists, a marine research expert from Thailand, a Swedish scientist and two Americans. The latter are Miss Grace Pickering of Yale University, who is leaving the expedition in the Philippines, having traveled with it from Europe, and Dr. Claude Zobell, science professor from Scripps Institute of Zoology at La Jolla, Calif., who joined the party here. The Galathea is on a two-year circumnavigation trip sponsored by Prince Axel of Denmark.

The New York Times July 27, 1951

Where ocean life has been found six miles down (cross).

July 27, 1951

SEA MUD BELIEVED 10,000 YEARS OLD

Columbia Scientists Obtain Core Specimens From the Deepest Atlantic Area

---

DATA ON ICE AGES SOUGHT

Samples Indicate the Flow of Gulf Stream Has Stopped in Some Past Periods

By ROBERT K. PLUMB

Columbia University scientists have snatched two precious scientific specimens, at least 10,000 years old, from Davy Jones' locker.

From the deepest point in the Atlantic Ocean, the great Puerto Rico Trench, near the island of Puerto Rico, two priceless samples of sea bottom mud and sand have been brought up. They are now at the Lamont Geological Observatory in Palisades, N. Y., where they are being studied with microscopes, mass spectrographs and other powerful sensing instruments of science. The announcement of their being obtained was made yesterday.

The paleontologists, geologists, chemists and physicists who are studying the specimens hope to learn the story they tell of the history of the planet earth since the close of Würm time, the 10,000 years since the last glacier covered half of the North American Continent.

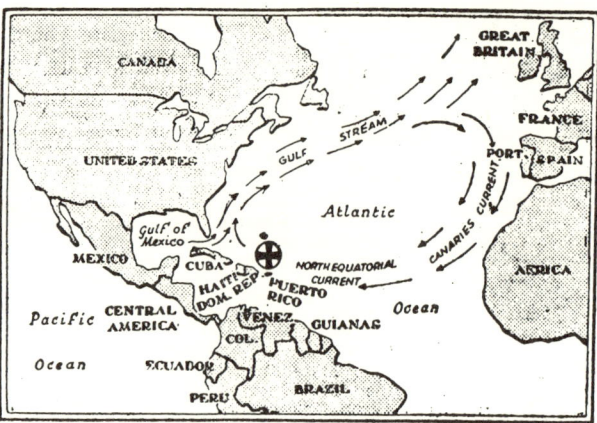

The New York Times   May 22, 1954
Puerto-Rico Trench (cross), where the samples were taken

A new concept of the history of the earth is being evolved from samples of oceanic floors by the Columbia University group. About 1,000 samples have been taken from the Gulf of Mexico to the Mediterranean, from the North Pole to the Equator. The leader of the group is Dr. Maurice Ewing, a geophysicist, who is now in Havana with the Vema, research ship of the Lamont Observatory, preparing for another Atlantic expedition.

Associated with him are Dr. David Ericson, a marine geologist; Dr. Lamar Worzel, geophysicist specializing in studies of gravity; Dr. Bruce Heezen, marine geologist specializing in the topography of ocean bottoms; Dr. Albert W. Bally, paleontologist studying ocean sediments, and Goesta Wollin, a marine geologist.

Deepest Core Sample Taken

The Puerto Rico Trench specimens are the deepest ocean core bottom samples ever obtained. The two are about two and one-half inches in diameter. One is ten feet long, the other is thirteen.

They were taken by lowering a sampling piston deep into the ocean on the end of a half-inch steel cable. Twelve feet from the bottom, a trigger is released and 1,000-pound lead weights drive the piston into the pristine sea floor. The piston and its contents are then hauled five miles to the surface. The substance obtained is a representative cross sample of the material that has fallen into the undersea deep.

Both Puerto Rican specimans are the same at the top, which represents the latest fall of material into the sea trough. That layer comprises six inches of red clay. It is assumed that it was deposited in the last 10,000 years.

Below this layer in both cores are three graded layers of sand of the type common in shallow waters. The finest is at the top of the sand layer as it would be if a river of sand flowed into the Trench.

The most startling find made so far in the study of the samples is that the shallow sands contain specks of an alga known as Halimeda. Halimeda needs sunlight. No sunlight can pierce the Trench. So the alga and the sand must have flowed into the Trench in a great sub-oceanic river, as the Lamont scientists have postulated other prominent features on the Atlantic bottom were formed.

Chemical Tests Give Data

These samples from the Trench, and others from other parts of the ocean, have been subject to extensive chemical analysis. It is known that the ratio of ordinary oxygen 16 to heavy oxygen 18 depends upon the temperature at the time the oxygen-containing particles were laid down. Also, some species of single-celled sea animals grow better in cold waters and some favor warm waters. The ratio of these creatures discovered in an oceanic specimen reveals something of the temperature when they died and became part of the sediment.

Oxygen measurements and counts of warm versus cold water animals come to surprising agreement when many sediment-temperature samples are studied together. They show the water to be cold in the last (Wisconsin) glacial stage and warm in the Sangamon inter-glacial.

Another phase of the studies has possibly shed light upon the cause of the glaciers. A tiny creature named Globorotalia menardii is fairly commonly spread through the path of the North Atlantic currents including the Gulf Stream. Studies of the occurance of this creature at spots on the sea bottom suggest that its appearance depends upon a supply of food carried by ocean streams and not on temperature alone. During the Ice Age, layers of oceanic sediment with few specimens of Globorotalia menardii were laid down.

From this observation the Columbia scientists speculate that the Gulf Stream stopped at one or more periods in earth history.

From these and similar studies the group hopes to report whether the earth is or is not to enter another Ice Age.

May 22, 1954

NEW DEVICE SPURS SEA FLOOR STUDY

Big Corer Built at Columbia Brings Up Diameter Foot of Sediment From Vast Depth

By ROBERT K. PLUMB

Ocean-bottom explorers at the Lamont Geological Observatory of Columbia University have constructed a new sampling device that will enable them to grasp a diameter-foot cross-section of sediment from the ocean bottom.

Effective to a depth of five miles, this device is expected to augment pioneering studies of ocean-bottom sediments conducted by the Columbia group over the last few years.

The scientists, working on the Vema, a three-masted schooner, have been taking samples with a piston device that obtains a sample only two and one-half inches in diameter. The new sampler will obtain specimens eleven and a half inches in diameter.

Ocean-bottom samples are analyzed by elaborate tests for geological origin, chemical content and the nature of tiny plants and animals entrapped in them.

Advantages Over Old Way

With a sample only two and a half inches in diameter, it was necessary to take a sediment layer at least one-half a centimeter (a fifth of an inch) thick to get enough material to provide for accurate analysis.

With the new eleven-and-one-half inch samples, a layer only millimeters thick (a millimeter is one twenty-fifth of an inch) will provide enough material for the essential analyses.

The long-used two-and-one-half inch piston on the Vema takes samples that have been as long as fifty to a hundred feet. Although the new device will be limited to a sample of one-foot depth on the ocean floor, there will be increased accuracy of analysis at millimeter intervals.

The new deep-sea equipment is a large diameter corer. It is lowered on a half-inch steel cable down thousands of feet into the ocean. A few feet off the bottom a trigger is released and a heavy lead weight drives the sampler into the sea floor. The corer and its contents are then hauled to the surface.

Successful Trial in Atlantic

Designed by Dr. W. Maurice Ewing, Professor of Geology at Columbia and director of the Lamont Observatory, the equipment was constructed in a machine shop at the observatory with funds from a Rockefeller Foundation grant for research in marine biology.

The new sampler was tried first in the Hudson River. Then the scientists took it to a point just off the Canary Islands in the Atlantic. In a recent test there, a precious sample of two gallons of undisturbed bottom ooze was obtained at a 13,000-foot depth.

The Vema is now in the Mediterranean, where both the new and the old samplers are being used in the research.

During past voyages of the Vema more than a thousand samples were obtained from the Gulf of Mexico to the Mediterranean, from near northern Arctic waters to the Equator. Many of the samples are at least 10,000 years old. From analysis of them is coming a new concept of the history of the earth and the origin of life in the deep sea.

July 13, 1956

A Huge Crack in the Floor of the Oceans Is Traced by Geologists

CRACK IN WORLD IS FOUND AT SEA

Columbia Teams Discover a 20-Mile-Wide, 2-Mile-Deep Fissure Circling Globe

By IRA HENRY FREEMAN

Geologists have traced a crack in the floor of the oceans that is twenty miles wide, two miles deep and runs around the world in a continuous line that is 45,000 miles long.

The line of the trench was located by teams of Columbia University scientists in five years of work in various parts of the world. Their success was announced yesterday by Prof. Maurice Ewing, director of the Lamont Geological Observatory, a unit of Columbia at Palisades, N. Y.

Dr. Ewing and Dr. Bruce Heezen, research associate, explained that the main line of the rift extended southward along the Atlantic from about the Greenwich Meridian in the Far North, bisecting Iceland, and running approximately midway between North and South America on the west and Europe and Africa on the east.

Splits Into 2 Branches

The great trench rounds the southern tip of Africa, then forks into two branches. The northern branch runs into the Arabian Sea and loops sharply south again through the Belgian Congo, Northern Rhodesia and other areas of the African continent. The southern branch continues eastward through the Indian Ocean, passes south of New Zealand and crosses the Southern Pacific until forking into two branches near Easter Island.

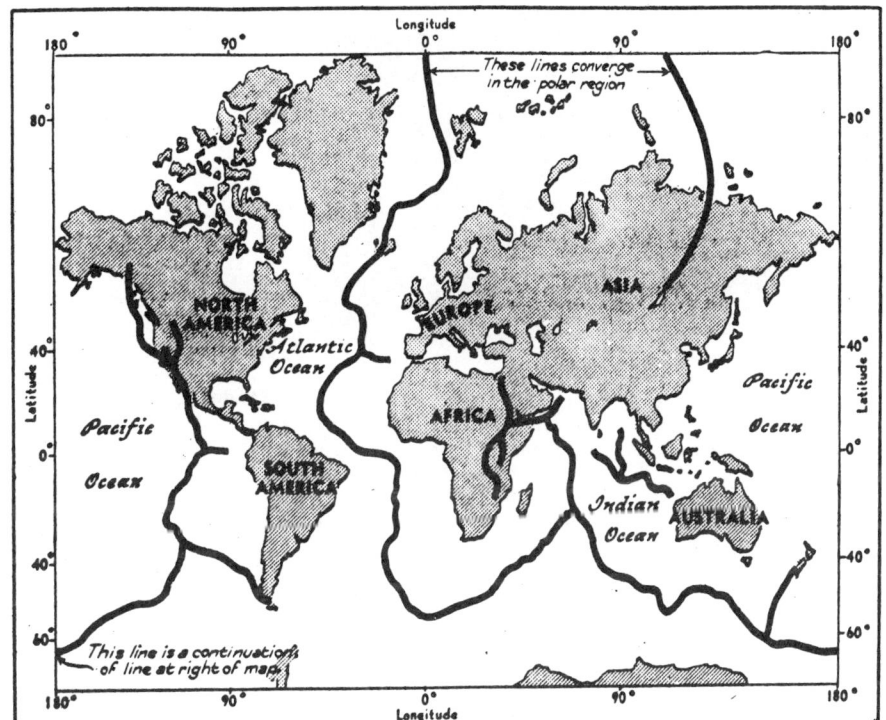

Heavy lines denote a 45,000-mile continuous trench and a shorter one in the Indian Ocean

The northern branch runs up the Gulf of California, while the southern branch curves down to the Straits of Magellan. There is "good evidence" of a connection in the Arctic Ocean, Dr. Ewing said, although this link has not been confirmed.

The rift was caused, the scientists theorized, by the pulling apart of the earth crust. Miss Marie Tharp, cartographer at the Lamont Observatory, had noticed that the locus of a great number of earthquakes in the North and South Atlantic in the last forty years coincided exactly with the great trench there.

Then the Columbia associates discovered that earthquake zones elsewhere followed the rift. On each side of the deep crack rise mountains from 6,000 to 12,000 feet tall, but the tallest peak is 3,600 feet below the surface of the sea. Everywhere the formation is remarkably similar—two belts of high, jagged mountains, each belt about seventy-five miles wide, separated by the trench about twenty to twenty-five miles wide and from two to four miles deep.

Rift Continues to Widen

"We became convinced that the undersea mountains and rift formed a world-wide system," Dr. Ewing said. "The earthquakes, which are shallow and still going on along the entire length of the rift, show that it is still being pulled apart."

Dr. Heezen expressed the opinion that the tracing of the rift "tended to weaken" the "theory of continental drift." This theory holds that North America, South America, Europe and Africa were one immense land mass eons ago but cracked roughly in the middle and pulled apart, forming the Atlantic Ocean basin.

"But notice that the rift rounds the Cape of Good Hope and runs up along the East Coast of Africa, too," Dr. Heezen commented. "If the Atlantic rift were caused by Africa moving eastward, how was the Arabian rift caused?"

The answer awaits further research. Columbia's three-masted schooner, the Vema, is exploring the rift in the South Atlantic now. Dr. Ewing will fly tomorrow to join her scientific crew in Buenos Aires.

"We believe," Dr. Ewing concluded, "the significance of these findings is that they may help determine the origin of the major surface features of the earth and changes that have taken place in its geological history."

February 1, 1957

Navy's Bathyscaph Dives 7 Miles in Pacific Trench

Special to The New York Times.

WASHINGTON, Jan. 23—In a feat of ocean diving that had scientific and military implications, the Navy's bathyscaph Trieste plunged more than seven miles to the bottom of the Marianas Trench of the Pacific Ocean today.

In announcing the feat as a world's record, the Navy said that the Trieste had gone down 37,800 feet in the deepest known hole in the world's oceans. This was considerably deeper than the record of 24,000 feet set by the same craft on Jan. 7.

It was also a greater distance downward toward the earth's core than Mount Everest in the Himalayas, the world's tallest mountain, rises above the surface of the earth—29,028 feet.

Jacques Piccard, the Swiss scientist, whose father, Auguste, designed and built the craft, was aboard the Trieste. With him was Navy Lieut. Don Walsh of San Diego.

No Difficulty in Dive

The Navy made the announcement at the Pentagon after having received an initial report that the ship had surfaced shortly before 1 A. M., Eastern Standard time, about 250 miles southwest of Guam.

The report said that the two men had encountered no difficulties during the ocean probe. At the depth of 7.15 miles, the hull of the undersea vessel sustained a pressure of 16.883 pounds a square inch.

The Navy, in a second announcement, said that the two scientists had found it "very cold" at the bottom. They emerged wet, because of moisture condensation inside their craft, and with teeth chattering.

They reported that they had spent a half hour at the bottom and could see living and moving objects. When the craft hit the bottom it shook up silt and "dust."

The Trieste submerged at 4:20 P. M. on Jan. 22, and reached the bottom at 9:10 P. M., the Navy reported. It thus took four hours and forty-eight minutes to sink. The two-man crew lost control half way down but regained it when resting at the bottom.

After half an hour's rest, it took three hours and seventeen minutes for the bathyscaph to return to surface. Half way up, the two scientists again lost control of their craft for a while. When they surfaced, Lieutenant Walsh released a plastic container with the American flag, the Navy reported.

"The purpose of today's dive is to demonstrate that the United States now possesses the capability for manned exploration of the sea down to the deepest part of its floor," the Navy announcement explained.

The series of deep-ocean plunges in which the Trieste is engaged, the Navy went on, is providing "scientific knowledge of sunlight penetration, underwater visibility, transmission of man-made sounds, and marine geological studies."

The Navy announcement did not mention that the information was also important to the operation of the United States' growing fleet of nuclear-powered submarines. It also is important to the development of anti-submarine warfare devices against the Soviet Union's huge fleet of submarines.

The descent proved that the depth of the Marianas Trench was greater than the 35,000 feet indicated in soundings by a Soviet oceanographic ship in 1957.

While the Trieste made her dive, two navy ships—the destroyer escort Lewis and the auxiliary ship Wandank—patrolled the surface.

January 24, 1960

Associated Press
Jacques Piccard

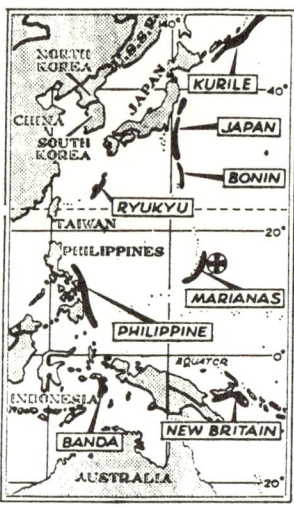
The New York Times Jan. 24, 1960
Scene of dive (cross). Other trenches in area also shown.

Sea Floor Drilled 2 Miles Down

Mohole Test Called a Success—Pacific Cores Recovered

By GLADWIN HILL
Special to The New York Times.

LOS ANGELES, March 31—A million-dollar scientific gamble, financed mostly by the nation's taxpayers, paid off early today when researchers brought up samples from the ocean floor more than two miles below the surface of the Pacific.

Bucking eight-foot waves, twenty-five-mile winds, and numerous other difficulties, a group of a hundred scientists, engineers and oil-field workers aboard an offshore oil rig 260 miles south of Los Angeles completed a sixty-hour operation by bringing up a cylindrical core of gray green clay millions of years old from 234 feet below the ocean bottom.

The operation, conducted by the National Academy of Sciences-National Research Council, was a test of feasibility of the Mohole project. This is a plan to bore into the ocean floor, through the earth's crust and into the underlying mantle.

Today's achievement was hailed by some of the nation's leading scientists and engineers as signaling a new era in oceanography, geology, and the exploitation of the earth's deeply buried resources. The operation was near Guadalupe Island, off Mexico's Lower California Peninsula. The project was financed by the National Science Foundation.

The scientists are from the University of California's Scripps Institution of Oceanography, the University of Southern California, and other institutions.

In earlier preliminary tests two weeks ago, the same rig took cores from the bottom near La Jolla, Calif., at a depth of about 3,000 feet.

The Mohole is a much greater undertaking, still several years off. Its aim is to reach the Moho or Mohorovicic discontinuity, marking the boundary between the earth's crust and mantle. It was named for its discoverer, Andrija Mohorovicic, a Yugoslav geologist.

PACIFIC STUDY: Site of sea floor drilling (cross).

EARTH SCIENCES

The Moho lies about six miles below the floor of ocean deeps. Under the continents it lies about three times as deep. For this reason the oceans offer the best prospects for success.

The oil rig, the Cuss I, towed by a tug and accompanied by several auxiliary vessels, left San Diego March 23 and arrived off Guadalupe Island late Sunday night.

The jointed drilling shaft, dropped through the center of the barge from a ninety-eight-foot derrick, touched bottom at 11,700 feet at noon on Tuesday. Nine hours later an initial core from 110 feet deep was brought up. Then drilling was resumed to the depth of 234 feet. The core from this was laid gently on the Cuss I's deck early this morning.

The results of the operation were announced at the project headquarters here by Gordon Lill, chairman of the National Academy of Sciences committee in charge of the project.

Working at such an extraordinary depth, the researchers had no way of knowing—aside from several years of calculations on paper—whether the drill would be controllable.

Two drill bits, each studded with $8,000 worth of diamonds, had been lost in the trials off La Jolla.

The drill was steadied by a metal framework "drill base" six feet square and one foot high resting on the ocean bottom.

The crux of the current tests is to see whether the 11,700-foot length of jointed four-and-one-half-inch steel drilling pipe, weighing 110 tons and unsupported by the surrounding earth of on-shore drilling, could be worked from an unanchored barge.

For the experiment the huge ship-like 260-foot oil exploration barge, Cuss I, chartered from the Global Marine Exploration Company of Los Angeles, was equipped with four 200-horsepower Diesel engines, one at each corner of the craft. These are activated by radio signals from small anchored buoys to keep the barge in place.

Boring to Continue

Experimental operations are scheduled to continue at the site for about another ten days, "until we run out of money," Mr. Lill said. The operation is costing about $6,000 a day.

The National Science Foundation's contract with the barge company totaled $735,800. Collateral expenditures have brought the total outlay for the operation probably to well over a million dollars.

The cores recovered will be taken to the Scripps laboratories at La Jolla, for study and apportionment among other scientific organizations.

They are expected to pre-date the Niocene geologic period of about 20,000,000 years ago. Mr. Lill said they might be from the Cretaceous period, which extended from 70,000,000 to 100,000,000 years ago.

The National Science Foundation's director, Dr. Alan T. Watermann, said in Washington:

"The pioneering deep sea drilling in nearly 12,000 feet of water places a radically new tool in the hands of scientists seeking to understand the secrets of the earth, and the history of the ocean enabling them for the first time to obtain samples from hundreds of feet within the earth's crust beneath the deep ocean."

Dr. Detlev W. Bronk, president of the National Academy of Siences-National Research Council, called the work "an extraordinary feat of ingenuity, skill and daring," and declared:

"The project has demonstrated once more what can be accomplished when the nation's scientists and engineers, the Federal Government and private industry unite in a common objective."

Robert F. Bauer, president of Global Marine, observed that the "unprecedented drilling achievement stemmed in great measure from the independent efforts of the oil industry in developing the techniques and equipment for offshore exploration from floating vessels."

The Cuss I is a World War II Navy cargo barge that was converted in 1956 into a sea-going drilling rig at a cost of more than a half-million dollars by four oil companies, Continental, Union, Shell and Superior, whose initials form its name.

When their original exploration program was completed in 1958, a group of oil company employes formed Global Marine to continue operating it. Mr. Bauer said the Shell company had deferred a charter claim on the barge to permit the scientific project to proceed.

The project was organized by a committee of Amsoc, an acronym for the American Miscellaneous Society. Amsoc set up by a group of scientists as an informal agency to pursue any projects they saw fit.

Amsoc's Willard Bascom is the project director. The scientific director is Dr. William Riedel, curator of cores at Scripps.

Auxiliary vessels on the expedition are the Scripps' research ships Spencer F. Baird and Horizon, and the Pacific Towboat Company's tug Ed V. Turner out of San Diego.

April 1, 1961

Earth sample was brought up from more than 2 miles below Pacific surface with drill operated from barge. The first bore hole was 234 feet into ocean floor.

OCEAN DRILL HITS LAYER OF BASALT

Scientists in Pacific Project Strike Lava-Like Rock 560 Feet Below Floor

DISCOVERY IS SURPRISE

Oceanographers Had Hoped Substance Would Fill Gaps in History of Evolution

By WALTER SULLIVAN

The floating rig drilling into the floor of the Pacific Ocean has reached the mysterious "second layer" and has found it to be basalt, a lava-like rock.

The discovery, described as a milestone in the exploration of the sea, is surprising to many oceanographers. They had hoped it would consist of solidified sediments containing the missing record of early evolution. The chances of finding such a record now seem somewhat dimmed.

Dr. Harry Hess, chairman of the Geology Department at Princeton University and a consultant to the project, noted last night that the basalt might be a thin layer. However, he doubts that the missing record will be found beneath it.

The grey-green clay penetrated to reach the second layer is believed to date back to the Miocene period, some 25,000,000 years ago. This is only a small fraction of the history of life, which is thought to be roughly a billion years old. The first half of this long history has never been unearthed in fossils.

Penetrated on 2d Try

The drilling that began last week near Guadalupe Island, off the west coast of Mexico, is the first ever attempted in the deep oceanic basins. These basins, lying under two or more miles of water, cover roughly half the earth's surface.

Last night word was received from California that the second layer had been penetrated on the second drilling attempt. For a number of years students of the ocean floor have speculated about this layer, which is observed beneath many ocean areas.

The only evidence of its existence has been in the analysis of shock waves from earthquakes or artificial explosions. These waves travel slowly through the ooze and other soft sediments deposited over millions of years by the rain of dead sea plants and animals. Layer upon layer, this rain has laid down a record of evolution.

The name "second layer" has been used because of uncertainty as to its composition. Beneath it is a third layer, which, it is widely agreed, consists of some form of basalt.

Moho Marks Boundary

Beneath that is the Moho (or Mohorovicic discontinuity) that marks the boundary between the earth's crust and the true interior of the planet. It is named for its discoverer, Andrija Mohorovicic, a Yugoslav geologist.

Dr. Hess believes the third layer is uniform everywhere, whereas the second layer may be heterogeneous. Thus, while it is basalt off Guadalupe, it may be consolidated sediments elsewhere, he said. The two types of rock "look" about the same in terms of shock-wave records.

The second layer was penetrated late Saturday after drilling for twenty and a half hours at that spot. Two cores of hard, fine-grained basalt were brought up from below 560 feet, one of them ten feet long. The water depth was 2.2 miles.

The drill rig, known as Cuss I, has been at the site for a week. The National Academy of Sciences is sponsoring the project.

April 3, 1961

Marine Mountains

The Mid-Ocean Ridge is the greatest geographical find since Columbus' day.

By LEONARD ENGEL

REYKJAVIK, Iceland.

SCIENTISTS from the far corners of the world are coming to Iceland to study the biggest and most remarkable geographical discovery in four centuries — a staggeringly huge, globe-girdling range of volcanic mountains beneath the sea called the Mid-Ocean Ridge.

The ridge meanders through all the world's oceans except the Central Pacific and dwarfs any range of mountains on land. It is 1,000 miles wide, 40,000 miles long, and covers an area nearly equal in size to Europe, Asia and Africa combined. Lava flows and earthquakes occur somewhere along its immense length every day.

Not many people have yet heard of the ridge. Parts of it haven't even been mapped. For most of the ridge is hidden beneath a mile or more of water.

There is, in fact, only one place in the world where you can really see what the Mid-Ocean Ridge is like without getting your feet wet. That place is Iceland. For Iceland and its hundreds of volcanoes, active and dead, are a segment of the ridge, risen out of the sea.

IN Iceland, moreover, one of the most striking and puzzling features of the ridge is evident. Along almost the entire length of the ridge there runs a great rift, ten to thirty miles wide and a mile deep, as though the earth's crust were being torn apart. This great rift runs right across Iceland, forming a giant trench through the mountains from the southwest corner of the island to the northern coast. Within the trench are scores of deep fissures showing that the earth's crust is indeed cracking apart along the line.

Icelanders call the fissures *gja* (pronounced "gyaw"). The largest and most famous is Almannagja, a four-and-a-half-mile-long, 120-foot-deep crack thirty miles from Reykjavik at the edge of a beautiful valley where the Althing the Iceland Parliament—met out in the open for hundreds of years. Speakers addressing the Althing used the Almannagja's lava walls as a natural sounding board.

Many *gja* are found elsewhere in the rift region of Iceland. Careful measurements have shown that most of the fissures are pulling apart at a rate of several inches a century. The Iceland rift as a whole is widening at a rate of two inches a year.

PARTS of the Mid-Ocean Ridge have been known for nearly a century. Key sections were found only two or three years ago, however, as the result of a bold deduction by two scientists at Columbia University's Lamont Geological Observatory, Maurice Ewing and Bruce C. Heezen.

During a research cruise shortly after World War II, Dr. Ewing discovered that the Mid-Atlantic Ridge, a section of the Mid-Ocean Ridge that was already known, had a huge rift in it. A further discovery was made a short time later. Earthquakes occurring in the Atlantic were all found to be centered in this crack in the Mid-Atlantic Ridge.

Additional earthquake charting soon revealed that quakes in other oceans were also centered along a line that went clear around the world. So, in 1956, Doctors Ewing and Heezen predicted that a Mid-Ocean Ridge would be found that went around the world and that it would have a great crack running along its entire crest.

DURING and after the International Geophysical Year, research ships, including the Columbia University schooner Vema, went out to look for missing sections of the ridge —and found them. One section, in the Arctic, was discovered by the atomic submarine Skate during its historic cruise under the ice.

Oceanographers hope some day to explore the ocean bottom in vehicles capable of cruising at great depths. In the meantime, they investigate it with depth recorders, underwater cameras and other instruments — and by studying Iceland.

The North Atlantic island, which is about the size of Ohio, has more active volcanoes than any other land area in the world of comparable size. More than forty have erupted one or more times since the island was first settled 1,100 years ago. One, Mount Hecla, has erupted no fewer than twenty-three times. Lava has also flowed from numerous vents in the ground. Volcanologists estimate that Iceland has produced a third of all the lava that has flowed in the land areas of the earth since 1500 A. D.

As a result of the volcanic activity (which is also responsible for the island's famous geysers and hot springs) Iceland is everywhere dotted by rugged, weirdly beautiful masses of lava and volcanic rock. Huge areas are entirely covered by bare volcanic rock; lava juts out in nearly every pasture.

"The Mid-Ocean Ridge must look much like Iceland, though it must be even more rugged

THE RIDGE — The heavy black line marks the Atlantic section of the Mid-Ocean Ridge, bisecting Iceland.

RIFT IN THE RIDGE—One of the scores of fissures that mark the course of the Mid-Ocean Ridge as it crosses Iceland.

because there are no winds and weather in the depths of the sea to round off peaks by erosion," declares Dr. Heezen. "Imagine millions of square miles of a tangled jumble of massive peaks, sawtooth ridges, cones, earthquake-fractured cliffs, valleys, lava formations of every conceivable shape—that is the Mid-Ocean Ridge."

IN spite of studies to date, the Mid-Ocean Ridge still conceals a central mystery. What is the meaning of the great rift in the ridge? Why is the earth's crust cracking along a line 40,000 miles long?

Many theories have been advanced, including the startling hypothesis that the earth is growing in size. Only two firm conclusions have been reached. The first is that the Mid-Ocean Ridge was formed by lava pouring forth from the great crack. The second is that the mighty ridge originated from forces within the earth, forces that also helped create the continents and oceans themselves.

November 12, 1961

DEEPEST ATLANTIC TEEMING WITH LIFE

Scientists in the Bathyscaph Surprised by Terracing in Puerto Rico Trench

By JOHN A. OSMUNDSEN

The first men to reach the deepest known spot in the Atlantic Ocean described the experience here yesterday.

On 10 occasions, teams of three men at a time climbed into the bright yellow French-owned bathyscaph, Archimède, and plunged downward three feet a second into the 27,500-foot-deep Puerto Rico Trench. The trench is an enormous undersea chasm that runs about 450 miles east and west, a hundred miles or so north of San Juan.

The pilot, engineer and scientist-passenger would chat freely throughout the two-to three-hour descent. All talk would cease when they hit bottom, however, and the silence inside the submarine-shaped vessel would match that of the cold dark water outside.

Found Action Everywhere

Peering through the inch-wide eyepieces of the bathyscaph's portholes, the three travelers — different on each trip — looked out on a scene illuminated by the vessel's 12 one-kilowatt searchlights. Not a square meter of the ocean floor they saw was without a suggestion of activity of some sort — a furrow, a hump, a tiny shrimp scurrying out of the light, a sea cucumber wriggling its tentacles expectantly.

The great abundance of life at a depth where pressures ran to 12,000 pounds a square inch surprised the scientists. They were also surprised that the number of particles — of what they did not know — suspended in the water did not, as expected, diminish with increasing depth, but remained largely uniform to the bottom.

What surprised them most so far, however, was the terracing they saw on both north and south walls of the trench. The bathyscaph bumped and glided down this fantastic undersea stairway, whose steps were hundreds of miles long, more than a hundred miles wide and about 10 feet high.

No Explanation Obtained

No explanation for this remarkable feature has yet been obtained, the French and American participants in the expedition told a news conference yesterday morning at the French Embassy Press and Information Division, 972 Fifth Avenue, at 79th Street.

The joint Franco-American scientific expedition began in April and ended this month. French participants arrived here on Tuesday to visit Columbia University's Lamont Geological Observatory and the university's Hudson Laboratories in Dobbs Ferry, one of the principals in the venture.

Other participants included the French Navy, the National Center of France, Woods Hole Oceanographic Institution in Massachusetts and the United States Navy Electronics Research Laboratory at San Diego, Calif.

August 21, 1964

Carpenter Leaves Pressure Tank

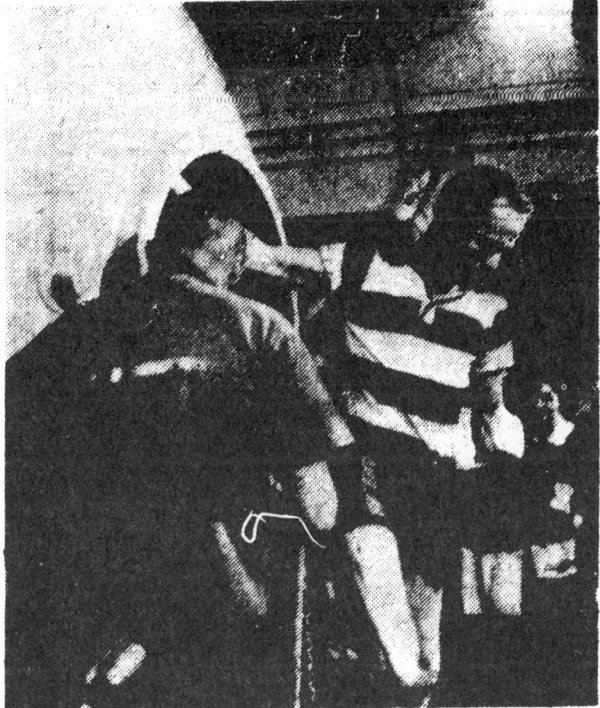

Comdr. Scott Carpenter emerges from decompression chamber off La Jolla, Calif., after 30 days below the sea.

LA JOLLA, Calif., Sept. 28 (UPI) — Comdr. M. Scott Carpenter said today his 30 days beneath the ocean's surface in Sealab 2 convinced him "man can stay under water indefinitely."

Comparing space to the underwater environment, however, the second United States astronaut to orbit the earth said he regarded "the ocean as a more hostile environment."

Commander Carpenter was reunited with his wife, Rene, yesterday after he and nine other Sealab 2 divers emerged from a decompression chamber. But the reunion was brief. The men were instructed to remain on the mother ship Berkone for 12 hours or more to guard against possible unexpected reactions.

Commander Carpenter told a news conference at the Scripps Institution of Oceanography today that man has good built-in defenses that make it possible for him to survive both in space and underwater.

"I think there is no difference in the length of time that a man can stay underwater and work effectively than if he were in a Dewline station (in the Arctic regions) or any other isolated stations," he said.

"If he can stand isolation anywhere else he can stand it underwater," Commander Carpenter said.

September 29, 1965

PLANT LIFE FOUND DEEP IN THE SEAS

Algae at 12,000 Feet Had Been Thought Impossible

By JOHN A. OSMUNDSEN
Special to The New York Times

MIAMI, Nov. 19 — Findings indicating that the amount of life in the oceans "may be many times greater than previously thought" were reported at a scientific meeting here today.

The report, by Prof. E. J. Ferguson Wood of the University of Miami's Institute of Marine Science, was made at an international meeting on tropical oceanography being held at Carillon Hotel. He told of finding algae at depths of 12,000 feet this summer.

The first indications that plants might live at extraordinary depths came in explorations in 1948 by Francis Bernard, a French scientist living in Algiers, Dr. Wood said. Professor Bernard reported finding plants as far down as 12,000 feet, Dr. Wood said, in the Mediterranean, the Atlantic and the Indian Ocean.

Those reports were largely

discounted Dr. Wood said, because of the "rule" that plant life should not extend beyond the depth to which 1 per cent of surface sunlight penetrates — about 240 feet.

Sunlight a Factor

Plant life much deeper than that was believed to be either impossible or a temporary and rare occurrence. Practically no light reaches deeper than 1,000 feet.

For those reasons, Dr. Wood's reports of finding living single-celled plants at depths of 21,000 to 30,000 feet in the Mindanao and Sunda Deeps in 1952 were similarly regarded very sceptically, he said.

It was suggested that plants brought up from such depths probably had been captured much shallower, on the way up or down. Now, however, improved equipment for collecting material at known great depths without the possibility of contamination elsewhere has ruled out that interpretation, Dr. Wood said.

Thus, the Miami scientist reported with confidence his collection of algae this summer east of the Isle of Rhodes in the Aegean and at other locations during an eight-months expedition of the Institute of Marine Sciences Research vessel John Elliott Pillsbury, which returned here Monday.

How energy from the surface gets down to those very deep plants was a matter not yet explained, Dr. Wood said.

Dr. Wood speculated that perhaps this was achieved partly through the "rain of death," by which expended organic material sinks slowly to the bottom of the ocean, and partly through a kind of cafeteria shuttle system.

The latter was described as the movement of animals from mid-depths to the surface for plant food some of which they later release undigested lower depths where it is used by the plants and animals there.

Dr. Wood said it was too early to say whether this newly discovered vast resource of organic matter could be harvested

November 20, 1965

The Valley Is Mostly Hidden

By WALTER SULLIVAN

One of the most remarkable features of the earth's surface is an almost continuous valley that snakes across it for some 40,000 miles. Its existence has become known only in recent years because most of it lies under water. It cleaves the mid-ocean ridges that bisect the great oceans, in particular the Atlantic, and is visible only in the few places where it appears above water.

Such is Iceland, which forms a portion of the Mid-Atlantic Ridge. The rift valley cuts directly across that island. Another exposed segment extends from the Dead Sea down through East Africa.

These valleys are areas of movement and frequently generate earthquakes. They are also marked, in places, by volcanic activity, as in Iceland. They manifest ferment deep within the earth that must be related to the existence of continents and oceanic basins, but the tale they have to tell is far from understood.

Ridge-Rift Charted

This summer some of the last uncharted portions of this ridge-and-rift system have been filled in on the maps, thanks largely to the work of American and Soviet oceanographers. What they seem to show is that the system was torn asunder by fracturing of the earth's crust east of Greenland. There, the southern sector of the system is displaced 300 miles southwest of the northern sector, which runs from north of Greenland across the Arctic Ocean toward the Lena River of Siberia.

In a recent study by two scientists of the Naval Oceanographic Office in Washington it is proposed that Spitzbergen may once have been north of Greenland. Crustal movement then shifted it southeast, parallel to the newly discovered fracture zone. What-

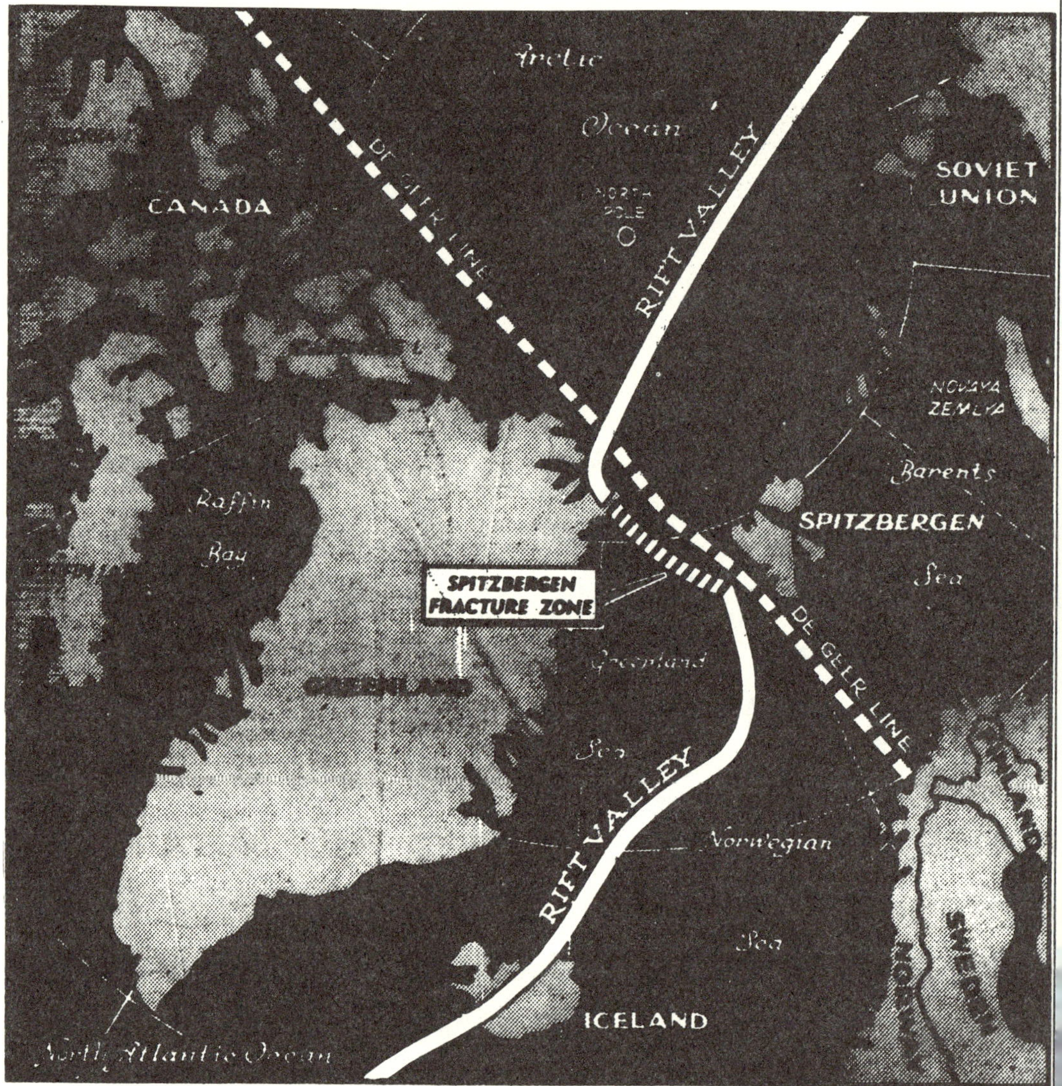

A new theory has been advanced to explain the eccentric pattern of the ocean floor between Greenland and Spitzbergen. Recent oceanographic studies of previously uncharted portions of the 40,000-mile worldwide mid-ocean ridge and rift system reveal that it extends across the Arctic Ocean. The studies also show that the Rift Valley shifted by some 300 miles (as shown by the striped pattern on map) between Spitzbergen and Greenland. This shift, a zone marked by frequent earthquakes, parallels the de Geer line, a hypothetical line of cleavage in the earth's crust extending from the Mackenzie River in Canada to the vicinity of Norway. It is thought that Norway once lay close to Greenland, and Spitzbergen lay north of Greenland until forces along the de Geer cleavage shifted them eastward. One of the major studies was carried out by two American scientists.

ever the nature of the fracturing process, it is not ended, for the fracture zone is still a frequent producer of earthquakes.

The de Geer Line

The authors, G. Leonard Johnson and Oscar B. Eckoff, note that this hypothetical movement parallels the so-called de Geer line, a geographic feature extending from the mouth of the Mackenzie River, along the Arctic rim of the Canadian islands and Greenland toward Norway. It has been proposed that the line represents a shearing of the earth's crust whereby Norway was shifted from an earlier position near the east coast of Greenland.

The Navy scientists have reconstructed the ocean floor structure between Greenland and Spitsbergen from soundings made by a long succession of expeditions. The earliest of importance, they say, was that carried out in 1937-38 by an American woman, Louise Boyd, in the ship Veslekari.

This was followed in recent years by the extensive journeys of Soviet research ships, some 5,400 miles of soundings by American icebreakers, plus work done by nuclear submarines — notably the U.S.S. Nautilus. Soundings farther north, in the pack ice of the Arctic Ocean, were made from stations established on the drifting ice. Further information from aerial magnetic surveys was reported this summer at the international oceanographic congress in Moscow.

Mrs. R. M. Demenitskaya of the Institute of Arctic Geology in Leningrad said these flights had shown a zebra pattern of magnetism on the floor of the Arctic Ocean parallel to the ridge system. This follows the patterns found in recent years on the floor of both the Atlantic and Pacific Oceans.

The pattern shows up if one charts the magnetic "anomalies." These are areas in which the earth's magnetism departs from normal. Some are positive, magnetically, and some negative. If one shows the positive anomalies as black and the negative as white, a zebra pattern emerges.

Ocean Pavement

Some believe the oceans are paved in this manner because hot, soft rock has been extruded from rifts in the mid-ocean ridges for millions of years, spreading slowly over the ocean floors on both sides of those rifts. This material, when it cools, captures the orientation of the earth's magnetic field at that time. It is argued that since the field periodically reverses its polarity, the oceans are paved with successive bands, each polarized in the opposite direction from its neighbors.

Mrs. Demenitskaya reported that the bands in the Arctic, like those elsewhere, were from 5 to 10 miles in width. Dr. Bruce C. Heezen of Columbia University's Lamont Geological Observatory termed these reports "very exciting." It has been he, as much as anyone, who has identified the worldwide system of mid-ocean ridges with their rift valleys.

August 21, 1966

4 Years of Sea Drilling Yields Vast Lore

By WALTER SULLIVAN

After four years of drilling into the ocean floor, the Glomar Challenger, an ungainly vessel with a giant drilling derrick amidships, has produced a vast treasure of important new information on the history of the earth.

Some of the findings are so revolutionary that there are scientists who still find them incredible. Much of the material, extracted from more than 350 holes bored deep into the sediments that have accumulated at the bottom of the sea for tens of millions of years, remains to be studied.

The saga of the drill ship —the stepchild of the ill-fated Mohole Project of the mid-nineteen-sixties — has been marked by dramatic encounters with fog-shrouded icebergs and storms, including a typhoon. There have also been bizarre mishaps, as when the ship-control computer ran amok.

But it is primarily a tale of discoveries, some probably of historic magnitude. Because they were made in remote areas and because the ship is almost constantly on the move, many of these findings have not until now been made public.

While some of the first drill holes confirmed the theory that sea floors are slowly but steadily moving away from the mid-ocean ridges, the more recent drilling has shown that the history of this activity is extremely complex. When that history is reconstructed, it should explain why the mountain ranges of the world formed where they did—and when they did.

The voyages of the Glomar Challenger, known as the Deep Sea Drilling Project, are financed by the National Science Foundation and are administered by the Scripps Institution of Oceanography. Interviews with participants and a study of the reports, published and unpublished, have brought to light, among others, the following discoveries and conclusions:

¶Although man is descended from creatures that lived in the sea, all of the oceans inhabited by his ancestors have vanished. None of the major ocean basins of today existed when the mammals began to evolve from certain reptilian species about 180 million years ago. The geography of the oceans has, so to speak, been completely recycled.

¶Some 5.5 million years ago, a waterfall on a scale far grander than any seen by modern man—with a flow at least 35 times that of Victoria Falls, today's largest — poured water from the Atlantic into the Mediterranean Basin, which, at the time, had completely dried up. Possibly man's slope-

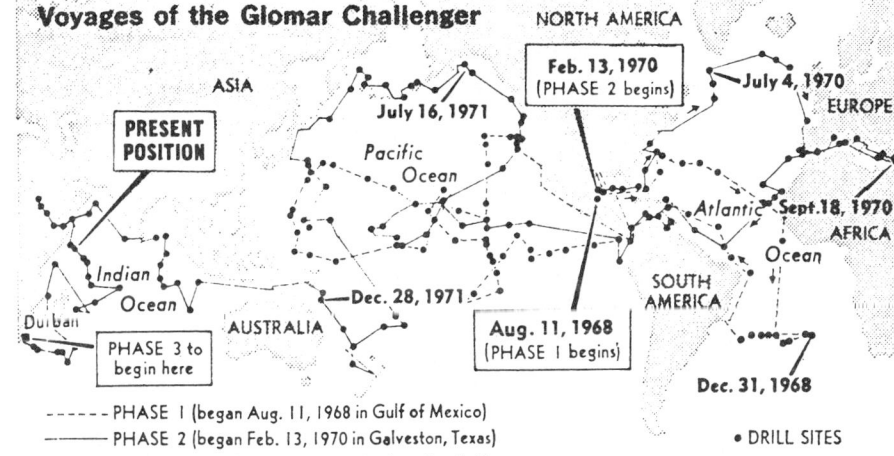

More than 350 holes have been drilled into the sea floor in the global treks of the Glomar Challenger.

browed ancestors gazed on this thundering spectacle.

¶A land larger than the combined areas of New Jersey, Connecticut and Massachusetts, which flourished in the days when early mammals and birds were evolving among the dinosaurs, has vanished. It has sunk almost a mile beneath the sea, midway between Ireland and Iceland, in a manner reminiscent of the mythical Atlantis. Such gross vertical movements, both up and down, as a by-product of the sea-floor spreading process, have been documented in many regions.

¶Gem stones from Brazil have been strewn across the mid-

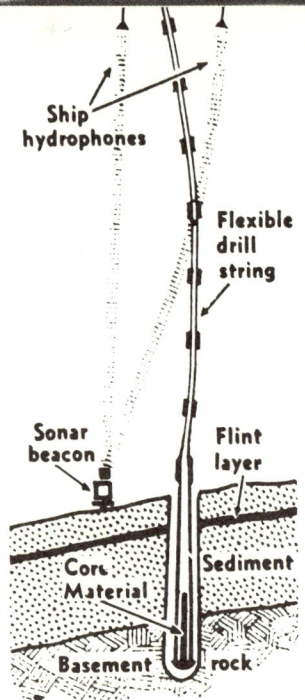

Glomar Challenger, above, with aid of beacon and hydrophones to keep position, can drill in three miles of water. Cores are hauled up through pipe.

Atlantic area by catastrophic "turbidity currents" sweeping hundreds of miles across the ocean floor from the mouth of the Amazon. They include topaz, tourmaline and aquamarine. Such currents have played a little recognized role in building the sea floor.

¶A metal-rich layer, in some areas consisting of almost pure oxides of iron and manganese, lies under the sediment across much of the Atlantic and Pacific. It is presumably being generated by volcanic activity as new sea floor is formed along mid-ocean ridges and is then carried away from the ridges. At the base of 800 feet of sediment on the continental margin off New York, for example, there is a deposit of iron, zinc and copper.

¶Great salt beds under the sediments of the Mediterranean, Gulf of Mexico and elsewhere give promise of immense new oil resources. In many places the salt has pushed up in the domes that typically mark oil accumulation. Natural gas (methane) was found at many sites and one core sample was "dripping with oil."

¶Drill holes west of Panama have documented a sequence of events in which movements of the sea floor away from a nearby submarine ridge pressed under Central America, lifting that region out of the sea some seven million years ago. This created the Isthmus of Panama and cut the equatorial link between the Atlantic and Pacific, revolutionizing circulation patterns in the world oceans.

¶In the floor of the Bering Sea, gravel deposited by melting icebergs was found in sediment laid down as early as seven million years ago. A similar time had been determined by other means for the onset of the South Polar glaciation. From two drill sites in the Labrador Sea it was estimated that glaciation in that region began three million years ago. It has been proposed from these observations that the onset of ice accumulation in both polar regions followed the change in ocean currents as Panama rose from the sea.

¶Another phenomenon that may be related to such changes is the occurrence of "tile floors" of chert—a hard type of rock including flint—that lie interbedded deep within the ocean sediments. Over large areas in both the Atlantic and Pacific they were laid down during the early Eocene, 50 million years ago, when dog-sized horses and primitive camels were roaming North America.

20 Million Years Skipped

Often, below this layer, the drillers found that the fossil record skipped a period of some 20 million years in the evolution of sea life. Among explanations advanced is that bottom currents became so pure — so lacking in the minerals of which marine fossils are formed—that those fossils quickly dissolved.

Another view is that some catastrophic change in the seas made them inhospitable to life. Since chert is formed from the silica-rich shells of diatoms, sponges and radiolaria, it is suspected that the missing link in the fossil record ended with an explosion of oceanic life that laid down the chert.

Still another view is that widespread volcanic activity rained ash on the seas, saturating the water with silica and thus preventing dissolution of silicate shells, which finally turned to chert under conditions of suitable pressure and temperature deep within the accumulating sediment.

Evidence of volcanic episodes on a scale unknown during recorded history was found in the sea floor: successions of volcanic ash layers far at sea and, in the Western Pacific, an apron of lava blanketing a region comparable to one of the dark "seas" of the moon.

It is the evidence for the total drying up of the Mediterranean, even where that sea is two miles deep, and for the vast waterfall replenishing it with water from the Atlantic that has created the greatest stir among oceanographers. The water, it is suggested, flowed over an escarpment east of Gibraltar that is now hidden beneath the sea.

The conclusion drawn from holes drilled in the Mediterranean is that the Strait of Gibraltar was repeatedly closed, starting some seven million years ago. Since the flow of rivers into the Mediterranean was insufficient to match the evaporation rate, the sea dried up. This deposited deep layers of salt on its floor and periodic incursions of Atlantic water that then also dried up added to this salt accumulation.

It was this salt deposit, thousands of feet deep, that was discovered beneath the sediments throughout the Mediterranean. Furthermore, the salt was in many areas in a form known as anhydrite, which occurs only where the salt has been baked in sunlight or otherwise exposed to heat. It was this that suggested the virtually complete drying up of the sea.

Letter From Russian Expert

One of the strongest pieces of evidence supporting the concept of a dry Mediterranean

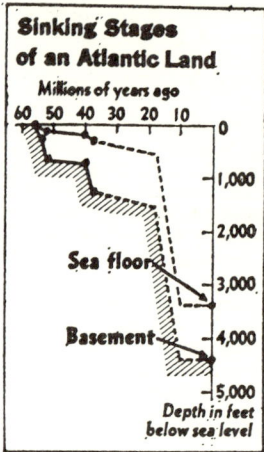

Sediment laid down on a submerged plateau that was once dry land shows 3 periods of rapid sinking.

basin became known in this country after Dr. I. S. Chumakov of the Geological Institute of the Soviet Academy of Sciences, in Moscow, read a New York Times account of the Mediterranean drilling. The article mentioned the possibility that the sea had, at times, been landlocked and dry, except for scattered lakes.

Dr. Chumakov wrote to Dr. William B. F. Ryan of Columbia University's Lamont-Doherty Geological Observatory:

"The article attracted my attention inasmuch as for some time I have been studying the development of the Mediterranean during the late Miocene-Early Pliocene, [five to seven million years ago] and your preliminary deductions coincide (admittedly not all of them!) or come very close to my conclusions obtained on the basis of work in some countries in the southern part of the Mediterranean."

In his research Dr. Chumakov had studied some 15 bore holes into the bottom of the Nile River just south of the Aswan High Dam, which Soviet engineers were helping to build. The holes were drilled to determine depths to bed rock and the nature of the overlying material.

This revealed that under the relatively flat bottom of the present river there was a sediment-filled canyon 950 feet deep. The lowest part was a narrow gorge with almost vertical walls. Most remarkable was the discovery that the bottom 500 feet of this canyon was filled with sediment of oceanic origin.

In other words, it had once been flooded by a sudden incursion of the sea, even though it lies 1,250 miles up the Nile. Dr. Chumakov concluded that the canyon was carved when the Mediterranean was at least 2,000 feet lower than it is to-

EARTH SCIENCES

day. Rapid filling of the sea sent sea water up the canyon and only when it silted up higher than the existing sea level did fresh-water fossils begin to appear in the accumulating sediment.

A similar sequence of events apparently took place in the Rhone Valley and along other rivers flowing into the Mediterranean Basin. Wells drilled in the Rhone Valley as far upstream as Lyon, almost 200 miles inland, have revealed marine sediment lying in the ancient river bed far below the present river floor.

Further evidence came to the attention of Dr. Ryan and his colleagues on the Mediterranean drilling venture—Dr. Kenneth J. Hsu, now at the Scripps Institution of Oceanography, and Dr. Maria B. Cita of the Institute of Geology and Paleontology in Milan, Italy. The new information was provided by Dr. F. T. Barr of the Oasis Oil Company of Libya.

Oil prospectors had discovered that beneath the Libyan sands, 600 miles west of the Nile, channels had been cut into the bedrock to 1,300 feet below the present sea level. Dr. Barr proposed that the channels had been carved when the Mediterranean was relatively empty of water in the late Miocene (about six million years ago). So unbelievable was his hypothesis then that he could not get it published in a scientific journal.

Other evidence indicates, too, that the floor of the Mediterranean itself was carved, as though by rivers, during this period.

In an account to be published in the annals of the Deep Sea Drilling Project, Drs. Ryan, Hsu and Cita state:

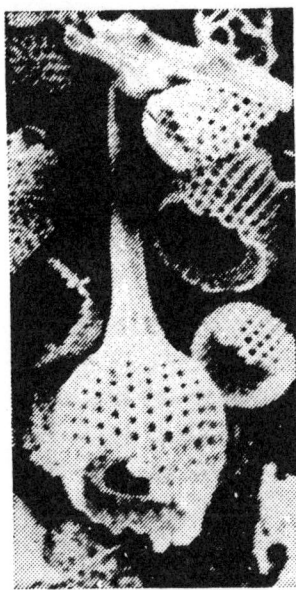

Radiolaria shells (magnified) from 429 feet below Pacific floor represent a stage in their evolution 30 million years ago.

"The idea that an ocean basin the size of the Mediterranean could actually dry up and leave behind a big hole thousands of meters below worldwide sea level is preposterous indeed. We ourselves were reluctant to adopt such an outrageous hypothesis until we were overwhelmed by many different lines of evidence."

When they first spoke of their idea, they said, "our venerated colleagues were vehement in their negative reactions; the idea was thought to be physically impossible." But then, they added, as more and more corroborative evidence came in, many of the skeptics were won over.

If all of the cores of sediment and rock extracted from the sea floor so far by the Glomar Challenger were laid end to end, they would reach more than 10 miles. Within these cores lies the history of oceanic changes and the evolution of oceanic life over the last 150 million years or more, since the present oceanic basins began forming.

Generations hence scientists will probably still be delving into this record.

The cores are collected as the drill bit, nine and a quarter inches in diameter, bores downward. A hole in the center of the bit allows a column of material to push up into the coring tube, which is periodically hauled up through the pipe. On deck the core is split, half being sealed and placed in refrigeration for eventual storage in repositories at the Scripps Institution in La Jolla, Calif., or the Lamont-Doherty Geological Observatory in Palisades, N. Y.

The other half of the split core is immediately examined in the ship's core laboratory for microscopic determination of its fossil shells. These tell the age of the sediment, constituting a milepost along the drill's journey into the past.

Scientists from most of the large maritime nations, including the Soviet Union, have taken part in the project. However, despite its international flavor, the project has also run into diplomatic constraints.

Off Brazil and Canada, drill sites had to be canceled because of disputes over territorial waters. Furthermore, the fear of an oil "blow-out," causing massive pollution of the sea and worldwide opprobrium, led to severe restrictions.

Whenever natural gas was detected in the cored material (which occurred in all holes drilled into the deep basin of the Gulf of Mexico), drilling was stopped and the hole was plugged with cement. Similar plugging was done in all Mediterranean holes, for a vast oil field may lie beneath that sea.

The only oceanographic expedition that has produced a comparable flood of new findings was that of the British steam corvette H.M.S. Challenger, from which the drill ship took its name. H.M.S. Challenger roamed the world for three and a half years in the eighteen-seventies, collecting a wealth of specimens and data.

The other part of the Glomar Challenger's name is derived from that of its owners, Global Marine Inc. of Los Angeles.

The four-year drilling effort has already cost $35-million, with another $33-million expenditure envisioned for the next three years. It was an offshoot of the ill-fated Mohole Project, which in the early nineteen-sixties fell victim to politics, escalating costs and scientific controversy that pitted those chiefly interested in a single hole through the earth's crust and its lower boundary — the "moho" — against those eager to drill many shallower holes like those drilled by the Glomar Challenger.

The project leaders at Scripps are Drs. William A. Nierenberg, who heads the institute, and Melvin N. A. Peterson.

May 29, 1972

Russians Will Help U.S. In Drilling in Sea's Floor

By WALTER SULLIVAN

The Soviet Union has agreed to help guide and finance the scientifically fruitful Deep Sea Drilling Project of the United States, contributing $1-million annually, or about one-tenth of its cost.

It is believed to be the first time another country has entered into such an arrangement with the United States. Final agreement came this week after the project's drill ship, the Glomar Challenger, reached Christchurch, New Zealand, with a new cargo of scientific discoveries from Antarctic waters.

These, as described at a news conference here yesterday, included the existence of natural gas in the floor of the Ross Sea and evidence that the Antarctic ice sheet, which now makes up more than 90 per cent of the world's ice, first formed 20 million years ago, some 15 million years earlier than had generally been believed.

The drilling of 16 holes in the ocean floor also showed evidence that five million years ago the bottom was scraped by an ice sheet that once extended 200 to 300 miles farther out into the Ross Sea than at present.

In contrast to the Ross Ice Shelf, an apron of continental ice, some 600 feet thick, that floats on the southern part of the Ross Sea, this former blanket of ice was far thicker, plowing across a sea floor that is from 1,500 to 2,000 feet below sea level.

The Glomar Challenger is the only ship that has ever been **able to drill through the sediments of the deep sea and sample the bedrock under them, often beneath water more than three miles deep. The resulting discoveries have vastly increased knowledge of the history of the ocean basins, their past climate and their inhabitants.**

The drill sites have been numbered consecutively, those of this portion of the project being numbers 264 through 274. It was the 28th segment of the ship's global operations. Beginning with the sixth, from Honolulu to Guam in 1969, Russians have occasionally taken part as visiting scientists.

Areas Out of Bounds

So have specialists from other countries, particularly those near the drilling areas. The Black Sea and the Baltic Sea, two regions of interest that border the Soviet Union, have been more or less out of bounds so far, but this may change with the new Soviet participation.

To date, policy for the project has been set by a consortium of five institutions on the East and West Coasts, known as JOIDES (for Joint Oceanographic Institutions for Deep Earth Sampling). The Institute of Oceanology of the

Soviet Academy of Sciences will now become a sixth member of this group.

Direct administration of the project is under the Scripps Institution of Oceanography at La Jolla, Calif.

Entry of the Soviet Union is an outcome of the Soviet-American agreement on cooperation in science and technology signed last May 24. In October Daniel Hunt and Dr. William E. Benson of the National Science Foundation and Arthur Maxwell, provost of the Woods Hole Oceanographic Institution, met in Moscow with Soviet scientists.

Details Worked Out

It was there that details of Soviet participation in the drilling project were worked out. From Monday through Wednesday this week the American-Soviet Joint Commission on Scientific and Technical Cooperation met in Washington and gave final approval to the arrangement.

Also agreed upon were joint projects in such research areas as pollution control and water resource management. The drilling project has been approved to 1975, and the agreement provides for Soviet participation if the project is continued.

It is hoped, if funds can be obtained, to modify the Glomar Challenger, or another drill ship, to penetrate considerably deeper into the rock crust beneath the sea. This may become feasible if more countries join in. West Germany, in recent weeks, has indicated an interest in becoming a member of the team.

Among the earliest holes drilled after the ship left Fremantle, Australia, on Dec. 20 were four at sites between the Antarctic coast and the ridge bisecting the sea that separates Australia from Antarctica. From other evidence it was suspected that material erupting from this ridge was driving the two continents apart, a few inches a year.

Evidence supporting this view was extracted from the bottom by the Glomar Challenger. The sea floor was found to be increasingly old at greater distances from the ridge. Fossils at the bottom of the sediment blanket, lying directly on the bedrock, at each site, were used to date its formation.

In the hole nearest the ridge the age was 13 million years.

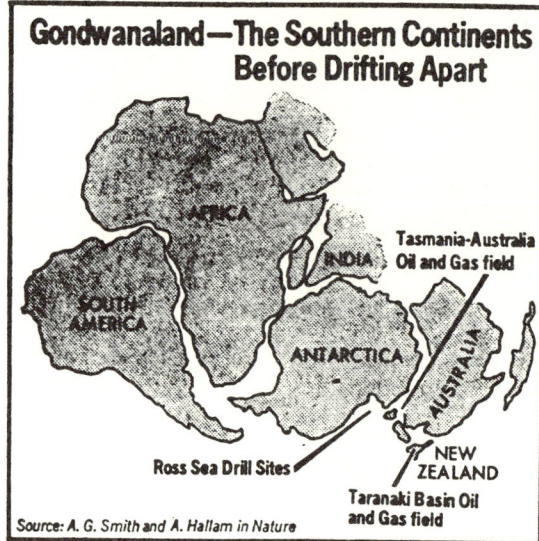

The drilling of holes deep into the sea floor off Antarctica by the ship Glomar Challenger has shown the ice sheet on that continent to have once been far larger than today, and has confirmed that spreading of the sea floor is pushing Australia and Antarctica apart. The drill sites have been numbered consecutively since the project began five years ago.

At holes successively farther south the ages were 24 million, 42 million and 50 million years, respectively. It is believed that Antarctica and Australia split apart about 55 million years ago with the breakup of the great southern continent, Gondwanaland.

Near Oil Reservoirs

Bubbles of gas — methane and, in some cases, ethane — were found in sediment hauled up at sites 271, 272 and 273, not many miles from the towering white cliffs of the Ross Ice Shelf. The operational doctrine of the project calls for an immediate halt to drilling whenever gas is detected lest the drill strike a high-pressure oil reservoir and produce a polluting blow-out.

At the news conference, Dr. Dennis E. Hayes of Columbia University's Lamont-Doherty Geological Observatory said that the existence of an important oil or gas reservoir there could not be ascertained under such circumstances.

However, he pointed out that, before the breakup of Gondwanaland, this region lay close to known reservoirs of oil and gas in New Zealand and beneath the sea between Australia and Tasmania.

There is evidence that, when the Atlantic first began forming, oil was laid down in narrow, shallow seas that separated the Americas from Europe and Africa. As the ocean grew, these reservoirs were split, and they now lie far apart, alongside opposite coasts of that ocean. The new discoveries suggest such a possibility in the Antarctic as well.

At the news conference, held at the Overseas Press Club in the Time-Life Building, Avenue of the Americas at 51st Street, Dr. Hayes said that the changing nature of the sediment, at different depths in the drill holes, testified to major changes in climate and oceanic circulation. These, presumably, developed as the continents moved apart and the geography of the Southern Hemisphere changed.

Menaced by Icebergs

The great growth of the polar ice sheet, some five million years ago, must have come about from greatly increased snowfall on the Antarctic, where precipitation today is meager. Signs that the sea floor had been scraped by such an ice sheet were found at all four drill sites in the Ross Sea.

Soundings of the sea floor sediments with sound waves indicate that this planed-off layer extends 200 to 300 miles north of the present ice shelf. The nearest drill site to the shelf was about 50 miles north of it.

Twice, when the ship's drill pipe was deep in the sea floor, making it immobile, an iceberg moved menacingly toward the ship. The largest of the bergs was about 10 times larger than the ship, but the Coast Guard icebreaker Burton Island pushed it away.

Sediment, brought up from directly above basement rock in this area, was laid down when the rock was at the surface, more than 20 million years ago. The sediment layers above it testified to slow subsidence of the rock, presumably as the increasing load of Antarctic ice down-warped the earth's crust.

Dr. Hayes, with Dr. Lawrence A. Frakes of Florida State University, was co-chief scientist on this segment of the project.

He believes the initial growth of the ice sheet created a great flow of icy water across the floors of the world oceans.

This, he said, could have swept away the top layers of sediment at that time, explaining the lack of a fossil record for the period from 20 million to 50 million years ago that was found at some drill sites in other ocean areas.

The findings seem to rule out the hypothesis of some students of ancient maps that Antarctica was all or largely free of ice early in human history. Much has been made, in this respect, of a map said to date from 1513 and attributed to the Turkish admiral Piri Re'is. It purportedly shows Antarctica as it would appear if its heavy blanket of ice was largely removed.

The indications from the drill cores, as well as from shallow ones obtained by driving a tube down into the sediment, are that there has been no major retreat of the ice for many millions of years.

March 23, 1973

Mid-Atlantic Rift Valley Spreading an Inch a Year

By WALTER SULLIVAN
Special to The New York Times

PONTA DELGADA, the Azores, July 31 — Intensive manned exploration of the Median Rift Valley that bisects the Atlantic Ocean has, according to the participants in dives, provided firm evidence that the valley is spreading away from its center line at the rate of about an inch a year.

They had been convinced of this by evidence obtained elsewhere indicating that eastern and western sides of the ocean are moving away from each other at that rate. However, the oceanographers said here today that the new discoveries should

EARTH SCIENCES

do much to convince those still skeptical.

While the descents far below depths tolerable by ordinary submarines are not without hazard, all of the dives — some of them more than 9,000 feet deep—have gone smoothly so far, according to the participants.

Using remotely controlled grasping tools, the divers have collected samples of rocks from the various formations to use in reconstructing the chemistry of sea floor production. Also, some 30,000 photographs of such formations have been made.

They show deep clefts, cliffs formed by abrupt uplifts and subsidences as well as other features peculiar to this region of oceanic growth. Thus, what seem among the youngest formations of the rift valley — Mount Venus and Mount Pluto half a mile apart from the valley center line — are partly encrusted with staghorn-like formations of volcanic glass.

The findings so far by two French and one American deep-diving craft were summarized at a news conference aboard the French vessel Le Noroit, mother ship to the "diving saucer" Cyana.

The Cyana has already made eight dives into a narrow cleft where the Median Valley is offset and has discovered there what seem to be twin geysers or hot springs that have spread manganese-iron deposits over the surrounding seabed.

Dr. Richard Holland of Harvard reported today on preliminary examination of specimens brought up from these features and said they contained large amounts of manganese and iron oxides in ratio of two to one.

The Netherlands is participating in the American phase of the Franco-American effort, known as FAMOUS, for "the French American Midocean Undersea Study."

New Undertakings

As one of the first tasks to be undertaken tomorrow after the French and American ships go to sea again with their submersibles, the Cyana will return to the geyser site to seek further rock and water samples and to get high precision data on temperatures.

Water within the throat of each vent is apparently not very hot, since fish were seen swimming in and out.

Latest series of dives into Mid-Atlantic Rift indicates that undersea valley floor is slowly spreading apart.

According to Dr. Xavier Le-Pichon, chief French scientist here, one of apparent areas of periodic hot water eruption is a narrow rift, and the other is small cone.

Both lie within an area some 130 feet long and 35 feet wide. Around them are what appear to be fragments of manganese-iron ejected from the vents as though from a volcano. The valley is more than 9,000 feet below the surface of the sea.

Dr. Tjeerd J. Van Andel of Oregon State University observed that if the features are active vents they will be of major importance in efforts to reconstruct how ore deposits of oceanic origin are formed. It is suspected that many deposits now on land originated at sea in this manner.

Vents Under Study

French and American scientists are conferring on the best means of determining the nature of these features. Dr. Le-Pichon believes occurrence of vents within one of the zones where the Median Valley is offset — a so-called transform fault known as Fracture Zone A — is significant. It may be within such zones, rather than the Median Valley itself, that the metal depositions occur, he said. The reason, he explained, is that in the Median Valley the frequent eruptions of lava that cause sea floor spreading from the center line prevent the persistence of deep cracks and maintain temperatures near the surface that are so high that water cannot percolate through the rock for prolonged periods.

In the fracture zones, on the other hand, the motions of opposing rock faces are lateral — caused by spreading from the Median Valley. Deep faulting can persist, allowing sea water to penetrate down three or four miles and remain long enough to pick up dissolved metals before carrying them up to the sea floor in submarine geysers.

Deformation of sediment layers in Fracture Zone A confirms side slippage in a manner predicted by the seafloor spreading concept, Dr. Le-Pichon said.

He added that examination of rocks from the Median Valley had shown that their ages increase at greater distances to either side of center line.

Along the center line, rocks have erupted within the last 10,000 years, he said, whereas in the valley walls the ages are about 150,000 years. This conforms to the predicted spreading one inch a year from the center, he said.

Whereas the area of rifting in the fracture zone is less than half a mile wide, forming a deep V-shaped valley, the area of disruption in the Median Valley extends across a width of some five miles. Thus, while lava intrusion occurs chiefly along the center line, there is also periodic rifting and intrusion for a few miles to either side of it.

Separation of Plates

This means the two great plates of the earth's surface that meet in the rift area under study some 220 miles southwest of here—one bearing the Americas and the other carrying Africa—are separated by a zone of rifting several miles wide rather than by a sharp demarcation line.

According to Dr. Van Andel, who has made five of the dives, the subjective feeling as one is carried along the rift valley by the Alvin, an American diving craft, is one of total isolation from the real world — from the world of air, sky and ships almost two miles overhead.

Visibility is limited by darkness and murkiness, and objects 60 feet away are already hazy. As a result, Dr. Van Andel said, "your eye doesn't believe it, though everything looks very large."

The strangely shaped boulders, clif-like scarps and other volcanic features, black and gray, are dusted with whitish sediment look, he added, "like the Rocky Mountains after a light snow, in moonlight — in a fog."

August 1, 1974

THE ATMOSPHERE

QUESTS IN UPPER AIR YIELD STRANGE FACTS

Our Knowledge of the Atmosphere Is Vastly Increased Through the Unending Work of Science

By WALDEMAR KAEMPFFERT.

Fourteen miles was the goal that Captain Albert Stevens and Captain Orvil Anderson intended to reach in the National Geographic Society's balloon Explorer II. If they failed to achieve precisely that height last Tuesday it must have been by only a very few hundred feet.

The breaking of the altitude record was but an incident in this voyage into the unknown. It was the carrying out of a scientific program that was all-important—a program which had as its five broad objectives: (1) measuring the intensity and direction of the cosmic rays, (2) determining the position of the ozone layer of the atmosphere, (3) collecting samples of air in the stratosphere for later analysis, (4) ascertaining if there is any life in the form of spores or bacteria at the highest altitudes, and (5) comparing the altitudes registered by barographs and thermographs with those computed from photographs.

Never was there a scientific balloon ascension more carefully prepared, never one more successful in achieving its end. Two-way telephone conversations were kept up between the stratosphere and the ground. Millions listened in by radio as the balloonists rose to new heights and gave terse accounts of their impressions—the intense cold, the steam that rose from the metal gondola, the bluish blackness of the sky, the seeming hollowness of the earth below, the working of the instruments as clicking cameras photographed their indicators.

An Ambitious Quest.

It is only in the last century or so that the wistful gaze of the explorer has turned upward to the clouds. With the invention of the balloon he ceased to be a two-dimensional animal in the sense that he was no longer limited to crawling over the surface of the globe.

Then came an important discovery: the atmosphere is not infinite in extent. Men gasp and die if they go high enough; yet far above the dying altitude, as it may be called, there is still an ocean of air. Probably the atmosphere reaches outward from the earth hundreds of miles. But even with oxygen it is doubtful for technical reasons if man can attain more than fifteen miles in a balloon.

Realizing this, small, free, unmanned balloons, freighted with featherweight automatic instruments, were sent aloft by Teisserenc de Bort in 1896 and later years. His successors have attained a height of somewhat more than twenty miles. There comes a time when the gas, constantly expanding with increasing altitude, bursts its rubber prison; a parachute or a second balloon which has remained intact, carries the instruments to the ground.

These instruments are artificial senses. They write down all that they feel—temperature, pressure and other facts of interest to scientists.

At first de Bort could hardly believe the scripts. More than six miles above sea level lies a strange stratum of air, he read, a layer as different from the air in which we live as the Arctic is from Panama. There are no clouds, no storms, nothing that we designate by the word "weather." One day is like another. Never is there even a mist. Perpetual sunshine. From the thinness of the air it can be inferred that the sun and stars blaze in a purplish-black sky—flickerless, hard. Here reign serenity, cold that must go to the very marrow, with the temperature steadier and lower over the tropics than over the Poles.

Naming the Strata.

Teisserenc de Bort and the meteorologists of his day spoke of the "isothermal" or "uniform temperature" layer. Later he coined the word "stratosphere" and designated by "troposphere" the thicker stratum that hugs the earth's surface. Between the stratosphere and the troposphere lies Sir Napier Shaw's "tropopause," a kind of No Man's Land.

How far the stratosphere extends no one knows exactly. About eighteen miles (some say thirty miles) above sea level is a layer of ozone which has been discovered by means of the spectroscope, by its ability to reflect sounds of heavy explosions on the earth and by the behavior of meteors.

If it hugged the earth that layer would be only an eighth of an inch thick. Upon it life on this earth depends for protection from the sun. Thin as a few sheets of paper, if compressed, it filters out an excess of ultra-violet rays, which, if they reached the earth, would strike us dead. It was a major object of the ascension made by Captain Stevens and Anderson to determine the position and distribution of this ozone layer in the stratosphere.

Beyond the ozone there must still be a little air. At forty-five miles there are signs of twilight, a phenomenon impossible in a vacuum. But what is the nature of this air? Nitrogen assuredly; perhaps some oxygen. Fifty miles high there are reflections to which the name "noctilucent clouds" has been given. Dust they are held to be by some. But how can dust manage to collect in a definite layer at such a height? And whence did it come?

Far-Away Auroras.

Far above these faëry clouds the aurora shimmers. The phenomenon implies a slight atmosphere, although the altitude may be at least 400 miles. Electrons shot from the sun strike the sparse atoms of air on the outer fringe of the earth's atmosphere and electrify them. The result is an aurora which, at the poles, may extend far down toward the earth and which has a definite connection with the earth's magnetism.

How sure can we be that the upper atmosphere is indeed electrified as bits of electricity called electrons, hurled into space by the sun, partly wreck atoms of air and thus excite them? Radio furnishes the answer.

The waves that are sent out by a powerful station encircle the earth. Yet they are a form of light—light that our eyes are not adapted to see. To expect radio waves to cross the ocean is much like expecting to see in New York a beacon blazing in London, despite the curvature of the earth. The physicists solved this riddle by presupposing and then discovering an invisible mirror in the sky. Between this mirror and the earth the waves are partly reflected, partly conducted. "Ionosphere" is the name now given to that reflecting layer.

Until recent years it was thought that there was but one reflecting layer, which drops as low as seventy miles above sea level. Now it is known that there is another below it, above it a third which begins at an altitude of 140 miles, and others.

The more recent attack on the upper air is identified with the study of the cosmic rays. It was Dr. Victor F. Hess of Innsbruck, Austria, who first recognized the extra-terrestrial character of the rays. He sent up unmanned balloons with measuring instruments and went up in balloons himself, and this before the World War. The rays grow stronger and stronger with increasing altitude. In order to discover the limit to this increase, mountain climbing and ballooning, not to mention high flights in airplanes, have been a part of cosmic-ray study ever since the war. Hence the stratosphere is as important to the physicist as it is to the meteorologist.

Big Expense Involved.

Balloon ascents of the kind that Stevens and Anderson undertook are expensive. Many different kinds of instruments can be carried aloft, and many different kinds of observations made, which is the reason why organizations like the National Geographic Society are willing to spend hundreds of thousands in building bags big enough to contain an office building. There will be more ascents into the stratosphere by adventurers imbued with the question-asking spirit of science. But physicists find it cheaper and for some purposes quite as effective to send up unmanned balloons carrying radio transmitters which wireless back messages about the state of the upper air.

Cosmic rays, auroras, reflecting layers, faëry clouds high in the sky—they are connected in some way not yet divined. Everywhere we see electrons at work—infinitesimal bits of matter or electricity, however we choose to regard them. The secret of the atmosphere may prove to be the secret of the atom, in which these electrons play an indispensable part.

New Concept Unfolds.

Meager as our knowledge is, a new conception of the earth has grown up since de Bort first re-

AS IT LOOKED ON THE WAY UP

Times Wide World.
Explorer II photographed from a plane over the Black Hills.

EARTH SCIENCES

INTO THE MYSTERIOUS STRATOSPHERE

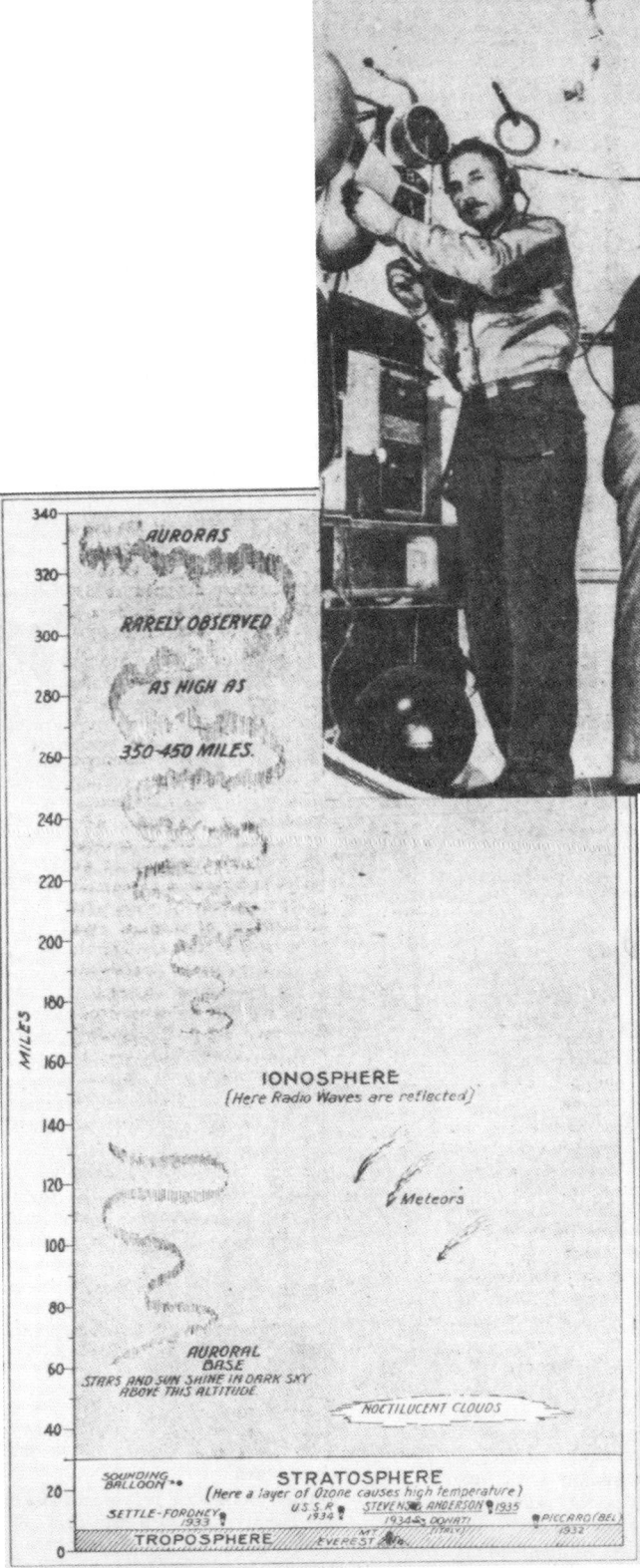

The photograph shows Captains Albert W. Stevens (left) and Orvil Anderson in the gondola of their balloon. The chart indicates regions of the upper air and flight records.

November 17, 1935

COSMIC RAY CEILING FOUND 10¾ MILES UP

Geographic Society Reports to Scientists Results of Stratosphere Flight.

CURVE OF EARTH PICTURED

Photo Made at 72,395 Feet Elevation Is Exhibited for the First Time.

By WILLIAM L. LAURENCE
Special to THE NEW YORK TIMES.

WASHINGTON, April 29.—Discovery of the existence of a cosmic ray ceiling at a height of ten and three-quarter miles above sea level, was announced here today by the National Geographic Society in its first official report on the scientific results of the stratosphere flight on Nov. 11, 1935. The report was written by Captain Albert W. Stevens, commander of Explorer II, which rose to an altitude of thirteen and three-quarter miles, the highest ever attained.

The society also gave a preview of the first photograph ever made of the top of the troposphere as well as of the bottom of the stratosphere, with a clear division line between them, which at the same time gives the first clear photographic view of the curvature of the earth. It was taken from an elevation of 72,395 feet, the highest camera shot ever made.

The division line between the troposphere, the lower part of the earth's atmosphere, and the stratosphere, the layer directly above the troposphere, was found to lie on the day of flight at a height of 37,000 feet.

From its vantage point far up in the stratosphere, the camera registered the horizon 350 miles away, sweeping like a great arch across the photograph. A straight black line has been ruled in to bring out clearly the curvature of the horizon, which is practically concentric with the earth's surface, and thus indicates the actual curvature of the earth.

The full report, including the photograph, appears in the May issue of The National Geographic Magazine. It outlines for the first time all the scientific results so far obtained, from a study by cooperating scientists, of the instrument readings made during the record stratosphere flight.

Results of Flight Shown

The balloon, sent up from the Black Hills of South Dakota under the auspices of the National Geographic Society and the Army Air Corps, was piloted by Captain Orvil A. Anderson. While some of the preliminary results have been published, this is the first time that a comprehensive report on the scientific results of the flight is being made, together with the first pub-

leased his sounding balloons. In the mind's eye we see it with all its aureoles—the concentric, fluttering halos that constitute its ionosphere, the auroras spreading out like a corona and testifying to powerful magnetic effects, and the atmosphere brilliantly blue.

If we could only transport ourselves to the moon! There we should see the misty, sapphire glitter of a planet that is unique in the solar system; unique, perhaps, in the vast universe of countless billion stars. Noctilucent clouds, ozone layer, stratosphere, troposphere—we should behold them as distinct shells. Deep down we should note a thick disturbed sediment. In this gaseous mud, stirred by winds, oceans and continents are visible through rifts in banks of clouds, life flourishes, airplanes fly, a race of thinking creatures gazes out and asks questions about the atmosphere without which it cannot live and becomes dimly aware that it has a cosmic destiny.

lication of the highest camera-view photographs ever made.

The cosmic ray ceiling was discovered with apparatus sent up by Dr. W. F. G. Swann of the Bartol Research Foundation of the Franklin Institute at Swarthmore, Pa. The ceiling has been located at a height of 57,000 feet above sea level.

At 40,000 feet above sea level (7½ miles) it was found that the rays coming straight down to the earth at terrific speeds from outer space were 40.1 times as numerous as at sea level. At 53,000 feet (10 miles) they were 51.5 times the sea-level number. At 57,000 feet (10¾ miles) they were 55 times the sea-level number. But at 72,395 feet, the highest point reached, the number of rays had decreased to 42 times the sea-level number.

The explanation of the decrease of the vertical rays above 57,000 feet, Dr. Swann believes, lies in the assumption that many, if not nearly all, of the rays entering the cunningly devised traps for the "interstellar imps," are secondary rays shot out from atoms of the air gases by the original, or primary, rays rushing in from outer space.

Above a certain level, where the air is very thin, it is believed, fewer secondaries are produced because there are fewer gas atoms for the primaries to tear apart.

Horizontal Rays Increase

A second fact about the cosmic rays brought to light by the flight, Captain Stevens reported, was that rays coming from the horizontal direction, which are negligible in number at sea level compared with the vertical rays, increase at higher levels, until at 13¾ miles up they are approximately equal in number to the vertical rays.

The rays traveling horizontally at the higher altitudes are believed, Captain Stevens stated, to have been swung into the horizontal plane from other planes by the earth's magnetic influence.

A remarkable correlation was also found between the cosmic-ray roof and the region in the atmosphere where the air's electrical conductivity is greatest. This spot was located at a height of eleven and a half miles, only a short distance above the cosmic-ray ceiling, which gave definite proof for the first time that the electrification in the upper atmosphere is due entirely to cosmic rays.

April 30, 1936

AIR HELD UNIFORM TO 35 MILES HIGH

New Mexico Experiments Show Make-Up to Be Constant in Most of Its Constituents

STRATOSPHERE ANALYZED

London Group Hears There Is Evidence of Layer of 'Lasting Stillness' Above Region

Special to THE NEW YORK TIMES.

LONDON, March 27—Experiments at the White Sands proving ground in New Mexico show that the composition of the air is uniform in most of its constituents up to a height of thirty-five miles, the Chemical Society was told tonight.

Prof. Frederick A. Paneth of the Radiochemistry Department of Durham University described how samples of stratospheric air sucked into the nose of rockets above the proving ground had been analyzed.

He said that his department, through this and other experiments, had reached the conclusion that the composition of the air up to thirty-five miles could be regarded as a geophysical constant.

But above that height, he said, there is evidence of a layer of "long lasting stillness" shown by the gravitational separation of the rarer stratospheric gases such as helium, argon and neon.

Such experiments are of considerable importance to meteorologists and physicists, but even for lay members of the audience there was a fascination in the details of the necessary microanalyses.

Relationship of Rarer Gases

Because of the difficulty of preserving oxygen in its gaseous state (it combines too readily with other substances), Professor Paneth and his co-workers relied on the relationship of the rarer gases—helium, neon and argon—to the nitrogen content of the air samples.

It was necessary first to establish the fact that the gases were evenly distributed throughout the atmospheric air. Helium, it was thought, might be more common over such places as the petroleum fields of Texas, where it occurs naturally. Accordingly, samples of surface air in bottles were sent to Durham University from about fifty meteorological stations in the major continents and oceans of the world.

Analyses showed that the helium content was as much a geophysical constant as atmosphere nitrogen, which could be determined with an accuracy of .001 per cent.

The next step was to examine the relative composition of the rare gases in the stratosphere. This was done through the cooperation of the United States Army Signal Corps and the Department of Aeronautical Engineering at the University of Michigan, which sent to Professor Paneth air samples collected in V-2 and Aerobee rockets at heights of between fifty and seventy kilometers (thirty to forty-five miles).

Taken at Current Altitude

To assure that the air samples were taken at the current altitude, three evacuated steel bottles were placed immediately behind the nose cones of the rockets. The bottles were soldered to thin copper tubes one inch in diameter. At a predetermined instant a steel knife was made to cut through the tube. Five seconds later the tube was squeezed shut at a lower point by a rocket-operated vise.

The second and third bottles were opened and shut at successive intervals as the rocket soared upward on its free climb.

The oxygen and carbon dioxide were bound and the argon was freed from the nitrogen by heating it with barium in a steel furnace. The neon and helium were then separated through charcoal and measured in a refinement of the Pirani gauge, which permits the determination of almost infinitesimally small amounts of gas.

March 28, 1952

Mystery of the Air We Explore

The wild black yonder high above the earth is challenged by man and his missiles.

By LEONARD ENGEL

MOST of us ordinarily take the vast sea of air around and above us for granted. The earth's atmosphere, however, clear out to the wildest black yonder, is in the news. Russian and American high-altitude sounding balloons have been making political as well as scientific headlines. Only a few thousand working hours from now, the first man-made moon will be hurled into space.

In 1714 the great astronomer Edmund Halley—discoverer of Halley's comet—could assert that the essential particulars of the atmosphere were

LEONARD ENGEL *is a freelance writer who specializes in scientific and related subjects.*

"perfectly well understood." In the light of the laws of natural science as then known, he was right.

Halley could not know that radio, rockets and Geiger counters were to reveal an immensely deeper, more complex and more mysterious atmosphere than he dreamed. He thought earth's blanket of air was forty-five miles thick. We know it extends at least 10,000 miles upward from the surface of the earth. He thought the atmosphere had a very simple structure: the density of the air merely diminished gradually as one went up until the air merged into the nothingness of space. He had no idea of the atmosphere's intricate, layered construction, or of the strange things that go on in some parts of it.

The atmosphere forms a quadruple shell around the earth. The bottommost layer is the troposphere, the layer of dense air in whose depths we live, ten miles thick at the equator, five miles thick at the poles. Next comes the stratosphere. The latter rises upward from the top of the troposphere to a height of about fifty miles.

Above the stratosphere is the third layer, the ionosphere. The ionosphere extends outward several hundred miles to the fourth and topmost atmospheric layer. This, the exosphere, is the layer in which the atmosphere does become indistinguishable from interplanetary space. The exosphere may go as far as 18,000 miles out.

LET us take up separately the four layers of the air, and what is known about each.

The troposphere represents less than one twenty-fifth of 1 per cent of the total volume of the atmosphere, but contains nearly 80 per cent of its gases. The air is dense enough to support life, however, only in the lower reaches of the troposphere. Supplementary

EARTH SCIENCES

oxygen is desirable for persons flying in unpressurized planes above 10,000 to 12,000 feet, and mandatory at 15,000 feet or more. The men who climbed Mount Everest, whose peak is well within the troposphere, had to carry oxygen with them, in tanks on their backs.

The word troposphere means turbulent sphere, and it is that. Hurricanes, tornadoes and the other great storms that buffet the surface of the earth are centered in the troposphere.

BY all odds, the most interesting recent discovery in the troposphere is the "jet stream," a river of wind circling the earth. In the northern hemisphere, the jet stream blows the year around from west to east at an altitude of 15,000 to 40,000 feet and a speed of 50 to 300 miles an hour.

The jet stream is obviously a product of the general circulation of the air around the world and has much to do with our weather, but its exact mechanics are unknown. The jet stream was uncovered with the aid of weather balloons. It is the jet stream that carries the propaganda balloons of the Free Europe Committee and United States Air Force weather balloons eastward over the Iron Curtain from Germany and Austria. The jet stream is also responsible for the recent appearance in Japan of Russian weather balloons launched in Siberia.

In the troposphere, the sky is blue by day because of the scattering of sunlight by dust particles and the like. In the stratosphere, the air is clear. There are no clouds, except some ice-crystal clouds (thirteen to eighteen miles up) and a rare form of dust cloud occasionally found fifty miles up. The stratosphere sky (as all the sky above that) is therefore brownish black. The stars may be seen by day. The sun has a fierce metallic glare and its flaming corona—visible from the surface of the earth only during eclipses—is constantly in evidence.

THE next layer of the atmosphere—the stratosphere—contains 20 per cent of the atmospheric gases, though it constitutes less than one-fifth of 1 per cent of the total volume of the atmosphere. Men have penetrated into the lower portion of the stratosphere both in planes and in balloons. But unmanned balloons have gone higher. The record for an unmanned balloon is 140,000 feet—twenty-six and a half miles above the earth, or almost twenty miles into the stratosphere.

Down where we live, near sea level, virtually all the air's life-giving oxygen is in the form of "molecular oxygen" —two atoms of oxygen joined in a single molecule. As one rises, another form of oxygen, ozone, containing three atoms of oxygen per molecule, appears. The concentration of ozone reaches a peak in the "ozone layer" of the lower stratosphere, at a height of twenty to forty miles.

The ozone

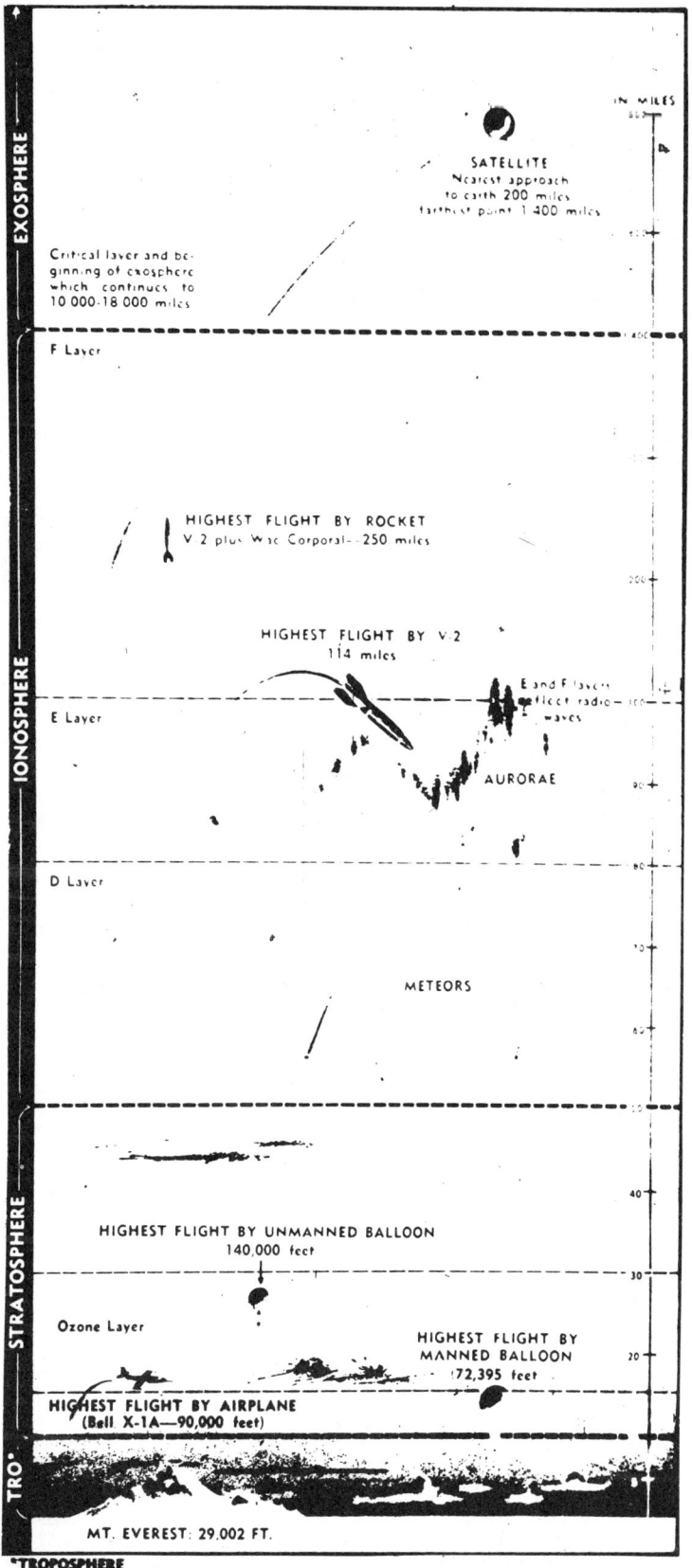

THE EARTH'S ATMOSPHERE is 10,000 to 18,000 miles thick and has an intricate layered construction. In the drawing here, the D, E and F layers are represented by lines for simplicity; they are actually bands many miles thick. (Other "layers," once called A, B and C, were found by scientists to be nonexistent.) The exosphere, the outermost portion of the atmosphere, will be penetrated in 1957 or 1958 by satellites to be launched for the International Geophysical Year.

layer absorbs a major part of the ultraviolet light from the sun. Because present forms of life on the earth's surface (including man) could not survive exposure to the full ultraviolet radiation of the sun, it it often said that the ozone layer is essential to life. This is true only for life on land (ultraviolet light does not penetrate the sea) and only for life as we know it. Had there been no ozone layer, other forms of life, resistant to ultraviolet light, might well have evolved.

NOW we come to the eerie, turbulent world of the ionosphere—primary target of the fantastic earth satellite project. Man has as yet hardly penetrated the ionosphere. In 1949, the Army mounted a WAC Corporal rocket in the nose of a V-2, took the WAC Corporal off the ground with the V-2, and launched the lighter rocket from high over the White Sands Proving Grounds. It reached an altitude of 250 miles. But most rockets go only 100 miles or so up. Radio studies and other indirect means have been the source of most of what has been learned about the ionosphere up to now.

The ionosphere is exposed to the practically undiluted radiation of the sun—visible light, ultra-violet rays, X-rays, infrared radiation, streams of hydrogen particles. Cosmic rays, from we know not where, beat upon the ionosphere and smash into occasional air atoms, causing "cosmic ray showers" (flying debris of the tiny terrible atomic explosions). Meteors (mostly dust particles the size of a pinhead, but sometimes much larger) whiz in from interplanetary space.

The air in the ionosphere is 10 million times rarer than air at sea level; the ionosphere is almost as empty as the "empty space" produced by the newest and the very best vacuum pumps.

Thin as it is, the air in the ionosphere is too thick for meteors. Virtually all are consumed in the lower part of the ionosphere—a region astronomers have dubbed the "crematorium of meteors"—by heat generated by the friction of their passage through the air.

IN the stratosphere, the main constituents of the air —oxygen and nitrogen—are the same (except for the modest percentage of ozone in the ozone layer) as at sea level. Two changes take place in the composition of the air, however, as we enter the ionosphere. First, oxygen molecules begin to break up into separate atoms of oxygen. Second, solar radiation electrifies some of the oxygen atoms and some of the oxygen and nitrogen molecules, converting them into what the physicist terms ions.

The over-all proportion of ions is certainly small, but their effects are enormous; the region is well named the sphere of the ion. To begin with, the ionosphere's ions make worldwide radio possible and, at the same time, are responsible for some of radio's worst headaches.

The ions are concentrated mainly in three layers. The lowest, located just above the start of the ionosphere, is the D layer. Thirty miles farther up (80 miles from the earth's surface) is the E layer. Another 20 miles up is the F layer, a broad band which is actually two layers by day and one layer at night.

The E and F layers reflect radio waves of the wave lengths used in regular broadcasting and in commercial shortwave and send them back to earth. That is why radio can reach long distances around the curvature of the globe.

The Army pierced the ionosphere, 250 miles up, with this rocket.

THE D layer, on the other hand, absorbs radio waves. Ordinarily it does not absorb enough radio waves to make trouble. But during solar flares —huge, flamelike eruptions of gas from the surface of the sun—the absorption of radio energy by the D layer is enormously increased and radio signals fade out. Solar flares may completely disrupt radio communications for days on end.

Ultrashort radio waves, like those used in radar, TV and FM broadcasting, are not bounced back by any of the ionized layers of the air, but pass right on through. It is for this reason that TV and FM stations, unlike regular broadcast stations, reach only to the horizon.

A few years ago, though, radio engineers discovered that even ultrashort waves occasionally went beyond the horizon, as though bounced off something up in the sky. The something proved to be turbulent masses of air. The discovery that turbulent air can bend radio waves back to the earth has led to an entirely new system of long-distance radio transmission called "radio scattering."

Radio scattering is not useful for home radio or TV, as expensive equipment is required. But it promises to be of great value for long-distance communications, particularly in regions like the Arctic, where relay stations are difficult to construct and where magnetic disturbances make conventional radio unreliable. Since the air is always turbulent, scatter broadcasts promise to be extremely reliable.

IF you were to step out into the ionosphere from some future space ship, the chances are you would feel nothing because of the thinness of the air. But the ionosphere is swept by winds and air currents that reach speeds of 300 miles an hour or more. They are caused by variations in temperature between day and night, and by the tidal pull of moon and sun on the air.

The great winds cause movement of the ions present in the ionosphere. The motion of the ions is responsible for the circulating electric current of the ionosphere, the so-called "dynamo current." The latter has a marked effect on the earth's magnetic field and the behavior of magnetic compasses.

In turn, the earth's magnetic field helps shape the most spectacular of all ionosphere phenomena — the polar aurorae. The polar lights have intrigued men since there were men near enough the poles to see them. An explanation of the aurora was found a few years ago, mainly by Dr. A. B. Meinel and his associates at the University of Chicago's Yerkes Observatory.

The auroral lights are caused by hydrogen particles from the sun. Since these particles are electrically charged, the earth's magnetic field diverts them to the poles. There they strike nitrogen and oxygen atoms in the ionosphere, inducing them to glow and producing the auroral displays.

There is but one more feature of the ionosphere that need be mentioned. In the lower part of the stratosphere, the air is very cold. In the mid-section it warms up to 32 degrees Fahrenheit, then the temperature drops again to 100 degrees below zero at the stratosphere-ionosphere boundary. Thereafter the temperature rises. No one knows yet why the temperature goes up and down that way.

It has been calculated that temperatures in the ionosphere may go as high as several thousand degrees. Upper-air visitors are nevertheless in no danger of burning up, provided they travel at a more modest speed than meteors. Solar radiation alone is not sufficient to make one feel hot; heat must also be transmitted by the surrounding air. Just as it is too thin to make one feel the ionospheric wind, the air in the ionosphere is too thin to make one feel the heat.

According to present plans, the artificial satellite is to be launched into an orbit 200 miles from the earth at its nearest point, and 1,400 miles at the farthest. Thus, the satellite is to journey in both the ionosphere and exosphere. In fact, the satellite will give man his first glimpse into the world beyond the ionosphere— the exosphere—and tell him just where it begins.

Upper-air physicists define the exosphere as that portion of the atmosphere where there is still air, but where the air is so thin and its molecules so far apart that the molecules can travel an indefinite distance without bumping into one other. They are then free, so to speak, to come and go as they please, save only for the influence of the earth's gravitational field.

PRESENT estimates indicate that the atmosphere reaches this state of "freedom" about 400 miles up. Thereafter, it is supposed, the air becomes thinner and thinner until oxygen molecules, nitrogen molecules and atoms cease to exist, and, in fact, nothing exists but the occasional wandering dust particles and hydrogen atoms of interplanetary space. That point is believed to be reached 10,000 to 18,000 miles up.

But much of this is very nearly pure speculation. Nothing is really known of the exosphere or of the upper half of the ionosphere or of how they may influence events far below on the surface of the earth, like radio broadcasting and the weather. Man will not know until he gets exploratory missiles and platforms up there— perhaps with man himself aboard one day—to find out.

April 15, 1956

EARTH SCIENCES

U. S. ROCKETS GIVE NEW SPACE DATA

116 I. G. Y. Firings Disclose Factors in Density Shifts in Upper Reaches

By WALTER SULLIVAN

Great tides and winds act upon the ocean of air that envelopes the earth, causing radical changes in the density of its upper reaches, depending on the season, latitude and time of day.

These and other "on-the-spot" discoveries are the fruit of 116 United States rocket firings during the first twelve months of the International Geophysical Year. They settle a number of theoretical controversies that could only be resolved by direct observation.

The results have been published by the National Academy of Sciences in thirty scientific papers, assembled as the first in a series of I. G. Y. rocket reports. Publication was timed to coincide with the conference of I. G. Y. participants now taking place in Moscow.

Fifty-four of the rockets were rockoons, a balloon-rocket combination launched from shipboard in otherwise inaccessible regions. Forty-one of the rockets were fired from Fort Churchill, on the edge of Hudson Bay; fifteen were launched from San Nicolas Island, off of California and six from White Sands, N. M.

70 Firings on Schedule

Another seventy have been scheduled for firing before the I. G. Y. is due to end, Dec. 31.

Three shots with Aerobee-Hi rockets, aimed at auroral displays over Fort Churchill last January and March, made it possible to determine, to the satisfaction of the experimenters, that the aurora is caused by high energy electrons showering on the atmosphere from space. At the same time ions seemed to be ruled out as the cause.

A rocket fired directly through an auroral display detected the electron shower when it was inside or above the curtains of light, but not elsewhere. Ions, on the other hand, were observed both inside and outside the display.

The air 125 miles above Hudson Bay was found to be twenty times denser on a summer's day than on a winter night. The sub-Arctic air was 6.5 times denser than the corresponding value over New Mexico.

The experimenters, associated with the Naval Research Laboratory, proposed the launching of an earth satellite in a pole-to-pole orbit to study this phenomenon. The new analyses of rocket results have produced air density figures that conform in general to the recent air drag studies of satellite flights.

The latter indicated densities, at satellite altitudes, that were many times greater than had been expected from earlier rocket studies. The I. G. Y. rocket examination of air density has been carried out by various methods.

Free-falling spheres the size of a melon have been ejected from rockets at high altitude. Inside each sphere was a device to observe and radio the speed of its acceleration—and hence the amount of air drag.

Pressure gauges and radio mass spectrometers were also used to determine densities. Rockets that successively fired nineteen grenades in flight were used to determine temperature and wind through the characteristics of sound travel from the rocket to ground stations. A wintertime west wind of 337.5 miles an hour was observed thirty-six miles above Fort Churchill.

The mass spectrometer reports the mass of the various components of the air at intervals along the flight path. The result was direct confirmation of some of the theories advanced to explain the radio-reflecting E and F layers of the ionosphere.

In the E-region, about seventy miles up, nitric oxide was found to be the predominant positive ion, although almost none of this gas was observed in neutral (non-ionized) form. In the F-region, roughly 150 miles up, atomic oxygen was found to be the predominant positive ion.

The only negative ion detected was nitrogen dioxide.

Other papers reported on rocket studies of cosmic rays, the earth's magnetic field, as well as the X-rays and ultraviolet rays that rain on the atmosphere from the sun. Some of the discoveries had already been announced and many of the results are preliminary, pending further study.

August 5, 1958

I. G. Y. EMPHASIZED WEATHER STUDIES

4 Jet Streams Discovered— First Data Gathered on Antarctica's Winter

By WALTER SULLIVAN

At exactly midnight, Greenwich Time, last Aug. 25, two Soviet scientists, looking something like divers with their moleskin face masks, special breathing apparatus and heavy furs, tried to launch a huge weather balloon.

At the same moment thousands of other weather men the world over were likewise engaged. At that hour, throughout the International Geophysical Year, they set loose special balloons designed to map the world's ocean of air up to eighteen miles in the sky.

For the Russians, however, this was a special day. The Diesel fuel that kept their little camp warm in the heart of Antarctica was thick as honey. The rubber-like covering of the balloon became brittle in the open air.

Lowest Reading Recorded

The temperature at their station, known as Vostok, had sunk to 125.3 degrees below zero, Fahrenheit—the lowest ever recorded on the face of the earth.

Probably the most thorough exploration carried out during the I. G. Y., an eighteen-month study that ended at midnight Wednesday, was that of the atmosphere. It only began the job, for weather is a four-dimensional problem.

The fourth dimension is time, a vital element, since it is the movement of air, sometimes gentle, sometimes destructive, that transports heat and produces weather.

The patterns of jet streams, seasonal wind reversals and other phenomena that were discovered made it clear that an intense and prolonged study was needed to gain the knowledge already much in demand for jet aircraft and rockets.

Four jet streams were discovered or explored: Two were so-called "polar night jets," blowing westward in the lower stratosphere roughly along the Arctic and Antarctic Circles. The other two were in the upper troposphere in middle latitudes.

Atmosphere's Lowest Layer

The troposphere is the lowest layer of the atmosphere, where the chief weather phenomena occur. It is marked by a steady decline in temperature with increasing altitude. At an elevation that varies from a few miles to more than a dozen the drop in temperature suddenly halts. This is the start of the stratosphere.

Preliminary study indicates a link between a wandering jet stream and major weather changes. From 1952 to 1956 Texas and more northern parts of the Great Plains were parched by extended droughts. In March and April of 1957 there were heavy rains and blizzards, followed by disastrous floods.

This bad weather, accompanied by tornadoes, moved gradually northward from Texas toward Canada in May and June. High-altitude weather mapping has shown that this northward movement was accompanied by a parallel shift of the middle-latitude jet stream.

In the spring months the jet stream was centered 40,000 feet above northern Mexico and southern Texas. It then began migrating north.

Weather Mapping Improves

The bigger, tougher balloons distributed to weather stations for the I. G. Y. made it possible to keep track of day-to-day weather changes at various levels of the stratosphere up to eighteen miles above the earth. This is high enough so that circulation patterns are not affected by mountain chains and the weather maps become smoother and less detailed.

It was hoped that this upward extension of weather knowledge would solve the mystery of "explosive warming," first observed over Berlin, Germany, in the winter of 1952.

At a time when one would expect the upper air to be coldest, a great mass of comparatively warm air would suddenly descend into the highest reaches of atmosphere within the range of weather balloons.

Some found in this direct evidence that streams of gas from the sun have a direct effect on weather. In some cases the temperature of the upper air near the Arctic rose eighty degrees in two days.

Data Compared

The daily, simultaneous release of the special balloons made it possible, beginning with the "rehearsal months" prior to the I. G. Y., to map the weather in the region of these "explosions." Comparison of Soviet and United States data showed that the warmings sometimes occurred at the same time over Siberia and Greenland.

Researchers in the United States Weather Bureau are doubtful that the heat comes from the sun. They believe that it derives from vertical movements of the earth's atmosphere that are not yet fully understood.

One part of the atmosphere was almost unexplored at the start of the I. G. Y. That lying over Antarctica. No man had spent the winter more than 100 miles inland and even in summer no interior station had been manned for more than a week.

The hinterland plateau of Antarctica, larger than Europe, was presumably the coldest region in the world, but nothing was

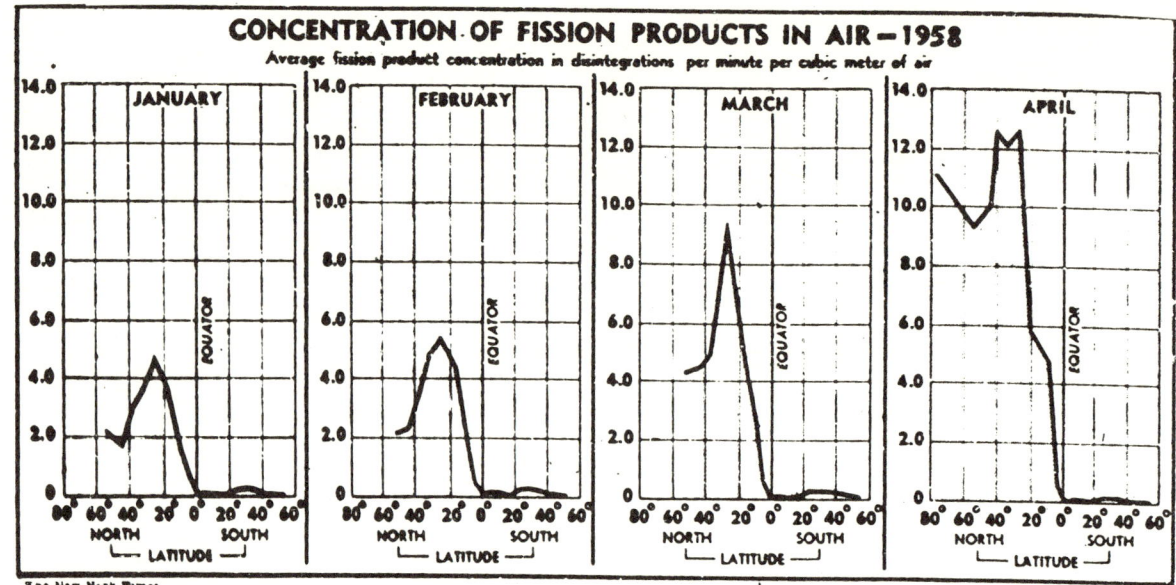

GROWING FALL-OUT: One of the projects of the International Geophysical Year was to study the interchange of atmosphere between the Northern and Southern Hemispheres by recording the extent to which nuclear bomb debris was carried across the equator. This series of graphs shows the amount of radioactive debris in the air for latitudes between the Arctic and Cape Horn. Although there was a marked increase of fission products in the Northern Hemisphere, almost none seem to have been carried south across the Equator.

known of its winter weather, when it lies in darkness for many months.

Discover Polar Jet

An international Weather Central at Little America pieced together information from observers at two score points on or near the continent. They discovered the Antarctic polar night jet, blowing westward around the coastline at as much as 150 miles an hour.

One of the chief surprises was the fact that the temperature at the South Pole, far inland, did not decline steadily until mid-winter, or soon thereafter. Instead it sank rapidly in early fall, at an average of one degree a day. Then, in May and June, it rose. It was the equivalent of December's being warmer than October in the Northern Hemisphere.

This effect continued until late winter. According to Dr. Harry Wexler, director of meteorological research at the United States Weather Bureau, this is due to the movement of warm oceanic air all the way to the heart of the continent.

Only late in the winter, when the halo of frozen ocean around Antarctica has grown out to 1,000 miles, is this effect curtailed. Then, in August, the mercury hits bottom, as evidenced by the experience of the Soviet scientists at Vostok.

Strangely, much of the time, the air warms rapidly with increasing altitude for the first 1,000 feet or so above the polar plateau. The rise may be as much as eighty degrees in that distance.

Above that level the trend reverses itself and the stratospheric air may be as cold as —135.4 degrees, a reading observed by a radio-equipped weather balloon thirteen miles above the United States station at the South Pole.

The staff of the Weather Central, while primarily American, included Argentine, Australian, French and Soviet scientists. Their work brought to light the basic patterns of weather circulation at the bottom of the world. Deep cyclones persisted over the Ross and Weddell Seas, whereas over the high plateau at the center of the continent there was a semi-permanent anticyclone.

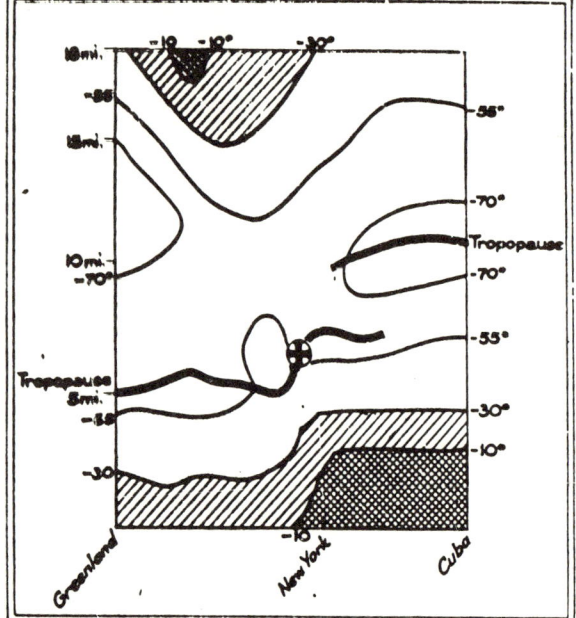

EXPLOSIVE WARMING: On Feb. 18, 1957, a great mass of comparatively warm air descended "from nowhere" into the winter sky over the Arctic. This is shown by the cross-section of the atmosphere from Cuba to Thule, Greenland. The warmth of the atmosphere is indicated by the degree of shading. The air high over Greenland [upper left] is comparable to air low over Florida [lower right]. The numbers mark lines of equal temperature. The cross represents the middle-latitude jet stream.

The Coldest Continent

Average temperatures in Antarctica are as follows:

Station	Elevation (in Feet)	Temperature (Fahrenheit)
Wilkes	50	+17
McMurdo Sound	100	+ 2
Little America	100	—10
Wilkes	4,000	—19
Byrd	5,000	—19
Station C (Victoria Land)	8,850	—49
South Pole	9,200	—59
Vostok (Geomagnetic Pole)	11,500	—69.2
Sovietskaya (Near center of continent)	12,200	—71.3

The lowness of the temperature is clearly related to the elevation. The figures for Wilkes, McMurdo Sound, Vostok and Sovietskaya are based on preliminary weather reports. Those for the other stations represent ice temperatures at a depth of twenty-five to thirty feet, which are thought to approximate the average air temperature. Sovietskaya lies near the Cold Pole, or coldest point in the world. The winters at Wilkes Station, the most northerly, are milder than those in Nebraska, due to the station's nearness to the sea.

EARTH SCIENCES

A cyclone is a wind system, centering on a low-pressure area, with a clockwise rotation in the Southern Hemisphere. It is usually marked by heavy precipitation.

An anticyclone rotates in the opposite direction, around a "high." It is usually marked by fair weather.

Classic Cyclone Noted

In the middle stratosphere there was a classic polar cyclone, reaching 150 miles an hour in the westerly jet. In October rapid heating of the stratosphere in returning sunlight caused a reversal to light easterly winds. This was followed by a slower warming at lower levels of the atmosphere.

I. G. Y. studies showed a similar reversal of winds between summer and winter in the stratosphere of the Arctic.

Whereas the interior of Wilkes Land produced the coldest temperatures recorded on earth, Wilkes Station on the coast proved to have a milder climate than Nebraska, due to the moderating influence of the sea.

The great reservoir of cold air over the continent was found to erupt, from time to time, into middle latitudes. Likewise warm air invaded the continent. "The battle between the two contrasting air masses may go on for many days," Dr. Wexler reported, "causing intense storms over large portions of Antarctica."

The upper air winds thus reach speeds of 200 miles an hour, he said. He noted that this knowledge would be of extreme importance in future air travel between continents at the bottom of the world.

Temperature readings at Little America taken at intervals since 1911 show a warming trend. The following average readings have been recorded there:

1911, —17 degrees Fahrenheit; 1929, —15.5; 1934, —14.3; 1940, —15.2; 1956, —13.2, and 1957, —11.6.

Another area of concentrated weather study was the South American segment of a pole-to-pole chain of I. G. Y. stations. There the United States Weather Bureau helped establish upper-air stations at Guayaquil, Ecuador; Lima, Peru, and Antofagasta, Quintero and Puerto Montt in Chile.

Local scientists were trained in the operation of the electronic equipment and were provided with manuals in Spanish. The result was the first extended observation of the upper atmosphere over South America, and led to the location of jet streams there. Specialists of Pan American-Grace Airways reported that this network had paid for itself in a single day: Dec. 8, 1957.

On that date it made possible the evaluation of a severe storm, enabling the airline to adjust its flight plans and continue operations safely.

It was said that, during the I. G. Y., forecasts of winds aloft improved by 50 per cent. As with much of the work initiated in smaller countries during the I. G. Y., it is hoped that these observations will be continued on a permanent basis.

Data Being Collated

Another aspect of I. G. Y. weather studies was the measurement of carbon dioxide in the air and oceans. This should provide a basis for future determination as to whether the burning of fuels is increasing the amount of this gas in the air.

The World Meteorological Organization is putting all I. G. Y. surface and upper-air observations on microcards. The data is on about 1,000,000 forms, but will be compressed onto 18,500 microcards that can be stored in fifteen or sixteen drawers of a standard card-catalogue cabinet.

A full set of cards will be sold for $5,990, making it possible for research institutions to own a record of world weather throughout the eighteen months of the I. G. Y. The meteorological organization reported in June that it had received data for these cards from ninety-eight countries and territories. Only four countries, all in Latin America, had sent none in.

It is hoped that intensive study of this data, making use of punch-card techniques, may reveal what causes major seasonal weather variations, such as the cold and snowy winter in the eastern United States last year.

Because 70 per cent of the atmosphere rests on the oceans, the heat exchange between these masses of air and water undoubtedly plays a major role in weather, as does the influence of the Antarctic ice sheet.

Some also believe, on the basis of I. G. Y. studies, that clouds of solar gas, which clearly affect the earth magnetically and electrically, also influence its weather.

Thus an understanding of weather—the I. G. Y. science that most intimately influences daily life—depends on most of the other I. G. Y. sciences. Dr. Wexler believes that the unraveling of these relationships may take decades, but that prolonged study of the data, and perhaps new I. G. Y.'s, should ultimately clarify the picture.

January 2, 1959

U.S. ORBITS WEATHER SATELLITE; IT TELEVISES EARTH AND STORMS; NEW ERA IN METEOROLOGY SEEN

2 CAMERAS IN USE

270-Pound Vehicle to Transmit Pictures for 3 Months

By RICHARD WITKIN
Special to The New York Times.

CAPE CANAVERAL, Fla., April 1— The first artificial satellite able to provide detailed photographs of the earth's weather was fired into orbit here today by the United States.

Two television cameras looking down from an altitude of about 450 miles made initial pictures of the earth's cloud patterns during the satellite's second orbital trip.

Four pictures, taken by the wider-viewing and therefore less-precise camera of the two, were proudly distributed this evening by the National Aeronautics and Space Administration. The space agency has over-all responsibility for the project.

The pictures showed the cloud cover that lay over the Northeast United States and the adjacent area of Canada this morning. They also showed an identifiable outline of the gulf of the St. Lawrence River. The curvature of the earth was clearly recognizable.

2 Storms Photographed

In Washington Dr. Harry Wexler, director of research of the Weather Bureau, said officials at Fort Monmouth, N. J., had reported that the satellite had taken photographs of a storm moving into the Middle West, a New England storm and the "cloud bridge" connecting the two.

Before being made public, the pictures had been taken to the White House by Dr. T. Keith Glennan, the head of the space agency.

The President told him:
"The earth doesn't look so big when you see that curvature."
He said also:
"I think it's a marvelous development."

Vast Gains Envisioned

It was understood that the second camera, with a narrow-angle lens for taking pictures with finer detail, was also working successfully. It may be that, for reasons of security, such pictures will not be released immediately.

For weather experts, today's successful launching held some of the promise that the discovery of the telescope must have held for astronomers in the seventeenth century.

The 270-pound satellite, named Tiros I, offered a means of illuminating vast areas of darkness in man's understanding of the weather.

With this illumination should come advances in the unsure

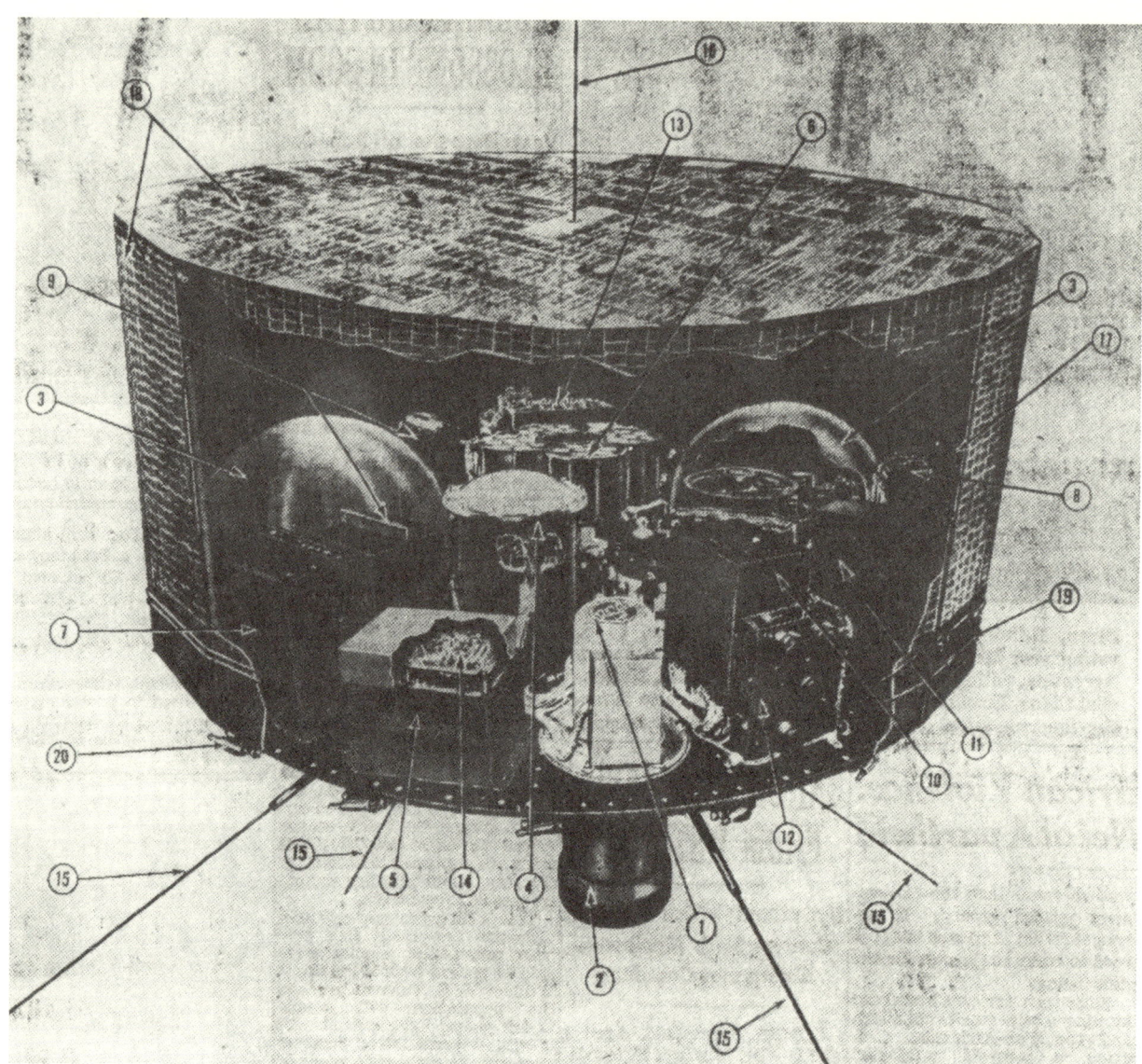

This diagram depicts articles in the satellite, some of which are more than one of a kind: (1) One of the two half-inch Vidicon TV cameras; (2) Wide-angle camera lens; (3) Tape recorders; (4) Electronic timer for operational sequencing; (5) TV transmitter; (6) Chemical batteries; (7) Camera electronics; (8) Tape recorder electronics; (9) Control circuits; (10) Auxiliary controls; (11) Power converter for tape motor; (12) Voltage regulator; (13) Battery charging regulator; (14) Auxiliary synchronizing generator for TV; (15) Transmitting antennas; (16) Receiving antenna; (17) Solar sensor to measure position of satellite with respect to sun; (18) Solar cells; (19) De-spin mechanism; (20) Spin-up rockets. Satellite, produced by R.C.A. affiliate at Princeton, N. J.

art of forecasting, with all that such advances imply economically, in crop planning; socially, in vacation planning, and militarily, in choosing an auspicious time to begin a campaign.

Eventually, scientists may understand weather well enough to manipulate it for peaceful or military advantage.

Tiros I is considered a prototype of satellites that are expected, in a few years, to provide twenty-four-hour, world-embracing data to forecasters who now must work with far less information.

Even some data radioed by Tiros I could be put to limited use in forecasts that will be made during the three months that scientists intend to operate the satellite's "weather eye."

But essentially, the launching of the new satellite is merely an experiment—a vital but early step on the way to practical advances.

It was at 6:40 Eastern Standard time, just four minutes before the expiration of a one-hour firing period, that Tiros I took off here on its scientific mission. It was propelled into space by a three-stage Air Force rocket, the Thor-Able.

Shaped Like a Hat Box

The 270-pound satellite was shaped like an enormous hat box. Each of its two television cameras was no larger than an ordinary water glass. They were built by the Radio Corporation of America under the direction of the Army's Signal Research and Development Laboratory at Fort Monmouth, N. J.

The camera with the wide-angle lens can view an area 800 miles square. The narrow-angle lens on the more precise camera was focused to peer at areas thirty miles square.

The more precise camera, it was hoped, would not merely show general cloud patterns but would also pick up individual types of clouds—cumulonimbus, stratus and so on.

The launching vehicle was aimed in such a way that the satellite, sweeping back and forth across the equator in 100-minute orbits, would cover an area between Lat. 50 degrees north and Lat. 50 degrees south. In the Western Hemisphere, this would take it as far north as Montreal and as far south as Santa Cruz, Argentina.

Stabilized in Space

A spinning motion was imparted to Tiros I—the name

EARTH SCIENCES

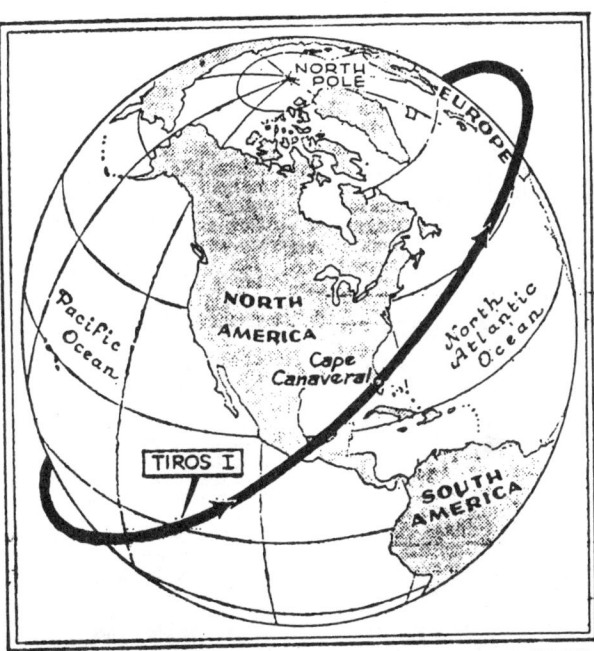

Schematic drawing of satellite orbit

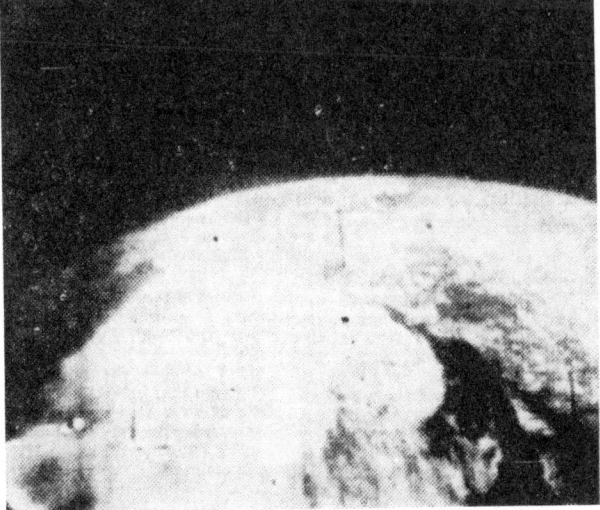

SENT BY SATELLITE: One of TV pictures from Tiros I. Dark area, lower right, is St. Lawrence River and Gulf.

stands for Television and Infra-Red Observation Satellite — at the burn-out of the second stage.

This was to stabilize it in space so that the camera eyes would not wander over the sky. It meant that the cameras, looking down from the bottom of the "hat box," would be pointed at the earth during brief but predictable periods in each orbit.

The equipment is designed to take pictures only during this period; to scan and store them on magnetic tape, and to relay them to earth as electronic signals when passing over one of two special receiving stations. These are at Fort Monmouth and Kaena Point, Hawaii.

When passage over a station coincides with the time the camera points at the earth, the intermediate step of storing the images on tape can be by-passed. It was understood that the pictures made public tonight had been acquired in this fashion.

Polar Orbits Planned

Eventual weather satellites will be built to keep their eyes riveted on the earth. They also will fly pole-to-pole orbits, so that, as the earth revolves beneath them, every inch will eventually pass before their lenses.

Little more than ninety minutes after today's firing the space agency confirmed in Washington what had appeared certain to the launch team here within minutes after the rocket disappeared to the northeast: Tiros I had gone into orbit.

The timing mechanism on board was set so that the camera shutters would not snap on the first orbit. This was because the third stage would not yet have dropped away when the cameras faced downward — over Mecca, incidentally — for the first time.

The plan was to cock the cameras for the first time as the satellite came within range of Fort Monmouth on the initial orbital circuit. The timer was to be set then to take a series of sixty-four photographs — thirty-two with each camera — when the satellite next pointed downward, over an area running approximately from Spain to Madagascar.

Informed sources said pictures stored on this first pass had been successfully radioed to the ground. But these pictures reportedly were of poorer quality than those that were made public.

The space agency plans to launch a second and more advanced Tiros satellite this summer or early in the fall. In addition to optical cameras it will use infra-red equipment to pick up atmospheric radiation invisible to the eye.

Tiros I, while not using infra-red for picture-taking, does employ it to help scientists match pictures with the orientation of the satellite at the time. So the name Tiros (the "i" and the "r" stand for infra-red) is not a misnomer.

Two years away are weather satellites of Project Nimbus, which are expected to open the era of practical forecasting from satellite data.

Tiros I is not the first satellite to obtain a picture of cloud cover. Crude pictures have been sent to earth by Vanguard II and Explorer VI. In addition, the Russians have shown the capability of space photography with their pictures of the formerly unseen side of the moon.

Orbit Nearly Circular

But Tiros I is the first space vehicle with equipment to obtain a comprehensive series of detailed weather pictures.

The injection of Tiros I into its nearly circular orbit was accomplished with a precision unequaled by any other of this country's fourteen satellites. The same goes for Russia's three Sputniks.

The scientists wanted the orbits to be as nearly circular as possible so that the photographs would be taken from nearly uniform distances.

The apogee, or high point, achieved was 468.28 miles. The perigee, or low point, was 435.5 miles — a remarkable spread of only thirty-five miles. The guidance used to obtain this accuracy was developed by the Bell Telephone Laboratories.

Orbit Slightly High

The orbit was a little higher than the 380 miles that had been aimed for.

In the three months during which it is planned to collect weather photographs, Tiros I will make 1,300 turns around the globe.

After three months, the relationship of the orbit and the position of the sun will have shifted in such a way that the down-pointing periods will not coincide with periods of daylight.

It was because of the need to establish the proper sun-orbit relationship that the launching team had to fire within a one-hour period.

The satellite is shaped somewhat like a cylinder, although it has eighteen flat sides instead of being round. It is forty-two inches wide and nineteen inches high. The sides are covered with 9,000 solar cells to provide electrical power.

The photo signals are coming in on 235 megacycles. The tracking signals are on 108 and 108.03 megacycles.

April 2, 1960

Upper Air Yields Secrets to New Methods of Research

Findings in Last 5 Years Surpass All in Past Studies

By WALTER SULLIVAN

Through the use of rockets, satellites, highly sensitive light detectors and pulses from the world's largest radar antenna, more has probably been learned of the upper air in the last five years than in all previous research.

This came out last week as scientists from East and West compared notes at the International Space Science Symposium in Washington.

There proved to be broad areas of agreement on the chemistry of the "chemosphere." The same was true of the manner in which components of sunlight produce the radio-reflecting layers, scores and hundreds of miles aloft.

Likewise added evidence was produced to support the contention of Dr. Marcel Nicolet that the higher portion of the atmosphere is a layer cake. From the top down, the layers are dominated, respectively, by hydrogen, helium, oxygen and nitrogen.

Until recently, Dr. Nicolet, a Belgian, was little known except in the scientific world. He was executive director of the body that ran the International Geophysical Year of 1957-58 and has emerged as probably the foremost authority on the upper air. He is now visiting Pennsylvania State University, in University Park, Pa.

A Surprise Discovery

Discovery of the helium layer is, perhaps, the most surprising development. As recently as a year ago it was generally thought that the upper air, or heterosphere, was divided into only two regions. The lower one, above fifty-five miles, was thought to be dominated by oxygen and the upper one, above 800 miles, by hydrogen.

However Dr. Nicolet deduced, from the drag effect of the upper air on the Echo I satellite, a balloon-like sphere 100 feet in diameter, that there must be a broad helium layer between the oxygen and hydrogen. Dr. N. N. Shefov of the Soviet Union then did optical measurements from the ground at twilight that seemed to confirm this.

Finally, this week, Dr. Kenneth L. Bowles of the National Bureau of Standards laboratory at Boulder, Colo., reported more evidence, based on observations with his gigantic radar. Its antenna system covers twenty-two acres near Lima, Peru, and is powered by a 4,000,000-watt pulse transmitter.

Because the earth's magnetic field, above Peru, is horizontal, the fine-grained structure of electrified components of the atmosphere is likewise horizontal. Hence a sufficient portion of the upward pulse is reflected to make possible the construction of a profile of atmospheric density out to 4,000 miles. By late June this may be extended to 15,000 miles.

The profile already obtained showed a helium layer in the region postulated by Dr. Nicolet. Furthermore, as reported last week by Robert R. Bourdeau and S. J. Bauer of the National Aeronautics and Space Administration, the layer has been detected in three rocket firings.

Theory Is Supported

Dr. Nicolet and others divide the atmosphere into two broad regions. Up to fifty-five miles is the homosphere, within which the primary components of air are uniformly mixed in their standard proportions by weight of 76 per cent nitrogen, 23 per cent oxygen and 1 per cent argon (apart from water vapor).

Above this is the heterosphere, reaching out to the limit of the region within which neutral atoms rotate with the earth—probably about 22,000 miles, or a tenth of the distance to the moon. The oxygen, helium and hydrogen layers lie within the heterosphere, whose name derives from the varying nature of its composition.

At such high elevations molecules of oxygen, hydrogen and nitrogen, in which atoms are paired, tend to be broken into individual atoms by ultraviolet light. Some are also ionized, generally by portions of the ultraviolet component of sunlight. This means that some electrons are dissociated from the atoms.

The more electrons there are, at any level of the atmosphere, the higher the frequency of radio signals (or shorter the wave length) that it will bend back to earth. The changing nature of these electron-rich regions is critical in planning long-range communications, but only recently has it become clear how they are formed.

Last week, in an interview, Dr. Nicolet described the various processes. A number of factors are involved. The density of air atoms decreases steadily with altitude and, hence, the density of electrons would likewise decrease if all the air were exposed to the same sunlight.

This, however, is not the case. The wave lengths of light that ionize various components of the upper air can penetrate only to certain depths. Furthermore, at high elevations, the air gases form in layers determined by their weight so that the composition of the air changes with height.

It is therefore the combination of these factors, plus the frequency of certain chemical reactions, that determines the existence of the "layers" constituting the ionosphere. the density of electronics within the ionosphere tends to increase, with height, but does not do so steadily. Rather there are regions of sharp increase and these are the so-called layers.

The highest, known as the F-2 region, is some 200 miles aloft, which means that it is well up in the region where oxygen atoms predominate. The peak electron density occurs where these atoms are ionized by light in the short-wave end of the ultraviolet.

A Zone of Transition

At about 125 miles the F-2 region is in the transition zone between the oxygen and nitrogen layers. The number of oxygen atoms and nitrogen molecules is roughly equal. Light at the short-wave end of the ultraviolet produces oxygen ions and some ions of nitric oxide.

Below this, at about sixty miles, is the E layer whose chief importance to radiomen is that it persists all night, despite the absence of sunlight. This is a region where nitrogen molecules are plentiful but where

---

REACTIONS: Dots indicate the electrons dissociated by ultraviolet light in the various layers of the heterosphere. Temperatures of 3 layers of homosphere are also given.

the Lyman-beta component of sunlight ionizes oxygen atoms. The two combine to form nitric oxide ions, plus nitrogen atoms.

It is the persistence of these nitric oxide ions that gives the E region its staying quality. Dr. Nicolet believes this discovery, largely by means of rockets, is one of the most important recent developments in this field.

The lowest, or D region, is formed by the action of the Lyman alpha component of sunlight in likewise producing oxygen ions and, hence, nitric oxide ions. However, the number of these reactions is low and the electron density in daytime is only about 1,000 a cubic centimeter (a thimbleful of air.).

Wide Range of Density

By contrast the electron density in the E region is 100,000 a cubic centimeter. In the F-1 region it is 200,000 and in the F-2 it is 1,000,000. Above that the gross density of atoms becomes so low that the electron density likewise drops off.

Below the ionosphere, between thirty and fifty miles aloft, is the chemosphere, a region in which chemical reactions are stimulated by sunlight.

These reactions involve substances such as ozone, hydroxyl and sodium, that are minor constituents of the atmosphere at those levels. Nevertheless, the reactions absorb much solar energy and transfer it to the upper atmosphere.

When sunlight is absent, at night, reverse reactions occur, with the emission of various forms of light detectable from the earth -- the so-called airglow.

Much has been learned and confirmed regarding these reactions by means of rockets.

Some of the most important reactions involve the three forms of oxygen: atomic oxygen (consisting of single oxygen atoms), molecular oxygen (the kind we breathe, with two atoms in a molecule) and ozone (with three atoms a molecule).

Ozone Molecule Split

In sunlight ozone in the chemosphere is split, by ultraviolet rays, into atomic and molecular oxygen. At night these substances recombine into ozone, emitting a characteristic glow.

The ozone also combines with atomic hydrogen to produce molecular oxygen and hydroxyl (a pairing of single oxygen and hydrogen atoms). This hydroxyl emits an infrared glow so intense that some nocturnal animals, whose eyes are sensitive to infrared light, may not be able to see any but the brightest stars.

According to Dr. Nicolet, the peak ozone absorption of ultraviolet is at the stratopause, or boundary between the stratosphere and the mesosphere, above it. This is some thirty miles aloft and is a comparatively warm layer.

From there on up the temperature drops until it reaches its minimum, for the entire atmosphere, at the mesopause, or top of the mesosphere, some fifty-five miles up. It is roughly 150 degrees below zero Fahrenheit.

Above that level the temperature rises but it does not melt space ships because the substance that is hot is so thin.

Dr. Nicolet and others lament the fact that, in many ways, the most important region of the atmosphere is the least known. This is the boundary region between the heterosphere and the homosphere.

Far more is known of the higher levels, where satellites have furnished masses of data, and of lower levels within reach of balloons. The intermediate region can be seen only in fleeting glimpses by means of rockets.

Dr. Nicolet points out that there have been only four measurements of pressure at this level. Calculations of what lies in the vast region above must rest upon such pressure data.

Wide Observations Urged

In a paper submitted to last week's meeting, Dr. V. I. Krassovsky of the Institute of Physics of the Atmosphere in Moscow, urged that upper air observations be made. He said these should be made at various hours, day and night, in different seasons, at different stages of the eleven-year sunspot cycle, and at various heights and latitudes. Only then, he said, would the picture be complete.

Dr. Nicolet argues that the upper air is a "laboratory" within which reactions can be observed that are unobtainable at lower levels. Even in the best vacuum in an earth-bound laboratory multitudes of atoms and molecules are constantly colliding in their random motions. In the upper air these "mean free paths" extend to tens or hundreds of miles and effects occur that cannot be observed elsewhere.

May 7, 1962

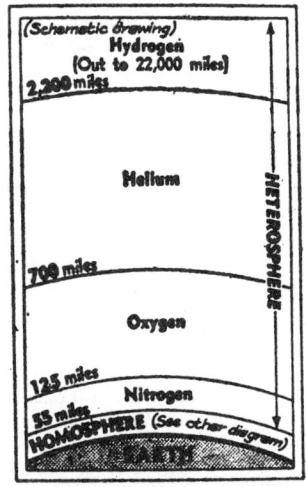

UPPER AIR: Discoveries indicate many-layered outer atmosphere, or heterosphere.

A Forecast: Not Much Change in Forecasting

By ALAN ANDERSON Jr.

In the last 30 years, sea ice has been building up in northern shipping routes, making navigation increasingly difficult. For the last six, drought in Africa has led to famine. And in the last three, the anchovy, critical to world food supply, disappeared, and then reappeared, off the Peruvian coast.

Is it possible to predict accurately changes in sea temperature and current, rainfall and wind, and other shifts, that when taken together make up what is called weather? Meteorologists, whose predictive vision is limited to days, and climatologists, whose primary concern is changes in the eons of the past, agree, not now.

Still, new technology and massive cooperative experiments are beginning to make it possible to record and analyze a variety of climatic change. Some will have an immediate effect for short-range weather prediction. The hope for the long-range is that the gap between the eons of the climatologists and the days of the meteorologists will be narrowed.

● In May, the first "Synchronous Meteorological Satellite" was launched by the United States to an altitude of 23,000 miles. The weather station is placed in synchronous orbit over South America, rotating exactly as fast as the earth. Because it is in fixed position, it is able to maintain constant surveillance over much of the Western Hemisphere; other weather satellites take as long as two hours to repeat observation of one region, too great an interval to warn of fast-developing storms like tornados.

● The first Global Atmospheric Research Program, the Atlantic Tropical Experiment, is now under way in an area that has largely been a meteorological blind spot, the tropics. The 100-day, 72-nation study runs from June 15 to Sept. 23, and involves 40 ships, 13 research aircraft, six kinds of satellites, nearly 1,000 land stations and around 4,000 people. Its purpose is to monitor every meteorological happening from a mile below the sea surface up to the ionosphere in a 20-million-square-mile swath from Latin America to the western Indian Ocean.

These regions receive about half of the solar energy that falls on the earth, and have been likened to the boiler that drives the planet's atmospheric engine. In the tropics, cyclonic storms are born; and there energy is transported to both mid-latitudes and polar regions via water and wind currents. This energy transport is now thought to be the key to accurate prediction of everything from hurricanes to weekend showers, everywhere in the world.

Data from the satellite and the South Atlantic experiment will feed a number of projects. At the National Oceanic and Atmospheric Administration's Geophysical Fluid Laboratory in Princeton, N. J., a group directed by Dr. Joseph Smagorinsky is trying to improve weather prediction through the use of mathematical models, the crystal balls of meteorology. They divide the earth's atmosphere into chunks, usually a few hundred kilometers on a side and several kilometers deep. To make a prediction, such as the familiar "five-day forecast," they feed into a computer every available piece of weather information about those chunks, along with all applicable physical laws. The complexity of the calculations is staggering; Dr. Smagorinsky is pressing the even capacity of the Princeton computer—the world's largest.

Current estimates for the range of prediction are around ten to 14 days, according to Dr. Donald Gilman of the National Weather Service. Even the best models become unreliable beyond about two weeks, he says, because minute errors or omissions become magnified.

Foreknowledge of such short-term, in climatological thinking, aberrations as five-year drought seem even more problematic. The immediate cause of drought is thought to be the immobilization of a large air mass which in turn traps another air mass beneath it. The downward pressure created by the upper mass slightly heats the lower air much as a bicycle pump becomes warm through use.

As temperature rises, this warm air holds water instead of dropping it as rain. Exactly what immobilizes such a mass of air is still unknown, says Dr. Jerome Namias of Scripps Institution of Oceanography.

The Oceanic 'Flywheel'

Weather researchers are pinning their hopes on identifying the controlling mechanism, or "flywheel," which seems to restrain the atmosphere from wild aberrations. "There seems to be some system on a longer term fluctuation than the normal daily and seasonal weather," says Dr. Gilman. "It is probably not the sun; people have looked for the simple relationships there already. Most people think it is the ocean."

The study of the oceanic flywheel and its crucial interactions with the atmosphere will continue after the Atlantic Experiment closes in the fall. Already planned are a Polar Experiment, a Monsoon Experiment, and, probably in 1977, a Global Experiment which promises to be the most complex scientific program ever attempted. By then the United States, Japan, the Soviet Union and the European scientific research organization plan to have launched a total of four more synchronous satellites spaced to keep most of the globe under constant surveillance.

Such surveillance, combined with improved models and ocean-atmosphere data from the Global Experiment should show whether it would be possible to predict long-term changes over a year or a century. One group studying how this might be done is CLIMAT, at Columbia University's Lamont-Doherty Geological Observatory. CLIMAT is analyzing the details of past climatic changes in hopes of using them to predict broad trends in the future. "Of course," acknowledges James Hays, project coordinator, "this may prove impossible to do. If so, we can all go back to the Farmers Almanac."

Alan Anderson Jr. writes frequently on science.

July 7, 1974

Issue and Debate

Concern Over the Ozone Layer

By WALTER SULLIVAN

Life on land became possible hundreds of millions of years ago when the atmosphere developed an ozone layer, cutting off those wavelengths of ultraviolet sunlight that break down the nucleic acids essential to life.

Today there is considerable concern that a variety of human activities are endangering the ozone layer. The concern has grown in formidable proportions in recent months as new alleged threats were identified.

The original alarm, sounded in 1971, was that exhaust gases from a large fleet of supersonic transport planes operating in the stratosphere would deplete the ozone there.

The effect, it was said, could be cumulative, leading initially to a rise in the incidence of skin cancer, but ultimately affecting climate, the production of food crops and, perhaps, threatening the very habitability of the earth.

Then it was suggested that high-altitude atomic bomb explosions could have a similar effect, making a nuclear war even more catastrophic than previously supposed. And it was argued, by others, that supposedly inert gases used in spray cans and for refrigeration were a far greater threat than SST exhaust.

Spokesmen for the various industries involved dismissed these claims as based chiefly on theoretical calculations, rather than direct measurements in the stratosphere.

The original SST debate led to the most ambitious attempt at technological assessment ever undertaken. The three-year effort, conducted at a cost of $20-million, was summarized last month at a wrap-up conference at the United States Department of Transportation research center in Cambridge, Mass.

Strikingly evident at the meeting was unanimity regarding factual findings of the study and deep disagreement as to their significance.

Those in favor of SST's found support for the view that, with certain safeguards, they could operate in the stratosphere without harmful effect.

Opponents emphasized the uncertainties, both with regard to development of low-emission engines, which was one of the proposed safeguards, and concerning the chemistry of the stratosphere.

Now, similar alignments are forming with regard to the atomic bomb and spray can arguments. The problem is being studied by a Federal study group headed by Carroll L. Pegler of the National Science Foundation and Dr. Warren R. Muir of the Council on Environmental Quality.

In a parallel effort, a panel is being formed by the National Academy of Sciences under Dr. Herbert S. Gutowsky, director of the School of Chemical Studies at the University of Illinois in Urbana. Today the academy is to issue a report on possible SST effects by its Climatic Impact Committee under Dr. Henry T. Booker of the University of California, San Diego.

The manner in which the alleged ozone threats are evaluated, including the investigation of economic as well as environmental considerations and alternative technologies, is bound to set precedents for a wide range of other issues such as the advisability of going all-out on atomic power.

Background

Ozone is a gas whose molecules consist of three oxygen atoms, whereas the oxygen molecules humans breathe are composed of two such atoms. Ozone is chiefly con-

EARTH SCIENCES

centrated between 8 and 30 miles aloft, which is within the stratosphere. The SST's operate in the lower part of this region, but their exhaust gases could diffuse upward.

The ozone is constantly broken down and replenished by a complex series of chemical reactions. Of key importance are those involving naturally occurring oxides of nitrogen, which would become more abundant, for example, from SST exhaust.

Not only do they react with the ozone, breaking it down, but they combine with the individual oxygen atoms needed to replenish ozone that has been lost.

The man chiefly credited with having sounded the alarm, with regard to the possible effects of oxides of nitrogen from SST exhaust, is Dr. Harold S. Johnston, an authority on atmospheric chemistry at the University of California, Berkeley.

In April, 1971, he circulated to fellow scientists, including members of President Nixon's Science Advisory Committee, a document setting forth his calculations in this regard. He pointed out that the oxides of nitrogen act as a catalyst in breaking down the ozone. That is, they bring about the chemical reaction but survive it intact.

The result, Dr. Johnston said, is that they continue destroying ozone "indefinitely." He and his colleagues then reported evidence that high-altitude atomic bomb explosions generate large amounts of nitric oxide. This, they said, apparently caused a marked depletion of stratospheric ozone from the heavy atomic weapons testing from 1952 to 1962.

This report was followed last year by a warning that nuclear war could have a catastrophic effect in this respect.

The SST Story

In 1970, following proposals that SST's might have a variety of environmental effects, Congress asked the Department of Transportation to complete an assessment by the end of 1974. The resulting Climatic Impact Assessment Program drew on nine Federal departments and agencies as well as seven foreign ones.

The results, delivered to Congress last Jan. 21, indicated that a small fleet of SST's, such as the 30 now scheduled to enter service, would cause effects so small that they could not be detected by present methods. But it was also found that a large fleet, using present

---

OZONE CHEMISTRY

Ozone in the stratosphere is chiefly formed by a two-stage process:

1. An oxygen molecule (formed of two oxygen atoms) is split by ultraviolet rays, producing two single oxygen atoms

$$O_2 + \text{ultraviolet ray} \rightarrow O + O$$

2. An oxygen atom then joins an oxygen molecule to form an ozone molecule (composed of three oxygen atoms)

$$O + O_2 \rightarrow O_3$$

Ozone can be depleted by two reactions involving oxides of nitrogen (of natural origin or from SST exhaust):

1. A nitric oxide molecule combines with one of ozone to produce nitrogen dioxide and an oxygen molecule

$$NO + O_3 \rightarrow NO_2 + O_2$$

2. Nitrogen dioxide reacts with a free oxygen atom to produce nitric oxide and an oxygen molecule, depriving the stratosphere of the free oxygen needed to replenish ozone

$$NO_2 + O \rightarrow NO + O_2$$

N = nitrogen atom O = oxygen atom O_2 = two oxygen atoms
O_3 = three oxygen atoms

The New York Times/March 31, 1975

---

engines and fuel, would present a hazard.

"Serious consequences" would follow, the report said, "if either supersonic or subsonic fleets are expanded to large numbers without imposing strict limitations on engine emissions." It noted that there are about 1,700 subsonic jets operating in the lower stratosphere, including more than 1,200 707's and DC-8's.

Jumbo jets, such as the 747, are more a threat than the smaller 707's because their exhaust emissions are much larger, the report said.

Thus, subsonic planes will present a problem as their numbers grow and their operating altitude rises.

It was also calculated that a moderately large fleet of SST's—125 planes producing emissions similar to those from the present Concorde and Soviet TU-144—could reduce the ozone by 0.5 per cent. This, it was believed, would increase ultraviolet radiation reaching the earth by 1 per cent and could therefore cause a 1 per cent rise in the less severe form of skin cancer.

Two kinds of skin cancer were dealt with in the study. One, a rare form known as melanoma, is fatal in about 40 per cent of the cases. Its relation to ultraviolet rays is unclear, for it occurs frequently in parts of the body normally protected by clothing. Yet its incidence varies in accordance with regional variations in ultraviolet exposure.

There are marked differences of exposure to ultraviolet rays depending on the latitude in which one lives. While this is due to various factors, participants in the study concluded that the chief one was the increased amount of ozone in the stratosphere at higher latitudes.

Day-to-day variations in one area may be as great as 25 per cent, but averaged over months and years, there is also a strong variation with latitude. Thus, on the average, there is 30 per cent more ozone over Minnesota than over Texas. National cancer surveys have shown markedly fewer cases of skin cancer in northern areas, where the ozone shield is denser.

The more common and less lethal type of skin cancer occurs in about 250 fair-skinned Americans per 100,000. Those with dark skins, including a sunburn tan, seem to be better protected whereas Celts, who rarely tan, are most susceptible. Melanoma cases occur in only about one in 100,000 Americans and affect all races.

At last month's meeting, SST opponents pointed out that from the Assessment Program findings it would appear that an over-all ozone depletion of only 0.5 per cent would lead to 6,000 additional cases of skin cancer yearly.

Dr. Alan J. Grobecker, manager of the Climatic Impact Assessment Program, replied that in terms of individuals this was comparable to the increased risk sustained by moving one's domicile from Baltimore to Washington, where the ozone layer would be slightly less dense.

Aerosol Challenge

Last June, Dr. F. Sherwood Rowland and Dr. Mario J. Molina of the University of California, Irvine, sounded an alarm more ominous than that related to SST's. It concerned the gases used as propellants in many aerosol spray cans.

Termed scientifically chlorofluoromethanes or fluorocarbons, they are best known as Freons—the trade name used by their chief producer, E. I. du Pont de Nemours & Co. The gases are also the prime ones employed in refrigeration and air-conditioning systems.

They were chosen as spray propellants because, being chemically inert, they do not alter the substance being sprayed, be it a deodorant, hair spray or pharmaceutical. The reason for concern was evidence that the Freons are no longer stable when exposed to ultraviolet rays such as those encountered in the stratosphere.

They apparently break down, releasing chlorine that can then act as a catalyst in breaking down ozone. The reaction is six times more rapid than that due to oxides of nitrogen.

While the amount of propellant in one aerosol can might seem inconsequential, their use has increased rapidly. World production of Freon gases in 1972 was about half a million tons a year and increasing about 8.7 per cent annually.

Being stable, the gases accumulate in the air. That this has been occurring on a worldwide basis has been observed in surveys conducted by the Naval Research Laboratory in Washington and by British researchers.

The level of alarm rose higher last September when Dr. Ralph J. Cicerone at the University of Michigan reported that, even if production of Freons was halted immediately, there would be significant ozone depletion, reaching its maximum in about 1990.

The lag would arise from the slow upward diffusion of those gases. It would thus appear that, by the time a long-term depletion of ozone became clearly evident, enough gas would be on its slow way upward to produce a perilous weakening of the shield.

Meanwhile, Dr. Michael B. McElroy and his colleagues at Harvard University were studying atmospheric effects of the projected space shuttle. On its way up into earth orbit, the shuttle would burn solid fuel that could leave chlorine in its wake.

Armed with computer calculations relating to the effects of such chlorine, the Harvard group turned to the Freon problem. Their initial finding was that, if Freon consumption continues to grow at 10 per cent yearly, the world's ozone could be depleted 16 per cent by the year 2000. The problem, they wrote, "must be addressed-as a matter of urgent priority."

While Freons used as refrigerants are also of concern, their use has not been growing so rapidly, although, according to an industry spokesman, because the pumps used to circulate them cannot be made leak-proof, some gas leaks out.

Industry Response

The makers of Freon-type gases and those manufacturers, like the Boeing Company, formerly involved in SST development, have emphasized the uncertainties involved in allegations of peril to atmospheric ozone.

Raymond L. McCarthy, product technical manager for the Freon Products Division of Du Pont, told a Congressional hearing that, in view of the paucity and difficulty of stratospheric observations, "There is no concrete evidence to show that the ozone depleting reaction with chlorine takes place."

While Du Pont is the prime maker of these gases, five other American companies also produce them, accounting for about half of world production. This year, according to the Du Pont Management Bulletin, industry dependent on Freon-type gases will contribute $8-billion to the economy and employ more than 200,000 workers.

This presumably includes production of air-conditioning systems and refrigerators.

"As refrigerants," said the bulletin, such gases "are at the base of our food processing, storage and distribution systems. Supply and distribution of medicines also rely heavily on fluorocarbon refrigerants."

Spokesmen for the industry also point out that a number of spray cans use propellants other than the fluorocarbons under suspicion.

Generally, they say, it is personal products that use fluorocarbons, such as hair sprays, deodorants, antiperspirants and pharmaceuticals.

Hydrocarbons are used as propellants in household products such as cleaners, paints, laundry products, waxes and polishes. While hydrocarbons are flammable, so are a number of these products. Hydrocarbons are used with shaving creams because they mix well with water.

Foods are generally propelled by nitrous oxide or carbon dioxide.

At the first meeting of the newly formed Federal Interagency Task Force on Inadvertent Modification of the Stratosphere, spokesmen for the industry said that, while other refrigerants and propellants could be used, they all had serious drawbacks.

Frank Bower, a division research head at Du Pont's Freon Products Laboratory, said that, in view of the constantly changing nature of the stratosphere, to "assess the validity" of the ozone depletion hypothesis, it will be necessary to measure the various substances suspected of taking part in the reactions "at the same time and the same place."

In its own research effort, he said, the industry plans to support balloon or rocket ascents into the stratosphere.

"It is possible," he added, "that the reactive chlorine atoms may undergo many reactions in the stratosphere, some of which could lead to stable, harmless products."

"With exceptional good fortune," he continued, "sufficient measurements and analyses to resolve the issue could be completed within three years. At the meeting, one who had sounded the alarm, Dr. McElroy of Harvard, said the more frightening calculations of his group had been refined and that a brief delay in a decision on the Freon issue seemed acceptable."

Assessment

Before long, the United States and other countries must decide what to do about the proposed threats to the ozone shield. To what extent does the Assessment Program's finding on the SST threat provide a model?

When the program's findings were announced in January, Dr. Grobecker told a news conference that the currently projected fleet of about 30 SST's represented essentially no threat.

Headlines across the country implied that the SST's had been exonerated. A Pennsylvania newspaper editorially chided the scientific community for having raised false alarms and having destroyed an important American industry—the SST opponents, citing the warnings within the findings, were furious.

The situation calls to mind other issues where extensive scientific evidence has been available, but its significance could be read in various ways. The debate on fallout from nuclear explosions would be an example. It came down to the question: Did keeping ahead in weaponry justify the risks involved? Likewise, nuclear power plants can never be made totally safe any more than can jet airliners, railroad trains or automobiles. Does the social benefit to be gained from nuclear power justify the risk?

The Assessment Program's study has been hailed as an excellent model for coping with other such problems, including the Freon dispute. Dr. Gordon J. F. MacDonald of Dartmouth College, an original member of President Nixon's Council on Environmental Quality, described the program as "the first example of what technology assessment really means."

But he noted that uncertainties inevitably characterize the conclusions reached by such studies. The public tends to expect yes-or-no, black-and-white answers from a scientific inquiry, he said, but "science is always probabalistic," and no one can say "for sure" what a large fleet of SST's would do.

The same applies to the Freon debate. While those who have sounded the alarm differ on the urgency of a halt in release of such gases, they are of one mind in arguing that, if there is a delay until it has been unequivocally demonstrated that the gases are depleting the ozone, it may well be too late.

March 31, 1975

CLIMATE

FOSSILS SHOW CHANGE IN CLIMATE ON PACIFIC

Scientists Find Remains of Tropical Plants Buried in Gravel of Placer Mines.

BERKELEY, Calif., Oct. 12.—Vast changes in the earth's climate during the last 100,000,000 years or so are revealed by fossils dug up by gold hunters in California.

The fossils, age-old but well preserved remains of plants that lived in California many tens of millions of years ago, reveal to Dr. Ralph W. Chaney of the University of California that America's West Coast had a tropical climate in those days.

Tropical forests such as now flourish only near the equator once grew all through the western part of California, Oregon and Washington, Dr. Chaney reports to the Carnegie Institution of Washington. Leaves of trees which now grow only in the most warm, moist regions of the earth and could not thrive in the present-day cool climate of the West Coast have been found preserved in very ancient gravel deposits.

They show that long before the human race appeared on earth the world's climate distribution was far different than today. They also suggest, says Dr. Chaney, that there are probably cycles in climate and that "the present epoch of relatively cold climate may be followed by one in which the air is sufficiently warm and humid to permit the return of forest trees which lived in California long ago and whose fossil remains, buried in the gold-bearing gravels, show us this early chapter of earth history."

October 13, 1934

ICE AGE IS BELIEVED NUMBER OF PERIODS

Studies Indicate It Was Broken by Warm Spells, Professor Tells Geological Society

Special to The New York Times.

DETROIT, Nov. 8—Studies of a prehistoric forest bed near Two Creek, Wis., disprove the belief that the Ice Age was a single, isolated event, the Geological Society of America was told today at the opening of its sixty-fourth annual meeting.

Dr. F. T. Thwaites, Professor of Geology at the University of Wisconsin, reported that the Ice Age was in fact divided into a number of glacial periods, separated by intervals in which the climate of what is now the northern United States warmed to something like its present temperature range.

The age of the Wisconsin forest bed was estimated at 11,400 years by Prof. W. F. Libby of the Institute for Nuclear Studies at Chicago.

A concurrent meeting of the Paleontological Society heard a report by Prof. Kenneth E. Caster of the University of Cincinnati and Erik N. K. Waering of Tulsa, Okla., on their discovery of large fossilized sea scorpions near Manchester, Ohio. Some of the skeletons are several feet long, they declared.

"These are the most primitive sea scorpions known," Professor Caster said. He estimated the age of the rock formation in which they were imbedded at 400,000,000 years. Although many fossil fragments of the same type have been found in more recent rock, he added, these were the first complete specimens reported.

The animals appeared to have been stopped in their tracks as they crawled over the floor of the vast inland sea that once covered much of the North American interior, the scientists said. The perfect preservation of the scorpion fossils was attributed to their sudden burial under a great fall of volcanic ash, which left no time for decay or scattering of the skeletons.

The Geological Society's annual Penrose Medal, in recognition of "eminent research in pure geology," was awarded tonight to Dr. Pentti Eskola, of the University of Finland. The medal was received in Dr. Eskola's behalf by Otso Vartiovaara, of the Finnish legation in Washington.

Dr. Martin J. Buerger, Professor of Mineralogy and Petrography at the Massachusetts Institute of Technology, received the Arthur L. Day Medal for his "outstanding contributions in the application of chemistry and physics to the solution of geologic problems."

November 9, 1951

Dating the Ice Age

Radiocarbon Method Indicates It Began 25,000 Years Ago

We are living at the end of an ice age. It began 25,000 or more years ago and reached a maximum between 20,000 and 18,000 years ago. The ice has been retreating for about 13,000 years. How do the scientists know? By the radio-carbon method which has been mentioned in this department over and over again.

The scientists who made the determination are Dr. Richard Foster Flint of Yale University and Drs. Meyer Rubin and Hans Suess of the United States Geological Survey. A report on their work appears in Science.

The radiocarbon dating method used is one developed by Dr. Suess. It is an improvement on the original method of Dr. W. F. Libby, now of the Atomic Energy Commission. Dr. Suess converts solid carbon in ancient wood into acetylene gas and works with this. Whereas Dr. Libby could go back only 25,000 years, Dr. Suess' gas method pushes back the calendar to 50,000 years.

June 26, 1955

I.G.Y. PUSHES HUNT FOR ICE-AGE CLUES

Mass of Data From Distant Posts Spurs New Studies

By WALTER SULLIVAN

Two months ago some twenty Americans, encamped on an ice floe drifting through winter darkness near the North Pole, were startled by an unbelievable sound, like that of an approaching express train. It ran right through their camp.

They hurried out of their clustered buildings with flashlights and hunted until they found a hair-like crack in the ice. It was evident that their home was doomed.

A few weeks later, on Dec. 21, thirty Russians similarly camped nearer the pole, found their community split in two.

These two groups of men had placed themselves in jeopardy to carry out some of the more dramatic research of the International Geophysical Year. That eighteen-month study of this planet ended Wednesday.

One of its fields of research was glaciology and other phenomena in the polar regions. It is with this part of the effort that this article deals.

Some of the scientists engaged in this program were at the opposite end of the world, at the South Pole itself. They also were on high mountain glaciers in Alaska and Siberia, having been carried there, with their huts and supplies, by ski planes, by helicopters, by hordes of reindeer or by horses. A dozen of the animals had been lost in mountain torrents or down glacier crevasses.

It was hoped that the combined efforts of these far-flung men would help solve the riddle of the ice ages. What causes them? Is a new one upon us? Will the Arctic Ocean soon be free of ice?

Of the 4,000 stations and observatories of the I. G. Y., all but four were fairly stationary, although some on glacier ice crept a few feet each day. The four exceptions were the drifting stations on the Arctic Ocean. Driven hither and yon by the whims of wind and current, they traveled as much as 4,000 miles. Two of the stations were American and two were Soviet.

Repeatedly cracks threatened the floe camps with disaster. Sometimes they dismembered the air strips, which were their only link to the outside world.

The perils they faced were often unexpected. Polar bears prowled the camps, even those hardly more than 100 miles from the North Pole.

The chief problem was not the cold, but summer melting. This sometimes flooded entire camps and made it impossible for aircraft to land for months at a time.

The scientists studied not only the weather above, but also the bottom of the ice beneath their feet, donning aqualungs to swim under the floes. Here are reports on some of these projects:

STATION A (UNITED STATES)

This was established in the spring of 1957 on an ice floe about seven feet thick and two miles square. It was then 925 miles north of Alaska and 375 miles from the North Pole. The Alaskan Air Command parachuted heavy equipment for smoothing a runway on the ice so that large planes could bring in the supplies.

All went well until last April, when part of the floe broke off and pressure ridges erupted through the camp. The Air Force flew out all but a skeleton crew of ten, leaving just enough men to keep the observations going. On May 2 it was decided to shift the camp to an adjacent floe a mile away.

The rest of the crew were flown in again, all twenty-one buildings were jacked up and placed on runners. Passes were cut through the pressure ridges. Power cable was dug out of the ice and everything dragged to the new site.

Few Observations Missed

Only two observations with radio weather balloons (rawinsondes) were missed. There was no interruption in the seismic, gravity, oceanographic and magnetic programs. By May 24 the move was completed and a new runway built, but hardly more than two weeks later the floe broke open across the runway in two places.

Likewise, the melting of summer snow had made the floe surface into a lake. Holes were bored through the ice to drain the water.

Twelve days later on June 22, a new runway had been laid out, but it cracked open almost immediately. Not until late September, when the weather cooled, was it possible to land an aircraft on the floe.

On Nov. 2, in the midst of the winter night, the floe, buffeted by a severe storm, split again, dividing the camp from its airfield. This time a rescue plane lifted out the entire crew and the camp was abandoned.

The under-ice swimming was done by T. Saunders English of the University of Alaska and two other men, wearing waterproof clothing and breathing apparatus. They measured the amount of light coming through the ice to support the plant life encrusted on the bottoms of the floes.

Likewise, they studied the small animals that graze on these plants. During their long drift they trawled the water at various depthss and scraped brittle stars and sea cucumbers from the deep ocean floor.

Norbert Untersteiner, an Austrian at the University of Washington, did ice studies. It was found that during the summer all of the snow cover and 11 per cent of the old ice melted. This formed ponds, which covered a third of the floe. This was less than the growth of the ice on the bottom the previous winter, that was determined, by cores drilled from the ice, to be a layer about two feet thick.

One of the stranger phenomena was the growth of ice from the bottom up in pools of fresh water under the floe. This water, drained from the floe, first froze nearest the underlying salt water. Hence the pools were sandwiched between the floe above and this layer of new ice below.

Station A moved as much as 150 miles in a month, from June 17 to July 17. Echoes from explosions set off once or twice a day brought to light a five-to-ten-thousand-foot ridge rising from the ocean floor and extending from off Ellesmere Island toward eastern Siberia.

STATION B (UNITED STATES)

This was placed on a heavy platter of ice, known as T-3 or Fletcher's Ice Island, north of Ellesmere Island. It is far more substantial than an ice floe, being about 150 feet thick, nine miles long and four and a half miles wide. It had been occupied before and is thought to have once been part of the apron of floating ice attached to Ellesmere Island.

Much of the work was similar to that at Station A. Air operations were impossible during summer months when the runway was flooded. Up to three feet melted off the top of the ice island. The water formed little rivers, which mystifyingly changed the direction of their flow from time to time. This, it was found, was caused by changes in tilt of the island in strong winds.

Skirts had been placed around the buildings to prevent melting of their ice foundations, but these were inadequate. Cargo parachutes were used to help keep off the sunlight, but even so, at the end of summer each building stood precariously on a pedestal of ice.

As had been predicted, the temperature extremes near the North Pole were far milder than deep in Siberia or Antarctica. The highest reading at Station A was 38 degrees Fahrenheit and the lowest 57 below zero. Station B, because of its size, acted to some extent like a land mass and the temperature fell to 65 below.

In Siberia the mercury falls to 90 below zero and in Antarctica to 125 below. The stratosphere over the drifting stations, nevertheless, became very cold. Temperatures as low as 121 below were recorded there by weather balloons.

NORTH POLE 6 (SOVIET)

This camp, on an ice island, was established about April, 1956. Up to last October it had drifted a straight line distance of 1,050 miles. In its meanderings it had covered 4,064 miles.

Its scientific program included three all-sky photographs of the aurora each minute, four radiosonde launchings a day to elevations between fourteen and sixteen miles, oceanic soundings several times a day and use of a searchlight at night to observe the height of the cloud ceiling.

The island was thinner than that of Station B. It was from twenty-one to forty-one feet thick, thinning toward the edges until it was no thicker than the surrounding floes. As at Station B, isolated rocks on the ice showed it had originated near land. The presence of salt throughout showed it had grown from the bottom, rather than from piled-up snow.

NORTH POLE 7 (SOVIET)

This ice floe camp was established in April, 1957. By last October it had moved a straight-line distance of more than 600 miles and a wandering course of almost 2,000 miles. As at the other stations, summer thaws made islands of the huts and holes had to be drilled.

The resulting drainage of several thousand tons of water caused the floe to rise. When it was only 112 miles from the Pole, a contingent of polar bears paid a visit and remained "some time," according to a Soviet account. At Station A a cub knocked out all the runway lights.

This last fall North Pole 7 moved an average of seven miles a day for a month. By December it had crumbled from an area of 4,200 square yards to a mere 900 square yards then, on Dec. 21, it split.

In addition to maintaining these stations, the Russians used planes to place automatic weather stations and radio beacons on ice floes in various parts of the Arctic Ocean to keep track of weather and ice movements.

They also planned, during a period of several days, to do an aerial survey of pack ice distribution in the entire ocean area north of the coasts of Europe and Asia.

Soviet plans call for an increased tempo of Arctic exploration in the next seven years, with twenty expeditions scheduled to take the field each year.

At the opposite end of the world eight nations joined forces to probe the secrets of Antarctica. It was the greatest assault ever launched on that continent and was widely publicized. Last winter, during what was summer in the Antarctic, parties with tractors or dog teams covered at least 14,000 miles of the ice sheet.

The operations will also be extensive during the current southern summer. Preliminary reports on the results have been received from scientists in the field and from Philip Benjamin, New York Times correspondent with the United States expedition.

Mr. Benjamin cites increasing evidence that Antarctica is cut in two by a trough between the Ross and Weddell Seas. The possibility of such an ice-filled link has long been postulated, but during the I. G. Y. such a feature was detected near the coast of the Weddell Sea. Its bottom averaged 3,500 feet below sea level.

At the opposite end of the supposed trough the ocean floor dipped beneath the floating Ross Ice Shelf to a point forty-eight miles south of Mount Discovery, where it was 4,400 feet below sea level. In the heart of Marie Byrd Land, midway between the two seas, it was found that the terrain dropped lower between Byrd Station and the mountains to the south.

The ice depths recorded by the many Antarctic trail parties indicated that there was about 40 per cent more ice in the world than had been supposed. It was enough to cover the entire earth to a depth of 100 feet.

There was some evidence of ice shrinkage. More than 1,000 square miles of Lady Newnes Ice Shelf had vanished since 1912.

P. Shumsky, head of ice studies of the Soviet expedition, reported that the surface of the ice sheet surrounding Mount Gauss had dropped an average of twenty-six feet in the fifty-five years since the Germans surveyed it. The peak lies close to the main Soviet base at Mirny.

On the other hand a survey of glaciers in the largely ice-free valleys near McMurdo Sound, carried out by Dr. Troy Péwé of the University of Alaska, indicated no change in the last forty-five years. It was

SCIENCE ADRIFT: Each of four drifting stations, whose meanderings in Arctic Ocean are shown, covered several thousand miles during International Geophysical Year. The broken lines are routes of Soviet stations. The solid lines show paths of United States camps. Alpha had to be abandoned in November after it had split. N. P. 7 split Dec. 21, but is still occupied. Final positions shown are for yesterday and were obtained from United States Weather Bureau, which receives reports from them daily. Although Alpha and N. P. 7 covered much of the same territory they never sighted one another.

EARTH SCIENCES

found that snowfall in Antarctica was surprisingly meager and melting almost non-existent. Hence any changes in the ice volume would be very gradual.

There was evidence that the climate in Antarctica had warmed, but not at the pace observable in the Atlantic sector of the Arctic. The average temperature at Little America was up five degrees since 1911.

Numerous expeditions were stationed on glaciers between the polar regions. They ranged from Mount Olympus in the State of Washington to the Brooks Range of Alaska, the Suntar-Khayata Mountains of Siberia, the Pamirs north of Afghanistan and even remote Franz Joseph Land in the Arctic Ocean. In the last-named place two girls of a Soviet party worked in a laboratory carved out of the glacier.

When the I. G. Y. ended, the basic questions about the ice ages still had not been answered. It had not been expected that they would be. Yet it seemed possible that the necessary information was in hand. It would have to be digested by business machines and studied by scientists in many fields.

Meanwhile, more clues are being sought. Consequently, sooner or later it seems likely that the world will know whether future generations will have to migrate southward, in flight from advancing ice, or inland to flee rising oceans, swollen by the melting of present ice sheets.

January 3, 1959

Scientists Ask Why World Climate Is Changing; Major Cooling May Be Ahead

By WALTER SULLIVAN

The world's climate is changing. Of that scientists are firmly convinced. But in what direction and why are subjects of deepening debate.

There are specialists who say that a new ice age is on the way—the inevitable consequence of a natural cyclic process, or as a result of man-made pollution of the atmosphere. And there are those who say that such pollution may actually head off an ice age.

Sooner or later a major cooling of the climate is widely considered inevitable. Hints that it may already have begun are evident. The drop in mean temperatures since 1950 in the Northern Hemisphere has been sufficient, for example, to shorten Britain's growing season for crops by two weeks.

As noted in a recent report of the National Academy of Sciences, "The global patterns of food production and population that have evolved are implicitly dependent on the climate of the present century."

Vulnerability to climate change, it says, is "all the more serious when we recognize that our present climate is in fact highly abnormal, and that we may already be producing climatic changes as a result of our own activities."

The first half of this century has apparently been the warmest period since the "hot spell" between 5,000 and 7,000 years ago immediately following the last ice age. That the climate, at least in the Northern Hemisphere, has been getting cooler since about 1950, is well established—if one ignores the last two winters.

It had been forecast by some specialists that last winter would be exceptionally cold, but as all ice skaters know, it was unusually mild in the New York area. In Boston it was the warmest in 22 years and in Moscow it was the second warmest in 230 years.

A major problem in seeking to assess the trend is to distinguish year-to-year fluctuations from those spread over decades, centuries and thousands of years.

Lack of agreement as to the factors that control climate change make it particularly difficult to assess current trends. Of major importance, therefore, is the debate as to the cause of such changes and the role of human activity in bringing them about. Among the major hypotheses are the following:

1. Solar Energy Variations

The amount of solar energy reaching the earth's surface at any one place and time of year varies because of changes in the earth's orbit and the tilt of its spin axis (The extent of that tilt determines the extent of seasonal changes).

There are also slight variations in the amount of energy radiated by the sun. They follow the 11-year sunspot cycle and relate chiefly to solar ultraviolet radiation.

Dr. Walter Orr Roberts, former head of the National Center for Atmospheric Research in Boulder, Colo., believes he has found a correlation between this cycle and weather phenomena such as jetstream behavior and droughts in the high plains east of the Rocky Mountains.

The droughts, he believes, tend to occur either in step with the 11-year cycle or with one of 20 to 22 years.

Such links are doubted by Dr. J. Murray Mitchell Jr., climatologist at the National Oceanic and Atmospheric Administration's Environmental Data Service. He sees no plausible explanation of how such slight variations in solar energy could affect the massive weather phenomena responsible for droughts and floods.

Tree-ring data from Nebraska and South Dakota, according to Dr. Mitchell, show that the pattern to which Dr. Roberts refers applies only to the last century, whereas earlier—as far back as the 16th century — major droughts occurred at irregular intervals generally longer than 20 years.

Triggering of the ice ages by cyclic changes of the earth's spin axis and orbit was proposed as early as in the nineteen twenties by a Yugoslav, Milutin Milankovitch. Because of tugging by the gravity of other planets, the orbit of the earth changes

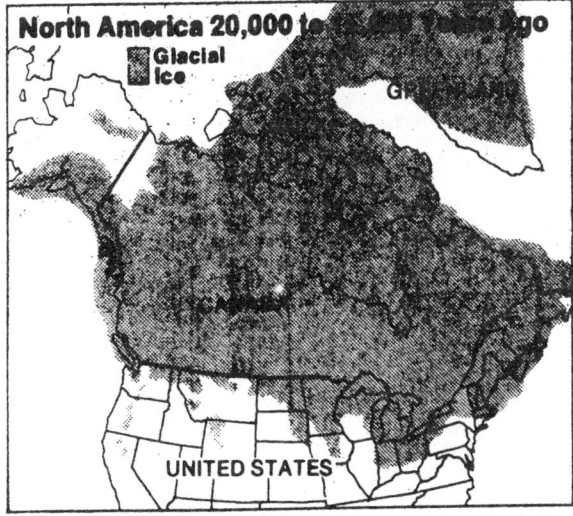

Shortly after its last major advance, the ice sheet covered almost all of Canada and much of the northern United States. Some scientists now believe, on the basis of recent findings, that ice resting on the sea floor extended far off the shore. The map is adapted from "The Earth and Its History" by R.F. Flint.

The New York Times/May 21, 1975

shape. Sometimes it is virtually circular. At other periods the earth's distance from the sun varies during each year by several million miles.

At present, 6 per cent more solar radiation reaches the earth on Jan. 14 than it does six months earlier or later, tempering northern winters. This variation in the shape of the orbit occurs in a cycle of about 93,000 years.

The tilt of the spin axis with respect to the earth's orbit around the sun varies from 22.0 to 24.5 degrees over a period of some 41,000 years. The aim of the axis with respect to the stars also rotates once every 26,000 years, causing precession of the equinoxes.

For many years the combined effects of these variations seemed too subtle to account for the ice ages, but recent discoveries have won converts for modernized versions of the Milankovitch thesis.

From the chemical composition of Pacific sediments, from studies of soil types in Central Europe and from fossil plankton that lived in the Caribbean it has been shown that in the last million years there have been considerably more ice ages than previously supposed.

According to the classic timetable, four great ice ages occurred. However, the new records of global climate show seven extraordinarily abrupt changes in the last million years. As noted in the academy report, they represent transition, in a few centuries, "from full glacial to full interglacial conditions."

Many scientists now consider it established that expansions of glaciers in the Southern Hemisphere coincided with the northern ice ages. Land areas, however, are meager in southern latitudes comparable to those that were heavily glaciated in the north.

Dr. George J. Kukla of Columbia University's Lamont-Doherty Geological Observatory has proposed a way in which small variations in solar energy falling on the middle latitudes—as in the Milankovitch concept—could affect the climate.

It is the extent to which northern seas and land areas become covered with snow and ice in the fall. When such cover is extensive, as in the fall of 1971, the white surface reflects sunlight back into space and there is a reduction in heating of the atmosphere.

This prolongs the northern winter and cools the globe. In 1971, according to images from earth satellites, autumn snow and ice cover increased by 1.5 million square miles.

The following year was one of freak weather throughout much of the world. The winter was exceptionally cold in North America, the Mediterranean and other areas. Severe drought struck many parts of Asia and Europe.

The implication was that a change in solar input that was slight, but sufficient to increase autumn snow and ice cover substantially, could eventually lead to a major climate change.

From a reworking of the Milankovitch calculations Dr. Kukla has found that solar energy falling on the atmosphere in the autumn hit a minimum 17,000 years ago, at the height of the last ice age. It reached a maximum some 6,000 years ago, when the world became warmest since the last ice age.

While the theory is, as yet, far from being a full explanation for climate changes it suggests, he said, that a trend toward cooling will continue for the next 4,000 years even though, since 1973, autumn snow cover has diminished somewhat.

2. Pendulum Swings

Some scientists believe that the ice ages are a product of cyclic phenomena affecting the flow of heat from the tropics to the polar regions through the sea and air.

Most of the solar energy that enters the oceans and drives the winds is received in the tropics and carried poleward. The polar regions radiate more energy into space than they receive from the sun, but ocean currents and winds bring in enough heat—or almost enough—to make up the deficit.

Until a few years ago some persons suspected that the presence or absence of pack ice covering the Arctic Ocean might play a key role in this delicately balanced process. An absence of pack ice, when ocean currents were carrying considerable heat into that ocean, would allow evaporation and the resulting moist winds would shed the snows of an ice age. Periodic freezing of the ocean would end the glaciation.

Recently, however, sediment samples extracted from the floor of the Arctic Ocean have shown that it was apparently never free of ice between the ice ages, even though before they began that ocean does appear to have been open.

In fact, according to Dr. F. Kenneth Hare, professor of geography at the University of Toronto, fossils from the Arctic islands of Canada, the Soviet Union and from Greenland all indicate an ice-free ocean with "luxuriant" forests along its shores.

Another proposal regarding built-in pendulum swings of climate is that of Dr. Reginald E. Newell, professor of meteorology at the Massachusetts Institute of Technology. He believes ice ages are initiated when energy losses at high latitudes exceed energy gains in the tropics—a state that may exist at present.

An ice age ends, in this concept, when enough of the ocean becomes ice covered to curtail the escape of heat being carried poleward by ocean currents. At the present stage of such a cycle, he said in a recent article, surface water in polar seas would be growing cooler, "in the slow process that will lead to the next ice age."

In a recent issue of the British journal Nature, Drs. Reid A. Bryson and E. W. Wahl of the Center for Climate Research at the University of Wisconsin cite records from nine North Atlantic weather ships indicating that from 1951 to the 1968-1972 period surface water temperatures dropped steadily.

The fall was comparable, they reported, to a return to the "Little Ice Age" that existed from 1430 to 1850. It was early in this period that pack ice apparently isolated the Norse colony in Greenland and led to its extinction. The temperature drop in the North Atlantic carried it one sixth of the way to the level of a full-fledged ice age, according to Drs. Bryson and Wahl.

Unfortunately, they said, several of these weather stations are being discontinued so that monitoring future trends will be difficult. Dr. Bryson attributes recent droughts in Africa and elsewhere to a southward displacement of the rain-bearing monsoons.

A similar change occurred in about 1600 B.C., he believes. The monsoon rains no longer reached northwest India. Fresh water lakes that had been there for 7,000 years dried into salt beds and the Indus Empire that had spread over the region for 1,500 years was destroyed.

3. Man-Made Influence

There is general agreement that introducing large amounts of smoke particles or carbon dioxide into the atmosphere can alter climate. The same would be true of generating industrial heat comparable to a substantial fraction of solar energy falling on the earth. The debate centers on the precise roles of these effects and the levels of pollution that would cause serious changes.

Carbon dioxide in the air acts like glass in a greenhouse. It permits solar energy to reach the earth as visible light, but it impedes the escape of that energy into space in the form of heat radiation (at infrared wave lengths).

Dr. Mitchell has pointed out that a variety of factors determine the role of carbon dioxide on earth. For example, the extent to which that gas, introduced into the atmosphere by smokestacks and exhaust pipes, is absorbed by the oceans depends on the temperature of surface waters.

This, in turn, is affected by climate, leading to so-called feedback effects. Plants consume carbon dioxide at rates that depend on temperature and the abundance of that gas in the air, complicating predictions of their role.

The observatory atop Mauna Loa, the great Hawaiian volcano, has recorded a steady rise in the annual mean level of carbon dioxide in the atmosphere, amounting to 4 per cent between 1958 and 1972. That, however, was a period of global cooling—not the reverse, as one would expect from a greenhouse effect.

The Mauna Loa observatory has also recorded a steady rise in atmospheric turbidity—the extent to which particles overhead dim the brightness of the sun. The academy study finds that human activity over the last 120 years has contributed more to this atmospheric dust than have volcanic eruptions.

However, it says, the present atmospheric load of man-made dust is perhaps only one fifth what was thrown into the stratosphere by the volcanic explosion of Krakatoa in 1883. The role of atmospheric dust is complex, for it cuts off sunlight from the earth, but is itself heated by that light, warming levels of atmosphere in which it resides.

Until recently the idea that ice ages are initiated by intense volcanic activity was unpopular for lack of evidence for such activity. The hypothesis has gained more credence from the analysis of sediment cores extracted from the ocean floors by the drill ship Glomar Challenger.

According to University of Rhode Island scientists, ash was far more common in layers laid down in the last two million years than in the previous 18 million years.

If worldwide energy consumption continues to increase at its present rates, catastrophic climate changes have been projected by M. I. Budyko, a leading Soviet specialist. He says that the critical level will probably be reached within a century.

This, he has written, will lead to "a complete destruction of polar ice covers." Not only would sea levels rise but, with the Arctic Ocean free of ice, the entire weather system of the Northern Hemisphere would be altered.

However, Dr. Mitchell has suggested, warming of the climate due to pollution might be enough to head off an ice age "quite inadvertently."

More precise knowledge of the past is certain to aid in choosing between various explanations for long-term climate changes. The Greenland Ice Sheet Program, with American, Danish and Swiss participants, is drilling a series of holes into the crest of the Greenland ice in the hope, ultimately, of reconstructing a year-by-year record of climate for the last 100,000 years.

So far the ice has been penetrated 1,325 feet, extending the record back 1,420 years. The yearly layers can be counted, like tree rings, in terms of summer and winter variation in the relative abundance of two forms of oxygen (oxygen 16 and oxygen 18). Their ratio indicates temperature at the time when the snow fell to form that layer of the ice sheet.

The isotopes also reflect the long-term climate changes. A remarkable finding, reported in the May issue of Nature, is that the trends in Greenland for the period 850 to 1700 A.D., closely match the British record for 1100 to 1950. California tree rings show a climate record similar to the one in Britain.

The implication is a lag of 250 years between climate variations in Greenland and those in regions east and west of the Atlantic.

If, in fact, the climatic cycles of Greenland precede those of Europe and North America by 250 years, a powerful means of prediction would be available. However, as noted in the Nature article, it is by no means certain that the effect is persistent.

The Academy of Sciences report notes that any assessment of climate trends is crippled by a lack of knowledge: "Not only are the basic scientific questions largely unanswered, but in many cases we do not yet know enough to pose the key questions."

The oceans clearly play an important—and little understood—role. Not only are they the chief source of water in the atmosphere but they harbor a vast reservoir of thermal energy. "When the dynamics of the ocean-atmosphere interaction are better known, according to the report, "we may find that the ocean plays a more important role than the atmosphere in climate changes."

The report, including a wide range of proposals for national and international programs of research, was prepared by the academy's Committee for the Global Atmospheric Research Program, headed by Dr. Verner E. Suomi of the University of Wisconsin.

In his preface Dr. Suomi notes that, by the end of this decade, space vehicles will be able, on a global scale to observe the sun's output, energy reflected from the earth, distributions of clouds, snow and ice, as well as ocean temperatures. With these and other inputs a better understanding of how and why the climate is changing should become possible.

May 21, 1975

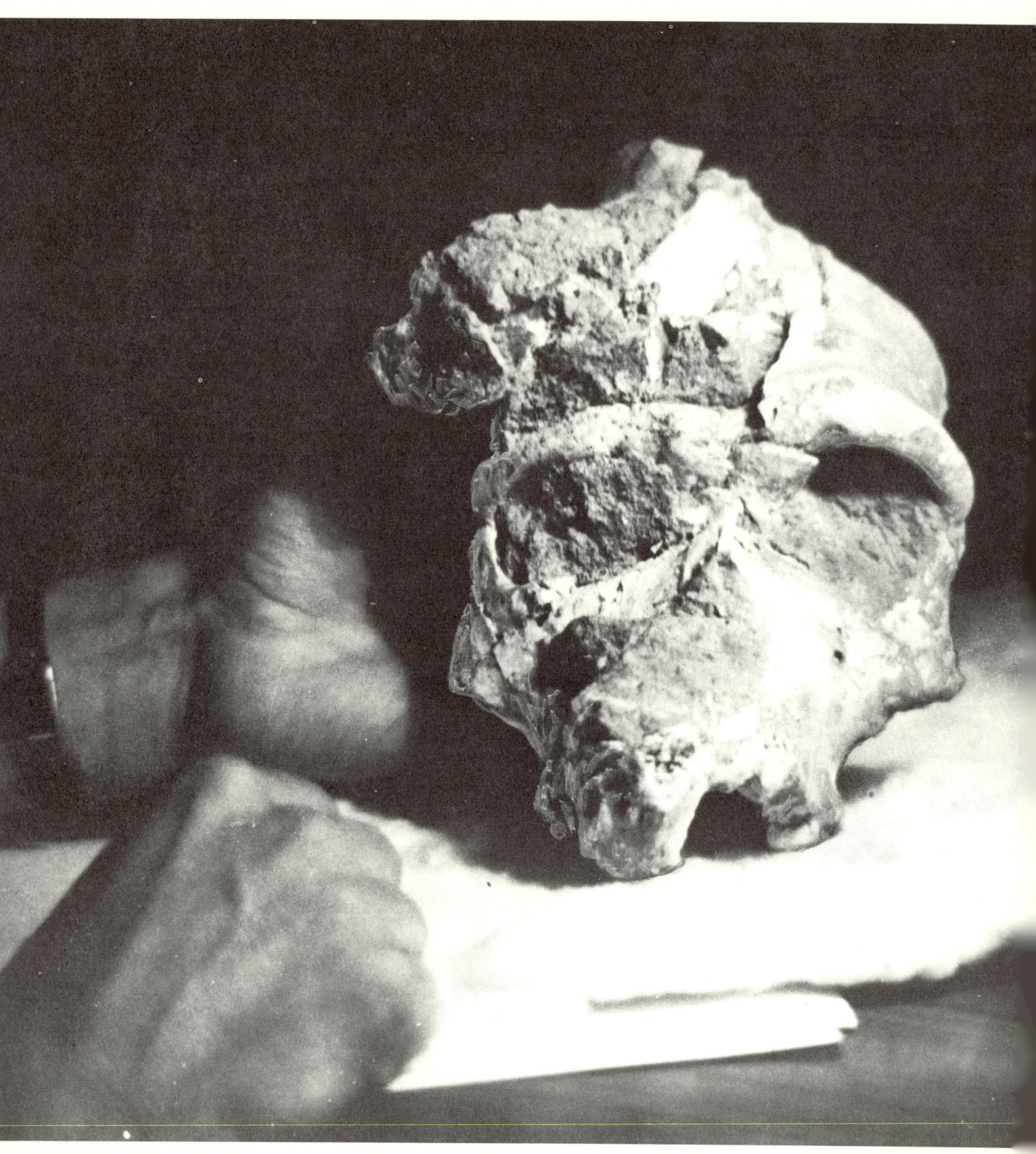

CHAPTER 4

Life

The skull of an ancient man discovered in Marseilles, France.

Walter Sullivan. The New York Times

GENESIS

"WHERE THE MICE GOT IN."

W. A. SHENSTONE, who has joined the reactionary movement led by Sir WILLIAM RAMSAY against the "mad" demonstrations by modern scientists of the origin of life from inanimate matter, presents an illuminating account of the history of abiogenetic investigation in The Cornhill Magazine for September. He shows how the idea that "the corruption of one thing is the life of another" was firmly held for centuries as axiomatic by disciples of religion and philosophy alike, and quotes the words of ST. PAUL, "Thou fool, that which thou sowest is not quickened, except it die," as characteristic of the belief of men until modern times. VAN HELMONT, chemist, physician, and mystic, whose authority was unquestioned in the early part of the seventeenth century, published receipts for the artificial production of mice and vipers. To make a pot of mice, he gravely asserted, "it suffices to press a dirty sheet into the mouth of a vessel containing a little corn, when after about twenty-one days the ferment proceeding from the linen, modified by the odor of the corn, effects the transformation of wheat into mice."

Since VAN HELMONT, and during cycles of experiment and investigation becoming more refined and accurate, the theory of life from inorganic matter has been successively set up and pulled down. Thus FRANCESCO REDI, the great Italian contemporary of HARVEY, demonstrated in 1670 to the satisfaction of his generation that there could be no life without antecedent life by the simple experiment of protecting from the inroads of flies by a gauze network strips of flesh exposed to the air, in which maggots had hitherto been supposed to generate spontaneously. But the revelations following the invention of the microscope overturned REDI's experiments, which depended altogether upon the testimony of the naked eye. Father NEEDHAM in the following century took similar strips of flesh, which he even heated to deprive of inhering eggs and germs, but found microscopic organisms slowly developing on the sterilized mass. It was then Father SPALLANZANI's turn to show "where the mice got in" in NEEDHAM's experiments, by proving that his organic infusions could not be protected from incursions of fresh germs from without, except the bottles containing them be hermetically sealed.

Again, when the theories of LIEBIG and PASTEUR concerning fermentation clashed in the nineteenth century, PASTEUR's biogenetic hypothesis triumphed. LIEBIG had proved to everybody that "putrescible matter," such as the concoction of beer, found its source of fermentation solely in its contact with the atmosphere. Of course the inference was inevitable that the living organisms of fermentation were in this way chemically produced. But PASTEUR demonstrated that if the beer and the air with which it came in contact were sterilized there could be no fermentation, and since his time no experimenter with sterilized "putrescible matter" has been able to show the contrary, or, until BURKE, to exhibit what might be construed as a symptom of organic life.

The world has been startled by BURKE's experiments with radium, tending to show that when small quantities of radium chloride are scattered upon the surface of carefully sterilized bouillon minute germlike objects appear, which finally lose their individuality and become resolved into crystals. In this case the abiogenetic "mice," as Dr. SHENSTONE and Sir WILLIAM RAMSAY aver, are really not in the liquid at all, but are running riot through the brains of their discoverer and his disciples. SHENSTONE emphasizes the fact that the radium cultures show no signs of reproduction after their initial appearance and subdivision, and that, unlike living organisms, they are readily soluble in water. To be sure, Prof. BURKE parries this objection by asserting that possibly he has discovered the "missing link" between animate and inanimate matter—that the radiobe crystal is a half-living organism, and like the lion in the Miltonic account of the Creation is

> pawing to get free
> Its hinder parts.

Yet the world need not too hastily accept a theory which progressive scientific experiment has hitherto negatived, and which now rests upon the slenderest of foundations.

September 8, 1906

SAYS LIFE CAME WITH LIGHT.

Germs Were Brought Here from Other Worlds, Prof. Arrhenius Asserts.

Prof. Soante Arrhenius discussed in the City College yesterday the question of how life came upon this earth.

"The theory of spontaneous generation having been proved impossible, we are reduced to believing that all life comes from preceding life," he said. "And since our earth was originally in a condition absolutely inimical to life, it must have come from other worlds.

"One theory maintains that meteors falling upon our earth brought germs of life. This idea, however, is not plausible. Meteors become white hot on passing through our atmosphere, which would destroy any life which might have existed upon them.

"About the only other thing which arrives on this earth from far away globes is light. This form of energy, rushing through space at 186,000 miles per second, became the carrier of the first life which took root on the earth. In the form of tiny spores, life germs, they were planted upon fertile soil, and through the process of evolution eventually became even man himself."

May 19, 1911

SEES IN PROTEIN ESSENCE OF LIFE

Dr. Vaughan Tells Chemists in Convention That All Organisms Rise From It.

FRIES DEFENDS WAR GAS

General Holds It Essential to Defense and Says Effects on Victims Are Exaggerated.

Special to The New York Times.
RICHMOND, Va., April 13.—Chemical agencies and environment were stressed as the dominant factors of life by Dr. Victor C. Vaughan of the National Research Council at Washington, in a paper presented at a meeting of the American Chemical Society here tonight.

"There is no life save in protein," asserted Dr. Vaughan, who for thirty years was Dean of the Department of Medicine and Surgery at the University of Michigan and who discussed "A Chemical Concept of the Origin and Development of Life."

There are protein molecules in nitrogenous bodies, that is, bodies containing nitrogen, and it is from such extremely small chemical units that all living matter is built up, according to Dr. Vaughan. Therefore, he held, even the bacteria and the bacteriaphage—the smaller fragments out of which bacteria are constructed—are protein molecules.

Dr. Vaughan pointed out that the difference between living and dead matter is that the living matter trades energy with its environment, and the character of that environment determines the form of the living protein matter. So long as the environment remains unchanged, the characteristics of this living matter are unchanged.

This is shown by the fact that the tubercular bacilli in the mummies of Egypt have been found to be just like those infesting human bodies today, and to cause the same kind of tissue destruction, Dr. Vaughan said. Throughout recorded time nothing has remained so little changed as the bacteria of that type, he declared.

Higher forms of life, including the life of man, have changed more through history than these bacteria have, he added. Notwithstanding all this, different kinds of tubercular bacilli have been produced by changing the environment.

The Origin of Species.

Dr. Vaughan explained that the origin of species, which most biologists attribute to change in the formation of cells, is in fact due to changes in the type of proteins, which are chemical compounds based on differences in environment.

"I can say with much confidence," he went on, "that the conversion of non-living into living matter is accompanied by increased molecular lability. By this I mean that the atoms or electrons within the molecule are energized; their orbits are enlarged. Within their orbits they move with greater speed. Their chemism is intensified so greatly that they are now able to drag into their orbits atoms, and possibly molecules which have hitherto been beyond their grasp. In other words, the molecules begin feeding on outside matter.

"All living things absorb, assimilate and aliminate. This means that metabolism or trading in energy begins. Such is the first evidence of life.

"Have we any idea of the nature of these primitive living molecules? Yes. They were and are protein molecules. There is no life save in protein.

"I do not regard bacteria as the simplest form of life. Their chemical structure is very complicated. They are essentially nucleins and their chief function is to multiply. Whether the individual consists of a single or many molecules I do not know. Probably their structure is multimolecular, but if so the chemism between the molecules must be very strong.

"In my opinion the assumption that bacteria and protozoa readily undergo mutation is not warranted by any fact, which can be gathered in a study of the history of infectious diseases. I am ready to assert that there has been less mutation in the tubercle bacillus or the virus of smallpox since the beginning of recorded times than there has been in man and the other higher animals.

"I find no difficulty in recognizing the action of chemical environment ever in the highest forms of life. Morphologists stress the stability of germ plasm, but some of them do admit that certain poisons, such as alcohol, lead and mercury, may deleteriously affect the reproductive cells. In my opinion, even more striking examples might be given.

"A boy and a girl born of healthy parents and raised to maturity under normal conditions may migrate into a goiterous district and after acquiring goiters may marry. Their children may be cretins. In this case it is the absence, according to the now accepted belief, of iodin in the food and drink which leads to this deterioration. It is only the absence of one chemical element which causes this disaster."

Favors Gas Warfare.

Addressing the chemical warfare group of the convention, Brig. Gen. Amos A. Fries, Chief of the American Chemical Warfare Service, made these points:

Surprises won the greatest battles in history. The worst fear man has is fear of surprise. To prevent surprise in case of attack the United States Government must have knowledge of the action of chemicals. Antidotes must be found for poisons.

He contended that chemical warfare was necessary for defense, and that in offensive war it was humanitarian and effective.

Referring to the great surprises of history he mentioned the attack on the Romans by the Goths in 378 A. D. and the gas attack by the Germans at Ypres. He said the war would have been won by the Germans had there been a sufficient number of troops on hand to be poured through the gaps made by the chemicals.

General Fries insisted that the gas of warfare was not poisonous and did not contain germs. It is false to say that gas leaves lingering after-effects, he asserted.

He said 70,552 soldiers were admitted to World War hospitals suffering from gas wounds. Two hundred died on the field from gas. Of the number taken to hospitals only 1,221 died. There were, he said, 155,000 soldiers brought to hospitals suffering from other injuries. 34,000 died on the field from other wounds. Of those admitted to hospitals 12,500 died. That is, the death rate from gas was about 2 per cent. and from other injuries 25 per cent.

Charts showed 838 soldiers were reported suffering from chlorine gas. Only thirty-seven of this number are dead today, he said, and twelve of them were killed later from other injuries after they had left the hospital.

Startling discoveries with regard to "polarized light," which may explain the influence of moonlight on the growth and deterioration of plants—and on moods of human beings—were announced today by Dr. David I. Macht of Johns Hopkins University. "Polarized light" is that light vibrations of which all lie in the same plane. Sea, snow and sky sometimes partially polarize daylight, and moonlight is largely polarized.

Dr. Macht learned from experiment that polarized light stimulates the growth of yeast and bacteria. Sick and poisoned rats die more quickly under this light.

G. Davis Buckner of the University of Kentucky stated egg laying was improved when hens were fed calcium carbonate, which is another name for marble.

E. G. Sharrard of the Forest Products Laboratory of Madison, Wis., reported that lactic acid could be made from waste in the manufacture of wood alcohol. Lactic acid has numerous industrial uses.

April 14, 1927

LABORATORY FRANKENSTEINS.

After the first organic compounds had been successfully synthesized, the great BERTHELOT wrote: "The objective of our "science is to banish 'life' from the "theories of organic chemistry." Those were the brave Victorian days when science saw only machinery in the solar system and in the wriggling of worms. Once the gears and rods of organized living matter had been laid bare, the chemists of a generation ago boldly spoke of creating life itself in the laboratory.

A change has come over the spirit of this scientific dream. Is it because the mathematical God of JEANS has dethroned the engineering God of KELVIN and revealed himself not in the interplay of mechanical forces but in probabilities and in nine-dimensional space? Gone for the moment is the old confidence in the possibility of discovering the secret of life—of constructing a primitive organism which will feed itself and reproduce its kind. Instead, we find a curious return to the eighteenth century doctrine of Vital Force, according to which minerals are the products of ordinary physical agencies but organisms of an inscrutable purposiveness.

When he received the Willard Gibbs Medal recently, Dr. LEVENE of the Rockefeller Institute made a stirring plea for a return to the romantic optimism of BERTHELOT and the Victorians. He admits the discouraging chemical complexity of a green leaf, an egg, a bit of steak. But there are new avenues of approach. Functions are now studied where once a knowledge of structure seemed sufficient. So it happens that the chemist is actually able to direct some of the forces at his disposal—notably the catalysts that control the body's chemical activities. Here we have under control a part of the purposiveness which underlies life. More important is the functioning of the organism as a whole. Some of the controlling agents are not complex. Thus the hormones and vitamins that determine whether we are Newtons or idiots prove to be remarkably simple when isolated—"definitely simpler than certain common drugs," Dr. LEVENE assures us. Such discoveries mark the beginning of a revolt against the vitalistic restriction of human curiosity, against the notion that life passeth all understanding.

The biologist may object to this hopefulness. Chemical compounds can be made to reproduce, to move spontaneously, to feed and even to convert what they feed upon into energy. Yet they are not alive. What, then, is life? No satisfactory definition has yet been framed. One test of life is the ability to evolve. An English physicist, L. L. WHYTE, has pointed out that we must reckon with time—nature's time—in our Frankensteinian attempts to fabricate life. Suppose we succeed in producing protein in an hour. Then it may require a month for primitive protoplasm to appear, ten years for a simple cell, a thousand for a flagellate and a million for a mammal. If science succeeds at all in creating life, it must content itself with the most elementary organism and let nature take her course.

June 7, 1931

BIOLOGISTS INSIST ON DISTINCT STATUS

Contend at London Congress Theirs Is an Independent Science in Study of Life.

Special Cable to THE NEW YORK TIMES.

LONDON, July 2.—Professor William E. Ritter of the University of California presided at today's session of the International Congress of the History of Science and Technology, at which Professor J. S. Haldane and other famous biologists contended that biology was an independent science in the study of life, having a different logical basis of interpretation from physical science.

"This view," said Professor Haldane, "is not what is known historically as vitalism, since the vitalists assumed a 'vital principle' controlling the phenomena of living organisms and interfering as a local opposing influence with physical phenomena. In my view there is no such interference but a coordinate maintenance which cannot be interpreted physically because in physical science phenomena are interpreted as if they consisted in separable events."

When Galileo took the first definite steps toward a physical interpretation of our experience, Professor Haldane declared, he entered an extraordinarily fruitful path but made the mistake of assuming that his interpretation was "objective" and represented a fundamental reality.

Dr. E. S. Russell of the British Board of Agriculture deplored the fact that the organismal method introduced by Aristotle had, "until recent years been overshadowed and inhibited by the success in the physical sciences of the methods, thought and procedure introduced by Galileo and Descartes." Great advances in parasitology and kindred forms of research, he urged, had not been based on mechanistic principles, but on biological observation. He insisted that it would be found that mechanistic explanations were insufficient.

July 3, 1931

Life Origin Traced to the Inanimate In '1936 Model' of Evolution Theory

Riddle, Carnegie Biologist, Credits Sun's Action on Inorganic Matter, Telling Scientists of New Clue to 'Twilight Zone'— Church Bars to Teaching Are Assailed at St. Louis.

By WILLIAM L. LAURENCE.
Special to THE NEW YORK TIMES.

ST. LOUIS, Jan. 1.—Evolution, model 1936, which brings Darwin up-to-the-minute and extends the picture of the origin of life and consciousness far back into its beginning from inanimate matter, by sunlight's action on water, carbon dioxide, sugar, nitrogen and similar inorganic substances, was presented tonight before the American Association for the Advancement of Science.

This version was offered by Dr. Oscar Riddle, internationally known biologist of the Station for Experimental Evolution, Carnegie Institution of Washington, Cold Spring Harbor, L. I.

Whereas Darwin, who first conceived of his epoch-making theory of evolution between October, 1835, and January, 1836, just a hundred years ago, traced the origin of life no further back than the lowest form of living matter, Dr. Riddle showed how the biological sciences had gradually been pushing the frontiers of life's origin further back into the inorganic world of substance.

One of the most important clues, the entity now believed to be the long-sought bridge between the living and the non-living, was traced only a few months ago in the Rockefeller Institute for Medical Research, he said, where Dr. W. M. Stanley succeeded last June in producing in the form of solid crystals one of the invisible plant-disease viruses.

Those mysterious entities elude even the most powerful microscope and are responsible for many of man's worst plagues, from the common cold to infantile paralysis.

From the study of these crystalline viruses, Dr. Riddle stated, it becomes evident that here science has at last found that which belongs neither to the living nor the non-living, a substance dwelling in the mysterious, shadowy borderland of the alive and the non-alive.

Step by step, like one piecing together a most intricate puzzle of innumerable parts, Dr. Riddle showed how such a "twilight creature" rose out of the pitch darkness of the primeval chaos of inanimateness and then proceeded upward on the scale of life, through bacteria, protozoa, and the first crawling things, out of the ooze out into the green pastures of mammalian life.

Then he took his audience through another labyrinthine maze, and traced the consciousness of animal life from its earliest manifestations in the form of primitive instincts, which, he said, modern science was now able to link definitely with the chemical processes of life.

Animals' Instincts Changed.

Only recently, he stated, it has been found possible to change the instincts of animals, formerly believed unalterable, by simply injecting into them certain hormones from the pituitary gland, or by depriving them of minute amounts of certain substances in their food.

In this manner, it has been possible to change by the act of a few seconds the behavior patterns nature has taken millions of years to evolve.

On the one hand, an animal's instinctive tendency to devour was changed into tender maternal solicitude, while on the other, the deeply rooted mother love of the animal was changed into an indifference that allowed its young to die.

In this manner consciousness is placed under a microscope and analyzed into its component parts. And the result of this scrutiny is, he stated, that science no longer accepts the doctrine of separation of mind from body, as it finds mind an essential function of the bodily processes, the outgrowth of physical, chemical and biological processes and no more.

At the end of this, the vice presidential address of the section on zoological sciences, Dr. Riddle launched a vitriolic attack against those who were instrumental, he charged, in keeping from our schools and colleges the true knowledge gained by modern biological sciences on the origin and nature of life and consciousness.

Holds Science Teaching "Garbled."

The blame must be laid, he said, against "traditional religion" which "has evoked the heavy hand of legislation," with a result that many States have passed laws forbidding the teaching of evolution, while in others the teaching of biological sciences is either curtailed or so garbled as to make such teaching worthless.

The net result of this legislative and other interference, Dr. Riddle said, has been that our schools teach much less of the biological sciences today than they did forty years ago. In presenting his view of evolution Dr. Riddle said, in part:

"Within the past twenty years it has been definitely proved that formaldehyde, sugars and other inorganic substances—some of them nitrogen-containing — are generated under conditions which must have prevailed on the earth's surface before, during and after the origin of living matter.

"The agents necessary to the building of such organic matter are none other than sunlight, ordinary temperatures, colored surfaces, water and carbon dioxide.

"Moreover, from one such sugar, glucose, in alkaline solution, plus a time factor, some hundreds of different organic substances have already been derived in laboratory experience.

"With only nitrate or nitrite added to the system already mentioned the very reactive nitrogenous substance formhydroxamic acid is formed; and with additional active formaldehyde this acid combines to form a whole series of compounds which are the commonest constituents of plants and animals—pyridine, purine and alpha amino acids.

"Natural Forces" Gave Sources.

"There exists scarcely a doubt that during long periods of earth history, preceding the appearance of living matter, many localized areas of the earth's surface provided suitable conditions for the synthesis of sugar and some amino acids.

"Thus the natural forces at work on the earth's surface prior to the appearance of living matter had almost certainly already provided a variety of both inorganic and organic compounds characteristically utilized by living matter, including the particular sugar, glucose, from which most living matter still derives an obligatory flow of free energy, and some alpha amino acids which must serve as building blocks for the protein which is the chief constituent of living matter.

"These syntheses of organic matter wholly apart from life are facts which must now be utilized in dealing with the question of the origin of life. Any general biological instruction that fails nowadays to direct attention to these facts is vitally deficient in its presentation of modern life-science.

"The longest single missing span may be that reaching from spontaneously formed sugars and amino acids to a protein-molecule like that of a virus.

"Studies of the past thirty years on the structure and functions of the brain, and on the relation of other parts of the body to brain function, have resolved but a fraction of the primary problem of the mind, but the mind has been firmly placed in an evolutionary frame, and several gaps in our knowledge formerly filled by myth or magic have been displaced by verifiable experience.

Development of Consciousness.

"The consciousness of dog and man has evolved, and redevelops in every ontogeny, in the same unbroken way that the function of the digestive or glandular system has evolved. The evidence on this important point today is probably stronger than that adduced in 1859—when Darwin published 'The Origin of Species'—for the principle of evolution.

"Consciousness, as well as some related subconscious phenomena, is associated with organization at a high biological level, and chiefly, though not solely, with the organization of the brain."

In discussing what he termed the difficulties encountered by competent teachers of the life sciences in providing the younger generation with the advances of knowledge gained every year in the research laboratories, Dr. Riddle said:

"In any consideration of this matter it is unquestionable that it was traditional religion that thus invoked the heavy hand of legislation. It is equally clear that elsewhere, without invoking the law but with its extended and varied influence, traditional religion is now effecting a widespread repression of the teaching of this central principle of biology in public schools through the United States and in practically all other civilized countries as well.

"In this country it sometimes forces the resignation of able zoologists even from college positions, and in high schools and late primary grades there are probably today few places where straightforward teaching of the unmitigated evolution principle can be done except at the peril of the teacher.

Parochial Schools Criticized.

"It is obvious that an eviscerated straw-man is set up in place of the reality for the younger students of denominational and parochial schools everywhere. In this country this means that many millions of our present and future citizens are robbed of a biological outlook, or they get one that is warped and unrecognizable, through direct responsibility of the church, while with somewhat less directness the same agency widely exercises a restraint upon effective biological teaching in the public schools.

"Biologists in nearly all countries, and particularly in our own, have tried a compromise with religious creeds. That compromise has failed. Most youth of 1935, like those of 1859, leave our schools without having opportunity to learn that the worthy facts concerning man's origin and destiny come not from religious traditions but from biological investigations made within the time of men now living. That compromise now robs most modern youth of opportunity to learn what is known concerning his or her place in nature.

"In what is said here I am not concerned with the question whether religion is important; nor whether one or another of the creeds of the earth has or has not sufficiently 'adjusted' its teachings to modern knowledge; nor whether one or another of them is good, bad, or quite indifferent. But whatever the answer to those questions the present restrictive influence of organized religion on the teaching of the best of biology is intolerable.

"Such an influence, from whatever source, is too highly harmful and dangerous to the well-being of man—to modern beehive aggregations of men who live under ever multiplied rules and laws which must wreck us if based on variegated tradition instead of upon a common knowledge.

"For moribund traditional beliefs to continue to exercise such influence over the educational program of a country is a confession and declaration either of the apathy, the cowardice, the impotence, or the intellectual bankruptcy of the enlightened leadership in that country.

Traditionalists Also in Schools.

"The tongues of the traditionalists are heard not merely from pulpits, but they echo also within our schools—the only possible home of science—and there they now curb or tie the tongues of biologic truth.

"This confusion is partly sustained—or at least the voices of tradition are prolonged and made more plausible—by the words of great authorities in one or another branch of learning. Today, as at Oxford in 1860, a professor can easily be had to support a bishop against a really good and far-reaching biological advance.

"The public cannot fix relative values to the words of different scientific men. The biologist who has learned the texture and ways of living stuff knows well enough, however, that when renowned physicists and astronomers elect to speak about life they really do this as laymen—and all too frequently their words are unconsciously filled with tradition, a thing which they also acquired as laymen.

"In addition to these volunteer voices from quite outside life-science, we are all aware that some high authorities in one or another branch of biological science persistently ignore the greater biologic accomplishment, and on some points they too still speak with tongues of a day that is gone. We may as well have it out with them.

"The issue today is whether the case and the course of civilization are to be guided by knowledge or by the dead hands of the past; whether the biological investigator of either yesterday or today may be permitted to give his best results to the world or whether he is to be more and more insulated by his own progress; whether, indeed, present man-in-the-mass has evolved sufficiently to prefer light to twilight, truth to tradition."

January 2, 1936

How Life Began

A Chance Chemical Union Which Could Not Occur Again

The theory that life on earth may have begun spontaneously through the chance chemical union of organic compounds floating in prehistoric seas well over a billion years ago was expounded by Dr. G. W. Beadle, geneticist of the California Institute of Technology, at a recent Mills College Centennial Symposium on evolution. Because there are no longer masses of organic compounds from which such life may have sprung, he thinks the process cannot be repeated again.

It may have taken as long as thousands or even millions of years for the first "living molecule" to produce a second like itself, in Dr. Beadle's opinion. "But somewhere, somehow," he said, "a chance combination of compounds must have acquired a new property it never had before—the ability to duplicate itself and to undergo mutation. This presumably was the precursor of all present living things."

Development of this first bit of life to even the simplest alga perhaps represents a greater step in evolution than the development from amoeba to man. The process may have take a billion years.

Viruses, according to Dr. Beadle, are at both the beginning and the end of the scale of living things. Not only are they like the first probable form of life, but they are also the final stage of degeneration of higher forms of living things under conditions of parasitism.

"All during the process during which living things have been evolving to give increasingly complex forms, evolution has been going on in the other direction as well," he said.

May 11, 1952

'INFINITE' WORLDS SEEN AS LIFE SITES

Dr. Urey Says Quadrillion in Known Universe May Have Produced Living Organisms

Dr. Harold C. Urey, Nobel-prize-winning atomic scientist of the Institute for Nuclear Studies of the University of Chicago, estimated here last night that there are 1,000,000,000,000,000 worlds in the known universe on which it is possible that life might have originated and survived.

Dr. Urey emphasized that he was not arguing that one out of a thousand of the bodies estimated to be in "space" are inhabited or even habitable. From chemical considerations that may apply to life on earth; however, it is possible that one quadrillion worlds might originate and support life, he said.

Dr. Urey gave the Rudolf Schoenheimer Memorial Lecture at a joint meeting of the American Society of European Chemists and Pharmacists and the Rudolf Virchow Medical Society. The meeting was at the New York Academy of Medicine.

Postulates Early Earth Conditions

He postulated that conditions of atmosphere, temperature and total water volume have not always been the same on earth as they are now. At some "starting point" in the two or three billion year history of the earth, he said, there was no oxygen in the earth's atmosphere. The earth's gaseous envelope consisted mostly of hydrogen, methane and ammonia gases. Seas around the continents were about 400 feet deeper than they are now. Ultra-violet rays from the sun provided the energy to change the gaseous methane, ammonia and water into more complex chemical compounds.

In this process, oxygen was liberated by the break-up of water and it accumulated in the atmosphere. Excess hydrogen escaped the earth and poured into space. As the water was used up, the level of the seas lowered and the concentration of elemental chemical "building blocks" in the oceans increased. Organic acids, aldehydes and other building blocks made up 1 per cent or more of the sea volume at a maximum.

A Vast Store of Chemicals

Thus it was, Dr. Urey said, that at some unknown time in the earth's far distant past the oceans were a vast pool of complex chemical compounds. Probably all the 500,000 compounds known to modern chemistry were present, perhaps many, many more, now unknown to chemists. This chemical store, concentrated in ocean water, served as a great world-wide test tube in which all reagents known to modern science were present. They interacted, perhaps countless billions of times over a billion years.

By chance, Dr. Urey continued, the milieu in some particular spot was conducive at some moment, to the natural development of a complex series of intermediaries. They led, perhaps, to the development of a protein molecule. After this molecule formed others like it were fashioned from the pool of raw materials. Eventually, by chance, the first bacterium, representative of the most elemental form of life, developed. It spread. Soon the seas were a concentration of one, two or more bacteria. Plant life flourished and animal life followed inevitably.

November 14, 1952

Origin of Life

Urey's Theory of Atmospheric Changes Supported by Tests

Of recent years Dr. Harold Urey has been lecturing on his conception of the earth's origin. According to him, the infant earth had an atmosphere made up of methane, (marsh gas), ammonia and hydrogen. In the course of time this atmosphere changed into the one we have—one composed largely of nitrogen and oxygen mixed with carbon dioxide, a little hydrogen and minute amounts of other gases. Dr. Urey's atmosphere would not support life as we know it. But it might provide the stuff out of which life could be formed.

It is agreed that the earth's atmosphere in its early days, billions of years ago, was highly electrified, so that there were countless flashes of lightning. These flashes would have some chemical effect on the primitive atmosphere. Dr. Stanley Miller, an organic chemist of Los Angeles, Calif., presented the Botanical Society of America, which recently met at Pasadena with the American Association for the Advancement of Science, an ingenious hypothesis, supported by experimental evidence, of what the electrical flashes may have accomplished.

Dr. Miller created an atmosphere in accordance with Dr. Urey's prescription. Through this atmosphere he passed electric sparks—miniature flashes of lightning. He got a large and strange mixture of organic compounds, among them amino acids.

Out of amino acids protein is formed. When we eat an egg (albumen) or drink milk we consume protein. This we break down into amino acids, whereupon we proceed to rebuild the amino acids into protein, that is into tissue. This process is indispensable in all living organisms. Without protein life is impossible.

Dr. Miller may not have proved beyond doubt that Dr. Urey's atmosphere is the one the earth had originally, but, given many terrific lightning flashes, he has shown that out of such an atmosphere the stuff required for the creation of life could have evolved.

July 3, 1955

FIRST LIFE TRACED TO SHORE, NOT SEA

2 Scientists Offer Theories of Cells' Genesis on Land 3 Billion Years Ago

By WALTER SULLIVAN
Special to The New York Times.

UNITED NATIONS, N. Y., Sept. 1—Two leading scientists said today that life might have begun ashore, rather than at sea, as has been supposed.

Prof. J. D. Bernal, physicist at the University of London, told the International Oceanographic Congress, meeting here, that the most likely starting place was in the mud of tidal river estuaries. The early steps took place, he said, at one of the earliest stages in the earth's history, when the seas were still hot, both in temperature and radioactivity.

Prof. G. Evelyn Hutchinson of Yale University said that he leaned toward shallow pools of fresh water, ashore, as a likely place for life's birth. He also suggested an answer to a mystery that has long puzzled students of evolution, namely: where are all the early fossils?

The oldest found to date represent the mid-point of evolution rather than the starting point. Virtually all the phyla, or basic classifications of plants and animals, had evolved by then. Dr. Hutchinson said that no earlier fossils would be found because there were none.

Until half way through evolution, he said, the world was utterly peaceful. Animals ate plants or dead organic matter. Then the first aggression was committed and meat-eaters appeared. The animals were forced to develop armour in a hurry. As a result, shells and skeletons appeared rapidly throughout the animal kingdom, providing the hard substance from which fossils are made.

This, he said, is a more acceptable explanation than the idea that earlier skeletons and shells were dissolved because sea water at that time was a better solvent.

Dr. Bernal spoke in place of Prof. I. G. Oparin of the Soviet Union, whom he described as the "originator" of modern theory on the origin of life. Dr. Oparin was reported unable to attend because of illness.

Life Foreseen on Planets

Because of what he described as the "fireplace principle," Dr. Bernal said he thought there was some form of life on one or more planets in every solar system. In front of an open fire, he said, you can achieve comfort if you stand at the proper distance from it.

By the same token, he explained, some planets would be the right distance from their source of heat and radiation to stimulate the building of the complex molecules that lie on the borderline of living matter.

Dr. Bernal said that laboratory experiments had left "no doubt" that organic compounds would form in the situation believed to have existed early in the earth's history and under the influence of sunlight. The resulting hydrocarbons or fatty molecules would float to the surface and be cast upon the beaches, thus concentrating, within shoreline clays, the building materials for life.

He told of work in Japan that had demonstrated that clays can serve as matrices on which complex polymer molecules form from amino acids. The production of such protein-like substances would overcome one of the chief hurdles on the path to life.

The proteins and nucleic acids, in themselves almost alive, banded, Dr. Bernal said, into colonies that could "live" more efficiently. Thus they created the first living cells.

All this took place, he said, at least 3,000,000,000 years ago.

The first life must have been anerobic, able to live without oxygen, Dr. Bernal said. Only later did the photosynthesis of carbonates begin, upon which most modern life is dependent. This final development came with the introduction of substantial quantities of oxygen into the atmosphere.

The congress is being held under auspices of the United Nations Educational Scientific and Cultural Organization and the Special Committee on Oceanic Research of the International Council of Scientific Unions. It was organized by the American Association for the Advancement of Science.

September 2, 1959

Life's Elements Found in Space

Discovery in meteorites of molecules resembling the basic constituents of genetic material on earth was reported recently by Dr. Melvin Calvin of the University of California. The discovery, which apparently provides the first concrete suggestion that conditions exist in space favorable to the development of living forms, was outlined by Dr. Calvin at a Compton lecture at the Massachusetts Institute of Technology.

"We have found very reasonable evidence of the presence of molecules of the aromatic heterocyclic type resembling the pyrimidines and purines present in terrestrial genetic material," Dr. Calvin said. Heterocyclic compounds are constituents of nucleotides which, in turn, are the basic structural units of nucleic acid, the stuff of which the heredity-transmitting genes are made. Purines and pyrimidines are compounds of hydrogen, carbon and nitrogen which combine with sugar and phosphoric acid to form the nucleotides.

Existed on Earth

Such primitive chemical forms, Dr. Calvin said, probably existed at one time on the earth, before biological forms arose. In this pre-biological age of the earth, it is theorized, simple chemicals like carbon and hydrogen were joined together into progressively more complex molecules by the energy of cosmic rays, ultra-violet light and electrical storms. Eventually nucleotides evolved, then the nucleic acids and the chromosomes, carriers of the genes.

The pre-biological heterocyclic forms, as such, no longer exist on earth, Dr. Calvin said. They have long since been used up in the formation of the more complex biological molecules.

Life Inevitable

The existence in meteorites of these chemicals suggests, Dr. Calvin said, that the formation of such species of molecules is going on by non-biological processes outside the earth. The findings support the view, recently advanced by Dr. Calvin, that, given the conditions that theoretically existed at the time the earth was formed, the development of living forms was inevitable.

Dr. Calvin obtained small pieces of the interiors of stone meteorites from the Smithsonian Institution and several other museums. He pointed out that while the exterior of a meteorite melts and evaporates under the enormous heat it is exposed to as it enters the earth's atmosphere, the interior remains cool and the molecules are unchanged. The carbon extracts of the meteorites were tested by spectroscopic methods.

December 20, 1959

TESTS OFFER CLUES TO ORIGIN OF LIFE

Basic Ingredients Made All Over Universe, Study Hints

By ROBERT K. PLUMB

Experimental evidence supporting the view that life may arise throughout the universe —and not just on earth—was presented here last week.

An experiment described here was the latest in a number of studies that seek clues to the secret of life's origin.

Dr. Rainer Berger of the Convair Scientific Research Laboratory in San Diego suggested that organic compounds, the basic ingredients of living material, are being synthesized now in space by cosmic radiation passing through frozen gas mixtures.

In his experiment, Dr. Berger took a mixture of methane, ammonia and water and froze it at the temperature of liquid nitrogen. That is about 383 degrees below zero Fahrenheit.

Hit By Beam of Protons

The frozen mixture was exposed to a beam of protons (the nuclei of hydrogen atoms) for two hundred seconds in the sixty-inch cyclotron of the Crocker Laboratory of the University of California at Berkeley.

Dr. Berger, a chemist, demonstrated that these conditions produced urea, acetamide and acetone from the original material. The temperature in the experiment was close to what might be expected in space.

Proton streams comprise some 85 per cent of all the cosmic radiation that pervades space. The number of proton impacts in the experiment is about what might be expected during the lifetime of a comet.

Thus, in Dr. Berger's view, the experiment demonstrates that the precursors of life can be created throughout space, and not just on earth, under the conditions of temperature, radiation and time that do exist.

Dr. Berger's scientific report has been published in the September issue of the Proceedings of the National Academy of Sciences. He discussed it and earlier experiments seeking to understand processes leading to the origin of life at a symposium here last week at the Sheraton-East Hotel.

The symposium was sponsored by Radiation Dynamics, Inc., of Westbury, L. I., a manufacturer of high power accelerators.

To this time, Dr. Berger said, there have been two simulation experiments bearing on the problem.

Dr. Stanley L. Miller, now at the University of California in La Jolla, Calif., reported that a mixture of amino acids (the precursors of protein molecules) can be achieved by exposing gases to an electrical discharge.

This experiment demonstrates the production of chemicals essential for the development of life under conditions that are believed to have existed on earth many years ago.

That is, the Miller experiment simulates a primitive earthly atmosphere in which bolts of lightning produce organic chemicals.

Dr. Berger's experiment goes a step farther to establish that conditions of space can lead to the production of organic chemicals essential to life.

Two other major studies bear upon the problem, he said. Dr. Melvin Calvin and associates of the University of California at Berkeley have found nucleic acid bases in meteorites. The bases are the elements that make up deoxyriboneucleic acid, the material of heredity. And a Fordham University team headed by Dr. Bartholomew Nagy has found more complex organic chemicals in another meteorite.

In all, Dr. Berger said, these and other studies lead to a master plan for the various types of evolution that have resulted in life as it is known on earth today.

In the beginning, he reported, came the genesis of the universe—an event that took place many billions of years ago and may, according to the "continuous creation" theory, be still going on. After the universe was formed came the evolution of the earth.

Chemical evolution occurred along with the evolution of the earth when first simple organic compounds were formed and then more complex ones.

Organic evolution during which life started, probably began more than two billion years ago. The earliest known fossils of earth, marking the beginning of evolution of plants and animals, date back to some half a billion years. The evolution of man probably started somewhat more than a million years ago.

All these types of evolution overlap, Dr. Berger said. Several evolutions thus were or are going on at the same time. A sentence of Aristotle describes the situation, Dr. Berger reported:

"Nature makes such a gradual transition from the inanimate to the animate that a boundary is doubtful or may not even exist."

Dr. Berger cautioned that there is an enormous step between the finding that simple chemical materials can be produced under conditions found throughout space (or on the primitive earth) and what is now regarded as "life."

The experimental conditions of temperature and radiation with which Dr. Berger produced urea, acetone and acetamide probably do prevail throughout the universe, he reported.

The mixture of methane, ammonia and water used in the experiment may be found on many comets, icy meteorites and probably on the interstellar dust that makes up half the mass of the universe. Thus it appears that organic compounds are formed on a large scale throughout the universe.

October 15, 1961

A KEY MOLECULE CREATED IN LAB

Ultraviolet Light Plays Role in Experiment on Coast

Special to The New York Times

SAN FRANCISO, Aug. 17—A scientist has reported creation in the laboratory of adenosine triphosphate, a highly complex molecule.

Dr. Cyril Ponnamperuma of the Ames Research Center of the National Aeronautics and Space Adminstration at Moffett Field said that the molecule was created by shining ultraviolet light on a solution of compounds thought to resemble the composition of the earth's oceans about 4,000,000,000 years ago.

"Such experiments," he said, "are lending significant support to the theory that biological molecules, which are the prerequisites of life, could have appeared by the interaction of forces and materials which existed on the earth before life did."

He said that ultraviolet light is now absorbed in the upper atmosphere, but that large amounts probably struck the earth billions of years ago.

Adenosine triphosphate (ATP) is now made by animal life by digestion of food, and by plant life through photosynthesis. Dr. Ponnamperuma suggested that his experiments indicated that in primitive times the molecule was created without any effort for the life forms that then existed.

The goal of Dr. Ponnamperuma's research is the synthesis of deoxyribonucleic acid (DNA), the basic determinant for hereditary traits. Previously he has shown that the purines in the nucleic acids, ribonucleic acid (RNA) and DNA, can be produced under prebiotic conditions. Also he has combined adenine with ribose to get the nucleoside adenosine, another key ingredient of the nucleic acids.

The significance of the discoveries in the search for life on other planets was cited by Dr. Harold Klein at Ames Research Center.

"Previously we thought that such chemicals were peculiar to living organisms, so that if we found them on other planets we could assume the existence of life," Dr. Klein said. "Now we know that such an assumption is incorrect — that these chemicals may be formed in the absence of life."

August 18, 1963

BUBBLES LINKED TO DEEP-SEA LIFE

Tiny Ocean Creatures Are Said to Feed on Particles That Form on Globules

BIOLOGICAL CLUE FOUND

New Hypothesis Is Thought to Explain the Survival of Zooplankton in Depths

By HAROLD M. SCHMECK Jr.

What may be an important missing link in the food chain that maintains life in the ocean depths has been discovered through laboratory and oceanographic research.

The discovery also offers hints concerning the origin of life on earth, its present abundance and the complex interdependence of its multitudinous forms.

The crux of the discovery was that particles or organic matter will be produced from dissolved organic chemicals in the sea water that gather on bubbles deep in the sea.

The dissolved organic substances are too dilute to be a useful source of food for zooplankton, the ocean's smallest animals. But it is believed that these microscopic creatures are able to harvest the bubble-produced particles.

More Food in Winter

The newly hypothesized source of food particles offers a handy explanation of the ability of populations of zooplankton to exist at depths where microscopic sea plants, the phytoplankton, may be scarce or absent.

Research shows that the particles are more abundant in the winter, when storms increase bubbling in the ocean, and that the abundance of particles also varies with the populations of sea organisms that discharge organic chemicals into the water.

The mechanism of particle formation is believed to be related to the electric charge properties of the bubble surface.

The same mechanism of particle formation offers a clue to the means by which naturally produced organic (carbon-containing) chemicals in the earth's early seas aggregated and combined over millions of years to give rise to the first self-replicating, and hence living, things.

Reports on Discovery

A statement on the discovery of the bubble phenomenon and the hypotheses that spring from it and related research was made public yesterday by the National Science Foundation. The foundation helps support the research at Yale University and the Woods Hole Oceanographic Institution.

The scientists principally involved are Dr. Gordon Riley and Dr. P. J. Wangersky of Yale, Dr. E. R. Baylor of Woods Hole and Dr. W. H. Sutcliffe, formerly of Woods Hole and now at Lehigh University.

There has long been a question as to how deep-sea creatures get enough food to survive, Dr. Riley said in a recent telephone interview. Phytoplankton, presumed to be the ultimate source of food, become scarce even on the surface in winter season and may not be found at all at great depth.

It has long been known that there is much nonliving organic material in ocean waters, perhaps 50 times the amount present in living creatures. But the dissolved organic material was thought to be useless to the sea's living things until it was broken down further into its inorganic constituents.

Laboratory studies at Woods Hole showing the effects of bubbling in generating organic particles offered an explanation of the puzzle. The sinking particles are probably the "marine snow" which has been observed by many oceanographers, but not heretofore satisfactorily explained, the scientists think. Oceanographic studies by Dr. Riley and others support the view that the particles are a source of food.

Students of the origin of life on earth are widely agreed that the large carbon-containing molecules that are today called "organic" were first generated from simpler inorganic substances in the early ocean and atmosphere.

Dr. Wangersky believes the bubbling phenomenon may be one of the crucial steps by which these essential raw materials of life first aggregated and concentrated into the huge, complex molecules that are the essence of all living things.

October 5, 1964

Berkeley Reports Clue to Chemistry Of Origin of Life

Another clue to the manner in which the chemistry of life may have evolved from non-living substances has been discovered, according to the University of California.

Research in recent years has brought to light a series of steps whereby the simple substances of the earth's original atmosphere could have been synthesized into more complex compounds.

This early atmosphere is thought to have consisted exclusively of hydrogen and its compounds, with little or no free oxygen.

The synthesis appears to have been activated by such energy sources as ultraviolet light from the sun, radiation and heat.

Substances vital to the life process, such as ATP (adenosine triphosphate) have been manufactured in this way. ATP delivers energy for movement in all living organisms.

However a number of the reactions along these suspected paths of chemical evolution were slow. Production rates were small and it was hard to see how significant amounts of the "life material" could have been synthesized.

Then it was found that certain substances that must have been present in the early days, such as ammonium cyanide, greatly accelerated the process.

A team at the Berkeley campus of the University of California that has long been working on this problem reported in yesterday's issue of the British journal Nature that the use of dicyanamide increased production rates of some long-chain molecules as much as tenfold.

Two weeks ago a group at Berkeley reported finding chemical evidence, in ancient rocks, that life existed on earth 2.7 billions years ago. This confirmed an earlier report of what seemed the patterns of primitive plants (algae) in limestone of comparable age.

Since the crust of the earth is believed to have formed some 4.7 billion years ago, these discoveries have led to the belief that life evolved within less than two billion years after the earth assumed its present form.

May 16, 1965

POISON HELD KEY TO ORIGIN OF LIFE

Hydrogen Cyanide Linked to Formation of Proteins

By HAROLD M. SCHMECK Jr.
Special to The New York Times

DURHAM, N. C., Oct. 20 — Hydrogen cyanide, one of the most potent of poisons, may have been the key to the origin of life on earth, according to a report here.

The same poison may, in the future, become an important sustainer of life, Dr. Clifford N. Matthews of the Monsanto Company suggested in a talk to the autumn meeting of the National Academy of Sciences which ended Wednesday at Duke University.

The scientist described chemical experiments that have led to a theory of the origin of life substantially different from that generally accepted today.

Synthetic Food Possible

The experiments offer at least the hope of developing efficient methods for making food materials synthetically, Dr. Matthews said in an interview. He emphasized that this latter idea had not yet been explored in detail but said it followed logically from the experiments on origin of life.

These experiments offer a view of the origin of life based on chemical events that seem fundamentally probable and likely to occur, he said.

In contrast, the other theories are based on some admittedly improbable events that are believed to have occurred simply because there were so many millions of years available in which they could happen. Even a most unlikely event could possibly happen a few times if the right conditions for it persist long enough, it was contended.

The theory reported here stems from experiments and calculations by Dr. Matthews and two colleagues, Dr. Robert M. Kliss and Dr. Robert E. Moser.

A Basic Premise

Like the other current theories of life's origin, this one starts with the premise that earth's early atmosphere was a noxious mixture of methane, ammonia and possibly water vapor that was laced with lightning flashes and bombarded by ultra-violet light and other radiations.

In the theory reported here, the effects of these energy sources on the gases was to produce huge quantities of hydrogen cyanide. From this gas larger more complicated molecules developed.

These molecules sank through the atmosphere to the early oceans where contact with water changed them into compounds very much like proteins or large pieces of protein, which are the key chemical ingredients of all living things.

Gradually over millions and perhaps billions of years the earth became covered with a scum of this material, Dr. Matthews suggested. In this layer were virtually all of the chemicals and basic substances that today make up living things.

The first living organisms were aggregations of these ingredients that somehow organized into self-reproducing units. These units in turn evolved into the first one-celled living creatures.

"From that scum we have emerged," said Dr. Matthews.

The experiment supporting this idea succeeded in producing in the laboratory substances that appear to be very much like large chunks of proteins.

This was done simply by subjecting hydrogen cyanide to an energy source—such as ultraviolet light—and then letting the resultant material react with water. By breaking down this laboratory-produced scum the scientists have found 15 of the roughly 20 smaller units, called amino acids, that are the ingredients of all natural proteins.

The ease with which these substances can be produced from such a simple compound as hydrogen cyanide suggests that some day synthetic proteins for food may be made the same way, Dr. Matthews said. This could be of some significance, he declared in the interview, in view of the possibility of a world food shortage within the next 10 to 20 years.

The other theories of the origin of life, also supported by experiments, postulate the existence of a dilute soup of amino acids in the earth's primordial oceans instead of the scum of more complex material that Dr. Matthews suggests.

Chemical theories to account for the origin of life on earth have been discussed since the 1920's. Laboratory experiments to test the ideas began about 15 years ago under the impetus of Dr. Harold Urey, then at University of Chicago.

Dr. Stanley Miller, a young colleague of Dr. Urey's, produced many of the amino acids in laboratory experiments in which the theorized conditions of an early earth atmosphere were reproduced. He and others have since made more complex substances through follow-up experiments.

October 21, 1966

Theories on Life Face Challenge

Chemistry of Ancient Rocks Said to Bar Accepted View

By WALTER SULLIVAN

A leading authority on the composition of the earth's rocks believes they have a story to tell about the origin of life that contradicts current theories on the subject.

At issue is the nature of the earth's early atmosphere.

A century ago Louis Pasteur persuaded the scientific world that life could not have arisen spontaneously on this planet because such a process was ruled out by the chemistry of our oxygen-rich air.

Pasteur pointed out that the amino acids and other substances that combine within a living cell to form the great molecules of life, such as proteins, could not survive in the open air.

Then, in the 1920's, J. B. S. Haldane of Britain proposed that, in its youth, the air of the earth was quite different from what it is today.

Oparin Theory of Life

Haldane noted that vast amounts of carbon were locked in the coal beds of the world. He reasoned that, before this carbon was converted to vegetable matter by plants, it must have been part of the air as carbon dioxide.

Like others who have sought to fathom how life began, Haldane recognized that the air must have contained the basic elements of all living matter: carbon, hydrogen, nitrogen and oxygen. Hence he proposed that the original air was composed of ammonia (which contains nitrogen) and water vapor, as well as carbon dioxide.

Today most scientists adhere to the theory advanced by Aleksander I. Oparin of the Soviet Union. His view was of an atmosphere of four components in which carbon was represented in the original air by methane (the lethal, odorless gas usually encountered in coal mines).

The new concept has been presented by Dr. Philip H. Abelson, a geochemist who is director of the Geophysical Laboratory at the Carnegie Institution of Washington.

Dr. Abelson believes the chemistry of ancient rocks is incompatible with an Oparin type of atmosphere. Instead, he has proposed that the early air, like that of today, contained nitrogen gas, plus hydrogen gas and carbon dioxide—but no methane or ammonia.

It is generally agreed that the earth, immediately after its formation, had no atmosphere. The gases of the air came from within the earth, probably through volcanic activity. This outgassing produced the water of the oceans and, according to Dr. Abelson, would envelop the earth in an atmosphere of nitrogen, hydrogen and carbon dioxide.

The chemical components in such an atmosphere, like that of the Haldane and Oparin versions, would be basically different from that of today's air, which contains oxygen. In the absence of oxygen there would be no rusting, no fire.

Theory by Abelson

Under these circumstances, any input of energy, such as that provided by ultraviolet light from the sun, would stimulate the synthesis of organic substances.

For more than a decade it has been known that such synthesis could occur in an Oparin type of atmosphere. Dr. Abelson has shown, however, that the same was true of a wide variety of gas mixtures.

His most recent work is described in the annual report of the Carnegie Institution, made public yesterday.

The injection of energy into an Abelson type of atmosphere by means of electric sparks synthesizes large amounts of hydrogen cyanide. This poisonous substance is composed of hydrogen, carbon and nitrogen. When mixed with water, it readily forms more complex organic substances under the influence of ultraviolet light.

In experimenting with this process, Dr. Abelson and his co-workers found that the products included such basic components of life's chemicals as glycine, serine, alanine, aspartic acid and glutamic acid. In some runs as much as 7 per cent of the carbon within the hydrogen cyanide came out in the form of glycine.

Dr. Abelson has also challenged the view that an ocean very rich in the building blocks of life was necessary to bring about the evolution of more complex substances. For a variety of reasons he believes that such a "thick soup" could not have existed.

Instead he proposes that the early ocean was a "thin soup" whose chemicals evolved step by step from the most simple organic compounds to those complex enough to participate in a primitive life process.

Many of those who have sought to reconstruct the history of life's origin have argued that a thick soup was needed to bring together the extraordinarily varied and complex molecules that take part in the life process. Dr. Abelson argues that the various components of such a soup would be broken down by radiation and other influences.

Furthermore, he believes a thick soup was necessary—that the chemistry life could evolve in steps without throwing together such a great variety of components.

THREE VIEWS OF COMPOSITION OF THE EARTH'S ORIGINAL ATMOSPHERE and its actual present composition

	Haldane	Oparin	Abelson	Present composition
CH_4 (Methane)		✓		
CO_2 (Carbon Dioxide)	✓		✓	
H_2 (Hydrogen gas)		✓	✓	
H_2O (Water)	✓	✓	✓	✓
N_2 (Nitrogen gas)			✓	✓
NH_3 (Ammonia)	✓	✓		
O_2 (Oxygen gas)				✓

December 20, 1966

Clues on Life's Beginning

By WALTER SULLIVAN

Almost two years ago, Dr. Charles C. Price, head of the American Chemical Society, proposed that this country set as "a national goal" the synthesis of a living organism. In an address to the society, he pointed out that sufficient steps in the evolution of life's chemistry have been reconstructed to make the remaining steps appear feasible.

Despite Dr. Price's call for action, present efforts in the field are still modest. The reason seems to be a suspicion that it will be extremely difficult to find out how the first living cell organized itself. We know now that even the simplest bacterium has a complex internal structure and lives by elaborate chemical processes.

Furthermore, it is known that this structure and these processes are controlled by inherited information, "tape recorded," so to speak, in the nucleic acids of the cell. The dilemma has called to mind the proverbial question: Which came first, chicken or egg? How could there be a living organism without nucleic acid to tell it how to live? And how could such nucleic acid evolve in the first place?

While the challenge of this problem has discouraged the less brave it has captured the attention of at least four recent winners of a Nobel Prize: Drs. Melvin Calvin, Joshua Lederberg, Fritz Lipmann and Edward L. Tatum. It has also led one researcher in the field to propose a solution, namely that the earliest pre-life forms were able to reproduce without the help of nucleic acids.

Possible Path

The researcher is Dr. Sidney W. Fox, head of the Institute of Molecular Evolution at the University of Miami. Dr. Fox is well known for his synthesis of "microspheres"—tiny spheres of protein-like material that, in his view, demonstrate a possible path whereby the first organisms could have evolved.

He has now shown that such microspheres can sprout buds. If these are knocked free, they can take on "food" from the surrounding fluid and grow to full microsphere size. The new microsphere can then grow its own sprouts. This, he argues, can be considered a primitive form of reproduction.

Once reproduction begins, the remarkably powerful process of evolution is under way. It is possible for such pre-life forms to "experiment" with various modes of operation—trials which could eventually have recruited simple nucleic acids as an aid in passing, to future generations, the lessons learned by the evolutionary process.

Dr. Fox has likened his microspheres to certain bacteria, notably Bacillus Cereus, which cause cream to turn bitter. He argues that they have spontaneously organized themselves internally, that they act, to some extent, like enzymes (proteins that are intermediaries in the chemical reactions of life), and that they can even move under certain chemical conditions.

His dramatic claims and his gift for popularizing his ideas have generated hostile reactions by various scientists. Yet some leaders in the field believe his proposals are highly significant. Among them is Dr. Aleksandr I. Oparin, father of modern research on the origin of life. Dr. Oparin, now 73, was interviewed recently in his Moscow laboratory.

First Big Break

The first great break in the origin-of-life inquiry came in the 1930's when Dr. Oparin proposed that the atmosphere of the young earth consisted of methane, ammonia, water vapor and hydrogen, with little or no free oxygen.

The Oparin hypothesis was tested at the University of Chicago in 1953 by Dr. Stanley L. Miller. For a week he fired electric sparks through a colorless mixture of the gases in Oparin's hypothetical atmosphere and found that a variety of organic compounds were synthesized, including building blocks of the sort that hook together to form the long and complex molecules of our bodies.

Dr. Fox has found that, by passing gases of an Oparin-type atmosphere through a hot tube containing silica and heating the products with water, a dozen or more of the amino acids found in protein are produced.

More remarkable is the fact that other forms of amino acid, not typical of protein, are almost completely absent.

This was a strong hint that the chemistry of life has arisen directly from non-living chemistry along paths built into the very fabric of non-living chemistry.

Dr. Fox and his colleagues use temperatures as high as 1,900 degrees Fahrenheit in their experiments, which has led some to doubt that his mode of synthesis could have occurred on a large scale. But he argues that volcanic activity was common on the young earth and that frequent rains could have furnished the water-washing effect that forms part of his process.

August 27, 1967

MILKY WAY YIELDS A CLUE ABOUT LIFE

Formaldehyde, a Chemical Building Block, Detected

By WALTER SULLIVAN

Radio astronomers have recorded evidence that in many parts of the Milky Way there are clouds of a material used as a laboratory preservative and thought, as well, to be a chemical precursor of life.

The substance is formaldehyde. The discovery reinforces the growing suspicion that evolution of the complex chemistry of life, as it exists on earth, began in space between the stars.

In recent months two other substances needed as chemicals for starting that evolution have been detected in distant space by similar means. They are water vapor and ammonia.

The formaldehyde observations were made with the 140-foot dish antenna of the National Radio Astronomy Observatory at Green Bank, W. Va. It was found that radio emissions passing through distant gas clouds on the way to earth were being absorbed at a wave length of roughly six centimeters, which is characteristic of formaldehyde.

The formaldehyde molecule consists of two hydrogen atoms together with an oxygen and a carbon atom. Recent experiments, seeking to find the paths by which simple gases may have evolved into the complex molecules essential to the life process, have suggested that a key role was played by formaldehyde. The other essential constituents are thought to have been methane, ammonia, water and hydrogen gas.

When stimulated by some form of energy input, these substances form such building blocks of life as alanine, one of the basic constituents of protein.

According to one participant in the formaldehyde discovery it may eventually be possible to look in the sky for the radio fingerprints of such amino acids, confirming that life's evolution began there, rather than on earth.

Such a discovery would strengthen the view that life has arisen on many other planets in the Milky Way.

The Green Bank observations were made by Drs. Lewis E. Snyder and David Buhl of that observatory, Benjamin Zuckerman of the University of Maryland and Patrick Palmer of the University of Chicago. Announcement was planned for a meeting of the American Astronomical Society late this month, but the finding was reported yesterday by a Boston newspaper.

March 21, 1969

Astronomers Detect a Substance That May Have Aided Evolution

By WALTER SULLIVAN

A substance widely considered to have been a key ingredient in the early evolution of life has been detected among the stars by a radio telescope atop Kitt Peak in Arizona.

It is hydrogen cyanide, a deadly poison to higher life forms. As a gas it is used, with special precautions, to fumigate pest-ridden ships and other confined areas.

However, current efforts to reconstruct the chemical paths whereby simple, starting substances evolved to the complexity of life have indicated that hydrogen cyanide was an abundant product of the early stages of the life-evolving process.

This process is thought to have begun with the synthesis of hydrogen cyanide and other more complex substances from the simplest compounds of hydrogen, carbon, nitrogen and oxygen.

Hydrogen cyanide is the second organic substance identified in radio emissions from space. Last year the National Radio Astronomy Observatory in Green Bank, W. Va., found evidence of formaldehyde. Both substances seem to be concentrated within the same well-defined clouds of dust and gas.

Recent observations have also disclosed the presence of water, carbon monoxide and cyanogen in some or all of these clouds, as well as ammonia in a cloud lying in the direction of the Milky Way core. All these substances are thought to have been associated, in one way or another, with the origin of life.

When the hydrogen cyanide findings were reported to a meeting of the American Astronomical Society in Boulder, Colo., last week, they evoked considerable interest.

It is becoming increasingly evident that the early stages of organic chemical evolution can take place in space, before a cloud of dust and gas draws together to form a new sun and orbiting planets.

Thus, some radio astronomers will be surprised if they detect evidence of amino acids —the building blocks of protein—in some of the condensing clouds.

They suspect now that when the earth formed in this manner, life's chemical evolution had already made considerable progress.

The new finding was discussed yesterday by Dr. David Buhl of the National Radio Astronomy Observatory, who was reached by phone at the observatory's field station on Kitt Peak.

The discovery was made with a radio dish, 36 feet in diameter, on that summit, by Dr. Buhl and Dr. Lewis E. Snyder of the University of Virginia.

The two astronomers, who participated earlier in the discovery of formaldehyde in distant space, have suggested that in some of the dust clouds responsible for these emissions —particularly those dense enough to shield the inner material from the destructive radiation of nearby stars, "primitive life forms may exist."

Hydrogen cyanide molecules are formed of single atoms of hydrogen, carbon and nitrogen. Formaldehyde is a mating of two hydrogen atoms with single atoms of carbon and oxygen. Cyanogen is a pairing of carbon and nitrogen atoms. Also found in many parts of the sky is hydroxyl—a pairing of hydrogen and oxygen atoms.

The observations are possible because all of these substances either absorb or emit radio waves at wave lengths characteristic of each. Thus, they can be identified from a distance, just as the chemical composition of stars can be deduced from the spectra of their light.

As noted by Dr. Buhl, other molecules associated with the genesis of life's evolution, such as methane and carbon dioxide, would be difficult to detect because they do not have easily identified "signatures" in the radio spectrum.

A striking feature of the evidence, he said, is the remarkably great abundance of large molecules, such as formaldehyde and hydrogen cyanide, relative to the simpler ones. "Something strange seems to be going on," he said.

One of the most startling discoveries has been that some of the molecules, such as those of water vapor and hydroxyl, emit the intense radiation characteristic of masers. The maser — precursor of the laser—is a device in which atoms or molecules of a certain kind are "pumped" to a higher energy state and then dump the energy in the form of intense emissions at some characteristic wave length.

It appears that nature performs what had been considered only a laboratory trick of intelligent man. However masering does not seem involved in emissions from the larger molecules.

Each of the emitting clouds has a characteristic motion, relative to the earth, and this is reflected in slight modification of the observed wave lengths.

It is evident that the hydrogen cyanide emissions are coming from the same clouds as those of formaldehyde, because emissions from the two substances observed at any one spot in the sky both exhibit this same modification.

June 17, 1970

Meteorite Hints Space Life Is Possible

By RICHARD D. LYONS
Special to The New York Times

WASHINGTON, Dec. 1— The freshest and strongest clues to the existence of conditions suitable for the possible evolution of life beyond the earth were reported today by a team of American scientists.

The group announced the recovery from the constituents of a meteorite that fell on Australia last year of 17 amino acids, including six that are normally found in living cells and are the precursors of life.

The discovery strongly supports the theory not only that life on earth grew from a primordial ooze of chemicals present after the crust solidified, but also that if it happened here it could have occurred elsewhere in the universe.

In the last several years, radio astronomers using increasingly more sophisticated equipment have detected such complex organic chemicals as methyl alcohol, formic acid and cyanoacetylene in the outer regions of the universe.

NASA
Dr. Cyril Ponnamperuma, report's principal author.

These chemicals are the building blocks of the far more complicated amino acids, whose existence in meteorites has previously been suspected but not proved because it was believed that they resulted from contamination of the specimens that were examined.

Dr. Cyril Ponnamperuma, the principal author of today's report, said the variety, the optical properties and the isotope content of the amino acids found in the so-called Murchison meteorite proved they were not of earthly origin.

Dr. Ponnamperuma, who was born in Ceylon, is chief of the chemical evolution branch in the Exobiology Division of the National Aeronautics and Space Administration's Ames Research Center at Mountain View, Calif.

The amino acid report, part of which will appear in the Dec. 5 issue of the British scientific journal Nature, also incorporates the work of Dr. Ian R. Kaplan of the University of California, Los Angeles, and Dr. Carleton Moore, director of the Center for Meteorite Studies at Arizona State University.

"No one really believed the previous reports that amino acids had been found in meteorites because you only have to make a thumbprint on a beaker and shake with water to obtain amino acids," Dr. Ponnamperuma said in a telephone interview. "We have a built-in way of showing that the amino acids that we have recovered are not earthly contaminants."

Amino acids, 21 of which compose the proteins that are needed for maintaining life, come in two forms, D and L, which are chemically indentical but have different optical properties.

When a beam of light is shined through the D form, its crystalline structure rotates the ray to the right. This is known as dextrorotation, hence the D.

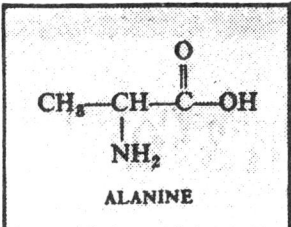

The New York Times Dec. 2, 1970
Complexity of amino acids in meteorite is shown by formula for one of them, alanine. Letters represent atoms of carbon, hydrogen, oxygen and nitrogen; numerals indicate more than one atom. Lines are single and double electric bonds.

247

The other form, the L form, for levorotation, means it turns light to the left.

Almost all amino acids found on earth are the L form, which is considered to be a criterion for life, although a few of the D form have been detected in bacteria and antibiotics. But D forms are relatively rare.

"The most significant finding is that these amino acids appear in both D and L forms n almost equal amounts," Dr. Ponnamperuma said, adding, "If the specimens had been contaminated by terrestrial amino acids they would almost certainly have been of the L form, which is not the case."

The amino acids found in the meteorite include glycine, the one with the simplest chemical structure, as well as, in increasing degrees of complexity, alanine, valine, aspartic acid, glutamic acid and proline. Eleven other amino acids that are not connected with the building of proteins, such as methylalomine and sarcosine, also were recovered.

Dr. Ponnamperuma noted that the types of amino acids found in the meteorite, together with their special optical properties, are similar to artificial amino acids that were created in laboratories during experiments in the nineteen-fifties and sixties.

These experiments, some of which were set up by Dr. Stanley Miller of the University of Chicago and Dr. Sidney Fox of the University of Miami, were performed to support the theory that amino acids could be created under the physical and chemical conditions that were believed to exist at the time of the formation of the earth and the solar system, or shortly thereafter.

Carbon From Asteroids

The meteorite that fell near the town of Murchison in Australia on Sept. 28, 1969, is of a type containing a large percentage of carbon that is believed to have stemmed from the belt of asteroids circling the sun between Mars and Jupiter.

The asteroids — several thousand are large enough to have been discerned by astronomers — are believed to date from the formation of the solar system, 4.5 billion years ago. The Murchison meteorite was found, from radioactivity decay rates, to be 4.5 billion years old.

This, Dr. Ponnamperuma said, would tend to indicate that the classes of chemicals necessary for the creation of life existed when the solar system was created, thus furthering the theory of the chemical evolution of life.

According to classical theory, the amino acids eventually formed themselves into proteins. When nucleic acids, such as deoxyribonucleic acid (DNA), interacted with the proteins, life was created.

Dr. Ponnamperuma said "radioastronomers have discovered molecules that are significant for the very early stages of chemical evolution, and we are now seeing the various stages of the processes that could have led to the creation of life."

"The picture seems to be emerging of a chemical evolutionary process that at some point under the right conditions leads to biological evolution," he added.

Early Atmosphere Theorized

The theory is that the early atmosphere of the earth contained a mixture of water vapor, methane, ammonia and hydrogen. When this blend of materials was acted upon by other chemicals, ultraviolet light, radioactivity, heat or some other form of energy, amino acids were formed, it is theorized.

This is exactly what happened in the laboratory experiments using a 60,000-volt high-frequency spark. Glycine, alanine and other amino acids resulted, and the acids were of both the D and the L forms.

Another indication that the amino acids are of extraterrestrial origin, Dr. Ponnamperuma said, is that their carbon isotopes differ from those normally found in similar chemical compounds known to have been created on earth. Carbon normally has an atomic weight of 12, but a small amount of the isotopes carbon 13 and 14 may be found in terrestrial materials.

But Dr. Ponnamperuma said the amino acids found in the meteorite contained about twice as much of the carbon 13 isotope as terrestrial materials.

"The variety of amino acids, their optical properties and their carbon 13 content indicate that this is the first case of well-defined evidence of extraterrestrial chemical evolution," Dr. Ponnamperuma said.

December 2, 1970

Chemicals Basic to Life Found Plentiful in Space

By WALTER SULLIVAN

Peering into the darkness between stars, radio astronomers have found that, instead of a void, there is a chemical ferment that is apparently the initial stage in the evolution of stars, planets and, ultimately, of life.

Recent discoveries, including one announced this month, bring to 23 the number of molecules that have been detected in the far reaches of the Milky Way.

Many are precursors to the chemistry of life, or biochemistry. Almost all are known as "organic" compounds because it was long thought that they could be produced only by living organisms.

A Chemical 'Trick'

While it was later shown that they could be synthesized, the fact that the simpler ones, at least, have formed in the deep vacuum of space has taken scientists by surprise.

According to Dr. Philip M. Solomon of the University of Minnesota, "The question of how far chemical evolution in the interstellar medium has proceeded toward biochemistry is not yet answered, but it has clearly gone much further than anyone would have estimated five years ago."

In the current issue of the journal Science it is suggested that a key factor in this evolution, during the final stages in formation of a new star and planetary system, is a chemical "trick" used in Germany during World War II to synthesize gasoline.

This is proposed by Drs. Edward Anders and Ryoichi Hayatsu at the University of Chicago and Dr. Martin H. Snider of the Argonne National Laboratory, near Chicago.

Found in Meteorites

In this process, known, for those who invented it in 1923, as the Fischer-Tropsch synthesis, intermediate substances, or catalysts, facilitate the assemblage of complex molecules.

The hypothesis has been advanced to account for the extraordinary variety of organic substances found in certain meteorites, known as carbonaceous chondrites. For example, fatty acids built on from 12 to 20 carbon atoms have been found in seven such meteorites.

These objects from the sky have yielded amino acids from which proteins are constructed, as well as purines and pyrimidines, substances of the type that form such key chemicals of life as DNA or deoxyribonucleic acid. Of 61 different kinds of hydrocarbons in the Murchison meteorite, 42 were synthesized in the laboratory by the Fischer-Tropsch method.

It is suggested that in the formative stage of the solar system such substances as magnetite and hydrated silicates served as the catalysts. The process is proposed as an alternative to the one suggested in the nineteen-fifties by Dr. Stanley L. Miller. It consisted of a straightforward synthesis using four gaseous starting substances — water, methane, ammonia and hydrogen—under stimulation by an energy source, such as electric sparks.

Among the telltale radio emissions detected from far out in the galaxy by radio astronomers are those of formaldehyde (precursor to the sugars), hydrocyanic acid (which can develop into the purines) and cyanoacetylene (precursor to the pyramidines). Thus, even in distant space, there is a start toward the evolution of the purine and pyramidine constituents of DNA.

The chemistry that is being observed with radio telescopes is unfamiliar to earthbound scientists, for it occurs under circumstances utterly remote from those that can be achieved in the laboratory. Both pressure and temperature are extremely low—the pressure being that of an almost total vacuum.

How Synthesis Occurs

Formation of the observed molecules cannot depend on random collisions between atoms, according to those who have analyzed the likelihood of such encounters. Even the entire age of the universe would not provide enough time to account for synthesis of the material in this manner.

Hence it is widely suspected

that the synthesis occurs on dust grains that collect the atoms and enable them to fuse. It appears that carbon is essential in this fusing, as it is in producing the molecules of life, such as the carbohydrates, which form a major part of the human diet.

According to Dr. Arno A. Penzias of the Bell Telephone Laboratories, carbon "seems to play a disproportionately large role, even as it does in the things we have for lunch." He and his colleagues have found that carbon monoxide is a thousand times more abundant in space between the stars than any molecule except that of hydrogen gas.

It has been proposed by Dr. David Buhl of the National Radio Astronomy Observatory that development of planets, stars and the chemistry of life begins in these dark clouds of dust, gas and evolving molecules. As the grains capture more molecules, they become heavy enough to exert a significant gravitational effect upon one another, like tiny planets.

This gently begins drawing them together to form larger and larger accretions of matter, forming a spinning cloud, or nebula, that finally condenses into a central star and planets. Surviving this process is enough of the organic matter to help in the evolution of life on one or more of the planets.

"The entire process," Dr. Buhl wrote earlier this year in the journal Sky and Telescope, "seems to be more than coincidental and suggests that the condensation of a star, the accumulation of the dust and molecules into planets and atmospheres, and even the subsequent evolution of life may all be part of an astronomical evolutionary cycle of very long time scale."

The headquarters of the observatory to which he is attached are at Charlottesville, Va., but its antennas are at Green Bank, W. Va., and atop Kitt Peak in Arizona. Their most recent achievement, reported this month, has been detection in space of the simplest—but highly elusive—organic molecule, the methylodine radical.

It is formed by the mating of single carbon and hydrogen atoms and is so reactive that, in the laboratory, it combines with something else and loses its identity before it can be studied. Yet its structure and its energy states, or quantum levels, are so fundamental to organic chemistry that its study in space is expected to yield important results.

Its changes in energy state account for its radio emissions.

Because of the difficulty of working with it in the laboratory, no precise predictions of the emitted wave lengths were possible (although its absorptions at optical wave lengths had been seen).

It is suspected by Dr. Anders and his coworkers that the carbon-rich meteorites are remnants of an intermediate region of the nebula from which the sun and planets were formed. The meteorites, they say, contain six "cosmothermometers" —constituents all of which indicate a temperature, during their formation, of about 160 degrees Fahrenheit.

It has also been determined that the pressure in the region where this material formed was some four million times lower than that of air on the earth's surface.

November 23, 1973

Experiments Indicate Chemical Reactions From Thunder or Sea Waves May Have Led to Origin of Life

By HAROLD M. SCHMECK Jr.
Special to The New York Times

WASHINGTON, April 21— Thunder and ocean waves may have helped set the stage for the emergence of life more than three and one-half billion years ago.

Chemists studying the origin of life reported today on recent experiments showing that the formation and collapse of bubbles in the clouds or oceans may have helped produce some of the huge stores of chemical building blocks from which life presumably emerged.

Cavitation, the formation and collapse of bubbles under extreme conditions of sound or turbulence, could have served as an energy source to make the simple chemicals, assumed to have been present, react to form more complex compounds.

Basic Building Blocks

The experiments, at the University of Maryland's Laboratory of Chemical Evolution, showed that cavitation in fluids containing the right simple chemicals could give rise to amino acids, the basic chemical building blocks of all living things. In related experiments, the research team has shown that the reactions likely to result from a comet head's explosive impact on earth might also have added to the store of chemicals needed for life to start.

In what were probably the most important in the series of recent experiments, the team used radiation to bombard a test-tube model of earth's early atmosphere and produced almost all the ingredients of a complex process vital to most forms of life.

The process, a series of chemical transformations called the Krebs cycle, is a vital part of the scheme of things by which living cells use the food available to them.

Dr. Cyril Ponnamperuma, professor of chemistry and director of the laboratory, said that the new evidence produced by his research team strengthened the view that life's emergence on earth might have been almost inevitable when the conditions for it became right. He spoke at a news briefing held by the American Chemical Society here.

He said that chemical evolution from the nonliving to the living might have taken place once on Mars and might perhaps be in progress now on Jupiter or one of its moons.

If life were ever extinguished on earth, however, he thinks it unlikely that it would reappear. The chemical conditions for such genesis are gone.

Many scientists believe life arose spontaneously in earth's early seas and estuaries using chemical ingredients from the atmosphere. The process is thought to have taken hundreds of millions of years, culminating about three and one-half billion years ago in the emergence of the first living organisms.

The planet's original atmosphere was a poisonous mixture of methane, ammonia and water vapor, according to specialists in chemical evolution. During hundreds of millions of years this atmosphere was presumably bombarded by ultraviolet light, X-rays and other radiations from the sun, and by lightning.

Thus cooked by electric discharges, heat and radiations, simple chemicals in the atmosphere were transformed into more complex forms that rained continuously into the oceans, creating a thin oceanic soup of those chemicals that are now called organic.

Many scientists believe that life emerged about three and one-half billion years ago, using as building blocks the chemicals in the water. For more than a decade, the theory has been tested in many laboratories by experiments in which the gases thought to be constituents of the early atmosphere were bombarded by energy sources such as X-rays, electric discharges and ultraviolet radiation.

The experiments have proved that the amino acids can be easily produced in this fashion, although previously it had been thought that they could arise naturally only from living things. As the experiments have continued, more and more of the necessary ingredients of living things have been produced in this way from simple chemicals. More sources of energy have been found capable of making the transformations.

Amino acids have also been found in meteorites from outer space.

Dr. Ponnamperuma has been one of the pioneers in this type of research, which has as its objective the understanding of the evolution of life and the possibilities of its existence elsewhere in the universe.

Up to now, he said, scientists have been looking mainly at individual building blocks of life, but the experiments concerning the Krebs cycle go significantly beyond that.

"Here we see practically an entire sequence," Dr. Ponnamperuma said at the news briefing.

The scientist said that his group's experiments showed that life did not have to invent all of the complex cycles of chemical transformation on which it now depends, but may have simply picked them up from natural nonliving processes already taking place. He said there were hints in the new results that parts of the actual sequence of events in the cycle might be taking place in the experiments.

Four reports by Dr. Ponnamperuma, Alicia Negron-Mendoza, Akira Shimoyama, Woo K. Park and Susan Burke are to be presented Wednesday at a regional meeting of the chemical society in Wilkes-Barre, Pa.

The Krebs cycle is a series of chemical transformations starting with acetic acid and ending up with complete burning of that acid to liberate energy. It was named after Sir Hans A. Krebs, a British chemist who first explained its complete role in living things in 1937.

April 22, 1974

GENETICS

ALTERED HEREDITY OF FLIES BY X-RAY

Dr. Muller, Reporting Experiments, Suggests Cosmic Rays May Shift Course of Nature.

BRAIN STUDY BAFFLING

Dr. Donaldson Tells Academy of Sciences at Washington of Examining That of Osler.

From a Staff Correspondent of The New York Times.

WASHINGTON, April 24.—A suggestion that the powerful cosmic rays, sometimes called the Millikan rays, and the rays from radioactive substances have played some part in bringing about revolutionary changes was made today to the National Academy of Sciences by Dr. H. J. Muller of the University of Texas.

His experiments showed heredity to be profoundly altered by the X-ray under certain conditions. What he accomplished in the laboratory by means of the X-ray might be caused in nature by the cosmic rays, which come from distant space, or by the radium rays which are everywhere present in some degree on the earth's surface.

For the purpose of his studies Dr. Muller used fruit flies whose pedigrees and hereditary characteristics were known for many generations back. Subjecting them to severe X-ray treatment he found that a great variety of changes appeared in their offspring—in the shape and color of their eyes, in the size and shape of their wings, in their hair antennae and in their general structure.

The X-rays were discovered to have directed effects on the germ cells within these germ cells. Within these germ cells are the sets of chromosomes or sets of small colored bodies, which contain the "genes" or the hereditary characteristics which are later to develop.

X-Ray Upsets Nature's Plans.

These "genes" which bear the same relation to the creature that plans and specifications do to a building, are arranged in a regular order, as in a filing system, in the chromosomes. Dr. Muller found that the X-rays sometimes broke the chromosomes in two; that the broken section wandered away and joined another chromosome. This disarranged nature's plans and specifications for the building of the individual. The specimens, which had their constitutions disordered before birth by the X-rays, grew up with a great variety of abnormalities.

The place which each hereditary unit or "gene" occupied in the order of arrangement had been surmised on previous evidence. The X-ray experiments confirmed in detail the previously worked out map or index of the hereditary units. They indicated also that each "gene" was a single big molecule.

In other words, each one of a series of large and complicated molecules contained one part of the personality or individuality which was to be unfolded later as the creature grew up. The X-rays thus caused a shuffling up of the preliminary arrangements for the future organism. Dr. Muller reported that others had found that radium rays produced a similar scrambling of the hereditary units.

Dr. Muller tried lead, arsenic and other poisons, in amounts less than fatal doses, to discover whether these would result in any rearrangements of the "genes" and in the production of abnormal offspring. Continuous poisoning, however, failed to achieve such a change. Dr. Muller reported that the cosmic rays and the radium rays were, as far as he knew, the only things occurring in nature which might disrupt the chromosomes and thus disturb the plan fixed by nature for the individual.

Says Brain Remains Mystery.

Some of the changes brought about by the X-ray remained fixed, that is, they occurred in generation after generation. The exact location of ninety-three of the heredity-carrying molecules has been found, according to Dr. Muller.

April 25, 1928

EVOLUTION DECLARED CHEMICAL PROCESS

Mechanistic Action in Cells Fashions Us, Dr. T. H. Morgan Tells Geneticists.

ITS CONTROL BEING STUDIED

X-Ray Is Peering Into Puzzle of Genes, Which Are Said to Sway Characteristics.

NEW SEX THEORY GIVEN

Ithaca Congress Head Holds No Single Gene but All Share in Determining It.

By WILLIAM L. LAURENCE.
Special to The New York Times.

ITHACA, N. Y., Aug. 25.—Evolution, of man, animal and plant, is entirely the result of the mechanistic workings of physical and chemical laws, determined by the inexorable working of the law of cause and effect, the International Congress of Genetics was told tonight by Dr. Thomas Hunt Morgan of the California Institute of Technology, president of the congress, former president of the American Association for the Advancement of Science, and one of the world's most eminent geneticists.

Dr. Morgan delivered the president's address on "The Rise of Genetics," in which he traced the development of man's knowledge of hereditary transmission of characteristics from its earliest beginnings with the discoveries of the sex of plants by Camerarius, in 1694, and Linnaeus, in 1760, through Mendel, whose original paper containing his famous Mendelian laws of inheritance, delivered first in 1865, had been forgotten for thirty-five years, and finally the rise of modern genetics during the last thirty years since the rediscovery of Mendel's paper by Devries.

Dr. Morgan's reaffirmation of the mechanistic, deterministic principle as applied to the evolution of man was regarded by the many eminent scientists present, who have come from many parts of the world to learn first hand of the progress made in the study of heredity of plants, animals and man, as a direct challenge to the tendency of modern scientists in the fields of physics and genetics to introduce a non-deterministic, mystical element in the workings of nature, notably Sir Arthur Eddington and Sir James Jeans in England, former South African Premier General Smuts and Dr. Henry Fairfield Osborn, president of the American Museum of Natural History.

Exact Laws Still Undetermined.

Genetics still has a long way to go, Dr. Morgan stated, before it will have learned the exact mechanistic, physicochemical laws that are responsible for what is known as the mutation of species; however, he added, we do know that these mutations are due to changes in the living cells, and that these changes are, in turn, due to physical, chemical laws.

Dr. Morgan reviewed the growth of man's knowledge of heredity through the study of the chromosomes, those strange rod-like entities in the cells, and the genes, those invisible, mysterious entities possessed by the chromosomes, which, like electrons and protons, have never yet been seen under any microscope but which are declared to be masters of man's heredity, the prime determinants of his characteristics long before he is born.

Modern genetics has discarded its original ideas about what determines sex, Dr. Morgan stated.

At the beginning, when chromosomes were first discovered, he stated, it was thought that a special chromosome controlled sex. Later it was believed that the sex-chromosome was merely the carrier of a special gene which controlled sex. Now the prevailing belief is that there is no single chromosome or gene controlling sex, but that either all, or at least many, genes enter into the determination of sex.

"The rapid expansion of genetics after 1900," Dr. Morgan said, "has been intimately connected with the applications of the chromosome theory to the experimental work in genetics. The integrity of the chromosomes and their continuity from one cell-generation to the next, the constancy in number of the chromosomes in each species, and the absence of mixing of the materials of the conjugating chromosomes at the time of meiosis, have furnished the basis on which genetics rests.

"I think we cannot overemphasize the significance of this relation between the theoretical side of genetics and the factual side, as observed in the known behavior of the material basis of heredity. To put the matter bluntly, the recognition that there is a mechanism to which genetic theory must conform, if it is to be productive, serves to keep us on the right track and acts as a check to irresponsible speculation, however attractive it might seem in print.

Theory of Genes Revised.

"Some one may reply it is not always an advantage to keep one's nose to the grindstone. Granted; but realizing how often ingenious speculation in the complex biological world has led nowhere, and how often the real advances in biology, as well as in chemistry, physics and astronomy, have kept within the bounds of mechanistic interpretation, we geneticists should rejoice, even with our noses to the grindstone (which means both eyes on the objectives), that we have at command an additional means of testing whatever original ideas pop into our heads."

When the existence of the gene was first discovered, Dr. Morgan stated, the belief prevailed that each gene was a particular unit of heredity, each one controlling a specific hereditary characteristic. This theory no longer holds in modern genetics.

"I need not labor the point more at this late date," he said, "that the characters of the individual are the product, both of its genetic make-up and its environment. The earlier, premature idea, that for each character there is a specific gene—the so-called unit-character, was never a cardinal doctrine of genetics, although some of the earlier popularizers of the new theory were certainly guilty of giving this impression. The opposite extreme statement, namely, that every character is the product of all of the genes, may also have its limitations, but is undoubtedly more nearly in accord with our conception of the relation of genes and characters.

Environment a Differential.

"A more accurate statement would be that the gene acts as a differential, turning the balance in a given direction, affecting certain characters more conspicuously than others. But let us not forget that the environment may also act as a differential, intensifying or diminishing, as the case may be, the action of the genes.

"The best illustration of this double relation is seen in the determination of sex. When an unpaired chromosome is present, in one or in the other sex, its genes determine, as a rule, whether a male or a female develops from each egg. Under environmental conditions, which, as we say, are normal, the differential acts almost perfectly; but under other unusual conditions and in a few special cases its power may be partially overcome, and even a reversal may take place.

"These unusual environmental conditions may be external agents, such as temperature or light. They may also be internal factors, such as hormones. Even 'age' itself may bring about a reversal of sex in certain types.

Only characters, or characteristics, that are inherited, Dr. Morgan continued, can take part in the process of evolution. The only characters that we know are inherited are those that arise as "mutants," by a change in a gene. Genetics has found out that all differences, normal and abnormal alike, follow the same laws of heredity. It is only the "old timers" who still cling to the idea that changes in species take place without regard to the laws of heredity, by a change in environment or by mere chance.

Progress by X-Ray Study.

We have not yet found out, Dr. Morgan stated, how to change any particular genes in any particular way, so as to be able to control heredity at will, but considerable progress has been made even in this direction. By work with X-rays and heat it has been possible to produce "mutants" that come up naturally without treatment, and new "mutants," similar to those which also appear spontaneously.

"There remains still the question of the casual origin of mutations," he said. "Here also some progress has been made, but the subject is admittedly by no means on the same footing as is our knowledge of the laws of inheritance. It behooves us then to be careful, for our progress in this respect has been slow and to some extent erratic.

"Even here, however, something has been done. In the work with X-rays and heat the same mutants appear that are already known, and that have come up without treatment. In addition, new mutants appear, as they do also without treatment. If it can be shown on a large scale that the same ratio for known mutations holds for X-ray and for spontaneous mutations, we may have found an opening for the further study of the causes of certain types of mutation. * * *

"Any change in a gene is almost certain to make some kind of change in the complex of physiological processes that lead to the development of the character of an individual, and a deficiency of effect would be, as a rule, the expected kind of result. * * *

"There still remains for consideration the theoretical conception that increasing complexity of structure in evolution means a corresponding complexity in genic composition.

Lists Problems for Geneticists.

"I have been challenged recently to state on this occasion what seemed to be the most important problems for genetics in the immediate future.
* * *
"First, the physical and physiological processes involved in the growth of genes and their duplication * * *. The ability of the new genes to retain the property of duplication is the background of all genetic theory. * * *

"Second, an interpretation in physical terms of the changes that take place during and after the conjugation of the chromosomes.

"Third, the relation of genes to characters. This is the explicit realization of the implicit power of the genes, and includes the physiological action of the gene on the rest of the cell. This is the gap in our knowledge.

"Fourth, the nature of the mutation process—perhaps I may say the chemico-physical changes involved when a gene changes to a new one.

"Fifth—The application of genetics to horticulture and to animal husbandry."

One of the most interesting and far-reaching recent discoveries in genetics, described by Dr. Morgan, is that not only do the genes influence the cells in which they are located but that they exert an influence outside the cells.

One of these extra-cellular activities of the genes results in the production of hormones, those tremendously important substances secreted by the endocrine, or ductless glands, such as the adrenal gland, the thyroid and the pituitary. Among some of the important hormones recently discovered are insulin, which has saved the lives of hundreds of thousands of diabetes sufferers; adrenalin, cortin, thyroxin, pituitrin, all of which have proved of large importance in saving lives and preserving health.

According to recent discoveries, Dr. Morgan said these hormones are an end-product of the genes. However, he added, it is not yet known whether the genes produce the hormone directly or after many intermediate steps.

August 26, 1932

SCIENTISTS TRACE HEREDITY SEEDS

Key to Long Sought Mystery of Evolution Is Credited to Carnegie Institution.

GIANT CHROMOSOME FOUND

Cold Spring Harbor Station Discovery Called Most Significant in Decade.

By The Associated Press.
COLD SPRING HARBOR, N. Y., Sept. 16.—The living equivalent for heredity of the Rosetta Stone is on demonstration here at the Station for Experimental Evolution.

Discovery of this heredity key, which promises to explain the mysteries of evolution much as the Rosetta Stone did the Egyptian hieroglyphs, was described by Dr. Calvin B. Bridges of the Carnegie Institution of Washington. He has been an associate of Dr. Thomas Hunt Morgan throughout the researches on heredity which recently won Dr. Morgan the Nobel Prize.

Among geneticists and cytologists the results reported on here are said to be one of the most significant steps in the past decade in advancement of the knowledge concerning the relation between hereditary units, the "genes," and the structure of the "chromosomes," on which they are strung in long rows.

Genes are so small that until the finding of this "Rosetta Stone" one has never been seen clearly. A few scientists have thought they discerned evidence of individual genes. Although this was in doubt, the existence of the mysterious things was clearly demonstrated.

They control development of such bodily characteristics as color of hair or eyes in human beings, or a mental trait or the predisposition to a certain disease.

Identify Separate Genes.

The modern "Rosetta Stone" is a giant chromosome, so large that the spots occupied by the separate genes along its length are thought to be identifiable. Actually a number of these spots have been proved to be separate genes, and it is expected that the whole gene "necklace" of heredity will be found the same as these first few.

The previously known ordinary chromosomes of the yeast fly with which this work was done are extraordinarily small. The largest is only fifteen hundred-thousandths of an inch long. It is only a dot under an excellent microscope. Yet it is composed of long strings of genes.

The giant chromosome is seventy times larger than ordinary. It is found in the cells of the yeast fly's salivary glands. Fully fifty years ago scientists knew that the salivary cells were different, with some parts greatly enlarged. But they were not examined critically until about four years ago by Heitz in Germany.

Last year Dr. Painter at the University of Texas studied the structure of these giant chromosomes. Upon them, somewhat like stripes around a stick of candy, were "cross striations." These he informed the scientific world appear to correspond with the positions of individual genes which had been mapped by Dr. Bridges and other fly workers.

Dot Contains 28 Bands.

Following Dr. Painter's announcement of pioneer work, an intensive study was begun by Dr. Bridges. This study showed that even the finest details of the giant chromosomes were constant and dependable.

Dr. Bridges has found and shown on accurate pictures twenty-eight distinct bands on the part of the "Rosetta Stone," which duplicates the tiniest of all the fly chromosomes. Hitherto that chromosome has never been visible as anything except a dot. In addition to the gene bands clearly seen are six others partly obscured by surrounding material.

Among the "genes" so identified to date are those which cause a black speck under the fly's wings, blistered wings and balloon wings. Another is "lethal," because it will if active cause a fly to suffocate in the larval stage.

These particular genes were previously known. They are among about 300 shown to exist by almost incredibly difficult work carried on by about 200 scientists in all parts of the world who are devoting their time to this particular problem.

Nature Provides Reserve.

The 300 genes are only a beginning, for this one small fly apparently possesses 2,000 to 3,000 individual genes. A full set of them seemingly is carried by every cell of the fly's body, and the giant "Rosetta" chromosome is proving itself to be a big-scale model of each and every one of the others.

Heretofore genes have been likened to beads on a string. Dr. Bridges's studies show that the giant chromosomes are really sixteen to thirty-two slender strands twisted together to form a cable. The cross bands are made of dots, each dot possibly a gene.

At present it appears possible that each dot in a band is a duplicate gene, indicating the pains nature has taken to have a reserve supply of each type of gene.

Hitherto these cross bands have been faintly discerned on the ordinary chromosomes, but their nature has been a puzzle.

September 17, 1934

Chemical Core of Life Is Isolated In Nucleoprotein, Kernel of Cell

By WILLIAM L. LAURENCE
Special to THE NEW YORK TIMES.

DALLAS, Texas, Dec. 31 — Drawn from various animal organs, the chemical core of living matter, the main constituent of the living cells serving as the building blocks of the animate world from amoeba to man, has been isolated in pure form for the first time.

The report of this advance in biology was made today to the Genetics Society of America at the annual meeting of the American Association for the Advancement of Science.

It was presented by Professor Arthur W. Pollister of Columbia University and Dr. Arthur E. Mirsky of the Hospital of the Rockefeller Institute for Medical Research, New York City.

The cell, basic unit of all things living, is composed of a nucleus, a tiny bit of protoplasm serving as its "yolk," surrounded by the cytoplasm, which may be compared with the white of an egg.

It has been known for more than seventy years that the main chemical constituent of the nuclei is a compound known as nucleoprotein, in which a protein is combined with phosphorus-containing nucleic acid.

In earlier demonstrations of the presence of nucleoprotein in spermatozoa it seemed certain that the nucleoprotein must have been derived from the nucleus and not from the cytoplasm. The lymphocytes (variety of white blood corpuscles) of thymus glands and lymph nodes are also mainly composed of nucleus, and from these cells nucleoproteins have been isolated in considerable amount.

The Pollister-Mirsky report stated that studies with ultraviolet light and a chemical test, the Feulgen nuclear reaction, had shown nucleoproteins to constitute nearly all of the chromatin, the more stainable portion of the cell nucleus that contains the heredity carriers, the chromosomes and the genes.

It thus becomes of prime importance for an understanding of the action of the genes to compare the nucleoproteins from different tissues and at different stages of development from any one tissue.

Indicating that the nucleoproteins might be greatly variable, especially in the protein component, the report cited considerable differences between those from spermatozoa and the thymus cells.

It was added that these differences, in turn, might account for the differences in heredity, in susceptibility and immunity to disease, the differences between the various tissue cells, and also those between normal and cancer cells.

Various difficulties, however, have hitherto prevented satisfactory extension of the study of nucleoproteins to a wide variety of tissues.

The methods of extraction that produced a good yield of nucleoprotein from sperm cells and lymphocytes provided a much smaller yield, in many cases practically no yield, from other tissues, much less than would be expected from the relative size of the nucleus.

Derivation From Nuclei

"Until we can study a yield that is large enough to indicate that a considerable part of the nucleoprotein has been removed from the nucleus," Drs. Pollister and Mirsky stated, "it will be impossible to know whether or not there are significant differences in the chemical content of the nuclei of the different tissues.

"Another difficulty has been that in other than sperm cells and lymphocytes there is much more cytoplasm than nucleus and the proof that the nucleoproteins isolated from these cells come from the nucleus has rested upon similarity to substances isolated from sperm cells and lymphocytes.

"Another difficulty of getting a high yield has been that some of the previous methods involved such drastic treatment of the tissues, such as drying, digestion with pepsin or prolonged maceration in water.

"In our researches we have used a simple method of extraction that gives a substantial yield of nucleoprotein, not only from sperm and thymus cells, but also from many other vertebrate organs, e. g., liver, pancreas, kidney and spleen.

"Furthermore, we have been able for the first time to make a direct cytological (cell study) demonstration that the nucleoprotein is derived from the nuclei.

"We feel that this simple and relatively much less drastic method of extraction of nucleoprotein from the nucleus is likely to open up a wide field of investigations of fundamental importance to an understanding of the role of the nucleus in cell differentiation (accounting for the differences in the cells of the various tissues and organs in embryonic development) and in adult tissues — a region in which research has up to now been able to do little more than grope tentatively."

The method used by Drs. Pollister and Mirsky was also used by Professor Robert R. Bensley of the University of Chicago in extracting a substance from cells which he considered to be of cytoplasmic, or non-nuclear, origin.

New Strains in Gene Research

The use of X-rays on pink bread mold to learn more about how the heredity-controlling genes exercise their control on vital functions in the organism was described by Drs. G. W. Beadle and E. L. Tatum of the Department of Biology, Stanford University.

The procedure comprises X-raying cultures of the bread mold and subsequently testing such cultures for induced loss of ability to synthesize specific substances.

The Stanford geneticists reported creating new strains (mutants) that differ from the original strain by a single gene factor.

Among the new strains were units unable separately to synthesize vitamin B-1, vitamin B-6 (Pyridoxin), nicotinic acid (the antipellagra vitamin), para-amino-benzoic acid (anti-gray-hair factor), methionine (an essential protein derivative or amino acid).

Two lost ability to synthesize vitamins as yet unidentified, and a ninth could no longer make what appears to be an as yet undiscovered essential amino acid.

"As a by-product the technic provides a general method of discovering new vitamins and other substances of vital significance to the organism," the report stated. "The mutants (new species) obtained are useful as test material for specific vitamin assays."

Physical Features of Mars

Dr. E. C. Slipher of the Lowell Observatory at Flagstaff, Ariz., discussing recent progress on planetary research, told the section on astronomy that Mars was the only planet that gave evidence of the presence of most of the three essentials to life—water, oxygen and sufficient warmth.

Stating that oxygen was the only one of the three about which there remained any serious doubt, he continued:

"Moreover, the face of Mars undergoes certain seasonal changes whose explanation is as yet impossible without invoking plant life.

"As to the canal network, I may say that we have recorded over and over again a large majority of the canals in the location and form (Percival) Lowell drew them.

"But, quite naturally, it is as yet impossible to say that his theory of the artificiality of the canals has been positively confirmed.

"Thus in Mars we look upon a very arid world, and doubtless view the inevitable future of the earth—a world where the cycle of life has nearly run its course and the final stages are being enacted across his face."

Professor Arthur H. Compton of the University of Chicago, who won the Nobel Prize in physics in 1927 for discoveries on the relationship of matter and energy, was elected president of the American Association for the Advancement of Science.

Dr. Compton, internationally known for his discovery of the Compton Effect and for his fundamental researches on the cosmic rays, was born in Wooster, Ohio, on Sept. 10, 1892, was graduated from the College of Wooster in 1913, and received his Ph. D. degree from Princeton in 1916.

January 1, 1942

NOTES ON SCIENCE

Genes Seen at Work Within a Cell

HEREDITY—

Genes, the submicroscopic units that determine how we inherit blue eyes, stub noses and blond hair from our parents, do not always govern alone, Prof. T. M. Sonneborn (University of Indiana) told the American Association for the Advancement of Science. He has found instances in which genes that determine certain factors or courses of events cannot operate except in the presence of at least traces of other substances which he calls "primers." What are these? Science simply doesn't know. This department wonders if they are not the same as Hans Spemann's "organizers," which determine just how a cell will develop.

W. K.

September 17, 1944

HEREDITY UNITS—

What is believed to be the first instance of genes, or heredity-determining units within a cell, actually being seen at work chemically influencing the course of physiology, is reported in Nature by Drs. J. F. Danielli and D. G. Catcheside, both of Cambridge University. The two were able to demonstrate a concentration of the enzyme phosphatase in zones of bands on a chromosome, one of the structures within the cell's nucleus credited with being the carriers of groups of genes. These zones of concentration correspond closely with the fixed positions assigned by geneticists to particular heredity-units. Phosphatase is an enzyme important in the life-economy of the cell because it influences the chemical transformation of phosphorus compounds. Drs. Danielli and Catcheside comment: "This apparent coincidence between sites of enzyme activity and of genetic activity suggests that we have here an indication of a process whereby genes influence cellular activity, and is, we believe, the first experimental indication of the nature of such processes."

November 18, 1945

HEREDITY CHEMICAL REPORTED ISOLATED

Prof. Stern of Brooklyn Puts Genoprotein Theory Before Stockholm Cytology Group

TRANSMISSION EXPLAINED

Scientist Ties Discovery to Individual Traits — He Also Offers New Gene Formula

By WILLIAM L. LAURENCE
Special to THE NEW YORK TIMES.

STOCKHOLM, July 14—The isolation from the nuclei of living cells of a chemical believed to be the substance transmitting heredity was reported today before the International Congress for Experimental Cytology.

The investigations, promising to open a new chapter in genetics, were described by Professor Kurt G. Stern of the chemistry department of the Brooklyn Polytechnic Institute.

The mechanism of heredity is controlled by genes, the invisible entities existing within pod-like bodies named chromosomes in the manner of peas inside a pod. These genes transmit the heredity of parents to offspring and determine the individual's physical characteristics together with his mental endowments that, along with environmental influences, shape the individual's destiny throughout life. The gene-bearing chromosomes are located within the nuclei of the cells composing the living body, the cells of all living things containing a specific number of chromosomes and genes.

Complex Substances Trailed

Within recent years scientists probing the mysteries of cells, locking within themselves the mystery of life, have been on the trail of a highly complex group of substances isolated from living cells which, because they are found in the nucleus, were named nucleic acids. These nucleic acids vary in form though all appear to have the same chemical structure and configuration.

The nucleic acids, which seem to play a vital role at the very basis of life, contain a rare form of sugar known as ribose, found in two forms, hence nucleic acids are classified under two distinct groupings, one known as ribo-nucleic acid and the second as desoxyribo-nucleic acid.

More recent investigations have revealed that the first exists within the cytoplasm—outer layer—of the cell, while the second exists within the nucleus.

Desoxyribo-nucleic acid in the nucleus is found combined with a protein substance, a complex of acid and protein being referred to as nucleo-protein. Many attempts have been made in recent years to isolate nucleo-protein in its native state, but the strong chemical solutions so far required for such isolation have resulted, Professor Stern states, in destroying the native state of the nuclear substance by dissociating the complicated large molecule.

Reports Isolation Success

Professor Stern reported that he and his Brooklyn Polytechnic co-workers have succeeded at last in isolating pure desoxyribo-nucleoprotein from the thymus gland of a calf in its native unspoiled form.

This was accomplished by a combination of physical and chemical methods involving high speed centrifuge and the inhibition of enzyme action inside the cells by an arsenate compound.

Electro-chemical and ultra-centrifugal tests have shown, Professor Stern reported, that the purified substance was homogeneous, namely a pure single substance. Particles appear to have a "fairly compact shape and a molecular weight order of one million."

Comparative studies of the nuclear acid-protein complex with electron microscope photographs, Professor Stern reported, provide strongly suggestive evidence that the substance is "intimately related to or identical with genic material contained in the chromosomes of the body's cells." He therefore proposed the name "Genoprotein" for the newly isolated substance.

New Theory on Gene Structure

Professor Stern also presented a new hypothesis offering a new theory for the structure of the gene, explaining how substances very closely resembling each other in chemical structure can, nevertheless, control entirely different hereditary traits.

Nucleic acids are composed of long chains of simpler units known as nucleotides, composed of still simpler substances known as purines and pyramidines. The chain of nucleotides is known as the polynucleotide chain.

According to Professor Stern's new picture, genetic substance is shaped in the manner of a long double comb, with "teeth" on one side composed of a long row of nucleotides while those on the other side consist of protein material.

The two rows of nucleotide and protein "teeth" are held together by a backbone of phosphorus-containing chemicals, named phosphate esters, according to Professor Stern.

"Individual genes," he said, "may differ, not with regard to their chemical composition, but to the modulation of the spatial arrangement of the nucleotide side chains projecting from the phosphate ester backbone of the polynucleotide chain.

"The role of the basic protein component of the nucleo-protein would then consist in the protection of the genic modulation of the nucleic acid chain against thermal and chemical disturbances."

July 15, 1947

GENES' PHOTOGRAPH TAKEN FOR 1ST TIME

USC Scientists Say They See Carriers of Heredity With Electron Microscope

LONG A RESEARCH DREAM

Particles One-Hundredth by One-Millionth of Centimeter Are Made Visible

By GLADWIN HILL
Special to THE NEW YORK TIMES.

LOS ANGELES, Jan. 7—Two scientists of the University of Southern California reported today that they had observed and photographed genes, the mysterious infinitesimal particles that transmit physical characteristics of living things from one generation to another.

The achievement, the university states, is the culmination of worldwide research going back fifty years. Observation of genes has been an ambition of scientists ever since Gregor Mendel indicated their existence in his sweet-pea breeding experiments a century ago.

The scientists, Drs. Daniel C. Pease and Richard F. Baker, described the genes as thin, double-pointed, "spindle-shaped" particles one one-hundredth of a centimeter long and one one-millionth of a centimeter wide.

The observations were made with a standard electronic microscope on unprecedentedly thin cross-sectional slices of animal tissue obtained by a new cutting technique. The two scientists developed and reported this technique some months ago.

By hardening the tissue specimen with paraffin, collodion and air chilled by dry ice, and using new methods of adjusting the cutting blade, the scientists said, they obtained slices only 1/250,000th of an inch thick, permitting for the first time the penetration of electrons necessary to disclose the genes.

Genes are carried in larger bodies called chromosomes. In these experiments, chromosomes of the fruit fly were used because they are relatively large.

The tissue sections were magnified 120,000 times. The genes appeared almost as specks, too small to be reproduced in ordinary printing, the scientists said.

The observation of genes at last is expected to be of great importance in medical and biological research, particularly in connection with germs and viruses.

Dr. Pease, assistant professor of anatomy at the university's School of Medicine, is 34 years old. A native of New York City, he was graduated from Yale University in 1936 and obtained his master of science degree at the California Institute of Technology in 1938 and his doctorate at Princeton in 1940. He has been at USC for four years.

Dr. Baker, assistant professor of experimental medicine, is 39 years old. He was born in Westfield, Pa., and was graduated from Pennsylvania State College in 1932. He received his doctorate at the University of Rochester. He was an electronic microscope expert at the Radio Corporation of America laboratories in New Jersey before coming to USC in 1947.

January 8, 1949

SCIENCE IDENTIFIES CELL CONSTITUENT

Discovery Indicates a Factor That May Help to Trace the Pathway of Evolution

For the first time a living cell constituent has been identified positively and discovered to be present in exactly the same amounts in all animals of a given species.

Further, this constant factor, a precise chemical measurement, has been shown to vary in an apparently explicable manner between different creatures so that it may soon be applied to tracing the evolutionary pathway chemically up from the amoeba to man.

These discoveries, representing a great stride in man's conquest of the secrets of life, were discussed informally yesterday afternoon before a meeting of the New York section of the American Chemical Society held in the Hunter College Playhouse, Sixty-eighth Street and Park Avenue.

The research project leading to them was described by Dr. Arthur

E. Mirsky of the Rockefeller Institute for Medical Research, where the work was performed. Dr. Mirsky, who has been at the institute for twenty-one years, was one of the principal investigators.

It has long been known that living cells contain nucleic acids in the chromosomes, Dr. Mirsky reported. Chromosomes are long strings made up of genes, half contributed by the male parent and half by the female parent of a given cell. An individual cell may contain up to 40,000 separate genes and their location on the chromosome string determines the cell function and characteristics that in turn encompass the entire variety of living creatures.

One Acid Identified

One acid was identified positively as a compound known to chemists as desoxyribonucleic acid, Dr. Mirsky said. After this step, it was found that the amount of desoxyribonucleic acid present within each cell of all creatures of a species was exactly the same.

This fundamental determination, which links all like creatures together and may prove to be the chemical difference between different species had been overlooked in previous biological work, Dr. Mirsky declared, because of difficulties in measuring the amount of acid present under different circumstances.

Intercellular bodies containing acid are located within the cell in a tiny sac, much as an egg yolk floats in the white protoplasm. The chromosomes are attached to molecules of protein. Previous efforts to measure the absolute amount of desoxyribonucleic acid present within each cell did not take into account the relative amounts of protein and surrounding material, known as cytoplasm, that might be present, Dr. Mirsky said.

The Rockefeller Institute scientist reported that the amount of protein and the amount of cytoplasm associated with each chromosome group could be varied by dietary regime, by the physical state of the living animal and by many other external considerations.

Factors Are Eliminated

When the varying factors were eliminated in recent laboratory tests, he said, the amount of desoxyribonucleic acid present proved to be exactly the same for all cells of any one creature, regardless of the organ or tissue from which the cell came.

Further tests with a variety of living creatures, including domestic fowl, shad, carp, trout, frogs, toads and turtles revealed that the constant acid content of the living cell was, in each case, peculiar to the species, he said.

Other work with higher mammals, including cells from the organs of cattle, is more complicated because chromosomal structures are more complex and the fundamental relationship between the amount of acid in each cell may be obscured, he asserted.

Application of the recent discoveries to heredity and the tracing of evolution is now under way, Dr. Mirsky said. In addition, the differences between cancerous and normal cells are being investigated on the basis of the apparently species-constant cell content of nucleic acid, he added, though this work is just beginning.

January 23, 1949

CLUE TO CHEMISTRY OF HEREDITY FOUND

American and Briton Report Solving Molecular Pattern of Vital Nucleic Acid

TESTS BY X-RAY PLANNED

Work Done in England, if It Is Confirmed, Should Make Biochemical History

Special to The New York Times.

LONDON, June 12—A scientific partnership between an American and a British biochemist at the Cavendish Laboratory in Cambridge has led to the unraveling of the structural pattern of a substance as important to biologists as uranium is to nuclear physicists. The substance is nucleic acid, the vital constituent of cells, the carrier of inherited characters and the fluid that links organic life with inorganic matter.

The form of nucleic acid under investigation is called DNA (desoxyribonucleic acid) and has been known since 1869.

But what nobody understood before the Cavendish Laboratory men considered the problem was how the molecules were grooved into each other like the strands of a wire hawser so they were able to pull inherited characters over from one generation to another.

Further Tests Slated

The two biochemists, James Dewey Watson, a former graduate student of the University of Chicago, and his British partner, Frances H. C. Crick, believe that in DNA they have at last found the clue to the chemistry of heredity. If further X-ray tests prove what has largely been demonstrated on paper, Drs. Watson and Crick will have made biochemical history.

Dr. Watson has now returned to the United States, where he intends to join Dr. Linus Pauling, of California, who has done most of the pioneer work on the problem.

[In Pasadena, Calif., Dr. Pauling said that the new Crick-Watson solution appeared to be somewhat better than the proposal for the structure of the nucleic acids worked out by Dr. Pauling and associates at the California Institute of Technology. The California solution was published in the February, 1953, issue of the Proceedings of the National Academy of Sciences.]

Dr. Crick may leave Britain, too, when he has done some more work on the problem. Right now, he said, it "simply smells right" and confirms research in many institutions, particularly the Rockefeller Foundation in the United States and at King's College in London.

The acid DNA, Dr. Crick explained is a "high polymer"—that is, its chemical components can be disentangled and rearranged in different ways.

DNA is the essential constituent of the miscroscopic life-threads called chromosomes that carry the genes of heredity like beads on a string.

In all life cells, including those of man, DNA is the substance that transmits inherited characters such as eye color, nose shape and certain types of blood and diseases. The transmission occurs at the vital moment of mitosis or cell division when a tangle of DNA containing chromosomes becomes thicker and the cell separates into two daughter cells.

Forming of Molecular Chain

Although DNA has never been synthesized, Drs. Watson and Crick knew it was composed of horizontal hook-ups of bases (sugars and phosphates) piled one above the other in chain-like formations. The problem was to find out how these giant molecules could be fitted together so they could duplicate themselves exactly.

By a method of scientific doodling with hand-drawn models of the molecules, Drs. Watson and Crick worked out which molecules could be joined together with regard to the fact that some molecules were more rigid than others and had critical angles of attachment. Some months ago they decided that the only possible interrelation of the molecules was in the form of two chains arranged in a double helix—like a spiral staircase, with the upper chain resembling the staircase handrail and the lower resembling the outside edge of the stairs.

New evidence for double DNA chains in helical form now has been obtained from the King's College Biophysics Department in London, where a group of workers extracted crystalline DNA from the thymus gland of a calf and bombarded it with X-rays.

The resulting X-ray diffraction photographs showed a whirlpool of light and shade that could be analyzed as the components of a double helix.

Dr. Crick emphasized that years of work still must be applied to the helical carriers of life's characteristics. But a working model to aid in the genetical studies of the future now has been laid out in blueprint form by Drs. Watson and Crick—or so most biochemists here believe.

Looks Good, Pauling Says

Reached by telephone in Pasadena, Dr. Pauling said last night that the Crick-Watson proposal for the structure of the nucleic acids "looks very good." Dr. Pauling has just returned from London where he talked with Dr. Crick and with Dr. Watson, who was formerly a student at California Institute of Technology.

Dr. Pauling said that he did not believe the problem of understanding "molecular genetics" had been finally solved, and that the shape of the molecules was a complicated matter. Both the California and the Crick-Watson explanations of the structure of the substances that control heredity are highly speculative, he remarked.

June 13, 1953

Making It Scientifically

THE DOUBLE HELIX. A Personal Account of the Discovery of the Structure of DNA. By James D. Watson. Illustrated. 226 pp. New York: Atheneum. $5.95.

By ROBERT K. MERTON

THIS is a wonderfully candid self-portrait of the scientist as a young man in a hurry. Chattily written with pungent and ironic wit and yet with an almost clinical detachment, it provides for the scientist and the general reader alike a fascinating case-history in the psychology and sociology of science as it describes the events that led up to one of the great biological discoveries of our time. I know of nothing quite like it in all the literature about scientists at work.

The bare facts of the case are public knowledge. In 1953, after two years of work in the famed Cavendish Laboratory, the 25-year-old American biologist, James D. Watson, and the 37-year-old English physicist-turned-biologist, Francis H. C. Crick, proposed a double-helical model of the molecular structure of *deoxyribonucleic acid* (DNA), the substance that transmits genetic information from one generation to the next, and observed that this suggests a copying mechanism for the genetic material. In 1962, they shared the Nobel Prize in physiology and medicine with Maurice Wilkins, their sometimes inadvertent collaborator at King's College (London) who had for years been engaged in X-ray studies of DNA.

Behind these sparse facts is the complex, absorbing story of how all this came to be. In "The Double Helix," Watson tells that story by adopting his heavily personalized version of the Rankean directive to write history *wie es eigentlich gewesen ist* (or, in the repulsive vernacular, to tell it like it was—or at least, as it seemed to the youthful Jim Watson). For this decision, he has ample precedent in principle, if not in practice. As far back as the early days of modern science, Francis Bacon was complaining that "never any knowledge was delivered in the same order it was invented." Ever since, men of science such as Leibniz and Mach, or to move swiftly to the present day, the physicist Richard Feynman, have periodically reminded us that the public record of science tends to produce a mythical imagery of scientific work in which disembodied intellects move toward discovery by inexorably logical steps, actuated all the while only by the aim to advance knowledge.

This is hardly the picture Watson paints, either of himself or of most of his colleagues. Instead, he depicts a variety and confusion of motives, in which the objective of finding the structure of DNA is intertwined with the tormenting pleasures of competition, contest and reward. Absorption in the scientific problem alternated with periodic idleness, escape, play and girl-watching. Friendship and hostility between collaborators was expressed in a nagging yet productive symbiosis in which neither could really do without the special abilities of the other. And all this engaged not only the passion for creating new knowledge but also the passion for recognition by scientific peers and the competition for place.

Watson makes no bones about it. In one of its aspects, the work on DNA was for him a race, principally against Linus Pauling, for the ultimate symbol of scientific accomplishment, the Nobel Prize. He tells all who will listen about the excitement of the race, takes unalloyed delight in learning that Pauling is apparently on the wrong track and, in his youthful enthusiasm, joins in a toast "to the Pauling failure. . . . Though the odds still appeared against us, Linus had not yet won his Nobel."

THOUGH it might surprise the outsider, this emphasis on competition in science will scarcely come as news to working scientists. They know from hardwon experience that multiple independent discoveries are one of their occupational hazards. Since discoveries are typically the temporary culmination of what has been found before, when several scientists are working independently on the same problem, they are apt to move toward the same conclusion. As a result, competition in science is as old as modern science itself. Almost everyone placed in the pantheon of science, from the days of Galileo and Newton, has been caught up in the consequent race for priority. But seldom before has a scientist so revealingly described for the general reader his own competitive motivation to get there first.

Watson's beautifully brash account serves to distinguish this competitive motive from the closely allied motive of contest. Competition involves the attempt to win out against the field for the rewards that come with victory; contest involves the directly sportive pleasure of beating particular others. Time and again, Watson records his youthful pleasure in testing his powers against the best there is. He is especially eager to outstrip the champions — Linus Pauling, "the world's greatest chemist," for one and Erwin Chargaff, "one of the world's leading authorities on DNA," for another.

And then there is the engagingly droll episode in which the energetic young Watson decides to match himself against the even more precocious *enfant terrible*, Joshua Lederberg. He reviews all of Lederberg's recent experimental work on the genetics of bacteria and finds, in true contest style, "particularly pleasing . . . the possibility that Joshua might be so stuck on his classical way of thinking that I would accomplish the unbelievable feat of beating him to the correct interpretation of his own experiments."

These elements of competition, contest and reward have made property rights an integral though still ambiguous part of the institution and ethics of science. For if the advancement of knowledge were the only institutionalized motive for scientists, then the concept of property rights would of course make little sense. What matters it who advances our knowledge, providing only that it gets done? Yet property rights have been a gray area in the mores of science for quite some time. More than a century ago, the nonpareil physicist, Clerk Maxwell was writing William Thomson: "I do not know the Game laws and Patent laws of science . . . but I certainly intend to poach among your electrical images."

IT is within this same context of property rights that Watson describes his own and Crick's initial hesitancy to move into work on DNA structure: ". . . this would create an awkward personal situation. At this time, molecular work on DNA in England was, for all practical purposes, the personal property of Maurice Wilkins. . . . It would have looked very bad if Francis [Crick] had jumped in on a problem that Maurice had worked over for several years."

In another of Watson's clinically described episodes, which reads like a paragraph drawn from Pepys, we see him ready to seize upon an odd expedient for gaining access to badly needed information from Wilkins. He experiences as a "tremendous stroke of good luck" the circumstance that Wilkins appears to have "noticed that my sister was very pretty. . . . Furthermore, if Maurice really liked my sister, it was inevitable that I would become closely associated with his X-ray work on DNA." The immediate outcome is anti-climactic: "Neither the beauty of my sister nor my intense interest in the DNA structure had snared him."

As is now often the case at the forefront of science, only a part of the information needed by Watson and Crick came through formal channels of publication. Some of the salient information traveled on grapevines of personal relations giving fact and rumor about who was doing what that might be pertinent to their own work. Here, too, kinship ties could occasionally be utilized to advantage. With temerity and self-mocking wit, Watson reports the occasions on which Linus Pauling's son, Peter, then a student at Cambridge, became a prime source of information about what his father was up to. This is the stuff that abounds in fiction but is rare in the proper histories of scientific ideas.

All this competition and jockeying for position might seem to suggest that science tends to recruit egotistic personalities, contentious and exceedingly hungry for fame. However that may be—I happen to doubt it—it does not explain these behavior patterns. For we know that even ordinarily modest and retiring men, such as the great 18th-century chemist Henry Cavendish himself, have been, however reluctantly, drawn into controversy over property rights in science. It appears rather, as we see in Watson's memoir, that the competitive behavior of scientists results largely from values central to the scientific enterprise itself. The institution of science puts an abiding emphasis on significant originality as an ultimate value, and demonstrated originality generally means coming upon the idea or finding first. Recognition and fame thus appear to be more than merely personal ambitions. They are institutionalized symbol and reward for having done one's job as a scientist superlatively well.

In the course of describing the behavior of his competitive and abrasive young self, Watson tells us much else about the workings of science at the frontiers. Some of this is just the sort of thing that scientists ordinarily take so much for granted that only the more reflective among them ever put it in so many words.

IN science as in every other field of human activity, taste is of prime importance. In one aspect, taste involves a capacity for distinguishing significant, that is, consequential problems from minor ones. What Watson describes as the chase for the Nobel Prize implies, of course, that Crick and he knew that they had hold of a problem of the first magnitude. Meanwhile, many of their able peers were busily and indispensably working on problems of far less consequence for biology.

Watson also alerts us to the functions of the basic self-confidence — even downright arrogance—of these young men of science as they entered upon a field of inquiry new to them. It must have required great ego-strength for them to take the plunge. For as Watson not merely admits but repeatedly insists, at the outset they were ignorant of much they needed to know in order to investigate the problem of DNA structure. The impressive inventory of this announced ignorance includes the techniques of X-ray diffraction, Pauling's work on the alpha-helix, Bragg's Law ("the most basic of all crystallographic ideas") and the chemistry of hydrogen bonds. Yet, despite occasional qualms, these newcomers had the adventurous fortitude to acquire much of the knowledge they needed and the good luck to have at their side the experts who could round out that knowledge enough for them to do the job of imaginative scientific carpentry that led to their momentous model.

From Watson's narrative, we learn as much about the micro-environments of these scientists as about their personalities. It soon becomes evident that Watson and Crick could not have accomplished what they did had it not been for the evocative environment in which they worked. Watson singles out five principals: Crick, Wilkins and Watson himself, of course, Pauling and Wilkins' associate, Rosalind Franklin.

But there are others who turn up in the story who were more than merely supporting members of the cast. And these were scientific minds of the first order. At the Cavendish itself, there were Max Perutz and John Kendrew (both destined to receive a Nobel Prize in the same year as Watson and Crick) and the director of the Laboratory (Sir Lawrence Bragg, who had been designated a laureate some forty years before). Important to the outcome perhaps above all else was the happy circumstance that placed the American crystallographer, Jerry Donohue, in the same office with Watson and Crick, for it was Donohue who put them on the right track by showing them where the textbooks of structural chemistry had gone wrong.

OUTSIDE the Cavendish, they were interacting with scientists of topmost caliber: Watson's teachers, S. E. Luria and Max Delbrück; the three laureates-to-be André Lwoff, Joshua Lederberg and Dorothy Hodgkin; Seymour Benzer, Gunther Stent and Erwin Chargaff (the man to whom Wilkins, in his Nobel Prize address, pays tribute for having laid "the foundations for nucleic acid structural studies and for generously helping us newcomers in the field of nucleic acids"). Each in his own way, Watson tells us in effect, played his part in making the outcome possible. This all adds up to the evident but often neglected fact that science is much more of a collaborative enterprise than is even hinted at by the lists of authors of scientific works.

Still, as Watson is the first to warn us, he is describing only his own, not necessarily representative, style of scientific inquiry. He emphasizes, moreover, that the entire narrative is only his distinctly personal version of how it all came about. The other participants in these events might see them differently. And he comes close to drawing the amply evident implication. If the other members of the cast would write their accounts, each from his own perspective, these could be collated to provide the fullest and most profoundly informative history we yet have of a basic contribution to science.

There remains only the question raised by some of Watson's fellow-scientists after this book was serialized in The Atlantic Monthly. Why did he decide to publish so intimate a history? Why has he conscientiously violated the mores that govern the public demeanor of scientists by reporting to all who would read what is ordinarily known only to the inner circle? The explicit reason for writing the book we have already noted: he wanted to give a full-blooded account of at least one style of scientific investigation. But in the prelude, not the preface, he intimates another reason for doing the book. This is how he reports an episode occurring on a walking trip in the Alps two years after the classical Watson-Crick paper had been published:

"We were only a few minutes out of sight of the hotel when we saw a party coming down upon us, and I quickly recognized one of the climbers. He was Willy Seeds, a scientist who several years before had worked at King's College, London, with Maurice Wilkins on the optical properties of DNA fibers. Willy soon spotted me, slowed down, and momentarily gave the impression that he might remove his rucksack and chat for a while. But all he said was, 'How's Honest Jim?' and quickly increasing his pace was soon below me on the path."

Placed within the context of the ambiguous norms of property in science, here, perhaps, is James Watson's *Apologia pro Vita Sua*.

February 25, 1968

Key Acid Substance in Heredity Charted in New X-Ray Studies

By PETER KIHSS

One of the fundamental substances of life has been laid bare in its living state through X-ray studies by British, French and American scientists.

The studies have shown that the three-dimensional structure of the substance — believed to make up the genes that control heredity—is identical in man, cattle, plants and bacteria.

Found in all living cells, the substance is deoxyribose nucleic acid, known as DNA. The new studies have been carrying forward research that began a dozen years ago. On the basis of theory and preliminary X-ray data they affirm the structure suggested tentatively for DNA in 1953.

The latest advances on this frontier of biochemistry have come to light through a report by nine scientists in the May 14 issue of the British scientific Journal Nature and through interviews in London and New York. The vital acid has been extracted, crystallized, X-rayed, analyzed and modeled to show how it hangs together.

In the long run, the report in Nature suggested, knowledge of the molecular structure of DNA and its linking with protein may lead to chemical explanation for the phenomena of inheritance.

In specific applications one hope is that the work may some day enable use of nucleic acids to treat cancer. The attempt would be to transfer the abnormal cells of that disease into normal cells through the processes of cellular reproduction.

The scientists making the studies include groups at King's College, London; the Genetical Institute, Paris, and the Sloan-Kettering Institute at 410 East Sixty-eighth Street here.

Studies with X-ray diffraction photography at King's College have been directed by Dr. M. H. F. Wilkins in the Wheatstone Physics Laboratory. Nucleic acid has been supplied by Dr. L. D. Hamilton and Dr. R. K. Barclay of Sloan-Kettering and by Dr. Harriet Ephrussi-Taylor of Paris.

The New York group has been providing DNA derived from many sources, such as human white blood cells—both healthy and cancerous—and the thymus gland of cattle. Dr. Ephrussi-Taylor has supplied nucleic acid from bacteria.

Over the years, studies by Dr. Wilkins and his associates have shown that nucleic acid has a helical structure—that is, its atoms of hydrogen, oxygen, carbon, nitrogen and phosphorus are bound in a double spiral twisting around an axis.

The acid chains are so small that the molecule can be viewed only—and then with difficulty—through an electron microscope. They measure about twenty angstroms in thickness and probably a thousand times more in length. An angstrom equals only one hundred millionth of a centimeter.

While the basic structure remains the same whether the nucleic acid is from a human being, a calf, a lettuce leaf or a bacterium, the studies show quantitative chemical differences in the DNA of various species. These are variations in the ratio of the nitrogenous building blocks that fit into the structure.

The nitrogenous chemicals—adenine and thymine, guanine and cytosine, in those pairs—have been found always present, but with different species having different proportions of those substances in pairs. The hypothesis of pairing was suggested in 1953 by Dr. J. D. Watson and Dr. F. H. C. Crick, working at Cavendish Laboratory in Cambridge, England.

Continuing X-ray diffraction studies have established precisely the double spiral structure of DNA and have corrected and added to the earlier picture in important respects. Detailed in the report in Nature, these include such technical problems as diameter and pitch of the spirals.

But, beyond this, the interviews brought out that Dr. Ephrussi-Taylor's bacterial nucleic acid was still alive—that is, its molecules were biologically able to join with other bacteria to produce new and altered generations of bacteria.

The mammalian-derived DNA supplied by Dr. Hamilton and Dr. Barclay from Sloan-Kettering has been shown by the X-ray studies to be identical in structure with the living Ephrussi-Taylor nucleic acid.

Thus far, the Hamilton-Barclay DNA from mamals, including man, has not yet been able to join and transform other cells in reproduction. It remains under study in this respect. But the ability of bacterial DNA to reproduce with other bacteria has given the strongest evidence yet available that DNA is actually the heredity-carrying material.

The King's College studies are being aided by a $35,000 grant from the Rockefeller Foundation for three years' work, continuing into 1958.

June 11, 1955

FRUIT FLY STUDIES TRACE GENE TREND

Development of Resistance to DDT Shows Heredity Is Not Result of Chance

By ROBERT K. PLUMB
Special to The New York Times.

COLD SPRING HARBOR, L. I., June 13—Studies of many generations of fruit flies, discussed here today, suggest that the genes, or carriers of heredity, are not just scrambled together by chance.

Rather, the experiments suggest, genes are organized into integrated patterns that are characteristic of entire populations.

In human terms, such a characteristic population might be a race or a smaller isolated community. Estimates place the number of human genes at about 40,000. Combinations of these elements of inheritance, contributed equally by male and female parents, determine physical characteristics attributed to heredity. Simple examples of these characteristics are hair type, eye color and skin pigmentation.

The fruit fly experiments were reported by two researchers of the Long Island Biological Association, Dr. James C. King and Dr. Bruce Wallace. Working with Dr. Wallace on one series of experiments was M. Vetukhiv.

Dr. King reported that fruit flies appeared to develop resistance to the insecticide DDT after twelve generations had been exposed to it. Fruit fly generations are about fourteen days long.

Resistance Is Explained

Dr. King said it was believed that perhaps five or six flies in each thousand or more are by chance slightly different in some unknown factor. The difference makes these particular insects more resistant to DDT. Insects sensitive to DDT are killed in the studies. This leaves to flourish the slightly-resistant flies. These insects mate. Again only a few—perhaps five or six in a thousand or more—will be a little more resistant than the rest.

This selection of a second degree of resistance among the offspring of the survivors of the first study may represent a different factor (still unknown) that contributes to resistance. After twelve such experiments have proceeded, Dr. King said, it is possible to demonstrate that strong DDT-resistance has been established in the surviving experimental insects.

The resistance to the insecticide that these insects demonstrate is unusual, he said, in that it persists in the insect populations. Usually flies selected for a particular train gradually die out because they have lost some other factor that contributes to their toughness. But this is not the case with DDT-resistance, Dr. King reported.

Dr. Wallace reported that when flies from different parts of the world are bred together, the hybrids that result at first are apparently tougher than were the original insects. But after a few further cross-matings, this effect disappears and the insects become similar to populations in which random matings from generation to generation establish a fairly uniform gene pool for the population. There is evidence that the genes form into organized patterns in these studies, he reported.

About 150 geneticists and others interested in evolution and breeding problems have gathered at the Biological Laboratory of the Long Island Biological Association here for eight days of meetings on "Population Genetics: The Nature and Causes of Genetic Variability in Populations."

June 14, 1955

SCIENCE IN REVIEW

DNA, Substance That Determines Heredity Of All Living Things, Is Synthesized

By WILLIAM L. LAURENCE

Synthesis from inert materials of a substance that "looks like and acts like DNA," the chemical that determines heredity in all living things from bacteria to man—and even in viruses, on the borderline between the living and non-living—was announced last week by a team of New York scientists at the International Congress of Biochemistry at Vienna. Further progress along these frontiers of life is expected to lead to a better understanding of life's most fundamental mechanisms, in health and in disease, as well as of the manner in which living things perpetuate themselves in their own image.

The studies were reported by Dr. Aaron Bendich and Herbert Rosenkranz of the Sloan-Kettering Institute at the Memorial Center for Cancer and Allied Diseases and Dr. S. M. Beiser of Columbia University's College of Physicians and Surgeons.

Of all the chemical compounds that play essential roles in the processes of life, DNA appears to be the most basic. It is found only in the chromosomes, the rod-like bodies in the nuclei of the cells of all living things that contain the genes, the agents that control heredity. All evidence strongly indicates that the active component of the genes consists mostly, if not entirely, of DNA.

Six Basic Units

Though it is composed of no more than six simple basic units named nucleotides, the DNA molecule can exist in billions of similar but unique forms, depending on the exact sequence of its basic units. It is now supposed that this varying sequence is a kind of code that contains the genetic information of the individual and the species. Each species has its own specific DNA, which determines whether any given germ cell is to develop into a bacterium, a plant, a mouse or a man. Furthermore, each individual in any given species has a specific pattern of DNA transmitted to him by his ancestors, which makes him different from any other individual in his species.

Moreover, recent studies now indicate that DNA, in addition to being the repository of genetic information, "the file of blueprints," so to speak, also participates in the construction and operation of a cell. Some believe that the DNA molecule serves as a kind of template upon which many of the biologically active molecules synthesized by the cell are constructed.

In their synthesis of a substance which "looks like and acts like DNA" the Sloan-Kettering and Columbia biochemists used nucleotides as their starting material. In all laboratory tests, they reported to the Vienna congress, the synthetic material resembled DNA removed from living organisms. In addition, and more important, the synthetic material appeared to have biological activity, in that it could affect the activity of DNA made by nature.

Previous synthesis of DNA and of its close relative, RNA, have been reported, but, in each of these cases, a small amount of natural DNA and RNA were necessary, and enzymes from living cells were used to build up the synthetic molecule. The present synthesis was achieved with wholly inert material.

Clue to Cancer

The achievement of the total synthesis of the chemical serving as the veritable powerhouse of heredity, and of other basic mechanisms of the living process, may be expected to open up a host of new approaches to unravelling the nature of some of life's fundamental chemical reactions and to a better understanding of the origin and possible prevention of some of the most serious and baffling diseases, such as cancer. Cancer cells, for example, are believed to have an altered, abnormal DNA, and it is hoped that synthetic DNA-like substances, which can interfere with the activity of natural DNA, might have a role in cancer treatment.

DNA contains an unusual sugar named desoxyribose. Other constituents are phosphorus, nitrogen, oxygen and hydrogen. The sugar molecule which constitutes as much as 48 per cent of the DNA molecule contains only five atoms of carbon, instead of six in the more common sugars. The full chemical name of DNA is desoxyribonucleic acid. It is a gigantic molecule made up of many thousands of atoms.

Recent pioneer studies at Cambridge University by Drs. J. D. Watson and F. H. C. Crick have revealed that the DNA molecule is a double helix in which the basic units (the nucleotides) are arranged on a sugar-like backbone strewn with phosphorus-containing groups. The phosphate-sugar backbones of the two halves of the molecule are coiled around each other in a long helix, each nucleotide in one coil being linked with a corresponding partner in the opposite coil.

Swiss Discovery

Nucleic acid, or DNA, was discovered as a constituent of the nucleus of living cells by the Swiss biochemist Friedrich Miescher in 1869, but it was not until recently that its true nature was identified. By far the most outstanding discoveries in this field were made during the course of the past fifty years by scientists of the Rockefeller Institute in New York City.

The story of these Rockefeller discoveries constitutes one of the brilliant chapters in the history of science. It began more than fifty years ago when Dr. P. A. Levene, one of the great chemists of his time, established that the sugar, desoxyribose, was a major constituent of nucleic acid. This work marked the beginning of a scientific super-detective story in which the true role of DNA as the chemical controlling heredity was gradually unravelled.

The long trail, in which scientists from many lands participated, began in 1928 when a British bacteriologist, Fred Griffith, reported that when he inoculated mice with a mixture of a harmless strain of live pneumonia germs and the dead remains of a harmful strain, the mice died from live pneumonia germs of the harmful strain. Here was a puzzle, indeed. Either the dead bacteria had been brought to life, or something in them was able to transform a harmless strain into a virulent one.

It was not until eighteen years later that Drs. O. T. Avery, Maclyn McCarthy and Colin MacLeod of the Rockefeller Institute, published a classic paper explaining the mystery. Their studies, later corroborated by more detailed experiments, disclosed that the DNA from the dead virulent bacteria was transferred to the living harmless species, transforming them into the virulent strain. What is more they found that the harmless strain, once transformed into a virulent one, transmitted its virulence to future generations by the transmission of the virulent type of DNA.

Pioneer Experiments

The next step was to prove that the DNA was actually the substance in the chromosomes that determines the transmission of heredity. Evidence that this was so was established in another series of pioneer experiments of the Rockefeller Institute by Dr. Alfred Mirsky and Dr. Hans Ris. They found that the amount of DNA in all the different cells of an organism is the same, even though the cells vary enormously in all other respects. This was suggestive because all the cells of a given organism contain the same number of chromosomes, the carriers of the genes. They also found that when sex cells have halved their number of chromosomes, in preparation for sexual reproduction, the amount of DNA they contain is also halved. They therefore concluded that DNA must be an essential part of the hereditary material of the chromosomes, the genes.

More recent work at the Rockefeller Institute by Dr. Norton Zinder, a young geneticist, working with Dr. Joshua Lederberg, a geneticist of the University of Wisconsin, made the significant discovery that viruses can transfer hereditary traits from one strain of bacteria to another by the transfer of the DNA of the first strain to the second. They call this process of transfer of heredity "transduction."

September 7, 1958

Nobel Given for Enzyme Study

For the second year in succession, American scientists have won the Nobel Prize in Medicine and Physiology. The winners of the 1959 Prize, the highest scientific honor in the world, are Dr. Severo Ochoa of New York University and Dr. Arthur Kornberg of Stanford University. They were cited for their fundamental discoveries on the biological synthesis of nucleic acids, chemicals that play a key role in the mechanism of life.

"All life is chemistry and the more we know of these chemical reactions the more we know of life," Dr. Ochoa said when he learned of the award last Thursday. What Dr. Ochoa and his former student, Dr. Kornberg, have done is to discover enzymes, or biological catalysts, with which they have been able to produce artificially in the test tube the two key chemicals instrumental in the origin of life and in its perpetuation ever since.

Synthesis of Protein

Dr. Ochoa, Spanish-born American

biochemist, discovered in 1955 a bacterial enzyme that can synthesize in the test tube, out of simpler constituents, the substance named ribonucleic acid, or RNA. This long-chain, double-coil-shaped molecule is present mostly in the cytoplasm, the protoplasmic envelope surrounding the nucleus, or "yolk," of the cells that constitute the bricks of living matter. It is the chemical believed to be crucial to the synthesis of protein, the basic stuff out of which all living matter, from lowest to the highest forms of life, is constituted. In viruses RNA is also believed to play a role in the transmission of heredity.

Brooklyn-born Dr. Kornberg discovered in 1956 a bacterial enzyme that has the ability to synthesize a close chemical relative of RNA named deoxyribonucleic acid, or DNA. DNA is the vital substance of which are made the genes, the heredity-transmitting units inside the chromosomes in the nucleus of living cells. These genes, different in each species, are nevertheless composed of the same essential ingredient, DNA. Nature creates its infinite variety of living things by merely shuffling the elementary units of the DNA molecule as one shuffles a deck of cards, the DNA in the various species differing largely in the architectural arrangements and the number of basic constituents of the highly complex molecule.

Master Pattern

The work of Dr. Kornberg has given experimental support to the concept that the DNA serves as a template, or master pattern, in each cell, allowing that cell to reproduce itself in its own image.

In Dr. Ochoa's laboratory it has been proved possible to promote synthesis of entirely unnatural chains of RNA-like material, such as an RNA molecule made up of only one of the four sub-components of which the molecule is normally made.

The two independent lines of research carried out by Dr. Ochoa and Dr. Kornberg have clarified many of the problems of cellular reproduction, the continuity of life and protein synthesis. Since cancer is a matter of abnormal growth and reproduction, it is hoped that the work will have eventual significance in solving the cancer problem. The work also promises to provide new understanding of the nature of viruses and their reproduction.

October 18, 1959

Scientists Find Clue To Heredity's Code

By JOHN A. OSMUNDSEN

Recently developed laboratory techniques and extraordinary luck have combined to produce a major scientific advance toward cracking the chemical code that governs the inheritance of all living things.

A report by two University of California scientists in today's issue of The Proceedings of the National Academy of Sciences describes the discovery of the first definite link between a mutation, or change in the inheritance code, and a specific alteration in the composition of a molecule manufactured according to that code.

With that much to go on, the scientists seem certain soon to come up with the first crude translation of that code. This would, in effect, define a single gene for the first time in terms of size and chemical composition. The work that has laid the foundation for such a future feat was done by Dr. A. Tsugita and Dr. Heinz Fraenkel-Conrat. Their achievement was largely made possible by techniques that Dr. Fraenkel-Conrat had developed over the last five years.

The new finding was described as "one of those rare breakthroughs" by Dr. Wendell M. Stanley, Nobel laureate and director of the University's Virus Laboratory in Berkeley, where the work was done.

The players in the microscopic drama of inheritance are the chromosomes of the living cell. They reside in the nucleus and carry the genes. Roughly speaking, each gene determines some specific character of the organism.

It used to be thought that genes existed as discrete units in a chromosome, not unlike beads on a string. Now, however, scientists have come to view a gene as a place on a very large and complex molecule underneath the protein coat of the chromosome.

The complex stuff of which genes are made is nucleic acid, and most generally deoxyribonucleic acid, or DNA. A molecule of DNA looks like a spiral ladder. The sides of the ladder are phosphate sugars, the rungs are organic bases of four types.

The arrangement of those four bases up and down the molecule constitutes a sort of four-letter code. Those four letters are presumably used to spell out words—which are the genes—in the language of life. The vocabulary of this language is virtually unlimited.

But how is that language translated? DNA can be dismantled piece by piece, but there is no way of knowing how many pieces make a gene, or word, and what such a genetic word might mean without a reference system of some sort.

Changes Looked For

In other words, something—a property of a cell, for example—must be found that expresses the effect of a gene in a meaningful way. A mutation or change in that gene should produce a detectable alteration in the property. Then, by chemically comparing the normal property with the changed one, an idea can be obtained about the size and character of that portion of the nucleic acid molecule—the gene—which determined the property.

The key to the reference system most scientists in this field are using to break the genetic code is protein, because its manufacture seems to be directed by nucleic acid. This is the way the process appears to work:

The double-stranded DNA molecule unwinds and separates into two single strands. Each single strand can either replicate itself, and so produce more DNA, or can direct the manufacture of another nucleic acid, called ribonucleic acid, or RNA. It is this single-stranded RNA that governs the production of protein.

Thus, a mutation or change in DNA is passed on as a change in RNA which transmits the alteration to the protein it makes. Scientists can then study a change in a protein resulting from a mutation and feel confident that this reflects an alteration in DNA and that the size and nature of the alteration defines a single gene.

This is what the Berkeley scientists have done, except that they have used a simpler system in which only RNA and protein are involved. With no DNA present, RNA governs its own replication as well as protein production.

The scientists produced a mutation in tobacco mosaic virus, by stripping the protein coat from the RNA and then treating the naked nucleic acid with nitrous acid.

The nitrous acid altered the RNA so that when the virus particles were reconstituted and rubbed on a tobacco plant, an effect of the viral protein different from the usual infection was produced. This reflected the mutation—a change in RNA that caused a corresponding change in the viral protein.

The scientists were particularly fortunate in their work on two counts. First, the mutation expressed itself in a way that was amenable to research with recently developed techniques. Second, the change in the protein occurred near the end of the molecular chain.

May 16, 1960

Synthesis of Proteins Achieved Under Controls Allowing Study

By JOHN W. FINNEY
Special to The New York Times.

WASHINGTON, March 29 — A major scientific advance was reported today in fathoming the complex biochemical and genetic processes going on in all living cells.

Through use of radioactivity as a research tool, scientists at the Oak Ridge National Laboratory in Tennessee have been able for the first time to observe in a test tube the genetic process by which proteins are synthesized. Protein synthesis—a central problem of modern biological research—is the process by which living cells convert food into energy and into new cell materials.

The research, which gives a new insight into living processes, was described by Dr. G. David Novelli, principal biochemist at the Oak Ridge Laboratory, in testimony before the Joint Congressional Atomic Energy subcommittee on research.

The Novelli testimony—the first extensive public description of his research work—caused an outburst of excitement among the scientists attending the subcommittee's hearings on the use being made of radio-isotopes in medical and agricultural research. The Novelli research was acclaimed by Dr. Charles L. Dunham, director of the Atomic Energy Commission's Division of Biology and Medicine, and Dr. H. Bentley Glass, geneticist at The Johns Hopkins University, as a significant advance in biological research.

Dr. Novelli predicted that the research should "open up whole new frontiers" in the exploration of the genetic process. It should now be possible, he said, to test in the laboratory many genetic theories, to isolate the gene, determine how big it is, to determine the role played by individual substances in genetic materials and to unravel the genetic code by which hereditary information is passed on to newly formed cells.

Through the research techniques developed by Dr. Novelli, scientists will have a way of dissecting and studying each step in the genetic process of protein synthesis. In the past, scientists have been restricted to observing the over-all genetic process and developing their theories and conclusions on a statistical, empirical basis.

Basic Questions Unanswered

Proteins, or enzymes, are large molecules built up of twenty building blocks called amino acids, and they act as the chemical machines that catalyze all the manifold processes occurring in living cells. The information on how the amino acids are arranged to form individual enzymes is contained in the genetic material of the cell.

For years scientists have been attempting to learn something about the mechanism by which cells synthesize proteins. Among the basic questions remaining unanswered are how does the information contained in the gene get to the place where the enzyme or protein is made? How are the amino acids arranged in the final structure? And what control mechanism does a cell utilize to regulate the rate at which a given protein is made?

Until recently, Dr. Novelli pointed out, little progress has been made in answering these questions. The reason has been that the problem had to be attacked by breaking open the cells and attempting to discover the stepwise fabrication of a protein. In the process, the internal structure of the cell becomes disorganized the enzymes are liberated that rapidly degrade and digest the proteins.

Through use of radiation and radio isotopes, Dr. Novelli has discovered a technique that permits study of gene action in a test tube in the absence of living cells.

Dr. Novelli came upon his discovery through experiments with the effects of ultraviolet radiation upon proteins. It was discovered that the irradiation inhibited the synthesis of a common enzyme known as galactosidase, which breaks down the sugar in milk. Further experimentation disclosed that the inhibition could be overcome by exposing the enzyme cells to white light.

Examination of the internal contents of the cells disclosed the presence of a hitherto unsuspected fraction of the enzyme. This fraction was then used as a cell-free extract for genetic reduplication and protein synthesis.

Through the addition of desoxynucleic acid—the genetic material of a cell—to the extract, new proteins were synthesized. Thus, Dr. Novelli said, a process was discovered for starting and stopping the process of protein synthesis and following it step by step.

Dr. Novelli was assisted by Dr. Tadanori Kameyama, a young biochemist sent by the Japanese Government to Oak Ridge for training in radiobiology. Dr. Kameyama is continuing the research now in Japan at the National Institute of Radiological Sciences in collaboration with Dr. Novelli.

March 30, 1961

NEW CLUE HINTED IN GENE FUNCTION

DNA Pictured in Key Role in Synthesizing Proteins

By ROBERT K. PLUMB

A model for the mechanism by which genetic information is used to construct proteins inside living cells was presented here yesterday at a conference at the Albert Einstein College of Medicine of Yeshiva University.

The model is the result of international teamwork between Dr. Sydney Brenner of the Cavendish Laboratory, Cambridge, England; Dr. Francois Jacob of the Institut Pasteur in Paris, and Dr. Matthew Meselson of the California Institute of Technology, Pasadena, Calif.

Dr. Brenner discussed the findings at a day-long conference on "The Molecular Basis of Biological Individuality" at the medical school in the Bronx.

The evidence is that the genes carry the information upon which proteins are constructed, Dr. Brenner pointed out. It is not clear, however, by what mechanism the information in the genes is converted to proteins built for particular purposes within the living body.

Carry Protein Code

In the model proposed by the group, the genes carry deoxyribonucleic acids (or DNA), which are the code for particular protein structures. The information encoded in the genes is transferred to an unstable intermediate or messenger form of nucleic acids known as ribonucleic acid (or RNA). This RNA serves as a "tape" containing instructions for the cell machinery or ribosomes.

The new theory postulates that the NA is a short-lived carrier of genetic information and that the ribosomes in the cell are prepared to receive the information and construct proteins in accordance with the information that is carried originally in the genes.

In the past, it was postulated that each gene controlled the operation of a specialized ribosome, which in turn directed the synthesis of a corresponding protein. Thus one gene acted on one ribosome to produce one protein.

In the new model, the ribosomes are seen as cell machinery, which acts under the instructions of the RNA intermediate to assemble amino acids into specific proteins. After an amino-acid sequence, or a particular protein, has been assembled, the message is destroyed and the ribosome is ready to follow new RNA instructions.

"We think," Dr. Brenner said, "that this is the way all genes act in viruses or in cells."

Tissue Study Presented

In another report, a model was presented for the construction of the collagen tissues, which are the major connective tissues of the body. The report was made by Dr. Alexander Rich of the Department of Biology of the Massachusetts Institute of Technology.

Collagen tissue is very strong, Dr. Rich pointed out. The model proposes that collagen might be constructed like a rope in which three helical strings of amino acids are twisted together. Every third unit in the helical strings is the amino acid glycine. The amino acids proline and hydroxyproline predominate at other points on the helical strings.

When collagen is stretched, it tends to compress itself like a rope, Dr. Rich said. This gives strength to the structure. He also suggested that the process of aging might alter collagen tissues through the accumulation of trace metals. This is a process akin to tanning of leather.

Collagen fibers are stable because hydrogen bonds hold the three strands together and because the three molecules tie into each other—as a rope is plaited—where they are held by the interaction of side chains on the amino acids.

Once the molecular basis of the stability of collagen tissues is understood, Dr. Rich said, it may be possible to understand the instabilities that arise in disease. The collagen diseases include rheumatic fever, other rheumatic diseases and lupus.

May 30, 1961

GAIN IS REPORTED IN HEREDITY STUDY

'Genetic Code' Partly Broken by U.S. Researchers

WASHINGTON, Dec. 20 (AP)—Government scientists said today they had succeeded in cracking partly the "genetic code," the key to the reproduction of all living matter.

Spokesmen for the National Institutes of Health hailed the work as of "major importance," and said the research might eventually make possible the following:

¶The creation of synthetic hereditary materials for use in treating certain diseases involving a hereditary defect, such as diabetes and gout.

¶A better understanding of cancer, a disease characterized by excessive growth of cells.

¶An increased understanding of normal hereditary processes, the mechanism, for example, by which similarity of facial and other physical characteristics are passed on in families. Also, the mechanism by which one group of chemicals in a developing embryo results in a heart, another in the formation of a kidney, and so on.

The "genetic code," it was explained in a report by scientists of the National Institute of Arthritis and Metabolic Diseases, is a system of messages between two chemicals contained in the nuclei or hearts of all living cells.

These chemicals are called, respectively, deoxyribonucleic acid, or DNA, and ribonucleic acid, also called RNA. The former is the stuff of which genes are made.

Together, these two nucleic acids work to bring about the manufacture of specific proteins which, in turn, are basic to the development of new cells from the original ones.

The theory has been that the DNA acts as a general storehouse of genetic information that determines the character of cells produced subsequently, and that the RNA serves as a kind of chemical messenger to transmit the information out of the original cell, and direct the manufacture of protein needed for life.

Like Morse Code

Another part of the theory

has been that the reason different kinds of protein develop from this process is that constituent chemicals in both DNA and RNA, arranged like beads on a string in a given molecule, are subject to a variety of differtri arrangements or sequences.

This, in turn, results in the selection of different combinations of the twenty different amino acids in much the same way that the sequence of the two alternates of Morse telegraph code—dots and dashes—spell out meaningful words from the twenty-six letters o fthe alphabet.

But, Dr. Marshall W. Nirenberg and Dr. J. Heinrich Matthaei reported, up to now there has been no experimental evidence that permits direct translation of such a code.

In their experiments, they prepared a laboratory system patterned after the genetic code.

They extracted genetic materials from bacterial cells and arranged things so that the DNA component could not come into play. Then they added different kinds of RNA of known composition. They found that different kinds of RNA selected different kinds of amino acids—apparently because of the particular "code" carried by the RNA. Ultimately, they were able to produce different kinds of proteins.

They have done that with fifteen of the twenty amino acids and are working toward doing it with the rest, they said.

Many Worked on Problem

Scientists all over the world have been working to break the "code of life" through which nucleic acids act.

The research has been largely independent at each laboratory working on the problem, though it is based on the same accumulated store of scientific knowledge.

Drs. Nirenburg and Matthaei reported their promising partial success in this huge endeavor last summer in Moscow at an international meeting of biochemists. They followed this with two detailed reports published in the October issue of The Proceedings of the National Academy of Sciences.

Recently, scientists at the New York University School of Medicine, led by the Nobel prize-winner, Dr. Severo Ochoa, have also made public important results in the same field.

Their disclosures appear to cover much the same territory of that of the scientists at the institute, even going beyond them in same respects.

In both cases, the accomplishment represents only a beginning, though an important one, in revealing details of the code.

December 21, 1961

Biologists Hopeful of Solving Secrets of Heredity This Year

By JOHN A. OSMUNDSEN

Biology is undergoing a revolution whose meaning and magnitude have become apparent only in recent weeks. The pace has been so swift, in fact, that many scientists do not fully appreciate the potential of the powder keg they are sitting on.

This was the clear import of a statement made by a participant in a recent symposium on genetics, who compared the state of that field today with the state of physics in 1927.

"They were saying then that physics was dead, that all of the major problems had been solved," he declared.

He did not have to mention that the first atomic bomb was detonated just about twenty years later.

In drawing his ironic comparison between the two fields, the scientist was calling attention to cries being heard today that genetics is dead.

That assessment is based upon the great likelihood that the chemical code of inheritance, which determines the form and function of every living thing and thereby provides the basis for genetics, will be cracked before the year is out. Scientific reports at the recent genetics meeting, moreover, only added conviction to that prediction.

There is no doubt that the breaking of the genetic code will rank among the greatest scientific achievements of all time. What some scientists apparently do not realize, however, is that, instead of killing genetics, breaking the code breathes new life into that field and into all of biological science.

Indeed, it is safe to say that some of the biological "bombs" that are likely to explode before long as a result of that achievement will rival even the atomic variety in their meaning for man.

Some of them might be:

¶Determining the basis of thought, if, as many scientists are beginning to expect, memory consists of information stored in giant molecules according to the genetic code.

¶Developing remedies for afflictions that today are incurable, such as cancer and many of the tragic inherited disorders.

¶Controlling the inheritance, and hence the destiny, of plants and animals and perhaps even men himself.

¶Creating life, of a sort, from chemicals that today are in the bottles on the scientists' laboratory shelves.

It may be hard to imagine that a magic cipher of any kind could be so powerful. Yet, this is certainly part of the attraction that the problem of deciphering the genetic code has had for physicists—both theoretical and experimental—mathematicians, chemists, biologists of all kinds, doctors and at least one astronomer and one political scientist.

Code a Vital Link

The question they are all trying ultimately to answer is how living things get to be the way they are. The problem reduces to a matter of deciphering the code because that is the first step in the process that ends with the fully developed organism, whether tree, horse or human.

This may be most easily appreciated if it is understood that the nature of the complete organism is determined by the interactions of all of its component cells, some acting one way, some another.

The character of each different cell, in turn, is determined by the cell's chemical composition and the chemical reactions that go on inside it.

Those reactions and the formation of structural elements of the cell are controlled primarily by substances called proteins. These are large complex molecules made up of numerous sub-units, called amino acids. And the function of each type of protein is determined specifically by the arrangement of the amino acids along the molecule.

Finally, the structure of every specific protein, that is, the order of the twenty-or-so kinds of amino acid in the chain-like protein molecule, is established by other cell chemicals called nucleic acids.

Structure Determines Nature

This is where the code comes in. The directions for hooking together amino acids into specific proteins, which will determine the nature of every cell, and thus, of the whole organ-

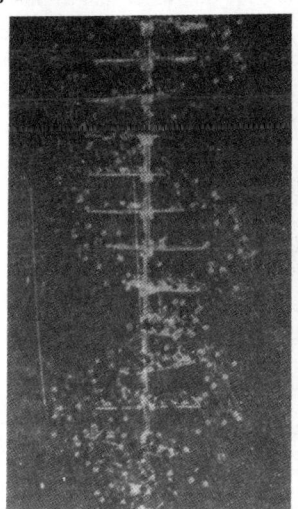
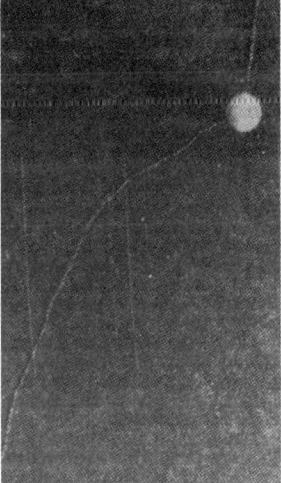
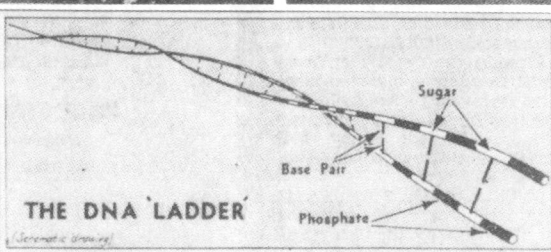

The New York Times — Feb. 2, 1962

DNA MOLECULE is visualized ideally in the plastic model of the Watson-Crick idea of its double helix structure (upper left), as it actually is in an electron micrograph of DNA strands (upper right), or schematically as a twisted ladder of sugar, phosphate and base groups (bottom).

ism, are encoded in the structure of the nucleic acids.

That code is embodied in the nucleic acids of the nuclei of egg and sperm when they unite to begin the history of a living thing.

The coded instructions for growth and development are then perpetuated throughout the lifetime of the organism when cells divide and specialize, giving bulk and identity to whatever they are part of.

This picture has been almost 100 years in the making. Almost all of the fine detail in it, however, has been supplied in the last decade and some of the most crucial strokes executed in recent weeks.

Thus has the pace in this fastest moving of all fields in science picked up as scientists near their goal of cracking the code.

A firm basis for the existence of such a code was not found until 1953, eighty-seven years after the "functional unit of heredity," later called the gene, had been discovered by Gregor Mendel, an Austrian abbot.

That winter at Cambridge University in England, two young scientists named Frances H. C. Crick and James D. Watson, an American now at Harvard University, puzzled out the molecular structure of a substance, deoxyribonucleic acid, or DNA, that exists in the nuclei of all cells and, just a few years earlier, had been shown conclusively to be the carrier of the hereditary information.

The structure for DNA that they arrived at was a double helix, a form that resembles a twisted ladder. The sides of the ladder, they found, consisted of alternate sugar and phosphate units. The rungs, between the sugars, were paired nitrogenous bases of four types: adenine, guanine, cytosine and thymine. Moreover, the bases paired in a very specific way, adenine with thymine only and cytosine only with guanine.

On the basis of that model, Dr. Crick and Dr. Watson were able to suggest mechanisms by which a DNA molecule performs its most notable functions: reproducing itself (replication) and carrying the genetic information.

The scientists speculated that the double-stranded DNA molecule might replicate by first unwinding and dividing into two long chains, in effect straightening out the ladder and splitting it lengthwise through the middles of the rungs.

Could Act as Template

Each half could then act as a mold, or template, on which a complementary strand could form. This might take place by the sticking of free nitrogenous bases in the cell to their corresponding ones on the halved-ladder mold, adenine sticking to thymine but bouncing off guanine and cytosine, for example.

In that way, a single DNA molecule would produce an exact copy of itself, resulting in two ladders with the same genetic information in both.

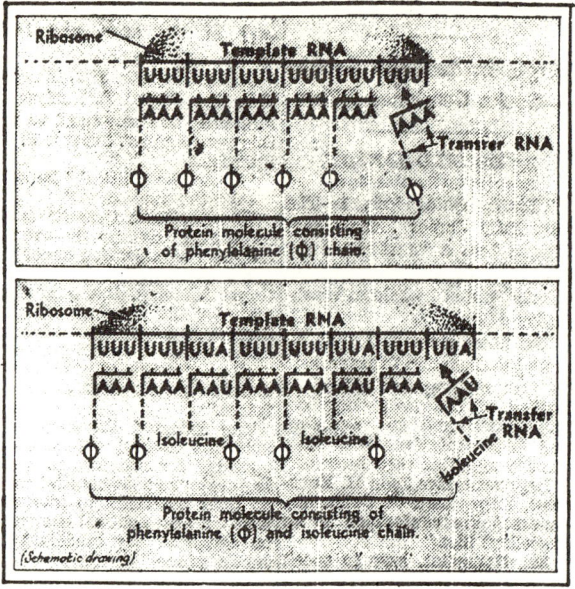

PROTEIN PRODUCTION with synthetic RNA's produced the biggest breaks in the genetic code. When all the bases of a synthetic template RNA were uracil (U), as in upper drawing, only those transfer RNA's whose attachment sequence of purely adenine (A) required that they carry phenylalanine, could match up, and a protein entirely of that amino acid resulted. This meant that the code letters for phenylalanine were U's only. "Doping" poly-U template RNA with small amounts of adenine, as in bottom drawing, incorporated isoleucine into the protein in a quantity indicating that the code for that particular amino acid would be two uracils to one adenine, or UUA.

This was demonstrated in the laboratory in 1957 by Dr. Arthur L. Kornberg at Washington University in St. Louis. For this work the scientist shared the 1959 Nobel Prize in physiology and medicine.

But in what form was that information carried by the DNA? The Cambridge scientists reasoned that the four bases, paired so that each one formed half a rung along the DNA ladder, might provide the basis for a four-letter code —A. G. C. T.—to which instructions for protein manufacture could be consigned.

That is, one base (or group of bases) might encode one particular amino acid, the adjacent base (or group of bases) might encode another amino acid, and so forth along the molecule.

The resultant chain of amino acids, complementing the sequence of bases, would then constitute the specific protein called for by the DNA.

Carrier Was Sought

That was all very well, but it had been known for some time that the site of protein manufacture is the cytoplasm, or cell sap, and not in the nucleus, where the cell's hereditary instructions are kept in the form of DNA. This suggested that another substance was needed to carry the instructions from the nuclear DNA into the cell sap.

The most likely candidate for that job was another nucleic acid called RNA or ribonucleic acid. RNA was known to exist in the cytoplasm and to be associated with protein synthesis. And its structure was very similar to DNA's.

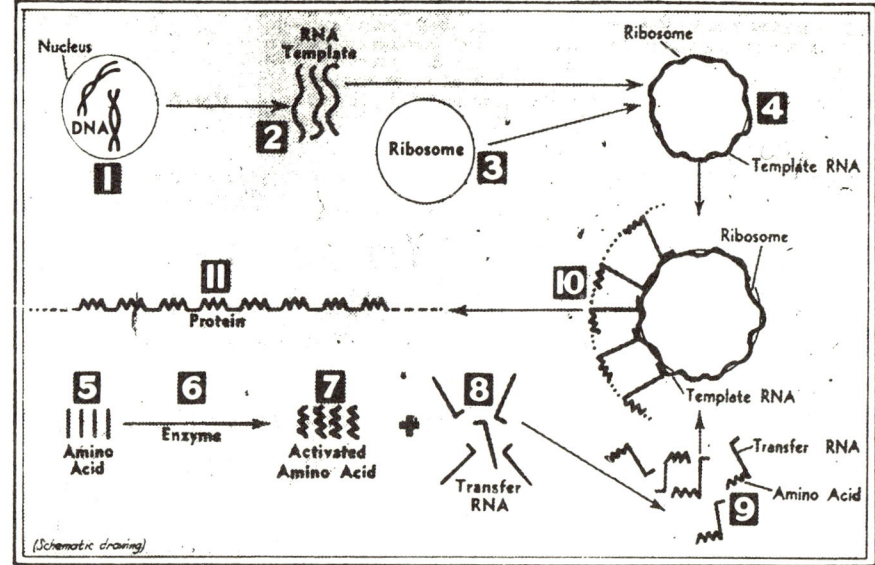

GENETIC INSTRUCTIONS for protein manufacture are coded in the structure of DNA in the cell nucleus (1), which directs manufacture of RNA strands (2) containing the same information. RNA carries the pattern to ribosomes (3) where the strands are laid down as a mold or template (4). Meanwhile, amino acids (5) are activated by special enzymes (6). The activated amino acids (7) are picked up by specific transfer RNA stands (8), and the complex (9) proceeds to the ribosomes (4) where each sticks to a special place on the template RNA according to the code. The dangling amino acids hook together (10) forming a chain that peels off ribosome as a completed protein (11).

The most important differences were that most RNA was single-stranded and that it contained a nitrogenous base called uracil in place of DNA's thymine. Otherwise, they were virtually identical.

This suggested a possible mechanism by which the genetic information could get to the ribosomes—minute particles in the cell sap—where proteins are made. DNA could direct the synthesis of RNA in much the same way that DNA reproduces itself.

In that way, the coded instructions for protein synthesis that are contained in the structure of the DNA molecule could be transmitted to RNA through the ordered arrangement of bases in the DNA-synthesized RNA strand.

Could Transmit Code

The RNA could then carry its instructions from the nucleus to the ribosomes in the cytoplasm. There, the long RNA copies of the DNA code could act as molds, or templates, on which amino acids could be laid down in chains of protein.

This theory — the role of a "template RNA" in protein synthesis — was put forth by Dr. Jacques Monod and Dr. Francois Jacob of the Pasteur Institute in Paris and has only recently been confirmed in experiments by Dr. Jerard J. Hurwitz and Dr. J. J. Furth of New York University and Dr. Paul Berg and Dr. William B. Wood of Stanford University.

That leaves three major questions: How do amino acids get to the ribosome, by what mechanism are they aligned there along the template RNA, and what sequences of bases correspond to which amino acids in the code?

The answers to the first two have been determined, and the answer to the last will almost surely be known this year.

As to how amino acids get to the ribosome, Dr. Crick and other scientists speculated a few years ago that an "adaptor" molecule of some sort might help. The discovery by several laboratories in 1957 of a type of RNA that was much smaller than template RNA, and recent experiments conducted on the basis of that finding, have identified the new RNA as the adaptor.

Called 'Transfer RNA'

Because of the adaptor's special function, it was dubbed "transfer RNA."

It works something like this: Special enzymes—protein regulators of biochemical reactions—were discovered by Dr. Mahlon B. Hoagland of Harvard Medical School in 1955. They activate amino acids. Transfer RNA molecules, one for each of the twenty-or-so known amino acids, then pick up these protein building blocks in the sap and carry them to the ribosome.

But transfer RNA is not finished there. It also provides part of the mechanism by which the amino acids are properly aligned on the template RNA.

It now appears that this is achieved by the mating of a particular group of bases on each transfer RNA strand with a complementary group of bases along the template RNA on the ribosome.

For example, the base sequence of uracil-adenine-guanine, or UAG, on the transfer RNA would mate with adenine-uracil-cytosine, or AUG, on the template RNA.

It was found that each amino acid attaches to one end of its particular transfer RNA. There are indications that the other end of that transfer RNA strand sticks to the template, like a foot. Thus, the amino acid is left dangling in a manner that allows it to join hands with similarly dangling ones on either side.

Once a full chain of amino acids has joined—in the sequence demanded by the template — the resultant protein peels off the ribosome ready to carry out its function in the cell sap, as an enzyme or hormone, for instance.

This seems to be exactly what happens, according to experiments conducted recently by Drs. Richard Schweet of the University of Kentucky and Howard M. Dintzis at the Massachusetts Institute of Technology. They found that some 150 amino acids of one type of protein lined up and linked together in about a minute and a half.

Thus, the sequential arrangement of amino acids in each specific protein reflects the genetic message encoded in the "master plan" of the DNA and passed on through the interaction of template and transfer RNA's.

This, however, does not say what the code is like—how many bases there are to an amino acid, what base sequences encode which amino acids, how this genetic information is "read out" of the base sequence, whether the code is universal or if different ones exist for different organisms.

It is on those problems that scientists lately have made the most spectacular headway.

Code Believed Universal

First, they have produced convincing evidence that the code probably is universal, greatly simplifying the task of deciphering it.

Experimental evidence for this was cited during the recent genetics symposium at Indiana University by Dr. Fritz Lipmann of the Rockefeller Institute. He told of manufacturing normal proteins from a mixture of RNA from one kind of organism and of the cell sap, containing ribosomes, amino acids and enzymes, from a completely unrelated organism.

The case for the universality of the code will be strengthened further if scientists can do that trick with normal cell sap and "unnatural," or synthetic, RNA's.

The ability to manufacture special RNA's of known composition was made possible by the discovery of an enzyme, called "polynucleotide phosphorolase," that hooks RNA parts together. Dr. Severo Ochoa shared a Nobel Prize for that achievement in 1959.

It was through the use of that enzyme that one of the biggest advances in breaking the code was made. This was announced last August at a biochemistry meeting in Moscow by Dr. Marshall W. Nirenberg and Dr. J. Henrich Matthaei, both of the National Institutes of Health in Bethesda, Md.

Manufactured RNA

They used the enzyme to make a synthetic RNA, all of whose bases were uracil, then put that nucleic acid to work making protein in the test tube. What they got back was a protein that consisted entirely of one kind of amino acid, phenylalanine. They knew immediately that they had determined the nucleic acid coding for the first amino acid.

That is, a group of bases

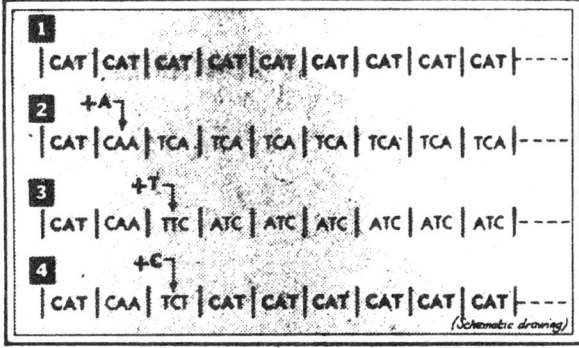

TRIPLET CODING of the genetic information in the structure of nucleic acids suggested by recent British experiments is illustrated above. Coding elements added one at a time to the message destroyed its meaning with one or two additions, restored the sense—"cat"—with a third.

The New York Times — Feb. 2, 1962

Amino Acid Code

Following is the RNA code for all twenty amino acids as determined by scientists of New York University and National Institutes of Health:

Amino Acids	RNA Bases*
Phenylalanine	UUU
Alanine	UCG
Arginine	UCG
Aspartic Acid	UAG
Asparagine	UAA, UAC
Cysteine	UUG
Glutamic Acid	UAG
Glutamine	UCG†
Glycine	UGG
Histidine	UAC
Isoleucine	UUA
Leucine	UUC, UUG, UUA
Lysine	UAA
Methionine	UAG
Proline	UCC
Serine	UUC
Threonine	UAC, UCC
Tryptophan	UGG
Tyrosine	UUA
Valine	UUG

* U = uracil; C = cytosine; G = guanine; A = adenine.

† Predicted, no experimental evidence.

Base triplets not containing U may exist.

Each sequence of bases has not yet been determined which explains why some appear with more than one amino acid.

consisting exclusively of uracil—UU, UUU, UUUU, or whatever number of bases the coded comprised—was the designation for phenylalanine. Wherever such a group occurs in the RNA molecule, a phenylalanine will appear in the corresponding protein.

The synthetic RNA — called "poly-U" — has since been used by Drs. Peter Lengyel, Joseph F. Speyer, Carlos Basilio and Ochoa at New York University and by the N. I. H. team as a "handle" to determine the base composition of the RNA code for all twenty amino acids.

This achievement was revealed for the first time at the Indiana meeting.

The experiments were done by "doping" the poly-U with small amounts of the other bases—adenine, guanine and cytosine—allowing the resultant synthetic RNA's to make protein and then analyzing the protein for amino acid content.

Variations Studied

For example, the New York group added a very small amount of adenine to the poly-U and recovered a protein that consisted mostly of phenylalanine with traces of another amino acid, isoleucine.

The proportion of uracil to adenine in the poly-U RNA indicated that the code for isoleucine was two U's (for uracil) to an A (for adenine). It could not be determined, however, whether the sequence was UUA, UAUU or AUU.

Similarly, when more adenine was put into the poly-U RNA, small amounts of asparagine were picked up in the protein in addition to the expected phenylalanine and isoleucine.

The proportions indicated this time that the code for asparagine was two A's to a U, though again the sequence could not be determined from the data.

The two-to-one composition of the base coding for the amino acids suggested that the code might be triplet in form, the four bases taken three at a time determining each amino acid.

Experimental evidence for this as well as other indications of how the code is read out were reported in the Dec. 30 number of Nature by Dr. Crick, Leslie Barnett and Drs. S. Brenner and R. J. Watts-Tobin of Cambridge University.

The picture of the code they favor is essentially this. It is a triplet, nonoverlapping code that reads out by starting at a fixed point and following the base groupings in one direction along the nucleic acid molecule.

Sequence Is Predicted

Thus, if the sequence of bases were "UUU-UUA-AAU-UUU," the sequence of amino acids would be, according to the N. Y. U. and N. I. H. findings, "phenylalanine, isoleucine, asparagine, phenylalanine."

An overlapping code, which has been proved impossible, would have produced, "phenylalanine, phenylalanine, isoleucine, isoleucine, asparagine, asparagine, isoleucine, phenylalanine, phenylalanine."

Dr. Crick's group produced experimental support for this triplet type of code with viruses that infect bacteria. They treated the viruses with certain chemicals that are believed to alter the base sequence of nucleic acids constituting the viral genes, by either adding or deleting a base at a time.

Those changes show up as mutations, alterations in the ability of offspring of the treated viruses to infect certain bacteria.

They found that the addition of one or two bases to a certain section of the virus' nucleic acid that controls that ability destroyed the function. Adding three bases to that section, however, restored the virus' infectivity for the particular bacterium.

The scientists interpreted this to mean that one base inserted in the nucleic acid threw off the message in the sequence of bases by one place in all that followed the addition. Addition of the second base threw off the message two places. The third base, however, moved the message ahead three places—the length of a whole amino acid code "word"—thereby restoring the meaning, which was to make the specific protein that would allow the virus to infect the test bacterium.

This would happen, Dr. Crick noted, only if the coding ratio was three bases to one amino acid or, less likely, a multiple of three to one. Therefore, the Cambridge scientists reasoned, the code is probably triplet and "degenerate," meaning that an amino acid might be coded by more than one base triplet. Both the N. I. H. and N. Y. U. groups have found evidence for this.

The problem remains, however, to determine the precise sequence of bases in each triplet for every amino acid and to ascertain that the code is, indeed, triplet in form. This might be achieved by statistical juggling and trial and error techniques applied to experiments such as those done by the N. I. H. and N. Y. U. groups.

The possibility of accomplishing both of those tasks more directly—the hard way—was reported at the Indiana meeting by Dr. Seymour Benzer of Purdue University.

With techniques he developed, and which were used by the Cambridge group to find the basic form of the code, Dr. Benzer has mapped individual genes in a bacterial virus.

Cistrons Are Located

That is, he determined the location of about 400 places on two sections of viral DNA, called "cistrons," which control the virus' ability to infect a certain bacterium. Each spot represents a pair of bases, a rung in the DNA ladder.

The Purdue scientist noted that certain chemicals were known to cause mutations in the virus by specifically affecting one or another kind of base in the DNA. Knowing the location of each mutation so created—from his gene map —and the kind of bases that the chemicals specifically affect has enabled him to specify the precise base pairings at several locations along the viral nucleic acid.

What he wants to do now, and believes he can do, is to find chemicals that specifically affect other bases in nucleic acid and then to isolate the protein whose synthesis is directed by the cistrons within the viral DNA he has mapped.

By analyzing the protein's amino acid sequence and comparing it with the order of the DNA bases, it will be possible to determine directly, unequivocally and precisely what the genetic code is.

With that done, to borrow a line from an article by Dr. Hurwitz in the current Scientific American:

"The dream of the gene has become the reality of the protein."

February 2, 1962

SCIENTISTS TELL OF GENETIC GAIN

'Template' RNA Is Isolated From Mammalian Cells

By JOHN A. OSMUNDSEN

A team of Rockefeller Institute scientists has reported the first isolation from mammalian cells of a chemical messenger that is believed to play a vital role in the process by which inheritance is determined.

The substance they have isolated was tentatively identified as "messenger" or "template" RNA, or ribonucleic acid.

That chemical is believed to act as a carrier of coded instructions from the genetic material, called deoxyribonucleic acid, or DNA, in the process that determines the nature of all living cells, and hence, of all living organisms.

The isolation of relatively large quantities of mammalian messenger RNA is not, by itself, considered as important as the promise that the techniques by which it was achieved hold for testing several basic biological theories and, perhaps, for empirically breaking the so-called "code of life."

Rapidly Expanding Field

At the same time, this achievement, as other recent advances in this rapidly expanding field, underlines what one member of the Rockefeller team called "problems that are too important to be left just to scientists."

The scientist Dr. Alfred E. Mirsky, was referring to the possibility that it is through such work that man may one day learn how to control his own heredity and thereby be able to practice a type of chemical eugenics.

Such prospects are probably a long way off, if, indeed they materialize at all. However, these matters are considered sufficiently important for a special symposium on them to be scheduled for the summer of 1963 in Washington, D. Mirsky noted. He and Dr. Vincent G. Allfrey, a co-author of the new report in the March number of The Proceedings of the National Academy of Sciences, then went on to discuss the significance of their findings in an interview at their laboratory.

May Lead to Brood Tests

They explained that the techniques they and their colleagues —Drs. A. Sibatani and S. R. De Koet—had developed for obtaining sizable amounts of messenger RNA might allow a specific chemical test of the widely held theory that all cells of an individual organism contain the same genetic instructions, only parts of which are acted upon by any given cell.

The nature of a cell is determined, largely, by the kinds of proteins it makes. The instructions for making each specific protein are encoded in the arrangement of structural elements of DNA within the cell's nucleus.

Those directions presumably are transmitted to sites of protein synthesis by messenger RNA which is made by DNA, the two kinds of nucleic acid having complementary structures.

According to the notion that all cells of an organism contain in each one's DNA the entire set of instructions for the whole organism, a segment of the DNA of a liver cell, for example, should contain the directions for making a certain protein found only in a skin cell.

Experiment Proposed

The Rockefeller Institute scientists propose to test this by extracting the messenger RNA from a cell that makes predominantly one kind of protein — hemoglobin is the one they have in mind—and seeing if they can get that nucleic acid to combine with a portion of the DNA from a cell that does not synthesize that protein.

Such an annealing of the hemoglobin messenger RNA with a DNA section would show that the cell had the information for making the protein but just was not using it.

Taking hemoglobin, whose structure is already known, may also allow them to determine empirically how the genetic instructions for protein synthesis are encoded in the arrangement of nucleic acid units called nucleotides.

They could do this by determining the exact sequence of nucleotides in the messenger RNA for hemoglobin and comparing that arrangement with the known sequence of amino acids in the hemoglobin molecule.

In addition, they could test the theory that the nucleic acid code is universal—that it is essentially the same in all living things. This might be accomplished by testing the compatibility of messenger RNAs among different species.

Contemplating the powerful

new tool for probing the chemistry of life that he and his colleagues had worked out, Dr. Mirsky said:

"There's just no end to the kinds of experiments that can be done now that we've actually got it [messenger RNA] in our hands."

Although the chemistry behind the way they "got their hands" on messenger RNA is somewhat complicated, their achievement reflects the capricious way science sometimes works: they found their quarry in a fraction of nuclear extract that "usually goes down the sink," Dr. Mirsky noted with amusement.

March 30, 1962

Second System of Genes Traced In Study by a Columbia Scientist

By JOHN A. OSMUNDSEN
Special to The New York Times

THE HAGUE, Sept. 7 — A vast unexplored region of genetics was opened here today with the description of the mechanism for a second system of inheritance.

The system does not involve chromosomes, the stringy structures in the cell nucleus on which the hereditary units called genes reside. Thus, the system is called nonchromosomal inheritance.

The findings were reported at a symposium here today at the 11th International Congress of Genetics.

According to findings reported by Dr. Ruth Sager, a geneticist from Columbia University in New York, the nonchromosomal system involves genes but follows rules completely different from the chromosomal one. This second genetic system also appears to influence a wide variety of vital cell traits, she said.

The chromosomal system of inheritance was discovered by Gregor Mendel, an Austrian monk, 98 years ago in his experiments with sweet peas. This system has been studied extensively since 1900, when the report of Mendel's findings was rediscovered.

Although Mendel never knew of the existence of chromosomes, he did postulate that inheritance was transmitted from parents to offspring in discrete particles. It was shown later that those little packages of posterity, which were later called genes, must be linked together as though they were on a linear structure of some sort. Further research revealed those structures to be the chromosomes.

In 1908, however, Carl Correns of Germany, one of three scientists who rediscovered Mendel's works, reported evidence for a kind of gene that was not linked to the ones on chromosomes. He suggested the existence of a second genetic system, a nonchromosomal, or non-Mendelian, one. Since then, more than 100 cases of this kind of inheritance have been described.

It was soon learned, however, that nonchromosomal genes would be extremely difficult, if not impossible, to study with the formal methods used with chromosomal genes.

One reason for this, and probably also for the failure of scientists to observe nonchromosomal inheritance in many more organisms than they have, was that the nonchromosomal genes did not seem to change spontaneously, or mutate, as chromosomal ones do.

Another reason was that the nonchromosomal genes, though possessed by both sexes, always seemed to be transmitted to offspring through the female. This would prevent scientists from studying nonchromosomal inheritance by performing experimental crosses, a fundamental tool for studying the chromosomal genetic system.

Although those properties of the nonchromosomal system presented serious obstacles to the study of it, Dr. Sager conjectured this morning that they might constitute a stabilizing mechanism that would endow the nonchromosomal genes with relative immunity to certain evolutionary forces to which chromosomal genes are sensitive.

The Columbia geneticist has nevertheless been able to get around those barriers to the study of the nonchromosomal genetic system and has produced the first description of how it works. Two findings were crucial to her achievement.

First, she and Yoshihiro Tsubo, who is now at Kobe University in Japan, developed a system for producing mutations in nonchromosomal genes and not in chromosomal ones.

Second, she found that males do sometimes transmit nonchromosomal genes, thus permitting the study of experimental crosses between maternal and paternal nonchromosomal genes.

The mutational method that they devised was based on the idea that there were probably many copies of each nonchromosomal gene in every cell, instead of just one, as there is on a chromosome. They therefore sought a way to knock out all of at least one kind of the nonchromosmal genes, so as to achieve an observable nonchromosomal gene change.

They did this by growing the micro-organism Chlamydomonas, in the presence of streptomycin under conditions that slowed the growth of the cells but did not kill them. After these cells divided several times, the offspring began to show new properties in their nonchromosomal genes. These properties were then proved to be the results of mutations.

So far, Dr. Sager has studied mutations in 20 different genes that govern such things as the requirements for certain chemicals and the growth rate.

Dr. Sager then overcame the problem of finding and using the rare cases of male transmission of nonchromosomal genes. She mated cells on a medium that would kill all those offspring that carried female nonchromosomal genes only, but not those that got nonchromosomal genes from the male as well.

The outsized cell that results from the mating is called a zygote. The ones that were in this way found to carry both parental traits, instead of just female ones, were termed exceptional zygotes.

It was through studies of progeny from those exceptional zygotes that Dr. Sager has now shown that nonchromosomal genes are, indeed, particles and that they folow rules of sorting that differ from the classical Mendelian ones that chromosomal genes obey.

In the first place, the sorting of genes among zygote progeny takes place a step later in the nonchromosomal system than in the chromosomal one. Sorting of nonchromosomal genes probably does not occur generally in nature, Dr. Sager said. There is little opportunity for it because the system operates mostly through maternal inheritance, she said.

The differences in sorting times found in this forced experimental situation was said by Dr. Sager to be further evidence that nonchromosomal systems are distinct from one another.

Secondly, the two systems produced different numbers of possible kinds of progeny.

Chromosomal genes from the two parents always sort out at a ratio of one to one among the offspring.

Ratios Are Different

Nonchromosomal genes, however, sort out at varying ratios that are probably characteristic for each type of gene. It appeared to be a consequence of a large number of copies that probably exist for each nonchromosomal gene, as opposed to one copy for each chromosomal gene.

The experimental techniques that Dr. Sager used to achieve these results are expected to be useful for finding nonchromosomal genes in another organisms and for studying them to learn their role in cell functions.

For example, Dr. Sager said that the nonchromosomal genetic system might be involved in coordinating chromosomal gene activity to promote the specialization or differentiation of cells. Or the system might direct the organization of molecules within the cell into membranes and other special structures.

Whatever their function, Dr. Sager is convinced that nonchromosomal genes represent an important element in inheritance that can now, for the first time, be studied on a systematic basis.

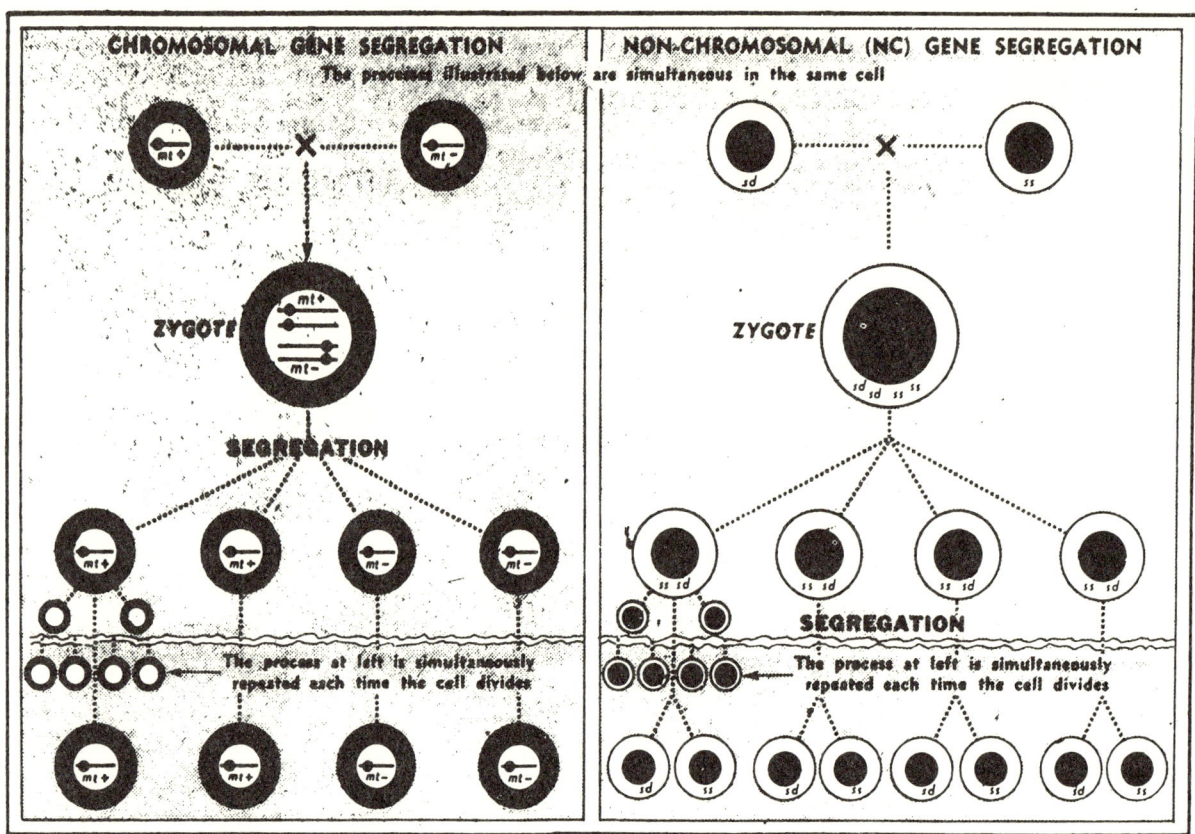

SEGREGATION: Diagram compares different ways segregation occurs in chromosomal and nonchromosomal genes

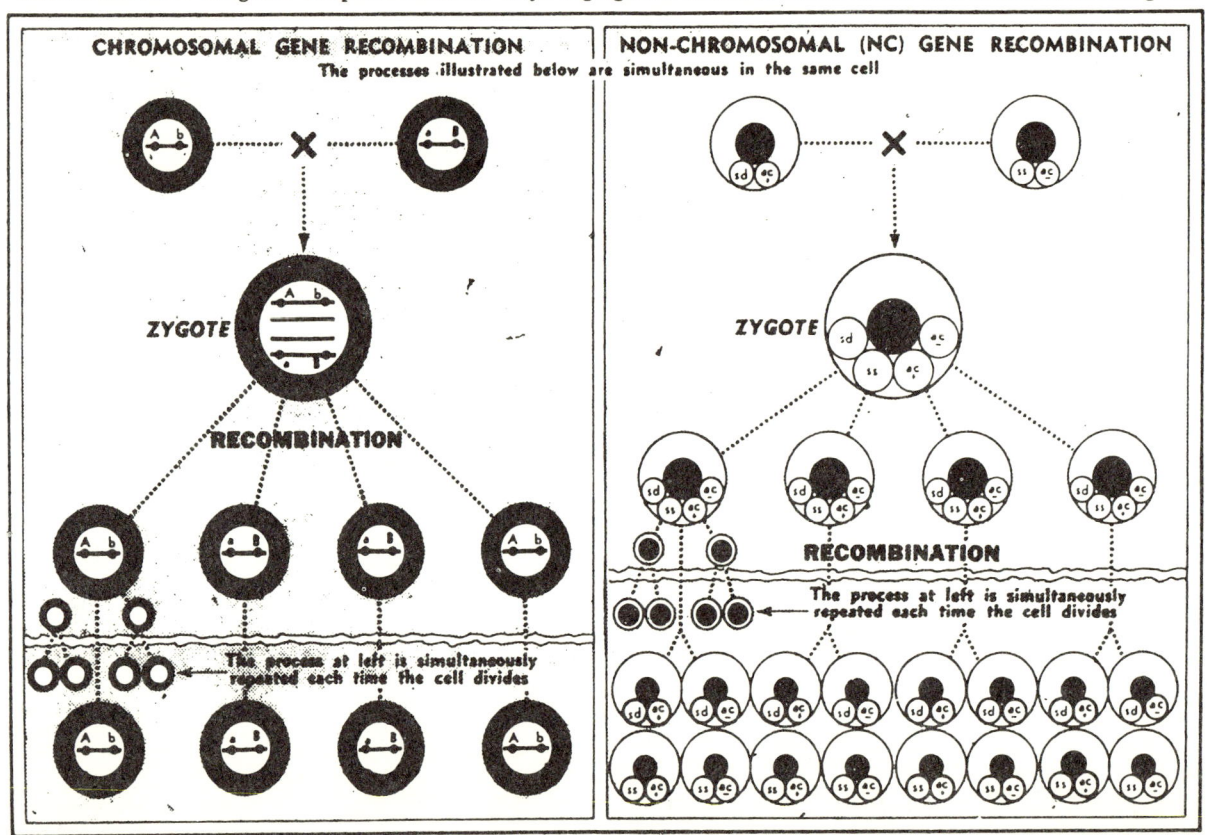

RECOMBINATION: Nonchromosomal genes recombine later than chromosomal genes do in Mendelian process

September 8, 1963

HOW DNA WORKS SHOWN IN TESTS

Only One Strand Generates the Genetic Message

By HAROLD M. SCHMECK Jr.

The message of heredity of all living things lies in the two twisted strands of the substance called deoxyribonucleic acid (DNA), but scientists are learning that only one of these two famous strands generates the message.

Experiments demonstrating this were described yesterday at a symposium of the American Chemical Society's national meeting here. The report was made by Dr. Sol Spiegelman of the University of Illinois.

DNA is generally agreed to be the essential stuff of heredity in all living things. Molecules of this substance take the shape of a double helix—a twisted spiral resembling in shape a spiral staircase.

The sub-units of which these large molecules are composed are arranged in a code that furnishes genetic instructions to each living, reproducing cell. These instructions ultimately determine the physical properties and the functions of the next generation of cells—and of all living creatures.

For some time scientists have suspected that only one of the two available strands in a given piece of DNA was generating the message, but only recently has there been proof of this.

Virus Is Defined

Dr. Spiegelman's experiments were done with viruses known to infect a common type of bacteria called Escherichia coli. In one of the viruses the active, infectious material was DNA of an unusual single-stranded form. The other was a virus containing no DNA at all, but the other form of nucleic acid called ribonucleic acid (RNA).

Viruses are minute particles that seem to occupy a position on the border between the living and the nonliving. They can reproduce only with the help of the living cells they infect.

In the experiments on DNA replication, Dr. Spiegelman showed the sequence of events that takes place when the unusual single-stranded DNA virus enters a living cell. First the virus converted itself into the normal double-stranded form of DNA by generating a second strand. Then the new strand was used to generate the genetic message for the production of new virus.

"These experiments, therefore, helped resolve a central issue of molecular genetics," Dr. Spiegelman said. "They established that only one of the two strands of the DNA helix generates translatable genetic information."

Heredity Process Traced

The experiments also showed that the single-strand DNA had to turn itself into the normal double-stranded form before it could start the process of replicating itself.

Recent research at several other institutions is also confirming, in slightly different ways, that only one of the two DNA strands is active in generating the genetic message, according to one of the leading specialists in this field. The point is considered important because it helps clarify the basic mechanics of heredity.

Dr. Spiegelman's research also showed that the nucleic acid of the RNA virus can act as a template, or instruction sheet, for reproduction of new viruses. RNA also is an essential component of all living cells. In these cells, however, its essential function is the manufacture of protein on instructions transmitted from the cell's DNA.

Since RNA viruses contain no DNA, the mechanics by which they reproduce posed something of a puzzle. Dr. Spiegelman's experiments showed that the RNA packages in these viruses have developed their own way of generating the needed genetic message and also give the pattern for manufacturing protein, including, of course, the enzyme necessary for the production of more RNA.

This basic difference between RNA viruses and DNA viruses may offer a new strategy for designing drugs to combat virus diseases, the speaker suggested in an interview following his talk.

The symposium on biochemical genetics was held at the Americana Hotel, one of several midtown hotels in which the American Chemical Society is holding its 145th national meeting this week.

September 12, 1963

NEW CLUE FOUND TO GENETIC CODE

One Organism's Genes Work in Another, Parley Told

By JOHN A. OSMUNDSEN

An Australian scientist has produced what many of her colleagues call "spectacular" new evidence that microbes, mice and men speak the same chemical language of inheritance.

The new evidence that the genetic code of life is universally the same in all living things was reported by Pamela Abel at the 29th Cold Spring Harbor symposium on quantitative biology last week. Dr. Abel did her research with Dr. T. A. Trautner at the genetics institute of the University of Cologne, Germany.

What Dr. Abel and Dr. Trautner did was to make the genes from one organism work in the strange environment of another, completely alien, organism.

The genes of all living things are believed to carry hereditary instructions for making all the different kinds of protein molecules that make each organism what it is. Those instructions are believed to be carried encoded in the structure of the stuff that genes are made of—nucleic acids. The code is translated by the protein-making machinery of the cell.

If the genetic code were universal, it would mean that in all organisms identical nucleic acid "words" would translate into identical protein sub-units. Thus, the nucleic acid from any organism should be able to direct the manufacture of its specific proteins when put together with the protein-making machinery of any other organism.

That such a natural law might be operating for the first time in 1961 when Rockefeller Institute scientists mixed genetic material from a rabbit with protein-synthesizing apparatus from a bacterium in a test tube and obtained protein that was similar, if not identical, to that from the rabbit.

The new work reported by Dr. Abel makes the argument for a universal genetic code even more convincing because not only protein was produced but also an entire biological entity—a virus—was assembled by a cell operating under the direction of just the genes from the virus that normally does not live in that cell.

The idea of trying this unlikely experiment occurred to Dr. Abel, she said, when colleagues found that a bacterium, called Baccilis subtilis, could be made infectable with the nucleic acid—the genes—from a virus whose natural host was bacteria and would produce new virus particles as a result of that infection.

Learning of that, she suggested that the same thing be tried with vaccinia virus, the kind used in smallpox vaccine.

At first, her co-workers scoffed at this idea, she said, because viruses are notoriously finicky about the kinds of cells they infect, most viruses having a very specific and narrow range of host cells they will multiply in. Moreover, the vaccinia virus's host is the mammal —notably cows and people—not bacteria.

Nevertheless, her experiment was tried, and it worked. A preparation of mostly the nucleic acid from vaccinia virus was put into the bacterial cells, and whole, ineffective virus particles came out, Dr. Abel reported.

Since this initial success, scientists in California and in Tübingen, Germany, have produced mature virus particles by infecting Baccilis subtilis with the genes from a mammalian cancer virus and by infecting tobacco leaves with the genes from a virus whose normal host is bacteria.

June 14, 1964

STRUCTURE FOUND FOR NUCLEIC ACID

Discovery of the Pattern of an RNA Hailed as a Step to Understanding Life

By JOHN A. OSMUNDSEN

American scientists, for the first time, have worked out the structure of a nucleic acid—a substance that helps direct the development of form and function of all living things.

The feat was hailed by authorities in the field as being of first rank importance in man's efforts to understand and ultimately control the evolutionary destiny of life on earth.

One leading biochemist, who is noted for his conservatism, termed the achievement "world shattering."

He said it marked a starting point from which scientists may hope eventually to write out the formulas of nucleic acids and designate in chemical symbols the specific parts of the molecules which direct particular actions within living cells.

This would amount to writing down the genetic code word by word and making sense out of it all.

Parallel to Rosetta Stone

Another biochemist drew an analogy between the contribution this first elucidation of the structure of a nucleic acid would make toward understanding the language of heredity and that which the deciphering of the Rosetta stone made toward the comprehension of Egyptian hieroglyphics.

It was made clear, however, that many fundamental and practical prospects for development of the new achievement may be years—perhaps decades —off. The reason for so long a delay after this "breakthrough" had to do with the enormous complexities of the problem, which further emphasized the difficulty of the feat.

The scientists who reported the new achievement in the March 19 number of Science were Robert W. Holley, Jean Apgar, George A. Everett, James T. Madison, Mark Marquisee, Susan H. Merrill, John Robert Penswick and Ada Zamir of the United States Plant, Soil and Nutrition Laboratory of the United States Department of Agriculture and the Department of Biochemistry of Cornell University in Ithaca.

4 Kinds of Material

The nucleic acid whose structure those scientists have worked out represents one of four kinds in a living cell.

One is deoxyribonucleic acid, or DNA. This is the genetic material in whose structure are encoded the hereditary instructions that make every living organism the way it is.

That is accomplished through the replication and transmission from cell generation to cell generation of the DNA and through the direction by DNA of the synthesis of three kinds of ribonucleic acid, or RNA. Those RNA's in turn direct the manufacture of the workhorse molecules of life, proteins.

One of the three RNA's is called messenger. It carries the hereditary instructions for protein synthesis to minute "protein factories" called ribosomes, in the cell. There amino acids are strung together into chain-like protein molecules. Ribosomes are made of the second kind of ribonucleic acid, ribosomal RNA.

The third form of ribonucleic acid is called "transfer RNA" because it transfers amino acids from the cell fluid to the ribosomes for assembly into proteins according to the instructions borne there from DNA by messenger RNA.

There are some 20 amino acids and perhaps three different transfer RNA's that are specific for each, or roughly 60 transfer RNA's. It was a transfer RNA specific for the amino acid, alanine, whose structure the scientific team, led by Dr. Holley, has elucidated.

He explained roughly how this was done in a press briefing at Cornell Medical College on Wednesday. The accomplishment was the culmination of about nine years of work in finding ways, first, of isolating alanine transfer RNA from more than 50 others in bakers yeast, and then of breaking up the molecule into fragments that could be pieced together like a linear jigsaw puzzle.

This was done, essentially, in four steps.

First, the RNA molecule was digested into small fragments with an enzyme, or biological catalyst, that selectively cleaved the chain of RNA subunits, called nucleotides, next to two particular ones, named cytidine and uridine. This produced 19 sequences of nucleotides.

Second, a similar digestion was carried out with another enzyme that cleaved the molecule next to the nucleotides, inosine and guanosine. This resulted in 17 such products.

Then, the results of the two digestions were compared for matchings and overlappings of nucleotide sequences among those groups of fragments. This produced 15 sequences of nucleotides. One of them was clearly from one end of the RNA molecule, another from the other end.

Finally, the gentle digestion of the transfer RNA molecule with one of the enzymes into a few large fragments furnished enough additional data with which to compare previously obtained nucleotide sequences to enable the scientists to determine the order of the 13 nucleotide sequences between the two end ones.

In that way, the scientists finally specified the complete sequence of 77 nucleotides in the alanine transfer RNA molecule.

Dr. Holley said that his group had also isolated the transfer RNA's specific for the amino acids, tyrosine and valine, and that techniques developed for determining the structure of the one that carries alanine would be useful for elucidating the structures of those next.

Comparisons of the structures may, he said, give clues to the activities of these vital molecules and may also provide the basis for attempts to synthesize them in the laboratory.

Thus, nucleic acid biochemists appear to be embarking on a course similar to the one launched about 15 years ago by Dr. Fred Sanger, the Nobel Laureate who, in fashion similar to Dr. Holley's, dissected and determined the structure of the first protein molecule, insulin.

Today, the structures of several other protein molecules are known, much is understood about how they act, and quite recently, the laboratory synthesis of biologically active insulin was achieved.

March 19, 1965

U.S. Scientists Decipher Structure of Key Enzyme

By WALTER SULLIVAN

After 16 years of intensive work and an expenditure of $2-million, scientists have deciphered the extremely complex structure of an enzyme that plays a key role in all living cells.

The substance, known as ribonuclease, contains more than 1,000 atoms. These atoms are arranged as a chain of 124 amino acid units twisted, coiled and cross-linked in intricate ways.

The feat is the first of its kind in this country, according to those who performed it at the Roswell Park Memorial Institute, in Buffalo.

The molecular structure was explored with the techniques of the science known as X-ray crystallography. Two other proteins—myoglobin and lysozyme —have been similarly completely deciphered in Britain. Like all enzymes, ribonuclease is a protein.

These British and American achievements are regarded as being of momentous importance, for they demonstrate the feasibility of determining how enzymes are built and how they work. Without the help of enzymes most of the body's chemical reactions would proceed at an extremely slow pace—or not at all.

Discovery of the complex structure of an enzyme, and particularly of the points in its structure that perform the chemical mission of the substance, opens the way to a host of possible applications. Chemists, it is believed, may be able to alter the structure to modify the role of the enzyme. Such tinkering with molecules is a standard pathway toward new life-saving drugs.

Simpler Versions Possible

Furthermore, knowledge of the active sites on a molecule has, in the past, made it possible to construct simpler man-produced versions that are just as effective—or even more so.

The role of ribonuclease is to break down another vital substance in the life process—ribonucleic acid, or RNA. The task of RNA is to carry, in its structure, the "blueprint" for manufacture of proteins by the cell. The RNA obtains this blueprint from another substance, DNA (for deoxyribonucleic acid), that serves as the archive of design. The DNA also passes the information on to future generations as the stuff of heredity.

There are stages in the life of a cell or an organism when it is important to deactivate RNA. This may be necessary to control cell growth or to dispose of the RNA when a cell dies. Such deactivation is the role performed by the newly deciphered enzyme.

Because this substance exerts control over cell growth, knowledge of its structure (and of ways in which it may become deformed) might help explain why cancer cells spread, unchecked, through the body. This, in turn, could provide clues for the development of ways to treat the disease.

Dr. David Harker, leader of the Roswell Park research team, began work on the deciphering of ribonuclease in 1950, while he was at the Polytechnic Institute of Brooklyn. He moved to Roswell Park in 1959. The institute there is a center for cancer treatment and research operated by the New York State Department of Health.

Some months ago this newspaper learned that the task of decipherment was virtually complete, but agreed to delay reporting on it until the results were sufficiently firm for presentation at a scientific meeting. They are being reported currently by Dr. Gopinath Kartha at an international meeting in Madras, India.

Dr. Kartha, who was educated at the University of Madras, has spent the last five years working on the project at Roswell Park. Dr. Jake Bello is also a member of the group.

The structure of the ribonuclease molecule was determined by scanning crystals of the substance with a beam of X-rays and analyzing the results on a computer. A crystal is formed when molecules of a given substance grow, one upon the other, into a symmetrical structure billions of times larger than the individual molecules.

When X-rays penetrate such a crystal they are diffracted by electrons in the molecules of the crystal, producing a characteristic pattern.

Because all of the molecules are aligned the same way, the effect is, in essence, that which would be produced by a single molecule. The latter would, of course, be unobservable, but the diffraction pattern generated by a whole crystal can be analyzed.

The first step, therefore, was to obtain pure crystals of ribonuclease. The most readily available source of the raw material was the pancreases of newly slaughtered cattle and the Roswell Park group turned to the Armour Company for that. The next problem was to grow crystals of the substance. This takes from two to six months and even then the crystals are hardly large enough to be visible.

Finally, the crystal had to be scanned with the X-ray beam and the results analyzed. This process is so laborious and difficult that it could only be achieved through extensive automation of the scanning process and long hours of work with computers.

During the last two years the IBM 7040 computer at Roswell Park has spent a third of its time on this project, and additional work was allocated to an IBM 7044 computer at the Buffalo campus of the State University of New York.

The core of the problem, as explained by Dr. Harker in a telephone interview, was to obtain a three-dimensional view. When one makes a chest X-ray the result is two-dimensional, but since the body's internal structure is already known, this is not much of an impediment. In this case, however, the goal was to learn the structure in depth.

X-rays, like other forms of light, have wavelike properties. When such waves are scattered from various points and depths within a crystal they interact with one another. Some waves merge to form bright spots; others cancel each other out. From these interactions it should theoretically be possible to calculate where the electrons are that have done the scattering.

However, in a molecule formed from more than 1,000 atoms the electrons are extremely numerous. It is therefore first necessary to outline the structure in general terms and this is done by sprinkling heavy atoms, such as those of uranium, through the molecules. The heavy atoms serve as survey beacons on which to hang the rest of the structure.

This technique was developed by Dr. Max F. Perutz of Britain's Cambridge University in the nineteen-fifties. He applied it to a partial analysis of horse hemoglobin, the respiratory pigment of red blood cells. Then his colleague, Dr. John C. Kendrew, used the method to decipher the structure of myoglobin (muscle hemoglobin) from sperm whales. It was the first complete mapping of a protein and won a Nobel Prize for both men.

This was followed in 1965 by the mapping of lysozyme, an enzyme that breaks down cell walls. It was achieved by a group at the Royal Institution in London under Dr. David C. Phillips. Lysozyme is formed from the 20 kinds of amino acid typical of all proteins arranged in twisted chain of 129 amino-acid units.

The heavy-atom "dyes" used in the British work to obtain three-dimensional information were, in general, inapplicable to the problem being attacked at Roswell Park. Scores of dyes were tried and seven were used in treating the crystals slated for analysis.

It was by comparing the diffraction patterns produced by these differently dyed crystals that the computer reconstructed the electron densities at various depths within the crystal.

The computer then generated a contour map of electron densities for a succession of cross-sections through the crystal. When these maps were inscribed on transparent sheets of plastic and stacked in the proper sequence, a three-dimensional picture of the molecule emerged.

It looks somewhat like a pile of worms—a tangle whose coils are locked together by four disulphide bonds. The amino acids themselves also interact through their hydrogen atoms to keep the structure in place.

The sequence of amino acids in this molecule and the locations of the disulphide bonds have been known for several years. The mystery concerned the manner in which this long rope was knotted and coiled.

As with other such substances, there are thought to be subtle differences in the structure of ribonuclease found in various organs of the body, as well as in various species of organisms. Also the molecule can stack into 14 different crystal patterns. Only one of these has been charted, but Dr. Harker does not believe this detracts from the validity of the results.

As with the British work, the technique was not sensitive enough to display the locations of individual atoms. However the atomic structure of the component amino acids was already well known. Roughly a half million diffraction points within the crystal were calculated. With the automation developed at Roswell Park Dr. Harker believes other proteins can now be analyzed with from 5 to 10 years' work, instead of 16.

Such analyses are already under way at a number of research centers.

January 22, 1967

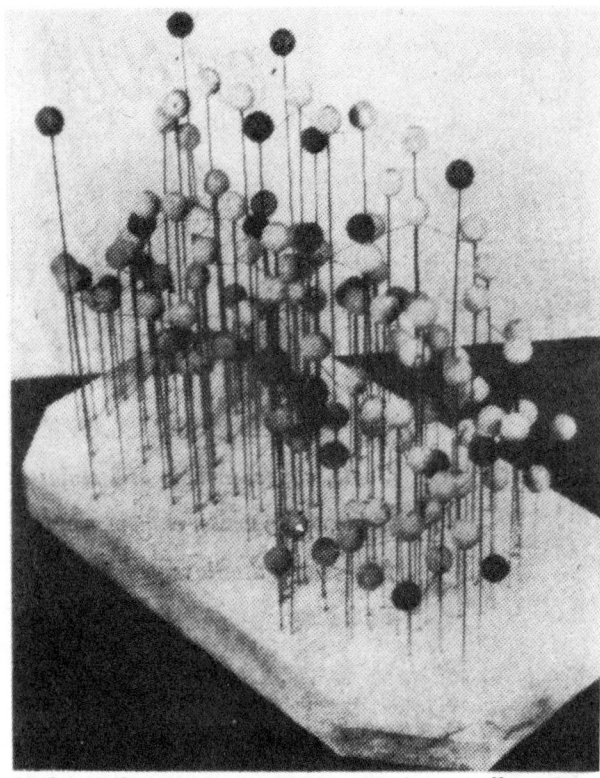

Model of the complex molecular structure of ribonuclease

2 KEY SUBSTANCES IN GENETICS FOUND

Proteins Are Believed to Control Gene Operation

Three Harvard scientists have isolated and identified for the first time two of the cell substances believed to control the process of turning genes "on" and "off."

Genes are the units within chromosomes that transmit hereditary characteristics and also govern the devlopment of cell structures.

The switching mechanism by which these units come into operation (on) or lie dormant (off) has been generally understood since 1961, but the precise biochemical process involved in the control mechanism has awaited isolation of the control substances.

The two substances, isolated from bacteria, were found to be protein molecules, one large and one small. Each acts in a system that is roughly analogous to a household thermostat that turns a furnace on and off.

The experiments were performed by Prof. Walter Gilbert, a 34-year-old biophysicist; Dr. Benno Müller-Hill, 34, a German-born biochemist, who is a research fellow; and Mark Ptashne, 26, a molecular biologist, and junior fellow at Harvard.

Their findings make possible the detailed study of the mechanism by which the expression of genetic characteristics are controlled. One geneticist said that such an understanding could have great medical significance as a step toward eventual control of diseases and disorders known to involve malfunctioning of genes.

"These findings confirm that a single gene can make a product that can turn off other genes," Mr. Ptashne said in a telephone interview. He added that, while the research dealt with bacterial cells, the knowledge gained could ultimately lead to elucidation of how the control mechanisms operate in complicated higher organisms.

Genes control cell development by directing the manufacture of enzymes, which are the proteins that govern the cell reactions that give each cell the special characteristics it needs to perform its job.

Each cell in the body contains an exact copy of every gene the organism possesses. Whether the cell becomes, say, a lung cell or a skin cell depends on which of its genes are turned off and which are turned on.

The general outlines of the control mechanism, in bacteria, were described in 1961 by two Frenchmen associated with the Pasteur Institute in Paris, Dr. Jacques Monod and Dr. François Jacob. For this work they won the Nobel Prize for Medicine in 1965.

According to the scheme proposed by the French scientists, the control mechanism is a negative one.

That is, a regular gene sends out a substance that represses an operator gene, an intermediary whose function it is to activate directly the structural genes that determine cell structure. When the repressor substance is blocked, however, the operator gene is free to do its job in activating the structural genes.

Drs. Monod and Jacob theorized that the represser substance passes, in effect, through a switch that is open or closed depending on the presence of a metabolite. A metabolite is a chemical substance needed by the cell for various interactions.

When the metabolite is present, the "switch" is closed and the repressor substance cannot "get through" to shut down the operator gene. Thus, the structural genes are free to function.

Conversely, when the metabolite is absent, the "switch" is open and the repressor substance does "get through" and shuts down the operator gene. Thus, the structural genes lie dormant.

This concept has won strong adherents among scientists.

The two substances isolated by the Harvard researchers are repressor substances.

One, isolated by Professor Gilbert and Dr. Müller-Hill, is the repressor involved in controlling the ability of a common intestinal bacterium called E. coli to use the sugar lactose for energy. It turned out to be a large protein.

Mr. Ptashne isolated the repressor that prevents the E. coli bacterium from being burst when attacked by infecting viruses. This repressor was found to be a small protein.

The results of the Harvard experiments were published in recent issues of the Proceedings of the National Academy of Sciences.

April 9, 1967

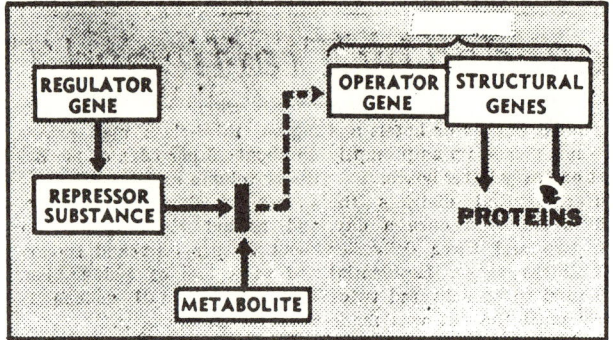

Schematic drawing of gene control mechanism. Operator gene activates structural genes, which produce proteins that control cell structure. Operator gene is turned off by repressor substance, sent out by regulator gene. Metabolite, a substance needed by cell, acts as switch, blocking repressor and allowing operator and structural genes to function. Harvard scientists isolated two repressors.

Core of Virus Is Made Artificially

Way Is Opened to Advances in Genetics

By HAROLD M. SCHMECK Jr.
Special to The New York Times

PALO ALTO, Calif., Dec. 14—Scientists have produced artificially in the laboratory the active, infectious inner core of a virus.

The achievement, by scientists at Stanford University, seems close to the laboratory production of life itself. It does not guarantee, however, that this will ever be achieved.

The active virus core material is DNA, or deoxyribonucleic acid, the master chemical of all life and the substance that determines the heredity of every living thing.

So far as is known, biologically active DNA had never before been produced artificially in the laboratory.

The accomplishment is believed to open the way to important progress in the study of genetics, the nature of virus infections and probably the nature of cancer.

Many Roads Open

The research, therefore, is believed to be significant for many of the illnesses and biological problems that beset mankind. It may be even more important in its potentialities for disclosing the most intimate processes of life itself.

When the artificially produced DNA was put into living cells it infected the cells just as a normal virus would. The infected cells stopped their own normal functioning to produce viruses instead. Then the cells ruptured and died and a host of new viruses spewed forth.

These viruses are indistinguishable from natural viruses, but their origin is the DNA produced artificially in the laboratory. In that respect, they might be called man-made viruses.

How close this achievement is to the test tube production of life is probably impossible to answer. The man-made DNA produces virus in the same way a natural virus does—by invading and infecting a cell.

But scientists are unsure whether to classify viruses among the living things. Most specialists say that viruses are on the borderline between the living and the inanimate.

Dr. Arthur Kornberg, Nobel Prize winner and senior member of the Stanford research team, believes the method can be used to synthesize other viruses, and to produce and modify the hereditary material of some living cells. He believes the method can also be used for the study and possibly the eventual control of some kinds of cancer.

It is widely believed that viruses are a factor in some human cancers, although this is yet to be proved. It is considered certain that cancers represent derangements in the cells' control system.

Landmark Is Praised

Dr. James A. Shannon, director of the National Institutes of Health, the Government's major institution for medical-biological research, said at Bethesda, Md., the accomplishment unquestionably will stand as one of the great landmarks of research in the life sciences."

It is the culmination of 11 years of work largely supported by the National Institute of General Medical Sciences, a unit of the National Institutes of Health, and by the National Science Foundation. The earlier work won the Nobel Prize for Dr. Kornberg, who is professor and executive head of the department of biochemistry at Stanford.

At a news conference today, Dr. Kornberg said that the achievements of his group rested on a solid background of research contributed by many scientists at many institutions.

Five research teams—two at Stanford, one at Harvard, one at the National Institutes of Health and one at Albert Einstein College of Medicine in New York—all contributed independently to the discovery of this enzyme that acts to join pieces of DNA together end to end.

While the Stanford research has produced the first artificially formed active DNA it is not the first project to help a virus be fabricated from such artificial precursors.

A similar achievement was reported in 1965 by Dr. Sol Spiegelman of the University of Illinois. He, too, produced living living viruses from off-the-shelf ratory materials.

The core material of the viruses he produced, however, was ribonucleic acid (RNA). This is the other master chemical of life, but its biological position is not quite so central as that of DNA.

DNA is the chemical substance of which the genes of all living things are made. These genes, which lie on the threadlike cell structures called the chromosomes, provided the chemical blueprints that determine the form and the function of every living thing.

The function of RNA is to translate these blueprints into action by guiding the manufacture of materials within the living cell.

Some viruses have a core of DNA. In others, it is RNA. When a virus infects a cell, the viral core material can subvert the genetic machinery of the cell and make it produce more viruses instead of its own normal materials. The cell then dies and the new generation of viruses emerge.

The achievement of the artificially produced infectious DNA at Stanford is reported in the forthcoming December issue of the proceedings of the National Academy of Sciences, one of the nation's most respected journals.

The co-authors of the report with Dr. Kornberg were Dr. Mehran Goulian, formerly a postdoctoral fellow at Stanford and now on the faculty of the University of Chicago, and Dr. Robert L. Sinsheimer of the division of biology at the California Institute of Technology.

Their artificially produced viruses are indistinguishable from natural viruses of a kind called Phi X 174. Dr. Sinsheimer has been a pioneer in the study of these viruses, which preyed naturally upon intestinal bacteria called Escherichia coli.

These viruses were chosen for the research at Stanford because they have been thoroughly studied and because they have several unusual properties, Dr. Kornberg said in an interview.

They are extraordinarily small, even for viruses, and are believed to contain only five or six genes in their little strand of DNA. Furthermore, the inner core of genetic material consists of only a single strand of DNA.

In most viruses and all known living cells, the DNA is normally double-stranded. Its form is roughly that of a ladder twisted into a long corkscrew or helix shape.

Another unusual characteristic of the DNA found in the Phi X 174 virus is that the genetic material is not in the form of a linear strand with two ends, but in the form of a closed ring.

The objective of the research was to produce reams of DNA like those found in the virus and to copy them so faithfully from the natural model that they would have the same biological properties as the natural virus.

This had never been done unequivocally with a DNA virus.

There were two essential tests of success—the production of viruses after the artificially made DNA invaded living cells and proof that no natural virus material was present.

The experiments required several complex steps.

First, the scientists took natural virus DNA that had been tagged with radioactive material so it could be identified.

To this almost natural material the scientists added off-the-shelf supplies of the four basic subunits or building blocks of DNA. They also added an enzyme called DNA polymerase.

The original virus DNA was to act as a template — a blueprint. The off-the-shelf building blocks of DNA lined themselves up against this template in a set complementary sequence. This is the normal way in which DNA replicates itself.

But the Phi X 174 virus DNA is in the form of a ring. The scientists had to close the ring of the complementary strand that formed around the original. To do this they used a second and crucial enzyme that they call a polynucleotide-joining enzyme.

This produced the desired effect. The ring closed. Now the scientists had what they call a duplex ring with the radioactively tagged but otherwise natural virus DNA forming the inner ring and the complementary strand formed from off-the-shelf chemicals forming the outer.

Chemically the outer strand was the same as one that would have been produced naturally in a virus infection except for one point. The scientists had substituted the chemical bromouracil for the chemical thymine that would be formed in nature.

This made no appreciable difference biologically but it made the synthetic product heavier than the natural rings. Consequently, the two could be separated by spinning them in a centrifuge.

To do this, however, each ring had to be broken in just one place to produce strands, rather than rings of DNA. This was done by adding a small amount of a third enzyme called DNase. The amount was enough to produce single breaks in about half the ring.

The delicacy of this work can be imagined by considering that the whole virus itself is too small to be seen with anything but a powerful electron microscope. Yet the scientists were manipulating just the inner core material of these viruses.

Once the rings had been broken, the centrifuge was used to separate and harvest the totally synthetic outer rings the scientists had made with their off-the-shelf building blocks.

The polynucleotide-joining enzyme was used again to reform these inner original ring shapes.

These synthetic rings were then put in cultures of living cells to see if they were infectious.

They were and they gave rise to viruses just like the natural virus.

This was a remarkable achievement in itself, but it was not quite the end of the scientists' quest. The synthetic rings of DNA were complementary to, not identical with, those of the natural virus.

To complete the story the scientists used these synthetic rings of DNA as templates to form another set of rings. These were complementary to the synthetic rings and therefore were identical to the original DNA of the natural virus. But those copies of the viral DNA had been made on an artificially prepared template and were built of off-the-shelf chemical building blocks.

These have also proved infectious in living cells and gave rise to complete viruses indistinguishable from the natural virus called Phi X 174.

The first actual production of the DNA that proved to be biologically active was done in June.

The four essential building blocks needed to produce the DNA are the nucleotides of adenine, thymine, guanine and cytosine. Each is available commercially.

The enzymes vital to the project, however, require sophisticated purification and extraction techniques.

The joining enzyme, an important key to the project, was discovered only this year.

Johnson Hails Feat
Special to The New York Times

WASHINGTON, Dec. 14—President Johnson hailed the Stanford announcement today as a great achievement and as evidence of the "creative partnership" that has developed over the years between science, the universities and the Federal Government. He expressed pride and congratulations on the achievement.

Addressing an audience gathered for the bicentennial of the Encyclopedia Brittanica at the Smithsonian Institution, Mr. Johnson told them of the biochemical achievement that was the "spectacular breakthrough in human knowledge" that was being announced "at this very moment."

"It's going to be one of the most important stories you ever read," he said, explaining that for the first time man had succeeded in manufacturing a synthetic molecule that displays the full biological activity of a natural molecule in a living organism.

"These men have unlocked a fundamental secret of life," the President said. "It is an awesome accomplishment. It opens a wide door to new discoveries in fighting disease and building healthier lives for mankind. It could be the first step toward the future control of certain types of cancer."

December 15, 1967

Viral Evolution Is Guided in Test Tube

By ROBERT REINHOLD
Special to The New York Times

TOKYO, Aug. 20 — The development of the first method of observing evolution under artificial conditions was reported here today by a leading authority on chemical genetics from the University of Illinois.

The achievement not only is of great theoretical value in understanding the underlying chemical mechanisms of heredity, but also suggests a new approach to curing viral diseases such as cancer by making the viruses incapable of infecting cells.

The Illinois scientist, Dr. Sol Spiegelman, described the development and its implications to an overflow audience of geneticists on the opening day of the 12th International Congress of Genetics at the Tokyo Prince Hotel.

His lecture stirred wide comment at the gathering and was hailed by many as an important new milestone in genetic studies. Dr. George W. Beadle, Nobel Prize-winning geneticist, who is honorary president of the congress, termed the report "very exciting."

'Test-Tube Evolution'

Others said it opened the possibility of "test-tube evolution," allowing scientists to observe, and control, the molecular events associated with evolutionary change under controlled laboratory conditions. Such studies are extremely difficult to perform in living organism.

"For the first time we can study the evolution and genetics of a molecule outside a cell," said Dr. Spiegelman in an interview.

The congress was opened earlier in the day in a ceremony at which Crown Prince Akihito and Princess Michiko greeted the 1,400 geneticists from 53 countries. Held only once every five years, this conference draws many of the world's leading authorities on heredity and is a kind of summit meeting of genetics.

Dr. Spiegelman's work was performed on the genetic material of a virus that infects bacteria. This material, called ribonucleic acid, or RNA, was first produced artificially in active form in Dr. Spiegelman's laboratory in 1965.

Catalyst Is Used

Normally, when the virus infects a cell, it produces more viruses. The RNA carries the information that directs this reproductive mechanism. All that is required to produce more of this RNA artificially is an enzyme, or biological catalyst, called replicase.

In the work reported today, viral RNA was placed in a test tube with replicase. As the RNA strands began to replicate themselves, only those that did so with unusual speed were allowed to continue. Eventually a strain of fast replicating viruses with a high affinity for the enzyme was obtained.

Thus, by setting up special conditions for survival, Dr. Spiegelman was able to produce genetic material that had no need for any genes except those dictating fast replication. Genes for other functions, such as cell infection, he theorized, were not needed and would ultimately be lost under the laws of natural selection.

This, in fact, was confirmed by experiment. What Dr. Spiegelman had witnessed was evolution at the most primitive and basic level.

For Killing Infection

Dr. Spiegelman believes the technique is of potential chemotherapeutic value in the destruction of infections, disease causing viruses. For example, a strand of fast replicating RNA could be introduced into the cell. Because of its high affinity for replicase, it would quickly monopolize the enzyme and leave the cell unable to produce new viruses.

Western scientists were impressed by the size of the Soviet delegation at the congress. Although not all had arrived by the opening session, 88 Russians had registered in advance.

This was considered a tribute to the recovery of Soviet genetics, which had stagnated for years under the rule of Trofim D. Lysenko. Mr. Lysenko, a geneticist who clung to the discredited notion that acquired characteristics could be inherited, was ousted soon after the fall of Nikita S. Khrushchev as Premier.

On the other hand, about 800 Americans had been expected, but more than 200 canceled their trips, largely because the Federal Government ordered cutbacks in travel grants to scientists this year to reduce expenditures abroad.

August 21, 1968

An Enzyme Is Synthesized for First Time

By WALTER SULLIVAN

For the first time an enzyme —one of the complex "master chemicals" of life—has been synthesized.

The feat, achieved virtually simultaneously by two laboratories using basically different methods, is expected to open new fields of research into the most intimate chemical processes of life.

"This is probably the beginning of a new generation of therapeutic agents," said Dr. Robert G. Denkewalter, vice president for exploratory research at the Merck Sharp & Dohme Research Laboratories, as the achievements were announced yesterday.

The enzyme ribonuclease has been synthesized not only by the Merck Laboratories, in Rahway, N.J., but also by the Rockefeller University, York Avenue and 66th Street, where yesterday's joint announcement was made.

Ribonuclease, which performs vital functions in almost all cells, has long been a target of research and analysis because of its comparatively simple structure.

Although the Merck researchers spoke of a new generation of drugs, neither they nor the scientists at Rockefeller University expected any immediate applications in medicine. Rather, they said, the door has been opened for research into how enzymes work. Once that has been learned, medical applications will inevitably follow.

As Dr. Denkewalter put it, enzymes are so important to life that "anything we learn about them is likely to be useful in medicine." They have long been a challenge — and "an embarrassment," he added.

They are an embarrassment because they enable the body to perform, in fractions of a second and at body temperature, chemical acrobatics that scientists, with all of their equipment and knowledge, can do only by means of extremely high temperatures, strong acids and other brute force methods.

Enzymes enable us to eat, breathe and exist. With them the body converts sugar and other basic foods into the countless substances needed for our bodily functions. The enzymes are essential to tissue-building, blood cell replacement and the release of chemical energy for muscle movement.

Ribonuclease consists of 124 molecular units, known as amino acids, linked chemically into a continuous chain like an 124-car train. As with others enzymes, ribonuclease belongs to the larger family of chemicals known as proteins.

While there are millions of different proteins, some of them with thousands of amino acids strung out in a chain, all are formed from the same 20 kinds of amino acid. In this sense, all life on earth is closely related, for the same 20 amino acids form the proteins and enzymes of bacteria and men.

The synthesis of such a substance can be likened to the assembly of a very long freight train. Ribonuclease is formed from 19 types of amino acid. Hence it is comparable to assembling a train made up of 19 kinds of freight cars—box cars, flat cars, tanker cars, cement cars and so forth.

It was doing so in the correct sequence that made the synthesis so difficult. This sequence determines the role of the enzyme and also its shape.

The three-dimensional shape of the assembled enzyme is controlled by the locations of the cysteine units along its chain. These units reach out with grasping "electrical" fingers, looking for other cysteine units. When they find one another, they lock together in a chemical bond that produces the tangled shape of the enzyme.

Until a few years ago, despite the basic functions of enzymes within the living cells, they played only a minor role in medical treatment. However, the situation is changing, as was pointed out yesterday by Dr. Max Tischler, senior vice president for research and development at Merck.

He noted that one enzyme, L-asparaginase, is a hopeful drug in the treatment of one kind of leukemia. Likewise, dextrinase is reportedly useful in removing the bacterial "plaque" on teeth that leads to cavities.

One difficulty in the medical use of such enzymes is obtain-

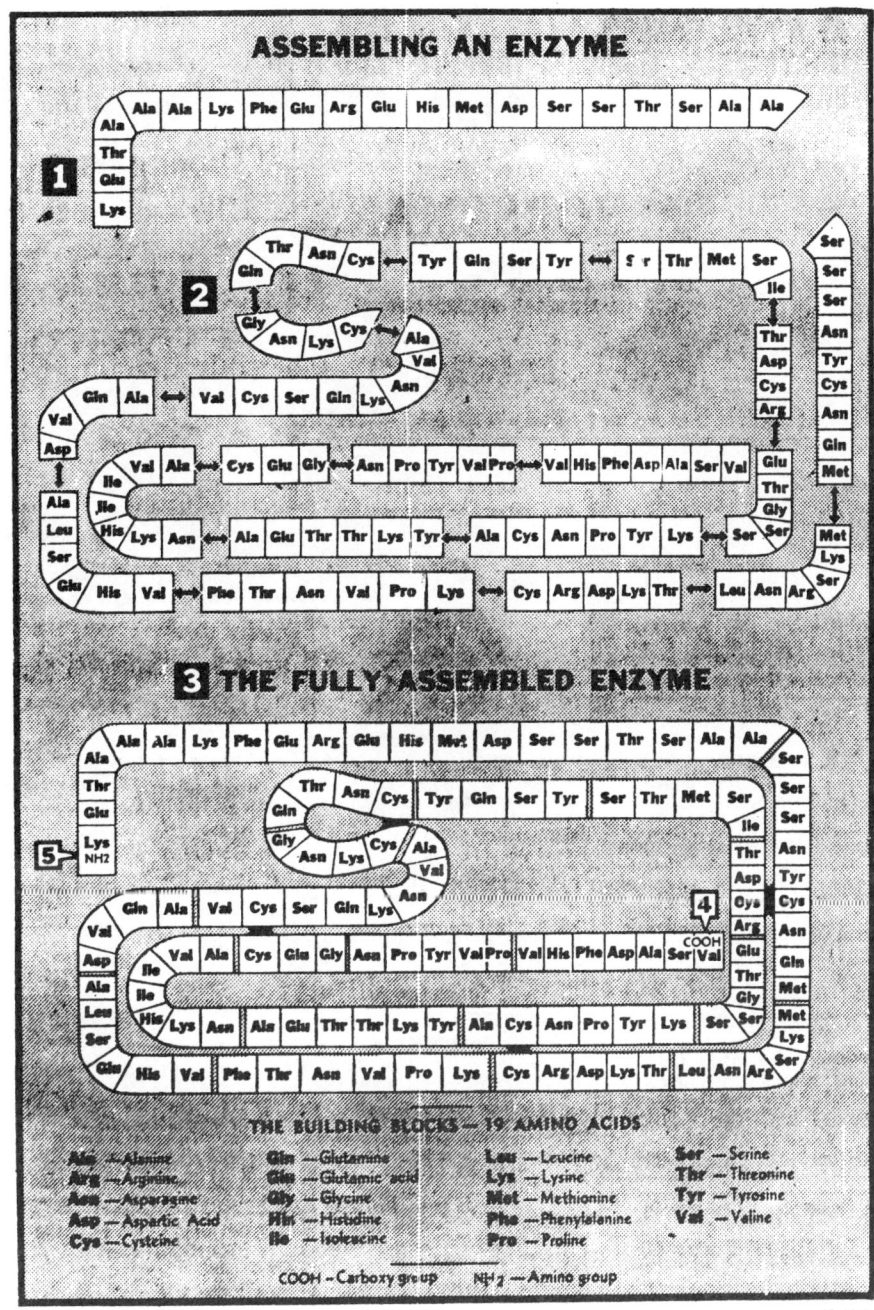

The first synthesis of an enzyme, ribonuclease, was carried out almost simultaneously by two laboratories. Merck & Co. scientists first synthesized two components from shorter chains of amino acid units (1 and 2 above). The two chains were then fused to produce the complete enzyme (3). At Rockefeller University the amino acids were added one by one, starting at the carboxy end (4). In nature the formation of ribonuclease begins at (5). Arrows indicate the fusing of small units, while dark bands indicate molecular bonds.

the case of phenylketonuria, a congenital disease that can lead to mental retardation. Some believe other diseases, including diabetes, certain forms of anemia and other blood disorders, may be related to improper enzyme function.

Until recently, a number of scientists believed that enzyme synthesis involved a mysterious process that could occur only within the living cell. How, it was argued, could scientists duplicate the can-of-worms structure of something so small it was not even visible in a microscope?

The structure of ribonuclease had been deduced by a variety of analytical techniques, but how could something that complex be put together?

Work at the National Institutes of Health in Bethesda, Md., suggested that enzyme molecules might be surprisingly "cooperative" in this respect. Their self-organizing ability has now been confirmed in both the Merck and Rockefeller work.

It has been found that it is only necessary to assemble the "freight train cars" in their proper sequence. If the chain is then allowed to float free in a solution, it twists and curls on its own, forming the bonds that give it the proper shape.

This is not only enlightening with regard to what goes on inside the cell. It makes it easier to understand how enzymes were formed in the beginning, when the chemistry of life was evolving on a lifeless world. It is further evidence that life's chemistry, despite its awesome complexity, has its origins in properties of the basic atoms and simpler molecules.

The Merck synthesis was reported by Dr. Denkewalter and Dr. Ralph F. Hirschmann, director of peptide research at the Rahway laboratories. It was achieved over an 18-month period in a multistage process in which a number of short amino acid chains were linked, to form two large units. These were then combined into a full-length molecule of ribonuclease.

The joining of the two parts was apparently not a full-fledged chemical bond, but the resulting material performed like its natural counterpart.

At Rockefeller University, under the direction of Dr. Bruce Merrifield, Dr. Bernd Gutte hooked on the 124 amino acids one by one. To this end he used an apparatus that automatically injected each amino acid, shook the solution for a given period, then sucked out the residue before injecting the next acid of the chain. The six-hour cycle for each amino acid was controlled much like that of a dishwasher.

ing them in quantity. It was proposed yesterday that both synthesis methods would help overcome this problem. They also open the way for made-to-order enzymes with properties not found in natural enzymes.

Another problem in the medical use of enzymes i delivering them to the right place in the body. Often they must function at some specific spot, at some specific time, deep in the mysterious interiors of certain cells.

When enzymes misbehave, the results can be tragic, as in

January 17, 1969

Scientists Isolate a Gene; Step in Heredity Control

By ROBERT REINHOLD
Special to The New York Times

BOSTON, Nov. 22—The basic chemical unit of heredity, the gene, has been isolated from an organism for the first time by a team of scientists at Harvard Medical School.

The feat, considered a major one by other scientists, paves the way for detailed study—and possibly eventual control—of the complex and little understood process by which genes determine tangible living traits. The team was headed by Dr. Jonathan Beckwith, 33 years old.

"This is a very significant achievement," said Dr. Philip C. Hartman of Johns Hopkins University, an authority in the field.

"It's one of those things other scientists take pleasure in reading about," remarked Dr. Sol Spiegelman of Columbia University, a leading microbiologist.

Despite the significance of the feat, scientists were hardly surprised. It was the logical next step in a series of dramatic events that has revolutionized the science of biology since the basic chemical stuff of heredity was identified 25 years ago as deoxyribonucleic acid, or DNA, a complex molecule present in the nuclei of all living cells.

It was a small segment of this substance that the Harvard group isolated from a common intestinal bacterium called Escherichia coli, or E. coli. The material represented the gene that controls the ability of the bacterium to metabolize, or use for fuel, a sugar known as lactose.

Because the gene came from such a lowly organism, the Harvard techniques cannot readily be applied to human or other complex organisms. However, the work has major implications for higher life because there is growing evidence that all living things, from the lowest single-celled bacterium to humans, receive their traits by fundamentally the same mechanism.

For this reason, the achievement probably brings much closer the day — less than 25 years off by some estimates—when it will be possible to cure human diseases or change inborn traits by injecting new genes. Because such methods could be misused, many scientists view this prospect with considerable ambivalence.

'Elegant' Experiment

The Harvard work, the experimental details of which are being published today in London in the journal Nature, was performed over two months last summer.

In an experiment that other scientists called "elegant," the term they apply to a sophisticated experiment that cuts to the heart of a complex conceptual problem with great simplicity, the Harvard team focused on the lactose gene, one of the largest and most thoroughly studied. It is one of 3,000 genes in the E. coli chromosome, the rod-shaped cell structure that bears the hereditary material.

They extracted the gene by using viruses as intermediary carriers. To do this they relied on the now classic techniques for which three American scientists—Max Delbruck, Alfred D. Hershey and Salvador E. Luria—were awarded a Nobel Prize last month.

These techniques are based on the fact that bacteria are infected by certain viruses, called phages. Phage particles enter the bacterial cells, pluck out small bits of the cell's genetic material (the DNA) and multiply.

The Harvard scientists used two such phages, known technically as lambda-plac5 and phi-80plac1, both of which contain the lactose gene picked up from E. coli cells, along with several other unwanted genes. The two phages differ mainly in that the sequence of chemical units that make up the lactose gene face in opposite directions, which was the key to the experiment.

The DNA in each of these phages is double stranded, shaped something like a twisted ladder, the rungs of which consist of chemically complementary pairs.

By treatment with special chemicals, these strands were unwound and separated. Then one strand from each phage was brought together in a test tube. Since only the lactose genes were opposite mirror images of each other, and therefor complementary, they immediately came together and formed a new double strand.

The remaining unwanted

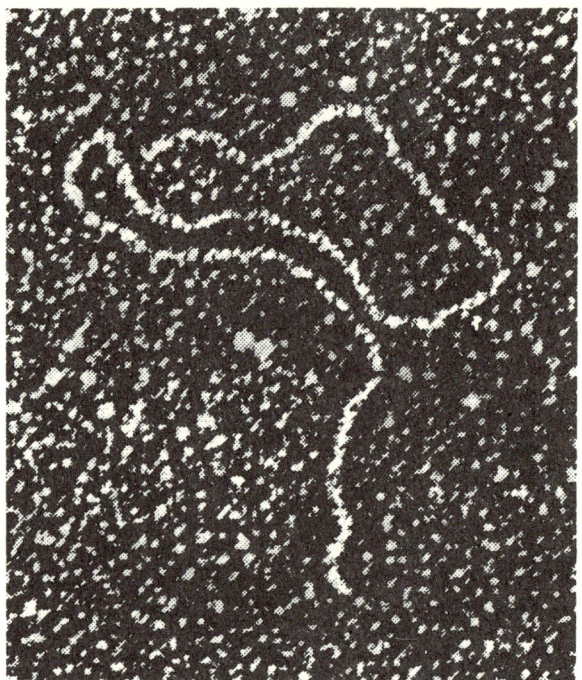

Electron microscope photograph of single gene, twisted strand in center of the picture. It is .000055 inches long.

gene segments could not find complementary partners and hung on as loose single-stranded ends. These ends were then dissolved by treating with a special enzyme, or biological catalyst, that affects only single-stranded DNA.

The result was a purified segment of bacterial DNA that was responsible for only one genetic function, which is the ultimate definition of a gene. Large enough to be photographed with an electron microscope, it measured 1.4 microns, or .000055 inches, in length.

Many scientists have been working on the gene purification problem for some time, using a variety of approaches. Until now, they had succeeded only in isolating three or four genes together.

It was important to obtain a single pure gene so that its action and the action of other cell components on it could be studied without other genes present to complicate the situation.

The most immediate practical significance of the feat is its implications for understanding the mechanisms by which genes are controlled. The broad outlines of these mechanisms were sketched by two French scientists, Francois Jacob and Jacques Monod, whose efforts were rewarded by a Nobel Prize. The inner biochemical and biophysical workings of the process, however, remain to be elucidated.

The French scientists advanced the theory, now substantially verified, that genes were controlled by a negative mechanism. That is, they are kept turned off by a "repressor" until another substance, an "inducer," comes along and disengages the repressor, allowing the gene to express itself.

Thus a gene consists of three elements, together called an operon. The elements are a "promoter" that produces the repressor, an "operator" that starts the gene operating but is normally dampened by the repressor and a "structural" portion, which is placed into action by the operator and does the main work of the gene. It does this by directing the production of cell proteins which in turn govern cell reactions and the formation of protoplasm, the cellular material. The proteins, thus, specify genetic traits.

Operon Isolated

Therefore, what Dr. Beckwith and his team did was to isolate the lactose operon. The "lac operon," as it is called, controls the bacteria's ability to metabolize lactose by governing the production of an enzyme, called beta-galactosidase, which chews up the sugar. Enzymes are proteins.

The achievement clears the path for a detailed study of the workings of the Jacob-Monod mechanism in the test tube under controlled conditions. A variety of experiments are possible, not only with the lac operon but with other bacterial genes, now that the techniques have been perfected.

For example, it should be possible to zero in on the mystery of where on the gene the repressor binds to prevent its function, as well as where and how the gene directs protein formation. Such knowledge may permit scientists to turn genes on and off at will.

In addition, the purified gene could be used to study the chemical products of a gene in action uncontaminated by the products of other genes. Ultimately, it may also be possible to pin down the exact sequence of chemical units that make up individual genes.

It is generally believed that the operator is the crucial site for the regulation of protein production. Recently, other scientists at Harvard University isolated the repressor substance. Now that the gene has also been isolated, the effect of the repressor can more readily be studied.

Genes operate by first producing a substance called ribonucleic acid, or RNA that acts as a messenger, carrying the genetic information of the DNA to the site of protein manufacture in the cells.

Other Steps Possible

With a purified gene now available, it should be possible in the test tube to measure with great precision the nature of the RNA produced and follow the complex steps of protein production.

Ultimately, it may also be possible to pin down the exact sequence of chemical units that make up individual genes. This may some day enable scientists to "manufacture" genes artificially to compensate for genetic deficiencies.

The genetic control mechanisms have attracted the attention not only of basic scientists like the Harvard team, but also medical men. It is widely thought that diseases like cancer are fundamentally cases of the genetic control mechanisms having broken down.

All of this has broad implications for "genetic engineering," the direct intervention in hereditary processes to instill desired new traits or to eliminate unwanted ones.

Misuse Feared

As yet, the methods are not perfected, but biologists speak confidently of some day infecting humans with viruses that carry new genes in order to cure hereditary diseases, such as hemophilia. But some also fear this same ability could be turned to destructive purposes by an unscrupulous government.

Such fears have caused much unease among scientists, including Dr. Beckwith and his associates, Dr. James Shapiro, Dr. Lorne A. MacHattie, Lawrence J. Eron, a student; Dr. Garret Ihler and Dr. Karen Ippen.

"The more we think about it, the more we realize that it could be used to purify genes in higher organisms," said Dr. Beckwith, a molecular geneticist who wears a dark beard, a Caesar-style haircut and flare-bottom trousers. "The steps do not exist now but it is not inconceivable that within not too long it could be used, and it becomes more and more frightening—especially when we see work in biology used by our Government in Vietnam and in devising chemical and biological weapons."

Dr. Shapiro concurred. "The work we have done may have bad consequences over which we have no control," he said, drawing a parallel to the development of atomic energy. "The use by the Government is the thing that frightens us."

Not all scientists agree with them. Dr. Joshua Lederberg, the Nobel-prizewinning geneticist at Stanford University, has consistently argued that the potential medical benefits of genetic manipulation outweigh the risks of misuse for political purposes.

The Harvard experiment was supported by the American Cancer Society, the Jane Coffin Childs Memorial Fund for Medical Research, the National Institutes of Health and the National Science Foundation.

November 23, 1969

Chicago Team Obtains Images of Atoms in Organic Compounds

By WALTER SULLIVAN

Scientists at the University of Chicago have obtained images of individual atoms within organic compounds, showing geometric structure of the compounds.

While previous attempts have disclosed spots that were suspected atoms, they say, this is the first time this has been done in a manner that could be verified.

he achievement represents a major step toward the day when scientists, using electron microscopes, will be able to "read" the coded genetic information in strands of DNA (deoxyribonucleic acid) and decipher the geometry of other large organic molecules.

The results were announced in Chicago yesterday by Dr. Albert V. Crewe of the University of Chicago, who for a number of years has led efforts there to develop such a technique. The pictures were made with a microscope whose scanning electron beam has 35 thousand volts.

The university is constructing a 100,000-volt electron microscope.

The present equipment can identify only atoms heavier mercury. One of the pictures displayed by Dr. Crewe showed a molecule of uranium salt resting on a film of carbon.

The two uranium atoms of the molecule appeared as two white blobs, side by side, but the connecting chain of carbon atoms could not be identified.

Another picture showed a chain of thorium atoms. Thorium has an atomic number of 90, which means it has 90 protons in its nucleus. Uranium has the highest atomic number of the naturally occurring elements—92.

While it is hoped that the technique will be used to sight smaller atoms, researchers are also developing methods whereby compounds containing heavy atoms can be inserted into organic compounds as tags. They could then be used to identify key elements of the molecular geometry.

Two of the graduate students working with Dr. Crewe discussed these efforts by telephone yesterday. They were Joseph S. Wall and John P. Langmore. They said that Dr. Michael Beer at the Johns Hopkins University was working on ways to tag the coding units, or "bases," in DNA for microscopic identification.

Thus it has been found that a compound of osmium (atomic number 76) reacts preferentially with thymine. Another compound seeks out guanine. These are two of the four coding bases of DNA.

In another experiement described by the Chicago scientists, the phosphate "backbone" of DNA is being tagged with heavy elements, such as thallium (atomic number 81), to study the twists and burns of the molecule.

The Chicago microscope, like others of this type, sweeps an electron beam back and forth across the specimen much as the beam from the electron gun of a television set sweeps the fluorescent screen to "paint" a picture.

The electrons that encounter the powerful positive charge of the atomic nucleus are deflected in a manner known as elastic scattering. Those that encounter the cloud of electrons surrounding the nucleus are not greatly deflected, but lose energy in what is called "inelastic scattering."

The specimen can be scanned, therefore, in a manner that brings out the nucleus and then in a way that shows the electron cloud. This facilitates positive identification of the atoms.

The Chicago effort to record single atoms has been under way about six months. Within another six months it is hoped that the larger microscope will be in operation.

Several years ago pictures were obtained showing the relative positions of tungsten atoms on the tip of a needle. The instrument used was a field ion microscope. The atoms appeared as a symmetrical pattern of spots.

According to the Chicago scientists, however, the new technique is the first that can show the relative positions of atoms within an organic compound.

Dr. Crewe, who was formerly head of the Argonne National Laboratory, near Chicago, told a news conference that the ultimate achievement of a method to identify all atoms in a molecule was the "holy grail" of those working in this field.

May 21, 1970

Complete Synthesis of Gene Reported

By WALTER SULLIVAN

A team at the University of Wisconsin reported yesterday the first complete synthesis of a gene — the inherited, coded "message" that tells a cell how to perform a certain chemical function.

While others have used a natural gene to make a new one, in this case a giant yeast molecule was pieced together, link by link. The gene in question consists of 77 such links, or nucleotides.

The synthesis was achieved by Dr. H. Gobind Khorana, who won a Nobel Prize in 1968 for his earlier work in this field, and his colleagues at the university in Madison. It marks a new step along the road toward manipulation of the hereditary material in plants, animals and, perhaps, man. However the road, as noted by the announcement, will probably be a long one.

For example, whereas the newly synthesized gene has 77 links, the genetic material within a human cell is estimated to have six billion such units, although not all are functional.

Last November, a group at Harvard reported another major development, the first isolation of a gene. Isolated at that time was the gene that controls the ability of a bacterium, Escherichia coli, to use its sugar fuel.

The term "gene" was devised at the start of this century by William Bateson at Cambridge University in England. He recognized that heredity is not transmitted in an amorphous mixing of material contributed by both parents, like paints on a pallete, but in discrete units, or "genes," contributed by each. From this evolved the science of genetics.

Some 25 years ago it was discovered that the genes consist of long molecules of DNA (deoxyribonucleic acide) formed of links, or nucleotides. Just as a computer does all of its "thinking" in terms of a binary code (one with only two numbers), so there are only four chemical "signals" within the DNA links.

They are paired, two to a link, in such a way that when the links are mated they form a twisted structure or double helix.

Dr. Khorana made use of the fact that Dr. Robert W. Holley, now at the Salk Institute in La Jolla, Calif., had determined the structure of another substance within yeast, known as alanine transfer RNA (ribonucleic acid).

Transfer RNA is a substance that is manufactured by DNA as a sort of template for the mass production of proteins that perform a needed function in the cell.

Just as one can deduce, from the shape of an automobile body, the structure of the stamping machine that produced it, so it was possible to deduce, the 77-link DNA molecule that made the substance deciphered by Dr. Holley.

With this knowledge, the Wisconsin group first assembled short strands formed of half-links of the projected chain. These strands of half-links were then mated with matching strands to form segments of the complete, twisted structure. Finally, using a natural enzyme (DNA ligase) as a chemical aid, they joined the segments into the full molecule.

The scientists felt reasonably sure of their results because each of the half-links (containing one of the chemical "signals") would only mate with a half-link containing one of the three other "signals" of the DNA code. Thus, if there was a mistake in the sequence, this would tend to become apparent.

Specialists in the field, commenting on this work, said Dr. Khorana's ingenuity in manipulating these strands in the face of various side reactions and other difficulties, was a major achievement. One likened it to figuring out how to assemble the many interlocking parts of an automobile.

The next step is to see if the artificial DNA manufactures its proper daughter product (alanine transfer RNA). The ultimate test would be to insert the DNA into living yeast cells to see if it worked properly. Preliminary chemical tests are already under way.

The Wisconsin group is also working on a new synthesis, that of the tyrosine transfer RNA that functions in E. coli. This was selected because there are strains of E. coli (a very common intestinal bacterium) that lack this substance, providing a ready means of testing any synthesis product.

The report said synthesis of the segments was almost finished and that the job should be complete within a few months. Meanwhile, Dr. Khorana hopes to use his technique to manipulate the link sequences in transfer RNA. This could lead to a better understanding of how this substance plays its vital role in all living cells.

June 3, 1970

GENE FUSING HELD EVOLUTION CLUE

Complex Enzyme Result of 2 Mutations in Bacteria

BERKELEY, Calif., Nov. 27 (UPI) — Two separate genes have been fused together inside bacteria to form a single enzyme-producing gene that performs the functions of both, a team of researchers reported today.

They described the experiment as an important clue to how evolution could occur at the most basic level of molecular activity.

The genes were fused when two successive mutations or sudden variations, blocked the operation of a chemical "punctuation mark" that ordinarily separates them. After fusion, the genes produced a single, large, complex enzyme that combined the distinctive functions of the two natural enzymes produced by the original genes.

"The main significance is that if you can do it in one case experimentally, events like this conceivably may have happened in nature," said John R. Roth, assistant professor of molecular biology at the University of California. "Complex enzymes may have evolved this way. It could be a step in the evolution of proteins."

The feat was reported in the scientific journal Nature by Professor Roth and Joseph Yourno and Tadaahiko Kohno, biochemists at Brookhaven National Laboratory in Upton, L. I.

Professor Roth said the fusion process had no apparent medical importance but might provide useful insight into the structure of the "punctuation" signals separating genes.

The researchers said DNA polymerase, a molecule that transfers the genetic code by serving as a template, is an example of a naturally occurring multifunctional enzyme that may have evolved through gene fusion.

"A class of proteins exists whose members have evolved as a consequence of gene fusion," they suggested. Prime candidates include enzymes with a single molecular chain performing several catalytic activities and those with extremely long chains, they added.

The scientists fused the second and third genes in the "histidine operon" of the Salmonella typhimurium, a bacterium blamed for some forms of food poisoning.

The histidine operon is a string of nine genes that perform related functions in controlling the production of histidine, one of the amino acids used to build cell proteins.

November 28, 1970

Researchers Complete Analysis of Make-Up of Gene

By JANE E. BRODY

A Belgian research team has for the first time completely analyzed the chemical make-up of a gene that dictates the production of a protein, an achievement that appears to bring closer the day when genes can be made to order and turned on and off at will.

The work is regarded by experts in genetics as a significant technological feat that, among other things, provides further proof that scientists have accurately translated the genetic code and shows which parts of the code are most often used in nature.

Previously, simple artificial genes with known structure have been synthesized, and two years ago the Nobel laureate Har Gobind Khorana manufactured a major part of a real gene using ingenious linking techniques to piece the molecule together. Dr. Khorana's gene did not direct the production of a protein.

The Belgian work, hailed this month in the British journal Nature as one of the year's most outstanding achievements in molecular biology, was made possible by sophisticated analytic techniques worked out by the Nobel laureate Frederick Sanger and colleagues in Cambridge, England, about five years ago.

These techniques, which showed how to determine in sequence the constituents of a molecule of ribonucleic acid (RNA), have not previously been applied to so complex and laborious a task as the complete analysis of a gene.

Genetics research is widely considered to be one of the most important avenues to understanding how living things are built at the most basic level.

This understanding may enable future doctors to correct inherited genetic defects and to treat more effectively a variety of degenerative diseases thought to be related to the breakdown of genes that control the function of cells.

Many scientists hope a better ability to manipulate genes in plants and animals will also lead to safer pesticides, new microbes to break down pollutants and more nutritious foods.

Like Mass Production

The achievement by the Belgian team, headed by Dr. Walter Fiers of the State University of Ghent, might be likened to the manufacture of the first mass-produced automobile after others had worked out the techniques of the assembly line.

"The work demonstrates how far science has come," said Dr. David Baltimore, a leading molecular biologist from the Massachusetts Institute of Technology. "It is a landmark of summation of many years of work, going back to many other groups. The Belgian team took on the hardest task. They brought Sanger's idea to fruition."

Dr. Sanger's team at the Medical Research Council Laboratory of Molecular Biology is currently trying to develop methods to analyze sequences of deoxyribonucleic acid (DNA) —the material that the genes of man and nearly all other living things are made of.

The RNA gene deciphered by the Belgian group came from a virus called MS2 that infects bacteria. The gene dictates the production of the protein that forms most of the outer coat of the virus. Although a number of isolated pieces of this gene had previously been determined, the Fiers group was the first to work out the entire gene structure and link it to the critical punctuation marks that tell where the gene starts and stops.

"If you ever want to manipulate a gene—to turn it on or off—you have to know where the signals are that say 'start here' and 'stop here'," Dr. Baltimore explained.

The ability to turn off (that is, make nonfunctional) an RNA gene might one day be important in cancer control, since most of the viruses known to cause cancer in animals consist of RNA genes.

Once the sequence of the MS2 gene was known, the Fiers group was able to suggest a probable model of how the gene exists when it is not working. The group concluded that the molecule must fold up on itself into hairpin-like loops, forming a structure that roughly resembles a flower. Hence, they dubbed it the "flower" model.

Dr. Norton Zinder of Rockefeller University, a leader in the field of RNA virus research who called the Belgian work "a real technological feat," said that knowledge of the folded structure "explains certain important biological properties of the molecule."

For example, the flower structure must first be opened up and the information of that gene "read" before the cell substances that translate the genetic code can recognize the start of the next attached gene.

Words of 3 Letters

The "letters" in the words of the genetic code are molecules called nucleotides. RNA consists of the nucleotides adenine, cytosine, guanine and uracil, commonly abbreviated A, C, G and U. Each word in the genetic code consists of three nucleotide letters, yielding 64 different possible words known as triplets.

Most of the triplets serve as code words for one or another of the 20 amino acids that, when strung together in order-

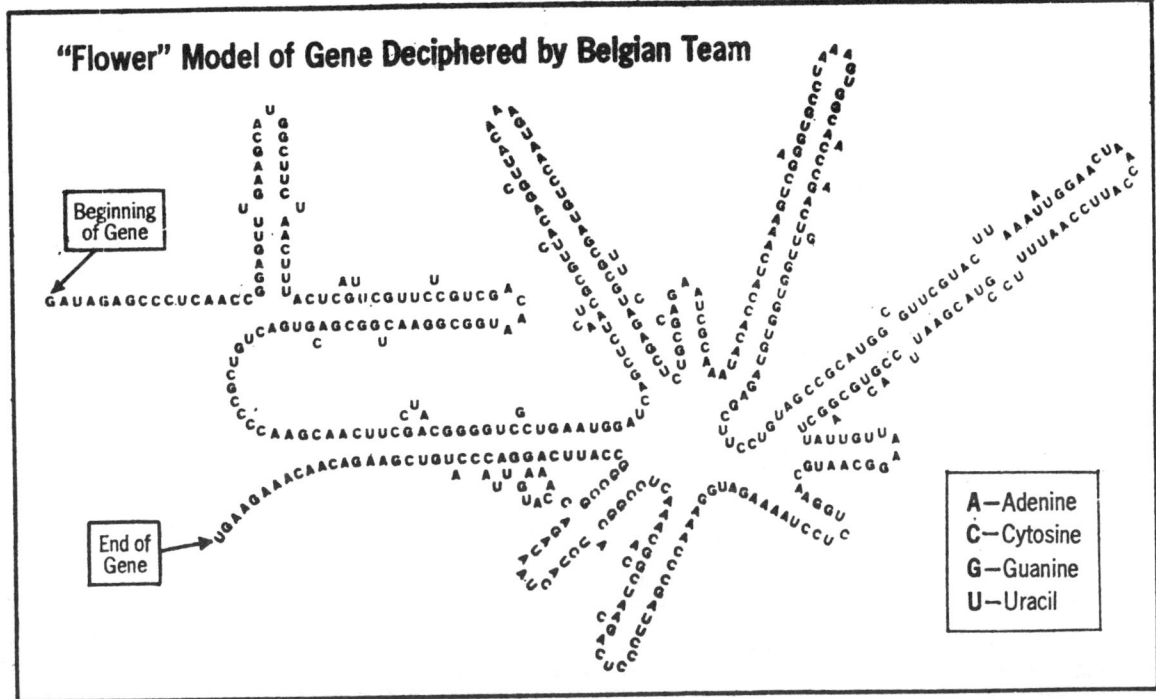

"Flower" Model of Gene Deciphered by Belgian Team

A—Adenine
C—Cytosine
G—Guanine
U—Uracil

The New York Times/Sept. 24, 1972

ly fashion, make up protein molecules. A few of the triplets act as punctuation marks.

Thus, the triplets GCU, GCC and GCA are translated as specifying the amino acid alanine, the triplets, UCU, AGC, UCA, UCG and UCC specify the amino acid serine, and so forth.

The MS2 coat protein, a rather small molecule as proteins go, contains 129 amino acid units, the exact sequence of which had been previously determined. Thus, the gene that codes for that protein would contain three times that number of nucleotides—or 387—plus a number of untranslated nucleotides that perform punctuation and other as yet unknown functions.

Enzymes Used

To determine precisely which code words are used and in what order, the Fiers team used enzymes to "digest" the RNA molecule in orderly fashion into ever smaller fragments. Then, by recognizing overlapping fragments, making educated guesses based on knowledge of the protein's structure and combining this new information with previously analyzed pieces of the gene, the team was able to deduce the precise nucleotide sequence of the complete gene.

In all, they found that 49 different code words were used in the gene.

It was "a real tedious job that took a lot of people two to three years to complete," Dr. Zinder said.

The authors of the original report of the achievement, published last May in Nature, were W. Min Jou, G. Haegeman, M. Ysebaert and Dr. Fiers.

The team's achievement may also be useful when science is ready to manufacture genes to perform certain desired functions, such as to supply a missing protein or neutralize the effects of an unwanted one.

Dr. Mark Ptashne, a geneticist at Harvard University, said that before a gene is made to order, it would be helpful to know which words in the genetic code are most frequently used in actual biological organisms.

Dr. Baltimore said there may also be important biological influences exerted by the "nearest neighbor" triplets.

"As we collect more and more knowledge of how genes are constructed, we can better unravel how they work," he added.

September 24, 1972

Structure of Protein-Building Molecule Determined

By WALTER SULLIVAN

The shape of a class of molecules that play an essential role in all life processes has been determined by a group at the Massachusetts Institute of Technology.

From knowledge of this structure, scientists hope it will be possible to learn how the molecule does its job normally and why it becomes altered in virus infections and cancer.

The molecule is transfer ribonucleic acid, or tRNA. Its task is to help assemble the proteins that are the chief structural material of all living things. It is said to be the first nucleic acid crystal whose three-dimensional structure has been determined by X-ray probing.

Proteins are like buildings formed of 20 different kinds of brick, known as amino acids. The tRNA molecules can be likened to bricklayers, each qualified to lay only one of the brick types.

The tRNA whose shape has been determined is the one that picks up a building block known as phenylalanine in constructing yeast proteins. It has been found to be L-shaped. Each arm of the L is a brief segment of a double helix. The latter is like two spiral staircases twisting around one another.

The best-known double helix in molecular biology is that of the DNA (or deoxyribonucleic acid) molecule. It forms an extremely elongated fiber whose structure has been determined not by X-ray probing of single crystals, but by analysis of fibers.

The structure of tRNA is a surprise to scientists, for the classic view of such molecules, based on theoretical considerations, is a clover-leaf shape, formed by a chain of substances (bases) that folds back on itself to make three hairpin turns. The two sides of each hairpin, except for the loop at its end, are held to one another by a succession of hydrogen bonds.

The configuration, as now deciphered, has three such loops and the predicted bonds, but its twists and turns form an L. While the tRNA molecules must differ from one another to perform their varying roles, it is known that structurally they are basically the same since in a mixture the various forms still order themselves into a single crystal.

The geometry of the molecule displays features whose function is known, as well as others whose role is still uncertain. They figure in a protein-making process whose starting point is DNA within the nucleus, or genetic archive, of a living cell.

The fiber-like DNA molecule "prints out" long molecules that, chemically speaking, carry complementary images of the DNA "message." These long molecules are known as messenger RNA.

The message consists of coded indicators, or codons, to muster the proper sequence of "bricklayers" to assemble the protein. The codons consist of a sequence of three chemical units, known as bases. Each such triplet, in key-and-lock fashion, matches a triplet, or anticodon, carried by the 20 different kinds of bricklayer (the tRNA molecules) for identification purposes.

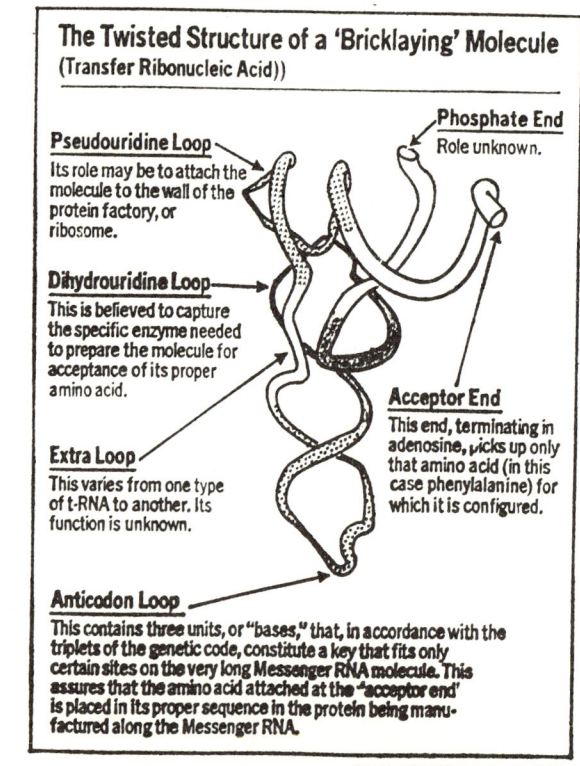

X-ray scanning of a crystal of transfer ribonucleic acid has shown its chain of 76 chemical components, or nucleotides, to be arranged in a twisted fashion.

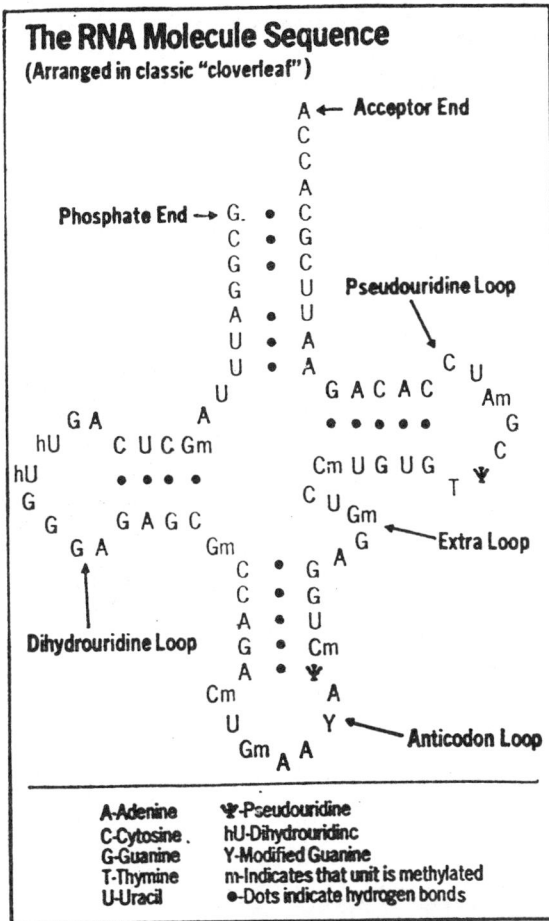

The sequence of 76 chemical constituents of transfer ribonucleic acid is shown above in its conventional, "cloverleaf" configuration. The methylated constituents are those that tend to become altered in tumor cells.

Like the perforated tape that can be run through a Teletype machine, the long messenger RNA molecule runs through a protein-making unit called a ribosome. The latter, with the aid of tRNA bricklayers, builds up the protein, amino acid by amino acid, according to the sequence indicated by the messenger RNA.

Within the ribosome the bricklayer molecules stand ready to do their job. As shown in the newly determined shape of these molecules, an "acceptor end" protrudes, standing ready, chemically and geometrically, to grasp onto the amino acid, which is this bricklayer's specialty.

However to do this the RNA molecule must be activated by a special enzyme. Another part of the molecule, its dihydrouridine loop, may be one of the sites where this enzyme attaches itself.

At the end opposite the acceptor site is the "anticodon"—the key of three bases that fits the proper triplet on the messenger RNA molecule. When this triplet, or codon, becomes exposed inside the ribosome, as the messenger RNA moves through that protein-making unit, the bricklayer molecule slips into place.

The messenger RNA moves up one notch to capture the next appropriate bricklayer. Then, apparently, the chemistry of the ribosome performs a remarkable trick. It lifts the brick (or amino acid) from the previously captured bricklayer and transfers it to the amino acid carried by the more recently captured one.

Growth of a Protein

Thus begins the gradual growth of a protein molecule. The bricklayer molecule that has been relieved of its load departs. Then the messenger RNA advances another slot and the process repeats itself, but this time, when the building blocks jump onto the most recently captured bricklayer, there are two such blocks.

As this process continues, along the length of the messenger RNA, this tail of amino acid building blocks gets longer and longer until, at the end, it is a fully grown protein with, typically, several hundred amino acids in their proper sequence.

Some of these amino acids have a strong affinity for one another, and this soon draws the molecule into its unique, three-dimensional geometry.

As explained by Dr. Alexander Rich of the M.I.T. group, a number of features of the TRNA structure remain to be explained. The role of the "extra loop," which differs markedly from one species of tRNA to another, is unknown.

Likewise, while the function of the "acceptor," or hydroxyl, end of the chain of 76 building blocks (nucleotides) forming this molecule is to capture the proper amino acid, the role of the other, or phosphate, end is uncertain. It may help identify the "right" activating enzyme.

Finally, while it is suspected that the pseudouridine loop briefly attaches the molecule to the ribosome, or protein factory, this is not certain.

Decipherment of the RNA structure, initiated about four years ago, developed into a race in which half a dozen laboratories, here and abroad, have been involved. In this respect it was reminiscent of the earlier race to determine the structure of DNA.

Early Results Published

Last month, the M.I.T. group published early results, from scanning down to a scale of 5.5 angstroms, (the entire molecule is 82 angstroms long). On Monday, Dr. Sung-hou Kim, formerly of the M.I.T. group and now at Duke University, will report on more detailed scanning, down to 4 angstroms, before the American Crystallography Association in Gainesville, Fla.

An account will also appear in Science next Friday. Meanwhile scanning at 3 angstroms resolution is under way.

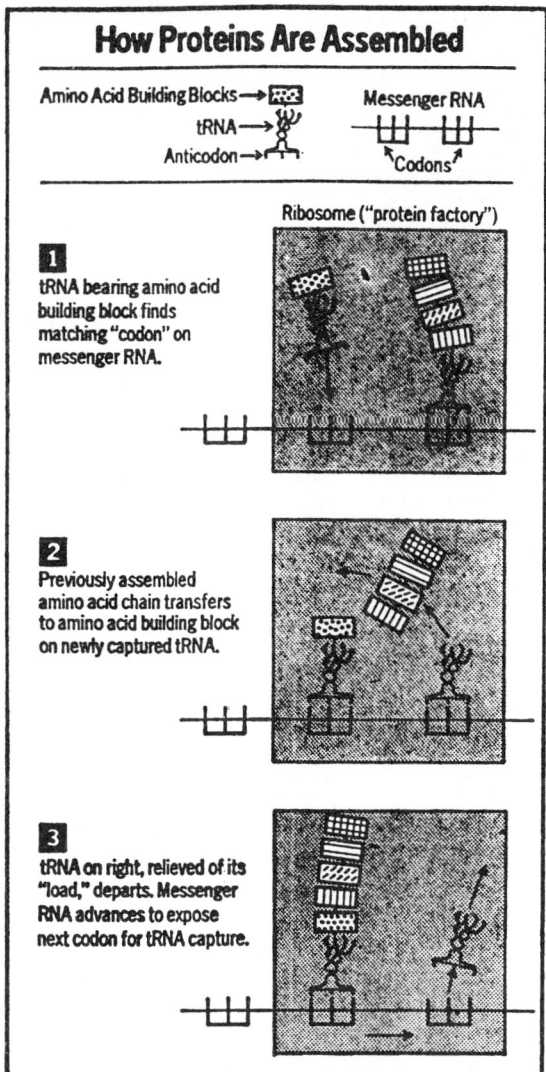

Three steps in the assembly of a protein molecule within the ribosome of a living cell are shown schematically above. Ultimately several hundred amino acid building blocks may be added to the protein chain in this manner.

The first sequence of tRNA building blocks (for the tRNA that carries alanine in yeast) was obtained eight years ago by Dr. Robert W. Holley.

Subsequently he received a Nobel Prize for this achievement. Since then about 40 different types of tRNA have been sequenced, and all fit into what has been diagrammed as a clover-leaf.

Because abnormal tRNA is found in tumor cells, the M.I.T. project has been supported by the American Cancer Society, as well as by the National Institutes of Health, the National Science Foundation and the National Aeronautics and Space Administration. Other participants included G.J. Quigley, F. L. Suddath, A. McPherson, D. Sneden, J.J. Kim and J. Weinzierl.

One of the most remarkable features of the triplet codin system, used in the protein-making process, in Dr. Rich's view, is its universality. The same code is used by all forms of life, from yeast to man.

A major puzzle is how nature devised this method to maintain continuity in the manufacture of proteins — a method that has made possible, as well, the continuity of life.

January 13, 1973

Columbia Biologists Determine Components of a Nucleic Acid

By WALTER SULLIVAN

Columbia University molecular biologists have determined the sequence of chemical "building blocks" in a nucleic acid capable of reproducing itself within a test tube, rather than in a living cell.

The finding, according to a report by the university's College of Physicians and Surgeons, will be of special value to those studying factors responsible for evolution. At its most basic level, evolution involves changes in hereditary information coded into nucleic acids of the cell.

The nucleic acid whose sequence of building blocks (or amino acid units) has been determined is a special form of ribonucleic acid, or RNA, developed through experiments of Dr. Sol Spiegelman, director of the university's Institute of Cancer Research. As nucleic acids go, its molecules are small, being composed of 218 amino acid units.

The most important part of the hereditary machinery is the deoxyribonucleic acid, or DNA, within the nuclei of cells, whose chains of amino acids often number in the hundreds of thousands. However RNA plays a number of key roles. The acid under study at the Columbia school was derived from the RNA in a virus that infects bacteria (a bacteriophage).

This modified RNA is known as MDV-1 and is single-stranded. Like other nucleic acids in single-stranded form it can assemble a complementary chain of amino acids to form a new molecule that, chemically speaking, is a mirror image of itself. In the case of MDV-1, the sequence of its complementary chains has also been determined.

These are not the first RNA molecules to be thus deciphered, but the finding, to be reported in the June 1 issue of Science, is important because of its applicability to laboratory studies of evolution.

According to Dr. Spiegelman, "A study of these nucleic acids and the changes that can be made in their molecular structure, now that we know what it is, offers the basis for conducting evolutionary experiments in the test tube."

May 26, 1973

Genetic Puzzle Solved, Scientists Say

By HAROLD M. SCHMECK Jr.
Special to The New York Times

WASHINGTON, Sept. 10— Scientists reported today that they had deciphered and synthesized most and perhaps all of the code that governs the turning on of the hereditary instructions in a gene.

The gene is the basic unit of heredity in all living things. It is made up of a twisted double strand of the master chemical deoxyribonucleic acid, or DNA. In any specific gene, the sequence of chemical subunits in its double helix of DNA is a code that spells out genetic instructions for the living cell. The sum of coded messages in all the genes of any cell gives it complete instructions, telling the cell what it can do and what it can become.

However, most of the genes are turned off most of the time in most cells. Thus, the puzzle of how genes are turned on and off is one of the key questions of modern biology.

M.I.T. Scientists

Today, Dr. Har Gobind Khorana and his colleagues at the Massachusetts Institute of Technology reported deciphering what they believe to be most and perhaps all of the "on" signal for a gene they had studied for several years, as well as a significant portion of the corresponding "off" signal. Their report was given at the American Chemical Society's national meeting in Atlantic City.

While the scientists believe they have correctly determined the chemical sequence for most or all of the "on" signal for the specific gene under study, conclusive proof depends on synthesizing the complete gene, together with its control sequence, and testing to see whether the control sequence can indeed function in turning on the gene.

Work toward this goal is now in progress. It is part of a major effort to understand how living cells control the transcription of the genetic instructions that are incorporated in their genes.

The gene under study is part of the hereditary endowment of a common bacterium called Escherichia coli, usually abbreviated as E. coli.

About a year ago, Dr. Khorana and his colleagues succeeded in synthesizing this gene. Since then, their research has included efforts to determine and synthesize the chemical sequence that makes up the gene's "on" and "off" signals.

The longterm objective is to test the totally synthesized gene and its control signals to see if it will function in living cells of E. coli.

Dr. Khorana, one of the major figures in study of the chemistry of genetics, shared a Nobel Prize in 1968 with two other Americans—Dr. Robert W. Holley and Dr. Marshall W. Nirenberg—for research involving the genetic code.

His co-authors in the current studies are Dr. Ramamoorthy Belagaje, Dr. Robert G. Lees, Dr. Dennis G. Kleid, Dr. Michael J. Gait and Dr. Kjeld E. Norris.

In previous studies, the researchers synthesized the complete 126-unit gene that has been the focus of the recent research. In its natural state, this gene and its control sequences are believed to constitute a double strand of DNA substantially longer than 126 of the units, which are called nucleotides.

In the newly reported work, the scientists determined a sequence of 29 nucleotide units that are thought to make up most, and perhaps all, of the "on" or "start" signal and a somewhat shorter sequence of the "off" or "stop" signal.

'Start' Signal Analyzed

Dr. Khorana believes the "start" signal for this gene probably consists of roughly between 30 and 50 nucleotide units. The nature and length of the "stop" signal is less clearly understood.

The specific gene under study is one that controls the bacterial cell's ability to incorporate an amino acid called tyrosine into the proteins the cell makes. The specific sequence of nucleotides that makes up this gene was first determined by a British research group in 1970.)

Dr. Khorana and his colleagues attempted to synthesize this gene and its control sequences because the characteristics of E. coli bacteria make it possible to detect the presence of the functioning gene readily in the living bacterial cells.

A long-range objective of the research at M.I.T. is to synthesize the gene and its controls, introduce them into a colony of the growing bacteria and show that the synthetic material works in the natural system.

September 11, 1974

Primate-to-Cat Gene Shift Laid to Virus of Long Ago

By HAROLD M. SCHMECK Jr.
Special to The New York Times

WASHINGTON, Nov. 28—A group of genes from early relatives of man and monkey was once transferred directly to cats by a virus infection, scientists have concluded after extensive chemical tests.

This seemingly bizarre case of genetic transfer between distant species is the first ever authenticated in animals, according to Dr. George J. Todaro, chief of the National Cancer Institute's Virus Leukemia and Lymphoma Branch.

Dr. Todaro believes the research has important implications for studies of evolution and the understanding of cancer. Such transfers of genetic information are known to occur in bacteria, but not heretofore in the higher animals, he said. How often it occurs in mammals is not known.

The event that caused the direct transfer of genetic information from primate to cat must have been a virus infection about 5 to 10 million years ago, Dr. Todaro and his colleagues believe.

The evidence that the transfer took place came from detailed chemical comparison of genetic material from the species studied. The conclusion that the event took place 5 to 10 million years ago is based on chemical differences in material from cat and primate assumed to have been caused by mutations over the long span of years.

The group of genes involved in the transfer had been native to the ancestors of the modern primates and was carried by the virus particles to the ancestors of the domestic cat, the scientists concluded.

Their studies show that, today, all breeds of domestic cats have these viral genes, which show close chemical relationships to comparable genetic material found in man, gorilla, baboon, chimpanzee and a dozen or more species of Old World monkeys. Details of the studies are to be published soon in several scientific reports.

Genes are the chemical determinants of heredity in all living things. In their chemistry, they carry instructions telling each new generation of living cells what it can do and what it can become. They govern the shape and nature of life, determining which creatures will sprout feathers, fur or hair as well as such crucial subtleties as the chemistry of the blood and the production of enzymes.

The chemical of which are the genes are made is the complex nucleic acid called deoxyribonucleic acid (DNA). Viruses, which some scientists have described as genes looking for a place to function, are minute packages containing either DNA or a closely related nucleic acid called ribonucleic acid (RNA).

Passed to Next Generation

The foreign genes that have become a part of cats' normal inheritance are presumed to be material from viruses that has become so intimately integrated with the animals' genetic machinery that it is passed from generation to generation with normal cellular genes.

These viral genes can nevertheless sometimes cause cells to manufacture complete virus particles that can then get out and act as infectious agents, according to the scientists' theory.

"Our data suggest that viral genes from one group of animals can give rise to infectious particles that not only can integrate into the DNA of another species, but can also be incorporated into the germ line and be transmitted as cellular genes," said a report to be published soon in the British Scientific journal Nature. The authors are Dr. Raoul E. Benveniste and Dr. Todaro.

The group at the Cancer Institute, which also includes Dr. Robert Callahan, Dr. Michael M. Lieber and Dr. Charles J. Sherr, has found that nucleic acid relates to the virus also present in man and all of the higher nonhuman primates.

The genetic material is believed to be that of viruses of a kind called C type viruses, which many scientists consider to be the key factors in the causation of cancer.

Selective Advantage Seen

In a recent interview, Dr. Todaro said the genetic material must also have important selective advantage to the species' harboring it, since it appears to have persisted in the store of genetic information for millions of years. He thinks the viral genes may possibly have a role in controlling cell growth and differentiation and perhaps in protecting calls from invasion by extraneous viruses.

While C type viruses are thought by many to be linked to cancer, the scientists think this must be a relatively rare effect of the genes' influence.

For example, the harmful effect allowing cells to turn cancerous might be switched on by damage from X-rays, chemicals or extraneous viruses, or simply by mistakes in internal body chemistry.

The scientists at the Cancer Institute made the discovery concerning cats while pursuing one of the key scientific objectives of modern research—the still unsuccessful search for human cancer viruses.

A virus, now known as RD-114, had been isolated in experiments with human cells growing in laboratory flasks and was first thought to be a possible human cancer virus. It proved instead to be a contaminant virus from cats, but chemically it was closely related to the viral genes that had long been part of the primates' fundamental genetic material.

Matching Studies

Painstaking studies were done in which the nucleic acid from a C-type virus of baboons was matched, chemical subunit by chemical subunit with DNA from man and other primates as well as with the cat virus.

The matching was done by a method called DNA hybridization in which comparable chemical subunits fit together in a manner suggestive of the fit between adjacent pieces in a jigsaw puzzles.

The degree of relatedness between the materials from the various species of primates, including man, matched the previously known evolutionary relationships among the species.

The virus material amounts to only a small fraction of the total genetic material in a normal human or primate cell. The scientists' theory is that the virus material became a part of primates genetic endowment many millions of years ago and changed slightly through multiple mutations as each species evolved.

Thus, the differences in the virus-related DNA of the several species could be used to estimate the amount of evolutionary separation between the species, Dr. Todaro explains. He believes studies of this kind may offer a powerful new tool for studying evolution.

November 29, 1974

Genetic Mutations Called Minor Evolutionary Factor

By BOYCE RENSBERGER

Two molecular biologists have challenged the conventional belief that mutations that alter the structure of individual genes are responsible for the major anatomical and behavioral differences between the various animal species.

Although they do not dispute the idea that such "point mutations" within genes account for minor changes, their evidence suggests that a second more powerful mechanism of evolution must exist.

The scientists contend that the major evolutionary changes, such as the obvious ones separating man from chimpanzee, are caused by new arrangements in the sequence of old genes that simply change the timing and duration of a gene's activity.

In other words, the scientists contend, the big steps in evolution are not the result of mutations creating new "words" in the inherited genetic message, but to a rearrangement of the old "words."

Such a rearrangement might, for example, change the time

during a creature's growth and development, when a certain gene—containing the blueprint for synthesizing a particular protein—is turned on, how long it operates and when it is turned off. Virtually all the organs, enzymes and other components of living creatures are built of various kinds of protein.

Potential Modification

Although the scientists' challenge is, at this time, no more than a theory designed to explain certain paradoxes that biologists have discovered in recent years, their arguments represent a significant potential modification of one of the most important concepts in biology, organic evolution.

The new theory has been put forth by Dr. Alan C. Wilson, a professor of biochemistry at the University of California, Berkeley, and Dr. Mary-Claire King, formerly a graduate student working under Dr. Wilson and now a research geneticist at the Hooper Foundation and the University of California, San Francisco. Their theory is set forth in the April 11 issue of Science.

The new theory is based on comparisons of the molecular composition of proteins, say a comparison between a hemoglobin in the blood from one species and the related hemoglobin from another species. The more alike the proteins are, the fewer mutations there have been to alter the gene that determines the protein's structure.

In comparisons of 44 different proteins, various scientists have found over the last decade that those from human beings are more than 99 per cent identical to those of chimpanzees.

A Minute Difference

This minute difference, less than 1 per cent, means that there have been very few mutations altering genes since the days of the common ancestor of man and ape, 20 to 30 million years ago. The scientists suggest that some other form of mutation must be operating to have produced the obviously vast differences between people and chimpanzees.

The difference between the proteins seems especially small when compared with the differences found to exist between proteins of many species known to be more closely related. For example, two species of frog that are similar in anatomy and behaviour may have protein differences 30 to 40 times greater than those between man and chimpanzee.

According to conventional evolutionary theory, relying only on mutations within genes, there should be fewer differences between species of frog than between creatures known to be more widely separated on the evolutionary family tree.

This paradox has been increasingly apparent over the last decade as various scientists, comparing protein after protein from chimpanzees and humans, consistently reported that there was hardly any difference between the two species on the biochemical level.

'Genetic Distance'

Dr. Wilson and his colleague at Berkeley, Dr. Vincent M. Sarich, have been pioneers in using the mutational differences in proteins between various species to construct evolutionary "family trees" based on the "genetic distance" calculated to exist between any two living species.

Thus, for example, if more differences in the composition of a particular protein are found when comparing humans with horses than in comparing humans with monkeys, it is assumed that monkeys are more closely related to people than are horses.

For the most part, family trees constructed on protein findings have agreed fairly well with the conclusions reached by biologists comparing gross anatomy.

Mutations within genes are generally believed to be caused by chemicals or natural radiation breaking up the structural units within a gene and leaving them in some random disarray. In most cases the altered structure constitutes a nonsense message in the genetic code, and the mutant individual fails to dveelop. In a few cases, however, the change creates a new "word" that makes biochemical sense, and the mutant bears a new trait.

Protein Compared

The point mutations that survive can be detected by comparing living proteins of one species with another. Some of the various kinds of hemoglobin proteins, for example, consist of a chain of 146 links called amino acids. There are some 22 kinds of amino acids, and each protein contains several kinds in a fixed linear sequence.

When most of the related hemoglobin chains from chimpanzee and human are compared, all 146 amino acid sites are found to be filled in identical fashion. In some other types of hemoglobin chains there may be a difference at only one of 146 sites.

That one difference is assumed to be the result of a point mutation that changed the code from specifying, say, the amino acid alanine to a code specifying the amino acid guanine.

Dr. Wilson and Dr. King believe It is unlikely that this kind of mutation accounts for major evolutionary changes. The method they postulate involves a different kind of mutation, such as the break-up of whole chromosomes, each of which contains many genes, and their reassembly into a different format.

This phenomenon can cause a gene, or set of genes that was originally on one chromosome, to be "translocated" to another. Sometimes chromosomes become fused together or break up permanently, both of which can change the gene arrangement.

Although the genes may become scrambled and may end up next to different regulator genes, which alter the timing and duration of their activity, the genes themselves remain unchanged.

A modification in the duration of a gene's activity might, for example, mean that bones grow longer or change their shape, that fur grows or does not, or that brains continue growing until they reach human size or do not. Bone remains bone, hair remains hair, and brain cells remain brain cells, but the size or number of any of these may change, and these are the differences that distinguish one species from another.

Once a species becomes well adapted to its lifestyle, it does not usually evolve much unless environmental conditions change.

Frogs, for example, became adapted to their stable aquatic habitats many millions of years ago and have not had to evolve in gross anatomy appreciably since. Whatever changes in gene regulation that have occurred have not conferred any advantage to survival.

April 18, 1975

MANIPULATING LIFE

LIFE'S REMOLDING TRACED BY SCIENCE

Step in Regenerating Organs From Cells Reported by Schotte of Amherst

HEAD GROWN ON TADPOLE

Tests, as Clue Toward Man's Revival of Himself, Told Before National Academy

By WILLIAM L. LAURENCE
Special to THE NEW YORK TIMES.

WASHINGTON, April 26.—Experiments on salamanders and tadpoles, in which the elemental clay of life was molded into eyes, nose, ears, mouth and other organs of the head, foreshadowing a time when man might regenerate lost organs, were described today before the annual Spring meeting of the National Academy of Sciences.

In these experiments, heads were made to grow on the tails of adult tadpoles out of undifferentiated connective tissue.

They revealed for the first time the residence within the body of all living things of a "sculptor of life," who, with the proper "clay" and favorable working conditions, could make new bodies out of old and regenerate the individual over and over again.

When man finally succeeded in definitely isolating this "sculptor" and in determining the sort of tools and working conditions he requires, goals toward which the experiments reported constitute an important step, man, to all intents and purposes, would have learned the secret of immortality, it was stated.

Once man learned how to put this sculptor to work, perhaps in another 50 or 100 years, men and women would possess the means for reversing the processes of life, to become embryos once more and thus start life all over again.

Duplicative Reproduction

In such a case the grown individual, living his second, third or successive lives, would be an exact reproduction of the original individual, for he would be regenerated out of the tissues of his original body.

Such a "resurrection," it was remarked, would result in an individual still possessing all the original genes, or units, of heredity; whereas in the case of the newly born child, it inherits half the genes of each parent, so that only half of an individual survives in his child.

The experiments, which pushed further the frontiers of knowledge on life's creation and organization opened up by Professors Hans Spemann of the University of Freiburg, Germany, and Ross G. Harrison of Yale University, were reported by Dr. Oscar E. Schotte, Professor of Comparative Anatomy, Embryology and Histology at Amherst College.

Professor Schotte studied under both Dr. Spemann and Dr. Harrison, and his paper was introduced by Dr. Harrison, one of the world's outstanding authorities in this field.

In the work of Dr. Spemann, for which he received the Nobel Prize, it was first shown that there exists in the early stages of embryonic development an "organizer" which molds the aboriginal clay of life into its own image, forming the various organs and tissues of the body.

Dr. Spemann showed the scientific world how this life-organizer works. Taking a microscopic bit of the early embryo of an amphibian—a bit that, if left undisturbed, was destined to form skin—and transplanting it to a place destined to occupy the still-unformed brain, he showed that the organizer existing in the spot of the brain-to-be transformed potential skin tissue into brain tissue.

Schotte's Extension of Work

However, Dr. Spemann could accomplish his brain-making out of potential skin tissue only in the very early stages of embryonic development. When transplanted at later stages the skin tissue could produce only skin tissue.

From these results, Dr. Spemann and other workers in the field concluded that the organizer existed only in the early stages of embryonic development and that, once the embryo had become more differentiated, the organizer bowed itself off the stage of life.

Professor Schotte succeeded for the first time, he reported, in transplanting unorganized, ordinary connective tissue into the tails of adult amphibians, and accomplished the feat of transforming this tissue not merely into more tail tissue, as would be expected from Dr. Spemann's results, but into organs from entirely different parts of the body, namely eyes, nose, mouth, ears—in fact, practically an entire head.

This shows for the first time that the organizer, or inductor, as Professor Schotte names the molder of life, does not step out on completing the work of embryonic development, but remains in the adult body throughout life, although with very little work to do because it has no more elemental clay to work with, and because the working conditions are no longer propitious.

Dr. Schotte, in discussion of his experiments, remarked on the "fundamental difference between the processes of induction in the embryo and in the regenerative tissue of the adult, although the same essential principles are involved."

Supporting Dr. Harrison's opinion of the wide possibilities of regeneration, under the right conditions, he said:

"Genetics and tissue culture and also experimental embryology have shown that every cell possesses potentially everything to produce any type of tissue or organ."

Kasner Develops Horn-Angle Study

Further studies on the horn angle by Professor Edward Kasner of Columbia University, who was the first to succeed in devising a means to measure it, are revealing an entirely new type of geometry, Professor Kasner reported to the academy. He named the new geometry "trihornometry."

A horn angle is the angle formed when two or more curved lines emerge from a common stem, thus forming the shape of a horn. Before the Harvard tercentenary conference last September Professor Kasner reported having found that the sum of the measure of its parts was greater than the measure of the whole. Today he reported some further anomalies.

The equivalent of the triangle in the horn angle is what Dr. Kasner calls a trihorn. Unlike the triangle, Professor Kasner found, the trihorn can be neither equilateral nor equiangular. Furthermore, he reported, if two sides of a trihorn are equal, the opposite angles are never equal, and the sum of the opposite angles is always equal to one.

"In general," he said, "any three of the six parts of a trihorn (except the three angles) determine the other three, but sometimes two distinct solutions exist.

"Two trihorns with sides respectively equal are not necessarily congruent. A full set of formulas of trihornometry is found. The medians are concurrent, but not the altitudes."

April 27, 1937

NUCLEI TRANSFERS OF CELLS ACHIEVED

Prof. J. F. Danielli Describes Two Experiments Involving Primitive Organisms

MULTIPLYING NOT BROKEN

British Cyto-Embryologist Hints, at Yale, Some Key to Whole Nature of Life

By WILLIAM L. LAURENCE
Special to THE NEW YORK TIMES.

NEW HAVEN, Conn., Sept. 5—Successful transfer of the nuclei of living cells from one species of primitive organism to another, so that they continue to multiply, was reported here today at Yale University before delegates at the Seventh International Congress of the International Society of Cell Biology.

The investigations into the basic units of living matter, which open the way to more fundamental problems into the nature of life and its development, were described by Prof. J. F. Danielli, an outstanding cytoembryologist of King's College, London, England. He also is affiliated with the Chester Beatty Research Institute of the Royal Cancer Hospital in London.

Dr. Danielli reported on two sets of investigations on cells, carried out in collaboration with Drs. I. J. Lorch at the Marine Biological Laboratories, Plymouth, England, and with Dr. Lorch and Prof. Sven Horstadius of the University of Uppsala, Sweden, at the Marine Biological Laboratories at Roscoff, Brittany, France.

Details of Experiments

In the first set of experiments the nuclei of one species of amoeba (single-celled organisms) were transferred into the cytoplasm of another species from which the nuclei had been removed. In the second set of experiments the nuclei of one-half of the cells of sea-urchin embryos were removed during the early stages of the embryo's development up the stage when the original fertilized egg had reached the eight-cell state.

The transplanted nucleus had a difficult time adjusting itself to the new environment, Dr. Danielli reported, and most of the cells failed to reproduce in mass culture. In one set of experiments, however, the "shot-gun marriage" led to the successful production of a mass culture, creating a new type of centaur of microscopic life.

Further work along these lines is expected to provide new knowledge on the roles played by the nucleus and the cytoplasm of the basic units of life and their interrelationship, knowledge essential for the understanding of the basic mechanisms involved in normal and abnormal manifestations of the living process.

The juggling of the embryonic cells of the sea urchin have already revealed the existence within the cytoplasm of mysterious forces that exert a profound influence on the nuclei of the developing embryo, Professor Danielli reported.

Several Changes Traced

When the embryo has reached the stage in which the fertilized egg had divided into eight cells, four of these cells are destined to form the outer half of the animal (known as the animal half) while the other four are destined to form the inner half (known as the vegetal half).

When the nuclei of four of these cells were removed and left in contact with the other four cells, Dr. Danielli and his collaborators found, the cytoplasm of the enucleated cells exerted a powerful influence on the nuclei of the other four cells.

The experiments revealed, Dr. Danielli declared, that each set of four cells exerted in inhibiting influence on the other set, this inhibiting effect being removed with the removal of the nuclei from one set. When the nuclei are removed from one set of four, the enucleated cytoplasm appears to exert a potent influence on the remaining nuclei.

This work may open the way to a new approach to the study of the cancer cell.

There are many types of cancer, and it is not known at present whether the action of the cancer cell lies in the cytoplasm or in the nucleus. This means that we do not know what part of the cancer cell to attack, whether it should be concentrated on the cytoplasm or on the nucleus.

Possible Assault on Cancer

It may also turn out that some types of cancer should be attacked one way, but directing it at the cytoplasm, while the attack against another type should be directed against the cancer cell's nucleus.

The studies are also expected to have an important bearing on genetics. Early geneticists had believed that the role of the nucleus in the transmission of heredity was absolute, the cytoplasm playing no part whatever.

More recent investigations have provided evidence that the cytoplasm does have a definite role to play in heredity by exerting an influence on the nucleus. These experiments provide new evidence that this is the case.

September 6, 1950

'GENETIC SURGERY' MAKING ADVANCES

'Artificial Selection' Is Cited at Science Conference

By JOHN A. OSMUNDSEN
Special to The New York Times

THE HAGUE, Sept. 10 — Nearly 2,000 scientists have spent the last nine days here laying the groundwork for what may be a dramatic change in the future of mankind.

Only occasionally did this note creep openly into the hundreds of technical papers that were presented at the 11th International Congress of Genetics, but no one with imagination could fail to see the implications in much of the work that is going on in this field.

After 1.7 billion years in which life has existed on this planet and evolved under the pressure of two forces—mutation, or spontaneous changes in the heredity material, and natural selection—man seems on the verge of adding a completely new evolutionary force.

This might be called "unnatural" or "artificial" selection, for it would be accomplished by the deliberate manipulation of the hereditary material in one way or another according to specified plans.

Scientists who talk about such things stress that this capability may be a very long way off or, indeed, may never come about.

Nevertheless, the kind of techniques that are envisioned for performing "genetic surgery," as scientists term the concept are in use on lower forms of life in laboratories. Some success has been reported with them on test-tube cultures of human cells.

Transduction Is Used

For example, geneticists use a technique known as transduction to transfer some of the genetic material of one microorganism into another. They do this by infecting a "donor" cell with a virus that attaches to the heredity-carrying structure called a chromosome and carries part of this into the recipient cell that the virus subsequently infects. The second cell then adopts some of the heredity traits of the first.

Milislav Demerec of the Brookhaven National Laboratory at Upton, L. I., described experiments during the closing session of the congress here this morning in which he used this technique to mix the inheritance of two species of bacteria so that he could compare the arrangement of the hereditary units, called genes, on each one's chromosome.

Another method that scientists have considered is called transformation. This technique consists of injecting naked molecules of the genetic material from one cell into another.

10 Papers Presented

The first successful transformation of living cells in this way was achieved a few years ago on bacteria. Earlier this year, Waclaw Szybalski, reported the transforming of human cells in tissue culture at the University of Wisconsin. At least 10 papers on transformation of microorganisms were presented at this congress.

As yet, neithr of these techniques nor others that are used to alter the genetic material in viruses and cells — such as as irradiation and treatment with certain chemicals — have produced directed mutations, or planned changes. The alterations have sofar been hit or miss. But the possibility of producing specified ones has been foreseen in advances made recently in deciphering the o-called code of life that is transcribed in the molecular structure of the genetic material and in the development of techniques for isolating specific strands of that substance.

September 11, 1963

Frogs Made From Single Body Cells

By WALTER SULLIVAN
Special to The New York Times

OXFORD, England, Oct. 6 — Experimenters at Oxford University believe they have removed any shadow of doubt as to the validity of a series of experiments in which frogs have been produced from single body cells extracted from another frog.

The experiments in what is known as vegetative reproduction bear on one of the most fundamental problems in biology—namely, what it is that turns on and turns off the genetic material buried within each cell of the body.

The frogs, according to the experimenters, have been produced from cells that line the intestine. This has demonstrated that even such highly specialized cells contain, within their nuclei, the information needed to construct an entire new individual. Normally such information lies dormant, but in the Oxford experiments it has been activated.

Until now, some scientists have believed that, in such specialized cells, the genetic information unrelated to that cell's function has been permanently erased. Those skeptical of the Oxford experiments, under way for several years, argued that the frogs raised here grew from nonspecialized cells that somehow made their way into the intestine.

In an interview, however, Dr. John Gurdon of Oxford, the zoologist in charge of the research, said it was now evident that this could not be so. With improved laboratory technique more than 30 per cent of the intestinal cells can be made to grow at least to the tadpole stage.

This, he said, is incompatible with the idea that these tadpoles have grown from rarely occurring nonspecialized cells. Only 1 or 2 per cent of the cells used in the Oxford experiments have grown to fully mature and fertile frogs, but this in Dr. Gurdon's view is because of subtle damage to the cell nucleus during manipulation.

The experiments have created a sensation here because they imply that, in theory if not in practice, it should be possible to mass produce identical twins of people gifted with exceptional ability or beauty. However Dr. Gurdon has struggled valiantly to dissociate himself from such speculation.

His goal, he explained during a visit to his laboratory last week, is to understand how the genetic information in a body cell is controlled. Thus, while the nucleus of a cell lining a man's intestine contains all the information needed to produce an identical twin, only one tiny bit of that information is active.

It says: "You are an intestinal liner; you must grow in a certain way and perform certain chemical functions."

Another Message

What Dr. Gurdon and his co-workers have done is to take such a nucleus from the intestine of a tadpole and implant it inside a frog egg whose own nucleus had been destroyed by the researchers. Something in the cytoplasm, or non-nuclear material, of the egg tells that nucleus: "You are no longer an intestinal cell nucleus; you are an egg cell nucleus; go to work." The result, if all goes well, is a new frog.

Dr. Gurdon knows no one, to date, who has been able to carry out a comparable transfer of nuclei in mammals. The procedure, he said, is bound to be extremely difficult, if even possible.

The frog egg is ideally suited to such work. It is large and its nucleus is on its surface, rather than buried deep within it.

This means that the old nucleus, containing the genetic information provided by the two parents of the egg, can easily be destroyed by ultraviolet light. The nucleus from the body cell of another frog can then be implanted by manipulation under a microscope.

The eggs of mammals, including human beings, are

about a hundredth as large. The low percentage of frogs that grow to maturity in the Oxford experiments presumably reflects the hazards of manipulation. The hazards in higher animals would be correspondingly greater.

A striking feature of vegetative reproduction is that the characteristics of the offspring are entirely predictable. In sexual reproduction each parent provides a set of bodily characteristics, or genes. These rival blueprints then compete at every level within the secret inner sanctum of the body's evolving cells and there is no predicting the outcome.

Experiments in which plants have been derived from single cells (not seed cells) have been reported in the United States. It is also often possible to grow new plants from cuttings. This process, however, is more comparable to the regeneration of limbs in newts and salamanders, according to Dr. Gurdon.

Mysterious Appearance

In such cases the new growth is not initiated by muscle cells, skin cells or other specialized cells in the stump that change their role. Instead a peculiar form of unidentified cell—like that in the earliest stages of embryo development—mysteriously appears at the site and begins to do the growing.

In the interview and in recent published reports, Dr. Gurdon emphasized that many workers in other laboratories laid the experimental foundations for the Oxford work. Notable among these were R. Briggs and T. J. King, who in the early 1950's achieved the first nucleus transplants, using the American frog Ramapipiens.

The Oxford work has been on the African clawed frog. Dr. Gurdon began research in this field in 1956 while working on his doctoral thesis here under Dr. Michail Fischberg, a pioneer in the field now at the University of Geneva.

Dr. Gurdon explained that intestinal cells were used because they subdivide frequently. In fact, those of growing tadpoles do so very often. This "eagerness" of the cells to subdivide makes them particularly prone to subdivide into entirely new individuals.

To demonstrate that frogs generated in this manner resemble the donor of the intestinal cells, rather than the donor of the egg, frogs of clearly distinctive types are used as the two donors in each case.

The Oxford researchers are now trying to find out what it is in the egg cytoplasm that controls activity in the nucleus. Because brain cells rarely, if ever, subdivide in maturity, the nuclei of frog brain cells have been used in these tests.

When such a nucleus is injected into egg cytoplasm it begins, within an hour, to manufacture deoxyribonucleic acid (DNA)—that is, to replicate its genetic information just as though it were getting ready to subdivide. Furthermore, the substance within frog eggs that performs this role has the same effect on mouse liver nuclei. The magic material is thus common to mammals and amphibians.

While its composition is still unknown, it appears in some cases, according to Dr. Gurdon, to contain the enzyme DNA-polymerase.

While salamanders can grow a new leg, man cannot grow new limbs or replace damaged parts of his brain or heart. Those seeking to understand these differences have long hoped to know what it is that turns genes on and off. They suspect, as well, that cancer may occur when this control system runs amok.

For these reasons the Oxford experiments are attracting worldwide attention.

October 7, 1968

Topics: *Genetic Manipulation and Morality*

By AMITAI ETZIONI

The acceleration of biological engineering has been urged before Congress by Nobel Laureate Dr. Joshua Lederberg. He has called for the establishment of a National Genetics Task Force to increase the momentum of efforts aimed at unlocking the genetic code of man. Such a breakthrough in biology could lead to the prevention of many illnesses whose origin is wholly or partially in the genetic code.

There is much to be said in favor of such a task force. But it ought to be accompanied by a task force on the social and moral consequences of genetic manipulation. The imminent breakthroughs in biology may affect man as much or more as he was affected by previous revolutions in engineering and physics: the imposition of a new set of capacities, of freedoms, of choices society must make, of evil it can inflict.

Gene manipulation may also allow man to tamper with biological elements which heretofore had to be accepted, including the sex of children to be conceived, their features and color, and ultimately their race, energy levels, and perhaps even their IQ's. Thus, what may start as the biological control of illnesses could become an attempt to breed supermen. While this may appeal to some, think about the agonizing problems if man has to act as the creator and fashion the image of man.

Shopping for Genes

What supermen will the national task force order? Blond or brown, white or black? Highly charged or low-keyed? More males? And, who will make all these decisions — the parents shopping for genes in the supermarket, again expecting society to pick up the bill for the aggregate effects of individual decisions? Or, a Government agency, a task force?

Fortunately, it seems we do not have to stop the genetic combat of illness to prevent genetic engineering for racist purposes. Contrary to widely held beliefs, studies show that the energy of science may be guided into one area to the relative neglect of others. It is generally thought that scientific work requires that the scientist follow any lead his investigating spirit encounters and which may take him any place. The findings of a sub-discipline of a field trickle freely into the others: hence, one kind of genetic manipulation will willy-nilly open the door to others.

Actually, most scientific findings are not readily transferable, and their application is affected by moral taboos. Next to no work is carried out in the psychology needed to develop subliminal advertising, and those scientists who sought to prove racist theories are starved for funds and academic recognition.

Before such guiding of scientific efforts can be effectively applied to the new genetics, we must have a clearer notion of the moral and social choices involved in the biological revolution and the mechanisms by which science can be guided without being stifled.

Explore the Options

Let us not again sail blindly into a storm unleashed by scientists anxious to unlock all of nature's secrets with little concern for who and what will be blown over in the resulting tidal waves.

To this end, I suggest that at least 1 per cent of the $10-million a year requested for National Genetics Task Force be set aside to explore the options genetic engineering is about to impose on us.

Amitai Etzioni is Chairman of the Department of Sociology at Columbia University and Director of the Center for Policy Research.

September 15, 1970

Buffalo Scientists Report Synthesis of Living Cell

By WALTER SULLIVAN

The State University of New York at Buffalo reported yesterday that scientists there had achieved "the first artificial synthesis of a living [and reproducing] cell."

The reference was to experiments at the university's Center for Theoretical Biology in which amoebas were partly dismembered, then put back together, using some components from other amoebas.

The reconstituted amoebas not only survived but also reproduced themselves and were indistinguishable from normal amoebas, it was reported. Amoebas are single-celled animals.

"This work," the report said, "opens up a new era for artificial life synthesis, now being explored."

For example, it continued, it helps clear the way "for the synthesis of new micro-organisms, new egg cells and an organism capable of living on Mars."

The environment of Mars appears so inhospitable that there are serious doubts among biologists that any form of life has evolved there. It has been suggested by some scientists, however, that organisms could be tailor-made to exist on that frigid planet by reconstituting or modifying existing life forms.

The experiment was reported to newsmen in a letter signed by Dr. Raymond Ewell, vice president for research, and Dr. James F. Danielli, the center director who led the team that has done the experiments.

The letter invited newsmen to a dinner on Dec. 7 at which further details would be reported. Its contents were made public by several radio stations and news services.

While some researchers in the field were not willing to endorse the somewhat sensational evaluation of the work as described in the report, specialists at the National Aeronautics and Space Administration, which is supporting the work, described it as "exciting" and "a big step."

They said it went considerably beyond the experiments of Dr. John Gurdon at Oxford University, who has implanted the nucleus of a body cell from one frog in the egg cell of another. The result was a frog identical to the one from whom the body cell was taken.

The work of Dr. Danielli and his colleagues, Dr. K. W. Jeon and Dr. I. J. Lorch, was described in preliminary form in the March 20 issue of the journal Science.

They removed the nucleus from an amoeba, as well as about three quarters of its cytoplasm, or "flesh." What remained was the rest of the cytoplasm, plus the cell membrane, or "skin."

All Parts Combined

They then inserted cytoplasm and a nucleus taken from one or more other amoebas. They found that if the inserted material came from amoebas of the same strain, about 80 per cent of the reassembled organisms lived and were able to reproduce by subdivision.

They found that in many cases where the inserted material came from other strains, survival was curtailed and the man-made amoebas divided only three or four times. Apparently material from one strain contained "lethal factors" that acted against another strain.

However, the Buffalo group was apparently able to overcome this effect, at least in some cases, producing amoebas that combined characteristics from strains that were not closely related. To the NASA specialists this was particularly significant.

Experimenters in the past have been able to modify the characteristics of an amoeba by injecting foreign cytoplasm into it. However, according to the Buffalo group, it should now be possible to remove various parts of the cell — the nucleus, or some other internal component, such as an organelle — and subject them to various tests.

After such tests — with radiation, drugs and the like — the units would be reinserted into cells to see if their function was affected.

As described in the Science report, the Buffalo group has not been able completely to dismember a cell. If more than three-quarters of the cytoplasm is removed, the reassembled cell will not live. Furthermore when reassembly is completed, the resulting cell is only three quarters as large as the original one.

Until now experimentation on the origin of life has focused largely on efforts to reconstruct the process whereby simple substances on the primitive earth interacted with one another. spontaneously synthesizing ever more complex compounds. It has thus been possible to reconstruct a plausible pathway for the early stages of such chemical evolution.

November 13, 1970

Embryos of Mice Are Developed To Heart-Beating Stage in Lab

By WALTER SULLIVAN

In what is regarded as an important step toward the gestation of mammals—possibly including human beings—outside the body, fertilized mouse eggs have been grown in the laboratory to the heart-beating stage.

According to specialists, the achievement is significant because it has carried embryo development past what had been considered a major obstacle to laboratory gestation. This is the stage at which the embryo attaches itself to the wall of the womb.

Such implantation is followed by the development of the primitive embryonic cells into specialized organs.

The culturing of mouse embryos through to this stage of initial organ development, has been achieved by Dr. Yu-Chih Hsu of the Johns Hopkins University School of Hygiene and Public Health. A report on the work was presented yesterday at the annual meeting of the Federation of American Societies for Experimental Biology in Chicago.

A number of laboratories in this country and abroad are seeking to develop techniques for laboratory culturing of embryos. The goal is not so much the production of "test-tube babies" as an understanding of factors controlling the gestation process.

This, in turn, would help explain what goes wrong in some cases and leads to birth defects. It would also enable researchers to test the effects on embryonic development of such things as drugs and food additives that might lead to birth malformations.

At present there is no way to monitor some of the most critical stages of human development—those that occur before birth.

In January, a group from Britain's Cambridge University and Oldham General Hospital in Lancashire reported culturing laboratory-fertilized human embryos to the "blastocyst" stage. At this stage the embryos were six days old, and the original egg cells had divided into 110 or more cells forming a hollow sphere. Furthermore, the cells had begun to develop specialized functions.

However, in these studies, as in earlier animal experiments, it appeared that the next step, in which the blastocyst normally implants itself in the uterus wall, stood as a barrier to further development. Something mysterious, it was thought, must occur in the mammalian womb to help the embryo past this critical stage.

If a mammalian embryo is removed from the womb shortly after implantation—or, as in one study, just as this is about to take place—it continues to develop. Only when it grows large enough to become critically dependent on the mother's blood for nourishment and removal of wastes is further growth impeded.

From interviews with a number of researchers in this field, it now appears that the development of an artificial placenta to perform this function is the chief obstacle to full laboratory gestation.

In Dr. Hsu's work, fertilized, subdividing eggs were removed from female mice and cultured in dishes coated with fibrous material (collagen) obtained from rat tails. This material provided a surface for implantation.

On the third day of incubation, some of the developing embryos attached themselves to the collagen; and cells that, in normal gestation, help constitute the placenta began spreading through the collagen.

After 10 to 14 days of incubation, primitive blood vessels and brilliant red blood cells began to develop in from 5 to 20 per cent of the implanted embryos.

In 7 of 25 embryos reaching this stage, rhythmic contractions, at 70 to 80 beats per minute, were evident.

Such apparent beating of the embryonic heart continued, usually, for two to four days, and by this action the red blood cells were pumped back and forth in the tiny blood vessels. However, the embryonic development was clearly abnormal and did not progress past this state.

This abnormality, Dr. Hsu said, may be due to an accumulation of waste products or a lack of sufficient nutrients. An artificial placenta would perform both waste removal and nourishment. In the womb, the placenta serves as the meeting ground where the blood of mother and embryo are able, through semipermeable membranes, to exchange such substances.

In the typical laboratory culturing of embryos, oxygen is bubbled into the culturing medium to keep it aerated, but there is no attempt to replicate the role of the placenta. Nourishment of the embryos in Dr. Hsu's work has apparently been largely by direct diffusion into the cells rather than via the primitive blood vessels.

According to Dr. Hsu it is not certain whether his implantations occur simply because the conditions of culture are right or because some vital substance is present. The culturing medium includes serum from fetal calves that, he notes, could contain some hormone-related compound.

Dr. Joseph C. Daniel at the University of Colorado in Boulder has found a protein that floods the uterine secretions at the time when implantation occurs and early organ formation begins. In studies with rabbits he has found that its release is apparently stimulated by the hormone progesterone.

In commenting this week on Dr. Hsu's report he noted that efforts several years ago to obtain funds for development of an artificial placenta had failed because such an effort seemed premature. It is now evident, he added, that this is no longer so.

Other specialists in embryo culture agreed that the lack of an artificial placenta stood in the way of further progress.

April 14, 1971

The Frankenstein myth becomes a reality—

We Have the Awful Knowledge To Make Exact Copies Of Human Beings

Studies In Revolution

This article, exploring ethical and social issues raised by one of the most ominous developments of the "biological revolution" —cloning—derives in large part from discussions at the Institute of Society, Ethics and the Life Sciences.

Based at Hastings-on-Hudson, N.Y., the institute was founded three years ago by a group who shared the conviction that too little time, money and interdisciplinary study was being devoted to problems that are already here. The institute, a research and teaching center, brings together biologists, physicians, social scientists, theologians and lawyers — 73 fellows in all. In groups, they are exploring problems of medical ethics, population control, treatment of the dying and the meaning of death, the impact of behavior control and genetic engineering.

By WILLARD GAYLIN

IN the winter of 1971, before a committee of the House of Representatives, the biologist J. D. Watson expressed dismay that the population had been insufficiently alerted to some of the profound implications of new technologies in genetic research. To the uninitiated, the fact that the statement came from a scientist whose own research is in that field may seem analogous to Dr. Frankenstein chastising the Swiss citizenry for failing to storm his laboratories. But this forceful testimony by the distinguished codiscoverer of DNA has been applauded by a growing group of scientists, social scientists and ethicists who sense that the people are shielded, by the complexity of genetic science, from an understanding of the nature and magnitude of threats it poses to their ways of life, their identities and their very existence as a species. The public attention to Watson's testimony—confirming their thesis—was minimal.

The Frankenstein myth has a viability that transcends its original intentions and a relevance beyond its original time. The image of the frightened scientist, guilt-ridden over his own creations, ceased to be theoretical with the explosion of the first atomic bomb. The revulsion of some of the young, idealistic men who were involved in the actual making of the bomb or in the theoretical work that led to it, had a demonstrable influence in the scientific community from the nineteen-fifties forward. Some biological scientists, now wary and forewarned, are trying to consider the ethical, social and political implications of their research before its

---

WILLARD GAYLIN, M.D., is president of the Institute of Society, Ethics and the Life Sciences, Hastings-on-Hudson, N.Y., and a professor of psychiatry and law at Columbia Law School.

use makes any contemplation merely an expiating exercise. They are even starting to ask whether some research ought to be done at all. With the serious introduction of questions of "ought," ethics has been introduced — and is beginning to shake some of the traditional illusions of a "science above morality," or a "value-free science."

Of course, in 1818 when Mary Shelley first created her story, the scientific domination of society was just beginning. The idea of one human being fabricating another was purely metaphorical. The process was presumed to be impossible, a grotesque exaggeration which cast in the form of a Gothic tale the author's philosophical concern about man's constant reaching for new knowledge and control over the forces of nature (the traditional Greek anxiety about *hubris*). It was, to use her words, "a ghost story," a fantasy to frame a poetic truth.

But the inconceivable has become conceivable, and in the 20th century we find ourselves, indeed, patching human beings together out of parts. We sew on detached arms, and fix shattered hips in place with metal spikes; we patch arterial tubing with plastic; we borrow corneas from the dead, and kidneys from the living or dead; automatic, rechargeable pacemakers placed under the skin regulate the heartbeat, and radio receivers placed in the brain case may shortly control behavior; there are artificial limbs, artificial lungs, artificial kidneys and artificial hearts; and respected scientific researchers — in a real-life parody of art — are publicly accused of stealing secret and mysterious devices from the laboratories of their rivals.

The issue which seemed most worrisome to Watson, and in his opinion called for a campaign to inform the world's citizens so that they might take part in planning possible control measures, was the cloning of human beings. Cloning is the production of genetically identical copies of an individual organism. Just as one can take hundreds of cuttings from a specific plant (indeed, the word *klon* is the Greek word for "twig" or "slip"), each of which can then develop into a mature plant—genetic replicas of the parent—it is now possible to clone animals. The possibility of human cloning seems to produce in nonscientists more titillation than terror or awe—perhaps because it is usually visualized as "a garden of Raquel Welches," blooming by the hundred, genetically identical from nipples to finger nails.

PEAS AND CARROTS

To understand the complications and implications of human cloning it is necessary to review some of the "facts of life" and to approach distressingly close to those bromides, the birds and the bees.

Every species of living organism has the capacity to reproduce its own kind. Indeed, this capability is so fundamental to the concept of being "alive," that it is part of the definition distinguishing animate from inanimate. The mechanism whereby species likeness is transmitted from one generation to the other was discovered by the Austrian priest Gregor Mendel in the 19th century, in some of the most amazing research of modern science. While it is not possible to do justice here to Mendel's genetic principles, it is necessary to recall a few of his conclusions. Working with common garden peas, taking such variables as the color of the flower, the size of the plant, the shape and texture of the seeds, Mendel defined the basic laws of heredity. Unlike previous vague conceptions of offspring as some loose amalgam of parental qualities (Darwin's *panmixis*) in which blood lines fused just as blue and yellow water colors blend to make green, Mendel established that the offspring inherit relatively discrete, independent traits which never mix nor modify each other, but maintain a segregated existence ready to be passed on in pure form to a future generation. He saw the instrument for transfer as a discrete body, later to be called a gene, and recognized that while one gene might dominate another, thus appearing as a particular property, they both existed and were ready to be shipped out to a next generation, again in pure and segregated form. Generally speaking, for example, if you inherit a gene for blue eyes and a gene for brown eyes, you do not usually get the muddy mixture of the two but will have the brown eyes of the dominant gene. Nonetheless, the gene for blue eyes is a part of your genetic potential ready to be handed down intact to your children.

As a corollary of this segregation principle, Mendel observed that the various traits are inherited relatively independently of each other. There may be a separate gene for the size of the pea, the color of the flower and the height of the plant, and in hybridization a variety of gene patterns is possible through chance combinations. It is obvious to even those who are not interested in gardening that this has now become a mechanism for controlling, in plants at least, the development of specialized and desired traits; for example, one could grow a large, fully double, high-centered, heavily scented, disease-resistant, thornless rose of a specific color.

Perhaps the most striking fact about Mendel's laws is that they are valid for *all* living beings. The principles of heredity discovered in the garden pea also apply to the prelate who discovered them. We have since discovered that not only is the principle the same, but that it works in all organisms by means of the identical chemical (DNA) mechanism!

Heredity, however, can be modified by the mechanisms of reproduction, which are not the same in all living beings. Humans, like most advanced life forms, reproduce sexually, which might not seem like the hottest piece of news, but the biological significance of this fact must be understood. Sexual reproduction does not always depend upon copulation. While that happens to be the means available to human beings, the variety of methods evolved in nature might overwhelm the imagination of man, whose sexual fantasy, even, is defined by the specific nature of his own rather limited sexual apparatus. Obviously, plants do not copulate, yet they have a variety of mechanisms for self-pollination or, more commonly, cross-pollination (those bees again), both of which are examples of *sexual* reproduction.

Animals also have a wide range of reproductive styles. Sexual reproduction can take place within a single organism — the hermaphrodite forms —but consider the case of the earthworm: Hermaphrodite though it be, producing both sperm and egg, the earthworm, like most higher plants, generally eschews self-fertilization; instead, it seeks out a partner and, in what seems to be a model of sexual courtesy and cooperation, inseminates the other while being inseminated itself. At the other extreme are forms in which fertilization of the egg by the sperm occurs outside of the organism, without contact between the parents.

What is essential to the definition of sexual, as distinguished from asexual, reproduction is that the new generation is formed by a combination of individual genes, half contributed by one parent and half by the other—the variability of the mix in the higher species being so complex as to almost guarantee the uniqueness of each individual. By contrast, in the asexual reproduction of lower forms such as the amoeba, there is a splitting of the organism and the genetic makeup of the two creatures derived from the original is identical, carrying the same undisturbed gene pattern as the "parent" organism.

THE genes in human beings are distributed among 46 chromosomes. These 46 chains of inheritance exist in the nucleus of every single body cell of the organism *except* for the sex cells. These cells, ova in women and sperm in men, contain only half the normal quantity—23—and are called haploid. When fertilization occurs, the nuclei of the sperm and the egg fuse, forming an egg cell with a full complement of chromosomes. The fertilized egg proceeds to undergo spontaneous division into two, then four, then eight, finally into the billions of cells that comprise the human body. In the meantime, the cells "differentiate," changing drastically in shape and function, thus forming the various tissues and organs of the body. The genetic code, embodied in that chance mixtures of genes from parental chromosomes, guides and contributes in some as yet unknown way to the ultimate form of the adult organism. Sexual reproduction with separate male and female forms guarantees a richness and a variability to the species. This process, combined with Darwinian principles, permits the evolution of individuals with enhanced adaptability and survival values. It is sexual reproduction which mandates continued change—and, therefore, ultimately, improved adaptive capacity.

The process of differentiation represents one of the great unsolved mysteries in biology. How can these cells, which are identical in early divisions with each containing the exact same nucleus (meaning the full potential to form the entire creature) evolve so differently? Lung tissue looks different from bone, skin from blood, muscle from cartilage, because the microscopic cells that make up the tissue have evolved into entirely different forms. The individual cell—ignoring most of its potential—becomes a specialist, and takes the form most suited to its function, which also has become specialized. Some cells will become chemical manufacturing units—reproducing, for example, insulin; some will be the wirelike cells of the nervous system that conduct impulses from other cells that have become pain receptors in the skin, to still others that have become "appreciators" of pain in the brain.

However it may have occurred, once differentiation develops it would seem that there is virtually no way back, short of regeneration itself. If this is true of an animal, it seems equally true of a vegetable.

IF man's heredity mechanism was first understood from common garden peas, it seems only equitable that the mechanism's undoing may be from the common garden carrot. Most of us have had the experience of growing vegetables from seeds. The seed is the equivalent of the fertilized egg ready to go, and, since the earth is its natural womb, the planter is merely a mechanical middleman. In a startling set of experiments during the early nineteen-sixties, Prof. F. C. Steward, a cellular physiologist at Cornell University, began agitating individual cells from carrot root in various nutritive media. Almost any mechanical or chemical stimulus can cause an egg or seed cell to begin dividing—heat, light, touching, shaking, or more exposure to a nutrient medium. Steward used differentiated cells, not seeds, yet amazingly these cells began to proliferate. Eventually, with patience and changing media and techniques, Steward was able to force the individual root cells to form clumps and organized masses; what is more, they began to differentiate again into other kinds of cells.

He finally succeeded in carrying one individual cell to the ultimate stage of a full-grown carrot plant — roots, stalk, leaves, flowers, seeds and all. Any cell can, conceivably, be thus forced, once the technology is understood, to grow into a full plant. And what is possible with a vegetable cell is, at least theoretically, just as possible with an animal cell. Animal cells, of course, have already been cultured in the laboratory. Tissue cultures are a basic medical research tool. But tissue cultures are not whole organisms — merely sheets of identical type cells —and the concept of growing a whole organism from one cell asexually in a laboratory would seem impossible. But that Cornell carrot confronts our incredulity. To a scientific mind, the leap from single cell to cloned carrot is greater than the leap from cloned carrot to cloned man.

MAN THE CREATOR

IS cloning a man foreseeable in any reasonable time? Years ago, J. B. S. Haldane, the brilliant British biologist and mathematician, confidently assumed the imminence of human cloning and eagerly anticipated its potential uses. Yet, to most people, such a development was inconceivable. One could imagine taking a single sloughed cell from the skin of a person's hand, or even from the hand of a mummy (since cells are neither "alive" nor "dead," but are merely intact or not intact), and seeing it perpetuate itself into a sheet of skin tissue. But could one really visualize the cell forming a finger, let alone a hand, let alone an embryo, let alone another Amenhotep?

There is an entirely different laboratory procedure, known for years, that also offers an alternative to sexual reproduction. When an egg cell is stimulated mechanically or chemically, it will start the division process which leads to the adult form even though it is unfertilized. This virgin birth, or parthenogenesis, occurs in nature, the typical example being the honey bee, whose fertilized eggs produce workers and queens and whose unfertilized eggs develop parthenogenetically into drones or males. Beginning with simple sea forms, laboratory parthenogenesis progressed up the evolutionary ladder to the point that in 1939 a whole rabbit was reported created from an unfertilized egg. However, since in most species the unfertilized sex cell, unlike all of the other cells of the body, is haploid, the individual formed is *not genetically* identical to its mother, or indeed genetically identical to anything.

It remained for Prof. John Gurdon, a biologist at Oxford, to perform the stunning experiment that bridged the technology of parthenogenesis and that of Steward's carrots. In the mid-sixties, Professor Gurdon, working with a frog's eggs, devised a technique, employing radiation, that destroyed the nucleus of an egg cell without damaging the body of the egg. Then, by equally complicated mechanisms, he managed to take the nucleus from an ordinary body cell of the frog (with its full complement of chromosomes) and intrude it into the egg cell. Until now, it was an unproved assumption that the nuclei of all cells, regardless of how different they might be, were identical in their genetic inheritance and contained the entire latent potential for reproduction of a differentiated, multicelled adult. If Gurdon's hypothesis was correct, the newly constructed egg cell was now the equivalent of a fertilized egg and should, on stimulation, be capable of producing an adult form. This is precisely what happened. Some of the cells, on division, formed perfectly normal tadpoles, some of which, indeed, became perfectly normal frogs genetically identical to the frog that donated the nucleus.

John Gurdon used an intestinal cell. He could have used any other body cell, and the cell could have been from a male or a female. The enucleated egg into which the nucleus was injected was also unimportant, genetically speaking; it was merely the environment. The means now exist to produce thousands of genetically identical offspring in the laboratory—at least in frogs.

What seemed like Haldane's immense and overvalued faith in scientific technology now sounds like a rational prediction. In 1969 Robert Sinsheimer, chairman of the division of biology at California Institute of Technology, stated that he assumed it would be possible to clone human organisms within 10 to 20 years. The way has thus been paved for the production of genetic copies of particularly prized individuals, in enormous quantities if desired—for whatever purposes.

THERE are still major obstacles to the cloning of human beings. Human ova and frog ova are vastly different in some respects — size, for

CLONE—This frog grew in the lab of Oxford biologist John Gurdon from an intestinal-cell nucleus of one "parent" that had been inserted in an egg cell of another. It has the exact genetic make-up of the nucleus donor.

one. Contrary to what one might guess, the frog egg is huge compared to the microscopic human ovum. This is because the frog egg, like a chicken egg, must contain all the nutrient to support the complete development of the embryo; in the human being the egg is implanted in the wall of the maternal uterus soon after fertilization, and a placenta forms which permits direct feeding of the fetus by the mother. The size of human ova, therefore, is incredibly small considering the size of the offspring. H. J. Muller, the great biologist, calculated that all the human eggs from the total population of the earth (then two and one-half billion) would occupy less than a gallon of space. Because of the minute size of human ova, further advances in microsurgery and laboratory techniques will be necessary before cloning becomes possible.

Gurdon has already supplied most of the technology for human cloning. Following the method he used on frogs, the nucleus of an egg cell from any donor would be destroyed. A nucleus (they are all alike) from any convenient cell of the person to be "replicated" would be inserted into the enucleated egg by microsurgical techniques (which have not yet been developed). On placing this new egg cell into an appropriate nutrient medium — a number of recipes have been devised—the "normal" process of division would commence. By the time it has divided into the 8- to 32-cell stage—four to six days—it would be ready for implantation.

A number of simple implantation devices have already been successfully worked out in animals. The developing egg can be injected directly into the uterine wall at the proper menstrual stage of receptivity. Or, more elegantly, it may be injected into the Fallopian tube and permitted to pass normally into the uterine cavity for self-implantation.

Many technical problems still remain, but given sufficient imperative they will be solved. Whether we will actually do human cloning involves other considerations.

OUGHT MAN?

THE types of questions that normally arise about any new and dramatic technological procedure fall into the categories of: can man, will man, and ought man. There is a tendency, particularly in antitechnology treatises, to lump the first two together and to consider the third an ndependent problem. This kind of reasoning usually assures us that what science can do, it will do. The facts are more complicated, as usual, than the polemics. There is much that man can do which he does not do — because he is aware that he ought not. We do not, for example, perform many behavioral experiments on babies, even though some research would unquestionably contribute to knowledge and the common good. Societal morality has traditionally disapproved of the use of the human beings as research animals. Their humanness protects them from certain kinds of destructive research. But even this rule is being violated in some instances. In at least one recent situation, for example, human fetuses that were about to be aborted were used as part of an experiment to determine the potentially harmful effects of ultrasound.

The typical scientist is a product of the culture's ethical system and reacts intuitively to its built-in values—even if he has never thought through its philosophical premises. In general, the culture-value system is one input into the broader psychological forces that drive men toward certain goals and tacitly discourage others.

In pure research, however, a goal may be pursued with no advance knowledge of its utility. Thus may a startling technique become available before we are prepared to consider all the implications of its application. Similarly, confusion can arise when the pursuit of one problem leads, accidentally, to the solution of another which, because unanticipated, was insufficiently evaluated. In these circumstances, the experimentalist is often tempted to do what can be done—merely for the excitement of doing it. The work on DNA of J. D. Watson and Francis Crick has opened the way to all sorts of experiments in genetic surgery that may be beyond the intent of the two pioneers.

WHAT would be the value of cloning? The most immediate answer comes from the field of animal husbandry, which would gain new breeding techniques on a par with those already available to the plant biologist. If a particular brand of rice or wheat is developed, a true line can be offered so that the genetically pure strain, and only that strain, can be propagated. With animals, we have been dependent on the chance fusion of chromosomes from a champion race horse or a prize steer with those of their respective mates. Cloning would give us the option of making 10,000 identical copies of the champion race horse. Of course, that might raise the question of why we would then bother racing horses at all!

This technique would also permit us to manipulate the massive genetic multiples involved in breeding the best cattle possible for meat. It would, however, stop the evolutionary process, for it is precisely the random combination and recombination of genes from one parent and another that produce not only lesser creatures than the parents but superior creatures, and thus permit the continuing expansion and enrichment of the gene pool itself. Cloning could also be used to augment the number of members of an endangered species to that critical level necessary for group survival.

Would there be legitimate uses for human cloning? Certainly the general speculation about multiple Mozarts, basketball teams composed of five Kareem Abdul-Jabbars (four more on the bench?) or an army of supersoldiers who are identical in every respect, with replaceable parts available for convenient transplants in case of injury, are insufficient to motivate scientific research.

They are not only insufficient, they are naive—as are the horror stories about the power-hungry dictator cloning his race of *Uebermenschen*. A human being is more (or less) than his genetic potential. It is the interaction of his genetic variables with the environment that produces the "person." The idea of seeing "yourself" born—as has been suggested—is a joke. The individual can be altered by the cytoplasm of the egg; by the biochemistry of the circulating blood through the placenta; by the diet and emotions of the woman carrying the child and by the trauma of the birth process. And all these environmental influences come to bear before what is usually visualized as life experience has yet begun!

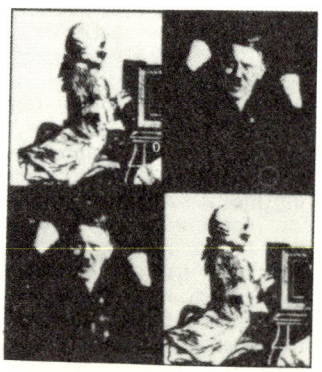

290

Take identical twins — nature's cloning. Even though they share "the same" pre-birth environment, they will emerge as different in such simple matters as weight and size. When identical twins are separated after birth, they develop remarkable variability while growing up—not just in achievement and psychological makeup, but in body weight and height.

The artificially cloned individual would have to be raised in a culture at least one generation apart from that of its donor (parent? sponsor?); this amount of time would be necessary to evaluate the worth and success of the prototype. To get an idea what a difference one generation makes, we need only look at our own bewildering offspring. Life experience pounds, pulls and shapes the same genetic clay into wondrous and ludicrous variations. If identical twins separated from birth show disparate form and personality, identical twins separated by a generation of time —clones—might not even recognize one another. We are not only what we are genetically given, but what we eat, hear, see, smell, learn, feel, touch, do and have done to. A genetic St. Francis clone could evolve into a tyrant. Or, more optimistically, a Hitler clone has the potential for sainthood.

The technical steps necessary to do human cloning are likely to be inspired not by the quest for a super race but by the need to solve compelling problems. Once developed to a point of predictable success, cloning will first be used as an eccentric application of a standard procedure, for a humanitarian end, as illustrated by a hypothetical case: A couple which has one adored infant and is incapable of having another learns that the child has been mortally injured. What possible harm would result, it may be asked, if one of the child's cells is taken so that he could be genetically reproduced (with the clone implanted in his mother's womb, or a substitute's) and nine months later "reborn" to the delight and comfort of his mourning parents?

Cloning—that most artificial of phenomena—would in this way be exploited to serve the most fundamental of human needs, bearing and raising children. Yet, on the other hand, it would totally cleave that need from related physiological and procreative behavior (sexual passion, tenderness and romantic love) which have traditionally initiated, accompanied and complemented parenthood.

CLONING AS METAPHOR

CLONING commands our attention more because it dramatizes the developing issues in bioethics than because of its potential threat to our way of life. Many biologists, ethicists and social scientists see it not as a pressing problem but a metaphoric device serving to focus attention on identical problems that arise from less dramatic forms of genetic engineering and that might slip into public use, protected from public debate by the incremental nature of the changes they impose.

All the issues have certain common features. The new technology will be motivated by the most humanitarian ends (with the exception of biological war research — another story). Its purpose will be to relieve suffering, to conquer disease, to restore normal capacities (as in conceiving or bearing a child). The difficulties of assessing the worth of this work vs. the cost are compounded because the benefits are immediate, concrete and tend to serve the individual, while the costs, if any, are perceived in abstractions ("humanness," "relatedness" and "quality of life"), are apparent only with time and are paid for by society as a whole or future generations.

The human being is the only species capable of systematically altering its "normal" biological system by use of its equally "normal" intellectual capacity. Cloning is but one example of such an intrusion into the reproductive area. The oldest is probably birth control of one sort or another. This capacity, when institutionalized, has a greater impact on a society than merely determining family size. It is a factor in defining the social roles available to women, levels of affluence of the society, and so on and on.

Abortion is also on its way to becoming institutionalized, in the sense of becoming accepted, relatively without question, as part of the normal order of things. It, too, will produce broad social effects, particularly when combined with amniocentesis, a technique which permits diagnosis of intrauterine fetal conditions (by withdrawing a sample of the amniotic fluid with needle and syringe) and would thus give couples the choice of aborting a defective fetus in its early stages.

ARTIFICIAL insemination as a solution for male sterility has been accepted, with some discussion of its psychological reverberations but little of its possible sociological impact.

A parallel problem exists when the husband produces adequate semen, but the wife either fails to produce eggs or has a blocked passageway from the ovary to the uterus. If the woman merely has the blocked passageway, it might indeed be possible to remove an egg from the ovary, fertilize it in a test tube, and replant it at the proper stage of division in her uterine wall. Each clause in that statement represents staggering technical problems — yet each problem has been solved, or is on the verge of solution. Drs. R. G. Edwards and P. C. Steptoe, who have been instrumental in much of the work in this area, are expected imminently to attempt the implantation in a uterus of an in-vitro fertilized egg, with a good chance that it will grow into a normal baby. They may have already tried it, but so far there has been no news about the results if they have. In-vitro fertilization offers an added advantage—or complication, depending on your moral position—over more traditional methods. In the laboratory, during the short interval of days between fertilization and implantation, the sex of the newborn baby will be determinable. There are now selective stains which when applied to a single cell can establish gender without even a chromosome smear and evaluation. Therefore, one could fertilize a number of eggs and offer the parents a choice of gender—as well as other options of genetic composition. On the other hand, it would probably be safer to allow implantation, wait a few weeks and then determine genetic composition by amniocentesis — aborting, if the fetus does not meet parental expectations or standards.

By utilizing the same technology, a woman with no ovaries but a healthy uterus can borrow an egg from a donor, just as semen may be obtained from a donor. It can then be fertilized by her husband's semen in the test tube and, when ready, implanted in her uterus. By the same token, if a woman has intact ovaries but has had her uterus removed (a not uncommon procedure), she can have a laboratory conceived baby that genetically is hers and her husband's, and ask that it be raised in the uterus of another woman for nine months. There is no technical problem here at all. It is just as simple to insert the fertilized egg into one uterus that has been prepared for it as another.

Once such procedures become accepted, need the reasons for utilizing them be limited to the biological?

A professional woman, for reasons of necessity, vanity or anxiety, might prefer not to carry her child. She might gladly pay for nine months' service from another woman. While certain liberationists might applaud the idea of freeing women from the nine-month pregnancy period, they might be appalled at the exploitation of another woman. This should provide the incentive for the development of an artificial placenta — doing away with the need for carrying the fetus in the womb—an undertaking that should not be immensely difficult.

A GREATER DANGER

THE artificial placenta is a long way from coitus interruptus, but in a definite continuum. Each step in the continuum offers potential for satisfying legitimate needs of individuals—and potentials for creating harm. Each step will incrementally influence society for good or bad or both beyond its meaning for those who are directly served. Some of the problems raised by tampering with reproduction may not be so obvious. To what degree will the procedure itself—independent of utility—

reduce man by altering the concept of the sanctity of life, birth and death? To what degree will it intrude on institutions and relationships traditionally deemed fundamental to human experience, perhaps to "humanness"? When might a technique that satisfies certain individual needs become a sociological or psychological problem?

We know now that simple conditioning can take place in the uterus. An embryo can be trained to move, for example, when exposed to specific sounds. The internal environment, then, is already operating on the embryo. Certainly, so is the more immediate environment of maternal emotion and physiology, in ways not yet known.

After birth, conditioning potential is increased enormously. Evidence in animals has demonstrated that when the neonate is given a "perfect" environment, but deprived only of affectual contact with the mother, monstrous psychological damage is done to the developing infant.

We must take particular care that we do not produce physiologically sound semblances of human beings who will turn out, at some later time, to contain deep psychological flaws. For this very reason, some scientists might argue, we should employ intelligent, systematic conditioning in-vitro, in-utero or neonatally — to guarantee some optimal future character or personality traits.

Who will determine what will be done and what will not? Who will determine what should be done and what should not? What controls should there be? How do we balance private rights and the general good? On what basis will we allocate decisions to either personal conscience or public policy?

Are there areas in which control of human development and behavior is bad *per se*, independent of the nature of the controlled things, the intention of the controller or the reasons for control? Are there processes which, once started, will bring irreversible changes so slight as to not be significant in one generation — but may, inexorably and incrementally, bring major changes to successive generations?

And if we do attempt human cloning, what will we do with the "debris," the discarded messes along the line? What will we do with those pieces and parts, near-successes and almost-persons? What will we call the debris? At what arbitrary point will the damaged "goods" become damaged "children," requiring nurture rather than disposal? The more successful one became at this kind of experimentation, the more horrifyingly close to human would be the failures. The whole thing seems beyond contemplation for ethical and esthetic, as well as scientific reasons.

Planned single alterations inevitably turn out to be package deals. The unpredicted complexities of environmental intervention, with the resulting ecological disasters, should serve as a warning model. Improvement is a form of substitution. The increasing capacity of man to reconstruct himself is, by definition, the capacity to destroy himself through transformation into another creature—perhaps better, but not man.

Any attempt at genetic engineering is bound to spark a public debate because it involves *physical* tampering with the substance of living things, and to most people physical tampering seems more permanent, more irreversible, and therefore more serious than mere psychological manipulation.

Scientists, however, are impressed with the similarities, rather than the differences, between the effectiveness of direct, physical control of behavior—for example, by electrical stimulation of the brain —and psychological conditioning. And it has been demonstrated in much research that one can successfully condition babies in the first days of life. For example, the neonate can be taught to switch on either a sound or light show on the basis of his sucking speed—and neonates do elect the entertainment of their choice with great consistency.

THE psychologist and psychiatrist have been aware that early imprinting on the mind and emotions has as profound a hold on an organism, and is as inflexible a determinant of adult behavior, as any genetic trait or physiological endowment. Now research is confirming that beyond even this, early environmental conditioning will actually produce organic, neurological changes in the brain.

That wise and sensitive geneticist Theodosius Dobzhansky has constantly emphasized that man's culture is not only his product, but his creator:

"Culture is, however, an instrument of adaptation which is vastly more efficient than the biological processes which led to its inception and advancement. It is more efficient among other things because it is more rapid — changed genes are transmitted only to the direct descendants of the individuals in whom they first appear; to replace the old genes, the carriers of the new ones must gradually outbreed and supplant the former. Changed culture may be transmitted to anybody regardless of biological parentage, or borrowed ready-made from other peoples.... In producing the genetic basis of culture, biological evolution has transcended itself—it has produced the superorganic."

Controlled culture has become a substitute for the natural selection of lower animals, and, with the homogenizing of the world's cultures, the variability and richness that randomness and chance produce are still further diminished.

If, in a time of anxiety, when the human species is unsure of its future and frightened by developments it does not understand, it is offered a planned environment, it may accept. If man is promised security and assured survival at the cost of his personal freedom and essential dignity, he may accept, particularly if he is told that the freedom he abandons is an illusion and the dignity only a conceit. Modern learning theory applied early through global television (the average American 4-year-old already watches 40 to 60 hours a week) and other teaching machines can program man beyond anything yet seen. The real danger of "pure strains" may come equally from conditioning and cloning. And both, as well as a frightening array of other problems not mentioned, demonstrate the fine line between promise and perdition in the new biotechnology.

WHEN Mary Shelley conceived Dr. Frankenstein, science was all promise. The technological age existed only in the excitement of anticipation, and there was leisure to philosophize. Man was ascending, and the only terror was that in his rise he would offend God by assuming too much and reaching too high, by coming too close. The scientist was the new Prometheus.

By the end of the 19th century, technology had surpassed even its own expectations. Man was too arrogant to recognize arrogance. Man did not have to fear God, he had replaced him. There was nothing that technology would not eventually solve. The whole of history seemed to be contrived to serve the purposes and glorify the name of Homo sapiens.

It seems grossly unfair that so short a time as the last 25 years should have produced so precipitous a fall. But then, the way down the mountain has traditionally been faster than the way up. Man has been handed the bill and he is not sure he has enough assets to pay up. We have destroyed much of our environment, exhausted much of our resources and have manufactured weapons of total destruction without sufficiently secure control mechanisms. The biological revolution may offer relief or hasten total failure. Unfortunately, things now move faster, and we are less sure of how to even recognize success or failure.

But technology has elevated man—and there is no going back. "Natural man" is the cooperative creation of nature and man. Antitechnology is self-hatred.

The tragic irony is not that Mary Shelley's "fantasy" once again has a relevance. The tragedy is that it is no longer a fantasy — and that in its realization we no longer identify with Dr. Frankenstein but with his monster. ■

March 5, 1972

Gene Transplants Seen Helping Farmers and Doctors

By VICTOR K. McELHENY

Biochemists working in California have developed a practical method of transplanting genes, the chemical units of heredity, from cells as complex as those of animals into the extremely simple, fast-multiplying cells known as bacteria.

The technique was worked out under the leadership of Dr. Stanley N. Cohen of the Stanford University Medical School and Dr. Herbert W. Boyer of the University of California Medical Center in San Francisco.

The new method, its discoverers say, gives promise of meeting some of the most fundamental needs of both medicine and agriculture, such as supplies of now scarce hormones, and nitrogen-fixing micro-organisms growing near the roots of wheat and corn plants, thus reducing requirements for fertilizers.

In addition, the ability to transfer genes—that is, particular stretches of deoxyribonucleic acid, the genetic chemical DNA, into bacterial "factories"—is expected to enable biologists to use bacteria, already the most closely studied cells, as laboratories for examining how medically significant animal genes are "turned on" and "turned off."

The gene transplantation is carried out by inserting the desired gene into a certain "plasmid," a small but complete ring of DNA that can multiply within the bacterium called Escherichia coli. The plasmid can be duplicated alongside the relatively huge circular chains of DNA that constitute the chromosome, the genetic endowment of the bacterium.

This new method was called "a major crossroads" in handling DNA, comparable to the technique of fusing different types of cells, by Dr. Joshua Lederberg, a Nobel Prize-winner at Stanford Medical School.

Dr. Lederberg said the work on gene transplantation had benefited from early support by the Government for research in fundamental biology, which he said had been sharply curtailed in recent years. He said it would be a mistake to conclude from achievements like those of Dr. Cohen and Dr. Boyer that the outlook for biological research was "rosy."

Because the method has already been used to transfer genetic characteristics from completely different types of micro organisms into Escherichia coli, Dr. Cohen said in a telephone interview he hoped the method could be used to transfer the ability to "fix" nitrogen from the air into species of microbes living next to the roots of corn and wheat.

Fertilizers Needed

Because the world's major food grains lack such nitrogen fixing microbes as the rhizobium, which grows next to the roots of the protein-rich soybean plant, they are heavy consumers of artificial nitrogen fertilizer.

Because such fertilizers derive from petroleum, their price is soaring after sharp price inceases imposed on oil by the petroleum - exporting nations. This, in turn, is creating a major food supply problem for populous underdeveloped nations such as India, Pakistan and Bangladesh.

The same ability to create what amounts to new species of bacteria, according to Dr. Cohen, could lead to colonies of Escherichia coli, equipped with the gene-carrying plasmids, growing large supplies of insulin for diabetics, who now depend on supplies obtained from beef and pork pancreases.

Another possibility on which Dr. Cohen and his associates are working is the transplanting of the genes responsibile for making the antibiotic Streptomycin from the streptomycete soil bacteria that now produce it. Dr. Cohen noted Escherichia coli were far easier to cultivate than streptomycetes.

Genes From Sea Urchins

Dr. Larry Kedes of Stanford is using the technique to transplant the genes that specify the composition of proteins called histomes of sea urchins into Echerichia coli.

Dr. Kay Ptashne of Dr. Cohen's laboratory and Dr. Oliver Richards of the University of Utah at Salt Lake City are using similar techniques with genes involved in the solar-energy-trapping process called photosynthesis in a water-dwelling organism called Euglena. Bacteria lack the genes for photosynthesis.

In San Francisco, in collaboration with Dr. Boyer, Dr. Brian McCarthy is using the technique to transplant particular genes from the fruit fly, drosophila, into Escherichia coli.

Also using the plasmid transplantation technique for drosophila genes is Dr. David Hogness of Stanford Medical School.

Last year, the method was used to transplant a particular gene from the frequently studied South African toad, Xenopus laevis, into Escherichia coli. The Xenopus gene is found in hundreds of identical or near-identical copies in female egg cells called oocytes. The gene embodies chemical instructions for making a substance called ribosomal RNA. This chainlike molecule, chemically similar to DNA, plays an important but still mysterious role in the assembly of the gobular structures called ribosomes.

A Stitching Platform

The ribosomes serve as the platforms for stitching together the thousands of different types of proteins that even a bacterium requires for vital functions.

A report of the work with the Xenopus ribosomal RNA gene is to be published in the forthcoming issue of the Proceedings of the National Academy of Sciences. The leading author of that paper is Dr. John F. Morrow, until recently at Stanford, but now at the Carnegie Institution of Washington's laboratory at the Johns Hopkins Medical School in Baltimore.

Co-authors of the paper are Annie C. Y. Chang, Dr. Cohen, Dr. Boyer, Dr. Howard M. Goodman and Dr. Robert B. Helling, all of whom figured in development of the plasmid transplantation technique.

The new method exploits the fact that even in bacteria, the most elementary of cells, the genes are not all grouped into the giant, closed-ended necklaces of DNA forming the chromosome. Bacteria generally have two identical copies of their chromosome, which must be duplicated in each hourlong "generation" of bacterial life before the cell divides into genetically identical "daughter" cells.

Fast Multiplication

Genes can also be found in the plasmids, which often multiply somewhat faster than the chromosome. Such plasmids often specify several special characteristics of a bacterial cell, such as resistance to such antibiotics as Streptomycin or tetracycline.

Another genetic characteristic embodied in a plasmid is the ability to form a bridge and inject part or all of a bacterial chomosome into a neighboring bacterium.

Earlier, biochemists had little success in transferring purified DNA from such drug-resistance plasmids into Escherichia coli. But then, Dr. Cohen's group tried using a calcium chloride treatment developed in another laboratory for a different purpose. Escherichia coli cells that were normally killed by various antibiotics took up drug-resistance plasmids and became resistant to the drugs.

Then, Dr. Cohen's group found a way to "shear" a plasmid into fragments containing only about 10 per cent of the original material but still viable, that is, still possessing resistance to the drug and the instructions for self-duplication.

One of these plasmids, called PSC-101, could be cut open by an enzyme called RI, described in Dr. Boyer's laboratory, in such a way that foreign genes could be inserted without impairing the plasmid's formal normal functions.

Unrelated Species

The California group found that they could add the genes of other plasmids from different varieties of Escherichia coli, and from such unrelated species of microorganism as staphylococcus. At almost the same time, the work of inserting the genes from the Xenopus toad was carried out.

Even though the work with animal genes might be drawing more interest from the scientific community, Dr. Cohen said, the success in assembling genes from different kinds of Escherichia coli and transferring genes from the "totally unrelated" staphylococcus was "the key" in suggesting that a general technique for gene transplantation had been found.

Thus, Dr. Cohen felt able to write in a staff paper for his colleagues, "It may be possible to introduce into E. coli genes specifying metabolic or synthetic functions such as photosynthesis or antibiotic production indigenous to other biological classes."

In describing the interest felt by many biologists in the new technique, Dr. Lederberg said that it provided "a completely new handle for studying regulatory phenomena" by isolating animal or even human genes from the enormous complexity of "higher" cells and putting these genes into "the well-defined context" of the much-studied Escherichia coli bacteria.

May 20, 1974

Genetic Tests Renounced Over Possible Hazards

By VICTOR K. McELHENY

In an action rare in the history of science, prominent American biologists, including one winner of the Nobel Prize, are voluntarily renouncing for the present two types of genetic experiments that they consider could be hazardous.

The classic approach in the sciences has been to pursue a trail of scientific inquiry wherever it might lead. However, in line with a growing scientific concern about some implications of modern molecular biology, the scientists warned against "indiscriminate application" of new techniques involved in the experiments.

The scientists are announcing this action in a letter that, within the next week, will reach much of the world scientific community. They said they were taking the step because the gene-transplantation experiments might accidentally increase the resistance of some micro-organisms to drugs, or lead to the spread of some types of cancer-causing virus.

The letter is being published this week in issues of the American journal Science and the British journal Nature. The recommendations of the scientists have been endorsed, in an equally unusual action, by the Assembly of Life Sciences of the National Academy of Sciences.

The scientists request in their letter that Dr. Robert Stone, director of the National Institute of Health, the United States Government's leading medical research agency, establish an advisory committee to oversee experiments to evaluate and minimize the possible hazards and set guidelines for researchers in the field.

Dr. Stone reportedly has already written to the National Academy to ask it to establish the committee. The National Academy has scheduled a news conference in Washington this morning for signers of the letter to discuss their action.

Among the signers are Dr. Paul Berg of Stanford University Medical School, Dr. David Baltimore of the Massachusetts Institute of Technology, and Dr. James D. Watson, the Nobel Prize-winner who directs the Cold Spring Harbor Laboratory of Quantitative Biology on Long Island.

The "potential rather than demonstrated risk" seen by the scientists rises out of a newly developed technique for transplanting certain animal genes into single-cell bacteria. The discovery of this technique was reported in The New York Times on May 20, and shortly thereafter in the May issue of the Proceedings of the National Academy of Sciences.

Genes Are Multiplied

The bacteria not only multiply the foreign genes, embodied in the genetic chemical called DNA, along with their own genes, but also provide an intensively studied context to observe how these foreign genes operate. The common bacterium of the human colon, Escherichia coli, is generally regarded as the organism that biologists have studied most completely.

The technique uses newly discovered enzyme proteins that act almost as surgical knives to cut DNA at precise points, and in a way that allows stretches of the stringlike, double-stranded DNA to be stitched together.

The technique also exploits the fact that bacteria like Escherichia coli contain small "satellite" rings of DNA, called "plasmids," which multiply alongside the much larger DNA ring containing most of the genes of the bacterium. Such plasmids can carry characteristics of a micro-organism suwch as resistance to one or more drugs.

The plasmids can be cut open with the enzymes, have genes from animals, viruses or other bacteria inserted in them, and then be stitched into a ring, which then can enter cells of Eschenchia coli to multiply.

Infectious Elements

Work planned by several groups, including those of the signers of the letter, would, according to the singers, create "novel types fo infectious DNA elements whose biological properties cannot be completely predicted in advance."

The signers said, "There is serious concern that some of thhese artificial DNA molecules could prove biologically hazardous."

Because Escherichia coli is found in the human intestinal tract, the scientists said, and because such innocuous bacteria can exchange DNA with other types, harmful to man, new DNA elements introduced into the bacteria might possibly become widely disseminated among human, bacterial, plant or animal populations with unpredictable effects.

This concern, based on a new tecnique, differs from the "genetic engineering" concepts, involving such futuristic ideas as producing whole armies of genetically identical human beings, or using viruses to inject missing genes into people, which biologists and the public have been discussing for the last decade.

The scientists asked others "throughout the world" to join them in "voluntarily deferring" two types of experiments. One would involve creating new "plasmids" containing a combination of drug-resistance not found in nature, or in using plasmids to give such resistance to bacteria now lacking it.

Other Experiments

The second class of experiments would involve attaching cancer-causing or other animal viruses to the plasmids, or to the DNA of other viruses.

The scientists said that such DNA molecules "might be more easily disseminated to bacterial populations in humans and other species and thus possibly increase the incidence of cancer and other diseases."

In addition, the scientists advised caution on experiments where animal DNA would be linked to bacterial plasmids.

The scientists urged that an international meeting be held early next year "to review scientific progress in this area and to further discuss ways to deal with the potential biohazards."

They said they realized that the risks were "potential rather than demonstrated" and that "scientifically worthwhile" research might be delayed. But they urged the delay "until attempts have been made to evaluate the hazards and some resolution of the outstanding questions has been achieved."

As publication of the letter neared, scientists searched for an exact precedent and found none. An agreement among non-German nuclear physicists in the early nineteen-forties to cease publication about the splitting of the atom was not intended to halt work thought hazardous but to deny advantages to Nazi Germany, where the phenomenon had been discovered.

The spur to the scientist's letter came last year at the annual Gordon Conference on Nucleic Acids. After the conference, two of the participants, Dr. Maxine Singer of the National Institutes of Health and Dr. Dieter Soll of Yale University, wrote a letter warning about the risks in joining DNA from "diverse sources." That letter was published in Science last Sept. 21.

A meeting to draft the current letter was held in Cambridge, Mass., in April.

Other signers included Drs. Stanley Cohen, Ronald Davis and David Hogness of Stanford University, Richard Roblin of Harvard Medical School, Norton Zinder of Rockefeller University, Daniel Nathans of Johns Hopkins University, Herbert Boyer of the University of California and Sherman Weissman of Yale University.

July 18, 1974

World Biologists Tighten Rules On 'Genetic Engineering' Work

By VICTOR K. McELHENY
Special to The New York Times

PACIFIC GROVE, Calif., Feb. 27—An international conference of biologists decided today on tighter professional standards governing research in so-called "genetic engineering."

The new guidelines, rare in the history of science, are designed to prevent medical risks to man while opening the door to many experiments intended to benefit medicine and agriculture.

The proposed rules have moral force only, but they were devised by a large fraction of the world's scientists now working with powerful new techniques for transplanting genes, the chemical units of heredity, from one type of living cell to another.

Scientists from 16 nations, including the Soviet Union, attended the conference, although more than half the participants were Americans. The conference's conclusions are expected to provide a blueprint for regulations to be written in months ahead by government health research agencies around the world.

The safeguards—designed to prevent the escape of potentially harmful organisms from the laboratory—are intended to replace a voluntary deferral of some kinds of genetic engineering research that went into effect last July.

The deferral was proposed by a group of American biologists, led by Dr. Paul Berg of Stanford University, who was the chairman of the conference.

Like the new safeguards, the deferral was a rare event in a scientific tradition that has resisted ruling out specific lines of inquiry as a violation of intellectual freedom.

The decision of the conference followed four days of struggle by more than 100 biologists to balance the potential hazards and benefits of the new techniques for manipulating genes.

Using enzymes to break up or stitch together pieces of the gene-containing chemical called D.N.A., the scientists can insert genes from complex cells like those of sea urchin eggs into simple, rapidly dividing cells like bacteria.

By doing this, they seek to build up large supplies of particular genes they want to study. They also hope someday to use simple organisms like bacteria to grow substances like human growth hormone for individuals who lack the ability to make their own.

Other hopes are to simplify the manufacture of disease-fighting antibiotics, open up another way to obtain supplies of insulin for diabetics, and to improve the capacity of microorganisms in the soil to "fix" nitrogen from the air and thus reduce demand for scarce fertilizer made from petroleum or natural gas.

Nobel Dissenters

The scientists agreed today that, for now, there were some classes of genetic engineering experiments involving dangerous viruses that they can see no safe way to perform.

They also agreed on an immediate collaborative effort to develop safer biological tools for gene manipulation. The idea is to restrict the usefulness of these tools to the laboratory, and make them unable to invade human beings.

Dissenting in part from the conference's conclusions were two Nobel Prize biologists. Dr. Joshua Lederborg of Stanford University and Dr. James D. Watson, director of the Cold Spring Harbor Laboratory on Long Island. Both said they regarded the safeguards as virtually unenforceable because of the difficulty of determining exactly the risk of specific experiments.

The conference was held at Asilomar, an oceanside conference center operated by the California park system.

The events leading up to the gathering began about five years ago, when molecular biologists discovered a special class of enzymes they called "restriction enzymes."

These proteins have the property of recognizing specific sequences of sub-units of the genetic chemical D.N.A., and cutting the double-stranded molecules in such a way that other strips of D.N.A. can be joined to them and re-form a closed circle.

In many types of single-cell bacteria, there are auxiliary rings of D.N.A. besides the large ones containing the genes of the bacteria. These auxiliary rings are called "plasmids."

It is into these plasmids that scientists have been inserting strips of D.N.A. from animal cells, such as those of the eggs of the South African toad Xenopus, in order to study them.

Bacteria multiplying with the animal genes in them can be regarded as man-made organisms.

The bacteria used most often are a special laboratory strain called K12 of Escherichia coli, a bacterium that commonly inhabits the human intestinal tract along with many others. Because the bacterium is common in human beings, scientists have feared that laboratory experiments with genetic engineering might lead to transfer of novel infections or even cancer viruses to man.

Two years ago, at a conference of biologists in New Hampton, N. H., reports of the potential of the new techniques so disturbed participants that the cochairmen of the conference wrote an open letter urging attention to possible hazards.

Drs. Maxine Singer of the National Institutes of Health and Bieter Soll of Yale University, wrote that hybrid D.N.A. molecules created by the new methods "may prove hazardous to laboratory workers and the public."

"Although no hazard has yet been established, prudence suggests that the potential hazard be seriously considered," they declared.

Last July, a committee of biologists headed by Dr. Berg published a letter in several scientific journals urging the voluntary deferral of work that might accidentally increase the resistance of some microorganisms to drugs or lead to the spread of some types of cancer-causing viruses.

'Enormous Progress' Cited

They said their concern was based on "potential rather than demonstrated risk" and added that "adherence to our major recommendations will entail postponement or possibly abandonment of certain types of scientifically worthwhile experiments."

In the following months, committees of American scientists worked on detailed safeguards for various aspects of the work with the bacterial "plasmids." Their reports were considered at the meeting, and broad conclusions from them were approved by the conference today.

The draft final report of the conference spoke of "enormous scientific progress already achieved in this field" and said that the new methods promised to "revolutionize the practice of molecular biology."

The draft, written by the organizing committee and amended at an often-heated session this morning, said, "The new techniques combining genetic information from very different organisms place us in an area of biology with many unknowns. It is this ignorance that has compelled us to conclude that it would be wise to exercise the utmost caution.

"Nevertheless," the report continued, "the work should proceed, but with appropriate safeguards. Although future experience may dispel many fears, standards of protection should be set high at the beginning. Each escalation, however small, should be carefully assessed."

February 28, 1975

EVOLUTION

NEW PUBLICATIONS.

The Descent of Man.*

THE PRESENT POSITION OF THE ARGUMENT.

It is generally understood among Mr. DARWIN'S friends in this country that the only terms which he made with the Messrs. APPLETON, in issuing his new book on the *Descent of Man*, were that they should print a thoroughly revised and corrected edition of his first great work, *The Origin of Species*, of which hitherto there had been no complete American edition. It is to be hoped, for the credit of American publishers, that the Messrs. APPLETON will be more generous than their bond toward the distinguished author, as undoubtedly Mr. DARWIN'S last work is destined to enjoy an enormous circulation for many years to come on this side of the Atlantic. He was wise, however, in the reported contract. The work on which his fame will rest is *The Origin of Species*; and it was of the utmost importance to him as a scientific thinker that every improvement which his well-known candor, and a wider investigation, suggested in the form of the celebrated argument should be engrafted on the American edition. The Messrs. APPLETON have accordingly issued a revised "Fifth Edition," in a not very handsome style, however, of this remarkable work.

Whatever coming generations may think of the Darwinian hypothesis; whether it will be classed with the Epicurean speculations on the origin of matter, with LAMARCK'S theories of development, or the ingenious suggestions of the *Vestiges of Creation*, captivating but unsound, or whether it will stand with the nebular hypothesis, as a stable working theory of the origin of the kingdoms of life—in any case the future historian of the progress of the human mind must testify that no one speculation has so affected every branch of scientific investigation in the present century, and so influenced the whole field of natural research, as the Darwinian theory of "Natural Selection." There is not a specialty in the whole domain of natural science, however remote, in which the student will not at once perceive the influence of "Darwinism" on all new investigators.

The mode of looking at nature is changed. BACON'S philosophy did not more entirely alter the process of investigation in all branches of science, to the English mind, than has the Darwinian theory among the students of nature in Europe and America. The supernatural method of accounting for phenomena is dropped. The duty of science—not ignoring the ultimate supernatural Cause—is solely with the immediate causes of all the phenomena of life on this earth. It is assumed that every form or appearance in organic or inorganic existence can be explained, though no present explanation may be possible. The task of science is simply to collect facts with the utmost patience and fidelity, and then, if possible, to group them under some general law or mode of action, and still again attempt, with untiring industry and a love of truth which is almost a religion, to simplify and generalize from these laws a still broader law. Always to be open for new facts and premises, to candidly confess if these be opposed to the theory, to begin and build again on a broader basis if the foundation be too narrow; but never for an instant to doubt that each phenomenon in this world has a direct and sufficient natural cause—this is the spirit of the new Darwinian philosophy. Under the influence of this school the former narrow *specialists* in the field of natural science are becoming broadened. Each investigator carefully collects facts, but he gathers them under the light of broad generalizations, which connect them with the whole world of nature, and give a wonderful attraction and dignity to his pursuit.

The essence of Darwinism is that nature—though originating from divinity—is an intricate, complicated, but explainable system of things, where the obligation of the student is not to say this is an ultimate fact—this "an idea of the Creator"—this a "final cause," or "plan of creation," or "ideal form of Nature," but "what is the cause of this appearance?" DARWIN, indeed offers a cause of specific appearances, an hypothesis of origin, to which his name will always be attached. This law he has poetically named "natural selection," or the "survival of the fittest," (as SPENCER has defined it.) Under this, with the unexplained law of variation, the well-known principle of inheritance, and the great facts of overproduction and consequent "struggle for existence," he has attempted to explain the origin of all the kingdoms of life, and all the natural phenomena of this world of ours.

The works, whose titles we have placed below, are only a small portion of the important treatises issued during the past ten years, illustrating or opposing this great hypothesis. German book-sellers and libraries already have an important division in

---

*THE ORIGIN OF SPECIES. Fifth edition, with additions and corrections. By CHARLES DARWIN. Pp. 447. D. APPLETON & CO., New-York. 1871.
THE DESCENT OF MAN. By same. 2 vols., 12 mo, pp. 406, 436. 1871. APPLETONS.
CONTRIBUTIONS TO THE THEORY OF NATURAL SELECTION. By A. R. WALLACE. Pp. 384. MACMILLAN & CO., New-York. 1870.
A REVIEW OF MR. DARWIN'S DESCENT OF MAN. By A. R. WALLACE. Pp. 36. BRENTANO, New-York. 1871.
THE GENESIS OF SPECIES. 12 mo, pp. 314. By ST. GEORGE MIVART. 1871. APPLETONS.
HEREDITARY GENIUS. By FRANCIS GALTON. 12 mo. 1871. APPLETONS.

---

their lists, headed "*Darwinismus*." In studying these and the writings of the co-discoverer of the law of natural selection, Mr. WALLACE, and such works as those of Sir JOHN LUBBOCK and the Duke of ARGYLL, the natural question arises, "How stands the great argument on origin?"

Many of the works we have mentioned, and such standard books of science as Sir CHARLES LYELL'S, are, in one aspect, only side illustrations of the application of the theory of natural selection. Mr. GALTON, in the volume on *Hereditary Genius*, has made, under this hypothesis, one of the most ingenious and scientific investigations on the subject of the transmission of mental and moral faculties in man which has ever been undertaken. We regret that we have no more space here to analyze its methods or quote its results. It is a book which should be in the hands of every scientific student of man. As a model of this close scientific research, we commend his ingenious explanation of the gradual extinction of the families of celebrated Judges in English history.

The strongest work by far which has appeared against Darwinism, pure and simple, is undoubtedly MIVART'S *Genesis of Species*. The Duke of ARGYLL'S writings are weak in comparison, as are nearly all arguments, especially the theological ones, on the other side. Next to this will come Mr. WALLACE'S writings, both in the article whose title we give, and in his volume on natural selection. The treatise of Mr. MIVART (whose name, unhappily, is not familiar to our scientific world) is strong from his remarkable candor and courtesy, and his thorough familiarity with points in natural history which bear on the discussion. He writes, too, with singular balance and clearness. Some of the points he makes seem almost unanswerable; and so confident are all scientific students of Mr. DARWIN'S unassailable integrity and candor, that they would not be surprised to see him at any moment abandoning some of his strongest positions under such a respectful and truth-loving assault.

The most important of MIVART'S objections are, briefly, that "natural selection" cannot account for the origin of beneficial variations; that the co-existence of closely-similar structures of diverse origin is inconsistent with it; that species sometimes develop suddenly and not gradually, and have very definite lines to their variability; and finally, that there are some phenomena in organic forms on which natural selection throws no light. His points in regard to the absence of certain fossil transitional forms, the peculiar distribution of animal life on the earth, and the physiological differences between "species" and "races," do not seem near so strong as the other, and are all capable of explanation under the Darwinian theory. Mr. WALLACE is a naturalist perhaps even superior to Mr. DARWIN. As a theorist, he early, in coincidence with the other, struck on the hypothesis of natural selection. He was one of the first—an American writer on ethnology being the first—to apply the theory to man. His explanation of the action of this principle on the formation or development of man's mental and moral faculties is not surpassed in ingenuity and delicacy of reasoning by anything which DARWIN has written.

Anything, therefore, which he could offer on the "great argument" is worthy of all attention. Both he and MIVART believe in evolution; both apparently believe in the development of man's body from a semi-human form. Indeed, MIVART has suggested, in a recent letter in Nature, that if the Darwinian hypothesis of man's origin be true, among the three Simian lines of descent—the Orang, Chimpanzee and Gibbon—the one which has been most parallel to man is that of the Gibbon.

But where they diverge from the great philosopher is in the extent of the working of the law of natural selection, and in his theories of the origin of the human faculties, and especially of the moral powers. They admit "natural selection" to be a *vera causa*, but not a *causa sufficiens*. It explains, but does not explain all. Mr. WALLACE is staggered by the hairless back of man, the uselessly large brain of the savage, and the ideal faculties of the lowest man, which apparently could not be formed under the law of the survival of the most useful capacities. From his long experience with savage tribes, moreover, and as a result of reflection, he has become an "intuitionist," and has no faith in the development dogma that conscience is the transmitted and accumulated result of experiences of benefit to the community in certain actions, or of the damage from the conquest of the higher social instinct by a lower instinct.

He has found a wild savage in the Malay Archipelago acting under an apparent intuition of truth and justice, where the action only brought loss both to the individual and the community. Mr. MIVART, again, is a devout orthodox believer, (though an evolutionist,) and, therefore, rejects the Darwinian explanation of the origin of the sense of right and wrong. The position of both on the great question is one which is more and more being adopted by scientific thinkers of a not too ultra stamp in both America and Europe; that the principle of natural selection is a law; that it underlies and explains the formation of varieties, and, perhaps, of nearly all species; that it has had an enormous influence in giving the whole organic world its present shape, appearance and variety, but that there are forms of life, organs, qualities and changes so peculiar and intricate, structures so complicated, features of so little benefit or injury, and sometimes of such apparent loss and disadvantage, that some other cause or law is necessary to explain them than merely the principles of "variation," "inheritance" and "struggle for existence."

Dr. ASA GRAY—than whom no American has written more lucidly on this theme, and himself a Darwinian—has somewhere said that, in the experience of naturalists, "Variation often seems to be *led* along some beneficial line." Mr. MIVART concurs with this, and we suppose Mr. WALLACE would agree; yet DARWIN himself could never accept it and hold his theory in its pure form.

In the particular argument on the "Descent of Man," Mr. WALLACE disagrees especially with his distinguished associate; and on this field his investigations and experience have probably been even greater than Mr. DARWIN'S. The immense chasm separating man from the highest form of monkey is what causes him here to doubt the universality of his own law, "Natural Selection." How a four-handed, creeping, hairy, speechless semi-human creature with small brain and corresponding capacities, could develop into an erect, smooth-skinned, two-handed, large-brained, fire-using, and above all, speaking man, at a time when the struggle for existence is severest, and remain in a limited area of tropical earth during the enormous interval of these changes, he cannot understand. "His absolute erectness of posture, the completeness of his nudity, the harmonious perfection of his hands, the almost infinite capacities of his brain, constitute a series of correlated advances too great to be accounted for by the struggle for existence of an isolated group of apes in a limited area." There must be, he conceives, other causes than the struggle for existence to produce such an immense contrast between man and the apes.

Leaving now Mr. DARWIN'S critics, it becomes necessary to ask, "How he himself stands in the great argument?" His last book, on the *Descent of Man*, will undoubtedly be by far the most generally read of all his writings. And yet—remarkable as it is in the candor of its discussions, in the extraordinary range of its facts, the ingenuity of its explanations, and clearness of the style—it is not by any means the greatest of his works. To his many admirers, it will be almost discouraging to note that the very first application of his theory to the simplest physical and supposed generic connection, has, after all, so little force or ground to rest upon. To the physicist, no two families would seem easier to connect, under the Darwinian hypothesis, than the human and simian. And yet the careful reader will finish DARWIN'S ingenious list of physical resemblances between man and the monkey, and find them, after all, extremely vague and remote, while the gap between the two, moral and mental, will appear, as it does to WALLACE, unexplainable. Mr. DARWIN has in this volume contributed nothing to the science of ethnology, and added little to the weight of his former arguments. His presentation, however, of the analogies or resemblances between the faculties of animals and men, though not new, is deeply interesting, and cannot fail to lead to more profound investigations in this attractive field.

In one point of view, it must have a deeply humane influence—in showing the many close points of sympathy between man and "the mutes," (as a fellow-Darwinian so respectfully terms "the brutes.") DARWIN'S explanation of "conscience," which was, of course, the great stumbling-block to the application of his theory, is highly ingenious and striking, but is odious and utterly unsatisfactory to the "intuitional" thinkers everywhere, and will without doubt revolt the religious world more than any other portion of his writings. We have not space here to discuss it. Singularly enough, there is in this last work a slight "change of front" by the great theorist. He is by nature such a lover of truth, that we are convinced he would to-morrow throw up his whole hypothesis, if he found unanswerable objections to it. The arguments of his opponents have shown that there are very many structures or organs, or appearances in the organic kingdom, which cannot be said to be either useful or injurious; and which accordingly have not come under the law of natural selection.

The remarkable fact, too, of Beauty—the wonderful and delicate combinations of colors and outlines of forms through the organic and inorganic world, where no advantage seems to result from these pleasing and exquisite appearances—has staggered him. Under pure Darwinism, Beauty has no place except as it is nullity. To meet these very weighty objections, Mr. DARWIN, in the volume on the Descent of Man, has presented a new hypothesis which is almost as original as his great one.

It was hinted at in his first volume, and undoubtedly led his German followers to work it out somewhat in advance of this presentation. It is the theory of "Sexual Selection" which occupies some two-thirds of these last volumes. Nothing that DARWIN has written is more ingenious or suggestive than the long, minute and careful investigation in this field, presented in his last work. The argument is too extended to be even analyzed here. It is sufficient to say that most of his followers, and the majority of students of science, will find it ingenious, but unsatisfactory. They will admit that it explains many phenomena, but they find that Mr. DARWIN accounts for too much by it. Many of the facts included under it would seem more easily explained by natural than sexual selection. Mr. WALLACE objects also that it does not explain the beautiful colors and appearance of insects, or the lowest forms of life. The "Intuitionists" everywhere will rejoice that it leaves Beauty as unexplained as ever, while it proves, to our surprise, that the birds have the same refined and cultivated tastes as the most cultured human beings. It cannot be admitted that DARWIN, by the new theory, has removed the main objections to the old. The great question of "Origin" is as unsettled as ever.

Shall we ever know more of the "Descent of Man"? At present there is an immense and unfilled gap between human beings and the highest animals. It is not at all probable, whatever may be our theory of the physical genealogy of man, that this chasm will ever be bridged. If there be an intermediate form, it must have perished from the earth millions of years ago; for recent investigations by Prof. WHITNEY in the Drift, and even the Pleiocene of the Sierras, show that man, in his perfect development, existed at that enormously distant period on the earth. A single link in man's long genealogical chain might easily perish from existence and never be seen again. On the other hand, unbiased naturalists like WALLACE and MIVART, believe that no known natural causes can explain the origin of man's mental and moral faculties; in other words, that in the beginnings of the first human soul or mind, a supernatural power intervened. If this be allowed, it would not be a potent supposition that to this spiritual existence a peculiar medicine was supernaturally adopted, in harmony with the nearest physical forms. DARWIN himself admits that somewhere in the vast line of human development, the soul, by Divine power, was made immortal. The student of physical science may equally believe that somewhere in that time it was created, and its brain and body adopted to it.

June 1, 1871

DEATH OF CHAS. DARWIN

THE LIFE AND WORK OF THE EMINENT NATURALIST.

HIS ANCESTRY AND EDUCATION—EARLIEST SCIENTIFIC WORK—HIS PUBLICATIONS—THE THEORY OF EVOLUTION AND THE USE HE MADE OF IT.

The announcement that Charles Robert Darwin died on Wednesday at his residence, Down House, near Orpington, will be read by very few individuals who have not some degree of acquaintance with the physical theories formulated and taught by this distinguished naturalist, however scanty may be their actual knowledge of his works. Darwin has been read much, but talked about more. Since the publication of his work "On the Origin of Species" in 1859, and particularly within the 11 years which have elapsed since his "Descent of Man" was given to the world, he has been the most widely known of living thinkers. Doctrines such as he set forth could not long remain the exclusive property of philosophers nor of educated people. They made their way at once into the reading and thought of the masses until the slightest allusion to Darwinism was sure of instant recognition from even the most illiterate individual or audience. It is not to be supposed that every country clergyman, with a library not extending far beyond Richard Baxter, Jortin, Bishop Berkeley, and a commentary or two, became profoundly versed in the doctrines of evolution, or that little laughing schoolgirls joking each other about a monkey ancestry had followed Mr. Darwin very far in his speculations on differentiation of species, but the ministers somehow all knew that evolution was an abominable heresy to be by precept and example thrust out of men's minds, and the school children intuitively understood that if man is descended from the ape he cannot be descended from Adam. All that part of the world which had never thought of such things before was aroused by the shock of a new idea. Previous speculations upon the origin of man had, it is safe to say, been the diversion of the learned; the people at large had no part in them. But here was a scientist, not a speculative philosopher, who dealt with facts and logical inferences in a new way, and he speedily had the whole world for an audience. Everybody saw that the history of living forms as his books taught it was widely at variance with the Mosaic account of the creation, and from the moment when the Darwinian theory of evolution was publicly stated the modern struggle between science and theological dogma took its rise. There had been skeptics and atheists and deists and what not before, but what grave essayists call scientific unbelief sprang primarily from works of Charles Darwin, and is fed chiefly from the writings of other scientists who are at work extending and completing the frame work he erected. Mr. Darwin, therefore, may be called a epoch-making man.

The qualities and natural bent of his clear mind were inherited. His father and grandfather were naturalists, though the latter, Dr. Erasmus Darwin, was a much more famous and productive man than his son, Dr. R. W. Darwin. Erasmus Darwin, a botanist of renown, is best known as the author of a remarkable poem called the "Botanic Garden," which, though destitute of the poetic feeling, shows its author to have been deeply versed in the Linnæan system of botany. Of Dr. R. W. Darwin little is known save that he was a member of the Royal Society. Charles Robert Darwin was born at Shrewsbury, England, Feb. 12, 1809. His early education was received in the Shrewsbury Grammar School, under the tuition of Dr. Butler. In 1825, when he was 16 years old, he entered Edinburgh University, and remained there two years, going then to Christ's College, Cambridge, where he received the degree of Bachelor of Arts in 1831. In December of the same year he was selected as a naturalist to make a voyage of scientific exploration around the world on board the ship Beagle. Five years were spent in this way. We may fairly suppose that Mr. Darwin was a naturalist of some competence and training when he set out on this voyage. The opportunities for research, experiment, and study which it gave him, particularly during his stay in South America, were fruitful in the material and hints out of which his later theories were evolved. Indeed, in geographical and geological distribution Mr. Darwin found the weightiest proofs of the truth of his system. Returning from this voyage in 1836, he began the preparation of a "Journal of Researches" into the geology and natural history of the countries visited by the expedition. This was published as a part of Capt. Fitzray's "Narrative of the Surveying Voyages of H. M. S. Beagle and Adventurer." In succeeding years he edited, in five parts, the "Zoology of the Voyage of the Beagle," the notes of the habits and range of mammalia being by his hand. Out of the material obtained on this cruise he prepared for publication in 1842 "The Structure and Distribution of Coral Reefs;" in 1874, "Geological Observations on Volcanic Islands," and in 1846, "Geological Observations on South America." Between 1844 and 1859 his publications were mostly brief monographs contributed to scientific publications or read before learned societies. But during this long period of slight literary productivity he occupied himself with untiring zeal and systematic regularity in the study of nature, making a series of observations upon the forms and habits of animals, plants, and minerals—for it is hard to say whether he was most eminent in zoology, botany, or geology—and slowly accumulating that vast mass of facts and registered phenomena to which he was later on to apply his theory of evolution.

The publication in 1859 of his work "On the Origin of Species by Means of Natural Selection; or, the Preservation of Favored Races in the Struggle for Life," was the announcement to his friends that he had at length passed over the sea of hypothesis to the firm ground of scientific assertion, and to the world that it must revise or fortify its opinions on biological subjects. After making in 1862 one of those excursions into the by-ways of scientific inquiry of which he was so fond, of which the outcome was his work on "The Various Contrivances by which Orchids are Fertilized by Insects," and still another in 1865 to publish the well-known book on "The Movements and Habits of Climbing Plants," he put forth in 1868 another important work on "The Variation of Animals and Plants Under Domestication," in two volumes. In 1871 appeared the best known of all his books, "The Descent of Man, and Selection in Relation to Sex," in two volumes. The following year saw the publication of "The Expression of the Emotions in Man and Animals;" in 1875 appeared "Insectiverous Plants;" in 1876, "The Effects of Cross and Self-fertilization in the Vegetable Kingdom;" in 1877, "The Different Forms of Flowers and Plants of the Same Species," and in 1881 "The Power of Movement in Plants." Each of these books has its place in the development of the theory which bears their author's name. All of them, even those which concern only a single order of the phenomena, abound in illustrations pertinent to his great theme, and supply those who wish to use or investigate his theories with the classified results of his accurate observations. But it is upon the "Origin of Species" and the "Descent of Man," in which the theory of evolution is made to tell the history of life upon the earth as we now see it, that his fame chiefly rests.

If asked to define Darwinism, the orthodox antagonist of the scientific unbelief of the day will reply that it is an attempt to show how blind matter became the seeing eye; the biologist of the Haeckel school will say that it is a description of the mechanical process by which the cosmic system was produced out of elementary matter acted upon by its own laws. Neither definition is correct, for Mr. Darwin made an extremely modest use of his great attainments. He did not construct a theory of the cosmos, and he did not deal with the entire theory of evolution. He was content to leave others to poke about in the original protoplasmic mire, and to extend the evolutionary law to social and political phenomena. For himself, he tried to show how higher organic forms were evolved out of lower. He starts with life already existing, and traces it through its successive forms up to the highest—man. The central principle—his opponents call it a dogma—of Mr. Darwin's system is "natural selection," called by Herbert Spencer "the survival of the fittest," a choice which results inevitably from "the struggle for existence." It is a law and fact in nature that there shall be the weak and the strong. The strong shall triumph and the weak shall go to the wall. The law, though involving destruction, is really preservative. If all plants and animals were free to reproduce their kind under like and equally favorable conditions, if all were equally strong and well equipped for obtaining sustenance and making their way in the world, there would soon be no room on the earth for even a single species. Thirty millions of men in less than 700 years of unchecked reproduction, under the conditions we have mentioned, would have living offspring enough to cover the whole earth at the rate of one for each square foot of its surface. The limit of subsistence and the power of reproduction are the bounds between which the conflict rages. In this struggle the multitudes are slain and the few survive. But the survivors do not owe their good luck to chance. Their adaptation to their surroundings is the secret of their exemption from the fate which overtakes those less happily circumstanced. A variety of squirrels, for instance, which is capable of wandering far afield in pursuit of its food, which is cunning and swift enough to evade its enemies, and has a habit of providing a store of nuts for Winter use, will naturally have a better chance of survival than a variety deficient in these qualities. But Mr. Darwin also discovered that natural selection created special fitness for given circumstances and surroundings. Climate, soil, food supply, and other conditions act in this way, and the result is the differentiation of species. A certain thistle grows in a kind of soil which is rich in the elements which go to produce the tiny hairs upon the surface of the plant. The seeds are thus furnished with downy wings longer than usual, and are wafted further off where they have plenty of space to grow, and they, in turn, reproduce and emphasize the changes to which they owe their existence. Seeds or nuts developing a thick covering for the kernel are thus protected from birds and animals, and live to germinate, producing also hard-shelled seeds, and thus the process goes on. Varieties which do not develop a high degree of special adaptation to their surroundings fall out of the race, unable to defend themselves against their innumerable aggressors. An infinitesimally minute variation of function or structure repeated and becoming more marked through many successive generations, results ultimately in the production of a variety, or even of a species, quite unlike the parent individual.

Mr. Darwin was by no means the discoverer of the theory of evolution. That is at least as old as Aristotle, who supposed individuals to be produced, not by a simultaneous creation of a minute copy of the adult, with all the different organs, but by epigenesis—that is, by successive acts of generation or growth, in which the rudiment or cell received additions. Other ancient philosophers, and in more modern times Descartes, Spinoza, Leibnitz, Bonnet, Lamarck, and Cuvier, have adopted and used this theory to a greater or less extent. But it never had a substantial basis of fact or a thoroughly scientific application until Mr. Darwin worked it out. Others, as we have said, and notably Mr. Spencer, have given it a more comprehensive scope, but within the limits he set for himself Mr. Darwin meets no rival claimant for the honors the scientific and thinking world would have accorded him. The dispatch announcing his death says that he had been suffering for some time from weakness of the heart, but continued to work to the last. He was taken ill on Tuesday night with pains in the chest, faintness, and nausea. The nausea lasted more or less during Wednesday and culminated in death in the afternoon. Mr. Darwin remained fully conscious until within a quarter of an hour of his death.

April 21, 1882

SCIENCE REBUILDS THE 'MISSING LINK'

Reconstructed Model of Ape-Woman's Skull Is Shown at South Kensington Museum.

HUGE TEETH, LITTLE BRAIN

Fragments of the Skull Found in a Sussex Pit May Now Be Viewed by Students.

By Marconi Transatlantic Wireless Telegraph to The New York Times.

LONDON, May 17.—A scientifically reconstructed skull of the oldest woman in the world is now on exhibition at the Natural History Museum in South Kensington.

It was the discovery of the fragmentary remains of the original skull that sent such a thrill of excitement throughout the scientific world last Autumn, when Charles Dawson unearthed it from a pit at Piltdown Common, Sussex, and a great company of distinguished professors gathered at the meeting of the Geological Society where the discovery was first made public.

The skull is not an object of beauty, even as skulls may be regarded as varying in gracefulness, but in her defense it may be pleaded that the woman was semi-simian, combining in herself traits of the human being with characteristics of the ape.

Scientists regard her as the one specimen extant of the "missing link." Her age eludes one even now. She may have lived 50,000 years ago, or 100,000, or even 200,000, for geologists agree to differ upon so delicate a subject, but it is believed that she belongs to the Pliocene period.

The discovery of her remains forms one of the most romantic incidents in the history of geological research, and, although a pilgrimage of scientists to England to inspect them is not predicted this year, no geologist from Europe or America who finds himself in London this Summer will be likely to miss the opportunity of having a call upon her at the museum.

The actual remains are not exhibited to the general public; savants and students may examine them, however. These remains comprised no more than a portion of the left side of the skull and a piece of the lower jaw, but with these as a guide Frank Barlow has succeeded in reproducing what is regarded as a faithful and trustworthy model of the whole, by careful observation and scientific deduction.

The task of making the model occupied Mr. Barlow many weeks. By noting the formation of the left half of the skull, he explained, it was possible to build up the right side with a considerable degree of certainty. On the same plan the lower jaw could be completed with a sure touch, and the conformation of the whole skull could thus be satisfactorily established.

The appearance of the facial bones and the upper jaw is largely conjecture, but the jaw bone is in every respect characteristic of that of the chimpanzee, and Mr. Barlow said that in constructing the model he followed the logical course, providing it with a dental equipment of the simian type. No modern human being possesses teeth of the size and shape of those seen in the model, and, more than anything else, the powerful teeth of the heavy under jaw serve to emphasize the ape-like characteristics of this primitive being.

A cast of the brain taken from the restored skull is also on view. While the brain cavity of the normal human being measures more than 90 cubic inches, this Pliocene skull has a capacity of no more than 64¾ cubic inches, showing that the brain development in modern woman is more than one-third greater than that of her semi-simian ancestor. From the greater development of brain at the back of the left lobe is judged that the individual was right handed, another item in the chain of evidence proving that the skull is of the human species.

There can be little doubt that the Piltdown woman's remains are the earliest yet uncovered, older than the fragments found at Neanderthal, Prussia, in 1856, or even the jaw found at Heidelberg in 1907.

Home scientists, Prof. Klaatsch among them, hold that this primitive type was driven back and extirpated by the higher race of man which existed contemporaneously with it on earth. There is, however, no evidence to support this view.

May 18, 1913

THINKS NEW SKULL LINKS MAN AND APE

Professor Dart Names His Discovery the Australopithecus Africanus.

OLDER THAN THE 'JAVA MAN'

London Expert Hopes It May Carry Back Human History to Period Hitherto Unknown.

Copyright, 1925, by The New York Times Company.
By Wireless to THE NEW YORK TIMES.

JOHANNESBURG, Feb. 4.—Professor Raymond Dart, discoverer of the fossil skull at Taungs, 1,000 miles south of Broken Hill, where three years ago a skull said to be of great antiquity was found, has made the following statement concerning the importance of the present find, to which he has given the name of Australopithecus Africanus:

"The geological records of the different species of man have been rendered fairly perfect. Where geological evidence has been lacking is in the specimens of that phase of pre-human existence between the most primitive and ancient of men and the most advanced of apes. This gap is now filled by the Australopithecus Africanus of Taungs.

"The individual, in brief, was not a human being and yet was a much more intelligent being than a gorilla or a chimpanzee, which is the highest of living apes. He was unable to talk, but the brain was advanced in the direction required in ancestors whose offspring were required to attain ultimately the power of communicating with their fellows by the symbolism of speech.

"He is therefore regarded not as an ape-like man but rather as a man-like ape and he reveals to us that period of human evolution more remote than the Pithecanthropus [the Java man] early in the Pliocene geological epoch, or even in the Miocene, when human stock had only very slightly advanced beyond that which led to the modern apes."

The Johannesburg Star makes the following addition to the above statement:

"The great importance of the Taungs skull lies in three things: it brings the record of the rocks down a thousand miles south of Broken Hill; it relates to a form of life really intermediate between the most advanced apes and primitive human beings, and it plainly reveals South Africa more clearly than ever as a mine of information regarding the dim past of the human race."

Found in a Limestone Mine.

Copyright, 1925, by The Chicago Tribune Co.

CAPE TOWN, Feb. 4.—The recent discovery at Taungs, Bechuanaland, of the skull of a creature with an intelligence rated as being between that of man and an ape, is considered by scientists here as the most important anthropological discovery ever made.

Professor Raymond Dart, who found the skull, told The Tribune today that once his attention had been drawn to the Taunga district, he enlisted the aid of Professor Young, a geologist, who had obtained permission to explore a limestone mine in the district. Professor Young eventually returned to Johannesburg with several monkey casts and also some rock fragments.

While examining these finds, Professor Dart discovered that the cast of the head fitted exactly with the fossil bones of the jaw. After cleaning a piece of rock the outline of a skeleton came into view. The limestone was then carefully chipped away and an almost complete skull was found.

This piece of limestone had been blasted from a cliff by dynamite. Professor Dart said that thousands of other fossils undoubtedly have been unearthed at Taungs by line workers with records far more significant than that just discovered. Most are either lying on the quarry dump or have gone in lime to the Natal sugar refineries.

Professor Dart intends making further researches in the same district and in the meanwhile he has decided that the skull just found shall remain in South Africa until he has completed his scientific examination.

Says It Backs a Darwin Theory.

Copyright, 1925, by The New York Times Company.
By Wireless to THE NEW YORK TIMES.

LONDON, Feb. 4.—Professor G. Elliot Smith, Professor of Anatomy at the University of London, informed THE NEW YORK TIMES representative today that if Professor Dart's discovery bore out the present interpretatiton upon fuller knowledge it certainly seemed that the skull of an intermediate creature between the ape and man had been discovered. In that case it would contribute to the knowledge of the stages and methods by which human evolution was effected.

Professor Smith, with whom Professor Dart had been working as an assistant before he went to Witwatersrand University, said:

"An interesting point which emerges is that the discovery supports Darwin's theory that Africa was probably the original home of the human family. That view has not been favorably regarded by many writers, though I have always inclined to that view, as both of those unenterprising relatives of the human family, the gorilla and the chimpanzee, happen to live in the forests of tropical Africa. This has always seemed to me a strong argument in favor of the Darwin view that Africa was the original home of the first creatures definitely committed to a human career.

"Great scientific interest, therefore, may attach to this discovery, which may put back the history of the human family to a much earlier period than known hitherto, but judging from a summary of Professor Dart's account of the discovery it seems that the skull represents a phase of simian development, which reveals a close affinity to the human family without having yet acquired distinctly human characteristics."

Professor Smith added that scientists were now for the first time grappling closely with the real problems of the differences between the respective mental capacities of man and the ape, and were now learning more accurately to associate them with certain definite localities in the brain.

It is interesting to note, he said, that parts of the brain concerned with touch, movement, hearing and vision were just as well developed in apes as in man, but where man was superior was in the development of those parts of the brain associated with the performance of skilled movements and discrimination of shape and size and with properties of objects—in short, those parts of the brain which made it possible to profit by experience.

Awaits Study of Brain.

What one hoped this Taungs skull would disclose, said Professor Smith, was that those particular parts of the brain would show greater development than that found in the ape, but a lesser development than that found in the Pithecanthropus skull, which, by the way, is regarded as only slightly older than the Piltdown skull found in England.

The pithecanthropus skull is one, which it is believed, belongs to a creature which had already been able to speak in some sort of way owing to a localized swelling of that portion of the brain in the region concerned with the appreciation of auditory symbolism. It is, however, not regarded by German scientists as being a human skull, but as that of a giant anthropoid ape, while Dutch scientists hold that it is actually the so-called "missing link."

British scientists, on the other hand, regard it as definitely human, and base this opinion among other things upon the relative size of the skull to other bones belonging to the creature, which upon the ape supposition would presuppose a creature some fifteen feet high, whereas on the man theory the normal proportion between the size of the skull and the other bones would give the creature about five feet ten inches in height.

Reverting to the fossil remains found in the South African hills, Professor Smith said that they had revealed an extremely interesting group of apes, some of whom seemed nearer to human characteristics than anything yet found, and while it is now claimed that they were human it afforded evidence that they were akin to the ancestors of the human family.

"Further Proof" of Evolution.

LONDON, Feb. 4.—"Australopithecus Africanus," as Professor Raymond Dart has named the original owner of the skull which he discovered at Taungs, Bechuanaland, has stirred lively interest among anthropologists and anatomists here. Their attention is all the more closely engaged because of Professor Dart's well-established reputation.

"If Professor Dart makes claim to a certain discovery, I am prepared to regard it as reliable," was the comment of Sir Arthur Keith, noted anthropologist. Sir Arthur, however, added cautiously that it would probably be more correct to regard the skull as one of hundreds of "missing links" rather than "the missing link"—in other words, as an additional piece of evidence in the already overwhelming proof of man's evolution.

Professor Dart was born in Queensland, studied medicine at the University of Sydney and served in the Australian Medical Corps during the war. He joined Professor Smith's staff here in 1919 and in 1920 visited the United States, where he worked in various American medical schools. He was appointed professor of anatomy in Witwatersrand University in 1922.

February 5, 1925

'DAWN FISH' FOSSIL THROWS NEW LIGHT ON ORIGIN OF MAN

Primitive Vertebra, More Than 50,000,000 Years Old, Is Found in Vermont.

THE FIRST OF ITS KIND

Scale of the Earliest Chordate, Size of the Head of a Match, Is Dug Up.

NOW IN PRINCETON MUSEUM

Dr. Howell's Discovery Extends History of Man's Predecessors Back Millions of Years.

Special to The New York Times.

PRINCETON, N. J., Dec. 25.—Professor B. F. Howell of the Department of Geology at Princeton will announce to the Paleontological Society at New Haven next week, it was learned today, the discovery of the first fossil of a vertebrate of the Cambrian period of geological history, which extends backward millions of years the history of man's predecessors.

The fossil is that of the scale of a primitive fish, the earliest chordate—the primitive form of the vertebrate—and as such is said to be, if not the direct antecedent, at least the earliest known species of the branch of this original antecedent.

The fossil has been named "Eoicthys Howelli," or "Howell's Dawn Fish," in honor of its discoverer. It is hardly larger than the head of a match. It was found by Professor Howell and Professor Charles Schuchert of the Department of Geology at Yale, at St. Albans, Franklin County, Vt., last Summer, on the final day of a field expedition which was one of a series conducted throughout thirteen years by Professor Howell, in an effort to find a fossil of a vertebrate in the earliest period of the Paleozoic era. Heretofore only the fossil of marine invertebrates and plants of a low order had been found in Cambrian rocks. Professor Howell dug up the fish scale with the fossils of trilobates, primitive shrimp-like creatures, and brachiopods, early types of shell animals, which were said to be unquestionably marine organisms of fifty to one hundred million years ago.

Corroborated by Dr. Brant.

Professor Howell did not know the nature of his find until he returned to Princeton and placed the fossil of the fish scale under a microscope and studied it. There he saw the significance of the discovery. Dr. William L. Brant, Director of the Park Museum of Providence, R. I., corroborated Professor Howell's description of it and named it in his honor. The fossil, which is described as a fish plate, is now in the Princeton Museum of Paleontology, in Guyot Hall.

Evidences of chordata have been found in every geological period except the Cambrian, which is the earliest period during which a connected and orderly assemblage of fossil forms has been exhibited in rock strata. For many years scientists believed chordata existed in that period, but none was able to find evidence of them until Professor Howell unearthed the fish plate, and he collected 40,000 or 50,000 fossils during the thirteen years he sought the chordate.

In the Devonian period, the age preceding the formation of coal deposits, fish were the dominant species, according to the fossils which have been found, and it is called the age of fish. A few chordata were found in deposits of the Silurian period, which preceded the Devonian and which is the age of invertebrates. In 1893, however, some fish plates were found of the Ordovician, or lower Silurian, period, and then scientists began to hope that evidence of chordata would be found in the preceding Cambrian period. Professor Howell's discovery is the first to make this hope realized.

Professor Howell said the fish plate he found corresponds to a scale on the extant species of fish. Dr. Bryant described it as being part of the armor which probably defended the head and foretrunk of some hitherto unknown species of fish.

Indicates Salt Water Origin.

This discovery may throw more light, it was said, on scientific belief as to the origin and early habitat of the fish. Hitherto it has been generally held that fish had their origin in fresh water deposits and that in course of time some were washed into the sea. The reason for this belief was that the oldest known fish heretofore had been found in deposits which were probably of fresh water origin. This fish plate, however, was found in a shale which contained trilobates and brachiopods and was taken to indicate a possibility that fish had their origin in salt water.

Professor Howell will discuss his discovery in a paper at New Haven while the Paleontologicol Society is in session next Monday, Tuesday and Wednesday and will also read a paper on the discovery written by Dr. Bryant.

Julian W. Fess of Cleveland, Ohio, a member of the junior class at Princeton, will read to the society a paper on the reproduction of fossils in sealing wax and type metal. He has conducted experiments in this respect in his independent work in the geology department, and the Princeton authorities said his accomplishment is one of the fruits of the new plan of upper-class study at Princeton which is based on the premise that undergraduates will do better work if they are permitted to develop original ideas, with assistance of professors, in special fields in which they are interested.

The Princeton Geological Alumni Association will hold its annual dinner at the Lawn Club in New Haven Monday evening. William Berryman Scott, '77, who is to make his retiring address as President of the Geological Society of America, will be the guest of honor. Some of the other scientists expected are Dr. E. C. Oelrich of the United States National Museum, Dr. R. S. Bass'er, the United States National Museum, Secretary of the Paleontological Society of America, and Dr. W. D. Matthey, curator of vertebrate paleontology in the American Museum of Natural History. Dr. W. M. Agar, '16, of the Yale geological department and President of the Princeton Geological Alumni Association, will preside at the dinner.

December 26, 1925

'MISSING LINK' SEEN IN FIND NEAR PEKING; SCIENTISTS STIRRED

Ten Skeletons and a Skull in Perfect Condition Hailed as Ancestors of Man.

HELD GREATEST DISCOVERY

Dr. G. Eliot Smith Sees Piltdown and Java Finds Outrivaled by That in Chinese Cave.

CANADIAN LED EXCAVATORS

Rockefeller Foundation Cooperated —Scientists Now Studying Bones Believed 1,000,000 Years Old.

Special Cable to The New York Times.

LONDON, Dec. 15.—The discovery in a cave near Peking of the fossilized bones of ten men, who possibly lived 1,000,000 years ago, as reported by scientists representing the Rockefeller Foundation and the Geological Survey of China, is held here to excel in interest all previous findings of this kind.

Of paramount importance is the discovery of a perfect skull, now in the possession of Dr. Davidson Black, a Canadian paleontologist, which, it is asserted, bears characteristics showing that even at the beginning of the ice age there existed men with the power of thinking and who, unlike the "ape men," walked erect.

From the fact that the ten skeletons lay huddled together in the cave, found in a field at Chou Outien, thirty miles from Peking, the scientists hold that they led a community life.

Believed the Ancestor of Man.

According to the Peking dispatches, the deduction made by scientists there from the discoveries is that the Java ape man is relegated definitely to a minor position in man's ancestry. The "Peking man," in fact, is believed to be a direct ancestor of the present human race.

"Every authority in Peking who has seen the skull," says a dispatch to The London Daily Telegraph, "unhesitatingly describes the discovery as epoch-making and outrivaling all previous discoveries, including the Piltdown man and the Neanderthal man."

The Chinese insisted that a Chinese excavator named Pei Wen-chung be permitted to accompany Dr. Black's party, and it was he who actually exposed the matrix containing the skull in the limestone cave. The field was first surveyed in 1927 by Dr. J. G. Anderson, Swedish geologist, in company with Dr. Grainger, paleontologist of the American Museum of Natural History in New York.

Dr. Anderson, continuing his investigation later with Dr. Bohlin, also of Sweden, who is now a member of Dr. Sven Hedin's expedition, discovered a tooth of the Peking man, the first finding in connection with this type of ancient man.

Calls Discovery the Greatest.

Dr. G. Elliot Smith, London anthropologist, in an interview tonight declares the discovery attributed to Dr. Black must be regarded as the most important discovery of the remains of ancient men ever made. He said:

"In the case of the ape man in Java there was recovered simply a tantalizing fragment of a skull, with three teeth and a thigh bone, and with no certain knowledge that the leg bone was part of the same individual as the fragment of skull.

"In the case of the Piltdown man there was simply a fragment of skull and a piece of jaw. Hence, the discovery in China of the remains of no fewer than ten individuals belonging to a primitive type which may be as old as the Piltdown man in England and the ape man in Java should make it possible to acquire a much fuller knowledge of the nature of the earliest known members of the human family than it seemed reasonable ever to expect."

Suggests Age of Fossils.

Sir Arthur Keith, another prominent English anthropologist, says anything found in Chou Outien cave and with which Dr. Black's name is associated is certainly of very great importance.

"The existence of an interesting and ney type of man has been revealed by a previous excavation there," he said tonight. "I should say the age of these relics is between 200,000 and 250,000 years and that the Peking man stands at the beginning of the Pleistocene period."

Sir Arthur smiled a little incredulously when told that the remains of ten men had been discovered.

"Discoveries are not made in this way," he said. "The other finds appear to be a continuation of the most important discoveries already made in this cave, which is well known to anthropologists."

Ten Skeletons, One Skull, Found.

PEKING, China, Dec. 15 (*Æ*).—Scientists today hinted that ten skeletons unearthed simultaneously with an unbroken skull found in the quarries at Chow Outien, thirty miles from here, may be the nearest approach yet made to finding the ancestors of the human race.

The skull was unearthed by a Chinese geologist, who said it belonged to a species of the famous Peking man, the Sinanthropus Pekinensis, said to be associated with the period of the Piltdown skull and the Java ape man.

The skull as well as the ten skeletons were said to be in a splendid state of preservation. The scientists continued their search in the hope of finding more skulls. Nine of the skeletons were headless.

The discoveries were made in the same limestone quarries where the Peking man jaw teeth were found in 1928. The location of the skeletons was said to have convinced the discoverers that the ancient home of a distinct type of primitive man had been uncovered.

It was understood the scientists believed that the various skeletons as well as the complete skull provided material enough to reconstruct the entire drama of the life of the prehistoric colony, or at least to sketch a portrait of man as he existed in the region of Peking more than a million years ago.

The scientists were unable to account for the fact that ten scattered skeletons were found, while only one skull was located. One theory advanced was that the limestone bed might have been a burial ground for persons who had been beheaded and that the heads had been deposited elsewhere.

Dr. J. C. Anderson, Swedish adviser to the Chinese Geological Survey, and other searchers are now seeking the other skulls.

December 16, 1929

'Extinct' Fish Reappears

Last December trawlers off East London, South Africa, hauled up half a ton of red fish and a ton and a half of sharks. In the flapping conglomeration was a living fish the like of which they had never seen. It was a fine steely blue, five feet long and it weighed 127 pounds.

The tales that came from South Africa about the catch were too weird to be accepted at their face value—too reminiscent of Loch Ness monsters and sea serpents. Accordingly, this department decided to wait for the authorities to speak.

They are now heard in the persons of Dr. J. L. B. Smith and Dr. Errol I. White. The one, an ichthyologist connected with Rhodes University College, Grahamstown, writes in Nature, and the other, deputy keeper in the British Museum's department of geology, in The Illustrated London News.

Theoretically Extinct

The discovery turns out to be of sensational importance. Theoretically, the strange fish was extinct. It had no more right to be caught alive than we have to see a dinosaur walking up Broadway, poking his little head into third-story windows. The dinosaurs were on the way to extinction when this fish was in its heyday and the first real plants were flowering. That was from fifty to a hundred million years ago.

The new fish belongs to the fringed-finned order (Crossopterygii, if you must have the scientific name), of which the earliest known specimens, nothing but stony fossils now, go back 300,000,000 years. These fringed fins spread all over the earth from Norway to South Africa. It was supposed that after swimming about for some 250,000,000 years they went the way of contemporary dinosaurs —why is not clear, though there are many hypotheses.

To an untrained eye the South African fish looks queer enough to attract the attention of any deep-sea angler, but not so queer as to raise a world-wide hullaballoo. Still the South African trawlers were astute enough to notify Miss M. Courtney-Latimer, curator of the East London Museum.

One glance was enough to convince the learned lady that here was indeed a new species. Whereupon Dr. Smith was summoned. By that time the fish was dead and the taxidermists had begun their work. While alive it had exuded twenty gallons of oil, some of it very fine oil from the spine.

These fringed fins had been painstakingly reconstructed from their fossil remains by such men as Sir A. Woodward, Professor E. A. Stensiö of Sweden and J. A. Moy-Thomas. As Dr. White describes them, they are all "heavy, rather wooden-looking fishes with large, boxlike heads having a prominent snout."

Fins Like Fans

The fins are their striking feature. Instead of rising directly from the body as in most fishes they were like fans. The tail, too, is remarkable in that it is double. There are other peculiarities, such as complex jaws, a tubular backbone that is not all bone but partly cartilage and scales covered with a shining enamel.

Dr. Smith sighs when he thinks of all the years spent in piecing together the facts about a supposedly extinct creature, when suddenly a living specimen turns up.

Now the naturalists are wondering what other odd life may be brought up from that still unplumbed lost world, the ocean.

April 2, 1939

Reports From Research Fields

THAT EXTINCT FISH —The scientists who wrote about the extinct fish which we described on this page last week must have made a mistake in their arithmetical calculations. We were told that the fish was five feet long, that it weighed 127 pounds and that twenty gallons of oil exuded from it. A reader calls our attention to the fact that twenty gallons of water (there is not a very great difference in weight between oil and water) weigh about 167 pounds. Possibly twenty "gallons" should have read twenty "pounds."

April 9, 1939

Scientist Stumbles Upon Method To Fix Age of Earth's Materials

Americanist Congress Hears Dr. W. F. Libby Tell How Special Geiger Counter May Make 'Prehistoric' Things Historic

Experiments with nuclear energy have accidentally uncovered a method of determining with relative accuracy the age of earth's materials by measuring their radiocarbon content, an associate of the Institute of Nuclear Studies, University of Chicago, reported yesterday.

Dr. Willard F. Libby, 40-year-old chemistry professor, who stumbled on the technique two years ago when studying cosmic ray action on the atmosphere, said it might enable scientists to fix dates of "living materials" up to 25,000 years old. He developed a special Geiger counter in connection with the "dating" technique.

Colleagues, anthropologists and archaeologists have greeted the discovery as a tremendous advance. It now will be possible, it was said, to date the early pre-Columbian civilizations in the Western Hemisphere as well as to explore earth's chemical composition in the Pleistocene Age (600,000 to 10,000 B.C.).

Dr. Libby emphasized, however, that the testing of materials with his Geiger counter would not be effective on anything beyond 25,000 years of age. He said he hoped to create a calendar to set the ages for such material as wood, charcoal, well-crystallized shell and teeth.

He reported his discovery at an opening session of the twenty-ninth International Congress of Americanists at the American Museum of Natural History before 300 scientists from thirty-five countries. Collaborating in the experiment were Dr. E. C. Anderson of the University of Chicago and Dr. A. V. Grosse, president of the Temple University Research Institute.

Their experimentation established that radiocarbon was generated some 30,000 feet up in the atmosphere and that a "steady state condition" has existed on earth for a long time. Dr. Libby estimated then that there must be present somewhere on earth a total of thirty-three metric tons of radiocarbon.

This amount, he explained, is several million times the amount of the isotope likely to be manufactured by the Atomic Energy Commission "over a period of many years."

Originating in the air, the carbon atoms produce carbon dioxide, the essential food for plants, hence are radioactive, Dr. Libby reasoned. Since animals and humans eat plants, they also are radioactive as are all things not isolated from the atmospheric life cycle.

Dr. Libby defined life as plants' assimilation of carbon dioxide from the air and the returning of carbon dioxide to the air by the animals that have eaten the plants.

"One then expects," he explained, "that death terminates the flow of radiocarbon atoms into the organism to replace those which are steadily disintegrating to form nitrogen."

Since this disintegration rate is immutable, he added, the elapsed time since death can be measured by determination of the radiocarbon "assay" and comparing it with the general world-wide assay which obtained at the time of death. The word "assay" is equivalent to the measurement of radioactivity.

He made the point that materials must contain exactly the same carbon atoms they had at time of death, and that their cosmic ray intensity must have remained constant over tens of thousands of years.

Dr. Libby reported that Dr. James Arnold of the University of Chicago collaborated during the last two years to "prove" the method. Using the special Geiger counter, they measured samples of wood that had their origins in antiquity. They were thus able to check for accuracy because a tree's age can be determined by its number or rings.

Wood samples collected from all parts of the world included fragments from a floor of a room in a palace of the "Syro-Hittite" period in northwestern Syria, dated at 625 to 725 B.C.; a California redwood tree felled in 1874, wood from a mummiform coffin from Egypt dated in the Ptolemaic period, 332 to 30 B.C. Two samples came from the tombs of Egyptian kings, whose coffins were dated 4,600 years ago.

Measurements Agree

Measurements were lumped and they agree with one another. In further investigations some eleven different major dating questions were selected and prominent scientists in each case were invited to collaborate. The venture was supported financially by the Viking Fund of New York.

Dr. Libby said that within the next year a "decisive" test of the method would be obtained through the other inquiries. Some twenty-five "unknown" samples have been measured with satisfactory results.

"We have reason to believe," Dr. Libby said, "that ages up to 15,000 to 20,000 years can be measured with some accuracy by the present method, and we hope to go to 25,000 years."

Dr. Helmut de Terra, who is associated with the Viking Fund, said at the symposium that he had supplied many samples to Dr. Libby that had been collected from the core of the Pyramid of the Sun at Teotiuacan in the Valley of Mexico, said to be one of the oldest pyramids in this hemisphere. The pyramid's age had never been known but estimates had ranged up to 15,000 years. The Geiger counter set the correct age as 2,951 years.

Wider information of scientific validity is looked for which may develop a historical record of the earth's physical formation, and in that sense make the pre-historic period historic.

September 6, 1949

Find Indicates 'Modern Man' Existed 75,000 Years Ago

Three Skeletons Unearthed Near Caspian Together With Crude Tools

Special to THE NEW YORK TIMES.

TEHERAN, Iran, April 27—Evidence that a modern form of man lived at the same time as previously known primitive types some 75,000 years ago was produced here today by Prof. Carleton S. Coon of the University of Pennsylvania.

Digging in a cave near Behshahr on the Caspian Sea, Dr. Coon, an anthropologist, and his assistant, Louis Dupree of Harvard University, a geologist, uncovered three human skeletons in what appeared to be Pleistocene Age gravels.

The skeletons were found by Dr. Coon and Mr. Dupree after they had dug through three gravel and three sand layers to a depth of about twelve meters (approximately thirty-nine feet). The two scientists concluded that their find dated back to somewhere near the end of the third interglacial period, thus making it roughly 75,000 years old.

Directly underneath the skeletons were about two meters of silt clays ending with a level of kaolin—evidence of a warm climate, probably in the third interglacial period. Crude flint tools, but no human remains, were turned up at this lower level.

On the basis of a preliminary field examination of the skeletons, Dr. Coon deduces that a true Homo Sapiens may prove to have been much older than subhuman types found elsewhere, such as the 50,000-year-old Neanderthal man. If this theory is borne out by subsequent studies, it presumably will modify the present concepts of man's earlier evolution.

As a corollary, Dr. Coon believes Northern Iran to be a "nuclear" region, one where cultures were evolved, and not a marginal region, where cultures lagged—in other words, a place where things began and probably witnessed some of the earliest agriculture in the world.

Consequently, modern man may not have evolved from the Peking Man (Sinanthropus Pekinensis) at all.

Dr. Coon and Mr. Dupree found the three skeletons below the surface of the cave called Hotu. The three prehistoric persons apparently had been sitting around a hearth when the roof caved in and crushed them. Near them were crude flint and bone instruments, animal bones and charcoal. Like other Paleolithic specimens, these were hunting and gathering folk.

Come Upon Pottery

Above their heads was an intensive rock fall. The next occupation level—the Neolithic one—began four meters eighty centimeters (approximately fifteen feet) below the surface and contained evidences of agriculture and domesticated animals. Here Dr. Coon and Mr. Dupree came upon a crude slow-fired pottery, bone implements and flint sickle blades polished through intensive use in reaping early cereals.

These remains date from 16,000 to 8,000 B. C., Dr. Coon believes, and may point to the very beginnings of agriculture.

A crude type of painted pottery began to appear near the top of the Neolithic level, which in turn gave way to thin and well-fired painted pottery.

The advent of the bronze or copper age was heralded by the appearance of well-fired pottery types in sophisticated shapes. One vase, for instance, had a handle in the form of a horse's head.

Cranium Described

The principal difference between the skeletons found by Dr. Coon and Mr. Dupree and modern man lies in the size of the cranium. Although the skull of the Hotu Man is human in shape, his cranial capacity, which Dr. Coon guesses to be in the neighborhood of 1,150 cubic centimeters (approximately seventy cubic inches), is appreciably smaller. The cranial capacity of modern man averages around 1,450 cubic centimeters.

In other respects, the Hotu Man is similar to some human types encountered today. He stood about 5 feet 8 inches tall, had a rather massive body structure with large hands and feet, long teeth, prominent molars, low eye orbits and a strong, perfectly human chin.

In the course of their diggings, Dr. Coon and Mr. Dupree collected charcoal from all prehistoric levels. Analysis of these charcoal specimens will aid in dating all known prehistoric sites in the Middle East, they believe.

The two scientists, who have been digging since January, are now on their way back to the United States.

Little Data Available From Area

Dr. Harry L. Shapiro, chairman of the department of anthropology of the American Museum of Natural History, reported that there had been little anthropological data available from the Iranian site.

The new find suggests that a markedly modern type of man lived there at a time in the earth's history when primitive men inhabited other parts of Asia and Africa, he added.

The description of agricultural implements confirms an earlier impression that an organized type of human living was practiced around 10,000 years ago, Dr. Shapiro said. It also suggests, he declared, that the practice of agriculture may have originated in Iran.

April 28, 1951

Dr. Carleton S. Coon
Associated Press

The New York Times   April 28, 1951
Area of discovery (cross)

14-Year Hunt Yields 'Missing Link' Fish

Special to THE NEW YORK TIMES.

DURBAN, South Africa, Dec. 29—Prof. J. L. B. Smith, Rhodes University ichthyologist, returned to Durban tonight after a 3,300-mile plane flight with what he described as "one of the most amazing discoveries ever made in science." It was a five-foot-long prehistoric fish called a coelacanth (pronounced SEE-la-kanth), long thought to have been extinct for millions of years.

When Professor Smith first saw the fish he broke down and wept. It was the culmination of fourteen years of search during which he tramped thousands of miles along the eastern seaboard of Africa with his wife. He distributed thousands of leaflets in English, French and Portuguese to enlist the aid of natives.

The body of the fish is intact and fairly well preserved, unlike a specimen caught in 1938 by a trawler from East London, South Africa. Of that fish, only the skeleton and skin could be preserved. It was also identified by Professor Smith.

"The coelacanth retained structures in its body unchanged over vast periods of time and so we hope in the soft parts of the fish to learn something of the early types of life," Professor Smith said.

The professor was returning by ship to his home at Rhodes University in Grahamstown with tons of fish specimens collected during a five-month research tour along the East African seaboard when in Durban Harbor he received a message from Eric Hunt, who trades in a small vessel between Zanzibar and the Comoro Islands off Madagascar. Professor Smith had previously briefed Captain Hunt on the coelacanth, which scientists estimate first appeared 300,000,000 years ago.

At Captain Hunt's word that he had a coelacanth, Professor Smith left Durban hurriedly in a military plane provided by Prime Minister Daniel F. Malan.

Landing at the islet of Dzaoudzi in the Comoros Islands, he found the coelacanth swathed in cotton wool on the deck of Captain Hunt's schooner nearby.

"I couldn't touch it," Professor Smith declared. "I asked them to open it."

When he saw the fish, Professor Smith knelt on the deck and wept, for, he said, "it was a coelacanth and, what's more, a species different from the 1938 specimen."

A native fisherman had caught the fish in sixty-five feet of water 200 yards offshore and taken it to market the next morning. He had been prevented from selling it by an excited native who recognized it from Professor Smith's pamphlet and said, "Plenty money." A reward of £100 ($280) had been offered by Professor Smith.

They carted the fish twenty-five miles over difficult mountainous country to Captain Hunt's vessel.

Captain Hunt sliced the fish open along the sides and packed it in salt to preserve it. Then he traveled to the nearest medical officer, miles away, obtained a syringe and five liters of formaldehyde and injected it into the fish.

The fish had a strong odor when Professor Smith arrived, but all the vital soft parts were intact and in good order.

The scientist preserved the fish properly and will fly to Capetown tomorrow to show the specimen to Dr. Malan, who made its recovery possible by providing the plane. Professor Smith proposes to call the fish Malania anjouanae in honor of Dr. Malan and Anjouan Island, near which it was caught.

A Link to Vertebrates

At the American Museum of Natural History, Dr. Bobb Schaeffer, associate curator of fossil fishes, said that the coelacanth found in the Comoro Islands would provide an opportunity for the first time to study the soft anatomy of a 300,000,000-year-old animal that is the closest living relative to a group of fishes that gave rise to the land-living vertebrates.

Fossil finds have given a good idea of the skeletal systems of the ancient species, Dr. Schaeffer said. But the soft organs (such as heart, stomach and intestines), the blood vessels and the nervous systems have never been preserved. By comparing the organization of the soft anatomy of the "living fossil" with that of the soft anatomy of modern vertebrates, Dr. Schaeffer said, scientists may determine more accurately the early pathway of evolution.

The African find (Latimeria chalumnae) is of a species about 300,000,000 years old, he estimated. The earliest known ancestor of living vertebrates is a cousin of Latimeria chalumnae. Both evolved from a common more distant fish ancestor. Until 1938 it was believed that Latimeria chalumnae and the cousin (ancestor of the first vertebrates) were both extinct 75,000,000 years ago.

Dr. Schaeffer said the sets of fins of the specimen found in 1938 were close to what must have been the appearance of the first sets of arms and legs.

December 30, 1952

Model of a coelacanth, of which a new species was found in the Comoro Islands (cross on map)

Piltdown Man Hoax Is Exposed; Jaw an Ape's, Skull Fairly Recent

By JOHN HILLABY
Special to THE NEW YORK TIMES.

LONDON, Nov. 21—Part of the skull of the Piltdown man, one of the most famous fossil skulls in the world, has been declared a hoax by authorities at the British Natural History Museum.

It is now stated that the jawbone associated with the skull is that of a modern ape, probably an orangutan, that has been "doctored" with chemicals to give it an aged appearance.

In addition, it is said that the cap of the skull is genuine but far more recent than had been believed—50,000 instead of 500,000 years old.

This declaration in the current issue of the museum's bulletin has been made after twenty years of rumors and uneasy speculation among European paleontologists about the authenticity of the bones.

The report, signed by Dr. J. S. Weiner, Dr. K. P. Oakley and Prof. W. E. Le Gros Clark, said: "The faking of the mandible [jawbone] is so extraordinarily skillful and the perpetration of the hoax appears to have been so entirely unscrupulous and inexplicable as to find no parallel in the history of paleontological discovery."

Paleontology is the science dealing with the life of past geological periods. It is based on the study of fossils.

The relics of the so-called first Englishman were unearthed in a gravel pit in the hamlet of Piltdown in Sussex, forty-five miles south of London, in 1911. The first bone was handed to Charles Dawson, a local lawyer and amateur geologist, by workmen who had apparently mistaken the skull for a "petrified coconut" and smashed it into pieces.

Mr. Dawson took the find home but apparently thought little about it. Several years later he found other portions of the skull some distance away and realized that he had found fragments of a "very thick-headed man."

The bones were taken to the Natural History Museum in London, where the Keeper of Geology, the late Sir Arthur Smith Wood-

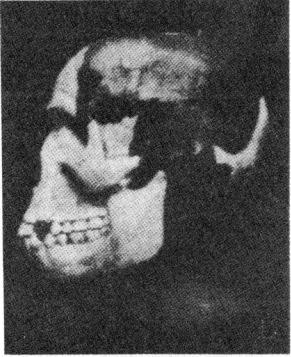

DEVALUED: Model in American Museum of Natural History of Piltdown skull in the British Natural History Museum, some parts of which have been declared to be spurious.

ward, pronounced them to be a portion of an ancient human parietal bone, part of the skull.

Within a year the Piltdown man was famous. The gravel pit was combed for other bones. Mr. Dawson found the right half of a jawbone at the same spot where the skull pieces had been turned up. Some teeth believed to be human were also found.

Sir Arthur Keith, famous British paleontologist, spent more than five years piecing together the fragments of what he called a "remarkable" discovery. He said the brain case was "primitive in some respects but in all its characters distinctly human."

The Piltdown man was named Eoanthropus dawsonii, or Dawn man, in honor of its discoverer, and paleontologists throughout the world handled it with reverence.

Although the fossil was generally accepted as the earliest known specimen of sapient man, as opposed to the apeman of China and Java, many research workers reserved their opinions about the disputatious jawbone.

Skeptical British paleontologists said it was probably a relic of another specimen that had been found in accidental conjunction with the Piltdown cranium.

The late Franz Weidenreich, who unearthed the Pekin man, said bluntly that the jaw was that of an orangutan, a statement that puzzled paleontologists here since no anthropoid apes are known to have inhabited Britain.

Although the name Piltdown became a landmark in the history of man's search for his earliest ancestors, the skull itself was placed in 1939 in what the British Museum politely called a "suspense account."

Meanwhile Mr. Dawson had died.

The first serious doubts about the authenticity of the skull were voiced in 1949. Chronologists reviewed the remaining evidence at the site and said that neither the skull cap nor the jaw was "particularly old."

These doubts were finally solved by Dr. Oakley, who tested all the bones for their relative content of fluorine absorbed from the soil. He found they contained "very little fluorine," a circumstance suggesting that the "relics" were relatively recent.

Now the opinion of comparative anatomists and further tests by chemists have established that the jaw is that of a modern ape treated with potassium bichromate and iron salt, giving it an aged appearance.

It has also been established that the teeth have been pared down so that they could have been associated with the jaw of a primitive man.

The cranium is believed to be genuine but about 50,000 years old. This age brings the "first English-man" into line with scores of early men found in Europe and elsewhere.

But it is assumed that the jaw was "planted" at the gravel pit. By whom? The writers of the museum bulletin are not prepared to say categorically that it was Mr. Dawson, the venerated lawyer and geologist of Sussex.

The Times of London, however, says that if the hoaxer were proved to be Mr. Dawson "it would be but one more instance of the desire for fame (since money was certainly not the object) bringing a scholar into dishonesty."

This abrupt devaluation of the Piltdown man means that the oldest skull of sapient man found in the world is the relic from Swanscombe, on the south bank of the Thames Estuary. It is about 200,000 years old and was found in conjunction with many "datable" flint hand axes. Paleontologists who have spoken eloquently about its antiquity are doubly thankful that it has been shown to be genuine by Dr. Oakley's telltale fluorine technique.

Hooton Jolted by the Fact

PROVIDENCE, R. I., Nov. 21 (AP)—One of the world's most famous anthropologists expressed surprise tonight at the new plight of the Piltdown man.

The Providence Journal quoted Prof. Ernest A. Hooton of Harvard as expressing shock and disbelief at the implications of the hoax.

As to the effect of the exposé on the study of anthropology, the Journal quoted Dr. Hooton as saying:

"It doesn't disturb our ideas of human evolution at all. If it is right that the head is a fake, it loses all its significance and removes a very puzzling link."

Professor Hooton told The Journal that the findings "impugn the honesty" of the late Sir Arthur Smith Woodward, formerly keeper of the natural history section of the British Museum and one of the most famous geologists of Britain.

"It is like implying that the Secretary of the Treasury is running a counterfeiting business on the side," Dr. Hooton commented.

At the same time Dr. Hooton described two of the scientists who collaborated on the report which branded the Piltdown man a fake as "very reputable and distinguished fellows."

"I know them both," he said of Dr. K. P. Oakley and Prof. Le Gros Clark.

November 22, 1953

African Cave Yields Fossil Proof Of Man's First Use of Tools, Fire

By ROBERT K. PLUMB

A discovery in a cave in South Africa has clearly established a point on the evolutionary ladder at which man's ancestors first learned to use tools.

Word of the discovery was received here yesterday afternoon by the Wenner-Gren Foundation for Anthropological Research, 14 East Seventy-first Street.

The discovery definitely establishes for the first time that a creature known as Australopithecus prometheus lived during the time of and at the place of a very primitive stone "pebble culture."

Australopithecus prometheus (the "fire-giver") has been known since 1925, when it was described by Dr. Raymond A. Dart, head of the Department of Anatomy of the Medical School of the University of the Witwatersrand in Johannesburg. Many specimens of this very early man-ape or ape-man have been uncovered in a series of remarkable caves in the valley of the Makapansgat River, thirteen miles northeast of Potgietersrust in the Transvaal.

The twenty-odd caves in the valley have served as a terrestrial yardstick for reaching into man's past, back perhaps 200,000 years. The caves were occupied by different creatures in era after era. Layers of bones and artifacts mark each era.

Crudely worked stone tools (relics of the "pebble culture") have often been found. But scientists have questioned whether the tools were really works of an intelligent creature or were merely unusual specimens of stones worked by river water.

They have questioned whether "smoke" marks in one of the caves, known as "the Cave of Hearths," were really caused by fire.

No good answer to these questions has been available, for no fragments of the creature have been found in a layer along with the pebble culture "tools."

Professor Dart has informed the foundation, which has financed extensive explorations of the caves, that such a find has just been made. Damaged specimens of the first and second molars from the right upper jaw of an adult Australopithecus prometheus have been found in a layer containing the crude tools.

The implication clearly is that the creature fashioned the tools. With this established, Professor Dart and Dr. Paul Fejos, director of research for the foundation, and his assistant, Dr. William L. Thomas Jr., are able to describe the life of the earliest known "man," the tool-using, fire-using, intelligent Australopithecus prometheus.

Lived Many Millennia Ago

This creature, a man-ape or ape-man, lived in the cave something over 100,000 years ago, perhaps 1,000,000 years. The age is not exact because it can only be determined by reference to geological epochs which left marks on the site. The geological epochs are not well dated.

Prometheus weighed eighty to 100 pounds, stood erect, had delicate hands and feet. He resembled modern man more than he did the great apes. His brain size, in proportion to his body weight, was not much less than that of modern man.

The creature was an aggressive hunter. The Makapansgat caves contain the bones of the hippopotamus, rhinoceros and elephant as well as other animals prometheus killed and ate. Skulls of these creatures are broken in a neat but forceful manner. This was the principal use of the tools of prometheus: he stood face to face with his prey or his adversary and slew it with a blow of a stone worked to meet the purpose.

Darwin's Explanation

The position of Australopithecus prometheus, the fire and tool user on the evolutionary ladder, was well predicted by Charles Darwin, who in 1871 in "The Descent of Man" postulated that:

"* * * as they (the early male forefathers of man) gradually acquired the habit of using stones, clubs or other weapons for fighting with their enemies or rivals they would use their jaws and teeth less and less.

"In this case the jaws, together with the teeth, would become reduced in size, as we may feel almost sure from innumerable analogous cases."

The significance of the find of a jaw and molar fragment is heightened by this prediction. One of the characteristics which to physical anthropologists distinguishes man from beast is shape, size, setting and structure of teeth. The two molars in the newly found specimen are damaged. But they appear to be sharp. This indicates that the creature ate flesh as well as vegetables, for animals with all-vegetable diets wear down their molars on the sand and dirt that they chew along with greens.

The caves along the river were preserved for thousands of years because the river periodically flooded them. Each inflow carried in stones and clay and sand. The silt formed a cement which sealed the successive layers and preserved them until this day. Removal of these rocks has revealed the find.

Australopithecus prometheus may be as much as 1,000,000 years old, according to studies of Dr. Dart and the late Sir Robert Bloom, British anthropologist. But no specimens earlier than the present bones are known.

June 4, 1955

AGE OF LIFE PUT AT 2,000,000,000

Fossil Dating Said to Show Primitive Forms May Run to 3,000,000,000 Years

By JOHN HILLABY
Special to The New York Times.

LONDON, April 25 — Radio-chemical methods of dating have established "beyond reasonable doubt" that primitive forms of life have existed on earth for at least 2,000,000,000 and probably 3,000,000,000 years, it has been reported here.

These figures are three to four times greater than the traditional estimates of the age of life (usually around 800,000,000 years). Observers have already brought forward some supporting theories about whether the earth would have been cool enough to support life in the distant past.

The estimates have been propounded by Prof. Arthur Holmes, of the University of Edinburgh, a fellow of the Royal Society and one of the most distinguished geologists in Britain. They have the support of Sir Gavin de Beer, the director of the Natural History Museum in London, and Dr. W. J. Pugh, director of the British Geological Museum and Survey.

They are based on discoveries in Canada, Britain and South Africa of the remains of algae, the simples of green plants, and also of fungus spores and protozoa in the most ancient rocks in the world.

The stones in which the fossils were found are declared to be sedimentary, that is, they were laid down under water in the form of sediment. Fortunately, they became sandwiched between radioactive rocks, called pegmatites, of later origin. The pegmatites can be dated by comparing the relationship of the infinitely small amounts of different forms of lead they contain.

Rock Sample From Ontario

One sample of rock extracted from the northern shore of Lake Superior, near Shreiber, Ont., has been dated by Prof. Patrick Hurley of the Massachusetts Institute of Technology as being 1,300,000,000 years old.

Several other samples from more ancient rock systems in Southern Rhodesia have been dated at the Government's Chemical Research Laboratory near London, with assistance from researchers in Prof. Alfred O. Nier's laboratory at the University of Minnesota. The most ancient, a granite pebble, was about 3,300,000,000 years old.

The next most aancient rock was a piece of pegmatite from the Bulawayan rock system of South Africa. It was judged to be 2,640,000,000 years old, with an estimated margin of error of 40,000,000 years either way.

Prof. Arthur Holmes, who is collating these results, said in an interview that "the Southern Rhodesian algae existed, say, about 3,000,000,000 years and, of course, life itself must have existed a long time [before that] for algae to be involved."

The presence in the rocks of the alga, now the oldest known organism in the world, was first detected by Dr. A. M. MacGregor of the Gelogical Survey of Southern Rhodesia. It was embedded in a piece of liimestone.

Other primitive fossils, including two kinds of fungus, two other algae aand what is believd to be a petrified flagellate, the whip-like motor-organ of a primitive protozoan, or "first animal," were found in Canada about three years ago by Prof. Elso S. Barghoorn of Harvard and Prof. Staanlley A. Tyler of the University of Wisconsin.

Data Held Indubitable

The geological evidence that the fossiliferous rocks are at least as old as, and probably older than, the associated strata that have been dated is said by Professor Holmes to be "indubitable."

Asked to comment on the possibility of recognizable forms of life more than half the estimated age of the earth (5,000,000,000 years), Sir Gavin de Beer said that the discovery was "not surprising." He added that a great deal of evolutionary activity must have taken place before relatively complex animals became fossilized in the well-known Cambrian and Silurian rocks that were usually considered to be ancient. They were laid down between 400,000,000 and 600,000,000 years ago.

Most British geophysicists are of the opinion that the earth had cooled sufficiently to maintain life when pegmatite rocks were being formed. An extreme view has been taken by Dr. Thomas Gold, assistant to the British Astronomer Royal, who doubts whether the earth was ever very hot at the surface.

"The process of earth formation was probably very slow, and it took place over such an enormous area that any heat generated by impacting particles would be strictly localized and quickly dissipated by radiation," he said.

He added that if any portion of the exterior of the earth was ever molten, which he doubted, a "thin skin would very soon be formed cool enough to support life."

May 6, 1956

GIGANTIC FOSSILS OF ANIMALS FOUND

East African Expert Cites Sheep Big as Horses, Hogs With Elephant-Like Tusks

By JOHN HILLABY
Special to The New York Times.

LONDON, May 10—The remains of gigantic animals, including sheep as big as present-day cart-horses and hogs with tusks like elephants, have been dug out of an ancient gorge at Olduvai in East Africa.

Two giant human teeth have also been found. The fossils are considered to be among the most remarkable paleontological discoveries of recent times.

Specimens have been brought to London for comparative studies by the man who found them. He is Dr. Louis S. B. Leakey, director of the Coryndon Museum at Nairobi, Kenya, a distinguished archeologist and fossil hunter.

The gorge, he said in an interview, this week, is in a deep cleft in the Serengeti steppe of Tanganyika, a region now famed for its abundant game herds. Dr. Leakey first visited the locality in 1931 with Hans Reck, a Berlin volcanologist, and within six hours of his arrival he found a magnificent stone axe showing that the stratified layers of the gorge were once the home of at least one very primitive race of man.

Since then he has located thousands of stone tools. Some are of the earliest type known. They are chipped pebbles called Chelleans from a site believed to be of similar age at Chelles-sur-Marne to the east of Paris. Others, fashioned with more care, have a double face and a razor-sharp cutting edge.

It was thought that the Chelleans lived about a quarter of a million years ago, when north Europe, Asia and North America were covered by enormous ice-sheets in the age known as the Pleistocene.

But no one could date the tools with much accuracy because no organic remains of Chellean Man were found and nobody knew whether the Ice Age of the north, the yardstick of Pleistocene dating, was represented in the tropics at the same time.

Light has now been thrown on the shadowy history of the East African Chelleans by Dr. Leakey's recent discovery of tons of material at a depth of about two hundred feet below the present ground level at Olduvai.

He has been able to assign relative dates to the Chellean culture by comparing the evolutionary stages of the development of the extinct animals found in profusion above and below the "working floors" of Chellean Man. He thinks it flourished about 400,000 years ago.

He also considers that the first Chelleans overlapped the last of the little South African ape-men called the Australopithecines. But the most remarkable character of the Chellean Age was, he says, the gigantic size of the animals existing at that time.

Dr. Leakey has found a complete skeleton of a Pleistocene sheep with a horn-span of fourteen feet. It was about as big as a big cart-horse, he says. He has also unearthed the bones of wild hogs the size of a rhinoceros with elephant-sized tusks. He has found a giant giraffe, baboons as big as modern gorillas, massive zebras and antelopes.

But his most important discovery, from the point of view of the evolution of man, is two giant teeth. They are either from a Chellean child or they are from another species of man hunted and eaten by cannibalistic Chelleans.

They are undoubtedly human and they are bigger than anything previously discovered. One molar is bigger than the mature molars of the Peking Man, one of the earliest human beings known to science.

Dr. Leakey has no doubts about the gigantic size of the Chellean animals. He has found complete skeletons. But he is not prepared to say that Chellean Man was a giant. He may have been merely a little man with huge teeth, he says. He points out that giant teeth of ancient men found in Java and South Africa apparently came from beings of normal size.

The Olduvaian animals of 400,000 years ago probably lived in the delta of a big river, in Dr. Leakey's estimation. As the erosion, or break-down, of the rocks in pluvial times was two or three times what it is today, vital salts known as trace-elements, necessary for growth, collected in unusually large quantities at the mouths of rivers and were taken up by the vegetation there.

The vegetation, in turn, was eaten by delta-haunting animals which grew to abnormal size, or at least so Dr. Leakey surmises.

May 11, 1958

Finder Says Fossil Links Ape and Man

Special to The New York Times.

NAIROBI, Kenya, Sept. 3 — Dr. Louis S. B. Leakey, a leading British anthropologist, reported today that a recently discovered skull was the link between man and the South African ape man.

He said he and his wife made the discovery July 17 in the Olduvai Gorge, Tanganyika, after an exploration that began there twenty-seven years ago. He believes the skull is between 600,000 and 1,000,000 years old.

Dr. Leakey said the skull, which was almost intact, was the oldest yet discovered of tool-making man and a convincing piece of evidence of truth of Darwin's view that man and apes evolved from common stock.

Features of the skull, he said, were its enormous teeth, small brain cavity and very large face. It was that of a youth of about 18 who had lived mainly on nuts and had just begun to eat mice, snakes and lizards, the anthropologist said.

The youth died of causes other than violence, possibly pneumonia, and had probably been covered with thornbushes by comrades to prevent his being eaten by hyenas.

Soon afterward, a lake rose in the gorge, covering and preserving him.

Dr. Leakey said at a news conference that he announced his discovery last week at the fourth Pan-African Congress of Prehistory at Leopoldville, Belgian Congo, but that the world outside scientific circles was still unaware of its importance. He said leading authorities at the congress had agreed that the skull was authentic and were "jubilant" at the discovery.

"For years," he said, "many scientists have been trying to find the connecting link between the South African near-men or ape men—Australopithecus and Paranthropus—and true man as we know him, from the primitive Pithecanthropus in Java and China on the one hand, to the more advanced humans—Atlanthropus of North Africa and the skulls from Steinheim and Swanscombe in Europe—on the other. Now at last we have got this link.

"We discovered the Olduvai skull on his living floor, with examples of the very primitive stone culture called Oldowan. On the same living floor were the bones of the animals, birds and reptiles that formed part of his diet."

Dr. Leakey said the characters of the skull separated it clearly from Australopithecus and Paranthropus. In some of these characters, the creature seemed even more primitive, while in others he was much closer to Pithecanthropus and Homo sapiens. The teeth and palate were far bigger than the largest previously known on a man or man-like fossil.

Find Hailed, 'Link' Doubted

Special to The New York Times.

YORK, England, Sept. 3 — British paleontologists are excited about the Tanganyika fossil discovery.

A distinguished one who is present at the British Association for the Advancement of Science meetings here said today:

"This seems to be a landmark in the history of the evolution of man."

But he was unwilling to be more specific until he learned more about the pattern on the cusps of the teeth of the lower jaw, which he said were critical in deciding whether a skull could be described as that of a very advanced ape or a very lowly hominid or member of the human family.

He felt "almost all professional paleontologists" reject the idea that the Olduvai skull was a "missing link"—not because the words have an "unscientific connotation" but because the whole series of Australopithecine (southern ape-man) skulls "have long been regarded as near-man.

"To be hailed as a link, a skull would have to connect the apes to the Australopithecines," he explained.

The latest specimen, he said, settles the long disputed question whether Australopithecus was an ape or a man. It seems to be a man because it was dug out of a layer in which primitive stone tools hve been found for more than thirty years, he said.

U. S. Experts to Run Tests

American scientists reacted much as the British ones did. They wanted more details, but agreed that proof of the assertion would be "very exciting."

Such proof may be six months or more in coming, according to one of the California scientists

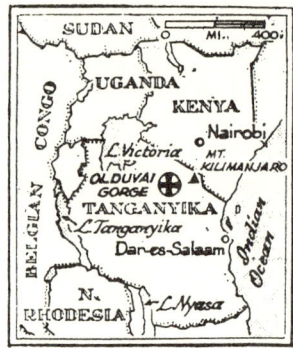

Site of discovery (cross)

who will run potassium-argon age tests on specimens shipped from Africa by Dr. Leakey.

Even then the scientists may not know any more than they do now. The reason is that mineral deposits much older than the skull may have contaminated the samples.

Such contamination would probably negate any results obtained from the new dating technique, which will be used by Dr. Jack Evernden and Dr. Garniss H. Curtis at the Berkeley campus of the University of California.

Dr. Evernden, in a telephone interview last night, said each sample would be dissolved in molten sodium hydroxide to liberate argon-40, a radioactive decay product of potassium-40, also contained in the rock. Because the decay occurs at a known rate—half the potassium-40 being converted to argon-40 in 1,300,000,000 years—the ratio of the two isotopes should give the rough age of the rock.

September 4, 1959

New Chronology of Geologic Time Is Fixed

'Datebook' Is Revised After Scientists Compare Notes on Decay in Rocks

By WALTER SULLIVAN

An assembly of specialists in the earth's history, gathered from many parts of the world, has pieced together a new chronology of geologic time.

It furnishes students of geology and evolution with a new "datebook" in which to record their discoveries. The timetable was constructed after the scientists had compared notes on the dating of rock specimens, based on measurements of their radioactive decay.

They emphasized that the new time scale was by no means final. As Dr. J. Laurence Kulp, the conference chairman, put it, many further "refinements" will be made.

Dr. Kulp, who is at the Lamont Geological Observatory of Columbia University, is a leading authority on radioactive-dating methods. He said they had doubled the estimated lengths of some past periods of geologic time and had halved others.

The pioneer in constructing a time scale for the past was Dr. Arthur Holmes of the University of Edinburgh. Using some of the earliest dates, based on studies of uranium decay into lead, he published a chornology forty-seven years ago. Its total length, back to the start of the Cambrian Period, was only about 15 per cent shorter than that proposed yesterday.

It was during the Cambrian Period that the earliest known

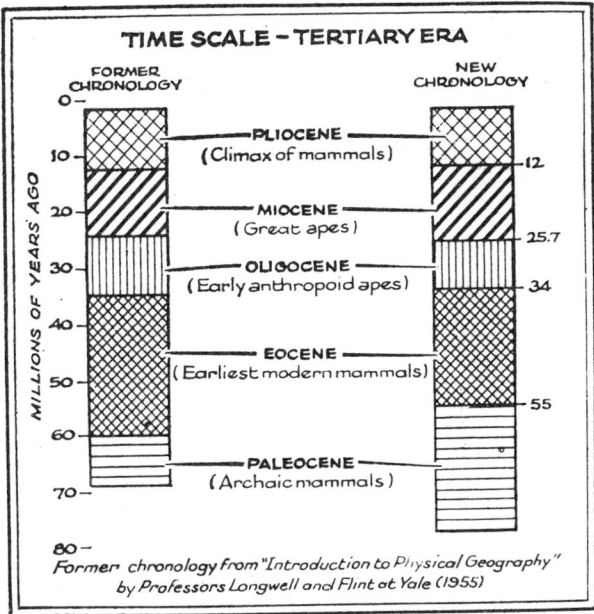

Timetable for period immediately prior to most recent period (the Quaternary), which lasted one million years.

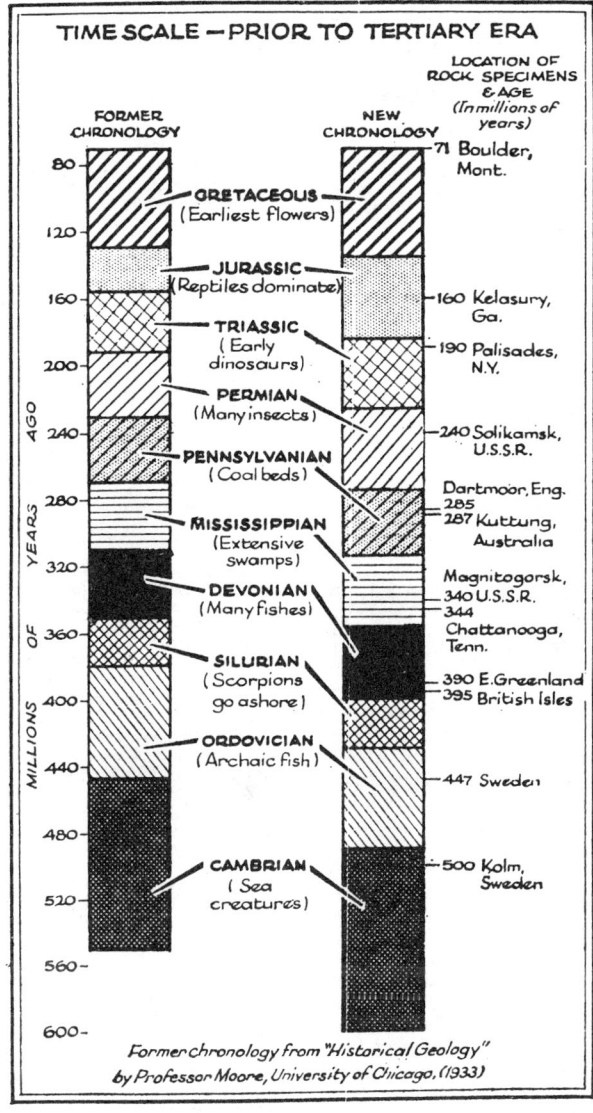

Table for period preceding that shown in chart at left

March 5, 1960

fossil beds were laid down. The conferees, meeting at the Barbizon-Plaza Hotel, sent a message of greeting to Dr. Holmes, now living in retirement in Scotland. The conference is being sponsored by the New York Academy of Sciences.

The ladder of earth's history has been put together, from the Cambrian to the Quaternary (the most recent period) on the basis of life's evolution, as revealed in the fossils, and the changes in climate and topography recorded in the rocks. The sequence of these periods has long been established, but their dates have been the subject of much debate.

The rock specimens upon which the new chronology has been based were found in deposits that could be clearly linked to a certain place on the ladder. They were collected in many parts of the world, from the Ural Mountains of the Soviet Union to New South Wales in Australia.

Their dating is still subject to margins of error that vary with each specimen.

It was hoped that a group of invited Soviet scientists would contribute additional information at the conference, but they did not come. Several Soviet dates, however, were available from published reports.

A RARE REPTILE FOUND IN BORNEO

Foot-Long Lizard Regarded as a 'Missing Link'

By JOHN HILLABY
Special to The New York Times.

LONDON, May 1—News has reached the British Museum of Natural History of the capture of what is considered one of the rarest reptiles.

This "missing link" among reptiles is called the Bornean earless monitor. It is believed to resemble the animal from which all snakes evolved at the time of dinosaurs 100,000,000 years ago.

The monitor, known to science as lanthanotus borneensis, is reported to be about a foot long. It was found recently at Niah in Sarawak, Borneo, after the Sarawak Museum had offered $50 for the capture of a live specimen.

A local collector spotted it buried six inches in soft yellow clay. The collector, Jommel Bin Bogol, is searching similar localities for more specimens.

Dead Specimens Known

In his book, "The Reptile World," Clifford H. Pope, formerly of the American Museum of Natural History, said that "the mere sight of one of these lizards alive and kicking would be the fulfillment of a dream no student has yet realized."

The skins of a few dead specimens have been known to reptile experts for more than fifty years. They all came from Sarawak. When no live specimen were found despite long searches, most experts thought the lizard had become extinct.

In appearance and some habits, the monitor is said to resemble the Mexican beaded lizard and the Gila monster, or heloderm, of Mexico, Arizona and Nevada, the only two poisonous lizards known.

The monitor may be poisonous, but the Sarawak Museum staff is said to be chiefly concerned about keeping it alive. It has so far refused all offers of food, including fish fry, insects, worms and small frogs.

Common Ancestor

Explaining the significance of the find, a member of the British Museum staff said tonight that snakes and lizards were newcomers in the evolutionary scale of development.

They had a common ancestor in the geological period known as the Cretaceous, when chalk was laid under the sea, he added. In that period, he said, the land was dominated by giant reptiles called dinosaurs, and the sea by giant lizards called mosasaurs.

The dinosaurs and mosasaurs died out suddenly. But the small lizards, the descendants of the mosasaurs, began to populate the land among newly evolving mammals. From one group of primeval lizards, snakes emerged.

Scientists have not been able to find a fossilized specimen of a snake-lizard, the expert said, "but in its structure it must have resembled the new monitor of Sarawak."

"It is a discovery of very great importance to zoologists and evolutionists," he emphasized.

Importance Confirmed

The Borneo lizard is the first live specimen of its type to fall into the hands of anyone who knew what it was, Charles M. Bogert of the Museum of Natural History said here last night.

He said the find was of considerable interest because, among living lizards, the monitor's ancestry was closest to that of snakes. He said there was almost no chance that it was venomous. There are indications, he added, that its normal diet may include earthworms.

He said the Sarawak Museum had recently informed specialists in this country of the discovery.

May 2, 1961

Million Years Added To Man's Evolution

By HAROLD M. SCHMECK Jr.

Primitive toolmaking man may have roamed the earth 1,750,000 years ago, according to a radioactive-dating measurement announced yesterday.

The new date would make the earliest toolmakers yet discovered just about three times as old as scientists had previously thought they were.

It pushes evidence for the emergence of toolmaking — one of the two most essential basic attributes of humanity — back, past the earliest fringes of the Pleistocene, into the Pliocene Age. A figure of roughly 1,000,000 years ago is commonly given for the beginning of the Pleistocene.

The dating was done by Drs. J. F. Evernden and Garniss H. Curtis, geologists at the University of California, Berkeley.

The ancient man whose age they have established is Zinjanthropus boisei found two years ago in East Africa by Dr. Louis S. B. Leakey of Coryndon Memorial Museum, Nairobi, Kenya. This anthropologist had estimated the age of Zinjanthropus as "more than 600,000 years." Even at that age the creature was considered the oldest for which clear evidence of toolmaking has been found.

The new date for Zinjanthropus was announced, for publication yesterday, by the National Geographic Society in Washington. An article on Dr. Leakey's finds and the new date are to be published in The National Geographic Magazine. The age figure is based on a potassium-argon radioactive dating process.

Importance Is Stressed

In a telephone interview, Dr. T. Dale Stewart of the Smithsonian Institution called the new date "a very important contribution" and one that would be a surprise to many.

The anthropologist declared, however, that the emergence of toolmaking 1,750,000 years ago seemed to him easier to fit into the picture of human evolution than the 600,000-year figure.

Zinjanthropus, of course, was not a man of the modern species. That is, he was not Homo sapiens.

"Everyone seems to see in Zinjanthropus a type that could be pretty close to the line, if not actually in the line, of human evolution," Dr. Stewart commented.

The creature was capable of walking erect. Its hands were freed for such tasks as making extremely primitive cutting tools from pieces of quartz rock. But Zinjanthropus had a small brain in comparison with that of modern man.

Course of Evolution

Anthropologists consider it probable that the larger brain and higher intelligence evolved as primitive man-like creatures learned to use tools and that other basic essential of humanity—a communication by speech.

Such a course of evolution seems more reasonable starting nearly 2,000,000 years ago than it would at the faster evolutionary pace required if the process started only about a half million years ago, Dr. Stewart reasoned.

Since no animal species other than man both makes and uses tools, Dr. Leakey defines Zinjanthropus as a man, albeit of a primitive type.

Dr. Evernden and Dr. Curtis established the date by radioactivity tests on volcanic rock samples taken from locations directly above and below the spot at which Dr. Leakey found fragments of the Zinjanthropus skull. The geologists also made several trips to the Olduvai Gorge in East Africa where Dr. Leakey made the Zinjanthropus and other important anthro- 7 per cent.

PRIMITIVE MAN: This portrait of a man believed to have lived 1,750,000 years ago was made by Peter Bianchi under the direction of Dr. Louis S. B. Leakey, anthropologist.

Six Rock Samples Tested

Reached by telephone at his laboratory, Dr. Curtis said that six samples of rock were tested by the potassium-argon dating process and that the 1,750,000-year figure represented the average. He estimated the probable range of experimental error in the tests at about 5 to 7 per cent. The range of dates obtained with the six samples was "a couple hundred thousand years," but the geologist said the most reliable of the six seem to be among the oldest. So Zinjanthropus may, if anything, be even more ancient than 1,750,000 years.

Some of the "youngest" dates in the series, Dr. Curtis said, were from rock samples that showed evidence of weathering, a process that could affect their apparent age as measured by radioactive dating.

The geologist said that evidence obtained earlier from Olduvai Gorge had made him and his colleagues suspicious that Zinjanthropus might be considerably older than the 600,000-plus figure. The new date confirms that earlier suspicion.

Dr. Curtis noted that the possibility always must be considered that erosion and like processes might have contaminated the rock samples with older minerals. This would distort the radioactive-dating results and would make the rocks appear older than they actually are.

Dating Process Explained

He said, however, that this possibility had been carefully considered and that all the pertinent criteria for such contamination were applied in the research. The evidence indicated that no such contamination was involved.

"My feeling is that these dates are good," he said.

The potassium-argon system, like other comparable dating methods, is based on the radioactive decay of elements in the rocks.

Over the ages, potassium 40 breaks down into two other elements—calcium 40 and argon 40. The rate is extremely slow so that half of any given sample of radioactive potassium would decay into its daughter products in about 1,200,000,000 years. The dating system, an extremely delicate process, depends on measuring the amount of argon 40 in a sample as compared with the amount of potassium.

The mineral tested was a feldspar called anorthoclase contained in volcanic ash.

The potassium-argon system is the only one of several available radioactive-dating process that offers a good chance to date specimens in this particular age range considered "young" by geologists and "old" by anthropological standards.

July 23, 1961

AFRICA BONES ADD LINK TO EVOLUTION

Man-Like Creature Existed 14 Million Years Ago

Special to The New York Times.

WASHINGTON, March 22— Remains have been discovered in Africa of a creature that lived 14,000,000 years ago and, in the broad stream of evolution, seems to lie midway between the early apes and man.

This was reported here today by Dr. Louis S. B. Leakey, the British anthropologist. Hitherto the earliest find on the branch of the evolutionary tree that had culminated in man, as opposed to the ape, was the Zinjanthropus, or East African man.

Zinjanthropus lived some 1,750,000 years ago. Its first remains were discovered by Mrs. Leakey in the Olduvai Gorge of Tanganyika in 1959. Since then 400 fossilized skull fragments have been found.

The new find was at Fort Ternan, in Kenya, about forty miles east of the eastern tip of Lake Victoria. The bone-rich deposit was discovered last year and dates from the lower (older) part of the Pliocene, which is far back in the history of two legged creatures.

In 1948, Dr. and Mrs. Leakey found in Kenya the skull of a primitive ape that lived from twenty-five to forty million years ago, in the period known as the lower Miocene. He called it "Proconsul" and believes it represents the root stock of the higher primates, including man.

Because the new discovery lies roughly midway between "Proconsul" and the earliest known man-like species, it is viewed as one of special importance. Dr. George Gaylord Simpson, Professor of Vertebrate Paleontology at Harvard University, termed the find "very significant." He said that

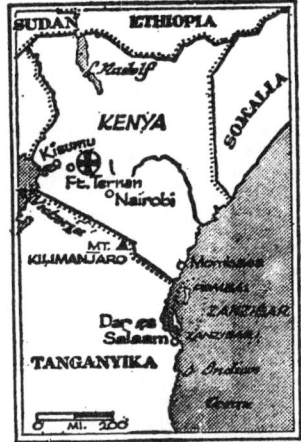

Remains of a 14,000,000-year-old manlike creature have been found at Fort Ternan (cross) in Kenya.

it fills an "enormous gap" in the history of the primates.

He was present last summer when the first fragment of the ancient skull was discovered. This was a portion of the roof of the mouth. Subsequently the other half of the roof was discovered, plus a lower tooth.

Dr. Simpson noted that excavation of the site has hardly begun and that further discoveries there may make it possible to piece together a more complete picture of this as yet unnamed creature.

Dr. Leakey believes the site is the first to reveal what sort of vertebrates lived in Africa, south of the Sahara, during the Pliocene. His finds include a giraffe the size of a donkey, an elephant no larger than a cow, five kinds of antelope and extinct forms of hippopotamus, pig and rhinoceros.

The African-born British anthropologist told a press conference at the National Geographic Society that he had suspected there might be pliocene outcrops in that area. He asked local farmers to watch for bones and, in 1959, received a sack of them from an orange grower named Fred Wicker.

In two months of digging on the Wicker ranch last summer.

he and his wife obtained 1,200 specimens.

The work has been supported by the National Geographic Society, which, tomorrow, will present to the husband-wife team its Hubbard Medal. The presentation is to be made by Chief Justice Earl Warren at Constitution Hall, prior to a lecture by Dr. Leakey.

The dates of Dr. Leakey's finds have been determined by Drs. Jack F. Evernden and Garniss H. Curtis at the University of California in Berkeley. They did so by dating rocks associated with the deposits. This was achieved by determining the extent to which radioactive potassium in the rock had decayed into argon.

Dr. Leakey said the California scientists had phoned yes-

Dr. and Mrs. Louis S. B. Leakey at news conference in Washington. Dr. Leakey holds fossil found in Africa.

terday to report that, after making four analyses, they had decided on an age of 14,000,000 years, with a probable error of no more than a few hundred thousand years.

Dr. Leakey said that, as in man, the canine teeth of the creature protrude only slightly below the level of its other teeth, whereas in the apes this protrusion is marked. Likewise, it has a depression in the cheek bone that is found in man but never in apes.

Nevertheless, he stressed that the find is not a species of man, as such. He would not, for example, expect to find that it had used tools, whereas Zinjanthropus apparently did so.

March 23, 1962

OLDEST ANCESTOR OF MAN GETS NAME

Homo Habilis of 1.7 Million Years Ago Made Tools

By JOHN HILLABY
Special to The New York Times

LONDON, April 2—The oldest ancestors of man, a race of upright but small-brained toolmakers that lived in East Africa about 1,750,000 years ago, have been given the official name of Homo habilis. The name was taken from the Latin, meaning "able, handy, mentally skillful, vigorous."

The bone fossil remnants of these men were found, mostly between 1961 and 1963, in the Olduvai gorge of Northern Tanganyika by Dr. Louis Leakey and his wife, Mary, also a well-known scientist. Reports of

these discoveries have been made previously.

A detailed description of the bones, the brain size and other physiological peculiarities of the "first handymen," will be published in the British science journal Nature on Saturday.

The reconstructions have been carried out by Dr. and Mrs. Leakey, Prof. Phillip T. Tobias of the University of Witwatersrand, Johannesburg, and by Dr. John Napier, of the Royal Free Hospital Medical School in London.

A Matter of Centimeters

Professor Tobias begins by explaining that most definitions of the border line between highly-advanced apes and primitive men (sometimes called homines or euhominids) hinge about what he calls a "cerebral Rubicon." That is the point at which the volume of the brain measured in cubic centimeters of cerebral tissue can be adjudged human.

He points out that between apes and men "there is almost

an overlap" since one large male gorilla was known to have a brain capacity of 752 cubic centimeters whilst the smallest human capacity hitherto reported (a specimen of Pithecanthropus) had no more than 775 cubic centimeters of brain.

From many casts of the shattered remnants of the upper skull bones (parientals) of the handymen of Olduvai, Professor Tobias says his estimates range between 642.7 to 723.6 cubic centimeters.

He seems convinced that when this brain size is considered in association with the human shape of the teeth, jaws and limb bones, most experts will agree that Habilis deserves the name of homo, a man.

Age Not Mentioned

There is no specific mention of the age of Homo habilis in Nature's preliminary account of the finds. But most investigators who have handled the bones consider they are probably 1,750,000 years old. More relics were found earlier this year at Lake Natron some 50 miles northeast of Olduvai.

Nearer Olduvai Dr. Leakey also discovered a rough circle of loosely piled stones in the same geological stratum as the bone site. He says:

"It seems that the early homindis were capable of making rough shelters or windbreaks and it is likely that Homo habilis may have been responsible."

The evolutionary picture from Olduvai tends to simplify man's emergence from his ape-like ancestors. There were fewer divisions between them than had been thought.

Palentologists agree that the primates sprang from long-tailed tree shrews long after the extinction of the great reptiles. That was about 90 million years ago.

These lemur-like animals developed prehensile hands and feet with opposable thumbs and great toes. They had digits topped with flattened nails, mobile arms, well-developed brains and binocular vision.

It is now thought that about 2,000,000 years ago the central evolutionary stem divided like the two prongs of an upturned hayfork.

The shaft represents common ancestors. One prong marks the ascent and gradual extinction of the little South African ape men, the Australopithecines. The other gives rise to Homo habilis (Olduvai man), Homo erectus (Java and Pekin man) and, eventually, Homo sapiens.

The latest report in Nature is one in a series in which Dr. Leakey and his colleagues have given detailed descriptions and analyses of the important fossil specimens found in East Africa in the last several years.

A publication about two weeks ago analyzed fossil hand and foot bones found at Olduvai Gorge.

This analysis of specimens called pre-Zinjanthropine indicated that the creature from which the specimens came was directly ancestral to man, yet older than any other direct ancestor previously named.

The pre-Zinjanthropine material was found at a level about two feet below that of Zinjanthropus, the man-ape discovered in Olduvai Gorge in 1959.

April 3, 1964

Fossil Hunter Supreme
Louis Seymour Bazett Leakey

Special to The New York Times

LONDON, April 2 — Dr. Louis Seymour Bazett Leakey, the noted paleontologist, is reckoned by most persons in the fossil-hunting business to be something of a human divining rod. They say that, faced by a stark gulley in the desert, he sniffs expectantly, stamps off with his spectacles pushed onto the top of his head and returns, triumphantly within a short time, with a prehistoric handax or a shattered jaw-bone.

Man in the News

Dr. Leakey, who is now touring the United States, is expected tomorrow to tell a news conference at the National Geographic Society in Washington of findings that, he belives, add about a million years to the ancestry of man.

Even Dr. Leakey's vigorous critics—and he has quite a few—admit that he seems to be able to smell human bones. His latest discovery, the relics of Homo habilis, the oldest race of men on earth, assures him of a high place in the paleontological halls of fame that include Eugene DuBois, the young Dutch doctor who unearthed Pithecanthropus in Java, and Raymond Dart, who found the remains of the Taungs child in South Africa. These two discoveries were the first of the so-called missing links between apes and men.

Son Of A Missionary

Dr. Leakey was born at Kabete, in Kenya, on Aug. 7, 1903. His father was a distinguished missionary.

Dr. Leakey learned the local dialects, especially Kikuyu, at a time when the British had made a peace pact with the Masai warriors and were driving the Bantu into the hills around Mount Kenya. Those were troubled times. The young paleontologist-to-be was sent to Britain to be educated at Weymouth College and St. John's College, Cambridge.

In 1924 he returned to East Africa as a member of the British Museum expedition. For the next 15 years he crisscrossed the country in search of the fossilized ancestors of man.

His greatest discoveries included the since-disputed Kanam jaw, a prehistoric flint "factory" at Olorgosailie near Nairobi and the remains of a race of unspecialized apes called Proconsul on the Rusinga Island of Lake Victoria.

After getting the job of curator at the Coryndon Museum in Nairobi, Dr. Leakey began to spend more and more time in a massive rift of the Serengeti plains of Tanganyika. This is Olduvai, a gorge discovered by a German zoologist in search of butterflies but renowned in latter years as "the cradle of mankind."

From the sun-cracked face of the Olduvai, Dr. Leakey and his wife, Mary, extracted tons of ancient bones. Among them were the skeletons of an extinct group of giant animals, pigs and sheep of their time, but as big as rhinoceros and elephant. More important they also found the remains of man.

At a point where the gorge drops sheer for 300 feet they found the living floors or hunting encampments where countless generations of prehistoric men and sub-men had lived for nearly two million years.

Working through the night with the aid of a kerosene lamp, the master bone-finder hastily reconstructed the skull of one of man's brutish cousins called Zinjanthropus or the nutcracker. It was flown to London and widely proclaimed.

Olduvai eventually yielded Homo habilis, "the first handyman."

Dr. Leakey then went back to fossil-hunting, taking his son, Richard, with him. Another son, Jonathon, became a professional snake collector.

Honors and awards have been showered on Dr. Leakey. The honors came from his own college, St. John's, which made him a fellow; from the National Geographic Societies of Sweden, Britain and the United States and from the Wenner-Gren foundation of New York.

April 3, 1964

EARLY MAN STIRS SCIENCE DISPUTE

African Fossils Challenged as a Distinct Species

By HAROLD M. SCHMECK Jr.

A sharp difference of opinion is developing over a creature that has been called the world's earliest known true man.

A British specialist doubts that the creature, named Homo habilis, should be classed as a distinct early species of man at all. He believes it should probably be classed as a sub-species of the African man-apes, or near-men, called the Australopithecines.

Homo habilis was named earlier this year by Dr. Louis S. B. Leakey, the internationally known scientist whose research in Africa has yielded many important clues to man's origins.

On the basis of many fossil bone fragments representing at least seven individuals, Dr. Leakey and a group of scientific colleagues concluded that Homo habilis represented a newly found species and the earliest true man. Radioactive dating indicates that some of the specimens may be two million years old.

The conclusions and the evidence were published earlier this year in the British science periodical Nature.

A conflicting viewpoint is to be published in the June issue of another British publication, Discovery, by Dr. Bernard Campbell of Cambridge University. He is a scientist highly regarded for his studies of the taxonomy, of classification, of early human and prehuman fossils. In the same issue Dr. John R. Napier, one of Dr. Leakey's co-authors in the earlier reports, defends the separate and human status of Homo habilis.

Large Brain Is Found

From detailed studies of fossils and their setting Dr. Leakey and his colleagues deduced that Homo habilis had a relatively large brain (though smaller than modern man's), was only about 4 feet tall, stood erect on feet that were almost human and probably, made crude stone tools with hands that were rather primitive in comparison with those of modern men.

The evidence supports the view that the first steps along the evolutionary path from near-man to man were the change to upright posture and the ability to make crude tools; and that thereafter the hand became more advanced and the brain larger.

Dr. Leakey's evidence from the sites in Olduvai Gorge where Homo habilis was found indicates the creature lived on large and small game, and may have built crude windbreaks of piled stones as protection against a cool wet climate.

The physical characteristics including brain size and the evidence of tool manufacture support the view that Homo habilis was a separate species on the human side of the boundary between men and near-men, Dr. Leakey, Dr. Napier and their colleagues have said.

Two Criticisms Voiced

In his article in Discovery, Dr. Campbell takes issue with this conclusion, although he agrees that the fossils are important. He has two general criticisms of the view that these creatures represent a separate species.

First, Dr. Campbell said, there is not enough basic difference between some of the more advanced Australopithecine man-apes and some specimens of Homo erectus, heretofore generally agreed to be the earliest true men, to leave evolutionary room for an entirely distinct species between them. He said it was more likely that Homo habilis was really synonymous with a subspecies of Australopithecus found in South Africa several years ago and called Telanthropus capensis.

The second general criticism, Dr. Campbell said, is that the specimens of the proposed new species have not been compared in detail with presumed closely related species.

"New species must be shown to be clearly distinguishable from their nearest neighbors in evolution," Dr. Campbell said. "This of course is common sense and is designed to prevent two names being given to one species—a situation which occurs all too often. To the zoologist's astonishment, he reads in Nature (April 4) that Dr. Leakey and his colleagues have not yet compared the new finds with what by general agreement are the most closely related forms, that is, those known as Telanthropus capensis."

Evidence Is Questioned

Dr. Campbell also questions the estimates made for Homo habilis' brain capacity and certain other details. On the significance of the creature's ability to make crude stone tools, the British scientist notes that stone tools have also been found in association with fossil remains of Australopithecines from several African sites.

He concludes "that the archeological evidence and published accounts of these new fossil discoveries are not sufficiently detailed to justify the creation of a new species of man [genus Homo]."

Dr. Campbell's doubts have found support among some other scientists on both sides of the Atlantic. One logical difficulty in calling Homo habilis a member of genus Homo is the great length of time the species seems to have persisted, said Dr. Elwyn Simons of Yale. If all the specimens are of one species and if the dates are correct, Homo habilis must have existed from about two million years to a time estimated at 750,000 years ago. This is a span of more than a million years, yet Homo erectus, hitherto generally agreed to be the earliest true humans, and Homo sapiens (modern man) have a history between them of only about 500,000 years.

June 7, 1964

MOLECULES HELP EVOLUTION STUDY

Scientists Get New Insight Into Life's Chemical Basis

By JOHN A. OSMUNDSEN

Studies of the evolution of living things at the molecular level are giving new insight into the chemical nature of life.

This was the conclusion that emerged from a two-day scientific symposium on "Evolving Genes and Proteins" that was held Thursday and Friday at the Institute of Microbiology of Rutgers, the State University of New Jersey in New Brunswick.

The scientists reported finding striking concordances between the structure of certain molecules or the course of certain chemical pathways of metabolism in organisms that had been regarded as closely related according to classical methods of classification.

But they also found some surprising disparities, indicating the possibility of unexpected evolutionary connections or misconnections.

What appeared to be most significant, however, were discoveries of extremely similar or identical features in the structures of vital molecules in a wide range of living things.

The great evolutionary stability of those features and their widespread appearance among a great diversity of organisms indicate that the features are probably very important to life. Scientists would like to find out why.

Enzyme Yields Clue

For example, Dr. E. Margoliash of Abbott Laboratories reported the results of studies on an enzyme, or biological regulator, called cytochrome-C. Like all proteins, cytochrome-C consists of a chain of subunits, called amino acids, of which there are about 20 kinds. Like all enzymes, this one has certain regions along its length that account for its predominant activity. Such a region is called an active site, the rest of the amino acid units presumably serving to give the enzyme molecule its characteristic, functional shape.

Dr. Margoliash reported that there were differences among the amino acid sequences of the active sites of different organism's forms of this enzyme. Yet, he said, there was a stretch of 10 amino acids in the cytochrome-C molecule that was not in a known active site but that was identical for all organisms studied, from yeast to man.

Interpretation Disputed

The question that this finding raised was: Why is that seemingly innocuous portion of the molecule so important as to be shared by so many different organisms, and what does it do?

The same question was asked, indirectly, about minor but important differences between the structures of molecules of hemoglobin, the red blood pigment from different organisms. A quite new look at the importance of the differences emerged from the symposium.

For example, there is an hereditary condition known as sickle cell anemia. The structure of the hemoglobin molecule is the root of the problem, that of those who suffer this anemia differing from the normal by a single, particular amino acid subunit.

That difference arose by mutation, or a change in the hereditary material that directs the production of the hemoglobin molecule. The question has been why such a deleterious mutation should survive instead of being selected against.

The answer to that question has been widely assumed to be that the sickle-shaped red blood cell is not a suitable home for the malaria parasite, giving persons who have this condition a selective advantage over those with normal hemoglobin in malaria-ridden regions of the world where sickle cell anemia is most prevalent.

At the Rutgers meeting, however, that interpretation was disputed by Dr. John Buettner-Janusch, a physical anthropologist from Yale. He said that the evidence for that theory was not only inconclusive, but that there was also evidence against it—mainly, the flourishing of sickle cell anemia in nonmalarial regions.

Acid Difference Explained

It was Dr. Buettner-Janusch's point that the single amino-acid difference between the normal and the sickle cell hemoglobin molecules must be vitally important for some survival advantage, probably not against malaria but possibly for something similar to fecundity. He noted that sickle-cell women tended to have slightly larger families than normal.

Dr. Margoliash had also said that the structures of the cytochrome-C molecules from pig, sheep and cow were identical. This finding was called remarkable by Dr. Ernst Mayr, Agassiz Professor of Zoology at Harvard.

Dr. Mayr said that since the pig was a quite distant relative to the two other animals, the genetic changes that produced the precise structure of the cytochrome-C molecule that the three share must have been very important to have become stabilized independently. Nor, he said, was this the only example. Man he explained, is the closest mammal to the rattlesnake in terms of the amino acid sequences of their cytochromes.

Other scientists reported using various chemical techniques to trace relationships between organisms. For example, Dr. N. O. Kaplan of Brandeis University presented data on similarities between forms of a particular enzyme that showed that the sturgeon had changed less than the halibut since they both diverged from the chicken; and that the ostrich and rhea — both primitive birds — were more closely related to the higher reptiles than chickens, pigeons and ducks.

Two other studies of the relationship between vital chemicals in different organisms tended to lend further support to new ideas about cancer and viruses.

Dr. B. H. Hoyer of the National Institutes of Health reported that the genetic material of a virus that caused cancer in animals was much more closely related to that of animals in which the tumors were produced than that from other animals.

Dr. Jules Marmur of the Albert Einstein Medical College told the meeting that the chemical composition of the hereditary material of viruses that infect bacteria varied widely from the bacterial cells they infected.

That of viruses that kill bacteria is most different from the infected cells, he said, whereas the genetic material of those that "get along" with their bacterial hosts is quite similar to the bacteria's.

This finding was considered interesting in connection with the virus-cancer relationship because tumor viruses are presumed able to live in -- but to change to malignancy — cells they infect, whereas other viral actions kill cells.

September 20, 1964

Earth Yields Clues to Past

By WALTER SULLIVAN

Significant new dimensions have been added to the history of life on earth. In recent days evidence has been presented indicating that our planet has been inhabited for at least three billion years. The known history of complex animals has been lengthened by 20 per cent. And excavations for a interstate highway have cast a beam of light into the murky seas of the Devonian Period, when giant sharks and other creatures marked the emergence of the vertebrates.

According to the Natural Science Museum of Cleveland, which is supervising the collection of fossils along Interstate 71, skirting the city, the slicing of Ohio hills to make way for the highway has doubled the number of known vertebrates from that period, some 400 million years ago.

The crust of the earth is thought to have been formed not much more than 4.5 billion years ago. The discovery that life was established 1.5 billion years later therefore means that the atmosphere stabilized, oceans formed, and the earth became habitable in a rather short time. It means, as well, that the complex chemistry of life evolved into self-replicating organisms far more rapidly than previously thought.

The discovery of fossil life three billion years old was reported last week to the geological Society of America in Kansas City by Dr. Elso S. Barghoorn, Professor of Botany at Harvard University. Dr. Barghoorn in recent years has been examining very ancient rock formations for signs of life.

He and others in the field have steadily pushed back the date of the earliest organisms. In 1961 he reported finding coal in northern Michigan that was 1.7 billion years old. Then it was reported that signs of ancient life had been found in rock 2.7 billion years ago.

One-Celled Fossils

The new discovery was made in rocks 20 miles southeast of Barberton, South Africa. Specimens were brought back to Harvard and studied with an electron microscope, the most powerful magnifying device of its kind. This revealed rod-like objects in the rock that looked strikingly like bacteria.

Presumably the earliest life consisted of one-celled organisms akin to bacteria that lived in water or mud. The remains of this life were subsequently preserved in sediment that was converted, by heat and chemical reactions, into solid rock.

Bacteria are classed as plants, although some of them are self-propelled. The first animals were presumably precursors of the protozoa. The time of their origin is still uncertain. What has puzzled scientists more than anything else, in the study of evolution, has been the sudden appearance of complex animals of many kinds at the start of the Cambrian Era, some 600 million years ago.

Most of the major divisions, or phyla, of the plant and animal kingdoms had been established by then, yet there is no record of how they evolved. A recent proposal has been that evolution, after plodding along at a snail's pace for two billion years, suddenly erupted when plant activity introduced oxygen as a free component of the air.

Oxygen made possible efficient energy production by oxidation, which is exploited by most modern life forms. However this manner of life also called for complex organs to handle ingestion, digestion respiration, locomotion and coordination. The result was a diversification of species that moved so rapidly it left little record in rocks.

Complex Organisms

At the Kansas City conference Dr. Andrew H. McNair of Dartmouth College reported finding fossil evidence of such creatures as clams and worms at least 720 million years old, adding 120 million years to the known record of such organisms. While their bodies seem simple by present day standards, they have many of the essential features of higher animals: intestines, muscles, a nervous system, and so forth.

The dating of these fossils was done by measuring the extent to which rubidium 87, a radioactive component of the rock, had decayed into strontium 87. Since the decay rate is known, it was possible to estimate the age at more than 720 million years. Dr. Barghoorn said the South Africans had also used radioactive decay measurements to date the ancient finds in that region.

While the Dartmouth discoveries do not explain how the clams and worms evolved, they may open the way for discoveries that will unravel some of the early history of the animals.

The Cleveland finds relate to the Devonian Period when oceanic activity laid down giant beds of limestone and sandstone across many areas of the world that are now above water. While the vertebrates originated earlier, they evolved rapidly during the Devonian. The sharks of today are relics of this time. Their spines, made of cartilage, are not true backbones and their stupidity is proverbial.

According to the Cleveland museum some 60 species of Devonian vertebrates were hitherto known and the excavations along Interstate 71 have added another 60.

SEARCHING FOR THE BEGINNINGS OF LIFE

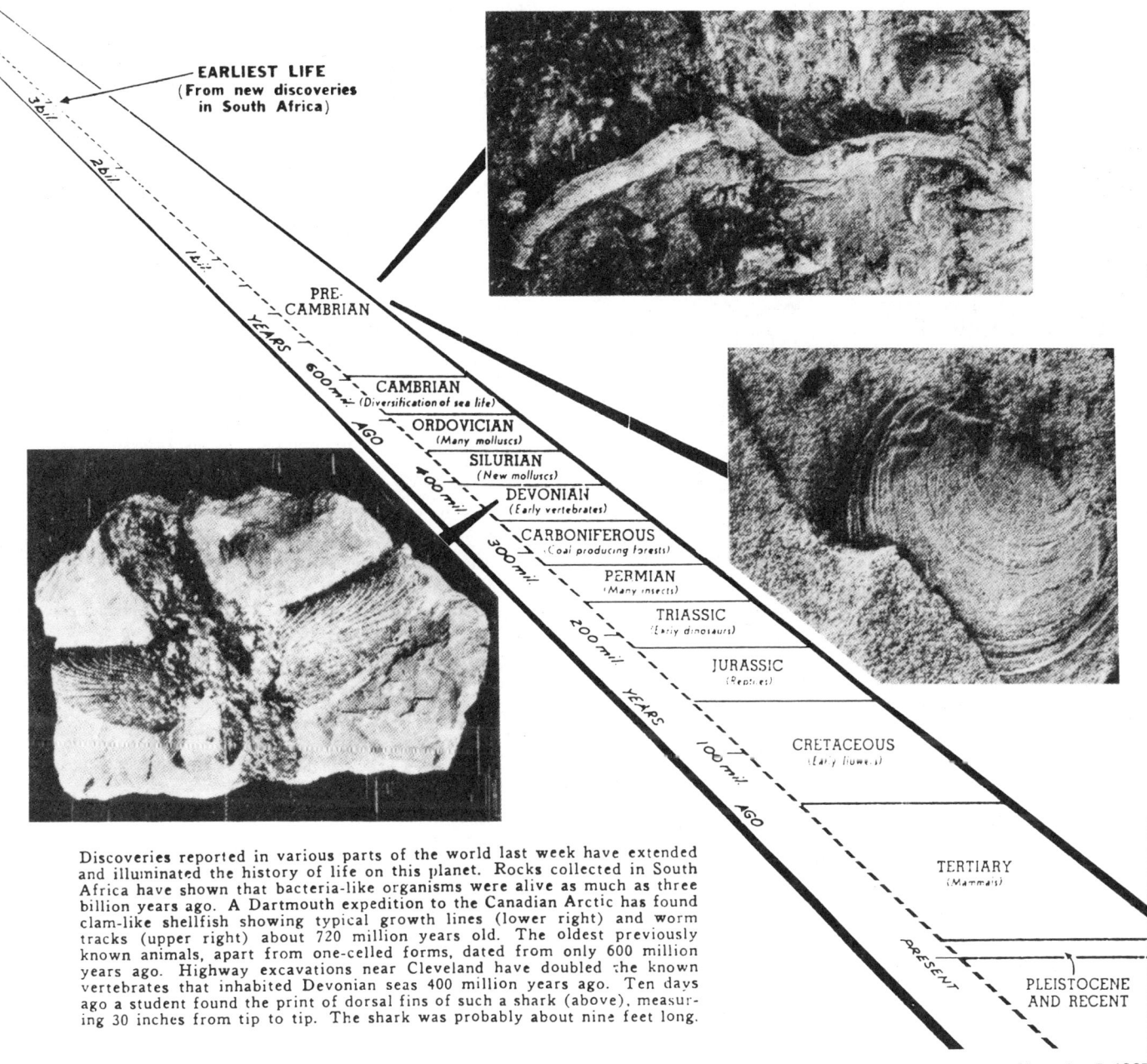

Discoveries reported in various parts of the world last week have extended and illuminated the history of life on this planet. Rocks collected in South Africa have shown that bacteria-like organisms were alive as much as three billion years ago. A Dartmouth expedition to the Canadian Arctic has found clam-like shellfish showing typical growth lines (lower right) and worm tracks (upper right) about 720 million years old. The oldest previously known animals, apart from one-celled forms, dated from only 600 million years ago. Highway excavations near Cleveland have doubled the known vertebrates that inhabited Devonian seas 400 million years ago. Ten days ago a student found the print of dorsal fins of such a shark (above), measuring 30 inches from tip to tip. The shark was probably about nine feet long.

November 7, 1965

Ancestor of Man and Ape Traced 5 Million Years Back by Study

By JOHN NOBLE WILFORD

A new evolutionary dating method that uses blood molecules indicates that man and the African ape are descendants of a common ancestor that lived no more than five million years ago, according to a study by two California scientists.

The new evidence could lead to the revision of earlier theories that the common ancestry of human beings, gorillas and chimpanzees dates as far back as 30 million years.

It also shows that gorillas and chimpanzees are apparently no more closely related to each other than either is to man.

These conclusions were published yesterday in the journal Science on the basis of investigations by Dr. Vincent M. Sarich and Dr. Allan C. Wilson, biochemists at the University of California at Berkeley.

If these conclusions are correct, the scientists said, it is likely that man and living apes evolved from a small chimpanzee-like animal that managed to survive for ages because it developed the ability to swing from branch to branch in search of food.

Blood Constituent

The method of dating evolutionary changes is based on comparisons of serum albumin, a common protein constituent of blood. Protein molecules are the basic substances of all living things, and their structure is known to closely reflect that of genes.

In analyzing the albumin in humans and a variety of apes, the scientists found an "extremely close structural similarity."

This in itself is not surprising, since it is generally agreed that the African ape is man's closest living relative. But based on theories of the rate of albumin evolutionary change, the California scientists concluded that "it seems likely that apes and man share a more recent common ancestry than is usually supposed."

In other words, if man had evolved at an earlier time, the differences between man's albumin and the ape's should have been more marked.

Dr. Sarich and Dr. Wilson suggested in their report that albumin molecules could serve as "an evolutionary clock or dating device."

Thus, according to their calculations, the time of divergence of man from the African apes occurred about five million years ago. The two, the scientist said, must have shared a common ancestor in the Pliocene period.

Lineage Separation

Moreover, they calculated that the lineage leading to the orangutan separated from that leading to the African apes eight million years ago, and the gibbon lineage separated 10 million years ago.

The scientists conceded that there were some "uncertainties" in their calculations but doubted that any errors would be of sufficient magnitude to invalidate their general conclusions.

For one thing, Dr. Sarich and Dr. Wilson said that their hypothesis would account for many of the physical similarities between apes and man that had previously been attributed to parallel evolution since they diverged from the common ancestor.

"In our opinion," the scientist concluded, "the albumin data definitely favor those who have postulated that man and the African apes shared a common ancestor in the Pliocene."

The scientists suggested, however, that their concept of evolutionary dating should be confirmed by investigating the albumin evolution in other mammals where an extensive fossil record is available for the sake of comparison.

December 2, 1967

JAWBONES' AGE PUT AT 8,000,000 YEARS

Yale Scientists Say Fossils Were of Human Species

By JOHN DARNTON
Special to The New York Times

NEW HAVEN, July 26—Two Yale scientists announced this week that they had identified two fossilized jawbones belonging to what they said was a human species that lived between eight and 15 million years ago.

The finding, according to Dr. Elwyn L. Simons, professor of paleontology, and Dr. David R. Pilbeam, assistant professor of anthropology, confirms the theory that man's ancestors began an evolutionary break from the apes 14 million years ago.

This estimate, while close to that accepted by many physical anthropologists, disputes a well-publicized statement of Dr. Louis B. Leakey of the Kenyan National Museum of Natural History. Two years ago Dr. Leakey, an archeologist, declared that he had found hominid, or human, fossils in Africa that were 20 million years old—by far the earliest time claimed for the appearance of man's family.

Dental Anatomy a Key

Dr. Simons considers the species cited by Dr. Leakey to be an ape because it "lacks any traces of the dental mechanism which we know to characterize animals related to the lineage of man."

In addition, the two Yale scientists said that the dental anatomy of the jawbones called into question a theory originally suggested by Charles Darwin and recently advocated by Dr. Sherwood Washburn of the University of California at Berkeley.

This theory argues that man's ancestors lost the need for large, stabbing canine teeth because they became toolmakers. Instead, pre-men may have had their canine teeth reduced by virtue of a change in eating habits long before they got around to producing tools.

The specimens were found in the British Museum in London and the Calcutta Museum in India. One was unearthed in the nineteen-thirties south of Rawalpindi, West Pakistan, and the other was found about 1928 by members of the geological survey of India in the northern state of Himachal Pradesh.

According to the two scientists, the specimens have been incorrectly identified until now. They were called Dryopithecus (a genus of extinct apes), when they should have been classified as Ramapithecus.

Controversial Creature

Ramapithecus is a controversial creature whose fossilized jawbone has turned up on the Indian subcontinent and in Africa. According to some scientists, Ramapithecus belongs to the Pongid, or monkey, line of evolution. According to others, he is sufficiently advanced to have traveled the fork toward the human side.

In the opinion of Dr. Simons, the two fossils show that Ramapitecus is in the hominid lineage and is linked to Australopithecus, which lived from 1 to 5 million years ago and is categorized as near-man.

"The two specimens which we have identified this year show functional resemblances to Australopithecus that are so extensive that there is little possibility of their having been more closely related to apes," Dr. Simons stated.

The fossil in the British Museum contains a part of the lower jaw, including roots of some of the front teeth, and is the most complete specimen of Ramapithecus recognized to date. A description of it was published this week by Dr. Pilbeam in Nature, a British scientific journal.

The structure of the specimen, Dr. Pilbeam said, indicated that Ramapithecut had strong back molars with powerful cheek muscles set in a robust jaw and a short face. The front of the jaw contains small canine teeth that instead of projecting forward, as in apes, joined the incisors to form a continuous cutting blade.

This strong dental mechanism—a slicing in front and a grinding in the back—is characteristic of hominids, in contrast to the up-and-down chomping of apes, who feed on soft vegetation.

"We have been led to the conclusion that the reduction of the canines is associated with a complex of changes in the front of the jaw, which in turn are correlated with changes in feeding behavior," Dr. Pilbeam said.

The new feeding behavior, Dr. Simons explained, was foraging for small objects such as roots, berries, nuts, grains and seeds, perhaps by a creature who had come out of the trees to the open ground.

Dr. Simons also pointed out that the lower jaw of Ramapitecus did not contain the ape-like gaps between the teeth to accommodate the overlapping canine teeth.

In the Calcutta specimen he noted a high gradient of decreasing molarware to the back, suggesting that the second molars and wisdom teeth came in much later than the first. This might indicate that Ramapithecus, like later hominids, had a long adolescence and infancy, which has been correlated with greater individual learning.

July 27, 1969

Bone Traces Man Back 5 Million Years

By ROBERT REINHOLD
Special to The New York Times

CAMBRIDGE, Mass., Feb. 18—A fragment of jawbone that appears to push early man back to five and a half million years ago has been discovered in Kenya.

The scholars at Harvard University who made the find say that the fragment represents the oldest member of the human family ever found. It is a million and a half years older than the earliest known specimen to date, an elbow bone found in the same region of Africa in 1965.

The new specimen, the right half of a lower jaw with one molar still in place, was found in deposits on Lothagam Hill in northern Kenya, in an area that has yielded many important finds over the years.

The leader of the Harvard expedition, Prof. Bryan Patterson of the Museum of Comparative Zoology, believes that the bone came

LIFE

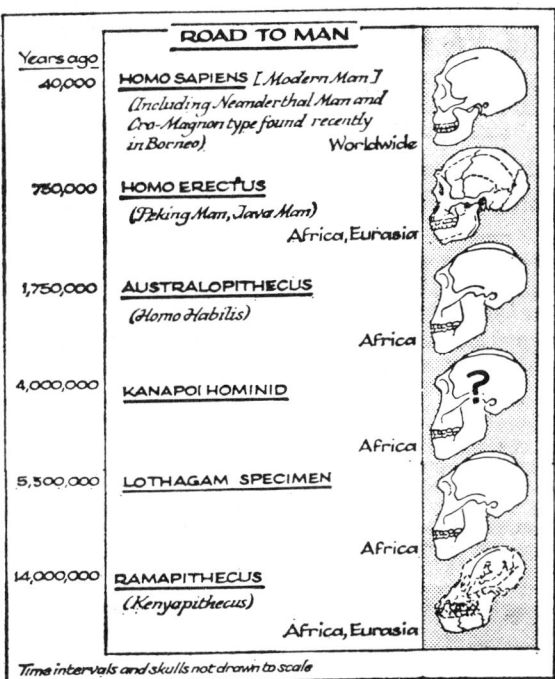

The New York Times — Feb. 19, 1971

Harvard group's find fills in part of the gap between the later forms of Australopithecus and Ramapithecus, a "man-ape" that lived 14 million years ago. Only an arm fragment bone of Kanapoi Hominid has been found.

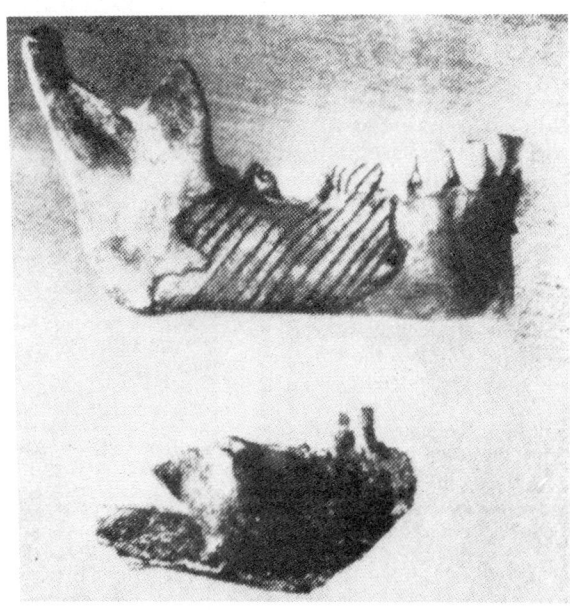

Jawbone fragment found in Kenya is shown below modern jawbone, marked to denote area similar to fragment.

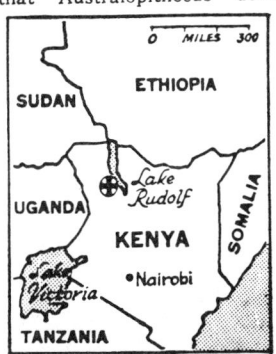

The New York Times — Feb. 19, 1971
Site of the find (cross)

from a creature closely related to Australopithecus, a genus of which many later examples have been found in South Africa and other parts of Africa.

It was Australopithecus, a five-foot high creature that walked upright and had a thick, heavy jawbone, that eventually evolved into Homo sapiens, or modern man.

The find is thought likely to spur some rethinking on a number of fundamental questions about the evolution of man. It suggests, for example, that man was evolving during the Pliocene period, which began about 13 million years ago.

It was previously believed that Australopithecus developed largely during the following period, the Pleistocene, which began about two million years ago and was marked by advancing and receding ice sheets.

The discovery also fills in part of the evolutionary gap between Australopithecus and Ramapithecus, a "man-ape" that lived about 14 million years ago and seems the most likely ancestor of man.

But it still leaves open the question of how that evolution took place and when the "man-apes" developed into hominids with upright posture and ground living habit. The hominids, which consist of modern man and his two immediate predecessors, Homo erectus and Australopithecus, are distinguished from apes in their upright walk, which freed the arms to use tools, and larger brains.

As recently as 1959, anthropologists were astonished when Dr. Louis S. B. Leakey found remains that were 1.75 million years old in Olduvai Gorge in Tanzania. He called that specimen Homo habilis, which many anthropologists classify as Australopithecus.

The Leakey find was three times as old as anything known at that time. The new Harvard discovery is more than three times as old as Dr. Leakey's find. In 1965, the Harvard group found an elbow bone at Kanapoi, Kenya, that was then judged to be an Australopithecus fossil 2.5 million years old. Recent work has shown the bone to be four million years old.

The new specimen was found on a "nice hot afternoon" in 1967 by Arnold D. Lewis, who heads the preparation laboratory at the museum. It was resting on the surface of the hill and Mr. Lewis immediately recognized it as a hominid. But it took about three years of laboratory work and research to establish its date and identity conclusively.

Professor Patterson, who is a vertebrate paleontologist, said this afternoon that he felt "rather pleased with this object" but added that "I wish it was rather more complete than it is."

From the condition of the tooth and the size of the bone, he offered a "guess" that the creature was a relatively mature female. But because no leg bones were found, it was impossible to judge its height or way of life.

The creature shared its habitat with elephants, and many tons of elephant bones were collected with the jaw in the same sediments, which lie between two old volcanic masses on the hill.

Dating of Fossil Bones

The elephant bones played a major role in the dating. Because elephants were evolving very rapidly at that time, it is relatively easy to date fossil bones. This work was performed by Dr. Vincent J. Maglio, a 28-year-old research fellow.

Together with radioactive dating techniques, Dr. Maglio was able to correlate the fossils in the Lothagam sediment to others of known date in the region. The age was placed at 5.5 million years ago.

A number of questions remained. For example, there was not enough bone to determine conclusively if the specimen was Australopithecus africanus or his somewhat larger relative Australopithecus robustus. Professor Patterson believes it was the africanus type.

The find is likely to generate some discussion among anthropologists as to its relationship to Paranthropus, an early hominid that lived at the same time as Australopithecus and had a much more robust jaw. The Harvard group, whose work was supported by the National Science Foundation, believes the new find is not closely related to the Paranthropus.

Australopithecus was first discovered in 1924 in South Africa. Until that time it was thought that man originated in Central Asia. Homo erectus made his appearance about 750,000 years ago in Africa and Eurasia and Homo sapiens emerged about 40,000 years ago.

February 19, 1971

200,000-Year-Old Skull of Man Is Found

By WALTER SULLIVAN

Digging in congealed sand in the floor of a Pyrenees cave, French paleontologists have found bony remains of a face that looked on the world some 200,000 years ago. It is like no face previously known.

While a few fragments of skull cap and jawbone have been found from the long period that ended with the emergence of Europe's Neanderthal Man, some 90,000 years ago, this find—an almost complete skull—is said to be the first clue to the facial appearance of the men who inhabited Europe during the 200,000 years that preceded the Neanderthal.

Little is known of this period, which was critical in the evolution of modern man. Stone implements found at sites from Britain to the Middle East have made it possible to reconstruct, to some extent, the development of toolmaking during this period.

A jawbone found near Heidelberg, in Germany, is thought by some paleontologists to be 400,000 years old, but no reliable dating has been possible. Without any complete skulls, it has been impossible to decide whether several races or just one, inhabited Europe during this prolonged period.

The Pyrenees skull is expected to help resolve this uncertainty.

The skull, which is still partly caked with sand, has massive eyebrow ridges, a remarkably flat forehead, more horizontal than vertical, and a narrow, elongated brain case. Yet despite his small brain, this creature was a potent hunter.

Surrounding the skull were numerous rhinoceros and horse bones as well as bones from bear, panther, a form of elephant, turtle, deer, a big archaic cow, wolf, rabbit and birds. Some bones had been broken and were clearly relics of ancient feasts. Yet these hunters had only the most primitive weapons — crude, stone axes and roughly chipped quartz points.

More than 100,000 pieces of worked stone have so far been found in the cave. Those im-

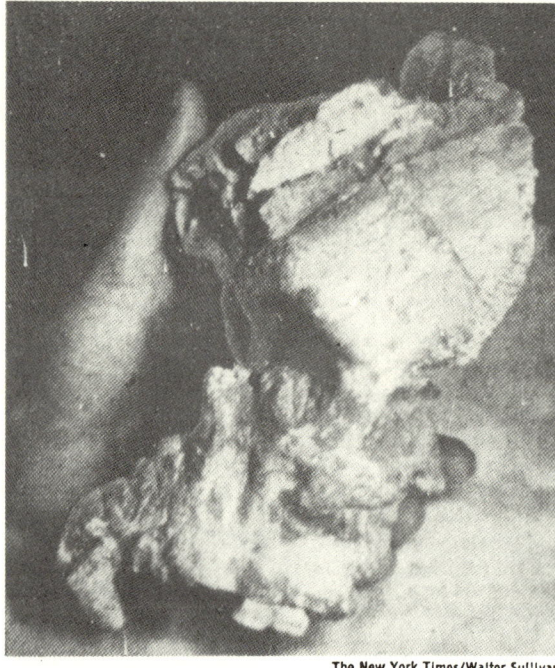

Remains found in Pyrenees. Diagram identifies features.

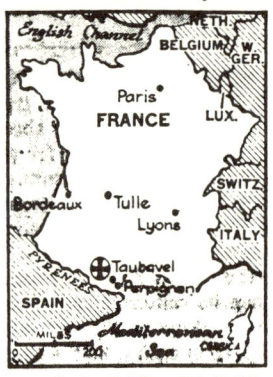

plements associated with the skull are classed by the French archaeologists as belonging to the so-called Tayacian culture.

Neanderthal Features

The skull has features that anticipate those of the Neanderthal and other characteristics reminiscent of the far more primitive Pithecanthropus, represented by Java Man and Peking Man, who lived more than 400,000 years ago.

However, its discoverers are inclined to believe that the skull will show that the European Neanderthal evolved quite independently from his contemporaries in Africa and Asia.

The skull was found this summer in a cave above the village of Tautavel, near Perpignan, in the eastern Pyrenees. This discovery was made by a group working under Henry de Lumley, a geologist at the Faculty of Sciences of the University of Aix-Marseilles, and his wife, Marie-Antoinette de Lumley, paleontologists at the same university.

Last July 22, systematic excavation of the cave disclosed two massive teeth—twice as large as those of modern man — protruding from the sand. Careful digging showed the teeth to be attached to a largely intact skull, which has now been extracted and taken to Marseilles.

Laboratory Work Set

At the end of this month it will be transferred to the laboratory of Jean Piveteau, professor of human archaeology at the University of Paris, who is an internationally known authority on early man. In recent interviews, Professor Piveteau and Mrs. de Lumley said there was no doubt that the skull dated from some 200,000 years ago.

That was the start of the next-to-last ice age, known in Europe as the Riss Glaciation. Excavation of the cave floor has penetrated a succession of layers reaching further and further into the past. Each period could be identified from the animal bones, tools, pollen grains and other indicators of climate and culture.

It was this layering that helped to date the skull. The cave is 15 feet deep and 33 feet wide at the maximum. As early as 1838 the finding there of "antediluvian" animal bones was reported. In 1964 the excavation began and it took three summers to remove a shepherd's shelter and other debris to reach the uppermost level of human relics.

Climate Was Cold

The level most recently excavated was laid down during a period of dry, cold climate when the cave was occupied only intermittently. When it was abandoned, sand blew in, covering the human relics and providing a clean floor for the next occupants.

More than 20 habitation levels were laid down during the period represented by the newly found skull.

During the summers of 1969 and 1970 parts of two human mandibles, or jawbones, were found at approximately the same level. Neither belongs to the newly found skull, which lacks its lower jaw, but they show that this creature was lantern-jawed and lacked the chin-knob characteristic of modern man.

A favorite indicator of primitiveness among paleontologists is thickness of that section of the lower jaw that supports the teeth. In one of the jaws from the Pyrenees cave, the thickness is 25 millimeters (almost one inch)—considerably thicker than the presumably far older Heidelberg jaw.

This could indicate that a different race inhabited the region.

One of the two mandibles is apparently that of a woman and the other, which is heavier, that of a man. The skull is thought to be that of a youth aged about 20 (deduced from the absence of tooth wear). What is striking about the two mandibles, according to Mrs. de Lumley, is the sharp structural difference between the sexes.

Whereas the Pithecanthropus skull was marked by a central ridge running back over the crown from the forehead, this skull seems to have a slight valley along the same route. It is extremely prognathous. That is, the jaws protruded far in front of the upper face.

October 13, 1971

Skull Pushes Back Man's Origin

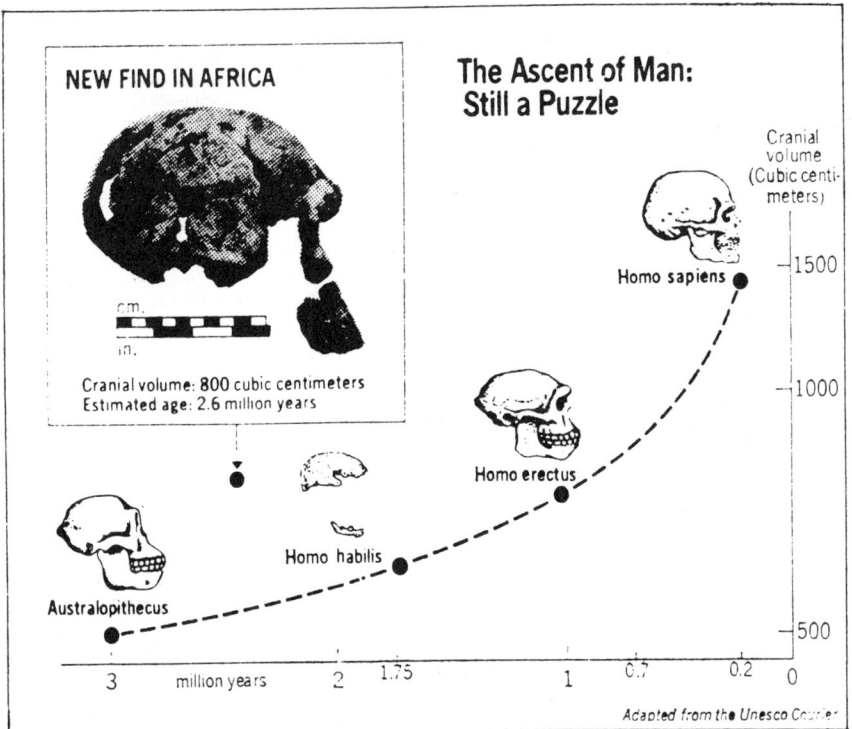

The gradual evolution of man's brain has been represented as a progression from the skull of Australopithecus to Homo sapiens. This progression is challenged by the recent discovery (inset) of an ancient skull near Lake Rudolf in Africa.

By WALTER SULLIVAN

Ancient bones found in Africa have been assembled into a skull that may extend man's immediate ancestry back more than one million years earlier than previously believed.

The fragments, making up a skull with striking resemblances to that of modern man, were found in a layer of material that had been deposited about 2.6 million years ago.

Richard Leakey, co-leader of the expedition that found the bones, said the skull seemed to displace two other man-like creatures widely thought to represent the early stages in man's development.

One of them, a beetle-browed type known as Homo erectus, lived far more recently—about a million years ago—yet is less like modern man than the newly found skull.

The other reputed ancestor, Australopithecus, an ape-like "man" that walked relatively erect, lived 2.5 to 3 million years ago. It now appears to have been a contemporary of the more modern-looking type, rather than ancestral to the men of today.

"There is now clear evidence that in eastern Africa, a truly upright and bipedal form of the genus Homo existed contemporaneously with Australopithecus more than 2.5 million years ago," Mr. Leakey said.

The new find was made in a desert region on the east side of Lake Rudolph, one of the chain of elongated lakes within the rifts of East Africa. In the last two years the site has produced a rich harvest of stone tools, considered to be the oldest known artifacts of man.

According to Dr. J. Desmond Clark, professor of anthropology at the University of California at Berkeley, whose former student, Dr. Glynn Isaac,

The New York Times/Nov. 10, 1972

was co-leader of the expedition with Mr. Leakey, the tools lay within a layer of congealed volcanic ash, or tuff, known to be about 2.6 million years old.

Dr. Isaac is in Kenya, but Dr. Clark was reached by telephone in Berkeley. The skull, he said, was originally buried by this ash layer and so must be at least as old.

The skull had been crushed to hundreds of fragments that were found, eroding on a slope, and have been pieced together by Dr. Maeve Leakey, Richard Leakey's wife. Mr. Leakey is the son of the late Louis S. B. Leakey, pioneer collector, in this region, of fossil remnants of man's earliest ancestors.

The find was announced yesterday by the National Geographic Society, which, with the National Science Foundation and the National Museums of Kenya, provided support for the Lake Rudolf excavations. It quoted Mr. Leakey as saying:

"While the skull is different from our own species, Homo sapiens, it is also different from all other known forms of early man and thus does not fit into any of the presently held theories of human evolution."

Cranial Volume Large

Mr. Leakey, who is administrative director of the National Museums of Kenya, said the cranial volume of this skull was large, indicating a brain size of 800 cubic centimeters.

By contrast the cranial volume of this early man's contemporary, Australopithecus, was less than 500 cubic centimeters, he said. That of modern

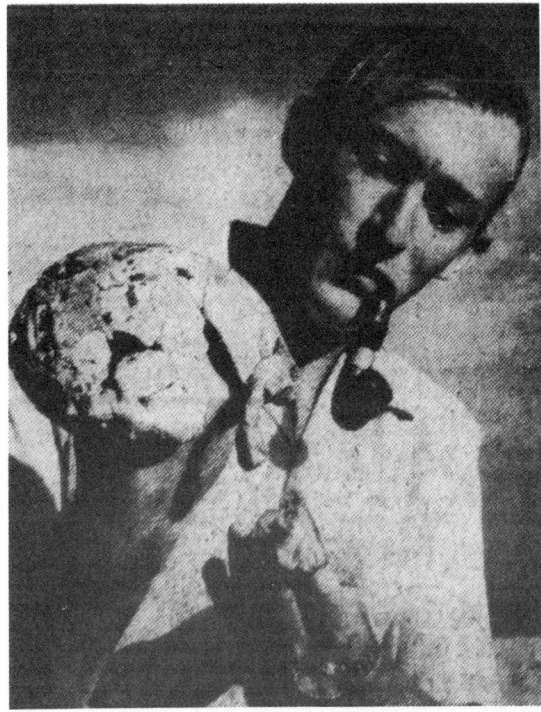

©National Geographic Society/Bob Campbell
Richard Leakey with the skull, found in fragments

man is some 1,500 cubic centimeters.

Unresolved is the relationship of the new find to Homo habilis, whose skull fragments were found in Olduvai Gorge, 500 miles south of the Lake Rudolf site, in a layer 1.75 million years old. The hand bones of this species suggested a dexterity approaching that of modern man, and Dr. Clark suspects the new find may be an earlier form of Homo habilis.

However, the brain chamber of the latter was only about 656 cubic centimeters, compared to Mr. Leakey's estimate of 800 for the newly pieced-together skull. The subsequent Homo erectus, represented by Java Man, Peking Man and various forms in Europe and Africa, had a brain volume from 935 to 1,225 cubic centimeters.

Whereas Homo erectus had the extremely heavy brow bone that continued to be typical of the early cave dwellers of Europe, as well as a thick jaw bone, these features seem to be far less prominent in the new, though much earlier, skull.

"The whole shape of the brain case is remarkably reminiscent of modern man, lacking the heavy and protruding eyebrow ridges and thick bone that are characteristic of Homo Erectus known from young deposits in both Africa and Asia," Mr. Leakey said.

Thigh Bones Found

Also found at the site were two intact thigh bones from other individuals as well as parts of the lower leg (tibula and fibula). They indicate that at this stage man had already graduated from the stooped, loping gait inherited from his arboreal ancestors. Previous studies, Mr. Leakey said, had led to the belief that this graduation occurred much later.

The leg specimens, he added, "have astounded anatomists and other scientists because they are practically indistinguishable from the same bones of modern man."

Dating of the African finds has been done by three methods. In one, a sample of volcanic debris or other material associated with the find is analyzed to determine the extent to which a radioactive form of potassium has decayed into argon gas. This indicates the time since the material erupted at high temperature.

Another method measures how long a specimen containing uranium has been in existence through the number of tracks left in it by the radioactive decay of that uranium.

The third technique uses the known timetable of reversals in the earth's magnetic field over the last few million years as a time scale. When lava cools, after eruption, it becomes imprinted with the direction of the earth's magnetism at the time of cooling. Under favorable circumstances determination of the magnetism in a specimen can be fitted to the timetable of field reversals to obtain the age of a lava flow associated with human remains.

Specimens from the African rift valleys have been dated with particular confidence because volcanic activity there has been frequent, burying the specimens in layers that could be dated by one or all three of these methods.

November 10, 1972

Fossils in Ethiopia Said to Show Man As Million Years Older Than Believed

By Reuters

ADDIS ABABA, Ethiopia, Oct. 25—Anthropologists said today that they had found fossilized human remains three million to four million years old that, they predicted, would revolutionize thinking on the origins of man.

Members of a joint American-French-Ethiopian expedition held a news conference to show parts of jawbones discovered this month in the central region of Ethiopia, near the Awash River.

Preliminary dating indicates the fossils may be as much as 1.5 million years older than those discovered by the American anthropologist Richard Leakey on the shores of Lake Rudolf, in Kenya. These were said to be the oldest relics of humans.

The latest finds consist of a complete upper jaw with all the teeth in place, half of an upper jaw and half of a lower jaw, both with teeth.

The expedition was led by Dr. Karl Johanson, an anthropologist from Case Western Reserve University and the Cleveland Museum of Natural History, and Dr. Maurice Taieb of the French Scientific Research Center.

A statement by the expedition said: "These specimens clearly exhibit traits which must be considered as indicative of the genus homo. Taken together they represent the most complete remains of this genus from anywhere in the world at a very ancient time. All previous theories of the origin of the lineage which lead to modern man must now be totally revised. We must throw out many existing theories and consider the possibility that man's origins go back to well over four million years."

The statement recalled that discoveries from Lake Rudolf and from Olduvai Gorge, in Tanzania—where Mr. Leakey's father, the late Dr. Louis S. B. Leakey, the archeologist, made historic fossil finds—had taken man's earliest remains back to over two million years. Two years ago Richard Leakey presented a manlike skull from Lake Rudolf that he said was 2.6 million years old.

The finds in Ethiopia are "perhaps the most provocative human fossils ever discovered on the African Continent," the statement said, adding, "It is certain that anthropologists from all over the world will meet these discoveries with extreme controversy and amazement."

It said the location of the finds suggested the "revolutionary postulate" that human origins lay outside Africa— Richard Leakey maintains that Africa was the cradle of humankind—but it conceded that this idea would be greeted with extreme skepticism.

Wealth of Recent Finds

If further study confirms that the fossils found in Ethiopia are indeed what their discoverers have asserted, they should prove to be of considerable significance in understanding how man evolved.

In recent years scientists have recovered a wealth of fossil bones of hominids (manlike creatures, including man) from half a dozen areas in Tanzania, Kenya and Ethiopia, multiplying several fold the quantity in hand only five years ago. Among the most productive areas has been a region east of Lake Rudolf explored under the direction of Richard Leakey, who is a museum director in Kenya.

Mr. Leakey's finds had already suggested that two million to three million years ago there were at least two and perhaps three or four distinct hominid lineages. His most provocative find, made two years ago, was a skull of nearly modern proportions uncovered in sediments below a volcanic ash layer dated at 2.6 million years of age. The skull itself may be closer to three million years old.

That find, once again extending man's antiquity, as have a succession of fossil discoveries in recent years, had already led many anthropologists to suspend the assignment of a date for man's origin.

Whether the new fossils will be accepted as further extending man's antiquity will depend on agreement by other specialists that the jawbones and teeth are sufficiently like modern man's and the more precise determination of their age. Both are points that frequently evoke substantial controversy among physical anthropologists.

The reported assertion that the discovery suggests that man arose outside Africa cannot be assessed without more information. Most experts contend that the fossil studied so far makes a strong case for man's having evolved in Africa.

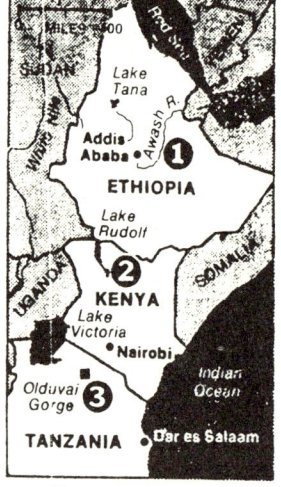

The New York Times/Oct. 26, 1974
Fossils were found near an Ethiopian river (1). Earlier finds were at a lake in Kenya (2) and a Tanzanian gorge (3).

October 26, 1974

Fossil of Giant Winged Reptile Is Found

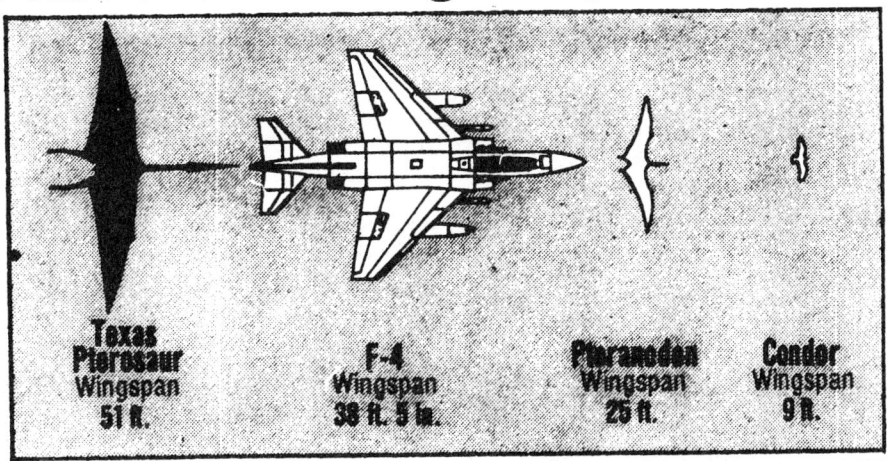

The New York Times/March 12, 1975

By BOYCE RENSBERGER

The largest known creature ever to have flown, an extinct winged reptile with an estimated wingspan of 51 feet—the length of an IRT subway car—has been discovered by fossil hunters in Texas.

The animal, which lived something more than 60 million years ago, had twice the wingspan of the biggest previously known pterodactyl, or winged reptile, and nearly six times the wingspan of the condor, the largest flying bird now alive.

The estimated size of the creature is derived from calculations based on the sizes of many fragmentary and some complete bones found in excavations during the last three years in Big Bend National Park in Brewster County, Texas.

Announcement of the discovery, in the March 14 issue of Science, is expected to rekindle a half-century-old debate among paleontologists over whether flying reptiles flapped their featherless, leathery wings like birds or merely climbed up onto high perches and leaped into the air currents to soar like gliders.

One scientist familiar with the discovery said that the great size of the newly found creature might make it seem improbable to some experts that it was able to rise into the air under wing power alone. He noted, however, that because no one had any reliable estimate of how much winged reptiles weighed, it was virtually impossible to calculate their aerodynamic properties.

The fossils were found by Douglas A. Lawson, a graduate student at the University of California at Berkeley, who began searching in the Big Bend area while a student at the University of Texas at Austin.

The New York Times/March 12, 1975

Mr. Lawson's continuing explorations and study of the fossils are being carried out under auspices of the university's Vertebrate Paleontology Laboratory.

Although the reptile clearly represents a new species, it has not yet been given a formal scientific name. There are many known species of flying reptiles. Scientists generally refer to all as pterosaurs, but the popular name pterodactyl is also considered correct. All are now extinct.

"The thing that's so extraordinary about this thing is its tremendous size," said Dr. Wann Langston Jr., director of the Vertebrate Paleontology Laboratory. "There's never been anything like this before."

Heretofore, the largest known pterosaur had been the species Pteranodon with a wingspread of about 25 feet. The largest flying creature alive today is the condor, a bird with a wingspan of about nine feet.

In his report Mr. Lawson said he had discovered the partial skeletons of three of the large pterosaurs, including the remains of four wings, a neck, the hind legs (forelimbs with claws frequently are part of the wing structure) and jaws. As is typical of pterosaurs, the jaws were toothless.

Dr. Langston, who is Mr. Lawson's academic supervisor, said in an interview that several more skeletons had been discovered and that Mr. Lawson was now on the way to resume their excavation.

Unlike most previously known Pterosaurs, the Big Bend creature was found in nonmarine sediments, suggesting that its habitat was away from oceans. Most pterosaurs are considered to have been fish eaters, scooping up their prey while gliding over the waves.

The Big Bend fossils were found in freshwater sediments far from the oceans of that time. In his report, Mr. Lawson suggested that the animal may have been a carrion-eater, feeding on dead dinosaurs much as the condors and other vultures of today consume dead animals.

This speculation is based on the big pterosaur's unusually long neck, an adaptation common to carrion-eating birds that enables the animal to reach deeply within a dead body to feed.

"This pterosaur," Mr. Lawson wrote, "had a neck long enough to probe a dinosaur carcass."

Whether the Big Bend pterosaur flew like a bird may never be answered satisfactorily, Dr. Langston said. He noted, however, that one factor in the speculation about the creature would be that a reconstruction of the topography of the fossil site indicates there were no mountains or high rocks for 20 to 30 miles around when it lived.

If the animal inhabited such a region, it probably would have been able to take off from level ground. It is possible, however, that the skeletons were washed far downstream from mountainous areas suitable for gliding where the reptiles may have lived and died.

March 12, 1975

East Africa Fossils Suggest That Man Is a Million Years Older Than He Thinks

By BOYCE RENSBERGER

Fossil remains of early man discovered over the last two years in East Africa are causing a major upheaval in the study of human evolution by suggesting that man's origins lie more than twice as far into the past as had been supposed from earlier evidence.

The new evidence is beginning to suggest that relatively intelligent human beings originated at least three million years ago in the plains of Africa, living in complex societies and inventing stone tools. Hitherto the earliest evidence of man was about 1.75 million years old.

The latest fossils also indicate that the early ancestors of modern man may not have been the only manlike creatures alive at that time. It now appears that early man coexisted with at least two and perhaps three or more other species of "near man" whose physical appearance may have been largely human but whose brains had remained apelike.

Until recently these "near men," called Australopithecus, were almost universally accepted as ancestors of modern man. Now there is evidence that they were merely smaller-brained contemporaries who died out as their larger-brained cousins continued to evolve.

A Major Rethinking

Some authorities also believe that for early man to have attained the level of advancement inferred from certain three-million-year-old fossils and stone tools found over the last two years in Kenya and Ethiopia, the human lineage must have begun sometime between four and six million years ago.

These emerging views of man's evolution and the fossils upon which they are based are forcing a major rethinking of what was once considered a fairly straightforward interpretation of human origins.

Although these new views are not widely accepted among paleoanthropologists, who study human evolution, they constitute a revolution in the effort to learn where man came from. The more conservative experts who do not yet accept the new views concede only that the latest fossils, which are significantly different from previously known finds, have thrown the field into a disarray that may take years to sort out.

Some paleoanthropologists, for example, insist that two forms of early manlike creature could not have coexisted for long because competition would have quickly eliminated one or the other.

One of the basic difficulties now confronting paleoanthro-

Sarah Landry, for "Sociobiology," by Edward O. Wilson/Harvard University Press

Artist's conception of Australopithecus hunters.

Finds by Mr. Leakey and others by Dr. Johanson occasion restudy of whether man evolved from these or from one of several types.

pologists is the lack of wide agreement as to what physical traits qualify a creature to be considered human. Where, in other words, is the fine line between a creature that is "almost human" and its descendant or contemporary that is "truly man?"

All manlike creatures, whether ancestral to modern man or not, are called hominids. "True man," given the genus name Homo, is considered but one branch of the hominid family tree.

Some authorities regard brain size as the key variable. Others stress tooth shape and size or erect posture. Because individual fossil finds almost never include evidence for all these traits, the decision is often made on very sketchy evidence or, increasingly these days, the decision is suspended pending further fossil discoveries.

The crucial issue that paleoanthropologists are grappling with now is an anatomist's version of the philosopher's ancient question: What is man?

At a recent meeting of fossil man experts in London, Dr. F. Clark Howell of the University of California at Berkeley said many new fossil finds possessed such perplexing anatomic features that it was almost impossible to put them into any coherent evolutionary sequence.

New Evolution Concept

"We ought to throw away a lot of ideas and try to start anew," said Dr. Howell, who has led numerous fossil-hunting expeditions in East Africa. "And then we might be able to understand a little more about human origins. In due course we might get a very, very shadowy picture about human emergence and what evolution was all about."

The current reassessment of man's evolution began in the spring of 1973 when Richard Leakey, a Kenyan scientist, announced the discovery of a 2.9-million-year-old form of man with a brain that was unexpectedly large for its antiquity. It has come to be known from its catalog number in the expedition's records as "1470."

That announcement, which is forcing a re-evaluation of whether Australopithecus was an ancestor of man, is leading to the view that the human race did not evolve simply from an isolated, linear succession of steadily advancing hominids. Rather, it seems more likely that modern man represents but one surviving lineage from among several lines of pre-human or nearly human creatures.

It is now generally agreed that there were at least two types of Australopithecus, a large "robust" form and a smaller "gracile" form. In addition, several of the most recently discovered hominid fossils, including some that have not yet been described in the scientific literature, seem not to belong to any of the known or hypothesized lineages and may represent still other near-man types.

To grasp the complexity of the current welter of fossils, it is necessary to understand something of what had been the accepted story of human evolution. Following is what most experts in the field used to think and what virtually all texts still repeat:

The evolution of man away from a hypothetical ancestral apelike creature is thought to have begun sometime between eight and 20 million years ago, during what is known as the Miocene period.

The main evidence for this

The Cleveland Museum of Natural History

Richard Leakey's discovery of the skull above, called "1470," suggests that man's development was not as has been thought. At left, Dr. Donald Carl Johanson shows most nearly complete fossil skeleton found, perhaps of Australopithecus. Squares on map note sites of major fossil finds.

consists of only a few jaws with teeth that are between 10 and 14 million years old. The only clue they give that the hominid lineage may have begun to separate from the ape line is that the teeth have certain shapes found only in modern human beings and in extinct undisputed hominids. This creature, which not all experts accept as hominid, is called Ramapithecus.

From Ramapithecus until the earliest Australopithecus appears about four to five million years ago, there are no fossils yet discovered that suggest what directions hominid evolution took.

However, from fossils dated at between five million and one million years of age, scientists have recovered the remains of hundreds of hominid, including skulls and body bones in East Africa and South Africa. Nearly all of these are called Australopithecus and had been considered a crucial transition species incorporating traits of both ape and man.

Most experts accept that there was one small species of Australopithecus, known as A. africanus, and possibly two large species known as A. robustus and A. boiseii.

Australopithecus is the creature Robert Ardrey made famous in his book, "African Genesis," which popularized the idea that these animals were "killer apes" from which modern man inherited a supposed propensity for violence. This view has never enjoyed much support among fossil experts.

From the bones it is known that Australopithecus "man-apes" had short-statured bodies proportioned very much like those of modern man, walked erect or nearly so, had teeth adapted for a mixed diet of meat and vegetation, but had a brain of only 450 cubic centimeters, approximating that of chimpanzees. In modern human beings brain size ranges from 1,200 to 1,800 cubic centimeters.

Sharing of Knowledge

The discovery and acceptance of Australopithecus as an ancestor of man forced the view, now being questioned, that man's brain and body did not evolve simultaneously. Rather, it was held, erect posture and the use of the hands came first, followed by the big brain.

From Australopithecus, it has generally been believed, hominids evolved into the first "true men" between one million and 1.5 million years ago. Fossils of this new creature, called Homo erectus, have been found in Africa, Asia and Europe, and the brain size has ranged from 800 to 1,000 cubic centimeters.

One major variation from this view was the opinion of Louis Leakey that some of the creatures that others called Australopithecus were, in fact, more advanced creatures with estimated brain sizes well into the 600 cc range.

Dr. Leakey called these 1.8 million-year-old creatures, which were found at Olduvai Gorge in Tanzania with a variety of stone tools, Homo habilis and insisted they were the first true human beings.

The existence of a "habilis" species distinct from the lineage of small Australopithecines has not been widely accepted.

Peking Man, who lived about 500,000 years ago, is now considered a Homo erectus although it was originally named Sinanthropus pekinensis. This creature and others from Asia known as Pithecanthropus erectus are now all called Homo erectus.

Branches of Evolution

This is the first stage of hominid evolution in which man is thought to have domesticated fire, built shelters and migrated out of Africa.

From Homo erectus, it is believed that Homo sapiens with a still larger brain and different skull shape evolved perhaps 200,000 years ago. Neanderthal Man was one form of early Homo sapiens who either died out or evolved into modern man, Homo sapiens sapiens, about 50,000 years ago.

Though the beginning and end of this evolutionary sequence are still generally accepted, the middle — dealing with Australopithecus and the emergence of the first true human beings ancestral to modern people—is now very much in doubt.

There is growing evidence that Australopithecus may not have been in man's ancestral sequence at all but rather only a contemporary of true man who may have come into being before nearly all the known Australopithecines lived. Some experts still hold that a hypothetical very early form of Australopithecus seems reasonable as the ancestor of an early Homo, scientific name for man.

The first suggestion of this revised picture came with Richard Leakey's announcement in early 1973 of a fossil hominid skull, found in northern Kenya, that had a brain size approaching that of Homo erectus but which lived as much as two million years earlier. Mr. Leakey is the 31-year-old son of Louis and Mary Leakey, pioneers in the study of man's evolution in Africa.

A Controversial Find

That skull, known as 1470, is still one of the most controversial and unsettling of hominid fossils.

"I still don't know what to make of 1470," said Dr. Howell, who has led numerous fossil-hunting expeditions in East Africa. "It isn't like any of the other fossils we've seen. If we had found 1470 in younger strata, we wouldn't have hesitated to call it Homo erectus."

Dr. Howell, who is regarded as one of the leading authorities on interpreting early man fossils, said that if it is accepted that the 1470 skull and two thigh bones of remarkably human proportions, also found by Richard Leakey, have been accurately dated, "it changes the whole picture of man's evolution."

When it first was announced, the 1470 skull was an isolated exception to the accepted sequence of evolution. Since then, however, several additional finds by Richard Leakey's expedition based on the eastern shore of Kenya's Lake Rudolf, have supported similar interpretations. Some of the finds have not yet been fully described in the scientific literature but have been seen by a number of experts.

Because the brain size of the 1470 skull was about 800 cubic centimeters, far surpassing that of the Australopithecus specimens, the vast majority of which lived later than 1470, Richard Leakey and other experts are suggesting that the smaller-brained hominids are less likely to have been the ancestors of modern human beings than were creatures like 1470.

Because of its brain size and the relatively modern-looking shape of the skull, Richard Leakey has called 1470 "Homo" —meaning he regards it as true man—but declined to say whether it should be considered a full-fledged "erectus" or a member of some more primitive species of Homo such as "habilis."

Dr. Leakey's Views

Although "habilis," meaning handy or able, has never been fully accepted among anthropologists as a distinct species, it conformed to Dr. Leakey's long-standing view that man arose as a distinct lineage very long ago and evolved independently of the Australopithecus "near men."

That view, long discounted by many other anthropologists, is now gaining support largely as a result of the findings of his son, Richard.

Young Leakey has speculated that the brain sizes of some of the "habilis" specimens which are fragmentary, have been underestimated and that they may belong to the same group as 1470.

One of the more spectacular fossil discoveries in recent months has been that of a partial skeleton recovered from deposits in Ethiopia believed to be between 3.0 and 3.5 million years old.

With some complete limb bones and fragments of the skull, jaw and other limb bones, the find comprises about 40 per cent of a complete skeleton and is easily the most nearly complete example of an extinct hominid yet found.

The skeleton was found by Dr. Donald Carl Johanson in the Afar region of Ethiopia near the Hadar River. He is a professor of anthropology at Case-Western Reserve University and curator of physical anthropology at the Cleveland Museum of Natural History.

Dr. Johanson has calculated that Lucy, as the skeleton is named, was a young adult female who stood about three-and-a-half to four feet tall. He and others agree that the skeleton looks like a small version of Australopithecus, which is generally considered to have stood about four-and-a-half feet tall.

While the discovery is spectacular in its relative completeness, it does not prompt any major change in thinking about human evolution. More startling was a 1973 discovery by Dr. Johanson of a knee joint that appears "very human" but which may be as much as four million years old.

A Startling Discovery

Dr. Johanson considers the shape of the ancient knee bones to indicate an upright posture, thus making the fossil the oldest evidence for the appearance of a key human trait. Without clues to of the creature's brain size or other crucial anatomic features, however, one can only speculate about whether the knee's original possessor merited the designation Homo.

While most of the new fossils from East Africa can be fairly well classified as either the large or small species of Australopitheous or Homo, a few cannot. One of the new skulls from East Rudolf, for example, has small teeth that would be considered Homo but a brain size that is more like Australopithecus. Another has a relatively large brain but, paradoxically, the heavy lower jaw and a crest atop the skull typical of the robust Australopithecus.

Years ago either of the fossils would have prompted its finder to announce the discovery of a new species and the find would be given a new scientific name. Now such seemingly anomalous fossils are, in Richard Leakey's words, "kept in the cupboard" until further finds make them easier to interpret.

Although only the briefest announcement of such a fossil is made in the scientific journals, the originals or casts are usually made available to other scientists for study. This practice is in marked contrast to the fierce competitiveness that prevailed among paleoanthropologists until less than a decade ago.

"Scientists are cooperating for the first time in the study of early man," Mr. Leakey said recently, "and I think this indicates a welcome degree of maturity in our field."

Although there are several fossils that cannot yet be explained, a few anthropologists have attempted to construct a hypothetical hominid evolutionary tree that takes into account the latest, well-described finds.

Man first arose, according to this tentative view, about four to six million years ago from a parent stock of hominids that may have been the hypothetical proto-Australopithecus. It was during this same period that many of the modern mammalian species arose from their now-extinct, more primitive ancestors.

This parent stock, the new view suggests, evolved into several lineages. By about three million years ago, it seems clear that there were at least three species of hominid living in East Africa. These would include the large and small Australopithecines and an early Homo of the 1470 type. Some experts believe there were two large australopithecines.

Although the australopithecines probably walked on two legs and, from the neck down were virtually human, their brain size seems to have remained small over a period of perhaps two million years. Homo, on the other hand, had a similarly advanced body but also a brain that was larger and still enlarging.

Eventually, it seems likely, the superior competitive ability of a steadily advancing Homo lineage pushed the australopithecines into extinction. Man became dominant on the earth.

This view, which does not explain several key fossils and which makes some assumptions that cannot yet be supported, is undoubtedly a vast oversimplification. However, given the rate at which scientists are turning up new fossils in East Africa, spurred by the earlier findings of the Leakey family in the nineteen-sixties, some experts are hopeful that a new, more accurate picture of man's evolution will appear.

"It seems clear," Richard Leakey said, "that within the next 5 to 10 years man will be able to look back, as a species, and see where he came from."

April 12, 1975

Sociobiology: Updating Darwin on Behavior

By BOYCE RENSBERGER

The tightly organized societies of bees and ants, the mating rituals of birds, the hunting tactics of lion prides, the social hierarchies of monkey troops—these and dozens of other examples of animal behavior have long fascinated people. But they have rarely been offered as anything more than intriguing evidence for the remarkable variety of nature.

Lately, however, some of the biologists who study animal ways have come to believe that their findings point to a far more profound conclusion. Beneath the superficial aspects of social behavior that vary so across the animal kingdom, the scientists assert, there lie common behavioral patterns governed by the genes and shaped by Darwinian evolution.

This belief—the product of a new field of scientific inquiry called sociobiology — carries with it the revolutionary implication that much of man's behavior toward his fellows, ranging from aggressive impulses to humanitarian inspirations, may be as much a product of evolution as is the structure of the hand or the size of the brain.

That implication is likely to be rejected by many students of humanity, especially psychologists and sociologists who have long held that man's behavior arises almost entirely from unique intellectual and emotional capacities.

Drawing shows dolphins helping wounded comrade, upper center, with harpoon protruding from tail, to swim to the surface and breathe—otherwise it would drown. The drawing, by Sarah Landry, is for a book on sociobiology by Dr. Edward O. Wilson, right.

But a growing number of scientists participating in the study of sociobiology are convinced that the ideas now emerging from their work are so powerful that every rational effort to understand the ways of man must someday take them into account.

The greatest impact outside of animal research is expected to be on sociology and psychology, which, some forecast, will eventually be forced to modify their theories and practices.

Sociobiology is the study of the biological basis for social behavior in every species from the lowliest ameba colony to modern human society. It seeks to explain the origin of that behavior in terms of how it improves an individual's or a society's fitness to survive.

Sociobiology's key contribution is the integration of Darwinian theory with the observations of animal behavior research, which have largely ben descriptive.

Much of the raw data has come from ethologists such as Konrad Lorenz and Hiko Tinbergen who study communication systems among animals and from skilled naturalists and biologists such as Jane Goodall and George Schaller who have devoted years to detailed observations of wild animal groups.

Definitive Book Due

Although the nascent field of sociobiology has been around for a few years, it is only now beginning to draw much attention from outside its own realm because a long-awaited definitive book on the subject is about to be published.

The volume, "Sociobiology, The New Synthesis" by Edward O. Wilson, a widely recognized zoologist at Harvard University, is not due to be published until late June, but word of its preparation has spread far among biologists. Among those scientists who have seen advance copies of the book, which comprises 700 oversize pages, it has already stimulated considerable excitement.

Perhaps the most startling assertions of sociobiology have to do with the origins of altruism, a type of social behavior once regarded as uniquely human but which biologists have observed in many animal species, even down to microorganisms.

In sociobiology, altruism is defined as any self-sacrificing behavior that benefits another individual.

For example, when a predator breaks into a nest of termites or ants, members of the colony's "soldier caste" instinctively rush to place themselves between the intruder and the rest of the colony.

Altruism in Animals

Like virtually all insect behavior, this act is believed to be governed by inherited instinct. Because soldiers often lose their lives defending the colony, however, students of evolution from Charles Darwin to now have regarded this and other altruistic acts among animals as puzzles.

How, they have asked, could behavior that reduces an individual's chances for survival have evolved if natural selection favors only traits that improve the ability to survive? Any animal exhibiting such risky behavior would be expected to die out while its more selfish fellows prospered.

The question has been raised about many other examples of self-sacrificing behavior among animals.

Several species of birds and mammals will give warning signals to others of their kind if they see predators approaching. In the process they draw the predator's attention to themselves, diminishing their own chances of survival.

Among birds, such as the Florida scrub jay, yearling offspring commonly remain with their parents for two or three more years to help feed the hatchlings of succeeding seasons.

That parents should help their offspring fits conventional evolutionary theories because they are ensuring the survival of those directly inheriting their own genes. The puzzle is why brothers and sisters should tax themselves to help those who are not their direct descendents.

Growing Following

Sociobiologists believe their theories give the answer. In science, theories gain credence by explaining puzzles. Much of sociobiology's growing following can be attributed to the fact that its theories help explain several previously baffling phenomena.

The central theory of sociobiology is that the social behavior of individuals evolves so as to maximize the chances of genes like the individual's own to survive in the greatest number.

The theory is based on the common observation that the more closely related individuals are, the more likely one is to act altruistically in the other's behalf. Geneticists know that the closer the relationship, the more genes the individuals have in common.

Thus, when a soldier ant sacrifices its life for other members of its colony—actually members of its own family—it is improving the chances that genes like many of its own will survive. The creatures who risk their lives to warn their families of danger or who help rear siblings do so for the same reason.

The scientists are not suggesting that animals consciously decide to help their relatives because they know anything about genetics. Rather, they believe the animals have inherited genes that produce the behavior automatically whenever environmental circumstances are right.

Evolution has favored such genes because, although some individuals are sacrificed, the sacrifice is advantageous to the society as a whole and the society has the capability to reproduce the lost individuals' genes many times over.

A Temporary Carrier

"In a Darwinist sense," Dr. Wilson wrote in his new book, the organism does not live for itself. Its primary function is not even to reproduce other organisms; it reproduces genes and serves as their temporary carrier."

Dr. Wilson paraphrases Samuel Butler's famous aphorism to the effect that a chicken is just an egg's way of making another egg by saying "the organism is only DNA's way of making more DNA." DNA—for deoxyribonucleic acid—is the abbreviation for the type of molecule of which genes are made.

Are acts of human kindness or heroism or brotherhood governed by the chemistry of DNA? Sociobiologists readily concede that they have no proof, but they believe the evidence from the rest of the animal kingdom is strong enough to suggest that human beings have inherited genes that influence social behavior.

The scientists are not saying that man and animals behave the same way. They are simply saying that their various behavioral patterns all evolved according to the same principles, just as common principles governed anatomic evolution.

Sociobiologists note that much human altruism is directed primarily toward one's closest relatives and only secondarily to more distant kin and, only after that, to larger groups such as nations.

Dr. Wilson wrote that further study in sociobiology might help "transform the insights of ancient religions into a precise account of the evolutionary origins of ethics and hence explain why we make certain moral choices instead of others at particular times.

In an interview, however, Dr. Wilson emphasized that he was not suggesting that all human social behavior is tightly controlled by the genes.

"I see a great deal of human behavior that is unique," he said. Man's intelligence, he continued, has enabled him to develop complex social relationships that lead to many kinds and degrees of moral commitments beyond those of other animals.

Sociobiologists note that man is virtually unique in expressing a type of altruism that extends beyond the immediate family to friends and sometimes to total strangers.

Dr. Robert Trivers of Harvard, another prominent sociobiologist, calls such behavior "reciprocal altruism because the motivation is not entirely selfless. For example, people who have risked their lives to help a stranger commonly say they did so because they would expect the other person to do the same for them.

Although altruism is among the most prominent types of behavior studied by sociobiologists, it is only one among many. Others include communication, territorial defense behavior, aggression, and parental behavior. All are thought by sociobiologists to have evolved by Darwinian principles.

Dr. Trivers, who calls himself a social theorist, said he felt the value of using evolutionary theory to understand social behavior would someday come to be accepted by scientists and academics who now attempt to understand human society largely without any biological concepts.

"Ten years from now," Dr. Trivers said, "the training of sociologists will have to include genetics and evolutionary theory. I'm sure sociobiology will come to have enormous impact on sociology and psychology."

Dr. Wilson said he thinks sociobiology is leading to a view of man as being under the influence of inherited "programs of behavior that are more strict than many psychologists would have us believe."

He said he believes the control mechanisms are more complex than the simple conditioning response suggested by B. F. Skinner, the psychologist, but that people are not as rigidly "pre-wired" for fixed behavioral patterns as followers of Konrad Lorenz often assert.

Dr. Wilson said he believes the view of man being developed by sociobiologists is most closely approximated in other disciplines by the work of Jean Piaget, the Swiss psychologist who has studied the remarkably predictable developmental stages of children's behavior, and by Claude Levi-Strauss, the Belgain-born anthropologist, who sees common, logical patterns underlying the structures of societies from the most "primitive" to the most "advanced."

Dr. David Barash of the University of Washington in Seattle is a psychologist who is also a sociobiologist. His work has included one of the most interesting facets of sociology—the ways in which environments can change the social system of a species.

For example, when marmots, which are relatives of woodchucks, live in regions where the growing season is long and the food plentiful, they establish and defend territories around their burrows. But where the climate is harsher and the food supply is tighter, marmots live in colonies and forage over common land.

Sociobiological theory suggests that there are Darwinian explanations for these differences, namely that each system offers the best advantages for survival in the specific environmental situation. It has been determined, for example, that where food is scarce and sparsely distributed, it can be found more readily by group foraging and subsequent sharing than by individual foraging, which would be adequate where food is well distributed.

Thus, it appears to sociobiologists that even among animals that may seem to be selfish individualists, evolution has built in a capacity for cooperation when the living conditions become bad enough.

In recent years observations of similar variations in how a species adapts to different environments have been made in other animals and have served to weaken the old notion that behavioral traits such as territoriality or the lack of it are necessarily fixed.

Accordingly, sociobiologists such as Dr. Barash emphasize that the genes do not necessarily encode any immutable behavioral pattern, but rather they contain instructions for a more basic repertoire : of behaviors and the ability to adapt them to suit differing environments.

May 28, 1975

CHAPTER 5

Science and Society

Werner von Braun, developer of the German V-2 rocket and prominent U.S. rocket scientist.

Courtesy The New York Times.

FUNDING AND ORGANIZING RESEARCH

GERMAN SCIENCE ADVANCES BY COORDINATING RESEARCH

Thirty-two Institutes Function as One in Three Important Fields of Effort

GERMAN advancement in science has resulted, it is contended, not by accident but by the thoroughly organized effort to conduct research in many fields. The movement received impetus in 1910 when Adolf von Harneck issued a memorial to Wilhelm von Humboldt, recalling that scientist's plan of a century earlier for the establishment of Hilfsinstitute, comprising auxiliary scientific organizations in which investigators could pursue their researches.

The result of Dr. von Harnack's plea for the carrying out of von Humboldt's plan was that the German people subscribed more than 15,000,000 marks to establish the Gesellschaft zur Förderung der Wissenschaften (Kaiser Wilhelm Institute for the Furthering of the Sciences) in 1911. The basic idea was to make "general staff officers of science" who could later, if they did not prefer to devote their whole lives to research, give to universities and industry the benefit of their work.

This extensive organization, which has thirty-two institutes in Germany and Austria, plays an important part in the scientific and industrial life of Europe. It has three main fields of effort. One division devotes itself to the building up of the theoretical science of chemistry, physics, botany and medicine; the second deals with the application of theory to practical applied science, and the third with history, jurisprudence and allied subjects.

Work of Various Institutes

The biological-physiological group includes the three institutes for biology, biochemistry, anthropology, human heredity and eugenics. Here are studied the mechanism in the development of plants and animals; physiological, physical and chemical problems of life; the alcohol question, cancer, the effects of foods on the different organs, and a multitude of other problems.

Dr. Fritz Haber is directing investigations into physical chemistry and electro-chemistry. In another institute experts are studying radiochemical problems. The institute for physics has the task of supplying to the other institutes a corps of able physicists.

In Göttingen twenty-four scientists are working on problems of aerodynamics, currents of air and electricity. Researches of great importance to aviation have been carried out, such as the measurements of the velocity and accompanying resistance of wind, the efficiency of propellers, the result of different arrangements of radiators for motors, and related subjects. Air pressure on buildings also has been studied and the results have been placed at the disposal of architects and builders.

The institute for experimental therapy and brain study in Berlin-Dahlem was established by a gift from the Krupp family. The institute welcomes foreign scientists. It maintains in Sao Paulo, Brazil, an important laboratory for micro-biology, which has made many discoveries to aid in battling tropical diseases. Research here into the anatomy of the brain, begun by Oskar and Cecilie Vogt, resulted in Dr. Oskar Vogt being summoned to Moscow by the Russian Government to make a microscopic study of Lenin's brain.

Rockefeller Assistance.

The Rockefeller Foundation has shown interest in medical research in Germany, giving $325,000 for a new building for the German psychiatric institute of research in Munich. All these institutes pay more attention to finding the foundations of scientific knowledge than to inventions. In like manner, the second great field of the Kaiser Wilhelm Gesellschaft, covered by the institutes for applied science, is purely scientific.

The first of these institutes was that for coal research in Mühlheim-on-the-Ruhr. Under the direction of Dr. Franz Fischer it deals with the entire chemistry of coal, and especially with the liquifying of coal. Since chemical light reactions played an important part in metamorphosing the basic substances into coal, a special laboratory for light synthesis was established. It also deals with the problem of making briquettes without any binding material.

In Düsseldorf an institute for iron research devotes itself to the metallurgy and metallography of iron and steel. The institute for metal research in Dahlem devotes itself mainly to problems in the production and manufacture of light metals. The institute for silicate research in Dahlem serves the glass manufacturers and the ceramic and cement industries.

The institute for leather research in Dresden deals with the chemistry of the different tanning substances. The institute for the chemistry of fibrous materials, which works in cooperation with the textile industry, devotes itself to discovering the tensile strength and make-up of those materials, especially cellulose.

Crop and Labor Inquiries.

The new institute for research into crop raising is at Müncheberg, near Berlin. The institute for the physiology of labor has been moved from Berlin to the Rhenish-Westphalian industrial district. This institute conducts research into the rationalizing of manual labor, symptoms of fatigue and adaptability to particular kinds of work. The aim is to find the maximum of efficiency with the minimum of labor.

The Kaiser Wilhelm Gesellschaft maintains a number of natural science stations. Among these are the hydrobiological institute in Plon, Holstein, for inland waters; the biological station in Lunz for flora and fauna of the Alpine waters and the bird observatory in Rossitten, where Johannes Thienemann studies the secrets of the migration of the different birds. There are Alpine meteorological stations on the Sonneblick and the Hochobir in Austria. In the entomological museum of Dahlem is an international library on insect research.

The third field of labor in the Gesellschaft covers what the Germans call "the intellectual sciences." The institute for history deals with the political and cultural history of the ecclesiastical principalities of the Middle Ages. The institute for foreign and international private law and the institute for foreign law and international law deal with juristic problems.

The Kaiser Wilhelm Gesellschaft maintains friendly relations with the scientists of other countries by an exchange of scientists and through research organizations abroad. It maintains zoological stations in Rovigno and Naples and at the Bibliotheca Hertziana in Rome, an important art-historical institute which is a gathering place for scholars from all over the world.

SCIENCE AND SOCIETY

SOVIET USES SCIENCES TO ADVANCE THE STATE

Groups Those With Related Aims to Save Repetition, Making A System That May Prove Powerful—Striking Results of New Researches

By J. G. CROWTHER.

WILL Soviet Russia become not the richest but the most powerful country in the world within the next twenty years? A tour of the scientific institutions in Leningrad and Moscow inspires this question. It is impossible to visit many laboratories there without realizing that all scientific work in the Soviet Union is planned from above.

This planning of the whole of the nation's scientific work arises quite naturally from the fact that nearly all of the nation's wealth is owned by the State. The capital necessary for the foundation of scientific laboratories is owned by the State. Since the State has to pay for this institution and takes into consideration the condition of all institutions doing the same kind of work, it tends to group them together and see that each receives its fair share of the available money.

In American and Western European countries scientific institutions are not grouped and planned to work together by an all-powerful State authority; they tend to be independent and follow those lines of work in which they are interested, irrespective of whether other institutions are doing the same kind of work. This independence leads to repetition; half-a-dozen laboratories may all be trying to solve one problem in the same way, instead of tackling the problem in six different ways. In Russia the State-planning departments would try to re-arrange this and save the waste of repetition.

This State planning of scientific research may perhaps prove very powerful. After a number of years the Soviet may be able to organize the country and scientific research into one huge integrated unit. It may be that no one will become rich, but the whole country may make up by organization what it lacks in wealth. Marshal Foch said that morale is the most powerful factor in establishing a nation's might. It appears to me that the Soviet may produce by organization a very high morale, even if it does not produce great wealth.

Generous to Science.

Next to the all-embracing system of scientific research, the most noticeable thing to a visiting scientist is the astonishing sums the Soviet is spending on science. Soviet Russia is at present a poor country, but her expenditure in this direction compares favorably with that of rich countries such as America and Britain. She probably spends a much greater percentage of her surplus wealth on science than any other country. For instance, she spends $1,500,000 a year on research in applied botany. On seismology she is spending $250,000 next year. An experimental electrical laboratory in Moscow is being erected at a cost of $7,500,000. If Soviet Russia can make efforts such as these now, when the country is still very poor, what sort of efforts will she make when she becomes more prosperous?

The next point observed was that scientists have great prestige in Soviet Russia. They are better paid and more respected than nearly all other citizens. A good scientist receives about $2,500 a year. A very good one may receive $5,000. Some engineers receive as much as $17,500 a year. These salaries may appear small, but they are larger than those of other people. Business men usually receive less; in fact, there are very few business men in Russia. There are experts who run factories, but all factories are owned by the State.

The researches on applied botany are among the most remarkable in progress in Russia. As Russia is predominantly an agricultural country this is not unexpected. The department of applied botany organizes and plans the applied botanical research done everywhere in the country. For instance, the study of plant breeding is directed by Professor Vavilov. The whole world is ransacked for new specimens of wild oats and other cereals. A new specimen is discovered, say, by a Russian expedition in Abyssinia. It is brought back to Russia, and plants of the new species are sent to plant-breeding grounds in different parts of the Union.

In this way the reactions of a plant to a variety of environments are discovered. The plants grown in the different parts of Russia are then sent to Leningrad, where they are placed in a huge herbarium of cultivated plants. In this herbarium there are from 10,000 to 20,000 specimens of each of the chief cereals—oats, wheat and barley. Immense numbers of specimens of cabbages, peas, beans, fruits, &c., are also filed. In the station for the study of plant genetics at Detskoi Selo, formerly the Czar's villa near Leningrad, some extraordinary plant-breeding experiments are in progress.

New Vegetables Produced.

It will be remembered that plants of different species do not in general give fertile crosses, that is, the offspring of such crosses are sterile. Professor Vavilov's workers have produced fertile crosses between entirely different species; for instance, between the cabbage and the radish. This new plant is quite distinct and produces offspring like itself. Another plant crosses cabbage, radish and mustard. It also breeds true. So the Russian plant geneticists are inventing entirely new strains of vegetables.

Not only can they produce new species of plants, they can explain how they arise on the modern chromosome theories of heredity. They are much interested in plant physiology. Sometimes plants occur whose cells contain twice the normal number of chromosomes. When this occurs, the plant is often a giant, twice the normal size. It may also breed true. The Russian geneticists are trying to discover cabbages with double chromosome numbers so as to produce new species of cabbage double the normal size. This problem particularly interests the Russians, because the cabbage is used in many dishes, cabbage soup being a staple in Russian cookery.

In another palace at Detskoi Selo, once owned by Prince Yusupoff—in whose Leningrad palace Rasputin was murdered—there is a laboratory for testing the baking qualities of flour. Specimens of corn are sent there from all parts of the Union and milled into flour which is baked into loaves. In this way the food qualities of new races of wheat and corn can be studied and also the qualities of flour made from mixtures of various corns.

In the Biochemical Institute in Moscow experiments are being conducted on injections and immunity. It has been found that if eggs are injected at a certain spot with certain amino-acids, they hatch out large chickens which grow into giant fowls. If injected at another spot, the chickens are dwarf, and grow into dwarf fowls. The Soviet Department of Agriculture is endowing further research on this discovery. In the same institute an investigator has found that certain pure amino-acids when injected into an animal confer on it some degree of immunity from certain diseases.

Work in Biophysics.

Work is being done in biophysics. Professor Lazareff, who directs the Institute of Biophysics in Moscow, has worked out a mathematical theory of biological stimulation. When a piece of living matter, such as a finger or the eye, is stimulated, chemical changes occur in it. Now chemical changes do not happen anyhow. They proceed at rates governed by mathematical laws and the quantities and qualities of the substances engaged in the reaction. Professor Lasareff considers he has shown that the rate of change in living cells, after they have been stimulated, must follow certain mathematical laws. He can show further that the sensitivity of living matter must change according to its age.

On his theory, the brain of man is most sensitive at the age of about 21. It increases up to that age and then declines steadily until it is normally non-existent at 80 or 90 years of age. Now it is possible to measure by experiments on the eye and ear the sensitivity of those parts of the brain associated with these organs. Professor Lasareff states that these experiments have been made, and shows that the sensitivity of the brain normally changes, as his theory would indicate. In fact, by experiments on a healthy person's eye or ear he can deduce their age to within a year or two; he can detect persons who conceal their age.

But he finds, he says, that persons in ill health depart from the normal curve. If he is told a person's age and finds that his brain sensitivity does not correspond to his age, Lasareff can predict with some confidence that that person is ill, even if he shows no other symptoms. So he appears to have discovered an additional method of diagnosis. Professor Lasareff asserts that his results have been confirmed by direct experiment on brain centres during operations on the brains of living persons. Apart from the general level of sensitivity of the brain, there are various secondary changes in its sensitivity.

Sensitivity of the Eye.

For instance, the eye is most sensitive at about 2 A. M., and least at about 2 P. M. This is to be expected, since the exposure of the eye during the morning must fatigue it. He also finds that the brain-centres connected with the eyes in Indians are less sensitive than those in Russians, but if the Indian settles in Russia, the sensitivity of his brain centres gradually rises to the Russian level. This obviously would appear to be connected with the fact that the sun's glare in India is much greater than in Russia.

To turn to the physical sciences. Russia has always been eminent in the study of earthquakes. One of the pioneers of this subject was Prince Galitzin, and his successors are carrying on his work energetically. Until recently seismology was only one of the subsections of the mathematical and physical department of the Academy of Sciences. Now it has been given a special department of its own and a new institute is being built for it. The work is directed by Professor Nikiforoff. The activity in seismology considerably surprised me. Why should Bolshevik Russia spend money on earthquake study when there are so many other things to be done?

This question leads immediately to the five-year industrialization plan. The rulers of Russia are trying to carry out the first instalment of the industrialization of Russia during the passing five years. The study of mining is becoming important. Seismology now has considerable importance in mineral prospecting. The Soviet desires to prospect Russia thoroughly. So the seismological department receives money to research on apparatus for mining prospecting. They have already invented and manufactured improved field seismographs and Etvos balances.

The latter are used for detecting local changes in the strength of gravity. When there are deposits of certain minerals in any district, the value of gravity, i. e., the weight of a pound, departs from the normal. These sensitive instruments can detect very slight changes in the weight of a pound, and hence the presence of the minerals.

Another seismological study of importance is concerned with the effect of vibrations on buildings. The vibrations caused by heavy traffic are like tiny earthquakes. This research provides data which help to design buildings immune from the disintegrating effects of earthquakes and heavy traffic. Models of buildings and embankments are mounted on platforms caused to tremble as if they were having earthquakes. The effects of the tremblings can be measured experimentally, then the theory of them can be worked out and applied to buildings to be erected in earthquake zones.

Earthquake waves are utilized in warfare. When a gun is fired, it gives a shock to the ground which travels as a small earthquake. If the time the shock takes to reach two observation stations is measured, it is easy to calculate how far the waves have traveled and hence the whereabouts of the gun. The development of instruments for this purpose is the job of one subsection of the seismological department.

Electrical Development.

Perhaps the most remarkable of the new Russian scientific developments are those in electrical engineering. The Soviet is building many large hydroelectric plants in various parts of Russia, in the mountains of the Caucasus and on the rapids of the great rivers. The river Dnieper rapids are being harnessed to a 700,000 horsepower hydroelectric plant now under construction. In light engineering, the telephone and radio are being rapidly developed. This extension of electrical service is creating a demand for electrical engineers, and laboratories where electrical engineering problems can be studied.

The Soviet Government has met the demand. It is building a vast experimental institute in Moscow in which all the main branches of electrical engineering—dynamos, radio, telephones, X-ray apparatus—can be studied and tested. The first installation is costing $7,500,000. There is an immense laboratory for a 1,500,000-volt transformer. This apparatus will be used for studying the properties of the insulators used in the transmission lines which will spread the electricity generated in the new power plants to the surrounding districts.

December 29, 1929

WASHINGTON NOW TAKES RANK AS THE CAPITAL OF SCIENCE

More Than 5,000 Scientists in Many Institutions Are Gathered There as Miracle Workers for the Entire Nation

By FRANK GEORGE,
U. S. Department of Agriculture.

A GROUP of men in the Cosmos Club in Washington were discussing the researches of Heyl, the man who weighed the earth; of Munroe, inventor of smokeless powder; of Howard, explorer of insect life; of Abbot, engaged now in studying the effect of solar radiation upon plant growth.

The conversation veered to the increasing part being played in Washington in scientific research. Some one suggested that Washington the last ten years had become, at least in scope of scientific investigation, the science capital of the world. Whatever the problem of science, he said, somewhere in Washington some one may be found working on it.

As he spoke, a rasping voice came through the open window. An airplane pilot circling the Washington Monument a mile away was testing a new type of long-distance radio speaker. Raked by a battery of searchlights, the pilot flew in and out of the darkness; then, suddenly, he darted behind a cloud and was gone.

Five thousand scientists, many of them world famous, in public and

SCIENCE AND SOCIETY

private laboratories in Washington, a modern city of miracle-makers, are continually exploring the unknown in search of new ways to promote the welfare of mankind; to improve the technique of industry and agriculture. These scientists, far from declaring a science holiday, are girding for even greater scientific endeavor as they are enlarging their facilities for scientific research.

Laboratories are being provided for fundamental research in water-power, in chemistry, and in the mechanization of industry and agriculture. There is being constructed a hydraulics laboratory in which waterfalls, dams and power plants are to be erected in connection with studies of hydro-electric power.

A large part of a building encompassing a city block is to be devoted to chemistry research in agriculture. Plans are being made for a Bureau of Engineering which will study automotive problems. The world's largest fisheries laboratory is to be installed in the new Department of Commerce Building.

In this capital of science there is located the underground workshop of Heyl, where the weight of the earth and other physical phenomena are studied; an earthly factory of "Thor," where lightning is manufactured by human hands; experimental chambers in which conditions ranging from undersea depths of 500 feet to heights of more than 20,000 feet in the upper air are produced artificially.

An entire building is used for geophysical research in the development of means of locating rare metals and precious gems in the earth; an entire building is devoted to studies in terrestrial magnetism; another to astrophysical research in an effort to perfect long-range weather forecasting; another to radio and television research; another to aviation research.

Hidden Laboratories.

The combined scientific activities in Washington are so extensive that even the long experienced science reporter is continually surprised at the discovery of research laboratories in unsuspected places. One may visit the Smithsonian Institution many times and learn only by chance that in the medieval towers of that ancient edifice there are hidden research laboratories in which stranger never sets foot; columns of one-room laboratories to which the only access is by means of automatic elevators.

Government scientific research was begun originally as a protective measure to test the quality of commodities bought by the Federal Government. Research in agriculture began when the first American Ambassadors abroad sent home specimens of foreign crops which might be grown profitably on American soil. The establishment of a United States Navy stimulated research in steel, and as a result there have been evolved many special alloys of iron with nickel, chrome, vanadium and other metals.

Industry and Research.

Scientific research increased rapidly as industry made demands upon the government departments for scientific investigations of all kinds, until now scientific explorations are made in practically every field of human endeavor. Latest efforts in this field are being directed toward the development of new metals, lightweight, durable and pliable, to satisfy modern needs for larger airships, ocean liners more gigantic than present-day leviathans, buildings one hundred and more stories in height, mile-long bridge spans.

United States Government scientists are making X-rays of the earth in all parts of the world in search of unknown metallic substances. This is being done by means of seismographs that register the vibrations of natural and artificial earthquakes through different kinds of metals. More than 10,000 specimens of rocks and rare metals have been analyzed spectroscopically, some of them having been extracted from depths 8,000 feet in the earth. Hitherto unknown properties of metals already known are being identified in the search for new metals.

Veterans of Science.

Some of the government scientists have spent decades in search of new food cultures; in study of disease-carrying insects; in search of the heavens for the elements that make human life possible. Some of them will spend a whole lifetime peering through microscopes in search of something they will never find; but in the course of their investigations they will, in the future as in the past, solve other problems that have long baffled science.

There are many instances where the government scientists in pursuit of one line of research have accidentally made discoveries in other fields, and of the unexpected commercial utility of the results of so-called "pure" science. Dr. Paul Bartsch, studying common slugs, accidentally discovered that the slug has an extremely delicate sense of smell and wilts in the presence of one-twelfth the concentration of lethal gases harmful to man. This discovery is being turned to practical use in studying gases in factories, mines and tunnels.

Government geologists in search of oil-bearing substances in abandoned Texas oil fields accidentally discovered potash at a depth of 8,000 feet. Scientists seeking to curb the damage of smelter fumes to farm crops in the West have learned accidentally that the same apparatus may be used effectively in the recovery of low-grade ores. The discovery of insect causation of disease in cattle led to the discovery of the mosquito as the bearer of yellow fever.

Scores of expeditions of government geologists set out this Spring to explore mineral and metal fields in the Western States. Airplane photography is being used in the mapping of areas which may contain hitherto undiscovered mineral and metal deposits. The maps are studied by the geologists who later explore by means of geographical prospecting fields of potential resources. Coincidentally, metallurgists are endeavoring to develop new methods for the economical recovery of low-grade ores.

The practicality of research in pure science is continually being demonstrated. The Federal Government recently was made the defendant in lawsuits filed by persons who claimed they were full-blooded Indians and therefore entitled to certain moneys for lands preempted by the government. Dr. Hrdlicka, famous anthropologist, was called into consultation and demonstrated that measurements of their skulls and other characteristics were dissimilar from those of full-bloods of the same tribe.

Wide Fields of Endeavor.

A chicle company desiring to reduce waste caused by natives who were harvesting kinds of chicle unsuited for chewing gum manufacture sought the advice of a government botanist, who identified the desired species of chicle and taught the peons to gather only that variety. The crab industry in Chesapeake Bay, alarmed over the steady depletion of crabs, has been informed by a government taxonomist of extensive crab fields off the coast of Cuba.

For many years the Smithsonian Institution has been collecting samples of sea-bottom mud from all parts of the world. Joseph Cushman, a specialist in the National Museum, described and named the microscopic shells of animals called foraminifora, which these muds contain. Oil companies discovered then that fossil shells of those same animals occurring in certain strata in Southwestern oil fields could be made to identify strata and so serve as a guide to the presence of oil, provided there was any way of identifying the foraminifora. Joseph Cushman had that knowledge.

Strange Research.

Strange, yet practical, are the ways of the Federal scientists and their implements of research. In one laboratory a man measures the holes in a Swiss cheese, for it has been discovered that upon the size and number of holes depends the quality of the cheese; in another laboratory a man is organizing a fly flight in order to learn how far disease carriers can fly; in still another laboratory a chap with a paint brush daubs gooey substances on cotton fabrics in an effort to produce weather-resisting materials.

There is a laboratory where a woman chloroforms clothes moths in order to slow down their actions for scientific study; there is another where a scientist pulls bits of cotton between his thumbs in search of reasons why the quality of the American crop has been deteriorating in recent years; another man spends all of his time exploding samples of dust in order to learn the causes of factory fires.

Destructive Tests.

There is a series of connecting laboratories in the Department of Commerce seemingly engaged in an orgy of destruction of every conceivable commodity from asphalt to zithers, but in which the scientists are really engaged in discovering the weaknesses of commodities in order to build them stronger; there is a plant where synthetic ammonia is literally made out of air.

One laboratory during the past year tested 1,843 different automotive fuels and lubricants; another made 3,120 tests of the physical properties of engineering materials; another made 12,070 tests of cement, concreting materials, lime, &c.; another tested 6,933 specimens of textiles; still another tested 2,009 samples of paints, varnishes and bituminous materials.

Washington is the headquarters, also, for scores of branch government laboratories in all parts of the world—under sea and on mountain top, in desert, in jungle. In Washington is centralized the research in solar radiation conducted on barren Mount Montezuma in Chile and on Mount Brukkaros in Southwest Africa; the research in the growing of rubber-bearing trees and other tropical plants in Florida and Arizona; research in aviation weather forecasting at Weather Bureau stations in all parts of the country; the gathering of foreign insect parasites which are the natural enemies of the pests that destroy American crops.

In Washington headquarters of the Bureau of Mines is directed research in safety mining practices, rock dusting of mines to prevent explosions, the development of gas masks and other apparatus in combating mine gases. In near-by Maryland the Chemical Warfare Service is conducting research in the development of new warfare gases and in the industrial utilization of gases and chemicals.

The appearance of the robin each Spring signalizes the exodus from Washington of government explorers to the four corners of the earth in quest of the answers to the thousand and one riddles of man and his universe; its disappearance each Fall announces the return of the scientists to study during Winter months the results of their researches.

June 14, 1931

SCIENCE IN REVIEW

Need of Permanently Organized Scientific Research Shown by Our War Experience

By WALDEMAR KAEMPFFERT

Long before we were embroiled in the global war which has taxed the resources and ingenuity of all belligerents, we knew that physicists, chemists, engineers, mathematicians and physicians had as much to contribute to victory as machine-gunners and generals. Following the example of Great Britain, we proceeded to mobilize our scientists and to create the Office of Scientific Research and Development, for the express purpose of solving anticipated technical problems of the Army and Navy.

The OSRD is no vast palace of research. A project is placed before a competent university professor or the director of an industrial research laboratory, and a contract signed which calls for the expenditure of a certain sum of money in a certain time. The necessary technicians are engaged, and the man in charge is sworn to secrecy and left alone. Thus the Government saves the expense of building and administering its own laboratories, and it gets the kind of scientific thinking and experimenting that it wants. So far as the evidence goes, the plan works well. Never in our history have science and engineering been so completely organized and directed to achieve a military end.

No Military Domination

At the outset it was decided that science is science, whether it deals with stars or guns, disease or range-finders. A problem in ballistics can be turned over to a mathematical physicist with the certainty that it can be solved by him as well as by an artillery expert. Nine-tenths of the professors and engineers who have concerned themselves with armor, tanks, range-finders and jet-propulsion never smelled gunpowder in action. Any industrial or university laboratory in the country can deal with any military project, provided it is broken up into its parts and each part assigned to a competent scientist or engineer.

The OSRD proceeded on the obvious principle that research demands organization, planning and competent direction. Military and naval experts were consulted, but they did not dominate the laboratories. This was something new. As a result more imagination and creative ability were marshalled by the OSRD than ever before in our military history. The military and naval expert is needed chiefly to specify what conditions must be met.

What is to become of this far-flung organization after the war? Will the OSRD be disbanded? Will we lapse into the easy ways that followed the last war and trust to the good intentions of Europe? The tentative agreement reached by Great Britain, Russia and the United States on the control of Europe to prevent aggression indicates that nothing like disarmament is contemplated.

All this being so, it is clearly the business of Congress to maintain the OSRD in some form on an adequate scale. We simply cannot afford to wait for the outbreak of another war before we mobilize science again. Germany almost won this war by blitz methods. The pace of the next war will be even swifter. Two oceans will not save us from attack in the future.

Case of the Airplane

We cannot assume that we shall have time enough to prepare ourselves for another conflict. It took ten years to develop the Flying Fortress, at a cost of millions. The same story is repeated by other bombers. Moreover, at the outbreak of war we had no satisfactory airplane armor and no self-sealing fuel tanks. The development of the liquid-cooled Allison engine began in 1930, and it is not yet ended. The Sperry gyroscope pilot made its appearance in 1913, but as late as 1935 it was not yet considered of military value.

The Wrights gave the world the first practical flying machine in 1903, yet it was not until they had flown abroad, after having been cold-shouldered here, that the Army showed much interest in their work. Even then our activities in aviation were so feeble that in World War I our airmen flew in machines built by our allies.

Radar repeats the story. A. Hoyt Taylor and Leo C. Young had worked on the principle of radio reflection to detect distant airplanes as early as 1922.

Similarly we launched a successful flying bomb in 1918—at Belport, L. I., in cooperation with the Sperry Gyroscope Corporation—a bomber which carried 1,000 pounds of T.N.T. and which had a range of 400 miles. Charles Kettering and his staff had begun the development of a radio-controlled flying bomb about the same time. Yet we did nothing with either invention.

Enemies Anticipated Us

The Germans and the Italians anticipated us in the development of the rocket-plane. Jet propulsion was brought to perfection in Great Britain. The bazooka and similar rocket guns, the only really effective weapons against tanks, were first used by the Germans and the Russians. We simply improved them.

In the light of past history we cannot afford to arrest military and naval research when the last shot is fired or leave it to Army and Navy laboratories which have never had either money enough or personnel enough to solve problems in the systematic way that they have been solved by the OSRD. How such technical negligence can end is shown by the history of France. After the last war the French had more planes and better planes than any other power. But, like us, they rested on their laurels, with the result that French aviation rapidly declined. The military authorities relied on the Maginot Line, with consequences that are now familiar.

If we are right in assuming that we will have a strong standing Army in the future, with the equipment of that force at least equal to that of any other power, it is the manifest responsibility of Congress and of our industrial leaders as well as of the Army and Navy to make the most of our scientific resources. The old policy of appropriating money which is to be spent for research by the Army and Navy is not good enough because wars are now total and because total wars embrace everything, from nuts and bolts to instruments of precision, from insulating materials to armor-piercing shells. This means that research, to give us the kind of military and naval equipment we should have, must be directed and controlled by scientists and not only by officers. It also means that the Army, the Navy and the research laboratories must be welded into a single organization which will see to it that our fighting equipment will meet the onslaught of an enemy who relies on speed and striking power to win a quick victory.

Joint Responsibility

If this country is to keep abreast of scientific and technological developments, research is clearly a joint responsibility of science, industry, the Army and the Navy. During the last period of peace, scientific leaders and technically trained officers thought more of the fundamental principles of pure science than of preparing this country for an all-out struggle for its very existence. If any plan is to be a success our leaders must recognize the responsibility which is theirs to maintain a free society at all costs and to protect it.

Lastly there is the matter of what may be called military insurance. If the United States had been fully prepared to cope with Hitler in 1939 it is possible that there would have been no war. Germany was feared not only because she had a huge and highly efficient military organization but also because she had organized and directed research for conquest. The country that keeps its military and naval establishments at a high level of efficiency through research is more likely to aid in maintaining peace than is one that is utterly unprepared for aggression.

October 29, 1944

CIVILIANS ASSUME ATOM RULE IN U. S.

Truman Signs the Executive Order Shifting Power—New Boards Will Meet Friday

By SIDNEY SHALETT
Special to The New York Times.

WASHINGTON, Wednesday, Jan. 1—At midnight last night the destinies of America's atomic energy production passed from military to civilian control.

President Truman yesterday afternoon affixed his signature to the Executive Order whereby the great atomic empire of the Army's so-called Manhattan District — the war-time secret "cover" name — passed out of the War Department's hands. It now will be run by the newly appointed United States Atomic Energy Commission, headed by David E. Lilienthal, heretofore the guiding genius of the Tennessee Valley Authority.

Significantly, the turn-over came at the beginning of a new year.

Robert P. Patterson, Secretary of War, pledged to the new civilian commission "the continuing wholehearted cooperation and support" of the Army.

Groves Looks On

Chairman Lilienthal and other members of the Commission were present as Mr. Truman signed the historic document. Maj. Gen. Leslie R. Groves, who has commanded the project since Sept. 17, 1942, and who now steps out, also looked on. Mr. Lilienthal commented that "the people of the United States" were turning over "to five civilians the most potent weapon of all time."

The Commission, Mr. Lilienthal said, will pursue development of "the peaceful and beneficial possibilities of this great discovery." It will meet Friday and Saturday for the first time with the advisory committee of nine leading atomic scientists, appointed by President Truman on Dec. 12.

With the notable exception that General Groves is being relieved by a 36-year-old civilian scientist, Carroll L. Wilson, there will be scant discernible physical change in the Manhattan District. Present personnel will be kept on, although military personnel eventually will be replaced.

There are indications, however, that Mr. Lilienthal's commission, which is a strong-minded one, composed of Dr. Robert F. Bacher, Sumner T. Pike, Lewis L. Strauss and William W. Waymack, has definite ideas as to how the atomic energy project should be run in the future.

Services Retain a Voice

The challenging—and, if successful, world-revolutionizing—problem facing the new civilian commission is how best to employ the resources and brains of the Manhattan District to solve the riddle of how to turn the power of the atom to civilian use. If the death-dealing properties of the uranium pile can be controlled and harnessed to power plants and machinery, it well may change the entire pattern of industrial civilization as we now know it.

The military and naval establishments will continue to have a voice in proposing what developments they wish from the atomic energy project. There is a military liaison committee that will pass on to the new civilian arbiters of atomic energy the ideas of the armed forces, and machinery is provided for the military to protest to the President if any serious impasses develop.

The Commission said the transferred personnel included 254 officers, of whom seventy-six were regular army and eight regular navy, and 1,688 enlisted men. The Government also has 3,950 civilian employees and contract operators have 37,800. All civilians will be transferred to the commission "with no change in status."

Staff Less Than 44,000

In contrast with the present personnel strength of 43,700, the $2,000,000,000 project employed 500,000 persons at its peak.

The empire to be taken over, extending over eighteen States, includes the great plant at Oak Ridge, Tenn.; the Hanford Engineer Works at Richland, Wash.; the Los Alamos (N. M.) laboratory, the Argonne National Laboratory in Chicago, and supervisory functions over work of various college and university laboratories.

The commission also is taking over unexpended funds of the Manhattan project and will maintain temporary offices in the new War Department Building, now General Groves' headquarters, until permanent quarters in Washington are available.

Highly praising General Groves for his accomplishments, Secretary Patterson commented:

"With the transfer of responsibility for the nation's atomic energy program from the War Department to the United States Atomic Energy Commission we will have carried out the long-range plans of President Roosevelt, President Truman, Secretary Stimson and General Marshall, who, months before Hiroshima, clearly recognized that Congress should create an independent agency of the Government to carry on this vital work.

"The War Department has consistently supported the principle of civilian control of atomic energy in its broad aspects, and we look forward to a relationship with the Atomic Energy Commission that will be of mutual advantage."

To this the Commission replied that its members also wished to join Secretary Patterson in expressing high praise to General Groves "for his outstanding service to the country."

It declared he had performed "an unprecedented feat of organization and management" and that he had borne "a responsibility not only unique for an Army officer but also vital to the security and welfare of the people of the United States."

January 1, 1947

TRUMAN SIGNS BILL FOR SCIENCE STUDY

National Foundation Is Created to Promote Basic Research to Maintain Leadership

Special to THE NEW YORK TIMES.

ABOARD PRESIDENT TRUMAN'S TRAIN IN IDAHO, May 10 —President Truman today signed the bill creating the National Science Foundation, asserting that insecurity engendered by the "cold war" made the measure more essential than heretofore.

The foundation, designed to promote "basic research and education in the sciences," including physical, biological and engineering phases, was needed, Mr. Truman said, not only so we might keep abreast of the rest of the world in science but also because "we must maintain our leadership."

In a statement he recalled that he had vetoed an earlier bill passed by the Eightieth Congress in 1947 because of "features which were undesirable from the standpoint of public policy and unworkable from the standpoint of administration."

Now, the President added, he was glad to sign it and expressed appreciation to members of the House of Representatives and Senate for "unselfishly reconciling divergent views."

TEXT OF STATEMENT

The text of Mr. Truman's statement follows:

I have today signed S. 247., an act creating the National Science Foundation.

The foundation will be an independent agency, in the Executive Branch of the Government, headed by a national science board and a director. It will be the function of the foundation to develop a national policy for the promotion of basic research and education in the sciences. The foundation will initiate and support basic research in the physical, biological, engineering, and other sciences. It will also grant scholarships and graduate fellowships in the sciences, and in other ways encourage scientific progress in this country.

The establishment of the National Science Foundation is a major landmark in the history of science in the United States. Its establishment climaxes five years of effort on the part of the Executive Branch, the Congress, and leading private citizens.

Three months after I assumed the Presidency in 1945, I received a report from Dr. Vannevar Bush and his colleagues, entitled "Science, the Endless Frontier." That report recommended the creation of an agency such as the National Science Foundation, to promote the development of new scientific knowledge and new scientific talent. It was assumed at that time that the world was close to an enduring peace. The foundation was to be an instrument in promoting reconstruction and in maintaining our wartime momentum in scientific progress.

The fact that the world has not found post-war security in no way lessens the need for the National Science Foundation. On the contrary it underscores this need.

Depend on Scientific Progress

We have come to know that our ability to survive and grow as a nation depends to a very large degree upon our scientific progress. Moreover, it is not enough simply to keep abreast of the rest of the world in scientific matters. We must maintain our leadership. The National Science Foundation will stimulate basic research and education in nearly every branch of science, and thereby add to the supply of knowledge which is indispensable to our continued growth, prosperity and security.

During the period that the National Science Foundation has been under consideration, there has never been any significant disagreement concerning the objective to be sought. Some differences of opinion have arisen concerning the means which should be employed in carrying the program forward. I was obliged to disapprove a bill which was passed by the Eightieth Congress in 1947, because it contained features which were undesirable from the standpoint of public policy and unworkable from the standpoint of administration. However, on that occasion I expressed my deep regret at the necessity of disapproving the bill, and I urged reconsideration by the Congress.

The present measure has satisfactorily met the objections which I expressed to the earlier bill. I appreciate the fact that members of both parties in the Senate and the House of Representatives have worked unselfishly to reconcile divergent views concerning the organization of the foundation and its relationship to the Executive and legislative branches of the Government.

The nation's strength is being tested today on many fronts. The National Science Foundation faces a great challenge to advance basic scientific research and to develop a national research policy. Its work should have the complete support of the American people.

May 11, 1950

DANGER TO SCIENCE FEARED IN U. S. AID

Dr. Bush Warns of 'Stifling Bureaucracy' in Trend to Control Research

Special to THE NEW YORK TIMES.

WASHINGTON, Dec. 14—A dangerous trend toward the bureaucratic control of scientific research was discerned today by Dr. Vannevar Bush in his annual report as president of the Carnegie Institution of Washington.

Dr. Bush, a central figure in the development of atomic fission as wartime Director of the Office of Scientific Research and Development, deplored Government centralization in general and expressed particular concern over its effect on science.

The Federal Government, he said, is channeling about seven times as many dollars a year into research and development as before the war. Although much of the support to research has been wisely conducted, he asserted, present dangers are great and the trends are far from reassuring.

"Many universities," Dr. Bush noted, "are carrying the bulk of their research and the salaries of their graduate facilities on Government funds. There is an inevitable trend toward an inflation within an inflation and toward bureaucratic control of research.

"Dependence on variable and uncertain yearly Government appropriations increases the dangers of control and could put our universities into very serious financial organizational difficulties."

Dr. Bush said a major condition for developing science was "the preservation of that freedom and initiative which made this country great."

The trend toward Government centralization, he added, "could build a stifling bureaucracy" and "has already gone far in that direction."

Another danger that Dr. Bush cited was undue emphasis on applied science and lack of support for fundamental science. He deplored what he called the conviction among those in authority in the Government that "research could be made to pay off."

Dr. Bush urged policies that avoid undue pressure on university groups. He recommended improved contracting methods, independent review committees to prevent support of mediocre projects and the tailoring of research to available trained personnel.

The Carnegie Institution, in supporting basic research in science since 1902, has given "a notable impetus to sound fundamental science in this country," especially through its example, the institution's president said.

Dr. Bush said the hope of avoiding a world war had increased in the last year and "patience, determination and collaboration" might prevent it for another generation. He said "the price of peace" was "continual vigilance and a heavy cost in maintaining sufficient defensive power among free nations."

December 15, 1951

KILLIAN NAMED TO SPUR SCIENCE

A NEW POST FILLED

President Tells Nation of Move to Step Up Missiles Program

By ALLEN DRURY
Special to The New York Times.

WASHINGTON, Nov. 7 —President Eisenhower announced tonight that he had named Dr. James R. Killian Jr., president of the Massachusetts Institute of Technology, to advance the nation's scientific defense program.

The President said that Dr. Killian had accepted appointment to the newly created post of Special Assistant to the President for Science and Technology. The announcement came in a radio-television address to the nation designed to allay public fears concerning scientific achievements by the Soviet Union.

[In Cambridge, Mass., Dr. Killian promised to use "every means" at his disposal and to move swiftly to fulfill his mission.]

Wide Changes Disclosed

The President also disclosed:

¶Sweeping administrative changes within the Defense Department to give top priority to missile and rocket development without regard to interservice rivalries.

¶A plan for greatly increased scientific education, research and sharing with America's allies.

General Eisenhower assured the country that although the Soviet Union was ahead in some fields, such as earth satellites, "the over-all military strength of the free world is distinctly greater than that of the Communist countries."

SCIENCE AND SOCIETY

He said that Dr. Killian would be assisted by a staff of scientists and experts who would report to him "and to me" in the drive to coordinate and speed up the American effort.

The President said also he had ordered that the Pentagon's missiles director, William M. Holaday, be clothed with the full authority of the Secretary of Defense to override interservice rivalries in pushing the missiles program.

Missile Manager

Any new missile or related program, the President said, would as far as practicable be placed under a single manager. It, too, he said, would be administered without regard to service frictions.

The President asserted that he would ask Congress to remove legal barriers to the exchange of "appropriate technological information" with friendly countries. He said he would support the special scientific committee that he and Prime Minister Harold Macmillan of Britain agreed should be set up within the North Atlantic Treaty Organization.

The President said he would discuss whatever legislation might be necessary to carry out his plan with a bipartisan meeting of Congressional leaders he has invited to the White House on Dec. 3.

General Eisenhower coupled his outlining of the stepped-up American efforts with a renewed appeal to the Soviet Union to join in United Nations disarmament efforts. He said that what the world needed more than "a giant leap into outer space" was "a giant step toward peace." He said the United States had repeatedly demonstrated its eagerness to take such a step.

The President said the country should feel a "high sense of urgency" but should not "mount our charger and try to ride off in all directions at once." He said some of the facts he laid before the nation were "sternly demanding" but that others were reassuring.

Among the reassurances, General Eisenhower noted that the United States now had weapons adapted to every kind of distance, launching and use.

Cites Warning System

General Eisenhower acknowledged that the successes of the Soviet earth satellites were a scientific achievement of the first importance. But he said that in themselves the satellites had no "direct present effect" on the country's security, although there was real military significance in their launchings since they implied advanced techniques and a powerful propulsion force of some type.

However, he said, this country and its Allies have developed a Distant Early Warning system, are maintaining ground and naval units in strategic spots around the world, and have a strong retaliatory power in case of attack. The combination of these, he said, provides the real deterrent to war.

The President warned that in spite of these things, the United States in coming years still could fall behind unless it now faced up to certain requirements and set out to meet them at once.

These, he said, include higher priority and aid to scientific education and higher priority, both public and private, to basic research. He said he would have more to say on these two subjects when he spoke Nov. 13 in Oklahoma City.

November 8, 1957

BASIC RESEARCH IN U. S. IS SPURRED BY SOVIET'S GAINS

Quest for New Knowledge, Long Neglected, Will Be Aided by More Funds

By HOMER BIGART

Basic scientific research, never adequately supported in the United States, is likely to be glutted with money as a result of the national humiliation over the Soviet earth satellites.

Overnight, the starved status of basic research has become a public issue. The National Science Foundation, the sole Government agency for the promotion of fundamental research, seems certain for the first time in its seven-year history to obtain from the Administration and from Congress all the money it wants.

Basic research is the quest for new knowledge. It involves slow and laborious methods of observation, hypothesis, deduction and experimental verification. New truths obtained from such investigation will be the basis of any major technological advance. But the evolution of new products and new weapons from this knowledge may take years of applied research and engineering.

Shock Leads to Action

The sudden awareness that this country may have a shortage of brainpower required for an increased effort in basic research, and that many creative scientists today are unable to carry on promising investigations for lack of support has produced major developments. Among them are the following:

¶The National Science Foundation will seek funds to triple its research grants. In fiscal year 1957 the foundation awarded 997 grants totaling $15,528,925.

¶To increase scientific brainpower, the Administration is considering a $200,000,000 to $250,000,000 program of grants to the states to stimulate education, particularly in science and mathematics. This project is controversial. Some education groups say it holds the threat of Federal control of the public schools.

¶The Defense Department has set up a new office—the Advance Research Projects Agency—where scientists and engineers will have greater freedom to explore ideas for weapons. However, this office will be concerned mainly with applied research and development rather than basic research.

¶To provide better intelligence on the research efforts of other nations, the State Department is under strong pressure to restore science attachés in the foreign service. There have been no science attachés in the field for more than a year. Obviously there would be no point in having science attachés unless there was also someone highly placed in the State Department with the wit to understand their reports. But the post of science adviser in the State Department has been vacant since mid-1953.

¶To hold 48,000 scientists and engineers now on the Federal payroll and to attract 4,000 more, the Government last week raised salaries in those fields by $135 to $1,080 a year in five of six Civil Service classifications. The new salary range is $4,890 to $14,835. The top category, paying $16,000, is not affected.

¶Private foundations, which have tended to neglect the physical sciences, are beginning to shift their attention to this field. Next year the Ford Foundation intends to begin a $20,000,000 program in science and engineering. Dr. Henry T. Heald, its president, says the foundation is unlikely to underwrite any research; the money will go to the improvement of education in science and engineering.

(However, the main effort of the Ford Foundation will continue to aim at strengthening higher education. The Rockefeller Foundation will continue to concentrate on the life sciences, and the Carnegie Corporation of New York will remain concerned chiefly with problems in higher education. Only the Alfred P. Sloan Foundation among the major philanthropic organizations has had a strong program of support for the physical sciences. This foundation's grants to postdoctorate research in chemistry, physics and mathematics now approach $1,000,000 a year.)

¶When Congress convenes, there will be renewed agitation for the establishment of a Federal Department of Science and Technology, at cabinet level. This department was proposed in a recent staff report of the Senate Government Operations Committee.

Program Is Controversial

The proposed Administration program for education is controversial in several aspects. The Department of Health, Education and Welfare, which would administer the program through its Office of Education, has been reluctant to accept any crash program that would give priority to science over other subjects.

But pressures on Secretary Marion B. Folsom are building up. Although he still insists that the Federal Government "cannot and should not dragoon our young people into fields of science; cannot and should not tell the schools what to teach and how to teach it," Secretary Folsom now accepts the need for "more and better science in our schools and colleges."

Also laden with controversy is the proposal for a Federal Science Department. Many leading scientists, including Dr. Vannevar Bush, oppose it. They argue that no single agency could manage research in such diverse fields as defense, atomic energy, agriculture and public health.

Jealous of Independence

Jealous of their independence, scientists protest that any attempt to coordinate research activities under one agency would regiment and stifle the search for new knowledge.

"You can't push scientists around," says Dr. Alan T. Waterman, director of the National Science Foundation. "Scientists know what they should do, and no one in Washington should try to direct them."

Dr. Waterman believes that basic research by its nature, is not amenable to central coordination or to crash programs.

In this field, brainpower is vastly more important than

335

money. Often the only materials an investigator needs are pencils and notebooks. Costly laboratories and equipment are for the "applied" scientist, whose job is to find practical application for the new knowledge supplied by basic research.

Applied research and development can be speeded by short-term crash programs. Development is the engineering of scientfic knowledge toward the production of useful items.

But applied research and development must feed on a constant flow of new basic knowledge. Thus no major advances in technology are possible without a vigorous program of basic research.

How to stimulate basic research is a baffling problem. In the judgment of many scientists, the problem can be solved only by creating in this country an intellectual climate conducive to genius.

They ask for an overhaul of the educational system, starting with the secondary schools. They call for more emphasis on mathematics, physics and chemistry.

They say that the standards of science teaching in high schools, colleges and universities are much too low.

They say that many high schools fail to provide basic science courses, and that those that do are using poor textbooks and poor laboratory equipment.

Eighteen per cent of the nation's high schools offer neither physics nor chemistry, according to the latest Office of Education survey. The figure is based on a random sampling of 10 per cent of all high schools in the country for the academic year 1956-57.

The schools that neglected these subjects were small schools. But they enrolled 5 per cent of all high school seniors in the country.

The survey showed that one of every five high schools failed to offer plane geometry, normally a sophomore subject. Only a little more than 40 per cent of high school pupils were studying this subject, an requirement for entrance in most colleges.

Compared with the last Office of Education survey in 1948-49, the new figures do reveal a considerable increase in the number of high school students taking chemistry and a slight rise in the number of those studying physics.

But in both cases the percentage of increase failed to keep pace with the 30 per cent rise in enrollments and with the rapidly growing demand for science-trained men and women.

Physics courses enrolled 310,000 in 1956-57, an increase of only 6.5 per cent over the 291,000 physics students in 1948-49. Chemistry enrolled 520,000, up 26 per cent from the 412,000 in 1948-49.

Need Is for Teachers

Pointing out that two out of every three high school graduates have not taken a full year of chemistry, and that three out of four have not taken a year of physics, Secretary Folsom recently said that the first and fundamental step to improve the situation was to provide more and better science and mathematics teachers.

The urgent nature of this problem was illustrated by the fact that last year, of the 5,000 college graduates prepared to teach mathematics and science in elementary and secondary schools, 2,000 went into industrial jobs instead of the classroom.

Secretary Folsom commented: "In the field of science, more than in many other fields, there is a dangerously wide gap between the rewards attached to teaching and the rewards attached to other jobs."

This argument will be used to justify pay increases for science and mathematics teachers.

The nation's research problems can be tackled only by persons of advanced scientific training. The critical barrier to the production of these scholars lies in the limited capacity of the graduate schools of the universities and colleges. These schools prepare most of the scientists who conduct the research programs; they prepare most of the teachers and professors in higher education.

Deficit Increasing

These schools today are producing far fewer than half the number of fully prepared college teachers needed simply to replace the 18,000 who leave the profession each year.

Scientists are alarmed that the number of doctor's degrees awarded in the physical sciences have remained almost on a plateau since 1951. By mid-1955, the Soviet Union had 15,000 more doctorate-level scientists than the United States.

Bleak evidence of the shortage of top-level scholars is found in the meager output of doctorates in critical fields in 1955-56:

Mathematics, 235; physics, 470; chemistry, 986; geology, 128; astronomy, 23; geophysics, 7; meteorology, 10; oceanography, 10; theoretical metallurgy, 1.

There were 596 doctorates in engineering.

These figures represented a slight decline from 1954-55 when 1,005 doctorates were earned in chemistry, 511 in physics and 250 in mathematics.

Some comfort was found in the continuing increase in the number of bachelor's and master's degrees awarded in science and engineering. But scientists warned that the lag in doctorates would persist unless graduates were motivated to go on to advanced studies.

Science and engineering groups have long complained about the lack of adequate guidance and counseling in the high schools. Only eight States have in their education departments a special director to foster the study of science and mathematics. These are New York, Connecticut, Pennsylvania, Virginia, Indiana, North Carolina, Texas and New Mexico.

Accordingly, the Administration program may contain grants to the states to help furnish consultants to science and mathematics teachers, and to provide better counseling to students.

In the past few months two Presidential committees have urged general increases in teachers' pay. Last July the President's Commission on Education Beyond the High School recommended that average faculty salaries in colleges and universities be doubled in the next five to ten years.

This commission, which expired in October after Congress refused to give it funds, also recommended revision of Federal revenue laws to give income tax deductions to students, parents and others paying for a student's education.

Short of the Dramatic

On Nov. 30 the President's Committee on Scientists and Engineers urged a "substantial rise in the social and economic status of teachers."

The committee also recommended greater emphasis on subject matter in teacher-training courses, provision for summer jobs for teachers in science-based industries or laboratories, exchange programs between industrial scientists and teachers, better laboratory equipment in the high schools and revision of the science and mathematics curricula.

But all these recommendations, including those in the proposed Administration program, are too long-term and undramatic to satisfy the clamor for an early scientific tour de force, such as a flight to the moon, to retrieve prestige lost to the Russians.

Consequently many scientists fear that the Administration, under Congressional pressure, will put such heavy emphasis on rocketry and space travel that the basic long-range problem of producing more scientists will remain neglected. Moreover the passion to beat the Russians in this field might divert brains and money from other vital fields of research.

Waterman Shares Fear

Dr. Waterman, who shares this fear, says:

"We must establish some sort of priority list of those fields on which we choose to concentrate. There are some limits both to our national income and our available manpower.

"In looking ahead to the next ten years, we should fully support basic research in all areas of science. Only by so doing will we discover the full possibilities upon which we may capitalize.

"As research results come in, we must then decide which useful undertakings will yield the greatest returns on our research investment.

"If we want to beat the Russians to the moon, do we want to do so badly enough to spend

Rise in Science Pupils

Changes in enrollments in mathematics and science in public secondary schools in the United States (grades 9-12) and related data, 1948-49 and 1956-57.

Item	Typical Grade	Enrollments 1948-49	Enrollments 1956-57	Per Cent of Increase
Subject				
General Science	9	1,074,000	1,518,000	41.3
Biology	10	996,000	1,430,000	43.6
Chemistry	11	412,000	520,000	26.2
Physics	12	291,000	310,000	6.5
Other Science	9-12	155,000	265,000	70.9
Total	9-12	2,928,000	4,043,000	38.1
Elementary Algebra	9	1,042,000	1,518,000	45.7
Intermediate Algebra	11	372,000	484,000	30.1
General Mathematics	9	650,000	976,000	50.2
Plane Geometry	10	599,000	788,000	31.6
Solid Geometry	12	94,000	160,000	70.2
Trigonometry	12	109,000	200,000	83.5
Other Mathematics	9-12	91,000	275,000	202.2
Total	9-12	2,957,000	4,401,000	48.8
Population				
Age 14		2,126,000	2,556,000	20.2
Age 15		2,140,000	2,393,000	11.8
Age 16		2,231,000	2,292,000	2.7
Age 17		2,206,000	2,300,000	4.3
Age 14-17		8,703,000	9,541,000	9.6
Enrollment				
Grade 9		1,641,000	2,254,000	37.4
Grade 10		1,491,000	1,933,000	29.6
Grade 11		1,242,000	1,513,000	21.8
Grade 12		1,026,000	1,263,000	23.1
Grade 9-12		5,399,000	6,963,000	29.0

Source: Offerings and Enrollments in Science and Mathematics in Public High Schools (Office of Education Pamphlet No. 120).

SCIENCE AND SOCIETY

millions of dollars that might instead be profitably invested in a half-dozen other promising lines of scientific inquiry? This is not a question that the Government can wholly decide, except as it reflects the will of the American people."

Questions for Research

Here are some questions that hold great promise for basic researchers, according to Dr. Waterman:

¶What is life?
¶How does radiation affect the laws of heredity and evolution?
¶Exactly how do hormones and enzymes regulate growth, aging and disease?
¶What is the basic structure of the atomic nucleus?
¶Why at temperatures approaching absolute zero do certain substances afford no resistance to electric currents?

In other areas, Dr. Waterman says the United States is neglecting research in oceanography, meteorology and radio astronomy. He says the Russians have forged ahead in these fields.

In oceanography our researchers are equipped with "a few out-of-date vessels and a couple of modified Navy tugs" while the Russians are employing "two first-class oceanographic vessels fitted up specifically for this purpose."

Oceanographic research involves the mapping and measuring of deep ocean currents. Out of this work could come one immediate practical use:

If it is proved that waters in the great ocean depressions do not rise to the surface in less than decades, or even hundreds of years, these depths may offer a partial solution to the problem of disposal of radioactive wastes.

Long-Range Forecasting

In meteorology, the country should be putting greater research effort on long-range forecasting and weather modification studies, Dr. Waterman believes. The Russians are making a strong effort in this field.

He finds it ironic that the Russians are now ahead of the United States in radio astronomy, a science founded scarcely twenty-five years ago by an American, the late Dr. Karl G. Jansky, of the Bell Telephone Laboratories.

The Russians already have radio telescopes larger than the 140-foot instrument that the National Science Foundation is sponsoring at Green Bank, W. Va. This observatory will not be ready until 1960.

The fact that the Russians are pushing ahead in so many fields of science disputes the notion prevalent in this country that the Soviet research effort is directed exclusively to military ends.

The Russians, for example, in their exploration of the upper atmosphere are putting much emphasis on research in crystallography. The study of crystal substance does not appear to have any practical application.

Scientists say that the rapid advances in Soviet science threaten the technological supremacy of the United States. They fear that the Russians may be close to important breakthroughs in the frontiers of scientific knowledge.

Perhaps the satellites were mere "engineering achievements" that incorporated no new knowledge. But the lingering fear of American scientists is that the Russians may be on the verge of discovering fundamental truths in physics, chemistry, metallurgy and electronics.

Maybe they aren't, and maybe the fear is exaggerated. But scientists in the Government believe that fear is the only wedge that can pry from Congress, industry and private foundations enough money to support adequate science research and adequate scientific education.

They complain that Congress has never seemed able to grasp the utter reliance of technology on basic research. Because the time lag between publication of research results and their exploitation often is ten to twenty years, Congress sees only a nebulous connection between theory and invention.

Attempts to evaluate a new conceptual theory in terms of military hardware or civilian products are always hopeless.

Theories Wait for Years

Such theories may rest for years in the storehouse of knowledge before they receive practical application. The development of the atomic bomb, for example, was based in part on the transformation equation that Albert Einstein wrote in 1905.

The German astronomer Johannes Kepler completed his third law of planetary motion in 1619. Nearly three and one-half centuries passed before the Russians put Kepler's laws to practical test by rocketing the first man-made satellites into their proper orbits.

Congress has always insisted on "practical" use of research funds. Agencies seeking appropriations for basic research are always sharply challenged by Congress to justify costs by proving immediate practical benefits.

This is a reason why only 8 cents of every research and development dollar spent by the Government goes to basic research. Sixty cents is for development, 32 cents for applied research.

3-Billion for Development

The Government is spending at least $3,377,000,000 in the current fiscal year for research and development. Twenty-three Government departments and agencies have research programs.

By far the biggest spenders are the Defense Department and the Atomic Energy Commission, which together account for 85 per cent of the outlay.

According to National Science Foundation estimates the Defense Department will spend in fiscal year 1958 about $2,110,-723,000 for research and development. But basic research will get only $31,905,000 of this. These are pre-sputnik figures; both amounts may go up sharply as a result of the policy reappraisal going on in the Pentagon.

The Atomic Energy Commission has obligated $671,740,000 for research and development, with basic research getting about $59,849,000.

Other Big Spenders

Other big spenders on basic research are the Department of Health, Education and Welfare, $44,444,000 (nearly all of which —$42,658,000—goes to the Public Health Service); Department of Agriculture, $19,693,-000; Department of Commerce, $6,235,000 (of which $5,047,000 goes to the National Bureau of Standards and $1,146,000 to the Weather Bureau); Department of Interior, $16,076,000 (mostly to the National Park Service, Geological Survey and Bureau of Reclamation); the National Advisory Committee for Aeronautics, $17,500,000, and the National Science Foundation, $35,401,000.

Total expenditures for government-sponsored basic research will reach about $232,-621,000 in the next fiscal year, compared to $217,609,000 in fiscal year 1957 and $157,189,-000 in fiscal year 1956.

Some scientists contend that a lot of the $2,788,171,000 that will be spent by the Government in applied research and development could well be diverted to basic research. They say basic research should get at least a 10 per cent slice of the research and development fund.

Could Triple Its Projects

As for the National Science Foundation, which supports basic research only, Dr. Waterman, its director, believes the foundation could wisely spend at three times its present rate.

Congress gave the foundation only $40,000,000 for 1958, the same amount as in 1957. The foundation had asked for $90,-000,000, but the economy-minded Administration cut the request to $65,000,000.

From its inception in 1950 the foundation has had to struggle for funds. It even had trouble getting born.

The agency was proposed by Dr. Vannevar Bush, the wartime director of the Office of Scientific Research and Development, in a report to President Roosevelt in 1945.

Dr. Bush warned that the United States could no longer rely on Europe as a major source of basic research and that a new agency was needed to promote the flow of scientific knowledge.

Against Congress's Notions

But Dr. Bush's dream of an agency for fundamental research guidance ran counter to Congress's notions of an office that would subsidize inventors and gadgeteers. When Congress finally passed a science foundation bill in 1948, President Truman killed it with a pocket veto. He favored the general aims of the bill. He objected to certain administrative provisions, insisting that the governing board of the foundation should be responsible to him.

Dr. Bush finally won Mr. Truman over by telling him that an independent board would be a handy buffer against a bombardment of requests for special grants.

The National Science Foundation Act finally became law in 1950. But one of the compromises that helped secure its passage was a statutory limitation of $15,000,000 for annual appropriations. In its early years the foundation received barely enough money to get itself organized and make a few modest grants.

Congress became more generous after the Russians produced a hydrogen bomb. The statutory ceiling was lifted in 1954.

Foundation Grows

The Foundation, although still a midget alongside those giant dispensers of research money, Defense and the A.E.C., grew in vigor.

Presidential recommendations for N.S.F. funds and the amounts voted by Congress are shown in the following table:

Fiscal Year	Appropriation	Presidential Recommendation
1951	$ 225,000	$ 475,000
1952	3,500,000	14,000,000
1953	4,750,000	15,000,000
1954	8,000,000	15,000,000
1955	12,250,000	14,000,000
1956	16,000,000	20,000,000
1957	40,000,000	41,300,000
1958	40,000,000	65,000,000

The foundation was able to increase its grants in support of research from $1,000,000 in 1952 to $15,528,925 in 1957. This amount was almost evenly divided between the biological and medical sciences, which received $7,620,925, and the mathematical, physical and engineering sciences, which received $7,908,000.

These grants—there were 997 last year—support investigations ranging from solar energy to the effect of elevation on the distribution of spiders, scorpions and ticks.

The average grant has been slightly higher than $12,000, for an average duration of two years. All were for individual projects, administered through educational institutions; the foundation has not, as a matter of policy, given broad program grants.

Supports Fellowships

In other activities the foundation supports four fellowship programs totaling 1,000 awards for graduate or post-doctoral studies; underwrites a $5,000,000 summer institute program for high school and college teachers in mathematics and science, and provides a $4,350,000 program under which about 850 high school science teachers study for a full academic year at seventeen colleges and universities to improve their knowledge of science subject matter.

Since 1956, the foundation has been providing funds for construction of badly needed research facilities. It has alloted $4,000,000 for the National Radio-Astronomy Laboratory at Green Bank, W. Va.; $500,000 toward the construction of a nuclear reactor at the Massachusetts Institute of Technology; $135,500 toward the construction of computation centers and for research in numerical analysis at five universities, and $750,000 to assist field stations for biological research.

Generally the National Science Foundation has been applauded for what it has done and criticized for what it has not done. Critics complain that the foundation has failed to promote a national science policy.

In setting up the foundation, Congress gave it a clear mandate "to develop and encourage the pursuit of a national science policy for the promotion of basic research and education in the sciences."

But scientists have always opposed Federal direction and control. Past attempts to bring about strong central coordination of the Federal scientific effort have failed.

Both Dr. Waterman and the National Science Board have rejected the belief "that government can and should direct the course of scientific development in this country."

Dr. Waterman believes that the foundation can best fulfill its role by quietly pointing out to American scientists the gaps that exist in the basic research effort. He believes it would be futile to attempt Federal direction of their work. In any case, he said, no major policy decision should be taken that does not have wide support from the scientific community.

Evaluates U. S. Programs

Another task that Congress had delegated to the N. S. F. was:

"To evaluate the scientific research programs undertaken by agencies of the Federal Government."

In 1954 an executive order by President Eisenhower defined more clearly the roles of the principal Federal agencies involved in research. Each agency was given authority to conduct and support basic research in fields closely related to its operating responsibilities. The N. S. F. received the wider responsibility of supporting general-purpose basic research.

It was made the chief adviser to the rest of the Government on research activities, with responsibility for surveying national research, manpower and resource needs and for making appropriate recommendations.

Critics say that if the National Science Foundation had acted more aggressively in its role as adviser on research to the rest of the Government, and especially to the Defense Department, there might have been no need for President Eisenhower's bringing in Dr. James R. Killian Jr., president of M. I. T., as top scientific adviser on defense matters.

Killian to Check Rivalries

But President Eisenhower, in outlining Dr. Killian's job, indicated that its prime purpose would be to "help see that such things as alleged interservice competition or insufficient use of overtime shall not be allowed to create even the suspicion of harm to our scientific and development program."

Further, Dr. Killian was to "make sure that our best talent and the full necessary resources are applied on certain high-priority, top-secret items."

No such powers were ever handed to the National Science Foundation, nor were they sought by Dr. Waterman. Advisory functions of the N. S. F. did not include knocking heads together at the Pentagon.

How extensively the foundation has exercised its advisory functions is not known. It is known, however, that Dr. Waterman helped persuade the Defense Department to rescind the drastic cuts made last September in research contracts to the universities. He told Neil H. McElroy, Secretary of Defense, that the colleges and universities had become heavily dependent on these contracts to support their research programs.

Points Out Research Needs

In its formal statements on policy, the N. S. F. has hammered at the need for more Government and private support of basic scientific research.

On Oct. 15, the foundation submitted a sixty-four-page report to the President. This report, called "Basic Research: A National Resource," attempted to convey in nontechnical language the importance of basic research to the Nation's economy, health and defense.

It recommended "positive steps" to aid basic research. These included tax relief for individuals in the low-income and middle-income brackets who contribute money to educational institutions.

Another would permit corporations to deduct as business expenses outlays for the training of basic research scientists at the universities. These recommendations are under study by Congressional committees.

The N. S. F. found that the nation's colleges and universities, in their increasing dependence on contract research for the Government, and especially for the Defense Department, were too heavily committed to applied research and development rather than to basic research. Universities, colleges and technical institutes are working on Government contracts totaling more than $441,000,000. This figure includes work done in highly specialized off-campus laboratories.

The foundation recommended a shift of Federal support away from applied and toward basic research in these institutions. It recognized the fear that a drastically reduced level of applied research would cripple some institutions and suggested a "significant increase in basic-support research" to make up for this.

Developmental Work

Apparently the foundation shared the opinion of many scientists that most of the work classed as "research and development" carried out by the universities was actually development work that made scientific hacks of the faculty and added nothing to the training of graduate students.

Colleges and universities always have contended that they are very selective in their acceptance of contract research, but the charge has been frequently made that the contracts were undertaken primarily to augment operating budgets.

In the words of the N. S. F. report, "The true functions of American institutions of higher education have tended to become clouded since World War II."

Colleges and universities are the traditional home of basic research. Yet the latest N. S. F. study of science activities in these institutions showed that of about 70,000 scientists and engineers employed, only 50 per cent were performing basic research in the natural and social sciences.

Industry's contribution to the total basic research activity always has been difficult to measure. A corporation executive may be acutely aware that basic research is vital to his business, yet he hesitates to take the long-shot gamble that such activity often involves.

Ventures Into Research

Ventures into basic research must hold promise of an early pay-off; consequently, what industry might regard as the purest of ivory-tower investigations might be dismissed by others as applied research because the work has an immediate or near-immediate commercial objective.

There are some industrial laboratories however, where the theoretical scientist finds encouragement. Foremost among these are the Bell Telephone Laboratories and the Laboratories of General Electric Company.

Corporate aid to higher education in 1956 reached a record level of about $100,000,000, an increase of $25,000,000 over 1954 and a gain of more than $60,000,000 since 1950. But this was still only one-fifth of one per cent of net profits reported by business and industry in 1956, according to figures compiled by the Council for Financial Aid to Education.

Industry performs about two-thirds of all the research and development in the natural sciences and engineering.

Plays Modest Role

Of a total estimated national expenditure of nearly $5,400,000,000 for research and development during a twelve-month period in 1953-54, industry's performance cost nearly $3,900,000. About one-third of the research and development work done by industry was conducted for the Federal Government.

Basic research played only a modest role in industry's effort, receiving only about $168,000,000, or less than 4 per cent of the total cost of industrial research and development.

The aircraft and electrical equipment industries far exceeded all others in the scale of their R. & D. programs, performing about two-fifths of the total for all industries. In basic research, the chemical industry outstripped the others by a wide margin, spending about $38,000,000.

Other industries with large basic research programs included electrical equipment, $19,000,000; aircraft, $18,000,000, and scientific instruments, machinery, and petroleum products, each of which had a basic research cost between $11,000,000 and $12,000,000.

Grants to Education

These figures are from an N. S. F. survey undertaken more than three years ago; amounts being spent today by industry on basic research may be considerably higher. For example, du Pont de Nemours & Co. alone said it was spending more than $15,000,000 on basic research this year.

Du Pont is one of several large corporations that make grants to educational institutions in support of basic research and teaching. More than half of du Pont's $1,000,000 aid-to-education fund this year goes to the improvement of teaching in universities, colleges and high schools. The company is granting $290,000 to universities for basic research and $165,000 for fellowships in science and engineering.

But scientists say it would be naïve to expect industry and private foundations to assume the major burden for fostering in the United States the intellectual atmosphere required for a general assault on the frontiers of science.

The important money — and the leadership — they say, can come only from the Federal Government. Only the President, scientists believe, can persuade Americans to realize that a set of values that exalts financial success over scholarship may be fatally immature in the present crisis of survival. They look to him in the hope that he may speak out against the traits of weakness in American society.

SCIENCE AND SOCIETY

Books of The Times

By ORVILLE PRESCOTT

WHEN Charles Percy Snow—a large man with an impressive forehead—is serious, he looks rather like a sternly virtuous Roman Senator—Cincinnatus perhaps. When he smiles a transformation takes place and he is revealed as one of the most affable, kind and engaging of living men of letters. His talents are many, his mind is brilliantly analytical and coldly objective, his energy is formidable.

A scientist and a civil servant as well as a novelist, he is best known for a series of related novels called "Strangers and Brothers," on which he has been at work for twenty-five years. But Sir Charles lectures occasionally and his lectures seem to arouse as much interest as his fiction. Last year, when a lecture delivered at Cambridge was published as "The Two Cultures and the Scientific Revolution," it provoked a lively debate. Today three more of his lectures, which were delivered at Harvard last autumn, are published in a briskly interesting book called "Science and Government."*

In these times of scientific mysteries incomprehensible to most people, crucial and secret decisions frequently have to be made. Politicians and government administrators are not likely to have the technical knowledge to make such decisions wisely. The scientists they consult may or may not have a good understanding of "what those choices depend upon or what their results may be." And what if the scientists disagree?

Matched in a Fateful Duel

To illustrate the importance of this problem Sir Charles has told the story of "two men and two choices." The men were Sir Henry Tizard and Frederic A. Lindemann, Lord Cherwell. Sir Charles knew them both, and Tizard, whose papers he has examined, was his friend.

Tizard, "the best scientific mind that in England has ever applied itself to war," was a chemist, "dazzlingly clever," brave, proud, conservative, "a high-level scientific administrator." Lindemann, "a very odd and very gifted man, a genuine heavyweight of personality," was a physicist, brave, proud, reactionary, a vegetarian and a fanatical ascetic, an intimate friend of Winston Churchill. The two scientists had been close friends for twenty-five years, but for some reason they became enemies.

Their enmity provoked rows and feuds. Tizard was chairman of a committee on air defense that was responsible for the development of radar. When Lindemann became a member of the Tizard committee at Churchill's insistence, he tried to stop the radar development and advocated instead the development of infrared detection and of parachute bombs and mines to be dropped in front of hostile planes. Fortunately, Tizard prevailed, but subsequently he was forced out of the committee.

The other decision was a victory for Lindemann, who urged strategic bombing of German cities to destroy workers' houses. Tizard opposed this and calculated that Lindemann's estimate of the damage that would be done was five times too high. "The

*SCIENCE AND GOVERNMENT. By C. P. Snow. 88 pages. Harvard. $2.50.

C. P. Snow — Walter Bird

bombing survey after the war revealed that it had been ten times too high."

Such secret decisions, one wise and one not, are examples of what Sir Charles calls "closed politics." There are three kinds of closed politics, he says: committee politics; hierarchical politics (bureaucratic or military chain of command), and court politics. The pro-radar decision was a committee choice. The decision in favor of wasteful strategic bombing was a court decision because Lindemann made it in his capacity as Churchill's personal adviser.

In "Science and Government" Sir Charles describes two scientists in government. One, he thinks, made a major contribution to his country's defense; the other was mistaken on two profoundly important scientific issues. And yet Sir Charles concludes his lectures by pleading for more scientists in government because they are accustomed to think in terms of the future with foresight and are also accustomed to think in terms of constant change.

A Confrontation on Choice

So far so good. But Sir Charles does not indicate how any government is to recruit wise scientists like Tizard rather than misguided ones like Lindemann. How is the career politician or civil servant to tell the difference? Even Churchill, as astute a statesman as democratic politics ever produced, relied on personal friendship and put his trust in a man regarded by his fellow scientists as not a very good physicist.

It would be interesting to know in just what departments of government Sir Charles would like to have more scientists. Presumably the technical posts, for which no non-scientist is qualified, are already occupied by them. Does Sir Charles recommend that scientists should be employed in purely political or administrative tasks? His earlier remarks in "The Two Cultures and the Scientific Revolution" suggest that scientists, because of their specialized education, are as ill-prepared to understand politics and government as politicians are to understand nuclear physics.

April 3, 1961

Science and Richard Nixon

By Daniel S. Greenberg

Brilliant scientific discoveries continue to pour out of the nation's laboratories, and the Nobel and other grand prizes continue to pour in, but probably not since Depression days, when a career in research usually involved a pact with poverty, has the American scientific community been so enveloped in despair or felt so ill-treated by its great patron, the Federal Government. In the reign of Richard M. Nixon, lawyer-President surrounded by the high achievers of conglometry, public relations, advertising, corporate law and burglary, the "scientific-technological élite" of President Eisenhower's farewell address has fallen from political grace as has no other group (except, curiously, the poor).

Today, for the first time since 1957, when Eisenhower summoned James R. Killian Jr., president of M.I.T., to serve as a full-time counselor on the mysteries of space and advanced weaponry, the post of Science Adviser to the President stands vacant, and come July it will be abolished. Federal funds for research and development—the lifeblood of university-based science and of a vast amount of science and technology elsewhere—have been virtually level for the past three years, which means that purchasing power is down substantially. The exact decline is difficult to figure since the ongoing revolution in scientific instrumentation makes it ever more costly simply to hold one's place in the competitive world of basic science. But Harvard's Paul Doty Jr., one of the nation's leading biochemists, believes that inflation has eroded away as much as 30 per cent of the constant dollar figure. And if Mr. Nixon's budgetary plans for the coming fiscal year are carried out, the drop will be even steeper, for the President has proposed to eliminate Federal traineeships for the support of graduate students.

Within the Federal budget for research and de-

Daniel S. Greenberg, author of "The Politics of Pure Science," publishes an independent Washington-based newsletter, Science & Government Report.

---

velopment, spending in some categories is reduced, but in others it is up—or appears to be. The Administration boasts about the increases—its celebrated War on Cancer, for example, and the newly conceived companion program for heart and lung disease. But even in these high-priority areas, the promises of fiscal growth have not been accompanied by any new outpouring of funds. The heart and lung program has not yet been organized, and the National Cancer Institute is operating on an annual budget that is practically identical to that of 1972. Meanwhile, the other research centers that make up the National Institutes of Health are slated for reduced spending in the forthcoming fiscal year.

The gloom in the amorphous network of institutions that make up the "scientific community" has also been deepened by more specific cutbacks: the termination of further manned exploration of the moon, the "stretchout" or cancellation of several major scientific space projects and a standstill spending plan for the academic science divisions of the National Science Foundation, the principal mainstay of university researchers outside the biomedical area. Even the one-time noblemen of the scientific hierarchy, the high-energy physicists, whose stadium-sized particle accelerators once commanded blank checks in Washington, are hard-pressed by their principal source of finance, the Atomic Energy Commission. Five of the nation's major accelerators will have their already tight budgets reduced next year. The sixth—the $250-million National Accelerator Laboratory now nearing completion at Weston, Ill.—will get more than last year, but considerably less than originally expected for its research debut.

To outsiders, the inhabitants of this disaster area are the "technocracy," the Strangeloves, beneficiaries of a fabled grant economy, generators and masters of esoteric knowledge that osmotically permeates our cultural and political processes, regardless of the *pro forma* rules. Don K. Price Jr. of Harvard dubbed American science "The Fourth Estate," and observed that "it has become the

SCIENCE AND SOCIETY

major Establishment in the American political system: the only set of institutions for which tax funds are appropriated almost on faith, and under concordats which protect the autonomy, if not the cloistered calm, of the laboratory." Science and technology's extraordinary postwar ascent to prominence and affluence gave rise in the nineteen-sixties to such works as "The New Brahmins," by Spencer Klaw, and "The New Priesthood," by Ralph Lapp, both muckracking jobs, but not without awe for their freewheeling subjects. But today in the leather-upholstered, muraled and chandeliered Cosmos Club, the mannerly gathering place for Washington's resident and commuting men of learning, you will find the "scientific - technological élite" radiating the mood of a déclassé set awaiting the next disaster.

When Mr. Nixon first took office, the Office of Science and Technology (O.S.T.), considered the research community's embassy in Washington, was a well-established part of his Executive Office family. Now the President has simply wiped out that operation. His similar designs on the Office of Economic Opportunity were predictable, given the President's belief in self-reliance, but his abolition of the O.S.T. has puzzled the scientists and technologists, who, after all, play a significant part in a high-technology society afflicted by foreign competition and a tide of domestic operations that seem to invite scientific and technical remedies.

Consider the circumstances of O.S.T.'s demise. Leaping before his office was scuttled, Presidential Science Adviser and O.S.T. Director Edward E. David Jr. suddenly announced his resignation on Jan. 2 and immediately departed for a job in industry, thus eliminating the research community's highest ranking member at court. One month later, Mr. Nixon informed Congress of his intention to abolish the 50-member O.S.T.—the decision that had precipitated Dr. David's departure—and, by extension, an assortment of science advisory groups that had grown up around it since Eisenhower's Sputnik-induced summons for expert help. By way of explanation, Mr. Nixon stated that since scientific expertise was now sufficiently available throughout Government agencies, a full-time scientific presence at the Presidential elbow was no longer necessary. He failed to note, though, that his post-election wave of forced resignations has decimated the upper ranks of research virtually throughout the Federal bureaucracy. (Among the casualties, for example, was the director of the multibillion dollar National Institutes of Health, Robert Q. Marston, whose post had heretofore been immune to political tides. Marston's resignation was ordered and accepted without explanation, and without a successor in the wings to take his place.)

In connection with the abolition of O.S.T., Mr. Nixon explained, he and his entourage, when the need arose, would solicit scientific advice from the director of the National Science Foundation, an agency that is, at best, one of the larger midgets in Federal research affairs. (N.S.F. is budgeted next year for $446-million for the "conduct of research and development," compared with $8.3-billion for the Department of Defense, $3-billion for the National Aeronautics and Space Administration, $1.8-billion for the Department of Health, Education and Welfare and $1.4-billion for the Atomic Energy Commission.) To perform this job, the N.S.F. Director, H. Guyford Stever, would take on the additional title of Science Adviser—though not to anyone in particular. He's just the Science Adviser and, furthermore, he is not permitted to provide advice on military research, an area into which the liberal, academic O.S.T. frequently sought to poke, to the outrage of the military services and their allies.

Unlike his predecessor, the newly created Science Adviser will not have direct access to the President; rather, his channel leads to Treasury Secretary George P. Shultz, Mr. Nixon's newly designated White House adviser for economic affairs. But then, it turns out, he will not even have access directly to Mr. Shultz, but only to his chief aide in the White House, Kenneth Dam, a lawyer-economist - budgeteer alumnus of the Office of Management and Budget—a rare triple personification of the professions that are least impressed by science's plea for faith in research. All of which deepened the despair of the elders of science, who had had easy access to the White House in earlier times and recognized that attempting to run science from the pint-sized National Science Foundation is akin to directing a major symphony orchestra from the seat of second oboist, with no authority over the brass.

In view of all this, what, then, is going on between Mr. Nixon and American science, and, in particular, what does it portend for the quality, viability and utilization of the nation's scientific resources? Do we face "the virtual dismantling of the foremost health sciences research program in the world," as Paul Berg, chairman of Stanford's department of biochemistry, proclaimed — and as specialists in other disciplines similarly prophesied for their own fields — when the new budget was announced? Or does the weeping simply reflect the ups and downs of bureaucratic skirmishing, infighting of vital concern to the participants and their friends, but of no particular consequence to the rest of us?

An essential part, but only a part, of the answer is that Mr. Nixon, who is demonstrably not above grudgery, does not like the academic world, including its substantial scientific component, probably for the well established reason that the academic world long ago decided that it did not like Mr. Nixon. It was academe, home base of the "campus bums" whom Mr. Nixon once angrily decried, that ignited and sustained the antiwar movement. Many of its eminent professors, scientists well represented among them, served in one brain trust or another for Kennedy when he beat Nixon by a whisker in 1960. In the following Presidential election, the scientific community came out in force against Goldwater with a nationwide Scientists and Engineers for Johnson-Humphrey. Matters were less clear-cut in 1968, and fewer came out in support of Humphrey's candidacy, but the dominant trend was anti-Nixon. After Nixon was installed in the Presidency, it soon became clear to his closest henchmen that a piranha was on the premises in the form of one of the proudest descendants of Eisenhower's quest for scientific advice—the 18-member President's Science Advisory Committee (P.S.A.C.), composed of distinguished scientists, engineers and other specialists who would normally meet monthly in Washington to dispense independent thought on whatever matters of science and technology engaged their interest. Chaired by the President's Science Adviser, P.S.A.C. was originally created to help the Science Adviser and the President squelch the military services' conflicting claims for independent missile forces of their own. Arms control, disarmament and the nuclear test ban tended to dominate its thinking—especially during the period when Mr. Nixon was behaving like a cold warrior. In its spare time, P.S.A.C. trumpeted the importance of heavy Government support

341

for academic science, whence most of its members commuted, and the message was heeded. Around election season, many of its members semicovertly deployed their nationwide connections and influence in behalf of the Democratic candidate, as is noted in an M.I.T. doctoral thesis by Anne H. Cahn entitled "Eggheads and Warheads: Scientists and the ABM." She states: ". . . the October, 1968, meeting of the P.S.A.C. Military Strategic Panel was converted into a working session of Scientists and Engineers for Humphrey-Muskie, to the chagrin of at least one panel member who was a Nixon supporter. He viewed with distaste the sight of his colleagues arranging calling and canvassing activities at the expense of the legitimate responsibilities of P.S.A.C." The account is disputed by Richard Garwin, a top I.B.M. researcher, who says he was there. Be that as it may, Garwin himself was the cause of an outburst of White House rage several years later when, after having studied the supersonic transport at the request of O.S.T., he concluded that it was a poor bargain — and publicly said so when Congress was engaged in its eventually fatal deliberations on the project. The Federation of American Scientists subsequently bestowed its first annual Public Service Award on Garwin "for courageous and effective testimony on the SST." The White House was livid on many grounds.

Viewed against this brief sketch of some of Mr. Nixon's formative encounters with statesmen of science, it is not unusual that, in contrast to our last half-dozen Presidents, he has rarely made a ceremonial gesture toward science. Of his few utterances on science, the most puzzling occurred in May, 1971, when he awarded the Medal of Science — highest award of its kind — to a group at the White House. Noting that he had read the citations accompanying the awards, Mr. Nixon then went on to say in part (according to the official White House transcript):

"I have read them, and I want you to know that I do not understand them, but I want you to know, too, that because I do not understand them, I realize how enormously important their contributions are to this nation. That to me is the nature of science to the unsophisticated people."

Whatever the nature of science to the unsophisticated people, Mr. Nixon and his aides have apparently concluded that, for example, the Medal of Science can be dispensed with or delayed. Normally awarded annually, the medal dropped off the White House agenda after that 1971 presentation and has not been heard of since. And in May, 1972, the White House announced that Mr. Nixon had established a separate set of Presidential Awards for Technological Innovation—to be accompanied by prizes of $50,000 each—with the first presentation scheduled for Sept. 15, 1972. An expert panel forwarded a list of nominees to the White House in comfortable time for that date, but the awards have never been made nor has any explanation been forthcoming. When Science Adviser David resigned at the beginning of this year, his office safe contained 10 $50,000 checks made out to the winners. They have been there for months awaiting the **White House's word on when to schedule the awards ceremony.**

Mr. Nixon's cool feelings toward the men of science does not, however, explain it all. He is no devotee of the arts and humanities, but the Government foundation responsible for subsidizing them has flourished even in these most difficult of budgetary times— its funding has risen from $72-million this year to $120-million scheduled for next year. What else is involved?

Closely related to the absence of favorable Presidential interest is the fact that, after nearly a decade of exuberant, often aimless growth, research and development in the U.S. was due for a collision with economic reality and social utility. From the late nineteen-fifties through the mid-nineteen-sixties, a growth mania, mysteriously set at an annual minimum of 15 per cent — though it often exceeded that — underlay research - and - development demands on the U.S. Treasury. If that figure were not met, if more bright youngsters were not subsidized into scientific careers, if more laboratories were not built and splendidly equipped—well, the statesmen of science assured Government and public, the Russians would get ahead, or the health of the American people would suffer, the culture would decline, or still other misfortunes might occur. As Federal expenditures for research and development rose from under $7-billion in 1959 to $16.5-billion in 1966, the academic scientists who dominated the top Government advisory councils argued that defense, space and atomic - energy activities took the lion's share; leaving academic science with a minor share of the money—between 10 per cent and 15 per cent— but most of the blame for the boundless fiscal appetite of "science." Well, there is justice to that plaint, but in the public mind, it's all "science," whether it's a space shot sent aloft by engineers to test a missile nose cone or true science, such as basic biochemistry questing for a better understanding of cell processes. The politicians began to balk, and soon relatively hard times set in for a generation that, from graduate school onwards, had become accustomed to more every year. At first the wails did not protest an actual regression of funds, but a deceleration in growth, often speciously referred to as "cuts" when science lamented aloud. But then came real cuts and accompanying them were basic inquiries from the budget makers: What, after all, is the true value of science, in terms of cost-effectiveness, for improving health, industrial productivity and social well-being? Mainly through special studies convened by the National Academy of Sciences, the high temple of science, the research community responded with many-paged, vaporous replies, generally to the effect that investment in research inevitably works out to the good, but there is no way of knowing how beforehand. Typical was the 1967 assertion of Philip Handler, then chairman of biochemistry at Duke, chairman of the National Science Board, member of P.S.A.C. and currently president of the Academy. "The edifice which is being created by science," he said, " . . . is fully comparable to the cathedrals of the Middle Ages or to the art of the Renaissance. . . ." More prophetic in terms of political sentiment, however, was the observation of Harry G. Johnson, professor of economics at the University of Chicago, who, when asked by his scientific brethren to join in a defense of science, observed that "insistence on the obligation of society to support the pursuit of scientific knowledge for its own sake differs little from the historically earlier insistence on the obligation of society to support the pursuit of religious truth, an obligation recompensed by a similarly unspecified and problematic payoff in the distant future."

In the fertile soil provided by Mr. Nixon's own particular attitude toward research and its practitioners, the difficulties that had sprouted during the preceding years began to flourish. Rising unemployment among scientists and engineers, never very high, but startling for being there at all in the face of the research community's repeated warnings of trained - manpower shortages, caused the Nixon Administration to query why the Federal Government should continue heavy subsidies for graduate training. No persuasive answer was forthcoming, and gradually the budgetmakers began to prune and then virtually eliminate this support, their reasoning being that if aspiring lawyers, architects and business managers are able to get educated without direct Federal assistance, there is no reason why aspiring chemists, physicists and mathematicians cannot do the same. And then, as attention increasingly focused on the na-

SCIENCE AND SOCIETY

tion's assorted ills, particularly the international competition that had begun to develop in the previously golden field of high-technology exports, questions began to arise about the return that was being realized from the Federal Government's investment in research and development. A key clue to Mr. Nixon's thinking on this matter appeared in September, 1970, when he appointed Dr. David, an engineer and psychologist, to be his Science Adviser. Traditionally, the Advisers had come from academe, but Dr. David's 20-year career had been in communications research and management at Bell Labs. Introducing him at a brief ceremony in the White House Rose Garden, Mr. Nixon several times said of Dr. David, "He is a very practical man."

Practical he may have been, but influential he was not, for the apparatus that Dr. David presided over was still tainted by its independent, liberal and academic reputation. The White House simply could not tolerate it. And though the elders of science had long since forsaken the "cathedral" and "Renaissance art" metaphor to justify requests for Government patronage, questions persisted as to what we were getting from these massive expenditures.

A further clue to Mr. Nixon's sentiments came in the fall of 1971, when the White House announced it had ordered a massive examination of "technological opportunities" related to industrial productivity and domestic social problems — and had assigned directorship of the study to William Magruder, chief of the very SST project that one of O.S.T.'s consultants, Dr. Garwin of I.B.M., had helped shoot down in Congress. If any doubt about O.S.T.'s place in White House esteem still existed, it was dissolved by that bureaucratic affront. Magruder energetically went about his task, and eventually produced a multi-billion-dollar list of promising research possibilities that the Federal Government might help pursue. The list was dutifully examined by the economists and lawyers of the Office of Management and Budget, who concluded that little or no firm evidence had been adduced to establish that the proposed expenditures would produce any reasonable payoff. When last year's budget was published, Administration officials contended that some $700-million of proposed expenditures reflected Magruder's proposals in such fields as energy research, pollution abatement, crime control and transportation. But the only clearly identifiable newcomer was a $40-million item, to be jointly administered by the National Science Foundation and the National Bureau of Standards, to conduct "experiments" on collaboration in innovation among Government, industry and academic research organizations. Explained an official of O.M.B.: "Frankly, we don't think anyone really understands how ideas get translated into marketable and socially useful products. We want to study it before we start paying for it heavily."

At Dr. David's prompting, Nixon subsequently issued the first Presidential Message on Science and Technology; its thrust was that, while basic research must remain an important Federal responsibility, the time had come to reorient the national research enterprise toward the solution of domestic problems. The rhetoric is, of course, commendable, but, in fact, the carve-up of Federal research and development expenditures still remains heavily weighted toward national security affairs, with the Department of Defense receiving not only approximately half of all Federal r. & d. funds—$8.3-billion out of a total of $16.7-billion budgeted for next year—but also receiving $460-million of the $904-million growth incorporated into the new budget.

All of this leaves two questions: First, is this a proper time to dispense with full-time science advice at the White House level? Second, will the budgetary jolts now being experienced by research produce serious harm or cause valuable opportunities to be missed?

In answer to the first, it is clear that advice cannot be provided to he who does not wish to be advised. The formal dismantling of O.S.T. simply reflects the nearly complete erosion of its influence during Mr. Nixon's first term. And yet that erosion could well prove unwise for the Administration as well as for science. If the Administration is sincere in its stated desire to reorient federally supported research activities toward domestic problems, then it is all the more important to attach Presidential authority and prestige to the task of coordinating the sprawling Federal research enterprise, selecting from among the far-too-many technological opportunities those that merit priority, and, in general, looking after the care and feeding of that delicate entity known as the "research community."

As for the future of research, it is true that cutbacks will have no immediate devastating impact. The Nobel Prizes continue to flock to American research workers, and this is often cited as a measure of sustained American quality. But complacency is dangerous. The prizes are often awarded for work performed a decade back, and it is over the past five years or so that American science has suffered its most serious financial and administrative shocks. It is undoubtedly still the most productive research community in the world, and research workers in many industrialized nations tend to regard with amusement the austerity complaints of their American colleagues. But as funds shrink and bright youngsters increasingly shy away from careers in science, the situation that is developing is not unlike that of a baseball team with a superb lineup of starters, a sparsely filled bench and a decaying farm system. The Einsteins and the Fermis are going to get started in research and win renown no matter what the Federal Government does or does not do about supporting the training of graduate students. But according to the most recently available figures, Ph.D. output in the sciences declined virtually across the board last spring for the first time since the post-World War II "hump" of G.I. students distorted the curves. Chemistry was down 8.8 per cent; physics and astronomy 6 per cent and agricultural sciences 5 per cent. And further declines are on the way, for in response to a mixture of factors, among them the shrinkage of Federal support for graduate students, enrollments at virtually all of the nation's top graduate centers are dropping sharply. Next fall, Harvard will admit 550 graduate students, compared with 900 just a few years ago; at the University of Wisconsin, graduate enrollment has dropped by nearly 1,000 over the past four years, while at the University of Illinois, similarly sharp cuts have occurred. And the declines in the sciences may be even more severe than these figures suggest, since enrollments in other fields have been increasing.

With the Ph.D. "pipeline" extending from three to six years, the effects on the quality of scientific output are difficult to ascertain. But according to a senior member of Harvard's élite chemistry department, "it's gradually becoming more and more difficult to find promising young faculty members. And whereas we used to be worrying about the 'brain drain' to the U.S., we now find that some outstanding foreign researchers are reluctant to come here because they fear difficulties in obtaining research funds. Maybe it's socially desirable that people are being diverted to other fields, medicine among them, but I think we should take notice of what's happening to science." ■

June 17, 1973

Ford to Seek Re-Establishment Of White House Science Office

Special to The New York Times

WASHINGTON, May 22—President Ford pledged today to key members of Congress that he would act to re-establish as a permanent part of the White House organization, the Office of Science and Technology that his predecessor abolished.

James M. Cannon, the President's assistant for domestic matters, said that Mr. Ford would propose legislation creating the post of science adviser, with a staff of 10 to 15 and an annual budget of $1-million to $1.5-million.

Two Democrats who head Congressional committees supervising scientific matters—Senator Frank E. Moss of Utah and Representative Olin E. Teague of Texas—welcomed the President's offer and said they would begin hearings on the proposal next month. They were among eight members of Congress who met with Mr. Ford this morning.

Scientists and their Congressional allies have been pressing for re-establishment of a formal White House system to provide the President with nonpartisan advice on science and technology.

Mr. Ford's decision to re-create the office by statute, rather than by Executive order as previous Presidents had done, signaled an intention to make it a permanent part of the White House establishment.

The Office of Science and Technology had a staff of about 50 when President Nixon dismantled it three years ago and transferred its functions to the National Science Foundation. Mr. Nixon contended that the action was meant to save money and avoid duplication of effort.

But the scientific community suspected that the decision was based more on the increasingly unwelcome advice that Mr. Nixon and, before him, President Johnson had received from science advisers on such controversial issues as the supersonic trasport and the anti-ballistic missile system.

The first such post was established by President Roosevelt during World War II as the Office of Scientific Research and Development.

May 23, 1975

The Brain Bank of America

An Inquiry Into the Politics of Science.
By Philip Boffey.
Introduction by Ralph Nader.
312 pp. New York: McGraw-Hill Book Company. $10.95.

The other scientific method

By STEPHEN JAY GOULD

No myth deserves a more emphatic death than the idea that science is an inherently impartial and objective enterprise; objectivity has, after all, been battered by everything from Thomas Kuhn to Watergate. Yet it continues to thrive among working scientists because it serves us so well. It works within our profession by inspiring our students and sustaining us through inevitable periods of self-doubt; more crucially, it is the hallmark of our effort in public relations— a self-serving statement that enhances the social prestige and political clout of scientists. It also provides the rationale for America's scientific priesthood: the National Academy of Sciences.

My strongest evidence for this assertion lies in a confident suspicion that most educated nonscientists have never heard of the N.A.S. Though its actions have touched all our lives it has not been subject to previous scrutiny—what could an investigative

Stephen Jay Gould teaches geology, biology and history of science at Harvard University.

reporter gain from studying a well-oiled computer? When Philip Boffey announced his intention, most Academy members reacted by asking "why us?" One academician stated: "If Ralph Nader's center for study of responsive law (whatever that may be) has nothing more useful to do, I suggest that it fold up its tent."

The N.A.S. is a quasi-governmental agency composed of the nation's 1,000 or so top scientists (as defined by themselves). The affiliated National Academy of Engineering has some 500 members. Like the nucleic acids, the N.A.S. has only two functions: replication and transcription. As the sole judge of its membership, the N.A.S. seems to spend an inordinate amount of time in perpetuating itself. The Byzantine procedures for nomination and election do guarantee the eminence of membership, but they also enforce a bias for conservative caution, élite institutions, white males, and advanced age (in 1970, the median age of members was 62). But the N.A.S. is much more than an honor society, for it was created by Congress in 1863 to serve as an official adviser to the Federal Government in scientific matters. In this role of transcribing wisdom, the Academy operates hundreds of committees producing reports on virtually every controversial issue through which science presses upon our lives.

Boffey has had to proceed in I.F. Stone's manner—from the public record and from interviews, since the Academy does not divulge its internal documents. I am always delighted to see how much this method can reveal, for it gives me hope that most enlightened opinion requires only basic intelligence and dogged persistence rather than privileged access. The book is largely a series of case studies detailing the Academy's approach to controversial questions: the storage of radioactive wastes, the S.S.T., defoliation in Vietnam, food safety, pesticides and airborne lead. These issues, to be sure, represent a non-random sample of committee reports, but the procedures they reveal are disturbing, especially when combined with Boffey's documentation of the influence of the reports in supporting Governmental actions against ecological and political critics.

Two aspects of N.A.S. committees seem to preclude their impartiality. First, they are funded directly by the very agencies whose policies they might oppose. Moreover, these agencies are not adverse to cutting off the funding of a rambunctious committee, and the N.A.S. brass, in such crises, has generally knuckled under. In 1967, for example, the Atomic Energy Commission cut off funds for the Academy's committee on radioactive waste—a courageous group, dominated by geologists who knew that Governmental plans for disposal were economical but unsafe. The Academy made no effort to obtain additional

SCIENCE AND SOCIETY

support, but rather reached a "compromise" with the A.E.C. They dismissed their original committee, gave the A.E.C. both an implied veto over new membership and the right to suppress reports not to its liking and established a new committee which was supervised by chemists who knew very little about the geological dangers of disposal and also, in certain cases, had an obvious bias in the A.E.C.'s favor. (The first chairman had served as an assistant director of the A.E.C.'s reactor division, the very division that had opposed the original committee.)

The second aspect precluding impartiality, the cult of the "expert," has led to the appointment of committees dominated by scientists who are indisposed to criticize the funding agencies. If the myth of impartial objectivity were true, it might be reasonable to weigh a committee on the S.S.T. with our greatest aeronautical engineers (who invariably work or consult for the aircraft industry), and it might be best to stock a pesticide committee with the people who know most about the chemical action of insect killers (and who invariably have ties to the chemical industry). It might even be expedient to allow an agency to fund its own critical reports—to let the Pentagon sponsor a fair account of the ecological and medical consequences of defoliation in Vietnam, for example.

But the myth of impartial expertise is false, and the committee reports have often been superficial, compromised or whitewashed. The Academy, in addition, is often caught with some of its other biases showing—as when it selected many American businessmen and not a single representative from a developing country for a study group to determine how American corporations might strengthen the technical capacity of developing nations. And when, in an embarrassing display of zealous devotion to their provider, the S.S.T. committee went beyond its stated task and advised the Government on tactics for a public relations campaign to win public approval for the fast but noisy aircraft. The committee suggested, for example, that "the military agencies would wish to show the necessity for the supersonic plane for national defense."

There is also a human problem quite beyond any general argument about objectivity. The Academy fills its committees with renowned men who are busy to the point of distraction with their own careers. They are asked to serve, without pay, and analyze some of the most difficult problems that science can pose. Is it any wonder that their reports are often superficial and sloppy, and that they accept too readily the "official" version of many tales?

Boffey closes with some sensible suggestions for reform—a wider net for committee membership and the public announcement of projects, for example. But no reform can guarantee the Academy's impartiality in a society where big science depends upon Governmental support. It is, rather, remarkable that, with all the procedures upon it, the Academy so often produces excellent and courageous reports. This is the gentlest and yet the most telling point. The members of the Academy are not evil schemers or conscious conspirators; they are, in general, good and fair men. But conformity and compromise too often characterize the illusory pursuit of influence—better to mute one's voice and retain a tiny office in the corridors of power. But do you exert influence this way, or are you merely used?

A system lies most nakedly exposed because the best and the brightest are willing to operate so uncritically within its constraints. ■

May 4, 1975

SCIENCE AND POLITICS

THE "HEREDITY COMMISSION."

According to a Washington dispatch to The Brooklyn Eagle, Mr. WILLIAM HAYES, Assistant Secretary of Agriculture, has chosen a committee of scientists from the American Breeders' Association, of which he is the President, to investigate all proper means of influencing heredity with the idea of encouraging the increase of families of good blood, and of discouraging the vicious elements in the cross-bred American civilization. This announcement follows Mr. LUTHER BURBANK's interesting statement in the current Century Magazine that in the hybridization of the fifty distinct nationalities in America a new species of human being may be consciously evolved.

Mr. HAYES would invite science and religion, co-operating with Government efforts, "in an investigation at once conservative, careful, and possibly constructive." There is nothing intrinsically objectionable in the notion of invoking afresh these institutional forces for the purpose of bettering the race. They are already at work to this end. Marriage, sanctified by religion, safeguarded by law, and attended in its crises by fostering science; the prescriptions of custom and the legal provisions for educating children, and the ordinary ways and barriers of human intercourse are designed for the uplift and improvement of the human type. But in coming years custom, law, and religion may themselves be profoundly modified by the advances in science.

In a plea for the conscious improvement of the species FRANCIS GALTON notes that man has affected the quality of organic life so widely that the changes on the earth's surface, merely through his deforestings and agriculture, would be recognizable from a distance as great as that of the moon. But less obvious and more subtle organic changes have been wrought. Man has gained radical control of the processes of natural selection in plant and animal life, producing within brief periods results that nature unaided can compass, if at all, only in the breadth of centuries. Mr. BURBANK, for example, has given to the world numberless strangely beautiful and useful types of plants. The American Breeders' Association and kindred bodies abroad have exhibited their skill in the improvement of the various types of domestic animals. It is natural, therefore, that practical inquiries prosecuted in these lines should extend to the improvement of the human species.

But a "heredity commission" can be influential only by placing the results of its investigations at the service of education, which, if history be not belied, will eventually become crystallized in law and religion. Pride of family, as anciently, may be successfully appealed to by the scientists. The people of the West have to-day a keener appreciation and consciousness of their descent from the hardy pioneers than had they in turn of their inherited attributes from the choice and adventurous spirits of the Old World. A refined directive consciousness is forming, which will have its fruits in the betterment of the future type, quite as subtly and as powerfully, too, we venture to predict, in the strains of the "effete East" as in the more exuberant Western stocks.

May 20, 1906

FOR MORE SCIENCE IN WAR.

Sir William Ramsay Advocates a Central Advisory Body.

Special Cable to THE NEW YORK TIMES.

LONDON, Friday, July 2.—The eminent scientists and chemists, Sir William Mather, Sir William Ramsay, Sir Roverton Redwood, Sir Philip Magnus, M. P.; Professor Perry, Sir Ronald Ross, Sir Archibald Geikie, and Sir Alexander Pedler, condemned, at the annual meeting yesterday of the British Science Guild, the attitude adopted by the Government toward science in connection with the war and demanded that in the future greater use should be made of opportunities afforded by scientific knowledge.

"The history of the relation of Government to science is a painful one," said Sir Archibald Geikie, "and as an old civil servant I have seen the true inwardness of it. There always has been a sort of antipathy in the official mind toward science, and it continues to this day."

Sir William Ramsay said this was a war in which pure and applied chemistry played an all important part. As early as Aug. 4 the French Academy of Sciences offered all its resources to the French Government. That offer was accepted and commissions were appointed on every phase of the subjects, whose reports and suggestions went direct to the chamber. In this country nothing of the sort has been done.

Sir William instanced refusal of the Government to declare cotton contraband against the advice of scientific experts. Had they made cotton contraband in January when they were implored to do so, he said, in all probability Germany would have been running short of propulsive ammunition. It was notorious there was very little intercommunication between the various Government departments. The prime necessity of the present moment was a central body of scientific men to whom the various Government departments should be compelled to apply for advice and assistance. Moreover, it should be within the province of such a central organization of science to propose new methods of circumventing the enemy. Above all there must be no delay. Government departments are proverbially deliberate, but now speed means life, delay death.

July 2, 1915

CHEMISTS BAR GERMANS.

Club Takes Questionnaire to Eliminate Enemies.

The Trustees of the Chemists' Club have sent a letter and questionnaire to each member of the club for the purpose of eliminating from the membership all persons not heartily in accord with the attitude of the United States in the war.

The letter says that the German language shall not be used in conversation in the club; that all disloyal citizens of the United States or its allies must avoid the club, and that any member of the club, of whatever descent, and whether an American citizen, whose sympathies favor the enemies of this country, must resign forthwith. The questionnaire, when the returns are in, will make a complete record of each member's birth, and other antecedents.

"Chemistry is something that has for so long been intimately connected with things German," an officer of the club said last night, "that it was natural that Germans or German-Americans should feel an interest in the club and apply for membership. When this country entered the war last year, however, many of these left the club. The recent letter of the Trustees resulted in a few resignations, but the Chemists' Club is, and always has been, in the front rank as a patriotic institution, this being emphasized especially at this time when so many of its members are performing invaluable assistance to the Government and American war industries."

May 5, 1918

Nobel Prizes That Go Begging.

Two of the Nobel prizes for 1919, it appears, will not be awarded. The prizes in economics and medicine were to be assigned to Frenchmen, but in advance of the formal announcement both refused to accept the honors because the Swedish body which gives the prize for chemistry had seen fit to honor Dr. HABER, inventor of the poison gases used by the German Army. No doubt there will be protests against this "narrowness" from those who profess to believe in the international character of science—a doctrine worthy enough, but somewhat beside the point. One recalls the recent appeal for the erasing of the hard feelings aroused by the war which was published in France, in Austria, and in America by some of those who call themselves intellectuals. The French intellectuals who were eager to forget about the war all at once seem to have been almost wholly of the group centring about the sufficiently known M. BARBUSSE—gentlemen, that is, who have no interest in any war except the Bolshevist crusade against the world, in the interest of which they toil so earnestly. As for the American signatories, nearly all of them bore German names or were known for friendship with Germany.

So, though Dr. HABER undoubtedly has many scientific achievements to his credit besides his work in poison gas, and though the Swedes who made the award had probably no invidious intention, general sympathy will be felt with the Frenchmen who did not care to be honored in such company. One may wonder, indeed, why the Nobel prize for idealistic and imaginative literature was not given to the man who wrote General LUDENDORFF's daily communiqués to the German people on the absorbing subject of strategic retirements.

January 27, 1920

WAR WORK SHOCKS BRITISH CHEMISTS

Prof. Soddy's Refusal of a Government Invitation Stirs a Controversy.

RESEARCH ON POISON GAS

War Office Defends Step—Maurice Hears of New American Shell.

Copyright, 1920, by The New York Times Company.
Special Cable to THE NEW YORK TIMES.

LONDON, Nov. 10.—In the current issue of Nature is printed a letter from Dr. Frederick Soddy, a well-known Professor of Chemistry at Oxford University, stating that he had received an invitation from the British War Office to become an "associate member of a committee now being constituted as part of the new peace organization for chemical warfare, research and experiment."

"The function of the committee," added Dr. Soddy, "was stated to be the development to the utmost extent of both the offensive and defensive aspects of chemical warfare."

SCIENCE AND SOCIETY

He intimated that he was against accepting the invitation, as he felt that universities and scientific men stood for something higher than anything that had yet found expression and representation in governments, particularly in their international relations.

In nonscientific quarters it was suggested that the War Office was running counter to the League of Nations, which at its Brussels conference passed a resolution against the use of poison gas. On behalf of the War Office it was asserted that other countries were pursuing research along this line and that this country could not afford to drop behind.

General Sir Frederick Maurice, writing in The Daily News, says:

"The invitation issued by the War Office to scientists to form committees to study the application of chemistry to warfare, both offensive and defensive, has shocked a great number of people, including some scientists. Personally I am glad we are acting in the open in this matter and not, as are some other nations, in the dark.

"The effect of such a declaration must be to make clearer what the powers and functions of the League of Nations really are. It will be found, I suggest, that the League is something quite different from that which is approved by a number of people and ridiculed by a great many more. The League acts by consent, not by compulsion. It cannot compel the abolition of poison gas or the reduction of armies and navies. The Council has the question of the use of poison gas under consideration and finds itself considerably embarrassed in the matter.

"The potentialities of poison gas are so terrible that the Council cannot take upon itself the responsibility of advising the members of the League to abjure its use unless and until it can protect them adequately. At present it cannot possibly do that. The League cannot prevent a scientist from working out a formula for a gas far more deadly than any used in the great war, nor can it alter the fact that many chemical industries, such, for example, as the dye industry, can be used for the rapid production of gas. The members of the League cannot pledge themselves not to manufacture and use poison gas unless they are assured that it will not be used against them.

"At the present time the United States, which is not a member of the League, is understood to be experimenting with armor-piercing shells filled with a very deadly gas. This adds a new terror to naval warfare which other nations dare not neglect."

November 11, 1920

EINSTEIN TO LEAVE BERLIN.

Is Aroused by Unfair Attacks on Relativity Theory.

Copyright, 1920, by The New York Times Company.
Special Cable to THE NEW YORK TIMES.

BERLIN, Aug. 27.—Local newspapers state that Professor Albert Einstein will leave the German capital on account of the many unfair attacks made against his relativity theory and himself. These attacks recently took the shape of public lectures by alleged scientists who under the name of the "coalition of German natural philosophers" combined to make it unpleasant for Einstein. Nevertheless, the latter attended the first of these lectures expecting to see his theory criticised in a scientific manner. Instead, the first lecturer, Dr. Weyland, who was styled "a well known philosopher," concentrated his attacks mainly on Einstein's person and accused him of a "businesslike booming of his theory and name," while the second lecturer, Professor Gehrke, seemed only superficially conversant with the relativity theory.

Replying to these attacks in the Berliner Tageblatt today, Einstein says he personally would not condescend to reply to either one of his attackers, but friends have pressed him to do so. Einstein intimates that neither of the men acted from the creditable motive of striving for the truth, but for quite different reasons.

"If I were a reactionary nationalist wearing an anti-foreigners badge instead of being a Jew of liberal international ideas I might have been treated quite differently by those gentlemen," he said.

There certainly is no denying that the first anti-Einstein lecture had a decided anti-Semitic complexion, which applies equally to the lecture and to the large part of the audience.

August 29, 1920

Einstein Theory 'Bourgeois' And Dangerous, Say Russians

Copyright, 1922, by The New York Times Co.
Special Cable to THE NEW YORK TIMES.

PARIS, Nov. 15.—A message from Moscow to the Echo de Paris says that Professor Albert Einstein has been solemnly excommunicated by the Russian Communists.

At a special council meeting held in order to examine the question the Russian Communist Party condemned the Einstein theory as being "reactionary of nature, furnishing support for counter-revolutionary ideas"; also as being "the product of the bourgeois class in decomposition."

Professor Timirazeff presented a long report to the council in which he discussed whether Einstein's theories could be reconciled with the theory of materialism. He decided that they could not, and because, in his opinion, they led to "pure idealism," the council pronounced condemnation.

November 16, 1922

SOVIET DECREES END OF ABSTRACT SCIENCE

Conference Is Now Being Held to Harness All Research to Socialist Ends.

BUKHARIN LEADS ATTACK

He Says Pure Science Is Cloak for Reactionaries, but Parley Sidesteps View.

By WALTER DURANTY.
Wireless to THE NEW YORK TIMES.

MOSCOW, April 8.—Not content with existing troubles—great enough in all conscience, though perhaps sometimes exaggerated—the Kremlin has now taken upon itself a new battle, namely, to harness science for Socialist service.

There is now being held in Moscow what is purposely entitled "A Conference for Planned Organization of Scientific Investigational Work." It is interesting to note that the principal speech was made by Nikolai Bukharin, co-member with Alexei Rykoff, who also was recently restored to grace, and Mikhail Tomsky of the Right Opposition "troika" [trio].

M. Bukharin proved worthy of the Kremlin's renewed trust. He declared:

"So-called pure science, that is science devoid of contact with practical life, is a figment. [Would Einstein agree to this?] The whole fabric of scientific investigational work in capitalist countries is a weapon in the hands of capitalist magnates and governments and their industrial and military organizations.

"We Bolsheviki, on the other hand, have demanded a gigantic increase of scientific effort in the whole system of Socialist construction in the Soviet Union. The problems before us require a decisive and categorical break with bourgeois traditions of old academism and their conversion to the task of solving immediate, practical difficulties."

M. Bukharin continued that Russian scientists must now choose whether to throw in their lot wholeheartedly with the Socialist system or "hide their hostility behind a screen of pure science and its abstraction from the national life."

In the discussion which followed, it appears from newspaper reports, the assembled scientists, whether of the old régime or Communists, were reluctant to accept M. Bukharin's challenge. The newspaper Economic Life states that the speakers without exception avoided the basic thesis which M. Bukharin had outlined and preferred to confine their remarks to matters of detail. It would thus seem that science is not much readier than Pegasus to be tied to the Soviet chariot.

Indeed, Economic Life rubs home this impression by a sarcastic article in an adjoining column about a department of the Leningrad Academy of Science "wasting a whole year" in abstruse investigation of the characteristics of an oakshoot grown in a dark cellar, and whether it is easiest to pull a round, a triangular or a square nail from a piece of wood.

However suitable the soil of Russia may be to Bolshevist doctrines, it is also congenial to the vaguer and more abstract manifestations of science and art. It will be interesting to see what success the Soviet Government will have in bending either or both to the utilitarian aims of Socialist construction.

April 9, 1931

SCIENTISTS FORCED TO QUIT GERMANY

Nazi Policy Is Robbing Nation of Famous Specialists in Research.

MANY TEACHERS 'ON LEAVE'

Professors of Renown Keep Away From Campus and Laboratory to Avoid Boycott.

Special Correspondence, THE NEW YORK TIMES.

BERLIN, Sept. 25.—National socialism's exaction of political and racial conformity has now definitely robbed Germany of the services of hundreds of scientists and scholars. They include men renowned the world over, like Albert Einstein, James Franck, Max Born, Ernst Cassirer; to say nothing of highly trained specialists difficult or impossible to replace. Some lines of research, notably in certain departments of organic chemistry, have been brought almost to a standstill.

Whether the new rulers fail to realize the consequences of their attitude or whether, aware of the consequences, they stand ready to pay the price, seems as yet uncertain. The first alternative seems the more probable.

It is also not yet possible to assay with any approach to exactitude the actual impairment inflicted on German science and scholarship. There are still too many cases "pending"—a host of men, officially "on leave"—whose final disposition remains unsettled.

Government Controls Schools.

It should be realized, first, that in Germany, even more than in the United States, research and teaching in a university or other institution of higher learning are intertwined. Second, all these institutions in Germany are State institutions. There is here no Harvard or Stanford or Johns Hopkins. Third, even the institutes for pure research dissociated from teaching and mostly gathered under the direction of the Kaiser Wilhelm Society, are under direct government control. Thus the German Government is able to dictate the living terms of practically the whole body of German science and scholarship.

Few of the men who have fallen under the National-Socialist—which means the government—ban have independent means. A few physicists and chemists have sufficiently lucrative connections with industry to keep them pecuniarily comfortable even after the loss of their university posts. Such a connection, for example, Professor Fritz Haber has with I. G. Farben. But pecuniary hardship is not the heart of the matter.

The "Aryan" wife of a German-Jewish physician, who lectured on the medical faculty of a German university but desisted after the political upturn because he was informed he would be boycotted, said to this correspondent: "We have quite enough to live on, and my husband still has a good practice. But he is pining away because he cannot work in the university laboratory, because he cannot teach any more. His heart was there."

Listed as "On Leave."

This man, one out of some hundreds, is still in the faculty register of his university for the coming semester, but opposite his name is printed "on leave." He was in the field during the World War. This should give him the right to be confirmed in his university post. Nor have the authorities ousted him from it. But none the less he does not lecture any more or show himself on the university grounds.

The "Aryan" senior in his special line—the man who would have been the head of his department were there departments after the American manner in German universities—advised him to stay away. He was full of sympathy, but said: "What can I—what can we do? If you attempt to lecture, you will only get into trouble. Try to move out of Germany as soon as you can."

This—getting out of Germany—is what many of the German scientists shelved by the Nazi conquest have been strenuously trying to do, but most of them in vain. The majority of these dispossessed seeking new working spheres are Jews, or persons with a Jewish strain in their ancestry.

Interested quarters have come to the conclusion that only international coordination can solve the problem of finding new working fields for the German scientists and scholars rendered homeless by the Nazi conquest. It is urged that the countries of the Occidental civilization—the United States, England, France, and so forth—draw together and agree on quotas of evicted, or self-evicted, German scholars and scientists whom they severally are ready to put to work.

October 8, 1933

Soviet, Fearing Code, Barred A Russian Scientific Chart

Officials of the Soviet Government recently refused to permit a copy of "The Periodic Chart of the Elements"—one of Russia's great contributions to chemistry—to pass through the mails entering that country in the belief that it might be some form of foreign code message, it was said yesterday by Professor Herbert R. Moody, director of chemistry laboratories at the College of the City of New York and a member of the American Chemical Society.

Professor Moody said that the University of St. Petersburg had only a photostatic copy of the chart. A City College student, who is a White Russian, therefore sent two copies of it to his sister, who is a student at the university. Both were returned by the Soviet authorities, Professor Moody said.

December 14, 1931

NAZIS WOULD JUNK THEORETIC PHYSICS

Einstein School Denounced for Trying to Impose a 'Measure of All Things.'

STUDENT STARTS DEBATE

His Attack on 'Jewish' Science Seized Upon as Material for Anti-Semitic Campaign.

By OTTO D. TOLISCHUS.
Wireless to THE NEW YORK TIMES.

BERLIN, March 8.—Six German professors, all winners of the Nobel Prize for physics, are now engaged in a public controversy on the issue, "German Physics vs. Jewish Physics."

This controversy which, significantly enough, is being fought out mainly in the Voelkische Beobachter, Chancellor Adolf Hitler's own newspaper, is part of great activity in all lines of "Kultur" stimulated by the highest authority. Nazi leaders, including Dr. Joseph Goebbels, Alfred Rosenberg and Bernhard Rust, have been addressing mass meetings, such as only Nazis can organize, on the general topic of the National Socialist "Weltanschauung" and "Kultur."

Vast sums have been spent on the production of classic drama in superlative style. Germany's best talents, even those formerly outlawed, such as Paul Hindemith and Wilhelm Furtwaengler, have been mobilized to restore the pre-eminence of German music and opera. And since nothing interests the German public more than a fight, a whole series of politically innocuous controversies has been launched to demonstrate that honest minds can still clash in the Third Reich despite regimentation.

Debates Cover Wide Range.

One such controversy deals with the merits of modern German adaptation of Shakespeare's plays, with respect to which Dr. Goebbels has reserved for himself the rôle of supreme arbiter. But the most interesting controversy is that of the physics professors, illustrating the extent of the confusion wrought, even in eminent minds, when science is combined with politics and racial mysticism.

The exponents of "German" physics in this controversy are Professor Philipp Lenard, discoverer of "Lenard's rays," Nobel Prize winner in 1905 and now head of the Philipp Lenard Institute of Physics at Heidelberg, and Professor Johannes Stark, discoverer of "the Stark effect," Nobel Prize winner in 1919 and president of the German Physics Institute and the German Research Association.

Their opponents are Professor Max Planck, Germany's most eminent physicist, creator of the quantum theory, on which modern physics is based, Nobel Prize winner in 1918 and director of the Institute for Theoretical Physics of the University of Berlin; Professor Max von Laue, Nobel Prize winner in 1914, and Professors Erwin Schroedinger and Werner Heisenberg, Nobel Prize winners in 1933.

The controversy started when a student of physics, Willi Menzel, whose scientific attainments are still to be revealed to the world, but whose party orthodoxy apparently is unchallenged, published a violent attack in the Beobachter against Professor Albert Einstein and against all theoretical physicists as Jews or products of the Jewish spirit.

No Nazi Physics, He Admits.

He modestly admitted there was no National Socialist physics, but maintained that there was a "German" physics, which he defined as "experimental research into reality in inorganic nature caused by the joy of observing its forms of reaction." "Jewish" physics, as he defined it, "aims to make physics a purely mathematical thought construction, propagated in a characteristically Jewish manner."

The idea of using the purely scientific contrast between theoretical and experimental physics for a National Socialist campaign against the Jews, however, is not original with Willi Menzel. It is based on the violent diatribes in the same direction uttered repeatedly by Professor Lenard, who maintains that science "is conditioned by race and blood," and by Professor Stark, who denies that theoretical science has any merit whatever and denounces "the Jewish propaganda that makes Einstein the biggest scientist of all times and seeks to impose Jewish views as a measure of all things."

Mr. Menzel's attack was answered by Professor Heisenberg, who declined to follow his opponent into the field of political anti-Semitism, but confined himself purely to the defense of theoretical physics, citing in particular Professor Planck as an authority, demonstrating how through it new experiments had been stimulated, and above all their results had been coordinated and explained.

This answer, however, was followed by a statement from Professor Stark, commending Mr. Menzel, expanding the attacks on Dr. Einstein to all those who support the Einstein ideas or methods, and concluding with a demand that their influence be excluded in deciding future university appointments.

In this controversy the weight of numbers and authority seems to be on the side of the theoretical physicists, but the "German" physicists are winning out because they have greater party orthodoxy on their side.

March 9, 1936

British Scientists in Rebellion Against War's Abuse of Finds

Leaders at Blackpool Meeting Condemn Government Officials' Attitude, Demanding Share in Directing the State—Reform of Schools Urged to Provide Studies of Technical Achievement.

By WALDEMAR KAEMPFFERT
Wireless to THE NEW YORK TIMES.

BLACKPOOL, England, Sept. 12.—Unlike the meetings of the British Association for the Advancement of Science held in the past decade this year's is restless, even rebellious. Fascism and the dread of war have caused chemists, biologists and engineers to break the bounds of scientific reserve.

The social effects of science thus far have been the main consideration at Blackpool because science has been under fire for the last decade. In 1927, when the association met at Leeds, Bishop Burroughs of Ripon appealed for a ten years' truce in research and the cessation of all discovery till the world could catch its breath and assimilate what had been achieved.

More recently Dean Matthews of St. Paul's Cathedral suggested that men of science should make it a point of honor to keep secret any discovery that would be useful in war. Others hold that science is responsible for overproduction and unemployment and charge it with indifference to the social consequences of its own progress.

Hence the resentment in the addresses delivered at Blackpool whenever the opportunity for discussing the social aspects of science is presented. The phenomenon is strange. Science usually goes on its way, serenely indifferent to applause, misunderstandings or opposition. Now it fights back.

The revolt takes the form of attacks on constituted authority for its failure to grasp the potentialities of science, to cultivate research, to adapt the educational system, to inculcate the cultural aspects of science and to recognize the fact that this is a scientific age. In addition scientists openly express the opinion that they have been exploited by industry.

Sir A. Daniel Hall, scientific adviser to the Ministry of Agriculture, said:

"Science means power. It has given no consideration to whom that power should be entrusted and to what end it should be used. The acceleration of productive means has led to overproduction and unemployment.

"New developments of war methods put appalling powers of destruction into the hands of governments, and the greatest of all dangers lies in the temptation now offered to those who wield this scientific power. Once having gained control of the machinery, they can wipe out any further exercise of the popular will.

"Of old every autocracy ended in revolution. What chance has an uprising today against machine guns and gas bombs?"

Because of the abuses of its discoveries, for which research itself is not responsible, he pleaded for the organization of men of science "to make their point of view prevail."

He said he realized that interference with governmental processes was alien to their temperament and that they preferred to remain in their laboratories to make discoveries and not concern themselves with the application of these same discoveries to achieve unsocial ends.

"To continue in this frame of mind is to accept slavery," he declared.

To illuminate the official attitude of the average statesman to science he told the story of an American politician who waved his hand symbolically over a group of university professors and remarked, "Brains! I can buy that lot for $20."

Chemist Welcomes Unrest

Professor J. C. Philip, brilliant president of the chemical section of the British Association for the Advancement of Science, was no less combative. He welcomed the signs of restlessness among scientific workers and declared:

"They are increasingly impatient at the extent to which their knowledge is made to serve inhuman ends."

He even dared to wonder whether the "My country, right or wrong" kind of patriotism was compatible with the spirit of science.

"Impelled by patriotic motives," he said, "most scientists have put themselves freely at the disposal of the State in time of need. But many are hesitating to admit that patriotism must always override considerations of humanity.

"Whatever be our individual attitude in this matter it is time for chemists and scientists in general to throw their weight into the scale against the tendencies now dragging science and civilization down and debasing our heritage of intellectual and spiritual values."

These fervid words were not received in silence but with resounding applause.

With Sir Richard Gregory, distinguished editor of "Nature," it was the same.

"Modern technical achievement and scientific thought," he said, "foreshadow a new economic structure and should be used to exercise the right influence on the major policies of the State. But what do we behold? Political economists seem unable to adjust this new force to national needs and human life."

Overproduction and technological unemployment are to him ironical consequences of the State's failure to apply scientific methods to solving social problems.

Like Sir A. Daniel Hall and Sir Richard Gregory, he protested against the abuse of discoveries that have military value.

"Science," he went on, "must repudiate the methods of cultivated barbarism manifested in modern warfare. If it does not, it must lose the right to be a spiritual influence and acknowledge with despair that man's ethical evolution has reached the culminating point."

Especially bitter was Professor Lancelot Hogben, who occupies the chair of social biology in University College, London. Though he did not spare statesmen in the seats of the mighty, it was the failure of the educational system that brought forth his more caustic comments.

For Popularization of Science

To Professor Hogben the study of what the average Englishman needs in the way of food to keep body and soul together ought to stir the national conscience more deeply than volumes on man's moral or prudential character.

"A restatement of the need for the cultivation of science in terms of modern society is necessary," he said, urging a popularization of science. He maintains the use and misuse of science intimately affects everyday life and should therefore be considered in teaching and cultivating science.

"The pivotal issues of modern education," he went on, "are the production of political leaders who realize the new potentials of human welfare and the training of citizens who will choose leaders with the necessary knowledge to deal constructively with the impact of science on social institutions."

Professor Hogben sees two dangers to the world. One is the failure to anticipate the penalties that must be paid if science is misused. The other is a possible revolt of society against science, hence a lowering of the standard of life.

"We have trained a generation of specialists to mind their own business," he said, "and a generation of statesmen to legislate in ignorance of the technical forces which control the character of social relations."

Another voice of protest was that of an eminent engineer, Professor William Cramp. To him "engineering is the greatest instrument of civilization the world has ever seen, in the sense that it continually tends to promote closer contact, greater intimacy and therefore profounder understanding between individuals and nations. Three-fourths of the work of the engineer is devoted to the development of communications."

Professor Cramp cited roads, canals, bridges, railways, harbors, ships, motor cars, airplanes, telegraphs, telephones and television. These he called "humanity's hyphens."

"Their natural effect," he went on, "is to foster friendliness and dissolve differences. Left undisturbed by the politician, scaremongers and patriot, the engineer would demolish the Tower of Babel and render war impossible. Build a Channel tunnel, then Calais and Dover become neighbors and Anglo-French understanding ensues in all senses. Place transmitters in the trenches with receivers and televisors at home, and war becomes unthinkable.

"The very first thing the government does on going to war is to seize and control every means of communication and every engineering device that must otherwise serve to unite the combatants."

He protested against the manufacture of instruments of war for most European nations.

"Verily for the promotion of peace and understanding engineering easily outclasses every religion; and for battle, murder and sudden death has no equal," he asserted.

As a practical engineer Mr. Cramp realized the hopelessness of overcoming the obstacle to universal peace in the present frame of national minds, but he did suggest how Britain at least might improve her own social conditions by national utilization of scientific resources. He demanded "proper representation of science upon all government bodies in industry and upon all technical departments in the State."

September 13, 1936

MOSCOW CANCELS GENETICS PARLEY

Nazi Racial Theories Ascribed to Some Scientists Causes the Dropping of World Congress.

AMERICANS WERE TO GO

Prof. N. I. Vaviloff, a Famous Plant Expert, Is Arrested—Others Under Attack.

Wireless to THE NEW YORK TIMES.

MOSCOW, Dec. 13.—The Seventh International Congress on Genetics, which was to have been held here next August with a thousand of the world's leading scientists in this field participating, has been canceled by order of the Soviet Government, it is learned unofficially. Several British scientists who had expected to attend have been informed by Moscow that the congress will not be held.

About 100 Americans had been expected to attend, about forty of whom, including such authorities as Drs. C. P. Bridges, T. S. Painter, Sewall Wright and G. H. Shull, were preparing papers. A score of British geneticists had been expected, including Julian Huxley and Viscount Haldane.

An interesting story of a schism among Soviet scientists, some of the most prominent among whom are accused by Communist party authorities of holding German Fascist views on genetics and even being shielders of "Trotskyists," lies behind the cancellation. The fact that so many of the Soviet Union's most distinguished geneticists are under fire is believed to be the motive for the government's action.

Attacks "Classical Geneticists"

In the past three months T. D. Lysenko, botanist, who has won great acclaim and high favor with the government for his experiments in the "vernalization" of wheat and other agricultural products to shorten the growing season, has been attacking the "classical geneticists" in the monthly scientific magazine, Socialist Reconstruction of Agriculture.

He challenged the validity of classical genetics, including the Mendelian laws and the chromosome theories, and stigmatized them as "formalistic" and of no practical value, whereas his work, he said, is producing useful results. Mr. Lysenko said, "Genetics is merely an amusement, like chess or football," and he attacked the All-Union Institute of Plant Industry at Leningrad, headed by Academician N. I. Vaviloff, as useless.

Americans had a special interest in the congress because Professor Herman J. Muller of the University of Texas, who during a four-year leave here has attracted world-wide attention with his experiments on mutations of the fruit fly, was chairman of the program committee. He is now head of the Department of Mutations and Genes of the Institute of Genetics, Academy of Sciences of the U.S.S.R. Dr. Muller is assisted here by Dr. Daniel Raffel, Johns Hopkins graduate and a nephew of Gertrude Stein.

Among Soviet geneticists now under fire are Professor S. G. Levit, head of the Medico-Genetical Institute here, who was general secretary of the organization committee for the congress, and Professor Agol, a member of the Ukrainian Academy of Sciences. Both have worked in Dr. Muller's laboratory in the University of Texas and both now have a high standing in the genetics world.

Vaviloff Is Under Arrest

Professors Agol and Vaviloff, who have traveled extensively in America, have been arrested at Kiev on charges understood to involve Trotskyism. Professor Agol has been a Bolshevik since before the revolution. Professor Levit was originally a Menshevik, who joined the Bolshevist party in 1918.

Recently Professor Levit has come under heavy attack in the Communist party press, proving he has come under the displeasure of party authorities, who rule every phase of Soviet life, including the sciences, literature and the arts, as well as economics and politics. This has culminated in Professor Levit's being accused by the science subcommittee of the Moscow City Communist party committee of permitting the development of scientific views hostile to Soviet theory and friendly to Nazis in his institute—which was surprising in view of his published theories.

At a meeting of physicians and biologists a party representative said some Soviet scientists were not only ready to admire false and anti-scientific theories of Nazi biologists but were imitating their methods in their own scientific work.

Now the complete mental equality of all races is as firm a dogma of Soviet faith as inequality is a dogma of Nazi faith. The party representative particularly attacked Professor Shtyvko of Professor Levit's staff for making deductions "resembling the racial nonsense of German Fascists" in a recent paper published in a German scientific journal.

Professor Shtyvko studied fifty-four skeletons of adult victims of the Russian famine during the civil war period and is alleged to have placed them somewhere between the Germans and the yellow race. He attributed this to the strain of famine and civil war.

In another paper, the party representative said, Professor Shtyvko classed the Buryat Mongols—a Siberian people—as mentally equal to 12-year-old Europeans.

Professor B. I. Lavrentyeff, who was appointed chairman of a committee to look into these charges, warned Soviet medical and biological experts that they must protect Soviet science against any anti-scientific theories that might be dragged in.

NAZIS LIFT ANATHEMA ON EINSTEIN THEORY

Rosenberg Says No Limitation Should Be Placed on Inquiry in the Natural Sciences

Wireless to THE NEW YORK TIMES.

BERLIN, Dec. 11.—Professor Albert Einstein's theory of relativity, once denounced by National Socialist scholars as Jewish and un-German, appears to have been restored to grace in the Third Reich by virtue of a decree issued by Dr. Alfred Rosenberg, supervisor of spiritual and ideological training in the National Socialist party.

In his decree Dr. Rosenberg said the National Socialist party could not assume a dogmatic ideological attitude toward various problems of cosmophysics, experimental chemistry and prehistoric geography because these represent problems of the natural sciences, earnest and impartial investigation which is free to every scientist.

Although the relativity theory is not specifically mentioned and the decree seems to concern itself mainly with the vigorous dispute over Nordic and Germanic origins now raging within the National Socialist party, nevertheless, Professor Einstein's famous theory would seem to be included under the rubric of cosmophysics.

One newspaper, the Beobachter, says that attempts to tie the Nazi party or party formations to specific cosmognonic and prehistoric theories have been lacking, but that time has come to proclaim, on the one hand, freedom of scientific research and, on the other, the independence of National Socialist ideology from scientific disputes that do not affect its kernel.

December 12, 1937

December 14, 1936

REICH EXPERTS IDENTIFIED

Scientists Brought Here Said to Have Worked on V-Weapons

WASHINGTON, Nov. 17 (P)—A group of German scientists brought to the United States in September has been identified as the development staff of the Nazis' V-weapon base at Peenemuende. A recent dispatch from Stockholm quoted an unnamed British officer as saying the entire German staff at the rocket-weapon base, about ninety men, was transferred to the United States.

Asked about this, War Department officials referred questioners to a cryptic statement issued several weeks ago that said "certain outstanding German scientists and technicians" were being brought here to help the United States "take full advantage of these significant developments, which are deemed vital to our national security."

November 18, 1945

IS IT SCIENTIFIC?

The first automobiles were saluted with "Get a horse!" The first airplanes weren't greeted at all—they were ignored. Were we wrong! But never again. Now we're "scientific." We watch a television show, and even if we're disappointed we do not think of saying, "Get a movie projector!" Not one of us in a thousand has ever seen a helicopter, let alone flown backward and sideways in one, but we accept as fact that after the war we'll be commuting in helicopters and the helicopter-parking problem will be a national issue.

No doubt about it, the pendulum has made its full arc. The physicist, the chemist, any white-jacketed man in any test-tube-filled laboratory—they all have us under their thumbs. Maybe that's where we belong, but the position has some disadvantages. We'll believe any yarn now that wears the label "scientific," sometimes even if it has a Goebbels label.

Suppose we hear this story next: The Nazis have a secret weapon that uses a newly discovered rocket principle to hurl an 8,000-pound projectile 850 miles at 700 miles an hour, said projectile guiding itself by a television set installed in its nose and, therefore, able to pick out any given target. Will we believe it? Well, we've been conditioned that way, like Pavlov's dogs, and even if it doesn't sound true, it certainly sounds scientific. You never can tell what science will do next.

November 21, 1943

Einstein Group Incorporates

Special to THE NEW YORK TIMES.

TRENTON, N. J., Aug. 12 — Prof. Albert Einstein, internationally known scientist of Princeton, and seven leaders in the study of atomic energy formally incorporated today as the Emergency Committee of Atomic Scientists.

The incorporation papers stated a principal objective would be the study of atomic energy that would produce results "beneficial to mankind."

Associated with Professor Einstein as trustees in filing the documents with the New Jersey Secretary of State, Lloyd B. Marsh, are R. F. Bacher and H. A. Berthe of Cornell University; Edward U. Condon of the United States Bureau of Standards, Washington; T. R. Hogness, Harold C. Urey and Leo Szilard of the University of Chicago, and V. F. Weisskopf of the Massachusetts Institute of Technology.

August 13, 1946

SCIENCE AND SOCIETY

WIENER DENOUNCES DEVICES 'FOR WAR'

M. I. T. Mathematician Rebuffs Bid to Harvard Symposium on Calculating Machinery

Special to THE NEW YORK TIMES.

CAMBRIDGE, Mass., Jan. 8 — Professor Norbert Wiener of Massachusetts Institute of Technology, one of the world's leading mathematicians, refused to address a symposium on calculating machinery at Harvard University today because, he said, the devices under discussion were "for war purposes."

Dr. Wiener, in a letter titled "A Scientist Rebels," in the current issue of The Atlantic Monthly, wrote:

"I do not expect to publish any future work of mine which may do damage in the hands of irresponsible militarists."

He had been scheduled to speak at the morning session of the symposium in the Harvard Computation Laboratory. Reporters discovered that his name had been crossed off the printed program. Reached by telephone, he confirmed that he had refused to appear because of his stand on use of scientific advancements for war purposes.

The Harvard Computation Laboratory is working exclusively on Navy contracts, although the symposium in progress there, marking the opening of the laboratory, is under Harvard sponsorship.

Dr. Wiener's letter in the January Atlantic was written to a research scientist of an aircraft corporation who had asked him for a technical account of a line of research which he had conducted in the war. He wrote:

"The practical use of guided missiles can only be to be to kill foreign civilians indiscriminately, and it furnishes no protection whatever to civilians in this country."

Devices for endowing machines with memory, patterned after the loop of electrical impulses in the human nervous system, were described by six scientists today to several hundred colleagues at the symposium. Just as a person jots down figures for future use, the mechanical calculator "traps" them in an electrical cycle and holds them for use at a later stage in a problem.

T. Kite Sharples told of a new calculator to be installed at the University of Pennsylvania which will "remember" 1,000 numbers. Impulses will pass through mercury tubes at the rate of a million a second, but the tank will hold only forty at a time. Electronic tubes can pick up those trapped at any time and hold them until needed in another part of the machine.

January 9, 1947

SCIENTISTS REPORT 'UNFAIR CLEARANCE'

CHICAGO, Sept. 15 (P)—At least seventy-six persons, including fifty-six scientists, have been victims of "allegedly unfair clearance procedures" involving loyalty, a committee of scientists said today it had been informed. The report was made in the Bulletin of the Atomic Scientists by the committee on secrecy and clearance of the Federation of American Scientists.

The committee mailed a general request to members of the federation last November asking information on any cases known to them involving clearance procedures alleged to be unfair.

Without identifying any persons the committee published a selected number of individual cases in detail. Most of the letters published contended that the scientists left regular jobs to take up secret work after being advised that they had been cleared. Subsequently, they said, they were told their clearance was only temporary and they were asked to resign. Some contended that they were not able to learn the reasons why they were not cleared or the charges placed against them.

In Washington the Atomic Energy Commission stressed that the article dealt with the whole Government and not merely atomic research.

September 16, 1948

Atom Board Orders Loyalty Oaths To Bar Reds From Study Awards

By WALTER H. WAGGONER
Special to THE NEW YORK TIMES.

WASHINGTON, May 21 — The Atomic Energy Commission ordered oaths of loyalty and non-Communist affidavits into effect today for all present and future holders of commission fellowships in medical, physical and biological science.

The action followed the assurance yesterday by David E. Lilienthal, chairman of the commission, to a Senate Appropriations subcommittee that oaths of allegiance to the United States Government would be required of applicants for fellowships. The fellowship program is financed by Government funds and administered by the National Research Council.

The oath requirement is the outcome of disclosure that at least one holder of a commission fellowship, Hans Freistadt of the University of North Carolina, is an avowed Communist.

The oath of loyalty and the sworn affidavit that the signer is not a Communist nor an advocate of overthrowing the Government by force are similar to those now required of Government employes. All applicants and the 497 present holders of commission fellowships, whether or not they are or will be working in secret or non-secret fields of study, will have to take the oath and furnish the affidavit.

Of the present fellows, 240 hold fellowships becoming effective in the 1949-50 academic year after July 1, and 257 are already at work under grants for the current year.

Of those presently working, 103 have been investigated by the Federal Bureau of Investigation and cleared by the commission for access to restricted data, since they are engaged in secret research and study. The remaining 154 are doing non-secret study and have not undergone the formal investigation or clearance procedures.

The forms of the oath and affidavit require the signer to swear or affirm the following:

"A. Oath.

"I will support and defend the Constitution of the United States against all enemies, foreign and domestic; that I will bear true faitht and allegiance to the same; that I take this obligation freely without any mental reservation or purpose of evasion.

"B. Affidavit.

"I am not a Communist. I do not advocate nor am I a member of any organization that advocates the overthrow of the Government of the United States by force or violence. I do further swear (or affirm) that this oath (or affirmation) will remain in effect during the period that I am a holder of an Atomic Energy Commission fellowship."

May 22, 1949

SCIENTISTS SCORE PHYSICS INSTITUTE

Szilard and Oppenheimer, in Atomic Bulletin, Attack Wire to McMahon on Loyalty

A telegram sent to Senator Brien McMahon, chairman of the Joint Congressional Committee on Atomic Energy, by five members of the executive committee of the American Institute of Physics, approving the requirement of a loyalty pledge and non-Communist affidavit for applicants of Atomic Energy Commission fellowships, is sharply criticized by Prof. Leo Szilard of the University of Chicago in a letter published in the Bulletin of the Atomic Scientists, out today.

Views in agreement with those of Prof. Szilard, one of the key scientists in the development of the atomic bomb, a expressed in a letter to Senator McMahon by Dr. J. Robert Oppenheimer, director of the Institute for Advanced Study, Princeton, N. J., wartime director of the Los Alamos (N. M.) Laboratory, where the atomic bomb was designed; as well as in an editorial and in a statement of policy by the Federation of American Scientists.

The telegram to Senator McMahon said that "the requirement of a loyalty pledge and non-Communist affidavit for AEC fellows is reasonable, but security clearance would be an unnecessary extension to the field of education of measures appropriate only in secret work."

Those who made this statement, comments Professor Szilard, are choosing a lesser evil to avoid a greater one, but "those who follow the principle of the lesser evil will have to retreat again and again. If asking for a non-Communist affidavit is reasonable, then it is also reasonable for the Government to refuse to take an applicant's word for his not being a Communist, and to investigate all applicants.

"And if it is reasonable to investigate holders of AEC fellowships," Professor Szilard continues, "why is it not equally reasonable to investigate holders of fellowships from the National Science Foundation? And if a university receives Federal aid to its educational and research program, is it not quite reasonable to investigate the members of the faculty and the students who benefit from such aid?"

Must Stand Firm, He Says

Our universities, Dr. Szilard declares, will not be able to resist Federal political control of education unless scientists take a stand based on the major principle involved, on which they are united, the principle that scientific ability be made the sole criterion for the selection of those who are given facilities for research of faculty appointments. "Once we give up this stand and retreat," he asserts, "there is no second line of defense behind which we can unite."

"If a student does not violate any laws," he adds, "then it is difficult to see how it is possible under the Constitution to discriminate against him or to bar him from educational opportunities provided at the taxpayers' expense. We are asked to sanction something that comes very close to the persecution of a politiacl minority.

"Justice and freedom have never been secure for very long in any one area of the world. None of us can say for sure what fate awaits them in the United States in the crisis through which we shall be going in the remainder of this century.

"Freedom and justice might survive this crisis; or they might not. They might perish, and the efforts of scientists might be of little avail. What we scientists can do is to resolve that they shall not be allowed to perish without a fight. And those of us who do not wish to fight can at least refuse to help dig the grave."

Dr. Oppenheimer points out that much of the scientific work done by the AEC is not secret, and that "basic work in science, in aspects which are not under the direct control of any Federal agency, is a major source of our scientific progress, of invention, discovery and technical leadership." Many of these discoveries, he adds, have been made by Communists or Communist sympathizers.

"The major—one might almost say the only—present peaceful application of atomic energy," Dr. Oppenheimer writes, "rests on the preparation and use of artificial radioactive materials, which were discovered by Joliot-Curie, who is a Communist, and by his wife, who is a Communist sympathizer.

Plea for Non-Conformists.

"It would be folly to suppose that the United States would be the stronger, or our science and industry the more vigorous, if this discovery had not been made. It would be contrary to all experience to suppose that only those who throughout their lives have held conformist political views would make the great discoveries of the future.

"The people and the Government of the United States have a stake in scientific discovery and invention; and it is for this stake, rather than as an act of benevolence to the recipient of the grants-in-aid, that one must look for justification for having a fellowship program at all," Dr. Oppenheimer writes.

The Bulletin's editorial says that the assent of the AEC chairman to a loyalty oath was a compromise forced by "the political hysteria aroused by a combination of the words 'Communist' and 'atomic energy.'" This compromise, the editorial adds, "will not avoid the fight for the preservation of independence and integrity of American education and scientific research."

The policy statement by the Federation of American Scientists declares "we believe that oaths, affidavits and clearance investigations are unnecessary and potentially dangerous to scientific progress. We endorse the original policy of the Atomic Energy Commission of granting these fellowships solely on the basis of scientific competence."

July 2, 1949

TRUMAN ORDERS HYDROGEN BOMB BUILT FOR SECURITY PENDING AN ATOMIC PACT

HISTORIC DECISION

President Says He Must Defend Nation Against Possible Aggressor

SOVIET 'EXPLOSION' CITED

His Ruling Wins Bipartisan Support on Capitol Hill—No Fund Request Due Now

By ANTHONY LEVIERO
Special to THE NEW YORK TIMES.

WASHINGTON, Jan. 31—President Truman announced today that he had ordered the Atomic Energy Commission to produce the hydrogen bomb.

The Chief Executive acted in his role of Commander in Chief of the Armed forces, ordering an improved weapon for national security. Thus, from the domestic standpoint, he removed the question of producing the super-weapon as an issue that might be argued on moral grounds.

As for international statecraft, Mr. Truman, by treating the hydrogen bomb as an addition to the American armory, also removed it as an issue that might be interpreted as an advanced threat or inducement in seeking international control of atomic weapons.

Nevertheless, Mr. Truman said that his perseverance in providing for national defense would be matched by his efforts to seek international control of atomic weapons.

New Phase of Atomic Age

In his announcement, Mr. Truman regarded the hydrogen bomb as a progressive outgrowth of United States production of the uranium-plutonium atomic bomb. He put it this way: the commission was "to continue its work on all forms of atomic weapons, including the so-called hydrogen or super-bomb."

His use of the word "continue" was understood to imply that with national security the over-riding consideration, the chief factor guiding his decision was whether it was practicable to make the weapon. Scientists have said that it is.

In effect, the President's decision, which won wide acclaim in Congress, marked the advent of a new phase of the atomic age and a surge ahead of Russia in the race to retain military ascendancy.

The bombs that visited destruction on Hiroshima and Nagasaki split the atom. The new bomb would fuse atoms instead, but with a power 100 to 1,000 times greater than the improved fission bombs that have been developed since the Japanese cities were struck.

The President's Statement

The President made his decision known in the following brief statement:

"It is part of my responsibility as Commander in Chief of the armed forces to see to it that our country is able to defend itself against any possible aggressor. Accordingly, I have directed the Atomic Energy Commission to continue its work on all forms of atomic weapons, including the so-called hydrogen or super-bomb. Like all other work in the field of atomic weapons, it is being and will be carried forward on a basis consistent with the over-all objectives of our program for peace and security.

"This we shall continue to do until a satisfactory plan for international control of atomic energy is achieved. We shall also continue to examine all those factors that affect our program for peace and this country's security."

On Capitol Hill when news of the Chief Executive's decision was received there, Republicans and Democrats joined in approving it. This bipartisanship boded well for Congressional backing of the new project, though it was said in informed quarters that Mr. Truman would not request funds for it at this time.

The Joint Congressional Committee on Atomic Energy held a previously scheduled meeting about an hour after the President's statement came out, and its chairman, Senator Brien McMahon,

Democrat, of Connecticut, said that it had approved Mr. Truman's decision. He added that the committee would now proceed with meetings in which the implementation of the hydrogen bomb program would be studied.

Louis Johnson, Secretary of Defense, who had been in Mr. Truman's office today, would say no more than that "the President's statement speaks for itself." The view of the professional soldier was expressed by an anonymous but high-ranking officer speaking in the absence of Gen. Omar N. Bradley, Chairman of the Joint Chiefs of Staff. He said:

"This is one of the gravest decisions the United States has ever had to make, but it had to be done."

Mr. Truman was as undramatic in making his announcement as he was last Sept. 23 when he disclosed that Russia had achieved an atomic explosion — a development that clearly showed that our absolute dominance in atomic weapons was virtually ended. The President was not in his office when the historic statement came out. He was lunching at Blair House, the official residence.

It was 1:55 P.M. when Miss Genevieve Irish of the White House staff walked through the lobby of the Executive Offices and into the press room, crying "press."

White House reporters hurried into the office of Charles G. Ross, the press secretary. He requested that none should leave the room until each had a copy of the mimeographed statement that he held in his hands. He does not make such a request unless the subject is momentous.

Truman Preferred Secrecy

Mr. Truman's decision was a direct result of the discovery in September of the Russian explosion. After it was established beyond doubt that the Soviet Union had accomplished atomic fission, he called in David E. Lilienthal, chairman of the Atomic Energy Commission, and asked what should be done about holding this country's lead in atomic weapons.

Mr. Lilienthal, who is to resign about Feb. 15, was reported to have reminded the President of the possibility of producing the hydrogen super-bomb and asked if he wished to go head with it. Thereupon Mr. Truman sought the advice of Mr. Lilienthal as well as of the three other leading officials concerned—Mr. Johnson, General Bradley and Dean Acheson, Secretary of State.

It was learned that the President would have produced the hydrogen bomb in secrecy, as most military weapons have been in the past, except for the great debate over it that erupted last November and has continued since.

Mr. Truman was represented as feeling that while the new weapon was particularly destructive, this country should have kept its development secret so as to retain the element of surprise as an additional measure of security.

The President sought to discourage discussion of the new type of bomb after Senator Edwin C. Johnson, Democrat, of Colorado, said in a television broadcast that this country was making considerable progress in developing an atomic bomb 1,000 times deadlier than the one dropped on Nagasaki.

J. Howard McGrath, Attorney General, and Senator McMahon were called to the White House on Nov. 25 and urged by Mr. Truman to prevent leaks on data so vital to national security.

Informed sources characterized as ludicrous published estimates that it would take $2,000,000,000 to $4,000,000,000 to produce the hydrogen bomb. The figure would be nearer $200,000,000, it was said, since this country's vast, well-developed atomic plants and "know-how" would be used in its production. In this respect, it was added, the United States retained a great advantage over Russia.

Because the work could be undertaken by the Atomic Energy Commission with its present resources, officials said, no new funds would be needed immediately. The work could be carried on with present appropriations and plants until advanced stages were reached.

February 1, 1950

BRITISH JAIL ATOM SCIENTIST AS A SPY AFTER TIP BY F. B. I.; HE KNEW OF HYDROGEN BOMB

TWO CHARGES MADE

First Alleges Betrayal of Information in U. S., 2d Site Not Named

COURT HEARING IS BRIEF

Klaus Fuchs, a Ministry Aide, Is Remanded in Custody to Reappear on Friday

By BENJAMIN WELLES
Special to The New York Times.

LONDON, Feb. 3—A senior British scientist who has worked on atomic projects in the United States and Britain was charged here today with having betrayed atomic research secrets.

The accused was Dr. Klaus Emil Julius Fuchs, aged 38, employed at the main British atomic research center. He was arrested yesterday on information passed on to the British Government by the United States Federal Bureau of Investigation.

One of the two charges leveled against him in the Bow Street Magistrate's Court was that "on a day in February, 1945, in the United States" he "communicated to a person unknown information relating to atomic research which was calculated to be, or might be, directly or indirectly useful to an enemy."

The second identical charge placed the date of the alleged offense in 1947 but made no mention of where it had taken place.

The action against Dr. Fuchs was taken under the Official Secrets Act. The penalty on conviction under the act is penal servitude for three to fourteen years.

Earlier Case Recalled

In a similar case another British atomic scientist, Dr. Alan Nunn May, was sentenced to ten years imprisonment in 1946.

German-born, Dr. Fuchs acquired British nationality in 1942. He has been employed at the Harwell atomic research establishment, from where he was summoned yesterday to the headquarters in London of the Ministry of Supply, which controls atomic research. There he was arrested.

Dr. Fuchs declined to make any answer to the charges.

Police officers testified that on his arrest Dr. Fuchs had asked immediately to see his superior, M. W. Perrin, Deputy Controller of Atomic Energy at the Ministry of Supply. When Mr. Perrin entered the office Dr. Fuchs was reported to have said to him:

"Do you realize the effect of this at Harwell?" (Atomic energy research establishment where Dr. Fuchs was employed.)

Mr. Perrin was said to have replied that he thought he understood.

Otherwise the court proceedings were brief. The presiding magistrate, Sir Laurence Dunne, set hearing for Feb. 10.

At the United States Embassy it was learned that no request for Dr. Fuchs' apprehension had been made through embassy channels.

British Government agencies declined to comment on the case or even to discuss the scientist's biographical background.

In United States 3 Years

It was learned, however, that Dr. Fuchs had been in the United States on atomic energy matters in 1943 and had stayed until 1946, when he returned to the Ministry of Supply. On his return he became head of the Ministry's Theoretical Physics Division.

Last September he was a member of the British delegation to the Anglo-American-Canadian talks here on hazards and safety factors connected with atomic piles "and other related matters." These talks did not include atomic weapons.

Members of the United States Atomic Commission's reactor safeguard committee who took part were Drs. Edward Teller, chairman; Manson Benedict, Joseph W. Kennedy, Abel Wolman and John A. Wheeler. Dr. Frederic de Hoffmann and Comdr. Joseph M. Dunford of the United States informed Senators, was to harden Congressional opinion against the Atomic Energy Commission also took part.

There is a feeling that the arrest

of Dr. Fuchs may embarrass the British Government on the eve of the tri-power talks scheduled to be held here Feb. 9 to 12 to discuss a greater exchange of atomic information among Britain, the United States and Canada.

While Britain is not believed to have access to the same degree of atomic information as the United States, she has announced her intention of building atomic weapons at the earliest possible time.

Britain's atomic installations officially reported are an atomic research "university" at Harwell, a former Royal Air Force field near an isolated village on the Berkshire Downs; a research center at Didcot, Berkshire; a uranium smelting factory at Springfields, Lancashire, and a bulk plutonium production center at Sellafield in Cumberland.

Remanded Till Friday

LONDON, Feb. 3 (UP) — Dr. Fuchs' appearance in court was brief.

Prosecutor Christmas Humphreys asked that he be required only to give formal evidence of Dr. Fuchs' arrest, and that Dr. Fuchs then be remanded for one week.

The only witness was Comdr. Leonard Burt of the special branch of Scotland Yard—the section that deals with espionage.

Commander Burt said that at 3:30 P. M. yesterday, with a Scotland Yard inspector, he called on Dr. Fuchs and told him that he was a police officer and had come to arrest him.

"I told him the nature of the charge and I cautioned him," Commander Burt told the court—in Britain arresting officers must warn that anything a defendant says may be used against him.

"He made no reply."

"Do you want to ask any questions?" said the magistrate.

"No," Dr. Fuchs replied quietly.

"Do you want the court to do anything about legal representation for you?" the magistrate asked.

"I don't know of anybody," Dr. Fuchs replied.

"Fuchs is a man of means," the prosecutor interposed. "I do not wish to be faced next Friday with a request for legal aid—which would mean further delay.

"Will you bear that in mind?" said the magistrate. "This case will be taken next Friday. If you wish legal representation, any person you desire will be informed and you will be put in charge with him. But the case will be taken next Friday morning, and I shall not listen to any representations unless they are of a very extraordinary nature that will justify a further remand."

The police agreed to give Dr. Fuchs his money and eye-glasses and he was ordered to Brixton for one week.

February 4, 1950

U. S. Censors H-Bomb Data; 3,000 Magazine Copies Burnt

By WILLIAM R. CONKLIN

Gerard Piel, editor of the Scientific American, attacked the censorship policies of the Atomic Energy Commission yesterday when he disclosed that it had ordered the burning of 3,000 copies of the magazine's April issue because of an article on the hydrogen bomb.

"Strict compliance with the commission's policies would mean that we could not teach physics," the editor said at his office, 24 West Fortieth Street.

Mr. Piel said the article was written by Dr. Hans A. Bethe, Professor of Physics at Cornell University, who from 1943 to 1946 was chief of the theoretical physics division of the Los Alamos Scientific Laboratory.

On March 15 Mr. Piel said the commission requested that technical portions, amounting to about one-half the article, be withheld from publication. Since the April issue was then on the presses the magazine asked the commission to specify its objections. These objections, Mr. Piel said, came down to about one column of type. In a further compromise, the editor said, the objectionable material was reduced to fractions of a column.

Conceding that the commission has broad powers to enforce suppression of information, Mr. Piel noted that the power of injunction was included among them. However, he said he was ready to join the issue with the commission on such material as had been widely published here and abroad, material which was declassified, and material that had never been classified.

"The Bethe article was the second in a series of four," the editor said. "There was no objection by the commission to the first H-bomb article published in March. That article was written by Louis N. Ridenour, a physicist who worked on radar development during the war. He is now dean of the Graduate College of the University of Illinois.

"The commission has given us no intimation of whether it will use its power of injunction against the magazine or not. However, we intend to proceed with two further articles on the hydrogen bomb, for which the authors have not yet been chosen. If the commission makes it an issue, this case may go before the United States Supreme Court."

Commission Aide Visits Press

On March 20, Mr. Piel continued, Alvin F. Ryan of the A. E. C.'s New York operations office at 70 Columbus Avenue met Donald H. Miller Jr., business manager of Scientific American, and Joseph Chanko, general manager of the Condé Nast Press. This meeting was held at the Condé Nast Press in Greenwich, Conn., Mr. Piel said. In addition to the magazine-burning, Mr. Piel said the following steps occurred:

"At Mr. Ryan's direction the complete file of proofs, galley, page and foundry, were given to Mr. Ryan. The additional foundry proofs from various departments were all assembled and given to Mr. Ryan. The 'Final OK for Press' proofs were given to Mr. Ryan.

"The proofs of the electrotypes were given to Mr. Ryan. The objectionable linotype slugs on Pages 19, 20 and 21 were removed and delivered to the smelting room as indicated in a marked set of pages given to Mr. Ryan. The four printing plates were destroyed in Mr. Ryan's presence.

"A total of thirty-two pages were cut apart and burned in the incinerator in order to get rid of the sheets which had been run on the press. These pages were 13-28 inclusive and 45-60 inclusive. Mr. Ryan took with him six sets of signatures of the complete book, including those which had been destroyed."

In a formal statement on the action, Scientific American, which has a circulation of 100,000, said it had always cooperated with the commission when any question of security on atomic or hydrogen bomb research was involved.

"In this article Dr. Bethe develops the moral and social implications of the super-weapon out of a discussion of technical information about the bomb which has been widely published and is well-known to nuclear physicists the world over," the statement said.

"After reviewing the article the commission requested deletion of certain statements from the technical part of the article. These included some statements which had already been widely quoted in the press and statements which have since been made by Robert Bacher, former member of the commission, in a speech in the Los Angeles Town Hall on March 27. We deleted these statements from the article."

Prohibitions Are Cited

On March 14, the statement noted, the commission had issued a "sweeping prohibition" to all its contractors, employes and consultants against public discussion "of the thermonuclear reactions involved in the thermonuclear weapons project." On March 20, it added, this order was expanded to cover unclassified, as well as classified material, while permitting "unclassified discussion of what might be called the classical thermonuclear reaction."

"We consider that the commission's action with regard to the Bethe article and the sweeping subsequent prohibition issued to the nation's atomic scientists raises the question of whether the commission is thus suppressing information which the American people need in order to form intelligent judgments on this major problem," the statement added.

"While there are certainly areas of information which must be protected for reasons of national security, there is a very large area of technical information in the public domain which is essential to adequate public participation in the development of national policy and on which the American people are entitled to be informed by such recognized authorities as Dr. Bethe."

Sumner T. Pike of Maine, acting chairman of the commission, signed the request to Mr. Piel to withhold the technical part of the Bethe article on the ground that a security question might be involved. Other members of the commission are Gordon Dean of California and Henry De Wolf Smyth of New Jersey.

David E. Lilienthal of Tennessee, former chairman, resigned effective Feb. 15. Thomas E. Murray of New York has been nominated for a place on the commission. Lewis L. Strauss of Virginia resigned from the five-man agency effective April 15.

Wilbur E. Kelley, manager of the commission's New York Operations Office, was not available for comment yesterday. Corbin Allardice, public information director there, was reported "in conference" and could not be reached.

April 1, 1950

SCIENCE AND SOCIETY

LEADING PHYSICISTS PROTEST VISA CURBS AS PERIL TO SCIENCE

Charge McCarran Act Sets Up 'Paper Curtain' That Stifles Intellectual Liberty in U. S.

26 PASSPORT CASES CITED

Einstein, Urey and Compton Join in the Denunciation of Immigration Laws

By RICHARD J. H. JOHNSTON
Special to THE NEW YORK TIMES.

CHICAGO, Oct. 12—Thirty-four of the world's leading scientists today directed a concerted and vigorous attack at the visa and passport policies of the United States Government.

Headed by Dr. Albert Einstein, the world's most renowned theoretical physicist, American, British, Italian, French and Mexican scientists, in a special issue of The Bulletin of the Atomic Scientists, which will be placed on sale tomorrow, condemned the Internal Security Act of 1950 and the Immigration and Nationality Act of 1952 as stifling to scientific and intellectual freedom and destructive of political liberties.

The scientists' protest against the measures, sponsored by Senator Pat McCarran, Nevada Democrat, is based on their contention that the legislation has made it possible for United States consular officers and the United States Attorney General to prevent, at will, admission to or egress from the United States of scientists and teachers on the basis of mere suspicion that they may be Communists or sympathetic toward Communists.

Visa Difficulties Described

The special issue of the publication, a magazine for science and public affairs, has been devoted to "American visa policy and foreign scientists." Dr. J. Robert Oppenheimer, a leader in the development of the atomic bomb, is chairman of the magazine's sponsors.

Among the writers of the issue's ten articles, covering fifty-one pages, are Dr. Einstein, member of the Institute for Advanced Study in Princeton, N. J.; Dr. Harold C. Urey of the University of Chicago; Arthur H. Compton, atomic fission pioneer; Victor Weisskopf, Professor of Physics at the Massachusetts Institute of Technology; Michael Polany, a British chemist and social philosopher, and Raymond Aron, a French sociologist.

Others are Bruno Ferretti, a leading Italian theoretical physicist, and Manuel Sandoval Vallarta, director of the National Institute of Scientific Research in Mexico. Five of the scientists are Nobel Prize winners.

A number of the writers, including Dr. Linus Pauling, an American and director of the Gates and Crellin Laboratory of Chemistry at the California Institute of Technology, reported personal experiences in being unable to secure passports and visas or in suffering delays and embarrassments in seeking them.

Dr. Edward A. Shils, Professor of the Social Sciences, University of Chicago, was special editor for this issue, which details the cases of twenty-six famous scientists who either were denied visas to attend scientific meetings or accept teaching posts in the United States, or who were put to much trouble to get visas.

An editorial written by Professor Shils, entitled "America's Paper Curtain," summarizes the case against restrictions on travel of scientists.

"The number of frustrated applicants is, of course, far greater than the number which we print here," Professor Shils wrote. "We tried to reach a sample large enough and representative enough to make American readers aware of the wrong-headedness of our present visa policy."

The attack on the immigration curbs described the case of Dr. Pauling, to show how the restrictions worked on American scientists seeking to go abroad. He applied for a passport Jan. 24, 1952, to go to Britain for a meeting of the Royal Society to discuss proteins.

Cases Are Discussed

His application was rejected. But after pressure the State Department reconsidered and issued a passport on July 14 after he had signed a statement that he never had been a Communist.

Among the twenty-six detailed "examples" were the following:

Dr. Polany, a leading anti-Communist, according to The Bulletin, who was refused a visa to accept a professorship at the University of Chicago.

Dr. M. L. Oliphant of the Australian National University and leading authority on microwave radar, who failed to get a visa to attend a conference on nuclear physics at the University of Chicago last September.

Dr. Jacques Monod of the Pasteur Institute of Paris, who was refused a visa by the American Consul.

Dr. Leonardo Guzman, Professor of Medicine, University of Santiago, and former Prime Minister and Minister of Education of Chile, who was refused a visa to come to the United States to study isotopes in their relation to cancer therapy.

The laws under attack provide that persons may be refused admission to this country if consular officers or the Attorney General "knows or has reason to believe" that their activities here "would be prejudicial to the public interest, or endanger the welfare, safety or security of the United States." Barred, under the terms of the legislation, are Communists or members of any other "totalitarian party."

Exceptions Provided

The law presumes that the Communist party and its members advocate the overthrow of the Government of the United States by force, violence or other unconstitutional means. It provides that persons who have been "involuntary" members of totalitarian organizations may be admitted, however.

The provisions of the Immigration Act of 1952, which was enacted last June 27 over the veto of President Truman, do not go into effect until Dec. 27. The scientists thus are attacking the working of the 1950 Internal Security Act and the immigration restrictions now in effect that the new law retains and extends.

Dr. Einstein wrote in part:

"The free, unhampered exchange of ideas and scientific conclusions is necessary for the sound development of science as it is in all spheres of cultural life. In my opinion, there can be no doubt that the intervention of political authorities in this country in the free exchange of knowledge between individuals has already had significantly damaging effects.

"First of all, the damage is to be seen in the field of scientific work proper, and, after a while, it will become evident in technology and industrial production.

"The intrusion of the political authorities into the scientific life of our country is especially evident in the obstruction of the travels of American scientists and scholars abroad and of foreign scientists seeking to come to this country.

"Interference with the freedom of the oral and written communication of scientific results, the widespread attitude of political distrust which is supported by an immense police organization, the timidity and the anxiety of individuals to avoid everything which might cause suspicion and which could threaten their economic positions—all these are only symptoms, even though they reveal more clearly the threatening character of the illness."

The editorial by Professor Shils declares in part:

"Through the two McCarran Acts—the Security Act of 1950, and the Immigration and Nationality Act of 1952—and their excessively rigid, indiscriminate application by the State Department, the United States Government and the American people are undoing with one hand what they are so laboriously and expensively accomplishing with the other.

Principles Traduced

"While one part of American policy generously and far-sightedly has sought to defend the free societies of the West through the Marshall Plan, the North Atlantic Treaty and other measures, these two acts, and particularly the clauses bearing on the entry into the United States of foreign scientists, scholars and educators, in conjunction with the sheer ignorance and unconcern for consequences in some sections of the State Department, alienate our allies, comfort our enemies, enfeeble our free institutions and traduce the principles of liberty.

"In the past few years a very large number of distinguished European scientists, almost all of them anti-Communists and deeply devoted to the freedom in which scientific truth is sought and discovered, have been frustrated in their efforts to come to the United States to share their knowledge with their American colleagues.

"Their applications for visas have in many cases been refused, usually after long delay; in other cases the visas have been finally granted, but only after delays so long that the scientific meetings to which they were invited had taken place or the teaching appointments for which they had been engaged had lapsed through their failure to arrive in time to fulfill them."

October 13, 1952

FOUNDATIONS' AID TO SCIENCE LAUDED

Bush Says at House Inquiry He Knows of No Socialistic Trend in Private Research

WASHINGTON, Nov. 20 (*P*)—Dr. Vannevar Bush praised private research foundations today as leading the United States toward world supremacy in fundamental science.

The president of the Carnegie Institution of Washington said that if there was any trend toward socialism in the operations, he was not aware of it.

Testifying before a special House committee studying tax-exempt charitable, philanthropic and educational foundations, Dr. Bush said that one of their foremost achievements had been the development of a strong effective system of medical training.

"Here is a trend that is in exactly the opposite direction from socialism," he asserted. "We have no need for the English system of socialized medicine. One reason is the work done by the foundations."

The House group before which he appeared in under instructions to find out whether any of the tax-free foundations are promoting un-American or subversive aims.

Dr. Bush testified there was no doubt that the Soviet Union had greatly expanded its scientific efforts, but that little was known in this country about the caliber of

355

those efforts. He said that it was known that Soviet Russia had been able to develop an atomic bomb and excellent jet planes, but he added:

"Whether Russia can break new ground, I do not know. Russians must follow the party line, even in science. Great science never prospers under such circumstances. I am very happy personally that Russia has that system."

Dr. Fred Middlebush, president of the University of Missouri, and Dr. William Myers, Dean of Agriculture at Cornell University, gave the tax-free foundations credit for worthwhile advances in the fields of education, research and economics.

Dr. Middlebush said that one of the big unfinished jobs of the foundations was to find ways to improve the quality of teaching in the arts and sciences. He added that they already had done "considerable construction work" along that line.

Dr. Myers told the committee that economic and statistical studies financed by the foundations had been of great help to both business and government.

November 21, 1952

OPPENHEIMER LOSES APPEAL TO A.E.C., 4 TO 1

CLEARANCE DENIED

'Defects in Character' and Red Associations Cited in Decision

By JAMES RESTON
Special to The New York Times

WASHINGTON, June 29—Dr. J. Robert Oppenheimer, the man who directed the making of the first atomic bomb, today lost his long fight for reinstatement as an adviser to the Government.

The Atomic Energy Commission announced this afternoon that it had voted 4 to 1 to deny him further access to secret Government information.

Three members of the commission—Rear Admiral Lewis L. Strauss, Chairman; Joseph Campbell and Eugene M. Zuckert—voted against him (1) because of "proof of fundamental defects in his character"; and (2) because his association with known Communists "extended far beyond the tolerable limits of prudence and self-restraint."

One other member, Thomas E. Murray, concurred in this judgment but went beyond it. Unlike the commission's Special Security Clearance Board (the Gray board), which voted 2 to 1 against clearance but praised Dr. Oppenheimer's loyalty and discretion in handling atomic secrets, Mr. Murray condemned him for failing to show "exact fidelity" and "obedience" to the Government's security regulations. Mr. Murray concluded: "He was disloyal."

Dr. Henry DeWolf Smyth, the senior member of the commission, wrote the lone dissent.

"I agree with the 'clear conclusion' of the Gray board," he said, "that he is completely loyal, and I do not believe he is a security risk."

Further Plea Unlikely

Dr. Oppenheimer could still appeal to President Eisenhower, but this is thought unlikely.

The decision was announced by the A. E. C. at 4 P. M. Representative W. Sterling Cole, Republican of upstate New York and chairman of the Joint Congressional Atomic Energy Committee, announced the decision on the floor of the House during the foreign aid debate. It was greeted by a smattering of applause.

[In a statement issued in New York Tuesday night, Dr. Oppenheimer said that Dr. Smyth's dissenting opinion "says what needs to be said." He refrained from commenting directly on the commission's decision, but said:

["Our country is fortunate in its scientists, in their high skill and their devotion. I know that they will work faithfully to preserve and strengthen this country. I hope that the fruit of their work will be used with humanity, with wisdom and with courage. I know that their counsel when sought will be given honestly and freely. I hope that it will be heard."]

Many Roles End

The situation now is as follows:

Dr. Oppenheimer's contract as a consultant to the A. E. C. ends tomorrow and will not be renewed. He no longer will be called upon, as he has been in the past, to advise the National Security Council on such things as continental defense; the Defense Department on its new weapons program; the State Department on international control of atomic energy.

Though he is in possession of most of the secrets of the atomic and hydrogen bomb programs, he will be placed, more or less permanently, behind the "blank wall" that was put between him and classified information last Dec. 3 on the personal order of President Eisenhower.

Finally, it was generally assumed that, because his reliability, character and loyalty to the security regulations of the Government had been officially challenged, he would resign as Director of the Institute for Advanced Studies at Princeton, N. J., and be refused clearance on all allied Government atomic projects.

The majority opinion (Strauss-Campbell-Zuckert) differed from the Gray board's majority (Gordon Gray, president of the University of North Carolina, and Thomas A. Morgan, former president of the Sperry Corporation) in two main respects.

It did not attest to his loyalty or discretion, as the Gray board majority had done; and it did not condemn him for his role in opposing the development of the hydrogen bomb in 1949.

In fact, all the commissioners emphasized that they had not reached their conclusions on the basis of anything he had said or done, or failed to say or do, in the controversy over that weapon.

The commission majority pointed out that it had a duty under the Atomic Energy Act of 1946 to reach a determination as to the "character" and "associations" of A. E. C. employes as well as on their "loyalty."

The majority then made these statements:

¶ "On the basis of the record before the commission * * * we find Dr. Oppenheimer is not entitled to the continued confidence of the Government and of this commission because of the proof of fundamental defects in his 'character.' * * *

¶ "In respect to the criterion of 'associations,' we find that his associations with persons known to him to be Communists have extended far beyond the tolerable limits of prudence and self-restraint which are to be expected of one holding the high positions that the Government has continuously entrusted to him since 1942. * * *

¶ "A Government official having access to the most sensitive areas of restricted data and to the innermost details of national war plans and weapons must measure up to exemplary standards of reliability, self-discipline and trustworthiness. Dr. Oppenheimer has fallen far short of acceptable standards."

6 Incidents Listed

The commission majority members listed six incidents that persuaded them of fundamental defects in Dr. Oppenheimer's character:

¶ He admitted he had lied to a military intelligence officer (Col. Borist T. Pash) who was trying to get a straight story about a Soviet agent who had sought secret information from Dr. Oppenheimer.

¶ He told the Gray board he would not have asked for the employment of one Giovanni Rossi Lomanitz at Los Alamos if he had known that Lomanitz was an active Communist who had given information to an unauthorized person. On Aug. 26, 1943, however, he had told Colonel Pash that he did not know that Lomanitz was a Communist and that he (Lomanitz) had given out information to an unauthorized person.

¶ In 1943 he testified that he did not know Rudy Lambert, a Communist party functionary, but recently under oath he testified that he had seen Lambert at least half a dozen times before 1943 and knew he was a Communist.

¶ In 1949 he testified before the House Un-American Activities Committee about the Communist party membership of Dr. Bernard Peters and later wrote a letter to The Rochester Times-Union which, in effect, contradicted that testimony.

¶ He testified before the Gray board that the General Advisory Committee of the A. E. C. had been "surprisingly unanimous" in its recommendation against the hydrogen bomb program, but did

SCIENCE AND SOCIETY

not testify that one of the members of the committee had in fact written him a letter—before the General Advisory Committee met - favoring production of the hydrogen bomb.

Finally, in 1950, he told an agent of the Federal Bureau of Investigation that he had not known Joseph Weinberg to be a member of the Communist party but, on Sept. 12, 1943, he told another Government official that Weinberg was a Communist party member.

"The catalogue does not end with these six examples," the majority said. "The work of military intelligence, the Federal Bureau of Investigation and the Atomic Energy Commission—all at one time or another have felt the effect of his falsehoods, evasions and misrepresentations."

"Chevalier Incident" Cited

There was reason for believing, however, that the so-called "Chevalier incident" was more damaging to Dr. Oppenheimer's fight for reinstatement than any other. Indeed, it can be stated on fairly reliable authority that at least one of the commission members who voted against reinstatement would have switched but for this incident.

In the Chevalier incident, Dr. Oppenheimer was questioned in 1943 by Colonel Pash, Col. John Lansdale Jr., and Lieut. Gen. Leslie R. Groves, all officials at Los Alamos, about the attempt to obtain information from him on the atomic bomb project in the interest of the Soviet Government.

He waited eight months before mentioning the occurrence to the proper authorities. Thereafter for almost four months he refused to name the individual (Haakon Chevalier) who approached him.

Moreover, on Feb. 25, 1950, Dr. Oppenheimer tried to help Chevalier to get a United States passport.

Later that year Chevalier stayed with Dr. Oppenheimer for several days at the latter's home. In December, 1953, Dr. Oppenheimer visited with Chevalier privately on two occasions in Paris, and lent his name to Chevalier's dealings with the United States Embassy in Paris on a problem that, according to Dr. Oppenheimer involved Chevalier's clearance.

Late Associations Scored

The majority seemed to forgive the scientist for his earlier Communist associations, but it was clearly critical, as was the Gray board and Dr. Smyth, the

The New York Times
Lewis L. Strauss

lone dissenter, of his continuing this association right up until December of 1953.

Incidentally, there is one aspect of this visit with Chevalier that is not mentioned anywhere in the testimony but is known to have worried some of the commissioners. This was that, though Dr. Oppenheimer was not sure, even in 1953, that Chevalier was not a Communist, he visited him in Paris and took the chance that, like many other nuclear scientists, he might have been forced at gun-point into a plane and taken behind the Iron Curtain.

Dr. Oppenheimer scoffs at this possibility, pointing out that the United States Government knew where he was, but others here cite this prospect as evidence that he was casual about security matters that affected not only his own safety but the safety of the country.

In his dissent from the positions taken by the other four commissioners, Dr. Smyth looked to the future rather than to the past,

Because Dr. Oppenheimer was one of the most knowledgeable and lucid physicists in the United States, Dr. Smyth observed, his services could be of great value to the country in the future. Therefore, he argued, the only question before the Atomic Energy Commission was whether there was a possibility that Dr. Oppenheimer intentionally or unintentionally would reveal secret information to persons who should not have it.

"To me," he said, "this is what is meant within our security system by the term 'security risk.' Character and associations are important only in so far as they bear on the possibility that secret information will be improperly revealed.

"In my opinion the most important evidence in this regard is the fact that there is no indication in the entire record that Dr. Oppenheimer has ever divulged any secret information."

Dr. Smyth took up the six incidents catalogued against the scientist by the majority and sought to show that these did not justify what he called the "severe" decision of the majority.

There was a marked contrast in the estimates of the commissioners about the effect on the commission of the loss of Dr. Oppenheimer.

Dr. Smyth said that failure to employ a man of such great talents might impair the strength and power of the country, but Commissioner Campbell, relying on the judgment of the general manager, felt that Dr. Oppenheimer was not "indispensable."

He observed that the Atomic Energy Commission was absolutely vital to the survival of the nation; that the security regulations were intentionally severe; that no violations could be countenanced and that, "where responsibility is highest, fidelity should be most perfect."

"The American citizen in private life, the man who is not engaged in governmental service," he said, "is not bound by the requirements of the security system. However, those American citizens who have the privilege of participating in the operations of Government, especially in sensitive agencies, are necessarily subject to this special system of law.

"Consequently, their faithfulness to the lawful Government of the United States, that is to say, their loyalty, must be judged by the standard of their obedience to security regulations. Dr. Oppenheimer was subject to the security system which applies to those engaged in the atomic energy program. The measure of his obedience to the requirements of this system is the decisive measure of his loyalty to his lawful Government. No lesser test will settle the question of his loyalty. Dr. Oppenheimer did

not meet this decisive test. He was disloyal."

Commissioner Murray was sharply critical of Dr. Oppenheimer's judgments about the hydrogen bomb project. He asserted flatly that the scientist's political and technical reasons for producing the bomb had "been proved wrong." He defended Dr. Oppenheimer's right to put forward "moral reasons" against producing the hydrogen bomb, but observed that he (Oppenheimer) was no "expert in morality."

It was not on this ground, however, that he reached the stark conclusion: "He was disloyal." He put the main test on the question of the scientist's "obedience" to the security regulations of the Government. Commissioner Murray put it this way:

"The American citizen recognizes that his Government, for all its imperfections, is a government under law, of law, by law; therefore, he is loyal to it. Furthermore, he recognizes that his Government, because it is lawful, has the right and the responsibility to protect itself against the action of those who would subvert it.

"The cooperative effort of the citizen with the rightful action of American Government in its discharge of this primary responsibility also belongs to the very substance of American loyalty. This is the crucial principle in the present case."

Commissioner Zuckert, whose term on the commission will end tomorrow, said it was "a source of real sadness" to him that his last act as a public official should involve the denial of clearance to a man who had made "a substantial contribution to the United States."

This thread of "sadness" ran through his separate opinion. As long as there were human emotions like love of family or human feelings like pain, he observed, everyone was to some degree vulnerable to influence, and thus a potential risk in some degree to our security.

"But," he concluded, "when I see such a combination of seriously disturbing actions and events as are present in this case, then I believe the risk to security passes acceptable bounds. All these actions and events and the relation between them make no other conclusion possible in my opinion than to deny clearance to Dr. Oppenheimer."

June 30, 1954

If Einstein Were Young Again, He Says, He'd Become a Plumber

Dr. Albert Einstein has declared that, if he had his career to fashion all over again, "I would not try to become a scientist or scholar or teacher."

In a letter to the editor to be published tomorrow in The Reporter magazine, the famous physicist said he would become a plumber or a peddler to seek the independence that these vocations afford.

The letter was written in response to a request from The Reporter for Dr. Einstein's comments on a recent series in the magazine by Theodore H. White, "U. S. Science: The Troubled Quest." The series said that centers of intellectual life were troubled by recent Federal actions concerning scientists.

Dr. Einstein has been an outspoken critic of these actions. When Dr. J. Robert Oppenheimer was denied security clearance by the Atomic Energy Commission, Dr. Einstein said: "The systematic, widespread attempt to destroy mutual trust and confidence constitutes the severest possible blow against society."

Early this year he advised witnesses to refuse to testify on their activities before the legislative committee headed by Senator Joseph R. McCarthy, Republican of Wisconsin.

The text of Dr. Einstein's letter:

"To the Editor:

"You have asked me what I thought about your articles concerning the situation of the scientists in America. Instead of trying to analyze the problem, I may express my feeling in a short remark: If I would be a young man again and had to decide how to make my living, I would not try to become a scientist or scholar or teacher. I would rather choose to be a plumber or a peddler in the hope to find that modest degree of independence still available under present circumstances."

In Princeton, Dr. Einstein's secretary declined to elaborate on this comment.

In publishing the letter, Max Ascoli, the editor of The Reporter, said that it was an honor but "hardly a pleasure to publish this letter from Albert Einstein." The comment will be freely used by enemies of the United States, he said.

But he added that the freedom to protest, which Dr. Einstein used in making his comment, can still be afforded here. Our country must maintain a good record on this score, not just a better record than do the totalitarian nations, Mr. Ascoli said in an editorial comment.

November 10, 1954

PURE SCIENTISTS CALLED RED PREY

Judge, in Sentencing Student for Contempt of Congress, Deplores Defections

Special to The New York Times.

WASHINGTON, Dec. 13—A Federal judge asserted here today that "the younger generation of pure scientists" had become a "fertile field for Communist propaganda."

Judge Alexander Holtzoff made the comment in sentencing Bernhard Deutch, a graduate student of the University of Pennsylvania, to ninety days in jail for contempt of Congress. He allowed Deutch to remain free on $500 bond pending an appeal.

Deutch was a witness before the House Un-American Activities Committee at a hearing in Albany in April, 1954. He admitted membership in a Communist group while attending Cornell University and answered questions about his personal activities. He refused, however, to name other members of the group.

Indictment Reinstated

In November, 1955, Federal Judge James R. Kirkland dismissed a contempt indictment against Deutch on the ground that it had failed to specify "willful" intent.

The indictment was reinstated last July by the Federal Court of Appeals. It ruled that Deutch's refusal to answer had been a "positive, affirmative act" and "by its very nature deliberate and willful."

Deutch was tried on the charges and found guilty yesterday. Before passing sentence today, Judge Holtzoff gave him an opportunity to answer the questions in court. He again refused.

Deutch's refusal to answer was based on moral grounds and an asserted desire not to cause "tribulation" for others. Judge Holtzoff commented that if every man decided for himself which laws he would obey anarchy would result.

"This court has gleaned the inference that the younger generation of scientists, particularly in the field of physics, has succumbed to Communist propaganda," Judge Holtzoff stated.

He explained that he was not referring to scientists in engineering, chemistry and similar fields, but to younger persons "engaged in pure science."

Judge Holtzoff attributed the defections among young scientists to a change in educational methods in recent years. He said that the newer generation had not been given "a proper cultural background" and displayed "an abysmal ignorance" of history and economics.

Deutch, a 27-year-old New Yorker, is studying at Pennsylvania, for a doctorate in physics.

December 14, 1956

An Indictment of the Men Who Made the Bomb

BRIGHTER THAN A THOUSAND SUNS. A Personal History of the Atomic Scientists. By Robert Jungk. Translated by James Cleugh from the German, "Heller als Tausend Sonnen." 369 pp. New York: Harcourt, Brace & Co. $5.

By WILLIAM L. LAURENCE

As the subtitle informs us, this volume purports to be "a personal history of the atomic scientists." Actually, however, it is largely an indictment rather than a history, the personalities involved being mainly American physicists. It is the author's thesis that in lending their knowledge and skills to the building of the atom bomb, they committed a grave sin against the moral order, as well as against the noblest traditions of science; that in doing what they did they perverted science, born of the strivings of the human spirit and its quest for eternal truths, into a technological monster that threatens to destroy mankind.

Mr. Laurence was the only journalist present at the first test of the atom bomb in 1945 and was also an eye-witness of the atom-bombing of Nagasaki.

Because the book is so highly interesting, it is likely to be taken by the average reader, unfamiliar with the facts, as authentic history and this is far from being the case. For the author was primarily interested not so much in presenting an unbiased, objective record of one of the great stories of our time—indeed, of all time—but rather in summing up the case for the prosecution.

The author, a German-born journalist now an American citizen, pictures modern technology as the incarnation of Mephistopheles and the American atomic scientists and their associates, led by J. Robert Oppenheimer, as a collective Faust who sold their souls to the devil. He talks of "the voices of millions of anxious persons who would have liked the men of science to answer such questions as: 'Are you the same sort of beings as we are? Do you still see any sense in moderation, in the dignity of man and the commands of his Creator? Won't you tell us what you are really after?'"

As every fair-minded person knows, or should know, it was just because the atomic scientists of the United States, Britain, France and the rest of the free world passionately believed in the dignity of man and the commands of his Creator that they found themselves compelled to concentrate their knowledge and skills to prevent Hitler from becoming the sole possessor of the atom bomb. What they were "really after," was, of course, to save the free world from enslavement, being fully convinced at the time that Hitler had gained a considerable lead in the development of nuclear weapons. Did that, as the author charges, mean "the loss of that deeply rooted set of ethical beliefs out of which all science had formerly grown?"

The atomic scientists who gave evidence at the security hearings of Mr. Oppenheimer in 1954, Mr. Jungk asserts, "were themselves, in fact, also before the bar of justice. The critical question they ought to have answered was not 'Have you been loyal to the state?' but 'Have you been loyal to mankind?'"

Here the answer of the scientists would, of course, be only too obvious. In being loyal to the state they were, most certainly, also loyal to the highest values of mankind. Or does Mr. Jungk mean to imply that the best interests of mankind would have been served by allowing Hitler to be the sole possessor of the atom bomb?

To bolster his charges of moral turpitude against the American scientists (he hardly mentions the highly important role played by the British scientists), Mr. Jungk presents the German atomic scientists as, indeed, placing their loyalty to mankind above their loyalty to the state. This would be understandable, in view of the fact that loyalty to the Nazi state would, most certainly, have been contrary to loyalty to mankind. And, indeed, this has been the impression the German atomic scientists have been trying to create since the end of the war. Unfortunately, this represents a distortion of well-established facts.

Mr. Jungk would have us believe that the German atomic physicists, led by the Nobel Prize winner Werner Heisenberg, "did not push for the construction of such a bomb. * * * On the contrary, these physi-

SCIENCE AND SOCIETY

cists were able successfully to divert the minds of the National Socialist Service Departments from the idea of so inhuman a weapon."

That this is wholly untrue was revealed in a tape recording made of the reactions of the German scientists, then prisoners in England, following the announcement of the dropping of the atomic bomb over Hiroshima. These recordings, as reported by Samuel Goudsmit, scientific head of the mission (Alsos) sent into Germany to learn what progress the Germans had made in their atomic project, revealed that, with the exception of Otto Hahn, the German scientists did not push for the construction of an atom bomb because they did not believe one could be made, not because they were opposed to it on moral and humanitarian grounds.

THE recordings (made without the scientists being aware that they were being overheard), as well as confiscated records and diaries, showed that the German scientists, in arrogant belief in their superiority, were confident that, since they could not make an atom bomb, no other nation could. It came to them as a severe shock when they learned that American, British and exiled scientists had accomplished what German scientists had regarded as impossible. And the record also shows that it was only after they had recovered from their shock, that they began developing the legend of their refusal to work on atom bombs because of moral scruples.

Mr. Jungk severely condemns the decision to use the bomb, presenting data purporting to show that Japan would have surrendered within a short time without its use, as well as the arguments against the bomb's use outlined in the historic "Franck Report" submitted to the Secretary of War in June, 1945. Yet the fact remains that at the time the decision was made the choice was believed to be between ending the war quickly or prolonging it at the risk of the loss of an estimated half-million young Americans and several million Japanese of all ages.

In his assignment of the role of a collective Faust to American atomic scientists, Mr. Jungk presents a most unflattering portrait of Mr. Oppenheimer as an individual, which is wholly at variance with the character and qualities of this highly respected man of science as known to the scientific community throughout the world. In this Mr. Jungk shows not only highly biased judgment but also lamentable poor taste.

October 12, 1958

Trackdown of the German Scientist

Nazism and defeat scattered the leaders of a once-great scientific establishment. Herewith a review of where they went and what some of them did when they got there.

By ARTHUR J. OLSEN

BONN.

IN the years 1933 to 1950 Germany, intending no benefit to any other nation, executed unwittingly a foreign-aid program of remarkable significance to the military, political and intellectual history of our time. The unrequested export was the highly trained, protean scientific mind. This was a priceless resource with which Germany was richly endowed in the first third of the 20th century. In a few catastrophic years it was irrecoverably squandered.

The period began with the National Socialist take-over in 1933 and ended about 1950 when the last of the Allied investigatory teams had finished picking over the wreckage of the German scientific community. During those years many hundreds of the country's ablest physicists, chemists, rocketeers, electronics engineers and specialists in other abstruse disciplines left Germany to resume their labors in foreign laboratories. The circumstances of departure varied. Some fled for their lives. Some eagerly accepted foreign "invitations." Some were kidnapped.

This unique dispersion of a scientific community is still making history, as current dispatches from points as diverse as Huntsville, and Cairo, Egypt, testify.

AT the Marshall Space Flight Center at Huntsville, Wernher von Braun, a creator of the German V-2 rocket and the builder of America's first satellite, is shaping a giant Saturn rocket to put a man on the moon. In the capital of the United Arab Republic, a dozen lesser known German emigrés are rubbing Israeli nerves raw with a rocket-and-aircraft program undertaken for President Nasser.

In Germany the achievements of the *Ausgewanderte Wissenschaftler* are followed with mixed emotions of pride and regret. The pride was summed up in a float seen a few seasons ago at a pre-Lenten *Fasching* parade. Two papier-mâché missiles reached for the skies, one marked with the Red Flag, the other with the Stars and Stripes. Both bore the legend, "German product."

In the years 1933-39 the German nation liberated itself of one-half of its physicists, two-thirds of its physical chemists and comparable proportions of specialists in related sciences in the name of racial and political purity.

Only a decade before, Germany had been the world center of nuclear re-

ARTHUR J. OLSEN has been chief of The Times bureau in Bonn since early this year.

359

TO EGYPT—Dr. Eugene Saenger, (left), rocketeer, and Ferdinand Brandner, aircraft engineer, established development programs in Cairo.

TO THE U. S.—Dr. Wolfgang Noggerath, German jet-engine specialist, worked for the U. S. Navy after the war; now, above, he is with Lockheed at Sunnyvale, Calif.

search, with the University of Göttingen as its capital. But, as anxious Allied scientists, hurrying in the wake of invading armies, quickly learned, the Nazis did an effective job of dismantling this great edifice.

In December, 1944, Strasbourg was occupied, and an American investigating team headed by the Dutch-born physicist, Samuel Goudsmit, swooped down upon the laboratory of Prof. Rudolf Fleischmann. A leading nuclear physicist, Fleischmann had preserved in his archives a complete and careful record of German work in atomic fission since Dr. Otto Hahn had published in 1939 his electrifying paper on the fission process. In a few days Dr. Goudsmit was able to flash the reassuring word back to Washington: no bomb, no reactors, no serious nuclear program.

MANY Germans did contribute mightily to the exploitation of Hahn's equations—but for the other side. The great scientist's first act upon completion of his calculations was to telegraph the news to his lifelong colleague, Dr. Lise Meitner. Dr. Meitner, a Jewess, had fled to Stockholm only weeks before. Her urgent relay of the information throughout the world community of physicists led directly to the Manhattan Project. And emigré Germans contributed heavily to the project's success, as is illustrated in a wry incident of 1943.

Three physicists from England had been posted to high positions in the Manhattan Project. Leading scientists already on the job were summoned to Washington to bring the newcomers up to date. The briefing group included, among other foreign-born scientists, Enrico Fermi, an Italian, and Leo Szilard, a Hungarian. Fermi and company discovered to their surprise that the "Englishmen" were each and all refugees from Germany. Diplomatically using the English language, they launched into their briefing. But the conferees stumbled over the unfamiliar tongue and soon lapsed into the then *lingua franca* of nuclear physics. An hour later an American security officer nervously reported to his superior that "the whole crowd is talking German."

The Nazis' dissolution of the international prestige that the briefing incident pointed up is symbolized in the melancholy annals of the Institute for Physical Chemistry of Hamburg. In January, 1933, this small research institution was headed by Dr. Otto Stern, presiding over a small talented group of eight research assistants. Six of them and Stern were Jews.

TEN months later the director and all his Jewish colleagues were gone from Germany. The institute, probing at one of the most exciting frontiers of modern science, never recovered from the blow. Stern, described by one of his sorrowing colleagues as "one of the most German Germans I ever knew," made his way to the Carnegie Institute in Pittsburgh and to a Nobel Prize in 1943. One of his ablest assistants, Dr. Robert Frisch, a nephew of Lise Meitner, followed an emigré path that led him to Los Alamos during the war.

There were groups of scientists, however, who basked in favor under the Third Reich. These were the specialists in rocketry, aeronautics, communications and metallurgy whose work promised immediate military applications.

There happened to be relatively few Jewish scientists in these fields in prewar Germany, and none was spared. But Nazi leaders were inclined to be generous about the political shortcomings of Aryans working in the weapons field.

THEY conceived the guns, the aircraft and, of course, the missiles that made the Wehrmacht an awesome military machine almost to the hour of its collapse. As the long struggle ended, the Allied powers reached for the men who had made that machine possible. Following the tanks of Bradley and Montgomery, special scientific teams swarmed into German universities and laboratories in a vacuum-cleaner operation.

The United States was the most vigorous proselytizer, with Britain running a good second. The Soviet Union was hampered at first by a wipe-out-the-war-criminals mentality.

In 1945 the Red Army seized Dr. Ernst Oskar Müller, a distinguished aeronautics scientist, and threw him into prison. He died two years later at Buchenwald, then under Soviet auspices, an ironic end for a man who had been in constant trouble with the Nazis. Not until 1946 and 1947 did Moscow belatedly undertake a comprehensive roundup—using the midnight-arrest technique—of available German scientific talent. By then

SCIENCE AND SOCIETY

American, British and French experts had pretty thoroughly picked over the supply.

IN 1946 the British Ministry of Supply established an aeronautical research station at Volkenroda in North Germany where scores of German engineers and theorists were "invited" to work. Volkenroda became a sort of recruit depot from which many of the German names that now grace Western aircraft and aerospace enterprises were drawn.

Not long ago the German Aeronautical Research Institute published an updated Who's Who of its 150 leading members in 1945. The record shows that 72 of them went abroad, almost all in 1946-50. Half of these left Germany after putting in a year or two at Volkenroda. Twenty-six of the emigrés settled eventually in the United States, 11 in Britain, nine in France. Three went to India and two to Argentina. The Institute reported that an undetermined but "considerable" number of its 1945 membership were taken to the Soviet Union.

The pattern of dispersal of the Aeronautical Institute's intellectual resources was repeated for scores of other scientific bodies. A characteristic case history is that of Dr Wolfgang Nöggerath, a specialist in jet-engine research before and during the war.

IMMEDIATELY after the surrender he was swept up in the War Department's Operation Paperclip and hurried to Washington in a shipload of demobilizing G.I.'s. He was offered and accepted work with the Navy and ultimately put in eight years on naval aircraft development, first at the Naval Ordnance Laboratory outside Washington and later at the Naval Test Station at Inyokern, Calif. In 1954 he quit the Government to join the Lockheed Aircraft Corporation. Today Dr. Nöggerath, now 55 years old, is a division chief at Lockheed's Sunnyvale, Calif., plant where the Polaris missile is manufactured.

A similar career pattern can be traced for the 125 aviation and missile men brought to the United States in Operation Paperclip. Almost the entire Peenemünde rocket staff, hiding out in an Alpine village at the end of the war, was picked up. Walter Dornberger, the chief of Germany's rocket program, was brought to the United States as an Air Force adviser on missile technology. After a time at Wright-Patterson Air Force Base, he spent 10 years as a consultant to the Bell Aircraft Co. He has now returned to West Germany and retirement.

Today, only two German missile men remain key figures in the United States States space program—von Braun at Huntsville and Kurt Debus, head of the launch operations center at Cape Canaveral. Virtually all the rest are now scattered among the laboratories and production lines of the aerospace industry.

THEIR contribution to the development of American rocket technology is still argued when missile men get together. Certainly they played the vital role of teachers in the infant stage of American rocketry, bringing late-starting native scientists abreast of the art. Now the knowledge gap has been closed, and their jealous peers concede them only one special distinction—an unmatched talent for thorough and precise engineering.

The world has also ceased to attribute the postwar scientific successes of the Soviet Union to "captured German scientists." Yet the contributions which the Soviets exacted from the Germans paralleled those given freely to the United States and were doubtless of even greater importance to Soviet technological advance.

BY rough estimate the Soviet Union rounded up about 200 German scientists and engineer-technicians of superior quality. The biggest haul was made in a midnight requisition, Oct. 22, 1946, of about 100 marked men then living in East Berlin and the Soviet Zone of Germany. Most were released 12 years later.

One of those abducted, Dr. Peter Lertes, later recounted his experiences in the Soviet Union. His story seems to be in broad outline that of most of his unlucky colleagues.

"The Russian purpose in the first five years was simply to pump us dry of everything we knew in our field," said Dr. Lertes, a topflight designer of guidance systems and electronic control devices for the Wehrmacht. "At that time they were well behind Western science practically everywhere. But they were determined to catch up and, by and large, they did."

By 1951 Dr. Lertes and his colleagues were relieved of their teaching assignments. They were invited then to accept four-year contracts for specific tasks of scientific research. There was no choice but to renew the contracts in 1955 for what turned out to be three years more.

THE abducted Germans, it seems, did not dominate the Soviet atom bomb and satellite efforts. But many were put to work on specific tasks they knew to be integrated into those programs. Significantly, no German was ever granted complete responsibility for a piece of work. Every job handed them was undertaken by at least one parallel all-Soviet team and sometimes by as many as 10, all working independently.

Stalin's most valuable catch was Dr. Gustav Hertz, a mathematician and student of wave mechanics who won a Nobel Prize in 1925. Hertz played probably the largest role of any German in the Soviet atomic program. For several years he directed a staff of 200 German and Soviet nuclear technicians at the Suchum Radiation Laboratory on the Caucasian coast of the Black Sea. In 1954 Professor Hertz was allowed to return to East Germany, where he remains at Leipzig University.

IN the Caucasus, Professor Hertz worked with Dr. Manfred von Ardenne, renowned as a *Konstructeur*, a designer of scientific instrumentation. In the prewar years von Ardenne, a child prodigy who grew to arrogant manhood, made major contributions to the development of television and radar. He built one of the world's first electron microscopes and one of the first cyclotrons. German physicists believe he had a major hand in the engineering of the first Soviet atomic bombs.

Today von Ardenne, dubbed "the Red Baron" in recognition of his expensive, privileged style of living on a Dresden estate, directs the Manfred von Ardenne Nuclear Research Institute in East Germany. He is a neighbor—and reputed bitter rival—of Dr. Klaus Fuchs, another prewar emigré who came home to Communist Germany after serving time in a British prison for atomic espionage.

German scientists who have returned from abroad to West Germany have not enjoyed the hero's homecoming accorded Hertz and von Ardenne in the East. Some find themselves ill at ease in the new postwar Germany. Others have felt snubbed as "rejects" from foreign scientific communities. For a few the answer to the readjustment problem has been emigration again—this time to obscure or bizarre havens.

The small group which accepted lucrative armaments-development contracts in the United Arab Republic is regarded in professional circles in this light. The director of the aircraft program in Cairo is Ferdinand Brandner, a former Junkers and Messerschmidt engineer who did eight years of penance in the Soviet Union. The leader of the rocket team for several years was Dr. Eugene Saenger, a veteran of postwar ballistics research in France.

THE rocketeers now in Egypt are mostly men recruited with Saenger by French military authorities in 1946 for work on French armament research. In 1954 Saenger led them back home to form the nucleus of a propulsion research laboratory he founded in Stuttgart. Four years later the United Arab Republic took a substantial number of the laboratory staff under contract.

West German scientists acquainted with their colleagues in Cairo doubt that many are unreconstructed Nazis or virulent anti-Semites. The attractions of the Nile seem to be two—a generous Egyptian treasury and an opportunity to work in research still shunned in the Bonn Republic. These motives would tend to fit the psychological pattern of the generation of German scientists who came to maturity during the Third Reich. Few were devoted followers of Hitler, and few were passionate anti-Nazis. Apolitical men, most gave their hearts only to their work.

In the case of the only German nuclear physicist now known to be working in the Near East, the motive for his emigration, however, was personal and political. Dr. Kurt Sitte was caught up in the Soviet net at the end of the war and returned to East Germany with his family a decade later. Deeply affected by Nazi persecutions, Dr. Sitte promptly went to Israel, offering his services to the Jewish state. He joined the Israeli atomic energy program.

It was his misfortune to be soon entrapped by a Soviet spy, highly placed in the Jerusalem Government. Blackmailed by threats of harm to his family, still in East Germany, Dr. Sitte delivered reports on Israeli nuclear projects. The German physicist served a two-year prison sentence for his crime. Released not long ago, he returned to his labors of atonement.

Thirty years have now passed since the first brilliant wave of German scientists fled the Third Reich. It is 15 years since the victor nations combed over Germany for valuable "recruits." The impact upon the world scientific community of the scattering of Germany's intellectual wealth is fading. A new generation is taking over the conquest of nature. But in Germany the consequences are still being felt.

In the Bonn Republic the recovery has been slow. Despite urgent pleas of the scientific community, far less effort has been put into the restoration of the nation's intellectual resources than into material recovery. Federal outlays for support of science and research facilities have crept up from $10 million in 1950 to $250 million this year. But scientists, who suspect the Adenauer regime of an anti-intellectual bias, consider even this a pittance when measured against what is needed to catch up.

"Thanks to our malice and stupidity," mused an elderly chemist of international repute the other day, "we are now a second- or third-rate nation in many areas; above all, in the new sciences of the postwar era. I won't be present at the recovery. It will take another generation to make up our loss."

September 22, 1963

U.S. Scientists Irresponsible, Report of Coast Group Charges

The Center for the Study of Democratic Institutions has published a report criticizing American scientists as irresponsible.

A majority of scientists, especially those connected with the arms program, are not acting in the best interests of a free society, according to the report.

The center, situated in Santa Barbara, Calif., is a nonprofit educational institution established by the Fund for the Republic. Contributors to its report, "Science, Scientists and Politics," were Dr. Robert M. Hutchins, president of the center; Chalmers Sherwin, vice president of Aerospace Corporation; Dr. Donald Michael, director of the Peace Research Institute; Dr. Lynn White Jr., professor of history, University of California at Los Angeles; Dr. Scott Buchanan, former dean of St. John's College, Annapolis, Md., and James Real, a management consultant.

Judgments Are Scored

Dr. Buchanan declared that, as a rule, scientists "are not able to take responsibility for their own strategic judgments in science, to say nothing of the uses to which their work will be put."

"If the scientist's concern is truth," he continued, "it's his responsibility to be sure that science is not misused so that something false comes out of it."

Mr. Real charged that many scientists had become technologists who spent their lives serving the military. They are not developing ways of controlling the consequences of the war machines they are building, he said.

Dr. Hutchins declared that scientists were educated to do nothing but collect facts and were so specialized that they had "no general ideas."

The solution of our main problems, he wrote, "depends on moral and intellectual virtues rather than on specialized knowledge."

Dr. White said:

"The modern tendency to regard science as somehow apart from or even dominant over the main human currents that surround it is dangerous to its continuance and can be harmful even to progress within science."

Scientists working in the government community were criticized by Dr. White.

"Frequently," he declared, "the powerful members of this group are self-assured to the point of arrogance about their own abilities, about the overriding rightness of scientific values and methods and about the validity of their views of how society operates and what it needs."

Mr. Sherwin said that "people running government often do not understand science and technology."

"Despite some notable exceptions," he said, "scientific ignoramuses usally handle scientific decisions."

The report is made up of some of the papers presented at a recent conference at Santa Barbara on the responsiblities of science executives in the government service.

September 16, 1963

Soviet Genetics Reborn After Lysenko Period

By WALTER SULLIVAN

LAST spring a delegation from the National Academy of Sciences went to Moscow to present the Kimber Genetics Award to Nikolai V. Timofeyev-Ressovsky.

George B. Kistiakowsky, Harvard chemist, designer of the explosion that fired the first atomic bomb, former adviser to President Eisenhower and former member of the White Russian army, read the citation, a bouquet was thrust into Timofeyev-Ressovsky's arms and Nikolai N. Blokhin, president of the Soviet Academy of Medical Sciences, was about to end the ceremony when a woman pressed forward.

She had come up from Timofeyev-Ressovsky's institute, at near-by Obninsk, to witness the presentation and now she launched into a eulogy of the old geneticist, who left the Soviet Union in the nineteen-twenties to become a leading figure at the Max Planck Institute for Brain Research in Berlin. When the Red Army took Berlin he was arrested and sent to Siberia, but later he was rehabilitated.

Tears Hidden Behind Champagne

Suddenly the Americans saw that tears were running down the cheeks of some of the Russians facing them. A tray of champagne glasses was brought in and the tears were quickly hidden behind the bubbling wine.

Why the tears? They were shed both for the personal tragedy of Timofeyev-Ressovsky and for the national

SCIENCE AND SOCIETY

tragedy of Soviet genetics. They were also tears of joy at the renaissance in the Soviet Union of this science, so vital to the relief of human suffering, to the conquest of cancer, to agriculture and to the future of the human race.

A one-month tour of Soviet laboratories and consultations with a number of leading American specialists showed that, after years of suppression during the rule of Stalin and his protégé, Trofim D. Lysenko, Soviet genetics and molecular biology are making a swift comeback. In some areas the work has already reached parity with research in American, British and West European laboratories.

The speed of recovery can be attributed in part to the existence of a strong genetics tradition. So discredited did Russian genetics become, when it was dominated by Lysenko, that many have forgotten the surge of intensive genetics research during the early years of the Soviet regime.

Many in the West fail to realize that science had deep and rigorous traditions in Russia long before the Bolshevik coup d'état of 1917 put the Communists in command.

The early years of Communism in Russia were marked by an explosion of talent in many scientific fields, including genetics.

Some enthusiastic Soviet geneticists of the nineteen twenties saw in their science the possibility of improving the inherited qualities of mankind. Soon after the Revolution a Bureau of Eugenics was formed and genealogical studies of the intelligentsia were made in search of material for improving the breed of man.

Emphasis on Agriculture

Then famine hit Russia and the emphasis shifted to agriculture. A young man named Nikolai I. Vavilov was chosen to lead an effort to improve food production through the development of better strains of crops and livestock.

Vavilov, a student of William Bateson in England, the man who gave genetics its name, believed that the great food crops—wheat, potatoes, corn and barley—could be invigorated if crossbred with strains obtained from regions where those crops originated. He traveled to all parts of the world. He and his colleagues brought home 25,000 living samples of wheat.

Despite these efforts food production by Stalin's new collective farms was disastrously low. The stage was set for a man who made extravagant promises —Trofim Lysenko, who with his deep-set, burning eyes and almost religious fervor has been called the "Savonarola of Soviet science." He was a plant-breeder from the Ukraine who said he had a method of treating the seeds of winter wheat so that they could be planted in the spring and produce a heavy crop. This would open vast eastern regions in Siberia to productive farming.

Lysenko said characteristics acquired by an organism in its lifetime could be passed on to future generations. This had ideological appeal. Charles Darwin had believed it and the fathers of Communism, Marx and Engels, had been ardent admirers of Darwin.

Ridiculed Hybrid Corn Work

Modern genetics had shown that Darwin was wrong. But Lysenko contended that genetics "is merely an amusement, like chess or football" and ridiculed efforts to seek the basic principles of heredity through studies of the fast-breeding fruit fly, work that led to several Nobel Prizes.

Lysenko denounced the American method of breeding hybrid corn, which was based on classical genetics. He ignored evidence that the blight that affected potatoes was a virus disease. His policies produced disasters in food

RESEARCHER: Vladimir A. Engelgardt, a pioneer in muscle chemistry during the 1930's, whose interest has shifted to the nucleic acids, such as DNA, and their role in genetics.
The New York Times (by Walter Sullivan)

production, but by this time Lysenko's drive against genetics had put Vavilov in jail. He was sent to Siberia where he died in 1943.

Despite the rise of Lysenko as a virtual czar of science there was a move within the Soviet Academy of Sciences after World War II to create an institute that would revive classical genetics. It was to be headed by a man named Nikolai P. Dubinin, whose work in cytogenetics had excited interest abroad. The plan was blocked and in 1948 a conference was organized by the Academy of Agricultural Sciences to assess the rival claims of the geneticists and the followers of Lysenko.

The geneticists leaned over backwards to compromise and appease their opponents, but they were doomed. Lysenko waited until the end to announce that Stalin had officially endorsed his theories.

Some of the geneticists rose and recanted as Galileo had done before the Inquisition. The Academy of Sciences published a letter to Stalin pledging to abolish the institutes doing work in classical genetics, including the one where Dubinin was working. It promised to purge from the field of biology all those adhering to heretical (non-Lysenko) views.

That was the low point in Soviet biology. Today the situation has altered completely. Dubinin and his colleagues are on top. Lysenko is living in relative obscurity near Moscow.

Dubinin Heads Genetics Research

Why have the tables turned so completely? What has restored freedom of inquiry to Soviet genetics and biology? A number of those involved in the turnabout told their stories to this correspondent.

Dubinin is installed as head of a new Institute of General Genetics much like the one envisioned for him by the Academy of Sciences 20 years earlier. The temporary quarters of his new institute, barely a year old, is in a suburb of Moscow torn up by the construction of new apartments and laboratories.

His conference table was heaped with current issues of the American journal Science and other American, European and Soviet magazines. He ticked off the recent developments. He knew them well, for he is now president of the Scientific Council on Genetic Problems of the Academy of Sciences, a coordinating panel on which various research and public health agencies are represented.

He pointed to a new Soviet journal, Genetika.

"It is our first review on genetic questions," he said. "We began monthly publication in 1965. During the last two years, 10 new laboratories have been organized in the Institute of Biological Problems.

"In the Byelorussian Academy of Sciences, at Minsk, an Institute of Genetics was formed a year and a half ago under Turbin, who is working on hybrids of corn and wheat. In Kiev, at the Ukrainian Academy of Sciences, there is a department of genetics with several new laboratories doing experimental work on mutagenesis and radiation chemistry.

"In the Academy of Medical Sciences there will soon be an Institute of Human Genetics. And in Obninsk there is Timofeyev-Ressovsky. He heads the department of radiation genetics in the new Institute of Radiobiology."

A special source of pride to Dubinin is the Institute of Cytology and Genetics that he himself established near Novosibirsk, in the heart of Siberia. At the time when Nikita S. Khrushchev still retained some loyalty to Lysenko, he is reported to have asked why, if Dubinin was a "bad scientist" in Moscow, he was not also a bad scientist in Novosibirsk.

The first public criticism of Lysenko came in 1952, close to the end of the Stalin era. But it was the death of Lysenko's chief sponsor, Stalin himself, that turned the tide. The next year, 1954, the party organ, Kommunist, encouraged free scientific discussion (although calling for it to be based on dialectical materialism) and it opposed the suppression of divergent views.

It was evident to the new leader, Khrushchev that Soviet agriculture was in a sorry state. The Central Committee of the Communist party decreed the intensive development of hybrid corn, and this offered to the bolder geneticists a chance to strike back.

Foremost among them was Dubinin, who had been out of sight since 1948. In a 1955 article he wrote:

"T. D. Lysenko caused the stoppage of work at the critical moment when hybrid corn began to emerge from the experiments in the fields of our kolkhozes and sovkhoses [collective farms and state farms]. Now 20 years later, after the United States, using the very same methods which were worked out in our country, has achieved the introduction of hybrid corn and the establishment of the foundation for production of animal fodder, the U.S.S.R. faces the problem of catching up in a very short space of time on that which we let slip."

Kurchatov Aided Geneticists

The word began to get around that the Soviet Union's leading nuclear physicists had helped keep genetics alive by quietly sheltering some of its most able practitioners. The great physicist Igor V. Kurchatov (who died in 1960) played a central role in providing the Soviet Union with its first atomic weapons. His Institute of Theoretical Physics in Moscow was almost sacrosanct, as far as government interference was concerned.

Since radiation, such as that from X-rays or radioactive materials, can cause hereditary changes, there was some logic to the conduct of genetic research at the institute.

Thus in the nineteen-fifties there was a large department of genetics in the Kurchatov Institute, and a Laboratory of Radiation Genetics was quietly organized by Dubinin at the Institute of Biophysics of the Academy of Sciences in 1958. It was this laboratory which evolved into a the full-fledged institute that he now heads—an institute that itself has a dozen laboratories devoted to such problems as evolutionary genetics, space genetics, viral genetics, immunogenetics, and so forth.

A sampling showed that research at one laboratory has, as its starting point, discoveries reported by Richard F. Kimball at the Oak Ridge National Laboratory in Tennessee. Kimball has been studying the manner in which the living cell can sometimes repair genetic damage caused by radiation. In Moscow the work is concentrated on repair processes where the genetic damage, or mutation, has been caused by a chemical, such as mustard gas or ethylenimine.

In another sophisticated line of research, David M. Goldfarb is investigating the manner in which bacterial cells protect themselves against incorrect genetic information that might be infiltrated into them by foreign DNA.

DNA, or deoxyribonucleic acid, is the long, twisted molecule that carries the hereditary information needed for the continuity of life, much as magnetic tape can store an entire television program. Apparently the enzymes that unwind the DNA molecule, making it active, "know" the DNA native to that cell and do not act on foreign DNA.

Strong Practical Element in Research

There remains a strong practical element in Soviet genetic research. On a visit to the Institute of Cytology and Genetics near Novosibirsk, its director, Dmitri K. Belyayev, stressed the economic benefits deriving from the work there, much as an American scientist might do in testifying before Congress on behalf of his program.

Belyayev said it had been found that, contrary to earlier belief, mutations or heredity changes induced by radiation were not always predominantly harmful. In fact, with some plants there seems to be an optimum dosage that produces many beneficial mutations. He cited wheat, for which the best dosage lies between 5,000 and 10,000 roentgens. Treatment with such dosages has generated strains that have short, thick stems, making them resistant to wind, yet have good baking qualities.

At the institute there is a Laboratory of Polyploidy, a field that was anathema to the Lysenkoists, because it was firmly rooted in classical genetics. Polyploidy is the occurrence of plants (and in some cases animals) whose cells have unusual multiples of a basic number of chromosomes. The latter are the bundles of genetic material visible during cell division.

The sugar beet normally has 18 chromosomes (nine kinds, occurring in pairs, one set contributed by each parent plant). But by treatment with a chemical, colchicine, it is possible to produce plants that have four of each kind, instead of two; that is 36 chromosomes.

SCIENCE AND SOCIETY

These are "tetraploids." By crossing them with ordinary sugar beets it is possible to obtain an odd strain—a "triploid"—with three of each chromosome type; that is, with a total of 27.

Such a strain, according to Belyayev and his colleagues, has been developed with dramatic results. Sugar production per acre, they say, is 15 per cent higher than with ordinary beets. The new strain has been in use for three years and is now growing on many farms in southern regions of the U.S.S.R.

Belyayev has been interested in the fact that female silver foxes on fur farms have several periods of heat in a year whereas wild foxes have only one. He has found evidence of a genetic relationship between multiple periods of heat and docility.

Foxes that are docile are the most likely to be caught and the least likely to escape or die young in captivity. This has led to a fox-farm population that has the twin characteristics of docility and multiple periods of heat.

Belyayev predicted that in the near future scientists would decipher the manner in which genetic information wrapped up in the nucleus of the cell actually controls development. "We are trying to direct as much as possible of the work in our laboratories toward that goal," he said.

Soviet science has largely shaken off the shackles of Lysenkoism, although the final step was taken only three years ago, when Lysenko was ousted as head of the old Institute of Genetics. Could it happen again, could another Savonarola, another man of passionate, arrogant and intolerant views impose his will on science in the Soviet Union?

Causes of Lysenko's Rise

To assess this possibility one must examine the causes of the past episode. A number of factors were involved:

1. The ideological attraction of Lysenko's ideas as presented by his more rhetorical backers.
2. The existence of a strong national tradition in empirical plant breeding.
3. The agricultural crisis of the nineteen-thirties that made the party and government responsive to extravagant promises.
4. The existence of a dictator deaf to the logic of Lysenko's opponents and powerful enough to impose his will without challenge.

In talking to those in the current mainstream of Soviet science one is reminded at every hand of their awareness of the disastrous sequence of events behind them. As early as 1957 this was related by Aleksandr N. Nesmeyanov, then president of the Academy of Sciences, when he said:

"We must frankly state that our biology had been acquiring the bad habit of solving debated scientific problems through the pressuring and suppression of scientific opponents, the use of disparaging labels, and other unscientific means. All this has had a negative effect on the development of a number of branches of biological science. . . .

In general, it must be stated that one-sided evaluation and attempts at arriving at official evaluations in science by a majority of votes or more vocal behavior are not fruitful."

Such candor is now commonplace. One is left with the impression that the Soviet establishment has learned its lesson and that a recurrence of suppressive tyranny in the physical or biological sciences is unlikely.

Advances in Molecular Biology

Closely linked to the new genetic research has been the explosive growth of molecular biology. As with genetics, this revival had its birth in the concern of Soviet nuclear physicists with the deplorable state of biology. The extent to which the country had fallen behind became evident as windows to the West were opened after Stalin's death in 1953.

The situation was a frequent subject of discussion at the weekly seminars known as "Kapishniks" because Peter Kapitsa was their central figure. Kapitsa had been one of the most brilliant pupils of Ernest Rutherford, the British physicist.

In 1934 Kapitsa made a visit to his Soviet homeland and was told he could not return to England. Not until 1966 was he allowed to do so and he promptly returned to Russia, where he had become a spokesman for freedom of scientific discussion.

It was this freedom that dominated the Kapishniks. The participants included such figures as Igor E. Tamm and Lev D. Landau, both Nobel laureates in physics. Tamm, a robust figure who wore the emblem of Master of Sport in mountaineering, gave a lecture about 1956 that added fuel to the fire.

Speaking in the biology department of the Kurchatov Institute, he told of exciting new developments in Britain and the United States. It was evident that by analysis of DNA (deoxyribonucleic acid) Western scientists were zeroing in on the chemical mechanism of heredity.

The prospects for genetics, for biology, for the future of mankind were awesome, for there now loomed the possibility of engineering heredity, not by Lysenko's manipulation of the environment but by ingeniously controlled mutations. It was obvious to Tamm that Soviet science was in danger of being hopelessly left behind.

A Pioneer in Muscle Chemistry

One of the more venerable opponents of Lysenkoism had quietly moved into the field. He was Vladimir A Engelgardt, who, in the nineteen-thirties pioneered in muscle chemistry. His interest has shifted to the nucleic acids (such as DNA) and their role in genetics.

About 1958 (which seems to have been a critical turning point) it was decided to form an Institute of Radiational and Physical-Chemical Biology, with Engelgardt as director. This cumbersome title served as a camouflage for the real thrust of the work there.

A visit to his institute found him overflowing with enthusiasm despite his 72 years. He explained that at his request the name of his establishment had been changed to the Institute of Molecular Biology, which more accurately reflects the nature of its research.

At his side was Aleksandr Y. Braunshtein, his prize pupil and perhaps the best-known biochemist in the Soviet Union. Braunshtein told of his current studies of vitamin B_6 chemistry, an enormously complicated — and vital — problem. The vitamin, in its various forms, joins with more than 60 enzymes to catalyze, or stimulate, reactions in the life process.

Some of the antibiotics inhibit certain of these reactions, and so a three-dimensional understanding of how the molecules twist and turn in their interactions with one another has important medical implications. To achieve this understanding, Braunshtein is using ultraviolet, visible, and fluorescence spectra.

One of the men at Engelgardt's institute, A. A. Bayev, had recently deciphered the structure of another key substance — the transfer-RNA (ribonucleic acid) that is responsible for the destiny of valine. The latter is one of the 20 components (amino acids) that are linked in an almost infinite variety of sequences to form the giant protein molecules.

The role of the transfer-RNA is to pick up a molecule of valine and see to it that it slips into the correct slot in the formation of a protein. The only other transfer-RNA charted so far is that for alanine, a feat completed in 1965 by Robert W. Holley, at Cornell University. It involved charting the correct sequence of the 77 chemical units forming that molecule.

It took Bayev five years to do the job, Engelgardt said. Now, he added, Bayev is seeking out the structural features common to all the transfer-RNA's (about 50 in number).

American specialists who have toured Engelgardt's institute consider it well equipped by Western standards, with automatic spectrophotometers, ultracentrifuges, and other analytical devices. They have also remarked on a practice that they found somewhat quaint and that is based on the belief that those working with poisonous or radioactive substances should drink milk. Every day there is a pause while milk is taken from the laboratory refrigerators and distributed to the researchers—a half-liter apiece.

Engelhardt said that, as in genetics, there is a new journal, Molekularnaya Biologiya, whose first issue came out early in 1967, and an interdepartmental council has been formed to coordinate research in this field. In a crash training program, he added, a "winter school" is held each January at Dubna, the atomic research center on the Volga River. About 200 promising youngsters are brought together to hear talks by leading men in the field. "Actually," he said, "there is more discussion than lecturing. And after the sessions they go skiing."

Physicists Drawn In

"I tried to get physicists and biologists into the program to avoid compartmentalization, although at first the physicists were not very enthusiastic," he said.

He added that a new brand of scientist was appearing on the scene — a sort of Renaissance man in the scientific sense, grounded in many fields: chemistry, physics, biology, statistics.

A major source of such talent is the new Physical-Technical Institute, established near Moscow by a group of leading scientists including Kapitsa and Nikolai N. Semenov, a Nobel laureate in chemistry.

A visit to Semenov's department in the Academy of Sciences and to various Moscow institutes showed many signs of the Soviet race to catch up with the West in genetics, molecular biology and biochemistry.

Semenov explained that the Academy of Sciences, a giant organization that towers over Soviet science, was now divided into three sections, one of which, dealing with chemistry and biology, was under his direction. Another deals with the physical and mathematical sciences, and the third is concerned with the social sciences.

Semenov, at 71, was full of energy and humor. In founding the Institute of Chemical Physics in 1931 he had brought about a certain synthesis of chemistry and physics, emphasizing application of the new discoveries in physics to the construction of theories to account for chemical reaction rates and energy releases, including those of explosions. It was for such work that he won his Nobel Prize.

"My goal, since 1950, has been to achieve a marriage between biology and chemistry," he said. "At first it was slowed by the difficulties of the time—the Lysenko problem. However, five years ago [in 1962] I was able to form a new Division of Biophysics, Biochemistry and Physiologically Active Compounds within the Academy."

He found, as Engelgardt had, that physicists, biologists and chemists, trained to limited fields of interest, did not mix easily. Of this new unit in the institute he said with a grin: "At first it was a mechanical mixture, but now it is nearly a chemical compound!"

He told of a proliferation of new institutes in the field. One of them, to study the manner in which proteins are formed, is being organized by Aleksandr S. Spirin at the new "science city" of Pushchino.

"He has invited four or five colleagues, young men of 35 and 36 years, to join him there," Semenov said. "He is a biologist and his associates will be physicists and chemists."

American scientists speak of Spirin as one of the most promising young biochemists in the Soviet Union. He works at the Bakh Institute of Biochemistry, headed by Aleksandr I. Oparin, father of modern scientific thinking on the origin of life.

Spirin is a dark-haired, diminutive man who looks a little like Frank Sinatra. He told of his efforts to dissect and reconstruct ribosomes. These are tiny structures within the cell, so small that they cannot be studied in any detail even under the most powerful electron microscope.

The ribosomes are thought to be the site where the essence of the life process, the assembly of amino acids into proteins, is carried out, following genetic instructions stored in the DNA of the nucleus and carried to the ribosomes by RNA.

Medical Implications Seen

Spirin explained that knowledge of ribosome structure could be of major medical importance because many antibiotics, such as the tetracyclines, alter the mechanism of protein synthesis in the ribosomes of bacteria, thus interfering with their life process. Such drugs can be used to fight bacteria because they affect bacterial ribosomes, but, apparently, not those of human cells. Intimate knowledge of this process and of ribosomal structure could make it possible to design antibiotics against ailments now beyond reach—even, Spirin said, against cancer.

Each ribosome is formed of two rounded structures of unequal size. These were separated by American researchers in the late nineteen-fifties, and then Spirin, in 1963, managed to unfold both the large and small component into a tiny filament.

It was then possible to do what he called in English a "strip tease" of each filament, to see what it was made of. It turned out that the large component consisted of an RNA molecule enveloped in 30 protein molecules, whereas the other was an RNA molecule wrapped in 15 proteins.

In 1966 the components were reassembled into ribosomes by both his laboratory and American workers. Now he is trying other ways to reassemble the ribosomes into bodies that act normally in the life process. In this way it may be possible to learn more about details of their invisible structure.

To what extent have government and party interference handicapped the development of Soviet science outside of genetics? What about physics, chemistry and other fields?

The most serious ideological invasions date from a speech given by Andrei A. Zhdanov of the Politburo on June 24, 1947. This policy statement, part of a period of repression known as the "Zhdanovshchina," set the tone for what followed, including the grim meeting of 1948 at which Trofim D. Lysenko delivered the coup de grace to his adversaries in genetics.

It is ironic that two of the most prominent Western scientists denounced as corrupting modern, materialist science with bourgeois "idealism" were Bertrand Russell and Linus Pauling, who were later to champion the Soviet point of view on such questions as the Vietnam war and nuclear-weapons testing.

Early in 1950 Russell and his coworker Alfred North Whitehead were denounced for their trail-blazing role in the development of symbolic logic. Pauling was found guilty of inventing the resonance theory of chemical bonds.

In both cases those attacked had made use of symbols or descriptions of things that did not necessarily exist. This was heretical to adherents of Lenin's teaching. Lenin had written an entire book to rebut what he considered errors in the physics of his day, particularly the ideas of Ernst Mach, an Austrian physicist and philosopher.

Although Mach did most of his important work late in the 19th century, his influence on contemporary science has only recently been widely recognized. In fact, to laymen his name is still chiefly familiar because of its use in expressing the speed of a supersonic aircraft: a vehicle flying at Mach 2 is traveling at twice the speed of sound in that medium.

Mach demanded the most rigorous experimental criteria in determining the reality of what he considered abstract concepts, such as "atoms," "molecules," "space," and "time." He thus brushed cobwebby concepts from the minds of scientists and helped clear the way for Einstein's relativity theory. His parallel belief that all existence is sensation led to Lenin's rebuttal in 1909.

Lenin dismissed this view as idealism and said that concepts were meaningful only if they represented something with a material existence. Furthermore, he said, matter can exist quite independently from mental processes.

Ideological Diseases of Chemistry

In 1951, a conference was held to diagnose the "ideological diseases" of Soviet chemistry. Gennadi V. Chelintsev, a professor of chemical warfare at the Voroshilov Military Academy, tried to play the role in chemistry that Lysenko had performed in biology, but failed.

The leaders of Soviet chemistry, including the Academy of Sciences president, Aleksandr N. Nesmeyanov, devised a theory similar to that of Pauling, but without its emphasis on artificiality. By 1961 Pauling was lecturing on his theory in Moscow although the subject of resonance was still handled gingerly in the Soviet literature.

The controversy in physics has followed parallel lines, but it has also been intertwined with a worldwide debate on the meaning of quantum theory, with roots in one of the basic paradoxes of science: the seemingly contradictory evidence that light consists of waves and that it is formed of discrete particles.

A mathematical method, known as quantum mechanics, has been devised to account for the behavior of light and atomic particles, but what is its meaning with regard to the actual nature of those phenomena?

Twin concepts, advanced by Niels Bohr, of Denmark, and Werner Heisenberg, of Germany, prior to World War II, sought to resolve the problem. They said that no observable property of a particle, whether it refers to its wavelike characteristics or its manifestations as a particle, has any reality until that property is measured. This came to be called "complementarity."

Uncertainty Principle Stated

Likewise, Heisenberg's uncertainty principle said that any such measurement intruded into the situation in a manner that made it possible to measure, with precision, only one property of the particle.

This "Copenhagen interpretation" of quantum theory had been accepted by a number of Soviet physicists by the time Zhdanov made the 1947 speech, in which he said: "The Kantian vagaries of modern bourgeois atomic physicists lead them to inferences about the electron's possessing 'free will,' to attempts to describe matter as only a certain conjunction of waves, and to other devilish tricks."

Like symbolic logic and the resonance theory, the Copenhagen interpretation seemed inconsistent with Lenin's view that ideas were valid only if they had a clear material base. The effects of the Zhdanov speech endured until about 1960.

Today the debates in Russia on quantum theory largely parallel those in the West. Furthermore it is clear that ideological issues have not handicapped Soviet physics in any manner comparable to the throttling of biology by the Lysenko affair.

The most dramatic product of quantum electronics is the laser, a device that produces an intense and narrow beam of light and has already figured as a weapon in a James Bond moving picture. It is noteworthy that the Nobel Prize given in 1964 to the inventors of this device was shared by one American and two Russians.

Advances in Cybernetics Made

One of the most remarkable developments in Soviet science is the current vogue for cybernetics, the science of control systems in machines and living organisms. In the early postwar years Soviet theoreticians dismissed cybernetics as a capitalist scheme to replace troublesome workers with machines.

There was also objection to the concept of Norbert Wiener, father of cybernetics, that it could become a universal science, helping man solve his social as well as his material problems. This was seen by Communist theoreticians as a new ideology, attempting to usurp the role of the Communist dogma, dialectical materialism.

But by the nineteen-fifties it became apparent that unless the Soviet Union delegated some of its control operations to machines, it would be paralyzed by its own bureaucracy. An academician in Moscow predicted that if the then current trend continued, by 1980 the entire population of the Soviet Union would be involved in planning and administration.

However, a major obstacle to rapid development of cybernetics in the Soviet Union has been the absence of an electronics industry as broadly based, competitive and diversified as that of the United States. The Russians' successes in the launching and control of spacecraft and missiles have shown that they can perform in a first-rate manner with computers, but, as with many aspects of the Soviet scientific and technological scene, they have had to ration their resources and only activities with top priority have the benefit of such sophisticated aids.

October 17, 1967

22 Scientists Bid Johnson Bar Chemical Weapons in Vietnam

Twenty-two American scientists, including seven Nobel laureates, are calling on President Johnson to halt the use of antipersonnel and anticrop chemical weapons in the Vietnam war.

Any weakening of worldwide restraints on the use of chemical and biological weapons, they say in a letter drafted to be sent to the President, could have disastrous long-term consequences for the United States.

The letter, a copy of which was made available here, urges the President to "re-establish and categorically declare the intention of the United States to refrain from initiating the use of chemical and biological weapons."

The Council of the Federation of American Scientists has given its support to the letter. Copies have been sent to thousands of members for their signatures, and the initiators of the move believe that most members will sign it.

The letter, which does not judge the Vietnam war, says the "large-scale use of anticrop and 'nonlethal' antipersonnel chemical weapons in Vietnam . . . sets a dangerous precedent, with long-term hazards far outweighing the probable short-term military advantage."

It urges the President to institute a White House study of Government policy regarding chemical and biological (CB) weapons, "with a view to maintaining and reinforcing the world-wide restraints against CB warfare."

In a statement, the 22 scientists said:

"Chemical and biological weapons could be far more dangerous as instruments of mass extermination than anything except nuclear weapons.

"The United States, along with other nations, recognizes that the use of even the smallest nuclear artillery shell in war would raise issues of extreme gravity. It would break down barriers to the use of more powerful nuclear weapons, and no one could tell where the escalation might end.

"The use of chemical or biological weapons, even relatively mild ones, involves similar dangers."

The move to send the letter to the President arose from informal discussions among chemists, biochemists and bacteriologists at Harvard University, according to one signer, Matthew Meselson, a Harvard bacteriologist. Mr. Meselson, 29 years old, is the youngest of the initial signers.

The signers include some of the most distinguished chemists, biochemists, bacteriologists, physicists, biologists and men of medical science in the American scientific community.

The seven Nobel Prize winners are Felix Bloch of Stanford University, Nobel laureate in physics, 1952; Konrad E. Bloch of Harvard, medicine and physiology, 1964; Robert Hofstadter of Stanford, physics, 1961; Arthur Kornberg of Stanford Medical School, medicine and physiology, 1959; Fritz Lipmann of Rockefeller University, medicine, 1953; Severo Ochoa of New York University School of Medicine, medicine and physiology, 1959, and E. L. Tatum of Rockefeller Universty, physiology and medicine, 1958.

In a leaflet sent to other scientists, the 22 signers solicited signatures to the letter, to be returned by Oct. 31. The leaflet contained the text of the letter and also reproductions of articles from The New York Times, The Boston Globe and The Wall Street Journal.

In the article from The Times of Feb. 22, 1966, Defense Department officials explained that a helicopter-borne tear gas attack in Vietnam was designed to flush Vietcong troops out of bunkers and tunnels before a raid by B-52 bombers.

The article in The Globe of May 15, 1966, reported an increase in the use of chemicals to destroy crops in areas occupied by the Vietcong. The Wall Street Journal article in the issue of Aug. 16, 1963, reported a sharp rise since 1960 in United States expenditures on chemical-biological weapons.

The letter as drafted reads as follows:

"We, the American scientists whose names appear below, wish to warn against any weakening of the world-wide prohibitions and restraints on the use of chemical and biological (CB) weapons.

"CB weapons have the potential of inflicting, especially on civilians, enormous devastation and death which may be unpredictable in scope and intensity; they could become far cheaper and easier to produce than nuclear weapons, thereby placing great mass destructive power within reach of nations not now possessing it; they lend themselves to use by leadership that may be desperate, irresponsible, or unscrupulous. The barriers to the use of these weapons must not be allowed to break down.

"During the Second World War, the United States maintained a firm and clearly stated policy of not initiating the use of CB weapons. However, in the last few years the U. S. position has become less clear. Since the late 1950's, Defense Department expenditures on CB weapons have risen several fold and there has been no categorical reaffirmation of the World War II policy.

"Most recently, U. S. forces have begun the large-scale use of anticrop and 'nonlethal' antipersonnel chemical weapons in Vietnam. We believe that this sets a dangerous precedent, with long-term hazards far outweighing any probable short-term military advantage. The employment of any one CB weapon weakens the barriers to the use of others. No lasting distinction seems feasible between incapacitating and lethal weapons or between chemical and biological warfare. The great variety of possible agents forms a continuous spectrum from the temporarily incapacitating to the highly lethal. If the restraints on the use of one kind of CB weapon are broken down, the use of others will be encouraged.

"Therefore, Mr. President, we urge that you.

"—Institute a White House study of over-all Governmental policy regarding CB weapons and the possibility of arms control measures with a vew to maintaining and reinforcing the worldwide restraints against CB warfare.

"—Order an end to the employment of antipersonnel and anticrop chemical weapons in Vietnam.

"—Re-establish and categorically declare the intention of the United States to refrain from initiating the use of chemical and biological weapons."

The scientists' statement, citing the dangers in a weakening of controls, said:

"Under the intense pressures of actual war, and without any carefully worked out and internationally recognized guidelines, it is difficult to keep even so mild a substance as tear gas from being used in ways that can set the stage for the introduction of lethal chemicals.

"For example, when, in Vietnam, we spread tear gas over large areas to make persons emerge from protective cover to face attack by fragmentation bombs or when we use tear gas so that a moving target cannot move so fast, we use gas to kill."

Word has spread through the scientific community that secret research, financed by the Government, is seeking to devise biological weapons of total destructive power.

According to Jonathan Beckwith, assistant professor in the department of bacteriology at Harvard Medical School, "work is being conducted in this country to purposefully construct highly pathogenic bacterial strains" resistant to antibiotics.

"In particular," he said, "it has come to our attention that researchers in one of the Government centers studying germ warfare have created, amongst others, a strain of Pasteurella pestis (plague bacteria), which has been infected with such a factor making it resistant to most of the common antibiotics.

"Thus, not only are we developing ways to spread plague amongst an enemy, but also we shall be able to do it in such a way that treatment may be impossible.

September 10, 1966

Scientists Halt Work for a Day, Troubled Over Role in Research

By ROBERT REINHOLD
Special to The New York Times

CAMBRIDGE, Mass., March 4—Scientists here and throughout the country put down their slide rules and test tubes, turned off their centrifuges and gathered today to ponder what they had wrought.

Here at the Massachusetts Institute of Technology, where a national movement to suspend scientific research temporarily as a symbolic gesture began about two months ago in the physics department, many scientists gathered for a full day's discussion on the uses and misuses of scientific knowledge.

Similar programs were held on as many as 30 campuses across the country. At one school, the University of Pennsylvania in Philadelphia, all undergraduate classes were canceled for the day.

The movement, which was not an official function of M.I.T., reflected a growing concern among scientists over the consequences of their work.

Few were willing to guess how many M.I.T. scientists had closed down their laboratories for the day. However, a shifting audience of professors and students often filled the 1,200-seat Kresge Auditorium on the campus to hear speeches on reconversion to nonmilitary research, the relationship of the university to the Government, disarmament and the responsibilities of intellectuals.

Other scientists, however, have dissented from the stoppage movement on the ground that it implies that even obviously socially desirable research at M.I.T. is bad and that, in any case, scientific research cannot be turned on and off at will.

The priorities and uses of research supported by the Federal Government, which finances three-quarters of all research done in American universities, formed the basis for much of today's discussions which often digressed from the topic.

At one session, four scientists discussed alternatives to military-oriented research. Prof. Ronald Probstein, a mechanical engineer, told the gathering that members of his department, who had previously worked on bombs, missiles and jet engines, had recently converted their laboratories to research on pollution, medical engineering and desalination.

SCIENCE AND SOCIETY

'To Redress Imbalance'

"It is not our intention to sever connections with the Defense Department," he said, "but to redress an imbalance."

Discussing the role of the university in society, Prof. Howard Zinn, a historian from Boston University, called on the academic community to use knowledge to combat what he called the "lawlessness of government."

A more conservative and somewhat amused view of the day's proceedings was taken by Prof. Thomas C. Schelling, an economist at Harvard. He urged scientists not to make a moral issue out of all questions because "both sides then become unable to compromise."

Hundreds at Columbia Join
By MURRAY ILLSON

Meetings and demonstrations in support of the M.I.T. stoppage were held at educational institutions across the country yesterday.

At Columbia University several hundred graduate students and faculty members in the sciences joined in what was termed "a nationwide one-day moratorium on research." They attended a day-long series of workshops and discussion groups to consider such topics as chemical and biological weapons research, military research on university campuses and the antiballistic missile system.

More than 200 students and faculty signed a policy statement circulated by the Columbia Scientists March 4 Committee

About 30 computer programmers demonstrated at Rockefeller Center to protest what they termed the "misuse" of science for military purposes. Asserting that "we will not program death," the technicians handed out literature and circulated petitions to noon-time shoppers on Fifth Avenue between 49th and 50th Streets.

Lectures and discussion on "Responsibility in Science" were held at Rockefeller University's Caspary Auditorium, York Avenue and 66th Street, under the auspices of an ad hoc committee.

Other Meetings Here

Other meetings were held at Fordham University, New York University's Washington Square Center, Brooklyn Polytechnic Institute and at the State University Center at Stony Brook, L.I.

However, at the Argonne National Laboratory in Chicago, where a counter-movement against the protest is based, work went on as usual yesterday.

"I guess we're all working," commented Dr. Jack Uretsky, one of the prime movers in the newly formed Federation of Responsible Scientists, which had fostered a stay-on-the-job move.

About 500 students and faculty at Cornell University attended a symposium in support of the protest. The university did not cancel classes, but some professors devoted class time to a discussion of the issues.

About 500 students at the University of Wisconsin at Madison attended a "teach-in" on the "misuse of science" at which four professors spoke out against the use of scientific research for destructive purposes. Science students and professors also conducted a one-day research stoppage in support of the issue.

At the University of California at Berkeley, 500 students and faculty members attended a day-long symposium on "The Use and misuse of Science and Technology." Similar meetings were held at the University's campuses at San Francisco and Irvine in Southern California, at Stanford University and at the University of Washington in Seattle.

The protest movement had only a small effect at Princeton University, where 180 students and 40 faculty members signed an advertisement in the campus newspaper supporting it. No classes were canceled.

March 5, 1969

PANEL URGES SHIFT IN M.I.T. RESEARCH

Retention of Defense-Allied Laboratories Backed, With Stress on Civilian Role

By ROBERT REINHOLD
Special to The New York Times

CAMBRIDGE, Mass., June 2—A special committee at the Massachusetts Institute of Technology recommended today that the institute retain its two controversial defense-connected laboratories but attempt to shift their emphasis toward socially oriented civilian projects.

The panel stopped short of recommending that weapons research be ended, as some students and professors have urged. The members made clear their belief, however, that the laboratories were too heavily engaged in military research and should "energetically explore new projects to provide a more balanced research program."

Appointed last month by M.I.T.'s president, Howard W. Johnson, the committee consisted of 22 professors, students, alumni, staff and trustees of the prestigious institution. At the time Mr. Johnson ordered that new classified research be declined while the panel was deliberating.

Mr. Johnson issued a statement today endorsing the committee's recommendations and lifting the ban on new classified projects. The recommendations are not binding on the Institute.

The final responsibility for decisions at M.I.T. rests with the president and corporation, or governing board. In practice, however, the institution attempts to run by consensus, with faculty and student groups often being involved in decision making.

In a long report released today the committee discussed the complex issue of just what type of research was appropriate in an academic institution and yet responsive to public need. It concluded:

"The country's scientific and technological base rests in large part in the universities and this base should be available to support advances in defense-related fields. It is therefore clear that the two special laboratories should continue research on defense problems."

However, the report set some limitations.

"The panel," it said, "agrees that M.I.T. should avoid projects involving the actual development of a prototype weapons system, except in time of grave national emergency."

This was taken to mean that it might be appropriate for M.I.T. to design weapons but not to participate in their actual production or deployment. In this connection, the panel deemed that the institution's present role in the Poseidon missile project was inappropriate.

The laboratories are the Instrumentation Laboratory in Cambridge and the Lincoln Laboratory in nearby Lexington. Together they employ 3,670 persons and receive about $120-million yearly in Government contracts, largely from the Department of Defense for work on weapons systems, radar and related technology.

M.I.T. has come under fire from within for sponsoring such research, as have other institutions such as Stanford University, which have continued to sponsor classified military research. Unlike M.I.T., Stanford recently decided to cut its ties with a subsidiary that performs military research, the Stanford Research Institute.

Campus reaction here was varied. The Science Action Coordinating Committee, a student group agitating against weapons research, applauded the decision to urge more civilian research, but assailed the committee for not dealing with the political aspects of weapons development.

Dr. Stark Draper, the outspoken head of the Instrumentation Laboratory, said he had not yet seen the report and hesitated to comment. But, he said by telephone from Milwaukee: "I have not heard how we are going to get more civilian oriented research. The basic trouble is that there just isn't any support that nears the Defense Department."

The panel, headed by William F. Pounds, dean of the Alfred P. Sloan School of Management at M.I.T., also recommended:

¶An expansion of the "educational interaction" between the special laboratories and the main M.I.T. campus. It suggested cooperative work-study projects and more direct communication between laboratory and academic staff.

¶"Intensive efforts" to reduce classification and clearance barriers in the laboratories to enable such greater interaction.

¶The formation of a standing committee on the special laboratories to guide the president. Mr. Johnson said today he would press for such a committee.

The panel did not recommend expansion of the two laboratories to achieve its long-term objectives of increasing research in such socially related areas as biomedical engineering and transportation. Rather it suggested that military research be cut.

This, the group acknowledged, may be difficult to achieve because of Defense Department needs and the interests of some laboratory workers. In that case, they said, new laboratories or affiliations could be created to reduce defense work relative to work in other areas. This alternative, they said, would necessitate cutting the size of the special laboratories or allowing them to spin off.

All Members Concur

The panel was a diverse one.

369

It included not only such men as Julius A. Stratton, president emeritus of M.I.T., but also Prof. Noam A. Chomsky, a leading radical, and Jonathan P. Kabat, a graduate student in biology who has been associated with radical students here.

Despite this diversity, all members concurred in the report. Some of them, however, added personal statements in which they disagreed with some of the recommendations.

Dr. Chomsky said he felt that no war-related research was appropriate on the campus. "Acceptance of such research," he said, "implies support for particular judgments on military and strategy policy."

June 3, 1969

Second H.E.W. Blacklist Includes Nobel Laureate

Institutes of Health Bar 48 More Researchers From Federal Panels

By RICHARD D. LYONS
Special to The New York Times

WASHINGTON, Oct. 19 — A new Federal blacklist of scientists, containing the names of 48 researchers and including one of the three Americans who won the Nobel Prize in medicine last week, has come to light.

Prepared within the National Institutes of Health, a division of the Department of Health, Education and Welfare, the list contains the names of other internationally known leaders in research, as well as seven members of the National Academy of Sciences.

Telephone interviews with eight whose names are on the new list confirmed that they had not served on study sections and review committees that the Institutes of Health established to oversee research activities and the investment of Federal funds in them.

The Nobel laureate, Dr. Salvador E. Luria, said at his home in Lexington, Mass., that he had not served on such Federal panels. Dr. Luria, who is a professor of microbiology at the Massachusetts Institute of Technology, said he did not want to discuss the issue further.

Persons close to Dr. Luria said, however, that he was aware he had been blacklisted.

These persons added that Dr. Luria had become so enraged with security procedures within the Department of Health, Education and Welfare that he told the Health Institutes several years ago he would refuse to sit on review groups, even if his name were removed from the list, unless the security arrangements were changed.

Ironically, Robert H. Finch, Secretary of Health, Education and Welfare, congratulated Dr. Luria by telegram last week after the Nobel selections had been announced.

While the reason for barring Dr. Luria from Health Institutes panels was a mystery, political beliefs or personal associations have appeared to be the reasons in other blacklistings.

The New York Times, which obtained its second blacklist over the weekend, reported on Oct. 9 that blacklists had been drawn up within the Department of Health, Education and Welfare for years, seemingly for reasons of "security" and "suitability."

While each of the lists contains 48 names, the names of five persons appear on both lists. Although the two blacklists in The Times' possession total 93 different names, it appears that the total number of blacklisted scientists is in the hundreds.

Opposes Vietnam War

Dr. Luria said last Thursday that he strongly opposed the Vietnam war, as did Dr. Max Delbrück of Pasadena, Calif. They shared the Nobel Prize with Dr. Alfred D. Hershey of Cold Spring Harbor, L. I., for virus research.

Dr. Luria has been awarded H.E.W. funds for 10 years and has received $55,266 in the current fiscal year. Many other blacklisted scientists also have received Federal research grants.

The blacklisting of Dr. Luria was condemned by Dr. Dane G. Prugh, professor of psychiatry and pediatrics at the University of Colorado's School of Medicine in Denver and one of the leaders of a group trying to convince the Department of Health, Education and Welfare to relax its security procedures.

Dr. Prugh said in a telephone interview that he believed it was "pointless" to keep a scientist of Dr. Luria's standing from serving as a Federal adviser.

"It simply shows that the Government is being seriously deprived of the contributions of very outstanding people who themselves are being harmed," Dr. Prugh said. "It also demonstrates the perniciousness and damage to the civil liberties of these people."

Statement Prepared

Dr. Prugh, who is chairman of the ad hoc Steering Committee on H.E.W. Clearance, said his group would send a position statement to Mr. Finch tomorrow that was drawn up two weeks ago by representatives of 27 scientific and legal organizations.

"These rejections are made on the basis of 'security and suitability' criteria often related to political or other private activities, without the knowledge of the individuals," the statement said. "No opportunity is afforded for such individuals to confront the 'record' inserted against them, and there are no procedural safeguards."

"The entire process of conducting security investigations for nonsensitive positions is an unwarranted practice carried over from the early 1950's," the statement continued. "The positions in question are not regarded by the Government as sensitive ones and yet the investigations carried out by H.E.W. often go beyond those employed by other Government agencies."

Boycott Threat

Dr. Prugh added that scientists within the Health Institutes were trying to mount a protest about blacklisting and there have been discussions by study section members of a threat of boycott if things do not change.

When blacklisting reports were made public earlier this month, an H.E.W. spokesman announced that the department had appointed a five-member committee to examine security precautions.

Another resolution criticizing the department's security practices became public yesterday. This was drawn up last week at a meeting of 75 members of the National Academy of Sciences at Dartmouth College in Hanover, N.H.

This resolution, which awaits action by the academy's full membership, said participation on Federal review panels should be based solely on criteria of scientific competence, integrity and judgement.

21 Former Officers

The names of academy members are prominent on the new blacklist, which included the names of at least 21 former armed forces officers. Persons listed are prominent in the fields of organ transplantation, cancer research, heart disease, molecular genetics, brain studies and other areas of biomedical research.

Yale University has nine persons among the 93 persons mentioned on the two blacklists, the largest representation of any institution.

The Albert Einstein College of Medicine in the Bronx is next with seven blacklisted faculty members, while Harvard University has five. Columbia, Pennsylvania, Stanford and the State University of New York have four each.

"I didn't know I had so many high-caliber colleagues," said a Yale professor whose name appeared on the new blacklist, when told that Yale was heavily represented.

Named Last June

He is Dr. Clement L. Markert, chairman of the biology department and a member of the National Academy of Scientists, who discussed the problem during a telephone interview today. Dr. Markert originally was named as a blacklisted scientist on June 27 in a report in Science, the weekly journal of the American Association for the Advancement of Science.

"Since that article appeared, I've had no feelers from the Government about being taken off the list," Dr. Markert said. "But then again, I have never been officially told that I was on one."

He said friends at the Health Institutes had told him that he had been blacklisted, although he has served on an advisory panel of the National Science Foundation.

Dr. Markert, who is 52 years old, fought with the International Brigade during the Spanish Civil War when he was hardly out of his teens. He believes that it was this association that caused him to be blacklisted.

Rejected 'Informer' Role

In 1953, Dr. Markert said, he was called before a Congressional committee investigating security risks and was "asked to name names."

"They wanted me to be an informer and supply names of persons I had served with in Spain, and others, and I refused because I consider it immoral, and I still do," he said.

Dr. Markert said that "most academy members like myself do not feel that political criteria are irrelevant to selection for panel service when these panels are concerned with secret or classified work."

"If the research were classified, we would not object to political criteria," he continued. "My objection to being on the list is that the research I would be asked to study is not classified."

"These lists have nothing to do with scientific merit," Dr. Markert concluded.

"I hope it may be possible to force the National Institutes of Health to rearrange the criteria and select people on merit rather than the political prejudices of the people making the selection."

October 20, 1969

NIXON RENOUNCES GERM WEAPONS, ORDERS DESTRUCTION OF STOCKS; RESTRICTS USE OF CHEMICAL ARMS

A UNILATERAL ACT

Use of Defoliants in Vietnam War Will Be Continued

By JAMES M. NAUGHTON
Special to The New York Times

WASHINGTON, Nov. 25 — President Nixon pledged today that the United States would never engage in germ warfare and renounced all but defensive uses of chemical warfare weapons.

However, the White House made it clear that Mr. Nixon would exempt the use of tear gas and chemical defoliants, which the United States has been using in Vietnam.

The President pledged unilaterally not to make any use of bacteriological weapons, even to retaliate against an enemy attack. He ordered existing American germ warfare weapons destroyed and asked the Defense Department for recommendations on the disposal of the stocks.

'Initiative Toward Peace'

Mr. Nixon reaffirmed United States policy against the first use of lethal chemical weapons and extended the policy to include first use of "incapacitating weapons." White House sources later said that phrase did not include tear gas, which the Administration classes as a "riot control" weapon.

Reliable sources reported, however, that the President intended to impose closer control on the use of tear gas in Vietnam. It was suggested that, by guidelines or in some other fashion, he would tighten the restriction on the use of the gas, to limit both the instances and the purposes for which it is used by American forces.

In his statement, the President described his decisions as "an initiative toward peace."

He said, "Mankind already carries in its own hands too many of the seeds of its own destruction. By the examples we set today, we hope to contribute to an atmosphere of peace and understanding between nations and among men."

Two Compacts Cited

In an apparently unrelated

STORING NERVE GAS: Worker at the Newport Chemical Plant near Terre Haute, Ind., moves artillery shells filled with agent. Use of certain gases will be barred under plan.

action at the United Nations, the Soviet Union called today for a new international pact barring production of chemical and bacteriological weapons.

The coincidental emphasis by both the United States and Soviet Union on the need to halt the proliferation of chemical and biological weapons served to underscore the apparent desire of the world powers to think more seriously about disarmament.

The declarations today came barely a day after the United States and Soviet Union completed ratification of the nuclear nonproliferation treaty. And they occurred as negotiations for a limitation on strategic arms were going on at Helsinki, Finland.

Senator Clifford P. Case, Republican of New Jersey, said he saw in the related developments "an initiative which could bring the world closer to the real security of general arms control."

Mr. Nixon's policy statement focused on two international compacts—the 1925 Geneva Protocol prohibiting the first use of "asphyxiating, poisonous or other gases and of bacteriological methods of warfare" and a new British proposal for a halt in production and stockpiling of germ warfare weapons.

The President discussed his statement with bipartisan Congressional leaders at the White House before making it public.

He said he would resubmit to the Senate the Geneva protocol,' which has been signed by 62 nations but was never adopted by the United States. The Senate Democratic leader, Mike Mansfield of Montana, said it should receive swift approval this year.

"I see no reason why there should be any controversy," Mr. Mansfield told reporters. "It's 44 years overdue."

Mr. Nixon also said the United States would "associate itself" with the British proposal at the Geneva disarmament talks. Only Canada has previously indicated support of the United Kingdom draft proposal. Mr. Nixon cautioned, however, that the United States would "seek to clarify" provisions of the draft "to assure that necessary safeguards are included."

The British draft proposes to "reinforce" the Geneva Protocol by describing more specifically a total ban on the use of germ warfare and destruction of existing bacteriolgical weapons.

The decision to retain a retaliatory arsenal of chemical weapons, but not of germ weapons, was apparently a result of Mr. Nixon's belief that "biological weapons have massive, unpredictable and potentially uncontrollable consequences." They could produce global epidemics and "impair the health of future generations," he said.

The President said neither the decision to support the British proposal nor his decision to limit American bacteriological efforts to research would "leave us vulnerable to surprise by an enemy who does not observe these rational restraints."

"Our intelligence community will continue to watch carefully the nature and extent of the biological programs of others," he said.

Early reaction from key members of Congress was generally favorable. Senator J. W. Fulbright, Democrat of Arkansas and chairman of the Senate Foreign Relations Committee, who had urged the President to resubmit the Geneva Protocol, said he was pleased. He also predicted swift approval.

Senator Charles E. Goodell, Republican of New York, called Mr. Nixon's action "a great decision for the future of mankind."

The House Republican leader, Gerald R. Ford, of Michigan, saw in the announcement a "highly salutary impact" on the strategic arms talks in Helsinki.

"In taking the United States out of the field of germ warfare," Mr. Ford said, "the President has made it abundantly clear to the American people and to peoples throughout the world the great devotion that this nation has to the objective of universal peace."

Representative Richard D. McCarthy, upstate New York Democrat who has led a yearlong campaign against United States research and production of chemical and biological weaponry, said in a news conference that Mr. Nixon's decision should be "hailed as a very significant thing."

But Mr. McCarthy said that tear gas used in conjunction with weapons that kill should also be banned. He referred to the technique in Vietnam of flushing enemy troops out of bunkers with tear gas and then firing at them."

Mr. McCarthy also said he believed defoliants should not be used "in Vietnam or elsewhere" until it was determined they have no adverse effect on human life and said that at some future point he would like to see a treaty outlawing the use of napalm and phosphorous bombs.

The White House sources contended that, "technically, tear gas is not considered an 'incapacitating agent' " — the term used by the Geneva Protocol—because "its effects are very much limited in time and it dissipates immediately."

According to the source, at least one of some 80 nations that have ratified the 1925 protocol, Australia, takes the same position, that tear gas is a riot control agent, not a weapon of chemical warfare.

It was unclear precisely what effect the President's announcement would have on the Defense Department. Testimony released today by the House Appropriations Committee disclosed that the Army has spent $203.8-million in chemical and biological weapons research since 1963.

Jerry W. Friedheim, a spokesman for the Defense Department, said that active materials for bacteriological weapons were generally not storable. He said it was assumed the President's order to destroy such weapons would cover also the means of their production and delivery.

Mr. Friedheim said it was possible that some facilities would be closed.

Ronald L. Ziegler, the White House press secretary, said in response to questions from reporters that the methods of destroying the bacteriological weapons would be determined after the Defense Department had studied the problem.

The White House sources said that all biological programs would be confined to research to find methods of immunizing persons against bacteriological attack. As much as possible, this research will be shifted from the Defense Department to the Department of Health, Education and Welfare, the sources said.

The President's action today followed six months of study by the Administration of all aspects of chemical and biological warfare. Mr. Nixon said it was the first such comprehensive review in 15 years.

He told reporters he could recall the days when he sat on the National Security Council as Vice President, when it was considered "taboo" even to discuss chemical and biological warfare.

At interagency staff meetings that began last March, the discussion reportedly was freewheeling, however. Initially, representatives of the Pentagon and the Joint Chiefs of Staff opposed any reduction in the American biological capability. But by late summer the outline of Mr. Nixon's policy began to take form. Secretary of Defense Melvin R. Laird recommended a halt in manufacture of biological weapons, and, by the time the National Security Council brought the issue to a decisive point in a meeting last week only the Joint Chiefs remained opposed.

Mr. Nixon then ordered Henry A. Kissinger, his National Security adviser, to put the decisions into writing.

The White House said there was no special significance in the timing of the President's announcement, although Mr. Nixon "hoped this demonstrates our interest in the control of arms," one source said.

November 26, 1969

3 Activists Score Scientists as Decadent

By NANCY HICKS
Special to The New York Times

CHICAGO, Dec. 29 — Three former doctoral chemistry professors, all of whom gave up science for social action, charged the scientific community with decadence and with indifference to and complicity in many of the problems in America today.

They called on scientists to be more aware of the potential political nature of technological research and to become more involved in social change.

The three — George Wiley, executive director of the National Welfare Rights Organization; John R. Froines, an acquitted defendant in the Chicago 7 conspiracy trial, and Douglas LaFollette, an unsuccessful candidate for Congress in Wisconsin — joined Pete Seeger, the folk singer, in an afternoon symposium entitled, "Science and Human Needs."

The session, organized and led by Dr. George Wald, the Harvard University biologist, was part of the 137th annual meeting of the American Association for the Advancement of Science being held here this week.

While not in the majority, a number of the more than 5,000 delegates at the meeting have been raising the same issues as the three panelists, and several symposium have been organized around such themes.

The radical delegates charged that research in technology was going into weaponry, research in biology was going into chemical warfare, and research in industry was contributing to pollution. They added that scientists must organize to have more control over the way their research is used.

Their opponents have said that science is a neutral instrument, neither good nor bad by itself, and must not be contaminated by politics.

"Most science has always occurred in a social setting from the time of Copernicus," Mr. Froines, former assistant professor of chemistry at the University of Oregon, told the 800 persons in the audience. "The science of the Renaissance came out of a feeling about man. Darwin's work in evolution was an attempt to push back the superstition that hampered inquiry at the time.

"Look at the growth of science in the United States, which most people will agree took place after Sputnik," he continued. It was not a growth that came out of the desire for more pure knowledge and more technology in the human sense. It was done for armament purposes. The same is true of the space race. And scientists are being used to do it."

'A New Elite'

"I am not saying that scientists should leave science," he said. "It is fine to stay and seek knowledge, but you must get into other things too. And I am not saying that we should take over Government because it is not doing its job. We run the danger there of setting up a new elite. We must talk to each other as human beings and try to demystify science."

"Society has raised scientists to such a state that we are totally complacent; we have too much prestige and unchecked authority," Mr. LaFollette said.

"The popular image of a scientist is one of a man who is sleeping in his laboratory next to his research, spending night and day watching the progress it makes," he continued. "Maybe we have to give up some of the super-dedication to science to spend some more time in the political world and let research progress a little more slowly."

Dr. Wiley said that he had made no personal indictments against science or the institutions that funded his research when he left the university six years ago. He had served as an assistant professor of chemistry at the University of California at Berkeley, and an associate professor at Syracuse University before he quit to join the staff of the Congress of Racial Equality and then the National Welfare Rights Organization.

"When I left science it was because I found no way in my 21 years of studying, learning, teaching and pursuing chemistry that had any relation to the liberation of myself and millions of other black people who are degraded and humiliated every day in this country," he said. "Since then I have slowly realized that this was the profession that made war and oppression possible."

"In the black community, we have a word for those who sell their soul for a pittance," he said. "We call them 'Uncle Toms.' I say that the vast majority of the scientific community are 'Uncle Toms.' They have sold their souls to the Defense Department and the Federal Government for small grants, status in the intellectual community. Others have sold their souls to industry. We have blindly exchanged our dignity for a house in the suburbs, a $15,000-a-year job and two cars in the garage. "There must be a renaissance, indeed there must be a revolution, in science where people with technical experience say to themselves that today my brain power, manpower, skills and energy will not be used any longer to exploit oppressed people in this country and throughout the world."

December 30, 1970

PROFESSORS ASSAIL DEBATE SUPPRESSION

An arm of the American Association of University Professors assailed yesterday attempts by college students and teachers to suppress research and debate on the question whether heredity is a major factor in intelligence.

Noting that some participants in the controversy maintain that genetic studies of intelligence foster theories of racial inferiority, the association's Committee on Academic Freedom and Tenure declared that in a statement that "some of its own members are undermining the integrity of the academic community by attempting to suppress unpopular opinions."

The statement named neither those who advance the heredity argument—Arthur Jensen of the University of California, William Shockley of Stanford and Richard Herrnstein of Harvard —nor those who feel it should not be debated. Students and faculty members on a number of campuses have tried to prevent debate on the question, succeeding at several institutions.

February 14, 1974

GOD AND EVOLUTION

Charge That American Teachers of Darwinism "Make the Bible a Scrap of Paper"

By WILLIAM JENNINGS BRYAN

I APPRECIATE your invitation to present the objections to Darwinism, or evolution applied to man, and beg to submit to your readers the following:

The only part of evolution in which any considerable interest is felt is evolution applied to man. A hypothesis in regard to the rocks and plant life does not affect the philosophy upon which one's life is built. Evolution applied to fish, birds and beasts would not materially affect man's view of his own responsibilities except as the acceptance of an unsupported hypothesis as to these would be used to support a similiar hypothesis as to man. The evolution that is harmful—distinctly so—is the evolution that destroys man's family tree as taught by the Bible and makes him a descendant of the lower forms of life. This, as I shall try to show, is a very vital matter.

I deal with Darwinism because it is a definite hypothesis. In his "Descent of Man" and 'Origin of Species" Darwin has presumed to outline a family tree that begins, according to his estimate, about two hundred million years ago with marine animals. He attempts to trace man's line of descent from this obscure beginning up through fish, reptile, bird and animal to man. He has us descend from European, rather than American, apes and locates our first ancestors in Africa. Then he says, "But why speculate?"—a very significant phrase because it applies to everything that he says. His entire discussion is speculation.

Darwin's "Laws."

Darwin set forth two (so-called) laws by which he attempts to explain the changes which he thought had taken place in the development of life from the earlier forms to man. One of these is called "Natural Selection" or "Survival of the Fittest," his argument being that a form of life which had any characteristic that was beneficial had a better chance of survival than a form of life that lacked that characteristic. The second law that he assumed to declare was called "Sexual Selection," by which he attempted to account for every change that was not accounted for by Natural Selection. Sexual Selection has been laughed out of the class room. Even in his day Darwin said (see note to "Descent of Man" 1874 edition, page 625) that it aroused more criticism than anything else he had said, when he used Sexual Selection to explain how man became a hairless animal. Natural Selection is being increasingly discarded by scientists. John Burroughs, just before his death, registered a protest against it. But many evolutionists adhere to Darwin's *conclusions* while discarding his *explanations*. In other words, they accept the line of descent which he suggested *without any explanation whatever* to support it.

Other scientists accept the family tree which he outlined, but would have man branch off at a point below, or above, the development of apes and monkeys intead of coming through them. So far as I have been able to find, Darwin's line of descent has more supporters than any other outlined by evolutionists. If there is any other clearly defined family tree supported by a larger number of evolutionists, I shall be glad to have information about it that I may investigate it.

The first objection to Darwinism is that it is only a guess and was never anything more. It is called a "hypothesis," but the word "hypothesis," though euphonious, dignified and high-sounding, is merely a scientific synonym for the old-fashioned word "guess." If Darwin had advanced his views as a *guess* they would not have survived for a year, but they have floated for half a century, buoyed up by the inflated word "hypothesis." When it is understood that "hypothesis" means "guess," people will inspect it more carefully before accepting it.

No Support in the Bible.

The second objection to Darwin's guess is that it has not one syllable in the Bible to support it. This ought to make Christians cautious about accepting it without thorough investigation. The Bible not only describes man's creation, but gives a reason for it: man is a part of God's plan and is placed on earth for a purpose. Both the Old and New Testament deal with man and with man only. They tell of God's creation of him, of God's dealings with him and of God's plans for him. Is it not strange that a Christian will accept Darwinism as a substitute for the Bible when the Bible not only does not support Darwin's hypothesis but directly and expressly contradicts it?

Third—Neither Darwin nor his supporters have been able to find a fact in the universe to support their hypothesis. With millions of species, the investigators have not been able to find *one single instance* in which one species has changed into another, although, according to the hypothesis, *all* species have developed from one or a few germs of life, the development being through the action of "resident forces" and without outside aid. Wherever a form of life, found in the rocks, is found among living organisms, there is no material change from the earliest form in which it is found. With millions of examples, nothing imperfect is found—nothing in the process of change. This statement may surprise those who have accepted evolution without investigation, as most of those who call themselves evolutionists have done. One preacher who wrote to me expressing great regret that I should dissent from Darwin said that he had not investigated the matter for himself, but that nearly all scientists seemed to accept Darwinism.

The latest word that we have on this subject comes from Professor Bateson, a high English authority, who journeyed all the way from London to Toronto, Canada, to address the American Association for the Advancement of Science the 28th day of last December. His speech has been published in full in the January issue of Science.

Professor Bateson is an evolutionist, but he tells with real pathos how every effort to discover the origin of species has failed. He takes up different lines of investigation, commenced hopefully but ending in disappointment. He concludes by saying, "Let us then proclaim in precise and unmistakable language that our faith in evolution is unshaken," and then he adds, "our doubts are not as to the reality or truth of evolution, but as to the origin of species, a technical, almost domestic problem. Any day that mystery may be solved." Here is optimism at its maximum. They fall back on faith. They have not yet found the origin of species, and yet how can evolution explain life unless it can account for change in species? Is it not more rational to believe in creation of man by separate act of God than to believe in evolution without a particle of evidence?

Fourth—Darwinism is not only without foundation, but it compels its believers to resort to explanations that are more absurd than anything found in the "Arabian Nights." Darwin explains that man's mind became superior to woman's because, among our brute ancestors, the males fought for the females and thus strengthened their minds. If he had lived until now, he would not have felt it necessary to make so ridiculous an explanation, because woman's mind is not now believed to be inferior to man's.

As to Hairless Men.

Darwin also explained that the hair disappeared from the body, permitting man to become a hairless animal because, among our brute ancestors, the females preferred the males with the least hair and thus, in the course of ages, bred the hair off. It is hardly necessary to point out that these explanations conflict; the males and the females could not both select at the same time.

Evolutionists, not being willing to accept the theory of creation, have to explain everything, and their courage in this respect is as great as their efforts are laughable. The eye, for instance, according to evolutionists, was brought out by "the light beating upon the skin;" the ear came out in response to "air waves;" the leg is the development of a wart that chanced to appear on the belly of an animal; and so the tommyrot runs on ad infinitum, and sensible people are asked to swallow it.

Recently a college professor told an audience in Philadelphia that a baby wiggles its big toe without wiggling its other toes because its ancestors climbed trees; also that we dream of falling because our forefathers fell out of trees 50,000 years ago, adding that we are not hurt in our dreams of falling because we descended from those that fell and were *not killed*. (If we descended from animals at all, we certainly did not descend from those that were killed in falling). A professor in Illinois has fixed as the great day in history the day when a water puppy crawled upon the land and decided to stay there, thus becoming man's first progenitor. A dispatch from Paris recently announced that an eminent scientist had reported having communicated with the soul of a dog and learned that the dog was happy.

I simply mention these explanations to show what some people can believe who cannot believe the Bible. Evolution seems to close the heart of some to the plainest spiritual truths while it opens the mind to the wildest of guesses advanced in the name of science.

Guessing Is Not Science.

Guesses are not science. Science is classified knowledge, and a scientist ought to be the last person to insist upon a guess being accepted until proof removes it from the field of hypothesis into the field of demonstrated truth. Christianity has nothing to fear from any *truth*; no *fact* disturbs the Christian religion or the Christian. It is the unsupported *guess* that is substituted for science to which opposition is made, and I think the objection is a valid one.

But, it may be asked, why should one object to Darwinism *even though it is not true?* This is a proper question and deserves a candid answer. There are many guesses which are perfectly groundless and at the same time entirely harmless; and it is not worth while to worry about a guess or to disturb the guesser so long as his guess does not harm others.

The objection to Darwinism is that it is *harmful*, as well as groundless. It entirely changes one's view of life and undermines faith in the Bible. Evolution has no place for the miracle or the supernatural. It flatters the egotist to be told that there is nothing that his mind cannot understand. Evolution proposes to bring all the processes of nature within the comprehension of man by making it the explanation of everything that is known. Creation implies a Creator, and the finite mind cannot comprehend the Infinite. We can understand some things, but we run across mystery at every point. Evolution attempts to solve the mystery of life by suggesting a process of development commencing "in the dawn of time" and continuing uninterrupted up until now. Evolution does not explain creation; it simply diverts attention from it by hiding it behind eons of time. If a man accepts Darwinism, or evolution applied to man, and is consistent, he rejects the miracle and the supernatural as impossible. He commences with the first chapter of Genesis and blots out the Bible story of man's creation, not because the evidence is insufficient, but because the miracle is inconsistent with evolution. If he is consistent, he will go through the Old Testament step by step and cut out all the miracles and all the supernatural. He will then take up the New Testament and cut out all the supernatural—the virgin birth of Christ, His miracles and His resurrection, leaving the Bible a story book without binding authority upon the conscience of man. Of course, not all evolutionists are consistent; some fail to apply their hypothesis to the end just as some Christians fail to apply their Christianity to life.

Evolution and God.

Most of the evolutionists are materialists; some admitting that they are atheists, others calling themselves agnostics. Some call themselves "Theistic Evolutionists," but the theistic evolutionist puts God so far away that He ceases to be a present influence in the life. Canon Barnes of Westminster, some two years ago, interpreted evolution as to put God back of the time when the electrons came out of "stuff" and combined (about 1740 of them) to form an atom. Since then, according to Canon Barnes, things have been developing to God's plan but without God's aid.

It requires measureless credulity to enable one to believe that all that we see about us came by chance, by a series of happy-go-lucky accidents. If only an infinite God could have formed hydrogen and oxygen and united them in just the right proportions to produce water—the daily need of every living thing—scattered among the flowers all the colors of the rainbow and every variety of perfume, adjusted the mocking bird's throat to its musical scale, and fashioned a soul for man, why should we want to imprison God in an impenetrable past? This is a living world. Why not a living God upon the throne? Why not allow Him to work now?

Theistic evolutionists insist that they magnify God when they credit Him with devising evolution as a plan of development. They sometimes characterize the Bible God as a "carpenter god," who is described as repairing his work from time to time at man's request. The question is not whether God could have made the world according to the plan of evolution—of course, an all-powerful God could make the world as He pleased. The real question is, Did God use evolution as His plan? If it could be shown that man, instead of being made in the image of God, is a development of beasts we would have to accept it, regardless of its effect, for truth is truth and must prevail. But when there is no proof we have a right to consider the effect of the acceptance of an unsupported hypothesis.

Darwin's Agnosticism.

Darwinism made an agnostic out of Darwin. When he was a young man he believed in God; before he died he declared that the beginning of all things is a mystery insoluble by us. When he was a young man he believed in the Bible; just before his death he declared that he did not believe that there had ever been any revelation; that banished the Bible as the inspired Word of God, and, with it, the Christ of whom the Bible tells. When Darwin was young he believed in a future life; before he died he declared that each must decide the question for himself from vague, uncertain probabilities. He could not throw any light upon the great questions of life and immortality. He said that he "must be content to remain an agnostic."

And then he brought the most terrific indictment that I have read against his own hypothesis. He asks (just before his death): "Can the mind of man, which has, as I fully believe, been developed from a mind as low as that possessed by the lowest animal, be trusted when it draws such grand conclusions?" He brought man down to the brute level and then judged man's mind by brute standards.

This is Darwinism. This is Darwin's own testimony against himself. If Darwinism could make an agnostic of Darwin, what is its effect likely to be upon students to whom Darwinism is taught at the very age when they are throwing off parental authority and becoming independent? Darwin's guess gives the student an excuse for rejecting the authority of God, an excuse that appeals to him more strongly at this age than at any other age in life. Many of them come back after a while as Romanes came back. After feeding upon husks for twenty-five years, he began to feel his way back, like a prodigal son, to his father's house, but many never return.

Professor Leuba, who teaches psychology at Bryn Mawr, Pennsylvania, wrote a book about six years ago entitled "Belief in God and Immortality" (it can be obtained from the Open Court Publishing Company, Chicago), in which he declared that belief in God and immortality is dying out among the educated classes. As proof of this he gave the results which he obtained by submitting questions to prominent scientists in the United States. He says that he found that more than half of them, according to their own answers, do not believe in a personal God or a personal immortality. To reinforce his position, he sent questions to students of nine representative colleges and found that unbelief increases from 15 per cent. in the freshman year to 30 per cent. in the junior class, and to 40 to 45 per cent. (among the men) at graduation. This he attributes to the influence of the scholarly men under whose instruction they pass in college.

Religion Waning Among Children.

Any one desiring to verify these statistics can do so by inquiry at our leading State institutions and even among some of our religious denominational colleges. Fathers and mothers complain of their children losing their interest in religion and speaking lightly of the Bible. This begins when they come under the influence of a teacher who accepts Darwin's guess, ridicules the Bible story of creation and instructs the child upon the basis of the brute theory. In Columbia a teacher began his course in geology by telling the children to lay aside all that they had learned in Sunday School. A teacher of philosophy in the University of Michigan tells students that Christianity is a state of mind and that there are only two books of literary value in the Bible. Another professor in that university tells students that no thinking man can believe in God or in the Bible. A teacher in the University of Wisconsin tells his students that the Bible is a collection of myths. Another State university professor diverts a dozen young men from the ministry and the President of a prominent State university tells his students in a lecture on religion to throw away religion if it does not harmonize with the teaching of biology, psychology, &c.

The effect of Darwinism is seen in the pulpits; men of prominent denominations deny the virgin birth of Christ and some even His resurrection. Two Presbyterians, preaching in New York State, recently told me that agnosticism was the natural attitude of old people. Evolution naturally leads to agnosticism and, if continued, finally to atheism. Those who teach Darwinism are undermining the faith of Christians; they are raising questions about the Bible as an authoritative source of truth; they are teaching materialistic views that rob the life of the young of spiritual values.

Christians do not object to freedom of speech; they believe that Biblical truth can hold its own in a fair field. They concede the right of ministers to pass from belief to agnosticism or atheism, but they contend that they should be honest enough to separate themselves from the ministry and not attempt to debase the religion which they profess.

And so in the matter of education. Christians do not dispute the right of any teacher to be agnostic or atheistic, but Christians do deny the right of agnostics and atheists to use the public school as a forum for the teaching of their doctrines.

The Bible has in many places been excluded from the schools on the ground that religion should not be taught by those paid by public taxation. If this doctrine is sound, what right have the enemies of religion to teach irreligion in the public schools? If the Bible cannot be taught, why should Christian taxpayers permit the teaching of guesses that make the Bible a lie? A teacher might just as well write over the door of his room, "Leave Christianity behind you, all ye who enter here," as to ask his students to accept an hypothesis directly and irreconcilably antagonistic to the Bible.

Our opponents are not fair. When we find fault with the teaching of Darwin's unsupported hypothesis, they talk about Copernicus and Galileo and ask whether we shall exclude science and return to the dark ages. Their evasion is a confession of weakness. We do not ask for the exclusion of any scientific truth, but we do protest against an atheist teacher being allowed to blow his guesses in the face of the student. The Christians who want to teach religion in their schools furnish the money for denominational institutions. If atheists want to teach atheism, why do they not built their own schools and employ their own teachers? If a man really believes that he has brute blood in him, he can teach that to his children at home or he can send them to atheistic schools, where his children will not be in danger of losing their brute philosophy, but why should he be allowed to deal with other people's children as if they were little monkeys?

We stamp upon our coins "In God We Trust"; we administer to witnesses an oath in which God's name appears; our President takes his oath of office upon the Bible. Is it fanatical to suggest that public taxes should not be employed for the purpose of undermining the nation's God? When we defend the Mosaic account of man's creation and contend that man has no brute blood in him, but was made in God's image by separate act and placed on earth to carry out a divine decree, we are defending the God of the Jews as well as the God of the Gentiles, the God of the Catholics as well as the God of the Protestants. We believe that faith in a Supreme Being is essential to civilization as well as to religion and that abandonment of God means ruin to the world and chaos to society.

Let these believers in "the tree man" come down out of the trees and meet the issue. Let them defend the teaching of agnosticism or atheism if they dare. If they deny that the natural tendency of Darwinism is to lead many to a denial of God, let them frankly point out the portions of the Bible which they regard as consistent with Darwinism, or evolution applied to man. They weaken faith in God, discourage prayer, raise doubt as to a future life, reduce Christ to the stature of a man, and make the Bible a "scrap of paper." As religion is the only basis of morals, it is time for Christians to protect religion from its most insidious enemy.

February 26, 1922

Tennessee Bans the Teaching of Evolution; Governor Says the Bible Disproves Theory

Special to The New York Times.

NASHVILLE, Tenn., March 23.—Enactment of a law to prohibit the teaching of the theory of evolution in State-supported schools of Tennessee was completed today with the signing of the measure by Governor Austin Peay. The Governor, quoting from the Federal Constitution "that all men have a natural and indefeasible right to worship Almighty God according to the dictates of their own conscience," said in a statement that the bill was a "distinct protest against an irreligious tendency to exalt so-called science and deny the Bible in some schools and quarters."

"Nobody will deny that the Holy Bible teaches that man was created by God in His own image," the Governor continued. "This bill is founded on the idea and belief that the very integrity of the Bible in its statement of man's divine creation is denied by any theory that man descended or has ascended from any lower order of animals. That such theory is at utter variance with the Bible story of man's creation is incapable of successful contradiction."

Governor Peay pointed out that the bill does not require that any particular theory or interpretation of the Bible regarding man's creation be taught and does no more, in fact, than provide that the integrity of the Bible "be not negatived in the minds of the children."

The Anti-evolution bill provides:

"That it shall be unlawful for any teacher in any of the universities, normal and all other public schools of the State which are supported in whole or in part by the public school funds of the State to teach any theory that denies the story of the divine creation of man as taught in the Bible, and to teach, instead, that man has descended from a lower order of animals."

"This bill," Governor Peay says, "is a distinct protest against an irreligious tendency to exalt so-called science and deny the Bible in some schools and quarters — a tendency fundamentally wrong and fatally mischievous in its effects on our children, our institutions and country.

"Freedom of religion and strict separation of Church and State are fixed principles in this country. This bill should be rejected if it contravenes either proposition. In my judgment, it does neither."

March 24, 1925

ARREST EVOLUTION TEACHER

Tennessee Authorities Start Test Case Under New Law.

NASHVILLE, Tenn., May 6 (A. P.).—A Dayton, Tenn., despatch to The Banner says that J. T. Scopes, science teacher in Rhea High School, was arrested yesterday on a charge of violating the new Tennessee law prohibiting the teaching of evolution in the State public schools. George W. Rappleyea, Dayton business man, was the complainant. It was stated that the defense would attack the constitutionality of the new law.

The case is designed to test the new Tennessee law and may go to the United States Supreme Court, it was stated. The test, it was understood, is to be made under the auspices of the American Civil Liberties Union of New York.

May 7, 1925

JUDGE SHATTERS THE SCOPES DEFENSE BY BARRING TESTIMONY OF SCIENTISTS; SHARP CLASHES AS DARROW DEFIES COURT

TEXT OF JUDGE'S RULING

He Holds Law Makes Clarification by Scientists Unnecessary.

INSISTS INTENT IS CLEAR

This, He Explains, Is to Prohibit Theory That Man Descended From Lower Animals.

SEES JURISDICTION LIMITED

Court Holds That Proof of Theory and Question of Policy Are Matters for Legislature.

Special to The New York Times.
DAYTON, Tenn., July 17.—The text of Judge Raulston's ruling today in the Scopes trial, barring expert testimony on evolution and the Bible, was as follows:

This case is now before the Court upon a motion by the Attorney General to exclude from the consideration of the jury certain expert testimony offered by the defendant, the import of such testimony being an effort to explain the origin of man and life. The State insists that such evidence is wholly irrelevant, incompetent and impertinent to the issues pending, and that it should be excluded.

Upon the other hand, the defendant insists that this evidence is highly competent and relevant to the issues involved, and should be admitted.

The first section of the statute involved in this case reads as follows:

"Be it enacted by the General Assembly of the State of Tennessee that it shall be unlawful for any teacher in any of the universities, normals and all the public schools of the State which are supported in whole or in part by the public school funds of the State to teach any theory that denies the story of the Divine creation as taught in the Bible, and to teach instead that man has descended from a lower order of animals."

Outlines State's Contentions.

The State says that it is both proven and admitted that this defendant did teach in Rhea County, within the limits of the statute, that man descended from a lower order of animals, and that with these facts ascertained and proven, it has met the requirements of the statute and has absolutely established the defendant's guilt, and with his guilt thus admitted and established, his ultimate conviction is unavoidable and inevitable, and that no amount of expert testimony can aid and enlighten the Court and jury upon the real issues or affect the final results.

In other words, the State insists that by a fair and reasonable construction of the statute the real offense provided against in the act is to teach that man descended from a lower order of animals, and that when this is accomplished by a fair interpretation and by legal implication the whole offense is proved. That is, the State says that the latter clause interprets and explains that the Legislature meant and intended by the use of the clause "any theory that denies the story of Divine creation as taught in the Bible."

But the defendant is not content to agree with the State in its theory, but takes issue and says that before there can be any conviction the State must prove two things:

First—That the defendant taught evolution in the sense used in the statute.

Second—That this teaching was contrary to the Bible.

That these are questions of fact, that the proof must show what evolution is, so that the jury may determine whether evolution as taught by the defendant conflicts with the Bible, that it is not merely what the defendant said, or what the Book taught, and that they cannot do this without evidence. That is, that the defendant must have taught the descent of man from a lower order of animals and a theory contrary to that of divine creation as taught by the Bible. That the teaching of either would not be a crime.

Not Within Province of Court.

It is not within the province of the Court, under these issues, to determine which is true, the story of divine creation as taught in the Bible, or the story of creation of man as taught by evolution.

If the State is correct in its insistence, it is immaterial, so far as the results of this case are concerned, as to which theory is true, because it is within the province of the legislative branch, and not the judicial branch, of the Government to pass upon the policy of statute, and, the policy of this statute having been passed upon by that department of the Government, this Court is not further concerned as to its policy, but is interested only in its proper interpretation, and, if valid, its enforcement.

Let us now inquire what is the true interpretation of this statute. Did the Legislature mean that before an accused could be convicted the State must prove two things:

First, that the accused taught a theory denying the story of divine creation as taught in the Bible;

Second, that man descended from a lower order of animals?

If the first must be specifically proven, then we must have proof as to what the story of divine creation is, and that a theory was taught denying that story. But if the second clause is explanatory of the first, speaks into the act the intention of the Legislature and the meaning of the first clause, it would be otherwise.

To illustrate: When the Legislature had provided that it shall be unlawful to teach a theory that denies the divine story as taught in the Bible, and then, by the second clause, merely clarified their intention, and that the real intention provided by the statute taken as a whole was to make it unlawful to teach that man descended from a lower order of animals, then there would be no such ambiguity and uncertainty as to the meaning of the statute and as to the offense provided against as to justify the Court in calling in expert testimony to explain.

The Court will seek the aid or opinion of expert evidence only when the issues involve facts of such complex nature that a man of ordinary understanding is not competent and qualified to form an opinion.

The Intent of the Legislature.

In Tennessee an act should be construed so as to make it carry out the purposes for which it was enacted.

The legislative intent will prevail over the strict letter, and in order to carry into effect its intent general terms will be limited and those which are narrow expanded.

In the act involved in the case at bar, if it is found consistent to interpret the latter clause, having any theory which denies the Biblical story of the divine creation of men as explanatory of the legislative intent as to the offense provided against, then why call experts? The ordinary non-expert mind can comprehend the simple language, "descended from a lower order of animals."

These are not ambiguous words or complex terms. But while discussing these words, by way of parenthesis, I desire to suggest that I believe evolutionists should at least show man the consideration to substitute the word "ascend" for the word "descend."

In the final analysis, this Court, after a most earnest and careful consideration, has reached the conclusion that under the provisions of the act involved in this case it is made unlawful thereby to teach in the public schools of the State of Tennessee the theory that man descended from a lower order of animals. If the Court is correct in this, then the evidence of experts would shed no light on the issues.

Therefore, the Court is content to sustain the motion of the Attorney General to exclude the expert testimony, the purpose of which is to explain the origin of man and life in this world.

July 18, 1925

SCOPES GUILTY, FINED $100, SCORES LAW; BENEDICTION ENDS TRIAL, APPEAL STARTS; DARROW ANSWERS NINE BRYAN QUESTIONS

FINAL SCENES DRAMATIC

Defense Suddenly Decides to Make No Plea and Accept Conviction.

BRYAN IS DISAPPOINTED

Loses Chance to Examine Darrow and His Long-Prepared Speech Is Undelivered.

HIS EVIDENCE IS EXPUNGED

Differences Forgotten in the End as All Concerned Exchange Felicitations.

Special to The New York Times.

DAYTON, Tenn., July 21.—The trial of John Thomas Scopes for teaching evolution in Tennessee, which Clarence Darrow characterized today as "the first case of its kind since we stopped trying people for witchcraft," is over. Mr. Scopes was found guilty and fined $100, and his counsel will appeal to the Supreme Court of Tennessee for reversal of the verdict. The scene will then be shifted from Dayton to Knoxville, where the case will probably come up on the first Monday in September.

But the end of the trial did not end the battle on evolution, for not long after its conclusion William Jennings Bryan opened fire on Clarence Darrow with a strong statement and a list of nine questions on the basic principles of the Christian religion. To these Mr. Darrow replied and added a statement explaining Mr. Bryan's "rabies." Dudley Field Malone also contributed a statement predicting ultimate victory for evolution and repeating that Mr. Bryan ran away from the fight.

The end of the trial came as unexpectedly as everything else in this trial, in which nothing has happened according to schedule except the opening of court each morning with prayer. It was reached practically by agreement between counsel in an effort to end the case, which showed signs of going on forever, although all the testimony offered before the jury took only two hours.

Young Scopes, in his shirt sleeves, his collar open at the neck, his carrot-colored hair brushed back, stood up before the bar with a gold epauletted policeman beside him, and Judge Raulston had pronounced sentence before his counsel could suggest that Mr. Scopes might have something to say.

"Oh," exclaimed Judge Raulston. "Have you anything to say, Mr. Scopes, as to why the Court should not impose punishment upon you?"

Scopes Calls Statute Unjust.

Mr. Scopes, who is hardly more than a boy and whose pleasant demeanor and modest bearing have won him many friends since this case started, was nervous. His voice trembled a little as he folded his arms and said:

"Your Honor, I feel that I have been convicted of violating an unjust statute. I will continue in the future, as I have in the past, to oppose the law in any way I can. Any other action would be in violation of my ideal of academic freedom, that is, to teach the truth as guaranteed in our Constitution, of personal and religious freedom. I think the fine is unjust."

No one had expected such a quick ending. Mr. Darrow came into court full of the pleasant anticipation of another "go" at Mr. Bryan, whom he questioned to the delight of hundreds the day before. But the court had no sooner opened than Judge Raulston decided that there would be no further questioning, and then ordered Mr. Bryan's testimony expunged from the record.

Mr. Bryan, who had contented himself with the thought that he would have an opportunity to put Mr. Darrow on the stand and tear into him, was somewhat chagrined at this turn of the case, and announced that he would have to appeal to the fairness of the press to give prominence to the questions which he would have asked Mr. Darrow.

"I had not reached the point where I could give my statement to answer the charges made by the counsel for the defense as to my ignorance and bigotry," he said, bitterly.

Sparrow Poses as Dove of Peace.

But before the day's session was over a dove of peace hovered over the court room in the form of a frightened sparrow, which had strayed in through an open window, and everybody exchanged felicitations except Mr. Bryan and Mr. Darrow. Judge John Raulston declared that the Word of God, "given to man, that man may use it as a waybill to the other world," was an indestructible thing, and prayed God that he had decided right the questions raised in the trial. A minister pronounced a benediction and court adjourned.

The general gratification of the people at the end of a good show was shown by their applause whenever any member of either side, or the visiting spectators, rose to thank Dayton for its hospitality and kindness. And there was a further manifestation of the remarkable change in sentiment which has taken place since this trial began.

The defense faced a unitedly hostile audience when they started. There were clamors against Clarence Darrow, the agnostic, and resentment that outsiders should come in and tell Tennessee how to run its schools.

The tide turned when Dudley Field Malone made his first speech and won the hearts of Dayton, for Tennesseans love a fiery speaker, and rounded, eloquent periods delight them. Mr. Darrow ended by winning their respect by his courage in the face of hostility, and when today both sides spoke briefly their appreciation of courtesies, and Mr. Bryan defended his position, it was a repudiation of Mr. Bryan's charges by Mr. Malone and the denunciation of bigotry by Mr. Darrow that won the most fervent applause.

That does not mean that the majority was with them. It was not. But they had caught the fancy of the crowd, which has learned a lot about evolution from the scattered fragments presented to them. Many people of the State crowded around Mr. Darrow after court was over to thank him for his defense of Mr. Scopes and to say that they were ashamed of the Anti-Evolution law.

There was no doubt that the ruling of the Court against further examination of counsel on either side was as much a disappointment to Mr. Bryan as it was to Mr. Darrow. Mr. Bryan was obviously full of wrath at the position in which he had been placed, with no opportunity of justifying himself on the court record, and Mr. Darrow had come into court with the pleasant anticipation of learning what else Mr. Bryan knew about the Bible.

There was a council of war by the forces of the State last night, and whatever desire Mr. Bryan had of going on with his examination so that he could rip into Mr. Darrow and his colleagues, and brand them as agnostics or infidels, was suppressed by Attorney General Stewart, who has maintained an indignant, though dignified, opposition to the events of the last few days. He was anticipated, however, by the Court.

"Since the beginning of this trial the Judge of this court has had some big problems to pass upon," said Judge Raulston as soon as court opened. "Of course, there is no way for me to know whether I have decided these questions correctly or not until the courts of last resort speak. If I have made a mistake, it was a mistake of the head and not the heart.

"There are two things that may lead a Judge into error. One is prejudice and passion, and another is an over-zeal to be absolutely fair to all parties. I fear that I may have committed error on yesterday in my over-zeal to ascertain if there was anything in the proof that was offered that might aid the higher courts in determining whether or not I had committed error in my former decrees. I have no disposition to protect any decree that I make from being reversed by a higher court, because, if I am in error, I hope to God that somebody will correct my mistake.

"I feel that the testimony of Mr. Bryan can shed no light upon any issues that will be pending before the higher courts. The issue now is whether or not Mr. Scopes taught that man descended from a lower order of animals. It isn't a question of whether God created man as all complete at once, or it isn't a question as to whether God created man by the process of development and growth. Those questions have been eliminated by this Court, and the only question we have now is whether or not this defendant taught that man descended from a lower order of animals.

"As I see it, after due deliberation, I feel that Mr. Bryan's testimony cannot aid the higher court in determining that question. If the question before the higher court involved the issue as to what evolution is, or as to how God created man, or created the earth, or created the universe, this testimony might be relevant, but those questions are not before the Court, and so, taking this view of it, I am pleased to expunge this testimony given by Mr. Bryan yesterday from the records of this court, and it will not be further considered."

"Of course I am not at all sure that Mr. Bryan's testimony would aid the Supreme Court, or any other human being," said Mr. Darrow, "but he testified by the hour there and I haven't got through with him yet."

General Stewart objected to an argument, and Mr. Darrow took an exception from the Court's ruling, announcing that he would try to get from the Supreme Court a writ certifying the testimony. Then Mr. Darrow threw up his hands and ended the case.

"We have all been here quite a while, and I say it in perfectly good faith, we have no witnesses to offer, no proof to offer on the issues that the Court has laid down here," he said, "that Mr. Scopes did teach what the children said he taught, that man descended from a lower order of animals. We do not mean to contradict that and I think to save time we will ask the Court to bring in the jury and instruct the jury to find the defendant guilty. We make no objection to that and it will save a lot of time and I think it should be done."

Hays Again Offers Proof.

Arthur Garfield Hays then made again his offer of proof for the record.

"We offer to prove by Mr. Bryan that the Bible was not to be taken literally," he said, "that the earth was a million years old, and we had hoped to prove by him further that nothing in the Bible said what the processes were of man's creation. We feel that the statement that the earth was a million years old, and nothing said about the processes of man's creation, that it was perfectly clear that what Scopes taught would not violate the first part of the act."

All Mr. Hays's arguments brought nothing but exceptions, and then Mr. Bryan rose to his feet, a great weariness in his voice and with the look of a tired and disappointed man.

"At the conclusion of your decision to expunge from the testimony the testimony given by me upon the record I didn't have time to ask you a question," he said. I fully agree with the Court that the testimony taken yesterday was not legitimate or proper. I simply wanted the Court to understand that I was not in position to raise an objection at that time myself, nor was I willing to have it raised for me without asserting my willingness to be cross-examined. Now the testimony has ended and I assume that you expunge the questions as well as the answers."

"Yes, sir," said Judge Raulston.

"That it isn't a reflection upon the answers any more than it is upon the questions," continued Mr. Bryan in his dispirited voice.

"I expunge the whole proceedings," said the Court.

"Now, I hadn't reached the point where I could make a statement to answer the charges made by the counsel for the defense as to my ignorance and bigotry," said Mr. Bryan, turning to glare at Mr. Darrow, who hunched up his shoulders and said:

"I object, your Honor! Now, what's all this about?"

"Why do you want to make this, Colonel Bryan?" asked the Court.

"I just want to finish my sentence."

"Why can't he go outside on the lawn?" growled Mr. Darrow.

Judge Raulston said he would hear what he had to say, and Mr. Bryan continued:

"I shall have to trust to the justice of the press which reported what was said yesterday to report what I will say, not to the Court, but to the press, in answer to that charge scattered broadcast over the world, and I shall also avail myself of the opportunity to give to the press, not to the Court, the questions that I would have asked had I been permitted to call the attorneys on the other side."

"I think it would be better, Mr. Bryan," said Mr. Darrow, "for you to take us out also with the press and ask us the questions, and then the press will have both the questions and the answers."

"The gentleman who represents the defense not only differs from me," continued Mr. Bryan, "but he differed from the Court very often in the matter of procedure. I simply want to make that statement, and say that I shall have to avail myself of the press without having the dignity of its being presented in this court, but I think it is hardly fair"—his voice rose as he lifted his clenched hand—"to bring into the limelight my views on religion and stand behind a dark lantern that throws light on other people but conceals themselves. I think it is only fair that the country should know the religious attitude of the people who come down here to deprive the people of Tennessee of the right to run their own schools."

"If your Honor please," interrupted Mr. Malone, who manages to get into every argument, "I wish to make a statement if statements are in order."

"If your Honor please," added Mr. Malone, who has taken joy in standing up for the religious principles of all those on the side of the defense, "the attorneys for the defense are hiding behind no screen of any kind. They will be happy at any time in any forum to answer any questions which Mr. Bryan can ask along the lines that were asked him yesterday."

Attorney General Stewart, whose Fundamentalism has been the cause of much embarrassment to him all during the case, stopped fidgeting and suggested that the jury be brought in and the case ended. Mr. Darrow and Mr. Stewart, with Judge Raulston, then agreed that the jury be brought in and instructed by both sides to bring in a verdict of guilty.

When the jury were present Judge Raulston charged them that all they had to do was to determine whether or not Mr. Scopes taught that a man descended from a lower form of animals and that if they so found beyond a reasonable doubt they should bring in a verdict of guilty.

"May I say a few words to the jury," said Mr. Darrow. He stood before them, as plain a looking person as any one among them, his suspenders standing out against his shirt, his arms folded so that he grasped his shoulders, and smiled benignantly at them.

"We are sorry we have not had a chance to say anything to you. We will do it some other time," he said. "Now, we came down here to offer evidence in this case and the Court has held under the law that the evidence we had is not admissible, so all we can do is take an exception and carry it to a higher court to see whether the evidence is admissible or not. As far as this case stands before this jury, the Court has told you very plainly that if you think my client taught that man descended from a lower order of animals you will find him guilty, and you heard the testimony of the boys on that question and heard read the books, and there is no dispute about the facts.

"Scopes did not go on the stand, because he could not deny the statements made by the boys. I do not know how you may feel, I am not especially interested in it, but this case and this law will never be decided until it gets to a higher court, and it cannot get to a higher court very well unless you bring in a verdict. I do not want any of you to think we are going to find any fault with you as to your verdict.

"We cannot explain to you that we think you should return a verdict of not guilty. We do not see how you could. We do not ask it."

"What Mr. Darrow wanted to say to you was that he wanted you to find his client guilty," said Mr. Stewart to the Judge, "but did not want to be in the position of pleading guilty because it would destroy his rights in the Appellate Court."

Bryan's Speech Is Undelivered.

And that's all there was to that. Mr. Bryan did not get his chance to make the summing up he had been working on so long, the speech that was to rival his "Cross of Gold" speech of many years ago, which made him a candidate for the Presidency. He knew he would not get this into the records soon after court opened, and it was one of his great disappointments. He is said to have been writing it for three months. He never would have gotten a chance to make it even if the verdict of guilty had not been found by agreement, because the defense had planned to let the State open the summing up, and then refuse to argue, which would have shut off anything except the verbal onslaught of Sue Hicks, who was scheduled to talk first for the prosecution.

After Mr. Scopes had been found guilty and fined, Judge Raulston set bail in $500, which was furnished by The Baltimore Sun. The case will come up before the State Supreme Court in September, and Judge Raulston allowed thirty days for preparation of the appeal.

Then followed the felicitations, started by Mr. Malone, who said:

"Your Honor, may I at this time say on behalf of my colleagues that we wish to thank the people of the State of Tennessee, not only for their hospitality, but for the opportunity of trying out these great issues here."

He was applauded liberally by the crowd. One of the visiting spectators got up to thank Dayton for its hospitality. Gordon McKensie of the prosecution, thanked the defense for bringing into Tennessee new ideas.

"We have learned to take a broader view of life since you came," he said. "While much has been said and much written about the narrow-minded people of Tennessee, we do not feel hard toward you for having said that, because that is your idea. But we people here want to be more broad-minded than some have given us credit for, and we appreciate your coming and we have been greatly elevated, edified and educated by your presense, and should the time ever come when you are back near the garden spot of the world we hope that you will stop off and stay a while with us here in order that we may chat about the days of the past when the Scopes trial was tried in Dayton."

Dr. John R. Neal of the defense spoke, and then Mr. Bryan rose again and said the people would decide this issue.

"I don't know that there is any special reason why I should add to what has been said, and yet the subject has been presented from so many viewpoints that I hope the Court will pardon me if I mention a viewpoint that has not been referred to," he said. "Dayton is the centre and seat of this trial largely by circumstance. We are told that more words have been sent across the ocean by cable to Europe and Australia about this trial than has ever been sent by cable in regard to anything else doing in the United States. That isn't because the trial is held in Dayton. It isn't because a school teacher has been subjected to the danger of a fine of from $100 to $500, but I think it illustrates how people can be drawn into prominence by attaching themselves to a great cause.

"Causes stir the world, and this cause has stirred the world. It is because it goes deep. It is because it extends wide and because it reaches into the future beyond the power of man to see. Here has been fought out a little case of little consequence as a case, but the world is interested because it raises an issue, and that issue will some day be settled right, whether it is settled on our side or the other side. It is going to be settled right. There can be no settlement of a great cause without discussion, and people will not discuss a cause until their attention is drawn to it, and the value of this trial is not in any incident of the trial, it is not because of anybody who is attached to it, either in an official way or as counsel on either side.

"Human beings are mighty small, your Honor. We are apt to mignify the personal element and we sometimes become inflated with our importance, but the world little cares for man as an individual. He is born, he works, he dies, but causes go on forever, and we who have participated in this case may congratulate ourselves that we have attached ourselves to a mighty issue.

"Now, if I were to attempt to define that issue I might find objection from the other side. Their definition of the issue might not be as mine is, and, therefore, I will not take advantage of the privilege the Court gives me this morning to make a statement that might be controversial, and nothing that I would say would determine it.

"I have no power to define this issue finally and authoritatively. None of the counsel on our side has this power, and none of the counsel on the other side has this power. Even this honorable Court has no such power. The people will determine this issue. They will take sides upon this issue, they will state the questions involved in this issue, they will examine the information—not so much that which has been brought out here, for very little has been brought out here, but this case will stimulate investigation and investigation will bring out information, and the facts will be known, and upon the facts as ascertained the decision will be rendered, and I think, my friends and your Honor, that if we are actuated by the spirit that should actuate every one of us, no matter what our views may be, we ought not only desire but pray that that which is right will prevail, whether it be our way or somebody else's."

Darrow Makes Final Retort.

His words brought a last retort from Mr. Darrow. He thanked Dayton for its hospitality and courtesy and liberality and thanked the Court for not sending him to jail, which aroused laughter.

"Of course there is much that Mr. Bryan has said that is true," he continued. "And nature—nature, I refer to—does not choose any special setting for mere events. I fancy that the place where Magna Charta was wrested from the barons in England was a very small place, probably not as big as Dayton. But events come along as they come along.

"I think this case will be remembered because it is the first case of this sort since we stopped trying people in America for witchcraft, because here"—and he thundered out the last words—"we have done our best to turn back the tide that has sought to force itself upon this modern world of testing every fact in science by a religious dictum. That is all I care to say."

Judge Raulston was moved to join the proceedings.

"I recently read somewhere what I think was a definition of a great man, and that was this: That he possess a passionate love for the truth and has the courage to declare it in the face of all opposition. It is easy enough, my friends, to have a passion to find a truth, or to find a fact, rather, that coincides with our preconceived notions and ideas, but it sometimes takes courage to search diligently for a truth that may destroy our preconceived notions and ideas.

"The man that only has a passion to find the truth is not a complete and great man; but he must also have the courage to declare it in the face of all opposition. It does not take any great courage for a man to stand for a principle that meets with the approval of public sentiment around him, but it sometimes takes courage to declare a truth or stand for a fact that is in contravention to the public sentiment.

"Now, my friends, the man—I am not speaking in regard to the issues in this case, but I am speaking in general terms—that a man who is big enough to search for the truth and find it, and declare it in the face of all opposition, is a big man.

"Dayton has been referred to, that the law—that something big could not come out of Dayton. Why, my friends, the greatest Man that has ever walked on the face of the earth, the Man that left the portals of Heaven, the Man that came down from Heaven to earth that man might live, was born in a little town, and he lived and spent his life among a simple, unpretentious people.

"Now, my friends, the people in America are a great people. We are great in the South, and they are great in the North. We are great because we are willing to lay down our differences when we fight the battle out and be friends. And, let me tell you, there are two things in this world that are indestructible, that man cannot destroy, or no force in the world can destroy.

"One is the truth. You may crush it to the earth, but it will rise again. It is indestructible, and the causes of the law of God.

"Another thing indestructible in America and in Europe and everywhere else, is the Word of God, that He has given to man, that man may use it as a waybill to the other world. Indestructible, my friends, by any force because it is the Word of the Man, of the forces that created the universe, and He has said in His Word that 'My word will not perish,' but will live forever.

"I am glad to have had these gentlemen with us. This little talk of mine comes from my heart, gentlemen. I have had some difficult problems to decide in this lawsuit, and I only pray to God that I have decided them right.

"If I have not the higher courts will find the mistake. But if I failed to decide them right, it was for the want of legal learning and legal attainments, and not for the want of a disposition to do everybody justice."

The meeting was about to break up when Arthur Garfield Hays rose and called out to the Judge, laughingly:

"May I, as one of the counsel for the defense, ask your Honor to allow me to send you the 'Origin of Species' and the 'Descent of Man,' by Charles Darwin?"

There was a roar of laughter as the Judge said he would be glad to receive them.

"We will adjourn," said the Court, "and Brother Jones will pronounce the benediction."

Benediction was pronounced, and it was followed by a rush to the doors.

July 22, 1925

378

SCIENCE AND SOCIETY

VATICAN CENSURES JESUIT'S WRITINGS

Deceased Scientist's Ideas on Theology Criticized

By PAUL HOFMANN
Special to The New York Times.

ROME, June 30—The Vatican warned today against "dangers" in works by the late Father Pierre Teilhard de Chardin, the French Jesuit paleontologist and theologian who attempted to reconcile religious faith with scientific discoveries.

Father Teilhard, who died in New York in 1955, was co-discoverer of the Peking man, a forerunner of modern-man whose fossilized remains were found in China in the 1920's and 1930's.

The Vatican made it clear today that it objected above all to the Jesuit writer's effort to introduce the theory of evolution into theology.

The Roman Catholic Church's censure of Father Teilhard was made public in a statement by the Supreme Sacred Congregation of the Holy Office, the Vatican's highest tribunal in matters of faith and morals. Proceedings before the Holy Office are secret.

Step Short of Index

Ecclesiastics said the "warning" was one step short of placing Father Teilhard's works on the Roman Catholic Church's 400-year-old index of prohibited books, which is kept up to date by a special section of the Holy Office.

The writings of Father Teilhard, some of which were published only after his death, are getting a wide reading, particularly in France. A recent volume of personal letters caused something of a literary sensation.

Father Teilhard was born May 1, 1881, at Orcine in central France. He entered the Society of Jesus in 1899 and was ordained a priest in 1912. During World War I he served as a chaplain in front-line trenches.

He won a doctorate of science in 1922, taught at the Catholic Institute in Paris and acted as geological adviser to the Government of China from 1929 to 1945. The discovery of Peking man was made on one of his trips there.

Father Teilhard's privately published philosophical speculations about the nature of man were disapproved by Church authorities. Ordered by his superiors to leave Paris, he went first to live in Rome and then to the United States. He died in a hotel in New York on April 10, 1955.

The Latin "warning" by the Holy Office today noted that Father Teilhard's works are meeting with "no small success."

The Vatican statement said it was "quite clear that these works present such ambiguities and even grave errors in the philosophical and theological fields as to offend Catholic doctrine."

The declaration urged all Bishops Superior of religious institutions and rectors of Catholic seminaries and universities "to defend the minds, especially of students, from the dangers inherent in the works."

1955 Book Criticized

Much of what the Church criticizes in Father Teilhard's writings is contained in "The Phenomenon of Man," a book that was published in Paris shortly after his death. The book was published in the United States in 1959.

In "The Phenomenon of Man" Father Teilhard wrote that as evolution had ascended toward higher forms of life "it should culminate forward in some sort of supreme consciousness." He identified this state as Christian love.

July 1, 1962

A Game of Cosmic Roulette

The following is an interview with Jacques Monod, 1965 Nobel Prize winner in biology and author of "Chance and Necessity."

Q. You write that man was the product of pure chance. How do you come to that conclusion?

A. Well, it's relatively simple in principle, unless we accept the pure creationist view of the origins of the universe. Unless we do that, we have to find some form of natural interpretation.

I think the basic attitude of science, as defined after Galileo, is what I call the postulated objectivity. It's a negative postulate. It says that no interpretation of whatever happens in terms of final causes is acceptable. Now that's a pure postulate because clearly you can't prove that this is right.

If you start with this, then you are faced with the advanced paradox that within the cosmos there is no intention. There is no project. There comes about the biosphere full of systems which behave as if they had a project. This is the core problem in biology—how do we account for the fact that purposeful systems could have grown out of a universe devoid of purpose?

The understanding of how this could have happened is that the whole business started with certain microscopic structures which had the unique capacity of reproducing not only their structure, but also any accident which had happened within that structure. And that's the unique property of human beings.

Evolution by itself is not a privilege of human beings.

Any microscopic structure undergoes evolution, changes with time, because of small mutations which occur in the structure, and therefore tend to modify it. In fact, as a general rule, they tend to bring more and more disorder into that structure. With this, you immediately set up a huge roulette which is going to throw out numbers at a fantastic rate, and from time to time a good number is going to come out.

Q. What is a good number?

A. A good number is one which gives even better opportunities for this invariant self-producing structure to reproduce even faster, or better, or under new conditions.

We know the nature of those accidents within the structure. We know these accidents to be microscopic.

We can write chemical formulae corresponding to a mutation in the genetic code and we therefore know, as well as one can know anything in science, that there is no source of novelty in the biosphere other than the microscopic accidents. We also know that these microscopic accidents cannot and will probably never be treated in any way other than as accidents — as chance events: for the very simple but at the same time profound reason that they are microscopic accidents.

Now if we come to man, we must accept the fact that within the biosphere the emergence of man is a unique event. I'm not saying that is dependent on a single mutation. It must have involved a great many mutations. Interestingly, we don't know how many.

Maybe we will some day. But there's no question that each of these accidents was a chance event. And what has created man is a combination of chance events, whose probability was even, of course, much lower than any of the individual chance events.

So I don't think any serious biologist today would dispute this statement: that before the appearance of man, had there been a first-rate scientific brain lurking around, this first-rate brain could not possibly have predicted this event.

There's one point which I have made in the book, perhaps not strongly enough, which I find extremely significant. While man, from a purely biological, physiological, morphological point of view, has nothing unique, and while his general biochemistry and physiology are exactly the same as mammals, he is absolutely unique in one respect, namely, language. And that language would be, of course, the vehicle that opens up the possibility of culture.

The mere fact that this event occurred only once by itself indicates that it was not written somewhere in the evolution of the biosphere that language should appear.

Q. If I believe this thesis . . .

A. You cannot *not* believe it! There's no alternative. Either accepting this thesis, or taking a mythical or religious attitude, which of course cannot be disallowed. It is, of course, far simpler to accept a universe with an intention and with someone to guide its evolution.

Q. You talk about the fact that Western man, especially, has a need for explaining his existence.

A. This is very important because it is precisely that need that causes our ethical attitudes.

If I psychoanalyze myself, I'm a mixture of existentialism and intellectual Puritanism. I don't see any other possibility except once more — some sort of religious belief.

Q. Do we have some type of very deep psychological need to think that we're here for a reason? Maybe a certain insecurity of Western man that he can't accept that he's here truly by chance. He's developed philosophies and religions and myths for so many years...

A. All of which gave him a place in the universe. It's something exceedingly difficult to accept.

Q. Does Jean-Paul Sartre agree with you?

A. I think Jean-Paul Sartre should agree with me on the ethical conclusions; because essentially they are what we define as an existentialist attitude.

Q. How does this affect a poet, or a novelist? Much of Western literary tradition had been an attempt to explain man's relation to the universe. Do you think that anyone writing literature today has to come to grips now with what you say in "Chance and Necessity"?

A. I definitely think so. But I didn't invent this, you see. All this has been said before. The only originality is the logical thread which runs through the book.

If we accept the scientific attitude, if we accept the postulated objectivity, if we say that we are not going to recognize as knowledge any statement which is not based on the postulated objectivity, then we are forced into these conclusions — there's absolutely no way of getting out of it. It's literally a prison that we have built for ourselves.

Q. The Western world seems to have suffered a nervous breakdown. People say society doesn't work. Life is becoming more and more difficult to understand and enjoy. Do you think you provide an answer to what people seem to be groping for?

A. I would say that I attempt to delineate the types of answers or the type of answer that would be logically acceptable. Now to say that it would be logically acceptable does not imply that it will be psychologically acceptable.

It seems to me that one of the causes, not the only one but, perhaps, the most profound, is the one that attacks the soul. We live in societies that have developed on the basis of strong and widely accepted systems of value which are a more or less harmonious blend of the ideas of the philosophers of the Enlightenment, particularly Rousseau. Man is good and there is something called the natural rights of man which have to be sustained. It's an absolute law; since these are natural rights, we are therefore bound to defend them. Of course, if you analyze this idea of natural rights of man, it doesn't stand for a minute. There's no such thing as the natural rights of man.

In what way are they natural? Nobody could answer that.

We cannot believe in natural rights of man, nor can we believe in the religious basis of the Christian value system. Mind you, I'm not attacking the Christian value system. What I'm attacking are the mythological or metaphysical grounds upon which these values were purported to stand.

Q. This seems to be, I think, the greatest misunderstanding, because I have the impression that we have to wipe the slate clean and start all over again. Is everything that we've been believing for 2,000 years bunk? If so, then the question arises: Are you trying to create a sterile society?

A. Of course not. What I'm saying is that we have values that we operate by, and certainly a scrutiny of these values individually must be gone through, but this is not to mean that we're going to give them all up—it's ridiculous—values of love, of creativity, of respect for freedom and dignity in man, and so on, are all things that we certainly want to keep if we can.

But if we pretend to base these values on philosophical or religious systems that cannot stand any more, then they are in danger, and in fact the only way of preserving them is to find some other foundation.

Once you have a system—and that is certainly what the average Marxist creed is—which is all-embracing, and where the history of societies is linked up with the history of evolution in the biosphere, which in turn is linked up with the history of the cosmos, then there is nothing that you could not justify on that basis, absolutely nothing.

Any total system which pretends to understand the real meaning of our lives and of humanity has this inherent danger. One has to distinguish between the operational value system that a society works by, which is more or less reflected in its constitution, its laws, and so on, and the logical and philosophical justification of this complex system. We have no acceptable justification today. And my interpretation of the *mal de l'âme* is just that.

Q. If Charles de Gaulle were alive today, how do you think he'd react to your thesis?

A. I would think that the conclusions would have been completely unacceptable to him. He was a man of the 17th century. His whole political action was based on mythological concepts of nations and a profound lack of respect for man—individual man. Institutions, for him, had to be saved. Institutions and nationry. And man was just a passing nothing at the service of these institutions.

November 8, 1971

Suggested Reading

General

Asimov, Isaac. *Asimov's Guide to Science.* 3rd ed. New York: Basic Books, 1972

McGraw-Hill Encyclopedia of Science and Technology. 15 vols. New York: McGraw-Hill, 1971.

Newman, James R., ed. *The Harper Encyclopedia of Science.* rev. ed. New York: Harper & Row, 1967.

Taton, Rene, ed. History of Science. 4 vols. New York: Basic Books, 1963-66.

Matter and Energy

Clark, Ronald W. *Einstein: The Life and Times.* New York: World Publishing Company, 1971.

Ford, Kenneth W. *The World of Elementary Particles.* New York: Blaisdell, 1963. Pb*

Gardner, Martin. *The Ambidextrous Universe.* New York: Basic Books, 1964. Pb

—. *Relativity for the Million.* New York: Macmillan, 1962.

Groueff, Stephane. *Manhattan Project: The Untold Story of the Making of the Atomic Bomb.* Boston: Little, Brown & Company, 1967.

Untold Story of the Making of the Atomic Bomb. Boston: Little, Brown & Company, 1967.

Hoffman, Banesh. *The Strange Story of the Quantum.* New York: Dover Publications, 1959.

Park, David. *Contemporary Physics.* New York: Harcourt, Brace & World, 1964. Pb

Riedman, Sarah R. *Men and Women Behind the Atom.* New York: Abelard-Schuman, 1958.

Shamos, Morris H. *Great Experiments in Physics.* New York: Henry Holt & Company, 1959.

Astronomy & Planetary Science

Asimov, Isaac. *The Universe.* rev. ed. New York: Walker & Company, 1971. Pb

Hoyle, Fred. *Galaxies, Nuclei & Quasars.* New York: Harper & Row, 1965.

Jastrow, Robert. *Red Giants and White Dwarfs: The Evolution Of Stars, Planets and Life.* New York: Harper & Row, 1967. Pb

*Pb Indicates available in paperback.

Motz Lloyd and Anneta Duveen. *Essentials of Astronomy.* New York: Columbia University Press, 1972.

Sullivan, Walter. *We Are Not Alone: The Search for Intelligent Life on Other Worlds.* New York: McGraw-Hill, 1964. Pb

Earth Science

Gamow, George. *A Planet Called Earth.* New York: Viking Press, 1963.

Hodgson, John H. *Earthquakes and Earth Structure.* Englewood Cliffs, N.J.: Prentice-Hall, 1964.

Silverberg, Robert. *The Challenge of Climate: Man and His Environment.* Des Moines, Ia.: Meredith, 1969.

Stewart, Harris B. Jr. *Deep Challenge.* New York: Van Nostrand, 1966.

Sullivan, Walter. *Assault on the Unknown: The International Geophysical Year.* New York: McGraw-Hill, 1961.

—. *Continents in Motion: The New Earth Debate.* New York: McGraw-Hill, 1974.

Life

Darwin, Charles. *On the Origin of Species.* Cambridge, Mass.: Harvard University Press, 1964.

Dobzhansky, Theodosius. *Genetics of the Evolutionary Process.* New York: Columbia University Press, 1971. Pb

Karlson, P. *Introduction to Modern Biochemistry.* New York: Academic Press, 1963.

Moore, Ruth E. *Man, Time and Fossils.* rev. ed. New York: Knopf, 1961.

Pauling, Linus and R. Hayward. *The Architecture of Molecules.* San Francisco: W.H. Freeman & Company, 1964. Pb

Watson, J.D. *The Double Helix: A Personal Account of the Discovery of the Structure of DNA.* New York: Atheneum, 1968. Pb

Science and Society

Commoner, Barry. *Science and Survival.* New York: Viking Press, 1966. Pb

Greenberg, Daniel S. *The Politics of Pure Science.* New York: New American Library, 1968.

Jungk, Robert. *Brighter Than a Thousand Suns.* New York: Harcourt, Brace & Company, 1958. Pb

Klaw, Spencer. *The New Brahmins: Scientific Life in America.* New York: William Morrow & Company, 1968. Pb

Morgenbesser, Sidney, ed. *The Philosophy of Science Today.* New York: Basic Books, 1967.

Snow, C.P. *The Two Cultures and a Second Look.* New York: Cambridge University Press, 1964. Pb

Wiesner, Jerome B. *Where Science and Politics Meet.* New York: McGraw-Hill, 1965.

Index

Abel, Pamela, 267
Abelson, Philip H., 36, 245
Adams, Walter S., 92, 95
Adel, Arthur, 128
adenosine triphosphate, 243
African Genesis (Ardrey), 322
age: radioactive determination of, 17, 179, 231, 301-302, 308, 312
Aldrin, Edwin E., Jr., 152-155
Allard, Gilles O., 194
Alpha Orionis, 93-94
alpha rays, 7
altruism, 324
Ambartsumian, Viktor A., 107
amino acids, 243, 245, 247-248, 263-264
ammonia, and earth's atmosphere, 245
Anders, William A., 148-150
Anderson, Carl D., 19-20, 71
Anderson, Dan L., 162
Anderson, Herbert L., 25, 36
Andesite line, 192
Andromeda galaxy, 97, 101-102
anode rays, 3
Antarctica, 190: and continental drift, 195; and fossils, 193-195; temperatures in, 221-222
anticyclone, 223
antideuteron, 55
antimatter, 53-55, 74
antineutron, 43
antiparticles, 55
antiproton, 42-43
anti-xi-zero particle, 48
aperture synthesis, 121
Apollo 8, 148-150
Apollo 11, 152-156
Apollo 12, 156-157
Apollo 15, 158
Appalachian Mountains, 199-200
archaeometry, 301-302
Arctic Ocean, 234
Ardrey, Robert (*African Genesis*), 322
ARIES (Astronomical Radio Interferometric Earth Surveying), 186
Armstrong, Neil A., 152-155
asteroids, 146
asthenosphere, 200-201

Atlantic Ocean, 194: evolution of, 241-242
atmosphere, 216-230: atom(s), experimental collisions of, 11; and the nuclear force, 60-61; nucleus of, 17-18, 38, 40; splitting of, 19-20; structure of, 11-12, 15, 19, 42, 45, 47-52; visual detection of, 275
atomic bomb, 26, 30-35, 80, 82, 229
atomic energy, 5, 16-18, 27-29, 35-36, 80-89, 333
Atomic Scientists, Emergency Committee of, 350
atom smasher. *See* particle accelerators and names of machines — e.g. Bevatron
atomic submarines, 208
auroral lights, 220
Australia, and continental drift, 195
Australopithecines (Australopithecus), 310-311, 315, 317-318, 320-323
Australopithecus africanus, 299
Australopithecus prometheus, 304

Baade, Walter, 97, 102-103, 107
Bacher, R. F., 350
bacteria, 238-239, 312
Bainbridge, Kenneth T., 18, 24-25, 34
Baker, Richard F., 253
balloons, 216, 221
Barash, David, 324-325
Barghoorn, Elso S., 312
Barnes, H. T., 4
Barrett, Peter J., 193
Barton, Henry A., 18
baryons, 76
basalt, 207
Bateson, William, 276
bathyscaph, 206, 209
Beacon satellites, 182
Beadle, George Wells, 241: quoted, 83
Beckwith, Jonathan, 274-275
Becquerel, Henri, 4-5
Bello, Jake, 269
Belyayev, Dmitri K., 364-365
Benioff Zone, 199
Benveniste, Raoul E., 281
Berger, Rainer, 243
Bernal, J. D., 242

Berthelot, Pierre E. M., 239
beta rays, 7
Betelgeuse, 93-94
Bethe, Hans A., 27, 39, 350, 354
Bevatron, 43
big bang theory, 105-106, 110-111; see also universe
Bird, John M., 199-200
Bjorken, James D., 60, 62
Black, Davidson, 300
Blackett, Patrick M. S., 20, 189
black holes, 58, 114-116
Bloembergen, Nicholaas, 104
Boffey, Philip, 344-345
Bohr, Niels, 10-12, 20-21, 25-26, 28, 366-367
Bondi, Herman, 101, 105-106
Booth, Eugene T., 25
Borman, Frank, 148-150
Bornean earless monitor, 307-308
bosons, 76
Bothe, Walther, 20
Boyer, Herbert W., 293
Bracewell, Ronald N., quoted, 116
Bradbury, Norris E., 81
Brenner, Sydney, 260
Brickwedde, F. G., 18
Brighter Than a Thousand Suns (Jungk), 358-359
British Association for the Advancement of Science, 348-349
Bryan, William Jennings, 374-375, 377-378
Bukharin, Nikolai, 347
Bullen, Keith E., 180
Bullock, Fred, quoted, 77
Burbank, Luther, 346
Burbidge, G. R., 107
Bush, Vannevar, 33, 334-336
Byerly, Perry, 177-178

Callisto, 169-170
Calvin, Melvin, 242-243
Cambrian age, 300, 306-307, 312
Cameron, Alastair G. W., 162
Campbell, Bernard, 310-311
cancer, and ultraviolet rays, 229, 312
carbon 14, 53-54
Carnap, Rudolf, 69
Carnegie Corporation of New York, 335
Carpenter, M. Scott, 209
Cartan, Elie Joseph, 69-70
Cassiopeia-A, 106
Catcheside, D. G., 252
cathode rays, 2
cavitation, 249
cell differentiation, 252
cells, nuclei transfer in, 283-284
censorship, 29
Center for the Study of Democratic Institutions, 362
Cepheid variables, 101

Chadwick, Sir James, 19, 34-35
Chance and Necessity (Monod), 379-380
charm, 62-63
charmonium, 62-63
Chellean man, 305
chemical warfare, 346-347, 371-372
chemosphere, 226-227
chlorofluoromethanes, 229-300
chordata, fossils of, 300
chromosomes, 250-251; see also DNA, gene, genetics, RNA
Chumakov, I. S., 212-213
climate, 191, 231-235
cloning, 288-292
Cockran, J. R., 200
Cockroft, J. D., 19-20
coelacanth, 303
Cohen, Stanley N., 293
collagen tissues, 260
Collins, Michael, 153
comets, 123-125, 132-133, 165
communications, extraterrestrial, 116, 121-122
complementarity, theory of, 20-21, 366-367
Compton, Arthur H., 13, 16-17, 31
computers, celestial mechanics and, 130-131
conditioning, psychological, 292
Condon, Edward U., 350
consciousness, evolution of, 240
continental drift, 187-202, 205, 214
continental shelves, 191
Continents in Motion (Sullivan), 201-202
continuous creation, 13-14, 101, 105-106, 243; see also universe
Coon, Carleton S., 302
cosmic phoenix, 97
cosmic rays, 12-14, 20, 56-57, 74, 100, 216-218
cosmotron, 101
Crab Nebula, 113-114
Cretaceous age, 307-308
Crewe, Albert V., 275
Crick, Francis H. C., 255, 257, 262
Crookes, William, 3
Crookes tubes, 2
Crossopterygii, 301
Curie, Marie, 3-4, 6, 176
Curie, Pierre, 3, 6
Curie-Joliot, Irene, 23
Curtis, Garniss H., 308
cybernetics, 367
cyclone, 223
cyclotron, 24, 35
Cygnus-A, 106
cytogenetics, 364

Danielli, J. F., 252, 283
Darrow, Clarence, 376-378
Dart, Raymond A., 299, 304

Darwin, Charles, 240, 296-298: and acquired characteristics, 363; and altruism, 324; quoted, 304; and tektites, 185
dating, radiocarbon. *See* age
David, Edward E., Jr., 341
Dawson, Charles, 303-304
Dean, Gordon, 82
Debus, Kurt, 361
Deep-Sea Drilling Project, 211-214
De Geer Line, 211
Delbruck, Max, 274
De Lumley, Henry, 316
De Lumley, Marie-Antoinette, 316
deoxyribonucleic acid. *See* DNA
Descent of Man, The (Darwin), 296-297, 374-375
deuterium, 21-22; interstellar, 120
deutons, 22
Deutsch, Bernhard, 358
Deutsch, Martin, 41
Devonian period, 312-313
Dewey, John F., 199-200
diamonds, 182-184
diatremes, 183-184
Dicke, Robert H., 55-56, 112
Dietz, Robert S., 195
dinosaurs, 307
Dintzis, Howard M., 263
Dirac, Paul A. M., 20, 67-68, 75-76
DNA (deoxyribonucleic acid): identification of, 253-254; isolation of, 252-253; mechanism of, 267; and protein synthesis, 260; structure of, 255-257, 262; synthesis of, 243, 258, 270-271
Dobzhansky, Theodosius, 292
doppler effect, 101
Double Helix, The (Watson), 255-256
Drake, Frank D., 104, 112-114, 116
drought, 228, 233
Dubinin, Nikolai P., 364
Dumond, Jesse, 17
Dunning, John R., 24-25, 27
Du Pont Company, 31
Dyson, Freeman J., 109

earth: age of, 175-179, 312-313; atmospheric evolution of, 241-242; climate of, 231-235; composition of, 177-178, 180; crust movement, 190; early atmosphere of, 244-245; magnetic field of, 171, 181-182, 185, 211; rotation of, 174; shape of, 181-182; and solar energy, 233; tides of, 178; time scale of, 306-307
earthquakes: early research on, 174-175; and earth's resonance, 181-182; and horizontal slippage, 194-195; monitoring of, 186; records of, 177-178; zones of, 199, 205
Eastman Kodak Company, 80
Echo I, 226
Eddington, Sir Arthur, 97, 250
education, and science, 336-338

Edwards, R. G., 291
eightfold way, 51-53
Einstein, Albert, 13-16: and anti-Semitism, 347-348, 350; article by, 64-66; and atomic energy, 350; and a finite universe, 94-95; and quantum theory, 10-11; and the U.S. communist scare, 357-358; and U.S. visa curbs, 355; *see also* Relativity, theory of; Unified Field Theory
Eisenhower, Dwight David, 334-335
ekauranium atom, 25
electric inertia, 3
electrons, wave vs. particle theory of, 14-15
element 93, 23-24
Elliot, David H., 196
Engelgardt, Vladimir A., 365-366
Eniwetok, 41-42
environment, and sociobiology, 324-325
enzymes; synthesis of, 272-273; restriction, 295
Euclid, 9
eugenics, 264, 275, 363
Evernden, J. F., 308
evolution, 296-325: artificial, 272; chemical, 244, 249-250; controversy over, 240-241; and genetic transfer, 281; organic, 243; and Jacques Monad, 379-380; and the Scopes trial, 374-378; theory of, 281-282
Ewing, Maurice, 208
exosphere, 218, 220
Explorer X, 182
explosive warming, 221-222

Fairbank, William M., 56, 59
Fairbridge, Rhodes W., 197-198
fall-out, radioactive, 84 85, 222
Faraday, Michael, 64
Fermi, Enrico, 23-26, 28, 30, 34
Feynman, Richard P., 60
Field, George, 112
Fiers, Walter, 277-278
Fink, G. A., 24
Fisher, Clyde, 95
Fischer-Tropsch synthesis, 248
fission, nuclear, 33, 39
Flannery, Kent, quoted, 116
Fleischmann, Rudolf, 360
Flint, Richard F., 180
Ford, Gerald R., 344
Ford Foundation, 335
forecasting, weather, 227-228
formaldehyde, presence in interstellar space, 246-247
fossils, 193-196, 242, 312
Fox, Sidney W., 246
Fraenkel-Conrat, Heinz, 259
Frankenstein myth, 287
Franzini, Paulo, 50
Freons, 229-300
Fresnel, August J., 64
Frisch, Robert, 360

Froines, John R., 372-373
fruit flies, 250, 253, 257
fusion, nuclear, 39-40

Gagarin, Yuri, 138-139
galaxy(ies), and speed of recession, 103; *see also* Milky Way
Galton, Francis, 346
Gamow, George, 40
Ganymede, 164
garnets, 183-184
Gell-Mann, Murray, 44, 51-53, 59-60
genes, 250: analysis of, 277-278; control mechanism of, 270; fusing of, 276; isolation of, 274-275; non-chromosomal, 265-266; photo of, 253; snythesis of, 276; transplants, 293 294
Genesis of Species (Mivart), 296-297
genetic code, 260-264, 267, 280
genetic distance, 282
genetic engineering, 291-292: guidelines for, 294-295
genetic manipulation, 283-295
genetics, 250-282: and T. D. Lysenko, 349-350, 362-367; and morality, 285
genetic surgery, 284
genetic transfer, 281
Germany: and the advancement of science, 328; and Einstein, 347-348, 350; and emigration of scientists, 359-362
Giacconi, Riccardo, 115
Gilbert, Walter, 270
glaciers, 204, 212
Glashow, Sheldon L., 62
Glasoe, G. Norris, 25
global plate techtonics, 199
Glomar Challenger, 211-213, 235
Goddard, Robert H., 126
Goedel, K., 69-70
Gold, Thomas, 104, 305
Gondwanaland, 187, 190, 214
Goodall, Jane, 324
Goposchkin, Cecilia Payne, 98
Goudsmit, Samuel, 360
Gould, Laurence M., 195-196
Gould, Thomas, 101
Goulian, Mehran, 271
Graves, Alvin C., 31
gravitation, generalized theory of, 70-71
graviton, 49
gravity, negative, 109
gravity waves, 57-58, 76
Great Crab Nebula, 102
Greenland, 187, 210-211, 234-235
Griffith, Fred, 258
Groves, Maj. Gen. Leslie R., 33-34
Gurdon, John, 284-286, 289-290

Haber, Fritz, 346-347

Hadley, George, 166
Hadley circulation, 166
Hahn, Otto, 25, 28, 360
Haldane, J. B. S., 240, 245, 289
Hale, G. E., 92
Haley, Thomas J., 84
Halley, Edmund, 218
Halley's comet, 123-125, 218
Hamilton, Warren, 191
Harker, David, 269
Harriman, W. Averell, 86-87
Harrison, Ross G., 283
Hayes, H. C., 203
heavy water, 21-22
Hebard, Arthur F., 59
Heezen, Bruce C., 208
Heisenberg, Werner, 13, 16, 21, 38, 73-75, 366-367
helium, 4, 7, 112
heredity: commission for, 346; and intelligence, 343; and X-ray, 250; *see also* DNA; **genes**
Herrnstein, Richard, 373
Hershey, Alfred D., 274
Hertz, Heinrich, 64
Hertzian waves, 70
Herzberg, Gerhard, 165
Hess, Victor F., 80, 216
heterosphere, 226-227
Hethe, H. A., 96-98
Hildebrand, Joel H., quoted, 83
Hill, Henry A., 56
Hiroshima, 32-33, 82
Hjellming, Robert M., 117
Hlavaty, Vaclav, 73
Hoagland, Mahlon B., 263
Hofstadter, Robert, 42
Hogness, David, 293
Hogness, T. R., 350
Holley, Robert W., 268, 280
Holmes, Arthur, 305-306
hominids, 315, 318, 321
Homo erectus, 311, 315, 317-318, 322
Homo habilis, 309-311, 315, 318, 322
Homo sapiens, 311, 315, 320-323
homosphere, 226-227
Hooker, John Daggett, 92-93
Horstadius, Sven, 283
Hotu man, 302
Howell, B. F., 300
Howell, F. Clark, 321-322
Hoyle, Fred, 101, 106, 117
Hubble, Edwin P., 95, 101, 103, 109
Humason, M. L., 103
Hurst, Vernon J., 194
Husband, H. C., 103
Hutchinson, G. Evelyn, 242
Huyghens, Christian, 64
hydrogen, 18-22: galactic, 100-101
hydrogen bomb, 39-42, 352-353
hydrogen cyanide, 244-247

ice ages, 231-235
Iceland, 200-201, 208-209: volcanic activity on, 191-192
implantation, uterine, 291
interferometry, 121
inheritance: Mendelian laws of, 250; nonchromosomal, 265-266; and a universal genetic code, 267
Inquiry into the Politics of Science, An (Boffey), 344-345
instincts, and evolution, 240
intermediate boson, 56
International Geophysical Year, 131-133, 221-223, 231-233
ionosphere, components of, 221
Isaac, Glynn, 317-318
isobars, 24-25
isostasy, 178
isotopes, 11
Jacob, François, 260, 263, 270
Jansky, Karl G., 100, 105
Java ape man, 300
Jeans, James Hopwood, 177, 250
Jeffreys, Harold, 177
Jensen, Arthur, 373
jet stream, 219, 221
Johanson, Karl, 318, 322
Johnston, Harold S., 229
Joliot, Frederic, 23
Jones, Sir Harold Spencer, 101
Jordan, Edward B., 24-25
Jungk, Robert (*Brighter Than a Thousand Suns*), 358-359
Jupiter: atmosphere of, 128; exploration of, 163-164, 168; life on, 249; magnetic field of, 171; photos of, 169-170

Kaiser Wilhelm Institute for the Furthering of the Sciences, 321
Kaluza, Theodore, 67
Kapitsa, Peter, 365
Kartha, Gopinath, 269
Kedes, Larry, 293
Keenan, P. C., 98
Kelvin, William Thomson, Lord, 3-4, 176
Kendrew, John C., 269
Kennedy, Joseph W., 36
Khorana, Har Gobind, 276, 280
Killian, James R., Jr., 334-335, 340
King, Mary Claire, 282
Kirkpatrick, Harry, 17
Klaw, Spencer (*The New Brahmins*), 341
Kohno, Tadashiko, 276
Kohoutek, 165
Kornberg, Arthur L., 258-259, 262, 270-271
Krebs, Hans A., 249
Krebs cycle, 249
Kukla, George J., 234
Kurchatov, Igor V., 364

labyrinthodonts, 193
LaFollette, Douglas, 372-373

Langer, R. M., 23
Langmuir, Irving, 10
Langseth, Marcus G., 200-201
Lanthanotus borneensis, 307-308
Laplace, Pierre Simon de, 9
Lapp, Ralph (*The New Priesthood*), 341
lasers, 46
Lawrence, Ernest O., 22, 24, 34, 36, 38
Lawson, Douglas A., 319
Leakey, Louis S. B., 305-306, 308-311, 314-315
Leakey, Mary, 309-310
Leakey, Richard, 321-322
Lederberg, Joshua, 258
Lederman, Leon, 60
Lee, Tsung Dao, 44, 47, 50-51
Lee-Franzini, Juliet, 50
Lemaitre, Georges, 101, 106
leptons, 76
Lertes, Peter, 361
Levene, P. A., 258
Levi-Strauss, Claude, 324
Lewis, Arnold D., 315
Libby, Willard F., 301-302
Lichtenberger, Herald K., 31
Lie, Sophus, 52
life: age of, 305; extraterrestrial, 116, 145-146, 156, 242-243; intelligent, 112-113; origin of, 238-249
light: amplification of, 46; particle theory of, 13; wave theory of, 64
Lilienthal, David E., 333
Little Ice Age, 234
Livingston, M. Stanley, 22
Lodge, Sir Oliver, 3, 5, 15
Loomis, A. L., 178
Lorch, I. J., 283
Lorenz, Konrad, 324
Lovell, A. B. C., 103
Lovell, James A., Jr., 148-150
Lowell, Percival, 127
loyalty oaths, 351-352
Luria, Salvador E., 274, 370
Lysenko, Trofim D., 272, 349-350, 362-367
Lystrosaurus, 196

McCarthy, Brian, 293
McCarthy, Maclyn, 258
McGetchin, Thomas R., 183-184
Mach, Ernst, 366
McKenzie, D. P., 198
MacLeod, Colin, 258
McMillan, Edwin M., 34, 36
Magellanic clouds, 101
magnetism, 167, 189
Maiman, Theodore H., 46
Malania anjouanae, 303
Malton, Charles E., 183
man, evolution of, 296-325

Manhattan Project, 360
Mariner 2, 140-142
Mariner 4, 144-146
Mariner 5, 146-147
Mariner 6, 155-156
Mariner 7, 156
Mariner 9, 159-160
Mariner 10, 163, 165-167
Markert, Clement L., 370-371
marmots, 324-325
Mars: atmosphere of, 126, 128, 151, 155-156; canals on, 123; life on, 249; photos of, 144-146; surface of, 159-160
Marshak, Robert E., 75-76, 96-97
Marston, Robert Q., 341
masers, 36, 104
mass, negative, 109
mass spectrograph, 18, 24-25
Matthaei, J. Heinrich, 261, 263
Matthews, Clifford N., 244-245
Maxwell, James Clerk, 5, 64, 70
Mayall, N. U., 103
Mediterranean Sea, 212-213
Meitner, Lise, 25, 28
Mendel, Gregor, 288
Mendeleyev, Dmitri Ivanovich, 351
Mendelian laws of inheritance, 250
Mercury, 162, 167-168
Meseison, Matthew, 260
meson, 27, 37, 71
meson theory, 37, 40, 71
meteorites, 179, 242-243, 247-248
meteorology, 227-228
meteors, presence in ionosphere, 220
methane, and earth's atmosphere, 245
Meyer, Walter, 67-68
Michelson, Albert A., 93-94, 175
microspheres, synthesis of, 246
Mid-Ocean Ridge, 200-201, 205, 210-211, 214-215
Miescher, Friedrich, 258
Milankovitch, Milutin, 233
Milankovitch thesis, 234
military research, opposition to, 368-370
Milky Way: dimensions of, 94, 101-102; formaldehyde in, 246; galactic plane of, 104; magnetic field of, 107; molecules in, 248-249; stars in, 97-98
Miller, Stanley L., 243, 245-246
Millikan, Robert A., 12-14, 17-18
Mindanao Trench, 203
Minkowsky, Rudolph L., 107
Mintz, Yale, 165
Mirsky, Alfred E., 251-254, 258, 264-265
Mitchell, D. P., 24
Mohorovicic discontinuity, 180-181, 206-207
Mohole project, 180-181, 184, 206-207, 211, 213
molecular biology, 365
molecule structure, and evolutionary dating, 311-314
Monod, Jacques, 263, 270, 379-380
moon: age of, 156-157; craters of, 181; exploration of, 134-136, 148-150; gravitational pull of, 178; heat of, 158-159; men on, 152-155; origin of, 160-162; photographs of, 136-138, 142-143; radar contact with, 129-130; see also names of space programs — e.g. Apollo
moonquakes, 162
Morgan, Thomas Hung, 250-251
Morgan, W. Jason, 198
Morrow, John F., 293
mosasaurs, 307-308
Mossbauer, Rudolf L., 46
Mossbauer effect, 56
Moulton, Forrest, quoted, 94
mountain ranges, formation of, 199-200
mountains, floating, 178
Mount Wilson Observatory, 92-93
Muller, Ernst Oskar, 360
Muller, Hermann J., 250, 290
Muller-Hill, Benno, 270
Murray, Grover, 195-196
mutants, 250-251
mutations, 239, 259, 265-266, 281-282
muons, 56-57, 78-79

Nagasaki, 82
National Academy of Sciences, 344-345
National Science Foundation, 334-338, 341, 344
natural selection, 296-297, 324
Neanderthal man, 300, 322
nebulae, extragalactic, 95
Negy, Bartholomew, 243
Needham, 238
Ne'eman, Yuval, 51-53
Neptune: atmosphere of, 128; discovery of, 126
neptunium, 36
neutrinos, and stellar energy, 182
neutron, 18, 35-36
neutron ray, 22
neutron star, 113-115, 118
Newell, Reginald E., 234
New Brahmins, The (Klaw), 341
New Priesthood, The (Lapp), 341
Newton, Sir Isaac, 8-9, 64
Nicolet, Marcel, 226-227
Nier, Alfred O., 27
Nirenberg, Marshall W., 261, 263
Nishijima, K., 44
nitrogen, and Martian atmosphere, 155
nitrogen fixing microbes, 293
Nixon, Richard M.: and chemical warfare, 371-372; and men on moon, 152-153; and science budget, 340-343
Noggerath, Wolfgang, 361
North Pole, temperatures in, 232
Novelli, G. David, 259-260
nuclear force, 75
nuclear reactors, 89
nuclear submarines, 211

nuclear tests: ban on, 83-84, 86-87; effects of, 88
nucleic acid, 242, 268
nucleoprotein, isolation of, 251-252

ocean: food chain in depths of, 243-244; and deep-sea exploration, 211-215; life in depths of, 209-210; and life's origin, 245
ocean floor, 180-181: crack in, 205; and the drifting plate theory, 198; drilling of, 206-207; life on, 203; magnetic mapping of, 191-192; sediment of, 204-205
oceanic flywheel, 228
Occhialini, G., 20
Ochoa, Severo, 258-259, 263
Office of Science and Technology, 344
Okubo, S., 52
Olduvai Gorge, 305-306, 308
omega-minus particle, 51-52
Onsager, L., 26
Oparin, Aleksander I., 242, 245-246
Oppenheimer, J. Robert, 33-34, 37, 67, 351-352, 356-359
Origin of Continents and Oceans, The (Wegener), 188-189
Origin of Species, The (Darwin), 296, 374-375
Orion, 93-94
oxidation, 312
ozone, 3
ozone layer, 179, 219-220, 228-230

Pacific Ocean, 192, 198
packing fraction, 18
Palomar Observatory, 98
Panofsky, Wolfgang K. H., 62, 78-79, 89
Paranthropus, 315
parity, principle of, 44, 47, 50
parons, *See* quarks
parthenogenesis, 289
particle(s), 42-45: asymetric behavior, 50-51; classification of, 76; periodic table for, 73; *see also* names of specific particles
particle accelerators, 24, 35, 38, 43; *see also* names of machines — e.g. Bevatron
partons, 60
Pasteur, Louis, 238, 245
Patterson, Bryan, 314-315
Pauling, Linus, 83, 85, 255, 366
Pease, Daniel C., 253
pebble culture, 304
Pegram, George B., 24-25
Peking man, 300-302, 322, 379
Periodic Chart of the Elements, 348
Perutz, Max F., 269
phages, 274
Phillips, David C., 269
photomultiplier, 99
photon, 49
photon gas, 112

Piaget, Jean, 324
Pickering, William H., 144
Pilbeam, David R., 314
Piltdown man, 299-300: hoax exposed, 303-304
Pioneer 10, 163-164, 168
Pioneer 11, 169-171
pions, 49
Pithecanthropus, 316
Pitman, Walter C., 3d, 201
Pitzer, Kenneth S., quoted, 83
Piveteau, Jean, 316
placenta, artificial, 291
Planck, Max, 74
plasmid transplantation technique, 293-294
Pleiades, 98
Pleistocene age, 305, 308-309, 315
Pluto, discovery of, 127-128
plutonium, 35-37
poison gas, 346-347
Pollister, Arthur W., 251-252
polonium, 3-4
polynucleotide phosphorolase, 263
polyploidy, 364-365
poly-U (synthetic RNA), 263
Pomeranchuk, Issak Y., 61
Ponnamperuma, Cyril, 243, 247-249
Pontecorvo, Bruno, 47-49
population genetics, 257
Populations I and II, 97-98
positronium, 41
Pound, Robert V., 56
President's Science Advisory Committee, 341-342
Price, Charles C., 246
proconsul, 309-310
protein(s), 238-239: assembly of, 279; synthesis of, 244-245, 259-260
proton, 15, 40: bombardment of, 78, interior of, 59;
Prugh, Dane G., 370
Ptashne, Kay, 293
Ptashne, Mark, 270
pterodactyl, 319
pterosaurs, 319
pulsers, 58, 60, 113-114, 118

quantum theory, 10-11, 13, 16, 66, 68-72
quarks, 52, 59-61, 78-79
quasar(s), 58, 60: age of, 120; distance of, 108-110, 119; and earthquake predictions, 186; pulsations of, 110
quasi-stellar blue galaxies, 111

radiation: effects of, 84-85; and the Mossbauer effect, 46; stellar, 96
radioactive dating, 17, 179, 231, 301-302, 308, 312
radioactive fall-out, 84-85
radioactivity, 4-8, 23, 80

radio astronomy, 99-100, 105
radio galaxies, 108
radio scattering, 220
radio stars, 108
radio-telescope, 103-104
radio waves, 220
radium, 3-5
Ramapithecus, 314-315, 322-323
Ramsay, Sir William, 4, 7-8, 102, 238
Ranger 6, 143
Ranger 7, 142-143
Rayleigh, John W. Strutt, Lord, 176
Rayton, W. M., 178
recession, law of, 101
red dwarfs, 97
Redi, Francesco, 238
red shift, 106, 109-110, 119
relativity, theory of, 8-11, 18, 55-56, 58, 64-65, 68-71, 75
Rentschler, Harvey C., 31
Retallack, J. Gordon, 40
Reykjanes Ridge, 200-201
ribonuclease: 268-269; synthesis of, 272-273
ribonucleic acid. *See* RNA
ribosomes, 260, 366
Rich, Alexander, 279
Richards, Oliver, 293
Richter, Burton, 61
Riddle, Oscar, 240
Ris, Hans, 258
RNA (ribonucleic acid): components of, 280; and mutation, 259; and protein synthesis, 260; synthetic, 258-259, 263; template, 263-265; transfer, 263, 268, 278-280
Rockefeller Foundation, 335
rockets, 126
Roentgen, Wilhelm Konrad, 2-3
Roth, John R., 276
Ruffini, Remo, 117
Rumsey, Howard C., Jr., 163
Russell, Bertrand, 366: quoted, 68, 111
Russell, Henry Norris, 96-97
Russell mixture, 97
Rutherford, Sir Ernest, 4, 7-8, 11, 19
Ryan, William B. F., 212-213
Ryle, Sir Martin, 106, 120-121

Sager, Ruth, 265-266
Sahara Desert, 197-198
Salam, Abdus, 78
San Andreas Fault, 186, 189, 194
Sandage, Allan R., 103, 110-111, 119
Sanger, Fred, 268
Sarich, Vincent M., 313-314
satellites, artificial: first Soviet, 131-132; and radiation layer, 132-133; Uhuru, 114-115; and weather, 223-225, 228; U.S. response to Soviet, 335-338

Saturn: atmosphere of, 128; exploration of, 169
scalar-tensor theory, 56
Science and Government (Snow), 339
Scientific American, 354
Schaller, George, 324
Schilling, Jean-Guy El, 200-201
Schroedinger, Erwin, quoted, 73
Schwartz, Melvin, 47-49
Schwarzschild, Karl, 109
Schwarzschild, Martin, 106
Schwarzschild radius, 109
Schwarzfield, Martin, 99
Schweet, Richard, 263
Schwinger, Julian, 38-39
Scopes, John Thomas, 375-378
Seaborg, Glenn T., 36
Segré, Emilio, 36
seismographs, 177
Semenov, Nikolai N., 366
sex, determination of, 250
sexual reproduction, 288-289
Shapiro, Irwin I., 56
Shapley, Harlow, 94-95, 101-102
Shenstone, W. A., 238
Shklovsky, Iosif S., 108-109
Shockley, William, 373
sickle cell anemia, 311
Simons, Elwyn L., 314
Simpson, John A., 144-145
Sinsheimer, Robert L., 271, 289
Skinner, B. F., 324
Slack, Francis G., 25
Slipher, V. M., 127-129
Sloan, Alfred P., Foundation, 335
Snow, C. P. (*Science and Government*), 339
sociobiology, 323-325
Sociobiology: The New Synthesis (Wilson), 324
Soddy, Frederic, 4-7, 346-347
solar energy, 39
solar system, 159, 164, 179
solar wind, 145, 165-166, 181 182
Somerfeld, Arnold, 10-11
sonic depth finder, 203
South Pole, 197-198
space, chemicals in, 248-249
spaceships, 138-139
spectrometer, multiple crystal, 17
spectroscopy, 11
Spedding, Frank H., 21
Spemann, Hans, 283
Spencer, Herbert, 4
Spiegelman, Sol, 267, 271-272
spinor theory, the, 73
Spirin, Aleksandr S., 366
Spitsbergen, 210
Sproll, Walter, 195
Stimson, Henry L., 32-33
stars: burned-out, 115; collisions of, 106-107; dwarf, 96; energy of, 97-98; evolution of, 99; recession of, 101

Stebbins, Joel, 99
Steptoe, P. C., 291
Stern, Kurt G., 252-253
Stetson, Harlan T., 178
Steward, F. C., 289
SST, 228-230
strangeness, 61-62
Strassmann, Fritz, 25
stratosphere, 216, 218-221
strontium 90, 88
Sullivan, Walter (*Continents in Motion*), 201-202
sun: photography of, 99-100; temperature of, 97
sun spots, 124, 233
supernovae, 102, 106-107
Swings, Pol, 112
symmetry group theory, 52-53
synchrocyclotron, 38
Szilard, Leo, 30, 36, 350-352

Taieb, Maurice, 318
Talwani, Manik, 200
Tayacian culture, 316
Taylor, Richard, 60
TCP symmetry, 55
tectonics, 199-200
Teilhard de Chardin, Pierre, 379
tektites, 184-185
Telanthropus capemsla, 311
telemeter, 130
Teller, Edward, 83-84
tensor analysis, 69
Thaddeus, Patrick, 112
Third Cambridge Catalogue of Radio Sources, 109
Thomson, Sir Joseph J., 3-4, 8, 18
thorium, 4, 6
Thorp, Willard L., 257
tillite, 190
Timofeyev-Ressovsky, Nikolai V., 362
Tinbergen, Hiko, 324
Ting, Samuel C. C., 61-63
Tiros I, 223-225
tissue transplants, 283
tobacco mosaic virus, 259
Tobias, Phillip T., 309-310
Todaro, George J., 281
toolmaking, 308
Townes, Charles H., 46, 104, 116
transduction, 258
trihornometry, 283
Trivers, Robert, 324
troposphere, 216, 218-219, 221
Truman, Harry S.: and the Atomic Energy Commission, 333; and the National Science Foundation, 334; and Hiroshima, 32-33; and the hydrogen bomb, 352-353
Trumpler, R. J., 98
Tsubo, Yoshihiro, 265-266

Tsugita, A., 259
Tucker, Wallace, 115
Tunguska meteorite, 53-55

Uhuru, 114-115
ultraviolet radiation, 12, 229, 243, 245
uncertainty principle, 16, 21, 72, 74, 367
unified field theory, 64-79
U.S.S.R.: and the advancement of science, 329-330, 347, 348; and artificial satellites, 131-132; and the Deep Sea Drilling Project, 213-214; and exploration of Venus, 157-158; and German scientists, 360-361; and genetics, 349-350, 362-367; and moon exploration, 134-136; and a nuclear test ban treaty, 86-87; and spaceships, 138-139
United Nations Scientific Committee on the Effects of Atomic Radiation, 84-85
United States and the advancement of science: 330-333, 336-338; Atomic Energy Commission, 82-83, 333, 344-345, 351-352, 354, 356-357; Department of Health, Education and Welfare, 370-371; Environmental Protection Agency, 89; Office of Science and Technology, 341-343; and the McCarran act, 355-356
Universe: age of, 102-103, 105-106; expansion of, 111; largest known object in, 120-121; origin of, 119; oscillating, 112; size of, 107-108
uranium atom, splitting of, 25-26
uranium 235, 26-29
Uranus: atmosphere of, 128; Neptune and, 126
Urey, Harold C., 18, 21-22, 241-242, 245, 350

V-A theory, 76
Van Allen, James A., 133, 144
Van Allen belt, 132-133
Van de Graaff, Robert J., 17-18
Vanguard satellite, 181
Van Helmont, 238
Vavilov, Nikolai I., 363-364
vegetative reproduction, 284-285
Venera 7, 157-158
Venus: exploration of, 140-141, 146-147, 157-158; magnetic field of, 141; origin of, 166-167; and radar, 162-163; weather of, 165-166
viruses, 241: and genetics, 258, 267, 281; and phages, 274; and synthesized DNA, 270-271
vitalism, 240
volcanoes, 208, 234-235
Von Ardenne, Manfred, 361
Von Braun, Werner, 359-361
Von Harneck, Adolf, 328

Walker, G. P. L., 201

Wallace, A. R., 296-297
Walton, E. T. S., 19-20
Walton, Matt, Jr., 180
water, interstellar, 108
Waterman, Alan T., 335-338
Watson, James D., 255-257, 262, 287-288
weak force, the, 61-62
weather, theory of, 221-223
weather forecasting, 227-228
Weber, Joseph, 57-58
Wegener, Alfred, 187-188
Weil, George, 36
Weinberg, Steven, 76-79
Weisskopt, V. F., 350
Wells, H. G., 5, 9
Westinghouse Company, 31
Wheeler, John A., 28, 109, 117
Whipple, Fred L., 132, 165
white dwarfs, 97, 115
white hole concept, 117
Wiener, Norbert, 351, 367
Wigner, Eugene, 52-53
Wiley, George, 372-373
Wilkins, Maurice, 255
Willins, T. R., 178
Wilson, Alan C., 282, 313-314
Wilson, C. T. R., 13

Wilson, Edward O. (*Sociobiology*), 324
Woodward, Sir Arthur Smith, 304
Woolf, Nicholas, 112
World War I, 346
W particles, 49-50

xi-zero (particle), 44-45
X-rays, 2-3, 22, 250
X-ray star, 115

Yukawa, Hideki, 27, 40, 71
Yang, Chen Ning, 44, 47
yeast fly, 251
Yourno, Joseph, 276

Zielinski, Gary, 200
Zinder, Norton, 258
Zinjanthropus boisei, 308
Zinn, Walter H., 36